HUMAN PHYSIOLOGY

THE MECHANISMS OF BODY FUNCTION

FIFTH EDITION

HUMAN PHYSIOLOGY

THE MECHANISMS OF BODY FUNCTION

ARTHUR J. VANDER, M.D.

Professor of Physiology
University of Michigan

JAMES H. SHERMAN, PH.D.

Associate Professor of Physiology
University of Michigan

DOROTHY S. LUCIANO, PH.D.

Formerly of the Department of Physiology
University of Michigan

McGraw-Hill Publishing Company

New York St. Louis San Francisco Auckland Bogotá
Caracas Lisbon London Madrid Mexico Milan
Montreal New Delhi Paris San Juan Singapore
Sydney Tokyo Toronto

HUMAN PHYSIOLOGY
THE MECHANISMS OF BODY FUNCTION

3 4 5 6 7 8 9 0 VNH VNH   9 4 3 2 1

ISBN 0-07-066969-4

This book was set in Caledonia by York Graphic Services, Inc.
The editors were Irene Nunes, Denise T. Schanck, and James W. Bradley;
the designer was Jo Jones;
the production supervisor was Diane Renda.
New drawings were done by Oxford Illustrators Limited.
Von Hoffman Press, Inc., was printer and binder.

The following illustrations were rendered by Deanne McKeown, Uppercase Graphics, © 1990: Figures 1-1; 8-1; 8-6; 8-7; 8-8; 8-10A, B; 8-11; 8-12; 8-14; 8-15; 8-18; 8-47; 9-3; 9-11; 9-12; 9-14; 9-18; 9-25; 9-32; 9-34; 9-35; 9-36; 9-37; 9-38; 9-39; 9-40; 9-41; 9-44; 9-45; 9-46; 10-5; 10-13; 11-1; 11-33; 11-34; 11-35; 11-36; 12-2; 12-4B; 12-6; 12-8; 12-9; 12-10; 12-11; 12-12; 12-13A, B; 12-15; 12-16; 13-8; 13-9; 13-13; 13-18; 13-19; 13-25; 13-31; 13-52; 13-54; 13-59; 13-66; 14-1; 14-3; 14-4; 14-13; 14-27; 15-1; 15-2; 15-3; 15-4; 16-1; 16-3; 16-13; 16-14; 16-15; 16-16; 16-19; 16-28; 16-32; 17-15; 18-4; 18-5; 18-11; 18-12; 18-14; 18-24; 18-25; 18-27; 18-29; 19-5; 20-7; 20-9; 20-14; and 20-15.

Library of Congress Cataloging-in-Publication Data

Vander, Arthur J., (date).
    Human physiology: the mechanisms of body function / Arthur J. Vander,
      James H. Sherman, Dorothy S. Luciano. —5th ed. p.   cm.
    Includes bibliographical references.
    ISBN 0-07-066969-4
    1. Human physiology.   I. Sherman, James H., (date).
  II. Luciano, Dorothy S.   III. Title.
QP34.5.V36   1990
612—dc20   89-12664

CONTENTS

To the student: You need read no further, for this preface is unashamedly aimed not at you but at your instructors.

To the instructor: Please do read further. Because a large number of competing books have appeared in the 20 years since our first edition, we feel obliged to dwell at some length upon our own goals and the profound changes made in this revision to enhance the achievement of these goals.

GOALS AND ORIENTATION

The purpose of this book remains what it was in the first four editions: to present the fundamental principles and facts of human physiology in a format that is suitable for undergraduate students, regardless of academic backgrounds or fields of study—liberal arts, biology, nursing, pharmacy, or other allied health professions. The book is also suitable for dental students, and many medical students have used previous editions to lay the foundation for the more detailed coverage they receive in their courses.

The most significant feature of this book is its clear, up-to-date, accurate explanations of mechanisms rather than the mere description of facts and events. Because there are limits to what can be covered in an introductory text, it is essential to reinforce over and over, through clear explanations, that physiology can be understood in terms of basic themes and principles. As evidenced by the very large number of flow diagrams employed, the book emphasizes understanding, based on the ability to think in clearly defined chains of causal links. Our goal is to integrate and synthesize information rather than simply to describe, so that students will achieve a working knowledge of physiology, not just a memory bank of physiological facts.

In this regard, we must acknowledge that even an introductory course in physiology must present a rather frightening number of facts, and our book is no exception. We have made every effort, however, not to mention facts simply because they happen to be known, but rather to use facts, whenever possible, as building blocks for general principles and concepts. Since our aim has been to tell a coherent story, rather than to write an encyclopedia, we have been willing to devote considerable space to the logical development of difficult but essential concepts, such as membrane potentials.

In keeping with our goals, the book progresses from the cell to the total body, utilizing at each level of increasing complexity the information and principles developed previously. Part One presents basic cellular and molecular biology, emphasizing that all phenomena of life are ultimately describable in terms of physical and chemical laws. Part Two analyzes the concept of the internal environment, the generalized components of the homeostatic control systems that regulate this environment, and the properties and organization of the major specialized cell types that comprise these systems. Part Three then describes the coordinated body functions, emphasizing how these functions result from the precise control and integration of specialized cellular activities and serve to maintain the internal environment.

One example of this approach is as follows: The characteristics that account for protein specificity are presented in Chapter 4, and this material is then used in Chapter 5 to explain the "recognition" process exhibited by enzymes, used again in Part Two for receptors, and finally used again in Part Three for antibodies. In this manner, the student is helped to see the basic foundations upon which more complex functions are built.

Another example: Rather than presenting, in a single chapter, a gland-by-gland description of all the hormones, we give a description of the basic principles of endocrinology in Chapter 10, but then save the details of the individual hormones for later chapters. For example, insulin is described in Chapter 17, "Regulation of Organic Metabolism, Growth, and Energy Balance," and aldosterone in Chapter 15, "Regulation of the Kidneys and Inorganic Ions." This permits the student to focus on the functions of the hormones in the context of the control systems in which they participate.

A MATTER OF BALANCE

The most challenging task for those who teach physiology to the groups for which this book is intended is to strike several difficult balances: (1) that between the enormously exciting new findings in cellular and molecular biology, on the one hand, and integrated whole-person physiology, on the other; and (2) that between breadth and depth of coverage in either of these two domains. In making the infinite number of decisions that go into striking these balances, we have been led by our goal of having the student understand, at a reasonable level, the mechanisms of the events described. Thus, for example, we cover systematically the various signal-transduction mechanisms used by plasma-membrane receptors because these mechanisms are so crucial to all cellular responses to chemical and electrical signals. Moreover, in this context, the overall function of G proteins is described, as is the specific role of G_s and G_i proteins in modulating adenylate cyclase. We have chosen, however, not to go into still greater detail and describe the interaction of these proteins with GTP, since these facts are not essential to an understanding of the adenylate cyclase system.

We also frequently use the technique of presenting in depth one example of a complex phenomenon and let this example serve to flesh out the general principle. Thus, we describe the molecular mechanisms by which norepinephrine increases myocardial contractility to illustrate a full sequence of events from receptor activation to cell response.

Another difficult decision, in a book aimed at students with diverse academic backgrounds, is how to deal with the basic biology, chemistry, and physics requisite for an understanding of the physiology. We have gathered much of the chemistry into two of the chapters of Part One but have presented other basic science where relevant in the text. Students with little or no background in the sciences will find this material essential, while others, more sophisticated in those areas, should profit from this review.

ALTERNATIVE SEQUENCES

Given the inevitable restrictions of time, our organization permits a variety of sequences and approaches to be adopted when using the book. Chapter 1 should definitely be read first since it introduces the basic themes that dominate the book. Depending on the time available, the instructor's goals, and the students' backgrounds in physical science and/or cellular biology, the chapters of Part One can either be worked through systematically at the outset or be used more selectively as background reading in the contexts of Parts Two and Three. To facilitate either approach, frequent references to specific pages of Part One are made in these later chapters.

In Part Two, the absolutely essential chapters are, in order, Chapters 7, 8, 10, and 11, for they present the basic concepts and facts relevant to homeostasis, intercellular communication, signal transduction, the nervous and endocrine systems, and muscle. The material in Chapters 9 and 12 are not as critical for an understanding of later chapters.

It would be best to begin the coordinated body functions of Part Three with the circulation, Chapter 13, but otherwise the chapters of Part Three, as well as Chapters 9 and 12 of Part Two, can be rearranged and used or not used to suit individual instructors' preferences.

REVISION HIGHLIGHTS

Our three major goals for this revision were: (1) to maintain and even improve upon the clarity of our explanations; (2) to update completely all material and assure the greatest accuracy possible; and (3) to provide a comprehensive set of study aids, including a new illustration program, that would enhance learning.

The first goal was achieved by an exhaustive rewriting, based largely on the criticisms and suggestions of our editor, Irene Nunes, who served as a naïve reader. Because we continued to rewrite until the material was clear to her, we are more convinced than ever that the book can serve as a primary source of information for the student.

Helped by the input from a survey of more than 400 teachers of physiology courses, we have also made significant changes in the organization of individual chapters and in the order of the chapters.

1. Chapters 2 and 3 have been transposed so that the molecular structures of carbohydrates, fats, and pro-

teins could be used in describing cell structures. The section on membrane structure that was formerly in Chapter 6 has been moved to Chapter 3.

2. "The Sensory Systems," formerly Chapter 18, is now Chapter 9, immediately following the chapter, "Neural Control Mechanisms." This juxtaposition will help the student gain greater understanding of the sensory components of the control systems described in Chapter 8. "Control of Body Movement," formerly Chapter 19, is now Chapter 12, directly following "Muscle" so as to provide greater understanding of the neural components of muscle control. The result of these two chapter shifts is to present most of the book's neurophysiology in Part Two rather than spreading it across Parts Two and Three. We have left "Consciousness and Behavior" as the last chapter in the book, however, so we could describe in it the interactions between "mind" and the coordinated body functions described in Part Three.

3. Hemostasis has been moved from the chapter on circulation to Chapter 19, to emphasize its role as a defense mechanism and its analogies to the immune system.

Updating was, as always, a huge task. Preparation of each new edition has required us to do more reading than for the previous one, as the amount of scientific information continues to expand explosively. A great number of physiologists reviewed the manuscript (see list in the acknowledgements) to help assure not only accuracy but balance. The number of revisions are far too numerous to list, but the following are examples of topics that have been either significantly expanded because of new information or added for the first time:

Categories and characteristics of plasma-membrane channels

Molecular mechanisms of ion transport

Feedforward in control systems

Biological rhythms and their contributions to homeostasis

New signal transduction mechanisms: G proteins, cyclic GMP, phospholipase C, PIP_2, IP_3, DAG, protein kinase C, and receptors as tyrosine kinases

Synaptic effectiveness

Coding in the visual system

Synthesis of peptide hormones (preprohormones and prohormones)

Candidate hormones

New hypotheses on mechanisms of summation in muscle contraction

Lengthening contractions in muscle

Hierarchical view of motor control systems

Regulation of blood cell production (colony-stimulating factors)

Functions of endothelial cells, including endothelium-derived relaxing factors

Molecular mechanisms of catecholamine effects on the myocardium

Angiogenic factors

Blood volume and long-term regulation of arterial pressure

Atrial natriuretic factor

Insulin receptors and signal transduction

Bone growth

Growth factors and growth-inhibiting factors, including insulin-like growth factor I

Control of food intake

Acclimatization to cold

Cytokines, including interleukin 1, tumor necrosis factor, interleukin 6, and gamma interferon

Sertoli cell functions

Antigen processing and presentation, including MHC restrictions

Anticlotting factors: the fibrinolytic system, protein C, antithrombin III, heparin, tissue plasminogen activator, anticlotting roles of thrombin and endothelial cells

Coverage of still other topics was expanded in response to suggestions both from the survey of physiology teachers mentioned above and from other reviewers. Examples include:

Summary tables describing the cranial nerves

Summary of arteriolar control in specific organs and tissues

Determinants of alveolar gas pressures

Graphical presentation of pulmonary circulation pressures

Lung compliance

Bicarbonate reabsorption by the kidneys

Creatinine clearance

Summary of all major liver functions

Meiosis

Chorionic villus sampling

Contraceptive methods

EEG patterns

Exercise physiology

Pathophysiology

A separate word is needed about these last two categories. We have chosen to continue our practice of presenting exercise physiology where relevant throughout the book, rather than in a separate chapter, because we feel that the integrated tack provides an excellent way of summarizing and applying much of the material used in the individual chapters. Pathophysiology, too, is important for illustrating basic physiology and demonstrating its relevance. Examples of such topics that have been either expanded or included for the first time include: cancer and oncogenes, cataracts and presbyopia, classification of endocrine disorders (primary and secondary hyposecretion, etc.), muscle diseases, anemia, therapy of hypertension and myocardial infarctions, chronic obstructive lung disease, asthma, ventilation-perfusion inequalities, diuretics, kidney diseases, peritoneal dialysis, hypoglycemia, obesity, osteoporosis, AIDS, late-phase allergic responses, psychoneuroimmunology.

With all these additions you might assume that the total text material has increased, but such is not the case. The careful editing described earlier and a ruthless removal of material not deemed essential has permitted us to maintain the text size essentially constant. Indeed, if one excludes the end-of-chapter study aids (to be described below) added to this edition, the actual text has been decreased. For example, the text of the following chapters has been shortened by 10 to 25 percent: Chapter 4, "Molecular Control Mechanisms: DNA and Protein"; Chapter 6, "Movement of Molecules across Cell Membranes"; Chapter 8, "Neural Control Mechanisms"; Chapter 16, "The Digestion and Absorption of Food"; Chapter 20, "Consciousness and Behavior."

STUDY AIDS

A variety of pedagogical aids are used in this edition, some, as noted, for the first time:

1. **Bold face key terms** appear throughout each chapter. chapter.
2. **The illustration program** has been totally redone for this edition. The total number of figures has remained at approximately 600, but many old ones have been dropped and many new ones added. Moreover, every figure retained in this edition has been redrawn, many with extensive revisions. For the first time, full color has been employed where appropriate (approximately half the figures), a change that greatly enhances these figures' educational value. The extensive use of flow diagrams, which we introduced in 1970 (and which has been so widely imitated in other texts) has been expanded. These diagrams have been

improved not only by the frequent use of full color but by the introduction of conventions that are used throughout the book and that should enhance learning. Look, for example, at Figure 19-18 on page 678. The beginning and ending boxes of the flow diagram are in light green (or in gray, in figures lacking full color), and the beginning is further clarified, when needed, by the use of a "Begin" logo. Blue three-dimensional boxes are used to denote events occurring inside organs and tissues (identified by boldface underlined labels in the upper right of the boxes), so that the reader can easily pick out the anatomic entities that participate in the sequences of events. Finally, in full-color figures, the participation of hormones in the sequences stand out by the placing of changes in their plasma concentrations in orange boxes.

3. **Summary tables.** We have doubled the number of reference and summary tables in this edition (a total of 120). Some summarize small amounts of information (for example, the summary of second messengers in Table 7-7, page 162), whereas others bring together larger amounts of information that may be scattered throughout the book (for example the summary of liver functions in Table 17-6, page 580). In several places, mini-glossaries are included as tables in the text (for example, the list of immune-system cells and chemical mediators in Table 19-13, page 684). Because the tables complement the figures, these two learning aids taken together provide a rapid means of reviewing the most important material in a chapter.

4. **End-of-chapter study aids.** These are all new for this edition:
   (a) Extensive **chapter summaries in outline form**.
   (b) Comprehensive **review questions in essay format**, including **key-word definition lists** of all boldface words in chapter. These review questions, in essence, constitute a complete list of learning objectives.
   (c) **Thought questions** that challenge the student to go beyond memorization of facts to solve problems, often presented as case histories or experiments. Complete **answers to thought questions** appear as Appendix A.

5. **Glossary.** For the first time, we have added **pronunciation guides** for many of the terms in this comprehensive glossary, appearing as Appendix D.

6. Other appendixes (B and C) present **English-metric interconversions** and **electrophysiology equations**.

7. **Suggested readings** are listed at the back of the book, as Appendix F.

SUPPLEMENTS

1. For this edition, a completely new **study guide** has been prepared by Dr. Sharon Russell (University of California at Berkeley). It contains a large variety of aids, including learning hints, many test questions with answers, and additional relevant contemporary material.
2. An **instructor's manual/test bank,** which contains, among other materials, a complete set of learning objectives, is available from the publisher. The test bank is also available on software for the IBM PC, Apple II, and Macintosh computers.
3. A set of 150 **overhead transparencies** representing the most important figures from the book is also available.

ACKNOWLEDGMENTS

First, we would like to thank the more than 400 instructors who participated in the survey that was taken prior to beginning this revision.

Secondly, we are very grateful to the many physiologists who read one or more chapters during various stages of this revision: Harold B. Falls, Southwest Missouri State University; Margaret M. Gould, Georgia State University; Bruce Grayson, University of Miami; John P. Harley, Eastern Kentucky University; Barbara J. Howell, SUNY–Buffalo; Fred J. Karsch, University of Michigan; Landis Keyes, University of Michigan; Steven Kunkel, University of Michigan; Esmail Meisami, University of Illinois, Sandy Milner-Brown, University of Colorado; Mary E. Murphy, Washington State University; Stephen R. Overmann, Southeast Missouri State University; Sharon Russell, University of California–Berkeley; Evelyn H. Schlenker, University of South Dakota, Medical School; Stanley A. Schwartz, The University of Michigan; George Simone, Eastern Michigan University; Gregory A. Stephens, University of Delaware; Jerry Stinner, University of Akron; Winton Tong, University of Pittsburgh; and George Veomett, University of Nebraska–Lincoln. Their advice was extremely useful in helping us to be accurate and balanced in our coverage. We hope they will be understanding of the few occasions when we did not heed their advice, and we are, of course, solely responsible for any errors that have crept in.

Many people participated in the development and production of this book, certainly the most gargantuan task we have all faced, not excepting preparation of the first edition. Our sponsoring editor, Denise Schanck, our production editor, Diane Renda, our editing supervisor James W. Bradley, and his assistant, Marie Gagliardi, and our designer, Jo Jones, all somehow managed to see this complex project through to completion with skill and efficiency. Deanne McKeown contributed the beautiful medical illustrations that make their appearance in this edition, and Oxford Illustrators completed the gigantic task of doing all the line art under the pressure of very tight deadlines. But one person, our basic-book editor, Irene M. Nunes, deserves the lion's share of credit for the creation of this edition. With great skill and wonderful good humor, she organized and supervised virtually every facet of the project, developed with us all the new features of this edition, and then served as a merciless reviewer of every sentence and figure in the manuscript. Irene, thank you.

Arthur J. Vander

James H. Sherman

Dorothy S. Luciano

A FRAMEWORK FOR HUMAN PHYSIOLOGY

One cannot meaningfully analyze the complex activities of the human body without a framework upon which to build, a set of viewpoints to guide one's thinking. It is the purpose of this chapter to provide such an orientation to the subject of **human physiology**—the mechanisms by which the body functions.

MECHANISM AND CAUSALITY

The **mechanist view** of life holds that all phenomena, no matter how complex, are ultimately describable in terms of physical and chemical laws and that no "vital force" distinct from matter and energy is required to explain life. The human being is a machine—an enormously complex machine, but a machine nevertheless. This view has predominated in the twentieth century because virtually all information gathered from observation and experiment has agreed with it. But **vitalism**, the view that some force beyond physics and chemistry is required by living organisms, is not completely dead, nor is it surprising that this viewpoint remains specifically in brain physiology, where scientists are almost entirely lacking in physicochemical hypotheses to explain such phenomena as thought and consciousness. Most physiologists believe that even this area will ultimately yield to physicochemical analysis, but it would be unscientific, on the basis of present knowledge, to dismiss the problem out of hand.

A common denominator of physiological processes is their contribution to survival. Unfortunately, it is easy to misunderstand the nature of this relationship. Consider, for example, the statement, "During exercise a person sweats because the body *needs* to get rid of the excess heat generated." This type of statement is an example of **teleology**, the explanation of events in terms of purpose, but it is not an explanation at all in the scientific sense of the word. It is somewhat like saying, "The furnace is on because the house needs to be heated." Clearly, the furnace is on not because it senses in some mystical manner the house's "needs" but because the temperature has fallen below the thermostat's set point and the electric current in the connecting wires has turned on the heater.

Is it not true that sweating serves a useful purpose because the excess heat, if not eliminated, might cause sickness or even death? Yes, it is, but this is totally different from stating that a need to avoid injury caused the sweating. The cause of the sweating was a sequence of events initiated by the increased heat generation: increased heat generation → increased blood temperature → increased activity of specific nerve cells in the brain → increased activity of a series of nerve cells → increased production of sweat by the sweat-gland cells. Each step occurs by means of physicochemical changes in the cells involved. In science, to explain a phenomenon is to reduce it to a causally linked sequence of physicochemical events. This is the scientific meaning of causality, of the word "because."

This is a good place to emphasize that causal chains can be not only long, as in the example just cited, but also multiple. In other words, one should not assume the simple relationship of one cause, one effect. We shall see that multiple factors often must interact to elicit a response. To take an example from medicine, cigarette smoking causes lung cancer, but the likelihood of the cancer developing in a smoker depends on a variety of other factors, including the way that person's body processes the chemicals in cigarette smoke, the rate at which damaged molecules are repaired, and so on.

That a phenomenon is beneficial to the person, while not explaining the mechanism of the phenomenon, is of obvious interest and importance. It is attributable to evolutionary processes that result in the selecting of those responses having survival value. Evolution is the key to understanding why most bodily activities do indeed appear to be purposeful. Throughout this book we emphasize how a particular process contributes to survival, but the reader must never confuse this survival value of a process with the explanation of the mechanisms by which the process occurs.

A SOCIETY OF CELLS

Cells: The Basic Units

The simplest structural units into which a complex multicellular organism can be divided and still retain the functions characteristic of life are called **cells**. One of the unifying generalizations of biology is that certain fundamental activities are common to almost all cells and represent the minimal requirements for maintaining cell integrity and life. Thus, a human liver cell and an amoeba are remarkably similar in their means of exchanging materials with their immediate environments, of obtaining energy from organic nutrients, of synthesizing complex molecules, and of duplicating themselves.

Each human organism begins as a single cell, the fertilized ovum, which divides to form two cells, each of which divides in turn, resulting in four cells, and so on. If cell multiplication were the only event occurring, the end result would be a spherical mass of identical cells. During development, however, each cell becomes specialized in the performance of a particular function, such as developing force and movement (muscle cells) or generating electric signals (nerve cells). The process of transforming an unspecialized cell into a specialized cell is known as **cell differentiation**. In addition to differentiating, cells migrate to new locations during development and form selective adhesions with other cells to form multicellular structures. In this manner, the cells of the body are arranged in various combinations to form a hi-

erarchy of organized structures. Differentiated cells with similar properties aggregate to form **tissues** (nerve tissue, muscle tissue, and so on), which combine with other types of tissues to form **organs** (the heart, lungs, kidneys, and so on), which are linked together to form **organ systems** (Figure 1-1).

About 200 distinct kinds of cells can be identified in the body in terms of differences in structure and function. When cells are classified according to the broad types of function they perform, however, four categories emerge: (1) muscle cells, (2) nerve cells, (3) epithelial cells, and (4) connective-tissue cells. In each of these functional categories, there are several cell types that perform variations of the specialized function. For example, there are three types of muscle cells—skeletal, cardiac, and smooth-muscle cells—all of which generate forces and produce movement but differ from each other in shape, in the mechanisms controlling their contractile activity, and in their location in the various organs of the body.

Muscle cells are specialized to generate the mechanical forces that produce force and movement. They may be attached to bones and produce movements of the limbs or trunk. They may be attached to skin, as for example, the muscles producing facial expressions, or they may enclose hollow cavities so that their contraction expels the contents of the cavity, as in the case of the pumping of the heart. Muscle cells also surround many of the tubes in the body—blood vessels, for example—and their contraction changes the diameter of these tubes.

Nerve cells are specialized to initiate and conduct electric signals, often over long distances. The signal may influence the initiation of new electric signals in other nerve cells, or it may influence secretion by a gland cell or contraction of a muscle cell. Thus, nerve cells provide a major means of controlling the activities of other cells. Their activity also underlies such phenomena as consciousness, perception, and cognition.

Epithelial cells are specialized for the selective secretion and absorption of ions and organic molecules. They are located mainly at the surfaces that either cover the body or individual organs or else line the walls of various tubular and hollow structures within the body. Epithelial cells, which rest on a homogenous noncellular material called the **basement membrane**, form the boundaries between compartments and function as selective barriers regulating the exchange of molecules across them. For example, the epithelial cells at the surface of the skin form a barrier that prevents most substances in the external environment from entering the body through the skin. Epithelial cells are also found in glands that form from the invagination of the epithelial surfaces.

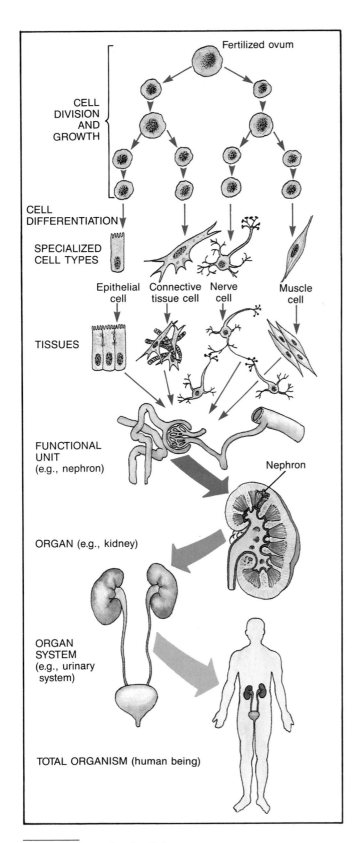

FIGURE 1-1 Levels of cellular organization.

Connective-tissue cells, as their name implies, have as their major function connecting, anchoring, and supporting the structures of the body. These cells typically have a large amount of extracellular material between them. The cells themselves include those in the loose meshwork of cells and fibers underlying most epithelial layers plus other types as diverse as fat-storing cells, bone cells, and red and white blood cells. Many connective-tissue cells secrete into the fluid surrounding them molecules that form a matrix consisting of various types of protein **fibers** embedded in a **ground substance** made of complex sugars, protein, and crystallized minerals. This matrix may vary in consistency from a semifluid gel, in loose connective tissue, to the solid crystalline structure of bone. These extracellular fibers include ropelike **collagen fibers**, which have a high tensile strength and resist stretching, rubber-band-like **elastin fibers**, and fine, highly branched **reticular fibers**.

Tissues

Most specialized cells are associated with other cells of a similar kind to form tissues. Corresponding to the four general categories of differentiated cells, there are four general classes of tissues: (1) **muscle tissue**, (2) **nerve tissue**, (3) **epithelial tissue**, and (4) **connective tissue**. It should be noted that the term "tissue" is frequently used in several ways. It is formally defined as just described, that is, an aggregate of a single type of specialized cell. However, it is also commonly used to denote the general cellular fabric of any organ or structure, for example, kidney tissue or lung tissue, each of which in fact usually contains all four classes of tissue.

Organs and Organ Systems

Organs are composed of the four kinds of tissues arranged in various proportions and patterns: sheets, tubes, layers, bundles, strips, and so on. For example, the kidneys consist largely of (1) a series of small tubes, each composed of a single layer of epithelial cells; (2) blood vessels, whose walls consist of an epithelial lining and varying quantities of smooth muscle and connective tissue; (3) nerve fibers with endings near the muscle and epithelial cells; and (4) a loose network of connective-tissue elements that are interspersed throughout the kidneys and also form enclosing capsules.

Many organs are organized into small, similar subunits often referred to as **functional units**, each performing the function of the organ. For example, the kidneys' 2 million functional units are termed nephrons, and the total production of urine by the kidneys is the sum of the amounts formed by the individual nephrons.

Finally we have the organ system, a collection of organs that together perform an overall function. For ex-

TABLE 1-1 ORGAN SYSTEMS OF THE BODY

System	Major Organs or Tissues	Primary Functions
Circulatory	Heart, blood vessels, blood (Some classifications also include lymphatic vessels and lymph in this system.)	Rapid flow of blood throughout the body's tissues
Respiratory	Nose, pharynx, larynx, trachea, bronchi, lungs	Exchange of carbon dioxide and oxygen; regulation of hydrogen-ion concentration
Digestive	Mouth, pharynx, esophagus, stomach, intestines, salivary glands, pancreas, liver, gallbladder	Digestion and absorption of organic nutrients, salts, and water
Urinary	Kidneys, ureters, bladder, urethra	Regulation of plasma composition through controlled excretion of organic wastes, salts, and water
Musculoskeletal	Cartilage, bone, ligaments, tendons, joints, skeletal muscle	Support, protection, and movement of the body
Immune	White blood cells, lymph vessels and nodes, spleen, thymus, and other lymphoid tissues	Defense against foreign invaders; return of extracellular fluid to blood; formation of white blood cells
Nervous	Brain, spinal cord, peripheral nerves and ganglia, special sense organs	Regulation and coordination of many activities in the body; detection of changes in the internal and external environments; states of consciousness; learning; cognition
Endocrine	All glands secreting hormones: Pancreas, testes, ovaries, hypothalamus, kidneys, pituitary, thyroid, parathyroid, adrenal, intestinal, thymus, and pineal	Regulation and coordination of many activities in the body
Reproductive	Male: Testes, penis, and associated ducts and glands	Production of sperm; transfer of sperm to female
	Female: Ovaries, uterine tubes, uterus, vagina, mammary glands	Production of eggs; provision of a nutritive environment for the developing embryo and fetus; nutrition of the infant
Integumentary	Skin	Protection against injury and dehydration; defense against foreign invaders; regulation of temperature

ample, the kidneys, the urinary bladder, the tubes leading from the kidneys to the bladder, and the tubes leading from the bladder to the exterior constitute the urinary system. There are ten organ systems in the human body. Their components and functions are given in Table 1-1.

To sum up, the human body can be viewed as a complex society of differentiated cells structurally and functionally combined and interrelated to carry on the functions essential to the survival of the organism. The individual cells constitute the basic units of this society, and almost all of these cells individually exhibit the fundamental activities common to all forms of life. Indeed,

many of the body's cells can be removed and maintained in test tubes as free-living organisms.

There is a paradox in this analysis. If each cell performs the fundamental activities required for its own survival, what contributions do the organ systems make? How is it that the functions of the organ systems are essential to the survival of the organism when each individual cell seems capable of performing its own fundamental activities? The resolution of this paradox is found in the isolation of most of the cells of a multicellular organism from the environment surrounding the organism—the **external environment**—and the presence of an internal environment.

THE INTERNAL ENVIRONMENT

An amoeba and a human liver cell both obtain their energy by breaking down certain organic nutrients. The chemical reactions involved in this intracellular process are remarkably similar in the two types of cells and involve the utilization of oxygen and the production of carbon dioxide. The amoeba picks up oxygen directly from the fluid surrounding it (its external environment) and eliminates carbon dioxide into the same fluid. But how can the liver cell and all other internal parts of the body obtain oxygen and eliminate carbon dioxide when, unlike the amoeba, they are not in direct contact with the external environment—the air surrounding the body?

Figure 1-2 summarizes the exchanges of matter that occur in a person. Supplying oxygen is the function both of the respiratory system, comprising the lungs and the airways leading to them, which takes up oxygen from the external environment, and of the circulatory system, which distributes oxygen to all parts of the body. In addition, the circulatory system carries the carbon dioxide generated by all the cells of the body to the lungs, which eliminate it to the exterior. Similarly, the digestive and circulatory systems working together make nutrients from the external environment available to all the body's cells. Wastes other than carbon dioxide are carried by the circulatory system from the cells that produced them to the kidneys and liver, which excrete them from the body. The kidneys also regulate the concentrations of water and many essential minerals in the plasma. The nervous and hormonal systems coordinate and control the activities of all the other organ systems.

Thus the overall effect of the activities of organ systems is to create within the body an environment in which all cells can survive and function. This environment surrounding each cell is called the **internal environment**. The internal environment is not merely a theoretical physiological concept. It can be identified quite specifically in anatomical terms: The body's internal environment is the **extracellular fluid** (literally, fluid outside the cells) that bathes each cell (Figure 1-2).

In other words, the internal environment in which each cell lives is not the external environment surrounding the entire body but the local extracellular fluid surrounding that cell. It is from this fluid that the cells receive oxygen and nutrients and into which they excrete wastes. A multicellular organism can survive only as long as it is able to maintain the composition of its internal environment in a state compatible with the survival of its individual cells. In 1857, the French physiologist Claude Bernard clearly described the central importance of the extracellular fluid: "It is the fixity of the internal environment that is the condition of free and independent life. . . . All the vital mechanisms, however varied they may be, have only one object, that of preserving constant the conditions of life in the internal environment." This concept of an internal environment and the necessity of maintaining its composition relatively constant is known as **homeostasis**. It is the single most important unifying idea to be kept in mind while attempting to understand the functions of the body's organ systems and their interrelationships.

Let us illustrate homeostasis with an example. When, for any reason, the concentration of oxygen in the blood significantly decreases below normal, the nervous system detects the change and increases its output to the

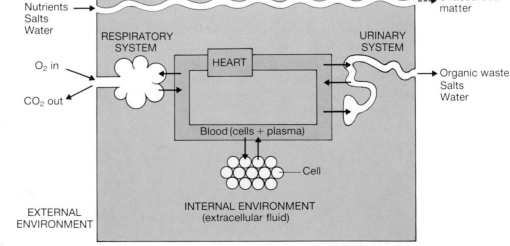

FIGURE 1-2 Exchanges of matter occur between the external environment and the circulatory system via the digestive, respiratory, and urinary systems. Extracellular fluid (plasma and interstitial fluid) is the internal environment of the body. The external environment is the air surrounding the body.

skeletal muscles responsible for breathing movements. The result is a compensatory increase in oxygen uptake by the body and a restoration of normal internal oxygen concentration.

To summarize, the total activities of every individual cell in the body fall into two categories: (1) Each cell performs for itself all those fundamental basic cellular processes—movement of materials across its membrane, extraction of energy, protein synthesis, and so on—that represent the minimal requirements for maintaining its individual integrity and life; and (2) each cell simultaneously performs one or more specialized activities that, in concert with the activities performed by the other cells of its tissue or organ system, contribute to the survival of the organism by helping maintain the stable internal environment required by all cells.

Plasma and Interstitial Fluid

Extracellular fluid exists in two locations—between cells and as the blood plasma. Approximately 80 percent of the fluid surrounds all the body's cells except the blood cells. Because it lies between cells, this 80 percent of the extracellular fluid is known as **intercellular fluid** or, more often, **interstitial fluid**. The remaining 20 percent of the extracellular fluid is the fluid portion of the blood, the **plasma**. As the blood (plasma plus suspended blood cells) is continuously circulated by the action of the heart to all parts of the body, the plasma exchanges oxygen, nutrients, wastes, and other metabolic products with the interstitial fluid surrounding the small blood vessels. Because of the exchanges between plasma and interstitial fluid, concentrations of dissolved substances are virtually identical in the two fluids, except for protein concentration.

With this major exception—higher protein concentration in plasma than in interstitial fluid—the entire extracellular fluid may be considered to have a homogeneous composition that is very different from that of the **intracellular fluid**—the fluid inside the cells.

With this introductory framework in mind, the overall organization and approach of this book should easily be understood. Because the fundamental features of cell function are shared by virtually all cells and because these features constitute the foundation upon which specialization develops, we devote the first part of the book to an analysis of basic cell physiology. Much of this cell biology is now referred to as **molecular biology** in recognition of the ultimate goal of explaining cellular processes in terms of interactions between molecules.

At the other end of the organizational spectrum, the final part of the book describes how coordinated functions (circulation, respiration, and so on) result from the precisely controlled and integrated activities of specialized cells grouped together in tissues and organs. The theme of these descriptions is that each function, with the obvious exception of reproduction, serves to keep some important aspect of the body's internal environment relatively constant.

The middle section of the book provides the principles and information required to bridge the gap between these two organizational levels, the cell and the body. Control systems are analyzed first in general terms to emphasize that the basic principles governing virtually all control systems are the same, but the bulk of Part 2 is concerned with the major components of the body's control systems (nerve cells, muscle cells, and gland cells). Once acquainted with the cast of major characters—nerve, muscle, and gland cells—and the theme—the maintenance of a stable internal environment—the reader will be free to follow the specific plot lines—circulation, respiration, and so on—of Part Three.

SUMMARY

Mechanism and Causality

  I. The mechanist view of life, the view taken by physiologists, holds that all phenomena are describable in terms of physical and chemical laws.

 II. Vitalism holds that some additional force is required to explain the function of living organisms.

A Society of Cells

  I. Cells are the simplest structural units into which a complex multicellular organism can be divided and still retain the functions characteristic of life.

 II. Cell differentiation results in the formation of four categories of specialized cells.
    A. Muscle cells generate the mechanical activities that produce force and movement.
    B. Nerve cells initiate and conduct electric signals.
    C. Epithelial cells selectively secrete and absorb ions and organic molecules.
    D. Connective-tissue cells connect, anchor, and support the structures of the body.

III. Specialized cells associate with similar cells to form tissues: muscle tissue, nerve tissue, epithelial tissue, and connective tissue.

IV. Organs are composed of the four kinds of tissues arranged in various proportions and patterns; many organs contain multiple small similar functional units.

 V. An organ system is a collection of organs that together perform an overall function.

The Internal Environment

  I. The extracellular fluid surrounding a cell is the cell's internal environment.

A. The function of organ systems is to maintain the internal environment relatively constant.

B. Each cell performs the basic cellular processes required to maintain its own integrity plus specialized activities that help maintain the internal environment relatively constant—homeostasis.

II. The extracellular fluid is composed of the blood plasma and the fluid between cells—the interstitial fluid.

A. Of the extracellular fluid, 80 percent is interstitial fluid and 20 percent is plasma.

B. Interstitial fluid and plasma have essentially the same composition except that plasma contains a much higher concentration of protein.

C. Extracellular fluid differs markedly in composition from the fluid inside cells—the intracellular fluid.

REVIEW QUESTIONS

1. Define:

human physiology	cells
mechanist view	cell differentiation
vitalism	tissues
teleology	organs

organ systems	epithelial tissue
muscle cells	connective tissue
nerve cells	functional units
epithelial cells	external environment
basement membrane	internal environment
connective-tissue cells	extracellular fluid
fibers	homeostasis
ground substance	intercellular fluid
collagen fibers	interstitial fluid
elastin fibers	plasma
reticular fibers	intracellular fluid
muscle tissue	molecular biology
nerve tissue	

2. Describe the levels of cellular differentiation and state the four types of specialized cells and tissues.

3. List the 10 organ systems of the body and give one-sentence descriptions of their functions.

4. Contrast the two categories of functions performed by every cell.

5. What two fluids constitute the extracellular fluid, what are their relative proportions, and in what way do they differ from each other in composition?

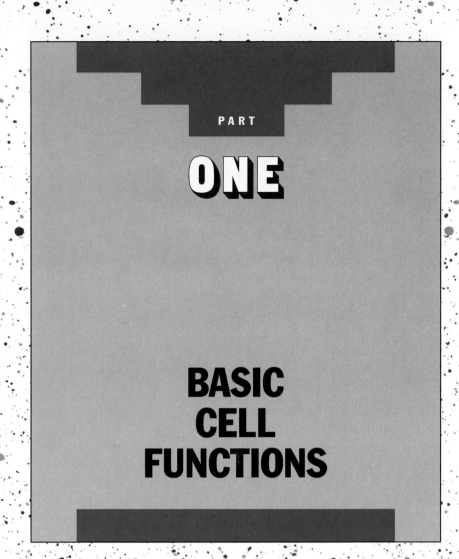

PART

ONE

BASIC CELL FUNCTIONS

CHEMICAL COMPOSITION OF THE BODY

Atoms and molecules are the chemical units of cell structure and function. In this chapter, we describe the distinguishing characteristics of the major chemicals in the human body. The specific roles of these substances will be discussed in subsequent chapters. This chapter is, in essence, an expanded glossary of chemical terms and structures, and like a glossary, it should be consulted according to need.

ATOMS

The units of matter that form all chemical substances are called **atoms.** The smallest atom, a hydrogen atom, is approximately 2.7 billionths of an inch in diameter. Each type of atom—carbon, hydrogen, oxygen, and so on—is called a **chemical element**. A one- or two-letter symbol is used as a shorthand identification of each chemical element (Table 2-1). Currently, 108 chemical elements (types of atoms) are known.

The chemical properties of atoms can be described in terms of three subatomic particles—**protons, neutrons, and electrons** (Table 2-2). The protons and neutrons are confined to a very small volume at the center of an atom, the **atomic nucleus**, whereas electrons revolve in orbits at various distances from the nucleus. This miniature-solar-system model of an atom is an oversimplification, but it is sufficient to provide a conceptual framework for understanding the chemical and physical interactions of atoms.

TABLE 2-1 ESSENTIAL ELEMENTS IN THE BODY

Element	Symbol
Major Elements: 99.3% Total Atoms	
Hydrogen	H (63%)
Oxygen	O (26%)
Carbon	C (9%)
Nitrogen	N (1%)
Mineral Elements: 0.7% Total Atoms	
Calcium	Ca
Phosphorus	P
Potassium	K (Latin *kalium*)
Sulfur	S
Sodium	Na (Latin *natrium*)
Chlorine	Cl
Magnesium	Mg
Trace Elements: Less than 0.01% Total Atoms	
Iron	Fe (Latin *ferrum*)
Iodine	I
Copper	Cu (Latin *cuprum*)
Zinc	Zn
Manganese	Mn
Cobalt	Co
Chromium	Cr
Selenium	Se
Molybdenum	Mo
Fluorine	F
Tin	Sn (Latin *stannum*)
Silicon	Si
Vanadium	V

Each of these subatomic particles has a different electric charge: protons have one unit of positive charge, electrons have one unit of negative charge, and neutrons are electrically neutral, that is, have no electric charge.

TABLE 2-2 CHARACTERISTICS OF MAJOR SUBATOMIC PARTICLES

Particle	Mass Relative to Electron Mass	Electric Charge	Location in Atom
Electron	1	−1	Orbiting the nucleus
Proton	1836	+1	Nucleus
Neutron	1839	0	Nucleus

Since the protons are located in the atomic nucleus, the nucleus has a net positive charge equal to the number of protons it contains. The entire atom has no net electric charge, however, because the number of negatively charged electrons orbiting the nucleus is equal to the number of positively charged protons in the nucleus.

Protons and neutrons are approximately equal in mass and are about 1800 times heavier than an electron (Table 2-2). Thus, almost all the mass of an atom is located in its nucleus.

Atomic Number

Every atom of each chemical element contains a specific number of protons, and this distinguishes one type of atom from another. This number is known as the **atomic number**. For example, hydrogen, the simplest atom, has an atomic number of 1, corresponding to its single proton, and calcium has an atomic number of 20, corresponding to its 20 protons. Since an atom is electrically neutral, the atomic number is also equal to the number of electrons in an atom of a given element.

Atomic Mass

Atoms have very little mass. A single hydrogen atom, for example, has a mass of only 1.67×10^{-24} g. The atomic mass scale indicates an atom's mass relative to the mass of other types of atoms. This scale is based upon assigning the carbon atom a value of 12. On this scale a hydrogen atom has an atomic mass of approximately 1, indicating that it has one-twelfth the mass of a carbon atom. A magnesium atom, with an atomic mass of 24, has twice the mass of a carbon atom. Since almost all the mass of an atom is due to its neutrons and protons, the atomic mass of each atom is approximately equal to the sum of the number of protons and neutrons it contains. The atomic mass of atoms is not exactly equal to this sum because electrons do have some mass and because protons and neutrons do not have exactly equal masses.

One **gram atomic mass** of a chemical element is the amount of the element in grams that is equal to the numerical value of its atomic mass. Thus, 12 g of carbon is 1 gram atomic mass of carbon, and 1 gram of hydrogen is 1 gram atomic mass of hydrogen. One gram atomic mass of any element contains the same number of atoms. Thus 1 g of hydrogen contains 6×10^{23} atoms, and 12 g of carbon, whose atoms have 12 times the mass of a hydrogen atom, also has 6×10^{23} atoms.

Atomic Composition of the Body

Of the 108 chemical elements, only 24 are known to be essential for the structure and function of the human body (Table 2-1). In fact, just 4 of these elements—

hydrogen, oxygen, carbon, and nitrogen—account for over 99 percent of the body's atoms.

The seven mineral elements listed in Table 2-1 are the most abundant substances dissolved in the extracellular and intracellular fluids. Most of the body's calcium and phosphorus atoms make up the solid matrix of bone tissue.

The 13 **trace elements** in Table 2-1 are present in extremely small quantities, but they are nonetheless essential. For example, iron plays a critical role in the transport of oxygen by the blood. Additional trace elements will likely be added to this list as the chemistry of the body becomes better understood.

Many other elements can be detected in the body in addition to the 24 listed in Table 2-1. These elements enter through the foods we eat and the air we breathe but do not have any known essential chemical function. Some elements, for example, mercury, not only are not required for normal function but can be toxic.

MOLECULES

Two or more atoms bonded together make up a **molecule**. For example, a molecule of water, which contains 2 hydrogen atoms and 1 oxygen atom, can be represented by H_2O, and the atomic composition of glucose, a sugar, is $C_6H_{12}O_6$. The formula for glucose indicates that the molecule contains 6 carbon atoms, 12 hydrogen atoms, and 6 oxygen atoms. Such formulas, however, do not indicate which atoms are linked to which in the molecule.

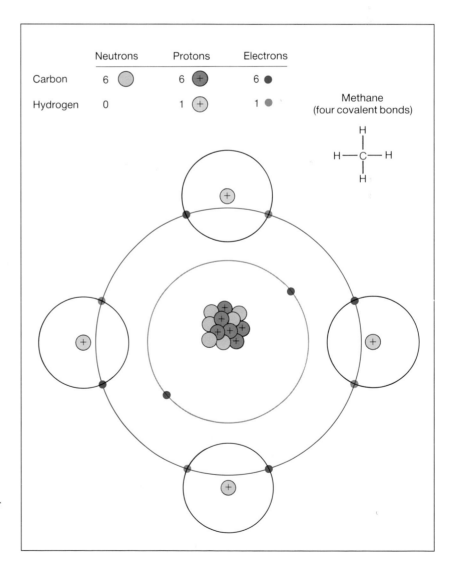

FIGURE 2-1 Each of the four hydrogen atoms in a molecule of methane (CH_4) forms a covalent bond with an atom of carbon by sharing its one electron with one of the electrons in the carbon atom. Each shared pair of electrons forms a covalent bond.

Covalent Chemical Bonds

The atoms in molecules are held together by chemical bonds, which are formed when electrons are transferred from one atom to another or are shared between two atoms. The strongest chemical bond between two atoms, a **covalent bond**, is formed when one electron in the outer electron orbit of each atom is shared between the two atoms (Figure 2-1). The atoms in most molecules found in the body are linked by covalent bonds.

The atoms of some elements can form more than one covalent bond and thus become linked simultaneously to two or more other atoms. Each type of atom has a characteristic number of covalent bonds it can form that depends on the number of electrons present in its outermost electron orbit. The number of chemical bonds formed by the four most abundant atoms in the body are: hydrogen, 1; oxygen, 2; nitrogen, 3; and carbon, 4. When the structure of a molecule is diagrammed, each covalent bond is represented by a line indicating a pair of shared electrons. The covalent bonds of the four elements mentioned above can be represented as:

$$H— \qquad —O— \qquad —\overset{\textstyle |}{N}— \qquad —\overset{\textstyle |}{\underset{\textstyle |}{C}}—$$

A molecule of water can be illustrated as:

$$H—O—H$$

In some cases, two covalent bonds—a double bond—are formed between two atoms when two electrons from each atom are shared. An example of this is carbon dioxide (CO_2):

$$O=C=O$$

Note that in this molecule the carbon atom still forms four covalent bonds and each oxygen atom only two.

Molecular Shape

When atoms are linked together, molecules with various shapes can be formed. Although we draw diagrammatic structures of molecules on flat sheets of paper, these molecules actually have three-dimensional shapes. When more than one covalent bond is formed with a given atom, the bonds are distributed about the atom in a pattern that may or may not be symmetrical (Figure 2-2).

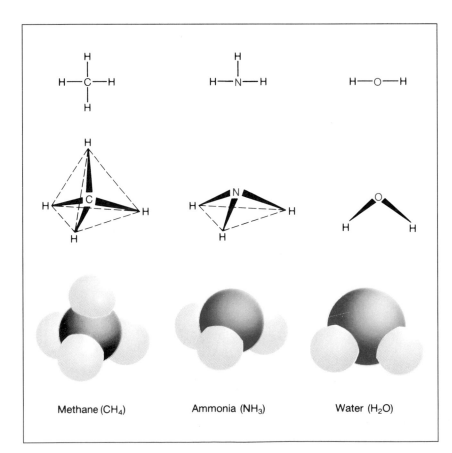

Methane (CH₄)   Ammonia (NH₃)   Water (H₂O)

FIGURE 2-2 Geometric configuration of covalent bonds around the carbon, nitrogen, and oxygen atoms bonded to hydrogen atoms.

the joined atoms can rotate. As illustrated in Figure 2-3, a sequence of six carbon atoms can assume a number of shapes as a result of rotations around various covalent bonds. As we shall see, the three-dimensional shape of molecules is one of the major factors governing molecular interactions.

IONS

An atom is electrically neutral since it contains equal numbers of negative electrons and positive protons. If, however, an atom gains or loses one or more electrons, it acquires a net electric charge and becomes an **ion**. For example, when a sodium atom, which has 11 electrons, loses 1 electron, it becomes a sodium ion (Na^+) with a net positive charge, that is, it now has only 10 electrons but 11 protons. On the other hand, a chlorine atom, which has 17 electrons, can gain an electron and become a chloride ion (Cl^-) with a net negative charge—it now has 18 electrons but only 17 protons. Some atoms can gain or lose more than 1 electron to become ions with two or even three units of electric charge, for example, calcium, Ca^{+2}.

Because of their ability to conduct electricity when dissolved in water, ions are collectively referred to as **electrolytes**. Hydrogen atoms and most of the mineral and trace element atoms readily form ions, whereas carbon, oxygen and nitrogen atoms do not. The number of electrons an atom may gain or lose in becoming an ion is a specific characteristic of each type of atom. Table 2-3 lists the ionic forms of some of these elements. Ions that have a net positive charge are called **cations**. Negatively charged ions are called **anions**.

The process of ion formation, known as ionization, can occur in single atoms or in molecules. Two commonly encountered groups of atoms within molecules that undergo ionization are the **carboxyl group** (—COOH) and the **amino group** (—NH_2). The formulas for molecules containing such groups are written R—COOH or R—NH_2, where R signifies the remaining portion of the molecule. The carboxyl group ionizes when the oxygen atom linked to the hydrogen atom captures the hydrogen atom's only electron to form a carboxyl ion (R—COO^-) and a hydrogen ion (H^+):

$$R—COOH \rightleftharpoons R—COO^- + H^+$$

The amino group can bind a hydrogen ion to form an ionized amino group (R—NH_3^+):

$$R—NH_2 + H^+ \rightleftharpoons R—NH_3^+$$

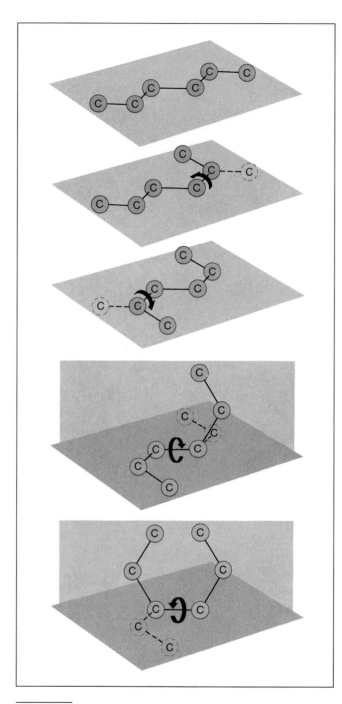

FIGURE 2-3 Changes in molecular shape occur as portions of a molecule rotate around different carbon-to-carbon bonds, transforming the molecule's shape from a relatively straight chain into a ring.

Molecules are not rigid, inflexible structures. Within certain limits, the shape of a molecule can be changed without breaking the covalent bonds linking its atoms together. A covalent bond is like an axle, around which

TABLE 2-3 MOST FREQUENTLY ENCOUNTERED IONIC FORMS OF ELEMENTS				
Atom	Chemical Symbol	Ion	Chemical Symbol	Electrons Gained or Lost
Hydrogen	H	Hydrogen ion	H^+	1 lost
Sodium	Na	Sodium ion	Na^+	1 lost
Potassium	K	Potassium ion	K^+	1 lost
Chlorine	Cl	Chloride ion	Cl^-	1 gained
Magnesium	Mg	Magnesium ion	Mg^{2+}	2 lost
Calcium	Ca	Calcium ion	Ca^{2+}	2 lost

The ionization of each of the above groups can be reversed, as indicated by the double arrows. The ionized carboxyl group can combine with a hydrogen ion to form an unionized carboxyl group, and the ionized amino group can loose a hydrogen ion and become an unionized amino group.

POLAR MOLECULES

As we have seen, when the electrons of two atoms interact, the two atoms may share the electrons equally, forming an electrically neutral covalent bond, or one of the atoms may completely capture an electron from the other atom, forming ions. Between these two extremes there exist covalent bonds in which the electrons are not shared equally between the two atoms but instead reside closer to one atom of the pair. This atom thus acquires a slight negative charge, while the other atom, having partly lost an electron, becomes slightly positive. Such bonds are known as **polar covalent bonds** since the atoms at each end of the bond have a different electric charge. Polar bonds do not have a net electric charge, as do ions, since they contain equal amounts of negative and positive charge. For example, the bond between hydrogen and oxygen in a **hydroxyl group** (R—OH) is a polar covalent bond in which the oxygen is slightly negative and the hydrogen slightly positive:

$$\overset{(-)\,(+)}{R—O—H}$$

The electric charge associated with the ends of a polar bond is considerably less than the charge of a fully ionized atom. For example, the oxygen in the polarized hydroxyl group has only about 13 percent of the negative charge associated with the oxygen in an ionized carboxyl group, R—COO⁻.

Atoms of oxygen and nitrogen, which have a strong attraction for electrons, form polar bonds with hydrogen atoms, whereas the bond between carbon and hydrogen atoms and that between two carbon atoms are electrically neutral (Table 2-4). A single molecule may contain nonpolar, polarized, and ionized bonds in different regions of the molecule. Molecules containing significant numbers of polar bonds or ionized groups are known as **polar molecules**, whereas molecules composed predominantly of electrically neutral bonds are known as **nonpolar molecules**. As we shall see, the physical characteristics of these two classes of molecules, especially their solubility in water, are quite different.

Hydrogen Bonds

The electrical attraction between the hydrogen atom in one polarized bond and the oxygen or nitrogen atom in a polarized bond of another molecule—or within the same molecule if the bonds are sufficiently separated from each other—forms a **hydrogen bond**. This type of bond is much weaker than a covalent bond; it has only about 4 percent of the strength of the covalent bonds linking the hydrogen and oxygen within a water molecule (H_2O). Hydrogen bonds between and within molecules play an important role in molecular interactions and in determining the shape of large molecules.

Water

Hydrogen is the most numerous atom in the body, and water is the most numerous molecule. Out of every 100 molecules, 99 are water. The covalent bonds linking the two hydrogen atoms to the oxygen atom in a water molecule are polar. Therefore, the oxygen in water has a slight negative charge, and each hydrogen atom has a slightly positive charge. The positively polarized regions

TABLE 2-4	EXAMPLES OF NONPOLAR AND POLAR BONDS, AND IONIZED CHEMICAL GROUPS	
Nonpolar bonds	$-\overset{\mid}{\underset{\mid}{C}}-H$	Carbon-hydrogen bond
	$-\overset{\mid}{\underset{\mid}{C}}-\overset{\mid}{\underset{\mid}{C}}-$	Carbon-carbon bond
Polar bonds	$R\overset{(-)}{-}O\overset{(+)}{-}\!\!\!-H$	Hydroxyl group (R—OH)
	$R\overset{(-)}{-}S\overset{(+)}{-}\!\!\!-H$	Sulfhydryl group (R—SH)
	$R-\overset{\overset{\textstyle H\ (+)}{\mid}}{\underset{}{N}}-R \quad (-)$	Nitrogen-hydrogen bond
Ionized groups	$R-\overset{\overset{\textstyle O}{\parallel}}{C}-O^{\ominus}$	Carboxyl group (R—COO$^-$)
	$R-\overset{\overset{\textstyle H}{\mid}}{\underset{\underset{\textstyle H}{\mid}}{\overset{\oplus}{N}}}-H$	Amino group (R—NH$_3^+$)
	$R-O-\overset{\overset{\textstyle O}{\parallel}}{\underset{\underset{\textstyle O^{\ominus}}{\mid}}{P}}-O^{\ominus}$	Phosphate group (R—PO$_4^{2-}$)

near the hydrogen atoms of one water molecule are electrically attracted to the negatively polarized regions of the oxygen atoms in adjacent water molecules forming hydrogen bonds (Figure 2-4).

At body temperature, the hydrogen bonds between water molecules are continuously being formed and broken, accounting for the liquid state of water. If the temperature is lowered, these hydrogen bonds are less frequently broken, and larger and larger clusters of water molecules are formed until at 0°C water freezes into a continuous crystalline matrix—ice.

Water molecules take part in many chemical reactions of the general type:

$$R_1-R_2 + H-O-H \rightleftharpoons R_1-OH + H-R_2$$

In this reaction the covalent bond between R_1 and R_2 and the covalent bond between a hydrogen atom and oxygen in water is broken and the hydroxyl group and hydrogen atom are transferred to R_1 and R_2, respectively. Reactions of this type are known as hydrolytic reactions or simply **hydrolysis**. Many large molecules in the body are broken down into smaller molecular units by hydrolysis.

SOLUTIONS

Substances dissolved in a liquid are known as **solutes**, and the liquid in which they are dissolved is the **solvent**. Solutes dissolved in a solvent form a **solution**. Water is the most abundant solvent in the body, accounting for 60 percent of the total-body weight. A majority of the chemical reactions that occur in the body involve molecules that are dissolved in water—either in the intracellular or extracellular fluids. However, not all molecules dissolve in water.

Molecular Solubility

In order to dissolve in water, a substance must be electrically attracted to water molecules. For example, table salt (NaCl) is a solid crystalline substance because of the strong electrical attraction between positive sodium ions and negative chloride ions. This strong attraction be-

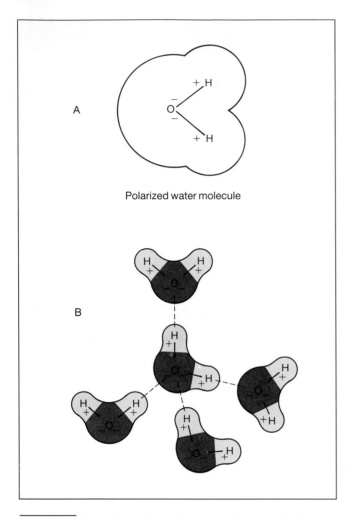

FIGURE 2-4 (A) Polarized covalent bonds link the hydrogen and oxygen atoms in a water molecule. (B) Hydrogen bonds between adjacent water molecules. Hydrogen bonds are represented in diagrams by dashed or dotted lines, and covalent bonds by solid lines.

tween two oppositely charged ions is known as an **ionic bond**. When a crystal of sodium chloride is placed in water, the polar water molecules are attracted to the charged sodium and chloride ions (Figure 2-5). The ions become surrounded by clusters of water molecules, allowing the sodium and chloride ions to separate from the salt crystal and enter the water, that is, to dissolve.

Molecules having a sufficient number of polar bonds and/or ionized groups will dissolve in water. Such molecules are said to be **hydrophilic**—water loving. Thus, in a molecule, the presence of ionized carboxyl and amino groups or of polar hydroxyl groups promotes solubility in water. In contrast, molecules composed predominantly of carbon and hydrogen are insoluble in water since their electrically neutral covalent bonds are not attracted to water molecules. These molecules are **hydrophobic**—water fearing.

When nonpolar molecules are mixed with water, two phases form, as occurs when oil is mixed with water. The strong attraction between polar water molecules "squeezes" the nonpolar molecules out of the water phase. Such a separation is never 100 percent complete, however, and very small amounts of nonpolar solutes remain dissolved in water. Although nonpolar molecules do not dissolve to any great extent in water, they will dissolve in nonpolar solvents, such as carbon tetrachloride.

Molecules that have a polar or ionized region at one end and a nonpolar region at the opposite end are called **amphipathic**—consisting of two parts. When mixed with water, amphipathic molecules form clusters, such that their polar (hydrophilic) regions are located at the surface of the cluster, where they are attracted to the sur-

FIGURE 2-5 The ability of water to dissolve sodium chloride crystals depends upon the electrical attraction between the polar water molecules and the charged sodium and chloride ions.

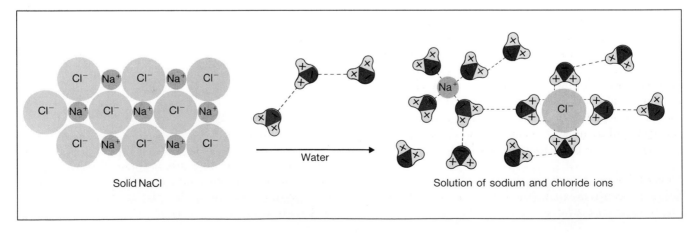

Solid NaCl

Water

Solution of sodium and chloride ions

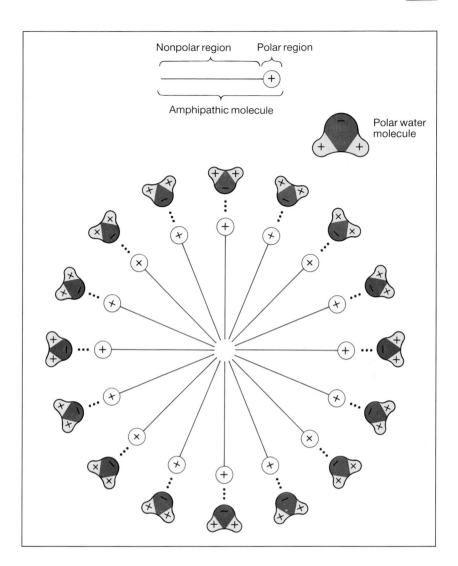

FIGURE 2-6 In water, amphipathic molecules aggregate into spherical clusters. Their polar regions form hydrogen bonds with water molecules at the surface of the cluster.

rounding water molecules, and the nonpolar (hydrophobic) ends are oriented toward the interior of the cluster (Figure 2-6). Such an arrangement provides the maximal interaction between the water molecules and the polar end of the amphipathic molecules. Nonpolar molecules can dissolve in the central nonpolar regions of these clusters and thus be carried in aqueous solutions in far higher amounts than would otherwise be possible based on their low solubility in water. As we shall see, the orientation of amphipathic molecules plays an important role in the structure of cell membranes and in both the absorption of nonpolar molecules from the gastrointestinal tract and their transport in the blood.

Concentration

Solute concentration is defined as the amount of the solute present in a unit volume of solution. The unit of volume in the metric system is a liter (L). One liter equals 1.06 quarts, or approximately three 12-ounce cans of soda pop. Smaller units are the milliliter (mL, or 0.001 liter) and the microliter (μL, or 0.001 mL) (see Appendix B). One way of expressing the amount of a substance is to give its mass in grams. Its concentration can then be expressed as the number of grams of the substance present in a liter of solution (g/L).

A comparison of the concentrations of two different substances on the basis of number of grams per liter of solution does not directly indicate how many molecules of each compound are present. For example, 10 g of compound X, whose molecules are heavier than those of compound Y, will contain fewer molecules than 10 g of compound Y. Concentrations in units of grams per liter are most often used when the chemical structure of the solute is unknown. When the structure of a molecule is known, concentrations are expressed as moles per liter, which provides a unit of concentration based upon the number of molecules of the solute in solution.

The **molecular mass** (or molecular weight) of a molecule is equal to the sum of the atomic masses of all the atoms in the molecule. For example, glucose ($C_6H_{12}O_6$) has a molecular mass of 180 [$(6 \times 12) + (12 \times 1) + (6 \times 16) = 180$]. One **mole** (or mol) of a substance is the amount of the substance in grams equal to its molecular mass. Accordingly, a solution containing 180 g of glucose in 1 L of solution is said to be a one-molar (abbreviated 1 M or 1 mol/L) solution of glucose. If 90 g of glucose were dissolved in enough water to produce 1 L of solution, the solution would have a concentration of 0.5 moles per liter (0.5 M). Just as 1 gram atomic mass of any element contains the same number of atoms, 1 gram molecular mass (1 mole) of any molecule will have the same number of molecules − 6×10^{23}. Thus, a 1 M solution of glucose contains the same number of solute molecules per liter as a 1 M solution of urea or any other substance.

The concentrations of solutes dissolved in the body fluids are much less than 1 M. Many have concentrations in the range of millimoles per liter (1 mM = 0.001 M), while others are present at even smaller concentrations— micromoles per liter (1 μM = 0.000001 M) or nanomoles per liter (1 nM = 0.000000001 M).

Hydrogen Ions and Acidity

As mentioned earlier, the hydrogen atom consists of a single proton orbited by a single electron. The hydrogen ion (H^+), formed by the loss of the electron, is thus a single free proton. Hydrogen ions are formed when the proton of a hydrogen atom in a molecule is released from the molecule, leaving behind its electron. Molecules that give rise to hydrogen ions in solution are called **acids**. For example,

$$HCl \longrightarrow H^+ + Cl^-$$

Hydrochloric acid   Chloride

$$H_2CO_3 \rightleftharpoons H^+ + HCO_3^-$$

Carbonic acid   Bicarbonate

$$
\begin{array}{c}
OH \\
|\\
CH_3-C-COOH \\
|\\
H
\end{array}
\rightleftharpoons H^+ +
\begin{array}{c}
OH \\
|\\
CH_3-C-COO^- \\
|\\
H
\end{array}
$$

Lactic acid   Lactate

Conversely, any substance that can accept a hydrogen ion is termed a **base**. In the reactions above, bicarbonate and lactate are bases since they can combine with hydrogen ions (note the double arrows in these two reactions above). These bases will remove hydrogen ions from solution, lowering the hydrogen-ion concentration. It must

be understood that the hydrogen-ion concentration refers only to the hydrogen ions that are free in solution and not to those that may be bound in lactic acid or to amino groups to form $R-NH_3^+$.

When hydrochloric acid is dissolved in water, 100 percent of its atoms separate to form hydrogen and chloride ions, which do not recombine in solution (note the one-way arrow above). In the case of lactic acid, however, only a fraction of the total number of lactic acid molecules in solution release hydrogen ions at any instant. Therefore, if a 1 M solution of hydrochloric acid is compared with a 1 M solution of lactic acid, the hydrogen-ion concentration will be lower in the lactic acid solution than in the hydrochloric acid solution. Hydrochloric acid and other acids that are 100 percent ionized in solution are known as **strong acids,** while carbonic and lactic acids, which do not completely ionize in solution, are **weak acids.** The same principles apply to the definition of **strong** and **weak bases.** Strong bases are 100 percent ionized in solution while weak bases are only partially ionized.

The acidity of a solution refers to the *free* (unbound), hydrogen-ion concentration in the solution. The higher the hydrogen-ion concentration, the greater the acidity. The hydrogen-ion concentration is frequently expressed in terms of the **pH** of a solution, which is defined as the negative logarithm to the base 10 of the hydrogen-ion concentration (the brackets around the symbol for the hydrogen ion below indicate concentration):

$$pH = -\log [H^+]$$

Thus, a solution with a hydrogen-ion concentration of 10^{-7} M has a pH of 7, whereas a more acidic solution with a concentration of 10^{-6} M has a pH of 6. Note that as the acidity *increases*, the pH *decreases* and that a change in pH from 7 to 6 represents a 10-fold increase in the hydrogen-ion concentration.

Pure water has a hydrogen-ion concentration of 10^{-7} M (pH = 7.0) and is a **neutral solution. Alkaline solutions** have a lower hydrogen-ion concentration (a higher pH than 7.0), while those with a higher hydrogen-ion concentration (a lower pH than 7.0) are **acidic solutions.** The body's extracellular fluid has a hydrogen-ion concentration of about 4×10^{-8} M (pH = 7.4), with a normal range of about 7.35 to 7.45, and is thus slightly alkaline. Intracellular fluids have a slightly higher hydrogen-ion concentration than extracellular fluid.

As we saw earlier, the ionization of carboxyl and amino groups in various molecules in the body involves the release and uptake, respectively, of hydrogen ions. These groups behave as weak acids and bases. Increasing the free hydrogen-ion concentration decreases the number of carboxyl groups that are ionized since it increases the probability that an ionized carboxyl group will bump into a hydrogen ion and combine with it:

$$R—COO^- + H^+ \rightleftharpoons R—COOH$$

Lowering the hydrogen-ion concentration has the opposite effect. Thus changes in the acidity of solutions containing molecules with carboxyl and amino groups will alter the net electric charge on these molecules by shifting the ionization reaction to the right or left as illustrated for the carboxyl group above. If the electric charge on a molecule is altered, its interaction with other molecules or with other regions within the same molecule is altered, and thus its functional characteristics are altered. In the extracellular fluid, hydrogen-ion concentrations beyond the 10-fold pH range of 7.8 to 6.8 are incompatible with human life if maintained for more than a brief period of time. Even small changes in the hydrogen-ion concentration can produce large changes in the activities of cells, as we shall see.

CLASSES OF ORGANIC MOLECULES

Because most naturally occurring carbon-containing molecules are found in living organisms, the study of these compounds became known as organic chemistry. (Inorganic chemistry is the study of non-carbon-containing molecules.) However, the chemistry of living organisms, biochemistry, now forms only a portion of the broad field of organic chemistry.

One of the properties of the carbon atom that makes life possible is its ability to form four covalent bonds with other atoms, in particular with other carbon atoms. Since carbon atoms can also combine with hydrogen, oxygen, nitrogen, and sulfur atoms, a vast number of compounds can be formed with relatively few chemical elements. Some of these molecules are extremely large (**macromolecules**), being composed of thousands of atoms. Such large molecules are formed by linking together hundreds of smaller molecules (subunits) and are thus known as **polymers** (many small parts). The structure of macromolecules depends upon the structure of the subunits, the number of subunits linked together, and the position along the chain of each type of subunit.

Most of the different types of organic molecules in the body can be classified into one of the following four groups: carbohydrates, lipids, proteins, and nucleic acids (Table 2-5).

Carbohydrates

Although carbohydrates account for only 3 percent of the body's organic matter, they play a central role in the chemical reactions that provide cells with energy. Carbohydrates are composed of mainly carbon, hydrogen, and oxygen atoms in the proportions represented by the

Category	Percent of Body Weight	Majority of Atoms	Subclass	Subunits
TABLE 2-5 MAJOR CATEGORIES OF ORGANIC MOLECULES IN THE BODY				
Carbohydrates	1	C, H, O	Monosaccharides (sugars)	
			Polysaccharides	Monosaccharides
Lipids	15	C, H	Triacylglycerols	3 fatty acids + glycerol
			Phospholipids	2 fatty acids + glycerol + phosphate + small charged nitrogen molecule
			Steroids	
Proteins	17	C, H, O, N		Amino acids
Nucleic acids	2	C, H, O, N	DNA	Nucleotides containing the bases adenine, cytosine, guanine, thymine and the sugar deoxyribose
			RNA	Nucleotides containing the bases adenine, cytosine, guanine, uracil and the sugar ribose

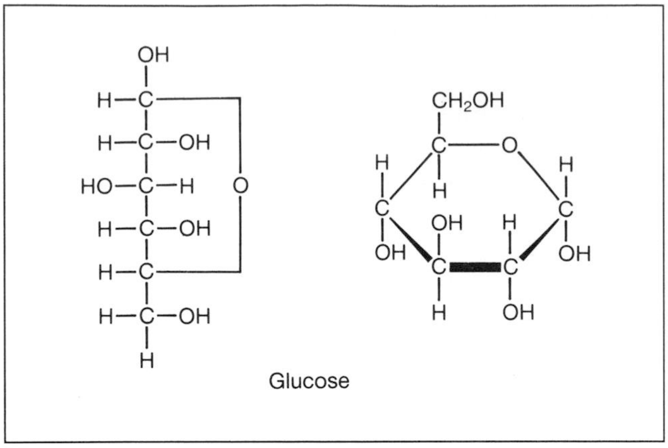

FIGURE 2-7 Two ways of diagramming the structure of the monosaccharide glucose.

general formula $C_n(H_2O)_n$, where n is any whole number. It is from this formula that this class of molecules gets its name, **carbohydrate**—water-containing (hydrated) carbon atoms. Linked to most of the carbon atoms in a carbohydrate are a hydrogen atom and a hydroxyl group:

$$H-\overset{\displaystyle |}{\underset{\displaystyle |}{C}}-OH$$

Chemical groups containing nitrogen and phosphorus atoms may also be linked to this basic carbohydrate structure. The presence of numerous hydroxyl groups, which are polar, makes carbohydrates readily soluble in water.

FIGURE 2-8 The difference between the monosaccharides glucose and galactose has to do with whether the hydroxyl group at the position indicated lies above or below the plane of the ring.

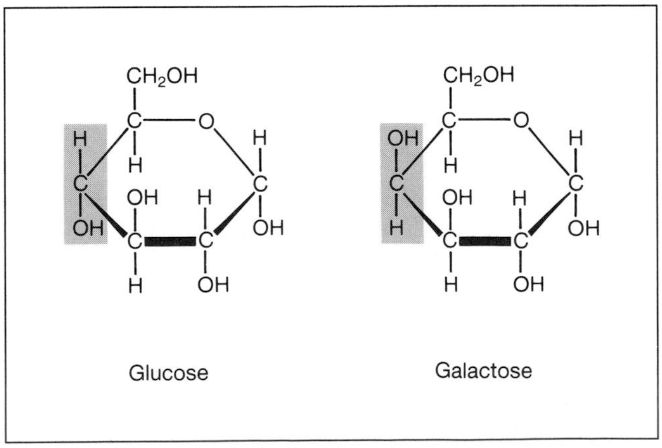

Glucose   Galactose

Most carbohydrates taste sweet, and it is among the carbohydrates that we find the substances known as sugars. The simplest sugars are the **monosaccharides** (single-sweet), the most abundant of which is **glucose**, a six-carbon molecule ($C_6H_{12}O_6$) often called "blood sugar" because it is the major monosaccharide found in the blood.

There are two ways of representing the structures of monosaccharides, as illustrated in Figure 2-7. The first is the conventional representation for organic molecules, but the second gives a better representation of the three-dimensional shape of monosaccharides. Five carbon atoms and an oxygen atom form a ring that lies in an essentially flat plane, the hydrogen and the hydroxyl groups on each carbon lying above and below the plane of this ring. If one of the hydroxyl groups below the ring is shifted to a position above the ring, as shown in Figure 2-8, a different monosaccharide is produced. Most monosaccharides in the body contain five or six carbon atoms and are called **pentoses** and **hexoses**, respectively.

Larger carbohydrate molecules can be formed by linking a number of monosaccharides together. Carbohydrate molecules composed of two monosaccharides are known as **disaccharides. Sucrose** (table sugar), Figure 2-9, is composed of two monosaccharides, glucose (a hexose) and fructose (a pentose). The linking together of most monosaccharides involves the removal of a hydroxyl group from one monosaccharide and of a hydrogen atom from the other, giving rise to a molecule of water and linking the two sugars together through an oxygen atom. Conversely, hydrolysis of the disaccharide breaks this linkage by adding back the water and thus uncoupling the two monosaccharides. Additional disaccharides are maltose (glucose-glucose) formed during the digestion of large carbohydrate molecules in the intestinal tract and lactose (glucose-galactose) present in breast milk.

When many monosaccharides are linked together to form large polymers, the molecules are known as **polysaccharides.** Starch found in plant cells, and **glycogen** (Figure 2-10), present in animal cells and often called "animal starch," are examples of polysaccharides. Both of these polysaccharides are composed of thousands of glucose molecules linked together in long chains. They differ only in the degree of branching along the chain. Hydrolysis of these polysaccharides leads to the release of the glucose subunits.

Lipids

Lipids are molecules composed predominantly of hydrogen and carbon atoms. Since these atoms are linked by neutral covalent bonds, lipids are nonpolar molecules and thus insoluble in water. It is this insolubility in water that characterizes this class of organic molecules. The

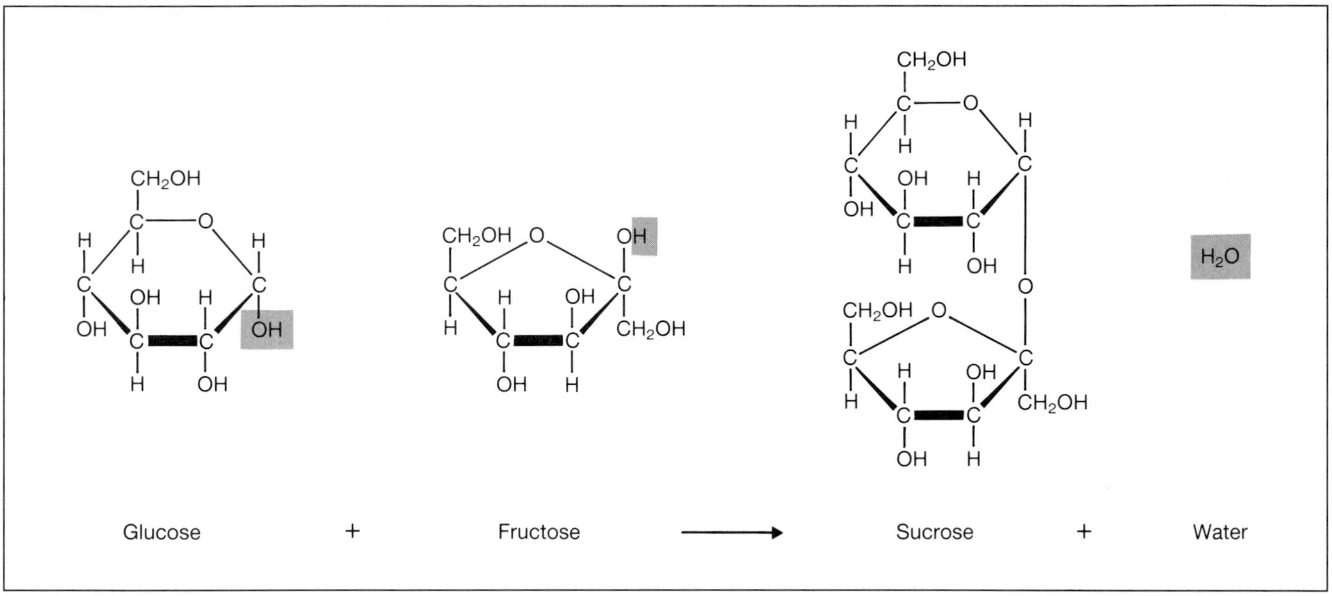

FIGURE 2-9 Sucrose (table sugar) is a disaccharide formed by the linking together of two monosaccharides, glucose and fructose.

lipids, which account for about 40 percent of the body's organic matter, can be divided into three major subclasses: triacylglycerols, phospholipids, and steroids. An additional class of lipids are the fatty acids, which are found either as subunits of triacylglycerols and phospholipids or as "free" molecules.

Triacylglycerols and fatty acids. The **triacylglycerols** (also known as triglycerides) constitute the majority of the body's lipids, and it is these molecules that are generally referred to simply as "fat." Triacylglycerols are formed by the linking together of **glycerol** and **fatty acids** (Figure 2-11). Glycerol is a three-carbon carbohydrate. A fatty acid consists of a chain of carbon atoms with a carboxyl group at one end. Each hydroxyl group in glycerol is linked to the carboxyl group of a fatty acid to form a triacylglycerol molecule.

Because fatty acids are synthesized in the body by the linking together of 2-carbon fragments, most fatty acids have an even number of carbon atoms, 16- and 18-carbon fatty acids being the most common. When all the carbons in a fatty acid chain are linked by single covalent bonds, the fatty acid is said to be a **saturated fatty acid**. Some fatty acids contain one or more double bonds between carbon atoms, and these are known as **unsaturated fatty acids.** If more than one double bond is present, the fatty acid is said to be **polyunsaturated** (Figure 2-11). Animal fats generally contain a high proportion of saturated fatty acids, whereas vegetable fats contain more polyunsaturated fatty acids.

The three fatty acids in a molecule of triacylglycerol need not be identical. Therefore, a variety of fats can be formed with fatty acids of different chain lengths and degrees of saturation. Hydrolysis of triacylglycerols releases the fatty acids from glycerol and these products can then be utilized to provide energy for cell functions.

Some fatty acids can be altered to produce a special class of molecules that regulate a number of cell functions. As described in more detail in Chapter 7, these modified fatty acids—collectively termed **eicosanoids**—are derived from the 20-carbon, polyunsaturated fatty acid **arachidonic acid.**

Phospholipids. The **phospholipids** are similar in overall structure to triacylglycerols, with one important difference. The third hydroxyl group of glycerol, rather than being attached to a molecule of fatty acid, is linked to phosphate. In addition, a smaller polar or ionized nitrogen-containing molecule is usually attached to this phosphate (Figure 2-11). These groups constitute a polar region at one end of the phospholipid molecule, while fatty acid chains provide a nonpolar (unionized) region at the opposite end. Therefore, phospholipids are amphipathic. In water they become organized into clusters, with their polar ends being attracted to the water molecules.

Steroids. The **steroids** have a distinctly different structure from that of the other two subclasses of lipid mole-

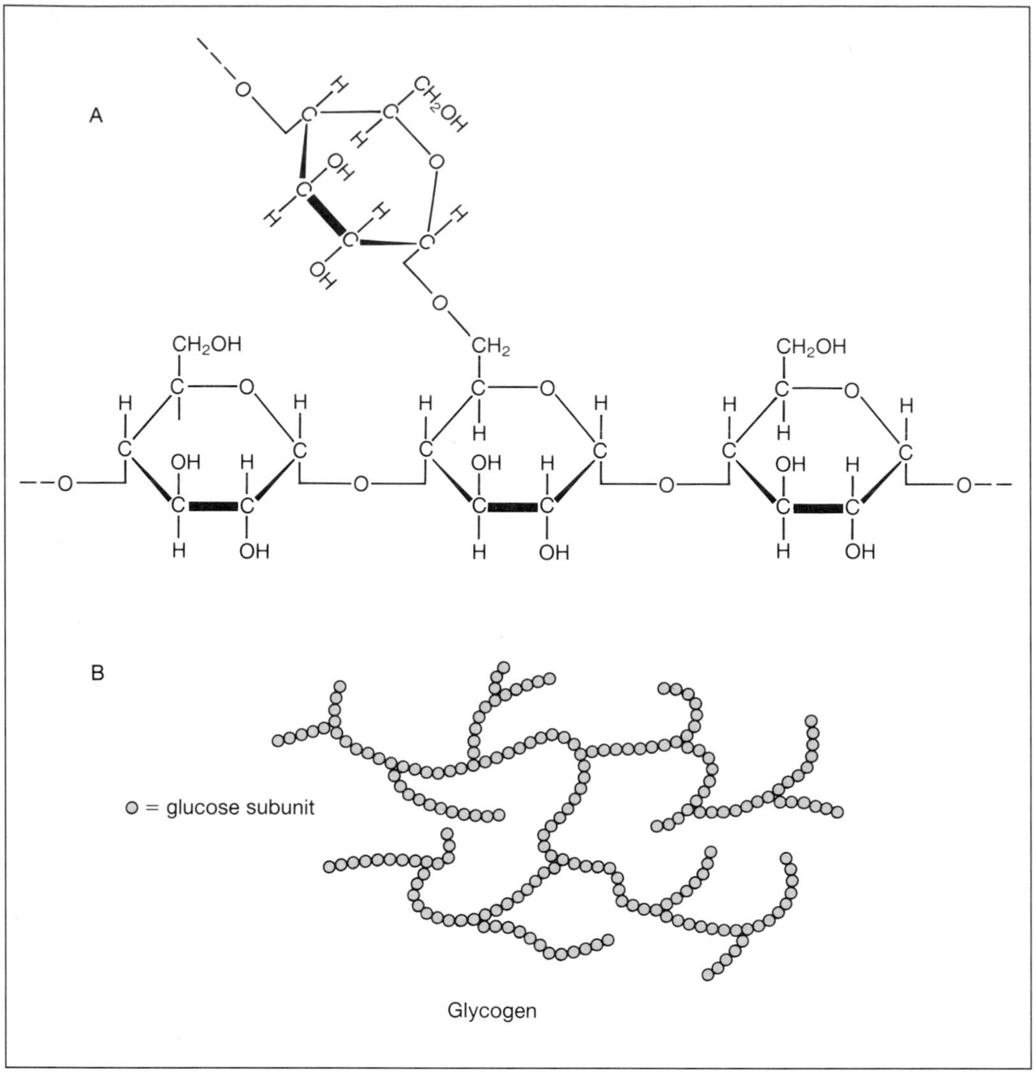

FIGURE 2-10 Many molecules of glucose linked end to end and at branch points form the branched chain polysaccharide glycogen (A), shown in diagrammatic form in (B).

cules. Four interconnected rings of carbon atoms form the skeleton of all steroids (Figure 2-12). A few polar hydroxyl groups may be attached to this ring structure, but they are not numerous enough to make a steroid water-soluble. Examples of steroids are cholesterol, cortisol from the adrenal glands, and female (estrogen) and male (testosterone) sex hormones secreted by the gonads.

Proteins

The term "protein" comes from the Greek *proteios* ("of the first rank"), which aptly describes their importance. These molecules, which account for about 50 percent of the organic material in the body, play critical roles in almost every physiological process. **Proteins** are composed of atoms of carbon, hydrogen, oxygen, nitrogen,

and small amounts of other elements, notably sulfur. They are macromolecules, often containing thousands of atoms, and like most large molecules, are formed by the linking together of a large number of small subunits to form long chains.

The subunit of protein structure is an **amino acid.** Thus, proteins are polymers of amino acids. Every amino acid except one—proline—has an amino ($-NH_2$) and a carboxyl ($-COOH$) group linked to the terminal carbon atom in the molecule (Figure 2-13). The third bond of this terminal carbon atom is linked to a hydrogen atom and the fourth to the remainder of the molecule, which is known as the **amino acid side chain.** The proteins of all living organisms, including human beings, are composed of the same set of 20 different amino acids, corresponding to 20 different side chains. The side

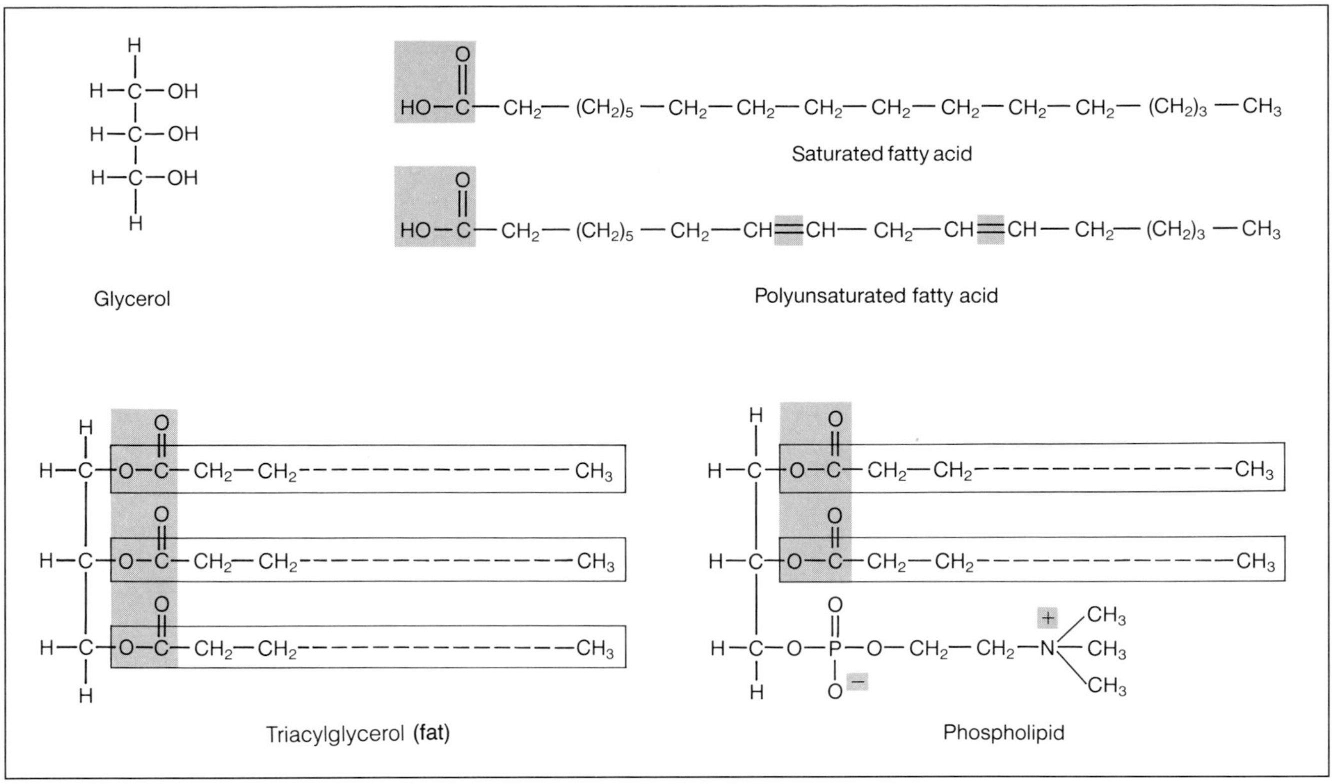

FIGURE 2-11 Glycerol and fatty acids are the major subunits that combine to form triacylglycerols and phospholipids.

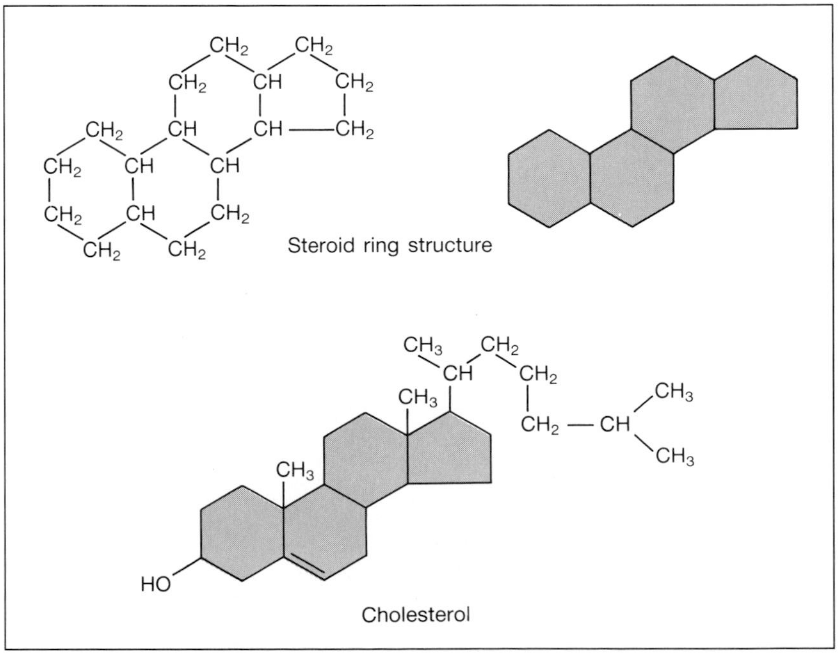

FIGURE 2-12 Steroid ring structure, shown with all the carbon and hydrogen atoms in the rings and again without these atoms to emphasize the overall ring structure of this class of lipids. Different steroids have different types and numbers of chemical groups attached at various locations on the steroid ring, as shown by the example of the steroid cholesterol.

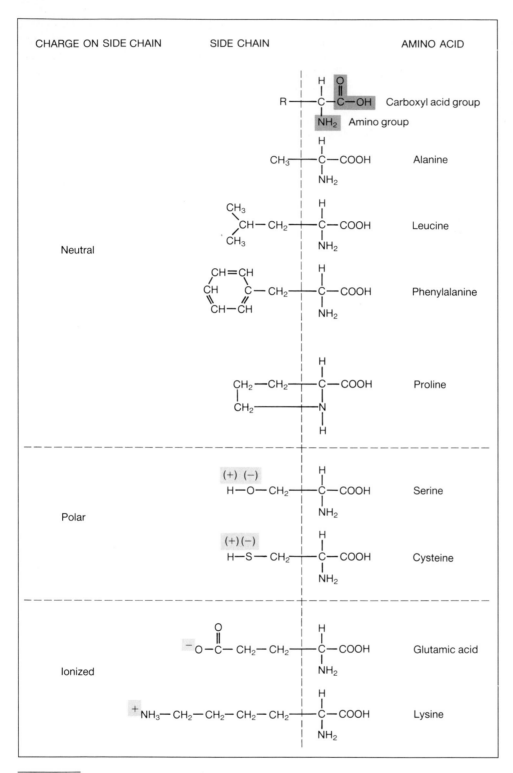

CHARGE ON SIDE CHAIN	SIDE CHAIN	AMINO ACID

Carboxyl acid group

Amino group

Alanine

Leucine

Neutral

Phenylalanine

Proline

Polar

Serine

Cysteine

Ionized

Glutamic acid

Lysine

FIGURE 2-13 Structures of 8 of the 20 amino acids found in proteins. Note that proline does not have a free amino group but it can still form a peptide bond. (The ionized form of glutamic acid is known as glutamate.)

chains may be nonpolar (8 amino acids), polar (7 amino acids), or ionized (5 amino acids).

Polypeptides. Amino acids are linked together in chains by a reaction between the carboxyl group of one amino acid and the amino group of the next amino acid in the sequence. In the process, a molecule of water is formed (Figure 2-14). The bond formed between the amino and carboxyl group is called a **peptide bond,** and it is a polar covalent bond. Note that when two amino acids are linked together, one end of the resulting molecule has a free amino group and the other has a free carboxyl group. Additional amino acids can be linked by peptide bonds to both ends of this chain. A sequence of amino acids linked by peptide bonds is known as a **polypeptide**. The peptide bonds form the backbone of the polypep-

tide, and the remainder of each amino acid sticks out to the side of the chain. If the number of amino acids in a polypeptide is fewer than 50, the molecule is known as a **peptide**; if the sequence is more than 50 amino acids units long, the polypeptide is known as a protein. The number 50 is arbitrary but has become the convention for distinguishing between large and small polypeptide chains.

After certain proteins have been synthesized, one or more monosaccharides are covalently attached to the side chains of specific amino acids (serine and theonine) to form a class of proteins known as **glycoproteins**.

Primary protein structure. There are only two variables that determine the primary structure of a protein or peptide: (1) the total number of amino acids in the

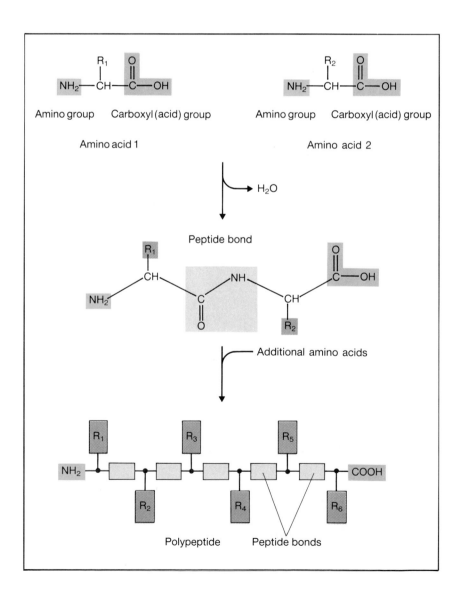

FIGURE 2-14 Linkage of amino acids by peptide bonds to form a polypeptide.

chain, and (2) the specific type of amino acid at each position along the chain (Figure 2-15). Each position can be occupied by any one of the 20 different amino acids. Let us consider the number of different peptides that can be formed with only three amino acids. Any one of the 20 different amino acids may occupy the first position in the sequence, any one of 20, the second position, and any one of 20, the third position, for a total of $20 \times 20 \times 20 = 20^3 = 8000$ possible sequences of 3 amino acids. If the peptide is 6 amino acids in length, $20^6 = 64,000,000$ possible combinations can be formed. Peptides that are only 6 amino acids long are still very small molecules compared to proteins, which may have sequences of 1000 or more amino acids. Thus, with 20 different amino acids an almost unlimited variety of polypeptides can be formed by altering both the amino acid sequence and the total number of amino acids in the chain.

Protein conformation. The structure of a polypeptide is analogous to a string of beads, each bead representing

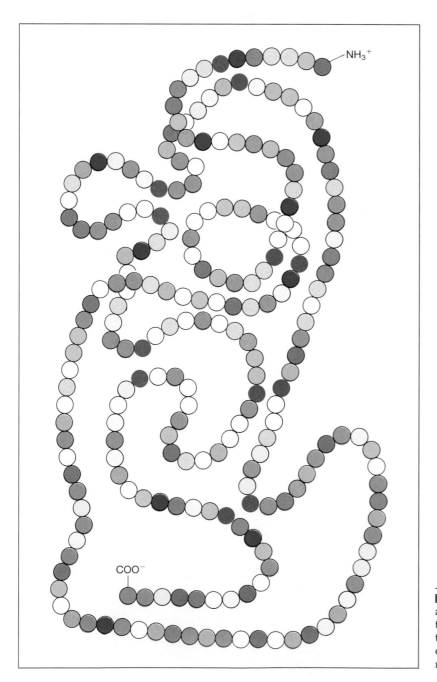

FIGURE 2-15 The position of each type of amino acid in a polypeptide chain and the total number of amino acids in the chain distinguish one polypeptide from another. The example shown contains 223 amino acids, each represented by a circle.

FIGURE 2-16 Conformation of a protein molecule (myoglobin). (*Adopted from Albert L. Lehninger.*)

a single amino acid (Figure 2-15). Moreover, since amino acids can rotate around their peptide bonds, a polypeptide chain is flexible and can be bent into a number of shapes, just as a string of beads can be twisted into many configurations. The three-dimensional shape of a molecule is known as its **conformation** (Figure 2-16). The conformations of peptides and proteins play a major role in the functioning of these molecules, as we shall see in Chapter 4.

Four factors determine the conformation of a polypeptide chain once the amino acid sequence has been formed: (1) attractions between polar and ionized regions along the polypeptide chain, (2) weak forces of attraction between nonpolar regions in close proximity to each

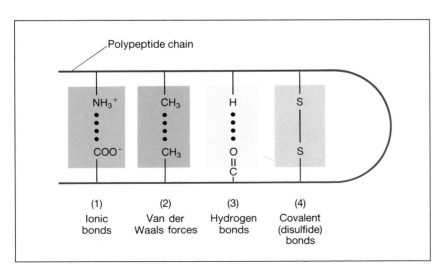

FIGURE 2-17 Factors that contribute to the folding of polypeptide chains and thus to their conformation are: (1) Ionic bonds between polar or ionized amino acid side chains, (2) weak van der Waals forces between nonpolar side chains, (3) hydrogen bonds between side chains or with surrounding water molecules, and (4) covalent bonds between side chains.

other and known as **van der Waals forces**, (3) attractions between the chain and surrounding water molecules, and (4) covalent bonds linking the side chains of two amino acids (Figure 2-17).

An example of the attractions between various regions along a polypeptide chain is the hydrogen bondings that can occur between the hydrogen linked to the nitrogen atom in one peptide bond and the double-bonded oxygen in another peptide bond (Figure 2-18). Since the peptide bonds occur at regular intervals along a polypeptide chain, the hydrogen bonds between them tend to force the polypeptide chain into a helical configuration known as an **alpha helix**. However, for several reasons, a given region of a polypeptide chain may not assume a helical shape. The sizes of the side chains and other types of side chain interactions in the region can interfere with the coiling and distort the helix, producing shapes with no regularity. These nonhelical regions are known as **random coil** configurations and occur in regions where the polypeptide is bent.

Water plays an important role in determining the conformation of a polypeptide chain. When polypeptides are dissolved in water, the polarized and ionized regions of the molecule are attracted to the surrounding polar water molecules. This tends to force the polar and ionized side chains toward the surface of the molecule, orienting them toward the water, while the hydrophobic side chains tend to become located in the interior of the molecule, removed from contact with water. When these hydrophobic side chains are close to each other,

they are attracted by van der Waals forces due to the weak electromagnetic field that surrounds electrically neutral atoms.

The fourth factor influencing the conformation of a polypeptide chain is the covalent bonding that can occur between certain amino acid side chains. The side chain of the amino acid cysteine contains a sulfhydryl group (R—SH), which can react with a sulfhydryl group in another cysteine side chain to produce a covalent **disulfide bond** (R—S—S—R) linking the two amino acid side chains together (Figure 2-19). Disulfide bonds form strong links between portions of a polypeptide chain, in contrast to the weaker hydrogen and ionic bonds, which are more easily broken. Table 2-6 provides a summary of the types of bonding forces that contribute to the conformation of polypeptide chains. These same bonds are also involved in other intermolecular interactions, which will be described in later chapters.

A number of proteins contain more than one polypeptide chain. The same factors that influence the conformation of a single polypeptide chain also determine the interactions between the polypeptide chains in a multichain protein. Thus, the separate chains can be held together by interactions between various polar, ionized, and nonpolar side chains, as well as by disulfide bonds between the chains.

The primary structures (amino acid sequences) of a large number of proteins are known, but three-dimensional conformations have been determined for only a few. Because of the multiple factors that can influence

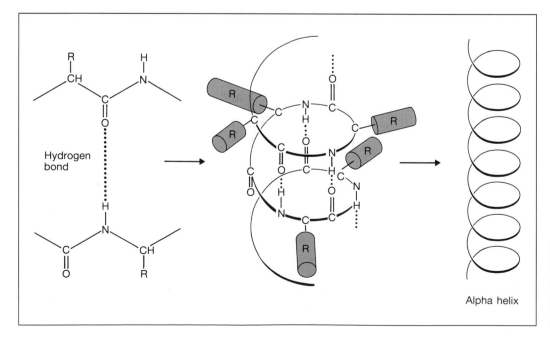

FIGURE 2-18 Hydrogen bonds between regularly spaced peptide bonds can produce a helical configuration in a polypeptide chain.

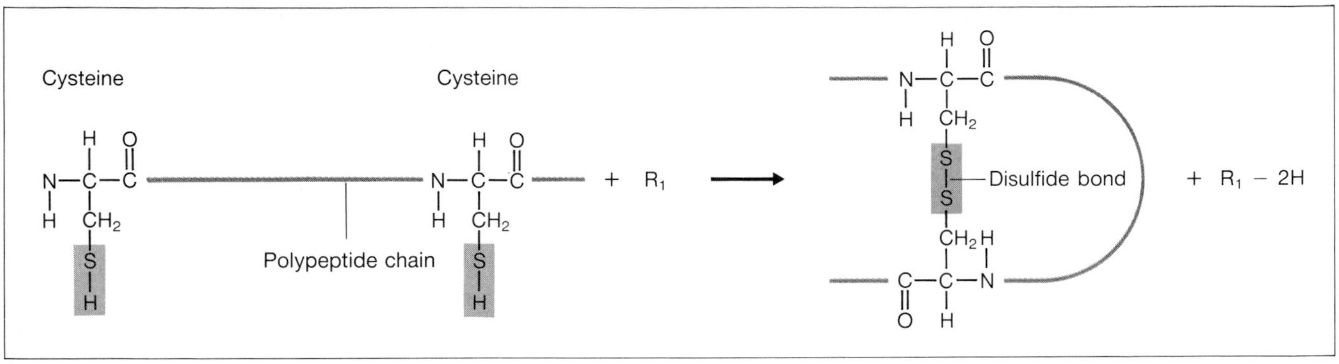

FIGURE 2-19 Formation of a disulfide bond between the side chains of two cysteine amino acids links two regions of the polypeptide together. The hydrogen atoms on the sulfhydryl groups of the cysteines are transferred to another molecule R_1 during the formation of the disulfide bond.

the folding of a polypeptide chain, it is not yet possible to predict accurately the conformation of a protein from its primary amino acid sequence.

Nucleic Acids

Nucleic acids account for only 2 percent of the body's weight, yet it is these molecules that ultimately determine the properties of all cells. Nucleic acids are responsible for the storage, expression, and transmission of genetic information. It is the expression of genetic information that determines whether one is a human being or a mouse, or whether a cell is a muscle cell or a nerve cell.

There are two classes of nucleic acids, **deoxyribonucleic acid (DNA)** and **ribonucleic acid (RNA)**. DNA molecules store genetic information coded in terms of their repeating subunit structure. RNA molecules are involved in the decoding of this information into instructions for linking together a specific sequence of amino acids to form a specific protein. Both types of nucleic acids are polymers composed of linear sequences of repeating subunits. Each subunit, known as a **nucleotide**, has three components: a phosphate group, a sugar, and a ring of carbon and nitrogen atoms often referred to as a base because it can accept hydrogen ions (Figure 2-20). The phosphate group of one nucleotide is linked to the

TABLE 2-6 BONDING FORCES BETWEEN ATOMS AND MOLECULES			
Bond	Strength	Characteristics	Examples
Covalent	Strong	Shared electrons between atoms	Most bonds linking atoms together to form molecules
Ionic	Strong	Electrical attraction between oppositely charged ionized groups	Attractions between ionized groups in amino acid side chains contributing to protein conformation; attractions between ions in a salt
Hydrogen	Weak	Electrical attraction between polarized bonds, usually hydrogen and oxygen	Attractions between peptide bonds forming the alpha helix structure of proteins and between polar amino acid side chains contributing to protein conformation; attractions between water molecules
Van der Waals	Very weak	Attraction between nonpolar molecules and groups when very close to each other	Attractions between nonpolar amino acids in proteins; protein conformation; attractions between lipid molecules

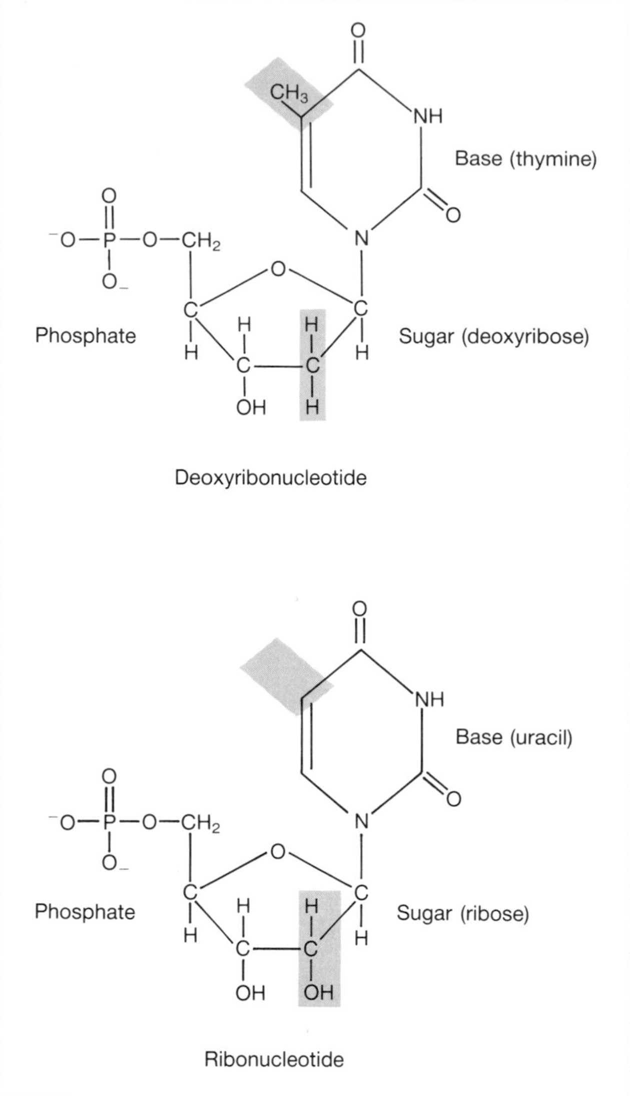

FIGURE 2-20 Nucleotide subunits of DNA and RNA. Nucleotides are composed of a sugar, a base, and phosphate. Deoxyribonucleotides present in DNA contain the sugar deoxyribose. The sugar in ribonucleotides, present in RNA, is ribose, which has an OH at the position that lacks this group in deoxyribose. One of the bases found in deoxyribonucleotides is thymine, which has a methyl group at the location where this group is lacking in the base uracil, one of the four bases present in RNA. The other three bases, adenine, guanine and cytosine, are the same in both DNA and RNA nucleotides.

sugar of the adjacent nucleotide to form a chain of nucleotides, with the bases sticking out to the side of the phosphate-sugar backbone (Figure 2-21).

DNA. The nucleotides in DNA contain the sugar **deoxyribose**—hence, the name deoxyribonucleic acid. Four different nucleotides are present in DNA, corresponding to the four different bases that can be linked to deoxyribose. The four nucleotide bases are divided into two classes: (1) the **purine** bases, **adenine** and **guanine**, which have double (fused) rings of nitrogen and carbon atoms, and (2) the **pyrimidine** bases, **cytosine** and **thymine**, which have only a single ring (Figure 2-21).

A DNA molecule consists of not one but two chains of nucleotides coiled around each other in the form of a double helix (Figure 2-22). The two chains are held together by hydrogen bonds between purine and pyrimidine bases. This pairing between purine and pyrimidine bases maintains a constant distance between the sugar-phosphate backbones of the two chains as they coil around each other. Specificity is imposed on the base pairings by the location of the hydrogen bonding groups in the four bases (Figure 2-22). Three hydrogen bonds are formed between the purine guanine and the pyrimidine cytosine (G—C pairing), while only two hydrogen bonds can be formed between the purine adenine and the pyrimidine thymine (A—T pairing). As a result, G is always paired with C, and A with T. In Chapter 4 we shall see how this specificity in base pairing provides the mechanism for duplicating and transferring genetic information.

RNA. The structure of RNA molecules differs in only a few respects from that of DNA (Table 2-7): (1) RNA consists of a single (rather than a double) chain of nucleotides; (2) in RNA, the sugar in each nucleotide is **ribose** rather than deoxyribose; and (3) the pyrimidine base thymine in DNA is replaced in RNA by the pyrimidine

TABLE 2-7 COMPARISON OF DNA AND RNA COMPOSITION		
	DNA	**RNA**
Nucleotide sugar	Deoxyribose	Ribose
Nucleotide bases		
Purines	Adenine	Adenine
	Guanine	Guanine
Pyrimidines	Cytosine	Cytosine
	Thymine	Uracil
Number of chains	Two	One

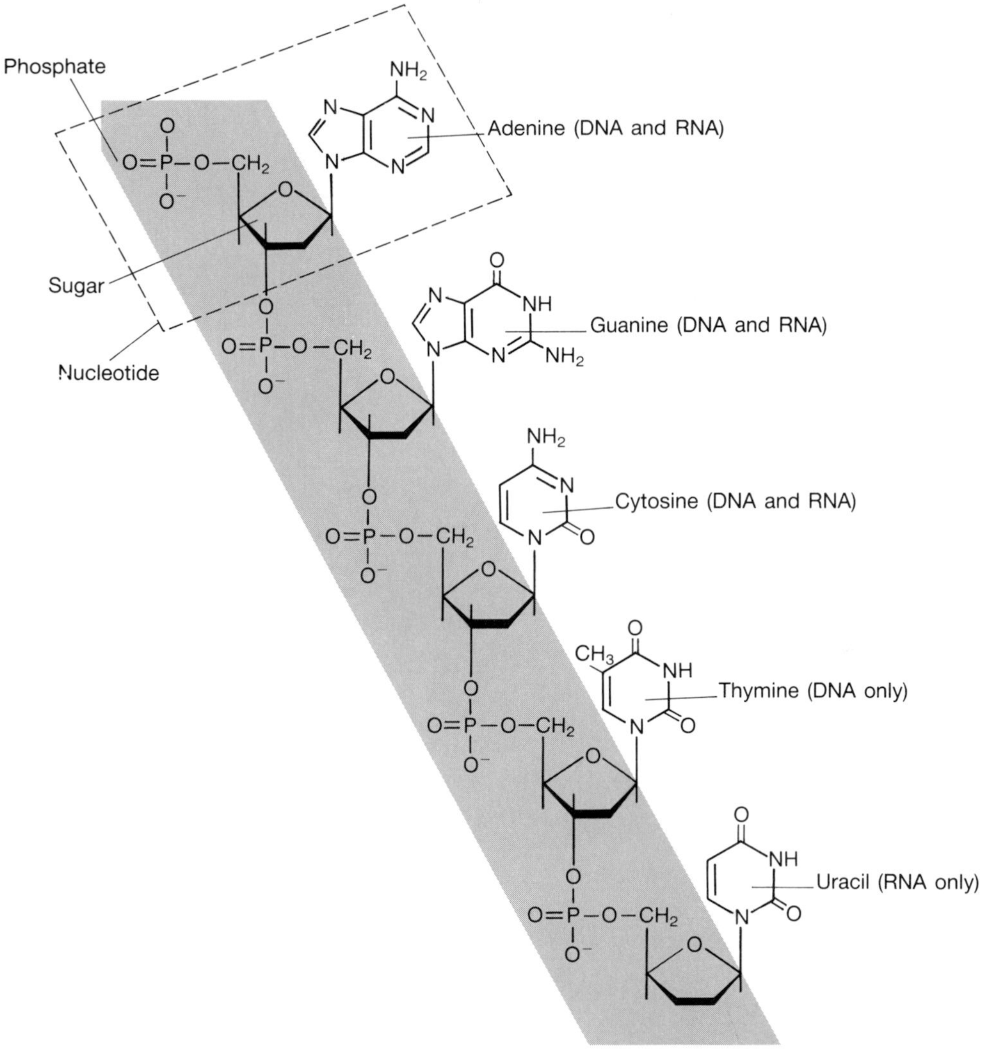

FIGURE 2-21 Phosphate-sugar bonds (color) link nucleotides in sequence to form nucleic acids.

base **uracil** (U) (Figure 2-20), which can base-pair with the purine adenine (A—U pairing). The other three bases, adenine, guanine, and cytosine, are the same in both DNA and RNA. Although RNA contains only a single chain of nucleotides, portions of this chain can bend back upon itself and base-pair with other nucleotides in the same chain.

SUMMARY

Atoms

I. Atoms are composed of three subatomic particles: positive protons and neutral neutrons, both located in the nucleus, and negative electrons revolving around the nucleus.

II. The atomic number is the number of protons in an atom, and because atoms are electrically neutral, it is also equal to the number of electrons.

III. The atomic mass is an atom's mass relative to that of a carbon-12 atom. It is approximately equal to the number of protons plus neutrons.

IV. One gram atomic mass is the number of grams of an element equal to its atomic mass.

V. The 24 elements essential for normal body function are listed in Table 2-1.

Molecules

I. Molecules are formed by linking atoms together.

II. A covalent bond is formed when two atoms share a pair of electrons. Each type of atom can form a characteristic

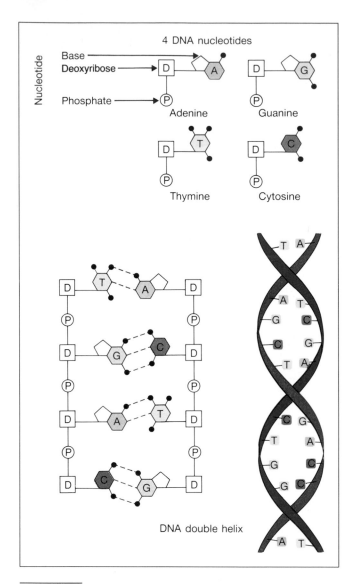

FIGURE 2-22 Base pairings between two nucleotide chains form the double-helical structure of DNA. D and P represent deoxyribose and phosphate, which form the backbone of each polynucleotide chain. The four bases are indicated by A, T, C, and G.

number of covalent bonds: hydrogen forms 1, oxygen 2, nitrogen 3, and carbon 4.

III. Molecules have characteristic shapes that can be altered within limits by the rotation of their atoms around covalent bonds.

Ions

I. When an atom gains or loses one or more electrons, it acquires a net electric charge and becomes an ion.

Polar Molecules

I. In polar covalent bonds, one atom attracts the bonding electrons more than does the other atom.

II. The electrical attraction between hydrogen and the oxygen or nitrogen atom in a separate molecule or different region of the same molecule forms a hydrogen bond.

III. Water, a polar molecule, is attracted to other water molecules by hydrogen bonds.

Solutions

I. Substances dissolved in a liquid are solutes, and the liquid in which they are dissolved is the solvent. Water is the most abundant solvent in the body.

II. Substances that have polar or ionized groups dissolve in water by being electrically attracted to polar water molecules.

III. In water, amphipathic molecules form clusters with the polar regions at the surface and the nonpolar regions in the interior of the cluster.

IV. The molecular mass of a molecule is the sum of the atomic masses of all its atoms. One mole of any substance contains 6×10^{23} molecules.

V. Substances that release a hydrogen ion in solution are called acids. Those that accept a hydrogen ion are bases.

A. The acidity of a solution is determined by its free hydrogen-ion concentration. The greater the hydrogen-ion concentration, the greater the acidity.

B. The pH of a solution is the negative logarithm of the hydrogen-ion concentration. As the acidity of a solution increases, the pH decreases.

C. Pure water has a pH of 7.0.

Classes of Organic Molecules

I. Carbohydrates are composed of carbon, hydrogen, and oxygen in the proportions $C_n(H_2O)_n$.

A. The presence of the polar hydroxyl groups makes carbohydrates soluble in water.

B. The most abundant monosaccharide in the body is glucose ($C_6H_{12}O_6$). Glucose is stored in cells in the form of the polysaccharide glycogen.

II. Lipids lack polar and ionized groups, a characteristic that makes them insoluble in water.

A. Triacylglycerols are formed when fatty acids are linked to each of the three hydroxyl groups in glycerol.

B. Phospholipids contain two fatty acids linked to two of the hydroxyl groups in glycerol, with the third hydroxyl linked to phosphate, which in turn is linked to a small charged or polar compound. The polar and ionized groups at one end of phospholipids make these molecules amphipathic.

C. Steroids are composed of four interconnected rings, often containing a few polar hydroxyl groups.

III. Proteins, macromolecules composed primarily of carbon, hydrogen, oxygen, and nitrogen, are polymers of amino acids.

A. Each of the 20 amino acids, except proline, has an amino ($-NH_2$) and a carboxyl ($-COOH$) group linked to the terminal carbon atom.

B. Amino acids are linked together by peptide bonds between the carboxyl group of one amino acid and the amino group of the next.

C. The primary structure of a polypeptide chain is determined by (1) the number of amino acids in sequence, and (2) the type of amino acid at each position.

D. The factors that determine the conformation of a polypeptide chain are summarized in Figure 2-17.

E. Hydrogen bonds between peptide bonds along a polypeptide chain tend to force the chain into an alpha helix.

F. The interactions of polar and ionized side chains with water tend to force these regions to the surface of the protein, leaving the nonpolar side chains to occupy the interior.

G. Covalent disulfide bonds can form between the sulfhydryl groups of cysteine side chains to hold regions of a polypeptide chain close to each other.

IV. Nucleic acids are responsible for the storage, expression, and transmission of genetic information.

A. Deoxyribonucleic acid (DNA) stores genetic information.

B. Ribonucleic acid (RNA) is involved in the decoding of the information coded in DNA into instructions for linking together amino acids to form proteins.

C. Both types of nucleic acids are polymers of nucleotides, each containing a phosphate group, a sugar, and a ring of carbon and nitrogen atoms known as a base.

D. DNA consists of two chains of nucleotides coiled around each other in a double helix. The chains are held together by hydrogen bonds between purine and pyrimidine bases in the two chains.

E. Base pairings in DNA always occur between guanine-cytosine and adenine-thymine.

F. RNA consists of a single chain of nucleotides, containing the sugar ribose and three of the four bases found in DNA. The fourth base is the pyrimidine uracil rather than thymine. Uracil base-pairs with adenine.

REVIEW QUESTIONS

1. Define:

atoms	carboxyl group
chemical element	amino group
protons	polar covalent bonds
neutrons	hydroxyl group
electrons	polar molecules
atomic nucleus	nonpolar molecules
atomic number	hydrogen bond
gram atomic mass	hydrolysis
trace elements	solutes
molecule	solvent
covalent bond	solution
ion	ionic bond
electrolytes	hydrophilic
cations	hydrophobic
anions	amphipathic

molecular mass	eicosanoids
mole	arachidonic acid
acids	phospholipids
base	steroids
strong acids	proteins
weak acids	amino acid
strong bases	amino acid side chain
weak bases	peptide bond
neutral solution	polypeptide
pH	peptide
alkaline solutions	glycoproteins
acidic solutions	conformation
macromolecules	van der Waals forces
polymers	alpha helix
carbohydrate	random coil
monosaccharides	disulfide bond
glucose	deoxyribonucleic acid
pentoses	(DNA)
hexoses	ribonucleic acid (RNA)
disaccharides	nucleotide
sucrose	deoxyribose
polysaccharides	purine
glycogen	adenine
lipids	guanine
triacylglycerols	pyrimidine
glycerol	cytosine
fatty acids	thymine
saturated fatty acids	ribose
unsaturated fatty acids	uracil
polyunsaturated fatty acids	

2. Describe the electric charge, mass, and location of the three major subatomic particles in an atom.

3. Which four atoms are the most abundant in the body?

4. Describe the distinguishing characteristics of the three classes of essential chemical elements found in the body.

5. How many covalent bonds can be formed by atoms of carbon, nitrogen, oxygen, and hydrogen?

6. What property of molecules allows them to change their three-dimensional shape?

7. Describe the process of forming an ion.

8. Draw the structures of an ionized carboxyl and amino group.

9. Describe the polar characteristics of a water molecule.

10. What determines a molecule's solubility or lack of solubility in water?

11. Describe the organization of amphipathic molecules in water.

12. What is the molar concentration of 80 g of glucose dissolved in sufficient water to make 2 L of solution?

13. What distinguishes a weak acid from a strong acid?

14. What effect does increasing the pH of a solution have upon the ionization of a carboxyl group? An amino group?

15. Name the four classes of organic molecules in the body.

16. Describe the three subclasses of carbohydrate molecules.

17. To which subclass of carbohydrates do each of the following molecules belong: glucose, sucrose, and glycogen?

18. What properties are characteristic of lipids?

19. Describe the three subclasses of lipids.

20. Describe the linkage between amino acids to form a polypeptide chain.

21. What is the difference between a peptide and a protein?

22. What two factors determine the primary structure of a polypeptide chain?

23. Describe the types of interactions that determine the conformation of a polypetide chain.

24. Describe the structure of DNA and RNA.

25. Describe the characteristic of base pairings between nucleotide bases.

CELL STRUCTURE

As we learned in Chapter 1, cells are the structural and functional units of all living organisms. The human body is composed of trillions of cells, each a microscopic compartment (Figure 3-1). In fact, the word "cell" means a small chamber (like a jail cell). In this chapter, we describe the structures present in most of the body's cells and state their functions. Subsequent chapters describe the mechanisms by which these structures perform their functions.

The cells of a mouse, those of a human being, and those of an elephant are all approximately the same size. An elephant is large because it has more cells, not because it has larger cells. Most cells in a human being have diameters of 10 to 20 μm, although cells as small as 2 μm and as large as 120 μm are present. A cell 10 μm in diameter is about one-tenth the size of the smallest object that can be seen with the naked eye. A microscope must therefore be used to observe cells and their internal structure.

MICROSCOPIC OBSERVATIONS OF CELLS

The smallest object that can be seen clearly through a microscope depends upon the wavelength of the radiation used to illuminate the specimen—the shorter the wavelength, the smaller the object that can be seen. With a **light microscope**, the smallest object that can be seen clearly is about 0.2 μm in diameter, and

so this instrument can be used to see individual cells as well as some intracellular structure. However, an **electron microscope** is necessary to make visible the fine details of cellular structure.

An electron microscope uses electrons instead of light rays to form an image of a specimen. A beam of electrons can be focused by electromagnetic lenses, just as glass lenses focus a beam of light in a light microscope. A greater resolution is achieved by an electron microscope because electrons behave as waves with much shorter wavelengths than those of visible light. Cell structures as small as 0.002 μm and very large molecules, such as proteins and nucleic acids, can be seen through an electron microscope, although the individual atoms that make up these molecules are still beyond the limits of effective resolution. Typical sizes of cells and cellular components are illustrated in Figure 3-2.

Although living cells can be observed with a light microscope, this is not possible with an electron microscope. To form an image with an electron beam, most of the electrons must pass through the specimen, just as light passes through a specimen in a light microscope. However, electrons can penetrate only a short distance through matter. Therefore, the observed specimen must be cut into very thin sections, like slices of salami. Each section is on the order of 0.1 μm thick, which is about one-hundredth of the thickness of a typical cell.

Because electron micrographs (such as Figure 3-3) are the images of very thin cell sections rather than the

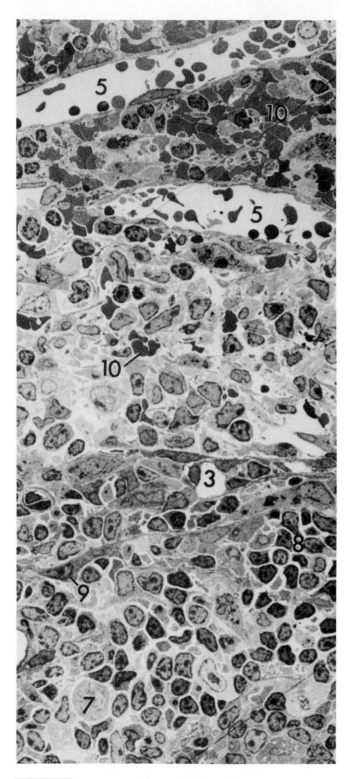

FIGURE 3-1 Cellular structure of the body's tissues, as illustrated by a portion of spleen. Circular and cylindrical spaces are blood vessels. The numbers in the figure are irrelevant. (*From Johannes A. G. Rhodin, "Histology: A Text & Atlas," Oxford University Press, New York, 1974.*)

whole cell, they can often be misleading. Structures that appear as separate objects in an electron micrograph may actually be continuous structures that are connected through a region lying outside the plane of the section. As an analogy, a thin section through a ball of string would appear as a collection of separate lines and disconnected dots even though the piece of string was originally continuous.

The microscopic exploration of the human body has revealed that cells have a variety of shapes. Most cells are surrounded by and anchored to neighboring cells, and such immobilized cells usually have a polyhedral shape (Figure 3-1). Specialized cells often have shapes that are related to the specific functions they perform. Nerve cells, for example, have extending from the cell one or more cylindrical processes, much like the branches of a tree, which are used to receive and transmit electrical signals.

CELL COMPARTMENTS

Compare an electron micrograph of a section through a cell, Figure 3-3, with a diagrammatic illustration of a typical human cell,[1] Figure 3-4. What is immediately obvious, from both the diagram and the electron micrograph, is the extensive structure inside the cell. A human cell is not simply a chemical solution surrounded by a limiting barrier, the **plasma membrane**, which covers the cell surface. The cell interior is divided into a number of compartments also surrounded by membranes. These membrane-bound compartments, along with some particles and filaments, are known as **cell organelles** (little organs). Each cell organelle performs a specific function that contributes to the cell's survival.

The interior of a cell is divided into two regions: (1) the **nucleus**, a spherical or oval organelle usually located near the cell's center and (2) the **cytoplasm**, the region that makes up all the interior of the cell except the nucleus. The cytoplasm contains two components: (1) cell organelles and (2) the fluid surrounding these

[1] There are three basic biological entities: prokaryotic cells, eukaryotic cells, and viruses. The cells in the human body, as well as other multicellular animals and plants, are eukaryotic cells (true-nucleus). These cells are distinguished by the presence of a nuclear membrane surrounding the cell nucleus (see below) and the presence of numerous membrane-bound organelles within the cell. Prokaryotic cells lack these membranous structures. The largest class of prokaryotic cells is the bacteria. Viruses have the simplest structure, consisting of only a nucleic acid molecule surrounded by a protein shell. Numerous infectious diseases are caused by the invasion of multicellular organisms by prokaryotic cells and viruses (Chapter 19). This chapter describes the structure of eukaryotic cells only.

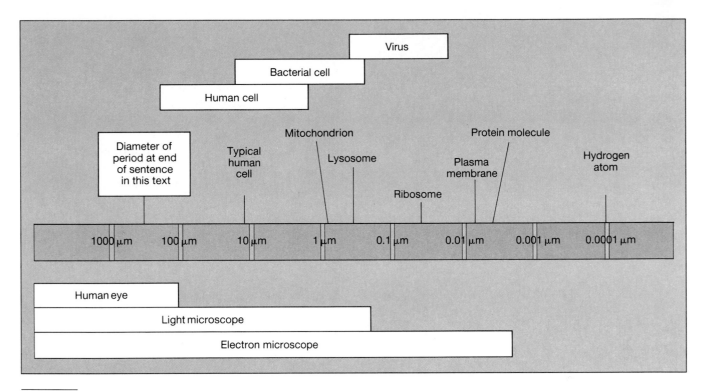

FIGURE 3-2 Size range of microscopic structures, plotted on a logarithmic scale. Typical sizes are given. Each component has a range of sizes around this typical value.

organelles known as the **cytosol** (Figure 3-5). The term **intracellular fluid** refers to all the fluid inside a cell—in other words, cytosol plus the fluid inside each organelle, including the nucleus. The chemical compositions of the fluids in these organelle compartments differ from that of the cytosol. Thus, intracellular fluid is not a fluid of uniform chemical composition.

Membranes

Membranes form a major structural element in human cells. They are found surrounding the entire cell (the plasma membrane) and enclosing most of the cell organelles. Although membranes perform a variety of functions, their most universal role is to act as a selective barrier to the passage of molecules, allowing some molecules to cross while excluding others. The plasma membrane regulates the passage of substances into and out of the cell, whereas the membranes surrounding cell organelles allow selective movement of substances between the organelles and the cytosol. One of the advantages of restricting the movements of molecules across membranes is that the products of chemical reactions can often be confined to specific cell organelles. As we shall see in Chapter 6, the resistance offered by a membrane to the passage of substances can be altered, so that

the flow of molecules or ions across the membrane is either increased or decreased.

The plasma membrane, in addition to acting as a selective barrier, plays an important role in detecting chemical signals from other cells, anchoring cells to adjacent cells, and providing sites for the attachment of protein filaments associated with the generation and transmission of force (Table 3-1).

Membrane structure.   All membranes consist of a double layer of lipid molecules in which proteins are embedded (Figure 3-6). The lipid layers prevent the movement of most molecules through the membrane, whereas the proteins provide the pathways for the selective transfer of certain substances through the lipid barrier. Cell membranes are 6 to 10 nm thick, too thin to be seen without the aid of an electron microscope. In an electron micrograph (Figure 3-6), a membrane appears as two dark lines separated by a light interspace. The dark lines correspond to the polar regions of the proteins and lipids, whereas the light interspace corresponds to the nonpolar regions of these molecules in the membrane.

The major membrane lipids are phospholipids. As described on page 23, these are amphipathic molecules.

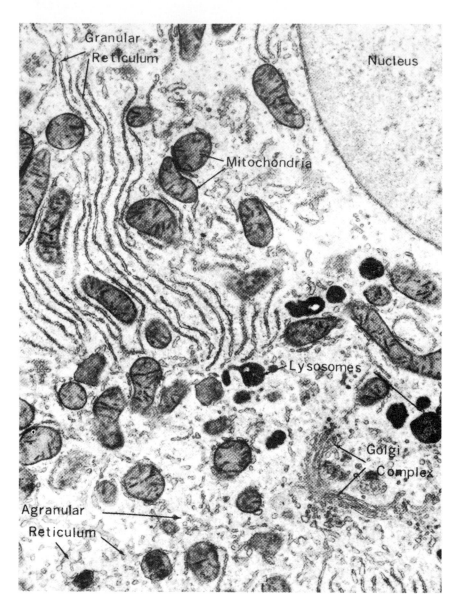

FIGURE 3-3 Electron micrograph of a thin section through a portion of a rat liver cell. [*From K. R. Porter in T. W. Goodwin and O. Lindberg (eds.), "Biological Structure and Function," vol. 1, Academic Press, Inc., New York, 1961.*]

One end has a charged region, and the remainder of the molecule, which consists of two long fatty acid chains, is nonpolar. The phospholipids in cell membranes are organized into a bimolecular layer, with the nonpolar fatty acid chains in the middle. In the bimolecular layer the polar regions are oriented toward the membrane surfaces due to their attraction to the polar water molecules in the extracellular and cytosolic fluids.

No chemical bonds link phospholipids to each other; therefore, each molecule is free to move independently. This results in considerable random lateral movement parallel to the surfaces of the bilayer. In addition, the long fatty acid chains can bend and wiggle back and forth. Thus, the lipid bilayer in the membrane has the characteristics of a fluid, much like a thin layer of oil on a water surface. This fluidity makes the membrane quite

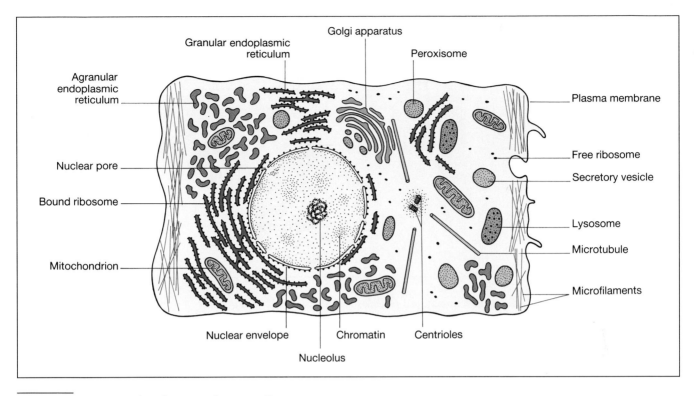

FIGURE 3-4 Structures found in most human cells.

flexible. Like a piece of cloth, membranes can be bent and folded but cannot be stretched without being torn. This flexibility, along with the fact that cells are filled with fluid, allows cells to undergo considerable changes in shape without disruption of their structural integrity.

In addition to phospholipids, the plasma membrane contains cholesterol, a steroid lipid (about one cholesterol molecule for each phospholipid molecule), whereas intracellular membranes contain very little cholesterol. Cholesterol is slightly amphipathic due to a single polar hydroxyl group (Figure 2-12) on its nonpolar ring structure. Therefore, cholesterol, like the phospholipids, is inserted into the membrane with its polar region at a bilayer surface and its nonpolar rings in the interior in association with the fatty acid chains. Plasma-membrane cholesterol appears to play an important role in maintaining the fluidity of the lipid bilayer.

FIGURE 3-5 Comparison of cytoplasm and cytosol. (A) Every point inside the cell except the nucleus interior is considered part of the cytoplasm. (B) Cytosol is a liquid that fills all the cytoplasmic region except the interior of the cell organelles.

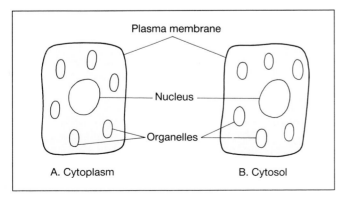

TABLE 3-1   FUNCTIONS OF PLASMA MEMBRANES
1. Regulate the passage of substances into and out of the cells
2. Detect chemical messengers arriving at the cell surface
3. Link adjacent cells together by membrane junctions
4. Anchor a variety of proteins, including intracellular and extracellular protein filaments involved in the generation and transmission of force, to the cell surface

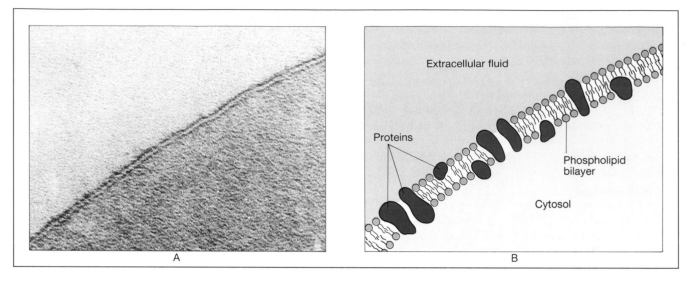

FIGURE 3-6 (A) Electron micrograph of a human red-cell plasma membrane. [*From J. D. Robertson in Michael Locke (ed.), "Cell Membranes in Development," Academic Press, Inc., New York, 1964.*] (B) Arrangement of the proteins and lipids in the membrane.

There are two classes of membrane proteins: integral and peripheral membrane proteins. **Integral membrane proteins** are closely associated with the membrane lipids and cannot be extracted from the membrane without disrupting the lipid bilayer. Like the phospholipids, integral proteins are amphipathic, having polar amino acid side chains in one region of the molecule while the nonpolar side chains are clustered in a separate region. Because they are amphipathic, integral proteins are arranged in the membrane with the same orientation as the amphipathic lipids—the polar regions at the surfaces in association with polar water molecules and the nonpolar regions in the interior in association with the non-

polar fatty acid chains (Figure 3-7). Like the membrane lipids, many of the integral proteins can move laterally in the membrane, but others are immobilized as a result of being linked to a network of peripheral proteins located primarily at the cytoplasmic surface of the membrane.

Most integral proteins span the entire membrane and are referred to as "transmembrane" proteins. These proteins have two polar regions connected by a nonpolar segment that associates with the nonpolar regions of the lipids in the membrane interior. The polar regions of these transmembrane proteins may extend far beyond the surface of the lipid bilayer. Some transmembrane proteins form channels through which ions or water can

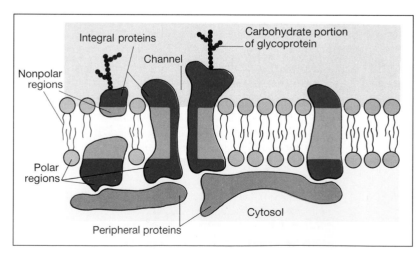

FIGURE 3-7 Arrangement of integral and peripheral membrane proteins in association with the biomolecular layer of phospholipids. Dark blue areas indicate the polar regions of proteins.

move across the membrane, while others are associated with the transmission of chemical signals across the membrane.

In contrast to transmembrane integral proteins, some integral membrane proteins do not cross the entire membrane and are present in either the outer or inner half, performing functions that are localized to only one membrane surface. These proteins are also amphiphatic and are oriented in the membrane in the same manner as the lipid molecules. Some of these proteins are anchored to the membrane by covalent bonds with phospholipid molecules.

Peripheral membrane proteins are not amphipathic and do not associate with the nonpolar regions of the lipids in the membrane interior. They are located at the membrane surface where they are bound to the polar regions of the integral membrane proteins (Figure 3-7). Most peripheral proteins are located on the cytoplasmic surface of the plasma membrane rather than on the extracellular surface and are associated with such properties as cell shape and motility.

In addition the plasma membrane contains small amounts of carbohydrate covalently linked to some of the membrane lipids and proteins. The carbohydrate portions of the membrane glycoproteins are all located at the extracellular surface. This is another example of the asymmetrical distribution of membrane components, the first being the location of most peripheral proteins at the inner surface. Also the lipids in the outer half of the

bilayer differ somewhat in kind and amount from those on the inner half, and, as we have seen, the proteins or portions of proteins on the outer surface differ from those on the inner surface. Many membrane functions are related to these asymmetries in chemical composition.

Although all membranes appear to have the general structure described above, which has come to be known as the **fluid-mosaic model** (Figure 3-8), the proteins and, to a lesser extent, the lipids in the plasma membrane are different from those in organelle membranes. Thus, the special functions of membranes, which depend primarily on the membrane proteins, will be different in the various membrane-bound organelles and plasma membranes of different types of cells. Most intracellular membranes are thinner than plasma membranes. This variation in thickness depends on the amount of protein at the membrane surfaces.

Membrane junctions.   In addition to providing a barrier to the movements of molecules between the intracellular and extracellular fluids, plasma membranes are involved in interactions between cells to form tissues. Some cells, particularly those of the blood, do not associate with other cells but remain independent, suspended in a fluid—the blood plasma in the case of blood cells. Most cells, however, are packaged into tissues and organs and are not free to move around the body. But even the cells in a tissue are not packed so tightly that the adjacent cell surfaces are in direct contact with each other. There usually exists a space of at least 20 nm between the opposing membranes of adjacent cells. This space is filled with extracellular fluid and provides the pathway for substances to pass between cells on their way to and from the blood.

The forces that organize cells into tissues and organs are poorly understood but appear to depend, at least in part, on the ability of binding sites on the plasma membranes to recognize and associate with specific proteins that bind cells together.

Some tissues can be separated into individual cells by gentle procedures, such as shaking the tissues in a calcium-free medium. The absence of calcium ions decreases the cells' ability to stick together, suggesting that the doubly charged Ca^{2+} may act as a link between cells by combining with negatively charged groups on adjacent cell surfaces. Many types of cells, however, require more drastic chemical procedures to separate them from their neighbors.

The electron microscope has revealed that many cells are physically joined by specialized types of membrane junctions known as desmosomes, tight junctions, and gap junctions.

Desmosomes (Figure 3-9A) consist of a region be-

FIGURE 3-8 Fluid-mosaic model of cell membrane structure. (*Redrawn from S. J. Singer and G. L. Nicholson, Science, 175:723. Copyright 1972 by the American Association for the Advancement of Science.*)

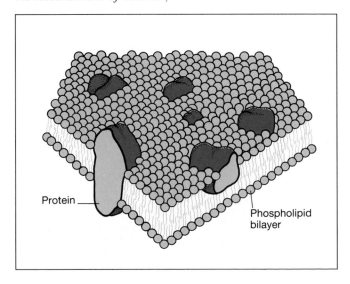

Protein

Phospholipid bilayer

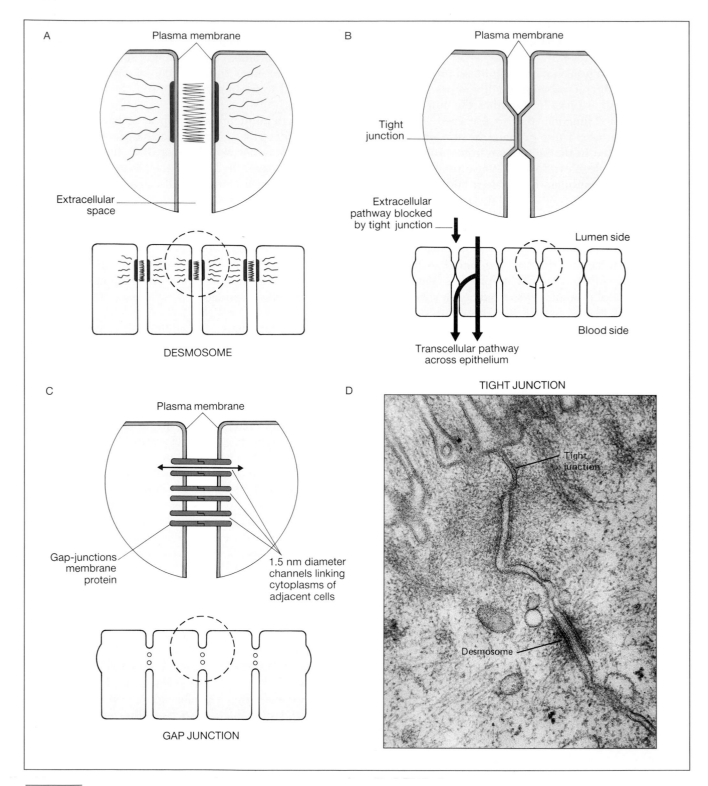

FIGURE 3-9 Three types of specialized cell junctions. (A) Desmosome, (B) tight junction, and (C) gap junction. (D) Electron micrograph of two intestinal epithelial cells joined by a tight junction near the luminal surface and a desmosome below the tight junction. [*Electron micrograph from M. Farquhar and G. E. Palade, J. Cell. Biol., 17:735–412 (1963).*]

tween two adjacent cells in which the two opposed plasma membranes are separated by about 20 nm and have a dense accumulation of protein at each cytoplasmic membrane surface and in the space between the two membranes. In addition, fibers extend from the cytoplasmic surface of each membrane into the cell and appear to be linked to other desmosomes on the opposite side of the cell. Desmosomes hold adjacent cells firmly together in areas that are subject to considerable stretching, such as the skin. The specialized membrane in the region of a desmosome is usually disk-shaped, and these membrane junctions could be likened to rivets or spotwelds that link two cells together at specific points.

A second type of membrane junction, the **tight junction** (Figure 3-9B), is formed when the extracellular surfaces of two adjacent plasma membranes are joined together so that there is no extracellular space between the adjacent cells in the region of the tight junction. Unlike the desmosome, which is limited to a disk-shaped area of the membrane, the tight junction occurs in a band around the entire circumference of the cell. Tight junctions greatly restrict the movement of molecules through the extracellular space between the joined cells.

Most epithelial cells are joined by tight junctions. For example, epithelial cells line the inner surface of the intestinal tract, where they come in contact with the digestion products in the **lumen**, defined as the cavity of a hollow organ. During absorption, the products of digestion move from the lumen across the epithelium and enter the blood. This transfer could take place by movement either through the extracellular space between the epithelial cells or across the epithelial cell. Movement through the extracellular space is blocked by the tight junctions that encircle the cells near their luminal border. Thus, absorption depends on the passage of molecules through, rather than between, cells. In this way the selective-barrier properties of the plasma membrane can control the types and amounts of absorbed substances. Tight junctions, which completely block the extracellular movements of most organic molecules, have a variable leakiness to small ions and water. Figure 3-9D shows both a tight junction and a desmosome located near the luminal border between two epithelial cells.

A third type of junction, known as a **gap junction**, contains protein channels linking the cytoplasms of adjacent cells (Figure 3-9C). In the region of the gap junction, the two opposing plasma membranes come within 2 to 4 nm of each other, allowing proteins from the two membranes to become joined, forming the walls of small channels. These channels extend across the gap between the cells, linking the cytoplasms. The small diameter of these channels (about 1.5 nm) limits what can pass between the connected cells to small molecules and ions,

such as sodium and potassium, and excludes the exchange of large protein molecules. A variety of cell types possess gap junctions, including the muscle cells of the heart and smooth-muscle cells. As we shall see in Chapter 11, these gap junctions play a very important role in the transmission of electrical activity between these muscle cells. In other cases, gap junctions are thought to coordinate the activities of adjacent cells by allowing chemical messengers to move from one cell to another.

CELL ORGANELLES

The Nucleus

Almost all cells contain a single nucleus. (A few specialized types of cells, for example, skeletal muscle cells, contain multiple nuclei, while the mature red blood cell has none.) The primary function of the nucleus is the transmission and expression of genetic information (Chapter 4). Surrounding the nucleus is a barrier, the **nuclear envelope**, composed of two membranes. At regular intervals along the surface of the nuclear envelope, the two membranes become joined to each other, forming the rims of circular openings known as **nuclear pores** (Figure 3-10). Through these pores, molecules that regulate the expression of genetic information move in both directions, between the nucleus and the cytoplasm.

The most prominent organelle within the nucleus is the **nucleolus**, a highly coiled structure associated with numerous particles but not surrounded by a membrane. The nucleolus is composed of DNA, RNA, and proteins and is the site at which the subunits of **ribosomes**, a cytoplasmic organelle, are assembled. These ribosomal subunits are transferred to the cytoplasm, where they combine to form ribosomes, the sites at which protein molecules are synthesized.

The nucleus also contains a fine network of threads known as **chromatin**, composed of DNA and protein, which are coiled to a greater or lesser degree, producing variations in the density of the nuclear contents seen in electron micrographs (Figure 3-10). These threads, which become highly condensed to form **chromosomes** when a cell divides, contain the DNA molecules that store genetic information and pass it on from cell to cell each time a cell divides.

Endoplasmic Reticulum

The most extensive cytoplasmic organelle is the system of membranes that forms the **endoplasmic reticulum** (Figure 3-11). It is found throughout the cytoplasm in the form of relatively flattened sacs or as a branching tubular network. The membranes enclose a space that is continuous throughout the network and connects with

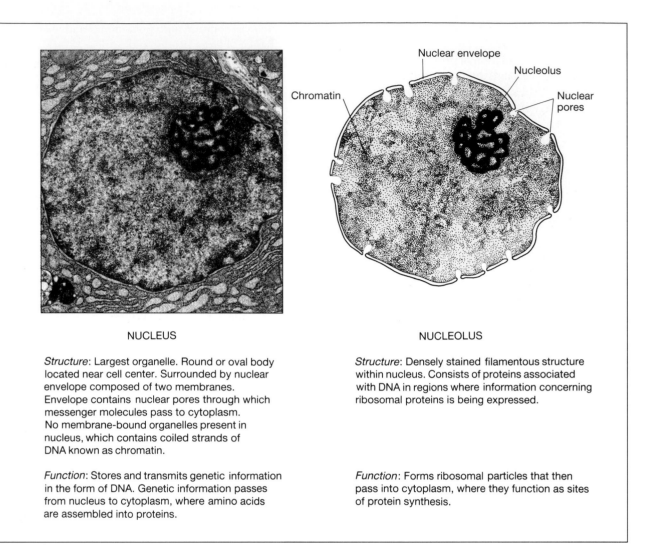

NUCLEUS

Structure: Largest organelle. Round or oval body located near cell center. Surrounded by nuclear envelope composed of two membranes. Envelope contains nuclear pores through which messenger molecules pass to cytoplasm. No membrane-bound organelles present in nucleus, which contains coiled strands of DNA known as chromatin.

Function: Stores and transmits genetic information in the form of DNA. Genetic information passes from nucleus to cytoplasm, where amino acids are assembled into proteins.

NUCLEOLUS

Structure: Densely stained filamentous structure within nucleus. Consists of proteins associated with DNA in regions where information concerning ribosomal proteins is being expressed.

Function: Forms ribosomal particles that then pass into cytoplasm, where they function as sites of protein synthesis.

FIGURE 3-10 Nucleus and nucleolus. (*Electron micrograph courtesy of K. R. Porter.*)

the space between the two membranes of the nuclear envelope (Figure 3-4). The continuity of the endoplasmic reticulum is not obvious when examining a single electron micrograph because only a portion of the network is present in any one section.

Two forms of endoplasmic reticulum can be distinguished: **granular** (rough-surfaced) and **agranular** (smooth-surfaced). The granular endoplasmic reticulum has ribosomes bound to its cytosolic surface and has a flattened sac appearance. The agranular endoplasmic reticulum has no ribosomal particles on its surface and has a branched, tubular structure. Both granular and agranular endoplasmic reticulum exist in the same cell. The relative amounts of the two types vary in different types of cells and even within the same cell during dif-

ferent periods of cell activity. Granular endoplasmic reticulum is involved in the packaging of proteins that are to be secreted by cells (Chapter 6). Smooth endoplasmic reticulum is the site at which lipids are synthesized (Chapter 5).

Ribosomes

Ribosomes are bound to the granular endoplasmic reticulum and are also found free in the cytoplasm. Each ribosome is composed of a large number of proteins and several RNA molecules (Chapter 4). Ribosomes synthesize proteins from amino acids using genetic information sent by messenger molecules from DNA in the nucleus.

The proteins synthesized on the free ribosomes are released into the cytosol, where they perform their func-

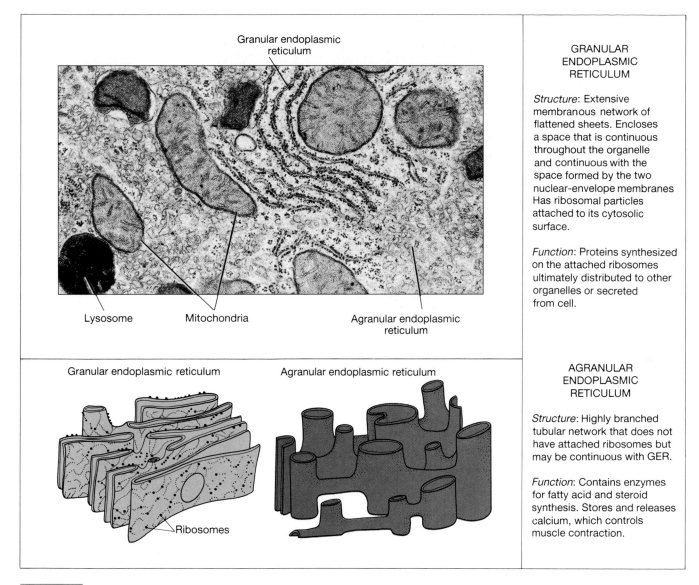

Granular endoplasmic reticulum

Lysosome   Mitochondria   Agranular endoplasmic reticulum

Granular endoplasmic reticulum   Agranular endoplasmic reticulum

Ribosomes

GRANULAR ENDOPLASMIC RETICULUM

Structure: Extensive membranous network of flattened sheets. Encloses a space that is continuous throughout the organelle and continuous with the space formed by the two nuclear-envelope membranes Has ribosomal particles attached to its cytosolic surface.

Function: Proteins synthesized on the attached ribosomes ultimately distributed to other organelles or secreted from cell.

AGRANULAR ENDOPLASMIC RETICULUM

Structure: Highly branched tubular network that does not have attached ribosomes but may be continuous with GER.

Function: Contains enzymes for fatty acid and steroid synthesis. Stores and releases calcium, which controls muscle contraction.

FIGURE 3-11 Endoplasmic reticulum. GER = granular endoplasmic reticulum. (*Electron micrograph from D. W. Fawcett, "The Cell, An Atlas of Fine Structure," W. B. Saunders Company, Philadelphia, 1966.*)

tions. The proteins synthesized by ribosomes attached to the granular endoplasmic reticulum pass into the lumen of the reticulum and onto another organelle, the Golgi apparatus.

Golgi Apparatus

The **Golgi apparatus**, consists of a series of closely opposed, flattened membranous sacs that are slightly curved, forming a cup-shaped structure (Figure 3-12). Most cells have a single Golgi apparatus located near the nucleus, although some cells may have several. Associated with this organelle, particularly near its concave surface, are a number of approximately spherical, membrane-enclosed sacs, referred to as vesicles. The Golgi

apparatus sorts the different types of proteins received from the granular endoplasmic reticulum into vesicles that will be delivered to various parts of the cell, including the plasma membrane where the protein contents of the vesicle are released to the outside of the cell. When proteins are secreted from cells, the vesicles that contain these proteins are known as **secretory vesicles**.

Mitochondria

Mitochondria (singular, *mitochondrion*) are spherical or elongated, rodlike structures surrounded by an inner and an outer membrane (Figure 3-13). The outer membrane is smooth, whereas the inner membrane is folded into sheets or tubules, known as **cristae**, that extend into

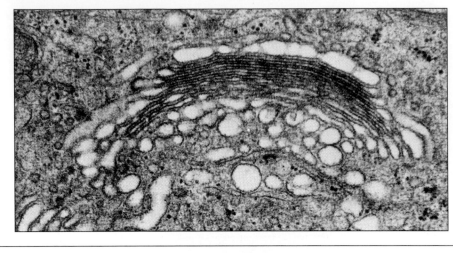

GOLGI APPARATUS

Structure: Series of cup-shaped, closely opposed, flattened, membranous sacs; associated with numerous vesicles. Generally, a single Golgi apparatus is located in central portion of cell near nucleus.

Function: Concentrates, modifies, and sorts newly synthesized proteins prior to their distribution, by way of the Golgi vesicles, to other organelles or their secretion from cell.

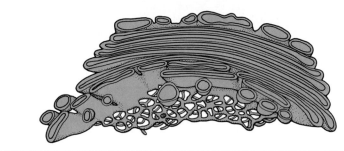

FIGURE 3-12 Golgi apparatus. (*Electron micrograph from W. Bloom and D. W. Fawcett, "Textbook of Histology," 9th edn. W. B. Saunders Company, Philadelphia, 1968.*)

the inner compartment (**matrix**). Mitochondria are found throughout the cytoplasm. Large numbers of them, as many as 1000, are present in cells that utilize large amounts of energy, whereas less active cells contain fewer.

Mitochondria are primarily concerned with the chemical processes by which energy is made available to cells in the form of adenosine triphosphate (ATP) (Chapter 5). Most of the ATP used by cells is formed in the mitochondria by a process that consumes oxygen and produces carbon dioxide.

Lysosomes

Lysosomes are spherical or oval organelles surrounded by a single membrane (Figure 3-11). A typical cell may contain several hundred lysosomes. They break down bacteria and the debris from dead cells that have been taken into the cell. They may also break down cell organelles that have been damaged and are no longer functioning normally. The lysosomes are thus a highly spe-

cialized intracellular digestive system. They play an especially important role in the various specialized cells that make up the defense systems of the body (Chapter 19).

Peroxisomes

The structure of **peroxisomes** is similar to that of lysosomes, that is, oval bodies enclosed by a single membrane, but their chemical composition is quite different. Like mitochondria, peroxisomes consume oxygen, although in much smaller amounts, and this oxygen is not used in the chemical reactions associated with ATP formation. Peroxisomes destroy certain products formed from oxygen, notably hydrogen peroxide, that can be quite toxic to cells—thus the organelles' name.

Filaments

In addition to membrane-enclosed organelles, the cytoplasm of most cells contains many filaments of various sizes. This cytoplasmic filamentous network is referred

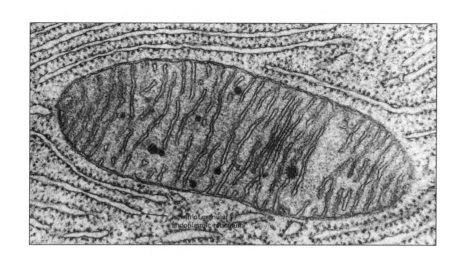

lumen of granular
endoplasmic reticulum

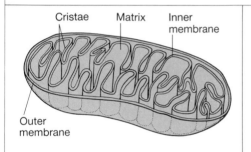

Cristae   Matrix   Inner
membrane

Outer
membrane

MITOCHONDRIA

Structure: Rod- or oval-shaped body
surrounded by two membranes. Inner
membrane folds into matrix of the
mitochondria, forming cristae.

Function: Major site of ATP production,
O_2 utilization, and CO_2 formation.
Contains enzymes of Krebs cycle and
oxidative phosphorylation.

FIGURE 3-13 Mitochondrion.
(*Electron micrograph courtesy of
K. R. Porter.*)

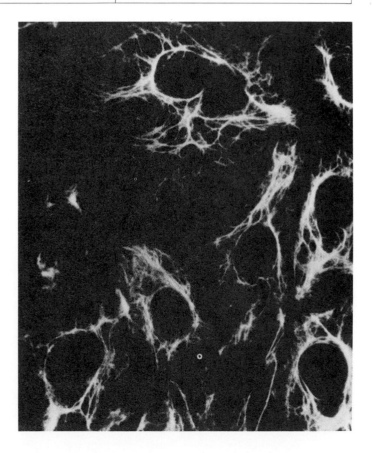

FIGURE 3-14 Cells stained to show the intermediate filament components of the cytoskeleton. (*From Roy A. Quinlan et al., "Annals
of the New York Academy of Sciences, vol.
455, New York, 1985.*)

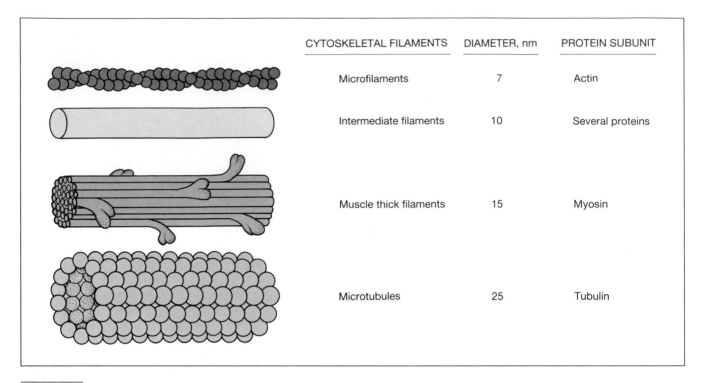

CYTOSKELETAL FILAMENTS	DIAMETER, nm	PROTEIN SUBUNIT
Microfilaments	7	Actin
Intermediate filaments	10	Several proteins
Muscle thick filaments	15	Myosin
Microtubules	25	Tubulin

FIGURE 3-15 Cytoskeletal filaments associated with cell shape and motility.

to as the cell's **cytoskeleton** (Figure 3-14), and, like the bony skeleton of the body, it is associated with processes that maintain and change cell shape and produce cell movements.

There are four classes of filaments grouped according to their diameter and the types of protein they contain (Figure 3-15). In order of size, starting with the thinnest, they are: (1) microfilaments, (2) intermediate filaments, (3) muscle thick filaments, and (4) microtubules. Microfilaments and microtubules can be rapidly assembled and disassembled, allowing a cell to change its cytoskeletal framework according to changing requirements. The other two types of filaments, once assembled, are less readily disassembled.

Microfilaments, which are composed of the contractile protein **actin**, make up a major portion of the cytoskeleton in all cells. **Intermediate filaments** are most extensively developed in regions of cells that are subject to mechanical stress. **Muscle thick filaments**, composed of the contractile protein **myosin**, are found only in muscle cells. However, myosin molecules not in filament form are also present in many, if not most, other cells where they interact with microfilaments to produce local forces and movements.

Microtubules, which are composed of the protein **tubulin**, are hollow tubes about 25 nm in diameter. Each

microtubule is composed of a helical arrangement of 13 tubulin subunits in each rotation of the helix. They are the most rigid of the cytoskeletal filaments and are present in the long extensions of nerve cells, where they provide the framework that maintains the cylindrical shape. Microtubules are also associated with movements of parts of cells. For example, when a cell divides, it duplicates its chromosomes and sends one copy to each new cell. The structure immediately responsible for the separation of the chromosomes, the spindle fibers, is composed of microtubules, as are the organelles that generate the spindle fibers at the time of cell division, the **centrioles** (Figure 3-4).

The hairlike extensions—**cilia**—on the surface of some epithelial cells have a central core of microtubules that produce the movements of the cilia. In hollow organs that are lined with ciliated epithelium, the cilia wave back and forth, propelling the luminal contents along the surface of the epithelium. Microtubules have also been implicated in the movements of organelles within the cytoplasm.

SUMMARY

I. All living matter is composed of cells.

Cell Compartments

I. Every cell is surrounded by a plasma membrane.

II. Within each cell are numerous membrane-bound compartments and nonmembranous particles and filaments, known collectively as cell organelles.

III. A cell is divided into two regions, the nucleus and the cytoplasm. The cytoplasm is composed of the fluid cytosol and cell organelles other than the nucleolus.

IV. The membranes that surround the cell and cell organelles regulate the movements of molecules and ions into and out of the cell and its compartments.

   A. Membranes consist of a bimolecular lipid layer, mainly phospholipids, in which proteins are embedded.

   B. The lipid layer prevents the movement of most molecules through the membrane, while the proteins provide pathways for the selective entry of substances.

   C. Integral membrane proteins are amphipathic proteins that often span the membrane, whereas the peripheral membrane proteins are confined to the surfaces of the membrane.

V. Three types of membrane junctions link adjacent cells:

   A. Desmosomes link cells that are subject to considerable stretching.

   B. Tight junctions, found primarily in epithelial cells, limit the passage of molecules through the extracellular space between the cells.

   C. Gap junctions form channels between the cytoplasms of adjacent cells.

Cell Organelles

I. The nucleus transmits and expresses genetic information.

   A. The nucleolus is the site at which ribosomes are formed.

   B. Threads of chromatin, composed of DNA and protein, form chromosomes when a cell divides.

II. The endoplasmic reticulum is a network of flattened sacs and tubules in the cytoplasm.

   A. Granular endoplasmic reticulum has attached ribosomes and is primarily involved in the packaging of proteins that are to be secreted from cells.

   B. Agranular endoplasmic reticulum is tubular, lacks ribosomes, and is the site of lipid synthesis.

III. Ribosomes, composed of RNA and protein, are the sites of protein synthesis.

IV. The Golgi apparatus sorts the proteins that are synthesized on the rough endoplasmic reticulum and packages them into secretory vesicles.

V. Mitochondria are the major sites at which chemical energy is transferred to ATP.

VI. Lysosomes digest particulate matter that enters the cell.

VII. Peroxisomes break down certain toxic products formed from oxygen.

VIII. The cytoplasm contains a network of four types of filaments known as the cytoskeleton: (1) microfilaments, (2) intermediate filaments, (3) muscle thick filaments, and (4) microtubules.

REVIEW QUESTIONS

1. Define:

light microscope	endoplasmic reticulum
electron microscope	granular endoplasmic
plasma membrane	reticulum
cell organelles	agranular endoplasmic
nucleus	reticulum
cytoplasm	Golgi apparatus
cytosol	secretory vesicles
intracellular fluid	mitochondria
integral membrane proteins	mitochondrial cristae
peripheral membrane	mitochondrial matrix
proteins	lysosomes
fluid-mosaic model	peroxisomes
desmosomes	cytoskeleton
tight junction	microfilaments
lumen	actin
gap junction	intermediate filaments
nuclear envelope	muscle thick filaments
nuclear pores	myosin
nucleolus	microtubules
ribosomes	tubulin
chromatin	centrioles
chromosomes	cilia

2. In terms of the size and number of cells, what makes an elephant larger than a mouse?

3. Identify the location of cytoplasm, cytosol, and intracellular fluid within a cell.

4. Identify the classes of organic molecules in cell membranes, and describe the major contribution of each to membrane function.

5. Describe the orientation of the phospholipid molecules in a membrane.

6. Which membrane component is responsible for membrane fluidity?

7. Describe the location and characteristics of integral and peripheral membrane proteins.

8. Describe the structure and function of the three types of junctions found between cells.

9. What function is performed by the nucleolus?

10. Contrast the structure and functions of the granular and agranular endoplasmic reticulum.

11. Describe the location and function of ribosomes.

12. What function is performed by the Golgi apparatus?

13. Describe the structure and primary function of mitochondria.

14. What functions are performed by lysosomes and peroxisomes?

15. List the four types of filaments associated with the cytoskeleton, and identify the structures in cells that are composed of microtubules.

4

MOLECULAR CONTROL MECHANISMS: DNA AND PROTEIN

The outstanding accomplishment of twentieth-century biology has been the discovery of the chemical basis of heredity and its relationship to protein synthesis. Whether an organism is a human being or a mouse, has blue eyes or black, has light skin or dark, is determined by the proteins it possesses. Moreover, within an individual organism, muscle cells differ from nerve cells or epithelial cells or any other type of cell because of the types of proteins they contain and the functions performed by these proteins. Most of the physiology described in this book is based upon the properties of the thousands of different proteins found in the human body.

Crucial for an understanding of protein function is the fact that each protein has a unique shape that determines its interaction with other molecules. In the first section of this chapter the relation between protein shape and function is described.

Since the functioning of the human body is determined primarily by its proteins, hereditary information consists of a set of instructions coded into DNA molecules that specify the types of proteins a cell can synthesize. The middle section of this chapter describes the chemical basis of the genetic code and the decoding mechanisms that lead to protein synthesis.

Given that different cell types have different proteins

and that the specifications for these proteins are coded in DNA, one might be led to conclude that different cell types contain different DNA molecules. However, such is not the case. All cells in the body, with the exception of sperm or ova, receive the same genetic information when DNA molecules are duplicated and passed on to daughter cells at the time of cell division. Cells differ in structure and function because only a portion of the total genetic information common to all cells is used by any given cell to synthesize proteins. In other words, different portions of the inherited genetic information are translated in different types of cells. The last section of this chapter describes some of the factors that govern the selective expression of genetic information, as well as the process of cell division.

SECTION A
PROTEIN BINDING SITES

BINDING-SITE CHARACTERISTICS

The ability to bind various molecules and ions to specific sites on the surface of a protein molecule forms the basis for the wide variety of functions performed by proteins. A **ligand** is any molecule or ion that is bound to the surface of a protein by forces other than covalent chemical bonds. These forces are either (1) attractions between oppositely charged ionic or polarized groups on the ligand and protein or (2) the weaker attractions, due to Van der Waals forces, between adjacent nonpolar regions on the two molecules. The region of a protein to which a ligand binds is known as a **binding site**. A protein may contain several binding sites, each specific for a different ligand.

Chemical Specificity

The force of electrical attraction between oppositely charged regions on a protein and a ligand decreases markedly as the distance between them increases. The even weaker Van der Waals forces act only between nonpolar groups that are very close to each other. Therefore, in order for a ligand to bind to the surface of a protein, it must be close to the protein. This occurs when the shape of the ligand is complementary to the shape of the protein binding site, such that the two fit together like pieces of a jigsaw puzzle (Figure 4-1).

The binding between a ligand and a protein may be so specific that a binding site may bind only one type of organic molecule or ion and no other. Such selectivity allows a protein to "identify" (by binding) one particular molecule in a solution containing hundreds of different molecules. This ability of a protein to bind specific ligands is known as **chemical specificity,** since the shape of the binding site determines the type of chemical compound that is bound.

In the second chapter we described how the conformation of a protein is determined by the location of the various amino acids along the polypeptide chain. Accordingly, proteins with different amino acid sequences have different shapes and therefore different binding sites, each with its own chemical specificity. As illus-

FIGURE 4-1 Complementary shapes of ligand and protein binding site determine the chemical specificity of binding.

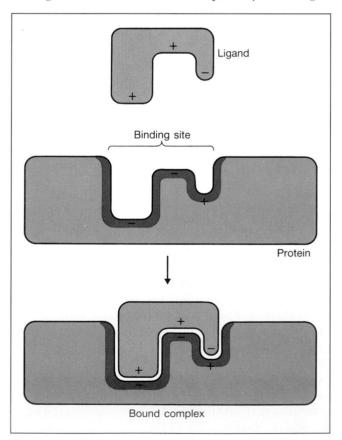

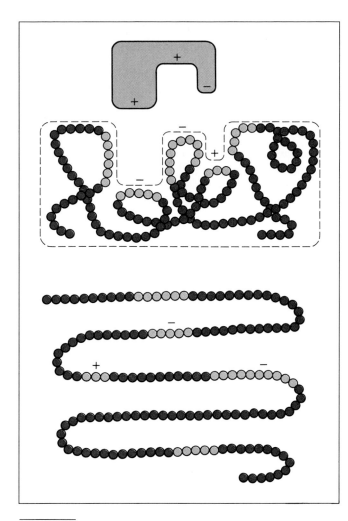

FIGURE 4-2 Amino acids that interact with the ligand at a binding site need not be at adjacent sites along the polypeptide chain, as indicated in this model showing the three-dimensional folding of a protein, with the unfolded polypeptide chain shown below it.

Affinity

The strength of ligand-protein binding is a property of the binding site known as **affinity**. The affinity of a binding site for a ligand determines how likely it is that a bound ligand will leave the protein surface and return to its unbound state. Binding sites that tightly bind a ligand are called high-affinity binding sites; those to which the ligand is weakly bound are low-affinity binding sites.

Affinity and chemical specificity are two distinct, although sometimes related, properties of binding sites. Chemical specificity depends only on the shape of the binding site, whereas affinity depends on the strength of the attraction between the protein and the ligand. Thus, different proteins may be able to bind the same ligand, that is, may have the same chemical specificity, but may

FIGURE 4-3 Protein X is able to bind all three ligands, which have similar chemical structures. Protein Y, because of the shape of its binding site, can bind only ligand C. Protein Y has a greater chemical specificity than protein X.

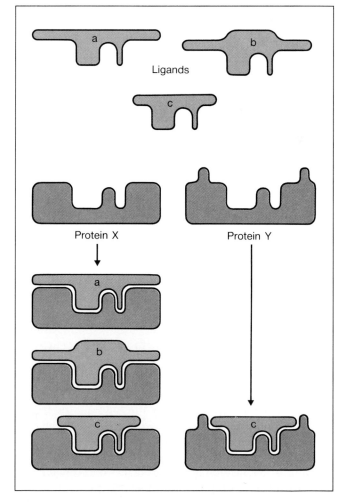

trated in Figure 4-2, the amino acids that interact with a ligand need not be adjacent to each other along the polypeptide chain since the folding of the protein may bring various segments of the molecule into juxtaposition.

Some binding sites have a chemical specificity that allows them to bind only one type of ligand, whereas other binding sites may be less specific and thus able to bind a number of related ligands. Three different ligands can combine with the binding site of protein X in Figure 4-3 since a portion of each ligand is complementary to the shape of the binding site. In contrast, protein Y has a greater, that is, more limited, specificity and can bind only one of the three ligands.

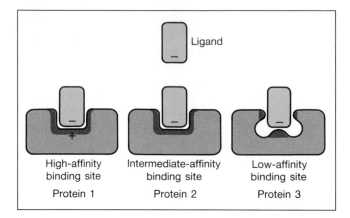

FIGURE 4-4 Three binding sites with the same chemical specificity for a ligand but different affinities.

have different affinities for it. For example, a ligand may have a negatively charged ionized group that would bind strongly to a site containing a positively charged amino acid side chain but would bind less strongly (Figure 4-4) to a binding site having the same shape but no positive charge. In addition, the closer the surfaces of the ligand and binding site to each other, the stronger the electrical attractions. Hence, the more closely the ligand shape matches the binding-site shape, the greater the affinity. In other words, shape can influence affinity as well as chemical specificity.

Saturation

In a solution containing both ligands and binding sites, some ligands will bind to unoccupied binding sites, while some of the bound ligands will be coming off binding sites. At any given time, a binding site will either be occupied or unoccupied. The term **saturation** refers to the fraction of binding sites that are occupied at any given time. When all the binding sites are occupied, the population of binding sites is said to be 100% percent saturated. When half of the available sites are occupied, the system is 50 percent saturated, and so on. A single binding site would also be said to be 50 percent saturated if it were occupied by a ligand 50 percent of the time.

The degree of saturation of a binding site depends upon two factors: (1) the concentration of unbound ligand in the solution and (2) the affinity of the binding site for the ligand.

The greater the ligand concentration, the greater the probability of a ligand encountering an unoccupied binding site and becoming bound. Thus, the percent satura-

tion of binding sites will increase with increasing ligand concentration until all the sites become occupied (Figure 4-5). If a ligand were a molecule that exerted a biological effect when it was bound to a protein, the magnitude of the effect would increase with increasing numbers of bound ligands until all the binding sites were occupied. Further increases in ligand concentration would produce no further effect since there would be no additional sites to be occupied. To generalize, a continuous increase in the magnitude of a chemical stimulus (ligand concentration), which exerts its effects by binding to proteins, will produce an increased biological response up to the point where the protein binding sites become saturated.

The second factor determining the degree of binding-site saturation is the affinity of the binding site. Collisions between molecules in a solution and those on a binding site can dislodge a loosely bound ligand, much as tackling a football player may cause a fumble. If a binding site has a high affinity for a ligand, even a low ligand concentration will result in a high degree of saturation since once bound to the site, the ligand is not easily dislodged. A low-affinity site, on the other hand, requires a much higher concentration of ligand to achieve the same degree of saturation since any given ligand remains bound for a much shorter period of time, requiring more frequent encounters between unbound ligand and the binding site to keep the site occupied. One measure of binding-site affinity is the ligand con-

FIGURE 4-5 Increasing ligand concentration increases the number of binding sites occupied, that is, increases the percent saturation. At 100 percent saturation, all the binding sites are occupied, and further increases in ligand concentration do not increase the amount bound.

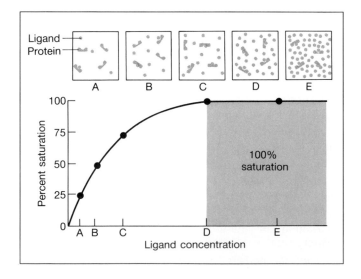

centration necessary to produce 50 percent saturation—the lower the ligand concentration at 50 percent saturation, the greater the affinity of the binding site (Figure 4-6).

Competition

As we have seen, more than one type of ligand can bind to certain binding sites (Figure 4-3). **Competition** occurs when two or more ligands compete for the same binding site. The presence of multiple ligands able to bind to the same binding site affects the percentage of binding sites occupied by any one ligand. If two competing ligands, A and B, are present, increasing the concentration of A will increase the amount of A that is bound and also decrease the number of sites available for the binding of B, thus decreasing the amount of B that is bound.

As a result of competition, the biological effects of one ligand may be markedly diminished by the presence of another. For example, many drugs produce their effects by competing with the body's natural ligands for binding sites. By occupying the binding sites, the drug decreases the amount of natural ligand that can be bound.

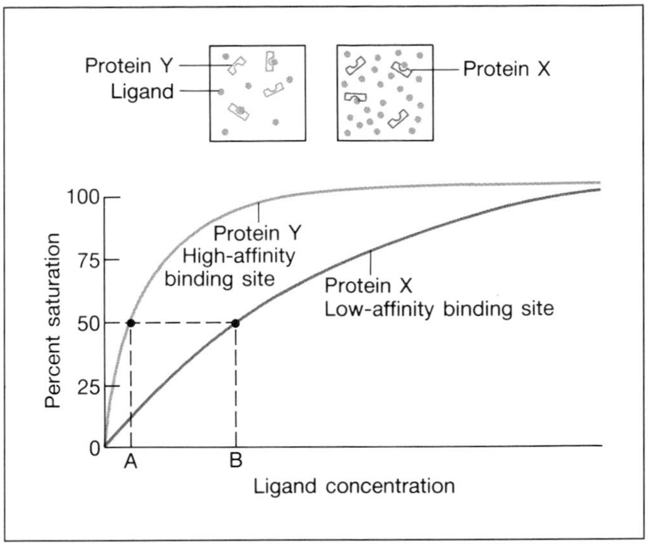

FIGURE 4-6 When two different proteins, X and Y, are able to bind the same ligand, the protein with the higher-affinity binding site (protein Y) is 50 percent saturated at a lower ligand concentration than is required to saturate 50 percent of the lower-affinity binding sites on protein X.

REGULATION OF BINDING-SITE CHARACTERISTICS

Because proteins are associated with practically every cell function, the mechanisms for controlling these functions center on the control of protein activity. There are two ways of controlling protein activity: (1) Changing protein shape alters its binding of ligands, and (2) regulation of protein synthesis determines the types and amounts of proteins in a cell. The first type of regulation—control of protein shape—is discussed in this section, and the second—protein synthesis—in later sections.

Since protein shape depends on electrical attractions between charged or polarized groups in various regions of a protein (page 29), a change in the charge distribution along a protein or in the polarity of the molecules immediately surrounding it will alter its shape.

Two nonspecific factors that alter protein shape are temperature, which alters the amount of kinetic energy available for breaking the electrostatic attractions between various regions of a protein, and acidity, which alters the ionization of amino acid side chains. Because both of these factors are maintained relatively constant in the body, they are not used to selectively regulate protein activity. Should large changes in temperature or acidity occur however, as in disease, they can produce marked alterations in protein shape and thus in protein activity.

The two mechanisms that are used by cells to selec-

tively alter protein shape are allosteric and covalent modulation.

Allosteric Modulation

Whenever a ligand binds to a protein, the attracting forces between ligand and protein alter the protein's shape. Very importantly, ligand binding can change the shape of the protein in regions that are not part of the binding site. Therefore, when a protein contains two binding sites, the binding of a ligand to one site can alter the shape of the second binding site and, hence, the binding characteristics of that site. This is termed **allosteric** (other shape) **modulation** (Figure 4-7), and such proteins are known as **allosteric proteins**.

One binding site on an allosteric protein, known as the **functional site**, carries out the protein's physiological function. The other binding site is the **regulatory site**, and the ligand that binds to this site is known as a **modulator molecule** since its binding to the regulatory site alters the shape and thus the activity of the functional site.

The regulatory site to which modulator molecules bind is the equivalent of a molecular switch that controls the affinity of the functional site. In some allosteric proteins, the binding of the modulator molecule to the regulatory site turns *on* the functional site by changing its shape so that it can bind the functional ligand. In other

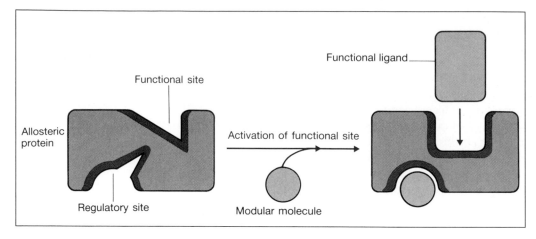

FIGURE 4-7 Allosteric modulation. Binding of a modulator molecule at a regulatory site alters the shape of the functional site.

cases, the binding of a modulator molecule turns *off* the functional site, preventing the functional site from binding its ligand, a state of zero affinity. In still other cases, binding of the modulator molecule may decrease or increase the affinity of the functional site so that the percent saturation of the functional site will be decreased or increased at any given concentration of the ligand that binds to the functional site.

A single cell contains not just one but many identical molecules of any particular allosteric protein. The total activity of these proteins depends on how many of the regulatory sites are occupied by modulator molecules. Therefore, the magnitude of a cell function can be regulated by varying the concentration of the modulator molecule and thus the number of allosteric proteins that have been altered by binding modulator molecules.

It should be emphasized that most proteins are not subject to allosteric modulation. Only certain key proteins associated with specific cell functions are regulated.

Covalent Modulation

A second way to alter the shape and therefore the activity of a protein is to covalently bind charged chemical groups to some of the protein's side chains. This is known as **covalent modulation**. In most cases a phosphate group, which has a net negative charge, is covalently attached by a chemical reaction called **phosphorylation**.

When one of the amino acid side chains in a protein is phosphorylated, a negative charge is introduced into this region of the protein. This alters the distribution of elec-

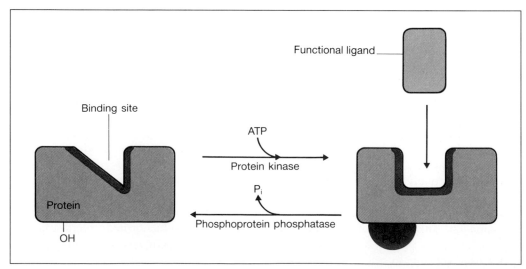

FIGURE 4-8 Covalent modulation. The shape of a protein binding site can be altered by the attachment, through a covalent chemical bond, of an ionized phosphate group to one of the amino acid side chains. The enzyme protein kinase catalyzes the phosphorylation of the protein, whereas a second enzyme, phosphoprotein phosphatase, removes the phosphate group, returning the protein to its original shape.

tric forces in the protein, producing a change in protein conformation (Figure 4-8). If this conformational change occurs in the region of a binding site, it will change the binding site's affinity. The effects produced by covalent modulation are the same as those of allosteric modulation, that is, phosphorylation may turn a binding site on or off or alter its affinity for ligand.

Unlike allosteric modulation, which involves simple binding of modulator molecules, covalent modulation requires chemical reactions in which covalent bonds are formed. Such reactions are mediated by a special class of proteins known as **enzymes**, whose properties will be discussed in Chapter 5. For now, suffice it to say that enzymes accelerate the rate at which one or more initial molecules called **substrates** are converted to new molecules called **products**. Any enzyme that mediates protein phosphorylation is called a **protein kinase**. These enzymes catalyze the transfer of phosphate from a molecule of adenosine triphosphate (ATP) (page 84) to a hydroxyl group present on the side chain of certain amino acids:

$$\text{protein} + \text{ATP} \xrightarrow{\text{protein kinase}} \text{protein}-\text{PO}_4^{2-} + \text{ADP}$$

The protein and ATP are the substrates for protein kinase, and the phosphorylated protein and ADP are the products of the reaction. There is also a mechanism for removing the phosphate group and returning the protein to its original shape. This dephosphorylation is accomplished by a second enzyme known as phosphoprotein

phosphatase:

$$\text{protein}-\text{PO}_4^{2-} + \text{H}_2\text{O} \xrightarrow{\text{phosphoprotein phosphatase}} \text{protein} + \text{HPO}_4^{2-}$$

Thus, two enzymes control a protein's activity by covalent modulation, one that adds phosphate and one that removes it.

Although most proteins contain amino acids capable of being phosphorylated, only those proteins that are substrates for specific protein kinases are phosphorylated. There are many protein kinases, each with specificities for different proteins, and several may be present in the same cell. The chemical specificities of the phosphoprotein phosphatases are broader, and a single enzyme can dephosphorylate many different phosphorylated proteins. Moreover, unlike protein kinases, phosphoprotein phosphatases tend to be continuously active.

When a protein kinase is active, the rate of phosphorylation exceeds that of dephosphorylation, and most of the proteins being covalently modulated are maintained in the phosphorylated state. Dephosphorylating such proteins requires only stopping the kinase reaction since the continuously active phosphoprotein phosphatase will remove the phosphate groups.

The kinase reaction can be controlled by modulator molecules since the protein kinases are allosteric proteins. Thus, the process of covalent modulation is itself regulated by allosteric mechanisms.

SECTION B
GENETIC INFORMATION AND PROTEIN SYNTHESIS

GENETIC INFORMATION

Molecules of DNA contain instructions, coded in the sequence of nucleotides, for the synthesis of proteins. There are many molecules of DNA in the cell nucleus, each containing different sets of instructions. A sequence of nucleotides containing the information that determines the amino acid sequence of a single polypeptide chain is known as a **gene**. A single molecule of DNA contains many genes. Genes are the units of hereditary information, and human cells contain between 50,000 and 100,000 genes.

As an example of the expression of genetic information, consider eye color, which is due to the presence of

pigment molecules in certain cells in the eye. A sequence of enzyme-mediated reactions is required to synthesize these pigments, and the genes that determine eye color do so by controlling the synthesis of the enzymes in this pathway. Note that these genes do not contain any information about the chemical structure of the eye pigments. All they contain is the information required to synthesize the enzymes that mediate the formation of the pigments.

Although DNA contains the information necessary for the synthesis of proteins, it does not itself participate *directly* in the assembly of protein molecules. Most of a cell's DNA is in the nucleus (a small amount is in the mitochondria), whereas most protein synthesis occurs in the cytoplasm. The transfer of information from DNA to

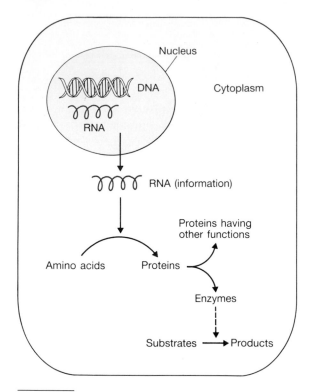

FIGURE 4-9 The expression of genetic information in a cell occurs through its transcription from DNA to RNA in the nucleus followed by the translation of the RNA information into protein synthesis in the cytoplasm.

the site of protein synthesis is the function of RNA molecules (page 32), whose synthesis is governed by the information in DNA. This mechanism for expressing genetic information occurs in all living organisms and has led to what is now called the "central dogma" of molecular biology: Genetic information flows from DNA to RNA and then to protein:[1]

$$DNA \rightarrow RNA \rightarrow protein$$

Figure 4-9 summarizes the general pathway by which the information stored in DNA influences cell activity via protein synthesis.

As described on page 32, a molecule of DNA consists of two polynucleotide chains coiled around each other to form a double helix (Figure 2-22). Each nucleotide

contains one of four bases—adenine (A), guanine (G), cytosine (C), or thymine (T)—and each of these bases is specifically paired, A to T and G to C, with a base on the opposite chain of the double helix. Thus, both nucleotide chains contain a specifically ordered sequence of bases, one chain being complementary to the other.

The genetic language is similar in principle to a written language, which consists of a set of symbols, such as α, β, σ, γ, forming an alphabet. The letters are arranged in specific sequences to form words, and the words are arranged in linear sequences to form sentences. The genetic language contains only four letters, corresponding to the four bases A, G, C, and T. The words are three-base sequences that specify particular amino acids, that is, each word in the genetic language is only three letters long. This is termed a triplet code. These words are arranged in a linear sequence along DNA, and the sequence of words specifying the structure of a single protein makes up a gene (Figure 4-10). A typical human gene is a sequence of approximately 20,000 base pairs. Thus, in our analogy, a gene is equivalent to a sentence, and the entire collection of genes in a cell is equivalent to a book.

How are the four bases in the DNA alphabet arranged to form at least 20 different three-letter code words that make up the **genetic code** for the 20 amino acids? Since the four bases can be arranged in 64 different three-letter combinations ($4 \times 4 \times 4 = 64$), a triplet code actually provides more than enough code words. It turns out that not just 20, but 61 of the 64 possible triplets are used to specify amino acids. This means that a given amino acid is usually specified by more than one code word. For example, the four triplets C-C-A, C-C-G, C-C-T and C-C-C all specify the same amino acid, glycine.

The three triplets that do not specify amino acids are known as **termination code words**. They perform the same function as does a period at the end of a sentence— they indicate that the end of a genetic message has been reached. The code word for the amino acid methionine does double duty by also indicating the beginning of the protein chain.

The genetic code is a universal language used by all living cells. For example, the code words for the amino acid tryptophan are the same in the DNA of a bacterium, an amoeba, a plant, and a human being. Although the same code words are used by all living cells, the messages they spell out—the sequences of code words (the "sentences") that determine the amino acid sequences in proteins—are different in each organism. The universal nature of the genetic code supports the concept that all forms of life on earth evolved from a common ancestor.

---

[1] In some viruses, genetic information can be transferred from RNA to protein in the absence of DNA, while in other cases the flow is from RNA to DNA to RNA to protein.

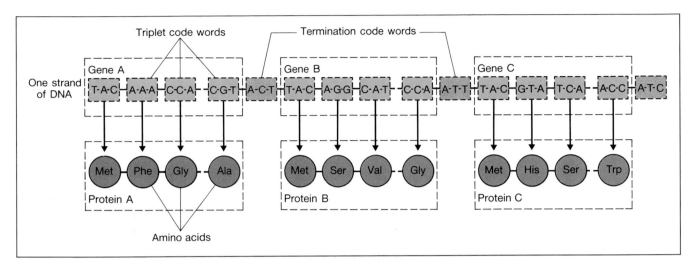

FIGURE 4-10 Relation between the sequence of triplet-code words in one strand of DNA and the amino acid sequences of proteins that correspond to them. The names of the amino acids are abbreviated. Note that all three genes are part of the same DNA molecule, which may contain hundreds of genes in a linear sequence, each gene being composed of hundreds of sequential bases.

PROTEIN SYNTHESIS

DNA molecules are too large to pass through the nuclear membrane into the cytoplasm. Yet it is in the cytoplasm, on ribosomes, that proteins are synthesized. Since DNA cannot leave the nucleus, a message carrying genetic information must pass from the nucleus to the cytoplasm. This message is carried by the RNA molecules known as **messenger RNA (mRNA)**. The transfer of genetic information from DNA to protein thus occurs in two stages: First, the genetic message is passed from DNA to mRNA in the nucleus (**transcription**); second, the message in mRNA passes from the nucleus into the cytoplasm where it is used to direct the assembly of the specific sequence of amino acids to form a protein (**translation**).

Transcription: mRNA Synthesis

During transcription, the sequence of nucleotides in DNA is used to determine the sequence of nucleotides in mRNA. As described on page 32, ribonucleic acids are single-chain polynucleotides whose nucleotides differ from DNA in that they contain the sugar ribose (rather than deoxyribose) and the base uracil (rather than thymine). The other three bases—adenine, guanine, and cytosine—occur in both DNA and RNA. The pool of subunits used to synthesize mRNA are free (uncombined) ribonucleotides, each containing three phosphate groups (nucleotide triphosphates): ATP, GTP, CTP, and UTP.

In DNA, the two polynucleotide chains are linked together by hydrogen bonds (page 32) between specific pairs of bases—A pairing with T and G with C. Transcription begins with the breakage of these hydrogen bonds so that a portion of the two chains of the DNA double helix separates (Figure 4-11). The bases in the exposed DNA nucleotides are then able to pair with the bases in the free ribonucleotide triphosphates. Free ribonucleotides containing adenine pair with any exposed thymine base in DNA. Likewise, free ribonucleotides containing G, C, or U pair with the exposed DNA bases C, G, and A, respectively. (Note that uracil, which is present in RNA but not DNA, pairs with the base adenine in DNA.) In this way, the nucleotide sequence in DNA acts as a template that determines the sequence of nucleotides in mRNA.

The aligned ribonucleotides are joined together by the enzyme **RNA polymerase**. This enzyme, which binds to DNA, catalyzes both the splitting-off of two of the three phosphate groups from each nucleotide and the covalent linkage of the RNA nucleotide to the next one in sequence. RNA polymerase is active only when bound to DNA and will not link free nucleotides to-

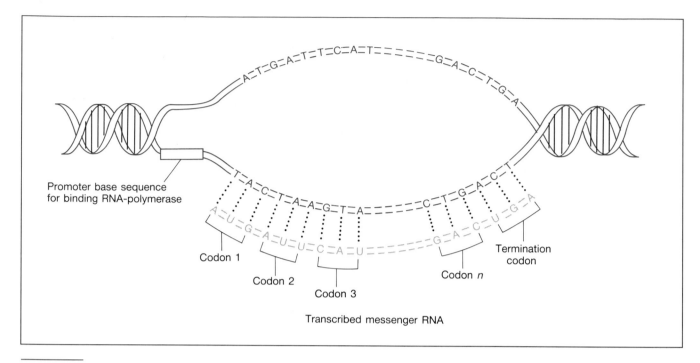

FIGURE 4-11 Transcription of a gene from a single strand of DNA to mRNA.

gether (in what would be a random sequence) when they are not base-paired with DNA. DNA acts as a modulator molecule that allosterically activates RNA polymerase.

Since DNA consists of two strands of polynucleotides, both of which are exposed during transcription, it should theoretically be possible to form two different mRNA molecules, one from each strand. However, only one of the two potential mRNAs is ever formed. Which of the two DNA strands is used as the template for mRNA synthesis is determined by a specific sequence of nucleotides in DNA, called the **promoter**, located at the beginning of each gene (Figure 4-11). The promoter, to which RNA polymerase binds, is present in only one of the two DNA strands. Beginning at the promoter end of a gene, the RNA polymerase separates the two DNA strands as it moves along one strand, joining one ribonucleotide at a time to the growing mRNA chain until it reaches a termination code word at the end of the gene, and this causes the RNA polymerase to release the newly formed mRNA.

In a given cell, the information in only a few of the thousands of genes present in DNA is transcribed into mRNA at any given time. Genes are transcribed only when RNA polymerase can bind to their promoter sites. Various mechanisms are used by cells either to block or to make accessible the promoter region of any particular gene. Such regulation of gene transcription provides a

means of controlling the synthesis of specific proteins and thereby the activities of cells.

It must be emphasized that the nucleotide sequence in mRNA is not identical to that in the corresponding strand of DNA, since its formation depends on the pairing between complementary, not identical, bases (Figure 4-11). A three-base sequence in mRNA that specifies one amino acid is called a **codon**. Each codon is complementary to a three-base sequence in DNA. For example, the base sequence T-A-C in DNA corresponds to the codon A-U-G in mRNA.

Although the entire sequence of nucleotides in a gene is transcribed into a corresponding sequence of nucleotides in mRNA, only selected portions of this sequence code for sequences of amino acids. These nucleotide sequences, known as **exons**, are separated from each other by noncoding sequences of nucleotides known as **introns**.

Before passing to the cytoplasm, a newly formed mRNA must undergo **RNA processing** (Figure 4-12) to remove the intron sequences. Nuclear enzymes that identify specific nucleotide sequences at the beginning and end of each intron remove the introns and splice the end of one exon to the beginning of another exon to form an mRNA with no intron segments. In some cases, the exons derived from a single gene can be spliced together in several different sequences, resulting in the formation

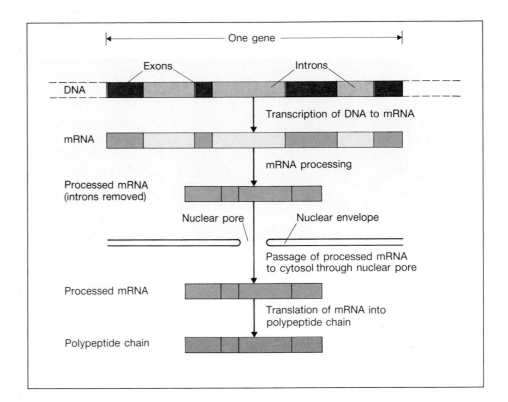

FIGURE 4-12 RNA processing removes the noncoding sequences (introns) before the mRNA passes through the nuclear pores to the cytosol. The length of the intron and exon segments represent the relative lengths of the base sequences in these regions.

of different mRNAs from the same gene and giving rise, in turn, to several different proteins.

The mRNAs formed as a result of RNA processing are 75 to 90 percent shorter than the originally transcribed mRNA, meaning that 75 to 90 percent of the nucleotide sequences in DNA are introns. What role such large amounts of "nonsense" DNA may perform is unclear.

Finally, it should be noted that, in addition to genes that give rise to mRNA and hence code for proteins, some genes form other types of RNA, as described in the next sections.

Translation: Polypeptide Synthesis

Once transcribed and processed, mRNA moves through the pores in the nuclear envelope into the cytoplasm where it binds to a ribosome, the cell organelle that contains the enzymes and other components required for the translation of mRNA's coded message into protein. Before describing this assembly process, we must first present the structure of a ribosome as well as the characteristics of two additional types of RNA involved in protein synthesis.

Ribosomes. As described in Chapter 3, ribosomes are small granules (diameter about 23 nm) located in the cytoplasm, either suspended in the cytosol (free ribo-

somes) or attached to the surface of the endoplasmic reticulum (bound ribosomes). Proteins synthesized on free ribosomes are released into the cytosol, whereas those synthesized on bound ribosomes are released into the lumen of the endoplasmic reticulum, from which they are either secreted from the cell or transferred to various organelles by processes described in Chapter 6.

Each ribosome consists of two subunits, a large 60S and a smaller 40S subunit. These subunits contain many proteins in association with a type of RNA known as **ribosomal RNA (rRNA)**. Ribosomal RNA is synthesized in the nucleus, with DNA again serving as a template for positioning the sequence of nucleotides in rRNA. The genes that code for rRNA are associated with the nucleolus, which is also the site at which the components of the two subunits of a ribosome are assembled. The ribosomal subunits then move into the cytoplasm (probably through the pores in the nuclear envelope).

When an mRNA molecule arrives in the cytoplasm, one end of it binds to the 40S subunit, and then this combination binds to the 60S subunit to form a fully functional ribosome with a portion of the mRNA lying in a groove between the two subunits (Figure 4-13). The mRNA is now ready to direct protein assembly. The sequence of codons in mRNA specifies the order of amino acids in the protein.

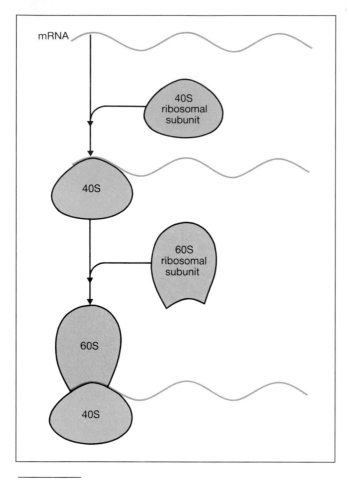

FIGURE 4-13 Ribosome-mRNA binding. First the 40S subunit binds to mRNA, and then the 60S subunit binds to the complex to form a functional ribosome.

Transfer RNA. How do individual amino acids identify the appropriate codons in mRNA during the process of translation? By themselves, free amino acids do not have the ability to bind to the bases in mRNA codons. This process of identification involves yet a third type of RNA, one known as **transfer RNA (tRNA).** Transfer RNA molecules are the smallest (about 80 nucleotides long) of the three types of RNA. The single chain of tRNA loops back upon itself, forming a structure resembling a cloverleaf with three loops (Figure 4-14).

Like mRNA and rRNA, tRNA is synthesized in the nucleus by base-pairing with DNA nucleotides, in this case at specific tRNA genes, then moves to the cytoplasm. The key to tRNA's role in protein synthesis is that it can combine with *both* a specific amino acid *and* a codon in mRNA specific for that amino acid. This permits tRNA to act as the link between an amino acid and the mRNA codon for that amino acid.

Transfer RNA is covalently linked to an amino acid by an enzyme known as aminoacyl-tRNA synthetase. There are at least 20 different aminoacyl-tRNA synthetases, each of which catalyzes the linkage of a specific amino acid to a particular type of tRNA. The next step is to link the tRNA, bearing its attached amino acid, to the mRNA codon for that amino acid. As one might predict, this is achieved by base-pairing between tRNA and mRNA. A three-nucleotide sequence at the end of one of the loops of tRNA can base-pair with a complementary codon in mRNA. This tRNA triplet sequence is appropriately termed an **anticodon.** Figure 4-14 illustrates the binding between mRNA and a tRNA specific for the amino acid alanine. Note that tRNA is covalently linked to alanine at one end and that its anticodon C-G-U is base-paired with the codon G-C-A in mRNA at the other end.

Protein assembly. The individual amino acids linked to mRNA by tRNA must now be bound to each other by peptide bonds. Several ribosomal proteins and rRNA interact in a complex manner to identify the initial codon sequence in mRNA and to catalyze the formation of a peptide bond between the first and second amino acids in the peptide chain being synthesized. This initial step is the slowest step in protein assembly, and the rate of protein synthesis can be regulated by factors that influence this initiation process.

Following the initiation of a protein's synthesis, the polypeptide chain is elongated by the successive addition of amino acids. The 60S subunit of the ribosome has two binding sites for tRNA: One holds the tRNA that is attached to the most recently added amino acid, and the other holds the tRNA containing the next amino acid to be added to the chain. Ribosomal enzymes catalyze the formation of a peptide bond between these two amino acids. Following the formation of the peptide bond, the tRNA at the first binding site is released from the ribosome, and the tRNA at the second site—now linked to the peptide chain—is transferred to the first binding site. The ribosome moves one codon space along the mRNA, making room for the binding of the next amino acid-tRNA molecule (Figure 4-15). This process is repeated over and over as each amino acid is added in succession to the growing peptide chain, at an average rate of 2 to 3 amino acids per second. When the ribosome reaches the termination codon in mRNA specifying the end of the protein, the link between the polypeptide chain and the last tRNA is broken, and the completed protein is released from the ribosome.

The same strand of mRNA can be used to synthesize many molecules of the protein because the message in mRNA is not destroyed during protein assembly. While one ribosome is moving along a strand of mRNA, a second ribosome may become attached to mRNA and begin the synthesis of a second protein molecule. Thus, a num-

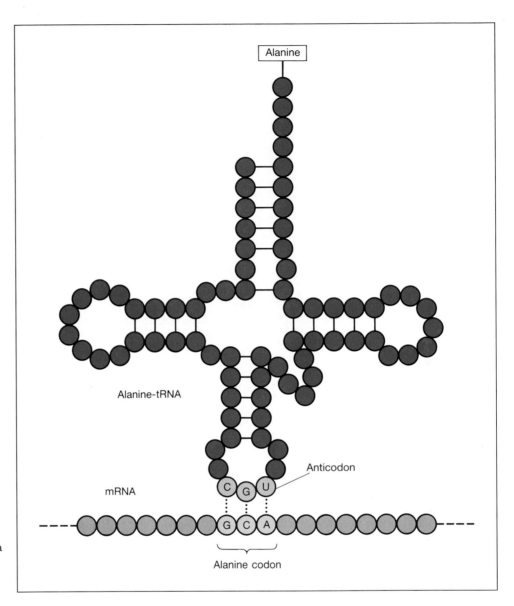

FIGURE 4-14 Base-pairing between the anticodon region of a tRNA molecule with the corresponding codon region of an mRNA molecule.

ber of ribosomes, as many as 70, may be attached to the same strand of mRNA at any one time. Ribosomes near the beginning of mRNA have short peptide chains representing the first few amino acids in the protein. Ribosomes near the end have protein chains that are almost completed (Figure 4-16).

Molecules of mRNA do not remain in the cytoplasm indefinitely. Eventually they are broken down into nucleotides by cytoplasmic enzymes. Therefore, if a gene corresponding to a particular protein ceases to be transcribed into mRNA, the synthesis of that protein will eventually slow down and cease as the mRNA is broken down.

Changes can occur in the structure of some polypeptide chains following their synthesis on a ribosome. Cer-

tain classes of proteins, the glycoproteins for example, are formed by the posttranslational addition of various carbohydrate groups to the protein. In other cases, specific peptide bonds in a large polypeptide are broken to produce a number of shorter polypeptides, each of which may perform a different function. For example, as illustrated in Figure 4-17, five proteins can be derived from the same mRNA as a result of posttranslational changes.

The steps leading from DNA to protein are summarized in Table 4-1.

Regulation of Protein Synthesis

As we mentioned earlier, certain proteins are present in most cells of the body, but they may be present at differ-

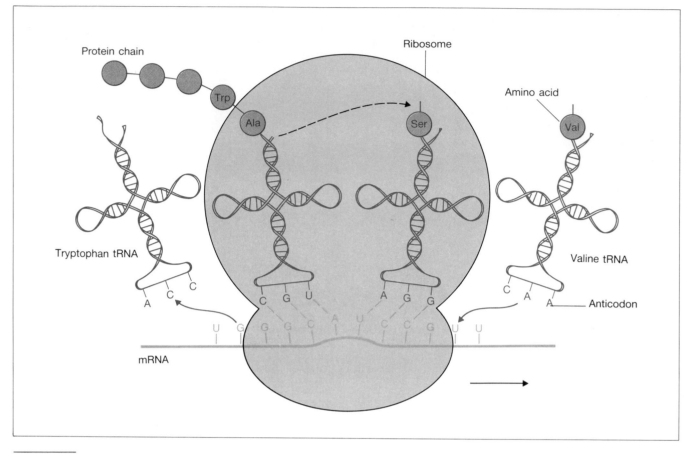

FIGURE 4-15 Sequence of events during the synthesis of a protein on the surface of a ribosome. The three-base anticodon of tRNA carrying one amino acid, for example, serine (Ser), binds to the corresponding codon of mRNA and transfers its amino acid to the growing polypeptide chain. As the ribosome moves along the strand of mRNA, successive amino acids are added to the polypeptide chain.

FIGURE 4-17 Posttranslational splitting of a protein can result in several proteins, each of which may perform a different function. All these proteins are derived from the same gene.

FIGURE 4-16 Protein assembly as multiple ribosomes move along a single strand of messenger RNA.

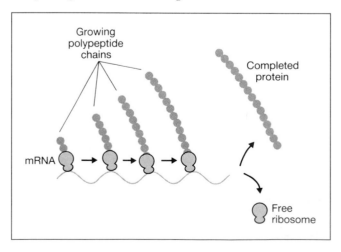

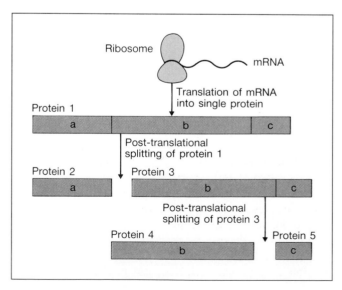

TABLE 4-1	EVENTS LEADING FROM DNA TO PROTEIN SYNTHESIS

Transcription

1. RNA polymerase binds to the promoter region of a gene and separates the two strands of the DNA double helix in the region of the gene to be transcribed.

2. Free ribonucleotide triphosphates base-pair with the deoxynucleotides in DNA.

3. The ribonucleotides paired with one strand of DNA are linked by RNA polymerase to form mRNA containing a sequence of bases complementary to one strand of the DNA base sequence.

4. mRNA processing removes the intron regions of mRNA, which contain noncoding sequences, and splices together the exon regions, which code for specific amino acids.

Translation

5. Processed mRNA passes from the nucleus to the cytoplasm, where one end of the mRNA binds to a ribosome.

6. Free amino acids are linked to their corresponding tRNAs by aminoacyl-tRNA synthetase.

7. The three-base anticodon in an amino acid-tRNA complex pairs with the corresponding codon in the region of the mRNA bound to the ribosome.

8. The portion of the peptide that has already been synthesized (and is still attached to a tRNA bound to the ribosome) is now linked by a peptide bond to the amino acid at the end of the tRNA next to it, thereby adding one more amino acid to the chain.

9. The tRNA that has been freed of the peptide chain is released from the ribosome.

10. The ribosome moves one codon step along mRNA.

11. Steps 7 to 10 are repeated over and over until the end of the mRNA message is reached.

12. The completed protein chain is released from the ribosome when the termination codon in mRNA is reached.

13. In some cases, the protein undergoes posttranslational processing in which various chemical groups are attached to specific side chains or the protein is split into several smaller peptide chains.

ent concentrations. Other proteins are found only in specific types of cells. Furthermore, cells do not synthesize their proteins at the same rate all the time. Thus, both the types and rates of protein synthesis are subject to physiological regulation.

Protein synthesis can be regulated at four steps in the process that leads from DNA to the final protein: (1) Regulation of gene transcription into mRNA can alter the rate of mRNA formation. (2) Regulation of mRNA processing can produce different combinations of exons or alter the ability of mRNA to associate with a ribosome. (3) Regulation of the stability of processed mRNA—the average time a molecule of mRNA remains in the cytoplasm before being broken down—can control the number of protein molecules formed from a molecule of mRNA. (4) Regulation of mRNA translation can alter the rate at which a protein is assembled on a ribosome. One or more of the above regulatory mechanisms may be involved in the synthesis of any given protein. Furthermore, these control systems may affect all the proteins that are being synthesized by a cell or only specific proteins.

It must be emphasized that in addition to regulating the synthesis of a particular protein, a cell may also regulate the activity of the completed protein by the allosteric and covalent mechanisms discussed earlier in this chapter.

The complex interactions that lead to the regulation of protein synthesis are only beginning to be worked out. In those cases where the mechanism is at least partially understood, the basis for regulation involves the binding of a molecule to DNA, various processing enzymes, or one of the ribosomal components. The result is either the inhibition or stimulation of the activity occurring at that site. The specific molecules that act upon these regulatory sites are for the most part still undiscovered. As we shall see in Chapter 10, hormones, chemical signals that travel from one cell to another by way of the blood, often stimulate or inhibit the synthesis of specific proteins.

REPLICATION AND EXPRESSION OF GENETIC INFORMATION

The development of the human body from a single fertilized ovum involves cell growth, cell division, and the differentiation of cells into specialized types such as nerve and muscle cells. The maintenance of structure and function in the adult body also requires these same processes. Each depends on the controlled synthesis of proteins. Cell division requires, in addition, the replication of DNA and the transmission of identical copies of this genetic information to each of the two resulting cells.

Replication of DNA

DNA is the only molecule in a cell able to duplicate itself without information from some other cell component. In contrast, as we have seen, mRNA can be formed only in the presence of DNA, protein can be formed only if mRNA is present, and all other molecules synthesized

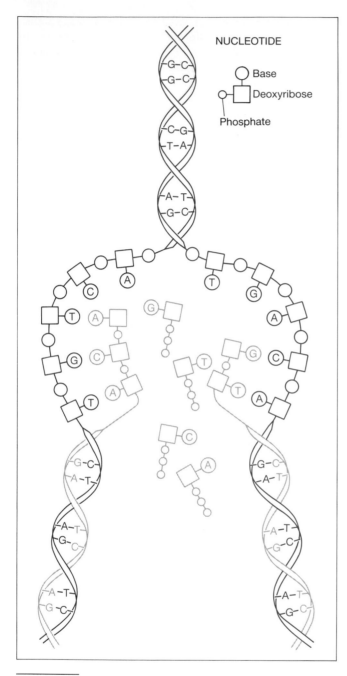

NUCLEOTIDE

◯ Base

◻ Deoxyribose

○ Phosphate

FIGURE 4-18 Replication of DNA involves the pairing of free nucleotides with the bases of each DNA strand, giving rise to two new identical DNA molecules, each containing one old and one new polynucleotide strand.

by a cell are formed by metabolic pathways that use proteins in the form of enzymes.

The replication of DNA is, in principle, similar to the process whereby mRNA is synthesized. During DNA replication (Figure 4-18), the two strands of the double helix separate, and the exposed bases in each strand base-pair with free deoxyribonucleotide triphosphates present in the nucleus. The enzyme **DNA polymerase** then links the free nucleotides together forming a new strand of DNA, a process very similar to that used to form mRNA. In contrast to the synthesis of mRNA, where only one strand of DNA was used as a template, here both strands of the original DNA act as templates for the synthesis of new strands. The end result is two identical molecules of DNA, each called a "copy." In each copy, one strand of nucleotides was present in the original DNA molecule and one strand has been newly synthesized.

Prior to cell division, the DNA molecules in the nucleus are replicated by the above process, and one copy will be passed on to each of the two new cells, termed **daughter cells**, when the cell divides. Thus, the daughter cells receive the same set of instructions present in the parent cell.

Cell Division

Starting with a single fertilized egg cell, the first cell division produces 2 cells. When these daughter cells divide, they each produce 2 cells, giving a total of 4 cells. These 4 cells produce a total of 8 cells, and so on. Thus, starting from a single cell, 3 division cycles will produce 8 cells (2^3), 10 division cycles will produce $2^{10} = 1024$ cells, and 20 division cycles will produce $2^{20} = 1,048,576$ cells. If the development of the human body involved only the cycle of cell division and growth without any cell death, it would require only about 46 division cycles to produce all the cells in the adult body. However, large numbers of cells die during the course of development, and even in the adult body many cells survive for only a few days and are continually being replaced by the division of existing cells.

Aside from the morphological description of the various stages of the division process as visualized with a light or electron microscope, surprisingly little is known about the underlying molecular events that occur during division or how the process is initiated and regulated. We will therefore limit our discussion of cell division to a brief description of the various stages of the process and focus our attention on the net result, which is the duplication, packaging, and distribution of DNA to the daughter cells.

Although the time between cell divisions varies considerably in different types of cells, the most rapidly growing cells divide about once every 24 h. During most of this period, there is no visible evidence that the cell will divide. For example, in a 24-h division cycle, visible changes in cell structure begin to appear 23 h after the previous division. The period between the end of one division and the appearance of the structural changes that indicate the beginning of the next is known as **inter-phase**. Since cell division takes only about 1 h, the cell spends most of its time in interphase, and most of the cell properties described in this book are properties of interphase cells. One very important event related to the subsequent cell division does occur during interphase, namely, the replication of DNA, which begins about 10 h prior to the first visible signs of division.

Cell division involves two processes: nuclear division (**mitosis**) and cytoplasmic division (**cytokinesis**). Although mitosis and cytokinesis are separate events, the term mitosis is often used in a broad sense to include the subsequent cytokinesis. Mitosis that is not followed by cytokinesis produces the multinucleated cells found in the liver, placenta, and in some embryonic cells and cancer cells.

During most of interphase, DNA is dispersed throughout the nucleus in association with proteins to form 46 extended nucleoprotein threads known as **chromatin** (Figure 4-19A). The DNA in each chromatin thread has a different nucleotide sequence and therefore carries a different set of genes. If a cell is to divide, as noted above, the DNA of each chromatin thread replicates during interphase, the result being two identical threads termed **sister chromatids**. In other words, there are now 46 pairs of sister chromatids. Each pair is joined together at one point known as the **centromere** (Figure 4-19B). As a cell enters mitosis (Figure 4-19C), each chromatid pair becomes highly coiled and condensed,

forming rod-shaped bodies known as **chromosomes** (colored bodies, so named because of their intense staining by the dyes used to visualize structures under a microscope).

As the chromosomes condense, the nuclear membrane breaks down, and the chromosomes become linked, at their centromeres, to spindle fibers. The **spindle fibers**, composed of microtubules, generate the forces that divide the cell. Some of the spindle fibers extend between two **centrioles** (Figure 4-19C) located on opposite sides of the cell, while others are connected to the chromosomes. Each centriole consists of two microtubular bodies oriented at right angles to each other. A single centriole will pass to each of the daughter cells during cytokinesis. During the preceding interphase, at the time that DNA was replicating, the second centriole formed, and the two centrioles moved to opposite sides of the nucleus. The spindle fibers and centrioles constitute the **mitotic apparatus**.

As mitosis proceeds, the sister chromatids in each chromosome separate at the centromere and move toward the opposed centrioles (Figure 4-19D). The spindle fibers act as though they were pulling the chromatids toward the poles although the actual molecular mechanism of this movement is still uncertain.

Cytokinesis begins as the sister chromatids separate. The cell begins to constrict along a plane perpendicular to the axis of the mitotic apparatus, and constriction continues until the cell has been pinched in half, forming

FIGURE 4-19 Mitosis and cytokinesis. (Only 4 of the 46 chromosomes in a human cell are illustrated.) (A) During interphase, chromatin exists in the nucleus as long, extended threads. (B) Prior to the onset of mitosis, DNA replicates, forming two sister chromatids that are joined at the centromere. A second centriole is also formed at this time. (C) As mitosis begins, the chromatids condense into chromosomes, which become attached to spindle fibers. (D) The two chromatids of each chromosome separate and move toward opposite poles of the cell as the cell divides (cytokinesis) into two daughter cells (E).

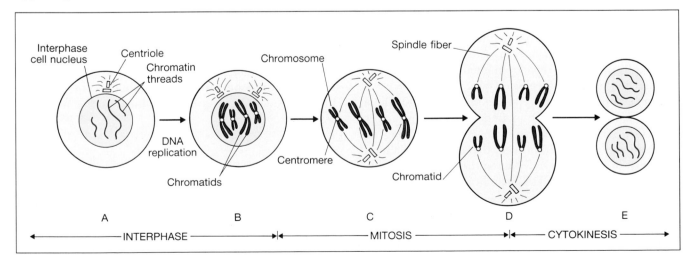

two daughter cells (Figure 4-19E). Following cytokinesis, the spindle fibers dissolve, a nuclear envelope forms, and the chromatids uncoil in each daughter cell.

Cell Differentiation

Since identical sets of DNA molecules pass to each of the daughter cells during cell division, every cell in the body, with the exception of the reproductive cells, contains the same genetic information as every other cell. How, then, is it possible for one cell to become a muscle cell and synthesize muscle proteins while another cell containing the same genetic information differentiates into a nerve cell and synthesizes a different set of proteins? We have already given a partial answer—different combinations of genes are active in the different cells. The genes that contain the information for synthesis of muscle proteins synthesize mRNA in muscle cells. These same genes are also present in nerve cells but do not form mRNA. Other genes are active in the nerve cell but are not active in muscle cells.

The problem of cell differentiation is thus related to the general problem of regulating protein synthesis. Certain genes are "turned *on*" or "turned *off*" during cell differentiation. The signals that control the transcription of specific genes are mostly unknown. In addition, there is the problem of timing. Some cells differentiate early during embryonic development, others later, or even after birth. What determines the timing of the signals that turn *on* certain genes and inhibit others during the course of development? To understand the complex process of cell differentiation and development, a great deal more must be learned about the processes that regulate protein synthesis.

Mutation

In order to form the 40 trillion cells of the adult human body, a minimum of 40 trillion cell divisions must occur. Thus, the DNA in the original fertilized ovum must be replicated at least 40 trillion times. Actually, many more than 40 trillion divisions occur during the growth of an ovum into an adult human being since, as noted earlier, many cells die during development and are replaced by the division of existing cells.

If a secretary were to type the same book 40 trillion times, one would expect to find some typing errors. Therefore, it is not surprising to find that during the duplication of DNA, errors occur that result in an altered sequence of bases and a change in the genetic message. What is amazing is that DNA can be duplicated so many times with relatively few errors. Any alteration in the genetic message carried by DNA is known as a **mutation**.

Factors in the environment that increase the mutation rate are known as **mutagens** and include certain chemicals and various forms of ionizing radiation, such as x-rays, cosmic rays, and atomic radiation. Most of these mutagens break chemical bonds in DNA, so that incorrect pairing between bases occurs or the wrong base is incorporated when the broken bonds are reformed. Even in the absence of specific agents that increase the likelihood of mutations, mistakes in DNA copying do occur, so that the mutation rate is never zero.

The simplest type of mutation occurs when a single base is inserted at the wrong position in DNA. For example, the base sequence C-G-T forms the DNA code word for the amino acid alanine. If guanine G is replaced by adenine A in this sequence, it becomes C-A-T, which is the code word for valine.[2] It is during the replication of DNA, when free nucleotides are being incorporated into new strands of DNA, that incorrect base substitution is most likely to occur.

In a second type of mutation, large sections of DNA are deleted from the molecule, or single bases are added or deleted. Such mutations may cause the loss of an entire gene or group of genes or may cause the misreading of a large sequence of bases. Figure 4-20 shows the effect of removing a single base on the reading of the genetic code. Since the code is read in sequences of three bases, the removal of one base not only alters the code word containing that base but also causes a misreading of all subsequent bases by shifting the reading sequence. Addition of an extra base would cause a similar misreading. Such mutations will result either in no protein being formed, if the gene has been deleted, or in the formation of a nonsense protein, a protein in which the amino acid sequence does not correspond to any functional protein, when single bases are added or deleted.

Assume that a mutation has altered a single code word so that it now codes for a different amino acid, say, for example, alanine C-G-T to valine C-A-T. What effect does this mutation have upon the cell? The answer depends upon both the type of gene and where in the gene the mutation has occurred. Although proteins are composed of many amino acids, the properties of a protein often depend upon only a very small region of the total molecule, such as the binding site of an enzyme. If the mutation does not alter the conformation of the binding site, there may be little or no change in the protein's properties. On the other hand, if the mutation alters the

---

[2] Because the genetic code has several different code words representing the same amino acid, a mutation in a single code word does not always alter the type of amino acid coded. We saw that substituting A for G in the alanine code word C-G-T produced the valine code word C-A-T. If, instead, the mutation caused cytosine C to be substituted for thymine T, the new codon C-G-C is one of several that correspond to alanine, and the protein formed by the mutant gene will not be altered in amino acid sequence in spite of the change in base sequence.

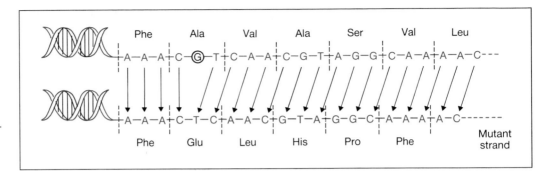

FIGURE 4-20 deletion mutation caused by the loss of a single base G in one of the two DNA strands causes a misreading of all code words beyond the point of the mutation.

binding site, a marked change in the protein's properties may occur. Thus, if the protein is an enzyme, a mutation may render it totally inactive or change its affinity for substrate. The mutated enzyme may even catalyze an entirely different type of reaction.

Let us assume that the mutation leads to an enzyme that is totally inactive. If the enzyme is in a pathway supplying most of the cell's chemical energy, the loss of the enzyme may lead to the death of the cell. On the other hand, the normal enzyme may be involved in the synthesis of a particular amino acid, and if the cell can obtain that amino acid from the extracellular fluid, the cell's functioning will not be impaired by the mutation.

To generalize, a mutation may have any one of three effects upon a cell: (1) It may cause no noticeable change in the cell's functioning, (2) it may modify cell function but still be compatible with cell growth and replication, or (3) it may lead to cell death.

With one exception—cancer, to be described below— the malfunction or death of a single cell, other than a sperm or ovum, as a result of a mutation usually has no significant effect because there are so many cells performing the same function in an organ. Unfortunately, the story is more complex when the mutation has occurred in a sperm or ovum, since these cells combine to form the fertilized ovum from which all the cells of a new individual will develop. In this case, the mutation will be passed on to all the cells in the body. Thus, mutations in the ovum or sperm do not affect the individual in which they occur but do affect, often catastrophically, the children produced by these cells. Moreover, these mutations will also be passed on to some individuals in future generations descended from the individual carrying the original mutant gene.

Inherited diseases resulting from gene mutation are termed **inborn errors in metabolism.** An example is phenylketonuria, a disorder that can lead to a form of mental retardation in children. Because of a single abnormal enzyme, these persons are unable to convert the amino acid phenylalanine to the amino acid tyrosine at a normal rate. Phenylalanine is therefore diverted into other biochemical pathways in large amounts, giving rise to products that interfere with the normal activity of the nervous system. These products are also excreted in large amounts in the urine, accounting for the name of the disease. Fortunately, the symptoms of the disease can be prevented if the content of phenylalanine in the diet is restricted during childhood, thus preventing accumulation of the toxic products formed from phenylalanine.

Cells possess mechanisms for protecting themselves against certain types of mutation. For example, if an abnormal base-pairing occurs in DNA, such as C with T (C normally pairs with G), a particular set of enzymes will cut out the segment containing the abnormal base T, allowing the normal strand to resynthesize the deleted segment by normal base-pairing.

Mutations contribute to evolution. Mutations may alter the activity of an enzyme in such a way that it is more, rather than less, active, or they may introduce an entirely new type of enzyme activity into a cell. If an organism carrying such a mutant gene is able to perform some function more effectively than an organism lacking the mutant gene, it has a better chance of surviving and passing the mutant gene on to its descendants. If the mutation produces an organism that functions less effectively than organisms lacking the mutation, the organism is less likely to survive and pass on the mutant gene. This is the principle of natural selection. Although any one mutation, if it is able to survive in the population, may cause only a very slight alteration in the properties of a cell, given enough time, a large number of small changes can accumulate to produce very large changes in the structure and function of an organism.

Cancer

In the adult body, the rates at which new cells in a tissue are formed and old cells die are in balance, producing a steady state in which the tissue does not increase in size. If the mechanisms regulating cell division are altered, however, the affected cells may produce new cells faster than old cells are removed, forming a growing mass of tissue known as a **tumor**. If the tumor cells remain localized and do not invade surrounding tissues, it is said to be a **benign tumor**. If, however, the tumor cells grow into the surrounding tissues, disrupting their functions, and/or spread to other regions of the body, it is said to be a **malignant tumor** and may lead to the death of the organism.

Malignant tumors are composed of **cancer cells**, which are cells that have lost the ability to respond to the normal control mechanisms that regulate cell growth. Cancer cells are characterized by their capacity for unlimited multiplication and for their ability to break away from the parent tumor and to spread by way of the circulatory system to other parts of the body, where they form multiple tumor sites, a process known as **metastasis**.

If cancer is detected in the early stages of its growth, before it has metastasized, the tumor may be removed by surgery. Once it has metastasized to many organs, surgery is no longer possible. Drugs and radiation can be used to inhibit cell multiplication and destroy malignant cells, both before and after metastasis. Unfortunately, these treatments also damage the growth of normal cells. Approximately 1200 Americans die each day from cancer.

Any normal cell in the body may at some point undergo a transformation to a malignant cell. A number of agents known as **carcinogens**, such as radiation, viruses, and certain chemicals, can induce the cancerous transformation of cells. These agents act by altering or activating various genes associated with cell growth.

It is important to note that mutagenic agents also tend to be carcinogenic. The mutation of certain genes in animal cells has been found to transform these cells into cancer cells, and similar genes have been found in human cells. Such cancer-producing genes are called **oncogenes** (Greek: *onkos*, mass, tumor—the branch of medicine that studies and treats cancer is known as oncology). Mutations produced by carcinogens are, however, only the first step in the development of many cancers. In addition, chemical agents, such as hormones or environmental chemicals, may be required to cause the enhanced cell replication leading from a transformed cell to a tumor.

At the present time it is unclear whether a single mutation or two or more mutations are necessary to transform a normal cell into a cancer cell. The multiple mutation hypothesis provides one explanation for the increased incidence of cancer with age. The first mutation does not produce cancer, only a precancerous state in the cell. Such a cell may then be transformed into a cancer cell when it later acquires the second mutation.

Some individuals may inherit one oncogene and thus be at increased risk for developing cancer if they should undergo a second mutation. As will be discussed in Chapter 19, the body's defense system is normally able to detect and destroy most cancerous cells. These defense systems become less efficient with age, which also contributes to the increased incidence of cancer with age.

Little is currently known about the molecular mechanisms that are altered by oncogenes or the regulatory systems that control normal tissue growth. Some of the proteins produced by oncogenes have been identified, however. These are either protein kinases or signal-detecting proteins that bind messenger molecules involved in regulating growth. In some cases it appears that the gene, prior to mutation, produces a protein that inhibits the transcription of another gene involved in growth. The loss of this inhibitory protein through mutation allows the unrestrained transcription of the growth-promoting gene.

It is hoped that by understanding the molecular mechanisms that produce cancer, we may be able to prevent or treat this disease, which is second only to heart disease as the major cause of death in the Western world.

Recombinant DNA

In the early 1970s, bacterial enzymes were discovered that split molecules of DNA in a unique manner. These enzymes, called **restriction nucleases**, identify specific nucleotide sequences, usually four to six nucleotides long, and in this region they cut DNA at a different site on each strand (Figure 4-21). The cut strands each have a short exposed sequence of bases that are complementary to each other and can bind together by base-pairing. A restriction nuclease breaks the DNA strands at many points, resulting in a number of DNA segments, each having an exposed base sequence at each end that can bind to the ends of other DNA segments split by the same restriction nuclease.

If the DNA from two organisms is cut by the same restriction nuclease, then a DNA fragment from one organism can base-pair with the complementary end of a DNA segment from a second organism. Another enzyme, known as a ligase, can then covalently link the two molecules of DNA together. By this procedure, DNA fragments (genes) from one organism can be inserted into the DNA of a second organism to form **recombinant DNA**. When the recombinant DNA is inserted into a living cell, its genetic message is transcribed into mRNA along with the messages from the host's genes and then translated into protein.

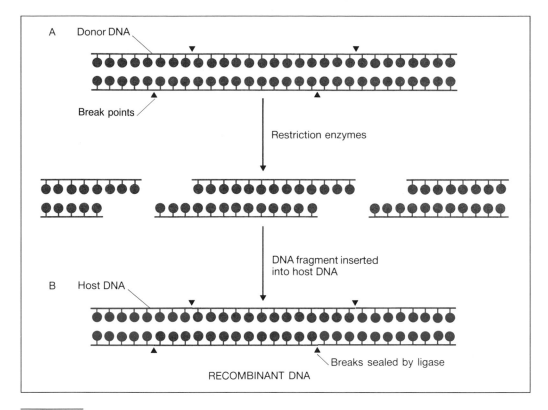

FIGURE 4-21 The basis of recombinant DNA. (A) Bacterial restriction enzymes break the two strands of DNA at different points producing ends with exposed bases that are complementary to each other. (B) A segment of DNA containing one or more genes from one organism (donor) can be inserted into the DNA of another organism (host) by using these restriction enzymes to produce DNA segments whose ends can bind together.

Recombinant DNA technology has greatly increased our knowledge of genes and how they work. It has led to the isolation of specific genes that can be transferred by recombinant DNA into bacteria, so that the multiplication of the bacteria will replicate the recombinant DNA along with the host DNA, producing large amounts of "cloned DNA." Determination of the nucleotide sequence of this cloned DNA can then be used to indirectly determine the amino acid sequence of the protein coded for by the gene. The amino acid sequences of many proteins are now being determined in this manner, a technial procedure that is easier than determining the amino acid sequence from a protein directly.

The potential benefits of this recombinant technique are many. For example, inserting the gene that codes for human insulin into bacterial DNA leads to the production of insulin by the bacteria. This hormone can then be extracted and used to treat diabetic patients who are unable to synthesize their own insulin. Although not currently possible, it is hoped that the techniques used to form recombinant DNA may one day provide the ability to selectively replace in humans mutant genes that cause inborn errors in metabolism with normal genes and thus provide a cure for these diseases.

On the other hand, the technology also bears potential hazards. For example, it would be theoretically possible to insert a gene that codes for a toxic protein into a harmless bacterium that commonly inhabits the human gastrointestinal tract and thereby produce a widespread epidemic. Only the future will tell whether the benefits of genetic "engineering" will outweigh the hazards of being able to manipulate an organism's genes.

SUMMARY

Section A. Protein Binding Sites

Binding-Site Characteristics

I. Ligands bind to proteins at sites that have shapes that are complementary to the ligand shape.

II. Protein binding sites have the properties of chemical specificity, affinity, saturation, and competition.

Regulation of Binding-Site Characteristics

I. Protein activity in a cell can be controlled by regulating either the shape of the protein or the amount of protein synthesized.

II. The binding of a modulator molecule to the regulatory site on an allosteric protein alters the shape of the functional binding site, thereby altering its binding characteristics and the activity of the protein. The activity of allosteric proteins is regulated by varying the concentrations of their modulator molecules.

III. Protein kinase enzymes catalyze the addition of a phosphate group to the side chains of certain amino acids in a protein, changing the shape of the protein's binding site and thus altering the protein's activity by covalent modulation. A second enzyme is required to remove the phosphate group, returning the protein to its original state.

Section B. Genetic Information and Protein Synthesis

Genetic Information

I. Genetic information is coded in the nucleotide sequence of DNA. A single gene contains either (a) the information that, via mRNA, determines the amino acid sequence in a specific protein, or (b) the information for forming rRNA or tRNA, which assist in protein assembly.

II. Genetic information is transferred from DNA to mRNA in the nucleus, and then mRNA passes to the cytoplasm, where its information is used to synthesize protein.

III. The code words in the DNA genetic code consist of a sequence of three nucleotide bases that specify a single amino acid. The sequence of triplet-code words along a gene determines the sequence of amino acids in a protein. There can be more than one code word specifying a given amino acid.

Protein Synthesis

I. Protein synthesis involves (a) transcription of genetic information from DNA to mRNA, which occurs in the nucleus, and (b) translation of the information in mRNA into protein, which occurs in the cytoplasm on the surface of ribosomes. Table 4-1 summarizes the steps leading from DNA to protein and the role of tRNA.

II. The rate of protein synthesis can be regulated by controlling: (1) transcription into mRNA, (2) processing of mRNA, (3) stability of mRNA and, (4) translation of mRNA by ribosomes.

Replication and Expression of Genetic Information

I. When DNA replicates, exposed bases in each of the two unwound strands base-pair with bases of free deoxyribonucleotide triphosphates. DNA polymerase joins the nucleotides together to form a new strand of DNA.

II. When a cell divides, one copy of DNA passes to each daughter cell, so that both receive the same set of genetic instructions.

III. Cell division, consisting of nuclear division—mitosis—and cytoplasmic division—cytokinesis—lasts about 1 h.

IV. During interphase, DNA replicates, forming two identical sister chromatids joined by a centromere.

V. The main events of mitosis are:
  A. The chromatids condense into highly coiled chromosomes.
  B. The centromeres of each chromosome become attached to spindle fibers extending from the centrioles that have migrated to opposite poles of the nucleus.
  C. The two chromatids of each chromosome separate and move toward opposite poles of the cell as the cell divides into two daughter cells. Following cell division, the condensed chromatids uncoil into their extended interphase form.

VI. The formation of specialized, differentiated cells involves the selective turning on or off of specific genes in the different types of cells, thereby controlling the types of proteins that are formed in these cells.

VII. Mutations, alterations in the genetic message, occur when incorrect bases are inserted into new strands of DNA during replication or when single bases or large sections of DNA are added or deleted.
  A. A mutation may (a) cause no noticeable change in a cell's function, (b) modify cell function but still be compatible with cell growth and replication, or (c) lead to the death of the cell.
  B. Mutations occurring in ova or sperm cells that combine to form a fertilized ovum will be passed on to all the cells in the body of a new individual and to some of the individuals in future generations.

VIII. Cancer cells are characterized by their capacity for unlimited multiplication and their ability to metastasize to other parts of the body, forming multiple tumor sites.
  A. The mutation of certain growth-regulating genes into oncogenes can transform a normal cell into a cancer cell. Additional chemical stimuli may be necessary to cause the enhanced cell replication of a cancer cell.
  B. More than one mutation may be necessary to cause the transformation of a normal cell into a cancer cell.

IX. Using bacterial enzymes, segments of DNA can be cut from the DNA of one organism and inserted into the DNA of another organism to form recombinant DNA. The recombinant DNA will then direct the synthesis of proteins coded by the donor DNA in the host organism.

REVIEW QUESTIONS

Section A. Protein Binding Sites

1. Define:

ligand	regulatory site
binding site	modulator molecule
chemical specificity	covalent modulation
affinity	phosphorylation
saturation	enzymes
competition	substrates
allosteric modulation	products
allosteric proteins	protein kinase
functional site	

2. List the four characteristics of a protein binding site.

3. List the types of forces that hold a ligand on a protein surface.

4. What characteristics of a binding site determine its chemical specificity?

5. Under what conditions can a single binding site have a chemical specificity for more than one type of ligand?

6. What characteristics of a binding site determine its affinity for a ligand?

7. What two factors determine the percent saturation of a binding site?

8. Describe the mechanism responsible for competition in terms of the properties of binding sites.

9. What are the two ways of controlling protein activity in a cell?

10. Name two nonspecific factors that can alter a protein's activity by altering its shape. Why are these not used to control protein activity in cells under normal conditions?

11. How is the activity of an allosteric protein modulated?

12. How does regulation of protein activity by covalent modulation differ from that of allosteric modulation?

Section B. Genetic Information and Protein Synthesis

1. Define:

gene	DNA polymerase
genetic code	daughter cells
termination code words	interphase
messenger RNA (mRNA)	mitosis
transcription	cytokinesis
translation	chromatin
RNA polymerase	sister chromatids
promoter	centromere
codon	chromosomes
exons	spindle fibers
introns	centrioles
RNA processing	mitotic apparatus
ribosomal RNA (rRNA)	mutation
transfer RNA (tRNA)	mutagens
anticodon	inborn errors in metabolism

tumor	carcinogens
benign tumor	oncogenes
malignant tumor	restriction nucleases
cancer cells	recombinant DNA
metastasis	

2. Summarize the direction of information flow during protein synthesis.

3. Describe the way in which the genetic code in DNA specifies the amino acid sequence in a protein.

4. List the four nucleotides used to form messenger RNA.

5. Describe the main events that occur during the transcription of genetic information into mRNA.

6. List the components of a ribosome, and identify the site of ribosomal subunit assembly.

7. Describe the role of tRNA in protein assembly.

8. Describe the events of protein translation that occur on the surface of a ribosome.

9. List the four ways in which regulation of protein synthesis can occur.

10. Describe the mechanism by which DNA is replicated.

11. Summarize the main events of mitosis and cytokinesis.

12. Describe in broad terms what makes a nerve cell different from a muscle cell even though they both contain the same genes.

13. How will the deletion of a single base in a gene affect the protein synthesized?

14. List the three effects that a mutation can have on a cell's functioning.

15. Describe the characteristics of a cancer cell.

16. What are some of the products formed by oncogenes?

17. Describe the mechanism by which recombinant DNA is formed.

THOUGHT QUESTIONS

(Answers are given in Appendix A.)

1. A variety of compounds that regulate acid secretion in the stomach bind to proteins in the membranes of the acid-secreting cells. Some of these binding reactions lead to increased acid secretion, others inhibit its secretion. In what ways might a drug that inhibits acid secretion be acting on these cells?

2. In one type of diabetes the plasma concentration of the hormone insulin is normal but the response of the cells to which insulin usually binds is markedly decreased. Suggest a reason for this observation in terms of the properties of protein binding sites.

3. Given the following substances in a cell and their effects on each other, predict the change in compound H that will result from an increase in compound A and diagram the sequence of changes.

Compound A is a modulator molecule that allosterically stimulates protein B.

Protein B is a protein kinase enzyme that stimulates protein C.

Protein C is an enzyme that converts compound D to compound E.

Compound E is a modulator molecule that allosterically inhibits protein F.

Protein F is an enzyme that converts compound G to compound H.

4. Shown below is the relation between the amount of acid secreted and the concentration of compound X, which stimulates acid secretion in the stomach by binding to a membrane protein.

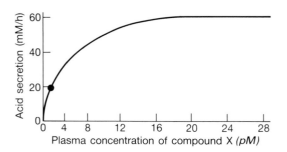

At a plasma concentration of 2 p*M*, compound X produces an acid secretion of 20 mmol/h.

A. Specify two ways in which acid secretion by compound X could be increased to 40 mmol/h.

B. Why will increasing the concentration of compound X to 28 p*M* not produce more acid secretion than increasing the concentration of X to 18 p*M*.

5. What would the consequences for protein regulation be of a mutation that led to the loss of phosphoprotein phosphatase from cells?

6. A base sequence in a portion of one strand of DNA is: A-G-T-G-C-A-A-G-T-C-T. Predict:

A. The base sequence in the corresponding strand of DNA.

B. The base sequence in a molecule of mRNA transcribed from the sequence shown.

7. The triplet code in DNA for the amino acid histidine is G-T-A. Predict the mRNA codon for this amino acid and the tRNA anticodon.

8. If a protein contains 100 amino acids, how many nucleotides will be present in the gene that codes for this protein?

9. Why will chemical agents that inhibit the polymerization of tubulin (Chapter 3, page 50) inhibit cell division?

10. Why are drugs that inhibit the replication of DNA potentially useful in the treatment of cancer? What are some of the limitations of such drugs?

ENERGY AND CELLULAR METABOLISM

Thousands of chemical reactions are occurring each instant throughout the body, the entire collection of reactions being termed **metabolism** (Greek: change). Metabolism includes the synthesis of the specific molecules required for cell structure and function, the breakdown of these molecules, and the provision of the energy for cell functions. The synthesis of organic molecules by cells is called **anabolism**, and the breakdown, **catabolism**.

The body's organic molecules undergo continuous transformations as some molecules are broken down while others of the same type are being synthesized. Chemically, no person is the same at noon as at 8 o'clock in the morning since during even this short period much of the body's structure has been torn apart and replaced with newly synthesized molecules. In adults the body's composition is in a state of dynamic equilibrium in which the anabolic and catabolic rates for most molecules are equal.

This chapter describes four aspects of metabolism: (1) the factors that determine the rate of chemical reactions, (2) the role of enzymes in accelerating chemical reactions and how enzyme activity can be controlled, (3) the biochemical pathways by which energy is obtained from the breakdown of carbohydrates, fats, and proteins, and (4) the pathways by which carbohydrates and fats are synthesized. (The pathway for protein synthesis was described in Chapter 4.) We begin by discussing the characteristics of chemical reactions because these concepts underlie all the other chapter topics.

CHEMICAL REACTIONS

Chemical reactions involve (1) the breaking of chemical bonds in reactant molecules followed by (2) the formation of new chemical bonds in the product molecules. In the chemical reaction in which carbonic acid is transformed into carbon dioxide and water, for example, two of the chemical bonds in carbonic acid are broken and the product molecules are formed by establishing two new bonds between different pairs of atoms.

$$\underset{\substack{\uparrow \\ \text{Broken} \quad \text{Broken}}}{H-O-\overset{\displaystyle O}{\underset{}{C}}-O-H} \longrightarrow \underset{\substack{\uparrow \qquad \uparrow \\ \text{Formed} \quad \text{Formed}}}{O=\overset{\displaystyle O}{\underset{}{C}} + H-O-H}$$

$$\underset{\substack{\text{Carbonic} \\ \text{acid}}}{H_2CO_3} \longrightarrow \underset{\substack{\text{Carbon} \\ \text{dioxide}}}{CO_2} + \underset{\text{Water}}{H_2O}$$

Since the energy content of the reactants and products are usually different, and because energy can neither be created nor destroyed, energy must either be added or released during a chemical reaction. For example, the reaction in which carbonic acid breaks down into carbon dioxide and water occurs with the release of 4 kcal of energy per mole of reactants since carbonic acid has a higher energy content (155 kcal/mol) than the sum of carbon dioxide and water (94 + 57 = 151 kcal/mol).

$$\underset{\substack{\text{Carbonic acid} \\ \\ \text{Reactants} \\ \\ 155 \text{ kcal/mol}}}{H_2CO_3} \longrightarrow \underset{\substack{\text{Carbon dioxide} \\ \\ \\ 94 \text{ kcal/mol}}}{CO_2} + \underset{\substack{\text{Water} \\ \\ \text{Products} \\ 57 \text{ kcal/mol}}}{H_2O} + \underset{\substack{\text{Heat energy}}}{4 \text{ kcal/mol}}$$

The energy that is released appears as heat energy, the energy of increased molecular motion, which is measured in units of calories. One **calorie** (1 cal) is the amount of heat energy required to raise the temperature of 1 g of water 1 C°. Energies associated with most chemical reactions are several thousand calories per mole and are reported as **kilocalories** (1 kcal = 1000 cal).

Determinants of Reaction Rates

The rate of a chemical reaction (in other words, how fast or slowly the reaction takes place) can be determined by measuring the change in the concentration of reactants or products per unit of time. The faster the product concentration increases or the reactant concentration decreases, the higher the rate of the reaction. Four factors (Table 5-1) influence the reaction rate: reactant concentration, activation energy, temperature, and the presence of a catalyst.

The lower the concentration of reactants, the slower the reaction simply because there are fewer molecules available to react. Conversely, the higher the concentration of reactants, the faster the reaction rate.

Given the same initial concentrations of reactants, however, all reactions do not occur at the same rate. Each type of chemical reaction has its own characteristic rate, which depends upon what is called the activation energy for the reaction. In order for a chemical reaction to occur, reactant molecules must acquire enough energy—the **activation energy**—to enter an activated state in which some of the chemical bonds can be broken. The activation energy does not alter the difference in energy content between the reactants and final products since the activation energy is released when the products are formed.

How do reactants acquire activation energy? In most of the metabolic reactions we will be considering, activation energy is obtained when reactants collide with other molecules. If the activation energy required for a reaction is large, then the probability of a given reactant molecule acquiring this amount of energy will be small, and the reaction rate will be slow. Thus, the higher the activation energy, the slower the rate of a chemical reaction.

Temperature is the third factor influencing reaction rates. The higher the temperature, the faster molecules move and thus the greater their impact when they collide. Therefore, one reason that increasing the temperature increases a reaction rate is that the reactants have a

TABLE 5-1 DETERMINANTS OF CHEMICAL REACTION RATES	
Reactant concentrations	Higher concentrations: faster reaction rate
Activation energy	Higher activation energy: slower reaction rate
Temperature	Higher temperature: faster reaction rate
Catalyst	Increases reaction rate

better chance of acquiring sufficient activation energy from a collision. In addition, faster-moving molecules will collide more frequently.

A **catalyst** is a substance that interacts with a reactant in such a manner that it alters the distribution of forces in the reactant and thereby decreases the activation energy for the reactant to be transformed into product. Since less activation energy is required, a reaction will proceed at a faster rate in the presence of a catalyst. The chemical composition of a catalyst is not altered by the reaction and thus a single catalyst molecule can be used over and over again to catalyze the conversion of many reactant molecules to products. Furthermore, a catalyst does not alter the difference in the energy contents of the reactants and products.

Reversible and Irreversible Reactions

Consider our previous example in which carbonic acid breaks down into carbon dioxide and water. The products of this reaction, carbon dioxide and water, can also recombine to form carbonic acid:

$$CO_2 \quad + \quad H_2O \quad + 4 \text{ kcal/mol} \longrightarrow H_2CO_3$$
$$\text{94 kcal/mol} \quad \text{57 kcal/mol} \quad \quad \quad \text{155 kcal/mol}$$

Since carbonic acid has a greater energy content than the sum of the energies contained in carbon dioxide and water, energy must be added to the latter molecules in order to form carbonic acid. (This 4 kcal of energy is *not* activation energy but an integral part of the energy balance.) This energy can be obtained, along with the activation energy, through collisions with other molecules.

Thus, in every chemical reaction there are two reactions going on simultaneously: a reaction (we will call it "forward") in which reactants are converted to products and one (we will call it "reverse") in which products are converted to reactants. The overall reaction is said to be a **reversible reaction**:

$$\text{Reactants} \underset{\text{Reverse}}{\overset{\text{Forward}}{\rightleftharpoons}} \text{products}$$

As a reaction progresses, the rate of the forward reaction will progressively decrease as the concentration of reactants decreases and simultaneously the rate of the reverse reaction will increase as the concentration of the product molecules increases. Eventually the reaction will reach a state of **chemical equilibrium** in which the forward and reverse reaction rates are equal. At this point there will be no further change in the concentrations of reactants or products. Their concentrations, however, need not be equal at equilibrium.

The ratio of product to reactant concentration at equilibrium varies with different reactions and depends upon the amount of energy released or added during the reaction. If there is no difference in the energy contents of reactants and products, their concentrations will be equal at equilibrium. If only a small amount of energy is released, on the order of 1 kcal or less, the product concentration at equilibrium will be between 1 and 10 × the reactant concentration.

When larger amounts of energy are released, the probability of the product molecules acquiring this energy and undergoing the reverse reaction is small, and almost all the reactant molecules are converted to products. For example, when 4 kcal of energy is released, the ratio of products to reactant molecules at equilibrium is about 1000 to 1. Thus, although all chemical reactions are reversible to some extent, those reactions that release large quantities of energy are said to be **irreversible reactions** in the sense that almost all of the reactant molecules have been converted to product molecules when chemical equilibrium is reached. A single arrow is used to indicate irreversible reactions and a double arrow for reversible reactions. The characteristics of reversible and irreversible reactions are summarized in Table 5-2.

Law of Mass Action

The concentrations of reactants and products play a very important role in determining not only the rates of the forward and reverse reactions but the direction in which the net reaction proceeds—whether products are accumulating or reactants are accumulating at a given time.

TABLE 5-2 CHARACTERISTICS OF REVERSIBLE AND IRREVERSIBLE CHEMICAL REACTIONS	
Reversible reactions	$A + B \rightleftharpoons C + D$ + small amount of energy At chemical equilibrium, product concentrations are only slightly higher than reactant concentrations.
Irreversible reactions	$E + F \longrightarrow G + H$ + large amount of energy At chemical equilibrium, almost all reactant molecules have been converted to product.

Consider the following reversible reaction that has reached chemical equilibrium:

$$A + B \underset{\text{Reverse}}{\overset{\text{Forward}}{\rightleftharpoons}} C + D$$

Reactants   Products

If at this point, the concentration of one of the reactants is increased, the rate of the forward reaction will increase and lead to increased product formation. In contrast, increasing the concentration of one of the product molecules drives the reaction in the reverse direction, increasing the formation of reactants. The direction in which the net reaction is proceeding can also be altered by *lowering* the concentration of one of the participants. Thus, lowering the concentration of one of the products will drive the net reaction in the forward direction since it will lower the rate of the reverse reaction without changing the rate of the forward reaction.

These effects of reactant and product concentrations on the direction in which the net reaction is proceeding is known as the **law of mass action**. Mass action is often a major determining factor controlling the direction in which metabolic pathways proceed.

ENZYMES

In order to achieve the high reaction rates observed in living organisms, catalysts are required to lower the activation energy of the reactions. These catalysts are called enzymes (meaning "in yeast," since the first enzymes were discovered in yeast cells.) Enzymes are protein molecules, and so an **enzyme** can be defined as a protein catalyst.[1]

To function, an enzyme must come into contact with reactants, which are called **substrates** in the case of enzyme-mediated reactions. The enzyme combines with substrate to form an enzyme-substrate complex, which breaks down to release products and enzyme (Figure 5-1). The reaction between enzyme and substrate can be written:

$$S + E \rightleftharpoons ES \rightleftharpoons P + E$$

Substrate Enzyme Enzyme-   Product Enzyme
                   substrate
                   complex

At the end of the reaction, the enzyme is free to undergo the same reaction with additional substrate molecules.

---

[1] Recently it has been found that some RNA molecules possess catalytic activity, but the number of reactions they catalyze is very small, and so we shall restrict the term "enzyme" to protein catalysts.

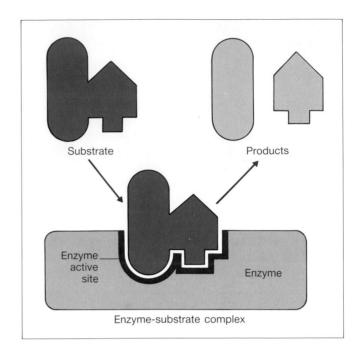

FIGURE 5-1 Binding of substrate to the active site of an enzyme catalyzes the formation of products.

The overall effect is to accelerate the conversion of substrate into product, the enzyme acting as a catalyst.

The interaction between substrate and enzyme has all the characteristics described in Chapter 4 for the binding of a ligand to a protein binding site—specificity, affinity, competition, and saturation. The region of the enzyme to which the substrate binds is known as the enzyme's **active site** (a term equivalent to "binding site"). The shape of the enzyme in the region of the active site provides the basis for the enzyme's chemical specificity since the shape of the active site is complementary to the substrate's shape (Figure 5-1).

Enzymes are generally named by adding the suffix *-ase* to the name of either the substrate or the type of reaction catalyzed by the enzyme. For example, the reaction in which carbonic acid is broken down to carbon dioxide and water is catalyzed by the enzyme carbonic anhydrase.

The catalytic activity of an enzyme can be extremely large. For example, a single molecule of carbonic anhydrase can convert about 100,000 substrate molecules to products in 1 s.

The major characteristics of enzymes are listed in Table 5-3.

Cofactors

Many enzymes are inactive in the absence of small amounts of other substances known as **cofactors**. In some cases, the cofactor is a trace metal, such as magne-

TABLE 5-3 CHARACTERISTICS OF ENZYMES

1. An enzyme undergoes no net chemical change as a consequence of the reaction it catalyzes.

2. The binding of substrate to an enzyme's active site has all the characteristics—chemical specificity, affinity, competition, and saturation—of ligand binding to a protein binding site.

3. An enzyme increases the rate of a chemical reaction but does not cause a reaction to occur that would not occur in its absence.

4. An enzyme increases both the forward and reverse rates of a chemical reaction and thus does not change the chemical equilibrium that is finally reached. It only increases the rate at which equilibrium is achieved.

5. An enzyme lowers the activation energy of a reaction but does not alter the net amount of energy that is added to or released by the reactants in the course of the reaction.

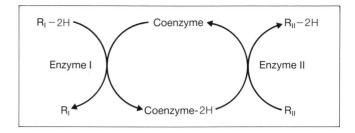

FIGURE 5-2 The transfer of two hydrogen atoms from R_I—2H to R_{II} by way of a coenzyme. Enzyme II regenerates the original coenzyme that can now repeat the cycle.

sium, iron, zinc, or copper, and its binding to an enzyme alters the conformation of the enzyme so that the enzyme can interact with substrate. Since only a few enzyme molecules need be present to catalyze the conversion of large amounts of substrate to product, very small quantities of these metals are sufficient to maintain enzymatic activity.

In other cases the cofactor is an organic molecule that directly participates as one of the substrates in the reaction being catalyzed and is termed a **coenzyme**. Enzymes that require coenzymes catalyze reactions in which a few atoms are either removed from or added to a substrate. For example:

$$R\text{—}2H + \text{coenzyme} \underset{}{\overset{\text{Enzyme}}{\rightleftharpoons}} R + \text{coenzyme—}2H$$

What makes a coenzyme different from an ordinary substrate is the fate of the coenzyme. In the example

above, the two hydrogen atoms that are transferred to the coenzyme can be transferred from the coenzyme to another substrate with the aid of a second enzyme. This second reaction converts the coenzyme back to its original form, which becomes available to accept two more hydrogen atoms (Figure 5-2). A single coenzyme molecule can be used over and over again to transfer molecular fragments from one reaction to another. Thus, as with the metallic cofactors, only small quantities of coenzymes are necessary to maintain those enzymatic reactions in which they participate.

Coenzymes are derived from a special class of nutrients known as vitamins. **Vitamins** are organic substances that are required, in small amounts, for metabolism but cannot be synthesized in the body from other nutrients and must therefore be supplied by the diet. Table 5-4 lists a few of the coenzymes, the vitamins from which they are derived, and the chemical groups they transfer.

REGULATION OF ENZYME-MEDIATED REACTIONS

The rate of an enzyme-mediated reaction depends mainly on substrate concentrations and on both the concentration and activity of the enzyme that catalyzes the reaction. Since body temperature is normally maintained nearly constant, changes in temperature are

TABLE 5-4 SOME COENZYMES

Coenzyme	Vitamin From Which Coenzyme Is Derived	Transferred Chemical Group
Nicotinamide adenine dinucleotide (NAD$^+$)	Niacin	Hydrogen (—2H)
Flavine adenine dinucleotide (FAD)	Riboflavin	Hydrogen (—2H)
Coenzyme A (CoA)	Pantothenic acid	Acetyl groups ($-\overset{\overset{\text{O}}{\|}}{\text{C}}-CH_3$)
Tetrahydrofolate	Folic acid	Methyl groups (—CH$_3$)

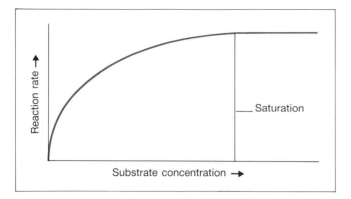

FIGURE 5-3 Rate of an enzyme-catalyzed reaction as a function of substrate concentration.

not used directly to alter the rates of metabolic reactions. Increases in body temperature can occur during fevers, however, and around muscle tissue during exercise, and such increases in temperature increase the rates of all metabolic reactions in the affected tissues.

Substrate Concentration

Substrate concentration may be altered as a result of factors that alter the supply of a substrate from outside a cell. Thus, there may be changes in its blood concentration due, for example, to changes in diet or rate of substrate absorption from the intestinal tract. In addition, a substrate's entry into the cell through the plasma membrane can be controlled by mechanisms that will be discussed in Chapter 6.

Intracellular substrate concentration can also be altered by reactions *in* a cell that either utilize the same substance, and thus tend to lower its concentration, or synthesize the substance, and thereby increase its intracellular concentration.

The rate of an enzyme-mediated reaction increases as the substrate concentration increases, as illustrated in Figure 5-3, until it reaches a maximal rate that remains constant despite further increases in substrate concentration. A maximal rate is reached because an enzyme can catalyze a reaction only when substrate is bound to its active site, and at high substrate concentrations the active site of every enzyme molecule is occupied.

Since coenzymes are substrates in enzyme reactions, changes in coenzyme concentration also affect a reaction rate by the same saturating mechanism as does the ordinary substrate.

Enzyme Concentration

At any substrate concentration, including saturating concentrations, the rate of an enzyme-mediated reaction can be increased by increasing enzyme concentration. If the number of enzyme molecules is doubled, twice as many active sites will be available and thus twice as much substrate will be converted to product (Figure 5-4). Certain reactions proceed faster in some cells than in others because more enzyme is present. In most metabolic reactions, the substrates are present at concentrations 10 to 100 times the enzyme concentration.

In order to change the concentration of an enzyme, either the enzyme synthesis rate or the enzyme breakdown rate must be altered. Since enzymes are proteins, this involves changing the rates of protein synthesis and breakdown. Little is known about the factors controlling the selective breakdown of enzymes. In Chapter 4 some of the mechanisms regulating protein synthesis were discussed. Regulation of gene transcription is the primary mechanism for altering most enzyme concentrations. The substrate or product molecules of the reaction in which the enzyme takes part are often the molecules that activate or inhibit the transcription of the gene for the enzyme. Changing the concentration of enzymes is a relatively slow process, generally requiring several hours.

Enzyme Activity

The rate of an enzyme-mediated reaction can be altered without changing the concentration of either substrate or enzyme and without altering temperature. Such a rate change occurs when the properties of the enzyme's active site are altered by allosteric or covalent modulation (Chapter 4). Such modulation alters the rate at which the binding site converts substrate to product or the affinity of the binding site for substrate, or both.

Figure 5-5 illustrates the effect of increasing the affinity of the enzyme's active site for substrate without changing the substrate concentration. Provided the sub-

FIGURE 5-4 Rate of an enzyme-catalyzed reaction as a function of substrate concentration at two enzyme concentrations A and B. Enzyme concentration B is twice enzyme concentration A and so produces twice the maximal reaction rate.

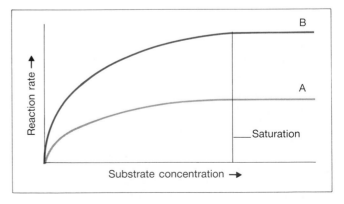

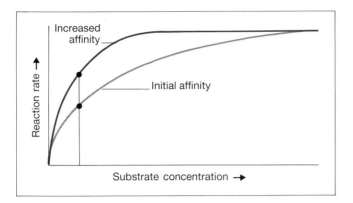

FIGURE 5-5 At a constant substrate concentration, increasing the affinity of an enzyme for its substrate by allosteric or covalent modulation increases the rate of the enzyme-mediated reaction. Note that increasing the enzyme's affinity does not increase the maximal rate of the enzyme-mediated reaction.

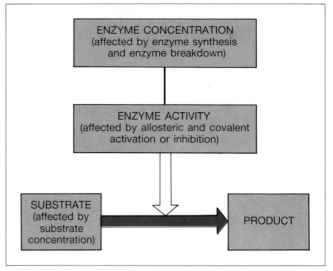

FIGURE 5-7 Factors that affect the rate of enzyme-mediated reactions.

strate concentration is less than the saturating concentration, the increased affinity of the enzyme's binding site will result in an increase in the number of active sites that contain bound substrate and thus in the probability that the substrate will be converted to product.

The major modulators of enzyme activity are the products of metabolism and chemical messengers such as hormones. The regulation of metabolism through the control of enzyme activity is an extremely complex process since, in many cases, the activity of an enzyme can be altered by more than one agent (Figure 5-6).

Figure 5-7 summarizes the factors that play a role in regulating the rate of an enzyme-mediated reaction.

FIGURE 5-6 On a single enzyme multiple sites can modulate the enzyme's activity and hence the reaction rate by increasing the affinity of the substrate-binding site—allosteric and covalent activation—or decreasing its affinity—allosteric and covalent inhibition.

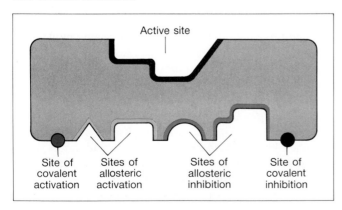

Active site

| Site of covalent activation | Sites of allosteric activation | Sites of allosteric inhibition | Site of covalent inhibition |

MULTIENZYME METABOLIC PATHWAYS

The sequence of enzyme-mediated reactions leading to the formation of a particular product is known as a metabolic pathway. For example, the 19 reactions that convert glucose to carbon dioxide and water constitute the metabolic pathway for glucose catabolism. Each reaction produces only a small change in the structure of the substrate (see, for example, Figures 5-12 and 5-15). By such a sequence of small steps, a complex chemical structure, such as glucose, can be transformed to the relatively simple molecular structures, carbon dioxide and water.

Consider a metabolic pathway containing four enzymes—e_1, e_2, e_3, and e_4—and leading from an initial substrate A to the end product E, through a series of intermediates, B, C, and D:

$$A \underset{}{\overset{e_1}{\rightleftharpoons}} B \underset{}{\overset{e_2}{\rightleftharpoons}} C \underset{}{\overset{e_3}{\rightleftharpoons}} D \overset{e_4}{\longrightarrow} E$$

(The irreversibility of the last reaction is of no consequence for the moment.) By mass action, increasing the concentration of A will lead to an increase in the concentration of B (provided e_1 is not already saturated with substrate) and so on until eventually there is an increase in the concentration of the end product E. Since different enzymes have different activities and concentrations, it would be extremely unlikely that the reaction rates of all these steps would be exactly the same. Thus, one step is very likely to be inherently slower than all the others. This step is known as the **rate-limiting reaction** in a met-

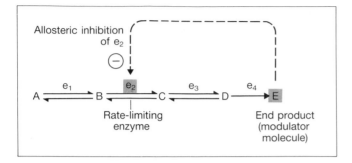

FIGURE 5-8 End-product inhibition of the rate-limiting enzyme in a metabolic pathway. The end product E becomes the modulator molecule that produces allosteric inhibition of enzyme e_2.

abolic pathway. None of the reactions that occur later in the sequence, including the formation of end product, can proceed more rapidly than the rate-limiting reaction since they depend upon the rate at which their substrates are being supplied by the previous steps. By regulating the activity of the rate-limiting enzyme, the rate of flow through the whole pathway can be increased or decreased. Thus, it is not necessary to alter the concentrations or activities of all the enzymes in a metabolic pathway to control the rate at which the final end product is produced.

Rate-limiting enzymes are often the sites of allosteric and covalent regulation. If enzyme e_2 is rate limiting in the pathway described above and is subject to allosteric inhibition by the end product E, we have a metabolic pathway that is able to regulate itself (Figure 5-8). This is termed **end-product inhibition** and is frequently found in synthetic pathways. It effectively shuts down the formation of end product when the end product is not being utilized and thus prevents its accumulation.

Control of enzyme activity or concentration also is critical for reversing a metabolic pathway. Consider the original pathway we have been discussing, ignoring the presence of end-product inhibition of enzyme e_2. It consists of three reversible reactions mediated by e_1, e_2, and e_3 followed by an irreversible reaction mediated by enzyme e_4. The last reaction is irreversible because a large amount of energy is released in the reaction. E can still be converted into D, however, if the reaction is coupled to the simultaneous breakdown of a molecule (X to Y), which releases large quantities of energy. In other words, an irreversible step can be "reversed" by an alternative route, using a second enzyme and its substrate to provide the required energy. Two such high-energy irreversible reactions are indicated by bowed arrows to emphasize that two separate enzymes are involved in the two directions:

$$A \rightleftharpoons_{e_1} B \rightleftharpoons_{e_2} C \rightleftharpoons_{e_3} D \underset{\overset{e_5}{\longleftarrow}}{\overset{e_4}{\longrightarrow}} E$$
$$Y \qquad X$$

By controlling the activities or concentrations of e_4 and e_5, the direction of flow through the pathway can be regulated. If e_4 is activated and/or e_5 inhibited, the flow will proceed from A to E, whereas inhibition of e_4 and/or activation of e_5 will produce flow from E to A.

Another situation involving the differential control of several enzymes arises when there is a branch in a metabolic pathway. A single metabolite may be the substrate for more than one enzyme, as illustrated by the pathway:

$$A \rightleftharpoons_{e_1} B \rightleftharpoons_{e_2} C$$
$$\overset{e_3}{\nearrow} D \rightleftharpoons_{e_4} E$$
$$\underset{e_5}{\searrow} F \rightleftharpoons_{e_6} G$$

Altering the activities or concentrations of e_3 and e_5 regulates the flow of metabolites through the two branches of the pathway.

When one considers the thousands of reactions that occur in the body and the permutations and combinations of possible control points, the overall result is staggering. The details of metabolic regulation at the enzymatic level are beyond the scope of this book. In the remainder of the chapter, we consider only the overall characteristics of the pathways by which cells obtain energy and the pathways by which carbohydrates, fats, and proteins are broken down and synthesized.

ATP AND CELLULAR ENERGY TRANSFER

The Role of ATP

The functioning of a cell depends upon its ability to extract and use the chemical energy in organic molecules. When a cell breaks down 1 mol of glucose, in the presence of oxygen, to carbon dioxide and water, 686 kcal of energy are released. Some of this energy appears as heat, but a cell cannot use this heat energy to perform work. The remainder of the energy is transferred to another molecule. That is, energy released when some of the chemical bonds of one molecule are broken is transferred to another molecule, thereby providing the energy needed to form new chemical bonds and increase the energy content of the second molecule. The energy transferred from one molecule to another, therefore, does not appear as heat during the reaction. The molecule to which the energy is transferred can in turn transfer it to yet another molecule or to energy-requiring processes.

In all cells, from bacterial to human, the molecule to which energy from the breakdown of fuel molecules is transferred and which then transfers this energy to cell functions is the nucleotide **adenosine triphosphate (ATP)** (Figure 5-9). For the moment we will disregard how ATP is formed from fuel molecules and focus on its energy release. When ATP is hydrolyzed, the removal of the terminal phosphate group is accompanied by the release of a large amount of energy, 7 kcal/mol.

$$\text{ATP} + \text{H}_2\text{O} \longrightarrow \text{ADP} + \text{P}_i + 7 \text{ kcal/mol}$$

Note that 7 kcal of energy are released when 1 mol (6×10^{23} molecules) of ATP are hydrolyzed, not just 1 molecule.

The energy derived from the hydrolysis of ATP is used by energy-requiring processes in cells for: (1) the production of force and movement as in muscle contrac-

tion (Chapter 11), (2) membrane active transport (Chapter 6), and (3) the synthesis of the organic molecules used for cell structure and function.

We must emphasize that cells use ATP not to *store* energy but rather to *transfer* it. ATP is an energy-carrying molecule that transfers relatively small amounts of energy from high-energy fuel molecules, such as carbohydrates, fats, and proteins, to the points in a cell that require energy. ATP is often referred to as the energy currency of the cell. By analogy, if the amount of usable energy released by the catabolism of glucose were equivalent to a 10-dollar bill, then the energy released by ATP hydrolysis would be worth about a quarter. The energy-requiring machinery of the cell uses only quarters—it will not accept 10-dollar bills. Transferring energy to ATP is the cell's way of making change.

Energy is continuously cycled through ATP molecules in a cell. Only about 40 percent of the energy released by the catabolism of fuel molecules can be transferred to ATP, the remaining 60 percent appears as heat. A typical ATP molecule may exist for only a few seconds before it is broken down to ADP and P_i, with the released energy used to perform a cell function. Equally

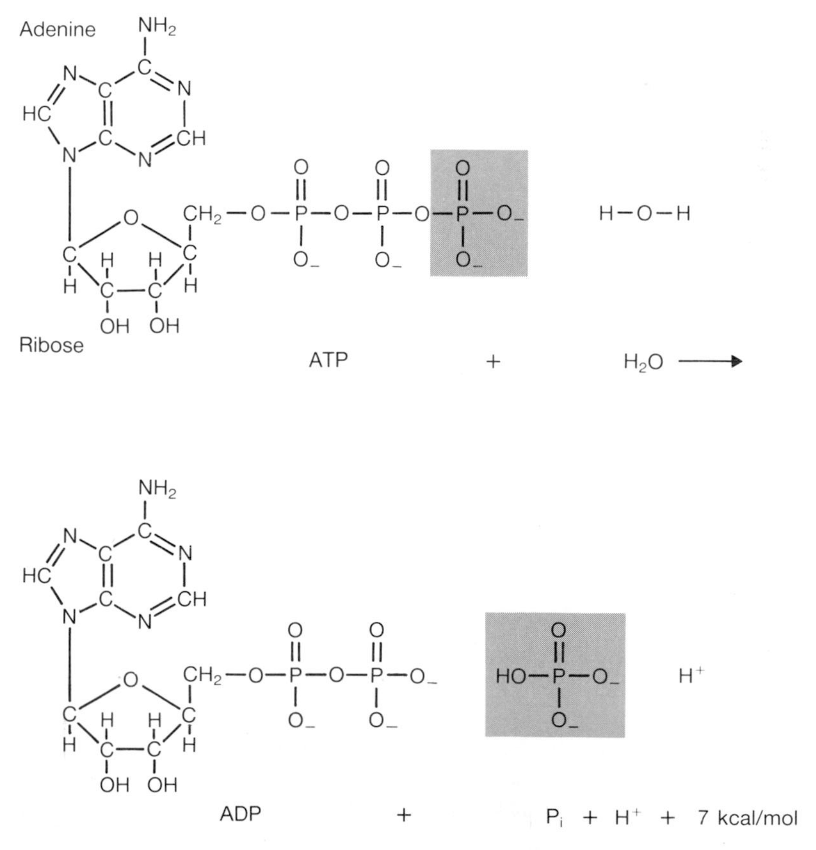

FIGURE 5-9 Chemical structure of ATP. Its breakdown to ADP and P_i is accompanied by the release of 7 kcal of energy per mole.

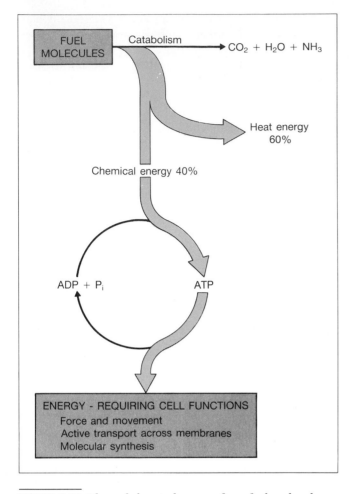

FIGURE 5-10 Flow of chemical energy from fuel molecules to ATP and heat, and then from ATP to energy-requiring cell functions.

rapidly, the products of ATP hydrolysis, ADP and P_i, are converted back into ATP through coupling to energy-releasing reactions, that is, the breakdown of carbohydrates, fats, or proteins (Figure 5-10). The total amount of ATP in the body is only sufficient to maintain the resting functions of the tissues for about 90 s. Thus, energy must be continuously transferred from fuel molecules to ATP.

Now we turn to ATP formation. In order to synthesize ATP from ADP and P_i, 7 kcal/mol of energy must be made available from the catabolism of fuel molecules:

$$ADP + P_i + 7 \text{ kcal/mol} \longrightarrow ATP + H_2O$$

There are two mechanisms by which energy can be transferred to ATP: (1) **oxidative phosphorylation** in which inorganic phosphate is coupled to ADP to form ATP, and (2) **substrate phosphorylation,** in which a phosphate group in an organic metabolic intermediate is

transferred directly to ADP to form ATP. Oxidative phosphorylation requires the presence of molecular oxygen, whereas substrate phosphorylation does not. Most of the ATP in the body is formed by oxidative phosphorylation.

Oxidative Phosphorylation

As we shall see in subsequent sections, the breakdown of carbohydrates, fats, and proteins involves the removal of hydrogen atoms from various intermediates and their transfer to coenzymes:

$$R—2H + \text{coenzyme} \underset{}{\overset{\text{Enzyme}}{\rightleftharpoons}} R + \text{coenzyme}—2H$$

During these reactions, some of the chemical energy in the intermediates is also transferred, along with the hydrogen atoms, to the coenzyme molecules. Oxidative phosphorylation then uses this energy to synthesize ATP. Since the hydrogen atoms are obtained from all three major types of fuel molecules, the energy obtained from all these sources can be funneled into the common energy carrier, ATP.

How does the energy transferred with hydrogen to the coenzyme get passed on to ATP? This involves a complex series of reactions that take place on the inner membranes of mitochondria. These membranes contain a group of iron-containing proteins known as **cytochromes** (because of their bright color) that accept a pair of electrons from two hydrogen atoms (Figure 5-11). These electrons pass through the various cytochromes and are eventually donated to molecular oxygen, which then reacts with the hydrogen ions, released at the beginning of the sequence of reactions, to form water. The overall reaction can be written:

$$\text{Coenzyme—2H} + \tfrac{1}{2} O_2 \xrightarrow{\text{Cytochromes}} \text{coenzyme} + H_2O + 52 \text{ kcal/mol}$$

Why must the cell proceed through a complex reaction sequence to achieve what is essentially the formation of water from hydrogen and oxygen? If a mole of coenzyme—2H reacted directly with oxygen, all 52 kcal of energy would be released, but since only 7 kcal of energy is used to form 1 mol of ATP from ADP and inorganic phosphate, the remaining 45 kcal of energy would appear as wasted heat. In contrast, by passing through the cytochrome chain, the same total energy is released in small steps during each electron transfer between successive cytochromes. The energy released at three of these steps is coupled to ATP synthesis. Either two or three molecules of ATP are formed from each pair of hydrogen atoms delivered to the cytochromes (Figure

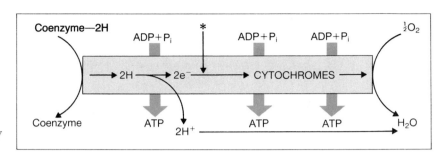

FIGURE 5-11 Oxidative phosphorylation. When coenzyme—2H donates its two hydrogens at the beginning of the electron-transport chain, three ATP are formed. When coenzyme—2H donation takes place at a later point in the chain (marked*), only two ATP are formed.

5-11), depending on which coenzyme delivers the hydrogen atoms and thus on the point at which the electrons enter the cytochrome chain. Of the total energy released during electron transport along the cytochrome chain, 40 percent is transferred to ATP, while the remaining 60 percent appears as heat.

The mechanism by which the energy released by the cytochrome reactions is coupled to the synthesis of ATP is complex and appears to involve the flow of hydrogen ions across the inner mitochondrial membrane.

It should be emphasized that, in addition to the formation of ATP, an important result of oxidative phosphorylation is the removal of hydrogen atoms from the coenzyme, returning the coenzyme to its original state. Without this regeneration, the enzymes that require these coenzymes to accept hydrogen from their substrates would be unable to act, and the flow of metabolites through these metabolic pathways would cease.

If oxygen is not available to the cytochrome system, ATP cannot be formed since the flow of electrons through the cytochrome chain depends on the ultimate donation of electrons to oxygen. The chemical cyanide can react with the final cytochrome in the chain, preventing electron transfer to oxygen and thus blocking the production of ATP by the mitochondria. The cause of death from cyanide poisoning is the same as that from a lack of oxygen—failure to form adequate amounts of ATP to maintain cell functions.

Oxygen-Independent ATP Synthesis: Glycolysis

Although most ATP is formed by oxidative phosphorylation, ATP can be formed by substrate phosphorylation, particularly in a pathway called **glycolysis** (derived from the Greek words *glycos*, sugar, and *lysis*, breakdown). The pathway is a series of either 10 or 11 enzymatic reactions beginning with glucose, a six-carbon molecule, and ending with either pyruvic acid or lactic acid, both three-carbon molecules (Figure 5-12). These reactions take place in the cytosol, in contrast to the mitochondrial location of the enzymes for oxidative phosphorylation.

The reactions in the glycolytic pathway can occur in either the presence or the absence of oxygen. In the presence of oxygen (**aerobic glycolysis**), the glycolytic pathway ends with pyruvic acid, whereas in the absence of oxygen (**anaerobic glycolysis**), pyruvic acid is converted in an additional step to lactic acid. In both cases the number of ATP molecules formed by substrate phosphorylation is the same.

The net chemical change that occurs during anaerobic glycolysis can be written:

$$\text{Glucose} + 2\ \text{ADP} + 2\ \text{P}_i \longrightarrow$$
$$2\ \text{lactic acid} + 2\ \text{ATP} + 2\ \text{H}_2\text{O}$$

In reactions 1 and 3 in Figure 5-12, two ATP molecules are used to add phosphate groups to glucose, but this energy debt is repaid with interest when four ATP molecules are formed later in the pathway, resulting in an overall net gain of two ATP molecules. The first synthesis of ATP during glycolysis occurs at reaction 7, in which one of the phosphate groups on 1,3-diphosphoglycerate is transferred to ADP to form ATP by substrate phosphorylation. A similar substrate phosphorylation occurs at reaction 10. Since reactions 4 and 5 convert one six-carbon molecule into two three-carbon molecules, and since each three-carbon intermediate forms one ATP molecule at reaction 7 and another at reaction 10, a total of four ATP molecules are formed by these steps.

Not only must the glycolytic enzymes be present for glycolysis to occur, but ATP, ADP, and inorganic phosphate must be available to participate as substrates in their appropriate reactions. In addition, a coenzyme to which hydrogen atoms are transferred in reaction 6 is required. Without this coenzyme, reaction 6 would not proceed and glycolysis would cease. Therefore, once the coenzyme has accepted a pair of hydrogen atoms, it must transfer them to another reaction in order to regenerate the coenzyme. In the absence of oxygen, the acceptor molecule is pyruvic acid, which is converted to the end product, lactic acid, by the addition of two hydrogen atoms from the coenzyme in reaction 11. Thus, during anaerobic glycolysis, coenzymes transfer hydrogen atoms from reaction 6 to 11 (Figure 5-13).

We have emphasized that glycolysis is a biochemical pathway that can form ATP in the absence of oxygen,

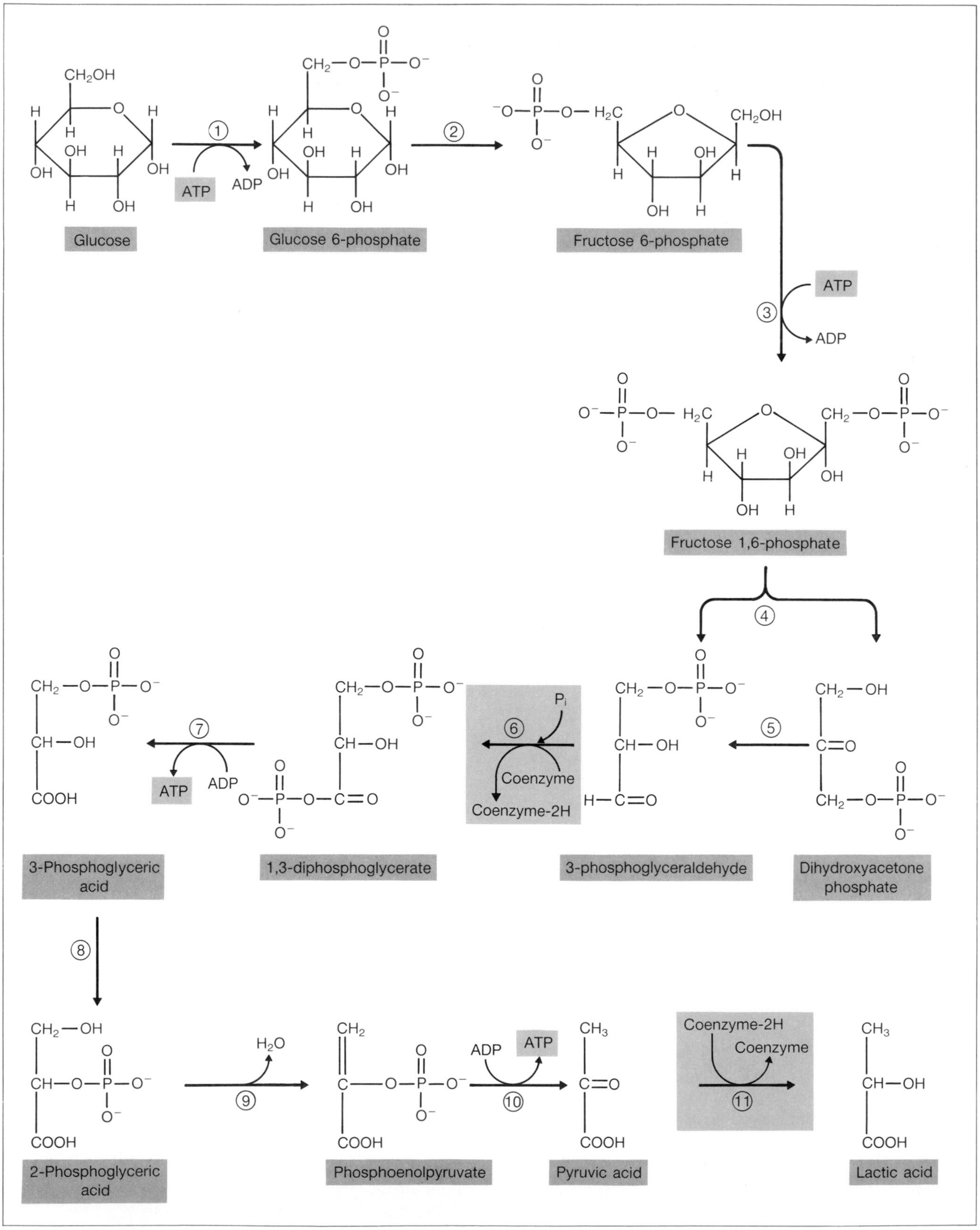
Glucose
ATP
ADP
Glucose 6-phosphate
Fructose 6-phosphate
ATP
ADP
Fructose 1,6-phosphate
Dihydroxyacetone phosphate
3-phosphoglyceraldehyde
Pi
Coenzyme
Coenzyme-2H
1,3-diphosphoglycerate
ADP
ATP
3-Phosphoglyceric acid
2-Phosphoglyceric acid
H2O
Phosphoenolpyruvate
ADP
ATP
Pyruvic acid
Coenzyme-2H
Coenzyme
Lactic acid

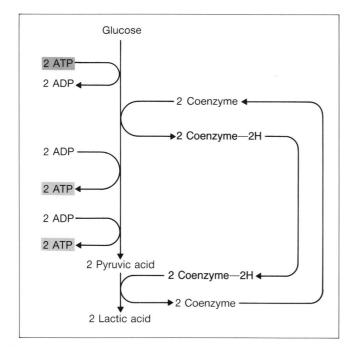

FIGURE 5-13 The coenzymes utilized in glycolysis under anaerobic conditions are regenerated when they transfer their hydrogen atoms to pyruvic acid during the formation of lactic acid.

but, as stated earlier, the 10 reactions from glucose to pyruvic acid also occur under aerobic conditions. In this case the coenzyme—2H that is formed by reaction 6 is regenerated by oxidative phosphorylation in the mitochondria. Thus, although substrate phosphorylation in glycolysis is identical whether oxygen is present or not, additional energy is gained under aerobic conditions when the hydrogen atoms generated are passed on to the cytochromes rather than to lactic acid.

To reiterate, the end product of anaerobic glycolysis is lactic acid, whereas that of aerobic glycolysis is pyruvic acid. As we shall see in the section on carbohydrate catabolism, this pyruvic acid goes on to be broken down to carbon dioxide and water.

In most cells, the amount of ATP produced by glycolysis is much smaller than the amount formed by oxidative phosphorylation. There are special cases, however, in which glycolysis supplies most, and in some cases all,

FIGURE 5-12 (*opposite*) Glycolytic pathway. Under anaerobic conditions there is a net synthesis of two molecules of ATP for every molecule of glucose that enters the pathway. Two molecules of ATP are used in reactions 1 and 3. Reactions 4 and 5 produce two molecules of 3-phosphoglyceraldehyde, each of which produces two molecules of ATP, one at step 7 and one at step 10. Thus, there is a total of four ATP produced and two consumed.

of a cell's ATP. For example, erythrocytes contain the enzymes for glycolysis but have no mitochondria. All of their ATP production occurs by glycolysis. Certain types of skeletal muscles contain considerable amounts of glycolytic enzymes and glycogen but have few mitochondria. During intense muscle activity, glycolysis provides most of the ATP in these cells. Despite these exceptions, however, most cells do not have sufficient concentrations of glycolytic enzymes or enough glucose to provide, by glycolysis alone, the high rates of ATP production required to maintain their energy requirements and thus are unable to function under anaerobic conditions.

The Krebs Cycle

As we have seen, a cell's ability to synthesize ATP by oxidative phosphorylation depends on a supply of hydrogen atoms that can combine with molecular oxygen. Some of these hydrogen atoms come from the coenzyme-linked hydrogen formed during aerobic glycolysis, and some also come, as we shall see, from the initial steps of fatty acid breakdown. However, the major source of hydrogen atoms for oxidative phosphorylation is the pathway known as the **Krebs cycle** (also called the **citric acid** or **tricarboxylic acid cycle**).

The Krebs cycle plays a central role in the breakdown of all three major classes of fuel molecules to carbon dioxide. In addition, ATP is synthesized by a substrate phosphorylation step in the Krebs cycle (an example of substrate phosphorylation in a pathway other than glycolysis). Thus, there are three products produced by the

FIGURE 5-14 Products produced by the Krebs-cycle reactions, which occur in mitochondria.

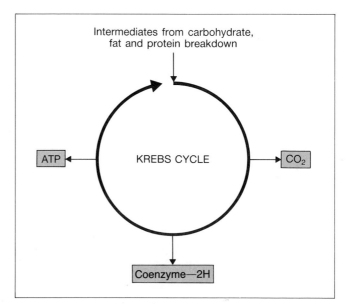

Krebs cycle reactions, (Figure 5-14): (1) hydrogen atoms bound to coenzyme molecules and available for oxidative phosphorylation to yield water and ATP, (2) carbon dioxide, and (3) ATP.

The net chemical change that occurs during the Krebs-cycle reactions can be written:

$$CH_3—\overset{\overset{O}{\|}}{C}—R + 3\ H_2O + 4\ coenzyme + ADP + P_i \longrightarrow$$
$$R—H + 2\ CO_2 + 4\ coenzyme—2H + ATP$$

The Krebs-cycle enzymes are soluble enzymes located in the inner compartment of mitochondria. Recall that the enzymes of oxidative phosphorylation are embedded in the inner mitochondrial membrane, which surrounds this compartment.

The seven reactions that constitute the Krebs cycle are illustrated in Figure 5-15. The starting point for the Krebs-cycle reactions is a molecule of **acetyl coenzyme A,** which is a carrier molecule for the two-carbon acetyl fragments:

$$—\overset{\overset{O}{\|}}{C}—CH_3$$

derived from the breakdown of carbohydrates, fats, and proteins. The Krebs cycle converts each two-carbon acetyl fragment into two molecules of carbon dioxide and transfers four pairs of hydrogen atoms to coenzyme molecules.

In reaction 1, acetyl coenzyme A donates its two-carbon acetyl fragment to the four-carbon molecule, oxaloacetic acid, to form the six-carbon molecule, citric acid. Carbon dioxide is formed from a carboxyl group (—COOH) in reaction 3 and again in reaction 4, leaving a four-carbon molecule, which after a few steps, becomes the four-carbon acceptor (oxaloacetic acid) of another acetyl group from acetyl coenzyme A to begin the cycle again.

The carbon dioxide formed in the Krebs cycle is the major waste product of metabolism. Thus, the two gases oxygen and carbon dioxide, which enter and leave the body by way of the lungs, take part in reactions that are located almost exclusively in mitochondria. Note, however, that the oxygen that appears in carbon dioxide is derived not from molecular oxygen but from the oxygen atoms in the carboxyl groups of the Krebs-cycle intermediates. As we have seen, molecular oxygen ends up in water molecules after combining with hydrogen during oxidative phosphorylation. Thus, water is a product not of the Krebs cycle per se, but of oxidative phosphorylation.

One molecule of ATP is formed in reaction 4 during each turn of the Krebs cycle. In this reaction a molecule of inorganic phosphate is linked to guanosine diphosphate (GDP) to form the high-energy molecule, guanosine triphosphate (GTP), a nucleotide similar to ATP but containing the base guanine rather than adenine. GTP, like ATP, is an energy-transferring molecule, and the energy in GTP is transferred to ATP by the following reaction:

$$GTP + ADP \longrightarrow GDP + ATP$$

In addition to this one molecule of ATP that is formed by substrate phosphorylation in the Krebs cycle, 11 ATP molecules are formed by oxidative phosphorylation from the four pairs of hydrogen atoms derived from each two-carbon fragment entering the cycle. Pairs of hydrogen atoms are transferred from intermediates to coenzyme molecules at reactions 3, 4, 5, and 7, producing a total of four coenzyme—2H molecules. As we have seen, there are three sites along the electron-transport chain at which ATP can be formed from ADP and inorganic phosphate. Three of the coenzymes from the Krebs cycle donate their hydrogen atoms to the cytochromes at the beginning of the chain, and thus each forms three ATP molecules. However, the type of coenzyme involved in reaction 5 donates its hydrogen at a point beyond the site where the first ATP is formed (Figure 5-11) and thereby gives rise to only two ATP molecules.

Combining the reactions of the Krebs cycle and oxidative phosphorylation, we see that for every two-carbon fragment that enters the mitochondria, 12 molecules of ATP are formed: 1 by substrate phosphorylation in the Krebs cycle and 11 during oxidative phosphorylation. At the same time, the carbon and oxygen atoms in the two-carbon fragment are converted into 2 molecules of carbon dioxide.

Although molecular oxygen is not a reactant in the Krebs-cycle reactions, it is required for the flow of metabolites through the Krebs cycle. Oxygen is necessary to regenerate the Krebs-cycle coenzymes by oxidative phosphorylation, and in the absence of oxygen these molecules will not be available to accept hydrogen atoms.

CARBOHYDRATE, FAT, AND PROTEIN METABOLISM

In this section, we outline the major pathways by which carbohydrates, fats, and proteins are catabolized to provide the energy for ATP formation. We also consider the synthesis of these fuel molecules and the pathways and restrictions governing the conversion

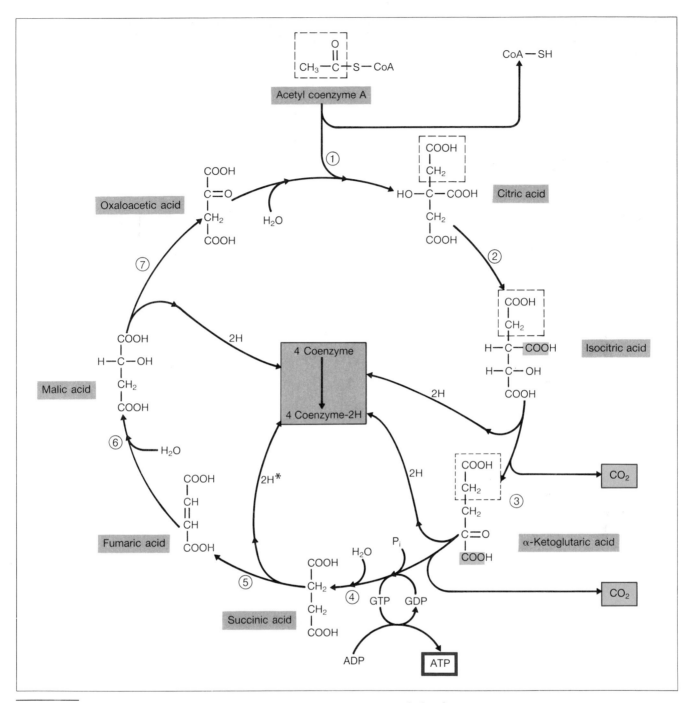

FIGURE 5-15 Krebs-cycle pathway. The acetyl group of coenzyme A is catabolized to CO_2, and its hydrogen atoms are transferred to coenzyme carrier molecules. The two hydrogen atoms (2H*) transferred to coenzyme in reaction 5 produce only two ATP during oxidative phosphorylation, whereas each of the other three pairs of hydrogen atoms gives rise to three ATP.

of them from one class to another. These anabolic pathways are also used to synthesize molecules that have functions other than the storage and release of energy. For example, the pathway for fat synthesis, with the addition of a few enzymes, is used for the synthesis of the phospholipids found in membranes.

Carbohydrate Metabolism

Carbohydrate catabolism. In the previous sections we described the major pathways of carbohydrate catabolism: the breakdown of glucose to pyruvic acid or lactic acid by way of the glycolytic pathway and the metabo-

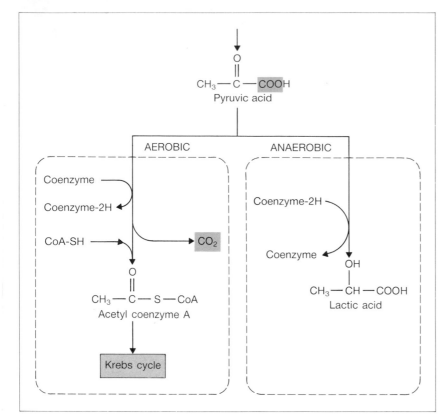

FIGURE 5-16 Anerobic and aerobic catabolism of glucose.

lism of acetyl fragments to carbon dioxide and hydrogen by way of the Krebs cycle. We have only to identify the link between these two pathways to provide the complete pathway for the catabolism of carbohydrate molecules.

We have emphasized that the breakdown of glucose to lactic acid occurs under anaerobic conditions. However, when the oxygen supply is adequate, the far more common condition, pyruvic acid is not converted to lactic acid to any great extent since the coenzyme molecules formed during glycolysis can donate their hydrogen atoms to oxygen. Instead, pyruvic acid enters the mitochondria, where enzymes transfer a two-carbon acetyl fragment from pyruvic acid to coenzyme A and release a molecule of carbon dioxide from the carboxyl group (Figure 5-16). (These reactions are also associated with the transfer of two hydrogen atoms to a coenzyme molecule, which will pass them on to the oxidative phosphorylation pathway.) The acetyl coenzyme A molecule can now enter the Krebs cycle, where the two carbon atoms in the acetyl group are degraded to carbon dioxide and its hydrogen atoms passed on to the oxidative phosphorylation pathway.

We have used glucose to illustrate carbohydrate catabolism because it is the most abundant carbohydrate in the body. Other sugars, such as fructose, found in table sugar, and galactose, found in milk, are catabolized by the same pathway after first being converted to glucose or to one of the early intermediates in the glycolytic pathway.

The amount of energy released during the catabolism of glucose to carbon dioxide and water is 686 kcal/mol of glucose:

$$C_6H_{12}O_6 + 6\ O_2 \longrightarrow 6\ H_2O + 6\ CO_2 + 686\ \text{kcal/mol}$$

How much of this energy can be transferred to ATP? Figure 5-17 illustrates the overall pathway of glucose catabolism and the sites at which ATP is formed. As we have seen, a net formation of two molecules of ATP occurs by substrate phosphorylation during glycolysis, and two more are formed by substrate phosphorylation in the Krebs cycle. Hydrogen atoms derived from glycolysis, the conversion of pyruvic acid to acetyl coenzyme A, and from the Krebs-cycle reactions provide the energy for an additional 34 ATP by oxidative phospho-

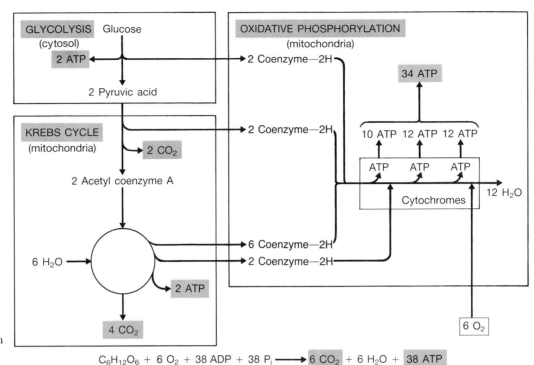

FIGURE 5-17 Pathways of aerobic glucose catabolism and their linkage to ATP formation.

$$C_6H_{12}O_6 + 6\ O_2 + 38\ ADP + 38\ P_i \longrightarrow 6\ CO_2 + 6\ H_2O + 38\ ATP$$

rylation.[2] Since 7 kcal is required to form 1 mol of ATP, and a total of 38 mol of ATP can be formed per mole of glucose catabolized, a total of 266 kcal (7 × 38) can be transferred from glucose to ATP. The remaining 420 kcal (686 − 266) is released as heat. Thus, 39 percent of the energy in glucose is transferred to ATP under aerobic conditions.

In the absence of oxygen, only two molecules of ATP can be formed by the breakdown of glucose to lactic acid. This yield represents only 2 percent of the energy stored in glucose. Thus, the evolution of aerobic metabolic pathways greatly increased the amount of energy available to a cell from glucose catabolism. For example, if a muscle consumed 38 molecules of ATP during a contraction, this amount of ATP could be supplied by the breakdown of 1 molecule of glucose in the presence of oxygen. However, 19 molecules of glucose would be required under anaerobic conditions to supply the same 38 mole-

cules of ATP. Accordingly, although only 2 molecules of ATP are formed per molecule of glucose under anaerobic conditions, large amounts of ATP can still be supplied by the glycolytic pathway if large amounts of glucose are broken down to lactic acid.

Glycogen storage.   A small amount of glucose is stored in the body to provide a reserve supply of fuel. It is stored in the form of the polysaccharide **glycogen,** and most of the body's glycogen is located in skeletal muscles and the liver.

Glycogen is synthesized from glucose by the pathway illustrated in Figure 5-18. The enzymes for both glycogen synthesis and glycogen breakdown are located in the cytosol. The first step, the transfer of phosphate from a molecule of ATP to glucose, forming glucose 6-phosphate, is the same as the first step in glycolysis. Thus, glucose 6-phosphate can either be broken down to pyruvic acid or be incorporated into glycogen.

Note that, as indicated by the bowed arrows in Figure 5-18, different enzymes are used to synthesize and break down glycogen. The existence of two pathways containing enzymes that are subject to both covalent and allosteric modulation provides a mechanism for regulating the flow of glucose to and from glycogen. When an excess of glucose is available to a liver or muscle cell, the enzymes in the glycogen-synthetic pathway are acti-

---

[2]The number of ATP molecules formed from the two coenzyme—2H molecules produced during glycolysis can be either 4 or 6, depending on the mechanism by which these coenzymes transfer their hydrogen atoms across the mitochondrial membrane. The other coenzymes in the Krebs cycle are already formed within the mitochondria and do not face the problem of having to pass through a membrane to reach the enzymes of oxidative phosphorylation. As a consequence, either 36 or 38 molecules of ATP can be formed from the catabolism of a single molecule of glucose.

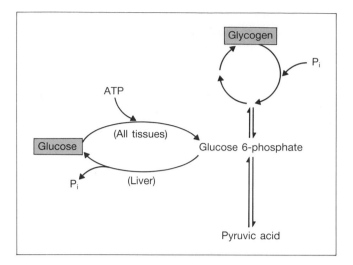

FIGURE 5-18 Pathways for the synthesis and breakdown of glycogen. Each bowed arrow indicates an irreversible reaction mediated by a different enzyme.

vated, by chemical signals described in Chapter 17, and the enzyme that breaks down glycogen is simultaneously inhibited, leading to the net storage of glucose in the form of glycogen. When less glucose is available, the reverse combination of enzyme stimulation and inhibition occurs, and there is a net breakdown of glycogen to glucose 6-phosphate.

Two paths are available to the glucose 6-phosphate formed from the breakdown of glycogen: (1) In most cells, including skeletal muscle cells, it enters the glycolytic pathway and, in the presence of oxygen, passes into the Krebs cycle to provide energy for ATP formation; and (2) in liver cells, glucose 6-phosphate can be converted to free glucose, which is able to pass out of the cell into the blood and become available to be used as fuel by other cells (Chapter 17).

Glucose synthesis. In addition to being able to form glucose from glycogen, liver and kidney cells can also synthesize it from intermediates derived from other metabolic pathways. Figure 5-19 illustrates the pathways used in the synthesis of glucose. The first important point to note is that the pathway leading from pyruvic acid to glucose 6-phosphate uses most, but not all, of the same cytosolic enzymes used in the glycolytic breakdown of glucose to pyruvic acid since these enzymes catalyze reversible reactions. However, two reactions in the glycolytic pathway (3 and 10) are irreversible. Reaction 3 and the separate reaction that achieves the reverse transformation are a major control point in the carbohydrate pathway because they involve two regulated enzymes, one mediating the flow of metabolites from glucose to pyruvic acid and the other the flow in the

opposite direction. By regulating the activities of these two enzymes, a liver or kidney cell can shift carbohydrate metabolism between net glucose breakdown and net synthesis, depending on the metabolic state of the cell and its energy requirements.

The second irreversible reaction in the glycolytic pathway, reaction 10, does not have a corresponding enzyme that can mediate the direct conversion of pyruvic acid back to phosphoenolpyruvic acid. Two additional enzymes accomplish this conversion by an indirect route, however. First, a molecule of carbon dioxide is linked to pyruvic acid to form a four-carbon intermediate oxaloacetic acid and then a second enzyme catalyzes the removal of carbon dioxide from oxaloacetic acid to form phosphoenolpyruvic acid. Thus because of these two enzymes, any organic molecule whose carbon atoms can be converted to pyruvic or oxaloacetic acid can give rise to glucose 6-phosphate. Regulation of these enzymes therefore provides another important mechanism for controlling glucose synthesis in liver and kidney.

The conversion of pyruvic acid to oxaloacetic acid not only provides a pathway that can eventually lead to glucose synthesis but also provides a means of forming this very important Krebs-cycle intermediate, which is the acceptor of acetyl fragments as they enter the Krebs cycle.

Having identified the enzymes and the pathways leading to glucose synthesis, we can now ask what types of molecules can be fed into this synthetic pathway. First, some amino acids can be converted into either pyruvic acid or one of the intermediates in the Krebs cycle that leads to oxaloacetic acid. Thus, proteins, by way of their amino acids, can be converted into carbohydrates. Second, lactic acid can be converted to pyruvic acid by the removal of hydrogen and thus can be used to synthesize glucose. The large quantities of lactic acid released into the blood by muscles during intense exercise can be taken up by liver cells and used to synthesize glucose, which is then released into the blood where it becomes available to other cells, including muscle. Third, glycerol, the three-carbon backbone to which fatty acids are attached to form triacylglycerols, can be used to synthesize glucose. Glycerol is converted into dihydroxyacetone phosphate, one of the three-carbon intermediates in glycolysis formed by reaction 4 (Figure 5-12), from which point it can be converted into glucose. Glycerol becomes available for glucose synthesis during periods when fat is being catabolized.

Fat Metabolism

Fat catabolism. Triacylglycerol (fat) consists of three fatty acids linked to glycerol (page 23) and constitutes

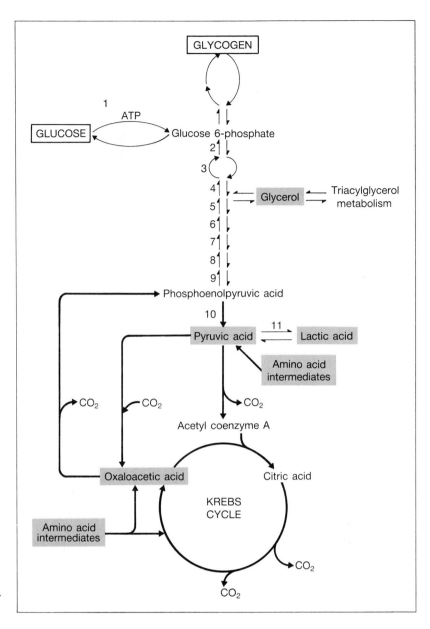

FIGURE 5-19 Pathways and intermediates leading to glucose or glycogen synthesis.

about 80 percent of the stored fuel in the body (Table 5-5). Under resting conditions, approximately half the energy used by such tissues as muscle, liver, and kidney is derived from the catabolism of fatty acids.

Although most cells store small amounts of fat, most of the body's fat is stored in specialized cells known as **adipocytes**. In these cells, almost the entire cytoplasm is filled with a single large fat droplet. Clusters of adipocytes form **adipose tissue**, most of which is located in deposits underlying the skin. The function of adipocytes is to synthesize and store triacylglycerols during periods of food uptake and to break them down between meals or during exercise. Following fat breakdown, the fatty acids are released into the blood and delivered to other

cells, which catabolize them to provide the energy for ATP formation. The glycerol portion of triacylglycerol is a carbohydrate and, as we have seen, can be used by the liver to synthesize glucose.

TABLE 5-5 FUEL CONTENT OF A 70-KG PERSON				
	Total-Body Content, kg	**Energy Content, kcal/g**	**Total-Body Energy Content**	
			kcal	**%**
Triacylglycerols	15.6	9	140,000	78
Proteins	9.5	4	38,000	21
Carbohydrates	0.5	4	2,000	1

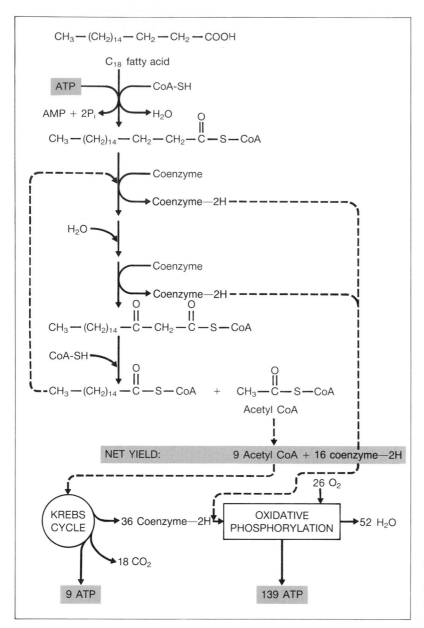

FIGURE 5-20 Pathway of fatty acid catabolism, which takes place in the mitochondria. The energy equivalent of two ATP is consumed at the start of the pathway.

What is the pathway for fatty acid catabolism? They are first catabolized to acetyl coenzyme A (Figure 5-20) by enzymes in the inner compartment of the mitochondria, where the Krebs-cycle enzymes are located. The breakdown of a fatty acid is initiated by linking a molecule of coenzyme A to the end of the fatty acid. This initial step is accompanied by the breakdown of ATP to AMP and two P_i. Since the terminal phosphates of both ATP and ADP are high-energy bonds, the energy released in this reaction is equivalent to that released when two ATP are hydrolyzed to two ADP and two P_i. Thus, although only one ATP is used in the reaction, the energy cost is equivalent to two ATP.

The coenzyme A derivative of the fatty acid then proceeds through a series of reactions known as **beta oxidation,** which split off a molecule of acetyl coenzyme A from the end of the fatty acid and transfer two pairs of hydrogen atoms to coenzymes. The two coenzyme—2H molecules enter the oxidative phosphorylation pathway, one forming 3 molecules of ATP, the other 2, for a total of 5 ATP for each pair of coenzyme—2H molecules formed.

When the terminal acetyl coenzyme A is split from the fatty acid, another is added (ATP is not required for this step), and the sequence is repeated. Each passage through this sequence shortens the fatty acid chain by two carbon atoms until all the carbon atoms have been transferred to coenzyme A. As we saw on page 90, the acetyl coenzyme A molecules then enter the Krebs cycle, each two-carbon fragment producing 2 molecules of CO_2 and ultimately 12 molecules of ATP.

How much ATP, then, is formed as a result of the total catabolism of a fatty acid? Most fatty acids in the body contain 14 to 22 carbons, 16 and 18 being most common. The catabolism of 1 18-carbon saturated fatty acid yields 146 ATP molecules (Table 5-6). In contrast, the catabolism of one glucose molecule yields a maximum of 38 ATP molecules. Thus, taking into account the difference in molecular weights of the fatty acid and glucose, it turns out that the amount of ATP formed from the catabolism of a gram of fat is about two and one-half times greater than the amount of ATP produced by catabolizing a gram of carbohydrate. If an average person stored most of his or her fuel as carbohydrate rather than as fat, body weight would have to be approximately 30 percent greater in order to store the same amount of usable energy, and the person would consume more energy moving this extra weight around. Thus, a major step in fuel economy occurred when animals evolved the ability to store fuel as fat. In contrast, plants store almost all their fuel as carbohydrate (starch).

Fat synthesis. The synthesis of fatty acids occurs by reactions that are almost the reverse of those that degrade them. The enzymes in the synthetic pathway are located in the cytosol, whereas (as we have just seen) the enzymes catalyzing fatty acid breakdown are located in the mitochondria. Fatty acid synthesis begins with acetyl coenzyme A, which, through a series of reactions requiring ATP and coenzyme—2H, transfers its acetyl group to another molecule of acetyl coenzyme A to form a 4-carbon chain. By repetition of this process, long-chain fatty acids are built up 2 carbons at a time, which accounts for the fact that all the fatty acids synthesized in the body contain an even number of carbon atoms. The overall reaction for the synthesis of an 18-carbon fatty acid can be written:

$$9 \ CH_3\overset{\overset{\displaystyle O}{\|}}{C}\!\!-\!\!S\!\!-\!\!CoA + 8 \ ATP + 16 \ coenzyme\!-\!2H \longrightarrow$$
$$CH_3(CH_2)_{16}COOH + 8 \ ADP + 8 \ P_i + 16 \ coenzyme$$
$$+ \ 7 \ H_2O + 9 \ CoA\!-\!SH$$

Once the fatty acids are formed, triacylglycerol can be synthesized by linking fatty acids to each of the three hydroxyl groups in glycerol, more specifically, in a phosphorylated form of glycerol called **glycerophosphate.** The synthesis of triacylglycerol is carried out by enzymes associated with the membranes of the smooth endoplasmic reticulum.

A comparison of the molecules required for both fatty acid and glycerol synthesis with those produced by glucose catabolism should reveal why sugar is so easily converted to fat. First, acetyl coenzyme A, the starting material for fatty acid synthesis, can be formed from pyruvic acid, the end product of aerobic glycolysis. Second, the other ingredients required for fatty acid synthesis—coenzyme—2H and ATP—are produced during carbohydrate catabolism. Third, glycerophosphate can be formed from a glucose intermediate. It should not be surprising, therefore, that much of the carbohydrate in food is converted into fat and stored in adipose tissue shortly after its absorption from the gastrointestinal tract. Mass action resulting from the increased concentration of glucose intermediates, as well as the specific hormonal regulation of key enzymes, promotes this conversion, as will be described in Chapter 17.

It is very important to note that fatty acids, or more specifically the acetyl coenzyme A derived from fatty acid breakdown, cannot be used to synthesize new molecules of glucose. The reasons for this can be seen by examining the pathways for glucose synthesis (Figure 5-19). First, because the reaction in which pyruvic acid is broken down to acetyl coenzyme A and carbon dioxide

---

TABLE 5-6   ATP FORMED FROM CATABOLISM OF ONE 18-CARBON FATTY ACID

	Molecules of ATP
9 molecules of acetyl coenzyme A	$9 \times 12 = 108$
(12 ATP formed per acetyl coenzyme A)	
8 pairs of coenzyme—2H	$8 \times 5 = 40$
(5 ATP formed per pair of coenzyme—2H)	———
	148
Energy equivalent of 2 ATP used at start of pathway	-2
	Total = 146

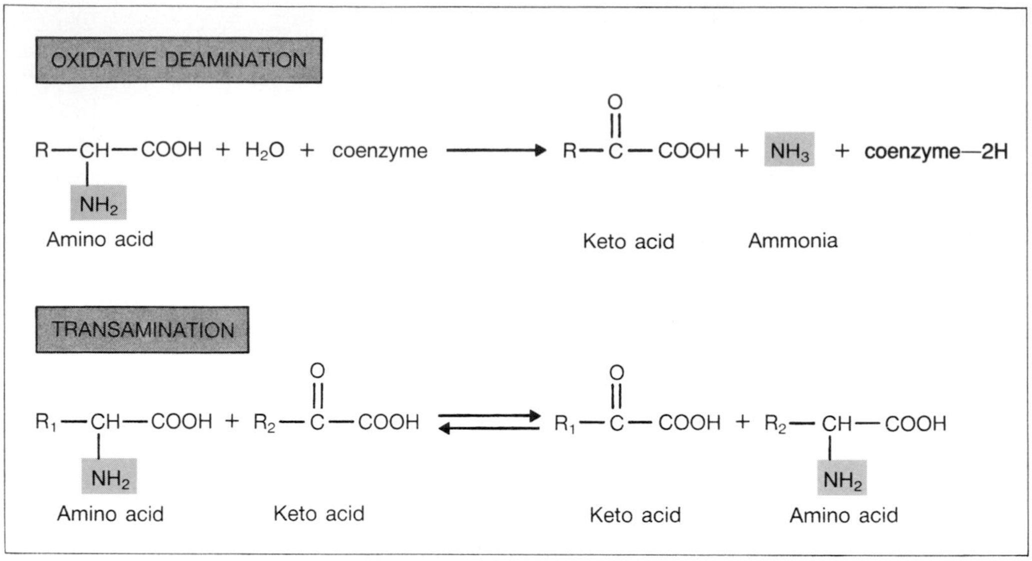

OXIDATIVE DEAMINATION

$$R\!-\!\underset{\underset{\text{NH}_2}{|}}{\text{CH}}\!-\!\text{COOH} + \text{H}_2\text{O} + \text{coenzyme} \longrightarrow R\!-\!\overset{\overset{\text{O}}{\|}}{\text{C}}\!-\!\text{COOH} + \boxed{\text{NH}_3} + \textbf{coenzyme}\!-\!\textbf{2H}$$

Amino acid   Keto acid   Ammonia

TRANSAMINATION

$$R_1\!-\!\underset{\underset{\text{NH}_2}{|}}{\text{CH}}\!-\!\text{COOH} + R_2\!-\!\overset{\overset{\text{O}}{\|}}{\text{C}}\!-\!\text{COOH} \rightleftharpoons R_1\!-\!\overset{\overset{\text{O}}{\|}}{\text{C}}\!-\!\text{COOH} + R_2\!-\!\underset{\underset{\text{NH}_2}{|}}{\text{CH}}\!-\!\text{COOH}$$

Amino acid   Keto acid   Keto acid   Amino acid

FIGURE 5-21 Oxidative deamination and transamination of amino acids.

is irreversible, acetyl coenzyme A cannot be converted into pyruvic acid, a molecule that can lead to the production of glucose. Second, the equivalent of the two carbon atoms in acetyl coenzyme A are converted into two molecules of carbon dioxide during their passage through the Krebs cycle before reaching oxaloacetic acid, another takeoff point for glucose synthesis, and therefore, cannot be used to synthesize net amounts of oxaloacetic acid.

Thus, glucose can readily be converted into fat, but the fatty acid portion of fat cannot be converted to glucose. The glycerol backbone of fat, as we mentioned earlier, can be converted to glucose.

Protein and Amino Acid Metabolism

In the last chapter we described how amino acids are linked together to form proteins and how DNA and RNA

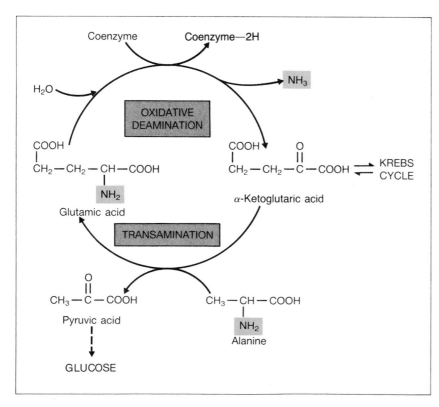

FIGURE 5-22 Oxidative deamination and transamination of the amino acids glutamic acid and alanine lead to keto acids that can enter the carbohydrate pathways.

are involved in directing the process. In contrast to the complexities of this synthetic pathway, protein catabolism requires only a few enzymes (**proteases**) to break the peptide bonds between amino acids. Some of these enzymes split off single amino acids, one at a time, from the ends of the protein chain, while others break peptide bonds between specific amino acids within the chain, forming peptides rather than free amino acids. The overall pathways for protein synthesis and degradation can thus be summarized as:

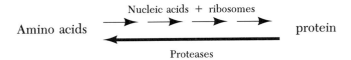

In addition to their incorporation into proteins, amino acids can be catabolized to provide energy for ATP synthesis and can provide the intermediates for the synthesis of a number of molecules. Since there are 20 different amino acids, there are a large number of intermediates that can be formed and many pathways for processing them. A few basic types of reactions common to most of these pathways can provide an overview of amino acid catabolism.

Unlike carbohydrates and fats, amino acids contain nitrogen atoms (in their amino groups) in addition to carbon, hydrogen, and oxygen atoms. Once the nitrogen-containing amino group is removed from most amino acids, the remainder of the amino acid can be metabolized to intermediates capable of entering either the glycolytic or Krebs-cycle pathways.

The two types of reactions by which the amino group is removed from an amino acid are illustrated in Figure 5-21. In the first reaction illustrated, **oxidative deamination,** the amino group gives rise to a molecule of ammonia (NH_3) and is replaced by an oxygen atom, derived from water, to form a **keto acid,** a categorical name, rather than the name of a specific molecule. The second means of removing an amino group is known as **transamination** and involves the transfer of the amino group from an amino acid to a keto acid. Note that the keto acid to which the amino group is transferred becomes an amino acid. The nitrogen derived from amino groups can also be used by cells to synthesize other important nitrogen-containing molecules, such as the purine and pyrimidine bases found in nucleic acids.

Figure 5-22 illustrates the oxidative deamination of the amino acid glutamic acid and the transamination of the amino acid alanine. Note that the keto acids formed are intermediates either in the Krebs cycle (α-ketoglutaric acid) or glycolytic pathway (pyruvic acid). Once formed, these keto acids can be metabolized to

produce carbon dioxide and form ATP, or they can be used as intermediates in the synthetic pathway leading to the formation of glucose. As a third alternative, they can be used to synthesize fatty acids after their conversion to acetyl coenzyme A by way of pyruvic acid. Thus, amino acids can be used as a source of energy, and some can be converted into carbohydrate or fat.

As we have seen, the catabolism of amino acids yields ammonia. This substance, which is highly toxic to cells if allowed to accumulate, readily passes through cell membranes and enters the blood, which carries it to the liver (Figure 5-23). Here, in a series of reactions, two mole-

FIGURE 5-23 Formation and excretion of urea, the major nitrogenous waste product of protein catabolism.

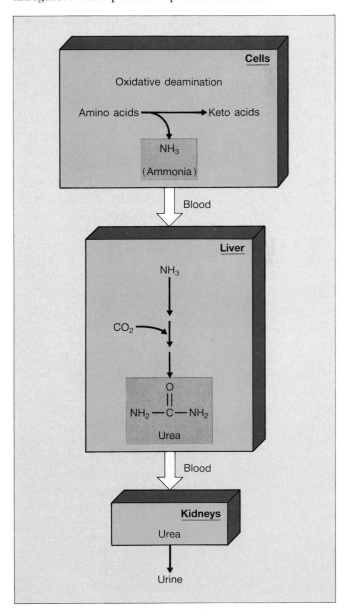

cules of ammonia are linked with carbon dioxide to form **urea.** Thus, urea, which is relatively nontoxic, is the major nitrogenous waste product of protein catabolism. It leaves the liver and is excreted by the kidneys into the urine. Two of the 20 amino acids also contain atoms of sulfur, which can be converted to sulfate, SO_4^{2-}, and excreted in the urine.

Thus far, we have discussed mainly amino acid catabolism. Figure 5-22 also demonstrates how amino acids can be formed. The keto acids, α-ketoglutaric and pyruvic acid, can be formed from the breakdown of glucose and then be transaminated, as described above, to form glutamic acid and alanine. Thus glucose can be used to form certain amino acids, provided other amino acids are available in the diet to supply amino groups for transamination. However, not all the amino acids can be formed by the body. Those that must be obtained from the food we eat (9 of the 20 amino acids) are known as **essential amino acids.**

Figure 5-24 provides a summary of the multiple routes by which amino acids are handled by the body. The amino acid pools, which consist of the body's total free amino acids, are derived both from ingested protein, which is degraded to amino acids during digestion in the intestinal tract, and from the continuous breakdown of body protein. These pools are the source of amino acids for resynthesis of body protein and a host of specialized amino acid derivatives. A very small quantity of amino acids and protein is lost from the body via the urine, skin, hair, fingernails, and in women, the menstrual fluid. The major route for the loss of amino acids from the body is not excretion of amino acids but rather their deamination, with ultimate excretion of the nitro-

gen atoms as urea in the urine. The terms **negative nitrogen balance** and **positive nitrogen balance** refer to whether there is net loss or gain, respectively, of amino acids in the body over any period of time.

If any of the essential amino acids is missing from the diet, negative nitrogen balance, that is, output greater than intake, always results. The proteins for which that amino acid is essential cannot be synthesized, and the other amino acids that would have been incorporated into the proteins are metabolized. This explains why a dietary requirement for protein cannot be specified without regard to the amino acid composition of that protein. Protein is graded in terms of how closely its relative proportions of essential amino acids approximate those in the average body protein. The highest-quality proteins are found in animal products, whereas the quality of most plant proteins is lower. Nevertheless, it is quite possible to obtain adequate quantities of all essential amino acids from a mixture of plant proteins alone.

Summary of Fuel Metabolism

Having discussed the metabolism of the three major classes of organic molecules, we can now briefly review how each class is related to the others and to the process of synthesizing ATP. Figure 5-25 illustrates the major pathways we have discussed and the relations of the common intermediates. All three classes of molecules can enter the Krebs cycle through some intermediate, and thus all three can be used as a source of chemical energy for the synthesis of ATP by oxidative phosphorylation. Glucose can be converted into fat or into some amino acids by way of common intermediates such as

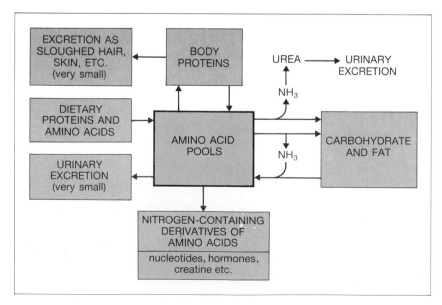

FIGURE 5-24 Pathways influencing the magnitude of amino acid pools in the body.

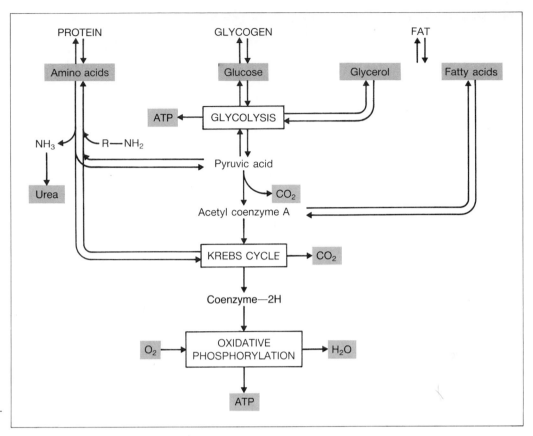

FIGURE 5-25 Interrelations between the pathways for the metabolism of carbohydrates, fats, and proteins.

pyruvic acid, α-ketoglutaric acid, and acetyl coenzyme A. Similarly, some amino acids can be converted into glucose and fat. Fatty acids cannot be converted into glucose because of the irreversibility of the reaction converting pyruvic acid to acetyl coenzyme A, but the glycerol portion of triacylglycerols can be converted into glucose. Fatty acids can be used to synthesize portions of the keto acids used to form amino acids. Metabolism is, thus, a highly integrated process in which all classes of molecules can be used, if necessary, to provide energy and in which each class of molecule can, to a large extent, provide the raw materials required to synthesize members of other classes.

The activities of many enzymes in these pathways can be altered by allosteric and covalent modulation. ATP and its products, ADP and P_i, can act as modulator molecules that are able to activate or inhibit various allosteric enzymes in these pathways. Changes in the concentrations of ATP, ADP, and P_i with altered rates of energy utilization can influence the rates at which ATP is being formed by these catabolic pathways. Metabolic intermediates, including glucose 6-phosphate, citric acid, and acetyl coenzyme A, also function as modulator molecules providing the cell with a large variety of interacting control systems.

ESSENTIAL NUTRIENTS

There are about 50 substances that are required for normal or optimal body function but that cannot be synthesized by the body at all or are synthesized in amounts inadequate to keep pace with the rates at which they are broken down or excreted. They are known as **essential nutrients** (Table 5-7). Because they are all removed at some finite rate, a continual new supply must be provided in the diet.

It should be emphasized that the term "essential nutrient" is reserved for substances that fulfill *two* criteria: (1) They must be essential for good health, and (2) they must not be synthesized by the body in adequate amounts. Thus, glucose, although "essential" for normal metabolism, is not classified as an essential nutrient because the body normally can synthesize all it needs, from amino acids, for example. Furthermore, the quantity of an essential nutrient that must be present in the diet in order to remain healthy is not a criterion for determining if the substance is essential. Approximately 1500 g of water, 2 g of the amino acid methionine, but only about 1 mg of the vitamin thiamine are required per day.

Water is an essential nutrient because far more of it is

TABLE 5-7 ESSENTIAL NUTRIENTS

Water
Mineral elements
    7 major mineral elements (Table 2-2)
    13 trace elements (Table 2-2)
Essential amino acids
    Isoleucine
    Leucine
    Lysine
    Methionine
    Phenylalanine
    Threonine
    Tryptophan
    Tyrosine
    Valine
Essential fatty acids
    Linoleic
    Linolenic
Vitamins
    Water-soluble vitamins
      B_1: thiamine
      B_2: riboflavin
      B_6: pyridoxine
      B_{12}: cobalamine
      Niacin
      Pantothenic acid
      Folic acid
      Biotin
      Lipoic acid
      Vitamin C
    Fat-soluble vitamins
      Vitamin A
      Vitamin D
      Vitamin E
      Vitamin K
Other essential nutrients
    Inositol
    Choline
    Carnitine

from the body in the urine, feces, and various secretions. The major minerals must be supplied in fairly large amounts, whereas only small quantities of the trace elements are required.

We have already noted that 9 of the 20 amino acids are essential. Two fatty acids that contain a number of double bonds and serve important roles in chemical messenger systems are also essential nutrients. Three additional essential nutrients—inositol, choline, and carnitine—have functions that will be described in later chapters but do not fall into any common category other than being essential nutrients. Finally, the class of essential nutrients known as vitamins deserves special attention.

Vitamins

Fourteen vitamins have been shown to be essential for normal growth and health in humans. The exact chemical structures of the first vitamins to be observed were unknown. Therefore, they were simply identified by letters of the alphabet. Vitamin B turned out to be composed of eight substances now known as the vitamin B complex. Plants and bacteria have the enzymes necessary for vitamin synthesis, and it is by eating plants or meat from animals that have eaten plants that we get our vitamins.

The vitamins as a class have no particular chemical structure in common, but they can be divided into the water-soluble vitamins and the fat-soluble vitamins (A, D, E, and K). The water-soluble vitamins function as coenzymes, as described earlier. The fat-soluble vitamins in general do not function as coenzymes. For example, vitamin A (retinol) forms the light-sensitive pigment in the eye, and lack of this vitamin leads to night blindness. The specific functions of each of the fat-soluble vitamins will be described in later chapters.

By themselves vitamins do not provide chemical energy, although they may participate as coenzymes in chemical reactions that release energy from other molecules. Increasing the amount of vitamins in the diet, beyond a certain minimum, does not necessarily increase the activity of those enzymes for which the vitamin functions as a coenzyme. Only very small quantities of coenzymes are necessary to saturate the binding sites for them on enzyme molecules, and increasing the concentration above this level does not increase the enzyme's activity.

Quantities of water-soluble vitamins in the diet that exceed the small amounts that are continually lost have not been proven to have beneficial effects. As the amount of these water-soluble vitamins in the diet is increased, so is the amount excreted in the urine, with the result that accumulation of these vitamins in the body is

lost in the urine and from the skin and respiratory tract than can be synthesized by the body. (Recall that water is formed as an end product of oxidative phosphorylation as well as from other metabolic reactions.) Therefore, to maintain water balance, an intake of water is necessary.

The mineral elements provide an example of substances that are not metabolized but are continually lost limited. On the other hand, fat-soluble vitamins can build up in the body because they are poorly excreted by the kidneys and because they dissolve in the fat stores in adipose tissue. Therefore, the intake of very large quantities of fat-soluble vitamins can produce toxic effects.

SUMMARY

I. In adults the rates at which molecules are synthesized (anabolism) and broken down (catabolism) are approximately equal, and so the chemical composition of the body does not change appreciably.

Chemical Reactions

  I. The difference in the energy content of reactants and products is the amount of energy that is released, or must be added, during a reaction.
 II. The energy released during a chemical reaction is either released as heat or is transferred to other molecules.
III. The four factors that can alter the rate of a chemical reaction are listed in Table 5-1.
 IV. The activation energy required to initiate the breaking of chemical bonds in a reaction is usually acquired through collisions with other molecules.
  V. Catalysts increase the rate of a reaction by lowering the activation energy.
 VI. The characteristics of reversible and irreversible reactions are listed in Table 5-2.
VII. The net direction in which a reaction proceeds can be altered, according to the law of mass action, by increases or decreases in the concentrations of reactants or products.

Enzymes

  I. Nearly all chemical reactions in the body are catalyzed by enzymes, the characteristics of which are summarized in Table 5-3.
 II. Some enzymes require small concentrations of cofactors for activity.
   A. The binding of trace metal cofactors maintains the conformation of the enzyme's binding site so that it is able to bind substrate.
   B. Coenzymes, derived from vitamins, transfer molecular fragments from one substrate to another. The coenzyme is regenerated in the course of these reactions and can be used over and over again.

Regulation of Enzyme-Mediated Reactions

  I. The rates of enzyme-mediated reactions can be altered by changes in temperature, substrate concentration, enzyme concentration, and enzyme activity. Enzyme activity is altered by allosteric and covalent modulation.

Multienzyme Metabolic Pathways

  I. The rate of product formation in a metabolic pathway can be controlled by allosteric or covalent modulation of the enzyme mediating the rate-limiting reaction in the pathway. The end product often acts as a modulator molecule, inhibiting the rate-limiting enzyme's activity.
 II. An irreversible step in a metabolic pathway can be reversed by the use of two enzymes, one for the forward reaction and one for the reverse direction via another energy-yielding reaction.

ATP and Cellular Energy Transfer

  I. In all cells, energy is transferred from the catabolism of fuel molecules to ATP. The hydrolysis of ATP to ADP and P_i then transfers this energy to cell functions. ATP is formed by oxidative phosphorylation and substrate phosphorylation.
 II. Formation of ATP by oxidative phosphorylation involves a series of reactions in which two hydrogen atoms are combined with an oxygen atom to form water.
   A. The enzymes for oxidative phosphorylation are located on the inner membrane of the mitochondria.
   B. The hydrogen atoms are obtained from the catabolism of fuel molecules and are transferred to the mitochondria as coenzyme—2H where the hydrogen atoms eventually combine with molecular oxygen to form water and the energy released is used for ATP synthesis.
III. In the formation of ATP by substrate phosphorylation, a phosphate group is transferred from a phosphorylated metabolic intermediate to ADP. Substrate phosphorylation occurs mainly in the glycolytic pathway, whose enzymes are located in the cytosol.
 IV. The end products of aerobic glycolysis are ATP and pyruvic acid, and those of anaerobic glycolysis are ATP and lactic acid.
   A. Carbohydrates are the only fuel molecules that can enter the glycolytic pathway.
   B. During anaerobic glycolysis, hydrogen atoms are transferred to a coenzyme that then transfers them to pyruvic acid to form lactic acid, thus regenerating the original coenzyme molecule.
   C. During aerobic glycolysis, coenzymes transfer hydrogen atoms to oxidative phosphorylation.
  V. The majority of the hydrogen atoms used in oxidative phosphorylation are derived from reactions in the Krebs cycle, whose enzymes are located in the matrix of the mitochondria.
   A. Acetyl coenzyme A, derived from all three classes of fuel molecules, is the major substrate entering the Krebs cycle.
   B. The two carbon atoms in the acetyl fragment of coenzyme A give rise to two molecules of carbon dioxide, and four molecules of coenzyme—2H are formed, along with one molecule of ATP by substrate phosphorylation.
   C. The 4 molecules of coenzyme—2H transfer their hydrogens to oxidative phosphorylation, producing 11 ATP.

Carbohydrate, Fat, and Protein Metabolism

I. The aerobic catabolism of carbohydrates proceeds through the glycolytic pathway to pyruvic acid, which enters the Krebs cycle and is broken down to carbon dioxide.

    A. Under aerobic conditions, 38 molecules of ATP can be formed from 1 molecule of glucose: 34 ATP by oxidative phosphorylation from coenzyme—2H, 2 ATP by glycolysis, and 2 ATP by substrate phosphorylation in the Krebs cycle.

    B. Under anaerobic conditions, 2 molecules of ATP are formed from 1 molecule of glucose.

    C. About 40 percent of the chemical energy in glucose can be transferred to ATP under aerobic conditions; the rest appears as heat.

II. Carbohydrates are stored as glycogen, primarily in the liver and skeletal muscles.

    A. Two different enzymes are used to synthesize and break down glycogen. The control of these enzymes regulates the flow of glucose to and from glycogen.

    B. In most cells the glucose formed by glycogen breakdown is catabolized to produce ATP. In liver cells, glucose can be produced and released from the cell into the blood.

III. Glucose can be synthesized from some amino acids and from lactic acid and glycerol via the enzymes that catalyze reversible reactions in the glycolytic pathway. Fatty acids cannot be used to synthesize glucose.

IV. Fat, stored primarily in adipose tissue, provides about 80 percent of the stored fuel in the body.

    A. Fatty acids are broken down in the mitochondrial inner compartment by beta oxidation, two carbon atoms at a time, to form acetyl coenzyme A, which is catabolized to carbon dioxide in the Krebs cycle.

    B. The amount of ATP formed by the catabolism of 1 g of fat is about two and one-half times greater than the amount of ATP that can be formed from 1 g of carbohydrate.

    C. Fatty acids are synthesized from acetyl coenzyme A by enzymes located in the cytosol and are linked to glycerophosphate to form triacylglycerols by enzymes in the smooth endoplasmic reticulum.

V. Proteins are broken down to free amino acids by proteases.

    A. The amino groups of amino acids are removed, forming keto acids that can be catabolized to provide the energy for the synthesis of ATP, or used to synthesize fatty acids.

    B. Amino groups are removed by (1) oxidative deamination, which gives rise to ammonia and (2) transamination, in which the amino group is transferred to a keto acid to form a new amino acid.

    C. The ammonia formed from the oxidative deamination of amino acids is converted into urea by enzymes in the liver and then excreted in the urine by the kidneys.

VI. Some amino acids can be synthesized from keto acids derived from glucose, while others cannot be synthesized by the body and must be provided in the diet.

Essential Nutrients

I. Approximately 50 essential nutrients are necessary for good health but cannot be synthesized in adequate amounts by the body and must therefore be provided in the diet. These are summarized in Table 5-7.

II. An excessive intake of water-soluble vitamins leads to their rapid excretion in the urine, while excessive intakes of fat-soluble vitamins leads to their accumulation in adipose tissue and may produce toxic effects.

REVIEW QUESTIONS

1. Define:

metabolism	cytochromes
anabolism	glycolysis
catabolism	aerobic glycolysis
calorie	anaerobic glycolysis
kilocalories	Krebs cycle
activation energy	citric acid cycle
catalyst	tricarboxylic acid cycle
reversible reaction	acetyl coenzyme A
chemical equilibrium	glycogen
irreversible reaction	adipocytes
law of mass action	adipose tissue
enzyme	beta oxidation
substrates	glycerophosphate
active sites	proteases
cofactors	oxidative deamination
coenzymes	keto acid
vitamins	transamination
rate-limiting reaction	urea
end-product inhibition	essential amino acids
adenosine triphosphate (ATP)	negative nitrogen balance
	positive nitrogen balance
oxidative phosphorylation	essential nutrients
substrate phosphorylation	

2. How do molecules acquire the activation energy required for a chemical reaction?

3. List the four factors that influence the rate of a chemical reaction and note whether increasing the factor will increase or decrease the rate of the reaction.

4. What characteristics of a chemical reaction make it reversible or irreversible?

5. List five characteristics of enzymes.

6. From what class of nutrients are coenzymes derived?

7. Why are small concentrations of coenzymes sufficient to maintain enzyme activity?

8. How can the rate of an enzyme-mediated reaction be altered?

9. How can an irreversible step in a metabolic pathway be reversed?

10. What is the function of ATP in metabolism?

11. Identify the two processes used to form ATP from ADP.

12. Approximately how much of the energy released from the catabolism of fuel molecules is transferred to ATP? What happens to the rest?

13. Identify the molecules that enter the oxidative phosphorylation pathway and the products that are formed.

14. What are the end products of anaerobic glycolysis? Of aerobic glycolysis?

15. To which molecule are the hydrogen atoms in coenzyme—2H transferred during anaerobic glycolysis? During aerobic glycolysis?

16. What are the major substrates entering the Krebs cycle and the products formed?

17. Where are the enzymes for the Krebs cycle located in a cell? The enzymes for oxidative phosphorylation? The enzymes for glycolysis?

18. Why will the Krebs cycle operate only under aerobic conditions even though molecular oxygen is not used in any of its reactions?

19. How many molecules of ATP can be formed from the breakdown of one molecule of glucose under aerobic conditions? Under anaerobic conditions?

20. Describe the pathways by which glycogen is synthesized and broken down by cells.

21. What molecules can be used to synthesize glucose?

22. Why can fatty acids not be used to synthesize glucose?

23. Describe the pathway used to catabolize fatty acids to carbon dioxide.

24. Why is it more efficient to store fuel as fat than as glycogen?

25. Describe the pathway by which glucose is converted into fat.

26. Describe the two processes by which amino groups are removed from amino acids.

27. What is the source of the nitrogen atoms in urea, and in what organ is urea synthesized?

28. Why is water considered an essential nutrient but glucose is not?

29. What are some of the consequences of ingesting large quantities of water-soluble vitamins? Fat-soluble vitamins?

concentration is A, what events could lead to an increase in the rate of the reaction to rate x? to rate y?

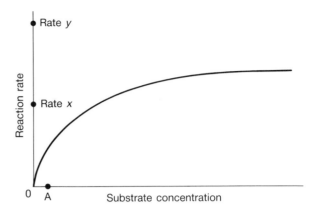

3. In the following metabolic pathway, in which the maximal rates of the individual steps are indicated, what is the rate at which the end product E is formed if substrate A is present at saturating concentrations?

$$A \xrightarrow{30} B \xrightarrow{5} C \xrightarrow{20} D \xrightarrow{40} E$$

4. Will breathing a gas containing a higher percentage of oxygen than normal air increase (by mass action) the rate of ATP production by cells? Explain the conditions when this may be true and when it would not.

5. During a prolonged period of starvation, which class of molecules—carbohydrates, fats, or proteins—are most likely to be used to provide the intermediates for the synthesis of blood glucose? Why?

6. Why does the catabolism of fatty acids occur only under aerobic conditions?

7. Why will certain forms of liver disease produce an increase in the blood levels of ammonia?

THOUGHT QUESTIONS

(Answers are given in Appendix A.)

1. How much energy is added to or released from a reaction in which reactants A and B are converted to products C and D if the energy content, in kilocalories per mole, of the participating molecules is: A = 55, B = 93, C = 62, and D = 87? Is this reaction a reversible or irreversible reaction? Explain.

2. The rate of an enzyme-mediated reaction as a function of substrate concentration is shown below. If the initial substrate

CHAPTER

6

MOVEMENT OF MOLECULES ACROSS CELL MEMBRANES

DIFFUSION
    Magnitude and Direction of Diffusion
    Speed of Diffusion
    Diffusion across Membranes
        Diffusion through the lipid bilayer
        Diffusion through protein channels
        Regulation of diffusion through membranes
MEDIATED-TRANSPORT SYSTEMS
    Facilitated Diffusion
    Active Transport
        Primary active transport

        Secondary active transport
OSMOSIS
    Extracellular Osmolarity and Cell Volume

ENDOCYTOSIS AND EXOCYTOSIS
    Endocytosis
    Exocytosis
        Protein secretion
        Membrane formation

EPITHELIAL TRANSPORT
    Glands

A s we saw in Chapter 3, the contents of a cell are separated from the surrounding extracellular fluid by a thin layer of lipids and protein—the plasma membrane. In addition, membranes associated with mitochondria, endoplasmic reticulum, lysosomes, the Golgi apparatus, and the nucleus divide the intracellular fluid into separate compartments.

The contributions of this array of membranes to cell function fall into two categories: (1) The membranes provide barriers to the movements of molecules and ions both between the various cell organelles and between the cell and the extracellular fluid, and (2) they provide a scaffolding to which various cell components are anchored, such as enzymes, contractile protein filaments, and binding sites for messenger molecules. This chapter focuses upon the barrier function of membranes, with emphasis on the plasma membrane.

Although cell membranes act as barriers, as just stated, they are not to be pictured as solid brick walls. Some molecules and ions do indeed pass from one side of these membranes to the other. Cells take in substances from the extracellular fluid and release other substances to it, all through the plasma membrane.

We say that a membrane is *permeable* to any substance that can pass through it regardless of the mechanism of permeation. If membranes were permeable to all substances, one would not expect the composition of the intracellular fluid to be much different from the extracellular fluid, while in fact (Table 6-1), there are large differences in the concentrations of substances on the two sides of the plasma membrane. These differences in intracellular and extracellular composition result from two properties of membranes: (1) Membranes are selectively permeable, allowing some substances to pass but not

TABLE 6-1	COMPOSITION OF EXTRACELLULAR AND INTRACELLULAR FLUIDS	
	Extracellular Concentration, mM	Intracellular Concentration,* mM
Na^+	145	15
K^+	4	150
Ca^{2+}	1	1.5
Mg^{2+}	1.5	12
Cl^-	110	10
HCO_3^-	24	10
P_i	2	40
Amino acids	2	8
Glucose	5.6	1
ATP	0	4
Protein	0.2	4

*The intracellular concentrations differ slightly from one tissue to another, but the concentrations shown above hold for most cells. Furthermore, the intracellular concentrations listed above may not reflect the free concentration of the substance in the cytosol since some may be bound to proteins or confined within cell organelles. For example, the free cytosolic concentration of calcium is only about 0.0001 mM.

others, and (2) the permeability of membranes to specific substances can be increased or decreased.

There are several mechanisms by which substances pass through membranes, and we begin with the physical process known as diffusion.

DIFFUSION

All molecules in any substance, be it solid, liquid or gas, are in a continuous state of movement at body temperature, due to the heat energy they contain. The average speed of molecular movement depends upon both the temperature and mass of the molecule. At body temperature, an average molecule of water moves at about 2500 km/h (1500 mi/h), whereas a molecule of glucose, which is 10 times heavier, moves at about 850 km/h. In solutions, such rapidly moving molecules cannot travel very far before colliding with other molecules. They bounce off each other like rubber balls, undergoing millions of collisions every second. Each collision alters the direction of the molecule's movement, so that the path of any one molecule becomes unpredictable. Since any molecule may at any instant be moving in any direction, such movement is said to be "random," meaning that it has no preferred direction of movement.

The random thermal motion of molecules in a liquid or gas will tend to redistribute them uniformly throughout the container, a process known as **diffusion**. Thus, if we start with a solution in which a solute is more concentrated in one region than another (Figure 6-1A), diffusion will redistribute the solute from regions of high concentration to regions of lower concentration until the solute reaches a uniform concentration throughout the solution (Figure 6-1B). Many properties of living organisms are closely associated with diffusion. For example, oxygen, nutrients, and other molecules enter and leave the blood by diffusion, and the movement of many substances across the plasma membrane and organelle membranes occurs by diffusion.

Magnitude and Direction of Diffusion

Any volume of solution can be looked upon as being composed of a number of small-volume elements joined together at their surfaces (Figure 6-2). The amount of material crossing a surface in a unit of time is known as a **flux**. If the number of molecules in a unit of volume is doubled, the flux of molecules across each surface of the unit will also be doubled, since twice as many molecules will be moving in any random direction at a given time. To generalize, the concentration of molecules (number of molecules in a unit of volume) in any region of a solution determines the magnitude of the flux across the "surfaces" of this region.

The diffusion of glucose between two compartments of equal volume and separated by a permeable barrier is

A

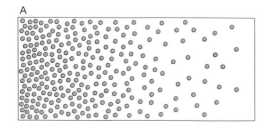

B

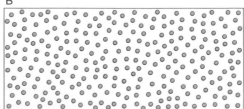

FIGURE 6-1 Molecules initially concentrated in one region of a solution (A) will, by their random thermal motion, become uniformly distributed throughout the solution (B).

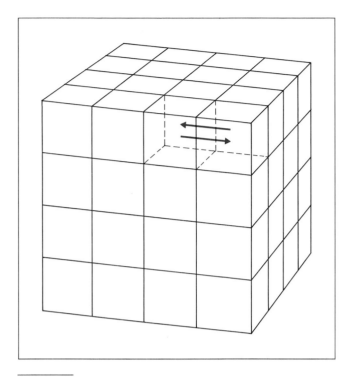

FIGURE 6-2 A large volume of solution is equivalent to a number of small-volume elements in which the movements of the molecules between the volume elements depend upon the concentrations in each volume element.

illustrated in Figure 6-3. Initially glucose is present in compartment 1 at a concentration of 20 mM, and there is no glucose in compartment 2. The random movements of the glucose molecules in compartment 1 carry some of them into compartment 2. This flux of glucose from compartment 1 to 2 depends on the concentration of glucose in compartment 1. After a short time, some of the glucose molecules that have entered compartment 2 will randomly move back into compartment 1 (Figure 6-3B). The magnitude of the glucose flux from compartment 2 to 1 depends upon the concentration of glucose in compartment 2 at any time. The **net flux** of glucose between the two compartments is the difference between the two one-way fluxes. It is the net flux that determines the net gain of molecules by compartment 2 and the net loss from compartment 1.

As the concentration of glucose in compartment 2 increases, the flux of glucose from 2 back into 1 increases. And as the concentration in compartment 1 decreases, the flux from 1 to 2 decreases. Eventually the concentrations of glucose in the two compartments will become equal at 10 mM. The two one-way fluxes will then be equal in magnitude but opposite in direction, and the net flux of glucose will be zero (Figure 6-3C). The system has now reached **diffusion equilibrium**, and no further

FIGURE 6-3 Diffusion of glucose between two compartments of equal volume separated by a permeable barrier. (A) Compartment 1 contains glucose at a concentration of 20 mM, and no glucose is present in compartment 2. (B) Some glucose molecules have moved into compartment 2, and some are moving at random back into compartment 1. The length of the arrows represents the magnitude of the one-way fluxes. (C) Diffusion equilibrium has been reached, the concentration of glucose is equal in the two compartments (10 mM), and the net flux is zero.

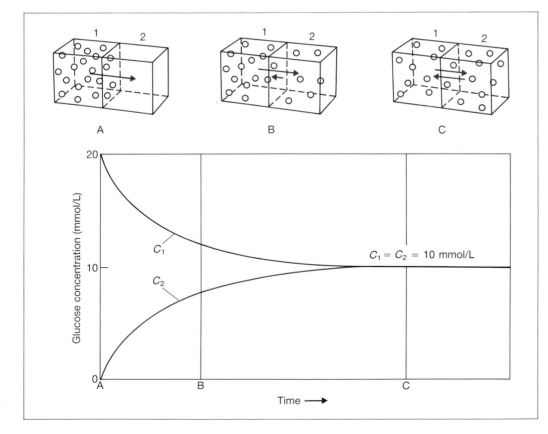

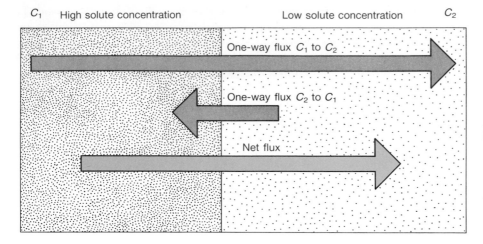

C_1 High solute concentration   Low solute concentration C_2

One-way flux C_1 to C_2

One-way flux C_2 to C_1

Net flux

FIGURE 6-4 The three fluxes occurring during the diffusion of solute across a boundary: two one-way fluxes and the net flux (the difference between the two one-way fluxes). The net flux always occurs in the direction from high to lower concentration.

change in the glucose concentration of the two compartments will occur, although thermal motion will continue to move equal numbers of glucose molecules in both directions between the two compartments.

Several important properties of diffusion can be re-emphasized using this example. Three fluxes can be identified at any surface—the two one-way fluxes occurring in opposite directions from one compartment to the other and the net flux, which is the difference between them (Figure 6-4). It is the net flux that is the most important component in diffusion since the net flux is the net amount of material transferred from one location to another. Although the movement of individual molecules is random, *the net flux always proceeds from regions of higher concentration to regions of lower concentration.* For this reason, we often speak of substances moving "downhill" by diffusion.

The greater the difference in concentration between any two regions, the greater the magnitude of the net flux. Thus, both the direction and the magnitude of the net flux are determined by the concentration difference.

At any concentration difference, however, the magnitude of the net flux depends on several additional factors: (1) temperature—the higher the temperature, the greater the speed of molecular movement and the greater the flux; (2) mass of the molecule—large molecules (for example, proteins) have a greater mass and lower speed than smaller molecules (for example, glucose) and thus have a smaller flux; (3) surface area—the greater the surface area between two regions, the greater the space available for diffusion and thus the greater the total net flux; (4) medium—molecules diffuse more rapidly in a gas phase than in water because collisions are less frequent in a gas phase.

Speed of Diffusion

The speed of diffusion is an important factor in determining the rate at which molecules can reach a cell from the blood or move throughout the interior of a cell after crossing the plasma membrane. Although individual molecules travel at high speeds, the number of collisions they undergo prevents them from traveling very far in a straight line. For example, it takes glucose approximately 3.5 s to reach 90 percent of diffusion equilibrium at a point 10 μm away from a source of glucose, such as the blood, but it would take over 11 years to reach the same concentration at a point that is 10 cm away from the source.

Thus, although diffusion can distribute molecules rapidly over short distances (for example, within cells or the extracellular regions surrounding a few layers of cells), diffusion takes a very long time when distances of a few centimeters or more are involved. For an organism as large as a person, the diffusion of oxygen and nutrients from the body surface to tissues located several centimeters below the surface would be far too slow to provide adequate nourishment. Accordingly, the circulatory system provides the mechanism for rapidly moving materials over large distances (by blood flow using a mechanical pump, the heart), with diffusion providing movement into and out of the blood and through the extracellular fluid.

The rate at which diffusion is able to move molecules *within* a cell is one of the factors limiting the size to which a cell can grow. The larger the cell, the greater is the volume of intracellular fluid that must be nourished by diffusion from a given unit of cell surface area. A cell would not have to be very large before diffusion fails to provide sufficient nutrients to the large volume of intracellular fluid. For example, the center of a 20-μm-diameter cell reaches diffusion equilibrium with extracellular oxygen in about 15 ms, but it would take 265 days to reach equilibrium at the center of a cell the size of a basketball.

Diffusion across Membranes

Some molecules are able to diffuse across cell membranes. The rate at which a substance diffuses across the plasma membrane can be measured by monitoring the rate at which its cytosolic concentration reaches diffusion equilibrium with its concentration in the extracellular fluid. If the volume of extracellular fluid is large, its solute concentration will remain essentially constant as the substance diffuses into the small intracellular volume (Figure 6-5). As with all diffusion processes, net flux F of material across the membrane is from the region of the high concentration (the extracellular solution in this case) to the region of low concentration. The net flux is the difference between the two one-way fluxes, the influx f_i into the cell and the efflux f_e out of the cell. The magnitude of the net flux is directly proportional to the difference in concentration across the membrane ($C_o - C_i$), the surface area of the membrane A, and the membrane **permeability constant** k_p:

$$\text{Influx:} \quad f_i = k_p A C_o$$
$$\text{Efflux:} \quad f_e = k_p A C_i$$

$$\text{Net flux:} \quad F = k_p A (C_o - C_i) \qquad (6\text{-}1)$$

FIGURE 6-5 The increase in intracellular concentration as a substance diffuses from a constant extracellular concentration until diffusion equilibrium is reached across the plasma membrane of a cell.

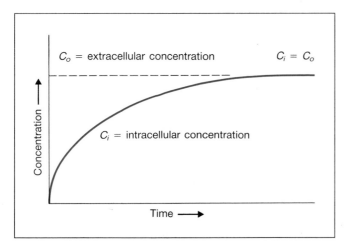

The numerical value of the permeability constant depends on the type of molecule, its molecular weight, the temperature, and the characteristics of the membrane through which the molecule is diffusing. It is an experimentally determined number for a given type of molecule at a given temperature, and it reflects the ease with which the molecule is able to move through a given membrane. In other words, the greater the permeability constant, the larger the net flux across the membrane for any given concentration difference and membrane area.

The above description has been in terms of molecules diffusing into a cell from a region of high extracellular concentration. Obviously, molecules can also undergo net diffusion out of a cell if the intracellular concentration is greater than the extracellular concentration, as occurs for substances that are being synthesized by cells. The same equation (Equation 6-1) applies, but now the net flux will be negative, indicating a net movement out of the cell.

The rates of diffusion across membranes are a thousand to a million times smaller than the diffusion of the same molecules in water. In other words, the molecules enter or leave a cell by diffusion at a much slower rate than if there were no physical barrier at the surface. Cell membranes are highly selective barriers, however, that allow some molecules to enter rapidly by diffusion while others that may diffuse at a similar rate in water enter more slowly or not at all. In other words, there is a different membrane permeability constant k_p for each substance.

Diffusion through the lipid bilayer. When the permeability constants of a number of different organic molecules are examined in relation to their molecular structures, a correlation emerges. Whereas most polar molecules diffuse into cells very slowly or not at all, nonpolar molecules rapidly diffuse across plasma membranes, that is, they have large permeability constants. The reason is that nonpolar molecules can dissolve in the nonpolar regions of the membrane—regions occupied by the fatty acid chains of the membrane phospholipids. In contrast, polar molecules have a much lower solubility in the nonpolar regions of the membrane. The more lipid-soluble the substance (the fewer its polar or ionized groups), the greater the number of its molecules that dissolve in the membrane lipids and the greater its flux across the membrane. Oxygen, carbon dioxide, fatty acids, and steroid hormones are examples of nonpolar molecules that diffuse rapidly through the lipid portions of membranes.

That it is the lipid-bilayer portion of the membrane that provides the selective barrier, allowing nonpolar molecules to pass but excluding most polar or ionized substances, is substantiated by experiments using artificial membranes consisting of a bimolecular layer of lipids but containing no protein. The rates of diffusion across these artificial lipid bilayers and their selectivity are very similar to those of cell membranes.

Diffusion through protein channels. Ions such as Na^+, K^+, Cl^-, and Ca^{2+} diffuse across plasma membranes at rates that are much faster than would be predicted from their very low solubility in lipid. Moreover, different cells have quite different permeabilities to these ions, whereas nonpolar substances, which diffuse through the lipid bilayer, have similar permeabilities when different cells are compared. The fact that artificial lipid bilayers containing no protein are practically impermeable to these ions, whereas cell membranes have a considerable permeability to them, suggests that it is the protein component of the membrane that is responsible for these permeability differences.

As we have seen (page 42), the integral membrane proteins span the lipid bilayer. Such proteins can form **channels** through which ions can diffuse across the membrane (Figure 6-6). A single protein may have a conformation similar to a doughnut, with the hole in the middle providing the channel for ion movement. In other cases, several proteins aggregate to form the walls of a channel. The diameters of such protein channels are very small, only slightly larger than those of the ions that pass through them. The small size of the channels pre-

vents larger, polar organic molecules from entering the channel. Only the mineral ions (and possibly water) are small enough to pass through these protein channels.

The variation in ion membrane permeability found in different cell membranes reflects differences in the number of ion channels in the membranes. The greater the number of channels, the greater the ion flux across the membrane for any given ion concentration difference. Furthermore, the ion channels show a selectivity for the type of ions that can pass through them. This selectivity is based partially on the channel diameter and partially on the charged and polar surfaces of the proteins that form the channel walls and that electrically attract or repel the ions. For example, some channels (K channels) will allow only potassium ions to pass, others are specific for sodium (Na channels), and still others allow both sodium and potassium ions to pass (Na,K channels). Two membranes that have the same permeability to potassium, because they have the same number of potassium channels, may have quite different permeabilities to sodium because of differences in the number of sodium channels.

Thus far we have described the direction and magnitude of the net flux in terms of the concentration difference across a membrane. When describing the diffusion of ions, however, one additional factor must also be considered: the electric forces acting upon ions. There is a separation of electric charge across plasma membranes

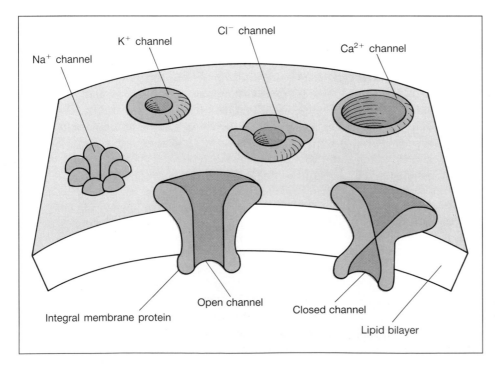

FIGURE 6-6 Hypothetical shapes of the integral membrane proteins that form the various channels through which ions diffuse across membranes. A single protein or a cluster of more than one protein forms the walls of a channel. A channel may be open or closed as a result of conformational changes in the channel proteins.

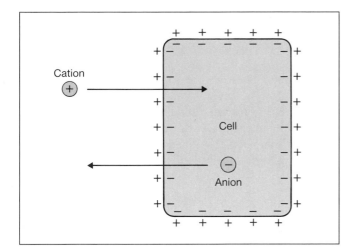

FIGURE 6-7 The separation of electric charge across a plasma membrane (membrane potential) provides the electric force that drives positive ions into a cell and negative ions out of a cell.

(Figure 6-7), the origin of which will be described in Chapter 8. Such a charge separation, known as a **membrane potential**, affects the diffusion of ions across the membrane because ions are charged. For example, if the inside of a cell has a net negative charge, as it does in most cells, there will be an electric force attracting positive ions into the cell and repelling negative ions. Even if there were no concentration difference across the membrane, there would still be a net flux of positive ions into and negative ions out of a cell because of the membrane potential. Thus, the direction and magnitude of ion fluxes across membranes depends on both the concentration difference and the electrical difference (the membrane potential.) These two driving forces are collectively known as the **electrochemical difference,** also termed the electrochemical gradient, across the membrane.

Regulation of diffusion through membranes. The net flux of a substance across a membrane may be altered by changes in one or more of the variables in Equation 6-1: (1) the surface area A, (2) the concentration difference for the substance $C_o - C_i$, and the electrical difference in the case of ions, and (3) the membrane permeability constant k_p for the substance. Of these three, the membrane permeability constant is the one over which cells have some direct control and the one that can be altered to effect rapid changes in ion fluxes across the membrane.

Changes in a membrane's permeability to ions can occur rapidly as a result of the opening or closing of ion channels. These openings and closings can occur randomly, suggesting that the channel protein is fluctuating between two (or more) conformations, possibly due to random collisions with molecules in the surrounding medium. Over an extended period of time, at any given electrochemical difference, the total number of ions that pass through a single channel depends on how frequently the channel opens and how long it stays open (Figure 6-8).

Three factors can alter the stability of the channel protein conformations, producing changes in the opening frequency or duration: (1) the binding of specific chemicals to the channel protein—**receptor-operated channels**; (2) changes in the membrane potential—**voltage-sensitive channels**; and (3) stretching the membrane—**stretch-activated channels**. A single channel may be affected by all three of these factors, only one, or none.

The same type of ion may pass through several different types of channels. For example, a membrane may contain receptor-operated potassium channels, voltage-sensitive potassium channels, and stretch-activated potassium channels. In addition, the same membrane may have several types of voltage-sensitive potassium channels, each responding to a different range of membrane voltage, or several types of receptor-operated potassium channels, each responding to a different chemical messenger. The roles of these various ion channels in determining the electrical activities of cells will be discussed in Chapter 8.

MEDIATED-TRANSPORT SYSTEMS

Although diffusion accounts for some of the transmembrane movement of ions, it does not account for all. Moreover, there are a number of other molecules, including amino acids and glucose, that are able to cross membranes yet are too polar to diffuse through the lipid bilayer or too large to diffuse through protein channels. The passage of these molecules and the nondiffusional movements of ions through membranes are mediated by integral membrane proteins known as **carriers** or transporters. Such **mediated-transport systems** are quite different from the passage of substances through membranes by diffusion.

A substance that crosses by mediated transport must first bind to a specific site on a carrier (Figure 6-9), a site that is exposed to the substance at one surface of the membrane. The carrier protein then undergoes a change in shape, exposing its binding site to the solution on the opposite side of the membrane. The dissociation of the substance from the carrier binding site completes the process of moving the material through the membrane. Using this carrier mechanism, molecules can move in

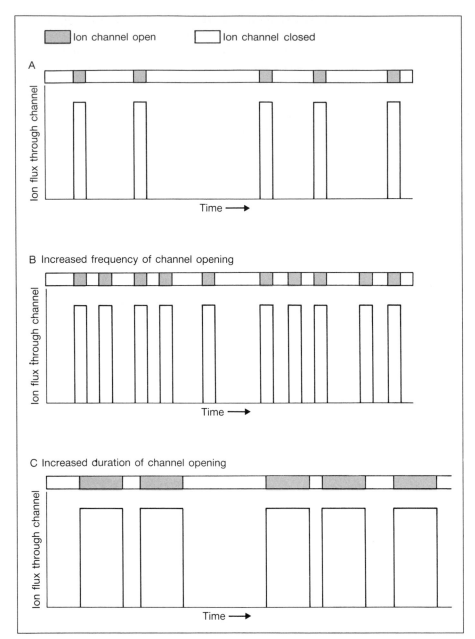

FIGURE 6-8 (A) Changes in the conformation of ion-channel proteins result in the fluctuating opening and closing of an ion channel. Increasing the frequency of channel openings (B) or the duration that a channel remains open (C) will increase the number of ions crossing the membrane.

either direction across the membrane, getting on the carrier at one side and off at the other.

At present, little is known about the specific shape changes that carrier proteins undergo during this process. Oscillations between the two conformations of a carrier probably occur spontaneously, whether or not the binding site is occupied as can occur with ion channel opening and closings.

There are many types of carrier proteins in membranes, each type having binding sites that are specific for a particular substance or a specific class of related substances. For example, although both amino acids and sugars undergo mediated transport, the carrier that transports amino acids does not transport sugars, and vice versa. Just as with ion channels, the plasma membranes of different cells may contain different types and numbers of carrier proteins and thus exhibit differences in the types of substances transported and their rates of transport.

Two factors determine the magnitude of the solute flux through a mediated-transport system: (1) the extent to which the carrier is saturated, which depends on both

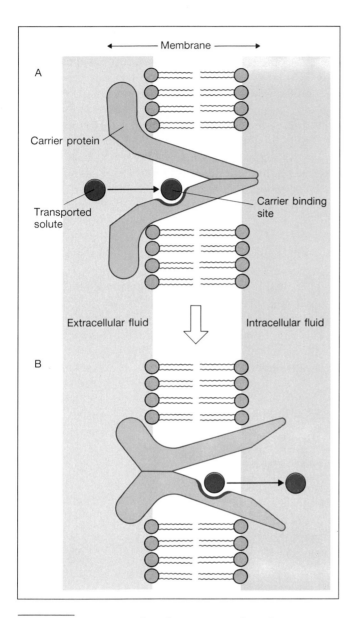

any binding site, as the concentration of the ligand (the solute to be transported in this case) is increased, the number of occupied binding sites on the carrier molecules increases until the system becomes saturated, that is, all the sites become occupied. Further increases in solute concentration will not increase the transport flux.

Contrast the saturation behavior of mediated-transport systems (Figure 6-10) with the expected fluxes produced by diffusion. A flux resulting from diffusion increases in direct proportion to the extracellular concentration, and there is no maximal limit since diffusion does not involve binding to a fixed number of sites.

The magnitude of a mediated-transport flux is subject to regulation by changing either the number of carrier proteins in a membrane, a process involving the insertion of new proteins into or removal of existing proteins from a membrane, or altering the affinity of the binding sites for the transported solute. Changing the affinity changes the number of occupied binding sites at any given solute concentration and thus changes the flux through the membrane.

There are two categories of carrier-mediated processes—**facilitated diffusion** and **active transport**. Facilitated diffusion uses a carrier protein to move solute from higher to lower concentration across a membrane,

FIGURE 6-9 Carrier-mediated transport. When the carrier protein is in conformation A, its solute-binding site is exposed to solute in the extracellular fluid. When the carrier protein changes its conformation from A to B, the solute-binding site becomes exposed to the intracellular fluid. These changes in carrier conformation move a solute molecule that is bound to the carrier-binding site through the membrane.

FIGURE 6-10 The flux of molecules diffusing across a plasma membrane continuously increases in proportion to the extracellular concentration, whereas the flux of molecules entering by a mediated-transport system reaches a maximal value. This maximal mediated-transport flux corresponds to the saturation of all the available carrier molecules by the transported solute.

the solute concentration and the affinity of the carrier protein for the solute, and (2) the number of carrier proteins in the membrane. The greater the number of carrier-proteins, the greater the flux at any level of saturation.

For any transported solute, there is a finite number of specific carrier proteins in a given membrane. As with

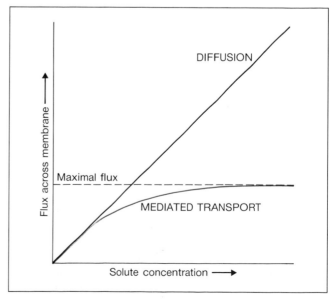

whereas active transport uses a carrier protein that is coupled to an energy source to move solute "uphill" across a membrane, that is, against its electrochemical difference.

Facilitated Diffusion

Facilitated diffusion is an unfortunate term since the process it denotes does not involve diffusion. The term arose because the end result of both diffusion and facilitated diffusion is the same. In both processes the net flux across a membrane always proceeds from higher to lower concentration and the concentrations on the two sides of the membrane are equal when equilibrium is reached. When the transported-solute concentration is increased sufficiently, however, facilitated-diffusion systems reach a maximal flux because their carriers become saturated, whereas fluxes due to diffusion do not saturate (Figure 6-11).

When a facilitated-diffusion carrier changes its shape, so that its binding site goes from facing one side of the membrane to facing the other, the binding properties of the site do not change. The flux from side 1 to side 2 of

a membrane depends on the number of carrier-binding sites accessible on side 1 that are occupied by solute, which in turn depends on the concentration of the transported solute on side 1. Movement in the opposite direction likewise depends on the number of carrier-binding sites available on side 2 and the concentration of solute on side 2. Since the binding-site characteristics are the same in the two carrier conformations, when the concentration of transported solute is the same on both sides of the membrane, equal numbers of molecules will move in opposite directions and the net flux will be zero. When the concentrations are not equal, the net flux will always proceed from higher to lower concentration since there will be more bound solute on the high-concentration side than on the low-concentration side, provided the system is not saturated.

One of the most important facilitated-diffusion systems in the body is the one that moves glucose across most plasma membranes. Without such glucose carriers, cells would be impermeable to glucose, which is a relatively large, polar molecule, and one of the most important nutrients would be unavailable to them. One might

FIGURE 6-11 Comparison of diffusion, facilitated diffusion, and active transport in terms of the level of intracellular concentration reached relative to the extracellular concentration and the presence or absence of a limiting maximal flux due to saturation. Note that the steady-state intracellular concentration does not distinguish between diffusion and facilitated diffusion, nor does the presence of saturation distinguish facilitated diffusion from active transport.

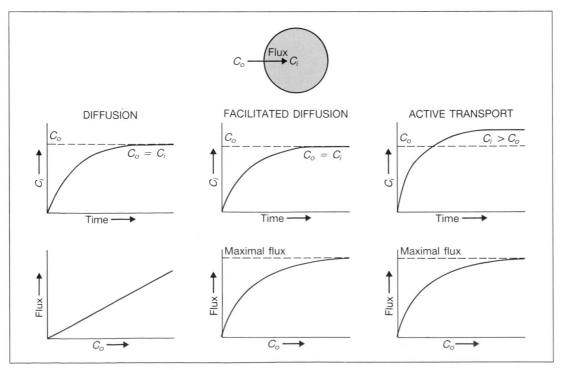

expect that as a result of facilitated diffusion, the glucose concentration inside cells would become equal to the extracellular concentration. This does not occur in most cells, however, because glucose is metabolized almost as quickly as it enters. Thus, the intracellular glucose concentration remains lower than the extracellular concentration (Table 6-1), and there is a continuous net flux of glucose into cells.

Active Transport

Active transport differs from facilitated diffusion in that the former uses energy to move substances "uphill," that is, from a region of lower concentration to one of higher concentration across a membrane (Figure 6-12). Because active-transport processes require energy, they are often referred to as "pumps." Recall that with ions, the electrical difference as well as the concentration difference influences fluxes across the membrane. Thus, the active transport of ions involves the movement of ions against an electrochemical difference. To simplify our discussion, however, we shall refer only to the concentration difference, by implication including the electrical difference when ions are involved.

Like facilitated diffusion, active-transport systems use carrier-protein molecules, exhibit chemical specificity, and reach a maximal flux when their binding sites are saturated. However, whereas the net flux becomes zero in a facilitated-diffusion system when the concentrations on the two sides of a membrane are equal, active-transport systems reach a zero net flux state when the concentration is higher on one side of the membrane than the other (Figure 6-11).

As in facilitated diffusion, an active-transport carrier is able to change its shape so that the binding site is accessible first to molecules on one side of a membrane and then to molecules on the other. However, in facilitated diffusion, as we saw, the characteristics of the binding sites are the same on both sides of the membrane. In contrast, active-transport carriers have binding sites that differ in affinity on opposite sides of the membrane, and it is this difference that is responsible for the ability of the carrier to move substances against a concentration gradient. An input of energy is required to produce the difference in affinity.

Figure 6-13A shows a membrane containing an active-transport carrier that separates two solutions initially having equal solute concentrations. If the binding sites exposed on side 1 have a greater affinity for the solute than do the sites on side 2, then the amount of solute bound on side 1 will be greater, and the flux from 1 to 2 will be greater than the flux from 2 to 1. The result is a net flux of solute from 1 to 2 even though initially there is no concentration difference across the mem-

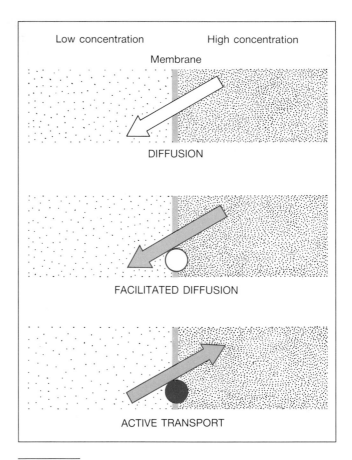

FIGURE 6-12 Direction of net solute flux crossing membrane by (1) diffusion—high to low concentration, (2) facilitated diffusion—high to low concentration, and (3) active transport—low to high concentration.

brane. Eventually the concentration on side 2 will increase to a level that will produce an occupancy of the low-affinity carrier sites on side 2 equal to the occupancy of the high-affinity sites on the opposite side of the membrane. At this point a steady state is reached (Figure 6-13B), and the net flux through the active-transport carrier is zero even though there is a concentration difference across the membrane.

The direction in which transport against a concentration difference takes place depends on which side of the membrane has the low-affinity binding sites, net "uphill" movement being directed toward the low-affinity side of the membrane. Some transport systems pump molecules into cells, others pump them out.

The role of energy in this process is to produce the difference in affinity of the carrier binding sites on the two sides of the membrane. Two means of linking energy to the active-transport process are known: (1) the direct use of ATP in **primary active transport** and (2) the use of an ion concentration difference across a membrane to

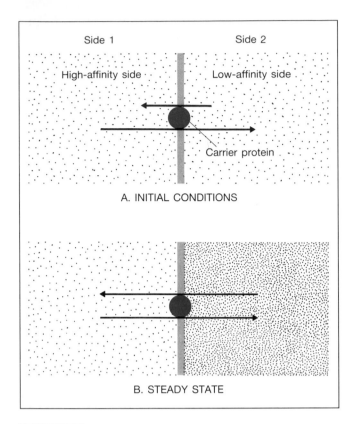

FIGURE 6-13 (A) When the concentration on the two sides of a membrane are equal, there is a net flux from side 1 to side 2 through an active-transport carrier from the high-affinity side of the membrane to the low-affinity side. (B) When the steady state is reached, there is no net flux through the active-transport carrier. (The symbol ⟶○⟶ will be used throughout this book to represent a carrier protein in a membrane. If a substance crosses a membrane by diffusion, no circle will be present.)

an ion concentration difference across a membrane to supply the energy in **secondary active transport**.

Primary active transport. Chemical energy is transferred directly from ATP to membrane carrier proteins during primary active transport. The carrier protein acts as an enzyme, an ATPase, that catalyzes its own phosphorylation. In this case an enzyme—the carrier protein—and one of its substrates—the carrier protein—are the same molecule. Phosphorylation of the carrier (Figure 6-14A) at one site alters the affinity of the solute-binding site located at a different point on the carrier protein. (Note that this is an example of covalent modulation, as discussed on page 58, in which the phosphorylation of a protein at one site modifies the binding-site characteristics of another site.)

When the carrier conformation changes and the binding site is exposed to the opposite side of the membrane (Figure 6-14B), the covalent bond between carrier and phosphate is broken, and the phosphate group is re-

FIGURE 6-14 Primary active transport. In this illustration, (A) phosphorylation of the carrier protein by ATP at the cytosolic surface increases the affinity of the solute-binding site that is exposed to the extracellular solution. (B) Removal of the phosphate group from the carrier decreases the affinity of the solute-binding site when it is exposed to the intracellular fluid. Since the saturation of carriers on the high-affinity side of the membrane will be greater than on the low-affinity side, there will be a net flow of solute into the cell from a low extracellular concentration to a higher intracellular concentration.

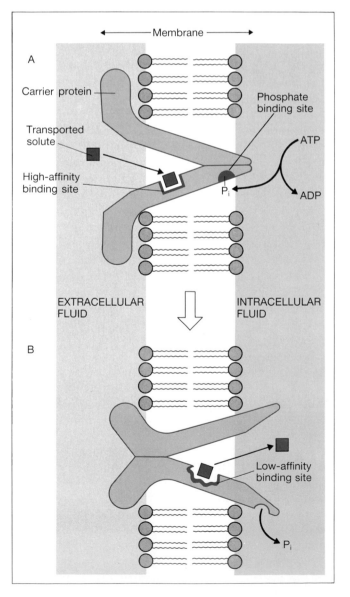

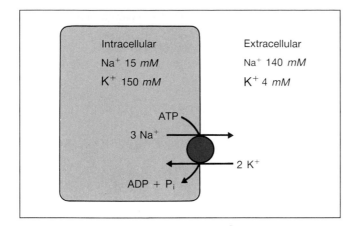

Intracellular

Na$^+$ 15 *mM*

K$^+$ 150 *mM*

Extracellular

Na$^+$ 140 *mM*

K$^+$ 4 *mM*

ATP

3 Na$^+$

2 K$^+$

ADP + P$_i$

FIGURE 6-15 The primary active transport of sodium and potassium ions in opposite directions by the Na,K-ATPase in plasma membranes is responsible for the low sodium and high potassium intracellular concentrations. For each ATP hydrolized, 3 sodium ions are moved out of a cell and 2 potassium ions are moved in.

leased, changing the affinity of the solute-binding site. The phosphorylation and dephosphorylation of the carrier protein thus produces binding sites of differing affinity on the two sides of the membrane, the condition necessary for active transport. Metabolic inhibitors that block ATP synthesis inhibit the primary active transport of solutes because they deprive the carrier of its energy source.

Only three primary active-transport carriers have been identified, each an ATPase and each involved in transporting ions: (1) one that simultaneously pumps sodium and potassium ions (Na,K-ATPase carrier); (2) one that pumps calcium ions (Ca-ATPase carrier); and (3) one that pumps hydrogen ions (H-ATPase carrier).

The Na,K-ATPase carrier is present in all plasma membranes. The pumping activity of this transport system leads to the characteristic distribution of high intracellular potassium and low intracellular sodium relative to their respective extracellular concentrations (Figure 6-15). This carrier moves three sodium ions *out* of a cell and two potassium ions *in* for each molecule of ATP that is hydrolyzed.

The Ca-ATPase carrier is located in the plasma membrane and in several organelle membranes, including the membranes of the endoplasmic reticulum and the mitochondria. In the plasma membrane, the direction of calcium transport is from cytosol to extracellular fluid. In organelle membranes it is from cytosol into the organelle lumen. The active transport of calcium out of the cytosol is one reason that the cytosol of most cells has a very low calcium concentration, about 10^{-7} mol/L compared with

an extracellular calcium of 10^{-3} mol/L, 10,000 times greater (a second reason will be given below).

The H-ATPase carrier is located in the plasma membrane and in several organelle membranes, including the inner mitochondrial and lysosomal membranes. In the plasma membrane the H carrier transports hydrogen ions out of cells.

Secondary active transport. The mechanism by which carrier affinity is altered on the two sides of a membrane distinguishes secondary active transport from primary active transport. In primary active transport the affinity of the solute-binding site on the carrier protein is altered by the phosphorylation of the carrier by ATP (covalent modulation), while in secondary active transport the affinity of the solute-binding site is altered by the binding of ions, often sodium, to the carrier. The latter is an example of allosteric modulation (page 57).

The carrier proteins for secondary active transport have two binding sites, one for, let us say, sodium ions and one for the actively transported solute. The binding of sodium at its site alters the affinity of the other site for the actively transported solute (Figure 6-16). If the concentration of sodium on the two sides of a membrane were equal, the same degree of sodium-binding-site saturation would occur and there would be no difference in the affinities of the transport sites on the two sides. However, as a result of the *primary* active transport of sodium out of cells by the Na,K-ATPase pumps, the extracellular concentration of sodium is much greater than the intracellular concentration. Therefore, more sodium is bound to the secondary transport carriers exposed on the outside of the cell than is bound to the sodium sites exposed to the lower intracellular sodium concentration. Because of this difference in sodium binding on the two sides of the membrane, the affinities of the binding sites for the transported solute will also differ. It is this difference in affinity, as is the case in primary active transport, that leads to the active transport of the solute from low to high concentration.

In addition to the actively transported solute that moves through the membrane when the carrier protein changes its conformation, the sodium bound to the carrier also crosses the membrane. The net movement of sodium by a secondary transport carrier is always from high extracellular concentration into the cell, where the concentration of sodium is lower. Thus, in secondary active transport, the movement of sodium is always "downhill," while the net movement of the actively transported solute on the same carrier is uphill, moving from lower to higher concentration. The movement of the actively transported solute can be either into the cell (in the same direction as sodium), in which case it is

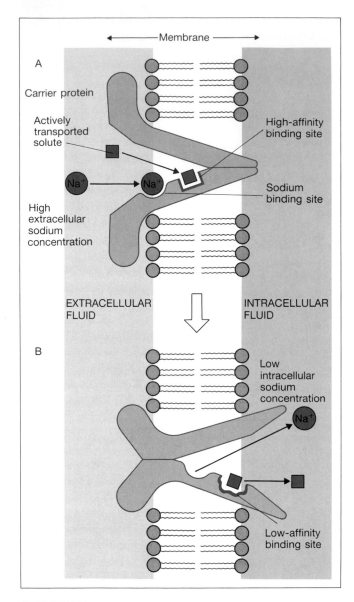

FIGURE 6-16 Secondary active transport. The binding of sodium ions to the carrier alters the affinity of the binding site for the actively transported solute. Differences in intracellular and extracellular sodium concentrations (the sodium gradient) lead, therefore, to low- and high-affinity sites for the actively transported solute on the two sides of the membrane.

transport pumps. Ultimately the energy for secondary active transport is derived from metabolism in the form of the ATP that is used by the primary Na,K-ATPase to create the sodium concentration gradient.[1] Between 10 and 40 percent of the ATP produced by a cell, under resting conditions, is used by the Na,K-pump to maintain the sodium gradient, which in turn drives a multitude of secondary active transport systems.

A variety of organic molecules and a few ions are moved across membranes by sodium-coupled secondary active transport. For example, amino acids, in most

---

[1]The energy stored in an ion concentration gradient across a membrane can also be used to synthesize ATP from ADP and P_i. Electron transport through the cytochrome chain produces a hydrogen-ion concentration gradient across the inner mitochondrial membrane. The movement of hydrogen ions down the concentration gradient provides the energy that is coupled to the synthesis of ATP during oxidative phosphorylation—the chemiosmotic hypothesis.

FIGURE 6-17 During secondary active transport involving sodium, this ion always moves *down* its concentration gradient into a cell, and the transported solute always moves *up* its gradient. Both sodium and the transported solute X move in the same direction through cotransport carriers but in opposite directions through countertransport carriers.

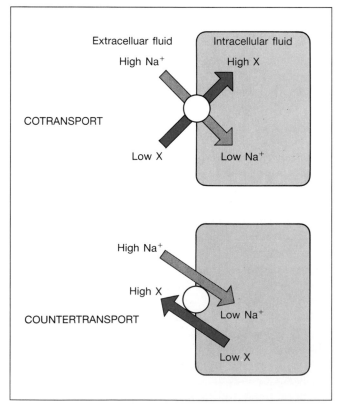

known as **cotransport**, or out of the cell (opposite to the direction of sodium), when it is called **countertransport** (Figure 6-17).

The creation of a sodium concentration difference across the plasma membrane by the primary active transport of sodium is a means of indirectly "storing" energy that can then be used to drive secondary active-

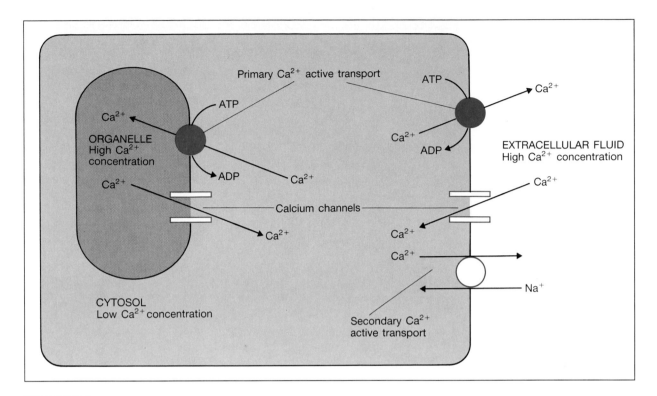

FIGURE 6-18 Pathways affecting cytosolic calcium concentration. The active transport of calcium, both by primary Ca-ATPase pumps and by secondary active calcium countertransport with sodium, moves calcium ions out of the cytosol. Calcium channels allow calcium to diffuse into the cytosol from either the extracellular fluid or cell organelles. Cytosolic calcium concentration is the resultant of all these processes.

cells, are actively transported into the cell by cotransport carriers, attaining intracellular concentrations 2 to 20 times higher than in the extracellular fluid.

An example of secondary active countertransport is provided by calcium. In addition to the previously described primary active transport of calcium from cytosol to extracellular fluid, there also exists in many membranes a Na-Ca countertransport carrier that uses the movement of sodium ions into a cell to move calcium ions out. In this system, should a decrease in the activity of the Na,K-ATPase pump occur, there would be a decrease in the sodium concentration difference across the plasma membrane leading to a rise in intracellular calcium. Figure 6-18 shows how cytosolic calcium concentration depends on calcium channels, a Ca-ATPase pump, and Na-Ca countertransport.

It must be emphasized that sodium is not the only ion whose movement down a concentration difference can achieve the secondary cotransport or countertransport of another substance. The less common participation of other ions will be described in subsequent chapters.

Table 6-2 provides a summary of the major characteristics of the different pathways by which substances move through cell membranes.

OSMOSIS

Water is a small, polar molecule about 0.3 nm in diameter that diffuses across most cell membranes very rapidly. One might expect that, because of its polar structure, water would not penetrate the nonpolar lipid regions of membranes. However, artificial phospholipid bilayers also have a high permeability to water, indicating that water can diffuse through the membrane lipid layer. The small size of the water molecule and the fact that it is not completely ionized, may contribute to its ability to penetrate the lipid region of membranes. Because of its small size, water can also diffuse through some protein channels but does not cross membranes by mediated-transport systems.

TABLE 6-2 MAJOR CHARACTERISTICS OF PATHWAYS BY WHICH SUBSTANCES CROSS MEMBRANES

	Diffusion		Mediated Transport		
	Through Lipid Bilayer	Through Protein Channel	Facilitated Diffusion	Primary Active Transport	Secondary Active Transport
Direction of net flux	High to low concentration	High to low concentration	High to low concentration	Low to high concentration	Low to high concentration
Equilibrium or steady state	$C_o = C_i$	$C_o = C_i$*	$C_o = C_i$	$C_o \neq C_i$	$C_o \neq C_i$
Use of integral membrane protein	No	Yes	Yes	Yes	Yes
Maximal flux at high concentration (saturation)	No	No	Yes	Yes	Yes
Chemical specificity	No	Yes	Yes	Yes	Yes
Use of energy and source	No	No	No	Yes; ATP	Yes; ion gradient (often Na)
Typical molecules using pathway	Nonpolar: O_2, CO_2, fatty acids	Ions: Na^+, K^+, Ca^{2+}	Polar: glucose	Ions: Na^+, K^+, Ca^{2+}, H^+	Polar: amino acids, glucose, some ions

*In the presence of a membrane potential, the intracellular and extracellular ion concentrations will not be equal at equilibrium.

Osmosis is the net diffusion of water from a region of high water concentration to a region of low water concentration when the movement of solute is prevented. There are thus two components to osmosis: (1) the diffusion of water and (2) a barrier, a membrane, for example, that will prevent solute movement but allow water movement.

As with any diffusion process, there must be a concentration difference in order to produce a net flux. How can a difference in water concentration be established across a membrane? The addition of a solute to water will lower the concentration of water in a solution compared to the concentration of pure water. A liter of pure water weighs about 1000 g. The molecular weight of water is 18. Thus, the concentration of pure water is 1000/18 = 55.5 M. However, if a solute such as glucose is dissolved in water, the concentration of water in the resulting solution is less than that of pure water. A given volume of a glucose solution will contain fewer water molecules than an equal volume of pure water since each glucose molecule in solution will occupy space formerly occupied by a

water molecule (Figure 6-19). The decrease in water concentration in a solution is approximately equal to the concentration of added solute. That is, one solute molecule will displace one water molecule.[2] The water concentration in a 1-M glucose solution is thus approximately 54.5 M rather than 55.5 M. Just as adding water to a solution will dilute the solute, adding solute to a solution will "dilute" the water. The greater the solute concentration, the lower the water concentration.

It is essential to recognize that the degree to which the water concentration is decreased by the addition of solute depends upon the *number* of particles (molecules or ions) of solute in solution (the solute concentration) and not upon the chemical nature of the solute. For example, 1 mol of glucose in 1 L of solution decreases the

---

[2] Because molecules have different sizes and may contain charged or polar groups that interact with the polar water molecules, there is not an exact one-to-one displacement of one water molecule for each solute molecule in solution, but for dilute solutions this generalization is approximately correct.

water concentration to approximately the same extent as does 1 mol of an amino acid, or 1 mol of urea, or 1 mol of any other molecule that exists as a single particle in solution. A molecule that ionizes in solution decreases the water concentration in proportion to the number of ions formed. Hence, 1 mol of sodium chloride gives rise to 1 mol of sodium ions and 1 mol of chloride ions, producing 2 mol of solute particles, which lowers the water concentration twice as much as does 1 mol of glucose. By the same reasoning, a 1-M $MgCl_2$ solution lowers the water concentration three times as much as does 1-M glucose solution.

Since water concentration depends upon the number of solute particles, it is useful to have a concentration term that refers to the total concentration of solute particles in a solution, regardless of their chemical composition. The total solute concentration of a solution is known as its **osmolarity**. One osmole is equal to 1 mol of solute particles. Thus, a 1-M solution of glucose has a concentration of 1 Osm (1 osmole per liter), but a 1-M solution of sodium chloride contains 2 osmoles of solute per liter of solution. A liter of solution containing 1 mol of glucose and 1 mol of sodium chloride has an osmolarity of 3 Osm. It is important to realize that although osmolarity refers to the concentration of solute particles in solution, it also determines the *water concentration* in the solution since the *higher* the osmolarity, the *lower* the water concentration.

The concentration of water in any two solutions having the same osmolarity is the same since the total number of solute particles per unit volume is the same. Such solutions are said to be isosmotic. A solution with an osmolarity of 3 Osm may contain 1 mol of glucose and 1 mol of sodium chloride, or 3 mol of glucose, or 1.5 mol of sodium chloride, or any other combination of solutes

as long as the total solute concentration is equal to 3 Osm.

Let us now apply these principles governing water concentration to the diffusion of water across membranes. Figure 6-20 shows two 1-L compartments separated by a membrane permeable to *both* solute and water. The concentration of solute is 2 Osm in compartment 1 and 4 Osm in compartment 2. This difference in solute concentration means there is also a difference in water concentration across the membrane: 53.5 M in compartment 1 and 51.5 M in compartment 2. Therefore, there will be a net diffusion of water from 1 to 2, and of solute in the opposite direction, from 2 to 1. When diffusion equilibrium is reached, both the solute and water concentrations will be equal in the two compartments, 3 Osm and 52.5 M, respectively. One mol of water will have diffused from compartment 1 to 2, and 1 mol of solute will have diffused from 2 to 1. Since 1 mol of solute has replaced 1 mol of water in compartment 1, and vice versa in compartment 2, there is no change in the volume of either compartment.

If the membrane is now replaced by one that is permeable to water but impermeable to solute (Figure 6-21), the same concentrations of water and solute will be reached at equilibrium as before, but there will also be a change in the volumes of the compartments. Water will diffuse from 1 to 2, but there will be no solute diffusion because the membrane is impermeable to solute. Water will continue to diffuse into compartment 2 until the water concentrations on the two sides become equal. The solute concentration in compartment 2 will decrease as it is diluted by the incoming water, and the solute in compartment 1 will become more concentrated as water moves out. When water has reached diffusion equilibrium, the osmolarities of the compartments will be equal

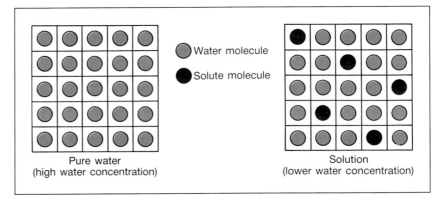

FIGURE 6-19 Decrease in water concentration resulting from the addition of solute molecules to pure water.

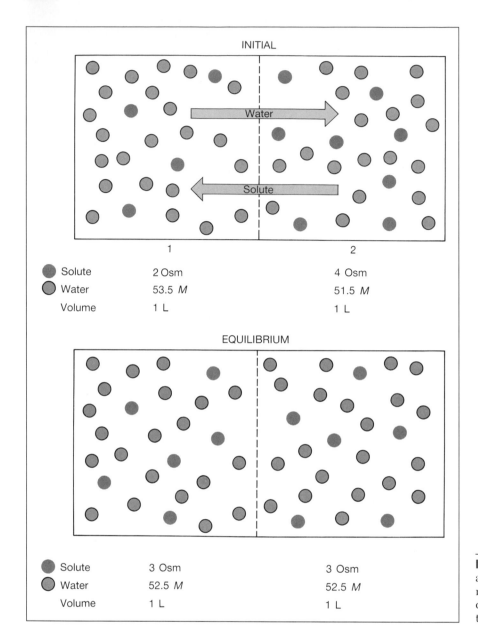

INITIAL

Water

Solute

1   2

	Solute	2 Osm	4 Osm
	Water	53.5 M	51.5 M
	Volume	1 L	1 L

EQUILIBRIUM

	Solute	3 Osm	3 Osm
	Water	52.5 M	52.5 M
	Volume	1 L	1 L

FIGURE 6-20 The net diffusion of water and solute in opposite directions across a membrane permeable to both leads to diffusion equilibrium with no change in the volume of either compartment.

and thus, the solute concentrations must also be equal. To reach this state of equilibrium, 18.5 mol[3] (0.33 L), of water had to be transferred by diffusion from compartment 1 to 2, increasing the volume of compartment 2 by one-third and decreasing the volume of compartment 1 by an equal amount. It is most important to realize that it is the presence of a membrane impermeable to sol-

ute that leads to the volume changes associated with osmosis.

We have treated the two compartments as if they were infinitely expandable, so that the net transfer of water does not create a pressure difference across the membrane. This is essentially the situation that occurs across plasma membranes. In contrast, if the walls of compartment 2 could not expand, the movement of water into compartment 2 would raise the pressure in compartment 2, which would oppose further net water entry. If a solution containing nonpenetrating solutes is separated from pure water by a membrane, the pressure that must be applied to the solution to prevent the net

---

[3] The 4 Osmol of solute initially present cannot leave compartment 1. The addition of 0.33 L of water to the liter of water already present will dilute the 4 Osmol of solute to 4 Osmol/1.33 L = 3 *Osm*. 0.33 L of water is equal (at .1 g/mL and 18 g/mol) to 333 g/18 = 18.5 mol of water.

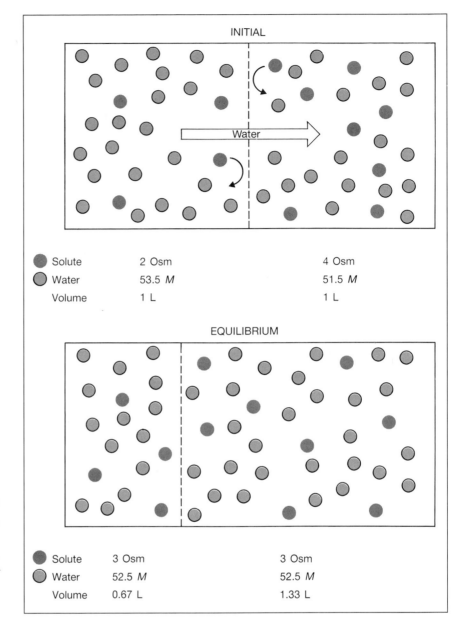

FIGURE 6-21 The movement of water across a membrane that is permeable to water but not permeable to solute leads to an equilibrium state in which there is a change in the volumes of the two compartments due to the net transfer of water (0.33 L in this case) from compartment 1 to 2. (The membrane in this example stretches as the volume of compartment 2 increases so that no significant change in intracellular pressure occurs.)

flow of water across the membrane is a measure of a property of the solution known as its **osmotic pressure**. The greater the osmolarity of a solution, the greater its osmotic pressure. It is important to note that the osmotic pressure of a solution does *not* produce the osmotic flow. Rather it is equal to the amount of pressure that must be applied to the solution to *prevent* osmotic flow.

Extracellular Osmolarity and Cell Volume

We can now apply the principles learned about osmosis to cells, which meet all the criteria necessary to produce an osmotic flow of water across a membrane. Both the intracellular and extracellular fluids contain water, and

cells are surrounded by a membrane that is very permeable to water but impermeable to many substances (**nonpenetrating solutes**).

About 85 percent of the extracellular solute particles are sodium and chloride ions, which can diffuse through protein channels in the plasma membrane. As we have seen, however, the membrane contains an active-transport system that pumps sodium ions out of the cell. Sodium thus behaves as if it were a nonpenetrating solute confined to the extracellular fluid. Secondary active-transport pumps and the membrane potential move chloride ions out of cells as rapidly as they diffuse in through chloride channels, with the result that chloride ions also behave as if they were nonpenetrating solutes confined to the extracellular fluid.

Inside the cell, the major solute particles are potassium ions and a number of organic solutes. Most of the latter are unable to cross the plasma membrane. Although potassium ions can leak out of a cell through potassium channels, they are actively transported back by the Na,K-ATPase pump. The net effect, as with sodium, is that potassium behaves as if it were a nonpenetrating solute, in this case in the intracellular fluid. Thus, sodium chloride outside the cell and potassium and organic solutes inside the cell represent the major nonpenetrating solutes on the two sides of the plasma membrane.

The osmolarity of the extracellular fluid is normally about 300 m*Osm*. Since water can diffuse across plasma membranes, the water in the intracellular fluid will come to diffusion equilibrium with the water in the extracellular fluid. At equilibrium, therefore, the osmolarity of the intracellular fluid is the same—300 m*Osm*—as the osmolarity of the extracellular fluid.

Changes in extracellular osmolarity can cause cells to shrink or swell due to the changes produced in the extracellular water concentration. As we shall see in Chapter 15, one of the major functions of the kidneys is to regulate the excretion of water in the urine so that the osmolarity of extracellular fluids remains nearly constant in spite of variations in salt and water intake and loss, thereby preventing damage to cells that would occur from excessive swelling or shrinkage.

If cells are placed in a solution of nonpenetrating solutes having an osmolarity of 300 m*Osm*, they will neither swell nor shrink since the water concentrations in the intra- and extracellular fluid are the same and the solutes cannot leave or enter. Such solutions are said to be **isotonic** (Figure 6-22)—having the same concentration of *nonpenetrating* solutes as normal extracellular fluid. Solutions containing less than 300 mOsmol of nonpenetrating solutes per liter (**hypotonic** solutions) will cause cells to swell because water will diffuse into the cell from its higher concentration in the extracellular fluid. Solutions containing greater than 300 mOsmol/L (**hypertonic** solutions) of nonpenetrating solutes will cause cells to shrink as water diffuses out of the cell into the fluid with a lower water concentration. Note that the concentration of *nonpenetrating* solutes in a solution, not the total osmolarity, determines its tonicity—hypotonic, isotonic, or hypertonic.

In contrast, another set of terms—**isosmotic**, **hyperosmotic**, and **hyposmotic**—denotes simply the osmolarity of a solution relative to that of cells without regard

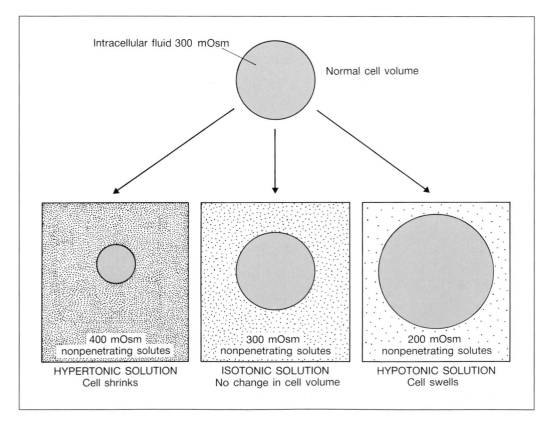

Intracellular fluid 300 mOsm

Normal cell volume

400 mOsm
nonpenetrating solutes

HYPERTONIC SOLUTION
Cell shrinks

300 mOsm
nonpenetrating solutes

ISOTONIC SOLUTION
No change in cell volume

200 mOsm
nonpenetrating solutes

HYPOTONIC SOLUTION
Cell swells

FIGURE 6-22 Changes in cell volume produced by hypotonic, isotonic, and hypertonic solutions.

TABLE 6-3	TERMS REFERRING TO THE SOLUTE OSMOLARITY AND TONICITY OF SOLUTIONS
Isosmotic	A solution having the same total solute concentration (osmolarity) as another solution regardless of its composition of membrane-penetrating and nonpenetrating solutes
Hyperosmotic	A solution containing greater than 300 mOsmol/L of solutes regardless of the composition of membrane-penetrating and nonpenetrating solutes
Hyposmotic	A solution containing less than 300 mOsmol/L of solutes regardless of the composition of membrane-penetrating and nonpenetrating solutes
Isotonic	A solution containing 300 mOsmol/L of nonpenetrating solutes, regardless of the concentration of membrane-penetrating solutes that may be present
Hypertonic	A solution containing greater than 300 mOsmol/L of nonpenetrating solutes, regardless of the concentration of membrane-penetrating solutes that may be present
Hypotonic	A solution containing less than 300 mOsmol/L of nonpenetrating solutes, regardless of the concentration of membrane-penetrating solutes that may be present

for whether the solute is penetrating or nonpenetrating. The two sets of terms are therefore not synonymous. For example, a 1-L solution containing 300 mOsmol of nonpenetrating NaCl and 100 mOsmol of urea, which can cross plasma membranes, would have a total osmolarity of 400 mOsm and would be hyperosmotic but would still be an isotonic solution (and so cells immersed in it would not shrink or swell). The reason is that urea will diffuse into the cells and reach the same concentration as the urea in the extracellular solution, and thus both the intracellular and extracellular solutions will have the same osmolarity (400 mOsm), and there will be no difference in the water concentration across the membrane and thus no change in cell volume.

Table 6-3 provides a comparison of the various terms used to describe the osmolarity of solutions. All hyposmotic solutions are also hypotonic whereas a hyperosmotic solution could be hypertonic, isotonic, or hypotonic.

The tonicity of solutions injected into the body is of great importance in medicine. Such solutions usually consist of a drug dissolved in an isotonic solution of NaCl (isotonic saline) or an isotonic solution of glucose (5%

dextrose solution). Injecting such solutions will not produce changes in cell volume, whereas injection of the same drug dissolved in pure water, forming a hypotonic solution, would produce cell swelling, perhaps to the point that plasma membranes would rupture, destroying cells.

ENDOCYTOSIS AND EXOCYTOSIS

In addition to diffusion and mediated transport, there is another pathway by which substances can enter or leave cells, one that does not require the molecules to pass through the structural matrix of the membrane. When living cells are observed under a light microscope, regions of the plasma membrane can be seen to fold into the cell, forming small pockets that pinch off to produce intracellular membrane-bound vesicles that enclose a small volume of extracellular fluid. This process is known as **endocytosis** (Figure 6-23). A similar process in the reverse direction, known as **exocytosis**, occurs when membrane-bound vesicles in the cytoplasm fuse with the

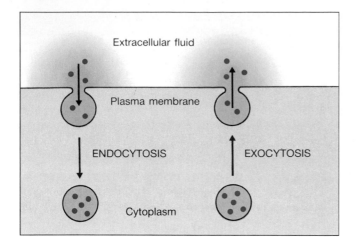

FIGURE 6-23 Endocytosis and exocytosis.

plasma membrane and release their contents to the outside of the cell.

Endocytosis

Several varieties of endocytosis can be identified. When the vesicle simply encloses a small volume of extracellular fluid, as described above, the process is known as **fluid endocytosis**. In other cases, specific molecules in the extracellular fluid bind to the plasma membrane and are carried into the cell, along with the extracellular fluid, when the membrane invaginates. This is known as **adsorptive endocytosis**. In addition to taking in trapped extracellular fluid, adsorptive endocytosis leads to the selective concentration in the vesicle of the material bound to the membrane. Both fluid—and adsorptive—endocytosis are often referred to as **pinocytosis** (cell drinking). A third type of endocytosis occurs when large particles, such as bacteria and debris from damaged tissues, are engulfed by the plasma membrane. In this form of endocytosis, known as **phagocytosis** (cell eating), the membrane folds around the surface of the particle so that little extracellular fluid is enclosed within the vesicle. While most cells undergo pinocytosis, only a few special cells carry out phagocytosis.

In most cells, fluid endocytosis occurs continuously, whereas adsorptive endocytosis and phagocytosis are stimulated by the binding of specific molecules or particles to the cell surface. Endocytosis of any kind requires metabolic energy in the form of ATP, although the molecular mechanisms by which this energy is coupled to membrane movements during endocytosis are unknown.

What is the fate of the endocytotic vesicles once they have entered the cell? In some cases the vesicles pass

through the cytoplasm and fuse with the plasma membrane on the opposite side of the cell, releasing their contents to the extracellular space (Figure 6-24). This provides a pathway for transferring large molecules, such as proteins, across the cells that separate two compartments, for example, blood and interstitial fluid. A similar process allows small amounts of protein to be moved across the intestinal epithelium.

In most cells, however, endocytotic vesicles fuse with the membranes of lysosomes (Figure 6-24), organelles that contain digestive enzymes that break down large molecules such as proteins, polysaccharides, and nucleic acids. The fusion of the endocytotic vesicle with the lysosomal membrane exposes the contents of the vesicle to these digestive enzymes. The endocytosis of bacteria and their destruction by the lysosomal digestive enzymes is one of the body's major defense mechanisms against germs (Chapter 19).

FIGURE 6-24 Fate of endocytotic vesicles. Pathway 1 transfers extracellular materials from one side of the cell to the other. Pathway 2 leads to fusion with lysosomes and the digestion of the vesicle contents.

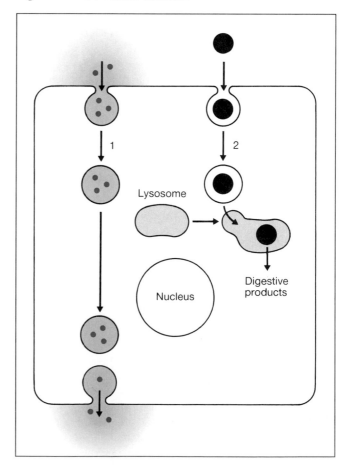

Endocytosis removes a small portion of the membrane from the cell surface. In cells that have a great deal of endocytotic activity, more than 100 percent of the plasma membrane may be internalized in an hour, yet the membrane surface area remains constant. This is because the membrane is replaced at about the same rate by vesicle membrane that fuses with the plasma membrane during exocytosis.

Exocytosis

Exocytosis performs two functions for cells: (1) It provides a way to replace portions of the plasma membrane that have been removed by endocytosis and in the process to replace or add new membrane components, and (2) it provides a route by which membrane-impermeable molecules synthesized by cells can be released into the extracellular fluid.

The secretion of substances by exocytosis is triggered in most cells by specific stimuli that lead to an increase in cytosolic calcium concentration by opening voltage-sensitive or receptor-operated calcium channels in the plasma membrane or in the membranes of intracellular organelles. The resulting increase in cytosolic calcium triggers, through unknown mechanisms, the fusion of secretory vesicles with the plasma membrane.

How are substances that are secreted by exocytosis packaged into vesicles? In some cases the enzymes synthesizing a particular exported product are located in the vesicles. In other cases, the vesicle membrane has mediated-transport systems that concentrate the product within the vesicle after it has been synthesized in the cytosol. Very high concentrations of certain products can be achieved within vesicles by a combination of mediated transport followed by their specific binding to protein binding sites within the vesicle. Material stored in secretory vesicles is available for rapid secretion in response to a stimulus, without delays that might occur if the material had to be synthesized after the arrival of the stimulus.

Protein secretion. The packaging of proteins into secretory vesicles creates special problems because of the complex enzymatic machinery associated with protein synthesis and the inability of these large molecules to pass through most membranes.

Upon arriving in the cytoplasm from the nucleus, a molecule of messenger RNA binds to a free ribosome and begins the process of translating its message into a polypeptide chain (Chapter 4). The first 15 to 30 amino acids to be linked together in the growing polypeptide chain determine whether the protein will be released into the cytosol to function as an intracellular protein or be packaged into a vesicle that will be transferred to a lysosome or secreted from the cell by exocytosis.

This initial portion of the protein chain is known as the **signal sequence.** If there is no signal sequence at the beginning of the protein, synthesis is completed on a free ribosome and the protein is released into the cytosol. If there is a signal sequence, however, it acts as a ligand that binds to a site on the surface of the granular endoplasmic reticulum. By this mechanism ribosomes become attached to the reticulum. Once the messenger RNA, with its attached ribosome and growing polypeptide chain, is bound to the reticulum, the polypeptide chain is inserted through the reticulum membrane as additional amino acids are added to the polypeptide (Figure 6-25).

At the completion of synthesis, the polypeptide is either released into the lumen of the endoplasmic reticulum or remains inserted in the reticular membrane as a membrane protein. The mechanism by which the growing polypeptide chain passes through the reticulum membrane is unknown. This is the only stage along its route to the outside of the cell at which a protein that is to be secreted passes through a membrane.

Within the lumen of the endoplasmic reticulum, enzymes remove the signal sequence, and so this portion of the protein, which was necessary for directing it to the endoplasmic reticulum, is not present in the final protein. In addition, carbohydrate groups are added to these proteins by enzymes in the endoplasmic reticulum. Almost all proteins that are secreted are glycoproteins.

Following these modifications, portions of the reticulum membrane that contain the newly synthesized proteins bud off, forming vesicles by a process similar to endocytosis. These vesicles migrate to the Golgi apparatus (Figure 6-25) and fuse with the Golgi membranes. Within the Golgi apparatus the protein undergoes still further modification. Some of the carbohydrates that were added in the granular endoplasmic reticulum are now removed and new groups added. These latter carbohydrate groups appear to function as labels that can be recognized when the protein encounters various binding sites during the remainder of its trip through the cell.

While in the Golgi apparatus, the many different proteins that have been funneled into this organelle become sorted out according to their final destination. How this sorting takes place is largely unknown, but it appears to involve the selective binding of the proteins to different regions of the Golgi apparatus.

Following modification and sorting, the proteins are packaged into vesicles that bud off the surface of the Golgi membrane. Some of the vesicles will travel to the plasma membrane where they release their contents to the extracellular fluid by exocytosis. Others will fuse

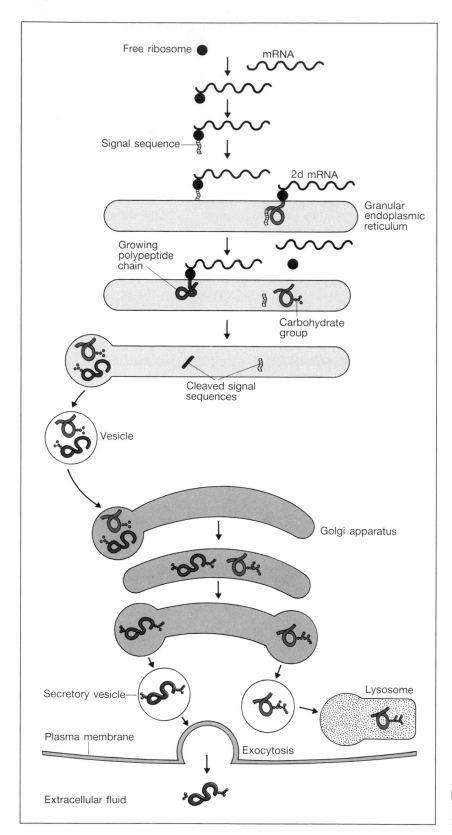

FIGURE 6-25 Pathway traveled by proteins that are to be secreted by cells or transferred to lysosomes.

with the lysosome membranes, delivering digestive enzymes to this organelle. Where a particular vesicle will go was determined by the sorting and "addressing" events that occurred in the Golgi apparatus.

Membrane formation. As we have seen, the budding and fusion of membranes provides the mechanism of adding or removing segments of membrane from the plasma membrane or cell organelle membranes. The lipid portions of these membranes are synthesized in association with the agranular endoplasmic reticulum where enzymes for phospholipid and cholesterol synthesis are located. Because of their amphipathic structure, lipid molecules become incorporated in the lipid bilayer of the endoplasmic reticulum membrane immediately following their synthesis.

The protein portions of most membranes (the mitochondrial membranes are the exception) are inserted into the membrane during protein synthesis on ribosomes attached to the granular endoplasmic reticulum. The membranes of the vesicles that bud off the endoplasmic reticulum then transport these newly synthesized membrane components to the membranes of the Golgi apparatus from which they eventually reach the plasma membrane or other cell organelle membranes.

EPITHELIAL TRANSPORT

Epithelial cells (page 2) line hollow organs or tubes and regulate the movement of substances across these surfaces. Most substances that cross an epithelium must pass from an extracellular compartment on one side of an epithelial cell, through a plasma membrane by diffusion or mediated transport into the cytosol, and then from the cytosol across a second plasma membrane on the other side of the epithelium into a second extracellular compartment, again by either diffusion or mediated transport. The presence of tight junctions (page 45) between epithelial cells prevents most substances from diffusing through the extracellular space between the cells. One surface of the epithelium generally faces a hollow fluid-filled chamber, and this side is referred to as the **luminal side** (also apical or mucosal side) of the epithelium. The opposite surface, which is usually adjacent to a network of blood vessels, is referred to as the **blood side** (also basolateral or serosal side).

Movements of molecules through the plasma membranes at the luminal and blood surfaces of an epithelial cell proceed through the pathways already described for movements across membranes in general. However, the characteristics of plasma membranes on the luminal and blood side of the cell are not the same. These membranes contain different channels for ion diffusion and different carrier proteins for mediated transport.

As a result of these differences in the two epithelial surfaces, it is possible for substances to undergo a net movement from a low concentration on one side of an epithelium to a higher concentration on the other side, that is, to undergo active transport across the epithelial layer of cells. Examples of such transport are the absorption of material from the gastrointestinal tract into the blood, the movements of substances between the kidney tubules and the blood during urine formation, and the secretion of salts and fluid by various glands.

Figure 6-26 and Figure 6-27 illustrate two examples of epithelial transport. Sodium is actively transported across most epithelia from lumen to blood side in absorptive processes and from blood side to lumen during secretion. During absorption, the movement of sodium from a lumen into an epithelial cell often occurs by diffusion through sodium channels in the luminal membrane (Figure 6-26). Diffusion occurs into the cell because the intracellular concentration of sodium is kept low by the active transport of sodium out of the cell across the plasma membrane on the blood side, where all of the Na,K-ATPase pumps are located. In other words, sodium moves downhill into the cell and then uphill out of it. The net result is that sodium can be moved from lower to higher concentration across the epithelium.

Figure 6-27 illustrates the absorption of organic molecules across an epithelium. In this case, entry of the organic molecule X across the luminal plasma membrane occurs through a secondary active-cotransport carrier that is linked to the downhill movement of sodium across the membrane. In the process, X moves from a lower concentration in the luminal fluid to a higher concentration in the cell. Exit of the substance on the blood side occurs by a facilitated-diffusion carrier that moves the material from its high concentration in the cell to a lower concentration in the extracellular fluid on the blood side. The concentration of the substance may be considerably higher on the blood side than in the lumen since the blood-side concentration can approach equilibrium with the high intracellular concentration achieved by the secondary active-transport process in the luminal membrane.

Although water is not actively transported across cell membranes, a net movement of water across an epithelium can be achieved by osmosis as a result of the active transport of solutes, especially sodium, across the epithelium. The active transport of sodium results in a decrease in the sodium concentration on one side and an increase in sodium on the other. These changes in solute concentration are accompanied by changes in the water concentration on the two sides since a change in solute concentration, as we have seen, produces a change in

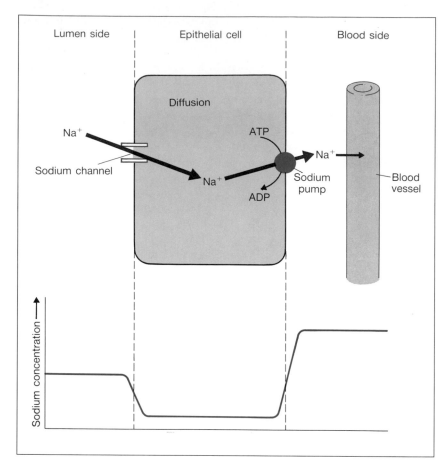

Lumen side | Epithelial cell | Blood side

FIGURE 6-26 Active transport of sodium across an epithelial cell. The transepithelial transport of sodium involves, in this case, diffusion through channels at one surface followed by active transport out of the cell, across the opposite surface. Shown below the cell is the concentration profile of the transported solute across the epithelium. In some epithelia downhill sodium movement across the first membrane is carrier-mediated rather than through channels.

water concentration. The water-concentration difference produced will cause water to move by osmosis from the low-sodium to the high-sodium side of the epithelium (Figure 6-28). Thus, net movement of solute across an epithelium will always be accompanied by a flow of water in the same direction provided the membrane is permeable to water.

Glands

The function of glands is to secrete substances into hollow organs, the blood, or onto the surface of the skin in response to various stimuli. The mechanisms of glandular secretion depend upon the principles of membrane permeability and transport described in this chapter.

Glands are formed during embryonic development by the infolding of the epithelial layer of an organ's surface. Many of these glands remain connected by ducts to the epithelial surfaces from which they were formed, while others lose this connection and become isolated clusters of cells. The first type of gland is known as an **exocrine gland**, and its secretions flow through ducts and are discharged into the lumen of an organ or, in the case of the skin glands, onto the surface of the skin (Figure 6-29). Sweat glands and salivary glands are examples of exocrine glands.

In the second type of gland, known as an **endocrine gland** or a ductless gland, secretions are released directly into the interstitial fluid surrounding the gland cells (Figure 6-29). From this point, the material diffuses into the blood, which carries it throughout the body. The endocrine glands secrete a major class of chemical messengers—the hormones.

In practical usage, the term endocrine gland has come to be synonymous with hormone-secreting glands. However, it should be noted that there are "ductless glands" that secrete nonhormonal, organic substances into the blood. For example, fat cells secrete fatty acids and glycerol, and liver cells secrete glucose, amino acids, fats, and proteins. These substances serve as nutrients for other cells or perform special functions in the blood, but

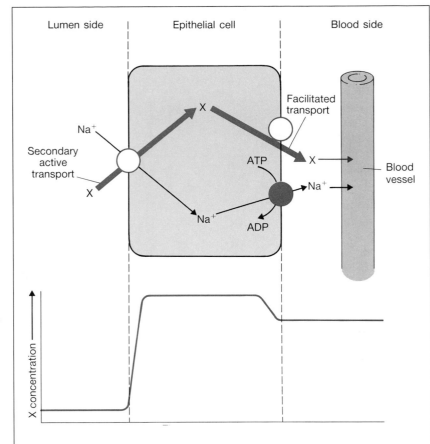

FIGURE 6-27 The transepithelial transport of most organic solutes (X) involves their movement into a cell through a secondary active-transport pump coupled to sodium, followed by movement out of the cell, down a concentration gradient, through a facilitated-diffusion carrier at the opposite surface. Shown below the cell is the concentration profile of the transported solute across the epithelium.

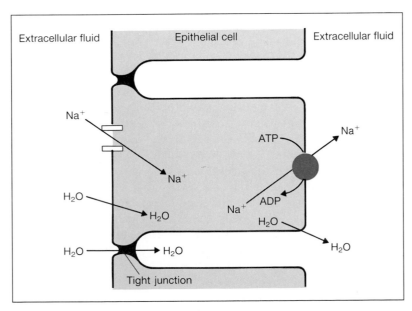

FIGURE 6-28 Movements of water across the epithelium. The active transport of sodium out of the cells, into the surrounding interstitial spaces, produces an elevated osmolarity in this region, leading to the osmotic flow of water across the epithelial membranes and tight junctions.

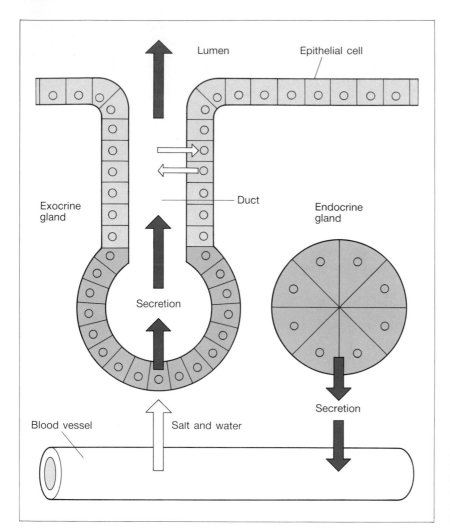

FIGURE 6-29 Glands are composed of epithelial cells. Exocrine gland secretions enter ducts whereas the hormones secreted by endocrine (ductless) glands diffuse into the blood.

they do not act as messengers and therefore are not hormones.

The substances secreted by glands fall into two chemical categories: (1) organic materials that are, for the most part, synthesized by the gland cells and (2) salt and water, which are simply moved from one extracellular compartment to another across the glandular epithelium.

Organic molecules are secreted by gland cells via all the pathways already described: diffusion in the case of lipid-soluble materials, mediated transport for some polar materials, and exocytosis for very large molecules such as proteins. Salts are actively transported across glandular epithelia, producing changes in the extracellular osmolarity, which in turn causes the osmotic flow of water. In exocrine glands, as the secreted fluid passes along the ducts connecting the gland to the luminal sur-

face, the composition of the fluid originally secreted may be altered as a result of absorption or secretion by the duct cells. Often the composition of the secreted material at the end of the duct will vary with the rate of secretion, reflecting the amount of time the fluid remains in the duct where it can be modified.

Most glands undergo a low, basal rate of secretion that can be greatly augmented in response to the appropriate signal, usually a nerve impulse or a hormone. The mechanism of the increased secretion is again one of altering some portion of the secretory pathway. This may involve: (1) increasing the rate at which a secreted substance is synthesized by activating the appropriate enzymes; (2) providing the calcium signal for the exocytosis of already-synthesized material; or (3) altering the pumping rates of a carrier-mediated process, or opening ion channels.

The volume of fluid secreted by an exocrine gland can be increased by directly increasing the sodium-pump activity or more often, by controlling the opening of sodium channels at the blood surface of an epithelial cell, allowing more sodium to enter the cell. This increases the amount of sodium pumped out of the cell (via the Na,K-ATPase pump), which in turn increases the flow of water across the epithelium since water osmotically follows the solute movement.

SUMMARY

Diffusion

I. Diffusion is the movement of molecules from one location to another by random thermal motion.
  A. The net flux between two compartments always proceeds from higher to lower concentration of the diffusing substance.
  B. Diffusion equilibrium is reached when the concentrations in the two compartments become equal, resulting in zero net flux.

II. The magnitude of the net flux F across a plasma membrane is directly proportional to the concentration difference across the membrane C_o - C_i, the surface area of the membrane A, and the membrane permeability constant k_p.

III. Nonpolar molecules diffuse through the lipid portions of membranes much more rapidly than polar or ionized molecules because nonpolar molecules can dissolve in the nonpolar lipids.

IV. Mineral ions diffuse across membranes by passing through ion channels formed by integral membrane proteins.
  A. The net diffusion of ions across a membrane depends on the concentration gradient and the membrane potential.
  B. The flux of ions across a membrane can be altered by opening or closing ion channels.

Mediated-Transport Systems

I. The mediated transport of molecules across a membrane involves the binding of the transported solute to a carrier protein in the membrane. Changes in the conformation of the carrier protein move the binding site to the opposite side of the membrane, where the solute dissociates from the carrier protein.
  A. The binding sites on carrier proteins exhibit chemical specificity and saturation.
  B. The magnitude of the flux through a mediated-transport system depends on the degree of carrier saturation and the concentration of carrier proteins in the membrane.

II. Facilitated diffusion is a carrier-mediated-transport process that moves molecules from higher to lower concentration across a membrane until the two concentrations become equal. Metabolic energy is not required for this process.

III. Active transport is a carrier-mediated-transport process that moves molecules against an electrochemical difference across a membrane and requires an input of energy.
  A. Active transport is achieved by altering the affinity of the carrier binding site on the two sides of the membrane such that net movement will proceed from the side of the membrane with the high-affinity binding site to the side with the low-affinity binding site.
  B. Primary active-transport carriers use the phosphorylation of the carrier by ATP (covalent modulation) to produce the change in binding-site affinity.
  C. Secondary active-transport carriers use the binding of ions (often sodium) to the carrier (allosteric modulation) to produce the change in binding-site affinity.
  D. In secondary active transport the downhill flow of an ion, often sodium, into the cell is linked either to movement of a second solute from lower extracellular to higher intracellular concentration—cotransport—or to solute movement from lower intracellular concentration to higher extracellular concentration—countertransport.

Osmosis

I. Osmosis is the diffusion of water from a region of high water concentration to a region of low water concentration when a barrier prevents the flow of solute. The osmolarity—the total solute concentration in a solution—determines the water concentration. The higher the osmolarity of a solution, the lower the water concentration.

II. Osmosis across a membrane that is permeable to water but impermeable to solute leads to an increase in the volume of the compartment on the initially high-osmolarity side and a decrease in the volume on the initially low-osmolarity side.

III. Application of a pressure equal to the osmotic pressure of the solution will prevent the osmotic flow of water into the solution from a compartment of pure water. The greater the osmolarity of a solution, the greater its osmotic pressure.

IV. The osmolarity of the extracellular fluid is about 300 mOsm. Since water is in diffusion equilibrium across cell membranes, the intracellular fluid also has an osmolarity of 300 mOsm.
  A. Na^+ and Cl^- ions are the major effectively nonpenetrating solutes in the extracellular fluids, while K^+ ions and the organic solutes are the major nonpenetrating solutes in the intracellular fluid.
  B. The terms used to describe the osmolarity of solutions containing different compositions of penetrating and nonpenetrating solutes are given in Table 6-3.
  C. Cells placed in hypotonic solutions swell, while cells placed in hypertonic solutions shrink.

Endocytosis and Exocytosis

I. During endocytosis, regions of the plasma membrane invaginate and pinch off to form vesicles that enclose a small volume of extracellular material.
  A. The three classes of endocytosis are: (1) fluid endocytosis, (2) adsorptive endocytosis, and (3) phagocytosis.

B. In most cells, endocytotic vesicles fuse with the membranes of lysosomes, in which the vesicle contents are digested by lysosomal enzymes.

II. Exocytosis, which occurs when intracellular vesicles fuse with the plasma membrane, (1) provides a way to insert new segments of membrane into the plasma membrane, and (2) provides a route by which membrane-impermeable molecules synthesized by cells can be released into the extracellular fluid.

III. Proteins that are secreted from cells by exocytosis pass through the sequence of steps illustrated in Figure 6-25.

IV. When membrane vesicles derived from the endoplasmic reticulum, and ultimately the Golgi apparatus, fuse with other membranes, they add the newly synthesized membrane components to these membranes.

Epithelial Transport

I. Molecules cross from the luminal side of an epithelial cell to the blood side by first crossing the luminal plasma membrane, passing through the cytoplasm, and then crossing the plasma membrane on the blood side of the cell.

II. The permeability and transport characteristics of the plasma membranes on the luminal and blood sides of an epithelial cell differ, resulting in the ability of epithelial cells to actively transport a substance between the extracellular fluid on one side of the cell and the extracellular fluid on the opposite side of the cell.

III. The active transport of sodium across epithelia increases the osmolarity on one side of the cell and decreases it on the other, causing water to move by osmosis in the same direction as the transported sodium.

IV. Glands are composed of epithelial cells that secrete water and solutes in response to stimulation.
A. There are two categories of glands: (1) exocrine glands that secrete into ducts, and (2) endocrine glands (ductless glands) that secrete hormones into the extracellular fluid from which they diffuse into the blood.
B. The secretions of glands consist of (1) organic substances that have been synthesized by the gland and (2) salt and water that have been transported across the gland cells from the blood.

REVIEW QUESTIONS

1. Define:

diffusion	carriers
flux	mediated-transport system
net flux	facilitated diffusion
diffusion equilibrium	active transport
permeability constant k_p	primary active transport
channels	secondary active transport
membrane potential	cotransport
electrochemical difference	countertransport
receptor-operated channels	osmosis
voltage-sensitive channels	osmolarity
stretch-activated channels	osmotic pressure
nonpenetrating solutes	fluid endocytosis
isotonic	adsorptive endocytosis
hypotonic	pinocytosis
hypertonic	phagocytosis
isosmotic	signal sequence
hyperosmotic	luminal side (of epithelium)
hyposmotic	blood side (of epithelium)
endocytosis	exocrine gland
exocytosis	endocrine gland

2. What determines the direction in which net diffusion will occur?

3. In what ways can the net solute flux between two compartments separated by a permeable membrane be increased?

4. Why are membranes more permeable to nonpolar than to most polar and ionized molecules?

5. Ions diffuse across cell membranes by what pathways?

6. When considering the diffusion of ions across a membrane, what driving force, in addition to the ion concentration gradient, must be considered?

7. What factors can alter the opening and closing of protein channels in a membrane?

8. Describe the mechanism by which a mediated-transport carrier moves a solute from one side of a membrane to the other.

9. What determines the magnitude of a mediated-transport flux across a membrane?

10. What characteristics distinguish diffusion from facilitated diffusion?

11. What characteristics distinguish facilitated diffusion from active transport?

12. What alteration in a carrier protein enables a solute molecule to move from a low concentration on one side of a membrane to a higher concentration on the opposite side?

13. Describe the mechanism by which energy is coupled to a carrier during (A) primary active transport and (B) secondary active transport.

14. Describe the direction in which sodium ions and a transported solute move during secondary active transport that occurs by cotransport and countertransport.

15. How can the concentration of water in a solution be decreased?

16. If two solutions having different osmolarities are separated by a water-permeable membrane, why will there be a change in the volumes of the two compartments if the membrane is impermeable to the solutes but no change in volume if the membrane is permeable to solute?

17. To which solution must a pressure be applied to prevent the osmotic flow of water across a membrane separating a solution of high osmolarity and a solution of lower osmolarity?

18. Why do sodium and chloride ions in the extracellular fluid and potassium ions in the intracellular fluid behave as if they were nonpenetrating solutes?

19. What is the osmolarity of the extracellular fluids of the body?

20. What change in cell volume will occur when a cell is placed in a hypotonic solution? In a hypertonic solution?

21. Under what conditions will a hyperosmotic solution be isotonic?

22. Endocytotic vesicles deliver their contents to which parts of a cell?

23. Describe the sequence of events that leads to the synthesis and secretion of proteins by cells.

24. What is the function of the signal sequence of a protein?

25. Where are membrane lipids and proteins synthesized in cells, and how are they inserted into the plasma membrane?

26. How do the mechanisms for actively transporting glucose and sodium across an epithelium differ?

27. By what mechanism does the active transport of sodium lead to the osmotic flow of water across an epithelium?

28. What is the difference between an endocrine gland and an exocrine gland?

THOUGHT QUESTIONS

(Answers are given in Appendix A.)

1. In two cases (A and B), the concentrations of solute X in two 1-L compartments separated by a membrane through which X can diffuse are

	Concentration, M	
Case	Compartment 1	Compartment 2
A	3	5
B	32	30

   a. In what direction will the net flux take place in case A and in case B?
   b. When diffusion equilibrium is reached, what will be the concentration of solute in each compartment in A and in B?
   c. Will A reach diffusion equilibrium faster, slower, or at the same rate as B?

2. When the extracellular concentration of the amino acid ala-nine is increased, the flux of the amino acid leucine into a cell is decreased. How might this observation be explained?

3. If an active-transport carrier protein has a lower affinity for the transported solute on the extracellular surface of the plasma membrane than on the intracellular surface, in what direction will there be a net transport of the solute across the membrane?

4. Why will an inhibition of ATP synthesis by a cell lead to a decrease in secondary active transport?

5. Given the following solutions, which has the lowest water concentration? Which two are isosmotic?

	Concentration, mM			
Solution	Glucose	Urea	NaCl	CaCl$_2$
A	20	30	150	10
B	10	100	20	50
C	100	200	10	20
D	30	10	60	100

6. If compartment 1 contains 200 mmol/L of NaCl and 100 mmol/L of urea and compartment 2 contains 100 mmol/L of NaCl and 300 mmol/L of urea, which compartment will have increased in volume when osmotic equilibrium has been reached if the membrane is permeable to urea but not permeable to NaCl?

7. What will happen to cell volume if a cell is placed in each of the following solutions?

	Concentration, mM	
Solution	NaCl (nonpenetrating)	Urea (penetrating)
A	150	100
B	100	150
C	200	100
D	100	50

8. Characterize each of the solutions in question 7 as to whether it is isotonic, hypotonic, hypertonic, hyposmotic, and/or hyperosmotic.

9. By what mechanism can an increase in intracellular sodium concentration lead to an increase in exocytosis?

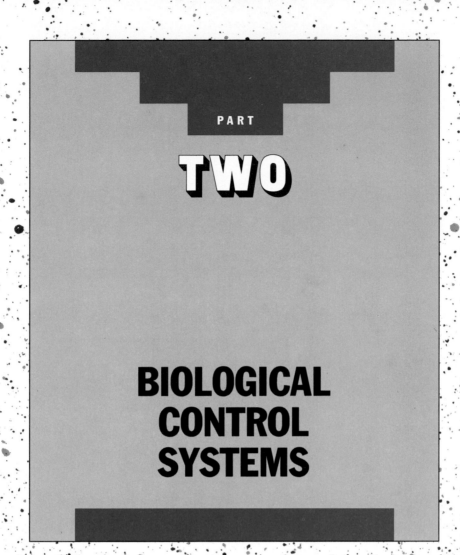

PART

TWO

BIOLOGICAL CONTROL SYSTEMS

7

HOMEOSTATIC MECHANISMS AND CELLULAR COMMUNICATION

As described in Chapter 1, the nineteenth-century French physiologist Claude Bernard was the first to recognize the central importance of maintaining a stable internal environment. This concept was further elaborated on and supported by the twentieth-century American physiologist W. B. Cannon, who emphasized that such stability could be achieved only through the operation of carefully coordinated physiological processes. The activities of cells, tissues, and organs must be regulated and integrated with each other in such a way that any change in the internal environment initiates a reaction to minimize the change. **Homeostasis** denotes the stable conditions of the internal environment that result from these compensating regulatory responses. Changes in the internal environment of the body do occur, but the magnitudes of these changes are small and are kept within narrow limits through multiple coordinated homeostatic processes, descriptions of which constitute the bulk of the remaining chapters of this book.

The concept of regulation has already been introduced in earlier chapters, in the context of a single cell. Each individual cell exhibits some degree of self-regulation, but the existence of a multitude of cells organized into tissues, which are further combined to form organs, imposes the requirement for overall regulatory mechanisms to coordinate the activities of all cells. To achieve this, communication between cells, often over relatively long distances, is essential.

GENERAL CHARACTERISTICS OF HOMEOSTATIC CONTROL SYSTEMS

We define a **homeostatic control system** as a collection of interconnected cells that functions to maintain a physical or chemical property of the internal environment relatively constant. Consider the regulation of body temperature. Our subject is a resting, lightly clad man in a room having a temperature of 20°C and moderate humidity. His internal body temperature is 37°C, and he is losing heat to the external environment because it is at a lower temperature. However, the chemical reactions occurring within the cells of his body are producing heat at a rate equal to the rate of heat loss. Under these conditions, the body undergoes no net gain or loss of heat and the body temperature remains constant. The system is said to be in a **steady state**, defined as a system in which a particular variable (temperature, in this case) is not changing but energy (in this case, heat) must be added continuously to maintain the variable constant. Steady states differ from *equilibria*, in which a particular variable is not changing but no input of energy is required to maintain the constancy. The steady-state temperature in our example is known as the **operating point** of the thermoregulatory system.

This example illustrates a crucial generalization about homeostasis: Stability of an internal environmental variable is achieved by the balancing of inputs and outputs. In this case, the variable—body temperature—remains constant because metabolic heat production (input) equals heat loss from the body (output).

Now we lower the temperature of the room rapidly, say, to 5°C, and keep it there. This immediately increases the loss of heat from our subject's warm skin, upsetting the dynamic balance between heat gain and loss. The body temperature therefore starts to fall. Very rapidly, however, a variety of homeostatic responses occur to limit the fall (Figure 7-1). First, the blood vessels to the skin narrow, reducing the amount of warm blood flowing through the skin and thus reducing heat loss. At a room temperature of 5°C, however, blood vessel constriction is not able to eliminate completely the extra heat loss from the skin. Our subject curls up in order to reduce the surface area of skin available for heat loss. This helps a bit, but excessive heat loss still continues and body temperature keeps falling, although at a slower rate. He has a strong desire to put on more clothing, for "voluntary" behavioral responses are often crucial components in homeostasis, but none is available. Clearly, then, if excessive heat loss (output) cannot be prevented, the only way of restoring the balance between heat input and output is to increase input, and this is precisely what occurs. He begins to shiver, and the chemical reactions responsible for the muscular contractions that constitute shivering produce large quantities of heat. The end result of all these responses is that heat loss from the body and heat production by the body are both high but are once again equal, so that body temperature stops falling. Indeed, heat production may transiently exceed heat loss so that body temperature begins to go back toward the value existing before the room temperature was lowered. It will eventually stabi-

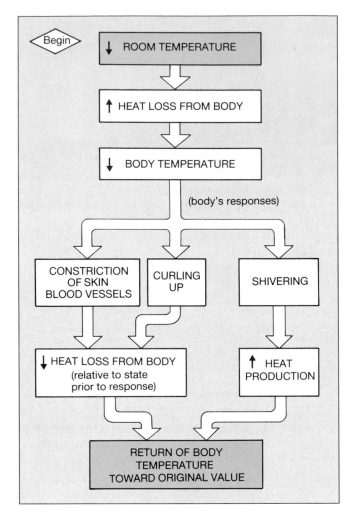

FIGURE 7-1 A homeostatic control system. This system maintains body temperature relatively constant when room temperature decreases. This diagram is typical of those used throughout this book to illustrate homeostatic systems, and several conventions should be noted. The "begin" sign indicates where to start. The arrows connecting any two boxes in the figure denote cause and effect; that is, an arrow can be read as "causes" or "leads to" (decreased room temperature "leads to" increased heat loss from body). The arrows next to each term denote increases or decreases. In general, one should add the words "tends to" in thinking about cause-and-effect relationships. For example, decreased room temperature tends to cause an increase in heat loss from the body, or curling up tends to cause a decrease in heat loss from the body. Qualifying the relationship in this way is necessary because variables like heat production and heat loss are under the influence of many factors, some of which oppose each other.

lize at a temperature a bit below this original value. Here is another crucial generalization about homeostasis: Stability of an internal environmental variable depends not upon the absolute magnitudes of opposing inputs and outputs but only upon the balance between them.

The type of thermoregulatory system just described is an example of a **negative-feedback system**, in which an increase or decrease in the variable being regulated brings about responses that tend to push the variable in the direction opposite ("negative") the direction of the original change. Thus, in our example, the *decrease* in body temperature led to responses that tended to *increase* the body temperature, that is, return it to its original value.

Negative-feedback control systems are the most common homeostatic mechanisms in the body, but there is another type of feedback known as **positive feedback** in which an initial disturbance in a system sets off a train of events that increases the disturbance even further. Thus positive feedback does not favor stability and often abruptly displaces a system away from its steady-state operating point. As we shall see, several important positive-feedback relationships occur in the body, blood clotting being one example.

Much of the sequence of events shown in Figure 7-1 is familiar, of course, since the ultimate responses—decreased skin temperature, curling up, and shivering—are visible, but what about the internal workings underlying them? First, what initiated the signals in the nerves to the various muscles mediating these responses? The nerves are controlled by nerve centers in the brain, which are in turn controlled by nerve signals coming from small groups of temperature-sensitive nerve cells in various parts of the body. The chain of events can be summarized: Decreased internal body temperature → increased activity of temperature-sensitive nerve cells → increased activity of a series of nerve cells in the central nervous system, the last of which stimulate the skin blood-vessel muscles and the skeletal muscles.

Another generalization about homeostatic control systems follows from the preceding description of their components and the nature of negative feedback: They cannot maintain *complete* constancy of the internal environment in the face of a persistent change in the external environment, but can only minimize changes (Figure 7-2). Imagine for the moment that the body's responses to a cold environment really did return body temperature completely to 37°C. What would happen to the output of the first component, the temperature-sensitive nerve cells—the "detectors"? Recall that it was the drop in internal body temperature that elicited the change in output from these detectors in the first place. If the body temperature were to return completely to its original value, so would the output of the detectors, and the entire sequence of events would stop. The blood vessels

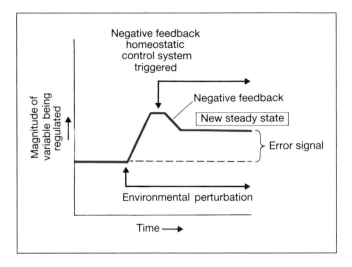

FIGURE 7-2 Negative-feedback control systems cannot return the regulated variable to its original value as long as the environmental perturbation persists. The deviation from the original value is called the error signal.

would stop constricting, shivering would cease, and body temperature would once again fall since the compensations would no longer be present. In other words, as long as the exposure to cold continues, some decrease in body temperature must persist to serve as a signal to maintain the responses. Such a signal is termed an **error signal**.

Just how big the error signal is in a control system depends on the magnitude of the external change, the sensitivity of the detectors, the efficiency of the connections between the system's components, and the responsiveness of the final apparatus in the control system. The temperature-regulating systems of the body are extremely sensitive, so that body temperature normally varies by only about 1 C° even in the face of marked changes in the external environment.

Inherent in the last sentence is another generalization about homeostasis: Even in reference to one individual, thus ignoring variation between persons, any regulated variable in the body cannot be assigned a single "normal" value but has a more or less narrow range of normal values, depending on external conditions. The more precise the homeostatic mechanisms for regulating a variable are, the narrower is the range. Thus, the concept of homeostasis is an ideal, for total constancy of the internal environment can never be achieved.

One more generalization: It is not possible for everything to be maintained relatively constant by homeostatic mechanisms. In our example, body temperature was maintained constant, but only because large changes

TABLE 7-1 SOME IMPORTANT GENERALIZATIONS ABOUT HOMEOSTATIC CONTROL SYSTEMS
1. Stability of an internal-environmental variable is achieved by balancing inputs and outputs. It is not the absolute magnitudes of the inputs and outputs that matter but the balance between them.
2. In negative-feedback systems, a change in the variable being regulated brings about responses that tend to push the variable in the direction opposite to the original change, that is, back toward the original value.
3. Homeostatic control systems cannot maintain complete constancy of any given feature of the internal environment. Therefore, any regulated variable will have a more or less narrow range of normal values depending on the external environmental conditions.
4. It is not possible for everything to be maintained relatively constant by homeostatic control systems. There is a hierarchy of importance, such that the constancy of certain variables is altered to maintain others relatively constant.

occurred in skin blood flow and skeletal muscle contraction. Moreover, because so many properties of the internal environment are closely interrelated, it is often possible to maintain one property relatively constant only by moving others farther from their usual values. These generalizations are summarized in Table 7-1.

Feedforward Regulation

Another type of regulatory process frequently used in conjunction with negative-feedback systems is feedforward. Let us give an example of feedforward and then define it. The temperature-sensitive nerve cells that trigger negative-feedback regulation of body temperature when body temperature begins to fall are located *inside* the body. In addition, there are temperature-sensitive nerve cells in the skin, and these cells, in effect, monitor *outside* temperature. When outside temperature falls, as in our example, these nerve cells immediately detect the change and relay this information to the brain, which then sends out signals to the blood vessels and muscles, resulting in heat conservation and increased heat production. In this manner compensatory thermoregulatory responses are activated *before* the colder outside temperature can cause the internal body temperature to fall. Thus, **feedforward** regulation anticipates changes in a regulated variable (internal body temperature in the example), improves the speed of the body's homeostatic responses, and minimizes fluctuations in the level of the variable being regulated.

Acclimatization

The term **adaptation** denotes a characteristic that favors survival in specific environments. Homeostatic control systems are inherited biological adaptations. An individual's ability to adapt to a particular environmental stress is not fixed, however. It can be enhanced, with no change in genetic endowment, by prolonged exposure to that stress. Such adaptation is known as **acclimatization**.

Let us take sweating in response to heat exposure as an example and perform a simple experiment. On day 1 we expose a person for 30 min to a high temperature and ask her to do a standardized exercise test. Body temperature rises, and sweating begins after a certain period of time. The sweating provides a mechanism for increasing heat loss from the body and thus tends to minimize the rise in body temperature in a hot environment. The volume of sweat produced is measured by weighing the subject before and after the test. Then, for a week, our subject enters the heat chamber for 1 or 2 h per day and exercises. On day 8, the sweating rate is again measured during the same exercise test performed on day 1. The striking finding is that the subject begins to sweat earlier and much more profusely than she did on day 1. Accordingly, her body temperature does not rise to nearly the same degree.

The subject has acclimatized to the heat; that is, she has undergone an adaptive change induced by exposure to the heat and is now specifically better adapted to heat exposure. In a sense, acclimatization may be viewed as an environmentally induced improvement in the functioning of an already existing genetically based homeostatic system. Up to some maximal limit, repeated or prolonged exposure to the environmental stress enhances the responsiveness of the system to that particular type of stress.

Heat acclimatization is completely reversible. If the daily exposures to heat are discontinued, the sweating rate of our subject will revert to the preacclimatized value within a relatively short time. If an acclimatization is induced very early in life, however, at the **critical period** for development of a structure or response, it is termed a **developmental acclimatization** and may be irreversible. For example, the barrel-shaped chests of natives of the Andes Mountains represent not a genetic difference between them and their lowland compatriots but rather an irreversible acclimatization induced during the first few years of their lives by their exposure to the low-oxygen environment of high altitude. The altered chest size remains even though the individual moves to a lowland environment later in life and stays there. Lowland persons who have suffered oxygen deprivation from heart or lung disease during their early years show precisely the same chest shape.

Biological Rhythms

A striking characteristic of many bodily functions is the rhythmical changes they manifest. The most common type is the **circadian rhythm**, which is approximately 24 h. The wake-sleep cycles, body temperature, the secretion of hormones into the blood, the excretion of ions into the urine, and many other functions undergo circadian variation (Figure 7-3). Other cycles have much longer periods, the menstrual cycle (approximately 28 days) being the most well-known.

A crucial point concerning most bodily rhythms is that they are *internally* driven. Environmental factors do not drive the rhythm but rather provide the timing cues important for **entrainment**, that is, setting of the actual hours, of the rhythm. A classic experiment will clarify this distinction.

Persons were put in experimental chambers that completely isolated them from the external environment.

FIGURE 7-3 Circadian rhythms of several physiological variables in a human subject with lights on (open bars at top) for 16 h and off (black bars at top) for 8 h. (*Adapted from Moore-Ede and Sulzman.*)

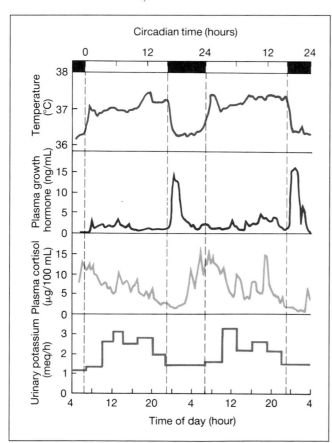

For the first few days, they were subjected to a 24-h rest-activity cycle in which the lights were turned on and off at the same time each day. Under these conditions, the persons' sleep-wake cycles were 24 h long. Then, all environmental time cues were eliminated, and the individuals were allowed to control the lights themselves. Immediately, their sleep-wake patterns began to change. On the average, bedtime got about 30 min later *each day* and so did wake-up time. Thus a sleep-wake cycle persisted in the complete absence of environmental cues, but the cycle was approximately 25 h rather than 24.

This experiment demonstrates that **free-running rhythms** can exist in the absence of environmental cues but that cues are required to entrain a cycle at 24 h. One more point: By altering the duration of the light-dark cycles, the sleep-wake cycles could be entrained to any value between 23 and 27 h, but shorter or longer durations could not be entrained, and the cycles simply became free-running. Because of this, the bodies of people whose work causes them to adopt sleep-wake cycles longer than 27 h are never able to make the proper adjustments and achieve stable rhythms. The result is symptoms similar to those of jet lag, to be described later.

The light-dark cycle is the most important environmental time cue in our lives but not the only one. Others include external environmental temperature, meal timing, and many social cues. Thus, if several people were undergoing the experiment just described in isolation from each other, their free-runs would be different, but if they were all in the same room, one person would be dominant and set the cycle for the others.

Environmental time cues also function to **phase-shift** rhythms, in other words, to reset the internal clock. Thus if one jets west or east to a different time zone, the sleep-wake cycle and other circadian rhythms slowly shift to the new light-dark cycle. These shifts take time, however, and the disparity between external time and internal time is one of the causes of the symptoms of jet lag—disruption of sleep, gastrointestinal disturbances, decreased vigilance and attention span, and a general feeling of malaise. Another, more subtle, factor is that the various rhythms do not all shift at the same rate (the reason for this is given below) and so become desynchronized from each other.

What is the neural basis of bodily rhythms? In the brain are several specific collections of nerve cells that function as the primary **pacemakers** (time clocks) for circadian rhythms. How they "keep time" independent of any external environmental cues is not really understood. The input the pacemakers receive from the eyes and many other parts of the nervous system mediate the entrainment effects exerted by the external environment.

The fact that there are several primary pacemakers accounts for the transient desynchronization of various rhythms when a phase shift occurs: when environmental cues change rapidly, the different pacemakers do not become reentrained at the same speeds. Under such conditions various specific rhythms, for example, those for sleep and for temperature, become temporarily out of phase with each other.

What have biological rhythms to do with homeostasis? They add a "predictive" component to homeostatic control systems, in effect a feedforward response operating without detectors. The negative-feedback homeostatic responses we described earlier in this chapter are corrective responses, in that they are initiated *after* the steady state of the individual has been perturbed. In contrast, the rhythms enable homeostatic mechanisms to be utilized immediately and automatically by activating them at times when a challenge is *likely* to occur but before it actually has occurred. For example, there is a rhythm in the urinary excretion of potassium such that excretion is high during the day and low at night. This makes sense since we ingest potassium in our food during the day, not at night when we are asleep. Therefore, the total amount of potassium in the body fluctuates less than if the rhythm did not exist.

It should not be surprising that rhythms have effects on the body's resistance to various stresses, such as that produced by bacteria or drugs. For example, when rats are given a large dose of a potentially lethal drug, their survival depends profoundly on the time the drug is given. Differences in sensitivity of this type unquestionably also exist in people, and the hope is that a fuller understanding of them will provide a more rational basis for the timing of drug therapies.

Aging and Homeostasis

The effects of aging on the body's homeostatic control systems are very important. At present, it is not possible to define aging in basic biological terms. Scientists agree, however, that aging represents the operation of a distinct process that should be distinguished from those diseases such as heart disease frequently associated with aging. Whether these diseases are, themselves, promoted by the aging process or simply occur over the same time frame is not known.

Aging is associated with a decrease in the number of cells in the body, due to some combination of decreased mitosis and increased cell death, and to malfunction of many of the cells that remain. The immediate cause of these changes is probably an interference in the function of the cells' macromolecules—not just DNA, but RNA

and cell proteins as well—and in the flow of information between these macromolecules.

The physiological manifestations of aging are a gradual deterioration in function and in the capacity of the body's homeostatic systems to respond to environmental stresses. However, it is difficult to sort out the extent to which any particular age-related change is due to aging per se and the extent to which it is secondary to disease and lifestyle changes. For example, until recently it was believed that the functioning of the nervous system markedly deteriorates as a result of aging per se, but this view is now being seriously questioned. It seems to have been based on studies of the performance of individuals with age-related diseases. More recent studies of people without such diseases do document changes—loss of memory, increased difficulty in learning new tasks, a decrease in the speed of processing by the brain, loss of brain mass—but these changes are modest, and most brain functions considered to underlie intelligence seem to remain relatively intact.

To take another example, a similar reevaluation of changes in heart function with age has suggested that much of the decrease in the ability of the heart to pump blood at rest and during exercise in older people may be the result of disease and lifestyle changes, decreased physical activity, for example, rather than to the aging process itself.

THE BALANCE CONCEPT AND CHEMICAL HOMEOSTASIS

Many homeostatic systems can be studied in terms of balance; some regulate the balance of a physical parameter (heat, pressure, flow, and so on), but most are concerned with the balance of a chemical component.

Figure 7-4 is a generalized schema of the possible pathways involved in the balance of a chemical substance. The **pool** occupies a position of central importance in the balance sheet. It is the body's readily available quantity of the particular substance and is frequently identical to the amount present in the extracellular fluid. The pool receives substances from and contributes them to all the other pathways.

The pathways on the left of the figure are sources of net gain to the body. A substance may be ingested and then absorbed from the gastrointestinal (GI) tract. The lungs offer a site of entry to the body for gases, particularly oxygen, and airborne chemicals. Finally, the substance may be synthesized by cells within the body.

The pathways on the right of the figure are sources of net loss from the body. A substance may be lost in the urine, feces, expired air, or menstrual fluid, as well as

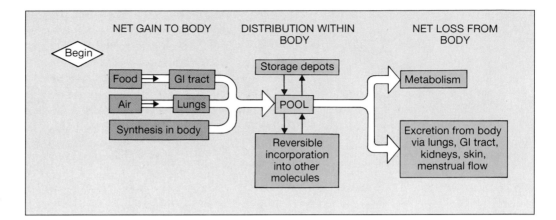

FIGURE 7-4 Balance diagram for a chemical substance.

from the surface of the body as skin, hair, nails, sweat, and tears. The substance may also be chemically altered and thus removed by metabolism.

The central portion of the figure illustrates the distribution of the substance within the body. The substance may be taken from the pool and accumulated in storage depots. Conversely, it may leave the storage depots to reenter the pool. Finally, the substance may be incorporated reversibly into some other molecular structure such as fatty acids into membranes or iodine into thyroxine. This process is reversible in that the substance is liberated again whenever the more complex molecule is broken down. Incorporation is also distinguished from storage in that the latter has no function other than the passive one of storage, whereas the incorporation of the substance into other molecules fulfills a specific function.

It should be recognized that not every pathway of this generalized schema is applicable to every substance. For example, mineral electrolytes such as sodium cannot be synthesized, do not normally enter through the lungs, and cannot be removed by metabolism.

The orientation of Figure 7-4 illustrates two important generalizations concerning the balance concept: (1) Total-body balance depends upon the rates of total-body net gain and net loss; and (2) the pool concentration depends not only upon total-body balance but also upon exchanges of the substance within the body.

For any chemical, three states of total-body balance are possible (Table 7-2): (1) Loss exceeds gain, so that the total amount of the substance in the body decreases and the person is said to be in **negative balance**; (2) gain exceeds loss, so that the total amount of the substance in the body increases and the person is said to be in **positive balance**; and (3) gain equals loss and the person is in **stable balance**.

Clearly a stable balance can be upset by alteration of the amount being gained or lost in any single pathway in the schema; for example, severe negative water balance can be caused by increased sweating. Conversely, stable balance can be restored by homeostatic control of water intake and output.

Let us take sodium as an example. The control systems for sodium balance have as their targets the kidneys, and the systems operate by inducing the kidneys to excrete into the urine an amount of sodium approximately equal to the amount ingested daily. In the example, we assume for simplicity that all sodium loss from the body occurs via the urine. Now imagine a person with a daily intake and excretion of 7 g of sodium and a stable amount of sodium in her body (Figure 7-5). On day 2 of our experiment, the subject changes her diet so that her daily sodium consumption rises to 15 g and remains there indefinitely. On this same day, the kidneys excrete into the urine somewhat more than 7 g of sodium but certainly not all the ingested 15 g. The result is that some excess sodium is retained in the body on that day, that is, the person goes into *positive* sodium balance. The kidneys do somewhat better on day 3, but it is probably not until day 4 or 5 that they are excreting 15 g. From this time on, output from the body once again equals input and sodium balance is once again *stable*.

But, and this is an important point, although again in stable balance, the woman has perhaps 2 to 3 percent

TABLE 7-2	THREE STATES OF TOTAL-BODY BALANCE OF A CHEMICAL
Stable balance:	Loss equals gain
Negative balance:	Loss exceeds gain
Positive balance:	Gain exceeds loss

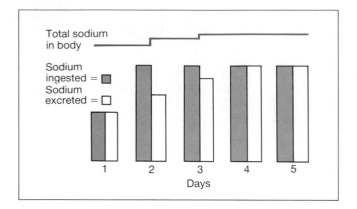

FIGURE 7-5 Effects of a continued change in amount of sodium ingestion on sodium excretion and total-body sodium balance. Stable sodium balance is reattained by day 4 but with some gain of total-body sodium.

more sodium in her body than was the case when she was in stable balance ingesting 7 g.[1] It is this 2 to 3 percent extra sodium that constitutes the continuous error signal to the control systems driving the kidneys to excrete 15 rather than 7g/day. An increase of 2 to 3 percent does not seem large, but it has been hypothesized that, over many years, this little extra might facilitate the development of high blood pressure in some persons.

COMPONENTS OF HOMEOSTATIC SYSTEMS

Reflexes

The thermoregulatory systems we used as an example earlier in the chapter and many of the body's other homeostatic systems belong to the general category of stimulus-response sequences known as "reflexes." Although in some reflexes we are aware of the stimulus and/or the response, many reflexes regulating the internal environment occur without any conscious awareness.

In the most narrow sense of the word, a **reflex** is an involuntary, unpremeditated, unlearned, "built-in" response to a stimulus. Examples of such reflexes include pulling one's hand away from a hot object or shutting one's eyes as an object rapidly approaches the face.

There are also many responses, however, that appear to be automatic and stereotyped but are actually the re-

sult of learning and practice. For example, an experienced driver performs many complicated acts in operating a car. To the driver these motions are, in large part, automatic, stereotyped, and unpremeditated, but they occur only because a great deal of conscious effort was spent learning them. We shall refer to such reflexes as **learned** or **acquired**. In general, most reflexes, no matter how basic they may appear to be, are subject to alteration by learning; that is, there is often no clear distinction between a basic reflex and one with a learned component.

The pathway mediating a reflex is known as the **reflex arc**, and its components are shown in Figure 7-6.

A **stimulus** is defined as a detectable change in the environment, such as a change in temperature, potassium concentration, or pressure. A **receptor** detects the environmental change; we referred to the receptor as a "detector" earlier. A stimulus acts upon a receptor to produce a signal that is relayed to an **integrating center**. The pathway traveled by the signal between the receptor and the integrating center is known as the **afferent pathway** (the general term "afferent" means "to carry to," in this case the integrating center).

An integrating center often receives signals from many receptors, some of which may be responding to quite different types of stimuli. Thus, the output of an

---

[1]Note that a large percentage change in intake (>100 percent) causes only a small change (2 to 3 percent) in total body balance before the new steady state is reached. In other words, the percentage change in the regulated variable is small compared to the percentage change in the altered input.

FIGURE 7-6 General components of a biological control system, the reflex arc. The response of the system has the effect of counteracting or eliminating the stimulus. This phenomenon of negative feedback is emphasized by the minus sign in the feedback loop.

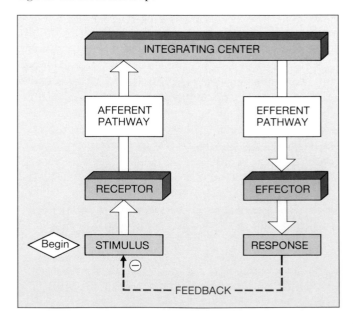

integrating center reflects the net effect of the total afferent input; that is, it represents an integration of numerous bits of information.

The output of an integrating center is sent to the last component of the system, a device whose change in activity constitutes the overall response of the system. This component is known as an **effector**. The information going from an integrating center to an effector is like a command directing the effector to alter its activity. The pathway along which this information travels is known as the **efferent pathway** (the general term "efferent" means "to bear away from," in this case away from the integrating center). In negative-feedback reflexes, the effector's response tends to counteract the original stimulus that triggered the sequence of events.

To illustrate the components of a homeostatic reflex arc, let us use Figure 7-7 to apply these terms to thermoregulation. The temperature receptors are the endings of certain nerve cells in various parts of the body. They generate electric signals in the nerve cells at a rate determined by the temperature. These electric signals are conducted by the nerve fibers—the afferent pathway—to a specific part of the brain—the integrating center—

which in turn determines the signals sent out along those nerve cells that cause skeletal muscles and the smooth muscles surrounding skin blood vessels to contract. The latter nerve fibers are the efferent pathway, and the muscles they innervate are the effectors.

Traditionally, the term "reflex" was restricted to situations in which the receptors, afferent pathway, integrating center, and efferent pathway are all parts of the nervous system, as in the thermoregulatory reflex. Present usage is not so restrictive, however, and recognizes that the principles are essentially the same when a blood-borne messenger known as a **hormone**, rather than a nerve fiber, serves as the efferent pathway, or when a hormone-secreting gland (**endocrine gland**) serves as integrating center. Thus, in the thermoregulation example, the integrating center in the brain not only sends signals by way of nerve fibers but also causes the release of several hormones that travel via the blood to many cells and produce an increase in the amount of heat produced in these cells. These hormones therefore also serve as an efferent pathway in thermoregulatory reflexes.

Accordingly, in our use of the term "reflex," we in-

FIGURE 7-7 Reflex for minimizing the decrease in body temperature that occurs on exposure to a reduced external environmental temperature. This figure provides the internal components for the reflex shown in Figure 7-1. The dashed arrow indicates the negative-feedback nature of the reflex.

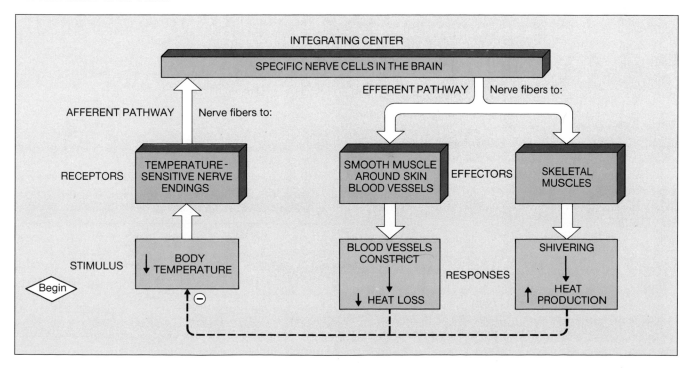

TABLE 7-3	QUESTIONS TO BE ASKED ABOUT ANY HOMEOSTATIC REFLEX

1. What is the variable (for example, plasma glucose concentration, body temperature, blood pressure) that is maintained relatively constant in the face of changing conditions?
2. Where are the receptors that detect changes in the state of this variable?
3. Where is the integrating center to which these receptors send information and from which information is sent out to the effectors, and what is the nature of these afferent and efferent pathways?
4. What are the effectors, and how do they alter their activities so as to maintain the regulated variable at the operating point of the system?

clude hormones as reflex components. Moreover, depending on the specific nature of the reflex, the integrating center may reside in either the nervous system or in an endocrine gland.

Finally, to complete our illustration of the components of a reflex arc, we must identify the effectors, the cells whose outputs constitute the ultimate responses in the reflexes. Almost all body cells can act as effectors in that their activity is subject to control by nerves or hormones. There are, however, two specialized classes of tissues—muscle and gland tissues—that are the major effectors of biological control systems. Muscle cells are specialized for the generation of force and movement, and their physiology is described in Chapter 11. Glands consist of epithelial cells specialized for the function of secretion (Chapter 6).

To summarize, many reflexes function, in a homeostatic manner, to keep a physical or chemical variable of the body relatively constant. One can analyze any such system by answering the questions listed in Table 7-3.

Local Homeostatic Responses

In addition to reflexes, another group of biological responses is of great importance for homeostasis. We shall call them **local homeostatic responses**. They are initiated by a change in the external or internal environment, that is, a stimulus, and they induce an alteration of cell activity with the net effect of counteracting the stimulus. Like a reflex, a local response is a sequence of events proceeding from stimulus to response. Unlike a reflex, however, the entire sequence occurs only in the area of the stimulus, and neither hormones nor nerves are involved.

Two examples should help clarify the nature and significance of local responses: (1) Damage to an area of skin causes cells in the damaged area to release certain chem-

icals that help the local defense against further injury; and (2) an exercising muscle liberates into the extracellular fluid chemicals that act locally to dilate the blood vessels in the area, thereby permitting the required inflow of additional blood to the muscle. The significance of such local responses is that they provide individual areas of the body with mechanisms for local self-regulation.

Chemicals as Intercellular Messengers in Homeostatic Systems

Essential to reflexes and local responses, and therefore to homeostasis, is the ability of cells to communicate with one another. When a hormone is involved in a reflex, the communication between hormone-secreting cell and effector is accomplished by a chemical agent, the hormone, with the blood acting as the delivery service. Most nerve cells also communicate with each other or with effectors by means of chemical agents called **neurotransmitters**. Thus, one nerve cell alters the activity of the next nerve cell by releasing from its ending a neurotransmitter that diffuses through the extracellular fluid separating the two nerve cells and acts upon the second (Figure 7-8). Similarly, neurotransmitters released from

---

FIGURE 7-8 Categories of chemical messengers. A neurohormone is the product of a nerve cell but is secreted into the blood; the latter characteristic makes it a hormone. An autocrine acts on the cell that secreted it. With the exception of autocrines, all other messengers act *inter*cellularly.

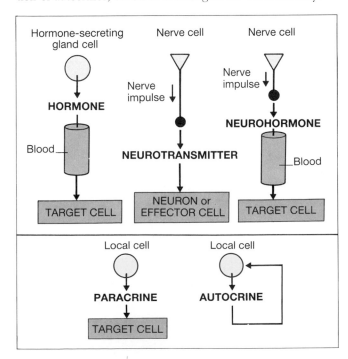

nerve cells going to effector cells constitute the immediate signal, or input, to the effector cells.

As described more fully in Chapter 10, the chemical messengers released by certain nerve cells do not act on adjacent nerve cells or effector cells but rather enter the bloodstream to act on cells elsewhere in the body. These messengers therefore are properly termed hormones, not neurotransmitters. They are often called **neurohormones** (Figure 7-8) in recognition of their secretion by nerve cells.

Chemicals act as intercellular messengers not only for reflexes but also for local responses, as the examples above illustrate. The chemical messengers, known as **paracrines**, involved in local responses are synthesized by cells and released, once given the appropriate stimulus, into the extracellular fluid. They then diffuse to neighboring cells and exert their particular effects.[2] Note that, given this broad definition, neurotransmitters could be classified as a subgroup of paracrines, but by convention they are not. Paracrines are generally inactivated rapidly by locally existing enzymes so that they do not enter the bloodstream in large quantities.

Paracrines are chemical substances synthesized and secreted *specifically* for the purpose of regulating local cells. During the course of metabolism, many substances, such as potassium and hydrogen ions, are released from cells and produce responses in neighboring cells, particularly smooth-muscle cells of blood vessels. Such substances serve to regulate local responses, therefore, but they are secondary by-products of cellular activity, not specific chemical messengers, and so are not classified as paracrines.

There is one category of local chemical messengers that is not an *intercellular* messenger at all. Rather, the chemical is secreted by a cell into the extracellular fluid and then acts upon the very cell that secreted it. The general term for such a substance is **autocrine**.

A point of great importance must be emphasized here to avoid later confusion: A nerve cell, endocrine gland cell, or other cell type may all secrete the same chemical messenger. Thus, a particular messenger may sometimes be referred to as a neurotransmitter, as a hormone, and as a paracrine or autocrine.

All the types of intercellular communication described in this section involve secretion of the chemical messenger into the extracellular fluid. However, there are two important types of intercellular chemical communication that do not require such secretion. In the first type, which occurs via gap junctions (page 46), chemicals move from one cell to an adjacent cell without

ever entering the extracellular fluid. In the second type, the chemical messenger is not actually released from the cell producing it but rather is located in the outer plasma membrane of that cell. In this case, the cell links up with its target cell via the membrane-bound messenger. This type of signaling is of particular importance in the functioning of the cells that protect the body against microbes and other foreign cells (Chapter 19).

Eicosanoids. The general approach of this text is to describe specific chemical messengers in the context of the functions they influence. However, one set of local chemical messengers exerts such a wide variety of effects that it is best described separately at this point. These are the **eicosanoids**, a family of substances at least some of which are synthesized in virtually all tissues of the body. This family includes the **prostaglandins** (formerly the general name loosely used to refer to all eicosanoids), **prostacyclin**, the **thromboxanes**, and the **leukotrienes** (Figure 7-9). They are all unsaturated fatty acids that have a carbon ring at one end. They are divided into main groups, each designated by a letter—for example, PGA and PGE for prostaglandins of the A and E families—on the basis of structural differences in their rings. Within each group, further subdivision occurs according to the number of double bonds in the fatty acid side chains; for example PGE_2 (Figure 7-10) has two double bonds.

The major chemical from which all eicosanoids are produced is the fatty acid **arachidonic acid** (Figure 7-10), which is present in plasma-membrane phospholipids. As shown in Figure 7-9, the appropriate stimulus—hormone, neurotransmitter, paracrine, drug, or toxic agent—activates an enzyme, **phospholipase A_2**, in the membrane of the stimulated cell. This enzyme splits off arachidonic acid from the phospholipid, and the arachidonic acid is then metabolized by several enzyme-mediated steps to yield the specific eicosanoids. There are two major branches from arachidonic acid, one catalyzed by the enzyme **cyclooxygenase** and leading to formation of the prostaglandins, prostacyclin, and thromboxanes and the other catalyzed by the enzyme **lipoxygenase** and leading to formation of leukotrienes.

Once synthesized in response to a stimulus, the eicosanoids are not stored to any extent but are released immediately and act locally. Accordingly, the eicosanoids are categorized as paracrines and autocrines. After they act, they are quickly metabolized by local enzymes to inactive forms. Whether certain eicosanoids can also function as hormones, that is, can be carried by the blood to exert effects on sites distant from their tissues of origin, remains unsettled.

The eicosanoids exert a bewildering array of effects.

---

[2] Paracrines are sometimes referred to as "local hormones."

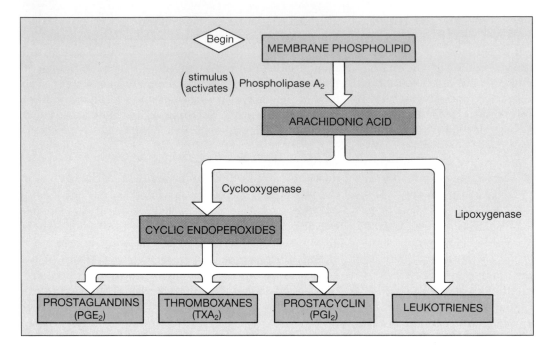

FIGURE 7-9 Pathways for the synthesis of eicosanoids. Phospholipase A_2 is the one enzyme common to the formation of all the eicosanoids; it is the site at which stimuli act. Different enzymes mediate each of the next steps, and because various cell types in the body possess different enzyme profiles, they produce different final substances in response to a stimulus. The step mediated by cyclooxygenase is inhibited by aspirin. The abbreviations in parentheses denote major examples of each of the families.

For example, they are important in blood clotting, in regulation of smooth-muscle contraction, in the modulation of neurotransmitter release and action, in multiple processes in the reproductive system, in the control of hormone secretion, and in the body's defenses against injury and infection. We shall describe many of these specific effects in future chapters.

FIGURE 7-10 Structure of arachidonic acid and PGE_2.

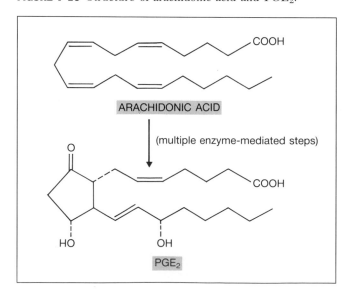

RECEPTORS

To reiterate, most homeostatic systems require cell-to-cell communication via chemical messengers. The first step in the action of any intercellular chemical messenger on its target cell is the binding of the messenger to specific protein molecules of the target cell. These molecules are known as **receptors**; in the general language of Chapter 4, the chemical messenger is a "ligand" and the receptor, a "binding site." The term "receptor" can be the source of confusion because the same word is used to denote the "detectors" in a reflex arc, as described earlier in this chapter. The reader must keep in mind the fact that "receptor" has two totally distinct meanings, but the context in which the term is used usually makes it quite clear which is meant.

What is the nature of the receptors with which chemical messengers combine? They are proteins located either in the cell's plasma membrane or inside the cell.

The plasma membrane is the much more common location, the exceptions being the receptors for steroid hormones, vitamin D, and the thyroid hormones; these substances all readily traverse the lipid-rich plasma membranes of cells, and their receptors are inside cells.

It is the combination of chemical messenger and receptor that initiates the events leading to the cell's response. The existence of receptors explains a very important characteristic of intercellular communication—**specificity** (see Table 7-4 for a glossary of terms concerning receptors). Although a chemical messenger—be it hormone, neurotransmitter, or paracrine—may come into contact with many different cells, it influences only certain cells and not others. The explanation is that cells differ in the types of receptors they contain. Accordingly, only certain cell types, frequently just one, possess the receptor required for combination with a given chemical messenger (Figure 7-11).

It is important to realize that, in the many cases where different cell types possess the same receptor type for a particular messenger, the responses of the various cell types to that messenger may differ from each other. For example, the neurotransmitter norepinephrine causes blood-vessel smooth muscle to contract, but via the same receptor type, norepinephrine causes cells in the pancreas to secrete less insulin. In essence, then, the receptor functions as a molecular "switch" that when "switched on" by the messenger binding to it, elicits the cell's response. Just as identical types of switches can be used to turn on a light or a radio, a single type of receptor can produce quite different responses in different cell types.

TABLE 7-4 A GLOSSARY OF TERMS CONCERNING RECEPTORS	
Receptor:	A specific protein in either the plasma membrane or interior of a target cell with which a chemical mediator combines to exert its effects.
Specificity:	Selectivity; the ability of a receptor to react with only one type or a limited number of structurally related types of molecule.
Saturation:	The degree to which receptors are occupied by a messenger. If all are occupied, the receptors are fully saturated; if half are occupied, the saturation is 50 percent, and so on.
Affinity:	The strength with which a chemical messenger binds to its receptor.
Competition:	The ability of different molecules very similar in structure to combine with the same receptor.
Antagonist:	A molecule that competes for a receptor with a chemical messenger normally present in the body. The antagonist binds to the receptor but does not trigger the cell's response.
Agonist:	A chemical messenger that binds to a receptor and triggers the cell's response; often refers to a drug that mimics a normal messenger's action.
Down-regulation:	A decrease in the total number of target-cell receptors for a given messenger in response to chronic high extracellular concentration of the messenger.
Up-regulation:	An increase in the total number of target-cell receptors for a given messenger in response to a chronic low extracellular concentration of the messenger.
Supersensitivity:	The increased responsiveness of a target cell to a given messenger, resulting from up-regulation.

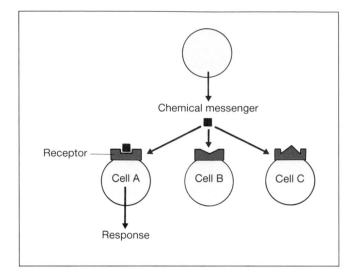

FIGURE 7-11 Specificity of receptors for chemical messengers.

Similar reasoning explains perhaps a more surprising phenomenon: A single cell may contain several different receptor types for a single messenger. Combination of the messenger with one of these receptor types may produce a cellular response quite different from, indeed sometimes opposite of, that produced when the messenger combines with the other receptor type. The degree to which the messenger molecules combine with the different receptor types in a single cell is determined by such factors as the affinity of the different receptor types for the messenger.

It should not be inferred from these descriptions that a cell has receptors for only one messenger. In fact, a single cell usually contains many different receptor types, that is, distinct receptors for different chemical messengers.

Other characteristics of messenger-receptor interactions are **saturation** and **competition**. These phenomena were described on page 56 for ligands binding to binding sites on proteins and are fully applicable here. In most systems, a cell's response to a messenger increases as the extracellular concentration of the messenger increases, because the number of receptors occupied by the messenger increases. There is an upper limit to this responsiveness, however, because only a finite number of receptors are available, and they become saturated at some point.

Competition is the ability of different messenger molecules very similar in structure to compete with each other for a receptor. Competition occurs physiologically with closely related messengers and also underlies much present-day use of drugs. If researchers or physicians

wish to interfere with the action of a particular messenger, they can administer competing molecules, if available, that bind to the receptors for that messenger. This prevents the messenger from doing so but does not trigger the cell's response. Such drugs are known as **antagonists** with regard to the usual chemical messenger. On the other hand, some drugs that bind to a particular receptor type do trigger the cells' response exactly as if the true chemical messenger had combined with receptor; such drugs are known as **agonists** and are used to mimic the messenger's action.

Regulation of Receptors

Receptors are themselves subject to physiological regulation. The number of receptors a cell has and the affinity of the receptors for their specific messenger can be increased or decreased, at least in certain systems. An important example of such regulation is the phenomenon of **down-regulation**. When a high extracellular concentration of messenger is maintained chronically, the total number of receptors for that messenger may decrease—be down-regulated. Down-regulation has the effect of reducing the target cells' responsiveness to high concentrations of messengers and thus represents a local negative-feedback mechanism. For example, a chronic high concentration of the hormone insulin, which stimulates glucose uptake by its target cells, causes down-regulation of its receptors, and this acts to dampen the ability of insulin to cause glucose uptake.

Change in the opposite direction (**up-regulation**) also occurs. Cells exposed chronically to very low concentrations of a messenger may come to have many more receptors for that messenger, thereby developing supersensitivity to it. For example, days after the nerves to a muscle are cut, thereby eliminating the neurotransmitter released by those nerves, the muscle will contract to amounts of experimentally injected neurotransmitter much smaller than those to which an innervated muscle can respond.

Up-regulation and down-regulation are made possible because there is a continuous degradation and synthesis of receptors. Membrane-bound receptors are internalized, that is, taken into the cell by means of endocytosis (page 127), and are either broken down or reinserted back into the membrane along with newly synthesized receptors during exocytosis (page 127). Often the cell contains stores of receptors bound in intracellular vesicles, and these are available for insertion into the membrane. Receptors located inside cells, rather than on plasma membranes, are also continuously being degraded and resynthesized.

Alteration of the rate of one or more of the catabolic and synthetic processes for receptors will raise or lower the number of cell receptors of a particular type. For example, some hormones induce endocytosis of the membrane receptors to which they bind. This increases the rate of receptor degradation so that at high hormone concentrations the number of plasma-membrane receptors gradually decreases. This is an example of down-regulation.

Down-regulation and up-regulation are physiological responses, but there also are many disease processes in which the number of receptors or their affinity for messenger becomes abnormal. The result is unusually large or small responses to any given level of messenger. For example, the disease called myasthenia gravis is due to a destruction of the skeletal muscle receptors for acetylcholine, the neurotransmitter that normally causes contraction of the muscle in response to nerve stimulation; the result is muscle weakness or paralysis.

SIGNAL TRANSDUCTION MECHANISMS FOR PLASMA-MEMBRANE RECEPTORS

We have moved, in this chapter, from the broad organismic concept of homeostasis to the mechanisms by which cells communicate with one another. Now we move into the cell and describe the mechanisms by which the binding of a chemical messenger to a receptor causes the cell to respond to the messenger.

The combination of messenger with receptor causes a change in the conformation of the receptor, and this step, known as **receptor activation**, is always the initial one leading to the cell's ultimate responses to the messenger, which can be: (1) changes in the cell's membrane permeability, in the rates at which it transports various substances, or in its electrical state; (2) changes in the rate at which a particular substance is synthesized or secreted by the cell; or (3) changes in the strength of contraction, if the cell is a muscle cell.

Despite the seeming variety of these ultimate responses, there is a common denominator: They are all due directly to alterations of particular cell proteins. Let us take a few examples of messenger-induced responses, all of which are described fully in subsequent chapters. Muscle contraction results from altered conformation of the contractile proteins in the muscle. Changes in the rate of glucose secretion by the liver reflect altered activity and concentration of enzymes in the metabolic pathways for glucose synthesis. Generation of electric signals in nerve cells reflects altered conformation of membrane proteins constituting ion channels through which substances can diffuse between extracellular fluid and intracellular fluid.

The subject of the remainder of this chapter is the mechanisms by which receptor activation leads to the changes in protein conformation, activity, or concentration that underlie a cell's ultimate response to a messenger. The mechanisms for messengers, notably the steroid hormones, that can diffuse through the plasma membrane and bind to intracellular receptors are given in Chapter 10 and so we shall restrict the present discussion to the much larger number of messengers that bind to plasma-membrane receptors. The main question we are concerned with is: How does the information that a receptor on the plasma membrane is occupied get relayed to and influence the cell's internal components, many of which are located far from the plasma membrane?

The relay mechanisms are termed **signal transductions**, and on the basis of the signal transduction pathways they use, plasma-membrane receptors can be classified into the five types listed in Table 7-5, which will all be described in subsequent sections. Before doing so, however, we must define three terms—G proteins, second messengers, and protein kinases—that will be used frequently in these descriptions.

G proteins are a family of plasma-membrane regulatory proteins that many activated receptors interact with and alter. The altered G protein then interacts with another protein—either an ion channel or an enzyme—*in the plasma membrane* to elicit the next step in the sequence of events leading to the cell's response.

Second messengers are substances that serve as the relay from the plasma membrane to the biochemical machinery *inside* the cell. In this terminology, **first messenger** denotes either the original chemical messenger combining with a plasma-membrane receptor or an electric message that arrives at a cell's plasma membrane. Note that because they do not themselves pass information directly from the plasma membrane to the inside of the cell, the G proteins are not second messengers, al-

TABLE 7-5	CLASSIFICATION OF RECEPTORS BASED ON THE SIGNAL TRANSDUCTION MECHANISMS THEY USE

(1) Receptors that affect membrane electrical activity and cytosolic calcium concentration by influencing ion channels

(2) Receptors that regulate cyclic AMP formation

(3) Receptors that regulate cyclic GMP formation

(4) Receptors that affect both cytosolic calcium concentration and protein kinase C activity by influencing the metabolism of membrane phosphatidyl inositol

(5) Receptors that possess protein kinase activity

though, as we shall see, they often contribute to the generation of second messengers.

Protein kinases are enzymes that phosphorylate other proteins by transferring to them a phosphate group from ATP. Introduction of the phosphate group changes the activity of the protein, often itself an enzyme.

Receptor-Operated Channels

In the first type of receptor signal transduction mechanism listed in Table 7-5, the alteration in membrane-receptor conformation caused by the binding of the messenger results in the opening or closing of ion channels in the plasma membrane. In some cases, the receptor is itself part of the protein complex that constitutes the ion channel, and the change in conformation of the receptor is the direct cause of the channel's opening or closing (Figure 7-12). In the remaining cases, instead of the receptor actually being part of the ion channel, the receptor interacts with an adjacent membrane G protein, which in turn opens or closes the channel.

In both cases, the channel that is opened or closed is termed, appropriately, a **receptor-operated channel** (or a ligand-operated channel).[3] The opening or closing results in an increase or decrease, respectively, in the net diffusion across the plasma membrane of the ion or ions specific to the channel. As we shall see in Chapter 8, such a change in ion diffusion is associated with an electric signal in the membrane, and this signal may be an

---

[3]In some terminologies, "receptor-operated channels" or "ligand-operated channels" refers only to channels that are part of the receptor. For simplicity, however, we also include in this category channels operated by G proteins.

FIGURE 7-12 Receptor-operated channels in the plasma membrane. The first messenger, which reaches the cell from the extracellular fluid, is a hormone, neurotransmitter, paracrine, or autocrine. The receptor may either be a part of the protein complex constituting the ion channel or it may affect the channel through the intermediation of an adjacent G protein in the membrane. In either case, some receptors open channels, whereas others close them.

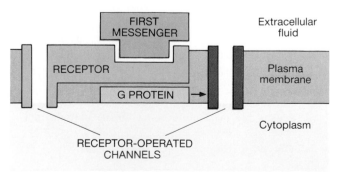

essential link in the cell's response to the original messenger. Indeed, the signal may constitute the entire response.

In addition to the electric signal, there may be a change in the cytosolic concentration of the ion(s) specific to the channel. In particular, when plasma-membrane calcium channels open, calcium ions move down their electrochemical gradient from the extracellular fluid through the plasma membrane and into the cytosol, thereby increasing cytosolic calcium concentration. Why an increased cytosolic calcium concentration is so important under these conditions will be described in the later section dealing with calcium as a second messenger.

Cyclic AMP

The second type of receptor mechanism listed in Table 7-5 is utilized by a remarkable number of chemical messengers and involves the generation, in the cytosol, of a particular second messenger known as cyclic AMP. The sequence of events is illustrated in Figure 7-13. The binding of the first messenger by the receptor induces a conformation change in the receptor that allows the receptor to bind to an adjacent membrane G protein known as G_s **protein** (the subscript s denotes "stimulatory"). This binding causes the G_s protein to activate the membrane enzyme called **adenylate cyclase**. The activated adenylate cyclase, whose catalytic site is located on the cytosolic surface of the plasma membrane, then catalyzes the conversion of cytosolic ATP to **cyclic 3′,5′-adenosine monophosphate**, called simply **cyclic AMP** (**cAMP**). Cyclic AMP can then diffuse throughout the cell to trigger the sequences of events leading to the cell's ultimate response to the first messenger.

In this manner, cAMP is acting as a second messenger. Its action is terminated by its breakdown to noncyclic AMP, a reaction catalyzed by the enzyme **phosphodiesterase** (Figure 7-14).

What does cyclic AMP actually do in the cell? It activates an enzyme known as **cAMP-dependent protein kinase**. In other words, the event common to all the sequences initiated by cAMP is activation of a protein kinase. As defined above, protein kinases phosphorylate other proteins—often enzymes—by transferring a phosphate group to them. The change in the activity of the protein owing to its phosphorylation by cAMP-dependent protein kinase brings about the response of the cell (secretion, contraction, and so on).

Thus, the activation of adenylate cyclase by G_s protein initiates a chain, or "cascade," of events in which proteins are converted in sequence from inactive to active forms. Figure 7-15 illustrates the benefit of such a cascade. While it is active, a single enzyme molecule is capable of transforming into product not one but many

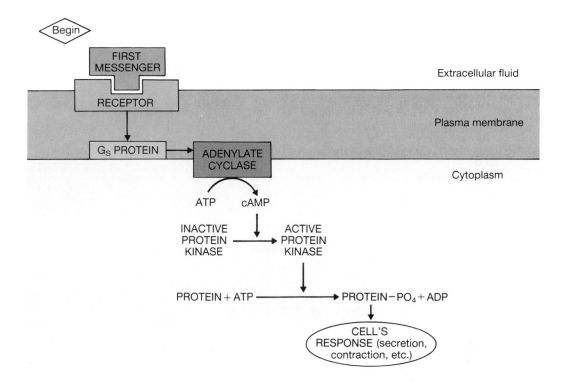

FIGURE 7-13 Cyclic AMP second-messenger system. Combination of the first messenger—hormone, neurotransmitter, or paracrine—with its specific receptor permits the receptor to bind to the membrane G protein termed G_s. This protein in turn binds to adenylate cyclase, activating it and causing it to catalyze the formation of cyclic AMP (cAMP) inside the cell. This cAMP activates cAMP-dependent protein kinase, which then phosphorylates specific proteins in the cell, thereby triggering the biochemical reactions leading ultimately to the cell's response. Not shown in the figure is the existence of another regulatory protein, G_i, with which certain receptors can react to cause inhibition of adenylate cyclase.

substrate molecules, let us say 100. Therefore, 1 active molecule of adenylate cyclase may catalyze the generation of 100 cAMP molecules. At each of two subsequent enzyme-activation steps in our example, another 100-fold amplification occurs. Therefore, the end result is that a single molecule of the first messenger could theoretically trigger the generation of 1 million product molecules. This fact helps to explain how hormones and other messengers can be effective at extremely low extracellular concentrations.

How can protein kinase activation by cAMP be an event common to the great variety of biochemical sequences and cell responses initiated by cAMP-generating messengers? The answer is that there are many distinct proteins that can be phosphorylated by cAMP-dependent protein kinase (Figure 7-16). For example, a fat cell responds to epinephrine (a hormone that utilizes the cyclic AMP pathway) not only with glycogen breakdown (mediated by one phosphorylated enzyme) but also with triacylglycerol breakdown (mediated by another phosphorylated enzyme).

Another source of variety is that, whereas phosphorylation mediated by cAMP-dependent protein kinase activates certain enzymes, it *inhibits* others. For example, the enzyme catalyzing the rate-limiting step in glycogen synthesis is inhibited by phosphorylation, which explains how epinephrine inhibits glycogen synthesis at the same time that it stimulates glycogen breakdown (Figure 7-17).

In summary, the biochemistry of cAMP-dependent protein kinase and its substrates (the proteins it phosphorylates) explains how a single molecule, cAMP, can produce so many different effects. It must be reemphasized that the ability of a cell to respond at all to a cAMP-generating first messenger depends upon the presence of specific receptors for that messenger in the plasma membrane.

Not mentioned previously is the fact that certain receptors, upon binding a messenger, *inhibit* adenylate cyclase and result in less, rather than more, generation of cAMP. This occurs because these receptors interact with a different regulatory protein, known as **G_i protein**

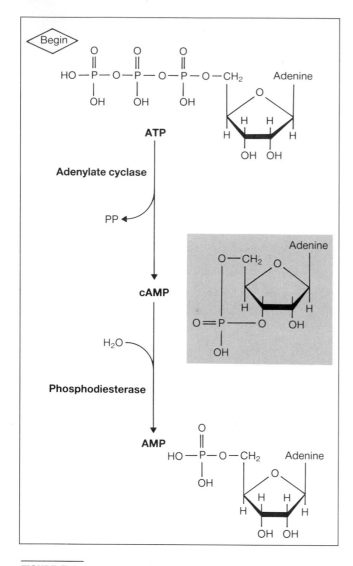

FIGURE 7-14 Structure of ATP, cyclic AMP, and AMP, the last resulting from enzymatic alteration of cyclic AMP.

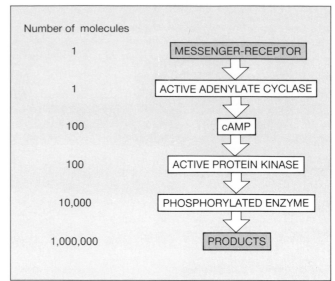

FIGURE 7-15 Example of amplification in the cAMP system. By means of an enzyme cascade, a single messenger-receptor complex ultimately leads to the generation of a very large number of product molecules.

turn, cGMP activates a **cGMP-dependent protein kinase**, which phosphorylates proteins. These proteins then elicit the cell's ultimate responses. The cGMP system is not nearly so widespread as the cAMP system.

Calcium

The calcium ion (Ca^{2+}) functions as a second messenger in a great variety of cellular responses to chemical messengers. The regulation of cytosolic calcium concentration is described on page 121 and should be reviewed at this point. By means of active-transport systems in the plasma membrane and cell organelles, this ion is maintained at an extremely low concentration in the cytosol. In response to a stimulus, its cytosolic concentration increases, and this increase in concentration triggers a sequence of reactions leading to the cell's responses. The physiology of calcium as a second messenger requires an analysis, therefore, of two broad questions: (1) How do stimuli elicit an increase in cytosolic calcium concentration? and (2) How does the increased calcium concentration elicit the cells' responses?

We answered part of the first question in the section on receptor-operated channels: When receptor activation by a first messenger causes plasma-membrane calcium channels to open, calcium ions move down their electrochemical gradient from the extracellular fluid through the plasma membrane and into the cytosol, thereby increasing cytosolic calcium concentration.

(the subscript i denotes "inhibitory"), in the adenylate cyclase system, and the binding of a receptor to a G_i protein causes an inhibition of adenylate cyclase. It is thought that several chemical messengers induce their target cells' responses by reducing the concentration of cAMP in this way and, thereby, decreasing the phosphorylation of key proteins inside the cell.

Cyclic GMP

Another cyclic nucleotide—**cyclic 3′,5′-guanosine monophosphate (cGMP)**—also serves as a second messenger. Some activated receptors activate or inhibit the enzyme **guanylate cyclase**, which catalyzes the generation of cGMP in a manner analagous to that for cAMP. In

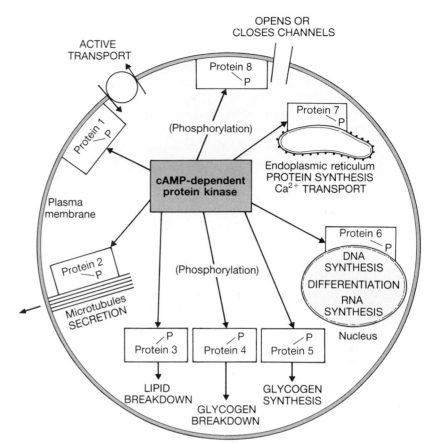

FIGURE 7-16 The variety of cellular responses induced by cAMP is due to the fact that cAMP-dependent protein kinase can phosphorylate many different proteins, activating them or inhibiting them. Our designations of the proteins being phosphorylated as 1, 2, and so on, are arbitrary. For clarity, the generation of cAMP and its activation of cAMP-dependent kinase is not shown in this figure (see Figure 7-13).

FIGURE 7-17 Coordinated stimulation of glycogen breakdown and inhibition of glycogen synthesis by epinephrine-induced generation of cAMP in liver cells. The cAMP activates cAMP-dependent protein kinase, which catalyzes the addition of phosphate to both the enzyme phosphorylase and the enzyme glycogen synthetase. The phosphorylation *activates* phosphorylase but *inhibits* glycogen synthetase.

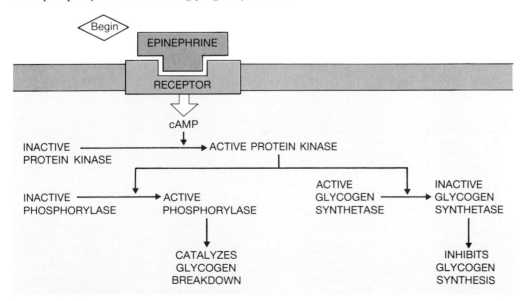

At this point, a very important digression must be made to distinguish between different kinds of regulatable calcium channels in the plasma membrane. The concept of such channels has been introduced in this chapter in the context of *chemical* signals—messengers that bind to and activate receptors. However, it must be emphasized that, as described in Chapter 6, there are other calcium channels in the plasma membrane that are not operated by receptors but rather open or close in response to *electric* signals in the membrane. These are known as **voltage-sensitive channels** (or voltage-gated channels). Thus, calcium can act as a second messenger in response not only to a chemical stimulus to the cell but to electric stimuli as well. Finally, there are still other regulatable calcium channels in the plasma membrane that are neither receptor-operated nor voltage-sensitive but instead respond to chemical reactions influencing them from the cytoplasmic side of the membrane.

Now let us return to answering our questions about calcium. The opening of plasma-membrane calcium channels is not the only way an activated receptor can cause cytosolic calcium concentration to increase. A quite different mechanism is the initiation of reactions that lead to the movement of calcium into the cytosol from the endoplasmic reticulum and possibly other cell organelles. In this sequence, the activated receptor stimulates, via the intermediation of a G protein, a plasma-membrane enzyme, **phospholipase C**. This enzyme catalyzes the breakdown of a plasma-membrane phospholipid known as **phosphatidylinositol bisphosphate**, abbreviated PIP_2, to **inositol trisphosphate (IP_3)** and **diacylglycerol (DAG)** (Figure 7-18). IP_3 enters the cytosol and acts upon the membranes of the endoplasmic reticulum to cause the movement of calcium from this structure into the cytosol. (The role of DAG is described in the next section.) Note that in this sequence IP_3 is acting as a second messenger to mobilize yet another second messenger—calcium.

These two mechanisms for increasing cytosolic calcium concentration—opening plasma-membrane calcium channels to allow calcium entry from the extracellular fluid and releasing calcium from the endoplasmic reticulum—are not mutually exclusive. Indeed, quite the opposite, they are often triggered simultaneously by the same messenger (Figure 7-19).

Now we turn to the second question posed earlier in this section: How does the increased cytosolic calcium concentration elicit the cells' responses? The common denominator of calcium's actions is its ability to bind tightly and specifically to various cytosolic calcium-binding proteins. One of the most important of these is a protein found in virtually all cells and known as **calmodulin** (Figure 7-20). On binding with calcium, calmodulin changes shape, and this allows it to combine with a large variety of enzymes and other proteins, activating them or inhibiting them.

At this point, part of the story for calcium-activated calmodulin becomes analogous to that for cyclic AMP, since several of the many proteins influenced by calmod-

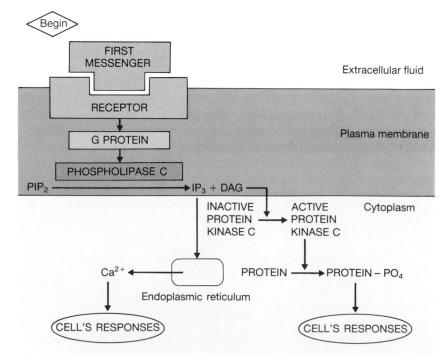

FIGURE 7-18 Mechanism by which an activated receptor stimulates the enzymatically mediated breakdown of PIP_2 to yield IP_3 and DAG. IP_3 then causes release of calcium ions from the endoplasmic reticulum, and DAG activates a particular protein kinase known as protein kinase C.

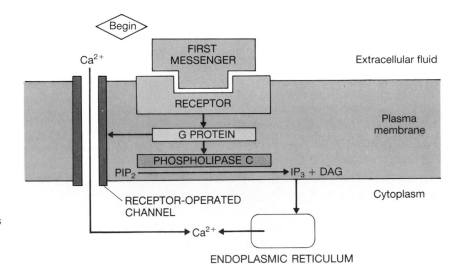

FIGURE 7-19 Summary of the two mechanisms by which a membrane receptor, after binding with a messenger, causes an increase in cytosolic calcium concentration. For simplicity, only one G protein is shown mediating both the opening of the channel and the activation of phospholipase C; it is more likely that two different G proteins would mediate these actions.

FIGURE 7-20 Calcium, calmodulin, calmodulin-dependent protein kinase system. The mechanisms for increasing cytosolic calcium concentration are summarized in Figure 7-19.

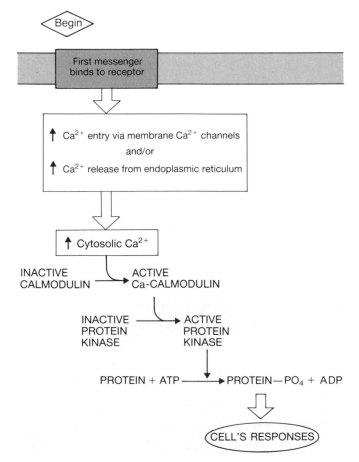

ulin are protein kinases. Activation or inhibition of **calmodulin-dependent protein kinases** leads, via phosphorylation, to activation or inhibition of proteins involved in the cell's ultimate responses—contraction, secretion, and so on—to the first messenger.

Calcium's second-messenger actions can involve proteins other than calmodulin. In some cases, the cytosolic calcium combines with calcium-binding proteins distinct from, but related to, calmodulin. These proteins then act as intermediaries, influencing other proteins in turn. In

---

TABLE 7-6   CALCIUM AS A SECOND MESSENGER

Mechanisms by which the binding of first chemical messengers to receptors leads to an increase in cytosolic Ca^{2+} concentration:

1. Plasma-membrane calcium channels open.
2. Calcium is released from endoplasmic reticulum (and possibly other organelles). This is mediated by inositol trisphosphate (IP_3) generated from the breakdown of membrane phospholipid.

Mechanisms by which an increase in cytosolic Ca^{2+} concentration induces the cell's responses:

1. Calcium binds to calmodulin. On binding calcium, the calmodulin changes shape, which allows it to combine with a large variety of enzymes and other proteins, activating them or inhibiting them. Many of these enzymes are protein kinases.
2. Calcium combines with calcium-binding intermediary proteins other than calmodulin. These proteins then act in a manner analagous to calmodulin.
3. Calcium combines with and alters response proteins directly, without the intermediation of any specific calcium-binding protein.

still other cases, the calcium directly alters cytosolic proteins immediately involved in the cell's responses, without the intermediation of any specific calcium-binding protein.

Calcium's function as a second messenger is summarized in Table 7-6.

DAG

As described in the previous section, certain receptors, following the binding of a messenger to them, stimulate the breakdown of PIP_2 to IP_3 and DAG. Both of these products act as second messengers: Whereas IP_3 enters the cytosol and acts on the endoplasmic reticulum, DAG activates a particular protein kinase known as **protein kinase C**. This enzyme then phosphorylates a large number of other proteins (Figure 7-18).

Protein kinase C is different from the other protein kinases we have described—those dependent on cAMP, cGMP, and calmodulin, and many of the proteins that it phosphorylates are also different. In particular, activated protein kinase C plays an important role, via phosphorylation, in the regulation of membrane proteins.

It must be emphasized that since DAG and IP_3 are generated by the same receptor-stimulated membrane reaction, any first messenger that triggers this reaction causes both the activation of protein kinase C—mediated by DAG—and an increase in cytosolic calcium concentration—mediated by IP_3's release of calcium from endoplasmic reticulum. Both pathways may interact in a variety of additive or synergistic ways. For example, protein kinase C is itself a calcium-dependent enzyme, and DAG and calcium can function synergistically to activate it.

Receptors as Protein Kinases

We have described four substances—cAMP, cGMP, calcium-activated calmodulin, and DAG—that have the ability to activate protein kinases. But there is at least one more important mechanism whereby a first messenger can trigger activation of protein kinases. When certain messengers involved in growth and development bind to their receptors, the receptors themselves become active protein kinases. The receptors then phosphorylate specific cytosolic and plasma-membrane proteins, including themselves. This, the fifth type of receptor mechanism listed in Table 7-5, provides yet another way of translating the first messenger's signal into cellular responses. The protein kinases that function in this way all belong to the family known as **tyrosine kinases**, because they phosphorylate specifically the tyrosine portions of proteins.

Interactions between cAMP, Calcium, IP₃, and DAG

Important interactions exist between the five second messengers we have described (Table 7-7). First, as we have seen, an obligatory link exists between receptor-induced elevation of protein kinase C levels and cytosolic calcium concentration. No obligatory link exists between the cAMP pathway and the other second messengers. Nevertheless, it is clear that they frequently function together. For example, in some cases, stimulation of a cell by a first messenger causes cAMP and cytosolic calcium concentrations to change in the same direction; in other cases, the concentrations of these two second messengers change in opposite directions. Moreover, they influence each other: In some cells calcium-activated calmodulin may itself activate adenylate cyclase, whereas in certain cells the cAMP pathway opens plasma-membrane calcium channels or influences the cell's calcium pumps.

An example of such interactions, as well as other concepts developed in the last sections of this chapter, is provided by the effects of the hormone epinephrine on the smooth-muscle cells surrounding blood vessels (Figure 7-21). There are two types of receptors (designated alpha and beta) for the messenger epinephrine on these muscle cells, and the combination of epinephrine with the two kinds produces diametrically opposed responses—contraction when epinephrine binds to alpha receptors and relaxation when it binds to beta receptors. The alpha-receptor effect is as follows: The binding of epinephrine to this receptor causes, in sequence, the

TABLE 7-7 SECOND MESSENGERS	
Substance	Effects
Cyclic AMP	Activates cAMP-dependent protein kinase
Cyclic GMP	Activates cGMP-dependent protein kinase
Calcium ion	Activates calmodulin and other calcium-binding proteins
Inositol trisphosphate (IP_3)	Releases calcium from endoplasmic reticulum
Diacylglycerol (DAG)	Activates protein kinase C

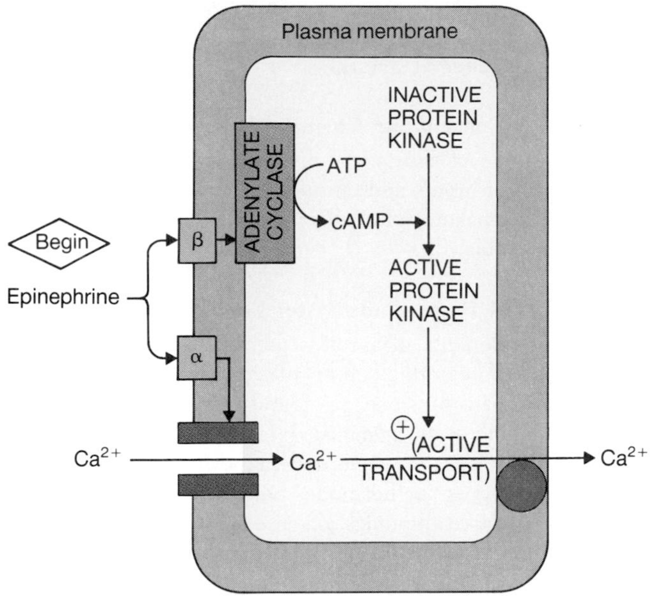

FIGURE 7-21 How epinephrine, via alpha and beta receptors, exerts opposing effects on cytosolic calcium concentration in the smooth-muscle cells of blood vessels. Binding to the alpha receptor results in the opening of membrane calcium channels and the diffusion of calcium into the cell (the links between the receptor and the channel, denoted by the direct arrow in the figure, are not yet known for smooth muscle). Binding to beta receptors results, via the cAMP system, in stimulation of the active-transport system that moves calcium out of the cell.

opening of receptor-operated calcium channels, an *increase* in cytosolic calcium concentration, activation of calmodulin, modulation of the contractile proteins, and *increased contraction*. The binding of epinephrine to a beta receptor results sequentially in activation of adenylate cyclase, generation of cAMP, activation of cAMP-dependent protein kinase, phosphorylation of plasma-membrane proteins that serve as calcium pumps, increased activity of the pumps, and increased active transport of calcium out of the smooth-muscle cells, thereby causing cytosolic calcium concentration to *decrease*, which favors *muscle relaxation*. Whether the alpha effect—contraction—or the beta effect—relaxation—predominates depends upon the relative numbers of the two receptor types on the smooth muscle and their affinities for epinephrine.

We noted in this example that beta receptors for epinephrine utilize the adenylate cyclase-cAMP system as their signal transduction mechanism. This relationship is true for beta receptors on any cell, not just a smooth-muscle cell. We point this out to emphasize that the linking of a certain receptor type with a particular signal transduction mechanism in a variety of cells is common.

SUMMARY

General Characteristics of Homeostatic Control Systems

I. Homeostasis denotes the stable conditions of the internal environment that result from the operation of compensatory homeostatic control systems.

A. In a negative-feedback control system, a change in the variable being regulated brings about responses that tend to push the variable in the direction opposite to the original change. Negative feedback minimizes changes from the operating point of the system, leading to stability.

B. In a positive-feedback system, an initial disturbance in the system sets off a train of events that increases the disturbance even further.

C. Homeostatic control systems minimize changes in the internal environment but cannot maintain complete constancy because of the need for error signals to drive the system.

D. Feedforward regulation anticipates changes in a regulated variable, improves the speed of the body's homeostatic responses, and minimizes fluctuations in the level of the variable being regulated.

II. An acclimatization is an improved ability to respond to an environmental stress.

A. The improvement is induced by prolonged exposure to the stress with no change in genetic endowment.

B. If acclimatization occurs early in life, it may be irreversible and is known as a developmental acclimatization.

III. Biological rhythms are internally driven by brain pacemakers.

A. The rhythms are entrained by environmental time cues, which also serve to phase-shift (reset) the rhythms when necessary.

B. In the absence of cues, rhythms free-run and may become desynchronized from each other.

IV. Aging is associated with a decrease in the number of cells in the body and a disordered functioning of many of the cells that remain.

A. It is a process distinct from the diseases associated with aging.

B. Its physiological manifestations are a deterioration in function and in capacity to respond homeostatically to environmental stresses.

The Balance Concept and Chemical Homeostasis

I. The balance of substances in the body is achieved by a matching of inputs and outputs.
II. Total-body balance of a substance may be negative, positive, or stable.

Components of Homeostatic Systems

I. The components of a reflex arc are: receptor, afferent pathway, integrating center, efferent pathway, and effector. The pathways may be neural or hormonal.
II. Local homeostatic responses are also stimulus-response sequences, but they occur only in the area of the stimulus, neither nerves nor hormones being involved.
III. Intercellular communication is essential to reflexes and local responses and is achieved by neurotransmitters, hormones, and paracrines. Less common is intercellular communication through either gap junctions or cell-bound messengers.
IV. The eicosanoids are a widespread family of messenger molecules derived from arachidonic acid. They function mainly as paracrines and autocrines in local responses.
  A. The first step in the production of the eicosanoids is the splitting off of arachidonic acid by the action of phospholipase A_2.
  B. There are two pathways from arachidonic acid, one mediated by cyclooxygenase and leading to formation of prostaglandins, prostacyclin, and thromboxanes, and the other mediated by lipoxygenase and leading to formation of leukotrienes.

Receptors

I. Receptors for chemical messengers are proteins located either inside the cell or, much more commonly, in the plasma membrane. The binding of a messenger by a receptor manifests specificity, saturation, and competition.
II. Receptors are subject to physiological regulation by their own messengers. This includes down-regulation and up-regulation.

Signal Transduction Mechanisms for Plasma-Membrane Receptors

I. Binding a chemical messenger activates a receptor, and this initiates one or more signal transduction pathways leading to the cell's response.
  A. The first step often involves receptor interaction with a G protein in the plasma membrane.
  B. Many of the pathways employ second messengers, substances that relay to the cell interior a chemical or electrical message that arrives at the cell's plasma membrane.
  C. Many of the pathways lead to the activation of protein kinases, enzymes that activate other proteins by phosphorylating them.

II. The receptor may open or close a membrane ion channel, resulting in an electric signal and changes in cytosolic calcium concentration.
III. The receptor may activate, via a G_s protein, or inhibit, via a G_i protein, the membrane enzyme adenylate cyclase, which catalyzes the conversion of cytosolic ATP to cyclic AMP.
  A. Cyclic AMP acts as a second messenger to activate intracellular cAMP-dependent protein kinase.
  B. This protein kinase phosphorylates proteins that mediate the cell's ultimate responses to the first messenger.
IV. Cyclic GMP, formed by the action of membrane guanylate cyclase, also functions as a second messenger through a protein kinase.
V. The calcium ion is one of the most widespread second messengers, and an activated receptor can increase cytosolic calcium concentration in several ways.
  A. The receptor may open a membrane calcium channel, which allows extracellular calcium to diffuse into the cell.
  B. The receptor may activate the membrane enzyme phospholipase C, which breaks down PIP_2 into two second messengers—IP_3 and DAG. IP_3 stimulates the release of calcium from the cell's endoplasmic reticulum.
  C. Calcium binds to one of several intracellular proteins, most often calmodulin. Calcium-activated calmodulin activates or inhibits many proteins, including calmodulin-dependent protein kinases.
VI. DAG activates protein kinase C, which phosphorylates many proteins.
VII. The receptor can itself act as a protein kinase termed a tyrosine kinase.
VIII. A single messenger may have multiple receptors, each triggering a different response.
IX. The various signal transduction mechanisms and second messengers often function together.

REVIEW QUESTIONS

1. Define:

homeostasis	entrainment
homeostatic control system	free-running rhythms
steady state	phase-shift
operating point	pacemakers
negative feedback	pool
positive feedback	negative balance
error signal	positive balance
feedforward	stable balance
adaptation	reflex
acclimatization	learned reflex
critical period	acquired reflex
developmental	reflex arc
acclimatization	stimulus
circadian rhythm	receptor (in reflex)

integrating center
afferent pathway
effector
efferent pathway
hormone
endocrine gland
local homeostatic responses
neurotransmitter
neurohormone
paracrine
autocrine
eicosanoids
prostaglandins
prostacyclin
thromboxanes
leukotrienes
arachidonic acid
phospholipase A_2
cyclooxygenase
lipoxygenase
receptors (for messengers)
specificity
saturation
competition
antagonists
agonists
down-regulation
up-regulation
receptor activation
signal transduction

G proteins
second messengers
first messengers
protein kinases
receptor-operated channel
G_s protein
adenylate cyclase
cyclic 3′,5′-adenosine
  monophosphate
cyclic AMP (cAMP)
phosphodiesterase
cAMP-dependent protein
  kinase
G_i protein
cyclic 3′,5′-guanosine
  monophosphate (cGMP)
guanylate cyclase
cGMP-dependent protein
  kinase
voltage-sensitive channels
phospholipase C
phosphatidyl inositol
  bisphosphate (PIP_2)
inositol trisphosphate (IP_3)
diacylglycerol (DAG)
calmodulin
calmodulin-dependent
  protein kinases
protein kinase C
tyrosine kinases

2. List four important generalizations about homeostatic control systems.

3. Contrast negative-feedback and positive-feedback systems.

4. Contrast feedforward and negative feedback.

5. How do error signals develop and why are they essential for maintaining homeostasis?

6. Describe the conditions under which acclimatization occurs. In what period of life might an acclimatization be irreversible? Are acclimatizations passed on to a person's offspring?

7. Under what conditions do circadian rhythms become free-running?

8. How do phase-shifts occur?

9. What are the important environmental time cues for entrainment of bodily rhythms?

10. What are the physiological manifestations of aging?

11. Draw a figure illustrating the balance concept in homeostasis.

12. What are the three possible states of total-body balance of any chemical?

13. List the components of a reflex arc.

14. What is the basic difference between a local homeostatic response and a reflex?

15. List the general categories of intercellular messengers.

16. Describe two types of intercellular communication that do not depend on chemical messengers.

17. Draw a figure illustrating the various pathways for eicosanoid synthesis.

18. What is the chemical nature of receptors? Where are they located?

19. Explain why different types of cells may respond differently to the same chemical messenger.

20. Describe how the metabolism of receptors can lead to down-regulation or up-regulation.

21. What is the first step in the action of a messenger on a cell?

22. Classify plasma-membrane receptors according to the signal transduction mechanisms they use.

23. What are the results of opening or closing membrane ion channels?

24. Draw a diagram describing the adenylate cyclase-cyclic AMP system.

25. What are the analogies between the cAMP and cGMP systems?

26. What are the two major mechanisms by which first messengers elicit an increase in cytosolic calcium concentration? What are the sources of the calcium in each mechanism?

27. Contrast receptor-operated and voltage-sensitive channels.

28. How does the calcium-calmodulin system function?

29. What are the two products released from PIP_2? Contrast their actions.

30. List the four types of molecules that activate protein kinases in signal transduction.

31. List five second messengers.

THOUGHT QUESTIONS

(Answers are given in Appendix A.)

1. A person's plasma potassium concentration (a homeostatically regulated variable) is 4 mM/L when she is eating 150 mmols of potassium per day. One day she doubles her potassium intake and continues to eat that amount indefinitely. At the new steady state, do you think her plasma potassium concentration is more likely to be 8, 4.4, or 4 mM/L? (The answer to this question requires no knowledge about potassium, only the ability to reason about homeostatic control systems.)

2. Eskimos have a remarkable ability to work in the cold without gloves and not suffer decreased skin blood flow. Does this prove that there is a genetic difference between Eskimos and other people with regard to this characteristic?

3. Patient A is given a drug that blocks the synthesis of all eicosanoids, whereas patient B is given a drug that blocks the synthesis of leukotrienes but none of the other eicosanoids. What are the enzymes most likely blocked by these drugs?

4. Certain nerves to the heart release the neurotransmitter norepinephrine. If these nerves are removed in experimental animals, the heart becomes extremely sensitive to the administration of a drug that is an agonist of norepinephrine. Explain why in terms of receptor physiology.

5. A particular hormone is known to elicit, completely by way of the cyclic AMP system, six different responses in its target cell. A drug is found that eliminates one of these responses but not the other five. Which of the following, if any, could the drug be blocking: the hormone's receptors, G_s protein, adenylate cyclase, or cyclic AMP?

6. If a drug that blocked all receptor-operated calcium channels were found, would this eliminate the role of calcium as a second messenger?

7. What are DAG and calcium-activated calmodulin analogous to in the cAMP system?

CHAPTER

8

NEURAL
CONTROL
MECHANISMS

The various parts of the nervous system are interconnected, but for convenience they can be divided into the **central nervous system** (**CNS**), composed of the brain and spinal cord, and the **peripheral nervous system**, consisting of the nerves extending from the brain and spinal cord out to all points of the body (Figure 8-1). For example, branches of the peripheral nervous system go from the base of the spine to the tips of the toes and, although they are not shown in Figure 8-1, from the base of the brain to the organs of the digestive system.

Along with the body's other communication system, the endocrine system, the nervous system regulates many internal functions and also coordinates the activities we know collectively as human behavior. These activities include not only such easily observed acts as

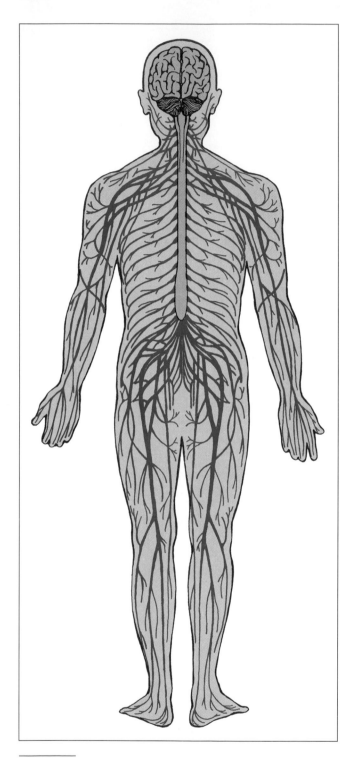

FIGURE 8-1 The central nervous system (violet) and the peripheral nervous system (dark blue). Some of the peripheral nerves connect with the brain (not shown), and others with the spinal cord.

smiling and walking but also experiences such as feeling angry, being motivated, having an idea, or remembering a long-past event. These phenomena that we attribute to the "mind" are believed to be related to the integrated activities of nerve cells.

In this chapter we are concerned with the components common to all neural mechanisms: the structure of individual nerve cells, the basic organization and major divisions of the nervous system, and the mechanisms underlying neural function.

SECTION A
STRUCTURE OF THE NERVOUS SYSTEM

FUNCTIONAL ANATOMY OF NEURONS

The basic unit of the nervous system is the individual nerve cell—the **neuron**. Nerve cells operate by generating electric signals and passing them from one part of the cell to another and by releasing chemical messengers to communicate with other cells.

Neurons occur in a variety of sizes and shapes; nevertheless, as shown in Figure 8-2, most of them consist of four parts: (1) the cell body, (2) the dendrites, (3) the axon, and (4) the axon terminals.

The **dendrites** form a series of highly branched outgrowths from the cell body. The dendrites and the cell body are the sites of most of the specialized junctions where signals are received from other neurons.

The **axon**, sometimes also called a **nerve fiber**, is a single long extension from the cell body. The portion of the axon closest to the cell body plus the part of the cell body where the axon is joined is known as the **initial segment**. It is at the initial segment that are initiated the electric signals that then propagate away from the cell body along the axon. The axon may have branches called collaterals along its course, and near their ends both the axon and the collaterals undergo further branching, each branch ending in an **axon terminal**. These terminals are responsible for transmitting chemical signals from the neuron to the cells contacted by the axon terminals. However, the axons of some neurons release their chemical messengers not from the axon terminals but from a series of bulging areas along the axon known as **varicosities**.

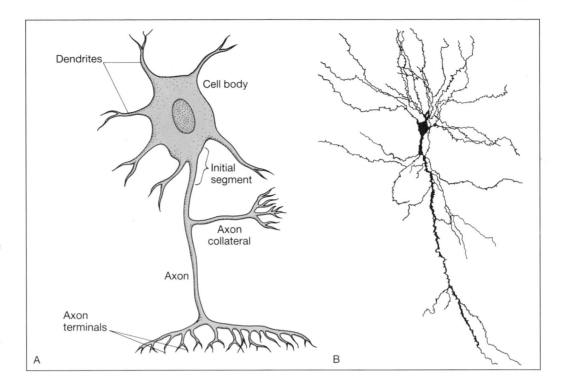

FIGURE 8-2 (A) Diagrammatic representation of a neuron. The proportions shown here are misleading because the axon may be 5000 to 10,000 times longer than the cell body is wide. Some neurons in the CNS may have no axons. (B) Drawing of a neuron as seen in a microscope.

The axons of some neurons are covered by **myelin** (Figure 8-3), a fatty, membranous sheath formed by nearby supporting cells that wrap their plasma membranes around the axon. The spaces between the myelin-forming cells where the axon's plasma membrane is exposed to extracellular fluid are **nodes of Ranvier**. Myelin speeds up passage of the electric signal along the axon.

As in other types of cells, a neuron's cell body contains the nucleus and ribosomes and thus has the genetic information and machinery necessary for protein synthesis. In order to maintain the structure and function of the cell processes, particularly when they are long, diverse organelles and materials must be moved to them from the cell body, where they are made. This is termed **axon transport**.

Axon transport of certain materials also occurs in the opposite direction, from the axon terminals to the cell body, and this permits growth factors and other chemical signals picked up at the terminals to affect the neuron's morphology, biochemistry, and connectivity. This is the route by which toxins, such as tetanus toxin, and herpes and polio viruses taken up by the nerve terminals enter the CNS.

Neurons can be divided into three functional classes: afferent neurons, efferent neurons, and interneurons.

Afferent neurons (Latin *ad*, to; and *ferre*, to carry) convey information from the tissues and organs of the body *into* the CNS, **efferent neurons** (Latin *ex*, out) transmit electric signals *from* the CNS out to effector cells (muscle or gland cells), and **interneurons** connect the afferent and efferent neurons (Figure 8-4, Table 8-1).

At the end of the cell farthest from the CNS, afferent neurons have receptors that, in response to various physical or chemical changes in their environment, cause electric signals to be generated in the neuron. The receptor may be a specialized ending of the neuron or a separate cell closely associated with it. The afferent neurons propagate these electric signals from the receptors into the brain or spinal cord. (Recall from Chapter 7 that the term "receptor" has two totally distinct meanings, the one as defined here and the other referring to the specific proteins with which a chemical messenger combines to exert its effects on a target cell; both types of receptors will be referred to frequently in this chapter. The meaning will be clear from the context.)

Afferent neurons are atypical in that they have no dendrites and only a single process, considered to be an axon. Shortly after leaving the cell body, the axon divides, one branch, the peripheral process, ending at the receptors and the other branch, the central process, entering the CNS to form junctions with other neurons.

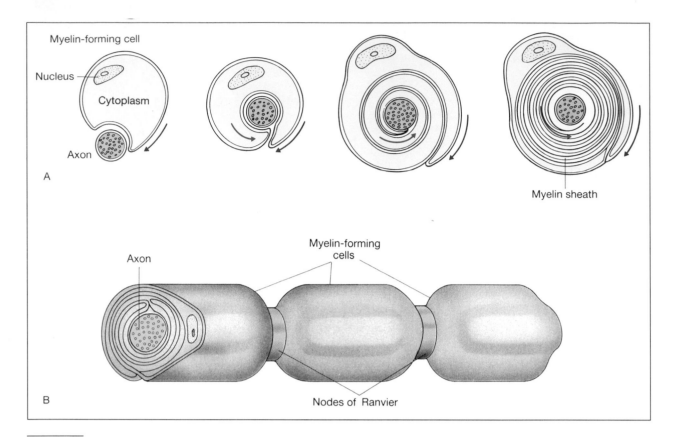

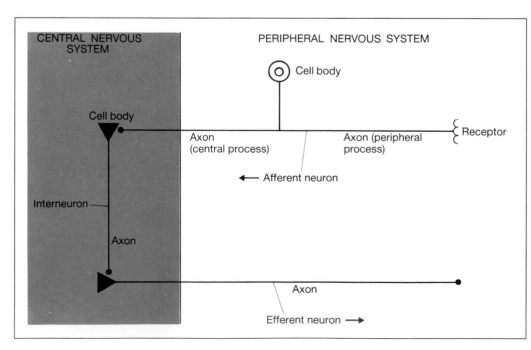

FIGURE 8-3 (A) Cross section of an axon in successive stages of myelinization. The myelin-forming cell may migrate around the axon, trailing successive layers of its plasma membrane (red arrows), or it may add to its tip, which lies against the axon, so that the tip is pushed around the axon, burrowing under the layers of myelin that are already formed (green arrows). The latter process must be used in the CNS where each myelin-forming cell may send branches to as many as 40 axons. (B) The myelin-forming cells are separated by a small space, the node of Ranvier.

FIGURE 8-4 Three classes of neurons. Note that the interneurons form the connections between afferent and efferent neurons and lie entirely within the CNS. The dendrites, which function as an extension of the neuron cell body, are not shown. The arrows indicate the direction of transmission of neural activity. The stylized neurons in this figure show the conventions that we will use throughout this book for the different parts of neurons.

TABLE 8-1 THREE CLASSES OF NEURONS

I. Afferent neurons
  A. Transmit information from receptors at their peripheral endings into CNS
  B. Most of cell (cell body and long peripheral process of axon) outside CNS; only short central process of axon enters CNS
  C. Have no dendrites
II. Efferent neurons
  A. Transmit information out of CNS to effector cells (muscles or glands)
  B. Cell body, dendrites, and a small segment of the axon in CNS; most of the axon outside CNS
III. Interneurons
  A. Entirely within CNS
  B. Account for 99 percent of all neurons

Note in Figure 8-4 that the cell body and the long peripheral process of the axon are outside the CNS and only part of the relatively short central process enters the brain or spinal cord.

The cell bodies and dendrites of efferent neurons are within the CNS, but the axons extend out into the periphery. The axons of the afferent and efferent neurons, except for the small part in the brain or spinal cord, form the nerves of the peripheral nervous system.

Interneurons lie entirely within the CNS. They account for about 99 percent of all neurons and have a wide range of physiological properties, shapes, chemistries, and functions. As a rough estimate, for each afferent neuron entering the CNS, there are about 10 efferent neurons and about 200,000 interneurons. The number of interneurons interposed between certain afferent and efferent neurons varies according to the complexity of the action. The reflex elicited by tapping below the kneecap has no interneurons, for here the afferent neurons end directly on the efferent neurons. In contrast, stimuli invoking memory or language may involve millions of interneurons.

The anatomically specialized junction between two neurons where one neuron alters the activity of another is called a **synapse**. At most synapses, a signal is transmitted from one neuron to another by chemical messengers known as **neurotransmitters**, a term that also includes the chemicals by which efferent neurons communicate with effector cells. The neurotransmitter released from one neuron alters the receiving neuron by binding with a specific membrane receptor on the receiving neuron. (The reader is cautioned one last time not to confuse this use of the term "receptor" with the receptors mentioned above that are on or associated with afferent neurons.)

Synapses generally occur between the axon terminal of one neuron and the cell body or dendrite of a second. In certain areas, however, synapses also occur between two dendrites, between a dendrite and a cell body, or between an axon terminal and a second axon terminal. A neuron conducting signals toward a synapse is called a **presynaptic neuron**, whereas neurons conducting signals away from a synapse are **postsynaptic neurons**. Figure 8-5 shows how, in a multineuronal pathway, a single neuron can be postsynaptic to one group of cells and presynaptic to another.

A postsynaptic neuron may have thousands of synaptic junctions on the surface of its dendrites and cell body so that signals from many presynaptic neurons can affect it. Certain neurons in the brain receive more than 100,000 synaptic inputs.

Only about 10 percent of the cells in the CNS are neurons; the remainder are **glial cells** (**neuroglia**). Glial cells branch but not as extensively as the neurons do,

FIGURE 8-5 A neuron postsynaptic to one group of cells can be presynaptic to another.

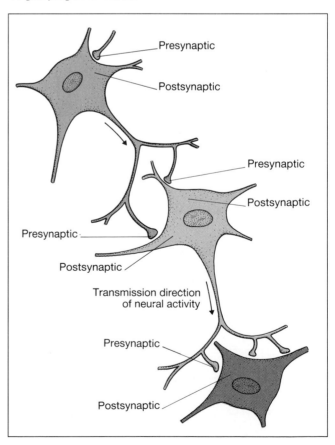

Presynaptic

Postsynaptic

Presynaptic

Postsynaptic

Presynaptic

Postsynaptic

Transmission direction of neural activity

Presynaptic

Postsynaptic

and the neurons occupy about 50 percent of the volume of the CNS. One type of glial cell, the oligodendroglia, forms the myelin covering of axons in the CNS.

A second type of glial cell, the astroglia, has multiple functions. Astroglia are thought to help regulate the composition of the extracellular fluid in the CNS because they remove potassium ions and neurotransmitters from the extracellular fluid around synapses. They also seem to help sustain the neurons metabolically—for example, by providing glucose and removing ammonia. In the embryo, astroglia guide neurons as they migrate during neural development, and they stimulate the neurons' growth and the formation of their axons and dendrites by secreting growth factors.

DIVISIONS OF THE NERVOUS SYSTEM

We shall now survey the anatomy and broad functions of the major structures of the nervous system; future chapters will describe these functions in more detail. First we must deal with some potentially confusing terminology. Recall that a long extension from a *single* neuron is called a nerve fiber. The term **nerve** refers to a group of *many* nerve fibers that are traveling together to the same general location in the peripheral nervous system. There are no nerves in the CNS. A group of nerve fibers traveling together in the CNS is called a **pathway**, **tract**, or, when it links the right and left halves of the CNS, a commissure.

The cell bodies of neurons having similar functions also are often clustered together. Groups of neuron cell bodies in the peripheral nervous system are called **ganglia** (sing. ganglion), but in the CNS they are usually called **nuclei** (singular *nucleus*).

The brain lies within the skull, and the spinal cord within the vertebral column. Between the soft neural tissues and the protective bones that house them are three protective membranous coverings, or **meninges**: the dura mater next to the bone, the arachnoid in the middle, and the pia mater next to the nervous tissue. A space, the subarachnoid space, between the arachnoid and pia is filled with **cerebrospinal fluid**.

Central Nervous System: Spinal Cord

The spinal cord (Figure 8-6) is a slender cylinder of soft tissue about as big around as the little finger. The central butterfly-shaped area of **gray matter** is composed of interneurons, the cell bodies and dendrites of efferent neurons, the entering fibers of afferent neurons, and glial cells. It is called gray matter because the structures there lack whitish myelin.

The gray matter is surrounded by **white matter**, which consists largely of the myelinated axons of interneurons, the fatty myelin giving the region a white appearance. These groups of axons in the CNS, that is, the

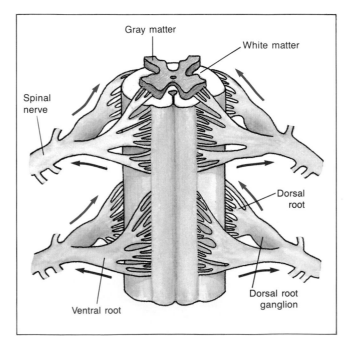

Gray matter

White matter

Spinal nerve

Dorsal root

Dorsal root ganglion

Ventral root

FIGURE 8-6 Section of the spinal cord, ventral view. The arrows indicate the direction of transmission of neural activity. (*Redrawn from Curtis et al.*)

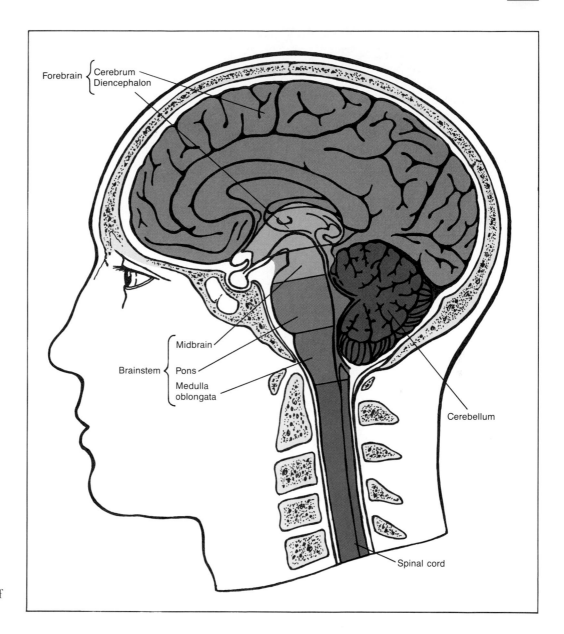

Forebrain { Cerebrum
Diencephalon

Midbrain
Brainstem { Pons
Medulla oblongata

Cerebellum

Spinal cord

FIGURE 8-7 The spinal cord and six divisions of the brain.

pathways, run longitudinally through the cord, some descending to convey information from the brain to the spinal cord or from upper levels of the spinal cord to lower levels, others ascending to transmit information to the brain.

Groups of afferent fibers that enter the spinal cord from the peripheral nerves enter on the dorsal side of the cord via the **dorsal roots**. Small bumps on the dorsal roots, the **dorsal root ganglia**, contain the cell bodies of the afferent neurons. The axons of efferent neurons leave the spinal cord on the ventral side via the **ventral roots**. A short distance from the cord, the dorsal and ventral

roots from the same level combine to form a pair of **spinal nerves**, one on each side of the spinal cord.

Central Nervous System: Brain

The brain is composed of six subdivisions: **cerebrum**, **diencephalon**, **midbrain**, **pons**, **medulla oblongata**, and **cerebellum** (Figure 8-7). The cerebrum and diencephalon together constitute the **forebrain**; the midbrain, pons, and medulla together form the **brainstem**.

The brain also contains four interconnected cavities, the **cerebral ventricles**, that contain cerebrospinal fluid (Figure 8-8).

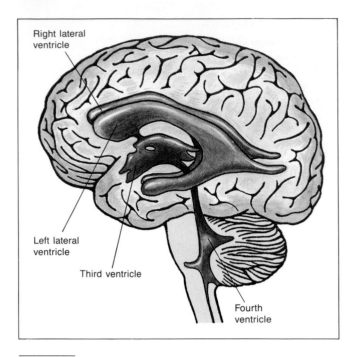

FIGURE 8-8 The four interconnected ventricles of the brain.

mental capacity. These ascending reticular pathways affect such things as wakefulness and the direction of attention to specific events.

The reticular formation also gives rise to connections with the cerebellum and the spinal cord. The fibers that descend to the spinal cord form the reticulospinal pathways. They influence activity in both the efferent and afferent neurons. There is considerable interaction between the ascending, descending, and cerebellar reticular pathways. For example, all three components function in controlling muscle activity.

Some reticular-formation neurons are clustered together, forming certain of the brainstem nuclei and integrating centers. These include the cardiovascular, respiratory, swallowing, and vomiting centers, all of which are discussed in subsequent chapters. The reticular formation also has nuclei important in eye movement control and the reflex orientation of the body in space.

The brainstem also contains nuclei involved in processing information for 10 of the 12 pairs of **cranial**

Brainstem. The brainstem is literally the stalk of the brain. Through it pass all the nerve fibers that relay signals between the spinal cord and the cerebrum or cerebellum. Information is transmitted between the brainstem and overlying cerebellum by three large bundles of nerve fibers, the **cerebellar peduncles**.

Running through the core of the brainstem and consisting largely of neuron cell bodies, that is, gray matter, is the **reticular formation**, which is the one part of the brain absolutely essential for life. It is composed of large numbers of loosely arranged interneurons that receive and integrate information from all regions of the CNS. The reticular formation is also responsible for the *output* of a great deal of neural information. Most reticular-formation neurons send axons for considerable distances up or down the brainstem and beyond, to most regions of the brain and spinal cord, which indicates the very large scope of influence the reticular formation has over other parts of the CNS.

The pathways that convey information from the reticular formation to the upper portions of the brain divide near the junction of the brainstem and forebrain; one branch goes to the thalamus, a large sensory way station in the forebrain, and a second branch goes to the base of the forebrain. Some of the fibers from these two branches continue to the cerebral cortex, which is thought to be that part of the brain having the greatest

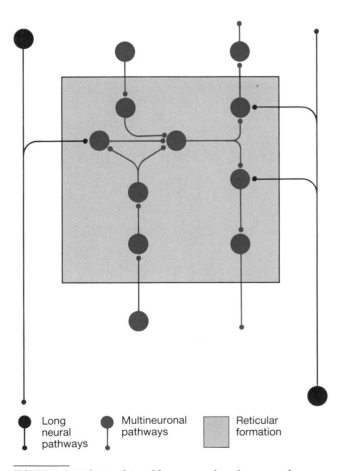

| ● | Long neural pathways | ● | Multineuronal pathways | ▨ | Reticular formation |

FIGURE 8-9 Relationship of long neural pathways and multineuronal (multisynaptic) pathways to the reticular formation.

nerves, the peripheral nerves that connect with the brain and innervate the muscles and glands of the head and many organs in the thoracic and abdominal cavities. The cranial nerves also innervate sensory receptors in these same areas.

Information can pass through the brainstem in two ways: (1) **long neural pathways**, which carry information directly between the brain and spinal cord or between the forebrain and brainstem, and (2) the **multineuronal**, or **multisynaptic**, **pathways** (Figure 8-9). As their name suggests, the multineuronal pathways are made up of many neurons and many synaptic connections. The long pathways, on the other hand, consist of few interconnected neurons and, therefore, contain few synapses, so there is little alteration in the information they transmit, but there are many opportunities for further neural processing in the multineuronal pathways.

Cerebellum. The cerebellum consists of an outer layer of cells, the cerebellar cortex, and several deeper cell clusters, the cerebellar nuclei. It is connected to the underlying brainstem by the cerebellar peduncles. Chiefly involved with skeletal muscle functions, the cerebellum is an important center for coordinating and learning movements and for controlling posture and balance. In order to carry out these functions, the cerebellum receives information from the muscles and joints, skin, eyes and ears, and even viscera, plus information from the parts of the brain involved in control of movement.

Forebrain. The forebrain consists of the central-core diencephalon and the right and left **cerebral hemispheres** that together make up the cerebrum.

The cerebral hemispheres (Figure 8-10) consist of an outer shell, the cerebral cortex, underlying subcortical nuclei, and the many nerve fibers that bring information into the cerebrum, carry information out, and interconnect the neurons in the cerebrum. The hemispheres, although largely separated by a longitudinal division, are

FIGURE 8-10 (A) Coronal (side-to-side) section of the brain, with the diencephalon shaded in violet. (B) The dashed line indicates the location of the cross section shown in A. (*Adapted from Nieuwenhuys et al.*)

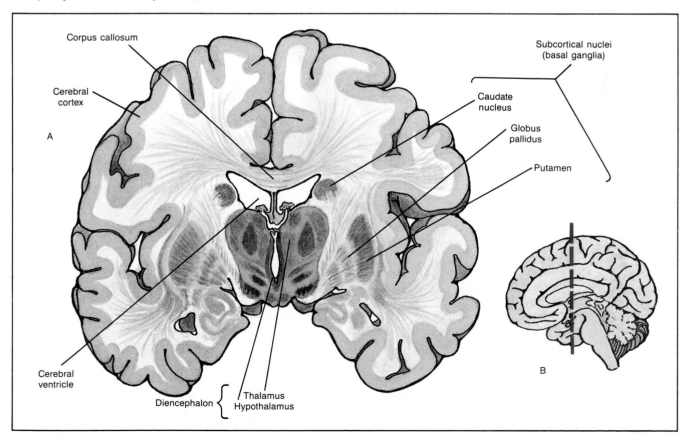

connected to each other by bundles of nerve fibers known as commissures, the **corpus callosum** being the largest (Figure 8-10). Areas within a hemisphere are connected to each other by association fibers, not shown in the figure.

The cortex of each hemisphere is divided into four **lobes**: the **frontal**, **parietal**, **occipital**, and **temporal** (Figure 8-11). The **cerebral cortex**, an area of gray matter, is about 3 mm thick, and its cells are arranged most often in six layers, a distinguishing trait of the cortex. The cortical neurons are of two basic types: pyramidal cells (named for the shape of their cell bodies) and nonpyramidal cells. The pyramidal cells form the major output cells of the cortex, sending their axons to other parts of the cerebral cortex and to other parts of the CNS. The cortex is highly folded, which increases the area available for cortical neurons without increasing appreciably the volume of the brain.

The cerebral cortex is the most complex integrating area of the nervous system and is necessary for the bringing together of basic afferent information into

TABLE 8-2 FUNCTIONS OF THE HYPOTHALAMUS
1. Regulation of anterior pituitary gland (Chapter 10)
2. Regulation of water balance (Chapter 15)
3. Regulation of autonomic nervous system (Chapters 8 and 17)
4. Regulation of eating and drinking behavior (Chapter 17)
5. Regulation of reproductive system (Chapters 10 and 18)
6. Reinforcement of certain behaviors (Chapter 20)
7. Generation and regulation of circadian rhythms (Chapters 7, 9, 10, and 17)

meaningful perceptual images and for the ultimate refinement of control over the **motor systems**, which control the movement of the skeletal muscles. Nerve fibers enter a particular site in the cortex from a variety of places but particularly from the thalamus, from other regions of the cortex, and from the brainstem reticular formation. Some of the input fibers convey information about specific events in the environment, whereas others are more concerned with controlling levels of cortical excitability, determining states of arousal, and directing attention to specific stimuli.

The **subcortical nuclei** form other areas of gray matter that lie deep within the cerebral hemispheres. Predominant among them are the **basal ganglia**, which play an important role in the control of movement and posture and in more complex aspects of behavior. In other parts of the forebrain, axons predominate, their whitish myelin coating distinguishing them as white matter.

The diencephalon, the second component of the forebrain, contains two major parts: the thalamus and the hypothalamus (Figure 8-10). The **thalamus** is a large cluster of nuclei that serves as a synaptic relay station and important integrating center for most inputs to the cortex.

The **hypothalamus** (Figure 8-10), which lies below the thalamus, is a tiny region whose volume is only 5 to 6 cm^3. Yet it is crucial to homeostatic regulation and is a principal site for regulating the behavior essential to the survival of the individual and the species. The integration achieved by the hypothalamus often involves correlation of neural and endocrine functions. Indeed, the hypothalamus appears to be the single most important control area for the regulation of the internal environment. It lies directly above the pituitary gland, an important endocrine structure, to which it is attached by a stalk. The functions of the hypothalamus are listed in Table 8-2.

FIGURE 8-11 A lateral view of the brain. The outer layer of the forebrain (the cortex) is divided into four lobes, as shown. (*Redrawn from Nieuwenhuys et al.*)

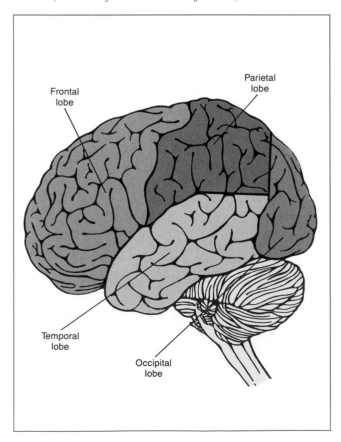

Frontal lobe

Parietal lobe

Temporal lobe

Occipital lobe

An area of the brain that includes both gray and white matter is the **limbic system**, an interconnected group of brain structures, including portions of frontal-lobe cortex, temporal lobe, thalamus, and hypothalamus, as well as the circuitous fiber pathways that connect them (Figure 8-12). The limbic system is associated with learning and emotional behavior. Besides being connected with each other, the parts of the limbic system are connected with many other parts of the CNS.

Peripheral Nervous System

Nerve fibers in the peripheral nervous system transmit signals between the CNS and all other parts of the body. As noted earlier, the nerve fibers are grouped into bundles called nerves. The peripheral nervous system consists of 43 pairs of nerves: 12 pairs of cranial nerves, and 31 pairs that connect with the spinal cord as the spinal nerves. The cranial nerves are listed in Table 8-3.

Each nerve fiber is surrounded by a **Schwann cell**, which is analogous to an oligodendroglial cell of the CNS (page 172). Some of the fibers are wrapped in layers of Schwann-cell membrane, and these tightly wrapped membranes form the myelin sheath (Figure 8-3). Other fibers are unmyelinated.

A nerve may contain nerve fibers that are the axons of efferent neurons or afferent neurons. Accordingly fibers may be classified as belonging to the **efferent** or the **afferent division** of the peripheral nervous system (Table 8-4). All the spinal nerves contain both afferent and efferent fibers, whereas some of the cranial nerves (the optic nerves from the eyes, for example) contain only afferent fibers.

Afferent division. Afferent neurons convey information from receptors at their peripheral endings to the CNS. The long part of their axon is outside the central nervous system and is part of the peripheral nervous system. Afferent neurons are sometimes called **primary afferents** or **first-order neurons** because they are the first cells entering the CNS in the synaptically linked chains of neurons that handle incoming information.

Efferent division. Efferent neurons carry signals from the CNS out to muscles or glands. The efferent division

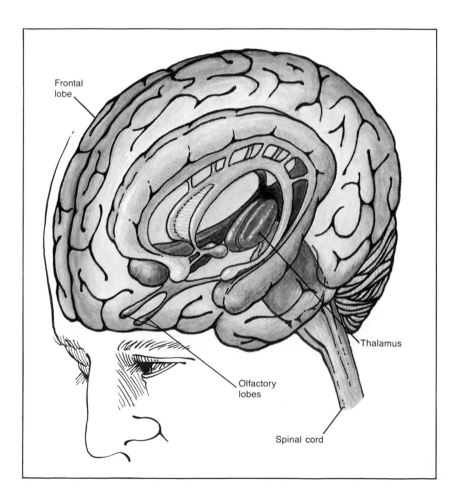

FIGURE 8-12 Structures of the limbic system are shown colored in this partially transparent view of the brain. (*Redrawn from Bloom, Lazerson, and Hofstadter.*)

TABLE 8-3 THE CRANIAL NERVES

Name	Fibers	Comments
I. Olfactory nerve	Afferent	Tract of brain, not true nerve. Carries input from receptors in olfactory (smell) epithelium.
II. Optic nerve	Afferent	Tract of brain, not true nerve. Carries input from receptors in eye.
III. Oculomotor nerve	Efferent	Innervates skeletal muscles that move eyeball up, down, and medially and raise upper eyelid; innervates smooth muscles that constrict pupil and alter lens shape for near and far vision.
	Afferent	Transmits information from receptors in muscles.
IV. Trochlear nerve	Efferent	Innervates skeletal muscles that move eyeball downward and laterally.
	Afferent	Transmits information from proprioceptors in muscles.
V. Trigeminal nerve	Efferent	Innervates skeletal chewing muscles.
	Afferent	Transmits information from receptors in skin and skeletal muscles of face, nose, and mouth and from teeth sockets.
VI. Abducens nerve	Efferent	Innervates skeletal muscles that move eyeball laterally.
	Afferent	Transmits information from proprioceptors in muscles.
VII. Facial nerve	Efferent	Innervates skeletal muscles of facial expression and swallowing; innervates nose, palate, lacrimal, and salivary glands.
	Afferent	Transmits information from taste buds in front of tongue and mouth.
VIII. Vestibulocochlear nerve	Afferent	Transmits information from receptors in ear.
IX. Glossopharyngeal nerve	Efferent	Innervates skeletal muscles involved in swallowing and parotid salivary gland.
	Afferent	Transmits information from taste buds at back of the tongue and receptors in auditory tube skin.
X. Vagus nerve	Efferent	Innervates skeletal muscles of pharynx and larynx and smooth muscle and glands of thorax and abdomen.
	Afferent	Transmits information from receptors in thorax and abdomen.
XI. Accessory nerve	Efferent	Innervates neck skeletal muscles.
XII. Hypoglossal nerve	Efferent	Innervates tongue skeletal muscles.

of the peripheral nervous system is more complicated than the afferent, being subdivided into a **somatic nervous system** and an **autonomic nervous system**. These terms are somewhat misleading because they suggest additional nervous systems distinct from the central and peripheral systems. Keep in mind that the terms refer simply to the efferent division of the peripheral nervous system.

The simplest distinction between the somatic and autonomic systems is that the neurons of the somatic division innervate skeletal muscle whereas the autonomic neurons innervate smooth and cardiac muscle, glands, and the neurons that form the enteric nervous system. This is a specialized nerve network in the wall of the gastrointestinal (GI) tract that regulates the glands and smooth muscles found there (Chapter 16). Other differences are listed in Table 8-5.

Somatic nervous system.   The somatic portion of the peripheral nervous system is made up of all the nerve fibers going from the CNS to skeletal muscle cells. The cell bodies of these neurons are located in groups in the brainstem or spinal cord. Their large-diameter, myelinated axons leave the CNS and pass without any synapses

TABLE 8-5 DIFFERENCES BETWEEN SOMATIC AND AUTOMATIC NERVOUS SYSTEMS

Somatic

1. Consists of single neuron between CNS and effector organ
2. Innervates skeletal muscle
3. Always leads to muscle excitation

Autonomic

1. Has two-neuron chain (connected by synapse) between CNS and effector organ
2. Innervates smooth and cardiac muscle, glands, and GI neurons
3. Can lead to excitation or inhibition of effector cells

to skeletal muscle cells. The neurotransmitter released by these neurons is **acetylcholine**. Because activity in the somatic neurons leads to contraction of the innervated skeletal muscle cells, these neurons are often called **motor neurons**. Excitation of motor neurons leads only to the *contraction* of skeletal muscle cells; there are no somatic neurons that inhibit the muscles.

Autonomic nervous system.   The innervation of all tissues other than skeletal muscle is by way of the autonomic nervous system. In the case of the intestinal tract, the autonomic nervous system innervates the neurons that are part of the enteric nervous system. The enteric nervous system actually fits the characteristics of neither the autonomic nor the somatic nervous system.

In the autonomic nervous system, the groups of axons between the CNS and the effector cells consist of two neurons and one synapse (Figure 8-13). (This is in contrast to the single neuron of the somatic system.) The first neuron has its cell body in the CNS. The synapse between the two neurons is outside the CNS, in a cell cluster called an **autonomic ganglion**. The nerve fibers passing between the CNS and the ganglia are called **preganglionic** autonomic fibers; those passing between the ganglia and the effector cells are the **postganglionic** fibers.

Anatomical and physiological differences within the autonomic nervous system are the basis for its further subdivision into **sympathetic** and **parasympathetic** components. The nerve fibers of these components leave the CNS at different levels—the sympathetic fibers from the thoracic and lumbar regions of the spinal cord and the parasympathetic fibers from the brain and the sacral portion of the spinal cord (Figure 8-14). The sympathetic division is also called the thoracolumbar division, and the parasympathetic, the craniosacral division.

The two divisions of the autonomic nervous system also differ in the location of ganglia. Most of the sympathetic ganglia lie close to the spinal cord and form the two chains of ganglia—one on each side of the cord—known as the **sympathetic trunks** (Figure 8-14). Other ganglia, called collateral ganglia—the celiac, superior mesenteric, and inferior mesenteric ganglia—lie far away from the spinal cord, closer to the innervated organs (Figure 8-14). In contrast, the parasympathetic ganglia lie within the effector organ.

The anatomy of the sympathetic nervous system can be confusing, partly because preganglionic sympathetic fibers leave the spinal cord only between the first thoracic and third lumbar segments whereas the sympathetic trunks extend the entire length of the cord—from the cervical levels high in the neck down to the sacral levels. The "extra" ganglia in the sympathetic trunks receive preganglionic fibers from the thoracolumbar re-

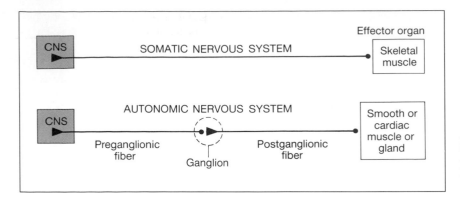

FIGURE 8-13 Efferent division of the peripheral nervous system. Overall plan of the somatic and autonomic nervous systems.

gions because some of the preganglionic fibers, once in the paravertebral chains, turn to travel upward or downward for several segments before forming synapses with postganglionic neurons (Figure 8-15, Parts 1 and 4). The ganglia at the cervical levels are the superior cervical, middle cervical, and stellate ganglia (Figure 8-14). Other possible paths taken by the sympathetic fibers are shown in Figure 8-15, Parts 2, 3, and 5.

The ganglia can act either as automatic relay stations with practically no change in the information they transmit to the effector organ or as important integrating centers capable of generating individualized responses. The anatomical arrangements in the sympathetic nervous system to some extent tie the entire system together so it can act as a single unit, although small segments of the system can still be regulated independently. The parasympathetic system, in contrast, is made up of relatively independent components. Thus, overall autonomic responses, made up of many small parts, are quite variable, and finely tailored to the specific demands of any given situation.

In both sympathetic and parasympathetic divisions, the major neurotransmitter released between pre- and postganglionic fibers is acetylcholine (Figure 8-16). In the parasympathetic division, the major neurotransmitter between the postganglionic fiber and the effector cell is also acetylcholine. In the sympathetic division, the major transmitter between the postganglionic fiber and the effector cell is usually **norepinephrine**.[1]

Because historically it was believed that epinephrine, rather than norepinephrine, was the major sympathetic postganglionic neurotransmitter and because epineph-

rine was called by its British name "adrenaline," nerve fibers that release norepinephrine came to be called **adrenergic** fibers. Fibers that release acetylcholine are called **cholinergic** fibers.

Recall that the somatic efferent nerve fibers are also cholinergic. Thus, the neurons in the entire efferent nervous system are either cholinergic or adrenergic (Figure 8-16).

Many of the drugs that stimulate or inhibit various components of the autonomic nervous system affect receptors for acetylcholine and norepinephrine. Indeed, experiments with these drugs have revealed that there are several types of receptors for each neurotransmitter (Table 8-6). Thus, acetylcholine receptors on all autonomic postganglionic neurons respond to low doses of the drug nicotine and are therefore called **nicotinic receptors**. In contrast, the acetylcholine receptors on the

---

[1] We stress "major" and "usually" because acetylcholine is released by some sympathetic postganglionic endings. Moreover, one or more substances known as cotransmitters are usually stored and released with the autonomic neurotransmitters; these include ATP, dopamine, and several of the neuropeptides. These substances, however, play a smaller role in sympathetic nervous system activity than norepinephrine does.

TABLE 8-6 CLASSES OF CHOLINERGIC AND ADRENERGIC RECEPTORS

I. Cholinergic receptors (receptors for acetylcholine)
  A. Nicotinic receptors: N_1 and N_2
      On postganglionic cell bodies in the autonomic ganglia
      At neuromuscular junctions of skeletal muscle
      On some neurons of CNS
  B. Muscarinic receptors
      On smooth muscle
      On cardiac muscle
      On gland cells
      On some neurons of CNS

II. Adrenergic receptors (receptors for norepinephrine and epinephrine): α_1, α_2 and β_1, β_2
      As we shall see, adrenergic receptors are found not only on smooth muscle, cardiac muscle, and glands but also in the CNS.

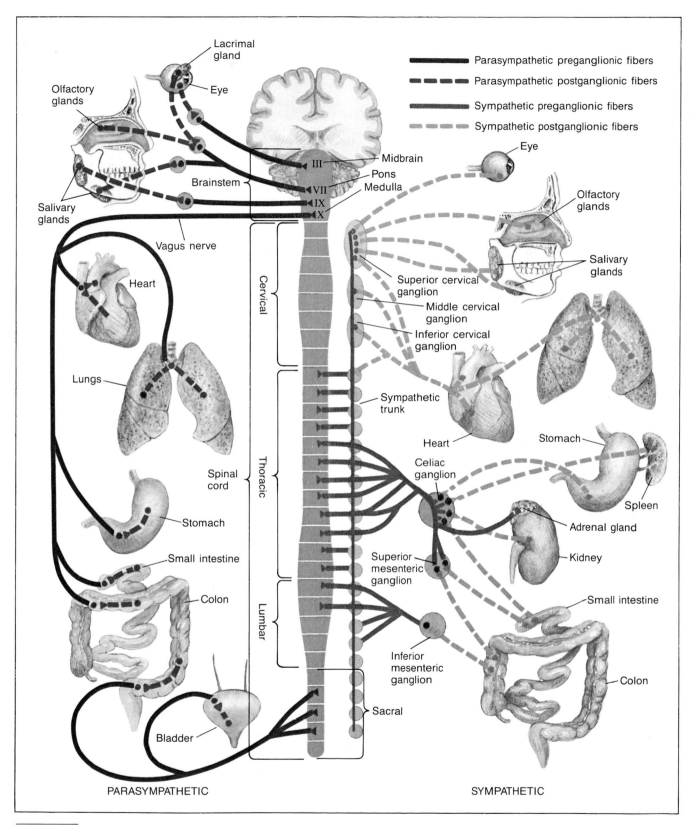

FIGURE 8-14 The parasympathetic (left) and sympathetic (right) divisions of the autonomic nervous system. The celiac, superior mesenteric, and inferior mesenteric ganglia are collateral ganglia. Only the sympathetic trunk on the right side of the spinal cord is shown, although another exists on the left. Not shown are the fibers passing to the liver, blood vessels, and skin glands.

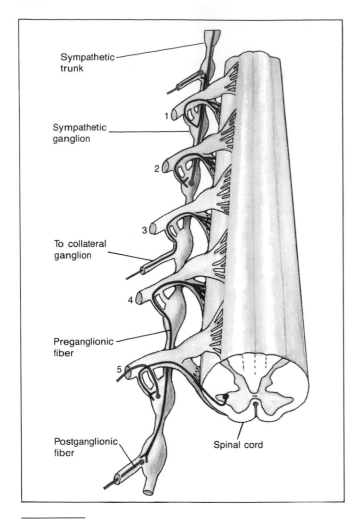

smooth-muscle, cardiac-muscle, and gland cells are not stimulated by nicotine but are stimulated by the mushroom poison muscarine; they are called **muscarinic receptors**. To complete the story of the peripheral cholinergic receptors, it should be emphasized that the cholinergic receptors on the neuromuscular junctions of skeletal muscle fibers, innervated by the *somatic* motor neurons, not autonomic neurons, are also nicotinic.

Similarly, there are two major classes of adrenergic receptors, also distinguished largely by the specific drugs that stimulate or block them: **alpha-adrenergic receptors** and **beta-adrenergic receptors**. One basic difference between the two is that activation of beta-adrenergic receptors results in the generation of a second messenger (cAMP, in this case), whereas activation of alpha-adrenergic receptors results in the opening of specific ion channels in the plasma membrane.

Both nicotinic and alpha- and beta-adrenergic receptors can be subdivided still further, again according to the drugs that influence them (Table 8-6).

One set of postganglionic neurons in the sympathetic division never develops axonal fibers. Instead, upon activation by preganglionic axons, the cells of this "ganglion" release their transmitters into the blood stream (Figure 8-16). This "ganglion," called the **adrenal medulla**, is therefore not really a ganglion at all; it is an endocrine gland whose secretion is controlled by the sympathetic preganglionic nerve fibers. It releases a mixture of about 80 percent epinephrine and 20 percent norepinephrine into the blood (plus small amounts of other substances, such as dopamine, ATP, and some of

FIGURE 8-15 Relationship between a sympathetic trunk and spinal cord. (1-5) Various courses that preganglionic sympathetic fibers (red lines) may take through the sympathetic trunk. Blue lines represent postganglionic fibers. A mirror image of this exists on the opposite side of the spinal cord.

FIGURE 8-16 Transmitters used in the various components of the peripheral nervous system. In a few cases (to be described later), sympathetic neurons release a transmitter other than norepinephrine. ACh, acetylcholine; NE, norepinephrine; Epi, epinephrine; DA, dopamine.

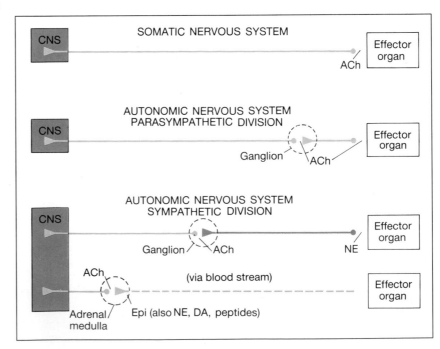

TABLE 8-7 SOME EFFECTS OF AUTONOMIC NERVOUS SYSTEM ACTIVITY*

Effector organ	Sympathetic Effect	Parasympathetic Effect
Eyes		
Iris muscle	Contracts radial muscle (widens pupil)	Contracts sphincter muscle (makes pupil smaller)
Ciliary muscle	Relaxes (flattens lens for far vision)	Contracts (allows lens to become more convex for near vision)
Heart		
S-A node	Increases heart rate	Decreases heart rate
Atria	Increases contractility	Decreases contractility
A-V node	Increases conduction velocity	Decreases conduction velocity
Ventricles	Increases contractility	Decreases contractility slightly
Arterioles		
Coronary	Constricts or dilates	Dilates
Skin	Constricts	—†
Skeletal muscle	Constricts or dilates	—
Abdominal viscera and kidneys	Constricts	—
Salivary glands	Constricts	Dilates
Penis or clitoris	Constricts	Dilates (causes erection)
Veins	Constricts or dilates	
Lungs		
Bronchial muscle	Relaxes	Contracts
Bronchial glands	Inhibits secretion	Stimulates secretion
Salivary glands	Stimulates secretion	Stimulates secretion
Stomach		
Motility, tone	Decreases	Increases
Sphincters	Contracts	Relaxes
Secretion	Inhibits	Stimulates
Gallbladder	Relaxes	Contracts
Liver	Glycogenolysis, gluconeogenesis	Glycogen synthesis
Pancreas		
Exocrine glands	Inhibits secretion	Stimulates secretion
Endocrine glands	Inhibits insulin secretion, stimulates glucagon secretion	—
Fat cells	Increases fat breakdown	—
Urinary bladder		
Bladder wall	Relaxes	Contracts
Sphincter	Contracts	Relaxes
Uterus	Pregnant: contracts; nonpregnant: relaxes	Variable
Reproductive tract (male)	Ejaculation	Erection
Skin		
Muscles causing hairs to stand erect	Contracts	—
Sweat glands	Localized secretion	Generalized secretion
Lacrimal glands	—	Secretion

*Adapted from Goodman and Gilman's, "The Pharmacological Basis of Therapeutics," Alfred Goodman Gilman, Louis S. Goodman, and Alfred Gilman, eds., 7th ed., Macmillan, New York, 1985.

† Dash means these cells are not innervated by this branch of the autonomic nervous system.

the neuropeptides). These substances, properly called hormones rather than neurotransmitters in these circumstances, are transported via the blood to receptor sites on effector cells sensitive to them. The receptor sites may be the same adrenergic receptors that are located near the release sites of sympathetic postganglionic neurons and that are normally activated by the norepinephrine released from these sites, or the receptor sites may be located at places that are far from the release sites and therefore activated only by the circulating epinephrine or norepinephrine.

Table 8-7 is a reference list of autonomic nervous system effects, which will be described in subsequent chapters. Note that the heart and many glands and smooth muscles are innervated by both sympathetic and parasympathetic fibers, that is, they receive **dual innervation**. Whatever effect one division has on the effector cells, the other division frequently (but not always) has just the opposite effect (Table 8-7).

Dual innervation by nerve fibers that cause opposite responses provides a very fine degree of control over the effector organ—it is like equipping a car with both an accelerator and a brake. One can slow the car simply by decreasing the pressure on the accelerator; with both accelerator and brake, however, the combined effects of releasing the accelerator and applying the brake provide faster and more accurate control. Analogously, the sympathetic and parasympathetic divisions are usually activated reciprocally; that is, as the activity of one division is increased, the activity of the other is decreased.

Autonomic responses usually occur without conscious control or awareness, as though they were indeed autonomous (in fact, the autonomic nervous system has been called the "involuntary" nervous system). However, it is wrong to assume that this is always the case, for it has been shown that discrete visceral or glandular responses can be learned and thus, to this extent, voluntarily controlled.

NEURAL GROWTH AND REGENERATION

The elaborate networks of nerve-cell processes that characterize the nervous system are remarkably similar in all human beings and depend upon controlled patterns of neuron development. The outgrowth of specific axons to specific targets is of utmost importance in the formation of this circuitry. The mechanisms that guide the development of the brain and spinal cord are today under intense investigation by neuroscientists.

Development of the nervous system in the embryo begins with the division of precursor cells (neuroblasts). After the last cell division, each cell, now called a neuron, migrates to its final location and begins to develop a spatial orientation, sending out processes that will become the axon and dendrites. A specialized enlargement, the growth cone, forms the tip of each extending process and is involved in finding the correct route and final target for the growing process.

As the process grows, it is guided along the surfaces of other cells, most commonly glial cells. Which particular route is followed depends in part on the presence of glycoproteins known as cell adhesion molecules (CAMs) on the membranes of early neurons and other embryonic cells that provide recognition markers.

Development is also influenced in part by growth factors, such as nerve growth factor, in the extracellular region immediately surrounding the growth cone. Once the target of the advancing growth cone is reached, synapses are formed. During these intricate early stages of neural development, alcohol and other drugs, radiation, and viruses can exert effects that cause permanent damage to the fetus.

A normal—although unexpected—aspect of development of the nervous system occurs after growth and projection of the axons: many of the neurons and synapses formed are destroyed. In fact, as many as 50 to 70 percent of normal neurons die in some regions of the developing nervous system! Why this seemingly wasteful process occurs is unknown. Although the basic shape of existing neurons in the mature CNS does not change, the creation and removal of synaptic contacts begun during fetal development continues at a slower pace throughout life as part of normal growth, learning, and aging.

Division of neuron precursors is essentially complete before birth, and no new neurons are formed to replace those that die. Damaged neurons can repair themselves, however, and significant function may be regained, providing the damage occurs outside the CNS and does not affect the neuron cell body. After such injury, the segment of the axon now separated from the cell body degenerates, an event known as Wallerian degeneration. The proximal part of the axon (the stump still attached to the cell body) degenerates back a few millimeters but then gives rise to a growth cone, which grows out to the effector organ so that, in some cases, function is restored.

In contrast, damage to axons within the CNS is followed by attempts at sprouting, but no significant regeneration occurs across the damaged site, and there are no well-documented reports of function return. Mature neurons of the CNS are capable of regrowing processes, but some property of their environment prevents this from happening. Researchers are attempting to create artificial tubes of Schwann-cell membrane or even fetal amniotic membrane to provide an environment that will support neuron regeneration in the CNS. Attempts are also being made to restore function to damaged or diseased brains by the implantation of pieces of fetal brain or of tissues, such as the patient's own adrenal medulla, that synthesize and secrete neurotransmitters.

BLOOD SUPPLY, BLOOD-BRAIN BARRIER PHENOMENA, AND CEREBROSPINAL FLUID

Usually, glucose is the only substrate metabolized by the brain to supply its energy requirements, and most of the energy from the oxidative breakdown of glucose is transferred to ATP. Its glycogen stores being negligible, the brain is completely dependent upon a continuous blood supply of glucose. Also essential is a blood supply of oxygen for oxidative phosphorylation. Although the adult brain is only 2 percent of the body weight, it receives 15 percent of the total blood supply to support its high oxygen utilization. If the oxygen supply is cut off for 4 to 5 min or if the glucose supply is cut off for 10 to 15 min, brain damage will occur. In fact, the most common form of brain damage is caused by a stoppage of the blood supply to a region of the brain, which results in a **stroke**. The cells in the region deprived of nutrients cease to function and die.

The exchange of substances between blood and neurons in the CNS is different from the more or less unrestricted diffusion of nonprotein substances from blood to cells in the other organs of the body. A complex group of **blood-brain barrier mechanisms** closely controls both the kinds of substances that enter the extracellular fluid of the brain and the rate at which they enter. These mechanisms regulate the chemical composition of the extracellular fluid of the brain and minimize the ability of many harmful substances to reach the neurons.

The blood-brain barrier resides within the endothelial cells that line the brain capillaries (capillaries are the smallest blood vessels and the major vessels across which substances pass between the blood and tissue fluids). In brain capillaries the edges of adjacent endothelial cells abut against each other, and the cells are completely sealed together by continuous tight junctions so that all substances entering or leaving the brain must pass through the two plasma membranes and cytoplasm of the endothelial cells (Figure 8-17). Capillaries in certain areas of the brain lack a blood-brain barrier, but these areas account for less than 1 percent of the brain volume.

The blood-brain barrier has both anatomical structures and physiological transport systems that handle different classes of substances in different ways; thus, the barrier is said to be selective. For example, substances that dissolve readily in the lipid components of the plasma membranes enter the brain quickly. This characteristic of the blood-brain barrier explains some drug

actions, as can be seen from the following scenario: Morphine differs chemically from heroin only in that morphine has two hydroxyl groups where heroin has two acetyl groups ($-COCH_3$). This small difference renders heroin highly lipid soluble and morphine highly lipid insoluble. Thus, heroin crosses the blood-brain barrier readily and morphine does not. As soon as heroin enters the brain, however, enzymes remove the acetyl groups from heroin and change it to morphine. The morphine, highly insoluble in lipid, is then effectively trapped in the brain where it continues to exert its effect. Some other drugs that have rapid effects in the CNS because of their high lipid solubility are the barbiturates, nicotine, caffeine, and alcohol.

Many substances that do not dissolve readily in lipids, such as glucose and other important substrates of brain

FIGURE 8-17 A diagram of the blood-brain barrier, which is formed by the endothelial cells that line the brain capillaries. (*Redrawn from Oldendorf.*)

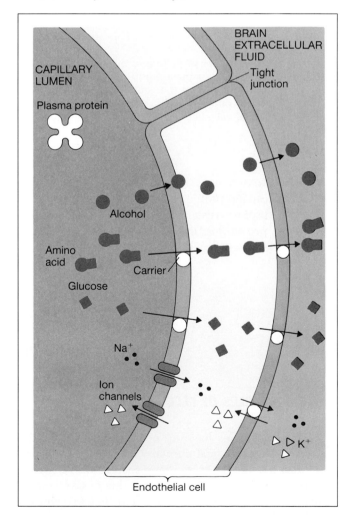

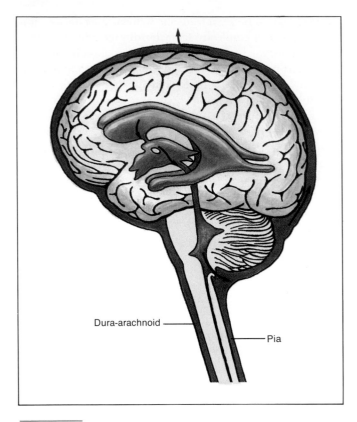

Dura-arachnoid

Pia

FIGURE 8-18 The ventricular system of the brain and the distribution of the cerebrospinal fluid, shown in color. Cerebrospinal fluid is formed in the ventricles, passes to the subarachnoid space outside the brain and spinal cord, and then through small valvelike structures into the large veins of the head.

metabolism, enter the brain rapidly by combining with membrane carrier proteins. Similar transport systems also move surplus substances out of the brain and into the blood, preventing the buildup of molecules that could interfere with brain function.

In addition to its blood supply, the CNS is perfused by a second fluid, the cerebrospinal fluid. This clear fluid fills the cerebral ventricles and the subarachnoid space, which surrounds the brain and spinal cord (Figure 8-18). Thus, the CNS literally floats in a cushion of cerebrospinal fluid. Since the brain and spinal cord are very soft, delicate tissues with about the consistency of gelatin, they are protected by the fluid from sudden and jarring movements.

Most of the cerebrospinal fluid is secreted into the ventricles by the **choroid plexuses**, which form part of the lining of the four ventricles. A barrier is present here, too, between the blood in the capillaries of the choroid plexuses and the cerebrospinal fluid that flows next to the neural tissue. Consistent with the barrier mechanisms, cerebrospinal fluid is a selective secretion, not a simple filtrate of plasma. For example, potassium and calcium concentrations are slightly lower in cerebrospinal fluid than in plasma, whereas the sodium and chloride concentrations are slightly higher.

The cerebrospinal fluid circulates through the interconnected ventricular system to the brainstem, where it passes through small openings out to the surface of the brain and spinal cord. Aided by circulatory, respiratory, and postural pressure changes, the fluid finally flows to the top of the outer surface of the brain, where most of it is reabsorbed into the bloodstream through one-way valves in large veins. If the flow is obstructed at any point, cerebrospinal fluid accumulates, causing hydrocephalus ("water on the brain"). In severe untreated cases, the resulting elevation of pressure in the ventricles may damage the brain and cause mental retardation.

This completes our survey of the organization of the nervous system. We now turn to the mechanisms by which neurons, synapses, and receptors on afferent neurons function, beginning with the electrical properties that underlie all these events.

SECTION B
MEMBRANE POTENTIALS

BASIC PRINCIPLES OF ELECTRICITY

Some molecules have no net electric charge since they contain equal numbers of electrons and protons (page 11). As we have seen, however, many molecules have a net electric charge due to such components as the negative carboxyl group $RCOO^-$ and the positive amino group RNH_3^+. Moreover, most of the mineral elements, such as sodium, potassium, and chloride, are present in solution as the charged particles known as ions (Na^+, K^+, and Cl^-).

With the exception of water, the major chemical substances in the extracellular fluid are, as we saw in Chapter 6, sodium and chloride ions, whereas the intracellular fluid contains high concentrations of potassium and ionized organic molecules, particularly proteins and phosphate compounds. Since many charged particles are found both inside and outside cells, it is not surprising that electrical phenomena resulting from the distribution of these charged particles play a significant role in cell function.

An electric force draws positively and negatively charged substances together. In contrast, like charges repel each other, that is, positive charge repels positive charge and negative charge repels negative charge. The strength of the force acting between electric charges increases when the charges are moved closer together and when the quantity of charge is increased.

When oppositely charged particles come together, the electric force of attraction moving them can be used to perform work. Conversely, to separate opposite charges, energy must be used to overcome this attractive force. Thus, separated electric charges of opposite sign have the potential of doing work if they are allowed to come together. This potential is called an **electric potential**, or, because it is determined by the difference in charge between two points, **potential difference**, which we shall often shorten to **potential**. The units of electric potential are volts, but since the total charge that can be separated in most biological systems is very small, the potential differences are small and are measured in millivolts (1 mV = 0.001 V).

The movement of electric charge is called a **current**. The electric force between charges tends to make them flow, producing a current. If the charges are of opposite sign, the current brings the charges toward each other; if the charges are alike, the current increases the separation between them. The amount of charge that moves—in other words, the current—depends on the potential difference between the charges and on the nature of the material through which they are moving. The hindrance to electric charge movement is known as the **resistance**. The relationship between current I, voltage E (for electrical potential difference), and resistance R is given by **Ohm's law**:

$$I = \frac{E}{R}$$

Materials that have a high electrical resistance are known as insulators, whereas materials having a low resistance are conductors.

Water is a poor conductor of electricity because it contains very few charged particles. When sodium chloride is added to water, however, the solution becomes a relatively good conductor with a low resistance because the sodium and chloride ions can carry the current. The intracellular and extracellular fluids contain numerous ions and can, therefore, carry current. Lipids, however, contain very few charged groups and cannot carry current, and they have a high electrical resistance. Therefore, the lipid layers of the plasma membrane are regions of high electrical resistance separating two water compartments of low resistance.

THE RESTING POTENTIAL

All cells under resting conditions have a potential difference across their plasma membranes oriented with the inside of the cell negatively charged with respect to the outside (Figure 8-19). This potential is the **resting membrane potential**; its magnitude varies from about −5 to −100 mV, depending upon the type of cell (in neurons, it is generally in the range of −40 to −75 mV).[2] As we shall see, the membrane potential of some cells can change rapidly in response to stimulation, an ability of key importance in their functioning.

The membrane potential can be accounted for by the fact that there is a net excess of negative ions inside the cell and a net excess of positive ions outside. The excess negative charges inside, not having positive charges to balance them, are electrically attracted to the excess positive charges outside the cell, and vice versa. Thus, the excess charges (ions) collect in a thin shell on the

---

[2] By convention, extracellular fluid is assigned a voltage of zero, and the polarity (positive or negative) of the membrane potential is stated in terms of the sign of the excess charge on the inside of the cell.

FIGURE 8-19 The potential difference across a cell membrane as measured by an intracellular microelectrode.

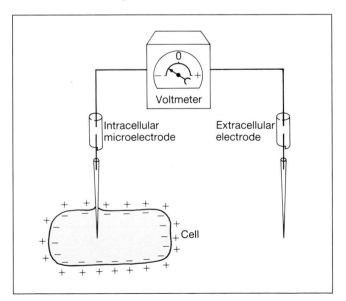

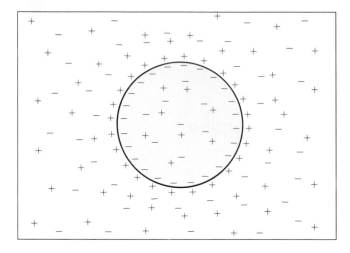

FIGURE 8-20 The excess of positive charge outside the cell and the excess negative charges inside collect close to the plasma membrane. In reality, these excess charges are but a very small fraction of the total number of the ions inside and outside the cell.

inner and outer surfaces of the plasma membrane (Figure 8-20), whereas the bulk of the intracellular and extracellular fluid is electrically neutral. Unlike Figure 8-20, the number of positive and negative charges that have to be separated across the membrane to account for the potential is an infinitesimal fraction of the total number of charges within the cell.

The magnitude of the resting membrane potential is determined mainly by two factors: (1) the difference in ion concentrations of the intracellular and extracellular fluids and (2) the permeabilities of the plasma membrane to the different ion species. The concentrations of sodium, potassium, and chloride ions in the extracellular and intracellular fluids of a nerve cell are listed in Table 8-8. There are many other ions, such as Mg^{2+}, Ca^{2+}, H^+, HCO_3^-, HPO_4^{2-}, SO_4^{2-}, amino acids, and proteins, in both fluid compartments, but sodium, potassium, and chloride ions are present in highest concentrations and therefore generally play the most important

roles in the generation of the resting membrane potential.

Note that the sodium and chloride concentrations are lower inside the cell than outside, and potassium concentration is greater inside. As we mentioned in Chapter 6, the concentration differences for sodium and potassium are due to the action of a plasma-membrane active-transport system, which pumps sodium out of the cell and potassium into it.

To understand how such concentration differences for sodium and potassium create membrane potentials, let us consider the situation in Figure 8-21. The assumption in this model is that the membrane contains open potassium channels but no sodium channels. Initially, there is no potential difference across the membrane because the two solutions are electrically neutral, that is, they contain equal numbers of positive and negative ions (the positive ions are different—sodium versus potassium—in the two compartments, but the numbers are the same, and each is balanced by an equal number of chloride ions). Because of the open potassium channels, potassium will diffuse down its concentration gradient from compartment 2 into compartment 1. After a few potassium ions have moved into compartment 1, that compartment will have an excess of positive charge, and

FIGURE 8-21 Generation of a diffusion potential across a membrane permeable only to potassium. Arrows represent ion movements.

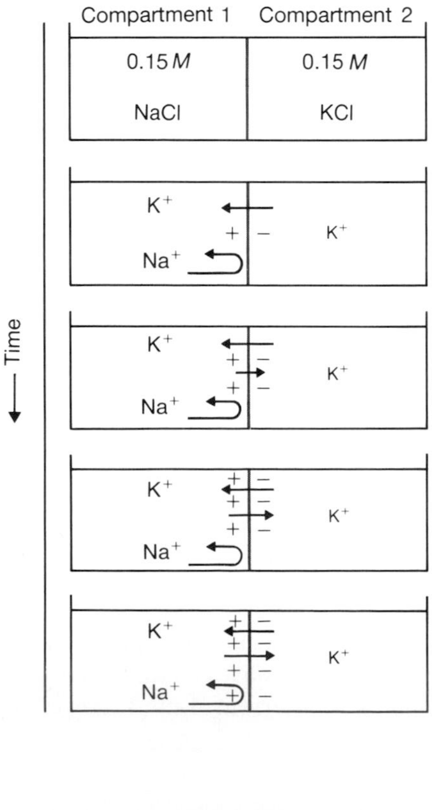

TABLE 8-8	DISTRIBUTION OF MAJOR IONS ACROSS THE PLASMA MEMBRANE OF A TYPICAL NERVE CELL	
	Concentration, mmol/L	
Ion	Extracellular	Intracellular
Na^+	150	15
Cl^-	110	10
K^+	5	150

compartment 2 will have an excess of negative charge, that is, a potential difference will exist across the membrane.

Now we introduce a second factor that can cause net movement across a membrane: an electrical potential. As compartment 1 becomes increasingly positive and compartment 2 increasingly negative, the membrane potential difference begins to influence the movement of the potassium ions. They are attracted by the negative charge of compartment 2 and repulsed by the positive charge of compartment 1.

As long as the force due to the concentration gradient is greater than the electric force, there will be net movement of potassium from compartment 2 to compartment 1, and the membrane potential will increase. Compartment 1 will become more and more positive until the *electric force* opposing the entry of potassium into that compartment equals the force due to the *concentration* gradient favoring entry. The membrane potential at which the electric force is equal in magnitude but opposite in direction to the concentration force is called the **equilibrium potential** for that ion. At the equilibrium potential there is no net movement of the ion because the opposing forces acting upon it are exactly balanced.

The value of the equilibrium potential for any ion depends on the concentration gradient for that ion across the membrane. If the concentrations on the two sides were equal, the force of the concentration gradient would be zero and the equilibrium potential would also be zero. The larger the concentration gradient, the larger the equilibrium potential since a larger electrically driven movement of ions would be required to balance the larger movement due to the concentration difference. For neurons, the equilibrium potential for potassium is close to -90 mV, the inside of the cell being negative with respect to the outside (Figure 8-22).

If the membrane separating the two compartments is replaced with one permeable only to sodium, a parallel situation will occur (Figure 8-23). A sodium equilibrium potential will eventually be established with compartment 2 positive with respect to compartment 1, at which point net movement will cease. For most neurons, the sodium equilibrium potential is about $+60$ mV, inside positive (Figure 8-24).

Thus, the equilibrium potential for each ion species is different in magnitude and, sometimes even in direction from those for other ion species, depending on the concentration gradients for each ion. The equilibrium potential for any ion can be calculated with the Nernst equation (Appendix C).

It is not difficult to move from these hypothetical examples to a nerve cell at rest where (1) the potassium concentration is much greater inside the cell than out-

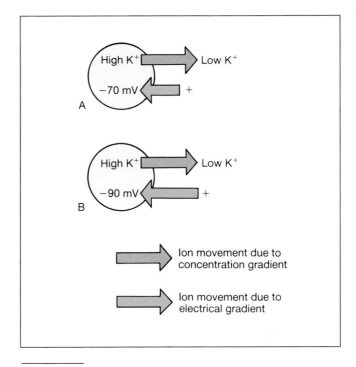

FIGURE 8-22 Forces acting on potassium when the membrane of a neuron is at (A) the resting potential (-70 mV, inside negative), and (B) the potassium equilibrium potential (-90 mV, inside negative).

side it and (2) the cell membrane is some 50 to 75 times more permeable to potassium than to sodium. A potential is generated across the plasma membrane largely because of the movement of potassium down its concentration gradient through open potassium channels, so

FIGURE 8-23 Generation of a diffusion potential across a membrane permeable only to sodium. Arrows represent ion movements.

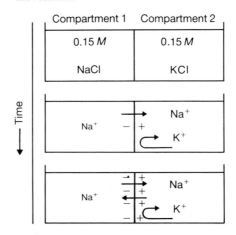

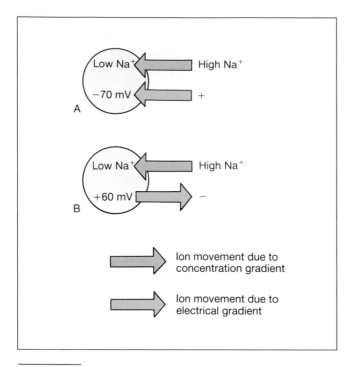

FIGURE 8-24 Forces acting on sodium when the membrane of a neuron is at (A) the resting potential (−70 mV, inside negative), and (B) the sodium equilibrium potential (+60 mV, inside positive).

dium into the cell and potassium out. If such net ion movements occur, why does the concentration of intracellular sodium not progressively increase and of intracellular potassium decrease? The reason is that active-transport mechanisms in the plasma membrane utilize energy derived from cellular metabolism to pump the sodium back out of the cell and the potassium back in. Actually, the pumping of these ions is linked because they are both transported by the Na,K-ATPase in the membrane (page 119).

In a resting cell, the number of ions moved by the pump equals the number of ions that diffuse through channels in the membrane down their concentration and/or electrical gradients (Figure 8-25), and the concentrations of sodium and potassium in the cell do not change. As long as the concentration gradients remain fixed and the ion permeabilities of the plasma membrane do not change, the electric potential across the resting membrane will remain constant. This constancy is not, however, an *equilibrium* state, because metabolic energy, in the form of the membrane pump, is required to maintain the status quo; rather, the constancy reflects a *steady-state* condition.

Thus far, we have described the membrane potential as due purely to the passive diffusion of ions down their electrical and concentration gradients, the concentration gradients having been established by membrane pumps. There is, however, another component to the membrane potential, reflecting the *direct* separation of charge and

that the inside of the cell becomes negative with respect to the outside. The experimentally measured resting membrane potential is not equal to the potassium equilibrium potential, however, because a small number of sodium channels are open in the resting membrane, and some sodium ions continually diffuse into the cell, canceling the effect of an equivalent number of potassium ions simultaneously moving out.

When more than one ion species can diffuse across the membrane, the permeabilities and concentration gradients for all the ions must be considered when accounting for the membrane potential. For a given concentration gradient, the greater the membrane permeability to an ion species, the greater the contribution that ion species will make to the membrane potential. Since the resting membrane is much more permeable to potassium than to sodium, the resting membrane potential is much closer to the potassium equilibrium potential than to that of sodium. The potential of a membrane permeable to several ion species can be calculated by the Goldman equation (Appendix C).

Thus, at a resting potential of −70 mV, a typical value for neurons, neither sodium nor potassium is at its equilibrium potential, and so there is a net diffusion of so-

FIGURE 8-25 Movements of sodium and potassium ions across the plasma membrane of a resting neuron. The passive movements are exactly balanced by the active transport of the ions in the opposite direction (*Redrawn from Kandel and Schwartz.*)

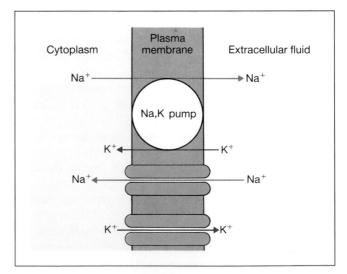

creation of a potential difference across the membrane by the transport of ions by the membrane Na,K-ATPase pump. If the pump were to move equal numbers of sodium and potassium ions across the membrane in opposite directions, it would not contribute directly to the separation of charge, and all the charge separation would result from ion diffusion. If equal numbers of ions are not moved and the pump does directly separate charge, however, it is known as an **electrogenic pump**.

The Na,K-ATPase pump is electrogenic since it moves three sodium ions out of the cell for every two potassium ions that it brings in. This unequal transport directly increases charge separation across the plasma membrane; that is, it makes the membrane potential larger (by transferring more positive charges to the outside) than it would be from the diffusion potential alone. In most cells (but by no means all), the electrogenic contribution to the membrane potential is quite small. It must be reemphasized that, even when the electrogenic contribution of the Na,K-ATPase pump is small, the pump makes an essential *indirect* contribution to the membrane potential because it maintains the concentration gradients down which the ions diffuse to produce most of the charge separation that makes up the potential.

We have not yet dealt with chloride ions. The plasma membranes of many cells are permeable to chloride ions and do not contain chloride-ion pumps. Therefore, in these cells, the membrane potential, set up by the interactions just described, acts on chloride ions. The inside negativity moves chloride out of the cell until a concentration gradient develops (the inside chloride concentration is lower than the outside). This concentration gradient produces a diffusion of chloride back into the cell that exactly opposes the diffusion out because of the electric force. The result is that the equilibrium potential for chloride ions is equal to the resting membrane potential, and chloride makes no contribution to the magnitude of the membrane potential.

In cells that do have an active-transport system that moves chloride out of the cell, the membrane potential is not at the chloride equilibrium potential, and net chloride diffusion contributes to the magnitude of the membrane potential. Since the net diffusion of chloride is into the cells, it contributes to the excess negative charge inside the cell; that is, it increases the magnitude of the membrane potential.

Most of the negative charge in neurons is accounted for, however, not by chloride ions but by negatively charged organic molecules, such as proteins. Unlike chloride, these molecules do not readily cross the plasma membrane but remain inside the cell, where their charge contributes to the total negative charge within the cell.

GRADED POTENTIALS AND ACTION POTENTIALS

Transient changes in the membrane potential from its resting level produce electrical signals that can alter cell activities. Such changes are the most important way that nerve cells process and transmit information. These signals occur in two forms: graded potentials and action potentials. Graded potentials are important in signaling over short distances, and action potentials are the long-distance signals of nerve and muscle membrane.

Adjectives such as "membrane," "resting," "action," and "graded" define the conditions under which the potential is measured or the way it develops (Table 8-9).

The terms "depolarize," "hyperpolarize," and "repolarize" will be used to describe the direction of changes in the membrane potential relative to the resting potential (Figure 8-26). The membrane is said to be **depolarized** when its potential is less negative (closer to zero) than the resting level. By convention, this includes states in which the polarity of the membrane potential reverses, that is, in which the inside of a cell becomes positive. The membrane is **hyperpolarized** when the potential is more negative than the resting level. When a membrane potential that has been either depolarized or hyperpolarized returns toward the resting value, it is said to be **repolarizing**.

Graded Potentials

Graded potentials are changes in membrane potential that are confined to a relatively small region of the membrane and die out within 1 to 2 mm of their site of origin. They can occur in either a depolarizing or hyperpolarizing direction (Figure 8-27A), and they are usually produced by some stimulus, that is, by a specific change in the cell's environment acting on a specialized region of the membrane. They are called "graded potentials" because the magnitude of the potential change is variable (graded) and is related to the magnitude of the stimulus (Figure 8-27B). We shall encounter a number of graded potentials that are given various names related to the location of the potential or to the function it performs: receptor potentials, synaptic potentials, end-plate potentials, pacemaker potentials.

Whenever a graded potential occurs, charge will flow between the place of origin of the potential and adjacent regions of the plasma membrane, which are at the resting potential. Thus, graded potentials create current in the intracellular and extracellular fluids in the region around them; the greater the potential change, the greater the current. By convention, the direction in which positive ions move is designated the direction of

TABLE 8-9   A MINIGLOSSARY

Potential = potential difference	The voltage difference between two points.
Membrane potential = transmembrane potential	The voltage difference between the inside and outside of a cell.
Diffusion potential	A potential created by ion diffusion.
Equilibrium potential	The voltage difference across a membrane that produces ion movement equal but opposite to the diffusion of that ion due to its concentration gradient.
Resting membrane potential = resting potential	The membrane potential of a cell that is not producing an electric signal.
Graded potential.	A potential change of variable amplitude that is conducted decrementally; it has no threshold or refractory period.
Action potential	A brief, all-or-none reversal of the membrane potential polarity; it has a threshold and refractory period and is conducted without decrement.
Postsynaptic potential	A graded potential change produced in the postsynaptic neuron in response to a release of a transmitter substance by a presynaptic terminal; it may be depolarizing (an *excitatory postsynaptic potential*) or hyperpolarizing (an *inhibitory postsynaptic potential*).
Receptor potential	A graded potential produced at the peripheral endings of afferent neurons (or in separate receptor cells) in response to an external stimulus.
Pacemaker potential	Spontaneously occurring graded potential change.

FIGURE 8-26 Depolarizing, repolarizing, and hyperpolarizing changes in membrane potential.

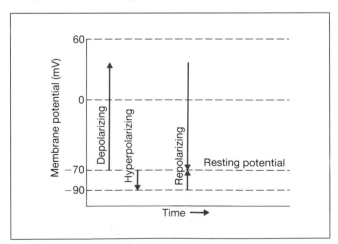

the current; negatively charged particles simultaneously move in the opposite direction.

In Figure 8-28A, a small region of a membrane has been depolarized by a stimulus and so has a potential less negative than that of adjacent areas. Inside the cell, positive charge will flow through the intracellular fluid away from the depolarized region and toward the more negative, resting regions of the membrane. Simultaneously, outside the cell, positive charge will flow from the more positive region of the resting membrane toward the less positive region just created by the depolarization (Figure 8-28B). This local current (movement of positive charge) removes positive charges from regions surrounding the depolarization site along the outside of the membrane and adds positive charge to regions surrounding the depolarization site along the inside of the membrane. Thus it produces a decrease in the amount of

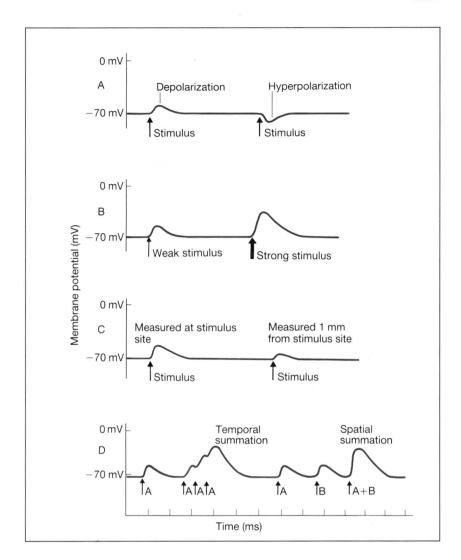

FIGURE 8-27 Graded potentials (A) can be depolarizing or hyperpolarizing, (B) can vary in size, (C) are conducted decrementally, and (D) can be summed. Temporal and spatial summation will be discussed later in the chapter.

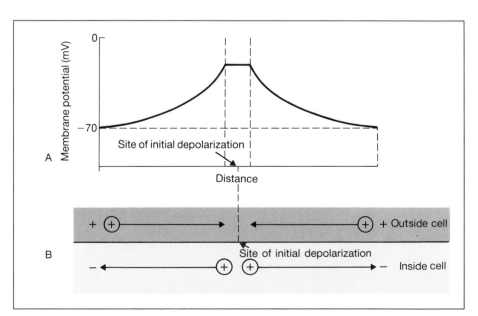

FIGURE 8-28 (A) Potential changes around a depolarization site. (B) Local current surrounding a depolarized region of a membrane produces a depolarization of adjacent regions.

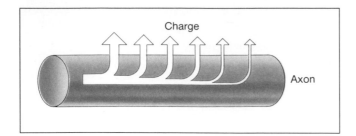

FIGURE 8-29 Leakage of charge across the plasma membrane reduces the local current at sites farther along the membrane.

charge separation (depolarization) in the membrane sites adjacent to the depolarized region. The local current is carried by ions such as K^+, Na^+, Cl^-, and HCO_3^-.

Local current flows much like water does through a leaky hose. Charge is lost across the membrane because the membrane is permeable to ions, just as water is lost from the hose, with the result that the magnitude of the current decreases with the distance away from the initial site of the potential change, just as water flow decreases the farther along the hose you are from the faucet (Figure 8-29). In fact, plasma membranes are so leaky to ions that local currents almost completely die out within a few millimeters of their point of origin. There is another

way of saying the same thing: Local current is **decremental**; that is, its amplitude decreases with increasing distance from the site of origin of the potential. The resulting change in membrane potential, therefore, also decreases with the distance of the potential (Figures 8-27C and 8-28A).

Because the electric signal decreases with distance, graded potentials (and the current they generate) can function as signals only over very short distances (a few millimeters). Nevertheless, graded potentials are the only means of communication used by some neurons and, as we shall see, play very important roles in the integration of signals by neurons and some other cells.

Action Potentials

Action potentials (**APs**) are very different from graded potentials (Table 8-10). They are rapid alterations in the membrane potential that may last only 1 ms, during which time the membrane potential may change 100 mV, from -70 to $+40$ mV, and then repolarize to its resting membrane potential (Figure 8-30). Only nerve, muscle, and some gland cells have plasma membranes capable of producing action potentials. These membranes are called **excitable membranes**, and their ability to generate action potentials is known as excitability. The propagation of action potentials is the mechanism

TABLE 8-10 DIFFERENCES BETWEEN GRADED POTENTIALS AND ACTION POTENTIALS	
Graded Potentials	**Action Potentials**
1. Graded response; amplitude varies with conditions of the initiating event	All-or-none response; once membrane is depolarized to threshold, amplitude independent of initiating event
2. Graded response; can be summed	All-or-none response; cannot be summed
3. Has no threshold	Has a threshold that is usually 10 to 15 mV depolarized relative to the resting potential
4. Has no refractory period	Has a refractory period
5. Is conducted decrementally; that is, amplitude decreases with distance	Is conducted without decrement; the amplitude is constant
6. Duration varies with initiating conditions	Duration constant for a given cell type
7. Can be a depolarization or hyperpolarization	Is a depolarization (with overshoot)
8. Initiated by environmental stimulus (receptor), by neurotransmitter (synapse), or spontaneously	Initiated by membrane depolarization

used by the nervous system to communicate over long distances.

How can an excitable membrane make rapid changes in its membrane potential? How does a stimulus interact with an excitable membrane to cause an action-potential response? How is an action potential propagated along an excitable membrane? These questions will be discussed in the following sections.

Ionic basis of the action potential. Action potentials can be explained by the concepts already developed for describing the origins of resting membrane potentials. We have seen that the magnitude of the resting membrane potential depends upon the concentration gradients of and membrane permeabilities to different ions, particularly sodium and potassium. This dependence is true for the period of the action potential as well: The action potential results from a transient change in membrane ion permeability whereas the concentration gradients remain unchanged. In the resting state the open channels in the plasma membrane are predominantly those that are permeable to potassium and chloride ions. Almost all the sodium-ion channels are closed, and the resting potential is therefore much closer to the potassium equilibrium potential than to the sodium equilibrium potential. During an action potential, however, the membrane permeabilities to sodium and potassium ions are markedly altered.

In the depolarizing phase of the action potential, sodium channels open, increasing the membrane permeability to sodium ions several hundredfold. This allows sodium ions to rush into the cell. During this period more positive charge enters the cell in the form of sodium ions than leaves in the form of potassium ions, and thus the membrane potential becomes less negative and eventually reverses polarity, becoming positive on the inside and negative on the outside of the membrane. In this phase the membrane potential approaches but does not quite reach the sodium equilibrium potential.

Action potentials in nerve cells last about 1 ms. (They may be much longer in certain types of muscle cells.) What causes the membrane potential to return so rapidly to its resting level? The answer to this question is twofold: (1) The sodium channels that were open during the depolarization phase close, and (2) a special set of potassium channels begin to open. The timing of these two events can be seen in Figure 8-30. Closure of the sodium channels alone would restore the membrane potential to its resting level since potassium flux out would then exceed sodium flux in. However, the process is speeded up by the simultaneous increase in potassium permeability. These two events, closure of the sodium channels and opening of potassium channels, lead to a potassium diffusion out of the cell that is much greater than the sodium diffusion in, rapidly returning the membrane potential to its resting level. In fact, after the sodium channels have closed, at least some of the potassium channels are still open, and there is generally a small hyperpolarizing overshoot of the membrane potential (**afterhyperpolarization**, Figure 8-30).

One might think that large movements of ions across the membrane would be required to produce such large changes in membrane potential. Actually, only about 1 of every 100,000 potassium ions in a cell diffuses out to charge the membrane potential to its resting level, and approximately the same number of sodium ions enters the cell during an action potential. These ion movements are so small that they produce only infinitesimal changes in the intracellular ion concentrations. Yet if this tiny number of ions crossing the membrane with each action potential were not eventually moved back across the membrane, the concentration gradients of sodium and potassium would gradually disappear and action potentials could no longer be generated. As might be expected, cellular accumulation of sodium and loss of potassium are prevented by the continuous action of the membrane Na,K-ATPase active-transport system.

The number of ions that cross the membrane during an action potential is so small that the pump need not keep up with the ion movements during each action potential, and even if the pump is stopped experimentally, thousands of action potentials can be produced before significant changes in the resting potential will occur. It

---

FIGURE 8-30 Changes in membrane potential and changes in the membrane permeability (P) to sodium and potassium ions during an action potential.

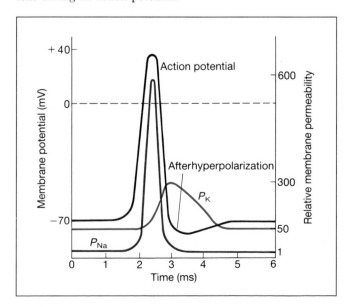

may seem foolish to let sodium move into the neuron and then pump it back out, but sodium movement into the cell creates the electric signal necessary for communication between parts of the cell, and pumping sodium out restores the concentration gradient so sodium will enter the cell and create another signal in response to a new stimulus.

Mechanism of permeability changes. Measurement of permeability changes reveals that the permeability to sodium is altered whenever the membrane potential changes. Specifically, hyperpolarization of the membrane causes a decrease in sodium permeability, whereas depolarization causes an increase in sodium permeability.

In light of our discussion of the ionic basis of membrane potentials, it is very easy to confuse the cause-and-effect relationships of the statements just made. Earlier we pointed out that an increase in sodium permeability *causes* membrane depolarization; now we are saying that depolarization *causes* an increase in sodium permeability. Combining these two distinct causal relationships yields the positive-feedback cycle (Figure 8-31) responsible for the rising phase of the action potential: Depolarization opens sodium channels so that the membrane permeability to sodium increases. Because of increased sodium permeability, sodium diffuses into the cell; this addition of positive charge to the cell further depolarizes the membrane, which, in turn, produces a still greater increase in sodium permeability, which, in turn, causes . . . .

Ion channels that can be opened or closed by changes in membrane potential are called voltage-sensitive ion channels (page 113). The potassium channels that open during an action potential are also voltage-sensitive.

Although we have discussed only sodium and potassium channels, there are other types of voltage-sensitive channels that have varying degrees of importance in dif-

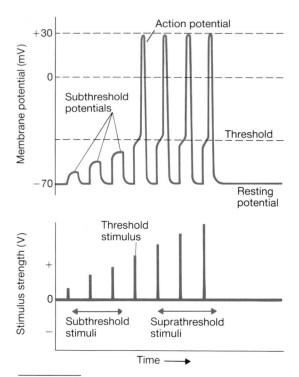

FIGURE 8-32 Changes in the membrane potential with increasing strength of depolarizing stimulus. When the membrane potential reaches threshold, action potentials are generated. Increasing the stimulus strength above threshold level does not cause larger action potentials. (The after-hyperpolarization has been omitted from this figure for clarity.)

ferent cell types and explain at least some of the special properties of these cells. For example, as we shall discuss later in the chapter, some regions of nerve-cell membrane have calcium channels that open in response to membrane depolarization. In some nonneural cells, voltage-sensitive calcium channels are so important that the action potentials of these cells are due entirely to calcium.

The generation of action potentials is prevented by local anesthetics such as procaine hydrochloride (Novocaine) and lidocaine (Xylocaine) because these prevent opening of the sodium channels in response to depolarization. Without action potentials, afferent signals generated in the periphery—in response to injury, for example—cannot reach the brain and give rise to the sensation of pain.

Threshold. Not all depolarizations trigger the positive-feedback relationship that leads to an action potential. Action potentials occur only when the membrane is depolarized enough (and therefore enough sodium channels are open) so that sodium entry exceeds potassium exit. In other words, action potentials occur only when

FIGURE 8-31 Positive-feedback relation between membrane depolarization and increased sodium permeability, which leads to the rapid depolarizing phase of the action potential.

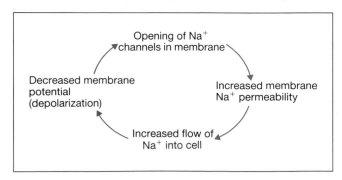

the *net* movement of positive charge is inward. The membrane potential at which the *net* movement of ions across the membrane first changes from outward to inward is the **threshold**, and stimuli that are just strong enough to depolarize the membrane to this level are **threshold stimuli** (Figure 8-32). The threshold of most excitable membranes is about 15 mV depolarized from the resting membrane potential. Thus, if the resting potential of a neuron is −70 mV, the threshold potential may be −55 mV. Stimuli depolarize the membrane, and in order to initiate an action potential in such a membrane, the membrane potential must be depolarized by at least 15 mV. At depolarizations less than threshold, outward potassium movement still exceeds the sodium entry and the positive-feedback cycle cannot get started despite the increase in sodium entry. In such cases, the membrane will return to its resting level as soon as the stimulus is removed, and no action potentials are generated. These weak depolarizations are **subthreshold potentials**, and the stimuli that cause them are **subthreshold stimuli**.

Just like threshold stimuli, stimuli of more than threshold magnitude (**suprathreshold stimuli**) elicit action potentials, but as can be seen in Figure 8-33, the action potentials resulting from such stimuli have exactly the same amplitude as those caused by threshold stimuli. This is explained by the fact that once threshold is reached, membrane events are no longer dependent upon stimulus strength. The depolarization continues on as an action potential because the positive-feedback cycle is operating. Action potentials either occur maximally or they do not occur at all. Another way of saying this is that action potentials are **all-or-none**.[3]

The firing of a gun is a mechanical analogy that shows the principle of all-or-none behavior. The magnitude of the explosion and the velocity at which the bullet leaves the gun do not depend on how hard the trigger is squeezed. Either the trigger is pulled hard enough to fire the gun, or it is not; one cannot fire a gun halfway.

Because of this all-or-none nature, a single action potential cannot convey information about the magnitude of the stimulus that initiated it. How then does one distinguish between a loud noise and a whisper, a light touch and a pinch? This information, as we shall see, depends in part upon the number of action potentials transmitted per unit of time, that is, frequency, and not upon their size.

---

[3]There is a qualification: The action potential fires maximally if it fires at all, but its maximal amplitude depends on the electrical and chemical conditions across the membrane and the state of the membrane. For example, if extracellular sodium is reduced, the action potential will be smaller than usual because of the reduced concentration force that drives sodium into the cell.

Refractory periods.   How soon after firing an action potential can an excitable membrane fire a second one? If we apply a threshold stimulus to a membrane twice in very close succession, the membrane does not respond to the second stimulus. Even though identical stimuli are applied, the membrane is unresponsive for several milliseconds, and is said to be refractory to a second stimulus.

Instead of applying the second stimulus at threshold strength, if we increase it to suprathreshold levels, we can distinguish two separate portions of the **refractory period** (Figure 8-33). During the action potential, a second stimulus—no matter how strong—will not produce a second action potential, and the membrane is said to be in its absolute refractory period. Following this period, there is an interval during which a second action potential can be produced but only if the stimulus strength is considerably greater than threshold. This is the relative refractory period, which can last 10 to 15 ms or longer.

The mechanisms responsible for refractory periods are related to the membrane mechanisms that alter so-

FIGURE 8-33 The magnitude of a second stimulus necessary to generate a second action potential during the refractory period is greater than that of the initial stimulus and decreases as the time between the first and second stimulus increases. During an action potential, the membrane is absolutely refractory to all stimulus strengths.

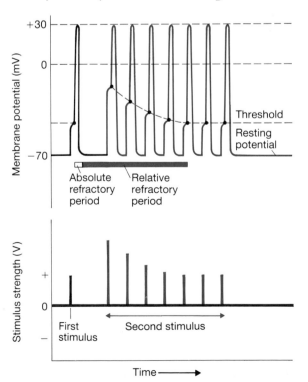

dium and potassium permeabilities. The absolute refractory period corresponds roughly with the period of sodium permeability changes, and the relative refractory period corresponds with the period of increased potassium permeability. A suprathreshold stimulus can cause an action potential during the relative refractory period when a threshold stimulus cannot because the suprathreshold stimulus opens more sodium channels so there is a greater chance that sodium entry will exceed potassium exit, triggering the positive feedback cycle pictured in Figure 8-31. After an action potential, time is required to restore the proteins of the voltage-sensitive ion channels to their original resting configuration.

The absolute refractory period limits the number of action potentials that can be produced by an excitable membrane in a given period of time. Most nerve cells respond physically at frequencies up to 100 action potentials per second, and some may produce much higher frequencies for brief periods of time.

Action potential propagation. Once generated, one particular action potential does not itself travel along the membrane. Rather, each action potential triggers, by creating a local current, a new one at an adjacent area of the membrane. The local current flow is great enough to depolarize the adjacent membrane site to its threshold potential so that the sodium positive-feedback cycle takes over, and a new action potential occurs there (Figure 8-34).

The new action potential is virtually identical to its predecessor, and it produces local currents of its own that depolarize the region adjacent to it, producing yet another action potential at the next site, and so on along the length of the membrane. Thus, no distortion occurs as this process repeats itself along the membrane, and the action potential arriving at the end of the membrane is identical in form to the initial one. In other words, action potentials are not conducted decrementally as are graded potentials.

Because local currents exist between regions of different electric potential, charge will also flow toward the original site of stimulation from newly stimulated adjacent sites (Figure 8-34). However, the membrane areas that have just undergone an action potential are refractory and cannot undergo another. Thus, the only direction of **action-potential propagation** is away from a region of membrane that has recently undergone an action potential.

Excitable membranes are able to conduct action potentials in either direction, the direction of propagation being determined by the stimulus location rather than by any intrinsic inability of the membrane to conduct in the opposite direction. For example, the action potentials in skeletal muscle cells are initiated near the middle of these cylindrical cells and propagate toward the two ends, but in most nerve cells, action potentials are initiated at one end of the cell and propagate toward the other end.

The velocity with which an action potential propagates along a membrane depends upon fiber diameter and whether or not the fiber is myelinated. The larger the fiber diameter, the faster is the action-potential propagation because a large fiber offers less resistance to local current, and adjacent regions of the membrane are brought to threshold faster.

Myelin, the fatty material surrounding axons of some

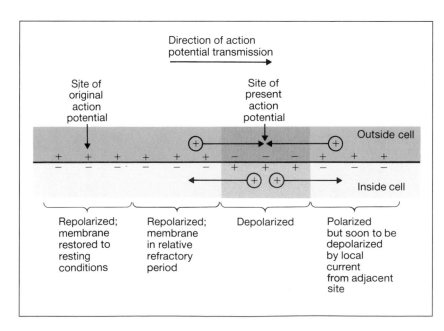

FIGURE 8-34 Propagation of an action potential along a plasma membrane.

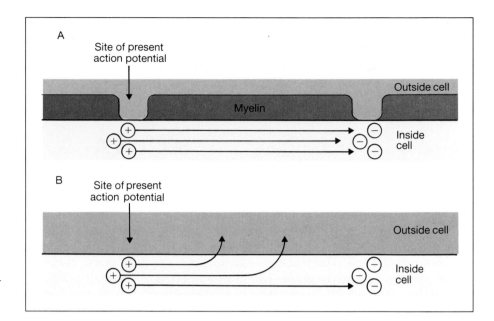

FIGURE 8-35 Current direction during an action potential in (A) a myelinated and (B) an unmyelinated axon.

neurons (page 169), is an insulator that makes it more difficult for charge to flow between intra- and extracellular fluid compartments. Action potentials do not occur along the sections of membrane protected by myelin. Rather, they occur only where the myelin coating is interrupted at the nodes of Ranvier at regular intervals along the axon. Thus, the action potential jumps from one node to the next as it propagates along a myelinated fiber, and for this reason this method of propagation is called saltatory conduction (Latin *saltare*, to leap).

Propagation via saltatory conduction is faster than propagation in nonmyelinated fibers of the same axon diameter because less charge leaks out through the myelin-covered section of the membrane (Figure 8-35). Thus more charge arrives in a given time at the node adjacent to the active node, and the node depolarizes faster and undergoes an action potential sooner than if the myelin were not present.

Conduction velocities range from about 0.5 m/s (1 mi/h) for small-diameter, unmyelinated fibers to about 175 m/s (400 mi/h) for large-diameter, myelinated fibers. At 0.5 m/s, an action potential will travel from the head to the toe of an average-size person in about 4 s; at a velocity of 175 m/s, it takes about 0.01 s.

Initiation of action potentials. In afferent neurons, the initial depolarization to threshold is achieved by a graded potential generated in the receptors at the peripheral ends of the neurons, in other words, at the ends farthest from the CNS where the nervous system functionally encounters the outside world. In all other neurons, the depolarization to threshold is due either to a

graded potential generated by synaptic input to the neuron or to a spontaneous change in the neuron's membrane potential, known as a **pacemaker potential**.

How synaptic and receptor potentials are produced is the subject of the next two sections.

Spontaneous generation by pacemaker potentials occurs in the absence of any identifiable external stimulus and is an inherent property of certain neurons (and other excitable cells, including certain smooth-muscle and cardiac cells). In all these cells, the ion channels in the plasma membrane appear to open and close spontaneously, the net effect being a graded depolarization of the membrane—the pacemaker potential. If threshold is reached, an action potential occurs (Figure 8-36); the

FIGURE 8-36 Action potentials resulting from pacemaker potentials.

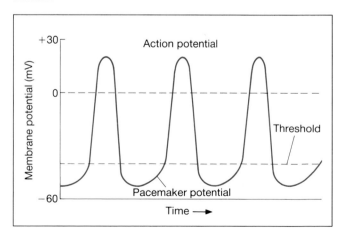

membrane then repolarizes and again begins to depolarize. There is no stable, resting membrane potential in such cells because of the continuous change in membrane permeability. The rate at which the membrane depolarizes determines the action-potential frequency. Pacemaker potentials are implicated in many rhythmic behaviors, such as breathing, the heartbeat, and movements of the stomach and intestines.

SECTION C
SYNAPSES

As defined earlier, a synapse is an anatomically specialized junction between two neurons, at which the electrical activity in one neuron, the presynaptic neuron, influences the electrical activity in the second, postsynaptic cell. The junction between a neuron and an effector—muscle or gland—cell is a **neuroeffector junction**. Anatomically, synapses include parts of the presynaptic and postsynaptic neurons and the extracellular space between these two cells.

When active, synapses can increase or decrease the activity in a postsynaptic neuron by producing a brief, graded potential there. The membrane potential of a postsynaptic neuron is brought closer to threshold at an **excitatory synapse** and is stabilized at the resting level or driven farther from threshold at an **inhibitory synapse**. "Stabilized" means not only kept at the resting level but also having a greater resistance to change from that level.

Thousands of synapses from many different presynaptic cells can affect a single postsynaptic cell (**convergence**), and a single presynaptic cell can send branches to affect many other postsynaptic cells (**divergence**, Figure 8-37).

The level of excitability of a postsynaptic cell at any moment, that is, how close its membrane potential is to threshold, depends on the number of synapses active at any one time and the number that are excitatory or inhibitory. If the membrane of the postsynaptic neuron reaches threshold, it will generate action potentials that are propagated along the axon to the terminal branches, which diverge to influence the excitability of other neurons or effector cells. Because the output of a neuron reflects the sum of all the incoming signals arriving in the form of excitatory and inhibitory synaptic inputs, neurons function as **integrators**.

FUNCTIONAL ANATOMY OF SYNAPSES

There are two types of synapses: **electric** and **chemical**. At electric synapses, the plasma membranes of the pre- and postsynaptic cells are joined by gap junctions (page 45), which allow the charges making up the local currents resulting from action potentials in the presynaptic neuron to flow directly into the postsynaptic cell, depolarizing the membrane to threshold and thus initiating an action potential in the postsynaptic cell. Only a minute percentage of the synapses in the mammalian nervous system are electric, and we shall discuss the much more common chemical synapse.

Figure 8-38 shows the structure of a single typical chemical synapse. Note that in actuality the size and shape of the pre- and postsynaptic elements can vary greatly (Figure 8-39). The axon of the presynaptic neuron ends in a slight swelling, the **axon terminal**. A narrow extracellular space, the **synaptic cleft**, separates the pre- and postsynaptic neurons and prevents *direct* propagation of the current from the presynaptic neuron to the postsynaptic cell. Instead, signals are transmitted across the synaptic cleft by means of a chemical neurotransmitter released from the presynaptic axon terminal or from axon varicosities. Sometimes, more than one neurotransmitter may be simultaneously released from an axon, in which case the additional neurotransmitter is called a cotransmitter.

FIGURE 8-37 Convergence of neural input from many neurons onto a single neuron, and divergence of output from a single neuron onto many others.

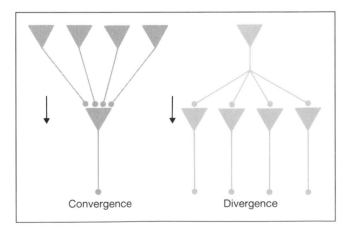

Convergence   Divergence

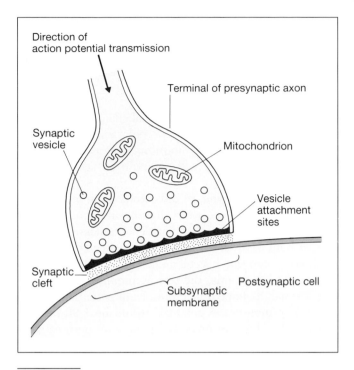

Direction of
action potential transmission

Terminal of presynaptic axon

Synaptic
vesicle

Mitochondrion

Vesicle
attachment
sites

Synaptic
cleft

Postsynaptic cell

Subsynaptic
membrane

FIGURE 8-38 Diagram of a synapse.

When an action potential in the presynaptic neuron reaches the end of the axon and depolarizes the axon terminal, small quantities of the neurotransmitter—and cotransmitter if there is one—are released from the terminal into the synaptic cleft. Calcium ions are the link between depolarization of the presynaptic membrane and neurotransmitter release. Depolarization of the terminal causes voltage-sensitive calcium channels in the membrane to open, and calcium diffuses into the axon terminal from the extracellular fluid, including that in the synaptic cleft.

The neurotransmitter in the terminals is stored in membrane-bound **synaptic vesicles**. The increase in calcium ions in the terminal causes some of these storage vesicles to fuse with specific vesicle-attachment sites on the inner surface of the presynaptic plasma membrane and liberate the vesicle contents into the synaptic cleft by the process of exocytosis.[4]

Once released from the axon terminal, the neuron transmitter molecules diffuse across the cleft, and a fraction of them bind to receptor sites on the plasma membrane of the postsynaptic cell (the fate of the others will

be described later). The combination of the transmitter with the receptor sites opens specific ion channels in the postsynaptic plasma membrane or initiates a second messenger cascade in the postsynaptic cell. If channels are involved, they belong to the class of receptor-operated channels discussed on page 113 and are distinct from the voltage-sensitive channels.

Although Figure 8-39 shows a few exceptions, in general the neurotransmitter is stored on the presynaptic side of the synaptic cleft and receptor sites are on the postsynaptic side so chemical synapses operate in only one direction. One-way conduction across synapses causes action potentials to be transmitted along a given multineuronal pathway in one direction.

There is a delay between the arrival of an action potential at a presynaptic terminal and the membrane-potential changes in the postsynaptic cell. It lasts about a

FIGURE 8-39 Variety of synaptic associations made upon a fiber of an ascending pathway in the spinal cord. Dendrites are shaded green. (A) Simple axodendritic connection, which may include (B) one or more presynaptic synapses or (C) an arrangement known as a synaptic triad. (D) Synaptic arrangement ending on a spine extending from the postsynaptic cell. Such spines are common in the central nervous system and are thought to play a role in learning. The arrows indicate the direction of synaptic activity. [*Redrawn from Maxwell and Rethelyi.*]

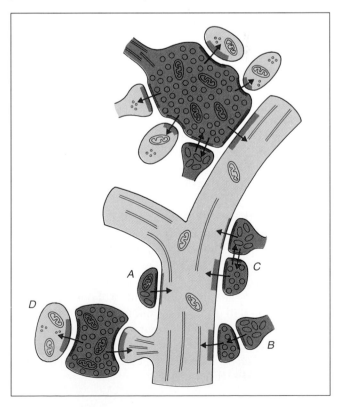

---

[4]Although the release of neurotransmitter by exocytosis of synaptic vesicles is generally accepted, it has been questioned by several investigators who believe the neurotransmitters are released directly from the cytosol.

millisecond and is called the synaptic delay. It can probably be accounted for by the time required for calcium to enter the axon terminal and for the synaptic vesicles to fuse with the plasma membrane and release their transmitter. The time required for the neurotransmitter to diffuse across the synaptic cleft is negligible.

The ion channels in the postsynaptic membrane close, stopping the currents responsible for the postsynaptic potential, when the activity of the neurotransmitter ceases. Neurotransmitter action may stop because of (1) chemical transformation into an ineffective substance, or (2) diffusion away from the receptor site, or (3) active transport back into the axon terminal or, in some cases, into nearby glial cells. The transmitters that never combined with the receptors in the first place meet the same fate as do the molecules that did combine with receptors.

The two kinds of chemical synapses, excitatory and inhibitory, are differentiated by the effects the neurotransmitter has on the postsynaptic cell. Whether the effect is inhibitory or excitatory depends on the type of receptor and on the type of channel the receptor controls. We emphasize that it is possible for the same neurotransmitter to produce different effects at different receptors; for example, a neurotransmitter may produce excitation at one synapse but inhibition at another. The response of a postsynaptic cell to a chemical depends on the type of receptor mechanism that is brought into operation when the chemical binds to a receptor. Thus, neurotransmitters are merely chemical messengers that trigger a response that is built into the target cell.

Excitatory Chemical Synapse

At an excitatory synapse, the postsynaptic response to the neurotransmitter is a depolarization, bringing the membrane potential closer to threshold. Here the effect of the neurotransmitter-receptor combination is usually to open postsynaptic-membrane ion channels that are permeable to sodium, potassium, and other small, positively charged ions. These ions then are free to move according to the electrical and chemical gradients across the membrane.

Recall that there is both an electrical and a concentration gradient driving sodium into the cell, whereas in the case of potassium, the electrical gradient is opposed by the concentration gradient. Opening channels to all small ions will therefore result in the simultaneous movement of a relatively small number of potassium ions out of the cell and a larger number of sodium ions into the cell. Thus, the *net* movement of positive ions is into the postsynaptic cell, and this slightly depolarizes it. This potential change is called an **excitatory postsynaptic potential** (**EPSP**), (Figure 8-40).

The EPSP is a graded potential that spreads decrementally away from the site of the synapse via local current. Its only function is to bring the membrane potential of the postsynaptic neuron closer to threshold.

Inhibitory Chemical Synapse

At inhibitory synapses, the potential change in the postsynaptic neuron is either hyperpolarizing or stabilizing. Thus, activation of an inhibitory synapse lessens the likelihood that the cell will generate an action potential.

At an inhibitory synapse the combination of the neurotransmitter with receptor sites on the postsynaptic membrane usually opens chloride or potassium channels, or both; sodium channels are not affected. On page 189 it was noted that if a cell membrane were permeable only to potassium ions, the resting membrane potential would equal the potassium equilibrium potential; that is, the resting membrane potential would be −90 instead of −70 mV. Thus, an increased potassium permeability at an inhibitory synapse would move the membrane potential closer to the potassium equilibrium potential. This hyperpolarization, also a graded potential, is referred to as an **inhibitory postsynaptic potential** (**IPSP**) (Figure 8-41).

In cells that actively transport chloride ions out of the cell, the chloride equilibrium potential (−80 mV) is more negative than the resting potential so that an increase in chloride permeability has a hyperpolarizing effect similar to the effect of increasing potassium perme-

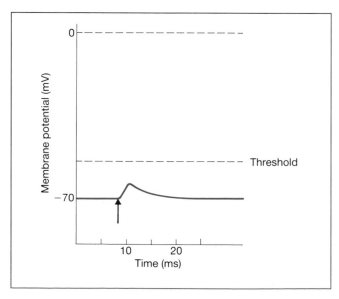

FIGURE 8-40 Excitatory postsynaptic potential (EPSP). Stimulation of the presynaptic neuron is marked by the arrow.

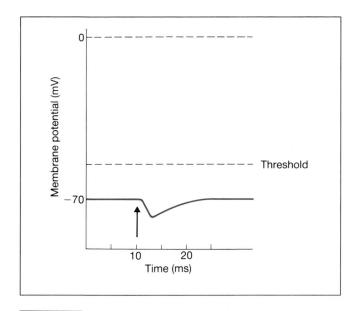

FIGURE 8-41 Inhibitory postsynaptic potential (IPSP). Stimulation of the presynaptic neuron is marked by the arrow.

ability; that is, such an increase generates an IPSP. In cells in which the equilibrium potential for chloride is equal to the resting membrane potential, a rise in chloride-ion permeability stabilizes the membrane at the resting level and lessens the likelihood that the cell will reach threshold during excitatory input. In other words, the increased chloride permeability increases the amount of sodium movement into the cell necessary to displace the membrane potential away from its resting value.

Activation of the Postsynaptic Cell

A feature that makes postsynaptic integration possible is that, in most neurons, one excitatory synaptic event by itself is not enough to reach threshold in the postsynaptic neuron, for example, a single EPSP in a motor neuron is estimated to be only 0.5 mV, whereas changes of 15 to 25 mV are necessary to depolarize the membrane to threshold. This being the case, an action potential can be initiated only by the combined effects of many excitatory synapses.

Of the thousands of synapses on any one neuron, probably hundreds are active simultaneously (or close enough in time so that the effects can add together), and the membrane potential of the neuron at any moment is the resultant of all the synaptic activity affecting it at that time. There is a depolarization of the membrane toward threshold when excitatory synaptic input predominates, which is known as facilitation, and a hyperpolarization when inhibitory input predominates (Figure 8-42).

Let us perform a simple experiment to see how two EPSPs, two IPSPs, or an EPSP plus an IPSP interact (Figure 8-43). Assume there are three synaptic inputs to the postsynaptic cell: The synapses from axons A and B are excitatory, and the synapse from axon C is inhibitory. There are laboratory stimulators on axons A, B, and C so that each can be activated individually. A very fine electrode is placed in the cell body of the postsynaptic neuron and connected to record the membrane potential. In Part 1 of the experiment, we shall test the interaction of two EPSPs by stimulating axon A and then, after a short while, stimulating it again. Part 1, Figure 8-43 shows that no interaction occurs between the two EPSPs. The reason is that the change in membrane potential associated with an EPSP is fairly short-lived. Within a few milliseconds (by the time we stimulate axon A for the second time), the postsynaptic cell has returned to its resting condition.

In Part 2, we stimulate axon A for the second time *before* the first EPSP has died away; the second synaptic potential adds to the previous one and creates a greater depolarization than from one synapse alone. This is called **temporal summation** since the input signals arrive at the same cell at different times. (The same effect could have been achieved by stimulating first A, then B a short time later.) In Part 3, axon B is stimulated alone to determine its response and then axons A and B are stimulated simultaneously. The two EPSPs that result also summate in the postsynaptic neuron; this is called **spatial summation** since the two inputs occurred at different locations on the same cell. The summation of multiple EPSPs can bring the membrane to threshold so that action potentials are initiated (Part 4).

FIGURE 8-42 Intracellular recording from a postsynaptic cell during episodes when (A) excitatory synaptic activity predominates and the cell is facilitated and (B) inhibitory synaptic activity dominates.

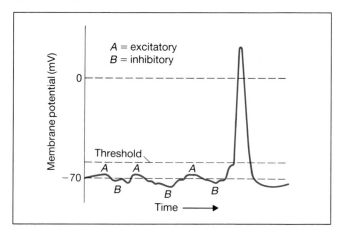

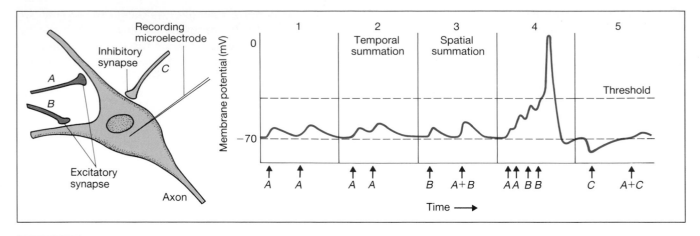

FIGURE 8-43 Interaction of EPSPs and IPSPs at the postsynaptic neuron. Arrows indicate time of stimulation.

So far we have tested only the patterns of interaction of excitatory synapses. What happens if an excitatory and inhibitory synapse are activated so that their effects occur in the postsynaptic cell simultaneously? Since EPSPs and IPSPs are due to oppositely directed local currents, they tend to cancel each other, and there is little or no change in membrane potential (Figure 8-43, Part 5). Inhibitory potentials can also show spatial and temporal summation.

As described above, the postsynaptic membrane is depolarized at an activated excitatory synapse and either stabilized or hyperpolarized at an activated inhibitory synapse. Via the local-current mechanisms described earlier, the plasma membrane of the entire postsynaptic cell body, including the initial segment, reflects these changes and becomes slightly depolarized during activation of an excitatory synapse and stabilized or slightly hyperpolarized during activation of an inhibitory synapse (Figure 8-44).

In the above examples, we referred to the threshold of the postsynaptic neuron as though it were the same for all parts of the cell. However, different parts of the neuron have different thresholds. The cell body and larger dendritic branches reach threshold when their membrane is depolarized about 25 mV from the resting level, but in many cells the initial segment has a threshold that is much closer to the resting potential. In cells whose initial-segment threshold is lower than that of the cell body and dendrites, the initial segment reaches threshold first whenever enough EPSPs summate, and the resulting action potential is propagated down the axon.

The fact that the initial segment has the lowest threshold explains why the location of individual synapses on the postsynaptic cell is important. A synapse

FIGURE 8-44 Comparison of excitatory and inhibitory synapses, showing current direction through the postsynaptic cell following synaptic activation. (A) Current through the postsynaptic cell is away from the excitatory synapse, depolarizing the cell. (B) Current through the postsynaptic cell is toward an inhibitory synapse, hyperpolarizing the cell.

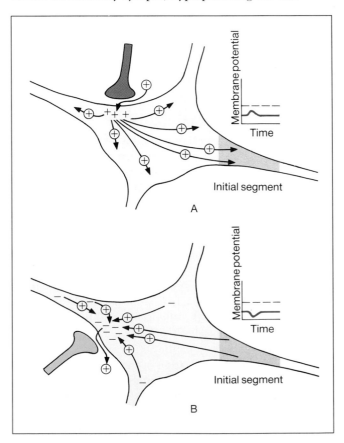

located near the initial segment will have much more impact than will a synapse on the outermost branch of a dendrite because it will expose the initial segment to stronger local current and there will be a greater chance that the threshold there will be reached.

Postsynaptic potentials last much longer than action potentials do. In the event that cumulative EPSPs cause the initial segment to still be depolarized above threshold after an action potential has been fired and the refractory period is over, a second action potential will occur. The greater the summed postsynaptic depolarization above threshold, the greater the frequency of action potentials. Neuronal responses almost always occur in bursts of action potentials rather than as single isolated events.

This discussion has no doubt left the impression that one neuron can influence another only at a synapse. This view requires qualification. Under certain circumstances, local currents from one neuron may directly affect the membrane potential of nearby neurons. This phenomenon is particularly important in areas of the CNS that contain a high density of unmyelinated neuronal processes.

SYNAPTIC EFFECTIVENESS

Synaptic events—whether excitatory or inhibitory—have been presented as though their effects are consistent and reproducible. Actually, the variability in postsynaptic potentials following any particular presynaptic input is enormous. The effect of a given synapse can be influenced by both presynaptic and postsynaptic mechanisms (Table 8-11).

The presynaptic mechanisms include a variety of factors that alter the amount of neurotransmitter (or neurotransmitters) released from the presynaptic neuron, because the greater the amount of neurotransmitter released, the greater the number of ion channels opened in the postsynaptic membrane, and the larger the amplitude of the EPSP or IPSP in the postsynaptic cell.

For example, calcium that has entered the terminal during previous action potentials is normally removed by being pumped out of the cell or (temporarily) into intracellular organelles. If the calcium removal systems are not able to keep up with entry, calcium concentration in the terminal and the amount of transmitter released will be greater than normal.

The amount of transmitter released is also altered by the activation of membrane receptors on the *presynaptic* axon terminal. These receptors may or may not be associated with a second synaptic ending known as an axon-axon or **presynaptic synapse**, in which axon terminals of

TABLE 8-11 FACTORS THAT DETERMINE SYNAPTIC EFFECTIVENESS
I. Presynaptic factors
A. Availability of neurotransmitter
1. Availability of precursor molecules
2. Amount (or activity) of the rate-limiting enzyme in neurotransmitter synthesis pathway
B. Axon terminal membrane potential
C. Axon terminal residual calcium
D. Activation of membrane receptors on presynaptic terminal
1. Presynaptic (axon-axon) synapses
2. Autoreceptors
3. Other receptors
E. Certain drugs and diseases
II. Postsynaptic factors
A. Immediate past history of electrical state of postsynaptic membrane, that is, facilitation or inhibition from temporal or spatial summation
B. Effects of other transmitter-type chemicals acting on postsynaptic neuron
C. Certain drugs or diseases

one neuron end on axon terminals of another. Figure 8-45 shows a typical synapse between a presynaptic neuron's axon terminal (B) and a postsynaptic neuron's cell body (C). What is new in this figure is that the axon terminal B itself receives synaptic input from yet another axon (A). The neurotransmitter released by A combines with receptor sites on B, resulting in a change in the

FIGURE 8-45 Axon-axon synapse between axon terminal A and axon terminal B. C is the final postsynaptic cell body.

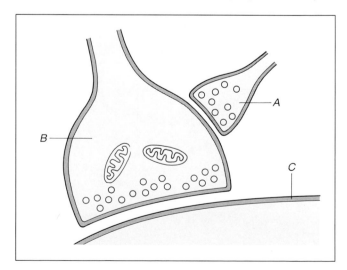

amount of transmitter released from B in response to action potentials. Thus, neuron A has no *direct* effect on neuron C but has an important influence on the ability of B to influence C. Neuron A is said to be exerting a presynaptic effect on the synapse between B and C, which can increase or decrease its effectiveness.

Depending upon the nature of the transmitter released from A and the type of receptor sites on B, the presynaptic effect may decrease the amount of transmitter released from B or increase it. If the final result is to inhibit neuron C, the synapse between A and B is called **presynaptic inhibition**; if it facilitates C, the synapse between A and B is known as **presynaptic facilitation**. Presynaptic synapses commonly occur, for example, in nerve fiber systems that transmit incoming afferent information.

FIGURE 8-46 Drug actions at synapses: (A) Increase leakage of neurotransmitter from vesicle to cytoplasm, exposing it to enzyme breakdown, (B) increase transmitter release, (C) block transmitter release, (D) inhibit transmitter synthesis, (E) block transmitter reuptake, (F) block enzymes that metabolize transmitter, (G) bind to receptor to block (antagonist) or mimic (agonist) transmitter action, (H) inhibit or facilitate second-messenger activity. (*Redrawn from Snyder.*)

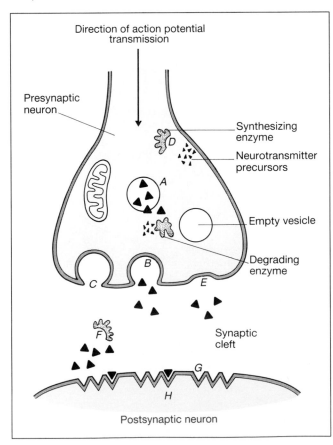

Receptors on the presynaptic terminal may not be associated with axon-axon synapses, in which case they are activated by specific chemicals released by nearby neurons or glia, distant tissues, or the presynaptic terminal itself. In the latter case, the receptors are called autoreceptors and provide a feedback mechanism by which the neuron can regulate its own output.

Modification of Synaptic Transmission by Drugs and Disease

The great majority of drugs that act on the nervous system do so by altering synaptic mechanisms and thus synaptic effectiveness. All the synaptic mechanisms labeled in Figure 8-46 are vulnerable.

The long-term effects of drugs are sometimes difficult to predict because the very processes affected by the drugs are dynamic, and the imbalances produced by the initial drug action are soon counteracted by feedback mechanisms that normally regulate the processes. For example, if a drug interferes with the action of norepinephrine by inhibiting the rate-limiting enzyme in the synthetic pathway for this neurotransmitter, the neurons may respond by increasing the rate of precursor transport into the axon terminals to maximize the use of any enzyme that is available.

Recall from Chapter 7 that drugs that, upon binding to a receptor on the pre- or postsynaptic membrane, produce a response similar to that caused by the neurotransmitter released at the synapse are called agonists, and drugs that bind to the receptor but are unable to activate it are antagonists. By occupying the receptors, antagonists prevent binding of the normal neurotransmitter when it is released at the synapse.

Diseases can also affect synaptic mechanisms. For example, a toxin produced by the tetanus bacillus interferes with inhibitory input to the motor neurons supplying skeletal muscles. This elimination of inhibitory input to the motor neurons leaves unchecked the excitatory inputs, resulting in excessive motor neuron activity, leading to contraction of the skeletal muscles and loss of coordination. The spasms of the jaw muscles appearing early in the disease are responsible for the common name of the condition, lockjaw.

NEUROTRANSMITTERS AND NEUROMODULATORS

We have emphasized the role of neurotransmitters in eliciting EPSPs and IPSPs. Some chemical messengers released by neurons elicit in neurons complex responses that cannot be simply described as EPSPs or IPSPs. The word "modulation" is used for these complex responses, and the messengers that cause them are called **neuromodulators**. The distinctions between neuromodulators and neurotransmitters are, however, far from clear.[5] In fact, certain neuromodulators are often synthesized by the presynpatic cell and released with the neurotransmitter. Neuromodulators often modify the postsynaptic cell's response to specific neurotransmitters, amplifying or dampening the effectiveness of ongoing synaptic activity.

In general, whereas neurotransmitters influence ion channels, neuromodulators bring about biochemical changes in neurons, often via a second-messenger system. The actions of both neuromodulators and neurotransmitters may result in changes in the membrane potential, but the changes induced by neurotransmitters are faster and more direct. Thus, neurotransmitter actions occur within milliseconds whereas neuromodulators tend to be associated with such events as learning, development, motivational states, or even some sensory or motor activities, all of which are measured in minutes, hours, or even days.

Like neurotransmitters, neuromodulators can activate both postsynaptic and presynaptic receptors. In the presynaptic effects, neuromodulators may alter the neurotransmitter's synthesis, release, reuptake, or metabolism.

There are many different substances known or suspected to be neurotransmitters or neuromodulators. Moreover, a given substance may act as a neurotransmitter in one area of the brain and as a neuromodulator in a different (or even the same) area.

Table 8-12 lists some substances thought presently to function as neurotransmitters or neuromodulators. In Chapter 10 you will see that some of these substances also function as hormones.

While writing this section, we have given in to the temptation to list behavioral functions for some of the substances, but we are fully aware that, in truth, the explanation of various behaviors, such as sleeping, learning, experiencing emotions, and eating, in terms of neurotransmitters and neuromodulators or the explanation of actions and experiences that follow use of a mind-altering drug are simply not available at this time. Even tracking down and understanding all the cellular events resulting from the use of a fairly simple drug such as LSD (lysergic acid diethylamide) is extremely difficult, and this level of understanding is far removed from explaining the behavior that results from use of this drug. Thus, we resort to the frequent use of the frustrating but honest terms "possibly," "may," "seems to," "perhaps," and so on.

Indeed, for simplicity's sake, in the following discus-

---

[5]To add to the complexity, it is clear that some hormones, paracrines, and agents of the immune system may serve as neuromodulators.

TABLE 8-12 CLASSES OF SOME OF THE CHEMICALS KNOWN OR PRESUMED TO BE NEUROTRANSMITTERS OR NEUROMODULATORS
1. Acetylcholine (ACh)

2. Biogenic amines
    Catecholamines
        Dopamine (DA)
        Norepinephrine (NE)
        Epinephrine (Epi)
    Serotonin (5-hydroxytryptamine, 5-HT)
    Histamine

3. Amino acid neurotransmitters
    Excitatory amino acids
        Glutamate
        Aspartate
    Inhibitory amino acids
        Gamma-aminobutyric acid (GABA)
        Glycine
        Taurine

4. Neuropeptides
    Endorphins (endorphin, enkephalins, dynorphin)
    Substance P
    Somatostatin
    Vasoactive intestinal peptide
    Cholecystokinin-like peptide
    Neurotensin
    Insulin
    Gastrin
    Glucagon
    Thyrotropin releasing hormone (TRH)
    Gonadotropin releasing hormone (GnRH)
    Adrenocorticotropin (ACTH)
    Angiotensin II
    Bradykinin
    Vasopressin (Antidiuretic hormone, ADH)
    Oxytocin
    Epidermal growth factor
    Prolactin
    Bombesin
    Motilin
    Peptide histidine isoleucine (PHI)
    Peptide histidine methionine (PHM)
    Secretin
    Neuropeptide Y (NPY)
    Peptide YY
    Pancreatic polypeptide
    Growth hormone releasing hormone (GHRH)
    Neurokinin A (Substance K)
    Neurokinin B
    Interferon
    Interleukin
    Thymosin

5. Miscellaneous
    Nucleotides
        Adenosine
        Adenosine triphosphate (ATP)
    Prostaglandins (Prostaglandin E)

sion of several major chemical messengers in the nervous system, we use the term neurotransmitter, realizing that sometimes the function may more appropriately be described as neuromodulator.

Acetylcholine

Acetylcholine (ACh) is synthesized from choline and acetyl coenzyme A in the cytoplasm of synaptic terminals and is stored in synaptic vesicles. After acetylcholine release and the activation of receptors on the postsynaptic membrane, the concentration of ACh at the postsynaptic membrane is reduced (thereby stopping receptor activation) by diffusion away from the receptors. The concentration is also reduced by an enzyme, **acetylcholinesterase** that is located on the pre- and postsynaptic membranes and rapidly destroys ACh, releasing choline. The choline is then actively transported back into the axon terminals where it is reused in the synthesis of ACh.

Acetylcholine is a major neurotransmitter in the efferent division of the peripheral nervous system (page 180) and is present in the brain. The cell bodies of the brain's cholinergic neurons are concentrated in relatively few areas, but their axons are widely distributed (Figure 8-47).

Cholinergic neurons in the brain cease to function normally in **Alzheimer's disease**, a degenerative brain disease that is usually age-related and is the most common cause of declining intellectual function in late life, affecting 10 to 15 percent of people over age 65. This disease is associated with a decreased release of ACh in certain areas of the brain and a loss of the postsynaptic neurons that would have responded to the ACh. These defects are presumably related to the mental impair-

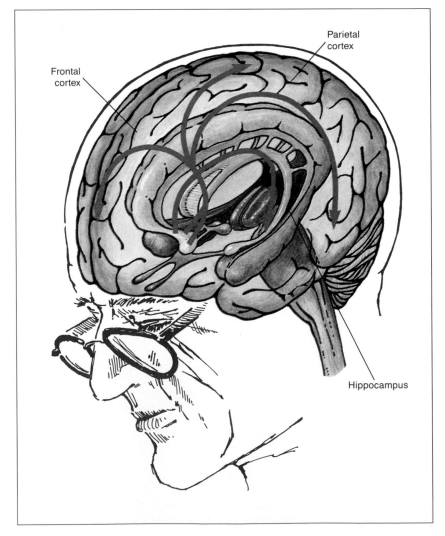

Frontal cortex

Parietal cortex

Hippocampus

FIGURE 8-47 The major cholinergic pathways implicated in Alzheimer's disease.

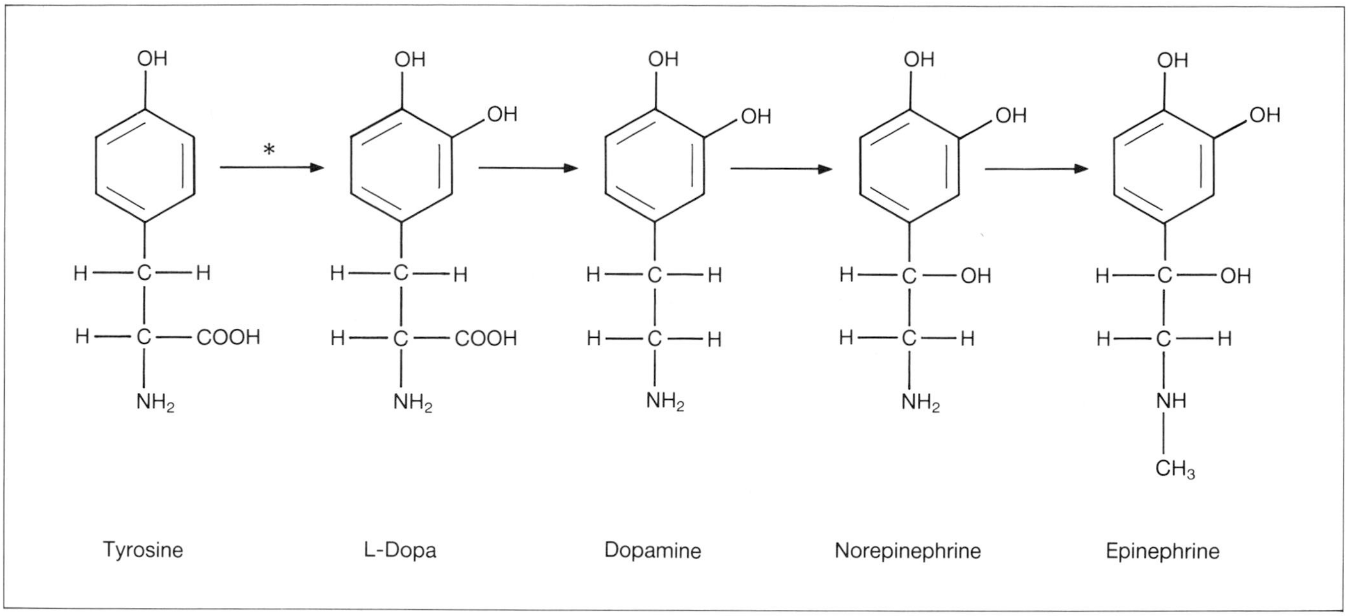

FIGURE 8-48 Catecholamine biosynthetic pathway. The asterisk indicates the site of action of tyrosine hydroxylase, the rate-limiting enzyme.

ment accompanied by the declining language and perceptual abilities, confusion, and memory loss that characterize Alzheimer's victims. The exact causes of this disease are unknown.

Biogenic Amines

The **biogenic amines** are of the basic type $R-NH_2$. (Amino acid neurotransmitters also fit this chemical type, but neurophysiologists traditionally put them into a category of their own.) The most common biogenic amines are dopamine, norepinephrine, serotonin, and histamine. Epinephrine, another biogenic amine, is not a common neurotransmitter in the brain or peripheral nervous system but is the major hormone secreted by the adrenal medulla.

Dopamine, norepinephrine, and epinephrine all contain a catechol ring (a six-sided carbon ring with two adjacent hydroxyl groups) and an amine group; thus they are called **catecholamines**. The catecholamines are formed from the amino acid tyrosine and share the synthetic pathway shown in Figure 8-48, which begins with the uptake of tyrosine by the axon terminals. Depending on the enzymes present in the terminals, the neurotransmitter finally formed may be any one of the three catecholamines. Release from the presynaptic terminals is strongly modulated by presynaptic receptors.

After activation of the receptors on the postsynaptic cell, the catecholamine concentration in the synaptic cleft declines mainly because the catecholamine is actively transported back into the axon terminal. It is also broken down by the enzymes **monoamine oxidase** and **catechol-O-methyltransferase**, which occur in the extracellular fluid and the axon terminal, but the intraneuronal enzymes play a larger role in catecholamine breakdown.

The catecholamines, particularly norepinephrine, in addition to their roles in the autonomic nervous system, are transmitters in the CNS. The cell bodies of the catecholamine releasing neurons lie in the brainstem, and although relatively few in number, their axons branch greatly and go to virtually all parts of the brain and spinal cord. The catecholamines exert a much greater influence than the number of neurons alone would suggest, possibly because of their neuromodulator-like effects on postsynaptic neurons. The time course of action of the biogenic amines is often much slower than that of other substances (their receptor action is linked to the second messengers cAMP, cGMP, or phosphatidylinositol). Thus, they may alter the actions of the postsynaptic neurons as these neurons respond to other transmitters that act more rapidly. The catecholamines are involved in the control of movement (particularly dopamine), mood, attention, and, possibly, certain endocrine, cardiovascular, and stress responses.

Serotonin (5-hydroxytryptamine, 5-HT) is produced from tryptophan, an essential amino acid, and is metabolized by monoamine oxidase. Its effects generally have a slow onset, indicating that it works as a neuromodulator. It is contained mainly in brainstem neurons that innervate virtually all other areas of the CNS. In general, serotonin has an excitatory effect on motor pathways and an inhibitory effect on the sensory pathways. The activity of serotonergic neurons is lowest or absent in sleep and highest during states of alert wakefulness when increased serotonergic activity enhances motor responsiveness and suppresses sensory systems to screen out distracting stimuli.

Serotonin may also be important in the neural pathways controlling mood, and several substances chemically related to it, for example, psilocybin, a hallucinogenic agent found in some mushrooms, have potent psychic effects. Moreover, LSD, the most potent hallucinogenic drug known, inhibits serotonergic neurons. Serotonergic pathways also function in the regulation of carbohydrate intake and hypothalamic releasing hormones, and they have been implicated in alcoholism and other obsessive-compulsive disorders. Serotonin is also present in many nonneural cells, for example, blood platelets, mast cells, and specialized cells lining the digestive tract. In fact, the brain contains only 1 to 2 percent of the body's serotonin.

Amino Acid Transmitters

As we have just seen, a number of neurotransmitters are synthesized from amino acids; in addition, several amino acids function directly as transmitters. In fact, they are the most abundant transmitter class in the CNS. **Aspartate** and **glutamate** are the major excitatory transmitters in the CNS, and **GABA** (gamma-aminobutyric acid), which is not really an amino acid but is synthesized from glutamic acid and is classified with the amino acids, is the major inhibitory transmitter, so widespread that most neurons in the CNS are believed to possess GABA receptors.

Neuropeptides

Neuropeptides are composed of two or more amino acids linked together by peptide bonds and are synthesized in neural tissue. More than 50 neuropeptides have been found, although the physiological roles are not known for many of them. It seems that evolution has selected the same chemical messengers for use in widely differing circumstances, and many of the neuropeptides had been previously identified in nonneural tissue where they are known to function as hormones, paracrine agents, or interleukins, depending on their site of release and target tissue. Thus, many of the peptides function in communication networks within the neural, endocrine, and immune systems. They generally retain the name they

were given when first discovered in the nonneural tissue.

The neuropeptides are formed differently from the other transmitters, which are synthesized in the axon terminals via a very few enzyme-mediated steps. The neuropeptides, in contrast, are derived from large precursor molecules called prehormones or preprohormones, which in themselves have little, if any, inherent biological activity. The synthesis of these precursors is directed by mRNA, which exists only in the cell body and large dendrites of the neuron, often a considerable distance from the axon terminals where the peptides are released.

In the cell body, the prohormone is packaged into vesicles, which are then moved by way of axon transport into the nerve terminal where their contents are cleaved by specific peptidases. Many of the prehormones and preprohormones are actually "polyproteins" in that they contain multiple peptides. These copies may be of the same peptide, related to it, or totally different. In many cases neuropeptides are cosecreted with another type of transmitter.

Certain neuropeptides, the **endorphins**, **dynorphins**, and **enkephalins**, have gained much interest since their discovery because receptors for these transmitters are the sites of action of the opiate drugs such as morphine and codeine. ("Endorphins" refers to a specific type of the opiate-like transmitters, but it is also used to refer to the entire class, including the dynorphins and the enkephalins.) No physiological function of any of the endorphins has been demonstrated conclusively, but there is some evidence that they play a role in regulating pain, in eating and drinking behavior, in central regulation of the cardiovascular system, and in cell development. There are at least three receptor types that respond to the enkephalins, and they occur in discrete locations in the CNS—in areas containing pathways that convey pain information, for example, and in parts of the brain involved in mood and emotions.

Substance P, another of the neuropeptides, is a transmitter for afferent neurons that relay sensory information into the CNS and is thought to be involved in the transmission of nociceptive (pain-producing) stimuli.

NEUROEFFECTOR COMMUNICATION

Thus far we have described the effects of neurotransmitters released from synapses. With the exception of the autonomic neurons that innervate the enteric nervous system of the gastrointestinal tract, the efferent neurons of the peripheral nervous system end not on neurons but on muscle and gland cells. The neurotransmitter released by the efferent neuron's terminals provides the link by which electrical activity of the nervous system is able to regulate effector-cell activity.

Recall that these junctions between efferent neurons and effector cells are called neuroeffector junctions, and the events that occur here are similar to those at a synapse. The neurotransmitter is released from the efferent neurons upon the arrival of an action potential at the axon terminals or varicosities, and the transmitter then diffuses to the surface of the effector cell, where it binds to receptors on that cell's plasma membrane. The recep-

tors may be directly under the axon terminal or varicosity, or they may be some distance away so that the diffusion path that must be followed by the transmitter is tortuous and long. The receptor of the effector cell may be associated with ion channels that alter the membrane potential of the cell, or it may be associated with an enzyme that results in the formation of a second messenger in the effector cell. The response (muscle contraction or glandular secretion) of the effector cell to these changes will be described in later chapters. As described earlier, the major neurotransmitters released by efferent neurons are acetylcholine and norepinephrine.

SECTION D
RECEPTORS

We have discussed the initiation of neural activity at synapses and by the spontaneous activity of neurons, and we now turn to the sensory receptors, which initiate neural activity at the border between the nervous system and the world outside it. Since some receptors respond to changes in the internal environment, the "outside world" in this sense can mean, for example, sound waves in the air around us or distension of a blood vessel inside our bodies.

Information about the external world and about the body's internal environment exists in different energy forms—pressure, temperature, light, sound waves, and so on. It is the function of receptors at the peripheral ends of afferent neurons to translate these energy forms into graded potentials that can initiate action potentials. (Again, by peripheral we mean the end farthest from the CNS.) The rest of the nervous system can extract meaning only from action potentials or, over very short distances, from graded potentials. Regardless of its original energy form, information from receptors linking the nervous system with the outside world must be translated into the language of action potentials or graded potentials. The energy that impinges upon and activates a receptor is known as a stimulus. The process by which a stimulus—a photon of light, say, or stretch of a muscle— is transformed into an electrical response at a receptor is known as **transduction**.

Receptors are either specialized peripheral endings of afferent neurons (Figure 8-49A) or separate cells that affect the peripheral ends of afferent neurons (Figure

8-49B). In the latter case, upon stimulation the receptor cell releases a neurotransmitter that diffuses across the extracellular cleft between the receptor cell and the afferent neuron and binds to specific sites on the afferent neuron. In both instances, the result is a graded potential in the afferent neuron, which leads to the generation of action potentials.

There are many types of receptors, each of which is specific; that is, each responds much more readily to one form of energy than to others, although virtually all receptors can be activated by several forms of energy if the

FIGURE 8-49 Afferent receptors. The sensitive membrane that responds to a stimulus is either (A) an ending of the afferent neuron or (B) on a separate cell adjacent to the afferent neuron (highly schematized).

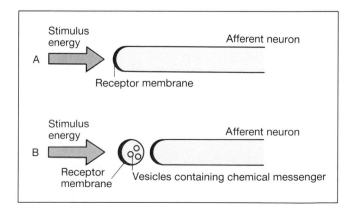

intensity is sufficiently high. For example, the receptors of the eye normally respond to light, but they can be activated by an intense mechanical stimulus, like a poke in the eye. Note, however, that one still experiences the sensation of light in response to the poke in the eye. Regardless of how the receptor is stimulated, any given receptor gives rise to only one sensation. This concept, that for every kind of sensation there is a special type of receptor whose activation always gives rise to that sensation, is stated as the **doctrine of specific nerve energies**.

Most receptors are exquisitely sensitive to their specific energy forms. For example, olfactory receptors can respond to as few as three or four odorous molecules in the inspired air, and visual receptors can respond to a single photon, the smallest quantity of light.

THE RECEPTOR POTENTIAL

The general mechanisms of receptor activation are believed to be the same for all types of receptors. The **stimulus** acts in some way to change the permeability of a specialized receptor membrane that receives information about the outside world. The permeability change produces a graded potential called a **receptor potential**. In describing the general mechanisms for receptor activation, we shall use as an example the pacinian corpuscle, which is the peripheral ending of an afferent neuron (Figure 8-50), because it is best-understood.

The pacinian corpuscle is a **mechanoreceptor**, which responds to mechanical deformation such as bending, stretching, or pressing on the receptor membrane. This occurrence in some unknown way causes ion channels in the membrane to open so that membrane permeability to small ions is increased. Because of the increased permeability, ions move across the membrane down their electrical and concentration gradients.

Just as at an activated excitatory synapse, the result of a nonselective increase in membrane permeability is a net depolarizing movement of positive charge (mainly sodium) into the cell. The movement of sodium into the receptor thus plays the major role in generating the receptor potential in the mechanoreceptor and very likely in most other receptor types as well. (A major exception is in the eye, as we shall see in Chapter 9.)

The receptor membrane of the afferent neuron ending in the pacinian corpuscle has a very high threshold, and the receptor potential is not large enough to cause an action potential there. Instead, local current flows from the receptor membrane a short distance to a region where the membrane threshold is lower. In myelinated afferent neurons, this region is usually at the first node of

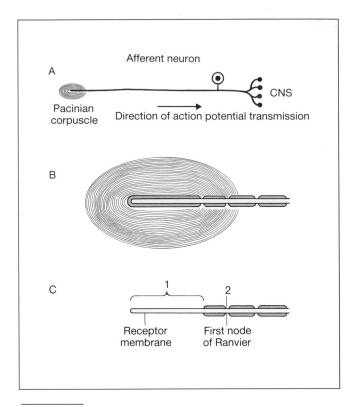

FIGURE 8-50 (A) An afferent neuron with a mechanoreceptor (pacinian corpuscle) ending. (B) A pacinian corpuscle showing how the nerve ending is modified by cellular structures. (C) The onion-like part has been removed to show the naked nerve ending of the same mechanoreceptor. The receptor potential arises at the nerve ending 1, and the action potential arises at the first node of the myelin sheath 2.

the myelin sheath. As is true of all graded potentials, the magnitude of the receptor potential decreases with distance from its origin; but if the amount of depolarization that reaches the first node is large enough to bring the membrane there to threshold, action potentials are initiated. The action potentials then propagate along the nerve fiber; the only function of the receptor potential is to trigger action potentials.

As long as the first node remains depolarized above threshold, action potentials continue to fire and propagate along the afferent neuron. The greater the depolarization, up to a point, the greater the frequency of action potentials. Therefore, the magnitude and duration of the receptor potential determine the action-potential frequency in the afferent neuron.

Although the receptor-potential amplitude is a determinant of action-potential frequency, it does not determine action-potential magnitude. Since the action po-

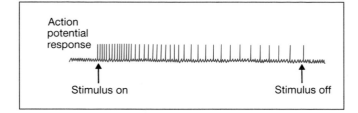

Action
potential
response

Stimulus on   Stimulus off

FIGURE 8-51 Action potentials in a single afferent nerve
fiber showing adaptation to a stimulus of constant strength.

tential is all-or-none, its magnitude is independent of the
strength of the initiating stimulus.

MAGNITUDE OF THE RECEPTOR POTENTIAL

Factors that control the magnitude of the receptor
potential include stimulus intensity (Figure 8-27B),
rate of change of stimulus application, summation
of successive receptor potentials (Figure 8-27D), and **adaptation**, which is a decrease in the frequency of action
potentials in an afferent neuron despite maintenance of
the stimulus at constant strength.

Adaptation of an afferent neuron can be seen in Figure 8-51. Some receptors adapt completely so that in
spite of a constantly maintained stimulus, the generation
of action potentials stops. This can be experienced with
the receptors for taste if the concentration of the solution
being tasted is moderate and the tongue is not moved. In
some pacinian corpuscles, the receptors fire only once at
the onset of the stimulus. In contrast to these rapidly
adapting receptors, slowly adapting types such as those
for skin temperature merely drop from an initial high
action-potential frequency to a lower level, which is then
maintained for the duration of the stimulus.

SUMMARY

Section A. Structure of the Nervous System

The nervous system is divided into two parts: The central nervous system comprises the brain and spinal cord, and the peripheral nervous system consists of nerves extending from the
CNS.

Functional Anatomy of Neurons
I. The basic unit of the nervous system is the nerve cell, or
neuron.
   A. Dendrites and the cell body receive information from
   other neurons.

B. The axon (nerve fiber), which may be covered with
sections of myelin separated by nodes of Ranvier,
transmits information to other neurons or effector
cells.
II. Neurons are classified in three ways:
   A. Afferent neurons transmit information *into* the CNS
   from receptors at their peripheral endings.
   B. Efferent neurons transmit information *out of* the CNS
   to effector cells.
   C. Interneurons lie entirely with the CNS and connect
   afferent and efferent neurons.
III. Information is transmitted across a synapse by a neurotransmitter, which is released by a presynaptic neuron
and combines with receptors on a postsynaptic neuron.
IV. The CNS also contains glial cells, which sustain the neurons metabolically, form myelin, and serve as guides for
the neurons during development.

Divisions of the Nervous System
I. Inside the skull and vertebral column, the brain and spinal cord are enclosed in and protected by the meninges.
II. The spinal cord is divided into two areas: a central gray
matter, which contains nerve cell bodies and dendrites,
and white matter, which surrounds the gray matter and
contains myelinated axons organized into ascending or
descending tracts.
III. The axons of afferent and efferent neurons join to form
the spinal nerves.
IV. The brain is divided into six regions: cerebrum, diencephalon, midbrain, pons, medulla oblongata, and cerebellum.
   A. The midbrain, pons, and medulla oblongata form the
   brainstem, which contains the reticular formation.
   B. The cerebellum plays a role in posture, movement,
   and some kinds of memory.
   C. The cerebrum, made up of right and left cerebral
   hemispheres, and the diencephalon together form the
   forebrain. The cerebral cortex forms the outer shell of
   the cerebrum and is divided into parietal, frontal, occipital, and temporal lobes.
   D. The diencephalon contains the thalamus and hypothalamus.
   E. The limbic system is a set of deep forebrain structures
   associated with learning and emotions.
V. The peripheral nervous system consists of 43 paired
nerves: 12 pairs of cranial and 31 pairs of spinal. Most
nerves contain axons of afferent and efferent neurons.
VI. The efferent division of the peripheral nervous system is
divided into somatic and autonomic parts. The somatic
fibers innervate skeletal muscle cells and release the neurotransmitter acetylcholine.
VII. The autonomic nervous system innervates cardiac and
smooth muscle, glands, and gastrointestinal-tract neurons. Each autonomic pathway consists of a preganglionic
neuron with its cell body in the CNS and a postganglionic
neuron with its cell body in an autonomic ganglion outside the CNS.

A. The autonomic nervous system is divided into sympathetic and parasympathetic components. The preganglionic neurons in both sympathetic and parasympathetic divisions and the postganglionic neurons in the parasympathetic division release acetylcholine; the postganglionic parasympathetic neurons release mainly norepinephrine.

B. The receptors that respond to acetylcholine (cholinergic receptors) or norepinephrine (adrenergic receptors) are further classified as nicotinic and muscarinic cholinergic and alpha- and beta-adrenergic types.

C. The adrenal medulla is a hormone-secreting part of the sympathetic nervous system and secretes mainly epinephrine.

D. Effector organs innervated by the autonomic nervous system generally receive dual innervation.

Neural Growth and Regeneration

I. Neurons develop from precursor cells, migrate to their final location, and send out processes to their target cells.

II. Cell division to form new neurons is completed by birth.

III. After degeneration of a severed axon, damaged peripheral neurons can regrow to their target organ. Damaged neurons of the CNS do not regenerate or restore function.

Blood Supply, Blood-Brain Barrier Phenomena, and Cerebrospinal Fluid

I. Brain tissue depends on a continuous supply of glucose and oxygen for metabolism.

II. The chemical composition of the extracellular fluid of the CNS is closely regulated by the blood-brain barrier.

III. The brain ventricles and the space within the meninges are filled with cerebrospinal fluid, which is formed by the choroid plexuses.

Section B. Membrane Potentials

The Resting Potential

I. Membrane potentials are due mainly to the diffusion of ions and are determined by (1) the membrane's relative permeabilities to ions and (2) the ionic concentration differences across the membrane.

A. Plasma membrane Na,K-ATPase pumps maintain intracellular sodium low and potassium high.

B. In almost all resting cells, the plasma membrane is much more permeable to potassium than sodium and so the membrane potential is close to the potassium equilibrium potential, that is, inside negative relative to the outside.

C. The Na,K-ATPase pumps also contribute directly to the potential because they are electrogenic.

Graded Potentials and Action Potentials

I. Neurons signal information by graded potentials and action potentials.

II. Graded potentials are local potentials whose magnitude can vary and that die out within 1 or 2 mm of their site of origin.

III. An action potential (AP) is a rapid change in the membrane potential that occurs when the potential reverses and the membrane becomes positive inside. APs provide long-distance transmission of information through the nervous system.

A. APs occur in excitable membranes because these membranes contain voltage-sensitive sodium channels, which open as the membrane depolarizes, causing a positive-feedback toward the sodium equilibrium potential.

B. The AP is ended as the sodium channels close and potassium channels open, which restores the resting conditions.

C. Depolarization of excitable membranes triggers action potentials only when the membrane potential exceeds threshold.

D. Regardless of the size of the stimulus, if the membrane reaches threshold, the APs generated are all the same size.

E. A membrane is refractory for a brief time even though stimuli that were previously effective are applied.

F. APs are propagated from one site to another along a membrane without any change in size.

G. In a myelinated nerve fiber, action potentials manifest saltatory conduction.

H. APs can be initiated by receptors at the ends of afferent neurons, by synapses, or in some cells, by pacemaker potentials.

Section C. Synapses

I. An excitatory synapse brings the membrane of the postsynaptic cell closer to threshold. An inhibitory synapse hyperpolarizes the postsynaptic cell (or stabilizes it at its resting level).

II. Whether a postsynaptic cell fires action potentials depends on the number of synapses that are active and whether they are excitatory or inhibitory.

Functional Anatomy of Synapses

I. The signal from a pre- to a postsynaptic neuron is a neurotransmitter stored in synaptic vesicles in the presynaptic axon terminal and released into the synaptic cleft when the axon terminal is depolarized, thereby raising the calcium concentration within the terminal.

II. The neurotransmitter diffuses across the synaptic cleft and binds to receptors on the postsynaptic cell, where they usually open ion channels.

A. At an excitatory synapse the electrical response in the postsynaptic cell is called an excitatory postsynaptic potential (EPSP). At an inhibitory synapse, it is an inhibitory postsynaptic potential (IPSP).

B. Usually at an excitatory synapse, channels in the postsynaptic cell that are permeable to sodium, potassium, and other small positive ions are opened; whereas at inhibitory synapses, channels to potassium and/or chloride are opened.

C. The postsynaptic cell's membrane potential is the result of temporal and spatial summation of the EPSPs and

IPSPs at the many active excitatory and inhibitory synapses on the cell.

Synaptic Effectiveness

Synaptic effects are influenced by pre- and postsynaptic events, drugs, and diseases (Table 8-11).

Neurotransmitters and Neuromodulators

I. In general, neurotransmitters cause EPSPs and IPSPs, and neuromodulators cause more complex metabolic effects in the postsynaptic cell.
II. The actions of neurotransmitters are usually faster than those of neuromodulators.
III. A substance can act as a neurotransmitter at one receptor type and as a neuromodulator at another.
IV. The known or suspected neurotransmitters and neuromodulators are listed in Table 8-12.

Neuroeffector Communication

I. The junction between a neuron and an effector cell is called a neuroeffector junction.
II. The events at a neuroeffector junction—release of transmitter into an extracellular space, diffusion of transmitter to the effector cell, and binding with a receptor on the effector cell—are similar to those at a synapse.

Section D. Receptors

I. Receptors translate information from the external world and internal environment into graded potentials, which then generate action potentials.
   A. Receptors may be either specialized endings of afferent neurons or separate cells adjacent to the neurons.
   B. Receptors respond best to one form of stimulus energy, but they may respond to other energy forms if the stimulus intensity is abnormally high.
   C. Regardless of how a specific receptor is stimulated, activation of that receptor always leads to perception of one sensation (the doctrine of specific nerve energies).
II. Application of a stimulus to a receptor opens ion channels in the receptor membrane. Ions then flow across the membrane, causing a receptor potential.
   A. Receptor-potential magnitude and action potential frequency increase as stimulus strength increases.
   B. Receptor potentials can be summed.

REVIEW QUESTIONS

Section A. Structure of the Nervous System

1. Define:

central nervous system
   (CNS)
peripheral nervous system
neuron
dendrites
axon
nerve fiber

initial segment
axon terminal
varicosities
myelin
nodes of Ranvier
axon transport
afferent neurons

efferent neurons
interneurons
synapse
neurotransmitters
presynaptic neuron
postsynaptic neurons
glial cells
neuroglia
nerve
pathway
tract
ganglia
nuclei
meninges
cerebrospinal fluid
gray matter
white matter
dorsal roots
dorsal root ganglia
ventral roots
spinal nerves
cerebrum
diencephalon
midbrain
pons
medulla oblongata
cerebellum
forebrain
brainstem
cerebral ventricles
cerebellar peduncles
reticular formation
cranial nerves
long neural pathways
multineuronal pathways
multisynaptic pathways
cerebral hemispheres
corpus callosum
lobes
frontal lobe
parietal lobe

occipital lobe
temporal lobe
cerebral cortex
motor systems
subcortical nuclei
basal ganglia
thalamus
hypothalamus
limbic system
Schwann cell
efferent division of the
   peripheral nervous
   system
afferent division of the
   peripheral nervous
   system
primary afferents
first-order neurons
somatic nervous system
autonomic nervous system
acetylcholine
motor neurons
autonomic ganglion
preganglionic
postganglionic
sympathetic division
parasympathetic division
sympathetic trunks
norepinephrine
adrenergic
cholinergic
nicotinic receptors
muscarinic receptors
alpha- and beta-adrenergic
   receptors
dual innervation
adrenal medulla
stroke
blood-brain barrier
   mechanisms
choroid plexuses

2. Describe the direction of information flow through a neuron; through a network consisting of afferent neurons, efferent neurons, and interneurons.

3. Contrast the two uses of the word "receptor."

4. Draw an organizational chart showing: central nervous system, peripheral nervous system, brain, spinal cord, spinal nerves, cranial nerves, forebrain, brainstem, cerebrum, diencephalon, midbrain, pons, medulla oblongata, and cerebellum.

5. Draw a cross section of the spinal cord showing the gray and white matter, dorsal and ventral roots, dorsal root ganglion, and spinal nerve. Indicate the general location of nuclei and pathways.

6. List three functions of the thalamus.

7. List seven functions of the hypothalamus.

8. Draw a peripheral nervous system chart indicating the relationships among afferent and efferent divisions, somatic and autonomic nervous systems, and sympathetic and parasympathetic divisions.

9. Contrast the somatic and autonomic nervous systems; mention at least three characteristics of each.

10. Name the neurotransmitter released at each synapse or neuroeffector junction in the somatic and autonomic systems.

11. Contrast the sympathetic and parasympathetic divisions; mention at least four characteristics of each.

12. Explain how the adrenal medulla can affect receptors on various effector organs despite the fact that its postganglionic cells have no axons.

13. The chemical composition of the CNS extracellular fluid is different from that of blood. Explain how this difference is achieved.

Section B. Membrane Potentials

1. Define:

electric potential	decremental
potential difference	action potential (AP)
potential	excitable membranes
current	afterhyperpolarization
resistance	threshold
Ohm's law	threshold stimuli
resting membrane potential	subthreshold potentials
equilibrium potential	subthreshold stimuli
electrogenic pump	suprathreshold stimuli
depolarize	all-or-none
hyperpolarize	refractory period
repolarize	action-potential propagation
graded potential	pacemaker potential

2. Describe the way that negative and positive charges interact and the potential that exists between separated charges.

3. Contrast the abilities of intracellular and extracellular fluids and membrane lipid to conduct electric current.

4. Draw a simple cell; indicate where the concentrations of Na^+, K^+, and Cl^- are high and low and the electric potential difference across the membrane when the cell is at rest.

5. Explain the conditions that give rise to the resting membrane potential. What effect does membrane permeability have on this potential? What is the role of the Na,K membrane pump on the membrane potential? Is this role direct or indirect?

6. Which two factors determine the magnitude of the resting membrane potential?

7. Explain why the resting membrane potential is not equal to the potassium equilibrium potential.

8. Draw a graded potential and an action potential on a graph of membrane potential versus time. Indicate zero membrane potential, resting membrane potential, and threshold potential; indicate when the membrane is depolarized, hyperpolarized, and repolarizing.

9. List at least eight differences between graded potentials and action potentials.

10. Describe the ionic basis of an action potential; include the role of voltage-sensitive channels and the positive-feedback cycle.

11. Explain threshold and the relative and absolute refractory periods in terms of the ionic basis of the action potential.

12. Describe the propagation of an action potential. Contrast this event in myelinated and unmyelinated axons.

13. List three ways in which action potentials can be initiated in neurons.

Section C. Synapses

1. Define:

neuroeffector junction	neuromodulators
excitatory synapse	acetylcholine (ACh)
inhibitory synapse	acetylcholinesterase
convergence	Alzheimer's disease
divergence	biogenic amines
integrators	catecholamines
electric synapse	monoamine oxidase
chemical synapse	catechol-*O*-
axon terminal	methyltransferase
synaptic cleft	serotonin
synaptic vesicles	amino acid transmitters
excitatory postsynaptic	aspartate
potential (EPSP)	glutamate
inhibitory postsynaptic	gamma-aminobutyric acid
potential (IPSP)	(GABA)
temporal summation	neuropeptides
spatial summation	endorphins
presynaptic synapse	dynorphins
presynaptic inhibition	enkephalins
presynaptic facilitation	substance P

2. Contrast the postsynaptic mechanisms of excitatory and inhibitory synapses.

3. Explain how synapses allow neurons to act as integrators; in your explanation include the concepts of facilitation, temporal and spatial summation, and convergence.

4. List at least eight ways in which the effectiveness of synapses may be altered.

5. Discuss the differences between neurotransmitters and neuromodulators. Why is it so difficult to assign chemical mediators clearly to one or the other category?

6. Discuss the relationship between dopamine, norepinephrine, and epinephrine.

Section D. Receptors

1. Define:

transduction	receptor potential
doctrine of specific nerve	mechanoreceptor
energies	adaptation
stimulus	

2. Describe the process of transduction for a pacinian corpuscle; include in your description the terms "specificity," "stimulus," "receptor potential," "graded potential," "action potential," and "summation."

3. List four ways in which the magnitude of a receptor potential can be varied.

THOUGHT QUESTIONS

(Answers are given in Appendix A.)

1. Neurons are treated with a drug that instantly and permanently stops the Na,K-ATPase pumps. What happens to the resting membrane potential immediately and over time?

2. Extracellular potassium concentration in a person is increased with no change in intracellular potassium concentration. What happens to the resting potential and the action potential?

3. A person has a severe blow to the head but appears to be all right. Over the next weeks, however, he develops loss of appetite, thirst, and sexual capacity but no loss in sensory or motor function. What part of the brain do you think may have been damaged?

4. A person is taking a drug that causes, among other things, dryness of the mouth and speeding of the heart but no impairment of the ability to use the skeletal muscles. What type of receptor does this drug probably block? (Table 8-7 will help you answer this.)

5. Some cells are treated with a drug that blocks chloride channels, and the membrane potential of these cells becomes less negative. From these facts, predict whether the plasma membrane of these cells actively transports chloride and in what direction.

6. If the enzyme acetylcholinesterase were blocked with a drug, what malfunctions would occur?

7. What would an action potential look like if the special potassium channels that open, after a slight delay, in response to depolarization were blocked with a drug?

CHAPTER

THE SENSORY SYSTEMS

A human being's awareness of the world is determined by the neural mechanisms that process afferent information. As we saw in Chapter 8, the initial step of this processing is the conversion of stimulus energy first into receptor potentials and then into action potentials in nerve fibers. The pattern of action potentials in particular nerve fibers is a code that provides information about the world even though, as is frequently the case with symbols, the action potentials differ vastly from what they represent.

A **sensory system** is a part of the nervous system that consists of sensory receptors that receive stimuli from the external or internal environment, the neural pathways that conduct information from the receptors to the brain, and those parts of the brain that deal primarily with processing the information.

Information processed by a sensory system may or may not lead to conscious awareness of the stimulus. Regardless of whether the information does or does not reach consciousness, it is called **sensory information**. If the information reaches consciousness, it can also be called a **sensation**. The understanding of the sensation's meaning is called **perception**. For example, pain is a sensation, but the awareness I have that my tooth hurts is a perception. Perceptions are acquired from the neural processing of sensory information. At present there is absolutely no understanding of the final stages in sensory information processing by which patterns of action potentials become sensations or perceptions.

Intuitively, it might seem that sensory systems operate like familiar electrical equipment, but this is true only up to a point. As an example, let us compare tele-

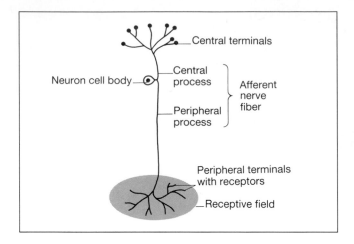

FIGURE 9-1 Sensory unit and receptive field.

phone transmission with our auditory (hearing) sensory system. The telephone changes sound waves into electrical impulses, which are then transmitted along wires to the receiver. Thus far the analogy holds. (Of course, the mechanisms by which electric currents and action potentials are transmitted are quite different, but this does not affect our argument.) The telephone then changes the coded electric impulses *back into sound waves*. Here is the crucial difference, for our brain does

not physically translate the code into sound. Rather, the coded information itself or some correlate of it is what we perceive as sound.

NEURAL PATHWAYS IN SENSORY SYSTEMS

In tracing sensory pathways, we shall begin at the receptor. A single afferent neuron with all its receptor endings makes up a **sensory unit**. In a few cases, the afferent neuron terminates at a single receptor, but generally the peripheral end of an afferent neuron divides into many fine branches, each terminating at a receptor. All the receptors of a sensory unit are preferentially sensitive to the same type of stimulus. For example, they are all sensitive to cold or all to pressure. The portion of the skin or other body surface that, when stimulated, leads to activity in the neuron is called the **receptive field** for that neuron (Figure 9-1).

The central processes of the afferent neurons terminate in the central nervous system, diverging to terminate on several, or many, interneurons (Figure 9-2A) and converging so that the processes of many afferent neurons terminate upon a single interneuron (Figure 9-2B).

The interneurons upon which the afferent neurons synapse are termed second-order neurons, which synapse with third-order neurons, and so on. These neural

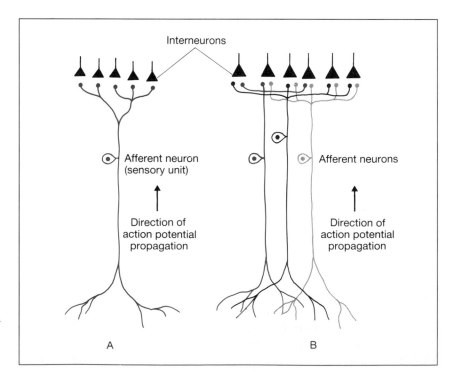

FIGURE 9-2 (A) Divergence of afferent neuron terminals. (B) Convergence of input from several afferent neurons onto single interneurons.

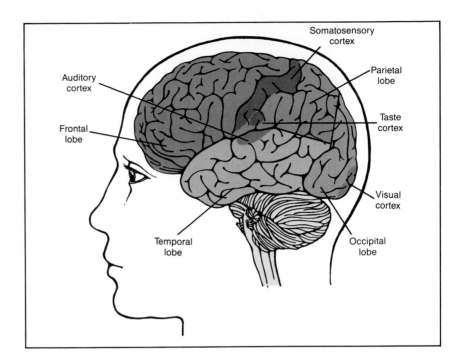

FIGURE 9-3 Primary sensory areas of cerebral cortex.

pathways, which go to the brain, are made up of parallel chains of interneurons grouped together. Each chain consists of three or more synaptically connected neuronal links.

Some of the pathways convey information about only a single type of sensory information. Thus, one pathway may be influenced only by information from mechanoreceptors, whereas another pathway may be influenced only by information from thermoreceptors. The ascending pathways in the spinal cord and brain that carry information about single types of stimuli are known as the **specific ascending pathways**. The specific pathways, except for the olfactory pathways, which go to parts of the limbic system, pass to the brainstem and thalamus, and the final neurons in the pathways go from there to different areas of the cerebral cortex.

The specific ascending pathways that transmit information from **somatic receptors**, that is, the receptors in the framework or outer walls of the body, including skin, skeletal muscle, tendons, and joints, synapse in the **somatosensory cortex**, a strip of cortex that lies in the parietal lobe just behind its junction with the frontal lobe (Figure 9-3).

The specific pathways from the eyes and ears go to other primary cortical receiving areas, the **visual** and **auditory cortex**, respectively (Figure 9-3), and those from the taste buds pass to cortical areas adjacent to the face region of the somatosensory cortex. The pathways serving olfaction have no representation in the cerebral cortex.

Finally, the processing of afferent information does not end in the thalamus or the primary cortical receiving areas but goes from these areas to association areas in the cerebral cortex (page 175).

FIGURE 9-4 Diagrammatic representation of the specific and nonspecific ascending pathways.

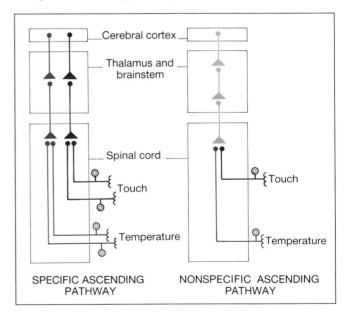

In contrast to the specific ascending pathways, neurons in the **nonspecific ascending pathways** are activated by sensory units of several different types (Figure 9-4) and therefore signal only general information. In other words, they indicate that *something* is happening, usually without specifying just what or where. A given neuron in the nonspecific pathways may respond, for example, to input from different afferent neurons activated by stimuli such as maintained skin pressure, heating, cooling, and skin stretch. Such neurons are called **polymodal**. The nonspecific pathways feed into the brainstem reticular formation and regions of the thalamus and cortex that are not highly discriminative.

BASIC CHARACTERISTICS OF SENSORY CODING

The sensory pathways provide information on stimulus type, intensity, and location. Somehow this information has to be coded in the language of action potentials conveyed over the nerve pathways.

Stimulus Type

As noted above, different receptors have different sensitivities for various types of stimuli. Therefore, the receptor type activated by a stimulus constitutes the first step in coding the different types of stimuli.

Stimulus type is indicated not only by which receptors are stimulated but also by the specificity of the pathways that convey the information to the brain and by the brain region. This means that stimulation of a specific nerve pathway at any point along its course will give rise to the sensation that would have occurred if the receptor

at the beginning of the pathway had been activated by its natural stimulus. For example, regardless of how or where the neural pathways from the eyes are stimulated, one always perceives some form of light, and activation of the pain pathway by the pressure of a growth in the spinal column results in the perception of pain.

Stimulus Intensity

How is information about different stimulus intensities relayed by action potentials of constant amplitude? Frequency of action potentials is one way since increased stimulus strength means a larger receptor potential and higher frequency of action-potential firing. A record of an experiment in which increased stimulus intensity is reflected in increased action-potential frequency in a single afferent nerve fiber is shown in Figure 9-5.

Moreover, as stimulus strength increases, more and more receptors on other branches of the afferent neuron begin to respond. (The receptors on different branches do not necessarily respond equally to a given stimulus.) The action potentials generated by these several receptors propagate along the branches to the main afferent nerve fiber and increase the frequency of action potentials there (Figure 9-6).

In addition to the increased firing frequency in a single neuron, similar receptors on the endings of other afferent neurons are also activated as stimulus strength increases, because stronger stimuli usually affect a larger area. For example, when one touches a surface lightly with a finger, the area of skin in contact with the surface is small, and only receptors in that skin area are stimulated. Pressing down firmly increases the area of skin stimulated. This "calling in" of receptors on additional nerve cells is known as **recruitment**.

FIGURE 9-5 Action potentials from the afferent fiber leading from a pressure receptor as the receptor was subjected to pressures of different magnitudes.

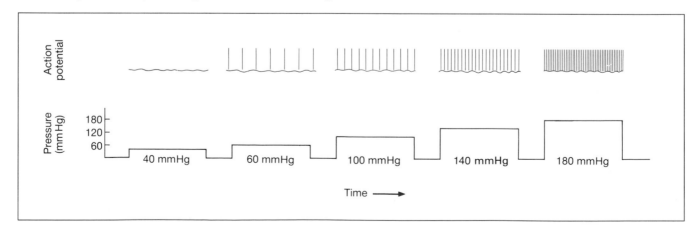

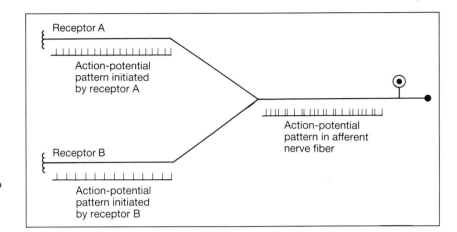

FIGURE 9-6 Simultaneous activation of two receptors in a sensory unit increases the number of action potentials transmitted along the afferent neuron.

Stimulus Location

A third type of information to be signaled is the location of the stimulus, in other words, where the stimulus is being applied. Since sensory units from only a restricted area of the body converge upon one interneuron of the pathways specific for that stimulation type, the pathway that contains that particular interneuron transmits information about only that restricted area. Thus, the specific pathway indicates stimulus location as well as type.

Adjacent receptive fields overlap, however, so that stimulation of only a single sensory unit almost never occurs. It is nevertheless possible for a person to pinpoint the location of a stimulus. Let us examine more closely how this is done.

The receptor density, that is, the number of receptors in a given area, varies within the receptive field of a sensory unit, usually being greatest at the geometric center. Thus, a stimulus activates more receptors and generates more action potentials if it occurs at the densest part of the receptive field (point A in Figure 9-7). The firing frequency of the afferent neuron is also related to stimulus strength, however, and a high frequency of impulses in the single afferent nerve fiber of Figure 9-7 could mean either that a moderate intensity stimulus was applied at A or that a strong stimulus was applied at B. Thus, neither the intensity nor the location of the stimulus can be detected precisely with a single neuron.

Since the peripheral terminations of different afferent neurons overlap, however, a stimulus can trigger activity in sensory units with overlapping receptive fields if the stimulus is in the overlap zone. In Figure 9-8, neurons A and C, stimulated near the edge of their receptive fields where the receptor density is low, fire at a lower frequency than neuron B, stimulated at the densest part of its receptive field. In the group of sensory units in Figure 9-8, a high action-potential frequency in neuron B

arriving simultaneously with a lower frequency in A and C permits more accurate locating of the stimulus near the center of neuron B's receptive field. Once this location is known, the firing frequency of neuron B can be used to indicate stimulus intensity.

The precision with which a stimulus can be located and differentiated from an adjacent stimulus depends both on the size of the receptive field covered by a single afferent neuron and on the amount of overlap of nearby receptive fields. For example, the ability to discriminate between two adjacent mechanical stimuli to the skin (two-point discrimination) is greater on the thumb, fingers, and lips, where the sensory units are small and overlap considerably, than on the back, where the sensory units are large and widely spaced. Locating sensations from the internal organs is less precise than from

FIGURE 9-7 Two stimulus points, A and B, in the receptive field of a single afferent neuron.

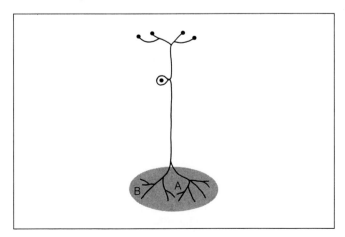

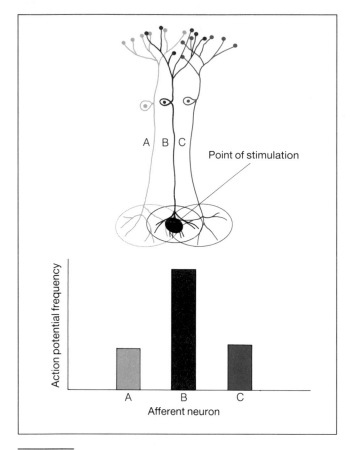

FIGURE 9-8 A stimulus point falls within the overlapping receptive fields of three afferent neurons. Note the difference in receptor response, that is, the action-potential frequency in the three neurons due to the difference in receptor distribution under the stimulus (low receptor density in A and C, high in B).

the skin because there are fewer afferent neurons in the internal organs and each has a larger receptive field.

Central Control of Afferent Information

All information coming into the nervous system is subject to extensive control at synaptic junctions before it reaches higher levels of the central nervous system. Much of it is reduced or even abolished by inhibition from collaterals from other afferent neurons, interneurons in the vicinity, and descending pathways from higher regions, particularly the reticular formation and cerebral cortex. The inhibitory controls are exerted via synapses mainly upon two sites: (1) the axon terminals of the primary afferent neurons and (2) the second-order neurons, that is, the interneurons directly activated by the afferent neurons. The afferent terminals are influ-

enced by presynaptic inhibition, whereas the second-order cells receive both pre- and postsynaptic inputs.

In some cases, for example pain pathways, the afferent input is continuously, that is, tonically, inhibited to some degree. This provides the flexibility of either removing the inhibition (disinhibition) so as to allow a greater degree of signal transmission or of increasing the inhibition so as to block the signal more completely.

In most afferent systems, the control is organized in such a way that stronger inputs are enhanced and the weaker inputs of adjacent sensory units are simultaneously inhibited (Figure 9-9). Such **lateral inhibition** can be demonstrated in the following way. While pressing the tip of a pencil against your finger with your eyes closed, you can localize the pencil point precisely, even though the region around the pencil tip is also indented and mechanoreceptors within this region are activated (Figure 9-10). This is because the information from the peripheral region is removed by lateral inhibition. Thus, lateral inhibition increases the contrast between "wanted" and "unwanted" information, thereby increasing the effectiveness of selected pathways and focusing sensory-processing mechanisms on "important" messages.

Lateral inhibition is utilized to the greatest degree in the pathways providing the most accurate localization. For example, movement of skin hairs, which we can locate quite well, activates pathways that have significant lateral inhibition, but temperature, which we locate only poorly, activates pathways that lack lateral inhibition.

This completes our general introduction to sensory system pathways and coding. We now present the individual systems.

SOMATIC SENSATION

The sensory function of the skin and body walls, that is, **somatic sensation**, is mediated by a variety of somatic receptors (Figure 9-11). Some respond to mechanical stimulation of the skin, hairs, and underlying tissues, whereas others respond to temperature changes and some chemical changes. Activation of skin receptors gives rise to the sensations of touch, pressure, heat, cold, pain, and the awareness of the position of the body parts and their movement. Each sensation is associated with a specific receptor type. In other words, there are distinct receptors for heat, cold, touch, pressure, joint position, and pain.

After entering the central nervous system, the afferent nerve fibers from the somatic receptors synapse on interneurons that form the specific ascending pathways going to the somatosensory cortex via the brainstem and

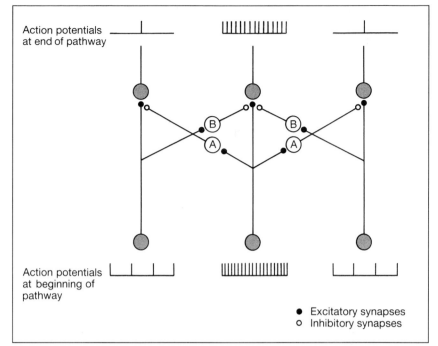

FIGURE 9-9 Afferent pathways showing lateral inhibition. The central fiber at the beginning of the pathway (bottom of figure) is firing at the highest frequency and inhibits, via inhibitory neurons A, the lateral neurons more strongly than the lateral pathways inhibit it, via inhibitory neurons B. Thus, because of lateral inhibition, the lateral pathways are inhibited more strongly than the central pathway and the contrast between the activity in the central and adjacent pathways is enhanced.

thalamus. They also synapse on interneurons that form the nonspecific pathways. The location of some important ascending pathways is shown in a cross section of the spinal cord (Figure 9-12A), and two are diagrammed in Figure 9-12B and 9-12C. Note that the pathways cross from the side at which the afferent neurons enter the central nervous system to the opposite side of the spinal cord (9-12B) or brainstem (9-12C). Thus, the sensory pathways from somatic receptors on the left side of the body go to the somatosensory cortex of the right cerebral hemisphere, and vice versa.

Because some pathways cross in the spinal cord at their point of entry whereas others delay crossing until they reach the brainstem, damage to one side of the spinal cord can result in a complex pattern of sensory deficiencies. Figure 9-13 shows how sensations derived from certain somatic receptors remain intact despite damage

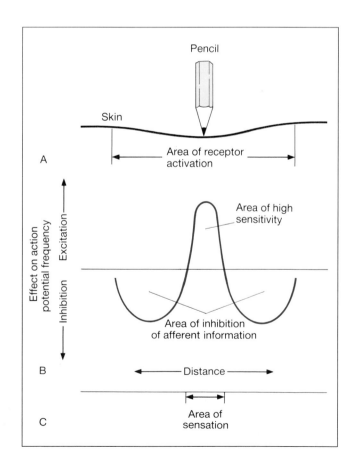

FIGURE 9-10 (A) A pencil tip pressed against the skin depresses surrounding tissue. Receptors are activated under the pencil tip and in the adjacent tissue. (B) Because of lateral inhibition, the central area of excitation is surrounded by an area where the afferent information is inhibited. (C) The sensation is localized to a more restricted region than that in which mechanoreceptors were actually stimulated.

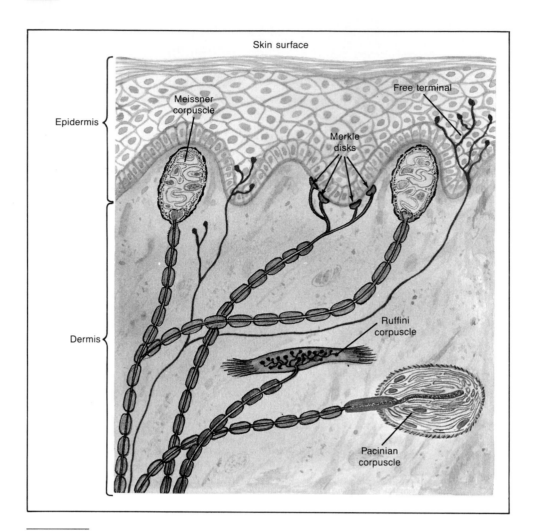

FIGURE 9-11 Skin receptors. Some nerve fibers, usually those with little or no myelin, have free nerve terminals not related to any apparent receptor structures. These terminals are sensitive to stimuli that give rise to pain and temperature. Thick, myelinated axons, on the other hand, end in receptors that can be quite complex, for example, the Pacinian, Meissner, and Ruffini corpuscles and the Merkel disks. These are all thought to be mechanoreceptors, sensitive to displacement of the skin, although the Pacinian corpuscles are especially sensitive to vibration and the Ruffini corpuscles and Merkel disks are more sensitive to sustained pressure. One afferent fiber is thought to connect to only one type of receptor so that the information coded by the different receptor types is kept separate.

to one side of the spinal cord, whereas sensations derived from other somatic receptors on the same side of the body are abolished.

In the somatosensory cortex, the axonal endings of the individual fibers of the specific somatic pathways are grouped according to the location of the receptors giving rise to the pathways (Figure 9-14). The parts of the body that are most densely innervated—fingers, thumb, and lips—are represented by the largest areas of the somatosensory cortex.

Touch-Pressure

The skin mechanoreceptors sensitive to touch-pressure are classified into two major categories according to their response to sustained pressure. About half the receptors are rapidly adapting and respond with a burst of action potentials only at the onset and removal of the stimulus, that is, when the stimulus is changing (Figure 9-15). The other receptors are slowly adapting and respond with a sustained discharge throughout the duration of the stimulus. Activation of rapidly adapting receptors gives rise

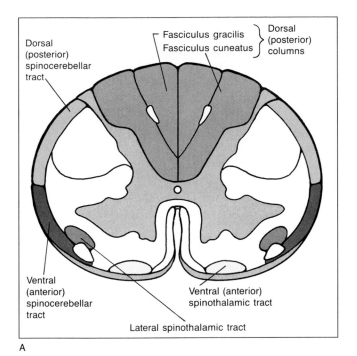

A

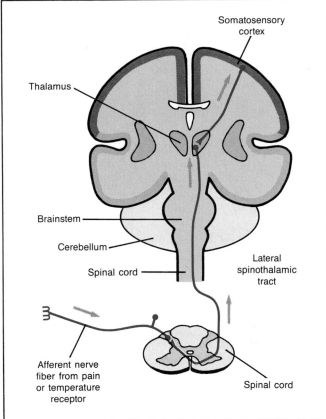

B

FIGURE 9-12 (A) A cross section of the spinal cord, showing the relative locations of the major ascending fiber tracts. Two pathways, the fasciculus gracilis and the fasciculus cuneatus, make up the dorsal columns. The dorsal (posterior) and ventral (anterior) spinocerebellar tracts carry information from the muscles or tendons. The ventral (anterior) spinothalamic tract conveys information associated with light touch. (B) The lateral spinothalamic tract, which conveys information associated with pain and temperature. (C) The posterior (dorsal) columns, which relay information related to vibration, joint position, and the ability to discriminate between two points placed near each other on the skin. (*Parts B and C adapted from Gardner.*)

to the sensations of touch, movement, vibration, and tickle, whereas slowly adapting receptors give rise to the sensation of pressure.

In both categories, some receptors have small, well-defined receptive fields and are able to provide precise information about the contours of objects indenting the skin. As might be expected, these receptors are concentrated at the fingertips. In contrast, other receptors have large receptive fields, with obscure boundaries sometimes covering a whole finger or a large part of the palm.

C

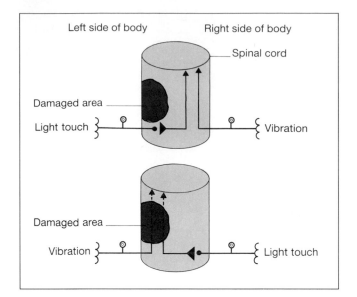

FIGURE 9-13 Damage to the left side of the spinal cord does not interfere with pathways that ascend on the undamaged right side. These include fibers that have crossed from the left to the right at their point of entry below the level of damage and fibers that remain on the undamaged side during ascent through the cord. Examples of crossed fibers are those that represent light touch, temperature, and pain, whereas uncrossed fibers relay information related to vibration, two-point discrimination, and joint position. Opposite effects are observed on the other side of the body because these pathways pass through the damaged area.

FIGURE 9-14 Location of pathway terminations for different parts of the body in the somatosensory cortex. The left half of the body is represented on the right hemisphere of the brain, and the right half of the body is represented on the left hemisphere.

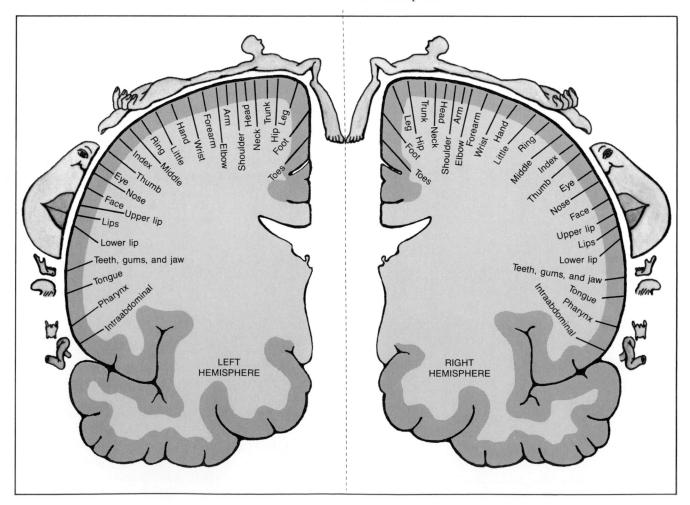

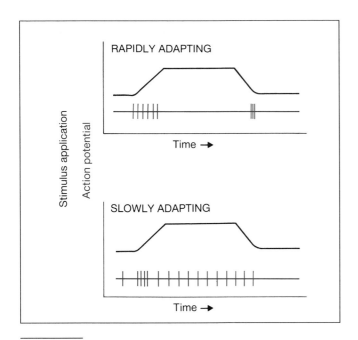

FIGURE 9-15 Rapidly and slowly adapting receptors. The top line indicates application of the stimulus, and the bottom line, the action-potential firing of the afferent nerve fiber from the receptor.

These receptors are not involved in detailed spatial discrimination but signal information about vibration, skin stretch, and joint movement.

Proprioception, Kinesthesia, and the Sense of Effort

The terms proprioception and kinesthesia are not always clearly distinguished, but we shall use **proprioception** to mean the sense of the body's position in space and **kinesthesia** to refer to the sensations associated with joint movement.

The major receptors responsible for these senses are the muscle-spindle stretch receptors that occur in skeletal muscles and respond both to the absolute magnitude of muscle stretch and to the rate at which the stretch occurs (Chapter 12). Mechanoreceptors in the joints, tendons, ligaments, and skin also play a role in proprioception and kinesthesia.

Sense of effort refers to an awareness of the amount of muscle contraction being exerted in a given situation. For this awareness, we rely on information in pathways that descend to the motor neurons from the parts of the brain controlling motor behavior. Collateral branches from these descending pathways synapse with parts of the brain involved in sensory function, which simultaneously leads to an awareness that one is exerting muscular effort.

Temperature

The thermoreceptors in the skin are classified according to their responses to cold and heat. The "warm receptors" are free nerve endings, but the structure of cold receptors is unknown. Warm receptors respond to temperatures between 30 and 43°C and increase their discharge rate upon warming, whereas cold receptors are stimulated by temperatures between 35 and 20°C and increase their discharge rate upon cooling. It is not known how temperatures alter the endings of the thermosensitive afferent neurons to generate action potentials.

Individual receptors signal constant temperatures by a steady discharge of action potentials whose rate depends on the temperature, and they indicate changes in temperature by altering the frequency of action potentials at the beginning of their response.

Pain

A stimulus that causes (or is on the verge of causing) tissue damage usually elicits a sensation of pain. Pain differs from the other somatosensory modalities in that such emotions as fear and anxiety and feelings of unpleasantness are experienced along with the perceived sensation. Also, a painful stimulus can evoke a reflex escape or withdrawal response as well as a gamut of physiological changes similar to those elicited during fear, rage, and aggression. These changes, mediated by the sympathetic nervous system, include increased heart rate and blood pressure, greater secretion of epinephrine into the bloodstream, increased blood glucose concentration, decreased gastric motility, decreased blood flow to the viscera and skin, dilation of the pupils, and sweating. In other words, the stimuli that give rise to pain result in a sensory experience *plus* a reaction to it.

Probably more than any other type of sensation, pain can be altered by past experiences, suggestion, emotions (particularly anxiety), and the simultaneous activation of other sensory modalities. Thus, the level of pain is not solely a physical property of the stimulus as, for example, temperature or shape would be. Instead, painfulness depends in part on the subjective opinion of the person receiving the stimulus.

The receptors, known as **nociceptors**, whose stimulation gives rise to pain, are located at the ends of small unmyelinated or lightly myelinated afferent neurons. Some of these receptors may respond preferentially to intense mechanical stimulation, others to mechanical and thermal stimulation, and still others to irritant chemicals as well. Chemicals such as histamine, bradykinin, and prostaglandins released from damaged tissue may depolarize nearby nociceptor nerve endings, initiating action potentials in the afferent nerve fibers.

The primary afferents coming from nociceptors synapse on interneurons after entering the central nervous system. Substance P is the transmitter released at some of these synapses. Information about pain is then transmitted to higher centers via both specific and nonspecific ascending pathways. The specific pathways convey information about where, when, and how strongly the stimulus was applied and about the sharp, localized aspect of pain. The nonspecific pathways convey information about the aspect of pain that is duller, longer lasting, and less well localized. Neurons in the reticular formation and thalamus are activated by the pain pathways and connect with the hypothalamus and other brain areas that integrate autonomic and endocrine stress responses and generate the behavioral patterns of aggression and defense.

The activation of interneurons by incoming nociceptive afferents may lead to the phenomenon of **referred pain**, in which case the sensation of pain is experienced at a site other than the injured or diseased part. For example, during a heart attack, pain is often experienced in the left arm instead of the heart. This type of referred pain occurs because both visceral and somatic afferents often converge on the same interneurons in the pain pathways. Excitation of the somatic afferent fibers is the usual source of afferent discharge, so we "refer" the location of receptor activation to the somatic source even though, in the case of visceral pain, the perception is incorrect.

Another type of referred pain is **phantom limb pain** in which the sensation of pain is experienced in a limb that has been lost by accident or amputation. Irritation of the damaged nociceptive afferent neurons at the stump of the limb or hyperexcitability of the interneurons contacted by these afferents causes phantom limb pain.

Electric stimulation of specific areas of the central nervous system can produce a profound reduction of pain, a phenomenon called **stimulation-produced analgesia**, by inhibiting pain pathways. Parts of the thalamus, central region of the brainstem, and spinal cord at the level of entry of the active nociceptor afferents have all been electrically stimulated with some success in alleviating pain.

Analgesia occurs from such stimulation because descending pathways that originate in these brain areas selectively inhibit the transmission of information originating on nociceptors. The descending axons end at lower brainstem and spinal levels on interneurons in the pain pathways as well as on the synaptic terminals of the afferent nociceptor neurons themselves. Some of the neurons in these inhibitory pathways release or are sensitive to certain of the endorphins. Thus, infusion of morphine, which stimulates the endorphin receptors, into the spinal cord at the level of entry of the active nociceptor fibers provides relief in many cases of intractable pain. This is separate from morphine's effect on the brain.

Transcutaneous electric nerve stimulation, or **TENS**, in which the painful site itself or the nerves leading from it are stimulated by electrodes placed on the surface of the skin, is often used in lessening pain. TENS works because the stimulation of nonpain, low-threshold afferent fibers, for example, the fibers from touch receptors, leads to inhibition of neurons in the pain pathways. We often apply our own type of TENS therapy when we rub or press hard on a painful area, utilizing the fact that pathways activated by the touch receptors add inhibitory synaptic input to the pain pathways.

Under certain circumstances **acupuncture** prevents or alleviates pain. During acupuncture analgesia, needles are introduced into specific parts of the body to stimulate afferent fibers, which causes analgesia by blocking, at a variety of sites, the transmission of neural activity perceived as painful.

VISION

The receptors of the eye are sensitive to only that tiny portion of the vast spectrum of electromagnetic radiation that we call visible light (Figure 9-16).

Light

Radiant energy is described in terms of wavelengths and frequencies. The **wavelength** is the distance between two successive wave peaks of the electromagnetic radiation (Figure 9-17). Wavelengths can vary from several kilometers at the long-wave radio end of the spectrum to minute fractions of a millimeter at the gamma ray end. The **frequency**, or the number of cycles per second, of the radiation wave varies inversely with wavelength. Those wavelengths capable of stimulating the receptors of the eye—the **visible spectrum**—are between 400 and 700 nm. Light of different wavelengths in this band is associated with different colors.

The Optics of Vision

A light wave can be represented by a line drawn in the direction in which the wave is traveling. Light waves are propagated in all directions from every point of a visible object. These divergent light waves must pass through an optical system that focuses them back into a point before an accurate image of the object is achieved. In the eye, the image of the object being viewed is focused upon the **retina**, a thin layer of neural tissue lining the back of the eyeball (Figure 9-18). The retina contains the light-sensitive receptor cells of the eye.

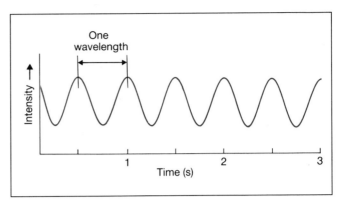

FIGURE 9-17 Properties of a wave. The frequency of this wave is 2 Hz (cycles/s).

The **lens** and **cornea** of the eye are the optical systems that focus an image upon the retina. At a boundary between two substances, such as the cornea and the air, light rays are bent so that they travel in a new direction. The cornea plays a larger role than the lens in focusing light rays because the rays are bent more in passing from air into the cornea than they are when passing into and out of the lens or any other transparent structures of the eye.

The surface of the cornea is curved so that light rays coming from a single point source hit the cornea at different angles and are bent different amounts, directing the light rays to a point after emerging from the lens (Figure 9-19). The image on the retina is upside down relative to the original light source (Figure 9-20), and it is also reversed right to left.

Light rays from objects close to the eye strike the cornea at greater angles and must be bent more by the lens in order to reconverge on the retina. Although the cornea performs the greater part quantitatively of focusing the visual image on the retina, all adjustments for distance are made by changes in lens shape. Such changes are part of the process known as **accommodation**.

The shape of the lens is controlled by the **ciliary muscle** and the tension it applies to the **zonular fibers**, which attach the muscle to the lens (Figure 9-21). The ciliary muscle is circular, like a sphincter, so that it draws nearer to the lens as it contracts and removes tension on the zonular fibers. The zonular fibers pull the lens to a flattened, oval shape, and when their pull is removed, the natural elasticity of the lens causes it to become more spherical. This more spherical shape provides additional bending of the light rays when near objects are viewed (Figure 9-22). The ciliary muscle is controlled by the parasympathetic nerves. Accommodation for viewing

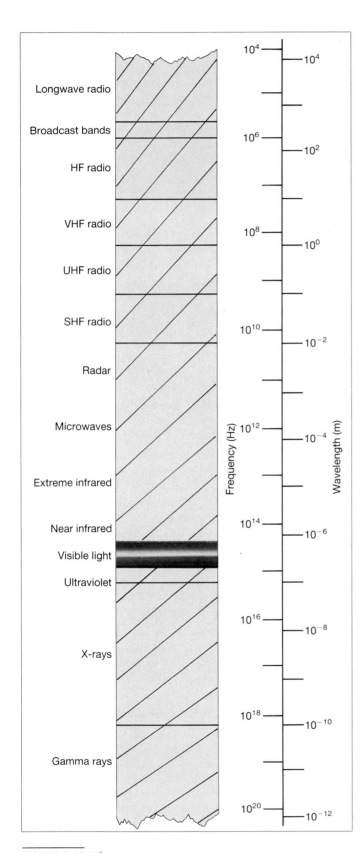

FIGURE 9-16 Electromagnetic spectrum.

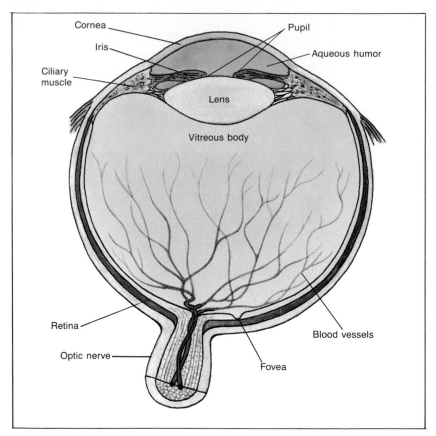

FIGURE 9-18 The human eye. The blood vessels depicted run along the back of the eye between the retina and the vitreous body, not through the vitreous body.

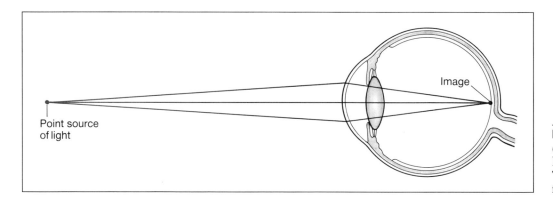

FIGURE 9-19 Refraction (bending) of light by the lens system of the eye. The focusing of light rays from a single point.

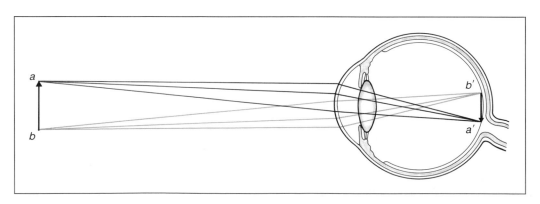

FIGURE 9-20 The focusing of light rays from more than one point to form an image on the retina. In reality, the image is not focused where the optic nerve leaves the eye, which is a "blind spot." Rather, the image is focused on the fovea, the area of the retina with greatest clarity, which is near the exit of the nerve.

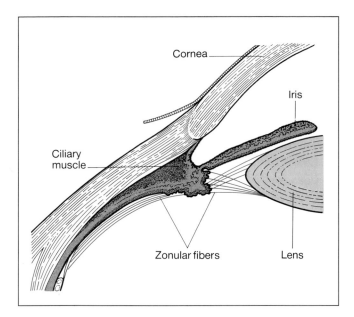

FIGURE 9-21 Ciliary muscle, zonular fibers, and lens of the eye. (*Redrawn from Davson.*)

compensating eyeglasses, contact lenses, or even an implanted artificial lens, effective vision can be restored, although the ability to accommodate will be lost.

Cornea and lens shape and eyeball length determine the point where light rays reconverge. Defects in vision occur if the eyeball is too long in relation to the lens size. In this case, the images of near objects fall on the retina but the images of far objects focus at a point in front of the retina. This is a **nearsighted**, or **myopic**, eye, which is unable to see distant objects clearly (Figure 9-23). If the eye is too short for the lens, distant objects are focused on the retina, but near objects are focused behind it (Figure 9-24). This eye is **farsighted**, or **hyperopic**, and near vision is poor. The use of corrective lenses for near- and farsighted vision is shown in Figures 9-23 and 9-24.

FIGURE 9-22 Accommodation of the lens for near vision.

near objects also includes moving the lens slightly toward the back of the eye, turning the eyes inward toward the nose (convergence), and constricting the pupil. The sequence of events for accommodation is reversed when distant objects are viewed.

The cells that make up most of the lens lose all their internal membranous structures early in life and are thus transparent but lack the ability to replicate. The only lens cells that retain the capacity to divide are on the surface of the lens, and as new cells are formed, older cells come to lie deeper within the lens. With increasing age, the central part of the lens becomes increasingly dense and stiff and acquires a coloration that progresses from yellow to black.

Since the lens must be elastic to assume a rounder shape during accommodation for near vision, the increasing stiffness of the lens that occurs with aging makes accommodation for near vision increasingly difficult. This condition, known as **presbyopia**, is a normal part of the aging process and is the reason that people around 45 years of age often begin wearing reading glasses or bifocals for close work.

The changes in lens color that occur with aging are responsible for **cataract**, which is an opacity of the lens and one of the most common eye disorders. Early changes in lens color do not interfere with vision, but vision is impaired as the process slowly continues. The opaque lens can be removed surgically. With the aid of

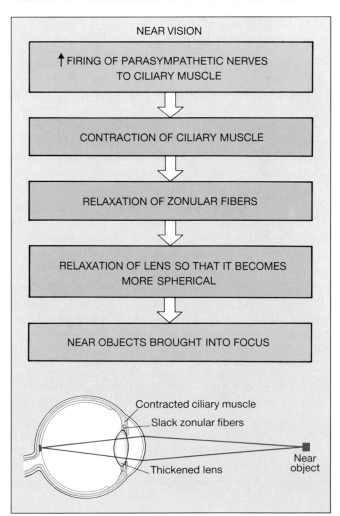

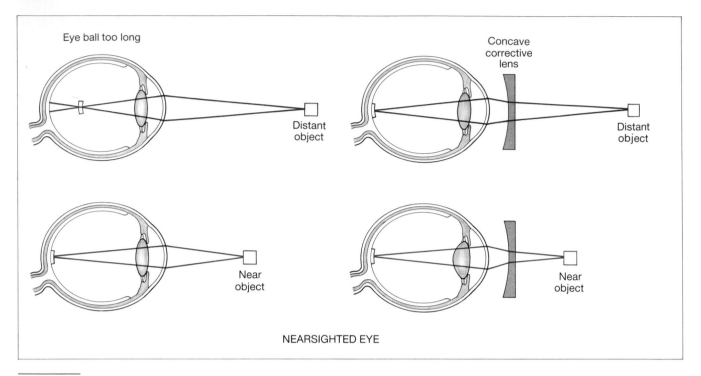

NEARSIGHTED EYE

FIGURE 9-23 In the nearsighted eye, light rays from a distant source are focused in front of the retina. A concave (cupping inward) lens placed before the eye bends the light rays out sufficiently to move the focused image back onto the retina. When near objects are viewed through concave lenses, the eye accommodates to focus the image on the retina.

Defects in vision also occur where the lens or cornea does not have a smoothly spherical surface, a condition known as **astigmatism**. These surface imperfections can usually be compensated for by eyeglasses.

The lens separates two fluid-filled chambers in the eye, the anterior chamber, which contains aqueous humor, and the posterior chamber, which contains the more viscous vitreous body, or vitreous humor (Figure 9-18). These two substances are colorless and permit the transmission of light from the front of the eye to the retina. The aqueous humor is formed by special vascular tissue that overlies the ciliary muscle. In some instances the aqueous humor is formed faster than it is removed, which results in increased pressure within the eye, a condition known as **glaucoma**.

The amount of light entering the eye is controlled by a ringlike pigmented muscle known as the **iris** (Figure 9-18), the color being of no importance as long as the tissue is sufficiently opaque to prevent the passage of light. The hole in the center of the iris through which light enters the eye is the **pupil**. The iris is composed of smooth muscles, which are innervated by autonomic nerves. Stimulation of sympathetic nerves to the iris causes the radially arranged muscle fibers to contract and enlarges the pupil, whereas stimulation of parasympathetic fibers to the iris causes the sphincter muscle fibers, which surround the pupil, to contract and makes the pupil smaller.

These neurally induced changes occur in response to light-sensitive reflexes. Bright light causes a decrease in the diameter of the pupil, which not only reduces the amount of light entering the eye but, more important, directs the light to the central part of the lens for more accurate vision. Conversely, the sphincter fibers relax in dim light, when maximal illumination is needed.

Receptor Cells

The photoreceptor cells in the retina are called **rods** and **cones** because of the shape of their light-sensitive tips (Figure 9-25). Photoreceptors contain molecules called **photopigments**, which absorb light. There are four different photopigments in the retina, one (**rhodopsin**) in the rods and one in each of the three cone types.

Each photopigment contains one of a group of proteins collectively known as **opsin**, which surrounds and binds a **chromophore** molecule (Figure 9-26). The chro-

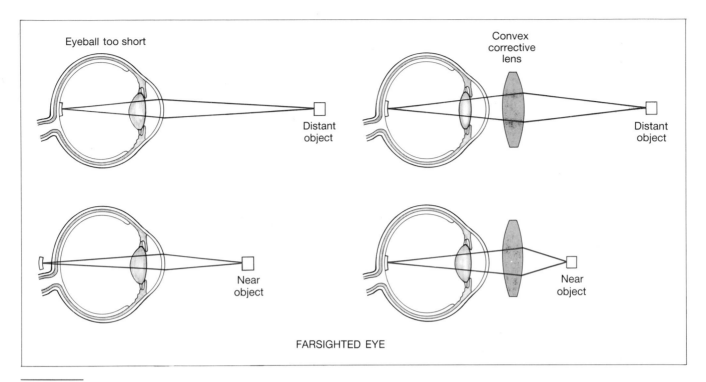

FIGURE 9-24 The normal eye views distant objects with a flat, stretched lens. The farsighted eye must accommodate to focus the image of distant objects on the retina. The accommodating power of the lens is sufficient for distant objects, and these objects are seen clearly, but the lens cannot accommodate enough to focus images of near objects on the retina, and they are blurred. A convex—bulging outward—lens converges light rays before they enter the eye and allows the eye's lens to work in a normal manner.

mophore, which is the actual light-sensitive part of the photopigment, is **retinal**, a slight variant of vitamin A, and it is the same in each of the four photopigments. The opsin, which filters the light reaching the retinal, differs in each of the four types. The different opsins cause each of the four photopigments to absorb light most effectively at a different part of the visible spectrum. For example, one photopigment absorbs wavelengths in the range of red light best, whereas another absorbs green light best.

The photopigments lie in specialized membranes within the receptor cells. The membranes are arranged in highly ordered stacks parallel to the retina (Figure 9-25). The repeated layers of membranes in each photoreceptor may contain over a billion molecules of photopigment, providing an effective trap for light.

Upon exposure to light, retinal rapidly changes its shape (Figure 9-27), which begins the complex process by which absorption of a photon of light leads to the formation of a receptor potential and alters the release of neurotransmitter from the photoreceptor. The change in

retinal shape causes an alteration in the opsin to which the retinal is bound. The opsin then activates a second-messenger cascade system, which results in a *decrease* in the concentration of cyclic GMP in the cytoplasm of the photoreceptor's outer segment.

In darkness, the intracellular concentration of cGMP in the photoreceptors is high. The cGMP opens receptor-operated plasma-membrane channels, permitting Na^+ and Ca^{2+} ions to enter the cell and *depolarize* the membrane. Thus, unlike most neurons, the photoreceptors have a large resting permeability to sodium, and the receptors are relatively depolarized in the dark. Transmitter release, which occurs with depolarization, is high in the dark. The decrease in cGMP that follows exposure to light closes the plasma-membrane ionic channels, and the plasma membrane *hyperpolarizes*. This causes a decrease in transmitter release, altering the responses of the postsynaptic neurons. Note that in the case of the photoreceptors, the receptor potential is a hyperpolarization.

After its activation by light, the retinal changes back

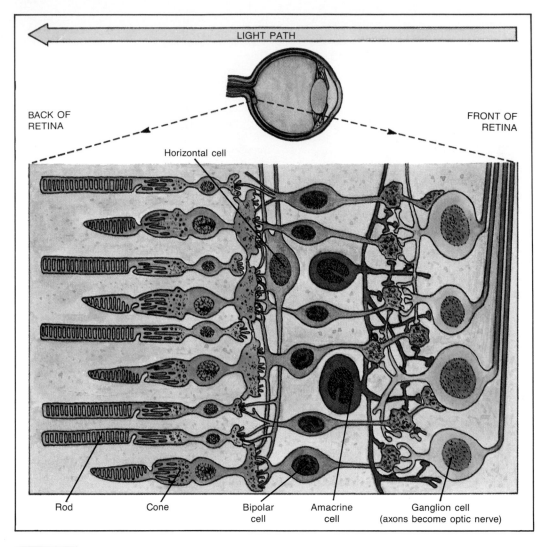

LIGHT PATH

BACK OF
RETINA

FRONT OF
RETINA

Horizontal cell

| Rod | Cone | Bipolar cell | Amacrine cell | Ganglion cell (axons become optic nerve) |

FIGURE 9-25 Organization of the retina. Light enters the eye through the lens and passes through the vitreous body and the front surface of the retina. Since the tips of the rods and cones are on the side of the retina opposite to the light entry, the light must pass through all the cell layers before reaching the photoreceptors and stimulating them. The pigmented layer of the eye (the choroid), which lies behind the retina, absorbs light and prevents light reflection back to the rods and cones. Reflection of this light would cause the visual image to be blurred. (*Redrawn from Kuffler, Nicholls, and Martin.*)

to its resting shape by several mechanisms that do not depend on light but are enzyme-mediated.

High sensitivity of the visual receptors is not always an advantage. If the receptor is saturated by high light levels, the receptor cannot respond to information about further small increases or decreases. Adaptation, however, reduces the sensitivity of the receptor back to a range that allows changes in the incoming stimulus to

again affect the receptor's output. Adaptation of the photoreceptors to light occurs by incompletely understood changes in the second-messenger cascade.

Neural Pathways of Vision

The neural pathways of vision begin with the rods and cones. These photoreceptors interact with each other and with second-order neurons, one of which is the **bi-**

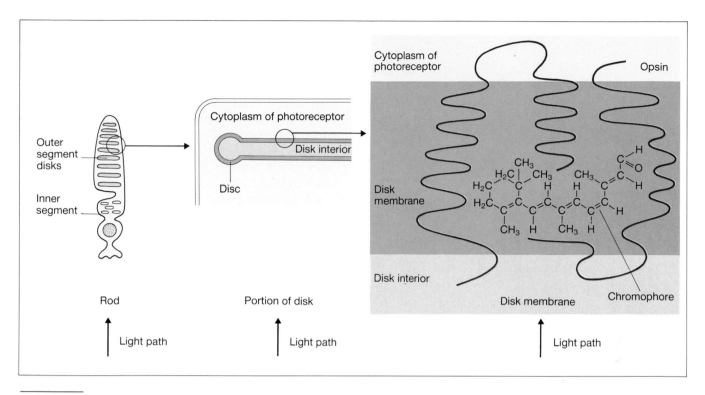

FIGURE 9-26 The arrangement of opsin and chromophore in the membrane of the photoreceptor disks. The opsin actually crosses the membrane seven times, not three as shown here. Note that the chromophore is perpendicular to the incoming light rays so that its ability to "catch" light is maximal.

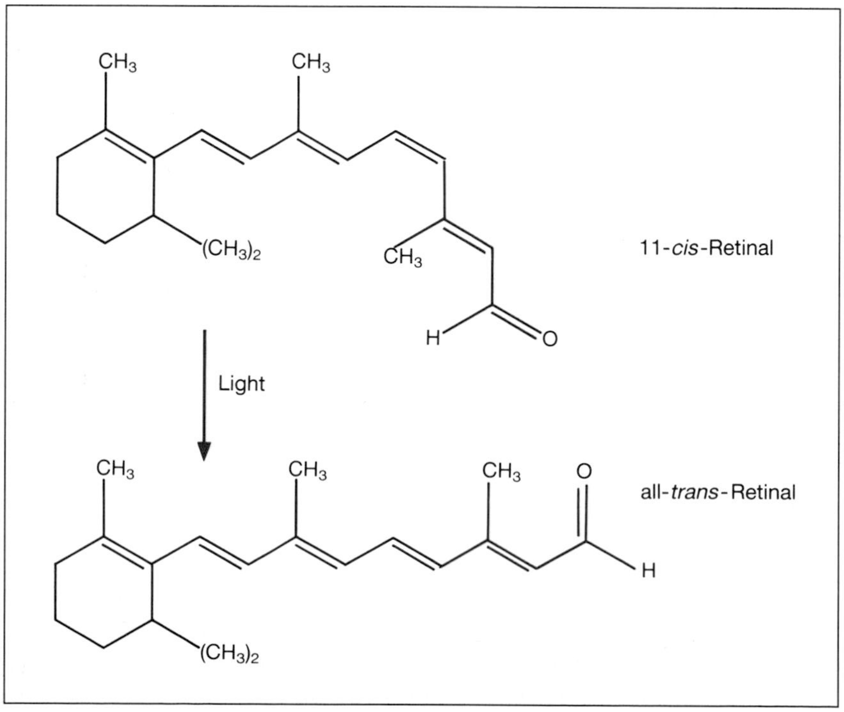

FIGURE 9-27 The change in retinal, the chromophore portion of the photopigment, in response to light.

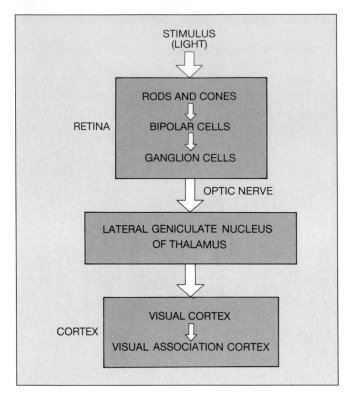

FIGURE 9-28 Diagrammatic representation of the visual pathway.

polar cell (Figures 9-25 and 9-28). The bipolar cells synapse (still within the retina) upon the **ganglion cells**. Other neurons in the retina pass information horizontally from one part of the retina to another, and a great deal of information processing takes place at this early stage of the sensory pathway.

Ganglion cells respond to activation by producing action potentials, whereas the rods and cones and almost all other retinal neurons produce only graded potentials. The axons of the ganglion cells form the output from the retina, the optic nerve, or cranial nerve II, which leads into the brain. The two optic nerves meet at the base of the brain to form the optic chiasm, where some of the fibers cross to the opposite side of the brain. This partial crossover provides both cerebral hemispheres with input from each eye.

Optic nerve fibers project to several structures in the brain, the largest number passing to the **lateral geniculate nucleus** in the thalamus (Figure 9-28). This nucleus transmits its information on to visual cortex, the primary visual area of cerebral cortex (Figure 9-3). The visual cortex has several subdivisions, each representing the complete **visual field**, that part of the visual world that stimulates the retina at any given time. The relay of information from the retina to the various areas of the visual cortex follows a precise point-to-point projection pattern so that adjacent retinal regions project to adjacent regions in each subdivision of the visual cortex.

Most of the neurons of the lateral geniculate nucleus do not receive afferents from the retina. Rather, they receive input from the brainstem reticular formation and input relayed back from the visual cortex. These nonretinal inputs can control the transmission of information from the retina to the visual cortex, and they are probably involved in the ability to shift attention between vision and the other sensory modalities or between different objects in the visual field.

We have mentioned that a substantial number of fibers of the visual pathway project to regions of the brain other than the visual cortex. For example, visual information is transmitted to the **suprachiasmatic nucleus**, which lies just above the optic chiasm and plays a major role in a variety of behaviors associated with biological clocks, such as eating, locomotion, sleep/wakefulness, and reproduction. In this nucleus the information about diurnal cycles of light intensity is used to synchronize the neuronal biological clocks. Other visual information is passed to the brainstem and cerebellum, where it is used in the coordination of eye and head movements, fixation of gaze, and constriction of the pupils.

Visual System Coding

This discussion is based on research done chiefly on cats and monkeys, but the results are consistent with results from perceptual experiments on people, and the results we describe almost certainly apply to people. In most of the experiments dealing with coding in the visual system, simple visual shapes such as white bars against a black background were projected onto a screen in front of an anesthetized animal while the activity of single cells in the visual system was recorded. Different parts of the retina, and therefore different populations of receptor cells, could be stimulated by varying the position of the bar on the screen.

We shall refer in the following discussion to the receptive fields of neurons within the visual pathway and to the responses of these neurons to light. It is essential to recognize that only the rods and cones respond directly to light. All other components of the pathway are influenced only by the synaptic input to them. Thus when we speak of the receptive field of a neuron in the visual pathway, we really mean the area of the retina that, when stimulated, can influence the activity of that neuron. Similarly the neuron's "response to light" is really its response to neural activity in the visual pathway initiated by light falling upon the rods and cones. We shall follow the information processing through the stages of the visual pathway, starting at the level of the bipolar cells (Figure 9-28).

Visual images have different characteristics, such as color, form, depth, movement, and texture, each processed by a separate channel in the visual system. This segregated, parallel processing of information begins in the retina, probably as early as the bipolar cells after their activation by the rods and cones, and continues in the brain at the highest stages of visual processing. In support of the concept of parallel processing, different classes of ganglion cells have been found to have different functional properties.

The receptive fields of bipolar cells and most ganglion cells are circular. In other words, any light falling within a specific circular area of the retina influences the activity of a given bipolar or ganglion cell by way of the receptor cells and, in the case of the ganglion cells, the bipolar cells that influence it. The responses of the bipolar and ganglion cells vary markedly, depending on the type of cell and the region stimulated in its receptive field. Some cells speed up their rate of firing when a spot of light is directed at the center of their receptive field and then slow down their firing when the periphery is stimulated. Such a cell is said to have an *on center*. The activity of a ganglion cell of this type is shown in Figure 9-29A. This type of cell is inhibited by light falling in the periphery of its receptive field (Figure 9-29B). Other ganglion cells, *off-center cells*, have just the opposite response, decreasing activity when the center of the cell's receptive field is stimulated (Figure 9-29C) and increasing activity when light stimulates the periphery

(Figure 9-29D). Notice the spontaneous activity of this *off*-center cell when the light is turned off and the abrupt inhibition of its activity when the light that is focused on the center of its receptive field is turned on (Figure 9-29C).

Although most ganglion cells have this circular arrangement of their receptive fields with the center and surround of the circle having opposite responses, the cells differ in their sensitivity to color, fineness of detail, response time—and, therefore, sensitivity to movement—and contrast in brightness between an object and its background. These major differences contribute to different aspects of vision.

These separate functions of cells are maintained in the visual cortex. An important aspect of lateral geniculate function is to separate and relay these different kinds of information from the retina to different cortical zones. Thus, most neurons in one subdivision of the visual cortex are responsive only to stimuli oriented in a particular direction in the visual field, a property important in the elaboration of the form of an object. The neurons in another subdivision are most responsive to movement of an object across the visual field. Furthermore, other neuronal groups may respond best to color, and others only to input from both eyes, an important clue to depth perception. Again, the functional differences between various areas of the visual cortex are determined by the input these areas receive from the distinct classes of ganglion cells.

Some of the information is kept separate even after it leaves the visual cortex. For example, pathways that transmit information dealing with the identification of objects in the visual field—analyzing the "what" of the visual stimulus—travel from the visual cortex to the association areas in the frontal lobe via the temporal lobe, whereas those dealing with the location of objects in the visual field—analyzing the "where" of the visual stimulus—travel to the frontal lobe via the parietal lobe. Thus, different aspects of visual information are carried in parallel pathways and are processed simultaneously in a number of independent ways in different parts of the cerebral cortex before they are reintegrated to produce the conscious sensation of sight.

Note that the cells of the visual pathways are organized to handle information about line, contrast, movement, and color. They do not, however, form a picture in the brain. Rather, they form a specifically coded electrical statement.

Color Vision

The colors we perceive are related to the wavelengths of light that are reflected, absorbed, or transmitted by the pigments in the objects of our visual world. For exam-

FIGURE 9-29 Recordings of the activities of two ganglion cells. (A) Response of an *on*-center ganglion cell increases when light stimulates the center of its receptive field. (B) Inhibition of the spontaneous firing of the same *on*-center cell when light stimulates the edge of its receptive field. (C) Activity of an *off*-center cell is suppressed when the center of its receptive field is stimulated. (D) The same *off*-center cell increases its activity when the light is restricted to the periphery. (*Adapted from Hubel and Wiesel.*)

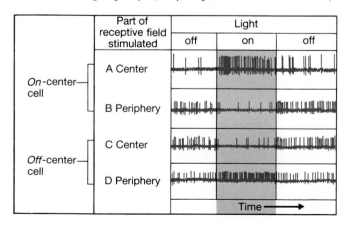

	Part of receptive field stimulated	Light		
		off	on	off
On-center cell	A Center			
	B Periphery			
Off-center cell	C Center			
	D Periphery			
			Time →	

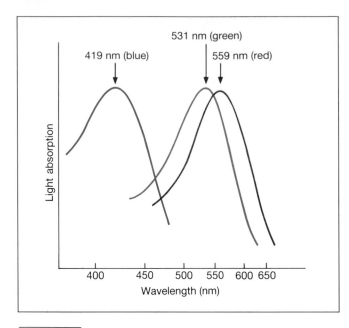

FIGURE 9-30 The sensitivities of the photopigments in the three types of cones in the normal human retina. Absorption of light by a photopigment is directly related to action-potential frequency in the optic nerve. (*Redrawn from Mollon.*)

ple, an object appears red because shorter wavelengths, which would be perceived as blue, are absorbed by the object while the longer wavelengths, perceived as red, are reflected to excite the photopigment of the retina most sensitive to red. Light perceived as white is a mixture of all wavelengths, and black is the absence of all light.

Color vision begins with activation of the photopigments in the cone receptor cells. Human retinas have three kinds of cones, which contain either yellow-, green-, or blue-sensitive photopigments that absorb and hence respond optimally to light of different wavelengths. Because the yellow pigment sensitivity extends far enough to sense the long wavelengths that correspond to red, this pigment is sometimes called the red photopigment.

Although each type of cone is excited most effectively by light of one particular wavelength, it responds to other wavelengths as well. Thus, for any given wavelength, the three cone types are excited to different degrees (Figure 9-30). For example, in response to a light of 530-nm wavelength, the green cones respond maximally, the red cones less, and the blue cones not at all. Our sensation of color depends upon the relative outputs of these three types of cone cells and their comparison by higher-order cells in the visual system.

The pathways for color vision follow those described in Figure 9-28. One type of ganglion cell responds to a broad band of wavelengths. In other words, it receives input from all three types of cones, and it signals not specific color but general brightness. The other type of ganglion cell codes specific colors. These latter cells are called **opponent color cells** because they have an excitatory input from one type of cone receptor and an inhibitory input from another. For example, the cell in Figure 9-31 increased its rate of firing when stimulated by a blue light but decreased it when a red light replaced the blue. The cell gave a weak response when stimulated with a white light because the light contained both blue and red wavelengths. Other more complicated patterns also exist. Opponent cells are also found in the lateral geniculate nucleus and visual cortex.

At high light intensities, as in daylight vision, most people—over 90 percent of the male population and over 99 percent of the female population—have normal color vision. People with the most common kind of color blindness lack either the red or green cone pigments and, as a result, have trouble perceiving red versus green.

Eye Movement

The cones are most concentrated in a specialized area of the retina known as the **fovea** (Figure 9-18), and images focused there are seen with the greatest acuity. In order to get the most important point in the visual image, the fixation point, focused on the fovea and keep it there, the eyeball must be able to move. Six skeletal muscles attached to the outside of each eyeball (Figure 9-32) control its movement. These muscles perform two basic movements, fast and slow.

The fast movements, called **saccades**, are small, jerk-

FIGURE 9-31 Response of a single opponent color ganglion cell to blue, red, and white lights. (*Redrawn from Hubel and Wiesel.*)

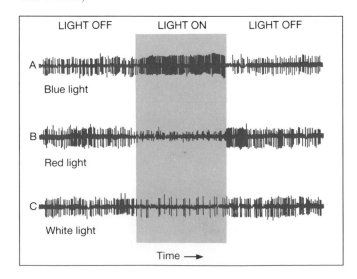

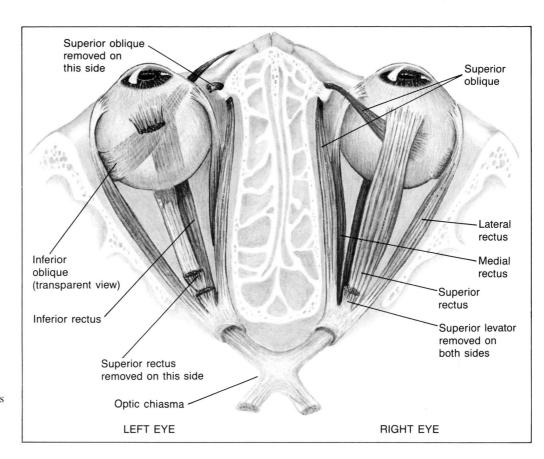

Superior oblique
removed on
this side

Superior
oblique

Lateral
rectus

Medial
rectus

Superior
rectus

Superior levator
removed on
both sides

Inferior
oblique
(transparent view)

Inferior rectus

Superior rectus
removed on this side

Optic chiasma

LEFT EYE

RIGHT EYE

FIGURE 9-32 The muscles that move the eye to direct the gaze and give convergence.

ing movements that rapidly bring the eye from one fixation point to another to allow a sweeping search of the visual field. In addition, saccades move the visual image over the receptors, thereby preventing adaptation. Saccades also occur during certain periods of sleep when the eyes are closed and may be associated with "watching" the visual imagery of dreams.

Slow eye movements are involved in tracking visual objects as they move through the visual field and during compensation for movements of the head. The control centers for these compensating movements obtain their information about head movement from the vestibular system, which will be described shortly. Control systems for the other slow movements of the eyes require the continuous feedback of visual information about the moving object.

HEARING

The sense of hearing is based on the physics of sound and the physiology of the external, middle, and inner ear, the nerves to the brain, and the parts of the brain involved in the analysis and comprehension of acoustic information.

Sound

Sound energy is transmitted through a medium by a movement of the molecules of the medium, but we shall restrict this discussion to situations when the medium is air because this is the most common medium for sound transmission. When there are no air molecules, as in a vacuum, there can be no sound. The disturbance of air molecules that makes up a sound wave consists of regions of compression, in which the air molecules are close together and the pressure is high, alternating with areas of rarefaction, where the molecules are farther apart and the pressure is lower (Figure 9-33A through D). A sound wave measured over time (Figure 9-33E) consists of rapidly alternating pressures that vary continuously from a high during compression of air molecules, to a low during rarefaction, and then back again.

Anything capable of creating such disturbances can serve as a sound source, but a common sound source is a vibrating object such as a tuning fork (Figure 9-33). The difference between the packing (or pressure) of air molecules in zones of compression and rarefaction determines the loudness of the sound. The frequency of vibration of the sound source determines the pitch we hear; the faster the vibration, the higher the pitch. The sounds

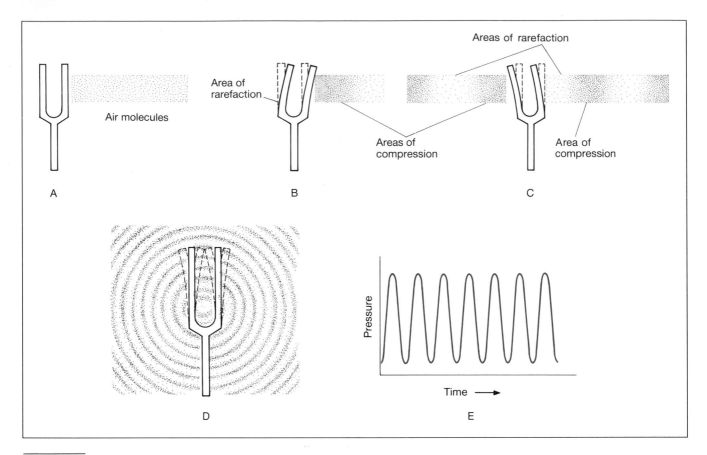

FIGURE 9-33 Formation of sound waves from a vibrating tuning fork.

heard most keenly by human ears are those from sources vibrating at frequencies between 1000 and 4000 Hz, but the entire range of frequencies audible to human beings extends from 20 to 20,000 Hz.

We can distinguish some 400,000 different sounds. For example, we can distinguish the note A played on a piano from the same note on a violin. We can also selectively *not* hear sounds, tuning out the babble of a party to concentrate on a single voice.

Sound Transmission in the Ear

The first step in hearing is the entrance of sound waves into the **ear canal** (Figure 9-34). The shapes of the outer ear (the pinna, or auricle) and the ear canal help to amplify and direct the sound. The sound waves reverberate from the sides and end of the ear canal, filling it with the continuous vibrations of pressure waves.

The **tympanic membrane** (eardrum) is stretched across the end of the ear canal, and the air molecules push against the membrane, causing it to vibrate at the same frequency as the sound wave. Under higher pressure during a wave of compression, the tympanic membrane bows inward. The distance the membrane moves,

although always very small, is a function of the force with which the air molecules hit it and is related to the loudness of the sound. During the following wave of rarefaction, the membrane returns to its original position. The exquisitely sensitive tympanic membrane responds to all the varying pressures of the sound waves, vibrating slowly in response to low-frequency sounds and rapidly in response to high-frequency ones. It is sensitive to pressures to which the most delicate touch receptors of the skin are totally insensitive.

The tympanic membrane separates the ear canal from the **middle-ear cavity**, a cavity in the temporal bone of the skull. The pressures in these two air-filled chambers are normally equal to atmospheric pressure. The middle ear is exposed to atmospheric pressure through the **auditory (eustachian) tube**, which connects the middle ear to the pharynx. The slitlike ending of this tube in the pharynx is normally closed, but muscle movements open the entire passage during yawning, swallowing, or sneezing, and the pressure in the middle ear equilibrates with atmospheric pressure. A difference in pressure can be produced with sudden changes in altitude (as in an ascending or descending elevator or airplane), when the

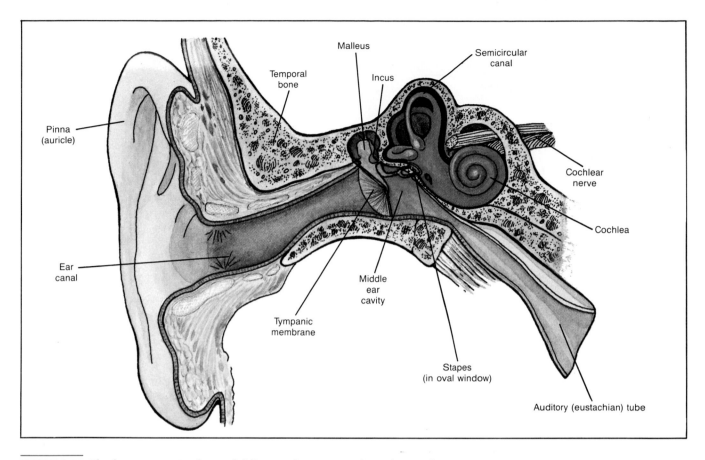

FIGURE 9-34 The human ear. In this and following drawings, violet indicates the outer ear, green the middle ear, and blue the inner ear.

pressure outside the ear and in the ear canal changes while the pressure in the middle ear remains constant because of the closed auditory tube. This pressure difference can excessively stretch the tympanic membrane and cause pain.

The second step in hearing is the transmission of sound energy from the tympanic membrane through the middle-ear cavity to the **inner ear**. The inner ear, called the **cochlea**, is a spiral-shaped passage in the temporal bone. The temporal bone also houses other passages, including the **semicircular canals**, which contain the sensory organs for equilibrium and movement, to be discussed later. The chambers of the inner ear are filled with fluid, and, because the liquid is more difficult to move than air, the pressure transmitted to the inner ear must be amplified. This is achieved by a movable chain of three small bones, the **malleus**, **incus**, and **stapes** (Figure 9-35), which acts as a piston and couples the motions of the tympanic membrane to a membrane-covered opening (the **oval window**) separating the middle and inner ear (Figure 9-36).

The *total* force of a sound wave applied to the tympanic membrane is transferred to the oval window, but because the oval window is so much smaller than the tympanic membrane, the *force per unit area*, that is, pressure, is increased 15 to 20 times. Additional advantage is gained through the lever action of the middle-ear bones. The amount of energy transmitted to the inner ear can be modified by the contraction of two small muscles in the middle ear that alter the tension of the tympanic membrane and the position of the stapes in the oval window. These muscles protect the delicate receptor apparatus of the inner ear from intense sound stimuli and possibly aid intent listening over certain frequency ranges.

The entire system described thus far has been concerned with the transmission of the sound energy into the cochlea, where the receptor cells are located. The cochlea is almost completely divided lengthwise by a fluid-filled membranous tube, the **cochlear duct**, which follows the cochlear spiral. One side of the cochlear duct is formed by the **basilar membrane** (Figure 9-37), to which is attached the **organ of Corti**, which contains the ear's sensitive receptor cells. These cells are usually referred to as **hair cells** because each contains hairlike **stereocilia** protruding from one end (Figure 9-37B). The

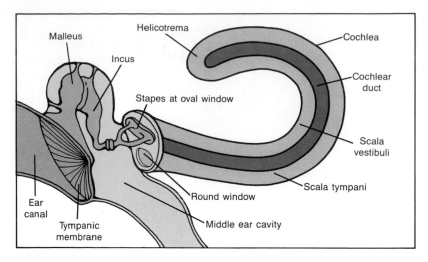

FIGURE 9-35 Relationship between the middle-ear bones and the cochlea. Movement of the stapes against the membrane covering the oval window sets up pressure waves in the fluid-filled scala vestibuli. These waves cause vibration of the cochlear duct and the basilar membrane. Some of the pressure is transmitted around the helicotrema directly into the scala tympani. (*Redrawn from Kandel and Schwartz.*)

hair cells are mechanoreceptors that transform the pressure waves in the cochlea into receptor potentials. Movements of the basilar membrane stimulate the hair cells because they are attached to the membrane.

On either side of the cochlear duct are fluid-filled compartments: the **scala vestibuli**, which is on the side of the cochlear duct that ends at the oval window, and the **scala tympani**, which is below the cochlear duct and ends at a second membrane-covered opening to the middle ear, the round window, (Figure 9-35). The scala vestibuli and scala tympani meet at the end of the cochlear duct at the helicotrema.

As the sound waves in the ear canal push in on the tympanic membrane, the chain of middle-ear bones presses the stapes against the membrane covering the oval window, causing it to bow into the scala vestibuli

and back out (Figure 9-38) creating waves of pressure there. The wall of the scala vestibuli is largely bone, but there are two paths by which the pressure waves can be dissipated. One path is to the helicotrema, where the waves pass around the end of the cochlear duct into the scala tympani and back to the round-window membrane, which is then bowed out into the middle-ear cavity. However, most of the pressure is transmitted from the scala vestibuli to the cochlear duct and thereby to the basilar membrane, which is caused to vibrate.

With each change in pressure in the inner ear, a wave of vibrations is made to travel down the basilar membrane. The region of maximal displacement of the basilar membrane varies with the frequency of the sound source. The properties of the membrane nearest the middle ear are such that this region vibrates most easily, that is, undergoes the greatest movement, in response to high-frequency (high-pitched) tones. The vibration of the basilar membrane in response to high-frequency sound waves soon dies out once it is past this region. This means that the hair cells attached to the region of the basilar membrane nearest the middle ear will be stimulated more in response to high-pitched tones than are hair cells in other regions.

As the frequency of the sound is lowered, the vibration waves travel out along the membrane for greater distances. Progressively more distant regions of the basilar membrane vibrate maximally in response to progressively lower tones.

The greatest stimulation of the hair cells occurs wherever displacement of the basilar membrane is greatest. Thus, individual hair cells develop a maximal receptor potential for a particular frequency of stimulation. The frequencies of the incoming sound waves are, in effect, sorted out along the length of the basilar membrane, and certain hair cells respond maximally to a particular frequency of stimulation.

FIGURE 9-36 Diagrammatic representation showing that the middle-ear bones act as a piston against the fluid of the inner ear. (*Redrawn from von Bekesy.*)

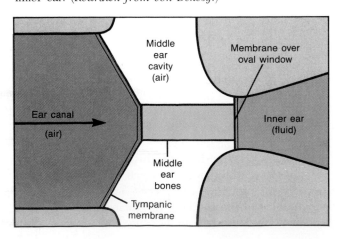

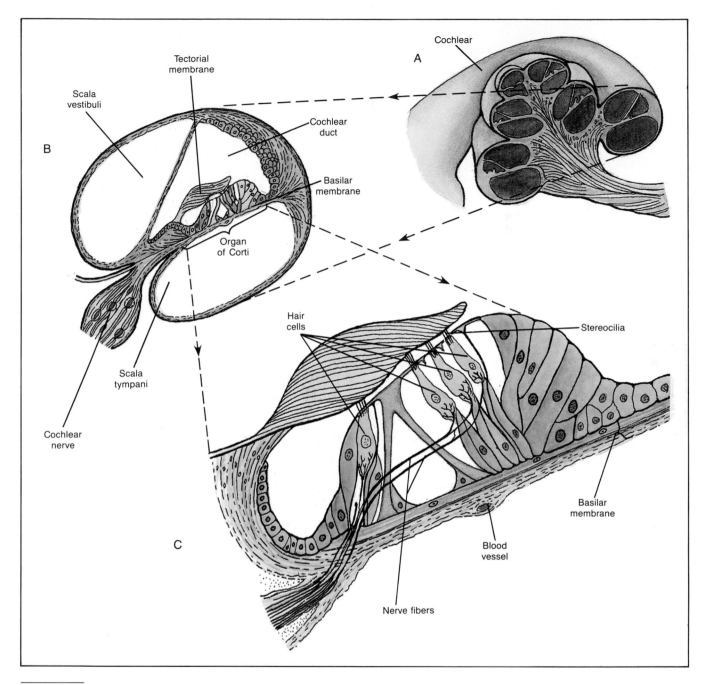

FIGURE 9-37 Cross section of the membranes and compartments of the inner ear with de-tailed views of the hair cells and other structures on the basilar membrane as shown with increasing magnifications in views (A), (B), and (C). (*Redrawn from Rasmussen.*)

Hair Cells of the Organ of Corti

The individual stereocilia that protrude from the top of the hair cells are actually hypertrophied microvilli. Each is a bundle of cross-linked actin microfilaments covered by plasma membrane. The stereocilia are in contact with the overhanging **tectorial membrane** (Figure 9-37), which projects inward from the side of the cochlea. As the basilar membrane is displaced by pressure waves, the hair cells move in relation to the tectorial mem-brane, and, consequently, the stereocilia are bent. Fi-bers at the roots of the stereocilia are coupled directly to ion channels in the plasma membrane of the hair cell so that the channels are opened whenever the stereocilia bend. When the channels are open, Na^+ and Ca^{2+} ions

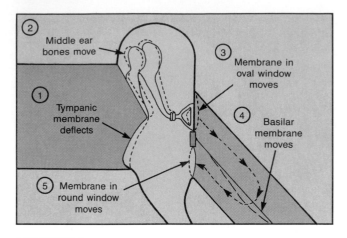

FIGURE 9-38 Transmission of sound vibrations through the middle and inner ear. (*Redrawn from Davis and Silverman.*)

enter the hair cell and depolarize it, creating a receptor potential.

The hair cells are easily damaged by exposure to high-intensity noises such as amplified rock music concerts, engines of jet planes, and revved-up motorcycles. The damaged sensory hairs form abnormal hair structures or are lost altogether. With chronic exposure to loud sounds, some hair cells completely degenerate. Much lesser noise levels also cause damage if exposure is chronic.

Although they have neither dendrites nor an axon, the hair cells nevertheless synapse with afferent neurons. Hair cell depolarization leads to release of a chemical transmitter that activates receptor sites on the terminals of the afferent neuron and causes the generation of action potentials in the neuron. The hair cells synapse with afferent fibers of the cochlear nerve, a component of cranial nerve VIII, one hair cell generally contacting one neuron. The greater the energy (loudness) of the sound wave, the greater the frequency of action potentials generated in the afferent nerve fiber. Each nerve fiber responds within a limited range of sound frequency and intensity, but it responds best to a single frequency, known as its best frequency.

Neural Pathways in Hearing

Cochlear nerve fibers enter the brainstem and synapse with interneurons there, fibers from both ears often converging on the same neuron. Many of these neurons are sensitive to the different arrival times and intensities of the input from the two ears, information that listeners use to determine the source of a sound. If, for example, a sound is louder in the right ear and arrives at the right ear slightly earlier than at the left, we assume that the sound source is on the right.

From the brainstem, the information is transmitted to the thalamus and on to the auditory cortex. Nerve pathways from different parts of the basilar membrane are connected to specific sites along the strips of auditory cortex in an orderly manner, according to sound frequency, in much the same way that signals from different regions of the body are represented at different sites in the somatosensory cortex.

VESTIBULAR SYSTEM

Changes in the motion and position of the head are detected by hair cells in the **vestibular apparatus** of the inner ear (Figure 9-39), a series of fluid-filled membranous tubes that connect with each other and with the cochlear duct. The vestibular apparatus consists of three **semicircular ducts** and two saclike swellings, the **utricle** and **saccule**, all of which lie in tunnels in the temporal bone on each side of the head.

The Semicircular Canals

The bony semicircular canals house the membranous semicircular ducts, which detect angular acceleration

FIGURE 9-39 A tunnel in the temporal bone contains a fluid-filled membranous duct system, part of which can be seen in this cutaway view. The semicircular duct, utricle, and saccule make up the vestibular apparatus. The vestibular duct is connected to the cochlear duct. The purple region on the ducts indicates the positions of hair (receptor) cells. (*Redrawn from Hudspeth.*)

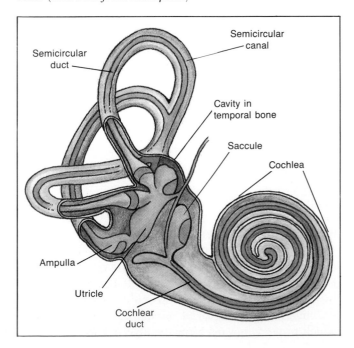

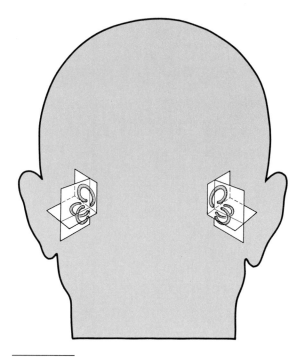

FIGURE 9-40 Relation of the two sets of semicircular canals.

during rotation of the head along three perpendicular axes. (The term semicircular canal is also used for the semicircular ducts.) The three axes of the semicircular canals are those activated while nodding the head up and down as in signifying "yes," shaking the head from side to side as in signifying "no," and tipping the head so the ear touches the shoulder on the same side (Figure 9-40).

The hairs of the receptor cells of the semicircular canals are closely ensheathed by a gelatinous mass, the **cupula**, which extends into the lumen of each membranous semicircular duct at the **ampulla**, a slight bulge in the wall of each duct (Figure 9-41). Whenever the head is moved, the bony semicircular canal, its enclosed membranous duct, and the attached bodies of the hair cells all turn with it. The fluid filling the duct, however, is not attached to the skull, and because of inertia, the fluid tends to retain its original position, that is, to be "left behind." Thus, the moving ampulla is pushed against the stationary fluid, and a pressure gradient is created across the cupula. This causes bending of the hairs, stimulation of the hair cells, and release of a chemical transmitter that activates the nerve terminals synapsing with the hair cells.

As the inertia is overcome and the duct fluid begins to move at the same rate as the rest of the head, the hairs slowly return to their resting position. For this reason, the hair cells are stimulated only during *changes* in the rate of motion, that is, during acceleration, of the head. The speed and magnitude of accelerating head move-

ments determine the way in which the hairs are bent and the hair cells stimulated. Each receptor has one direction of maximum sensitivity, and when its hairs are bent in this direction, the receptor cell depolarizes. When the hairs are bent in the opposite direction, the cell hyperpolarizes (Figure 9-42). The frequency of action potentials in the afferent nerve fibers that synapse with the hair cells is related to both the amount of force bending the hairs on the receptor cells and to the direction in which this force is applied.

The Utricle and Saccule

The utricle and saccule provide information about linear acceleration and changes in head position relative to the forces of gravity. The receptor cells here, too, are mechanoreceptors sensitive to the displacement of projecting hairs. The patch of hair cells in the utricle is nearly horizontal in a standing person, and that in the saccule is vertical.

The hairs are covered by a gelatinous substance in which tiny stones, or otoliths, made up of calcium carbonate crystals are embedded, making the gelatinous substance heavier than the surrounding fluid. In response to any linear acceleration, the gelatinous-otolith material changes its position, pulled by gravitational forces against the hair cells so that the hairs are bent and the receptor cells stimulated.

Vestibular Information and Dysfunction

Information about hair cell stimulation is relayed from the vestibular apparatus to the brainstem via cranial nerve VIII, the same nerve that carries acoustic information. This information is used in three ways. One is to control eye muscles so that, in spite of changes in head position, the eyes can remain fixed on the same point.

The second use of vestibular information is in reflex mechanisms for maintaining upright posture. The vestibular apparatus plays a role in the support of the head during movement, orientation of the head in space, and reflexes accompanying locomotion. Very few postural reflexes depend primarily on vestibular input, however, despite the fact that the vestibular organs are sometimes called the sense organs of balance.

The third use of vestibular information is, after relay via the thalamus to the cortex, in conscious awareness of the position and acceleration of the body.

Motion sickness can be induced in most people from an unexpected combination of inputs from the vestibular system and other sensory systems. For example, when one is on a boat in rough weather, linear and rotational accelerations occur in an unpredicted way. **Ménière's disease**, a disorder involving the vestibular system, is associated with episodes of abrupt and often severe diz-

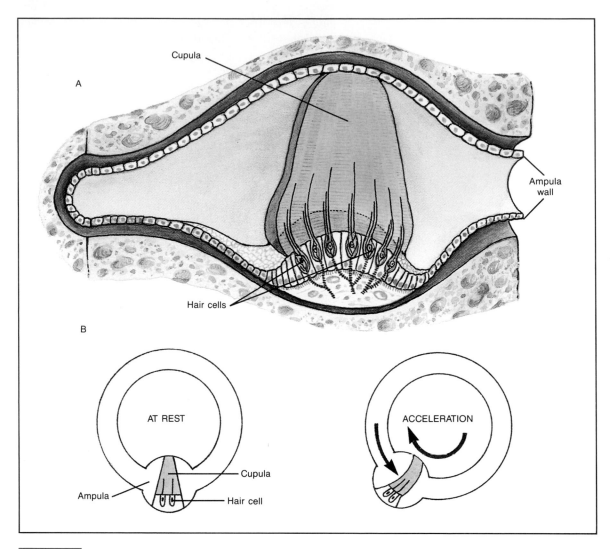

FIGURE 9-41 (A) A cupula and (B) the relation of the cupula to the ampulla when the head is at rest and when it is accelerating.

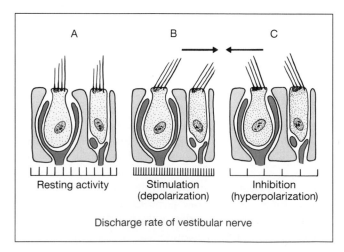

Resting activity

Stimulation (depolarization)

Inhibition (hyperpolarization)

Discharge rate of vestibular nerve

ziness, ringing in the ear, and bouts of hearing loss. It is due to an increased pressure of the fluid in the membrane duct system of the inner ear. The dizziness occurs because the inputs from the two ears are not balanced, either because only one ear is affected or because the process begins in one ear sooner than the other.

FIGURE 9-42 Relationship between position of hairs and activity in afferent neurons. (A) Resting activity. (B) Movement of hairs in one direction increases the action-potential frequency in the afferent nerve activated by the hair cell. (C) Movement in the opposite direction decreases the rate relative to the resting state. (*Redrawn from Wersall, Gleisner, and Lundquist.*)

CHEMICAL SENSES

Receptors sensitive to specific chemicals are **chemoreceptors**. Some of these respond to chemical changes in the internal environment, two examples being the oxygen and hydrogen-ion receptors in certain large blood vessels. Others respond to external chemical changes, and in this category are the receptors for taste and smell, the least understood of the major senses. Taste and smell affect a person's appetite, flow of saliva, gastric secretions, and avoidance of harmful substances.

Taste

The specialized sense organs for taste are the 10,000 or so **taste buds** that are found primarily on the tongue and roof of the mouth. Smaller numbers are also found in other areas of the mouth and pharynx. In the taste buds, the receptor cells and the adjacent cells that support them are arranged like the segments of an orange (Figure 9-43). The multifolded upper surfaces of the receptor cells extend into a small pore at the surface of the taste bud, where they are bathed by the fluids of the mouth.

Taste sensations are traditionally divided into four basic groups: sweet, sour, salty, and bitter (Figure 9-44). Despite these traditional four taste categories, single receptor cells can respond in varying degrees to many different chemical substances and to substances that fall into more than one taste category.

FIGURE 9-43 Structure and innervation of a taste bud. (*Redrawn from Elias, Pauly, and Burns.*)

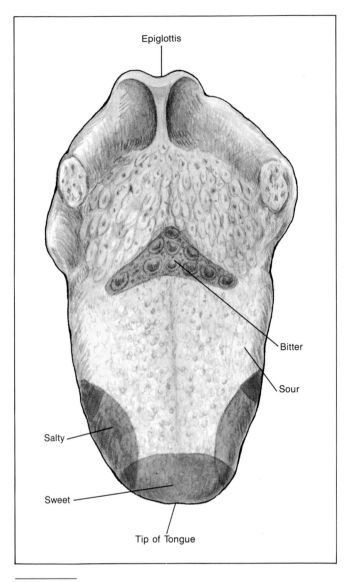

FIGURE 9-44 The four taste zones on the tongue. (*Redrawn from Elias, Pauly, and Burns.*)

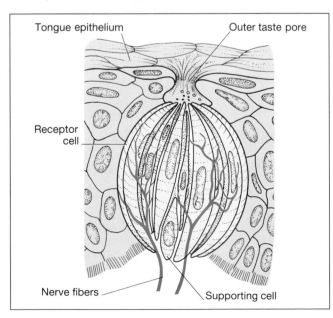

Afferent nerve fibers enter the taste buds to synapse with the receptor cells and are thought to be activated by a chemical transmitter released from the receptor cell. There is clearly no one-to-one relationship by which each receptor cell has a direct line into the nervous system. Awareness of the specific taste of a substance probably depends upon the pattern of firing in a group of neurons rather than firing of a specific neuron. The pathways for taste in the central nervous system project to parietal cortex, near the "mouth" region of the somatosensory cortex.

It is not known how we can distinguish so many different taste sensations when the receptor cells lack specificity both in terms of the kind of chemical to which they

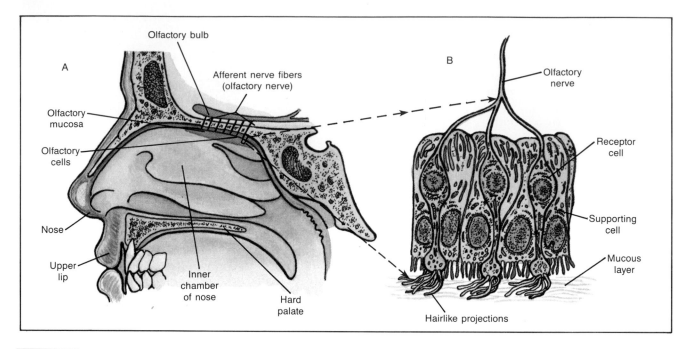

FIGURE 9-45 (A) Location and (B) structure of the olfactory receptors.

respond and the way in which they are connected to the brain. The odor of the substance clearly helps to identify a substance as is attested by the common experience that food lacks taste when one has a head cold.

Smell

The olfactory receptor cells, which give rise to the sense of smell, lie in a small patch of mucus-secreting membrane (the **olfactory mucosa**) in the upper part of the nasal cavity (Figure 9-45A). The olfactory receptors are not separate cells but are specialized afferent neurons. The cell bodies of these neurons have an enlarged extension from which several long hairlike processes, analogous to dendrites, extend out to the surface of the olfactory mucosa (Figure 9-45B). The axons of these neurons project to the brain as the olfactory nerve, which is cranial nerve I.

For an odorous substance, that is, an **odorant**, to be detected, molecules of the substance must diffuse into the air and pass into the nose to the region of the olfactory mucosa. Once there, the odorant molecules first dissolve in a layer of mucus that covers the receptor cells and then bind to receptor sites on the hairlike processes. **Odorant-binding proteins**, which are secreted into the mucus covering the olfactory receptors, may interact with the odorant molecules, transport them to the receptors, and facilitate the binding of the odorant to specific receptors on the olfactory neurons.

The family of protein receptors that underlies odor

recognition has not been identified, and the physiological basis for discrimination between the tens of thousands of different odor qualities is speculative. It is known, however, that olfactory discrimination varies with attentiveness, state of the olfactory mucosa—the sense of smell decreases when the mucosa is congested, as in a head cold; hunger—sensitivity is greater in hungry subjects; gender—women in general have keener olfactory sensitivities than men; and smoking—decreased sensitivity has been repeatedly associated with smoking.

Olfactory pathways in the central nervous system go to the olfactory bulbs, which lie on the undersurface of the frontal lobes of the cerebral cortex and are part of the limbic system (see Figure 8-12). It is not surprising that olfactory information is relayed directly to the limbic system, the part of the brain most intimately associated with neuroendocrine regulation and emotional, food-getting, and sexual behavior.

ASSOCIATION CORTEX AND PERCEPTUAL PROCESSING

Information from primary sensory areas in the cortex is processed further in the **cortical association areas** (Figure 9-46). These are brain areas that lie outside the primary cortical sensory or motor areas but are connected to them. The association areas play a role in the

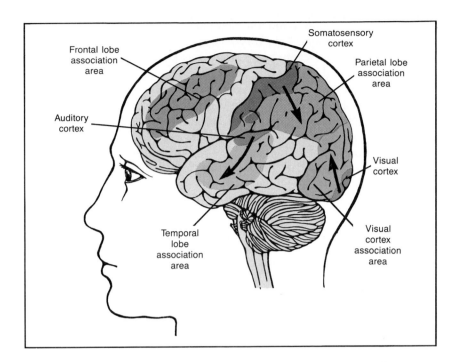

FIGURE 9-46 Areas of association cortex.

progressively complex analysis of incoming information. They also serve integrative functions and are implicated in many forms of behavior.

The primary cortical areas for visual, somatosensory, and auditory information each have lying next to them an association area, and each of these can be divided into two regions. The portions of the association cortex closer to the primary sensory areas receive input directly from the primary areas and serve relatively simple sensory-related functions, such as the recognition of patterns in the incoming information.

The more distant regions of association cortex receive information only after it has undergone initial processing by the region closer to the primary receiving area. Some of the neurons in the more distant regions receive input concerning two or even three of the different types of stimuli. These regions of association cortex are assumed to serve more complex functions. Thus, a neuron receiving input from both the "head" region of the somatosensory cortex and the visual cortex might be concerned with integrating visual information with sensory information about head position. For example, a tree is understood to be vertical even though the viewer's head is tipped sideways.

Neurons of these farther regions go to association areas in the frontal lobes that are part of the limbic system, and through these connections highly processed sensory information can be invested with emotional and motivational significance. For example, removal of part of the limbic association areas may abolish the emotional response to painful stimuli.

Further perceptual processing involves not only arousal, attention, learning, memory, language, and emotions but also comparing the information presented via one sensory modality with that of another. For example, we may hear a growling dog, but our perception of the event varies markedly, depending upon whether our visual system detects that the sound source is an angry animal or a loudspeaker.

Factors That Distort Perception

We put great trust in our sensory-perceptual processes despite the inevitable modifications we know to exist. Some factors known to distort our perceptions of the real world are as follows:

1. Afferent information is distorted by receptor mechanisms, for example, by adaptation, and by the processing of the information along afferent pathways.
2. Factors such as emotions, personality, experience, and social background can influence perceptions so that two people can witness the same events and yet perceive them differently.
3. Not all information entering the central nervous system gives rise to conscious sensation. Actually, this is a very good thing because many unwanted signals are generated by the extreme sensitivity of our receptors. For example, under ideal conditions the rods of the eye can detect the flame of a candle 17 mi away. The hair cells of the ear can detect vibrations of an amplitude much lower than those caused by blood flow through the ear's blood vessels and can even detect

molecules in random motion bumping against the tympanic membrane. Olfactory receptors respond to the presence of only four to eight odorous molecules. It is possible to detect one action potential generated by a pacinian corpuscle. If no mechanisms existed to select, restrain, and organize the barrage of impulses from the periphery, life would be unbearable. Information in some receptors' afferent pathways is not cancelled out—it simply does not give rise to a conscious sensation. For example, stretch receptors in the walls of some of the largest blood vessels effectively monitor both absolute blood pressure and its rate of change, but people have no conscious awareness of their blood pressure.

4. We lack suitable receptors for many energy forms. For example, we cannot directly detect ionizing radiation and radio or television waves.

However, the most dramatic examples of a clear difference between the real world and our perceptual world can be found in illusions and drug- and disease-induced hallucinations, when whole worlds can be created.

In conclusion, for perception to occur, the two processes—transmitting data through the nervous system and interpreting it—cannot be separated. Sensory information is processed at each synapse along the afferent pathways and at many levels of the CNS, with the more complex stages receiving input only after it has been processed by the more elementary systems. In addition to this hierarchical processing of afferent information along individual pathways, the information is processed by parallel pathways, each of which handles a limited aspect of the neural signals generated by the sensory transducers. Every synapse along the afferent pathways adds an element of organization and contributes to the sensory experience.

SUMMARY

Sensory systems process information that may lead to a sensation or to perception of an event within the body or in the outside world.

Neural Pathways in Sensory Systems

I. A single afferent neuron with all its receptor endings is a sensory unit.

II. The area of the body that, when stimulated, causes activity in a sensory unit or other neuron in the afferent pathway is called the receptive field for that neuron.

III. The specific ascending pathways convey information about only a single type of information to the cerebral cortex.

IV. Nonspecific ascending pathways convey information from more than one type of sensory unit to the brainstem reticular formation and regions of the thalamus not part of the specific ascending pathways.

Basic Characteristics of Sensory Coding

I. The type of stimulus perceived is determined by the type of receptor activated, the specific pathways activated, and the part of the brain in which these pathways terminate.

II. Stimulus intensity is coded by the rate of firing of individual sensory units and by the number of sensory units activated.

III. Location of the stimulus depends on the size of the receptive field covered by a single sensory unit and on the overlap of nearby receptive fields.

IV. Information coming into the nervous system is subject to control by ascending pathways and by descending pathways. Lateral inhibition is a means by which ascending pathways emphasize wanted information and increase the precision of location of the stimulus.

Somatic Sensation

I. Sensory function of the skin and underlying tissues is served by a variety of receptors sensitive to one (or a few) stimulus types.

II. Information about somatic sensation enters both specific and nonspecific ascending pathways. The specific pathways cross to the opposite side of the brain.

III. The somatic sensations include touch-pressure, proprioception and kinesthesia, temperature, and pain.

A. Some mechanoreceptors of the skin are rapidly adapting and give rise to sensations such as vibration, touch, movement, and tickle, whereas others are slowly adapting and give rise to the sensation of pressure.

B. Skin receptors having small receptive fields are involved in fine spatial discrimination, whereas receptors having larger receptive fields signal less precise touch/pressure sensations.

C. The major receptor type responsible for proprioception and kinesthesia is the muscle-spindle stretch receptor.

D. Cold receptors are sensitive to decreasing temperatures; warm receptors signal information about increasing temperature.

E. Specific receptors give rise to the sensation of pain, which may induce emotional and reflex responses as well as the perception of pain.

F. Stimulation-produced analgesia and TENS control pain by blocking transmission in the pain pathways.

Vision

I. Light is described by its wavelength or frequency.

II. The light that falls on the retina must be focused by the cornea and lens.

A. Lens shape is changed in response to viewing near or distant objects so that both are focused on the retina.

B. Presbyopia interferes with accommodation. Cataract prevents the passage of light to the retina.

C. An eyeball too long or too short relative to the focusing power of the lens causes nearsighted or farsighted vision, respectively.

III. The photopigments of the rods and cones are made up of a protein component (opsin) and a chromophore.

    A. The rods and each of the three cone types have different opsins that make each of the four receptor types sensitive to a different wavelength of light.

    B. When light falls upon the chromophore, the photic energy causes the chromophore to change shape, which triggers a cascade of events involving the second-messenger cyclic GMP and Ca^{2+}.

    C. When exposed to darkness, the rods and cones are depolarized and release their neurotransmitter. When exposed to light, they become hyperpolarized, and transmitter release is reduced.

IV. The rods and cones synapse on bipolar cells, which synapse on ganglion cells.

    A. Ganglion cell axons form the optic nerves, which lead into the brain.

    B. Half the optic nerve fibers cross to the opposite side of the brain in the optic chiasm. Fibers from the optic nerves terminate in the lateral geniculate nuclei of the thalamus, which send fibers to visual cortex.

    C. Visual information is also relayed to areas of the brain dealing with biological rhythms.

V. Coding in the visual system occurs along parallel pathways, in which different aspects of visual information, such as color, form, movement, and depth, are kept separate from each other.

VI. The colors we perceive are related to the wavelength of light. Different wavelengths excite one of the three cone photopigments most strongly.

    A. Certain ganglion cells are excited by input from one type of cone cell and inhibited by input from a different cone type.

    B. Our sensation of color depends on the output of the various cone opponent-color cells and the processing of this output by brain areas involved in color vision.

VII. Six skeletal muscles control eye movement to scan the visual field for objects of interest, keep the fixation point focused on the fovea despite movements of the object or the head, and move the eyes during accommodation.

Hearing

I. Sound energy is transmitted by movements of pressure waves.

    A. The sound wave frequency determines pitch.

    B. The sound wave amplitude determines loudness.

II. The sound transmission sequence is as follows:

    A. Sound waves enter the ear canal and press against the tympanic membrane, causing it to vibrate.

    B. The vibrating membrane causes movement of the three small middle-ear bones, and the stapes presses against the oval-window membrane.

    C. Movements of this membrane set up pressure waves in the fluid-filled scala vestibuli, which cause movements in the cochlear duct wall, setting up pressure waves in the fluid there.

    D. These pressure waves cause vibrations in the basilar membrane, located on one side of the cochlear duct.

    E. As this membrane vibrates, the hair cells of the organ of Corti move in relation to the tectorial membrane.

    F. Movement of the hair cells' stereocilia stimulates them to release neurotransmitter that activates receptors on the peripheral ends of the afferent nerve fibers.

III. Each part of the basilar membrane vibrates maximally in response to one particular sound frequency.

Vestibular System

I. A vestibular apparatus lies in the temporal bone on each side of the head and consists of three semicircular ducts, a utricle, and a saccule.

II. The semicircular ducts detect angular acceleration during rotation of the head, which cause bending of the stereocilia on the hair cells.

III. Otoliths in the gelatinous substance over the utricle and saccule hair cells move in response to changes in linear acceleration and the position of the head relative to gravity and stimulate the stereocilia on the hair cells.

Chemical Senses

I. The receptors for taste lie in taste buds throughout the mouth, principally on the tongue, and can respond to many different substances.

II. Olfactory receptors, which are part of the afferent olfactory neurons, lie in the mucosa in the upper nasal cavity.

    A. Odorant molecules, once dissolved in the mucus that bathes the olfactory receptors, are transported to the receptors by odorant binding proteins.

    B. Olfactory pathways go to the limbic system.

Association Cortex and Perceptual Processing

I. Information from the primary sensory cortical areas is elaborated after it is relayed to a cortical association area.

    A. The region of association cortex closest to the primary sensory cortical area develops the information in fairly simple ways and serves basic sensory-related functions.

    B. Regions of association cortex farther from the primary sensory areas process the sensory information in more complicated ways.

    C. Association cortical processing includes input from areas of the brain serving arousal, attention, memory, language, and emotions.

REVIEW QUESTIONS

1. Define:

sensory system	receptive field
sensory information	specific ascending pathways
sensation	somatic receptors
perception	somatosensory cortex
sensory unit	visual cortex
	auditory cortex

nonspecific ascending
  pathways
polymodal
recruitment
lateral inhibition
somatic sensation
proprioception
kinesthesia
sense of effort
nociceptors
referred pain
phantom limb pain
stimulation-produced
  analgesia
transcutaneous electric
  nerve stimulation
  (TENS)
acupuncture
wavelength
frequency
visible spectrum
retina
lens
cornea
accommodation
ciliary muscle
zonular fibers
presbyopia
cataract
nearsighted
myopic
farsighted
hyperopic
astigmatism
glaucoma
iris
pupil
rods
cones
photopigments
rhodopsin
opsin
chromophore

retinal
bipolar cell
ganglion cells
lateral geniculate nucleus
visual field
suprachiasmatic nucleus
on-center cell
off-center cells
opponent color cells
fovea
saccades
ear canal
tympanic membrane
middle-ear cavity
auditory (eustachian) tube
inner ear
cochlea
semicircular canals
malleus
incus
stapes
oval window
cochlear duct
basilar membrane
organ of Corti
hair cells
stereocilia
scala vestibuli
scala tympani
tectorial membrane
vestibular apparatus
semicircular ducts
utricle
saccule
cupula
ampulla
Meniere's disease
chemoreceptors
taste buds
olfactory mucosa
odorant
odorant-binding protein
cortical association areas

2. How does the nervous system distinguish between stimuli of different types?

3. How is information about stimulus intensity coded by the nervous system?

4. Diagram a specific ascending pathway that relays information from nociceptors. Diagram a nonspecific ascending pathway that relays information from mechanoreceptors and nociceptors.

5. Describe the similarities between pain and the other somatic sensations. Describe the differences.

6. List the structures through which light must pass before it reaches the photopigment in the rods and cones.

7. Describe the events that take place during accommodation for far vision.

8. What changes take place in neurotransmitter release from the rods or cones when they are exposed to light?

9. Beginning with the ganglion cells of the retina, describe the visual pathway.

10. List the sequence of events that occur between entry of a sound wave into the ear canal and the firing of action potentials in the cochlear nerve.

11. Describe the anatomical relationship between the cochlea and the cochlear duct.

12. What is the relationship between head movement and cupula movement in a semicircular canal?

13. What causes the release of neurotransmitter from the utricle and saccule receptor cells?

14. In what ways are the sensory systems for taste and olfaction similar? In what ways are they different?

15. Describe the relationship between sensory information processing in the primary cortical sensory areas and in the cortical association areas.

16. List four ways in which sensory information can be distorted.

THOUGHT QUESTION

(Answer is given in Appendix A.)

Describe six mechanisms by which pain could theoretically be controlled.

HORMONAL CONTROL MECHANISMS

The endocrine system is one of the body's two major communication systems, the nervous system being the other. The **endocrine system** consists of all those glands, termed **endocrine glands**, that secrete hormones. **Hormones**, as noted in Chapter 7, are chemical messengers that are carried by the blood from endocrine glands to the cells upon which they act. The cells influenced by a particular hormone are termed the **target cells** (or effector cells) for that hormone.

Table 10-1 summarizes, for reference and orientation, the endocrine glands, the hormones they secrete, and the major functions the hormones control. The endocrine system differs from most of the other organ systems of the body in that the various glands are not in anatomical continuity with each other. However, they do form a system in the functional sense. The reader may be puzzled to see listed as endocrine glands some organs—the heart, for instance—that clearly have other functions.

TABLE 10-1 SUMMARY OF THE HORMONES

Site Produced (Endocrine Gland)	Hormone	Major Function* Is Control of:
Hypothalamus	Releasing hormones	Secretion of hormones by the anterior pituitary
	Oxytocin	See posterior pituitary.
	Vasopressin	See posterior pituitary.
Anterior pituitary	Growth hormone (somatotropin, GH)†	Growth: secretion of IGF-I; organic metabolism
	Thyroid-stimulating hormone (TSH, thyrotropin)	Thyroid gland
	Adrenocorticotropic hormone (ACTH, corticotropin)	Adrenal cortex
	Prolactin	Breast growth and milk synthesis; permissive for certain reproductive functions in the male
	Gonadotropic hormones: Follicle-stimulating hormone (FSH) Luteinizing hormone (LH)	Gonads (gamete production and sex hormone secretion)
	β-lipotropin	Unknown.
	β-endorphin	Unknown.
Posterior pituitary	Oxytocin‡	Milk "let-down"; uterine motility
	Vasopressin (antidiurectic hormone, ADH)‡	Water excretion by the kidneys: blood pressure
Adrenal cortex	Cortisol	Organic metabolism; response to stresses; immune system
	Androgens	Sex drive in women
	Aldosterone	Sodium, potassium, and acid excretion by kidneys
Adrenal medulla	Epinephrine Norepinephrine	Organic metabolism; cardiovascular function; response to stress
Thyroid	Thyroxine (T$_4$) Triiodothyronine (T$_3$)	Metabolic rate; growth; brain development and function
	Calcitonin	Plasma calcium
Parathyroids	Parathyroid hormone (parathormone, PTH, PH)	Plasma calcium and phosphate
Gonads: Female: ovaries	Estrogen Progesterone	Reproductive system; breasts; growth and development
	Inhibin	FSH secretion
	Relaxin	Relaxation of cervix and pubic ligaments
Male: testes	Testosterone	Reproductive system; growth and development
	Inhibin	FSH secretion
	Müllerian-inhibiting hormone	Regression of Müllerian ducts
Pancreas	Insulin Glucagon Somatostatin Pancreatic polypeptide	Organic metabolism; plasma glucose

TABLE 10-1 *(continued)*

Site Produced (Endocrine Gland)	Hormone	Major Function* Is Control of:
Kidneys	Renin (→angitonsin II)§	Aldosterone secretion; blood pressure
	Erythropoietin	Erythrocyte production
	1,25-dihydroxyvitamin D_3	Calcium absorption by the intestine
Gastrointestinal tract	Gastrin	Gastrointestinal tract; liver; pancreas; gallbladder
	Secretin	
	Cholecystokinin	
	Glucose-dependent insulinotropic peptide (GIP)	
	Somatostatin	
Liver (and other cells)	Insulin-like growth factors (IGF-I and II)	Growth
Thymus	Thymosin (thymopoietin)	T-lymphocyte function
Pineal	Melatonin	? Sexual maturity; body rhythms
Placenta	Chorionic gonadotropin (CG)	Secretion by corpus luteum
	Estrogens	See ovaries
	Progesterone	
	Placental lactogen	Breast development; organic metabolism
Heart	Atrial natriuretic factor (ANF, atriopeptin, auriculin)	Sodium excretion by kidneys; blood pressure
Monocytes and macrophages	Interleukin-1 (IL-1)	See Chapter 19.
	Tumor necrosis factor (TNF)	
	Other monokines	
Multiple cell types	Growth factors (e.g., nerve growth factor)	Growth of specific tissues

*This table does not list all functions of the hormones.
†The names and abbreviations in parentheses are synonyms.
‡The posterior pituitary stores and secretes these hormones; they are made in the hypothalamus.
§Renin is an enzyme that initiates reactions in blood that generate angiotensin II

The explanation is that, in addition to the cells that carry out the organ's other functions, the organ also contains cells that secrete hormones.

Note also in Table 10-1 that the hypothalamus, a part of the brain, is considered part of the endocrine system too. This is because the chemical messengers released by certain neuron terminals in both the hypothalamus and its extension, the posterior pituitary, do not function as neurotransmitters affecting adjacent cells but rather enter the blood and are carried to sites of action elsewhere. As pointed out in Chapter 7, hormones released from neuron terminals are referred to as **neurohormones**.

Table 10-1 demonstrates that there are a large number of endocrine glands and hormones. One way of describing the physiology of the individual hormones is to present all relevant material, gland by gland, in a single chapter. In keeping with our emphasis on hormones as messengers in homeostatic control mechanisms, however, we have chosen to describe the physiology of specific hormones and the glands that secrete them in subsequent chapters, in the context of the control systems in which they participate. For example, the pancreatic hormones are described in Chapter 17 on organic metabolism, the parathyroid hormones in Chapter 15 in the context of calcium metabolism, and so on.

The aims of the present chapter are, therefore, limited to presenting: (1) the general principles of endocrinology, that is, a structural and functional analysis of hormones in general that transcends individual glands; and (2) an analysis of the hypothalamus-pituitary hormonal system. The control systems for the hormones of this particular system are so interconnected that they are best described as a unit to lay the foundation for subsequent descriptions in other chapters.

Before turning to these analyses, however, several additional general points should be made concerning Table 10-1. One phenomenon evident from this table is that a single gland may secrete multiple hormones. The usual pattern in such cases is that a single cell type secretes only one hormone, so that multiple hormone secretion reflects the presence of different endocrine cell types in the same gland. In a few cases, however, a single cell may secrete more than one hormone.

Another point of interest illustrated by Table 10-1 is that a particular hormone may be produced by more than one type of endocrine gland. For example, somatostatin is secreted by endocrine cells in both the gastrointestinal tract and the pancreas and is also one of the hormones secreted by the hypothalamus.

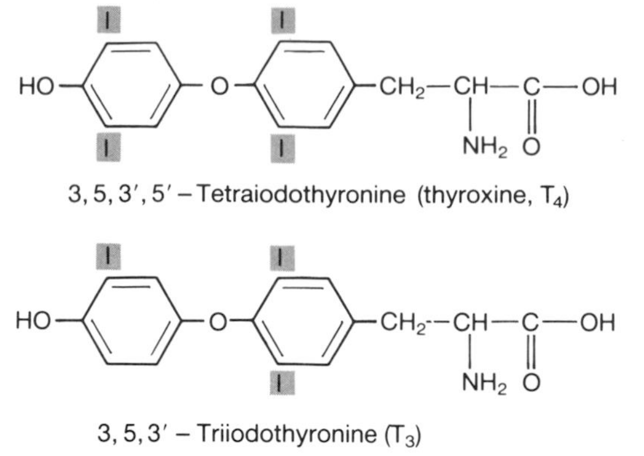

HORMONE STRUCTURES AND SYNTHESIS

Hormones fall into three general chemical classes: (1) amines, (2) peptides and proteins, and (3) steroids.

Amine Hormones

The **amine hormones** are nitrogen-containing derivatives of the amino acid tyrosine. They include the thyroid hormones and the hormones produced by the adrenal medulla.

Thyroid hormones. The **thyroid gland** is located in the lower part of the neck wrapped around the front of the trachea (windpipe). It secretes three hormones, but the two iodine-containing amine hormones shown in Figure 10-1—**thyroxine** (**T_4**) and **triiodothyronine** (**T_3**)—are collectively known as the **thyroid hormones** (**TH**). The third hormone, a peptide called calcitonin, is not included in subsequent references to "thyroid hormones."

Iodine is an essential element that functions as a component of T_4 and T_3. Most of the iodine ingested in food is transported into the blood from the gastrointestinal tract, which converts it to the ionized form, iodide. Iodide then leaves the blood and is actively transported into the thyroid cells. Once in the gland, the iodide is converted back to iodine, which is then used, along with tyrosine, for hormone synthesis.

The normal thyroid gland stores several weeks' supply of thyroid hormones bound to a protein known as thyroglobulin. Hormone secretion occurs by enzymatic splitting of T_4 and T_3 from the thyroglobulin and the entry of the freed hormones into the blood. T_4 is secreted in much larger amounts than is T_3. However, a

FIGURE 10-1 Chemical structures of the thyroid hormones, thyroxine and triiodothyronine. The two molecules differ by only one iodine atom, a difference noted in the abbreviations T_3 and T_4.

variety of tissues, including the hormone's target cells, convert most of this T_4 into T_3 by enzymatic removal of one iodine atom.

Virtually every tissue in the body is affected by the thyroid hormones. These effects, which are described in Chapter 17, include regulation of metabolic rate, brain development and function, and growth.

Adrenal medullary hormones. There are two adrenal glands, one on the top of each kidney. Each **adrenal gland** constitutes two distinct endocrine glands, an inner **adrenal medulla**, which secretes amine hormones, and a surrounding **adrenal cortex**, which secretes steroid hormones. As described in Chapter 8, the adrenal medulla is really like a sympathetic ganglion whose cell bodies do not send out nerve fibers but instead release their secretions into the blood, thereby fulfilling a criterion for an endocrine gland.

The adrenal medulla secretes two amine hormones, **epinephrine** (**E**) and **norepinephrine** (**NE**), which belong to the chemical family of catecholamines. In this chapter we shall use the terms "adrenal medullary hormones" and "catecholamine hormones" interchangeably. The structures and pathways for synthesis of the catecholamines were described in Chapter 8, when these substances were dealt with as neurotransmitters. In humans, the adrenal medulla secretes approximately four times more epinephrine than norepinephrine.[1] Epi-

---

[1]Yet under basal conditions, the plasma concentration of norepinephrine is much higher than that of epinephrine; the source of this norepinephrine is sympathetic neurons, not the adrenal medulla.

nephrine and norepinephrine exert actions similar to those of the sympathetic nerves. These effects are described in various chapters and summarized in Chapter 19 in the section on stress.

The adrenal medulla also secretes several substances other than catecholamines, but the functions, if any, of these secretions are unknown.

Peptide Hormones

The great majority of all hormones are either peptides or proteins. They range in size from small peptides having only three amino acids to small proteins, but for the sake of convenience, we shall follow a common practice of endocrinologists and refer to them all as **peptide hormones**.

They are initially synthesized (Figure 10-2) on the ribosomes of the endocrine cell as larger proteins known as preprohormones, which are then cleaved to **prohormones** by proteolytic enzymes in the granular endoplasmic reticulum of the cell. The prohormone is then packaged into secretory vesicles by the Golgi apparatus of the cell. In this process, the prohormone is cleaved to yield the hormone itself and other peptide chains found in the prohormone. Therefore, when the cell is stimulated to release the contents of the secretory vesicles by exocytosis, the other peptides are cosecreted with the hormone. In certain cases they, too, may exert hormonal effects. In other words, instead of just one peptide hormone, the cell may be secreting multiple peptide hormones that differ in their effects on target cells.

The direct stimulus of secretory-vesicle exocytosis in peptide-producing endocrine cells, as in neurons, is an increase in cytosolic calcium concentration. The source of the calcium during cell stimulation is either the extracellular fluid or the endoplasmic reticulum. Many of the stimuli for secretion directly or indirectly cause some degree of plasma-membrane depolarization—sometimes even action potentials—which opens voltage-sensitive calcium channels in the plasma membrane and allows diffusion of calcium into the cell.

One more point about peptide hormones: As mentioned in Chapters 7 and 8, many peptides serve as both neurotransmitters (or neuromodulators) and as hormones. For example, most of the hormones secreted by the endocrine glands in the gastrointestinal tract are also produced by neurons in the brain.

Steroid Hormones

The third family of hormones is the steroids, the lipids whose ringlike structure was described on page 31. **Steroid hormones** are produced by the adrenal cortex and the gonads (testes and ovaries), as well as by the placenta during pregnancy. Examples are shown in Figure 10-3.

Cholesterol is the precursor of all steroid hormones. Although steroid-producing endocrine glands synthesize some of their own cholesterol, their major source is the cholesterol delivered to the cells by plasma lipoproteins that are made in the liver and circulate in the blood. The many biochemical steps in steroid synthesis beyond cho-

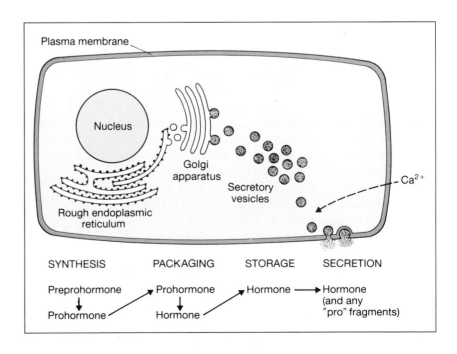

FIGURE 10-2 Synthesis and secretion of peptide hormones. *(Adapted from Hedge et al.)*

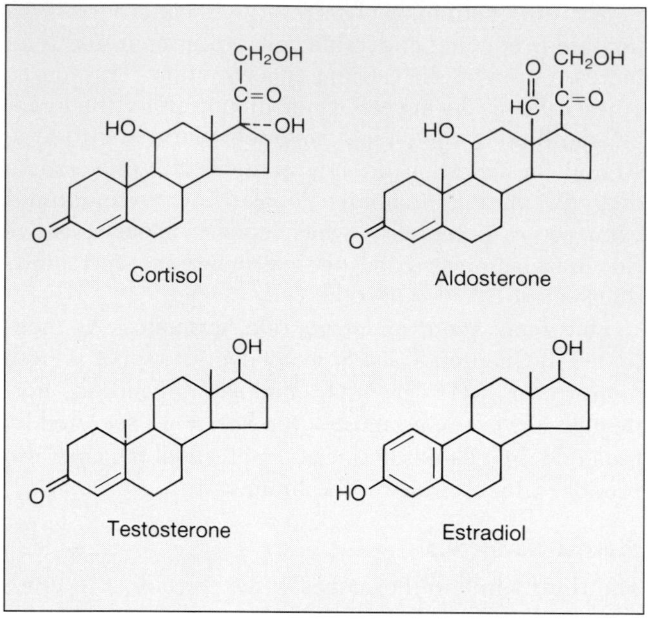

FIGURE 10-3 Structures of representative steroid hormones.

lesterol involve small changes in the molecules and are mediated by specific enzymes. The steroids a particular cell produces depend, therefore, on the types and concentrations of the enzymes the cell has. Because steroids are highly lipid-soluble, once they are synthesized they simply diffuse across the plasma membrane of the steroid-producing cell and enter the blood.

The next sections describe the pathways for steroid synthesis by the adrenal cortex and gonads. Those for the placenta are somewhat unusual and are discussed in Chapter 18.

Hormones of the adrenal cortex. Steroid synthesis by the adrenal cortex is illustrated in Figure 10-4. The five hormones normally secreted in physiologically significant amounts by the adrenal cortex are aldosterone, cortisol, corticosterone, dehydroepiandrosterone, and androstenedione. **Aldosterone** is known as a **mineralocorticoid** because its effects are on salt (mineral) metabolism, mainly on the kidney's handling of sodium, potassium, and hydrogen ions. **Cortisol** and corticosterone are called **glucocorticoids** because they have important effects on the metabolism of glucose and other organic nutrients;[2] cortisol is by far the more important of

---

[2] In very high concentrations glucocorticoids also exert some mineralocorticoid effects.

the two glucocorticoids in humans, and so we shall deal only with it in future discussions. In addition to its effects on organic metabolism (described in Chapter 17), cortisol exerts many other effects, including facilitation of the body's responses to stress (Chapter 19).

Dehydroepiandrosterone and androstenedione belong to the general class of hormones known as **androgens**, which includes the male sex hormone, testosterone, produced by the testes. All androgens have actions similar to testosterone. Because the adrenal androgens are much less potent than testosterone, they are of little physiological significance in the male; they do, however, play roles in the female, as described in Chapter 18.

The adrenal cortex is not a homogeneous gland but is composed of three distinct layers (Figure 10-5). The outer layer—the zona glomerulosa—possesses very high concentrations of the enzymes required to convert corticosterone to aldosterone but lacks the enzymes required for the formation of cortisol and androgens. Accordingly, this layer synthesizes and secretes aldosterone but not the other major adrenal cortical hormones. In contrast, the zona fasciculata and zona reticularis have just the opposite enzyme profile: They secrete no aldosterone but much cortisol and androgen.

In certain disease states, the adrenal cortex may secrete decreased or increased amounts of various steroids. For example, an absence of the enzymes for the formation of cortisol by the adrenal cortex can result in the shunting of the cortisol precursors into the androgen pathway. In a woman, the result of the large increase in androgen secretion would be masculinization.

Hormones of the gonads. Compared to the adrenal cortex, the gonads have very different concentrations of key enzymes in the steroid pathways. Endocrine cells in both testes and ovaries lack the enzymes needed to produce aldosterone and cortisol. They possess high concentrations of enzymes in the androgen pathways leading to androstenedione, as in the adrenal cortex. In addition, however, the cells in the testes contain a high concentration of the enzyme that converts androstenedione to **testosterone**, which is therefore the major androgen secreted by the testes. The ovarian cells that synthesize the major female sex hormone, **estradiol**, have high concentration of the enzyme required to go one step further, that is, to transform testosterone to estradiol. Accordingly, estradiol, rather than testosterone, is secreted by the ovaries. Ovarian cells also secrete progesterone.

Very small amounts of testosterone do leak out of ovarian endocrine cells, however, and very small amounts of estradiol are produced from testosterone in the testes. Moreover, following their secretion into the blood by the gonads and the adrenal cortex, steroid hor-

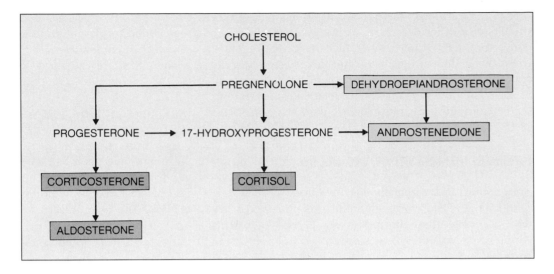

FIGURE 10-4 Simplified flow sheet for synthesis of steroid hormones by the adrenal cortex; many intermediate steps have been left out. The five hormones in boxes are the major steroid hormones secreted. Dehydroepiandrosterone and androstenedione are androgens, that is, testosterone-like hormones.

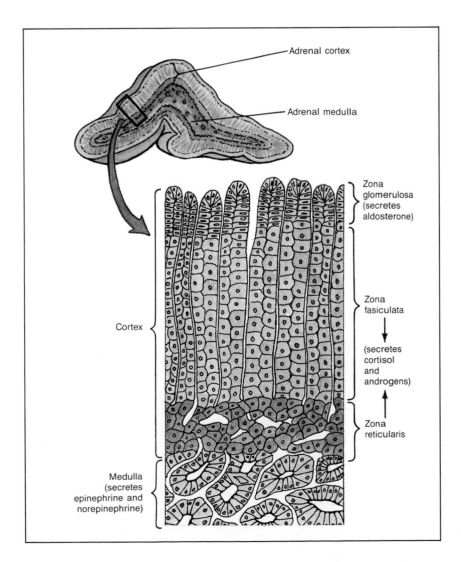

FIGURE 10-5 Section through an adrenal gland showing both adrenal medulla and adrenal cortex.

mones may undergo further interconversion in either the blood or other organs. For example, testosterone is converted to estradiol in some of its target cells. Thus, the major male and female sex hormones—testosterone and estradiol, respectively—are not unique to males and females, although, of course, the relative concentrations of the hormones are quite different in the two sexes.

HORMONE TRANSPORT IN THE BLOOD

Peptide and catecholamine hormones are water-soluble. Therefore, with the exceptions of a few peptides, these hormones are carried in plasma simply dissolved in the plasma water (Table 10-2). In contrast, the steroid hormones and the thyroid hormones are poorly soluble in water and circulate in the blood largely bound to plasma proteins. Some of these proteins are specialized in that they bind only certain hormones and have no other function, whereas other plasma proteins bind almost any water-insoluble hormone.

Even though the steroid and thyroid hormones exist in plasma mainly bound to proteins, small concentrations of these hormones do exist dissolved in the plasma. The dissolved, or free, hormone is in equilibrium with the bound hormone:

Free hormone + binding protein $\rightleftharpoons$
$$\text{hormone-protein complex}$$

The total hormone concentration in plasma is the sum of the free and bound hormone. It is important to realize, however, that only the free hormone can cross capillary walls and encounter its target cells. Accordingly, the concentration of the free hormone is what is physiologically important rather than the concentration of the total hormone, most of which is bound. For example, less than 1 percent of plasma thyroid hormone is free, that is, not bound to plasma proteins. As we shall see, the degree of protein binding also influences the rate of metabolism and the excretion of the hormone.

HORMONE METABOLISM AND EXCRETION

A hormone's concentration in the plasma depends not only upon its rate of secretion from the endocrine gland but also upon its rate of removal from the blood, either by excretion or by metabolic transformation. The liver and the kidneys are the most important organs that excrete or metabolize hormones. Because for many hormones and their metabolites the rate of urinary excretion is directly proportional to the original rate of secretion, the excretion rate can be used as an indicator of the secretory rate.

The liver and kidneys are not the only routes for eliminating hormones. Sometimes the hormone is metabolized by the cells upon which it acts. For example, endocytosis of plasma-membrane hormone-receptor complexes enables cells to remove peptide hormones rapidly from their surfaces and catabolize them intracellularly—the receptors are often recycled to the plasma membrane.

In general, catecholamine and peptide hormones, which circulate mainly free in the plasma, are comparatively easily excreted or attacked by enzymes in the blood and tissues. These hormones therefore tend to remain in the bloodstream for only brief periods—minutes to a few hours. In contrast, the circulating steroid and thyroid hormones are mainly bound to plasma proteins, and bound hormone is less vulnerable to excretion or metabolism by enzymes. Accordingly, removal of these hormones from plasma generally takes many hours.

In some cases, metabolism of the hormone after its

TABLE 10-2 CATEGORIES OF HORMONES				
Types	Form in Plasma	Location of Receptors	Signal Transduction Mechanisms	Rate of Excretion
Peptides and catecholamines	Free	Plasma membrane	Receptor-operated channels Cyclic AMP Cyclic GMP Ca^{2+}, IP_3, DAG	Fast
Steroids and thyroid hormones	Protein-bound	Cell interior	Activates genes	Slow

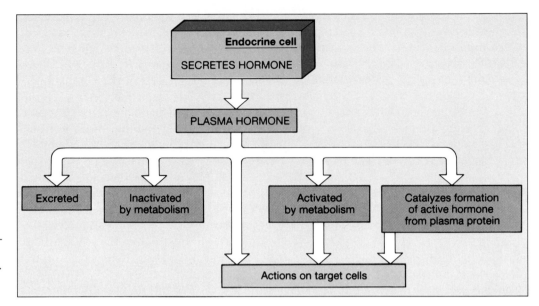

FIGURE 10-6 Possible fates and actions of a hormone following its secretion by an endocrine cell. Not all paths apply to all hormones.

secretion *activates* the hormone rather than inactivates it. In other words, the secreted hormone may be relatively or completely unable to act upon a target cell until metabolism elsewhere in the body transforms it into a substance that can act. We have already seen one example of hormone activation—the conversion of T_4 to the far more active T_3 by organs other than the thyroid. Another example is provided by testosterone, which is converted to dihydrotestosterone in certain of its target cells. This latter steroid, rather than testosterone itself, then elicits the response of the target cell.

There is another kind of activation that applies to a few hormones. Instead of the hormone itself being activated after secretion, it acts enzymatically on another plasma protein to split off a peptide that functions as the active hormone.

Figure 10-6 summarizes the fates of hormones after their secretion.

MECHANISMS OF HORMONE ACTION

Hormone Receptors

Because they travel in the blood, hormones are able to reach virtually all tissues. Yet the body's response to a hormone is not all-inclusive but highly specific, involving only the target cells for that hormone. The ability to respond depends upon the presence on (or in) the target cells of specific receptors for those hormones.

As emphasized in Chapter 7, the response of a target cell to a chemical messenger is only the final event in a sequence that begins when the messenger binds to specific receptors on the cell. Where in the target cells are the hormone receptors located? The receptors for steroid hormones and the thyroid hormones are proteins inside the target cells. Because these hormones are lipid-soluble, they readily cross the plasma membrane and combine with their specific receptors.

In contrast, the receptors for the peptide hormones and catecholamines are proteins located on the plasma membranes of the target cells. However, many peptide hormones do in fact gain entry to the interior of cells by endocytosis of the plasma-membrane hormone-receptor complex. It is likely that most of the hormone molecules internalized in this way simply undergo catabolism, but it is also possible that either they or the peptide fragments resulting from their catabolism exert additional effects upon the target cell.

Basic concepts of receptor modulation (up-regulation and down-regulation) were described in Chapter 7, in the context of the receptors for all types of chemical messengers, because we wished to emphasize the general applicability of these concepts. However, it is in the field of endocrinology that these phenomena have been most intensively studied. For hormones, **up-regulation** is an increase in the number of a hormone's receptors on a target cell resulting from prolonged exposure to a low concentration of the hormone. **Down-regulation** is a decrease in receptor number resulting from exposure to high concentrations of the hormone.

Hormones can alter not only their own receptors but the receptors for other hormones as well. If one hormone induces a loss of a second hormone's receptors, the result will be a reduction of the second hormone's effectiveness. On the other hand, a hormone may induce an increase in the number of receptors for a second hor-

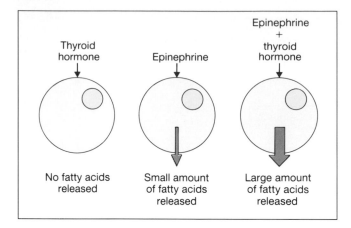

FIGURE 10-7 Ability of thyroid hormone to "permit" epinephrine-induced liberation of fatty acids from adipose-tissue cells. The thyroid hormones exert this effect by causing an increased number of epinephrine receptors on the cell.

mone. In this case the effectiveness of the second hormone is increased. This last phenomenon underlies the important hormone-hormone interaction known as permissiveness.

In general terms, **permissiveness** means that hormone A must be present for the full exertion of hormone B's effect. A low concentration of hormone A is usually all that is needed for this permissive effect, which is due to A's positive effect on B's receptors. For example, as shown in Figure 10-7, epinephrine causes marked release of fatty acids from adipose tissue, but only in the presence of permissive amounts of thyroid hormone, because thyroid hormone facilitates the synthesis of receptors for epinephrine in adipose tissue.

The ability of one hormone to influence the receptors of a second hormone also underlies certain situations in which a response requires actions of hormones in a sequence. For example, ovulation requires the action of two different hormones secreted by the anterior pituitary; the first hormone paves the way for the second hormone by increasing the number of the latter's receptors in the relevant ovarian cells.

Events Elicited by Hormone-Receptor Binding

Effects of steroid and thyroid hormones. The common denominator of the effects of the steroid and thyroid hormones is an increased synthesis of specific proteins—enzymes, structural proteins, and so on—by their target cells. These hormones increase protein synthesis by

stimulating the production of mRNA via the following sequence of events. First, a steroid or thyroid hormone enters a target cell and binds to specific receptors for it (Figure 10-8). The receptors for these hormones are synthesized in the cytoplasm but transported into the nucleus. In most cases, the hormone enters the nucleus and combines with the receptor already there. However, in other cases, for example, cortisol, the hormone combines with the receptor still in the cytoplasm, and the entire complex is transported into the nucleus. In all cases, the receptor, activated by its binding of the hormone, interacts with segments of DNA specific for it to trigger transcription of those segments. That is, the activated receptor stimulates the synthesis of mRNA, which then enters the cytoplasm and serves as a template for synthesis of a specific protein.

Effects of peptide hormones and adrenal catecholamines. As described above, the receptors for peptide hormones and the hormones of the adrenal medulla are located on the outer surface of the target cell's plasma membrane. Most of these receptors utilize one or more of the signal transduction mechanisms and second messengers described in Chapter 7: receptor-operated channels, cyclic AMP, cyclic GMP, calcium, and the phosphatidylinositol system. The receptors of at least three peptide hormones—insulin, growth hormone, and prolactin—probably utilize other, as yet unidentified, second messengers. Again as described in Chapter 7, all these signal transduction mechanisms result in alterations of protein activity, leading to the cell's ultimate response to the hormones.

Pharmacological Effects of Hormones

Administration of very large quantities of a hormone for medical purposes may have effects that are never seen in a normal person. They are called **pharmacological effects**, and they can also occur in diseases when excessive amounts of hormones are secreted. Pharmacological effects are of great importance in medicine since hormones are often used in large doses as therapeutic agents. Perhaps the most common example is that of the adrenal cortical hormone cortisol, which is administered in large amounts to suppress allergic and inflammatory reactions.

TYPES OF INPUTS THAT CONTROL HORMONE SECRETION

Most hormones are released in short bursts, with little or no release occurring between bursts. Accordingly, the plasma concentrations of hormones may fluctuate rapidly over brief time periods. Hormones also manifest 24-h cyclical variations in their

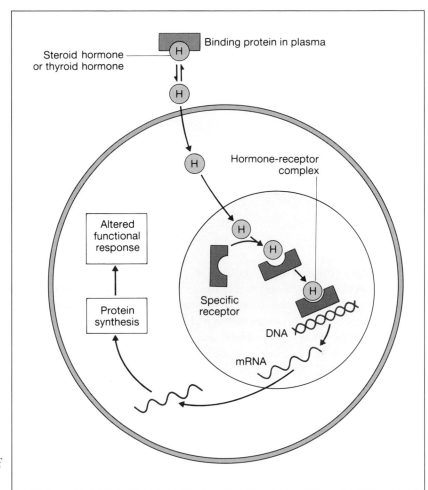

FIGURE 10-8 Steroid and thyroid hormones enter the target cell and bind to their specific receptors, usually in the nucleus. (For some hormones, not shown, the binding takes place in the cytoplasm, rather than the nucleus, and the hormone-receptor complex is then transported into the nucleus.) The activated receptor interacts with specific segments of DNA and stimulates transcription of new mRNA, which directs the synthesis of the proteins that mediate the cell's response.

secretory rates, the circadian patterns being different for different hormones. Some are clearly linked to sleep. For example, Figure 10-9 illustrates how growth hormone's secretion is markedly increased during the early period of sleep. The mechanisms underlying these cycles are ultimately traceable to cyclical variations in the

activity of neural pathways involved in the hormone's release.

Hormone secretion is controlled mainly by four types of inputs to endocrine cells (Figure 10-10): (1) changes in the plasma concentrations of mineral ions; (2) changes in the plasma concentrations of organic nutrients; (3) neu-

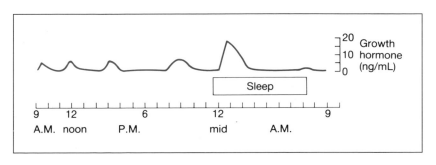

FIGURE 10-9 Plasma growth hormone concentrations in a normal 23-year-old male over 24 h. [*Adapted from Sassin et al. (1972).*]

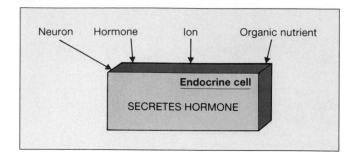

FIGURE 10-10 Types of inputs that act directly on endocrine-gland cells to stimulate or inhibit hormone secretion.

rotransmitters released from neurons impinging on the endocrine cell; and (4) another hormone acting on the endocrine cell. There is actually a fifth type of input, but it applies only to the hormones secreted by the gastrointestinal tract and will be described in Chapter 16.

Before we look more closely at each category, it must be stressed that, in many cases, hormone secretion is influenced by more than one input. For example, insulin secretion is controlled by the extracellular concentrations of glucose and other nutrients, by both sympathetic and parasympathetic neurons to the endocrine cells, and by several hormones acting on the endocrine cells. Thus, endocrine cells, like neurons, may be subject to multiple, simultaneous, often opposing inputs, and the resulting output—the rate of hormone secretion—reflects the integration of all these inputs.

Control by Plasma Concentrations of Mineral Ions or Organic Nutrients

There are at least five hormones whose secretion is directly controlled, at least in part, by the plasma concentrations of specific mineral ions or organic nutrients. In each case, a major function of the hormone is to regulate, in a negative-feedback manner, the plasma concentration of the ion or nutrient controlling its secretion. For example, insulin secretion is stimulated by an elevated plasma glucose concentration, and the additional insulin then causes, by several actions, the plasma glucose level to decrease (Figure 10-11).

Control by Neurons

The adrenal medulla behaves like a sympathetic ganglion and thus is stimulated by sympathetic preganglionic fibers. In addition to its control of the adrenal medulla, the autonomic nervous system has other influences on the endocrine system (Figure 10-12). Various endocrine glands receive a rich supply of sympathetic and, in some cases, parasympathetic neurons, whose

activity influences the glands' rates of hormone secretion. Examples are the secretion of insulin and the gastrointestinal hormones. In most cases studied, the autonomic input serves not as the sole regulator of the hormone's secretion but rather as one of multiple inputs.

Thus far, our discussion of neural control of hormone release has been limited to the role of the autonomic nervous system. However, one large group of hormones—those secreted by the hypothalamus and its extension, the posterior pituitary—are under the direct control not of autonomic neurons but of neurons in the brain (Figure 10-12). This category will be described in detail later in this chapter.

Control by Other Hormones

In many cases, the secretion of a particular hormone is directly controlled by the blood concentration of another hormone. As we shall see, there are often complex sequences in which the only function of the first few hormones in the sequence is to stimulate the secretion of

FIGURE 10-11 Example of how the direct control of hormone secretion by the plasma concentration of a substance, either organic nutrient or mineral ion, results in the negative-feedback control of that substance's plasma concentration.

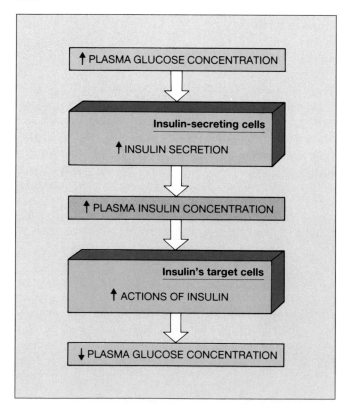

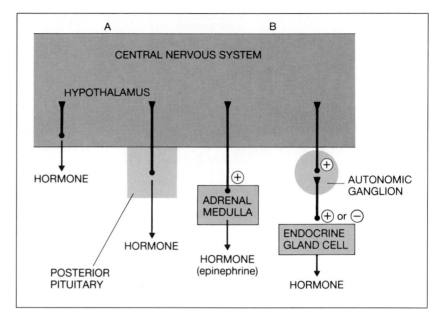

FIGURE 10-12 Pathways by which the nervous system influences hormone secretion. (A) Neurons in the hypothalamus, some of which terminate in the posterior pituitary, themselves secrete neurohormones. (B) The autonomic nervous system controls hormone secretion by the adrenal medulla and many other endocrine glands; both parasympathetic and sympathetic inputs to these other glands may occur, some inhibitory and some stimulatory. Each neuron in the central nervous system is, of course, under the influence of the many other neurons synapsing upon it.

the next one. A hormone that stimulates the secretion of another hormone is often referred to as a **tropic hormone**.

CONTROL SYSTEMS INVOLVING THE HYPOTHALAMUS AND PITUITARY

The **pituitary gland** lies in a pocket of bone at the base of the brain, just below the hypothalamus (Figure 10-13), to which it is connected by a stalk containing nerve fibers and small blood vessels. In adult human beings, the pituitary gland is composed of two adjacent lobes—the **anterior pituitary** (toward the front of the head) and the **posterior pituitary** (toward the back of the head)—each of which is a more or less distinct gland.

Posterior Pituitary Hormones

The posterior pituitary is an outgrowth of the hypothalamus and is neural tissue. The axons of two well-defined clusters of hypothalamic neurons pass through the connecting stalk and end within the posterior pituitary in close proximity to capillaries (the smallest of blood vessels) (Figure 10-14). The two peptide hormones released from the posterior pituitary are **oxytocin** and **vasopressin** (the latter is also known as **antidiuretic hormone**, or **ADH**). They are synthesized in the cell bodies of the hypothalamic neurons, only one hormone by any particular neuron. Enclosed in small vesicles, the hormones move down the neural axons to accumulate at the axon

terminals in the posterior pituitary. In response to an action potential in the axon, the hormones are released into the capillaries and are carried away by the blood returning to the heart. Therefore, the term "posterior pituitary hormones" is somewhat of a misnomer since the hormones are actually synthesized in the hypothalamus, of which the posterior pituitary is merely an extension.

Action potentials are generated in these hypothalamic hormone-secreting neurons as a result of the synaptic input to them, just as in any other neuron. The reflexes controlling vasopressin release, and the actions of this hormone on blood pressure and kidney function are described in Chapters 13 and 15. Oxytocin, which acts on the breasts and uterus, is described in Chapter 18.

Vasopressin and probably oxytocin are released by axon terminals in addition to those in the hypothalamus-posterior-pituitary complex. In these other locations the molecules serve as neurotransmitters or neuromodulators.

The Hypothalamus and Anterior Pituitary

The hypothalamus, in addition to producing oxytocin and vasopressin, secretes hormones that control the secretion of all the *anterior* pituitary hormones. With one exception, the basic pattern for this overall system is sequences of three hormones. The sequences begin with (1) a hypothalamic hormone, which controls the secretion of (2) an anterior pituitary hormone, which controls the secretion of (3) a hormone from some other endocrine gland (Figure 10-15). This last hormone then acts

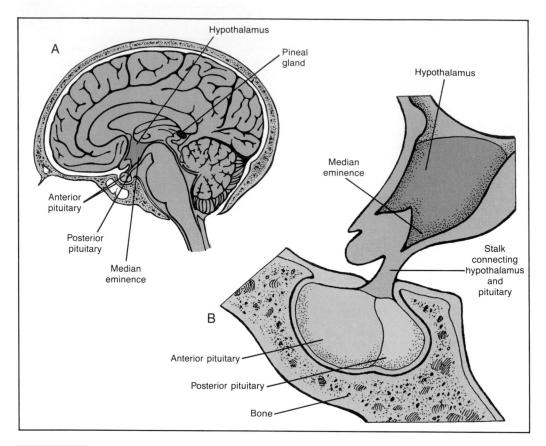

FIGURE 10-13 (A) Relation of the pituitary gland to the brain and hypothalamus. (B) An enlargement of the area including the hypothalamus and the pituitary.

FIGURE 10-14 Relationship between the hypothalamus and posterior pituitary.

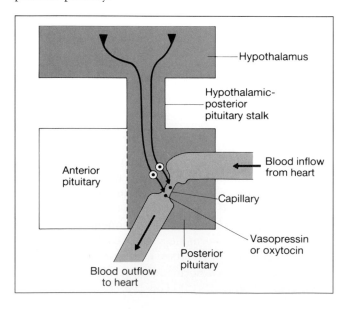

on its target cells. We begin our description of these sequences in the middle, that is, with the anterior pituitary hormones, because the names and actions of the hypothalamic hormones are all based on the names of the anterior pituitary hormones.

Anterior pituitary hormones. As shown in Table 10-1, the anterior pituitary secretes at least eight hormones, but only six have well-established functions. All peptides, these six "classical" hormones are: **follicle-stimulating hormone (FSH)**, **luteinizing hormone (LH)**, **growth hormone (GH)**, **thyroid-stimulating hormone (TSH, thyrotropin)**, **prolactin**, and **adrenocorticotropic hormone (ACTH, corticotropin)**. Each of the last four is probably secreted by a distinct cell type in the anterior pituitary, whereas FSH and LH, collectively termed **gonadotropic hormones** because they stimulate the gonads, are both secreted by the same cells.

What about the other two peptides—**β-lipotropin** and **β-endorphin**—secreted by the anterior pituitary? They are initially synthesized as part of the large **pro-**

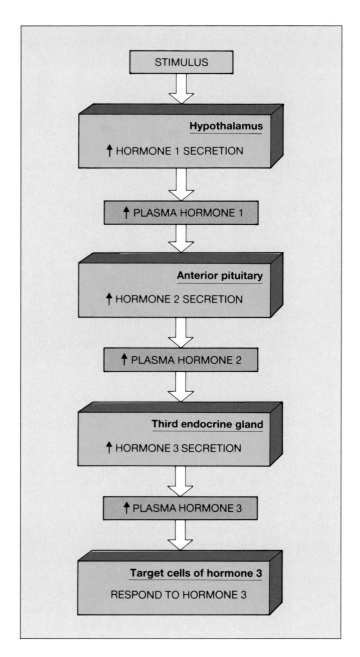

opiomelanocortin molecule (Figure 10-16), which is then cleaved in the cell to form ACTH and other peptides, including β-lipotropin. The latter molecule is cleaved, in turn, to form the powerful opioid β-endorphin. Then β-lipotropin and β-endorphin, and perhaps other cleavage products as well, are cosecreted from the anterior pituitary along with ACTH. Their functions are not presently known, but some possibilities are described in Chapters 9 and 19.

The target organs and major functions of the six classical anterior pituitary hormones are summarized in Figure 10-17. Note that the only major function of two of the six is to stimulate secretion of other hormones: Thyroid-stimulating hormone induces secretion of thyroid hormones—thyroxine and triiodothyronine—from the thyroid; adrenocorticotropic hormone, meaning "hormone that stimulates the adrenal cortex," stimulates the secretion of cortisol by that gland.

Three others also stimulate secretion of another hormone but have an additional function as well: Follicle-stimulating hormone and luteinizing hormone stimulate the secretion of the sex hormones—estrogen, progesterone, and testosterone—by the gonads and, in addition, regulate the growth and development of sperm and ova. Growth hormone both stimulates the liver and many

FIGURE 10-16 After its synthesis in the anterior pituitary, the pro-opiomelanocortin molecule is successively cleaved to yield ACTH and other peptides, the most important of which is β-endorphin, an opiate. These other peptides are cosecreted with ACTH. For simplicity, most of the peptides resulting from the cleavages have not been named in the figure and they are represented by empty boxes.

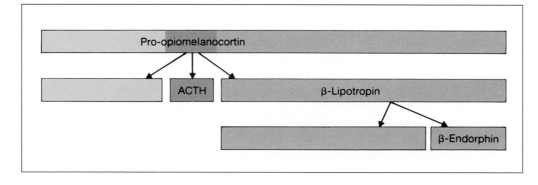

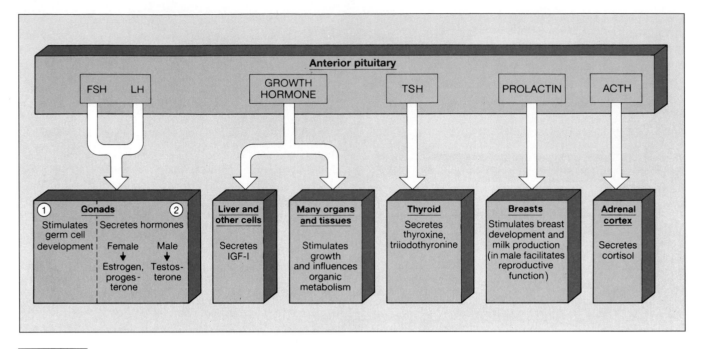

FIGURE 10-17 Targets and major functions of the six classical anterior pituitary hormones.

other body cells to secrete a growth-promoting peptide hormone known as **insulin-like growth factor I (IGF-I,** also called **somatomedin C)** and exerts direct effects on the organic metabolism of various organs and tissues (Chapter 17).

Prolactin is unique among the anterior pituitary hormones in that it does not exert major control over the secretion of a hormone by another endocrine gland. Rather, its most important action is to stimulate development of mammary glands and milk production by direct effects upon the breasts. In the male, prolactin facilitates several components of reproductive function.

It should be emphasized that many, perhaps all, of the anterior pituitary hormones are also secreted by cells other than those in the anterior pituitary and may, in those other sites, exert functions (as neurotransmitters, neuromodulators, or paracrines) quite different from those summarized in Figure 10-17.

Hypothalamic releasing hormones. Secretion of the anterior pituitary hormones is largely regulated by still other hormones, produced by hypothalamic neurons and termed **hypothalamic releasing hormones** (the reason for this name will be apparent later).

An appreciation of the anatomical relationships between the hypothalamus and the anterior pituitary is essential for understanding how the hypothalamic re-

leasing hormones control the anterior pituitary. In contrast to the neural connections between the hypothalamus and posterior pituitary (Figure 10-14), there are no important neural connections between the hypothalamus and anterior pituitary. There is, however, an unusual blood-vessel connection. The capillaries at the base of the hypothalamus (the **median eminence**) recombine into the **hypothalamo-pituitary portal vessels**—the term "portal" denotes vessels that connect distinct capillary beds. The portal vessels pass down the stalk connecting the hypothalamus and pituitary and enter the anterior pituitary where they drain into a second capillary bed, the anterior pituitary capillaries. Thus, the portal vessels offer a local route for blood flow directly from the hypothalamus to the anterior pituitary.

The axons of neurons that originate in diverse areas of the hypothalamus and secrete the releasing hormones terminate in the median eminence around the capillaries that are the origins of the portal vessels. The generation of action potentials in these neurons causes them to release their hormones, which enter the capillaries and are carried by the portal vessels directly to the anterior pituitary. There, they act upon the various pituitary cells to control their hormone secretions (Figure 10-18).

Thus, these hypothalamic neurons secrete hormones in a manner identical to that described previously for the hypothalamic neurons whose axons end in the posterior

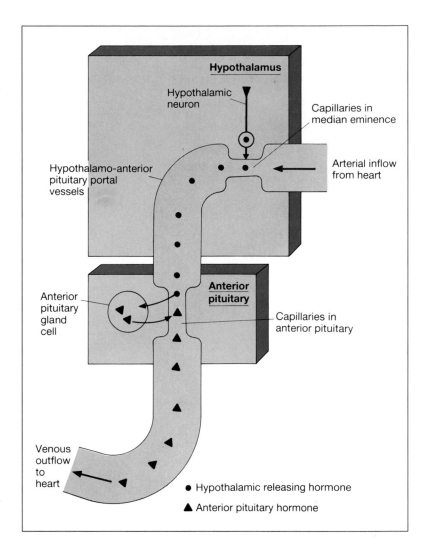

FIGURE 10-18 Relationship between the hypothalamus and anterior pituitary. Hormone secretion by the anterior pituitary is controlled by hypothalamic releasing hormones.

pituitary. In both cases the hormones are made in hypothalamic neurons, pass down the axons to the neuron terminals, and are released in response to action potentials in the neurons. The crucial differences, however, between the two systems need emphasizing: (1) The axons of the hypothalamic neurons that secrete vasopressin and oxytocin leave the hypothalamus and end in the posterior pituitary, whereas those that secrete the releasing hormones remain in the hypothalamus, ending in its median eminence. (2) The posterior pituitary capillaries, into which vasopressin and oxytocin are secreted, immediately join the main bloodstream, which carries the hormones to the heart to be distributed to the entire body. In contrast, the capillaries entered by the releasing hormones in the median eminence of the hypothalamus do not join the main bloodstream but empty into the portal vessels, which provide a short direct route to

the anterior pituitary. Thus, the anterior pituitary is exposed to very high plasma concentrations of the releasing hormones.

There are multiple discrete hypothalamic releasing hormones, each secreted by a particular group of hypothalamic neurons and influencing the release of one or more of the anterior pituitary hormones. For simplicity, Figure 10-19 summarizes only the major releasing hormones that perform this function. Each is named according to an anterior pituitary hormone whose secretion it controls; for example, the secretion of growth hormone is stimulated by **growth hormone releasing hormone (GHRH)**, and the secretion of ACTH is stimulated by **corticotropin releasing hormone (CRH)**.

The general term "releasing hormone" is somewhat unfortunate since, as shown in Figure 10-19, at least two of the hypothalamic releasing hormones do not cause

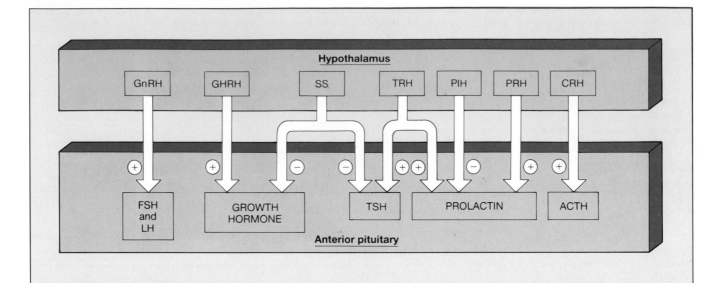

Hypothalamic releasing hormone	Effect on anterior pituitary
Corticotropin releasing hormone (CRH)*	Stimulates secretion of ACTH
Thyrotropin releasing hormone (TRH)†	Stimulates secretion of TSH and prolactin
Growth hormone releasing hormone (GHRH)	Stimulates secretion of GH
Somatostatin (SS, also known as growth hormone release inhibiting hormone, GIH)	Inhibits secretion of GH and TSH (and possibly several other hormones)
Gonadotropin releasing hormone (GnRH)	Stimulates secretion of LH and FSH
Prolactin releasing hormone (PRH)†	Stimulates secretion of prolactin
Prolactin release inhibiting hormone (PIH)‡	Inhibits secretion of prolactin

*CRH is the major stimulator of ACTH release, but vasopressin and possibly other messengers from the hypothalamus also stimulate ACTH release.

†Note that TRH stimulates the secretion not only of TSH but of prolactin as well; therefore, TRH is also a "PRH." However, there is a second PRH, which acts only on prolactin, and that is why PRH is listed separately here.

‡PIH has been identified as dopamine, a catecholamine. All the other hypothalamic releasing hormones are peptides.

FIGURE 10-19 Major hypothalamic releasing hormones and their effects on the anterior pituitary.

release of anterior pituitary hormones but rather inhibit their release.[3] One of them inhibits secretion of growth hormone and is most commonly called **somatostatin (SS)**. The other inhibits secretion of prolactin and is termed **prolactin release inhibiting hormone (PIH)**.

As also shown in Figure 10-19, there is not a strict one-to-one correspondence of hypothalamic releasing hormone and anterior pituitary hormone. For one thing, in at least three cases, a hypothalamic hormone influences the secretion of more than one anterior pituitary hormone: (1) **gonadotropin releasing hormone (GnRH)** stimulates the secretion of both FSH and LH; (2) **thyrotropin releasing hormone (TRH)** stimulates the secretion not only of TSH but of prolactin as well; and (3) somatostatin inhibits the secretion not only of growth hormone but of TSH. Indeed, somatostatin can also inhibit several other anterior pituitary hormones, although whether these inhibitory effects are physiological is still not clear.

The cases just described are of a single hypothalamic

[3]To avoid the awkwardness of having inhibitory hormones in the generic category of "releasing" hormones, some endocrinologists refer to the entire group as "hypophysiotropic hormones" rather than "releasing hormones."

hormone controlling more than one anterior pituitary hormone, but Figure 10-19 also shows that in at least three cases—growth hormone, TSH, and prolactin—secretion of a single anterior pituitary hormone is controlled by more than one hypothalamic releasing hormone. Moreover, the several releasing hormones involved may exert opposing effects. Thus, as we have seen, the secretion of growth hormone is controlled by two releasing hormones—somatostatin, which inhibits its release, and growth hormone releasing hormone, which stimulates it. In such multiple-control cases, the anterior-pituitary response depends upon the relative amounts of the opposing hormones released by the hypothalamic neurons, as well as upon the relative sensitivities of the anterior pituitary to them.

Prolactin offers an excellent example of this generalization. There are at least three hypothalamic hormones—TRH, PIH, and **prolactin releasing hormone (PRH)**—influencing prolactin secretion. Under basal conditions, the inhibitory hormone (PIH) is dominant over the two stimulatory hormones—TRH and PRH—and so prolactin secretion is quite low.

With one exception, all the hypothalamic releasing hormones listed in Figure 10-19 are peptides. The peptides all also occur in locations other than the hypothalamus, particularly in other areas of the central nervous system, where they function as neurotransmitters or neuromodulators, and in the gastrointestinal tract.

The one hypothalamic releasing hormone that is not a peptide is PIH, which has been identified as **dopamine**.[4] This substance, a third member of the catecholamine family, is an example of a chemical messenger that serves multiple roles. Thus, it is secreted not only by hypothalamic neurons but in small amounts by the adrenal medulla, and its most important role is that of a neurotransmitter in various areas of the central nervous system (Chapter 8).

Figure 10-20 summarizes the information presented

---

[4]There may be another hypothalamic hormone, in addition to dopamine, that inhibits the secretion of prolactin, that is, there may be two different PIHs.

FIGURE 10-20 A combination of Figures 10-17 and 10-19 summarizes the hypothalamic-anterior-pituitary system.

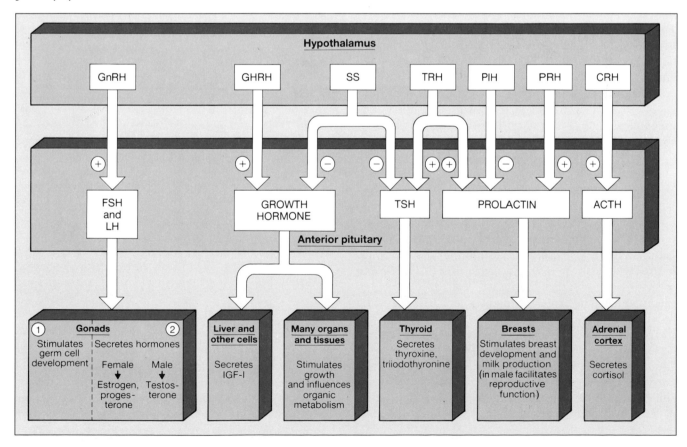

in Figures 10-17 and 10-19 to demonstrate the full sequences controlled by the hypothalamus.

Given that the hypothalamic releasing hormones control anterior-pituitary function, we must now ask: What controls secretion of the releasing hormones? Some of the neurons that secrete releasing hormones may possess spontaneous activity, but the firing of most of them requires neural and hormonal input. First, let us deal with the neural input.

Neural control of hypothalamic releasing hormones. The hypothalamus receives synaptic input, both stimulatory and inhibitory, from virtually all areas of the central nervous system, and specific neural pathways control, in large part, the secretion of the individual releasing hormones. A large number of neurotransmitters, including the catecholamines, acetylcholine, and β-endorphin, are used by the synapses on the hormone-secreting hypothalamic neurons, and this explains why the secretion of the releasing hormones can be altered by drugs that influence these neurotransmitters.

Figure 10-21 illustrates one example of the role of neural input to the hypothalamus. Corticotropin releasing hormone, secreted by the hypothalamus, stimulates the anterior pituitary to secrete ACTH, which in turn stimulates the adrenal cortex to secrete cortisol. A wide variety of stresses, both physical and emotional, act, via neural pathways to the hypothalamus, to increase CRH secretion and hence, ACTH and cortisol secretion, markedly above basal values. Thus, stress is the common denominator of reflexes leading to increased cortisol secretion. Cortisol then functions to facilitate an individual's response to stress. Even in an unstressed person, however, the secretion of cortisol varies in a highly stereotyped manner during a 24-h period, because neural rhythms within the central nervous system also impinge upon the hypothalamic neurons that secrete CRH.

Hormonal control of hypothalamic releasing hormones. A prominent feature of each of the hormonal sequences initiated by the hypothalamic releasing hormones is negative feedback exerted upon the hypothalamic-pituitary system by one or more of the hormones in the sequence. For example, in the CRH-ACTH-cortisol sequence, the final hormone—cortisol—acts upon the hypothalamus to reduce secretion of CRH by causing a decrease in the frequency of action potentials in the neurons secreting CRH. In addition, cortisol acts directly on the anterior pituitary to reduce the response of the ACTH-secreting cells to CRH. Thus, by a double-barreled action, cortisol exerts a negative-feedback control over its own secretion (Figure 10-21).

Such a system is effective in dampening hormonal

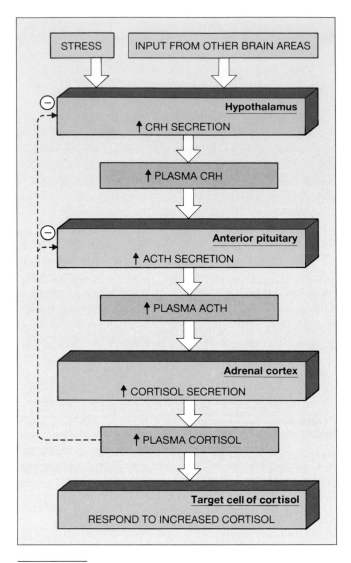

FIGURE 10-21 CRH-ACTH-cortisol sequence. The "stress" input to the hypothalamus is via neural pathways from other brain areas. Cortisol exerts a negative-feedback control over the system by acting on (1) the hypothalamus to inhibit CRH secretion and (2) the anterior pituitary to reduce responsiveness to CRH. Not shown is the fact that vasopressin and other messengers released both by hypothalamic neurons and by peripheral tissues also act on the anterior pituitary to stimulate ACTH secretion. The common denominator of many of the effects of cortisol is to increase plasma glucose concentration and to facilitate a person's responses to stress (Chapters 17 and 19).

responses, that is, in limiting the extremes of hormone secretory rates. For example, when a painful stimulus elicits increased secretion of CRH, ACTH, and cortisol, the resulting elevation in plasma cortisol concentration feeds back to inhibit the hypothalamus and anterior pituitary. Therefore, cortisol secretion does not rise as much as it would without these negative feedbacks.

Negative feedback also serves to maintain the plasma concentration of the final hormone in a sequence relatively constant whenever a primary change occurs in the metabolism of that hormone. An example of this is shown in Figure 10-22 for cortisol. Another example, which was quite common in the United States before the introduction of iodized salt and is still common in many areas of the world, is provided by the TRH-TSH-TH

system. The thyroid hormones exert a feedback inhibition on the hypothalamo-pituitary system, mainly by decreasing the response of TSH-secreting cells to the stimulatory effects of TRH. Since iodine is essential for the synthesis of TH, persons with iodine deficiency tend to have deficient production of TH. The resulting decrease in plasma TH concentration relieves some of the feedback inhibition TH exerts on the pituitary. Therefore, more TSH is secreted in response to the TRH coming from the hypothalamus, and the increased plasma

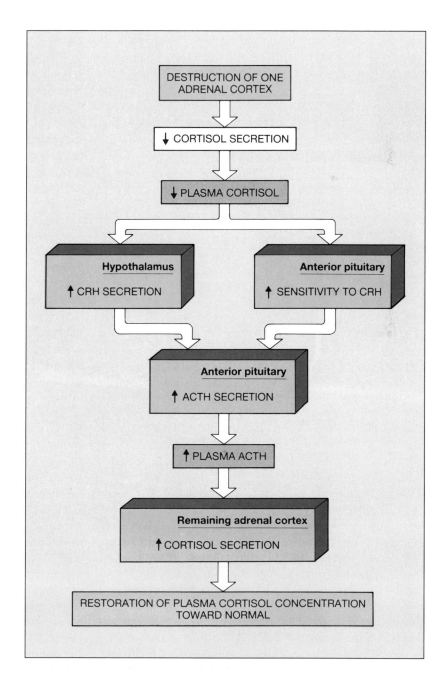

FIGURE 10-22 Example of how negative feedback in hormonal sequences helps maintain the plasma concentration of the final hormone when disease-induced changes in hormone secretion occur. The same analysis would apply if the original reduction in plasma cortisol were due to excessive metabolism of cortisol rather than deficient secretion.

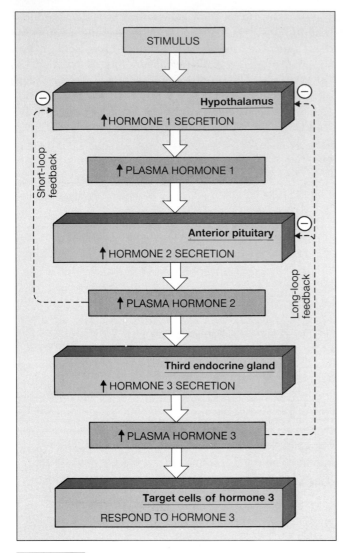

STIMULUS

Hypothalamus

↑HORMONE 1 SECRETION

Short-loop feedback

↑ PLASMA HORMONE 1

Anterior pituitary

↑ HORMONE 2 SECRETION

Long-loop feedback

↑ PLASMA HORMONE 2

Third endocrine gland

↑ HORMONE 3 SECRETION

↑PLASMA HORMONE 3

Target cells of hormone 3

RESPOND TO HORMONE 3

FIGURE 10-23 Short-loop and long-loop feedbacks. Long-loop feedback is exerted on the hypothalamus and/or anterior pituitary by the third hormone in the sequence. Short-loop feedback is exerted by the anterior-pituitary hormone on the hypothalamus.

TSH stimulates the thyroid gland to enlarge and to utilize more efficiently whatever iodine is available (the enlarged gland is known as iodine-deficient goiter). In this manner, plasma TH concentration can be kept quite close to normal.

The situation described above for cortisol and the thyroid hormones, in which the hormone secreted by the third endocrine gland in a sequence exerts a negative-feedback effect over the anterior pituitary and/or hypothalamus, is known as a **long-loop negative feedback** (Figure 10-23). This type of feedback exists for each of

the five three-hormone sequences initiated by a hypothalamic releasing hormone.

Long-loop feedback cannot exist for prolactin since this is the one anterior-pituitary hormone that does not have major control over another endocrine gland, that is, it does not participate in a three-hormone sequence. Nonetheless, there is negative feedback in the prolactin system, for this hormone itself acts upon the hypothalamus to influence the secretion of the hypothalamic hormones that control its secretion. This influence of an anterior-pituitary hormone on the hypothalamus is known as a **short-loop negative feedback** (Figure 10-23). Several anterior-pituitary hormones (growth hormone, for example) other than prolactin also exert such feedback on the hypothalamus.

It must be emphasized that there are many hormonal influences, both stimulatory and inhibitory, on the hypothalamus and/or anterior pituitary other than those that fit the feedback patterns just described. In other words, a hormone that is not itself in a particular hormonal sequence may nevertheless exert important influences on the secretion of the hypothalamic or anterior-pituitary hormones in that sequence. For example, estrogen markedly enhances the secretion of prolactin by the anterior pituitary, even though estrogen secretion is not controlled by prolactin. Thus, one should not view these sequences as isolated units.

A summary example: Control of growth hormone secretion. The three-hormone sequences beginning in the hypothalamus can be extremely complex, incorporating multiple sites of feedback, both long loop and short loop, as well as other hormones not in the sequence. Purely for the sake of illustrating this complexity, we describe here the control of growth hormone secretion (Figure 10-24), building upon the information already presented in this chapter.

Recall from Figure 10-19 that the secretion of GH by the anterior pituitary is controlled mainly by two hormones from the hypothalamus: (1) somatostatin, which inhibits GH secretion; and (2) GHRH, which stimulates it. With such a dual control system, the rate of GH secretion at any moment reflects the relative amounts of simultaneous stimulation by GHRH and inhibition by somatostatin. For example, as shown in Figure 10-9, very little growth hormone is secreted during the day in nonstressed persons who are eating normally, and whether this reflects a very low secretion of GHRH or a very high secretion of somatostatin is not yet clear. Similarly, a large number of physiological states (exercise, stress, fasting, a low plasma glucose concentration, and sleep) stimulate growth hormone secretion by decreasing the secretion of somatostatin and/or increasing that of GHRH.

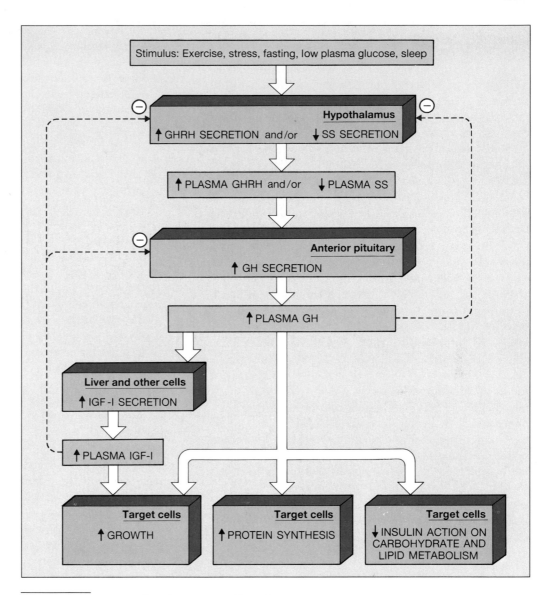

FIGURE 10-24 Hormonal pathways controlling the secretion of growth hormone and insulin-like growth factor I. At the hypothalamus, the minus sign (−) denotes that the input inhibits the secretion of GHRH and/or stimulates the release of somatostatin.

As shown in Figures 10-17 and 10-24, growth hormone not only exerts direct actions on peripheral tissues but also stimulates the secretion of another hormone, IGF-I, from the liver and many other target cells of GH. This makes possible a variety of feedbacks by which an increase in plasma GH results in inhibition of GH secretion (Figure 10-24): (1) A short-loop negative feedback, exerted by GH on the hypothalamus, and (2) a long-loop negative feedback, exerted by IGF-I on the hypothalamus. Both of these feedbacks operate by inhibiting secretion of GHRH and/or stimulating secretion of somatostatin, the result in either case being less stimulation of GH secretion. In addition, (3) IGF-I probably also acts directly on the pituitary to inhibit the stimulatory effect of GHRH on GH secretion; again, the result is decreased GH secretion.

Finally, the secretion of growth hormone is influenced by several other hormones not in the sequence, for example, by the sex hormones. Some hormones may affect the sequence indirectly by altering the release of somatostatin and/or GHRH from the hypothalamus; others may act directly upon the anterior pituitary.

CANDIDATE HORMONES

Many substances, termed **candidate hormones**, are suspected of being hormones in humans but are not considered classical hormones for one of two reasons: Either (1) their functions have not been conclusively documented, or (2) they have well-documented functions as paracrines and autocrines, but it is not certain that they ever reach additional target cells via the blood, an essential criterion for classification as a hormone. This second category includes certain of the prostaglandins (Chapter 7) and a large number of what are called growth factors (Chapter 17) that are secreted by multiple cell types and stimulate specific cells to differentiate and undergo mitosis.

A substance that fits the first category described above is the amino acid derivative **melatonin**, which is synthesized from serotonin. This candidate hormone is produced by the **pineal gland** (also known as the pineal body), an outgrowth from the roof of the diencephalon (Figure 10-13). The function of melatonin and the pineal gland in humans, in contrast to several other animals, is unknown. Its secretion is stimulated by sympathetic postganglionic neurons that constitute the last link in a neuronal chain primarily controlled by the prevailing light-dark environment. Melatonin secretion, therefore, undergoes a marked 24-h cycle, being high at night and low during the day, and melatonin probably influences a variety of bodily rhythms. Its relationship to "winter depression," its ability to reduce the symptoms of jet lag, and its possible role in the onset of puberty are all being studied.

TYPES OF ENDOCRINE DISORDERS

Most endocrine disorders fall into one of three categories: (1) too little hormone (**hyposecretion**); (2) too much hormone (**hypersecretion**); and (3) reduced response of the target cells (**hyporesponsiveness**). The phrases "too little hormone" and "too much hormone" here means "too little" or "too much" for any given physiological situation. For example, as we shall see, insulin secretion decreases during fasting, and this decrease is an adaptive physiological response, not too little insulin. In contrast, insulin secretion should increase after eating, and if its increase is less than normal, this is too little hormone secretion.

Hyposecretion

An endocrine gland may be secreting too little hormone because the gland is not able to function normally. This is termed **primary hyposecretion**. Examples of primary

hyposecretion include (a) genetic absence of a steroid-forming enzyme, leading to decreased cortisol secretion, and (b) dietary deficiency of iodine, leading to decreased secretion of thyroid hormones. There are many other causes—infections, toxic chemicals, and so on—all having the common denominator of damaging the endocrine gland.

In contrast to primary hyposecretion, a gland may be secreting too little hormone not because it is abnormal but because there is not enough of its tropic hormone. This is termed **secondary hyposecretion**. For example, there may be nothing wrong with the thyroid gland but it may be secreting too little thyroid hormone because the secretion of TSH by the anterior pituitary is abnormally low. Thus, the hyposecretion by the thyroid gland in this case is secondary to inadequate secretion by the anterior pituitary.

This example raises the next question applicable to any of the other anterior pituitary hormones as well: Is the hyposecretion of TSH primary, that is, due to a defect in the anterior pituitary, or is it secondary to a hypothalamic defect causing too little secretion of TRH? If the latter were true, then we would have the following sequence: primary hyposecretion of TRH leading to secondary hyposecretion of TSH leading to secondary hyposecretion of TH.

In diagnosing the presence of hyposecretion, a basic measurement to be made is the concentration of the hormone in either plasma or urine. The finding of a low concentration will not distinguish between primary and secondary hyposecretion, however. To do this, the concentration of the relevant tropic hormone must also be measured. Thus, in our example, if the hyposecretion of TH is secondary to hyposecretion of TSH, then the plasma concentrations of both will be decreased. If the hyposecretion of TH is primary, then TH concentration will be decreased and TSH concentration will be increased because of less negative-feedback inhibition by TH over TSH secretion.

Another diagnostic approach is to attempt to stimulate the gland in question by administering either its tropic hormone or some other substance known to elicit increased secretion. The increase in hormone secretion elicited by the stimulus will be normal if the original hyposecretion is secondary but less than normal if primary hyposecretion is the problem.

The most common means of treating hormone hyposecretion is to administer the hormone that is missing or present in too small amounts. In cases of secondary hyposecretion, there is a choice since at least two hormones are involved. In our example, the TH deficiency could be eliminated by administering either TH or TSH. It must be stressed, however, that while administering hormones to remedy deficiencies is theoretically straightforward, it is often very difficult in practice.

Hypersecretion

A hormone can also undergo either **primary hypersecretion** (the gland is secreting too much of the hormone on its own) or **secondary hypersecretion** (there is excessive stimulation by its tropic hormone). One of the most common causes of primary hypersecretion is the presence of a hormone-secreting endocrine-cell tumor.

The diagnosis of primary versus secondary hypersecretion of a particular hormone is analogous to that of hyposecretion. The concentrations of the hormone and, if relevant, its tropic hormone are measured in plasma or urine. If both concentrations are elevated, then the hormone in question is being secondarily hypersecreted. If the hypersecretion is primary, there will be a decreased concentration of the tropic hormone because of negative feedback by the high concentration of the hormone being hypersecreted.

When an endocrine tumor is the cause of hypersecretion, it can be removed surgically or destroyed with radiation. In many cases hypersecretion can also be blocked by drugs that inhibit the hormone's synthesis. Alternatively, the situation can be treated with drugs that do not alter the hormone's secretion but instead block the hormone's actions on its target cells.

Hyporesponsiveness

In some cases, the endocrine system may be dysfunctioning even though there is nothing wrong with hormone secretion. The problem is that the target cells do not respond normally to the hormone. This condition is termed hyporesponsiveness. One cause is either a lack or deficiency of receptors for the hormone. For example, certain men have a genetic defect manifested by the absence of receptors for dihydrotestosterone, the form of testosterone active in many target cells. In such men, these cells are unable to bind dihydrotestosterone, and the result is lack of development of certain male characteristics, just as though the hormone were not being produced.

In a second type of hyporesponsiveness the receptors for a hormone may be normal, but some event occurring after the hormone binds to a receptor may be defective. For example, the activated receptor might be unable to stimulate formation of cyclic AMP or open a plasma-membrane channel.

A third cause of hyporesponsiveness applies to hormones that require metabolic activation by some other tissue after secretion. There may be a lack or deficiency of the enzymes that catalyse the activation. For example, some men secrete testosterone normally and have normal receptors for dihydrotestosterone but are missing the enzyme that converts testosterone to dihydrotestosterone. The result is the same as the case in which there are no dihydrotestosterone receptors.

In situations characterized by hyporesponsiveness to a hormone, the plasma concentration of the hormone in question is normal or elevated, but the response to administered hormone is diminished.

Finally, in a few cases, *hyperresponsiveness* to a hormone can occur and cause problems.

SUMMARY

I. The endocrine system is one of the body's two major communications systems. It consists of all those glands that secrete hormones, which are chemical messengers carried by the blood from the endocrine glands to target cells elsewhere in the body.

Chemical Structures and Synthesis

I. The amine hormones are the iodine-containing thyroid hormones—thyroxine and triiodothyronine—and the catecholamines (mainly epinephrine) secreted by the adrenal medulla.

II. Most hormones are peptides and are synthesized as larger molecules that are then cleaved.

III. Steroid hormones are produced from cholesterol by the adrenal cortex and the gonads, and by the placenta during pregnancy.

   A. The most important steroid hormones produced by the adrenal cortex are the mineralocorticoid aldosterone, the glucocorticoid cortisol, and two androgens.

   B. The ovaries produce mainly estradiol and progesterone, and the testes mainly testosterone.

Transport of Hormones in the Blood

I. Peptide hormones and catecholamines dissolve in the plasma water, but steroid and thyroid hormones, being water-insoluble, circulate mainly bound to plasma proteins.

Hormone Metabolism and Excretion

I. The liver and kidneys are the major organs that remove hormones from the plasma by metabolizing or excreting them.

II. The peptide hormones and catecholamines are rapidly removed from the blood, whereas the steroid and thyroid hormones are removed more slowly.

III. After their secretion, some hormones are metabolized to more active molecules in their target cells or other organs.

Mechanisms of Hormone Action

I. The receptors for steroid and thyroid hormones are inside the target-cells; those for the peptide hormones and catecholamines are on the plasma membrane.

II. Hormones can cause up-regulation and down-regulation of their own receptors and those of other hormones. The induction of one hormone's receptors by another hormone increases the first hormone's effectiveness and may be essential to permit the first hormone to exert its effects.

III. Intracellular receptors activated by steroid and thyroid hormones combine with DNA in the nucleus and induce the transcription of DNA into mRNA; the result is increased synthesis of particular proteins.

IV. Receptors activated by peptide hormones and catecholamines utilize one or more of the signal transduction mechanisms available to plasma membrane receptors; the result is altered activity of proteins in the cell.

V. In pharmacological doses, hormones can have effects not seen under ordinary circumstances.

Types of Inputs that Control Hormone Secretion

I. The secretion of a hormone may be controlled by the plasma concentration of an ion or nutrient that the hormone regulates, by neural input to the endocrine cells, and by a tropic hormone.

II. The autonomic nervous system is the neural input controlling many hormones, but the hypothalamic and posterior-pituitary hormones are controlled by neurons in the brain.

Control Systems Involving the Hypothalamus and Pituitary

I. The pituitary gland, comprising the anterior pituitary and the posterior pituitary, is connected to the hypothalamus by a stalk containing nerve fibers and blood vessels.

II. The nerve fibers, which originate in the hypothalamus and terminate in the posterior pituitary, secrete oxytocin and vasopressin.

III. The anterior pituitary secretes growth hormone (GH), thyroid-stimulating hormone (TSH), adrenocorticotropic hormone (ACTH), prolactin, and two gonadotropic hormones—follicle-stimulating hormone (FSH) and luteinizing hormone (LH). The functions of these hormones are summarized in Figure 10-17.

IV. The anterior pituitary also produces β-lipotropin and β-endorphin along with ACTH, all being portions of the parent pro-opiomelanocortin molecule.

V. Secretion of the anterior pituitary hormones is controlled mainly by hypothalamic releasing hormones secreted into capillaries in the median eminence and reaching the anterior pituitary via the portal vessels connecting the hypothalamus and anterior pituitary. The actions of the hypothalamic releasing hormones on the anterior pituitary are summarized in Figure 10-19.

VI. The secretion of each releasing hormone is controlled by neuronal and hormonal input to the hypothalamic neurons producing it.

A. In each of the three-hormone sequences beginning with a hypothalamic releasing hormone, the third hormone exerts a long-loop negative-feedback effect on the secretion of the hypothalamic and/or anterior-pituitary hormone.

B. The anterior-pituitary hormone may exert a short-loop negative-feedback inhibition of the hypothalamic re-

leasing hormone(s) controlling it.

C. Hormones not in a sequence can also influence secretion of the hypothalamic and/or anterior-pituitary hormones in the sequence.

Candidate Hormones

I. Substances that are suspected of functioning as hormones but have not yet been proven to do so are called candidate hormones.

II. Melatonin is a hormone secreted with a 24-h rhythm by the pineal gland, an outgrowth of the diencephalon, but its functions in humans are not known.

Types of Endocrine Disorders

I. Endocrine disorders may be classified as hyposecretion, hypersecretion, and target-cell hyporesponsiveness.

A. Primary disorders are those in which the defect is in the cells that secrete the hormone.

B. Secondary disorders are those in which there is too much or too little tropic hormone.

C. Hyporesponsiveness is due to an alteration in the receptors for the hormone, to disordered postreceptor events, or to failure of normal metabolic activation of the hormone in those cases requiring such activation.

II. These disorders can be distinguished by measurements of the hormone and any tropic hormones under basal conditions and during experimental stimulation or suppression of the hormone's secretion.

REVIEW QUESTIONS

1. Define:

endocrine system	testosterone
endocrine glands	estradiol
hormones	up-regulation
target cells	down-regulation
neurohormones	permissiveness
amine hormones	pharmacological effects
thyroid gland	tropic hormone
thyroxine (T_4)	pituitary gland
triiodothyronine (T_3)	anterior pituitary
thyroid hormones (TH)	posterior pituitary
iodine	oxytocin
adrenal gland	vasopressin
adrenal medulla	antidiuretic hormone (ADH)
adrenal cortex	
epinephrine (E)	follicle-stimulating hormone (FSH)
norepinephrine (NE)	
peptide hormones	luteinizing hormone (LH)
prohormones	growth hormone (GH)
steroid hormones	thyroid-stimulating hormone (TSH)
aldosterone	
mineralocorticoid	thyrotropin
cortisol	prolactin
glucocorticoids	adrenocorticotropic hormone (ACTH)
androgens	

corticotropin
gonadotropic hormones
β-lipotropin
β-endorphin
pro-opiomelanocortin
insulin-like growth factor 1
    (IGF-1)
somatomedin C
hypothalamic releasing
    hormones
median eminence
hypothalamo-pituitary
    portal vessels
growth hormone releasing
    hormone (GHRH)
corticotropin releasing
    hormone (CRH)
somatostatin (SS)
prolactin release inhibiting
    hormone (PIH)

gonadotropin releasing
    hormone (GnRH)
thyrotropin releasing
    hormone (TRH)
prolactin releasing
    hormone (PRH)
dopamine
long-loop negative feedback
short-loop negative
    feedback
candidate hormones
melatonin
pineal gland
hyposecretion
hypersecretion
hyporesponsiveness
primary hyposecretion
secondary hyposecretion
primary hypersecretion
secondary hypersecretion

2. What are the three general chemical classes of hormones?

3. What essential nutrient is needed for the synthesis of the thyroid hormones?

4. Which catecholamine is secreted in the largest amount by the adrenal medulla?

5. What are the major hormones produced by the adrenal cortex? By the testes? By the ovaries?

6. Which classes of hormones are carried in the blood mainly as dissolved hormone? Mainly bound to plasma proteins?

7. Can bound hormones cross capillary walls?

8. Which organs are the major sites of hormone excretion and metabolic transformation?

9. How do the rates of metabolism and excretion differ for the various classes of hormones?

10. What must some hormones undergo after their secretion to become activated?

11. Contrast the locations of receptors for the various classes of hormones.

12. How do hormones influence the concentrations of their own receptors and those of other hormones? How does this explain permissiveness in hormone action?

13. Describe the sequence of events when steroid or thyroid hormones bind to their receptors.

14. Describe the sequence of events when peptide or catecholamine hormones bind to their receptors.

15. What are the four types of direct inputs to endocrine glands controlling hormone secretion?

16. How does control of hormone secretion by plasma mineral ions and nutrients achieve negative-feedback control of these substances?

17. What roles does the autonomic nervous system play in controlling hormone secretion?

18. For what groups of hormones is the important neural input neurons in the brain rather than in the autonomic nervous system?

19. Describe the anatomical relationship between the hypothalamus and the posterior pituitary and name the two posterior-pituitary hormones.

20. List the six classical anterior pituitary hormones and their functions; list two other anterior pituitary hormones and their biochemical origin.

21. Describe the anatomical relationships between the hypothalamus and anterior pituitary.

22. List the hypothalamic releasing hormones and the effects on the hormone(s) whose release each controls.

23. What kinds of inputs control the secretion of the hypothalamic releasing hormones?

24. Diagram the CRH-ACTH-cortisol system.

25. What is the difference between long-loop and short-loop negative feedback in the hypothalamic anterior-pituitary system?

26. How would you distinguish between primary and secondary hyposecretion of a hormone? Between hyposecretion and hyporesponsiveness?

THOUGHT QUESTIONS

(Answers are given in Appendix A.)

1. In an experimental animal the sympathetic preganglionic fibers to the adrenal medulla are cut. What happens to the plasma concentration of epinephrine at rest and during stress?

2. During pregnancy there is an increase in the production (by the liver) and the plasma concentration of the major plasma-binding protein for the thyroid hormones (TH). This causes a sequence of events involving feedback that results in an increase in the plasma concentration of TH but no evidence of hyperthyroidism. Describe the sequence of events.

3. A child shows the following symptoms: deficient growth; failure to show sexual development; decreased ability to respond to stress. What is the most likely cause of all these symptoms?

4. If all the neural connections between the hypothalamus and pituitary were severed, the secretion of which pituitary hormones would be affected? Which pituitary hormones would not be affected?

5. An antibody to a peptide combines with the peptide and renders it nonfunctional. If an animal were given an antibody to somatostatin, the secretion of which anterior pituitary hormones would change and in what direction?

6. A drug that blocks the action of norepinephrine is injected directly into the hypothalamus of an experimental animal, and the secretion rates of several anterior pituitary hormones are observed to change. How is this possible since norepinephrine is not a releasing hormone?

7. A person is receiving very large doses of a cortisol-like drug to treat her arthritis. What happens to her secretion of cortisol?

8. A person with symptoms of hypothyroidism, for example, sluggishness and intolerance to cold, is found to have abnormally low plasma concentrations of T_4 and T_3 and TSH. After an injection of TRH, the plasma concentrations of all three hormones increase. Where is the site of the defect leading to the hypothyroidism?

CHAPTER

11

MUSCLE

The ability to use chemical energy to produce force and movement is present to a limited extent in most living cells. It is, however, in muscle cells that the chemical apparatus for these activities has its greatest development. The primary function of these specialized cells is to generate the forces and movements that are used by multicellular organisms in the regulation of the internal environment and to produce the movements of the entire organism in its external environment. In the case of humans, speech and the ability to manipulate objects in the environment are also the result of muscle contractions. Indeed, it is only by controlling the activity of muscles that the human mind ultimately expresses itself.

On the basis of structure, contractile properties, and control mechanisms, three types of muscle can be identified: (1) **skeletal muscle**, (2) **smooth muscle**, and (3)

cardiac muscle. Most skeletal muscle, as the name implies, is attached to bone, and its contraction is responsible for supporting and moving the skeleton. The contraction of skeletal muscle is initiated by impulses in the motor neurons to the muscle and is under voluntary control.

Sheets of smooth muscle surround various hollow organs and tubes, including the stomach, intestinal tract, urinary bladder, uterus, blood vessels, and the airways in the lungs. Single smooth-muscle cells are also found distributed throughout organs and in small bundles of cells attached to the hair in the skin and the iris of the eye. Contraction of the smooth muscle surrounding hollow organs may propel the luminal contents through the organ, or it may regulate internal flow by changing the tube diameter. Smooth-muscle contraction is controlled by the autonomic nervous system, hormones, paracrines, and other local chemical signals. Some smooth muscles contract spontaneously, however, even in the absence of such signals.

Cardiac muscle is the muscle of the heart. Its contraction propels blood through the circulatory system. Like smooth muscle, it is regulated by the autonomic nervous system and hormones, and certain portions of it can undergo spontaneous contractions.

Although there are significant differences in these three types of muscle, the force-generating mechanism is similar in all of them. Skeletal muscle will be described first, followed by a description of smooth muscle. Cardiac muscle, which combines some of the properties of both skeletal and smooth muscle, will be described in Chapter 13 in association with cardiac muscle's role in the circulatory system.

STRUCTURE OF SKELETAL MUSCLES AND MUSCLE FIBERS

A single skeletal-muscle cell is known as a **muscle fiber.** Each muscle fiber is formed during development by the fusion of a number of undifferentiated, mononucleated cells, known as myoblasts, into a single, cylindrical, multinucleated muscle fiber. This stage of muscle differentiation is completed around the time of birth, after which new fibers are not normally formed. The existing fibers continue to increase in size with the growth of the child. If skeletal-muscle fibers are destroyed after birth due to injury, they cannot be replaced by the division of other existing muscle fibers. However, new fibers can be formed from undifferentiated cells known as satellite cells, which are located in the extracellular compartment adjacent to muscle fibers. This capacity for forming new skeletal-muscle fibers after birth is considerable but will not restore a severely damaged muscle to full strength. Much of the compensation for a loss of muscle tissue occurs through increased growth in the size of the remaining muscle fibers.

Each fiber has a diameter between 10 and 100 μm and a length that may extend up to 20 cm. The term **muscle** refers to a number of muscle fibers bound together by connective tissue (Figure 11-1) and usually linked to bones by bundles of collagen fibers, known as **tendons,** located at each end of the muscle. The relation between a single muscle fiber and a muscle is analogous to that between a single neuron and a nerve, which is composed of the axons of many neurons.

In some muscles, the individual fibers extend the entire length of the muscle, but in most, the fibers are shorter, often oriented at an angle to the longitudinal axis of the muscle. These short fibers are anchored to the connective-tissue network surrounding the muscle fibers. The transmission of force from muscle to bone is

FIGURE 11-1 Organization of cylindrical muscle fibers in a muscle that is attached to bones by tendons.

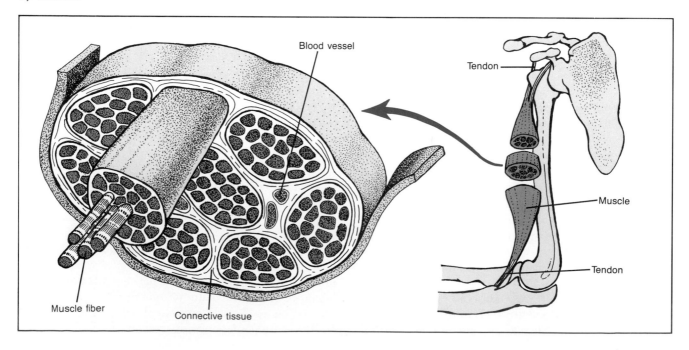

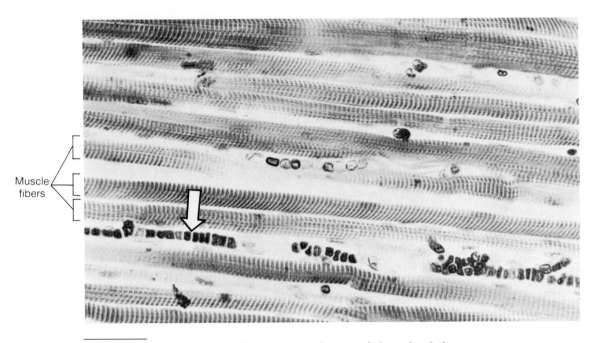

FIGURE 11-2 Skeletal-muscle fibers in a muscle viewed through a light microscope. Arrow indicates a blood vessel containing erythrocytes. (*From Edward K. Keith and Michael H. Ross, "Atlas of Descriptive Histology," Harper & Row, New York, 1968.*)

like a number of people pulling on a rope, each person corresponding to a single muscle fiber and the rope corresponding to the connective tissue and tendons.

Some tendons are very long with the site of tendon attachment to bone far removed from the muscle. For example, some of the muscles that move the fingers are in the forearm, as one can observe by wiggling one's fingers and feeling the movement of the muscles in the lower arm. These muscles are connected to the fingers by long tendons. If the muscles that move the fingers were located in the fingers themselves, we would have very thick fingers.

The most striking feature seen when observing a muscle fiber with a light microscope (Figure 11-2) is a series of light and dark bands oriented perpendicular to the long axis of the fiber. Both skeletal- and cardiac-muscle fibers (Figure 11-3) have this characteristic banding and are known as **striated muscles.** Smooth-muscle cells show no banding pattern. This striated pattern is due to the presence in the fiber cytoplasm of approximately cylindrical elements (1 to 2 μm in diameter) known as **myofibrils** (Figure 11-4). Most of the cytoplasm is filled with myofibrils, each of which extends from one end of a fiber to the other.

Each myofibril is composed of thick and thin filaments (Figure 11-5) arranged in a repeating pattern along the length of the myofibril. One unit of this repeating pattern is known as a **sarcomere** (little muscle). The

FIGURE 11-3 The three types of muscle fibers. Note the difference in their sizes. (The single nucleus of the cardiac-muscle fiber is not shown.)

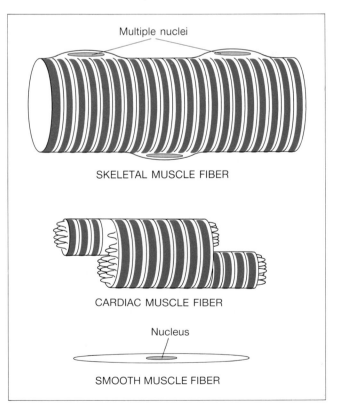

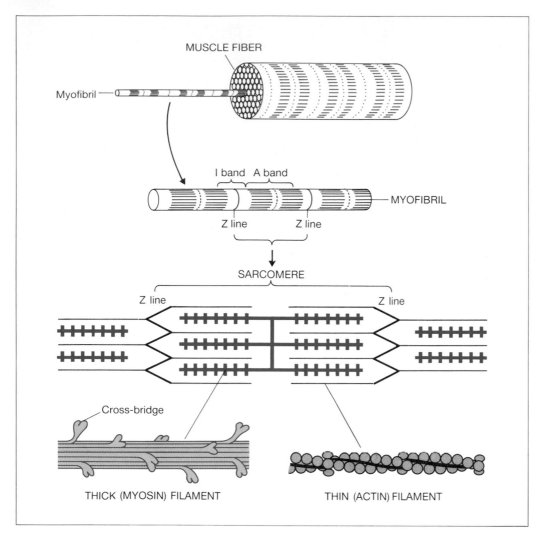

MUSCLE FIBER

Myofibril

I band A band

MYOFIBRIL

Z line   Z line

SARCOMERE

Z line   Z line

Cross-bridge

THICK (MYOSIN) FILAMENT

THIN (ACTIN) FILAMENT

FIGURE 11-4 Arrangement of skeletal-muscle fiber filaments that produce the striated banding pattern.

thick filaments are composed almost entirely of the contractile protein **myosin,** and the **thin filaments** (about half the diameter of the thick filaments) contain the contractile protein **actin** as well as two other proteins troponin and tropomyosin, that play important roles in regulating contraction.

The thick filaments are located in the middle of each sarcomere, where their orderly parallel arrangement produces the wide, dark band known as the **A band** (Figure 11-4). In contrast, each sarcomere contains *two* sets of thin filaments, one at each end. One end of each thin filament is anchored to a network of interconnecting proteins known as a **Z line,** whereas the other end overlaps a portion of the thick filaments. Two successive Z lines define the limits of one sarcomere. Thus, thin filaments from two adjacent sarcomeres are anchored to each Z line.

The **I band** (Figure 11-4) lies between the ends of the A bands of two adjacent sarcomeres and contains those

portions of the thin filaments that do not overlap the thick filaments. It is bisected by the Z line.

Two additional bands are present in the A-band region of each sarcomere (Figure 11-5). The **H zone** is a relatively light band in the center of the A band. It corresponds to the space between the ends of the two sets of thin filaments in each sarcomere; hence, only the central parts of thick filaments are found in the H zone. Finally, the narrow, dark band in the center of the H zone is known as the **M line** and is produced by proteins that bind all the thick filaments in a sarcomere together. Thus, neither the thick nor the thin filaments are free floating, since the thin filaments are anchored to the Z line and the thick filaments are linked together by the M line.

A cross section through the A bands of six adjacent myofibrils shows the regular, almost crystalline, arrangement of the thick and thin filaments (Figure 11-6). Each thick filament is surrounded by a hexagonal array

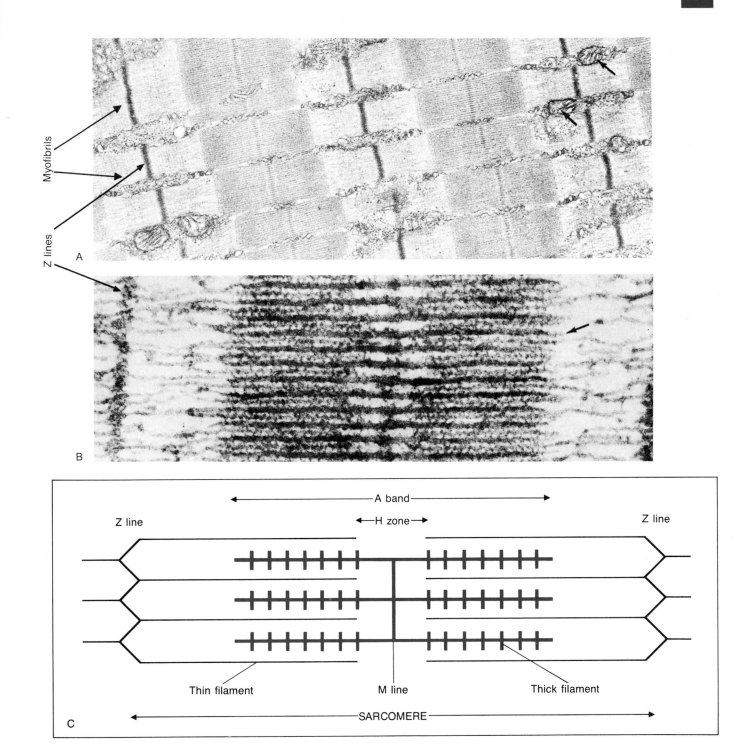

FIGURE 11-5 (A) Numerous myofibrils in a single skeletal-muscle fiber (arrow indicates a mitochondrion located between the myofibrils). (B) High magnification of a single sarcomere within a single myofibril (arrow indicates end of a thick filament). (C) Arrangement of thick and thin filaments in a single sarcomere.

of six thin filaments, and each thin filament is surrounded by a triangular arrangement of three thick filaments. Altogether there are twice as many thin as thick filaments in the region of filament overlap.

The space between adjacent thick and thin filaments is bridged by projections known as **cross bridges,** portions of myosin molecules that extend from the surface of the thick filaments toward the thin filaments (Figure

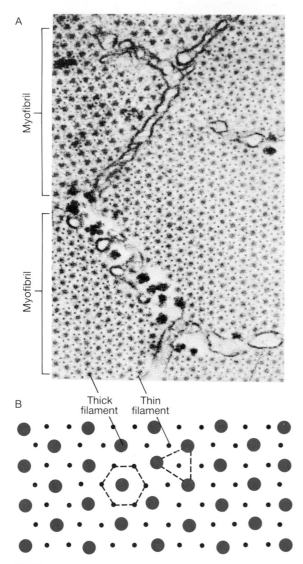

B

Thick filament   Thin filament

FIGURE 11-6 (A) Electron micrograph of a cross section through six myofibrils in a single skeletal-muscle fiber. [*From H. E. Huxley, J. Mol. Biol., 37:507– 520 (1968).*] (B) Hexagonal arrangements of the thick and thin filaments in the overlap region in a single myofibril.

FIGURE 11-7 High-magnification electron micrograph in the filament overlap region near the middle of a sarcomere. Cross bridges between the thick and thin filaments can be seen at regular intervals along the filaments. [*From H. E. Huxley and J. Hanson, in G. H. Bourne (ed.), "The Structure and Function of Muscle," Vol. 1, Academic Press, New York, 1960.*]

11-4 and 11-7). During muscle contraction, these cross bridges make contact with the thin filaments and exert force on them. The cross bridges are the force-generating sites in muscle cells.

MOLECULAR MECHANISMS OF CONTRACTION

When a skeletal muscle fiber is activated by a nerve impulse, the cross bridges bind to the thin filaments and exert a force on them. In order for shortening to occur, the forces exerted on the thin filaments must be greater than the force opposing shortening. The term **contraction,** as used in muscle physiology, does not necessarily mean shortening; rather, it refers only to the turning on of the force-generating sites—the cross bridges—in a muscle fiber. Following contraction, the mechanisms that initiate force generation are turned off, and tension generation declines, producing **relaxation** of the muscle fiber.

Sliding-Filament Mechanism

When contraction produces shortening of a skeletal-muscle fiber, the overlapping thick and thin filaments in each sarcomere move past each other, propelled by movements of the cross bridges. During this movement, there is no change in the lengths of either the thick or thin filaments (Figure 11-8). This is known as the **sliding-filament mechanism** of muscle contraction.

During shortening, each cross bridge attached to a thin filament moves in an arc much like the oars on a boat. This swiveling motion forces the thin filaments at either end of the A band toward the M line, thereby shortening the sarcomere (Figure 11-9). One stroke of a

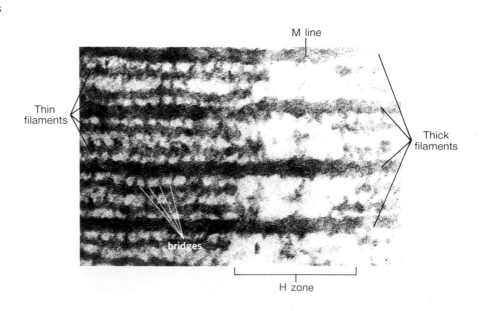

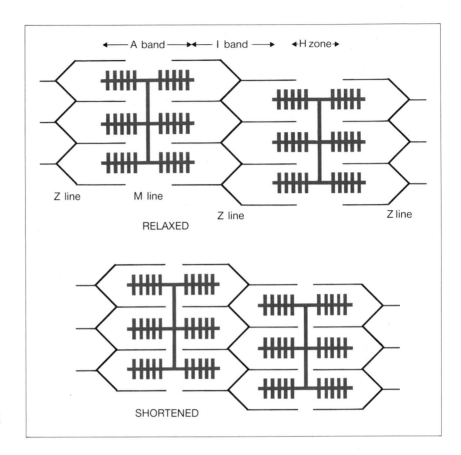

FIGURE 11-8 Changes in thick- and thin-filament alignment as a result of myofibril shortening.

cross bridge produces only a small movement of a thin filament relative to a thick filament. While the force-generating mechanism remains active, the cross bridges will repeat their swiveling motion many times, producing large movements that are made up of a series of very small steps.

The sequence of events that occurs between the time a cross bridge binds to a thin filament and the time it again binds to a thin filament to repeat the process is known as the **cross-bridge cycle.** Each cycle consists of four steps: (1) attachment of the cross bridge to a thin filament, (2) movement of the cross bridge, producing movement of the thin filament, (3) detachment of the cross bridge from the thin filament, and (4) movement of the cross bridge into a position where it can again reattach to a thin filament and repeat the cycle. Each cross bridge undergoes its own cycle of movements independently of the other cross bridges, so that at any one instant during contraction only about 50 percent of the cross bridges are attached to the thin filaments and are producing movement.

Let us look more closely at these events. A muscle fiber's ability to generate force and movement depends on the interactions of the two contractile proteins, myo-

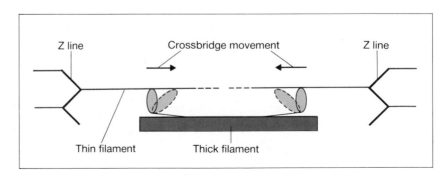

FIGURE 11-9 Cross bridges in the thick filaments bind to actin in the thin filaments and undergo a conformational change that propels the thin filaments toward the center of a sarcomere. (Only 2 of the approximately 200 cross bridges in each thick filament are shown.)

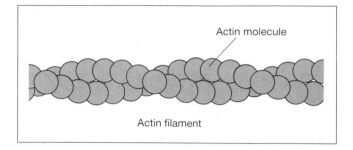

Actin molecule

Actin filament

FIGURE 11-10 Two helical chains of actin molecules form the primary structure of the thin filaments.

sin in the thick filaments and actin in the thin filaments, and the energy provided by ATP. Actin is a globular protein that polymerizes to form the two intertwined helical chains (Figure 11-10) that make up the core of the thin filaments. Myosin has a globular head attached to a long tail (Figure 11-11). The tails lie along the axis of the thick filament, and the globular heads extend out to the sides, forming the cross bridges. Each globular head contains a binding site for actin and an enzymatic site—an ATPase—that catalyzes the hydrolysis of ATP to ADP and inorganic phosphate, releasing the chemical energy stored in ATP.

The myosin molecules in the two halves of each thick filament are oriented in opposite directions, such that all their tail ends are directed toward the center of the filament (Figure 11-11). Because of this arrangement, the power strokes of the cross bridges in the two halves of each thick filament are directed toward the center, thereby moving the attached thin filaments at the two ends of the sarcomere toward the center during shortening (Figure 11-9).

The chemical and physical events occurring during the four steps of a cross-bridge cycle are illustrated in

Figure 11-12. At the conclusion of the preceding cycle (step 4), the ATP bound to myosin was split, releasing chemical energy. This energy was transferred to myosin (M), producing an energized form of myosin (M*) to which the products of ATP hydrolysis, ADP and inorganic phosphate (P$_i$), are still bound.

Step 4:

$$M{\cdot}ATP \xrightarrow[\text{ATP hydrolysis}]{} M^*{\cdot}ADP{\cdot}P_i$$

When a muscle fiber is stimulated to contract, the energized myosin cross bridge binds to an actin (A) molecule in a thin filament (step 1):

Step 1:

$$A + M^*{\cdot}ADP{\cdot}P_i \xrightarrow[\text{Actin binding}]{} A{\cdot}M^*{\cdot}ADP{\cdot}P_i$$

The binding of this energized myosin to actin triggers the discharge of the energy stored in myosin, with the resulting movement of the bound cross bridge (step 2). The ADP and P$_i$ are released from myosin during the cross bridge movement:

Step 2:

$$A{\cdot}M^*{\cdot}ADP{\cdot}P_i \xrightarrow[\substack{\text{Bridge}\\\text{movement}}]{} A{\cdot}M + ADP + P_i$$

This sequence of energy storage and release by myosin is analogous to the operation of a mousetrap: Energy is stored in the trap by cocking the spring (by ATP hydrolysis) and released by springing the trap (by binding to actin).

During the cross-bridge movement, myosin binds very firmly to actin, and this linkage must be broken in order to allow the cross bridge to reattach to a new actin

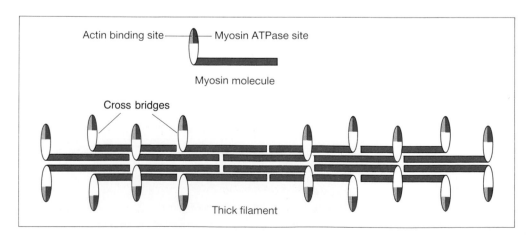

Actin binding site —— Myosin ATPase site

Myosin molecule

Cross bridges

Thick filament

FIGURE 11-11 Orientation of myosin molecules in one thick filament. The globular heads of myosin form cross bridges and the myosin tails are located in the core of the filament.

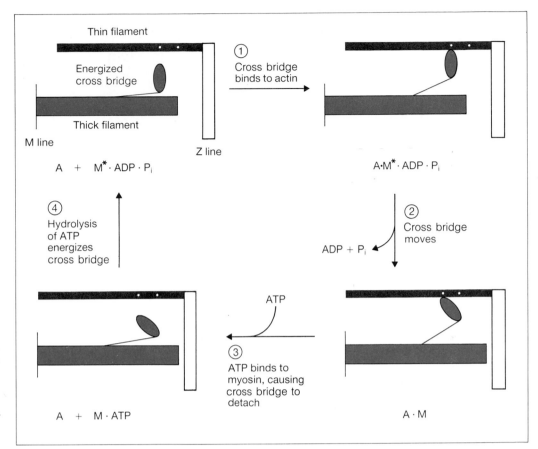

FIGURE 11-12 Chemical and mechanical changes during the four stages of a cross bridge cycle. In a resting muscle fiber, contraction begins with the binding of a cross bridge to actin in a thin filament—step 1. (M* represents an energized myosin cross bridge.)

molecule and repeat the cycle. The binding of a molecule of ATP to myosin is responsible for breaking the link between actin and myosin (step 3):

Step 3:

$$A \cdot M + ATP \xrightarrow{\hspace{3cm}} A + M \cdot ATP$$

Cross-bridge
dissociation from actin

The dissociation of actin and myosin by ATP is an example of allosteric regulation of protein activity. The binding of ATP at one site on myosin decreases the affinity of myosin for actin bound at another site on myosin. Thus, ATP is acting as a modulator molecule controlling the binding of actin to myosin. Note that ATP is not split in this step.

Following the dissociation of actin and myosin, the ATP bound to myosin is split (step 4), thereby re-forming the energized state of myosin, which can now reattach to a new site on the actin filament and repeat the cycle as long as the fiber remains activated.

To summarize, ATP performs two distinct roles in the cross-bridge cycle: (1) The energy released from ATP hydrolysis ultimately provides the energy for cross bridge movement, and (2) the binding (not hydrolysis) of

ATP to myosin breaks the link formed between actin and myosin during the cycle, allowing the cycle to be repeated. Note that the release of energy by the hydrolysis of ATP (step 4) and the movement of the cross bridge (step 2) are not simultaneous events.

The importance of ATP in dissociating actin and myosin during step 3 of a cross-bridge cycle is illustrated by **rigor mortis,** the stiffening of skeletal muscles, which begins several hours after death and is complete after about 12 h. The ATP concentration in cells, including muscle cells, declines after death because the nutrients and oxygen required by the metabolic pathways to form ATP are no longer supplied. In the absence of ATP, cross bridges can bind to actin, but the subsequent movement of the cross bridge and the breakage of the link between actin and myosin does not occur because they require the binding of ATP. The thick and thin filaments become bound to each other by immobilized cross bridges, producing the rigid condition of a dead muscle. The stiffness of rigor mortis disappears some 48 to 60 h after death due to the disintegration of muscle tissue.

Role of Calcium in Contraction

Since every muscle fiber contains all the ingredients necessary for cross-bridge activity—actin, myosin, and

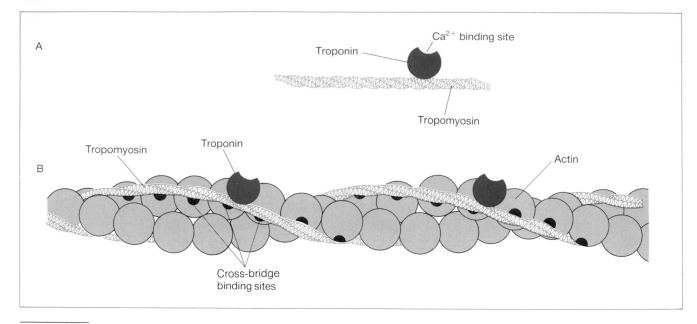

FIGURE 11-13 (A) Molecule of troponin bound to a molecule of tropomyosin. (B) Two chains of tropomyosin in a thin filament block cross-bridge binding sites on actin.

ATP—the question arises: Why are muscles not in a continuous state of contractile activity? The answer is that in a resting muscle fiber, the cross bridges are unable to bind to actin. This inhibition is due to two regulator proteins, **troponin** and **tropomyosin,** which, as noted earlier, are located on the thin filaments along with actin (Figure 11-13). Tropomyosin is a rod-shaped molecule composed of two intertwined polypeptides with a length approximately equal to that of seven actin molecules. Two chains of tropomyosin molecules are arranged end to end along the two strands of the actin double helix, where they partially cover the myosin-binding site on each actin, thereby preventing the cross bridges from making contact with actin. Each tropomyosin molecule is held in this blocking position by troponin, a smaller, globular protein that is bound to both tropomyosin and actin. One molecule of troponin binds to each molecule of tropomyosin.

Having described the system that prevents cross bridge activity and thus keeps a muscle fiber in a resting state, we can now ask what triggers a contraction, that is, what allows cross-bridge activity to occur? In order for the cross bridges to bind to actin, the tropomyosin molecules must be moved away from their actin-blocking positions. This occurs when calcium binds to specific binding sites on troponin. The binding of calcium produces a change in the shape of troponin such that it pulls the tropomyosin, to which it is bound, out of its blocking position, uncovering the cross-bridge binding sites on actin (Figure 11-14). Conversely, removal of calcium

from troponin reverses the process, and tropomyosin moves back into its blocking position so that cross-bridge activity is prevented.

Thus, cytosolic calcium-ion concentration determines the number ot troponin sites occupied by calcium, which in turn determines the number of cross bridges that can bind to actin and exert force on the thin filaments. These changes in cytosolic calcium concentration are controlled by electrical events occurring in the muscle plasma membrane, to which we now turn.

Excitation-Contraction Coupling

Excitation-contraction coupling refers to the sequence of events by which an action potential in the plasma membrane of a muscle fiber leads to cross-bridge activity by increasing cytosolic calcium concentration.

The skeletal-muscle plasma membrane is an excitable membrane capable of generating and propagating action potentials by mechanisms similar to those described for nerve cells (Chapter 8). An action potential in a skeletal-muscle fiber lasts 1 to 2 ms and is completed before any signs of mechanical activity begin (Figure 11-15). Once begun, the mechanical activity following a single action potential may last 100 ms or more. Hence, the electrical activity in the plasma membrane does not directly act upon the contractile proteins but instead produces a state (increased cytosolic calcium concentration) that continues to activate the contractile apparatus long after the electrical activity in the membrane has ceased.

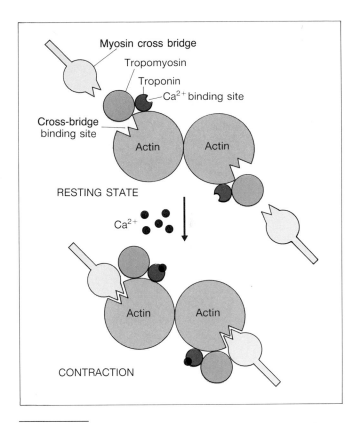

FIGURE 11-14 Cross section through a thin filament in the region of troponin. In the absence of calcium, tropomyosin blocks the cross-bridge binding sites on actin. Binding of calcium to troponin moves the tropomyosin to one side, exposing the binding sites and allowing myosin cross bridges to bind to actin.

cium that is released following membrane excitation. A separate tubular structure, the **transverse tubule** (t tubule) crosses the muscle fiber at the level of each A-I junction, passing between adjacent lateral sacs and eventually joining the plasma membrane. The lumen of the t tubule is continuous with the extracellular medium surrounding the muscle fiber.

The membrane of the t tubule, like the plasma membrane, is able to propagate action potentials. Once initiated in the plasma membrane, an action potential is rapidly conducted over the surface of the membrane and into the interior of the muscle fiber by way of the t tubules. As the action potential in a t tubule passes the sarcoplasmic reticulum, it triggers the opening of calcium channels in the lateral sacs, and calcium diffuses into the cytosol. The mechanism by which an electrical event in one membrane—the t tubule—triggers the opening of calcium channels in another membrane—the sarcoplasmic reticulum—remains a major unanswered question in muscle physiology.

As mentioned above, contraction is turned *off* when calcium is removed from troponin, and this is achieved by a lowering of the calcium concentration in the cytosol. The membranes of the lateral sacs contain primary active-transport proteins (Ca-ATPases) that pump calcium ions from the cytosol into the lumen of the reticulum. Calcium is rapidly released from the reticulum upon arrival of an action potential, but the pumping of the released calcium back into the reticulum requires a much

In a resting muscle fiber, the concentration of free calcium in the cytosol surrounding the thick and thin filaments is very low, about 10^{-7} mol/L. At this low calcium concentration, very few of the calcium binding sites on troponin are occupied, and thus cross-bridge activity is blocked by tropomyosin. Following an action potential, there is a rapid increase in cytosolic calcium concentration, calcium binds to troponin, removing the blocking effect of tropomyosin and allowing contraction. The source of the increased calcium following an action potential is the **sarcoplasmic reticulum** within the muscle fiber.

Sarcoplasmic reticulum. The sarcoplasmic reticulum in muscle is homologous to the endoplasmic reticulum found in most other kinds of cells and forms a series of sleevelike structures around each myofibril (Figure 11-16), one portion surrounding the A band and another around the I band. There are two enlarged regions of the sarcoplasmic reticulum known as **lateral sacs,** located at each end of the reticulum and connected by a series of smaller tubular elements. The lateral sacs store the cal-

FIGURE 11-15 Time relations between a skeletal-muscle fiber action potential and the resulting shortening and relaxation of the muscle fiber.

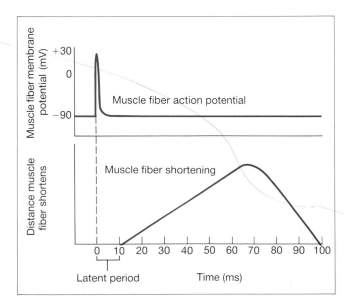

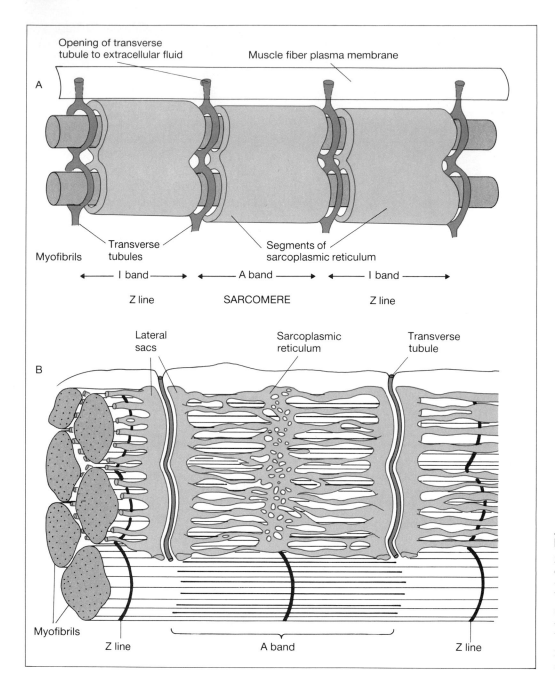

Opening of transverse
tubule to extracellular fluid

Muscle fiber plasma membrane

A

Myofibrils

Transverse
tubules

Segments of
sarcoplasmic reticulum

← I band → ← A band → ← I band →

Z line   SARCOMERE   Z line

Lateral
sacs

Sarcoplasmic
reticulum

Transverse
tubule

B

Myofibrils

Z line

A band

Z line

FIGURE 11-16 (A) Diagrammatic representation of the sarcoplasmic reticulum, the transverse tubules, and the myofibrils. (B) Three-dimensional view of transverse tubules and sarcoplasmic reticulum in a single skeletal-muscle fiber.

longer time. Therefore, the cytosolic calcium concentration remains elevated and the contraction continues for some time after each action potential.

To reiterate, just as contraction results from the release of calcium ions stored in the sarcoplasmic reticulum, so contraction ends and relaxation occurs as calcium is pumped back into the reticulum (Figure 11-17). ATP is required to provide the energy for the calcium pump, and this is the third major role of ATP in muscle contraction (Table 11-1).

Membrane Excitation: The Neuromuscular Junction

We have just seen that an action potential in the plasma membrane of a skeletal-muscle fiber is the signal that triggers contraction. The next question we must ask, then, is how are these action potentials initiated? Stimulation of the nerve fibers to a skeletal muscle is the only mechanism by which this type of muscle is normally activated. As we shall learn, there are additional mechanisms for activating cardiac- and smooth-muscle contraction.

FIGURE 11-17 Release and uptake of calcium by the sarcoplasmic reticulum during contraction and relaxation of a skeletal-muscle fiber.

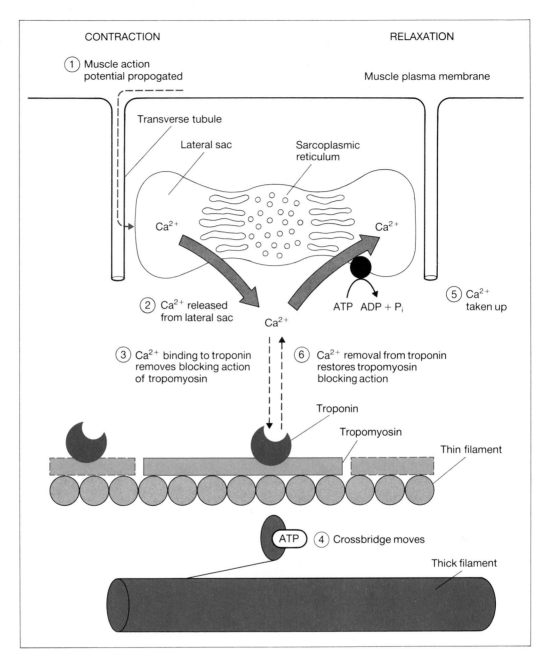

TABLE 11-1	FUNCTIONS OF ATP IN SKELETAL-MUSCLE CONTRACTION

1. Hydrolysis of ATP by myosin energizes the cross bridges, providing the energy for force generation.
2. Binding of ATP to myosin dissociates cross bridges bound to actin, allowing the bridges to repeat their cycle of activity.
3. Hydrolysis of ATP by the Ca-ATPase in the sarcoplasmic reticulum provides the energy for the active transport of calcium ions into the lateral sacs of the reticulum, lowering cytosolic calcium, ending the contraction and allowing the muscle fiber to relax.

The nerve cells whose axons innervate skeletal-muscle fibers are known as **motor neurons** (somatic efferent), and their cell bodies are located in either the brainstem or the spinal cord. The axons of motor neurons are myelinated and are the largest-diameter axons in the body. They are therefore able to propagate action potentials at high velocities, allowing signals from the central nervous system to be transmitted to skeletal-muscle fibers with minimal delay.

Within a muscle, the axon of a motor neuron divides into many branches, each branch forming a single junction with a muscle fiber. Thus, a single motor neuron innervates many muscle fibers, but each muscle fiber

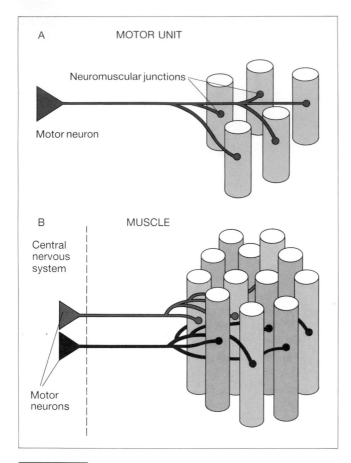

FIGURE 11-18 (A) Single motor unit consisting of one motor neuron and the muscle fibers it innervates. (B) Two motor units and their intermingled fibers in a muscle.

synaptic junctions, that is, junctions between two neurons. The vesicles contain the chemical transmitter acetylcholine (ACh). When an action potential in a motor neuron arrives at an axon terminal, it depolarizes the nerve plasma membrane, opening voltage-sensitive calcium channels, and thus allowing calcium ions to diffuse into the axon terminal. This calcium triggers the release, by exocytosis, of acetylcholine from the vesicles into the extracellular cleft separating the axon terminal and the motor end plate.

The acetylcholine diffuses across this cleft and binds to receptor sites [of the nicotinic type (page 180)] on the motor end plate. The binding of ACh activates the receptor, which opens ion channels in the end-plate membrane. Both sodium and potassium ions can pass through these channels. Because of the differences in electrochemical gradients across the plasma membrane (Chapter 8), more sodium moves in than potassium out, producing a local depolarization of the motor end plate known as an **end-plate potential** (**EPP**). Thus, an EPP is analogous to an EPSP (excitatory postsynaptic potential) at a synapse (page 202).

The magnitude of a single EPP is, however, much larger than that of an EPSP because in the former much larger amounts of neurotransmitter are released over a larger surface area, opening many more ion channels.

FIGURE 11-19 Neuromuscular junction. The motor axon terminals are embedded in grooves in the muscle fiber's surface.

has only one nerve junction and therefore is controlled by only one motor neuron. A motor neuron plus the muscle fibers it innervates is called a **motor unit** (Figure 11-18A). The muscle fibers in a single motor unit are all located in one muscle, but they are scattered throughout the muscle and do not all lie adjacent to each other (Figure 11-18B). When an action potential is produced in a single motor neuron, all the muscle fibers in its motor unit contract.

The myelin sheath surrounding the axon of a motor neuron ends near the surface of a muscle fiber, and the axon divides into a number of short processes that lie embedded in grooves on the muscle-fiber surface. The region of the muscle-fiber plasma membrane that lies directly under the terminal portion of the axon has special properties and is known as the **motor end plate**. The junction of an axon terminal with the motor end plate is known as a **neuromuscular junction** (Figure 11-19).

The axon terminals of a motor neuron contain membrane-bound vesicles resembling the vesicles found at

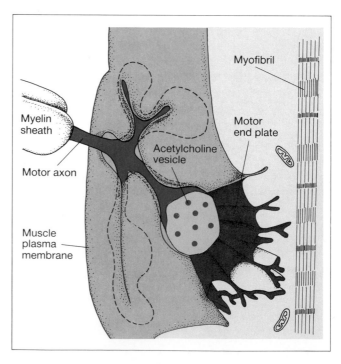

For this reason, an EPP is normally sufficient to depolarize the adjacent plasma membrane, by local current flow, to its threshold potential, initiating an action potential in the muscle plasma membrane. This action potential is then propagated over the surface of the muscle fiber by the same mechanism described in Chapter 8 for the propagation of action potentials along axon membranes (Figure 11-20). Most neuromuscular junctions are located in the middle of a muscle fiber, and it is from this region that a newly generated muscle action potential is propagated in both directions toward the ends of the fiber.

Because each EPP has a sufficient magnitude to depolarize the muscle membrane to threshold, every action potential in a motor neuron produces an action potential in each muscle fiber in its motor unit. Thus, there is a one-to-one transmission of an action potential from the motor neuron to the muscle fiber. This is quite different from synaptic junctions, where multiple EPSPs must occur, undergoing temporal and spatial summation, in order for threshold to be reached and an action potential elicited in the postsynaptic membrane.

A second difference between synaptic and neuromuscular junctions should be noted. As we saw in Chapter 8, at some synaptic junctions, inhibitory postsynaptic potentials (IPSP's) are produced. They hyperpolarize the postsynaptic membrane and decrease the probability of

its firing an action potential. Such inhibitory potentials are not found in human skeletal muscle; all neuromuscular junctions are excitatory.

In addition to receptor sites for ACh, the motor end plate on a muscle fiber contains the enzyme acetylcholinesterase at its surface. This enzyme breaks down ACh, just as occurs at ACh-mediated synapses in the nervous system. ACh bound to receptor sites is in equilibrium with free ACh in the cleft between the nerve and muscle membranes. As the concentration of free ACh falls because of its breakdown by acetylcholinesterase, less ACh will bind to the receptor sites. When the receptor sites no longer contain bound ACh, the ion channels in the end plate close, and the depolarized end plate returns to its resting potential and can respond to the arrival of a new burst of ACh released by another nerve action potential.

Table 11-2 summarizes the sequence of events that lead from an action potential in a motor neuron to the contraction and relaxation of a skeletal-muscle fiber.

There are many ways events at the neuromuscular junction can be modified by disease or drugs. For example, the deadly South American Indian arrowhead poison **curare** binds strongly to the ACh receptor site, but it does not open ion channels, nor is it destroyed by acetylcholinesterase. When a receptor site is occupied by curare, ACh cannot bind to the receptor. Therefore, al-

FIGURE 11-20 Events occurring at a neuromuscular junction that lead to an action potential in the muscle fiber plasma membrane.

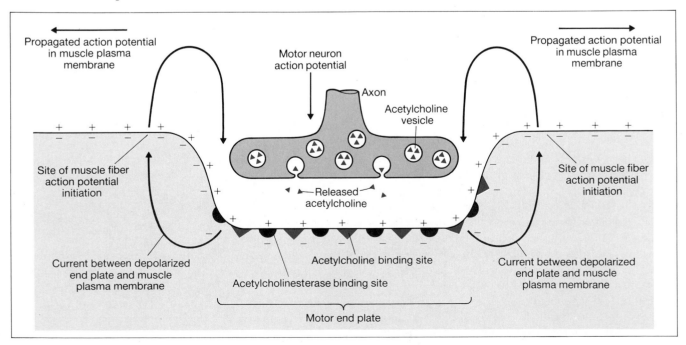

TABLE 11-2 SEQUENCE OF EVENTS BETWEEN A MOTOR NEURON ACTION POTENTIAL AND SKELETAL-MUSCLE FIBER ISOTONIC CONTRACTION

1. Action potential initiated and propagates in motor neuron axon.

2. Action potential triggers release of ACh from axon terminals at neuromuscular junction.

3. ACh diffuses from axon terminals to motor end plate in muscle fiber.

4. ACh binds to receptor sites on motor end plate, opening Na^+, K^+ ion channels.

5. More Na^+ moves into the fiber at the motor end plate than K^+ moves out, depolarizing the membrane, producing the end-plate potential (EPP).

6. Local currents depolarize the adjacent plasma membrane to its threshold potential, generating an action potential that propagates over muscle fiber surface and into the fiber along the transverse tubules.

7. Action potential in transverse tubules triggers release of Ca^{2+} from lateral sacs of sarcoplasmic reticulum.

8. Ca^{2+} binds to troponin on the thin filaments, causing tropomyosin to move away from its blocking position and thus uncovering cross-bridge binding sites on actin.

9. Energized myosin cross bridges on the thick filaments bind to actin:

$$A + M^* \cdot ADP \cdot P_i \longrightarrow A \cdot M^* \cdot ADP \cdot P_i$$

10. Cross-bridge binding triggers release of energy stored in myosin, producing an angular movement of each cross bridge:

$$A \cdot M^* \cdot ADP \cdot P_i \longrightarrow A \cdot M + ADP + P_i$$

11. ATP binds to myosin, breaking linkage between actin and myosin and thereby allowing cross bridges to dissociate from actin:

$$A \cdot M + ATP \longrightarrow A + M \cdot ATP$$

12. ATP bound to myosin is split, transferring energy to myosin cross bridge:

$$M \cdot ATP \longrightarrow M^* \cdot ADP \cdot P_i$$

13. Cross bridges repeat steps 9 to 12, producing movement of thin filaments past thick filaments. Cycles of cross-bridge movement continue as long as Ca^{2+} remains bound to troponin.

14. Cytosolic $[Ca^{2+}]$ decreases as Ca^{2+} is actively transported into sarcoplasmic reticulum by Ca-ATPase.

15. Removal of Ca^{2+} from troponin restores blocking action of tropomyosin, the cross-bridge cycle ceases, and the muscle fiber relaxes.

though the motor nerves still conduct normal action potentials and release ACh, there is no resulting EPP in the motor end plate and hence no contraction. Since the skeletal muscles responsible for breathing, like all skeletal muscles, depend upon neuromuscular transmission to initiate their contraction, curare poisoning can lead to death by asphyxiation. Curare and similar drugs are used in small amounts to prevent muscular contractions during certain types of surgical procedures when it is necessary to immobilize the surgical field. Such patients are artificially ventilated in order to maintain respiration until the drug has been removed from the system.

Neuromuscular transmission can also be blocked by inhibiting acetylcholinesterase. Some organophosphates, which are the main ingredients in certain pesticides and "nerve gases" (the latter developed for biological warfare), inhibit this enzyme. In the presence of such agents, ACh is released normally upon the arrival of an action potential at the axon terminal and binds to the end-plate receptors. The ACh is not destroyed, however, because the acetylcholinesterase is inhibited. The ion channels in the end plate therefore remain open, producing a maintained depolarization of the end plate and the muscle plasma membrane adjacent to the end plate. A skeletal-muscle membrane maintained in a depolarized state cannot generate action potentials because the voltage-sensitive sodium channels in the membrane have entered a refractory state that requires repolarization to remove. Thus, the muscle does not contract in response to nerve stimulation, and the result is skeletal-muscle paralysis and death from asphyxiation.

A third group of substances, including the toxin produced by the bacterium *Clostridium botulinum*, block the release of acetylcholine from nerve terminals, preventing the transmission of the chemical signal from the nerve to the muscle fiber. This toxin, which produces botulism, a type of food poisoning, is one of the most deadly poisons known.

MECHANICS OF SINGLE-FIBER CONTRACTION

The force exerted on an object *by* a contracting muscle is known as muscle **tension**, and the force exerted *on* the muscle by the weight of an object is the **load**. Muscle tension and load are opposing forces. Whether or not a contraction leads to fiber shortening depends on the relative magnitudes of the tension and the load. In order for muscle fibers to shorten, and therefore to move a load, muscle tension must be slightly greater than the opposing load.

When a muscle develops tension but does not shorten or lengthen, the contraction is said to be **isometric** (constant length). Such contractions occur when the muscle supports a load in a constant position or attempts to move a load that is greater than the tension developed by the muscle. A contraction occurring under conditions in which the load on a muscle remains constant but the muscle length is shortening is said to be **isotonic** (constant tension). A muscle undergoes an isotonic contraction when moving a load.

A third type of contraction is a **lengthening contraction**. This occurs when the load on a muscle is greater than the tension being generated by the cross bridges. In this situation, the load pulls the muscle to a longer length in spite of the opposing force being produced by the cross bridges. Such lengthening contractions occur when an object that is being supported by muscle contractions is lowered, such as occurs when you sit down from a standing position or walk down a flight of stairs. It must be emphasized that in these situations the lengthening of muscle fibers is not an active process produced by the contractile proteins but a consequence of the external forces being applied to the muscles. In the absence of external forces, a fiber will only shorten when stimulated; it will never lengthen. All three types of contractions—isometric, isotonic, and lengthening—occur in the natural course of muscle activity in the body.

During each type of contraction the cross bridges repeatedly go through the four steps of the cross-bridge cycle illustrated in Figure 11-12. During step 2 of an isotonic contraction, the cross bridges bound to actin move to their angled positions, causing shortening of the sarcomeres. In contrast, during an isometric contraction, the bound cross bridges are unable to move the thin filaments because of the load on the muscle fiber, but they do exert a force on the thin filaments—isometric tension. During a lengthening contraction, the cross bridges in step 2 are pulled toward the Z lines by the load while still bound to actin. The events of steps 1, 3, and 4 are the same in all three types of contraction. Thus, the chemical changes in the contractile proteins during each type of contraction are the same. The end result of shortening, no length change, or lengthening is determined by the magnitude of the load on the muscle.

Figure 11-21 illustrates the general method of recording isotonic and isometric contractions. During an isotonic contraction, the distance the muscle shortens as a function of time is recorded. The shortening distance is measured by a pen attached to the muscle and leaves a

FIGURE 11-21 Method of recording isotonic and isometric muscle contractions.

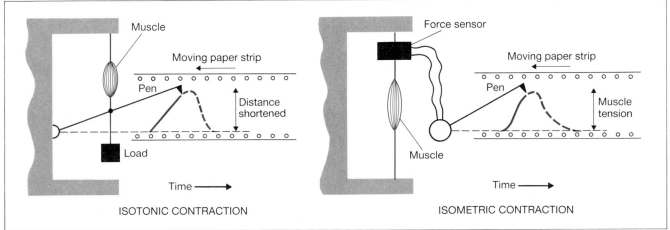

trace on a moving strip of paper as the muscle contracts and relaxes. During an isometric contraction, the force (tension) generated by the muscle as a function of time is measured by attaching the muscle at one end to a rigid support and at the other to an electronic-force-sensing device that controls the movement of a pen in proportion to the force exerted. Contraction terminology and recording methods apply to both single fibers and whole muscles. We first describe the mechanics of single-fiber contractions and later discuss the factors controlling the mechanics of whole-muscle contraction.

Twitch Contractions

The mechanical response of a muscle fiber to a single action potential is known as a **twitch**. Figure 11-22 shows the main features of an isometric twitch. Following the action potential, there is an interval of a few milliseconds, known as the **latent period,** before the tension in the muscle fiber begins to increase. During this latent period, the processes associated with excitation-contraction coupling (page 292) are occurring. The time interval from the beginning of tension development at the end of the latent period to the peak tension in an isometric contraction is the **contraction time**. Not all skeletal-muscle fibers have the same contraction times. Some fast fibers have contraction times as short as 10 ms, whereas slower fibers may take 100 ms or longer.

Comparing an isometric twitch with an isotonic twitch in the same muscle fiber, one can see from Figure 11-23 that the latent period is longer in an isotonic twitch, but the duration of the mechanical event—shortening—is briefer than the duration of isometric tension generation. The characteristics of an isotonic twitch depend

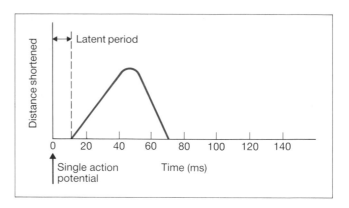

FIGURE 11-23 An isotonic twitch of a skeletal muscle fiber following a single action potential.

upon the magnitude of the load being lifted (Figure 11-24). At heavier loads the latent period is longer, but the velocity of shortening (distance shortened per unit of time), the duration of the twitch, and the distance shortened all decrease.

Let us look more closely at the sequence of events in an isotonic contraction. During the latent period of an isotonic contraction, the cross bridges begin to develop force, but shortening does not begin until the muscle tension just exceeds the load on the fiber. Thus, prior to shortening, there is a period of isometric contraction during which the tension increases. A similar period of isometric contraction occurs at the end of an isotonic contraction, as the tension declines to zero. The heavier the load, the longer it takes for the tension to increase to a value just exceeding the load. If the load on a fiber is

FIGURE 11-22 An isometric twitch of a skeletal-muscle fiber following a single action potential.

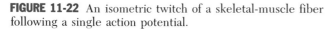
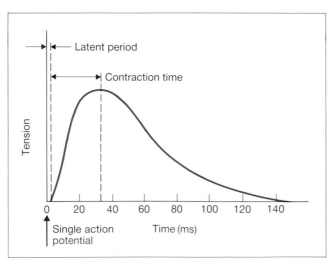

FIGURE 11-24 Isotonic twitches with different loads. The distance shortened, velocity of shortening, and duration of shortening all decrease with increased load, whereas the latent period increases with load.

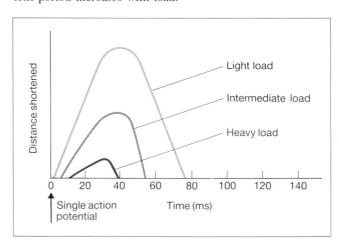

increased, eventually a load will be reached that the muscle is unable to lift, the velocity and distance of shortening will be zero, and the contraction becomes completely isometric. At still higher loads, the muscle fiber will lengthen, in spite of maximal cross-bridge force—a lengthening contraction.

Frequency-Tension Relation

Since a single action potential in a skeletal-muscle fiber lasts 1 to 2 ms but the twitch may last for 100 ms, it is possible for a second action potential to be initiated during the period of mechanical activity. Figure 11-25 illustrates the tension generated during isometric contractions of a muscle fiber in response to three successive stimuli. In Figure 11-25A, the isometric twitch following the first stimulus S1 lasts 150 ms. The second stimulus S2, applied to the muscle fiber 200 ms after S1 when the fiber has completely relaxed, causes a second identical twitch, and a third stimulus S3, equally timed produces a third identical twitch. In Figure 11-25B, the interval

between S1 and S2 remains 200 ms, but a third stimulus is applied 60 ms after S2, when the mechanical response resulting from S2 is beginning to decrease. Stimulus S3 induces a contractile response whose peak tension is greater than that produced by S2. In Figure 11-25C, the interval between S2 and S3 is further reduced to 10 ms, and the resulting peak tension is even greater. Indeed, the mechanical response to S3 is a smooth continuation of the mechanical response already induced by S2.

The increased mechanical response of a muscle fiber to a second action potential is known as **summation**. Contractile activity can be sustained, in the absence of fatigue, if a fiber is repeatedly stimulated at a frequency sufficient to prevent complete relaxation between stimuli. A maintained contraction in response to repetitive stimulation is known as a **tetanus**. At low stimulation frequencies, the tension may oscillate as the muscle fiber partially relaxes between stimuli, producing an unfused tetanus. A fused tetanus, with no oscillations, is produced at higher stimulation frequencies (Figure 11-26).

As the frequency of action potentials increases, the level of sustained tension also increases until a maximal tetanic tension is reached, beyond which tension no longer increases with further increases in stimulation frequency. This maximal tetanic tension is about three to five times greater than the isometric twitch tension. Since different muscle fibers have different contraction times, the stimulus frequency that will produce a maximal tetanic tension also differs from fiber to fiber.

The increased tension produced by increased stimulus frequency can be explained by events occurring in the muscle fiber. The tension produced by each sarcomere at any instant in time depends upon the number of cross bridges bound to actin and undergoing step 2 of the cross-bridge cycle. This number depends on two factors: (1) the rate at which unbound cross bridges can attach to actin and (2) the number of unblocked binding sites available on the actin.

The binding of energized cross bridges to actin (step 1 in Figure 11-12) takes a finite amount of time to occur. When energized myosin is exposed to actin, the tension will increase as more and more bridges become bound and exert tension. The time course of this reaction is reflected in the rise in tension during a maximal isometric tetanic contraction and includes the time to stretch elastic elements in series with the cross bridges.

The total number of cross bridges that eventually become bound to actin during a given time period depends on the number of unblocked binding sites available on the thin filaments during that time, which in turn depends on the number of troponin molecules that have bound calcium and thereby uncovered myosin-binding sites on actin. As we have seen, the number of troponin molecules that bind calcium depends on the cytosolic

FIGURE 11-25 Summation of isometric contractions produced by shortening the time between stimuli S2 and S3.

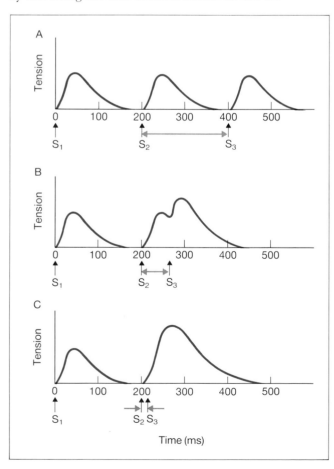

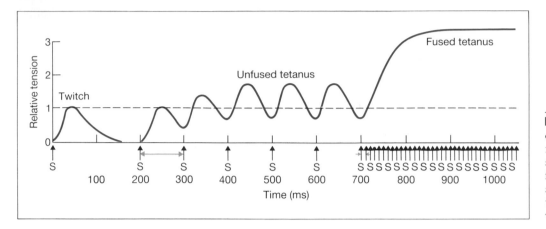

FIGURE 11-26 Isometric contractions produced by multiple stimuli of 10 stimuli per second (unfused tetanus) and 100 stimuli per second (fused tetanus), as compared with a single twitch.

concentration of calcium achieved by the release of calcium from the sarcoplasmic reticulum during excitation.

A single action potential in a skeletal-muscle fiber releases enough calcium to saturate troponin, and all the myosin-binding sites on the thin filaments are therefore initially available. As soon as calcium is released from the sarcoplasmic reticulum, however, it begins to be pumped back in, and the calcium concentration begins to fall from its elevated value, causing more and more of the myosin-binding sites on actin to be blocked by tropomyosin and become unavailable for cross-bridge binding. The tension reached during a single twitch is limited, therefore, by the rate at which the cross bridges can bind to actin and the declining number of available binding sites.

During a tetanic contraction, the successive action potentials each release calcium from the sarcoplasmic reticulum before all the calcium from the previous action potential has been pumped back into the reticulum. This results in a maintained elevated cytosolic calcium concentration and prevents any decline in the number of available binding sites on the thin filaments. Under these conditions, the maximum number of binding sites is available while the cross bridges are binding to actin, and the tension rises to its maximal level. At lower frequencies of stimulation, fewer binding sites are available, and the tension reached in summation is lower.

Length-Tension Relation

The amount of tension developed by a muscle fiber and thus its strength can be altered not only by changing the frequency of stimulation but also by changing the length of the fiber prior to contraction. A muscle fiber can be stretched to various lengths and the magnitude of the maximal isometric tetanic tension generated during subsequent contraction measured at each length (Figure 11-27). The length at which the fiber develops the greatest tension is termed the **optimal length,** l_o. When muscle-fiber length is 60 percent of l_o, the fiber develops no tension when stimulated. At longer lengths the isometric tension is increased up to a maximum at l_o, and further lengthening leads to a drop in tension. At lengths of 1.75 l_o or beyond, the fiber develops no tension when stimulated.

The lengths of most fibers, when all the skeletal muscles in the body are relaxed, is near l_o and thus near the optimal length for force generation. The length of a relaxed fiber can be altered by the load on the muscle or the contraction of other muscles, but the extent to which the relaxed length can be changed is limited by the muscle's attachments to bones. It rarely exceeds a 30 percent change from the optimal length and is often much less. Over this range of lengths, the ability to develop tension never falls below about half of the tension that can be developed at the optimal length (Figure 11-27).

The relationship between fiber length and the fiber's capacity to develop tension can be explained in terms of the sliding-filament mechanism. Stretching a relaxed muscle fiber pulls the thin filaments past the thick filaments, changing the amount of overlap between them. Stretching a fiber to 1.75 l_o pulls the filaments apart to the point where there is no overlap. At this point there can be no cross-bridge binding to actin and no development of tension. Between 1.75 l_o and l_o, there is more and more filament overlap, and the tension developed upon stimulation increases in proportion to the increased number of cross bridges in the overlap region. Filament overlap is greatest at l_o, allowing the maximal number of cross bridges to bind to the thin filaments, thereby producing maximal tension. At lengths less than l_o, tension declines because the overlapping of the two sets of thin filaments from opposite ends of the sarcomere interfere with cross-bridge binding in the regions of double overlap. In addition, for reasons that are un-

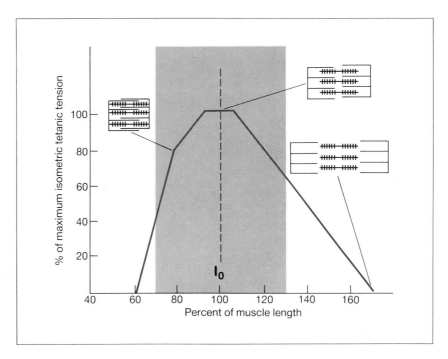

FIGURE 11-27 Variation in isometric tetanus tension with muscle fiber length. The shaded band represents the range of length changes that can occur physiologically in the body while muscles are attached to bones.

known, less calcium is released from the sarcoplasmic reticulum at these shorter fiber lengths.

Load-Velocity Relation

It is a common experience that light objects can be moved faster than heavy objects. The velocity at which a muscle fiber shortens while undergoing maximal tetanic stimulation decreases with increasing loads (Figure 11-28). The maximal shortening velocity is achieved at zero load and is zero when the load is equal to the maximal isometric tension. At loads greater than the maximal isometric tension, the fiber will lengthen at a velocity that increases with load, and at very high loads the fiber will break.

The shortening velocity is determined by the rate at which individual cross bridges undergo their cyclic activity. For complex reasons, increasing the load on a cross bridge decreases the rate at which the bridge proceeds through step 2, the force-generating step in the cross-bridge cycle. The slower the rate of cross-bridge cycling, the slower the shortening velocity.

SKELETAL-MUSCLE ENERGY METABOLISM

As we have seen, ATP performs three functions directly related to muscle fiber contraction and relaxation (Table 11-1). In no other cell type does the rate of ATP breakdown increase so much from one moment to the next as in a skeletal-muscle fiber when it goes from rest to a state of contractile activity. If a fiber is to sustain contractile activity, molecules of ATP must be supplied by metabolism as rapidly as they are broken down by the contractile process. The small supply of

FIGURE 11-28 Velocity of skeletal-muscle fiber shortening and lengthening as a function of load. Note that the force being sustained by cross bridges during a lengthening contraction is greater than the maximum isometric tension.

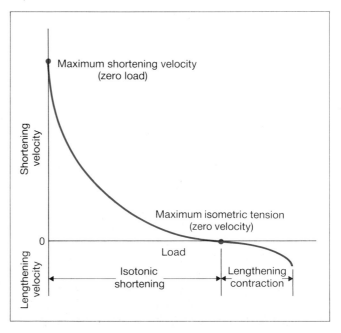

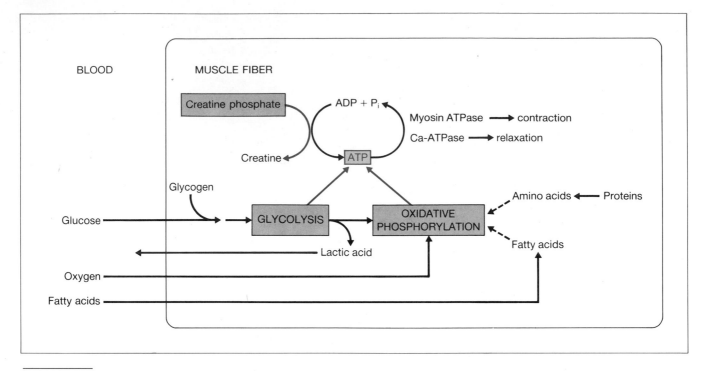

BLOOD   MUSCLE FIBER

FIGURE 11-29 Metabolic pathways producing the ATP utilized during muscle contraction.

preformed ATP that exists at the start of contractile activity is consumed within a few twitches. If this were the only source of energy, the muscle cross bridges would rapidly become locked in a state equivalent to rigor mortis as the ATP was depleted.

There are three ways a muscle fiber can form ATP during contractile activity (Figure 11-29): (1) phosphorylation of ADP by **creatine phosphate**; (2) oxidative phosphorylation of ADP in mitochondria, and (3) substrate phosphorylation of ADP, primarily by the glycolytic pathway in the cytosol.

Phosphorylation of ADP by creatine phosphate (CP) provides a very rapid means of forming ATP at the onset of contractile activity. When the chemical bond between creatine (C) and phosphate is broken, the amount of energy released is about the same as that released when the terminal phosphate bond in ATP is broken. This energy, along with the phosphate group, can be transferred to ADP to form ATP in a reversible reaction catalyzed by creatine kinase:

$$CP + ADP \xrightleftharpoons[\text{Creatine kinase}]{} C + ATP$$

In a resting muscle fiber, the concentration of ATP is always greater than ADP, leading, by mass action, to the formation of creatine phosphate. During periods of rest,

muscle fibers build up a concentration of creatine phosphate to a level approximately five times that of ATP. At the beginning of contraction, when the concentration of ATP begins to fall and ADP to rise owing to the increased rate of ATP breakdown, mass action favors the formation of ATP from creatine phosphate. This transfer of energy from creatine phosphate to ATP is so rapid that the concentration of ATP in a muscle fiber changes very little at the start of contraction, whereas the concentration of creatine phosphate falls rapidly.

Although the formation of ATP from creatine phosphate is very rapid, requiring only a single enzymatic reaction, the amount of ATP that can be formed by this process is limited by the initial concentration of creatine phosphate. If contractile activity is to be continued for more than a few seconds, the muscle must be able to form ATP from sources other than the limited creatine phosphate stores. The use of creatine phosphate at the start of contractile activity provides the few seconds necessary for the slower, multienzyme pathways of oxidative phosphorylation and glycolysis to increase their rates of ATP formation to levels that match the rates of ATP breakdown.

At moderate levels of muscular exercise (moderate rates of ATP breakdown), most of the ATP used for muscle contraction is formed by oxidative phosphorylation. During the first 5 to 10 min of exercise, the muscle's own glycogen is the major fuel consumed. For the next 30 min or so, blood-borne fuels become dominant, blood glucose and fatty acids contributing approximately equally to the oxygen consumption of the muscle. Beyond this period, fatty acids become progressively more important as glucose utilization decreases.

A number of factors may limit the production of ATP by oxidative phosphorylation in a muscle fiber: (1) the quantity of oxygen delivered by the blood; (2) the quantity of fuel molecules delivered by the blood; and (3) the rates at which the enzymes in the metabolic pathways can process the fuel molecules. Factors 1 and 2 depend upon the ability of the respiratory and cardiovascular systems to deliver these substances to a muscle (see Chapters 13 and 14 for the roles of the circulatory and respiratory systems in limiting maximal exercise). Factor 3 reflects the fact that the oxidative phosphorylation pathway is relatively slow and may not be able to produce ATP as rapidly as it is broken down even when oxygen and fuel molecules are available.

When the level of exercise exceeds about 70 percent of the maximal rate of ATP breakdown, glycolysis begins to contribute an increasingly significant fraction of the total ATP generated by the muscle. The glycolytic pathway, although producing only small quantities of ATP from each molecule of glucose metabolized, can produce large quantities of ATP rapidly when enough enzymes and substrate are available, and it can do so in the absence of oxygen. The glucose for glycolysis can be obtained from two sources; the blood or the stores of glycogen within the muscle fibers. As the intensity of muscle activity increases, more and more of the ATP is formed by the anaerobic breakdown of muscle glycogen, with a corresponding increase in the production of lactic acid.

At the end of muscle activity, creatine phosphate and glycogen levels in the muscle are decreased. To return a muscle fiber to its original state, the glycogen stores and creatine phosphate must be replaced. Both processes require energy, and so a muscle continues to consume increased amounts of oxygen for some time after it has ceased to contract (as evidenced by the fact that one continues to breathe deeply and rapidly for a period of time immediately following intense exercise). The longer and more intense the exercise, the longer it takes to restore the glycogen and creatine phosphate in a muscle fiber to their original concentrations.

Muscle Fatigue

When a skeletal-muscle fiber is continuously stimulated at a frequency that produces maximal tetanic contraction, the tension developed by the fiber eventually declines. This failure of a muscle fiber to maintain tension as a result of previous contractile activity is known as **muscle fatigue**. Although low-frequency twitches can be repeated indefinitely in certain types of fibers with no signs of fatigue (Figure 11-30A), other fibers undergo fatigue if twitches are repeated frequently enough over an extended period of time (Figure 11-30B). The onset of fatigue and its rate of development depend on the type of skeletal-muscle fiber and on the intensity and duration of contractile activity.

If a muscle is allowed to rest for several minutes after the onset of fatigue, it can recover its ability to contract upon restimulation (Figure 11-30C). The extent of recovery depends upon the duration and intensity of the previous activity. Some muscle fibers will fatigue fairly rapidly if stimulated at a high frequency but will also recover rapidly. This is the type of fatigue that accompanies short-duration, high-intensity types of exercise, such as weight lifting. Fatigue that develops more slowly with low-intensity, long-duration endurance exercise, such as long-distance running, requires much longer periods of rest, often up to 24h, before the muscle achieves complete recovery.

It might seem logical that depletion of available ATP would account for fatigue, but such depletion is not the immediate cause. The ATP concentration in fatigued muscle is only slightly lower than in a resting muscle. The mechanisms producing fatigue appear to be different in short-duration, high-intensity exercise and in long-duration, low-intensity exercise, although conclusive evidence for these mechanisms has yet to be achieved. Short-duration, high-intensity fatigue is caused in part by the increased intracellular acidity that accompanies the rise in lactic acid. This increased hydrogen-ion concentration affects the contractile proteins, decreasing the force generated by the cross-bridge movements. In addition, there is a decrease in the amount of calcium released by the sarcoplasmic reticulum in this type of fatigue. The factors responsible for the fatigue following long-duration, low-intensity activity are less well understood and appear to be related to glycogen metabolism since the onset of this type of fatigue is correlated with the depletion of muscle glycogen. The larger the glycogen content of a muscle, the longer the duration of activity before fatigue sets in.

Although not directly caused by a decrease in ATP concentration, fatigue does occur at a time when the ATP breakdown rate is beginning to exceed its rate of production. If contractile activity were to continue without entering a state of fatigue, the ATP concentration would decrease to the point that the cross bridges would become linked in a rigor configuration. Thus, muscle fatigue may have evolved as a mechanism for preventing the onset of rigor, which is very damaging to muscle fibers.

Failure of the appropriate levels of the cerebral cortex to send excitatory signals to the motor neurons can

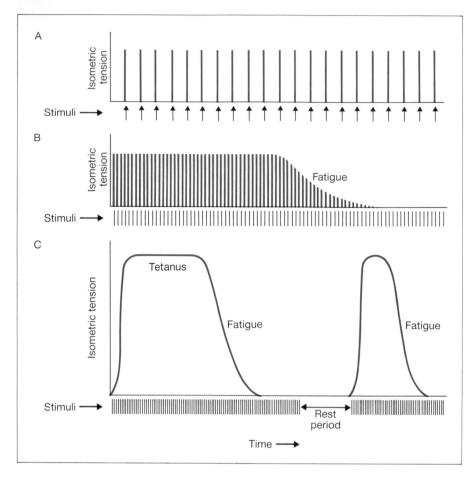

FIGURE 11-30 Muscle fatigue. (A) Lack of fatigue during repeated twitch contractions at low frequency. (B) Fatigue during repeated twitch contractions at high frequency. (C) Fatigue during a maintained tetanus and recovery following a period of rest.

also lead to the failure of muscle fibers to contract. This type of **psychological fatigue** causes an individual to stop exercising even though the muscles are not fatigued. An athlete's performance depends not only on the physical state of the appropriate muscles but also upon the "will to win," the ability to overcome psychological fatigue.

TYPES OF SKELETAL-MUSCLE FIBERS

All skeletal-muscle fibers do not have the same mechanical and metabolic characteristics. Different types of fibers can be identified on the basis of (1) their maximal velocities of shortening—fast and slow fibers—and (2) the major pathway used to form ATP—oxidative and glycolytic fibers.

Fast and slow fibers contain myosin isozymes[1] that differ in the maximal rates at which they split ATP, which in turn determines the maximal rate of cross-bridge cycling and hence the fibers' maximal shortening velocity. Fibers containing myosin with high ATPase activity are classified as **fast fibers,** and those containing myosin with a lower ATPase activity are **slow fibers.**

The second means of classifying skeletal-muscle fibers is according to the type of enzymatic machinery available for synthesizing ATP. Some fibers contain numerous mitochondria and thus have a high capacity for oxidative phosphorylation. These fibers are classified as **oxidative fibers.** Most of the ATP produced by such fibers is dependent upon blood flow to the muscle to deliver oxygen and fuel molecules, and these fibers are surrounded by numerous small blood vessels. They also contain large amounts of an oxygen-binding protein known as **myoglobin,** which increases the rate of oxygen movement into them in addition to providing a small store of oxygen. The large amounts of myoglobin present in oxidative fibers give the muscle a dark red color, and thus oxidative fibers are often referred to as **red muscle** fibers.

---

[1] Enzymes that catalyze the same reaction but have slightly different amino acid sequences, and therefore different affinities for substrate, are called isozymes. There is a separate gene for each isozyme.

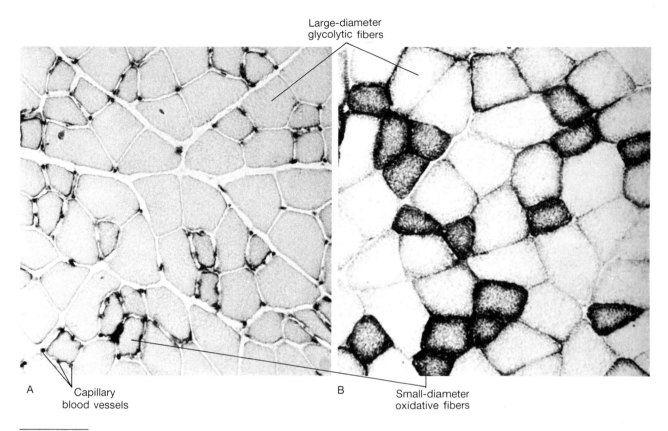

Large-diameter
glycolytic fibers

A Capillary
blood vessels

B Small-diameter
oxidative fibers

FIGURE 11-31 Cross sections of skeletal muscle. (A) The capillaries surrounding the muscle fibers have been stained. Note the large number of capillaries surrounding the small-diameter oxidative fibers. (B) The mitochondria have been stained. Fibers containing a high concentration of stained material indicate the large numbers of mitochondria in the small-diameter oxidative fibers. (*Courtesy of John A. Faulkner.*)

In contrast, **glycolytic fibers** have few mitochondria but possess a high concentration of glycolytic enzymes and a large store of glycogen. Corresponding to their limited use of oxygen, these fibers are surrounded by relatively few blood vessels and contain little myoglobin. The lack of myoglobin is responsible for the lack of color in glycolytic fibers and their frequent designation as **white muscle** fibers.

On the basis of these characteristics, three types of skeletal-muscle fibers can be distinguished:

1. **Slow-oxidative fibers** combine low myosin-ATPase activity with high oxidative capacity.
2. **Fast-oxidative fibers** combine high myosin-ATPase activity with high oxidative capacity.
3. **Fast-glycolytic fibers** combine high myosin-ATPase activity with high glycolytic capacity.

In addition to these biochemical differences, there are also size differences, glycolytic fibers generally having much larger diameters than oxidative fibers (Figure 11-31). This fact has significance for tension develop-

ment. The density of thick and thin filaments per unit of cross-sectional area is about the same in all types of skeletal-muscle fibers. The larger the diameter of a muscle fiber, therefore, the greater the maximum tension it can develop (the greater its strength) because there are more thick and thin filaments acting in parallel to produce force. Accordingly, a glycolytic fiber develops more tension when it contracts than does an oxidative fiber.

These three types of fibers also differ in their capacity to resist fatigue. Fast-glycolytic fibers fatigue rapidly, whereas slow-oxidative fibers are very resistant to fatigue, which allows them to maintain contractile activity for long periods with little loss of tension. The fast-oxidative fibers have an intermediate capacity to resist fatigue.

The characteristics of these three types of skeletal-muscle fibers are summarized in Table 11-3.

Most muscles are composed of all three fiber types interspersed with each other (Figure 11-31). Depending on the proportions of the three fiber types present, muscles can differ considerably in their maximal contraction

TABLE 11-3   CHARACTERISTICS OF THE THREE TYPES OF SKELETAL MUSCLE FIBERS

	Slow-Oxidative Fibers	Fast-Oxidative Fibers	Fast-Glycolytic Fibers
Primary source of ATP production	Oxidative phosphorylation	Oxidative phosphorylation	Anaerobic glycolysis
Mitochondria	Many	Many	Few
Capillaries	Many	Many	Few
Myoglobin content	High (red muscle)	High (red muscle)	Low (white muscle)
Glycolytic enzyme activity	Low	Intermediate	High
Glycogen content	Low	Intermediate	High
Rate of fatigue	Slow	Intermediate	Fast
Myosin-ATPase activity	Low	High	High
Contraction velocity	Slow	Fast	Fast
Fiber diameter	Small	Intermediate	Large
Motor unit size	Small	Intermediate	Large
Size of motor neuron innervating fiber	Small	Intermediate	Large

speed, strength, and fatigability. For example, the muscles of the back and legs must be able to maintain their activity for long periods of time without fatigue while supporting an upright posture. These muscles contain large numbers of slow-oxidative fibers, which are resistant to fatigue. In contrast, the muscles in the arms may be called upon to produce large amounts of tension over a short time period, as when lifting a heavy object, and these muscles have a greater proportion of fast-glycolytic fibers.

All the muscle fibers in a single motor unit are of the same fiber type. Thus, one can apply the fiber type designation to the whole motor unit and refer to slow-oxidative motor units, fast-glycolytic motor units, and so forth.

WHOLE-MUSCLE CONTRACTION

Thus far we have considered the biochemical and mechanical properties of single muscle fibers. As described earlier, however, whole muscles are made up of many muscle fibers organized into motor units. We will now use the characteristics of single fibers to a describe whole-muscle contraction and its control.

Control of Muscle Tension

The total tension a muscle can develop depends upon two factors: (1) the amount of tension developed by each fiber and (2) the number of fibers contracting at any given time. By controlling these two factors, the nervous system controls whole-muscle tension and shortening velocity. The factors that determine the amount of tension developed in a single fiber have been discussed previously and are summarized in the top of Table 11-4.

The second factor determining total muscle tension—the number of fibers contracting at any time—depends on: (1) the number of fibers in each motor unit and (2) the number of active motor units.

The number of muscle fibers in a motor unit (motor unit size) varies considerably from one muscle to another. The muscles in the hand and eye, which produce very delicate movements, contain small motor units. For example, one motor neuron innervates only about 13 fibers in an eye muscle. In contrast, in the more coarsely controlled muscles of the back and legs, each motor unit contains hundreds and in some cases several thousand fibers. When a muscle is composed of small motor units, the total tension produced by the muscle can be increased in small steps by activating additional motor units in the muscle. If the motor units are large, large increases in tension occur as each additional motor unit is activated. Thus, finer control of muscle tension is possible in muscles with small motor units.

The force produced by a single fiber, as we have seen (page 307), depends, in part, on the fiber diameter—the greater the diameter, the greater the force. We have also noted that fast-glycoltic fibers have the largest diame-

TABLE 11-4 FACTORS DETERMINING MUSCLE TENSION

I. Single muscle fibers
  A. Action-potential frequency (frequency-tension relation)
  B. Fiber length (length-tension relation)
  C. Fiber diameter
  D. Fatigue
II. Whole muscle
  A. Tension developed by each active fiber
  B. Number of active fibers
    1. Number of fibers per motor unit
    2. Number of active motor units

intensity, short-duration types of contraction, such as weight lifting.

To summarize, the neural control of whole-muscle tension involves both the frequency of action potentials in individual motor neurons (to vary the tension generated by each fiber) and the recruitment of motor units (to vary the number of active fibers). Most motor neuron activity occurs in bursts of action potentials, which produce tetanic contractions rather than single twitches. Recall from page 301 that the tension of a single fiber increases only three- to fivefold when going from a twitch to a maximal tetanic contraction. Therefore, varying the frequency of action potentials in the neurons supplying them provides a way to make only relatively small adjustments in the tension of the recruited motor

eters. Thus, a motor unit composed of 100 fast-glycoltic fibers would produce more force than a motor unit composed of 100 slow-oxidative fibers. In addition, fast-glycolytic motor units tend to have more muscle fibers. For all these reasons, activating a single fast-glycolytic motor unit will produce more force than activating a single slow-oxidative motor unit.

The process of increasing the number of motor units that are active in a muscle at any given time is called **recruitment**. It is achieved by increasing the excitatory synaptic input to the motor neurons. The greater the number of active motor neurons and hence motor units recruited, the greater the muscle tension.

Motor neuron size plays an important role in the recruitment of motor units. The size of a motor neuron refers to the diameter of the nerve cell body, which is usually correlated with the diameter of its axon and does not refer to the size of the motor unit it may control. Given the same number of ions entering a cell during synaptic activity in a large and a small motor neuron, the small neuron will undergo a greater depolarization because these ions will be distributed over a smaller membrane surface area. As a consequence, the recruitment of motor units during a contraction that gradually proceeds from very weak to very strong begins with the recruitment of the smallest neurons and proceeds to the larger neurons. Since the smallest motor neurons are part of the slow-oxidative motor units (Table 11-3), these motor units are recruited first, followed by fast-oxidative motor units and finally, during very strong contractions, by the fast-glycolytic motor units (Figure 11-32).

Thus, during moderate-strength contractions, such as are used in most endurance types of exercise, relatively few fast-glycolytic motor units are recruited, and most of the activity occurs in oxidative fibers, which are more resistant to fatigue. The large fast-glycolytic motor units, which rapidly fatigue, are recruited only during high-

FIGURE 11-32 (A) Diagram of a cross section through a muscle composed of three types of motor units. (B) Tetanic muscle tension resulting from the successive recruitment of the three motor units. Note that motor unit 3, composed of fast-glycolytic fibers, produces the greatest rise in tension because it is composed of the largest-diameter fibers and contains the largest number of fibers per motor unit.

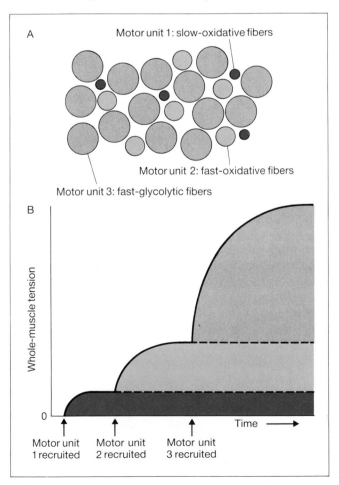

units. The force a whole muscle exerts can be varied over a much wider range than three- to fivefold, from very delicate movements to extremely powerful contractions. This wide range of force is controlled by the recruitment of motor units, which provides the primary means of varying tension in a whole muscle.

Control of Shortening Velocity

As we saw earlier, the velocity at which a single muscle fiber shortens is determined by (1) the load on the fiber (page 303) and (2) whether the fiber is a fast fiber or slow fiber (page 306). In a whole muscle, the shortening velocity can be varied from very fast to very slow while the load on the muscle remains constant. This range of velocities is controlled by the nervous system in the same way that tension is varied, namely, by the recruitment of additional motor units.

Consider, for the sake of illustration, a muscle composed of only two motor units of the same size and fiber type lifting a 4-g load. When both units are active, each motor unit is bearing only half the load and its fibers will shorten as if each fiber were lifting only a 2-g load. Thus, the muscle will lift the 4-g load at a higher velocity when both motor units are active.

Muscle Adaptation to Exercise

The frequency with which a muscle is used, as well as the duration and intensity of its activity, affects the properties of the muscle. If the neurons to a skeletal muscle are severed or otherwise destroyed, the denervated muscle fibers become progressively smaller in diameter, and the amount of contractile proteins they contain decreases. This condition is known as **denervation atrophy**. A muscle can also atrophy with its nerve supply intact if the muscle is not used for a long period of time, as when a broken arm or leg is immobilized in a cast. This condition is known as **disuse atrophy**.

In contrast to the decrease in muscle mass that results from a *lack* of neural stimulation, increased amounts of contractile activity—in other words, exercise—can produce a considerable increase in size (**hypertrophy**) of certain muscle fibers as well as other changes in their chemical composition. Since the number of fibers in a muscle remains essentially constant throughout adult life, the changes in muscle size with atrophy and hypertrophy do not result from changes in the number of muscle fibers but in the metabolic capacity and size of each fiber.

Two types of change can occur in muscle fibers as a result of exercise, or the lack thereof: (1) alterations in their ATP-forming capacity as a result of increases or decreases (a) in the synthesis of enzymes in the various ATP-producing pathways and (b) in the blood flow to the

muscle; and (2) changes in the diameters of the muscle fibers as a result of the loss or formation of myofibrils.

Exercise that is of relatively low intensity but of long duration, such as long-distance running and swimming, produces increases in the number of mitochondria in the fast- and slow-oxidative fibers that are recruited in this type of activity. In addition, there is an increase in the number of capillaries around these fibers. All these changes lead to an increase in the capacity for endurance activity with a minimum of fatigue. Fiber diameter decreases slightly, and thus there is a small decrease in the strength of muscles as a result of endurance exercise. As we shall see in later chapters, endurance exercise produces changes not only in the skeletal muscles but also in the respiratory and circulatory systems, changes that improve the delivery of oxygen and fuel molecules to the muscle.

In contrast, short-duration, high-intensity exercise, such as weight lifting, affects primarily the fast-glycolytic fibers, which are recruited when the intensity of contraction exceeds about 40 percent of the maximal tension produced by the muscle. These fibers undergo an increase in fiber diameter because increased synthesis of actin and myosin filaments forms more myofibrils. In addition, the number of glycolytic enzymes increases. The result of such high-intensity exercise is to increase the strength of the muscle, producing the bulging muscles of a conditioned weight lifter. Such muscles, although very powerful, have little capacity for endurance, and they fatigue rapidly.

Because different types of exercise produce quite different changes in the strength and endurance capacity of a muscle, an individual performing regular exercises to improve muscle performance must be careful to choose a type of exercise that is compatible with the activity he or she ultimately wishes to perform. Thus, lifting weights will not improve the endurance of a long-distance runner, and jogging will not produce the increased strength desired by a weight lifter. Most exercises have suboptimal effects on both strength and endurance.

If a regular pattern of exercise is stopped, the changes in the muscle that occurred as a result of the exercise will slowly revert, over a period of months, to their state before exercise began.

The signals responsible for the changes in muscle with different types of activity are unknown. They appear to be related to the frequency and intensity of the contractile activity in the muscle fibers and thus to the pattern of action potentials produced in the muscle over an extended period of time.

Exercise produces little change in the types of myosin formed by the fibers and thus little change in the proportions of fast and slow fibers in a muscle. As described above, however, exercise does change the rates at which metabolic enzymes are synthesized, leading to changes in the proportion of oxidative and glycolytic fibers within a muscle.

The maximum force generated by a muscle decreases by 30 to 40 percent between the ages of 30 and 80. This decrease in tension-generating capacity is due primarily to a decrease in average fiber diameter. In addition, the ability of a muscle to adapt to exercise decreases with age. The same intensity and duration of exercise in an older individual will not produce the same amount of change (adaptation) as in a younger person. This decreased ability to adapt to increased activity is seen in most organs as one ages (page 146). The precise cause is unknown but probably involves impairment of the mechanisms by which genes are transcribed and translated into proteins. Some of the change is simply the result of diminishing physical activity.

Lever Action of Muscles and Bones

A contracting muscle exerts a force on bones through its connecting tendons. When the force is great enough, the bone moves as the muscle shortens. A contracting muscle exerts only a pulling force, so that as the muscle shortens, the bones to which it is attached are pulled toward each other. **Flexion** refers to the bending of a limb at a joint, thereby decreasing the angle around the joint, whereas **extension** is the straightening of a limb —increasing the angle around a joint (Figure 11-33). These opposing motions require at least two muscles, one to cause flexion and the other extension. Groups of muscles that produce oppositely directed movements at a joint are known as **antagonists**. From Figure 11-33 it can be seen that contraction of the biceps causes flexion of the arm at the elbow, whereas contraction of the an-

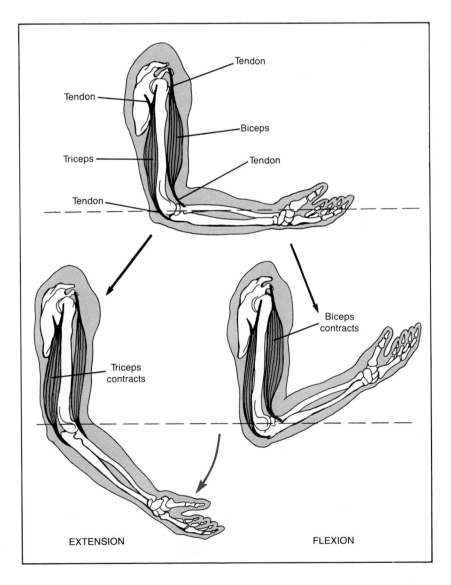

FIGURE 11-33 Antagonistic muscles for flexion and extension of the forearm.

tagonistic muscle, the triceps, causes the arm to extend. Both muscles exert only a pulling force upon the forearm when they contract.

Sets of antagonistic muscles are required not only for flexion-extension, but for side-to-side movements or rotation of a limb. The contraction of some muscles leads to two types of limb movement depending on the contractile state of other muscles acting on the same limb. For example, contraction of the gastrocnemius muscle in the leg causes a flexion of the leg at the knee, as in walking (Figure 11-34). However, contraction of the gastrocnemius muscle with the simultaneous contraction of the quadriceps femoris, which causes extension of the lower leg, prevents the knee joint from bending, leaving only the ankle joint capable of moving. The foot is extended, and the body rises on tiptoe.

Extensive exercise by an individual whose muscles have not adapted to the particular type of exercise leads to muscle soreness the next day. This soreness is the result of a mild inflammation of the muscle (Chapter 19), which occurs whenever tissues are damaged. The most severe inflammation occurs following a period of lengthening contractions, suggesting that the lengthening of a muscle fiber by an external force produces greater muscle damage than do either isotonic or isometric contractions. Thus, exercising by gradually lowering weights will produce greater muscle soreness than an equivalent amount of weight lifting.

The muscles, bones, and joints in the body are arranged in lever systems. The basic principle of a lever is illustrated by the flexion of the arm by the biceps muscle (Figure 11-35), which exerts an upward pulling force on the forearm about 5 cm away from the elbow. In this example, a 10-kg weight held in the hand exerts a downward force of 10 kg about 35 cm from the elbow. A law of physics tells us that the forearm is in mechanical equilibrium (no net forces acting on the system) when the product of the downward force (10 kg) and its distance from the elbow (35 cm) is equal to the product of the isometric tension exerted by the muscle (X), and its distance from the elbow (5 cm); that is, $10 \times 35 = 5 \times X$. Thus $X = 70$ kg. Note that this system is working at a mechanical disadvantage since the force exerted by the muscle (70 kg) is considerably greater than the load (10 kg) it is supporting. However, the mechanical disadvantage under which most muscle lever systems operate is offset by increased maneuverability. In Figure 11-36, when the biceps shortens 1 cm, the hand moves through a distance of 7 cm. Since the muscle shortens 1 cm in the same amount of time that the hand moves 7 cm, the velocity at which the hand moves is seven times greater than the rate of muscle shortening. The lever system amplifies the velocity of muscle shortening so that short, relatively slow movements of the muscle produce faster movements of the hand. Thus, a pitcher can throw a baseball at 90 to 100 mi/h even though his muscles shorten at only a fraction of this velocity.

Skeletal-Muscle Disease

A number of diseases can affect the contraction of skeletal muscle. Many of them are due to defects in the parts of the nervous system that control the contraction of the muscle fibers, however, rather than to defects in the muscle fibers themselves. For example, polio is a viral disease that destroys motor neurons, leading to the paralysis of skeletal muscle, and may result in death due to respiratory failure.

FIGURE 11-34 Depending on the activity of the quadriceps femoris muscle, either flexion of the leg or extension of the foot follows contraction of the gastrocnemius muscle.

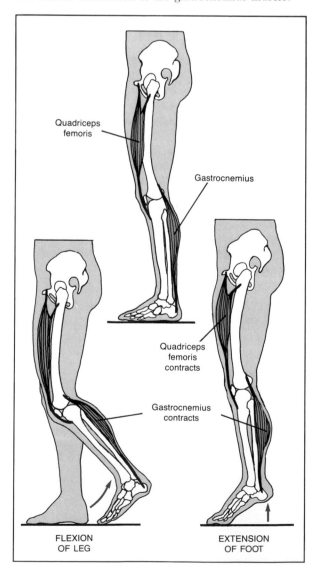

Quadriceps femoris

Gastrocnemius

Quadriceps femoris contracts

Gastrocnemius contracts

FLEXION OF LEG

EXTENSION OF FOOT

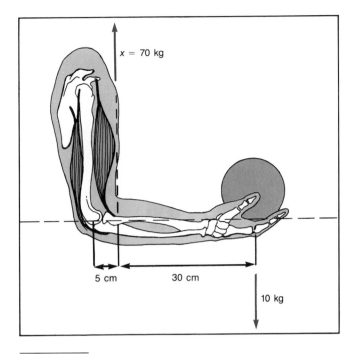

FIGURE 11-35 Mechanical equilibrium of forces acting on the forearm while supporting a 10-kg load.

FIGURE 11-36 Velocity of the biceps muscle is amplified by the lever system of the arm, producing a larger velocity of the hand. The range of movement is also amplified (1 cm of shortening by the muscle produces 7 cm of movement by the hand).

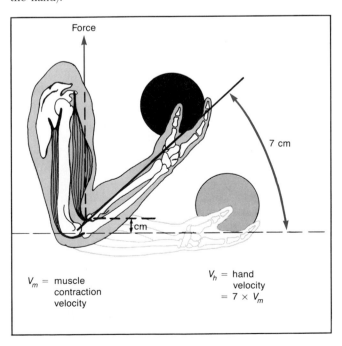

Muscle cramps. Involuntary tetanic contraction of skeletal muscles produces **muscle cramps**. During cramping, action potentials fire at rates as high as 300/s, a much greater rate than occurs during maximal voluntary contraction. The specific cause of this high activity is unknown but appears to be related to electrolyte imbalances in the extracellular fluid surrounding both the muscle and nerve fibers and changes in extracellular osmolarity, especially hyposmolarity.

Hypocalcemic tetany. Similar in symptoms to muscular cramping is the involuntary tetanic contraction of skeletal muscles that occurs when the extracellular calcium concentration falls to about 40 percent of its normal value. This may seem surprising since we have seen that calcium is required for muscle contraction. The effect of altered extracellular calcium is exerted not on the intracellular proteins, however, but on the plasma membrane. Low extracellular calcium increases the opening of sodium channels in excitable membranes, leading to membrane depolarization and the spontaneous firing of action potentials. The mechanisms controlling the extracellular concentration of calcium ions are discussed in Chapter 15.

Muscular dystrophy. This disease is one of the most frequently encountered genetic diseases, affecting one in every 4000 boys born in America. **Muscular dystrophy** is a degenerative disease that leads to the progressive wasting and loss of force-generating capacity in skeletal muscles. The symptoms become evident between 2 to 6 years of age, and most affected individuals do not survive much beyond the age of 20. The specific cellular defect responsible for muscular dystrophy is unknown, although the defective gene is known to be located on the X chromosome, (Chapter 18), which accounts for its expression in boys rather than girls, (with a few exceptions).

Myasthenia gravis. This condition, characterized by muscle fatigue and weakness that progressively worsens as the muscle is used, affects about 12,000 Americans. The symptoms result from a decrease in the number of ACh receptors on the motor end plate. The release of ACh from the nerve terminals is normal, but the magnitude of the muscle end-plate potential is markedly reduced because of the decreased number of receptor sites. As described in Chapter 19, the destruction of the ACh receptors is brought about by the body's own defense mechanisms gone awry, specifically because of the formation of antibodies to the ACh receptor proteins.

SMOOTH MUSCLE

Having described the properties and control of skeletal muscle, we now examine the characteristics of the second of the three types of muscle found in the body—smooth muscle. Two characteristics are common to all smooth muscles: They lack the cross-striated banding pattern found in skeletal and cardiac fibers (hence the name "smooth muscle"), and the nerves to them are derived from the autonomic division of the nervous system rather than the somatic division. Thus, smooth muscle is not under direct voluntary control.

Smooth muscle, like skeletal muscle, uses crossbridge movements between actin and myosin filaments to generate force and calcium ions to control crossbridge activity. However, the organization of the contractile filaments and the process of excitation-contraction coupling are quite different in these two types of muscle. Furthermore, there is considerable diversity among various smooth muscles with respect to the mechanism of excitation-contraction coupling.

Smooth-Muscle Structure

Each smooth-muscle fiber is a spindle-shaped cell with a diameter ranging from 2 to 10 μm, as compared to a range of 10 to 100 μm for skeletal-muscle fibers (Figure 11-3). Unlike skeletal-muscle fibers, the precursor cells of smooth-muscle fibers do not fuse during embryological development. In addition, skeletal-muscle fibers are multinucleate cells that are unable to divide, whereas smooth-muscle fibers contain only a single nucleus and are able to undergo cell division.

Two types of filaments are present in the cytoplasm of smooth-muscle fibers (Figure 11-37): thick myosin-containing filaments and thin actin-containing filaments. The latter are anchored either to the plasma membrane or to cytoplasmic structures known as **dense bodies**, which have characteristics similar to the Z lines in skeletal-muscle fibers. Note that the axis of the filaments is oriented slightly diagonally to the long axis of the cell. When the fiber shortens, the regions of the plasma membrane between the points where actin is attached to the membrane balloon out. The thick and thin filaments are not organized into myofibrils, as is the case in striated muscles, and there is no regular alignment of these filaments into sarcomeres, which accounts for the absence of a banding pattern in smooth-muscle fibers (Figure 11-38).

The amount of myosin present in smooth muscle is only about one-third that in striated muscle, whereas the actin content can be as much as twice as great. In spite of these differences, the maximal tension per unit of cross-

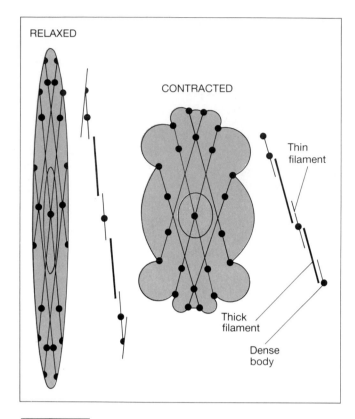

FIGURE 11-37 Arrangement of thick and thin filaments in a smooth-muscle fiber at rest and in contracted state.

sectional area developed by smooth muscles is similar to that developed by skeletal muscle.

The tension developed by smooth-muscle fibers varies with muscle length in a manner qualitatively similar to skeletal muscle. There is an optimal length at which tension development is maximal, and less tension is generated at lengths shorter or longer than this optimal length. The range of muscle lengths over which smooth muscle is able to develop tension is greater than it is in skeletal muscle, however. This property is highly adaptive since most smooth muscle surrounds hollow organs that undergo changes in volume with accompanying changes in the lengths of the smooth-muscle fibers in their walls. Even with relatively large increases in volume, as during the accumulation of large amounts of urine in the bladder, the smooth-muscle fibers in the wall retain some ability to develop tension, whereas such distortion would have stretched skeletal-muscle fibers beyond the point of thick- and thin-filament overlap.

The presence of actin and myosin, overlapping thick and thin filaments with cross bridges, and a length-tension relationship provide evidence that smooth-muscle contraction occurs by a sliding-filament mechanism similar to that in skeletal and cardiac muscle.

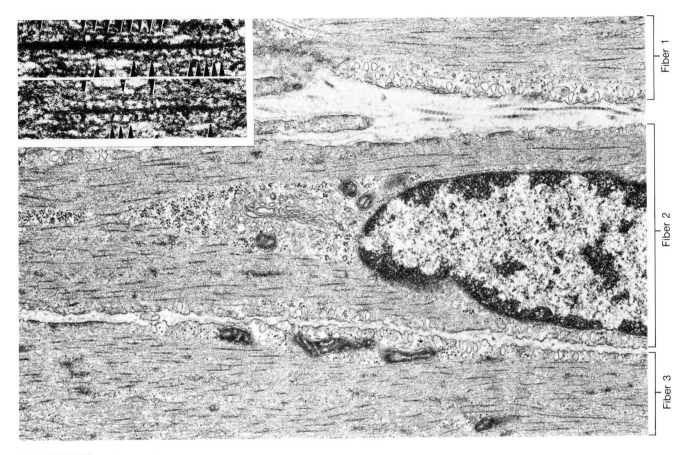

FIGURE 11-38 Electron micrograph of portions of three smooth-muscle fibers. Higher magnification of thick filaments (insert) with arrows indicating cross bridges connecting adjacent thin filaments. [*From A. P. Somlyo, C. E. Devine, Avril V. Somlyo, and R. V. Rice, Phil. Trans. R. Soc. Lond. B, 265:223–229, (1973).*]

Excitation-Contraction Coupling

Changes in cytosolic calcium concentration control the contractile activity in smooth-muscle fibers as in striated muscle. However, there are significant differences between the two types of muscle in the way in which calcium exerts its effects on cross-bridge activity and in the mechanisms by which the muscle fiber controls the cytosolic concentration of calcium.

The thin filaments in smooth muscle do not have the calcium-binding protein troponin that is present in both skeletal and cardiac muscle. In smooth muscle, calcium controls cross-bridge activity by regulating the enzymatic phosphorylation of myosin, an event necessary for smooth-muscle cross bridges to bind to the actin in thin filaments. The following sequence of events occurs after a rise in cytosolic calcium in a smooth-muscle fiber (Figure 11-39): Calcium binds to calmodulin, a calcium-binding protein that is present in most cells (Chapter 7) and whose structure is related to troponin; the calcium-calmodulin complex binds to a protein kinase, **myosin**

light-chain kinase, thereby activating the enzyme; the active protein kinase then uses ATP to phosphorylate myosin. The myosin isozymes in smooth muscle are different from those in striated muscle, and only the phosphorylated form of smooth-muscle myosin is able to bind to actin and undergo cross-bridge cycling. Hence, cross-bridge activity in smooth muscle is turned on by calcium-mediated changes in the thick filaments, whereas in skeletal muscle, calcium mediates changes in the thin filaments.

The smooth-muscle myosin isozyme also has a very low maximal rate of ATPase activity, on the order of 10 to 100 times less than that of skeletal-muscle myosin. Since the rate of ATP splitting determines the rate of cross-bridge cycling and thus shortening, the velocity of smooth-muscle shortening is much slower than that of skeletal muscle. The overall low rate of ATP utilization may account for the fact that smooth muscle does not undergo fatigue during prolonged periods of activity.

In addition, if the cytosolic calcium concentration

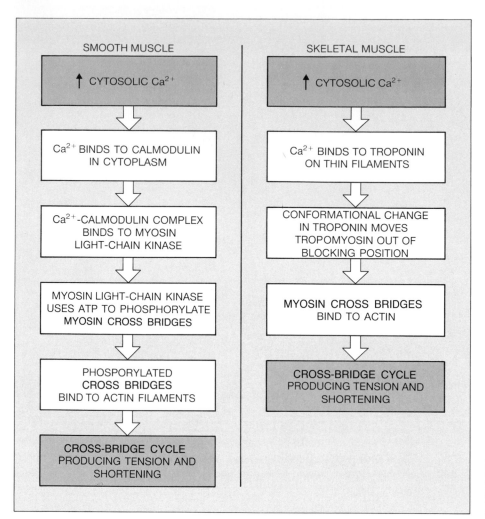

FIGURE 11-39 Pathways leading from increased cytoplasmic calcium to cross-bridge cycling in smooth- and skeletal-muscle fibers.

remains elevated, the rate of ATP splitting by the cross bridges declines even though tension is maintained. The cross bridges enter a state similar to rigor, in which they are attached to the thin filaments, maintaining tension, but are cycling at a very slow rate. This provides a very economical means of maintaining tension since it uses only small amounts of ATP.

To relax a contracted smooth muscle, the phosphorylated myosin must be dephosphorylated by a phosphatase enzyme. This enzyme is always present in its active form, even during contraction. During the rise in tension, the rate of myosin phosphorylation by the activated kinase exceeds the rate of dephosphorylation, and the amount of phosphorylated myosin in the cell increases. Lowering the cytosolic calcium relaxes the muscle by turning off the myosin light-chain kinase. Under these conditions, the rate of dephosphorylation will exceed the rate of phosphorylation, and the amount of phosphorylated myosin will decrease, producing relaxation.

Two sources of calcium contribute to the rise in cytosolic calcium that initiates smooth-muscle contraction: (1) the sarcoplasmic reticulum and (2) extracellular calcium. The amount of calcium contributed by each of these two sources differs among various smooth muscles, some being more dependent on extracellular calcium than the internal stores of calcium in the sarcoplasmic reticulum and vice versa.

Let us look first at the sarcoplasmic reticulum. The total quantity of this organelle in smooth muscle is smaller than in skeletal muscle, and it is not arranged in any specific pattern in relation to the thick and thin filaments. There are no transverse tubules connected to the plasma membrane in smooth muscle. The small fiber diameter and the slow rate of contraction do not require a rapid means of getting an excitatory signal into the muscle fiber in contrast to the case in skeletal muscle. Portions of the sarcoplasmic reticulum are located near the plasma membrane, forming associations similar to those between t tubules and the lateral sacs in skeletal muscle. Action potentials in the plasma membrane may be coupled to the release of calcium from those portions of the sarcoplasmic reticulum in close association with the plasma membrane. In addition, second messengers released in response to the binding of chemical messengers to plasma-membrane receptors can trigger the release of calcium from the sarcoplasmic reticulum.

What about extracellular calcium in excitation-contraction coupling? Since the concentration of calcium in the extracellular fluid is 10,000 times greater than cytosolic calcium concentration (page 108), the opening of calcium channels in the plasma membrane results in an increased flow of calcium into the cell. There are both voltage-sensitive and receptor-operated calcium channels in the plasma membrane of smooth muscles.

Removal of calcium from the cytosol, to bring about relaxation, is achieved by the active transport of calcium back into the sarcoplasmic reticulum as well as out of the cell across the plasma membrane. The rate of calcium removal in smooth muscle is much slower than in skeletal muscle, with the result that a single twitch lasts only a fraction of a second in skeletal muscle but may last several seconds in smooth muscle.

Unlike skeletal muscle, in which a single action potential releases sufficient calcium to turn on all the cross bridges in a fiber, only a portion of the cross bridges are activated in a smooth muscle fiber in response to most stimuli. Therefore, the tension generated by smooth muscle can be graded by varying the magnitude of the changes in cytosolic calcium concentration. The greater the increase in calcium concentration, the greater the number of cross bridges activated and the greater the tension.

In some smooth muscles, the cytosolic calcium concentration is sufficient to maintain a low level of cross-bridge activity in the absence of external stimuli. This low level of contractile activity is known as **tone**. The intensity of this tonic activity can be varied by factors that alter the cytosolic calcium concentration. Increasing the cytosolic calcium levels will cause an increase in tension on top of ongoing tonic activity.

In a sense, we have approached the question of excitation-contraction coupling in smooth muscle backward by first describing the coupling (the changes in cytosolic calcium), but now we must ask what constitutes the excitation that elicits these changes in calcium concentration.

In contrast to skeletal muscle, in which contractile activity is dependent on a single input—the somatic neurons to the muscle—many inputs affect the contractile activity of smooth muscle fibers (Table 11-5). All these inputs have the ultimate effect of increasing cytosolic calcium concentration, either by releasing calcium from

TABLE 11-5 INPUTS INFLUENCING SMOOTH-MUSCLE CONTRACTILE ACTIVITY
1. Spontaneous electrical activity in the fiber plasma membrane
2. Neurotransmitters released by autonomic neurons
3. Hormones
4. Locally induced changes in the chemical composition (paracrines, acidity, oxygen, osmolarity, and ion concentrations) of the extracellular fluid surrounding fiber
5. Stretch

the sarcoplasmic reticulum or by increasing the permeability of the plasma membrane to extracellular calcium by opening voltage-sensitive or receptor-operated calcium channels.

Note that we have said nothing thus far of action potentials in smooth-muscle membranes. Although most smooth muscles are capable of producing action potentials, some are not. In those smooth muscles in which action potentials occur, calcium ions, rather than sodium ions, carry positive charge into the cell during the rising phase of the action potential, that is, depolarization of the membrane opens voltage-sensitive calcium channels, producing calcium action potentials rather than sodium action potentials.

Spontaneous electrical activity. Some types of smooth-muscle fibers generate action potentials spontaneously in the absence of any neural or hormonal input. The plasma membrane in these fibers does not maintain a constant resting potential. Instead, it gradually depolarizes until it reaches the threshold potential and produces an action potential. Following repolarization, the membrane again begins to depolarize (Figure 11-40), so that a sequence of action potentials occurs, producing a tonic state of contractile activity. The potential change occurring during the spontaneous depolarization to threshold is known as a **pacemaker potential**. (Some cardiac-muscle fibers and a few neurons in the central nervous system also have pacemaker potentials and can spontaneously generate action potentials in the absence of external stimuli.)

Nerves and hormones. The contractile activity of smooth muscles is influenced by neurotransmitters released by autonomic nerve endings. Unlike skeletal muscle, smooth-muscle fibers do not have a specialized motor-end-plate region. As the axon of a postganglionic autonomic neuron enters the region of smooth-muscle fibers, it divides into numerous branches, each branch

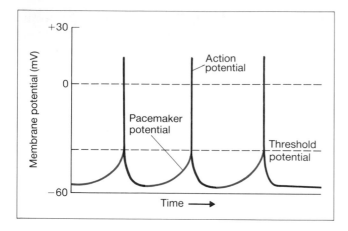

FIGURE 11-40 Generation of action potentials in a smooth-muscle fiber resulting from spontaneous depolarizations of the membrane (pacemaker potentials).

containing a series of swollen regions known as varicosities. Each varicosity contains numerous vesicles filled with neurotransmitter, which is released when an action potential conducted along the axon passes the varicosity. Several varicosities from a single axon may be located along several muscle fibers, and a single muscle fiber may be located near varicosities belonging to postganglionic fibers of both sympathetic and parasympathetic neurons (Figure 11-41). Therefore, a number of smooth-muscle fibers are influenced by neurotransmitter released by a single nerve fiber, and a single smooth-muscle fiber may be influenced by neurotransmitter from more than one neuron.

Whereas some neurotransmitters depolarize smooth-muscle membranes, others produce a hyperpolarization

that leads to a decreased cytosolic calcium concentration, producing a lessening of contractile activity. Thus, in contrast to skeletal muscle, which receives only excitatory input from its motor neurons, smooth-muscle tension can be either increased or decreased by neural activity.

Moreover, a given neurotransmitter may produce opposite effects in different smooth-muscle tissues. For example, norepinephrine, the neurotransmitter released from most postganglionic sympathetic neurons, depolarizes certain types of vascular smooth muscle, thereby initiating action potentials and contraction of the muscle. In contrast, sympathetic stimulation inhibits spontaneous activity in intestinal smooth muscle and produces a relaxation. Thus, the type of response (excitatory or inhibitory) depends on events initiated by the binding of the chemical messenger to receptor sites in the membrane.

In addition to receptor sites for neurotransmitters, the plasma membrane of smooth-muscle cells contains receptors for a variety of hormones. Binding of a hormone to its receptor may lead to a change in membrane potential and/or to the release of a second messenger. The result can be either increased or decreased contractive activity.

Although most changes in smooth-muscle contractile activity are accompanied by a change in membrane potential, this is not always the case. Second messengers, released in response to the binding of a chemical messenger to a receptor site, can cause the release of calcium from the sarcoplasmic reticulum, producing a contraction, without a change in membrane potential. Inositol triphosphate, which can release calcium from the endoplasmic reticulum of most cells (page 160), including smooth-muscle sarcoplasmic reticulum, is a prime candi-

FIGURE 11-41 Innervation of smooth muscle by postganglionic autonomic neurons. Neurotransmitter is released from the varicosities along the branched axons and diffuses to receptors on muscle fiber plasma membranes.

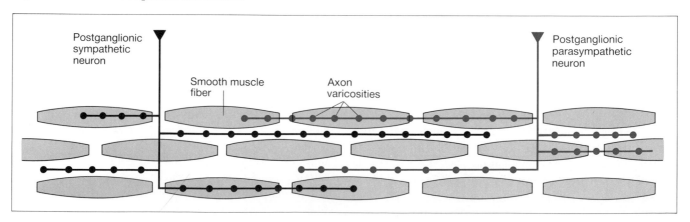

date for mediating such potential-independent contractions.

Local factors. Local factors, including paracrines, acidity, oxygen concentration, osmolarity, and the ion composition of the extracellular fluid can also alter smooth-muscle tension by affecting cytosolic calcium concentration. The specific mechanisms by which many of these local factors act on a smooth-muscle fiber to bring about changes in contraction are unknown.

Some smooth muscles respond by contracting when they are stretched. Stretching opens stretch-sensitive ion channels leading to membrane depolarization that induces contraction. The resulting contraction opposes the forces acting to stretch the muscle.

All these responses to local factors provide a means for altering smooth-muscle tension that can lead to local regulation independent of nerves and hormones.

Types of Smooth Muscle

The great diversity in the factors that can influence the contractile activity of smooth muscles from various organs has made it difficult to classify differences in smooth-muscle fibers. Many smooth muscles can be placed into one of two groups, however, based on the excitability characteristics of the muscle membrane and on the conduction of electrical activity from fiber to fiber

within the muscle tissue: (1) **single-unit smooth muscles**, whose membranes are capable of propagating action potentials from cell to cell and may manifest spontaneous action potentials and (2) **multiunit smooth muscles**, which exhibit little, if any, propagation of electrical activity from fiber to fiber and whose contractile activity is closely coupled to the neural activity to them.

It must be emphasized that most smooth muscles will not show all the characteristics of either single-unit or multiunit smooth muscles. A particular smooth muscle will, in general, have more of one class of characteristics than the other. These two prototypes should be considered to represent the two extremes in smooth-muscle characteristics, with many smooth muscles having characteristics that overlap the two groups.

Single-unit smooth muscle. The smooth muscles of the intestinal tract, the uterus, and small-diameter blood vessels are examples of single-unit smooth muscles. All the muscle fibers in a single-unit smooth muscle undergo synchronous activity, both electrical and mechanical. That is, the whole muscle responds to stimulation as a single unit. Each muscle fiber is linked to adjacent fibers by gap junctions through which small ions can move from cell to cell, carrying electric current. Thus, action potentials occurring in one cell are propagated by local currents through gap junctions into adjacent cells.

FIGURE 11-42 Innervation of a single-unit smooth muscle is often restricted to only a few fibers in the muscle. Electrical activity is conducted from fiber to fiber throughout the muscle by way of the gap junctions between the fibers.

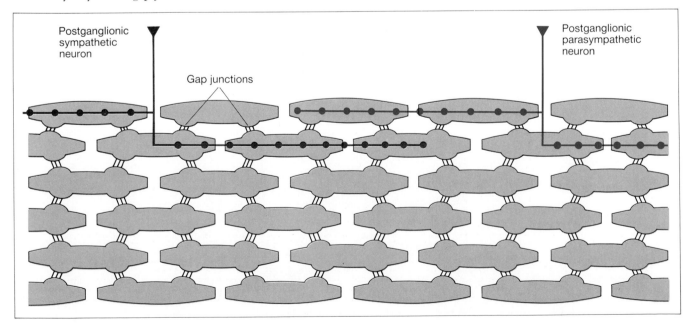

Therefore, electrical activity occurring anywhere within a group of single-unit smooth-muscle fibers can be conducted to all the other cells that are connected by gap junctions (Figure 11-42).

Some of the fibers in a single-unit muscle are pacemaker cells that spontaneously generate action potentials, which are conducted by way of gap junctions into fibers that do not spontaneously generate action potentials. The majority of cells in the muscle are nonpacemaker cells.

The contractile activity of single-unit smooth muscles can be altered by nerves, hormones, and local factors using the variety of mechanisms described previously for smooth muscles in general. The extent to which these muscles are innervated varies considerably in different organs. The nerve terminals are often restricted to the small regions of the muscle that contain pacemaker cells. By regulating the activity of the pacemaker cells, the activity of the entire muscle can be controlled.

One additional characteristic of single-unit smooth muscles is that a contractile response can often be induced by stretching the muscle. In several hollow organs—the uterus, for example—stretching the smooth muscles in the walls of the organ as a result of increases in the volume of material in the lumen initiates a contractile response that opposes the increase in volume.

Multiunit smooth muscle. The smooth muscle in the large airways to the lungs, in large arteries, and attached to the hairs in the skin are examples of multiunit smooth muscles. Because these smooth muscles have few gap junctions, each muscle fiber responds independently of its neighbors, and the muscle behaves as multiple units. Multiunit smooth muscles are richly innervated by branches of the autonomic nervous system. The contractile response of the whole muscle depends on the number of muscle fibers that are activated and on the frequency of nerve stimulation. Although stimulation of

TABLE 11-6 CHARACTERISTICS OF MUSCLE FIBERS

Characteristic	Skeletal Muscle	Smooth Muscle Single-Unit	Smooth Muscle Multiunit	Cardiac Muscle
Thick and thin filaments	Yes	Yes	Yes	Yes
Sarcomeres—banding pattern	Yes	No	No	Yes
Transverse tubules	Yes	No	No	Yes
Sarcoplasmic reticulum (SR)*	++++	+	+	++
Gap junctions between fibers	No	Yes	Few	Yes
Source of activating calcium	SR	SR and extracellular	SR and extracellular	SR and extracellular
Site of calcium regulation	Troponin	Myosin	Myosin	Troponin
Speed of contraction	Fast-slow	Very slow	Very slow	Slow
Spontaneous production of action potentials by pacemakers	No	Yes	No	Yes
Tone (low levels of maintained tension in the absence of external stimuli)	No	Yes	No	No
Effect of nerve stimulation	Excitation	Excitation or inhibition	Excitation or inhibition	Excitation or inhibition
Physiological effects of hormones on excitability and contraction	No	Yes	Yes	Yes
Stretch of fiber produces contraction	No	Yes	No	Yes

*Number of plus signs (+) indicates the relative amount of sarcoplasmic reticulum present in a given muscle type.

the nerve fibers to the muscle leads to a depolarization and a contractile response, action potentials do not occur in most multiunit smooth muscles. Circulating hormones can also increase or decrease contractile activity in multiunit smooth muscle, but stretching does not induce contraction in this type of muscle nor does the membrane spontaneously generate action potentials.

Table 11-6 compares the properties of the different types of muscle. Cardiac muscle has been included for completeness; its properties are discussed in Chapter 13.

SUMMARY

I. Three types of muscle, skeletal, smooth, and cardiac, are found in the body. Skeletal muscle is attached to bones and moves and supports the skeleton. Smooth muscle surrounds hollow cavities and tubes. Cardiac muscle is the muscle of the heart.

Structure of Skeletal Muscle and Muscle Fibers

I. Skeletal muscles, composed of cylindrical muscle fibers (cells), are linked to bones by tendons at each end of the muscle.

II. Skeletal-muscle fibers have a repeating, striated pattern of light and dark bands, due to the location of the thick and thin filaments within the myofibrils.

III. Actin-containing thin filaments are anchored to the Z lines at each end of a sarcomere, while their free ends partially overlap the myosin-containing thick filaments in the A band.

Molecular Mechanisms of Contraction

I. When a skeletal-muscle fiber actively shortens, the thin filaments are propelled toward the center of each sarcomere by movements of the myosin cross-bridges that bind to actin on the thin filaments.

A. The globular head of each myosin molecule contains a binding site for actin and an enzymatic site that splits ATP.

B. The four steps occuring during each cross-bridge cycle are summarized in Figure 11-12. The cross bridges undergo repeated cycles during a contraction, each producing only a small increment of movement.

C. The three functions of ATP in muscle contraction are summarized in Table 11-1.

II. In a resting muscle, attachment of cross bridges to actin is prevented by rod-shaped tropomyosin molecules that lie on the actin filaments and block the myosin-binding sites on actin.

III. Contraction is initiated by an increase in cytosolic calcium concentration. The calcium ions bind to troponin, producing a change in its shape that moves tropomyosin out of its blocking position, allowing cross bridges to bind to actin.

A. The rise in cytosolic calcium concentration is triggered by an action potential in the plasma membrane that is propagated into the interior of the fiber along the transverse tubules to the region of the sarcoplasmic reticulum, where it produces a release of calcium ions from the reticulum.

B. Relaxation of a contracting muscle fiber is achieved by actively transporting the cytosolic calcium ions back into the sarcoplasmic reticulum.

IV. The axon of a motor neuron forms a neuromuscular junction with many muscle fibers in a muscle. Each muscle fiber is innervated by only one motor neuron. A motor unit consists of a single motor neuron and the fibers it innervates.

A. Acetylcholine released by an action potential in a motor neuron binds to receptors on the motor end plate of the muscle membrane, opening ion channels that allow the passage of both sodium and potassium ions, and depolarizing the end-plate membrane.

B. A single action potential in a motor neuron is sufficient to produce a single action potential in a skeletal-muscle fiber.

V. Table 11-2 summarizes the molecular events leading to the contraction of a skeletal-muscle fiber.

Mechanics of Single-Fiber Contraction

I. Contraction refers to the turning on of the cross-bridge cycle. Whether there is an accompanying change in muscle length depends upon the external forces acting on the muscle.

II. Three types of contractions can occur following activation of a muscle fiber: an isometric contraction—the muscle generates tension but does not change length; an isotonic contraction—the muscle shortens, moving a load; and a lengthening contraction—the external load on the muscle is greater than the muscle tension, causing the muscle to lengthen during the period of contractile activity.

III. Increasing the frequency of action potentials in a muscle fiber increases the mechanical response (tension or shortening), up to the level of maximal tetanic tension.

IV. Maximum isometric tetanic tension is produced when there is a maximal overlap of thick and thin filaments, that is, at the optimal length l_o. Stretching a fiber beyond its optimal length decreases the filament overlap and decreases the tension produced, while decreasing the fiber length below l_o also decreases the tension generated due to interference with cross-bridge binding.

V. The velocity of muscle fiber shortening decreases with increases in load. Maximum velocity occurs at zero load.

Skeletal-Muscle Energy Metabolism

I. Muscle fibers form ATP by the transfer of phosphate from creatine phosphate to ADP, by oxidative phosphorylation of ADP in mitochondria, and by anaerobic substrate phosphorylation of ADP by the glycolytic pathway.

II. At the beginning of exercise, muscle glycogen is the major fuel consumed. As the exercise proceeds, glucose and fatty acids from the blood provide most of the fuel consumed, fatty acids becoming progressively more important during prolonged exercise. When the intensity of exercise exceeds about 70 percent of maximum, glycolysis begins to contribute an increasing fraction of the total ATP generated.

III. Muscle fatigue is caused by internal changes in acidity, glycogen depletion, and excitation-contraction coupling failure, not by a lack of ATP.

Types of Skeletal-Muscle Fibers

I. Three types of skeletal-muscle fibers—slow-oxidative, fast-oxidative, and fast-glycolytic fibers—can be distinguished based on their maximal shortening velocities and the predominate pathway used to form ATP.

    A. Differences in maximal shortening velocities are the result of different myosin isozymes with high or low ATPase activities, giving rise to fast and slow fibers.

    B. Fast-glycolytic fibers have a larger average diameter than oxidative fibers and therefore produce greater tension, but they also fatigue more rapidly.

II. All the muscle fibers in a single motor unit are the same fiber type, and most muscles contain all three fiber types.

III. Table 11-3 summarizes the characteristics of the three types of skeletal-muscle fibers.

Whole-Muscle Contraction

I. The tension produced by whole-muscle contraction depends on the amount of tension developed by each fiber and the number of active fibers in the muscle (Table 11-4).

II. Muscles that produce delicate movements have a small number of fibers per motor unit, whereas postural muscles have much larger motor units.

III. Fast-glycolytic motor units not only have large-diameter fibers but also tend to have large numbers of fibers per motor unit.

IV. Increases in muscle tension are controlled primarily by increasing the number of active motor units in a muscle—a process known as recruitment. Slow-oxidative motor units are recruited first during low levels of activity, then fast-oxidative motor units, and finally, fast-glycolytic motor units are recruited at high levels of activity.

V. Increasing motor unit recruitment increases the velocity at which a muscle will move a given load.

VI. The strength and susceptibility to fatigue of a muscle can be altered by exercise.

    A. Long-duration, low-intensity exercise increases a fiber's oxidative capacity for ATP production by increasing the number of mitochondria and blood flow to the muscle, resulting in increased endurance.

    B. Short-duration, high-intensity exercise increases fiber diameter due to increased synthesis of actin and myosin, resulting in increased strength.

VII. Movement around a joint requires two antagonistic groups of muscles: One flexes the limb at a joint and the other extends the limb.

VIII. The lever system of muscles requires muscle tensions far greater than the load in order to sustain a load in an isometric contraction. This disadvantage is offset because the lever system produces a shortening velocity at the end of the lever arm that is greater than the muscle-shortening velocity.

Smooth Muscle

I. Smooth-muscle fibers are spindle-shaped cells lacking striations, have a single nucleus, and are capable of cell division. They contain actin and myosin filaments, and contraction occurs by the sliding-filament mechanism.

II. An increase in cytosolic calcium leads to the binding of calcium by calmodulin. The calcium-calmodulin complex then binds to myosin light-chain kinase, activating the enzyme, which uses ATP to phosphorylate smooth-mucle myosin. In smooth muscle only phosphorylated myosin is able to bind to actin and undergo cross-bridge cycling.

III. Smooth-muscle myosin isozymes have a low rate of ATP splitting, and as a result the velocity of smooth-muscle contraction is much slower than that of striated muscles.

IV. Relaxation of smooth muscle occurs when the cytosolic calcium level is lowered by the active transport of calcium out of the cell across the plasma membrane or into the sarcoplasmic reticulum.

V. The two sources of the cytosolic calcium ions that initiate smooth-muscle contraction are the sarcoplasmic reticulum and extracellular calcium. The opening of both voltage-sensitive and receptor-operated calcium channels in the smooth-muscle plasma membrane allows extracellular calcium ions to enter the cell.

VI. The increase in cytosolic calcium resulting from most stimuli does not activate all the cross bridges in smooth muscle. Therefore smooth-muscle tension can be increased by agents that increase the concentration of cytosolic calcium ions.

VII. Table 11-5 summarizes the types of stimuli that can initiate smooth-muscle contraction.

VIII. Most, but not all, smooth-muscle cells can generate action potentials in their plasma membrane upon membrane depolarization. The rising phase of the smooth-muscle action potential is due to the influx of calcium ions into the cell through open calcium channels.

IX. Some smooth muscles generate action potentials spontaneously in the absence of any external input due to pacemaker potentials in the plasma membrane that repeatedly depolarize the membrane to threshold.

X. Smooth-muscle cells do not have a specialized end-plate region. A number of smooth-muscle fibers may be influenced by neurotransmitters released by a single nerve ending, and a single smooth-muscle fiber may be influenced by neurotransmitter from more than one neuron. Neurotransmitters may have either excitatory or inhibitory effects on smooth-muscle contraction.

XI. Smooth muscles are classified according to the degree to which they show the characteristics of single-unit or multiunit smooth muscle (Table 11-6).

REVIEW QUESTIONS

1. Define:

skeletal muscle	isometric contraction
smooth muscle	isotonic contraction
cardiac muscle	lengthening contraction
muscle fiber	twitch
muscle	latent period
tendons	contraction time
striated muscles	summation
myofibrils	tetanus
sarcomere	optimal length (l_o)
thick filaments	creatine phosphate
myosin	muscle fatigue
thin filaments	psychological fatigue
actin	fast fibers
A band	slow fibers
Z line	oxidative fibers
I band	myoglobin
H zone	red muscle
M line	glycolytic fibers
cross bridges	white muscle
contraction	slow-oxidative fibers
relaxation	fast-oxidative fibers
sliding-filament mechanism	fast-glycolytic fibers
cross-bridge cycle	recruitment
rigor mortis	denervation atrophy
troponin	disuse atrophy
tropomyosin	hypertrophy
excitation-contraction coupling	flexion
	extension
sarcoplasmic reticulum	antagonists
lateral sacs	muscle cramps
transverse tubule (t tubule)	hypocalcemic tetany
motor neurons	muscular dystrophy
motor unit	myasthenia gravis
motor end plate	dense bodies
neuromuscular junction	myosin light-chain kinase
end-plate potential (EPP)	tone
curare	pacemaker potential
tension	single-unit smooth muscles
load	multiunit smooth muscles

2. List the three types of muscle cells and where they are located.

3. Draw a diagram of the arrangement of thick and thin filaments in a striated-muscle sarcomere, and label the major bands that give rise to the striated pattern.

4. Describe the four steps of one cross-bridge cycle.

5. Describe the organization of myosin and actin molecules in the thick and thin filaments.

6. Describe the physical state of a muscle fiber in rigor mortis and the conditions that produce this state.

7. What three events in skeletal-muscle contraction and relaxation are dependent on ATP?

8. What prevents cross bridges from attaching to sites on the thin filaments in a resting skeletal muscle?

9. Describe the role of calcium ions in initiating contraction in skeletal muscle.

10. Describe the location, structure, and function of the sarcoplasmic reticulum in skeletal muscle fibers.

11. Describe the structure and function of the transverse tubules in skeletal muscle fibers.

12. Describe the events that result in the relaxation of skeletal muscle.

13. Define a motor unit and describe its structure.

14. Describe the sequence of events by which an action potential in a motor neuron produces an action potential in a skeletal muscle fiber plasma membrane.

15. What is an end-plate potential, and what ions are involved in producing it?

16. Compare and contrast the transmission of electrical activity at a neuromuscular junction with that at a synaptic junction.

17. Describe the characteristics of isometric, isotonic, and lengthening contractions.

18. What factors determine the duration of an isotonic twitch in skeletal muscle? An isometric twitch?

19. What effect does increasing the frequency of skeletal muscle fiber action potentials have upon the force of contraction? Explain the mechanism responsible for this effect.

20. Describe the length-tension relationship in striated muscle fibers.

21. Describe the effect of increasing the load on a skeletal muscle fiber on the velocity of shortening.

22. What is the function of creatine phosphate in skeletal muscle contraction?

23. What fuel molecules are metabolized to produce ATP during skeletal muscle activity?

24. What factors are responsible for skeletal muscle fatigue?

25. What component of skeletal muscle fibers accounts for differences in maximal shortening velocities between fibers?

26. Summarize the characteristics of the three types of skeletal-muscle fibers.

27. Upon what two factors does the amount of tension developed by a whole skeletal muscle depend?

28. Describe the process of motor unit recruitment in controlling: (a) whole-muscle tension and (b) velocity of whole-muscle shortening.

29. During voluntary increases in the force of skeletal muscle contraction, what is the order of recruitment of different types of motor units?

30. What happens to skeletal-muscle fibers when the motor neuron to the muscle is destroyed?

31. Describe the changes that occur in skeletal muscles following a period of: (a) long-duration, low-intensity exercise training; (b) of short-duration, high-intensity exercise training.

32. How are skeletal muscles arranged around joints so that a limb can push or pull?

33. What are the advantages and disadvantages of the muscle-bone-joint-lever system?

34. How does the organization of thick and thin filaments in smooth-muscle fibers differ from that in striated-muscle fibers?

35. Compare the mechanisms by which a rise in cytosolic calcium initiates contractile activity in skeletal- and smooth-muscle fibers.

36. What are the two sources of calcium that can lead to the increase in cytosolic calcium that triggers contraction in smooth muscle?

37. What types of stimuli can trigger a rise in cytosolic calcium in smooth-muscle fibers?

38. What effect does a pacemaker potential have on a smooth-muscle cell?

39. In what ways does the neural control of smooth-muscle contractile activity differ from the neural control of skeletal-muscle activity?

40. Describe how a stimulus may lead to the contraction of a smooth-muscle cell without a change in the plasma membrane potential.

41. Describe the distinguishing characteristics of single-unit smooth muscles and multiunit smooth muscles.

THOUGHT QUESTIONS

(Answers are given in Appendix A.)

1. Myosin will be in which of the following states: $M \cdot ATP$; $M^* \cdot ADP \cdot P_i$; $AM \cdot ADP \cdot P_i$; $A \cdot M$
   (a) under resting conditions?
   (b) in rigor mortis?

2. If the transverse tubules of a skeletal muscle were disconnected from the plasma membrane, would action potentials trigger a contraction? Give reasons.

3. If a small load is attached to a muscle that is then tetanically stimulated, the muscle will lift the load in an isotonic contraction over a certain distance and then stop shortening and enter a state of isometric contraction. If a heavier load is used, the distance shortened before entering an isometric contraction is shorter. Explain these shortening limits in terms of the sliding-filament model of contraction and the length-tension relation.

4. What conditions would produce the maximum tension from any given skeletal muscle fiber?

5. A skeletal muscle can often maintain a moderate level of active tension for long periods of time, even though many of its fibers become fatigued. Explain.

6. If the blood flow to a skeletal muscle were markedly decreased, which types of motor units would have their ability to produce ATP for muscle contraction severely reduced? Why?

7. As a result of an automobile accident, 50 percent of the muscle fibers in the biceps muscle of a patient were destroyed. Ten months later, the biceps muscle was able to generate 80 percent of its original force. Describe the changes that took place in the damaged muscle that enabled it to recover.

8. If the extracellular fluid surrounding a skeletal muscle contains no calcium ions, will the muscle contract when it is stimulated (1) directly by depolarizing its membrane or (2) by stimulating the nerve to the muscle? What would happen if this were a smooth muscle?

9. The following experiments were performed on a single-unit smooth muscle in the gastrointestinal tract:
   (a) Stimulating the parasympathetic nerves to the muscle produced a contraction.
   (b) Injecting a nonmetabolized form of acetylcholine into the blood vessels to the muscle produced a contraction.
   (c) Applying a drug that blocks the voltage-sensitive sodium channels in most plasma membranes led to a failure to contract upon stimulating the parasympathetic nerves but did not prevent the muscle from contracting in response to the nonmetabolized form of acetylcholine.
   (d) Applying a drug that binds to muscarinic receptors, Chapter 8, page 182, blocking the action of ACh at these receptors, prevented the smooth muscle from contracting when the nonmetabolized form of ACh was injected but did not prevent the muscle from contracting when the parasympathetic nerve was stimulated.

   From these observations, what might one conclude about the mechanism by which parasympathetic nerve stimulation is producing a contraction of the smooth muscle?

CONTROL
OF BODY
MOVEMENT

Carrying out a coordinated movement is a complicated process involving nerves, muscles, and bones. Consider the events associated with reaching out and grasping an object. The fingers are first *extended* (straightened) to reach around the object and then *flexed* (bent) to grasp it, the degree of extension depending upon the size of the object (Is it a golf ball or a soccer ball?) and the force of flexion depending upon its weight and consistency (A bowling ball or a balloon?). Simultaneously, the wrist, elbow, and shoulder are extended, and the trunk is inclined forward, the exact movements depending upon the object's position relative to the person doing the lifting. The shoulder, elbow, and wrist are also stabilized to support first the weight of the arm and hand and then the added weight of the object. Through all this, upright posture is maintained despite the body's continuously shifting center of gravity.

The building blocks for this simple action—as for all movements—are **motor units,** each comprising one motor neuron together with all the skeletal muscle fibers that neuron innervates (motor unit, page 296). It is through these motor units that all skeletal muscle contraction in the body is controlled.

All the motor neurons for a given muscle make up the **motor neuron pool** for the muscle. While all neurons in a pool can be activated at one time, as when trying to lift a very heavy object, this leads to a coarse movement. The more usual case is for only some of the neurons in a pool to be activated in order to carry out a movement. This partial activation gives much finer degrees of control, as when lacing and tying a shoe. Page 309 describes how motor neurons are recruited from a given pool.

Many neurons synapse on the motor neurons to control their activity. No one input source to a motor unit is

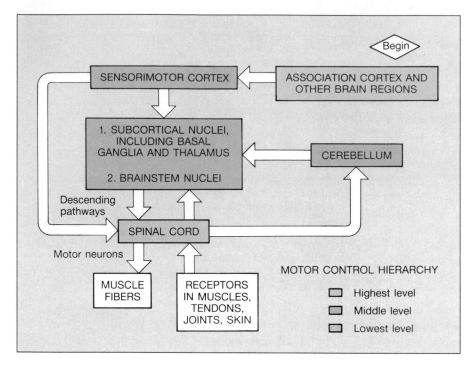

FIGURE 12-1 The hierarchical organization of the neural system controlling body movement. All the skeletal muscles of the body are controlled by motor neurons.

essential for movement, but a balanced input from all sources is necessary to provide the precision and speed of normally coordinated actions. For example, if an inhibitory system synapsing on the neuron of a given motor unit is damaged so that inhibitory input is decreased, the still-normal excitatory input to the unit will be unopposed, the motor neuron will fire excessively, and the muscle it controls will be hyperactive. This, you will recall from page 206, is what happens in the disease tetanus (lockjaw).

Each of the myriad coordinated body movements a person is capable of is achieved by many motor units from various motor pools activated in a precise temporal order. This chapter deals with the interrelating neural inputs that converge upon the neurons of these units to control the neuron's activity. We present first a summary of a model of motor-system functioning (Figure 12-1 and Table 12-1) and then describe each component of the model in detail.

Throughout the central nervous system, the neurons that are involved in controlling movement are organized in a hierarchical fashion—the **motor control hierarchy**—each level having a limited task in motor control. It is important to realize that we are speaking of a *functional* hierarchy, not an anatomical one. For example, parts of the cerebral cortex play roles in two levels of the hierarchy.

To begin a movement, a general "intention" such as

TABLE 12-1	MOTOR CONTROL HIERARCHY FOR VOLUNTARY MOVEMENTS

I. The highest level
   A. Structures: Areas involved with memory and emotion, supplementary motor area, and association cortex, which, in turn, receives and correlates input from many other brain structures.
   B. Function: forms complex plans according to person's intention.

II. The middle level
   A. Structures: sensorimotor cortex, cerebellum, parts of basal ganglia, some brainstem nuclei.
   B. Function: converts complex plans to a number of smaller motor programs, which determine the pattern of neural activation required to perform the movement. These programs are broken down into subprograms that determine the movements of individual joints. These programs and subprograms are transmitted via the descending pathways to the lowest control level.

III. The lowest level
   A. Structures: levels of brainstem or spinal cord from which motor neurons exit.
   B. Function: specifies tension of particular muscles and angle of specific joints necessary to carry out the programs and subprograms transmitted from the middle control levels.

"pick up sweater" or "write signature" or "answer telephone" is generated at the highest level of the hierarchy. This highest level encompasses many regions of the brain, including those involved in memory and emotions and areas of association cortex.[1]

Information[2] is relayed from the "command" neurons of the highest hierarchical level to parts of the brain that make up the middle level of the motor control hierarchy. These structures specify the postures and movements needed to carry out the desired action. In our example of picking up a sweater, structures of the middle hierarchical level coordinate the commands that tilt the body and extend the arm and hand toward the sweater and shift the body's weight so balance is maintained. The middle hierarchical structures are located in parts of the cerebral cortex, notably sensorimotor cortex (page 333), cerebellum, basal ganglia, and brainstem (Figure 12-2A and B).

As the neurons in the middle level of the hierarchy are receiving input from the "command" neurons, they simultaneously receive (afferent) information from re-

ceptors in the muscles, tendons, joints, skin, vestibular apparatus, and eyes about the starting position of the body part that is to be "commanded" to move. This afferent information is integrated by neurons of the middle level of the hierarchy with the signals from the "command" neurons, and the middle level of the hierarchy creates the **motor program,** that is, the pattern of neural activity required to perform the desired movement.

The information determined by that motor program is then transmitted to the lowest level of the motor control hierarchy, the local level, at which the motor neurons to the muscles exit the brainstem or spinal cord. It is this information that determines exactly which motor neurons will be activated.

The information is transmitted from the middle to the lowest level in two ways. The axons from some of the sensorimotor-cortex neurons descend directly to the level of the motor neurons. These direct descending pathways are called the corticospinal pathways. A second parallel set of descending pathways, known as the multineuronal pathways, also convey action potentials to the region of the motor neurons. Both pathways are involved simultaneously in virtually all actions, and both are described more fully later in the chapter.

After the initial motor program is implemented and the action has begun, those brain regions concerned with motor control continue to receive a constant stream

---

[1] It may be necessary for the student to review at this time brain anatomy (pages 173 to 177).

[2] Recall that all information in the nervous system is transmitted in the form of graded potentials or action potentials.

FIGURE 12-2 (A) Side view of brain showing three of the four components of the middle level of the motor control hierarchy. (B) Cross section of brain showing the parts of the basal ganglia, another important component of the hierarchy's middle level.

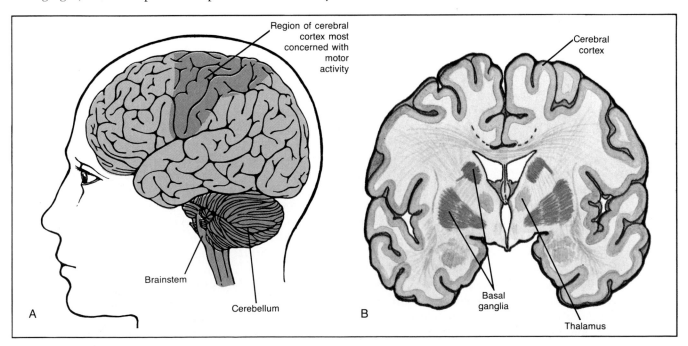

of updated information from receptors in the periphery about the movements taking place. Say, for example, that the sweater being picked up is wet and heavier than expected so that the initially determined amount of muscle contraction is not sufficient to lift it. Any discrepancies between the intended and actual movements are detected, program corrections are determined, and the corrections are relayed via the lowest level of the motor control hierarchy to its final output stage—the motor neurons. Thus, the motor program is continuously adjusted during the course of most movements.

When learning has taken place and a movement becomes "skilled," the initial information from the middle hierarchical level is more accurate and fewer corrections need be made. Also, movements performed at high speed without concern for fine control are made solely according to the initial motor program.

VOLUNTARY AND INVOLUNTARY ACTIONS

Given such a hierarchical model filled with information transmission, feedback, fine-tuning, and so on, it is difficult when speaking of body movement to use the phrase **voluntary movement** with any real precision. We shall use it, however, to refer to those actions that have the following characteristics: (1) The movement is accompanied by a conscious awareness of what we are doing and why we are doing it rather than the feeling that it "just happened," and (2) our attention is directed toward the action or its purpose.

The term "involuntary," on the other hand, describes actions that do not have these characteristics. "Unconscious," "automatic," and "reflex" are often taken to be synonyms for "involuntary."

Despite our attempts to distinguish between voluntary (conscious) and involuntary (unconscious) actions, almost all motor behavior involves both components, and the distinction between the two cannot be made easily. Even such a highly conscious act as threading a needle involves the unconscious postural support of the hand and arm and inhibition of the antagonistic muscles—those muscles whose activity would oppose the intended action, in this case, the muscles that straighten the fingers.

Most motor behavior is neither purely voluntary nor purely involuntary but falls at some point on a spectrum between these two extremes. Moreover, actions shift along the spectrum according to the frequency with which they are performed. When a person first learns to drive a car with a standard transmission, for example, stopping is a fairly complicated process and requires a great deal of conscious attention. With practice, the same actions become automatic, and a complicated pattern of muscle movements is shifted from the highly conscious end of the spectrum toward the involuntary end by the process of learning.

We now turn to an analysis of the individual components of the motor-control hierarchy, beginning with local control mechanisms because their activity serves as a base upon which the pathways descending from the brain exert their influence.

LOCAL CONTROL OF MOTOR NEURONS

In describing local control, we must add one more element to our motor-control system. So far we have the picture of the higher levels of the motor-control hierarchy sending messages down from the brain to the lowest hierarchical level—the level of local control—where the motor neurons are activated or inhibited. This is not the only path for information flow to the motor neurons, however, for afferent fibers in the muscles, tendons, joints, and skin of the body part to be moved enter the central nervous system and also transmit information not only to higher levels of the hierarchy, as noted earlier, but to the local level as well.

Such information from the periphery can influence body movements. For example, the pathways descending from the brain may be signaling that the motor neurons that control the extensor muscles of one leg are to fire so the body weight can be shifted to that leg. If nociceptors are simultaneously signaling that a sharp object is under that foot, however, these same motor neurons may be inhibited and the movement not made.

Interneurons

Most of the synaptic input to motor neurons arises not *directly* from the afferent neurons or the descending pathways emphasized in our hierarchical model but from interneurons, which are of several types. Some interneurons are confined to the general level of the motor neuron upon which they synapse and thus are called local interneurons. Others have processes that extend up or down short distances in the spinal cord and brainstem, whereas still others have very long processes that can extend throughout much of the length of the central nervous system. The interneurons with longer processes are important in movements that involve the coordinated interaction of, for example, an arm and a leg, whereas the interneurons with shorter processes are activated in movements that involve a smaller region of the body, for example, only a shoulder and arm.

The local interneurons are important integrative elements of the lowest level of the motor-control hierarchy.

Varied inputs converge upon them, not only from higher centers and peripheral receptors but from other local interneurons as well (Figure 12-3A). The functions of interneurons are complex. For example, an interneuron that receives excitatory input from descending pathways from the brain but that itself ends at inhibitory synapses on motor neurons serves to convert an excitatory influence into an inhibitory one (Figure 12-3B).

Interneurons can act as "switches" that enable a movement to be turned on or off under the command of higher motor centers. Thus, as we suggested earlier, reflexes do not necessarily have the obligatory nature we commonly associate with reflex behavior but rather are subject to permissive, modulatory influences from higher motor centers. For example, if we picked up a hot object, a reflex arc would be activated, resulting in our dropping the object. But if the object were something we had spent a great deal of time and trouble to prepare, we would continue to hold onto it until we could put it down safely. In other words, because of the convergence of local and descending inputs onto interneurons, descending commands are subject to local influence and vice versa.

Afferent Systems

Afferent fibers impinge upon the local interneurons, or in one case, directly on the motor neuron. These fibers bring information from receptors in three areas: (1) those in the very muscles controlled by the motor neurons, (2) those in other nearby muscles, and (3) those in the tendons, joints, and skin surrounding the muscle.

The receptors at the peripheral ends of these afferent fibers monitor the length and tension of the muscles and movement of the joints. This input provides negative-feedback control over muscle length and tension and is also transmitted by way of interneurons to motor areas of the brain, where, as we mentioned earlier, it is integrated with input from other types of receptors.

Length-monitoring systems and the stretch reflex. Muscle length and changes in length are monitored by **stretch receptors** embedded within the muscle. These receptors consist of endings of afferent nerve fibers that are wrapped around modified muscle fibers, several of which are enclosed in a connective-tissue capsule. The entire structure is called a **muscle spindle** (Figure 12-4A). The modified muscle fibers within the spindle are known as **spindle** (or **intrafusal**) **fibers,** whereas the skeletal muscle fibers that form the bulk of the muscle are the **skeletomotor** (or **extrafusal**) **fibers** (Figure 12-4B).

The muscle spindles are parallel to the skeletomotor fibers such that stretch of the muscle by an external force pulls on the spindle fibers, stretching them and activating their receptor endings (Figure 12-5A). The greater the stretch, the faster the receptors fire. In contrast, contraction of the skeletomotor fibers and the resultant shortening of the muscle slow down the rate of firing of the stretch receptor (Figure 12-5B). For example, when your arm is hanging loosely at your side, the biceps muscle (see Figure 11-33) and the muscle spindles within it are stretched by the weight of the lower arm. When you flex your arm, however, the biceps muscle shortens and the pull on the spindles is lessened.

Within a given spindle, there are two kinds of receptors: One responds best to how much the muscle has been stretched, the other to both the magnitude of the stretch and the speed with which it occurs. Although the two kinds of stretch receptors are separate entities, they will be referred to collectively as the **muscle-spindle stretch receptors.**

When the afferent fibers from the muscle spindle enter the central nervous system, they divide into

---

FIGURE 12-3 (A) Convergence of axons onto a local interneuron. (B) An inhibitory interneuron reverses the action of an excitatory descending pathway.

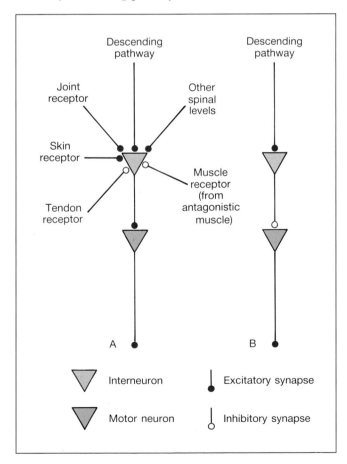

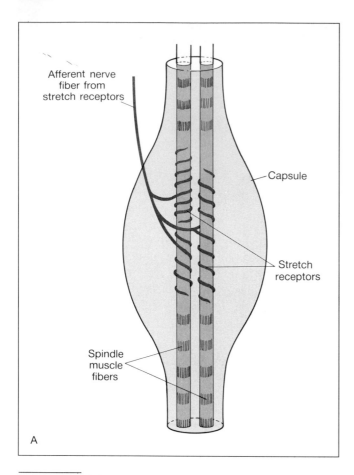

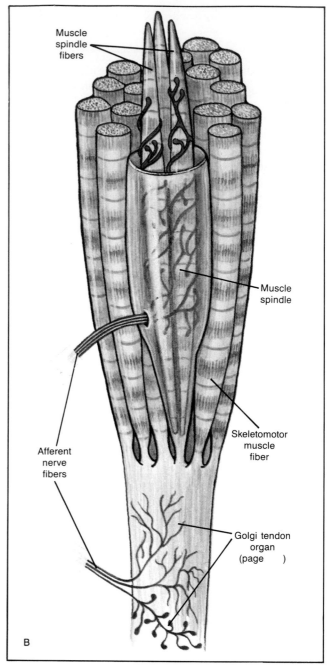

FIGURE 12-4 (A) Diagrammatic and (B) anatomic views of a muscle spindle. Note in B that the muscle spindle is parallel to the skeletomotor muscle fibers. [(B) *Adapted from Elias, Pauly, and Burns.*]

branches that can take several different paths. One path (A in Figure 12-6) directly stimulates motor neurons going back to the muscle that was stretched, thereby completing a reflex arc known as the **stretch reflex.**

This reflex is probably most familiar in the form of the **knee jerk,** part of a routine medical examination. The physician taps the patellar tendon (Figure 12-6), which passes over the knee and connects extensor muscles in the thigh to the tibia in the lower leg. As the tendon is pushed in by tapping it, the thigh muscles to which it is attached are stretched, and all the stretch receptors within these muscles are activated. Information about the change in muscle length is fed back by the afferent nerve fibers from the stretch receptors to the motor neurons that control these same muscles. The motor units are stimulated, the thigh muscles shorten, and the patient's lower leg is extended to give the knee jerk.

The proper performance of the knee jerk tells the physician that the stretch receptors, the afferent fibers, the balance of synaptic input to the motor neurons, the motor neurons themselves, the neuromuscular junctions, and the muscles are all functioning normally. In usual circumstances, of course, the stretch receptors are not all activated at the same time nor are they activated so strongly. Consequently, the response to the stretching that occurs all the time physiologically is a smooth movement rather than a jerk.

Because the afferent nerve fibers mediating the stretch reflex synapse directly with the motor neurons without the interposition of any interneurons, the stretch reflex is called **monosynaptic.** Stretch reflexes are the only known monosynaptic reflex arcs. All other

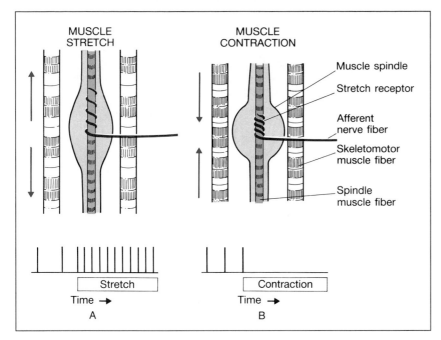

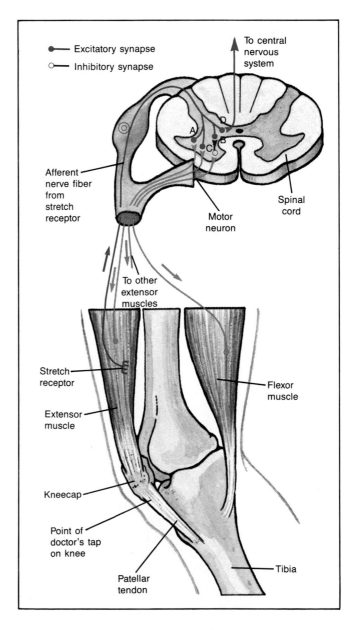

FIGURE 12-5 (A) Passive stretch of the muscle activates the spindle stretch receptors and causes an increased rate of action-potential firing in the afferent nerve. (B) Contraction of the muscle removes tension on the stretch receptors and lowers the rate of action-potential firing. Arrows indicate the direction of force on the muscle spindles.

reflex arcs are **polysynaptic,** having at least one inter-neuron, and usually many, between the afferent and efferent neurons.

A second path for the branches of the afferent nerve fibers from stretch receptors (B in Figure 12-6) ends on interneurons that, when activated, inhibit the motor neurons controlling antagonistic muscles whose contraction would interfere with the reflex response (in the knee jerk, for example, the flexor muscles of the knee are the antagonists). The activation of one muscle and the simultaneous inhibition of its antagonistic muscle is called **reciprocal inhibition,** which is characteristic of many movements, not just the stretch reflex.

A third path (C in Figure 12-6) ends on interneurons that, when activated, activate the motor neurons of **synergistic muscles,** that is, muscles whose contraction assists the intended motion (in our example, other leg extensor muscles). The muscles activated are on the same side of the body as are the receptors, and the response is therefore **ipsilateral** (a response on the opposite side of the body would be **contralateral**).

A fourth path (D in Figure 12-6) synapses with interneurons that convey information about the muscle length to areas of the brain dealing with motor control. Information from the muscle stretch receptors that reaches the cerebral cortex contributes to the conscious

FIGURE 12-6 Neural pathways involved in the knee jerk reflex. Hitting the patellar tendon stretches the extensor muscle causing (A and C) compensatory contraction of this and other extensor muscles, (B) relaxation of the flexor muscle, and (D) information about muscle length to be sent to the brain. Arrows indicate direction of action-potential propagation.

perception of the position of a limb or joint. This provides information about muscle action important during small, slow movements such as the performance of an unfamiliar action and for small adjustments in posture.

Alpha-gamma coactivation. As indicated in Figure 12-5B, stretch of the spindle fibers is decreased when the muscle shortens. If the spindle stretch receptors were permitted to go slack at this time, they would stop firing action potentials and there would be no indication of changes in muscle length the whole time the muscle was contracting. To prevent this loss of information, the two ends of each spindle muscle fiber are made to contract during the shortening of the skeletomotor fibers, thus maintaining tension in the central region of each spindle fiber, where the stretch receptors are located (Figure 12-7). Spindle fibers are not large or strong enough to shorten whole muscles and move joints. Their sole job is to maintain tension on the spindle stretch receptors.

FIGURE 12-7 As the muscles at the two ends of a spindle fiber contract in response to gamma motor neuron activation, they pull on the center of the fiber and stretch the receptor. The thin arrows beside the nerve fibers indicate the direction of action-potential propagation.

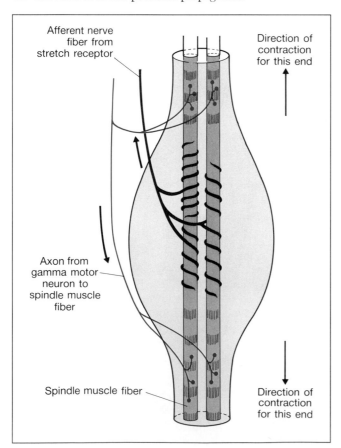

The spindle fibers contract only in response to activation by a motor neuron. The motor neurons that activate these fibers, however, are usually not those that activate the skeletomotor muscle fibers. The motor neurons controlling the skeletomotor muscle fibers are larger and are classified as **alpha motor neurons,** whereas the smaller motor neurons whose axons innervate the spindle fibers are known as **gamma motor neurons** (Figure 12-7).

Alpha and gamma motor neurons are both activated by interneurons in their immediate vicinity and by neurons of the descending pathways. In fact, they are often coactivated, that is, fired at almost the same time during many voluntary and involuntary movements. It seems, however, that during most voluntary actions, the gamma-efferent activity is of limited usefulness in invoking the stretch reflex and controlling alpha motor neuron firing since the reflex effect of the spindle afferents on the alpha motor neurons is relatively weak and few additional motor units are recruited in response to spindle stretch. Coactivation does ensure, however, that information about muscle length will be continuously available to the higher motor centers.

Tension-monitoring systems. A second component of the local motor-control apparatus monitors how much tension is being exerted by the contracting muscle or imposed on the muscle by external forces if the muscle is being stretched. At any given set of inputs to a group of motor neurons, the tension developed by a contracting muscle depends on various factors—muscle length, load on the muscle, and degree of muscle fatigue (page 308). Therefore, feedback is necessary to inform the motor-control systems of the tension actually achieved.

The receptors employed in the tension-monitoring system are the **Golgi tendon organs,** which are located in the tendons near their junction with the muscle (Figure 12-4B). Endings of afferent nerve fibers are wrapped around collagen bundles in the tendon, bundles that are slightly bowed in the resting state. When the attached skeletomotor muscle fibers contract, they pull on the tendon, and this action straightens the collagen bundles and distorts the receptor endings associated with them. Thus, the Golgi tendon organs discharge in response to the tension in the contracting muscle and initiate action potentials that are transmitted into the central nervous system.

The afferent neuron's firing activity supplies the motor-control systems (both locally and in the brain) with continuous information about the muscle's tension. In addition, branches of the afferent neuron inhibit, via polysynaptic reflexes, the motor neurons of the contracting muscle (A in Figure 12-8). They also activate the motor neurons of the antagonistic muscle (B in Figure 12-8). Thus when muscle contraction strains the tendon,

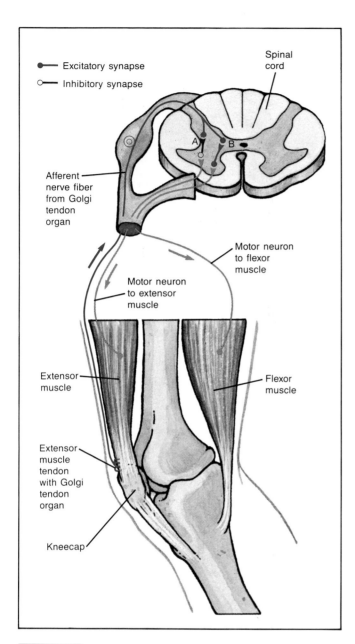

Excitatory synapse
○— **Inhibitory synapse**

Spinal cord

A   B

Afferent nerve fiber from Golgi tendon organ

Motor neuron to flexor muscle

Motor neuron to extensor muscle

Extensor muscle

Flexor muscle

Extensor muscle tendon with Golgi tendon organ

Kneecap

FIGURE 12-8 Neural pathways underlying the Golgi tendon organ component of the local control system. In this diagram, contraction of the extensor muscle causes tension in the Golgi tendon organ and increases the firing rate of action potentials in the afferent nerve fiber. By way of interneurons, this increased activity results in (A) inhibition of the extensor muscle's motor neuron and (B) excitation of the flexor muscle's motor neuron. Arrows indicate direction of action-potential transmission.

the Golgi tendon organs discharge and bring about an inhibition of the muscle's motor neurons, thus stopping the contraction and releasing the strain on the tendon. At high muscle tensions, this inhibition protects the muscle from damage due to the strong contraction. Note

that this effect is the opposite of the muscle-spindle activation.

Other local afferent systems. In addition to the afferent information from the spindle stretch receptors and Golgi tendon organs of the activated muscle, other input is fed into the local motor-control systems. For example, stimulation of the skin, particularly painful stimulation, activates the ipsilateral flexor motor neurons and inhibits the ipsilateral extensor motor neurons so that the body part is moved away from the stimulus. This is called the **flexion,** or **withdrawal, reflex** (Figure 12-9). The same stimulus causes just the opposite response on the contralateral side of the body—activation of the extensor motor neurons and inhibition of the flexor motor neurons (the **crossed-extensor reflex**). In the example of Figure 12-9, the strengthened extension of the contralateral leg means that this leg can support more of the body's weight as the hurt foot is raised from the ground.

The local motor-control systems are also affected by information from joint receptors. These inputs are all polysynaptic.

DESCENDING PATHWAYS AND THE BRAIN CENTERS THAT CONTROL THEM

The various motor centers in the brain influence the motor neurons by descending pathways that originate in either the cerebral cortex or the brainstem (Figure 12-1).

Cerebral Cortex

Many areas of the cerebral cortex function in the middle level of the hierarchy and give rise to the two types of descending pathways described on page 339. A large number of these nerve fibers come from two areas on the posterior part of the frontal lobe: the **primary motor cortex** and the **premotor area** (Figure 12-10). The primary motor cortex is sometimes called simply the motor cortex. Another term, **sensorimotor cortex,** is used to include all those parts of the cerebral cortex that act together in the control of muscle movement, including primary motor cortex, the premotor area, somatosensory cortex, and parts of the parietal-lobe association cortex adjacent to somatosensory cortex (Figure 12-10).

In general, the location of the neurons in the motor cortex varies with the parts of the body they serve. As one starts at the top of the brain and moves down along the side (A to B in Figure 12-11), the cortical neurons affecting movements of the toes and feet are at the top, followed by neurons controlling leg, trunk, arm, hand, fingers, neck, and face.

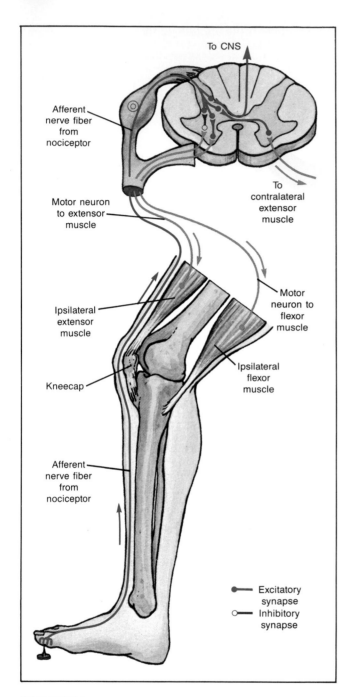

FIGURE 12-9 In response to pain, the ipsilateral flexor's motor neuron is stimulated (flexion reflex). In the case illustrated, the opposite limb is extended (crossed extensor reflex) to support the body's weight. Arrows indicate direction of action-potential propagation.

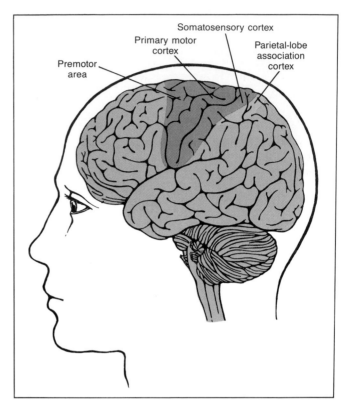

FIGURE 12-10 The major motor areas of cerebral cortex. The premotor area and the primary motor, somatosensory, and parietal-lobe association cortex together make up the sensorimotor cortex.

FIGURE 12-11 Primary motor cortex and the premotor area. The arrows A-B and C-D are explained in the text.

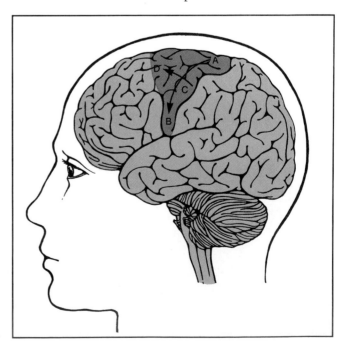

The size of the representation of each of the individual body parts in Figure 12-12 is proportional to the amount of cortex devoted to its control. Clearly, the cortical areas representing hand and face are the largest. The great number of cortical neurons for innervation of the hand and face is one factor responsible for the fine degree of motor control that can be exerted over those parts. The more delicate the function of any muscle, the more nearly one to one is the relationship between cortical neurons and the muscle's motor units.

Cortical neurons also contribute to different aspects of motor function as one explores from the back of the motor cortex forward (C to D in Figure 12-11). Neurons in the primary motor cortex, that is, the area closest to the junction of the frontal and parietal lobes, mainly contribute to movements involving the use of individual muscle groups, such as those that move a single finger. Neurons in this area of cortex are thought to enhance the accuracy of motor control.

Moving anteriorly (forward) in the cortex, primary motor cortex gives way to the premotor area, whose neurons are involved in such complicated motor functions as a change in the force or velocity of a movement, a change from one task to another, a movement made in response to visual input or a spoken command, two-handed coordination, and postural support for a wide variety of detailed movements.

The premotor area is also an important gateway for relaying processed information to the primary motor cortex or directly to the descending pathways from other regions of cerebral cortex and from other parts of the brain. Some of this relayed information comes, for example, from the parietal lobe association areas that process signals from all the sensory systems to indicate the body's position in space and the direction in which the body must be moved to achieve the desired goal. In other words, the input to the cortical neurons must select from a variety of ways of achieving a purpose (how to

FIGURE 12-12 Representation of the body in motor cortex. Note the disproportionately large areas devoted to control of the hands and facial muscles.

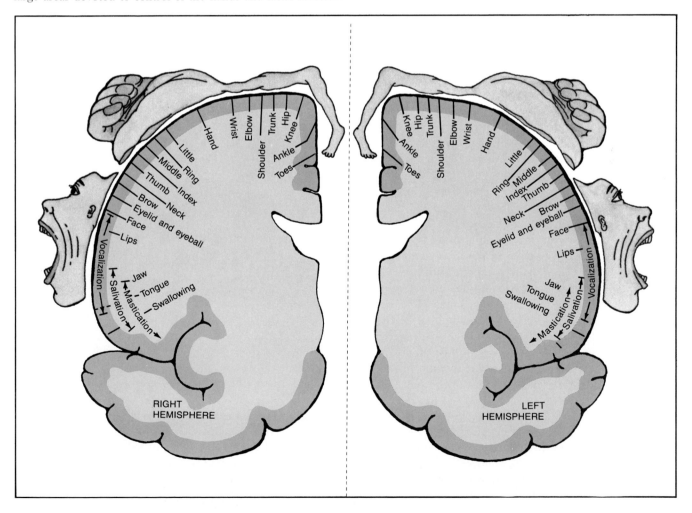

do it), since motor acts, even learned ones, can be executed in many ways with many sets of muscles. A simple example of this is a rat trained to press a lever whenever a light flashes. The rat will press quite consistently, but depending upon its original position in the cage, its movements can be quite varied. If the rat is to the left of the lever, it moves to the right; if it is to the right, it moves to the left. If its paw is on the floor, it raises the paw; if its paw is above the lever, it lowers the paw.

We have described the motor cortex and premotor area as giving rise, either directly or indirectly, to the pathways descending to the motor neurons. As we have stated, however, neurons of these areas are part of the middle hierarchical level and are not, therefore, the prime initiators of movement. They are simply a tremendously important relay station. We turn now to consider in more detail the synaptic input from the highest hierarchical level, which determines the activity of motor cortex and the premotor area.

About 800 ms before voluntary movements, that is, before any activity can be recorded in the motor cortex and premotor neurons, a distinct wave of electrical activity, known as the **readiness potential,** can be recorded from electrodes on the scalp. It is the earliest detectable sign of "getting ready" to make a voluntary motion, even something as simple as a slight finger movement. The readiness potential arises in a region—the **supplementary motor area** (Figure 12-13)—that lies near the limbic system on the surface of the brain where the cortex folds down between the two hemispheres. In fact, blood flow increases in this region earlier than any other brain area when one just thinks about making a sequence of finger movements. Other brain regions, particularly the association areas of the parietal and frontal lobes, also contribute to the readiness potential.

The readiness potential is followed by a **motor potential,** which occurs within the motor cortex about 55 ms before muscle activity begins. In the short time between the termination of the readiness potential and the beginning of the motor potential, the middle hierarchical level comes into play. Subcortical mechanisms become active, particularly in the basal ganglia, cerebellum, and some brainstem nuclei. This subcortical activity, as mentioned earlier, determines the spatial and temporal sequences of cortical and local motor neuron activation needed to accomplish the intended movement.

But the question remains: What makes neurons in the supplementary motor area become active? As part of the decision-making process for any voluntary movement, the questions "what to do" and "when" must be resolved. It appears that these matters are dealt with in parts of the cerebral cortex different from the sensorimotor regions. The question "what to do" may be deter-

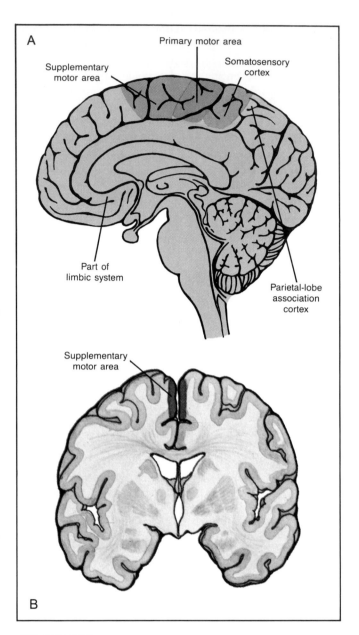

FIGURE 12-13 (A) A midline view of the brain showing the supplementary motor area. (B) This area lies in the part of cerebral cortex that is folded down between the two cerebral hemispheres.

mined in the lower (orbital) parts of the frontal cortex (Figure 12-10), although some investigators suggest that parts of parietal lobe association cortex play this role. Regardless of where the determination takes place, however, to arrive at an answer to this question, there must be a general attentiveness (input from the brainstem reticular system, page 174). There is also input from motivation, that is, a reason to make the movement, which implies that input from the limbic system signifies that a desired reward is available (Chapter 20). This, in turn, implies that memory-storage systems (Chapter 20) play a role, too.

The supplementary motor area, which also receives information from the motivational systems, seems to deal with the question "when."

We sense that at this point the reader will share with physiologists who study the motor system the frustration of trying to pin down a clear explanation of motor control. The problem is that the complete control network for any movement—complex or simple—is presently understood only imperfectly. Some features of motor control that are, however, becoming clear are listed in Table 12-2.

Subcortical and Brainstem Nuclei

Recall that the subcortical and brainstem nuclei are part of the middle level of the motor-control hierarchy and that this level converts the overall plan or goal of an action into programs that determine the specific movements needed to accomplish that action, that is, "how to do it." A very important feature of the subcortical and brainstem nuclei is the feedback effect they exert on the motor cortex through the thalamus (Figure 12-14).

One example of how such a feedback influence might work is as follows: During normal motor activity, motor cortex neurons fire sustained bursts of action potentials due to an excitatory loop from the thalamus. It is necessary, then, to have inhibitory mechanisms capable of modulating the positive influence (for example, to stop ongoing activities) and of suppressing unwanted, conflicting behaviors. This function seems to be played by neurons of the basal ganglia, which end in inhibitory synapses upon the thalamic cells (Figure 12-14). Cells of the basal ganglia thus seem to be responsible for selecting and maintaining desired patterns of motor activity while suppressing antagonistic or unwanted behaviors.

This theory explains the defects in motor control associated with persons suffering from **Parkinson's disease** who tend to maintain ongoing activity and have difficulty establishing new behaviors. Thus, if walking, they tend to keep walking; if seated, they tend to remain seated. One particular subcortical nucleus implicated in this disease is the substantia nigra (black substance), which gets its name from the dark pigment in its cells. The neurons of the substantia nigra project to the basal ganglia, where they liberate the transmitter dopamine from their axon terminals. Dopamine inhibits neurons in the basal ganglia.

When the substantia-nigra neurons degenerate, as they do for unknown reasons in Parkinson's disease, the amount of dopamine delivered to the basal ganglia is reduced. Consequently, excitatory activity in the basal

TABLE 12-2	SOME CHARACTERISTICS OF MOTOR-CONTROL SYSTEMS

1. The brain operates pretty much as a whole in producing movement, that is, many brain structures are involved.
2. The motor cortex seems to be unable to generate voluntary movements without input from the subcortical motor centers.
3. Not all voluntary movements are channeled through the motor cortex. For example, tongue movements during speech are controlled by special speech areas (Chapter 20).
4. A given set of muscles can be controlled by several neural systems, and the central nervous system can select from them to execute a movement.
5. Movements that involve learning, movements to be made in response to an expected signal, highly skilled movements that have become almost automatic—all these are controlled by different components of the motor-control systems.
6. The limbic system (emotion and motivation) plays an important role in motor control (we are all aware that it is usually difficult to make well-controlled, detailed movements when in a highly emotional state). The limbic system is also important in determining which control system will be used for a given set of muscles, depending upon the reasons for the movement, for example, escape, feeding, courtship, or curiosity.
7. The motor cortex is more important in controlling skilled, accurate movements than in producing automatic or rhythmic ones.
8. The movements performed to achieve a specific task are variable and seem almost inconsequential. For example, it usually does not matter whether you comb your hair with the right or left hand. It is the end result that matters, but only the intervening acts or movements can be programmed by the nervous system.
9. In certain circumstances a spinal reflex, for example, the flexion reflex can override a command from a higher hierarchical level, for example, extend leg to bear the body's weight.

ganglia predominates, the thalamus is overactive, and the positive input to the cortex runs unimpeded, resulting in a diminished ability to select specific behaviors.

A deficiency in the dopamine delivered to the basal ganglia occurs in Parkinson's disease, although other transmitter abnormalities may be involved as well. A common set of symptoms, in addition to difficulty in initiating and changing motor behavior, evolve, including a tremor, which is most prominent at rest, a change in facial expression resulting in a masklike, unemotional

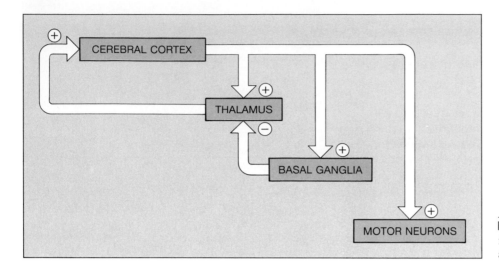

FIGURE 12-14 Neuronal loops linking activity in the cortex, thalamus, and basal ganglia.

appearance, muscular rigidity, a shuffling gait with loss of arm swing, a stooped posture, and loss of postural reflexes (see below).

One treatment of parkinsonism is the drug L-dopa, a precursor of dopamine. L-dopa enters the bloodstream, crosses the blood-brain barrier, and is converted to dopamine (dopamine is not used as medication because it cannot cross the blood-brain barrier). The newly formed dopamine relieves the symptoms of the disease.

In addition to the role played by the basal ganglia in posture and movement, they serve in intellectual function. Thus the basal ganglia participate in many of the same functions as the cerebral cortex and are not, as previously thought, limited to motor programming.

Cerebellum

The cerebellum sits on top of the brainstem, as can be seen in Figure 12-2A. Its most important function is to help coordinate the muscle activity involved in posture and movement, but it does not exert its influence *directly* by pathways that descend to the motor neurons. Rather, the cerebellum affects motor behavior by means of input to brainstem nuclei and (by way of the thalamus) to the motor cortex. In other words, the cerebellum does not initiate movements but modulates movements initiated by other parts of the brain. In this respect, the cerebellum is like the basal ganglia, but the cerebellum and basal ganglia project to different regions of the motor cortex and follow completely separate paths through the thalamus.

The cerebellum's role in motor functioning, at least in part, is to perform the critical task of comparing information about what the muscles should be doing with information about what they are actually doing. To achieve this, the cerebellum receives information from the brain centers involved in programming and executing a movement and compares the information with sensory information it simultaneously receives about motor performance during the course of the movement. If there is a discrepancy between the two, the cerebellum adjusts the activity in the descending pathways to correct the ongoing movement, and it sends an error signal to the motor cortex and subcortical centers to modify the central motor programs so that future movements of the same kind will be performed more accurately. In fact, the cerebellum plays an important role in the learning of motor skills.

The afferent inputs to the cerebellum come from the vestibular system, eyes, ears, skin, muscles, joints, and tendons, that is, from the major receptors affected by movement. Input from receptors in a single small area of the body end in the same region of the cerebellum as do the inputs from the higher brain centers controlling the muscles in that same body area. This permits the cerebellum to compare motor commands with muscle performance.

The intricate neural "wiring diagram" of the cerebellum allows it to serve another crucial function, that of regulating the timing of the complex muscle actions required for even the simplest motor act.

The symptoms of persons with cerebellar damage generally include the following:

1. Such persons cannot perform movements smoothly. If they reach for an object, their movements are jerky and accompanied by oscillating, to-and-fro tremors that become more marked as the hand approaches the object.
2. They walk awkwardly, with the feet well apart. They

have such difficulty maintaining balance that their gait appears drunken.

3. They cannot start or stop movements quickly or easily. If, for example, they are asked to rotate the wrist over and back rapidly, their motions are slow and irregular.

4. They cannot easily combine the movements of several joints into a single smooth, coordinated motion. To move the arm, they might first move the shoulder, then the elbow, and finally the wrist.

Descending Pathways

By three mechanisms the descending pathways alter the balance of excitatory and inhibitory synaptic input that converges upon the alpha motor neurons: (1) by synapsing directly upon the alpha motor neurons, a pathway that has the advantages of speed and specificity; (2) by synapsing directly on the gamma motor neurons, which invokes the stretch reflex; and (3) by synapsing on interneurons, often, as we have mentioned, the same ones serving the local reflexes. Although the interneuron route is not as fast as direct activation of the motor neurons, it has the advantage of the coordination built into the interneuron network (for example, reciprocal inhibition and recruitment of synergistic muscles). The degree to which each of these three mechanisms is employed varies depending upon the nature of the motor task to be performed and the descending pathway that is utilized.

As mentioned earlier, the descending pathways can be divided into two major categories: the corticospinal pathways and the multineuronal pathways.

Corticospinal pathway.

The nerve fibers of the **corticospinal pathways,** as the name implies, have their cell bodies in the cerebral cortex (specifically, in the sensorimotor regions), and end close to the region of the motor neurons in the brainstem and spinal cord. In the medulla, near the junction of the spinal cord and brainstem, most of the corticospinal fibers cross to descend on the opposite side of the spinal cord (Figure 12-15). Thus, the skeletal muscles on the left side of the body are controlled largely by neurons in the right half of the brain, and vice versa. The corticospinal pathways are also called the **pyramidal tracts,** or **pyramidal system** (so-called perhaps because of their shape as they pass along the surface of the brainstem medulla or because they were formerly thought to arise solely from the pyramidal cells of the motor cortex).

As the corticospinal fibers descend through the brain from the cerebral cortex, they are accompanied by fibers of the **corticobulbar pathway** (bulbar means "pertaining to the brainstem"), a pathway that begins in the cerebral

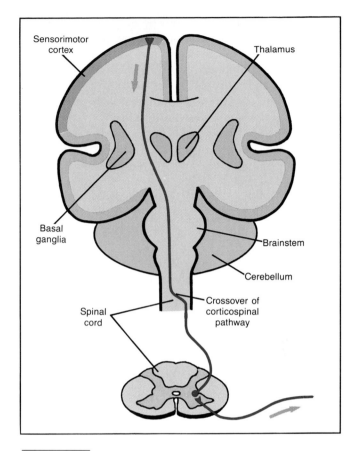

FIGURE 12-15 The corticospinal pathway. Most of these fibers cross in the brainstem to descend in the opposite side of the spinal cord. (*Adapted from Gardner.*)

cortex and ends in the brainstem. The corticobulbar fibers end near the motor neurons that innervate muscles of the eye, face, tongue, and throat. These fibers are the main source of control for the voluntary movement of the muscles of the head and neck, whereas the corticospinal fibers serve this function for the muscles of the rest of the body. [For convenience, we shall include corticobulbar pathway in the term corticospinal pathways.]

The corticospinal pathways serve fine, precise movements, such as those made when an object is manipulated by the fingers. Some corticospinal fibers end at excitatory synapses on alpha and gamma motor neurons, whereas others end on interneurons that may excite or inhibit the alpha motor neurons. Thus, the effect of the corticospinal pathway on the alpha motor neurons may be excitatory or inhibitory. In either case, the corticospinal fibers do not end diffusely. Rather, they generally affect motor neurons in a single motor neuron pool, that is, that innervate single muscles.

Surprisingly, some of the corticospinal fibers transmit information from the brain to *afferent* neurons and so can affect afferent systems. They do this by ending either (1) presynaptically on the axon terminals of afferent neurons as these fibers enter the central nervous system or (2) directly on the dendrites or cell bodies of neurons in the ascending pathways. The overall effect of this descending input to afferent systems is to limit the areas of skin, muscle, or joints allowed to influence the cortical neurons, thereby sharpening the focus of the afferent signal. Because of this descending (motor) control over ascending (sensory) information, there is clearly no real functional separation of the motor and sensory systems.

The descending fibers may also convey the information that a certain motor command is being delivered and give rise to the sense of effort (page 229).

Multineuronal pathways. The other axons that descend into the lower brainstem and spinal cord to influence motor neurons in the head and body originate in the upper brainstem. Yet these upper-brainstem cells are themselves only the final link in the **multineuronal pathways.** Thus, the multineuronal pathway is formed of many neurons that are synaptically linked together, one after another, as in a chain. At each successive neuron, information is altered according to the balance of excitatory and inhibitory input to that neuron.

The "chains" of interconnected neurons do not, however, travel in an orderly, unidirectional manner. Some multineuronal pathways form complicated circuits, looping back to earlier way stations and to the cerebral cortex (Figure 12-14). These loops feed back to the cortex the programs and program corrections (described in our original model) from subcortical brain areas. These corrections then influence the activity of the neurons that give rise to the descending pathways.

From several points in the circuits, descending fibers emerge that may, however, enter other nuclei rather than descending directly to the brainstem or spinal cord. The circuits that form the multineuronal pathways are so complex that we have omitted them in Figure 12-16, indicating only the major brain areas that contribute to them.

"Multineuronal pathways" is plural because in the spinal cord the fibers form distinct clusters, segregated according to their sites of origin. For example, the vestibulospinal pathway descends to the spinal cord from the vestibular nuclei in the brainstem, whereas the reticulospinal pathway descends from neurons in brainstem reticular formation. The multineuronal pathways are also collectively called the **extrapyramidal system** to distinguish them from the corticospinal (pyramidal) pathways.

General comments on the descending pathways. In general, the corticospinal neurons have greater influ-

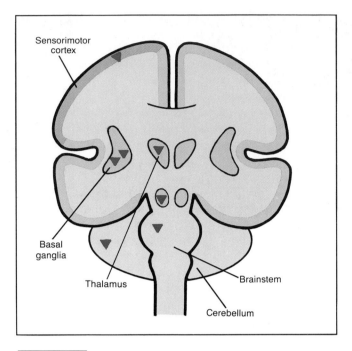

FIGURE 12-16 The symbol for a neuron cell body indicates the brain areas that contribute fibers to the multineuronal pathways. For clarity, we have not drawn in the complex circuitry of fibers that connect virtually all these brain areas with each other (and some of them with the spinal cord), yet it is these fibers that make up the multineuronal pathways. (*Adapted from Gardner.*)

ence over motor neurons that control muscles involved in fine movements, particularly those of the fingers and hands. The multineuronal pathways are more involved with the coordination of the large muscle groups used in the maintenance of upright posture, in locomotion, and in head and body movements when turning toward a specific stimulus.

There is actually much interaction between the various multineuronal pathways, however, and even between the multineuronal and corticospinal pathways. For example, the corticospinal pathway (Figure 12-15) is the most prominent output from the multineuronal circuits that loop back to the cortex. Moreover, some fibers of the corticospinal pathway end on interneurons that play important roles in posture, whereas fibers of the multineuronal pathways sometimes end directly on the alpha motor neurons to control discrete muscle movements. Because of this redundancy, loss of function resulting from damage to one system may be compensated for by the remaining system, although the compensation is generally not complete.

Thus, the distinctions between the corticospinal and multineuronal pathways are not clear-cut. It is incorrect to imagine a total separation of function between them since all movements, whether automatic or voluntary, require the continuous coordinated interaction of both pathways.

MUSCLE TONE

The resistance of muscle to continuous passive stretch, the resistance to someone else straightening your arm, for example, is known as **muscle tone.** It is usually tested as the examiner moves the relaxed limb of the subject. Under such circumstances in a normal person, the resistance to passive movement is moderate and uniform, regardless of the speed of the movement.

Muscle tone is due to the viscoelastic properties of the muscles and joints and to any ongoing contractile activity of the muscle. When a person is deeply relaxed, muscle tone is low, and the stretch reflexes and alpha motor neuron activity probably make no contribution to the tone. As the person becomes alert, however, activity in the descending pathways and activation of the alpha (and gamma) motor neurons increase, which, in turn, increases muscle tone. With increasing tone, active motor neurons play an ever-increasing role until during maximal voluntary contractions, they—or rather the myofibrillar cross-bridge activity they stimulate—account for much of the resistance to stretch. Evaluation of muscle tone provides important information.

Abnormal Muscle Tone

Abnormally high muscle tone, seen particularly clearly when a joint is moved passively at high speeds, is called **hypertonia.** The hypertonia can often be felt as a "catch" or unevenness in the resistance. The increased resistance is due to a steady state of motor neuron activity keeping a muscle contracted despite the individual's attempts to relax the muscle.

Spasticity is hypertonia accompanied by increased responses of motor reflexes such as the knee jerk reflex and by distorted voluntary actions (decreased coordination and strength of the movements). **Spasms** (sustained involuntary contractions) of flexor muscles are common in spasticity. The condition of **rigidity** is also a form of hypertonia but is characterized by normal local muscle reflexes and by an abnormality of muscle tone that is related less to the movement velocity than in spasticity. The rigidity of a person with Parkinson's disease is the most familiar example.

Hypertonia is usually found when there are disorders of the descending pathways that result in an imbalance in the facilitory and inhibitory inputs exerted by them on the motor neurons and interneurons. Clinically, the descending pathways, primarily the corticospinal pathways, and neurons of the motor cortex are often called the **upper motor neurons,** and abnormalities due to their dysfunction are classed as upper motor neuron disorders. (In this classification, the **lower motor neurons** are the alpha motor neurons.)

Hypotonia is a condition of abnormally low muscle tone despite voluntary attempts to contract the muscle. It is accompanied by weakness, atrophy (a decrease in muscle bulk), and decreased or absent reflex responses. Dexterity and coordination are generally preserved unless profound weakness is present. While hypotonia may develop after cerebellar disease, it more frequently accompanies disorders of the alpha motor neurons, neuromuscular junctions, or muscles. The term flaccid, which means weak or soft, is often used to describe hypotonic muscles.

MAINTENANCE OF UPRIGHT POSTURE AND BALANCE

The skeleton supporting the body is a system of long bones and a many-jointed spine that cannot stand erect against the forces of gravity without the support given by coordinated muscle activity. The muscles that maintain upright posture, that is, support the body's weight against gravity, are controlled by the brain and by reflex mechanisms that are "wired into" the neural networks of the brainstem and spinal cord. Many of the reflex pathways previously introduced, for example, the flexion and crossed-extensor reflexes, are used in posture control.

Added to the problem of maintaining upright posture is that of maintaining balance. A human body is a very tall structure balanced on a relatively small base, and its center of gravity is quite high, being situated just above the pelvis. For stability, the center of gravity must be kept within the base of support provided by the feet (Figure 12-17). Once the center of gravity has moved beyond this base, the body will fall unless one foot is shifted to broaden the base of support. Yet people often operate under conditions of unstable equilibrium and would topple easily if their balance were not protected by postural reflexes.

The maintenance of posture and balance is accomplished by means of complex interacting **postural reflexes,** all the components of which we have met previously. The afferent pathways of the reflex arcs come from three sources: the eyes, the vestibular apparatus, and the proprioceptors (page 229). The efferent pathways are the alpha motor neurons to the skeletal muscles, and the integrating centers are neuron networks in the brainstem and spinal cord.

There are many familiar examples of postural reflexes. The stretch reflex is activated, for example, when the knees begin to buckle, stretching the spindle receptors

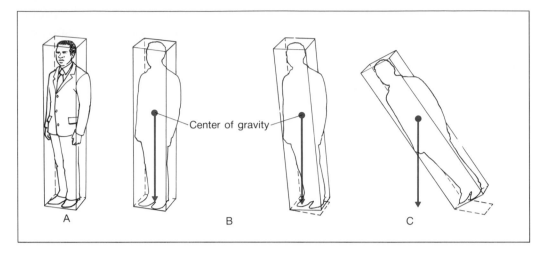

FIGURE 12-17 The center of gravity is the point in an object at which, if a string were attached to the object at this point and pulled up, all the downward force due to gravity would be exactly balanced. (A) The center of gravity must remain within the upward vertical projections of the object's base (the tall box outlined in the drawing) if stability is to be maintained. (B) Stable conditions: Box tilts a bit, but the center of gravity remains within the base area and so the box will return to its upright position. (C) Unstable conditions: The box tilts so far that its center of gravity is not above any part of the object's base—the dashed rectangle on the floor—and the object will fall.

in the thigh muscles. Although the input from the muscle-spindle afferent fibers onto the alpha motor neurons does not exert a powerful effect, it is thought to be useful for small adjustments of posture. The crossed-extensor reflex is clearly used to maintain postural support. As one leg is flexed, the other is extended more strongly to support the added weight of the body, and the positions of various parts of the body are shifted to move the center of gravity over the single, weight-bearing leg. This shift in the center of gravity, demonstrated in Figure 12-18, is an important component in the stepping mechanisms of locomotion.

It is clear that afferent input from several sources is necessary for effective postural adjustments, yet interfering with any one of these inputs alone does not cause a person to topple over. Blind people maintain their balance quite well with only a slight loss of precision, and people whose vestibular mechanisms have been destroyed have very little disability in everyday life as long as their visual system and proprioceptors are functioning.

The conclusion to be drawn from such examples is that the sensory information received by postural control mechanisms is redundant and that one particular type of information becomes critically important only when all the other inputs have been lost. This redundancy allows

postural control mechanisms to perform under a wide variety of environmental conditions.

WALKING

The cyclic, alternating movements of walking are controlled by **central pattern generators** within the central nervous system—some in the spinal cord at the level of the motor neurons and (probably) some in the brainstem.

Walking is initiated by allowing the body to fall forward to an unstable position and then moving one leg forward to regain equilibrium. The legs may be activated out of phase with each other, as in walking or running. In such cases, spinal pattern generators utilize reciprocal inhibition so that when the extensor muscles are activated on one side of the body, to bear the body's weight, the contralateral extensors are inhibited to allow that limb to flex and swing forward.

Although the central pattern generators can control the rhythmic movement of a limb in the absence of feedback information from proprioceptors, vestibular apparatus, or eyes, such afferent input normally contributes substantially to locomotion by acting on the generators and indirectly on the motor neurons themselves. Such

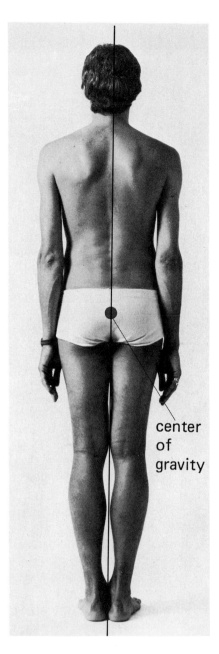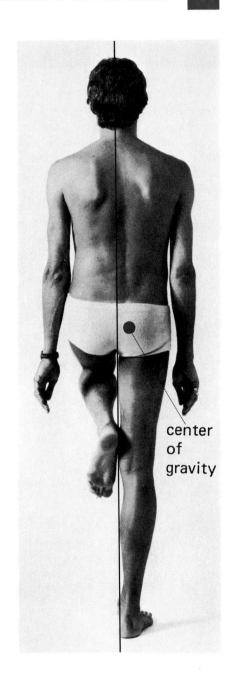

center
of
gravity

center
of
gravity

FIGURE 12-18 Postural changes with stepping. (Left) Normal standing posture. The line of the center of gravity falls directly between the two feet. (Right) As the left foot is raised, the whole body leans to the right so that the center of gravity shifts over the right foot.

afferent information assures that the output of the central pattern generator is optimally adapted to the environment. For example, if the limb movement should be slowed by external factors, as in wading through water, the afferent signals prevent a premature flexion of the limb.

The central pattern generators in the spinal cord are also affected by input from the descending pathways, both during the initiation of activity and during ongoing activity. Such input helps adjust the movement to the environment and to the person's goals. For example, depending upon commands from the descending path-

ways, the two limbs may be operated together, as in jumping, instead of reciprocally. One attractive hypothesis is that the brain can control stereotyped behaviors such as locomotion simply by activating the appropriate central program generators by way of the descending pathways. And then the brain can forget about it and do other things.

In summary, the picture that is emerging in the neural control of locomotion is that central pattern generators provide input to the motor neurons. The pattern of this input may be modified through a complex group of motor systems that act on the motor neurons, interneurons, and central pattern generators. The cerebellum is one such system. Information from peripheral receptors reaches the cerebellum along with information from other brain centers about the neural commands that are being delivered to the motor neurons. The cerebellum acts by way of multineuronal descending pathways on the local motor regions to modify their output. Furthermore, the basic locomotor patterns can be altered to adapt each movement to the goals of the person and to the environment.

The reader will have noticed our frequent use of such terms as "possibly," "it has been suggested," "hypothesis," and so on, which indicates how little is known for certain about locomotion and, indeed, motor behavior in general.

SUMMARY

Overview of the Motor Systems

I. The neural systems that control body movements are arranged in a motor-control hierarchy.
   A. The highest level determines the general intention of an action.
   B. The middle level specifies the postures and movements needed to carry out the intended action and, taking account of sensory information that indicates the body's position, establishes a motor program.
   C. The lowest level determines which motor neurons will be activated.
   D. As the movement progresses, information about what the muscles are doing is fed back to the motor-control centers, which make any needed program corrections.
II. Actions are voluntary when we are aware of what we are doing and why, or when we are paying attention to the action or its purpose.
III. Almost all actions have conscious and unconscious components.

Local Control of Motor Neurons

I. Most input to motor neurons arises from local interneurons, which themselves receive input from peripheral receptors, descending pathways, and other interneurons.
II. Muscle length and changes in length are monitored by muscle-spindle stretch receptors.
   A. Activation of these receptors initiates the stretch reflex, in which motor neurons of ipsilateral antagonists are inhibited and those of synergists are activated.
   B. Tension on the stretch receptors is maintained during muscle contraction by gamma-efferent activation of the spindle muscle fibers.
   C. Alpha and gamma motor neurons are often coactivated.
III. Muscle tension is monitored by Golgi tendon organs, which inhibit motor neurons of the contracting muscle and stimulate ipsilateral antagonists.

IV. The flexion reflex excites the ipsilateral flexor muscles and inhibits the ipsilateral extensors. The crossed-extensor reflex excites the contralateral extensor muscles during excitation of the ipsilateral flexors.

Descending Pathways and the Brain Centers that Control Them

I. The location of the neurons in motor cortex varies with the part of the body the neurons serve.
II. A readiness potential is fired from the supplementary motor area 800 ms before any electrical activity in motor cortex begins. This is followed by a motor potential in motor cortex.
III. The basal ganglia help determine the direction, force, and speed of movements.
IV. The cerebellum coordinates posture and movement.
V. The corticospinal pathways pass directly from the sensorimotor cortex to motor neurons in the spinal cord (in the brainstem, in the case of the corticobulbar pathways) or to interneurons near the motor neurons.
   A. In general, neurons on one side of the brain control muscles on the other side of the body.
   B. Corticospinal pathways serve predominantly fine, precise movements.
   C. Some corticospinal fibers affect the transmission of information in afferent pathways.
VI. The multineuronal pathways consist of chains of neurons that carry information from the sensorimotor cortex, basal ganglia and other subcortical nuclei, cerebellum, and brainstem nuclei to motor neurons in the brainstem or spinal cord or interneurons near them.
   A. Some fibers loop back to alter earlier pathway components.
   B. The multineuronal pathways are involved in the coordination of large groups of muscles used in posture and locomotion.
VII. There is some duplication of function between the two descending pathways.

Muscle Tone

I. Muscle tone is due to the viscoelastic properties of muscles and joints and to any ongoing contractile activity.
II. Hypertonia, as seen in spasticity or rigidity, usually occurs with disorders of the descending pathways.
III. Hypotonia can be seen with cerebellar disease or, more commonly, with alpha motor neuron or muscle disease.

Maintenance of Upright Posture and Balance

I. To maintain balance, the body's center of gravity must be maintained over the body's base.
II. Postural reflexes depend on inputs from eyes, vestibular apparatus, and proprioceptors.
III. The stretch and crossed-extensor reflexes are postural reflexes.

Walking

I. Central pattern generators generate the cyclic, alternating movements of locomotion.

II. The pattern generators are affected by feedback and motor programs.

REVIEW QUESTIONS

1. Define:

motor units	withdrawal reflex
motor neuron pool	crossed-extensor reflex
motor control hierarchy	primary motor cortex
motor program	premotor area
voluntary movement	sensorimotor cortex
stretch receptors	readiness potential
muscle spindle	supplementary motor area
spindle fibers	motor potential
intrafusal fibers	Parkinson's disease
skeletomotor fibers	corticospinal pathways
extrafusal fibers	pyramidal tracts
muscle-spindle stretch	pyramidal system
receptors	corticobulbar pathways
stretch reflex	multineuronal pathways
knee jerk	extrapyramidal system
monosynaptic	muscle tone
polysynaptic	hypertonia
reciprocal inhibition	spasticity
synergistic muscles	spasms
ipsilateral	rigidity
contralateral	upper motor neurons
alpha motor neurons	lower motor neurons
gamma motor neurons	hypotonia
Golgi tendon organs	postural reflexes
flexion reflex	central pattern generators

2. Describe motor control in terms of the motor control hierarchy, using the following terms: highest, middle, and lowest levels; motor program; descending pathways; and motor neuron.

3. List the characteristics of voluntary actions.

4. Picking up a book, for example, has both voluntary and involuntary components. List six components of this action, and indicate whether they are voluntary or involuntary.

5. List four inputs that can converge on the interneurons active in local motor control. List three types of neurons that can receive the synaptic output of these interneurons.

6. Draw a muscle spindle, labeling the spindle, spindle fibers and skeletomotor muscle fibers, stretch receptors, afferent fibers, and alpha- and gamma-efferent fibers.

7. Describe the components of the knee jerk reflex (stimulus, receptor, afferent pathway, integrating center, efferent pathway, effector, and response).

8. Describe the major function of alpha-gamma coactivation.

9. Distinguish among the following areas of cerebral cortex: primary motor, premotor area, sensorimotor, and supplementary motor area.

10. Contrast the two major descending pathways in terms of structure and function.

11. Describe the hypothesis concerning the events that may be taking place during the readiness potential.

12. Describe the roles that the basal ganglia and cerebellum play in motor control.

13. Explain how hypertonia might result from upper motor neuron disease.

14. Explain how hypotonia might result from lower motor neuron disease.

15. Explain the role played by the stretch reflex in postural stability.

16. Explain the role played by the crossed-extensor reflex in postural stability.

17. Explain the role of the central pattern generators in walking, incorporating in your discussion the following terms: interneuron, reciprocal inhibition, synergist, antagonist, and feedback.

THOUGHT QUESTIONS

(Answers are given in Appendix A.)

1. What changes would occur in the knee jerk reflex after destruction of the gamma motor neurons?

2. What changes would occur in the knee jerk reflex after destruction of the alpha motor neurons?

3. Draw a cross section of the spinal cord and a portion of the thigh (similar to Figure 12-6), but "wire up" the neurons so the leg becomes a stiff pillar, that is, the knee does not bend.

4. We have said that hypertonia is usually considered a sign of upper motor neuron disease, but how might it result from a lower motor neuron disorder?

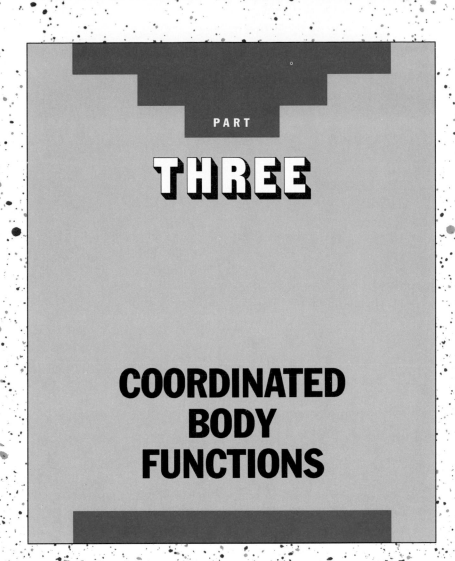

PART

THREE

COORDINATED BODY FUNCTIONS

CHAPTER

13

CIRCULATION

Beyond a distance of a few cell diameters, diffusion—the random movement of substances from a region of high concentration to one of low concentration—is not sufficiently rapid to meet the metabolic requirements of cells. For multicellular organisms to evolve to a size larger than a microscopic cluster of cells, some mechanism other than diffusion was needed to transport molecules rapidly over the long distances between internal cells and the body's surface and between the various specialized tissues and organs. This problem was solved in the animal kingdom by the **circulatory system**, which comprises the blood, the set of interconnected tubes (**blood vessels** or **vascular system**) through which the blood flows, and a pump (the **heart**) that produces this flow. The heart and blood vessels together are termed the **cardiovascular system**.

SECTION A
BLOOD

Blood is composed of cells and a liquid, plasma, in which they are suspended. The cells are the **erythrocytes** (red blood cells), the **leukocytes** (white blood cells), and the **platelets**. More than 99 percent of the cells are erythrocytes, which are the oxygen-carrying cells of the blood. The leukocytes protect against infection and cancer. The platelets function in blood clotting. In the cardiovascular system, the constant motion of the blood keeps all the cells well dispersed throughout the plasma.

The **hematocrit** is defined as the percentage of total blood volume that is erythrocytes. It is determined by centrifuging a sample of blood, the erythrocytes moving to the bottom of the tube and the plasma to the top, the leukocytes and platelets forming a very thin layer between them (Figure 13-1). The normal hematocrit is approximately 45 percent in men and 42 percent in women.

The total blood volume of an average person is approximately 5.5 L. If we take the hematocrit to be 45 percent, then:

$$\text{Total erythrocyte volume} = 0.45 \times 5.5 \text{ L} = 2.5 \text{ L}$$

Therefore:

$$\text{Plasma volume} = 5.5 \text{ L} - 2.5 \text{ L} = 3.0 \text{ L}$$

This last calculation ignores the volume occupied by the leukocytes and platelets because it is normally so small.

PLASMA

Plasma, the liquid portion of the blood, consists of a large number of organic and inorganic substances dissolved in water (Table 13-1). The proteins account for approximately 7 percent of the total weight of plasma. Most **plasma proteins** can be classified, according to certain physical and chemical reactions, into three broad groups: the **albumins** and **globulins**, which have many overlapping functions summarized in Table 13-1 and described in relevant chapters, and fibrinogen,

In addition to the organic solutes—proteins, nutrients, and metabolic waste products—plasma contains a variety of mineral electrolytes. Note in Table 13-1 that these ions constitute much less of the weight of plasma than do the proteins but have much higher molar concentrations. This is because molarity is a measure not of *weight* but of *number* of molecules or ions per unit volume. Thus, there are many more ions than protein molecules, but the protein molecules are so large that a very small number of them greatly outweighs the much larger number of ions.

THE BLOOD CELLS

Erythrocytes

Erythrocytes have the shape of a biconcave disk, that is, a disk thicker at the edge than in the middle, like a doughnut with a center depression on each side instead of a hole (Figure 13-2). Their shape and small size (7 μm in diameter) impart to them a high surface-to-volume ratio, so that oxygen and carbon dioxide can rapidly diffuse to and from the interior of the cell. The plasma membrane of erythrocytes contains specific proteins that differ from person to person, and these confer upon the blood its so-called type, as described in Chapter 19.

The outstanding characteristic of erythrocytes is the presence of the protein **hemoglobin**, which binds oxygen taken in by the lungs and constitutes approximately one-third of the total weight of the erythrocyte. Each hemoglobin molecule is made up of four subunits, each subunit consisting of an organic molecule known as **heme** attached to a polypeptide. The four identical polypeptides of a hemoglobin molecule are bound together and collectively called **globin**. Each of the four heme portions of a hemoglobin molecule (Figure 13-3) contains one atom of iron (Fe), which binds oxygen.

The site of erythrocyte production is the soft interior of bones called **bone marrow**, specifically the "red" bone marrow. The erythrocytes are descended from bone-marrow cells that contain no hemoglobin but do have nuclei and are therefore capable of cell division. After several divisions, cells emerge that are identifiable as immature erythrocytes because they contain hemoglobin. As maturation continues, these cells produce increased amounts of hemoglobin but then ultimately lose their nuclei and organelles. However, newly produced erythrocytes still contain a few ribosomes, which produce a weblike (reticular) appearance when treated with special stains, an appearance that gives young erythrocytes the name **reticulocyte**. Normally, only mature erythrocytes leave the bone marrow and enter the gen-

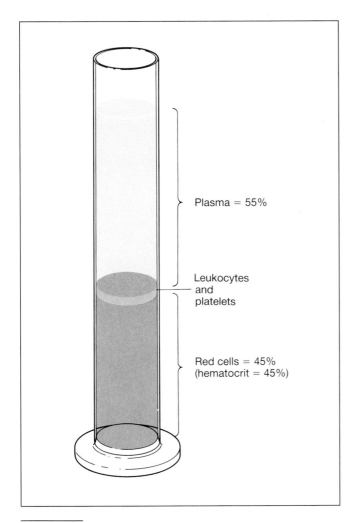

Plasma = 55%

Leukocytes and platelets

Red cells = 45% (hematocrit = 45%)

FIGURE 13-1 If a blood-filled capillary tube is centrifuged, the erythrocytes become packed in the lower portion, and the percentage of blood that is erythrocytes can be determined. This is called the hematocrit. The leukocytes and platelets form a very thin layer, known as a buffy coat, between the plasma and red cells.

which functions in blood clotting. The albumins are by far the most abundant of these groups and are synthesized by the liver.

It must be emphasized that the three major groups of plasma proteins normally do not leave the plasma to be taken up by cells. Cells use plasma amino acids, not plasma proteins, to make their own proteins. Accordingly, plasma proteins must be viewed quite differently from most of the other organic constituents of plasma, which use the plasma as a vehicle for transport but are used by cells. The plasma proteins function in the plasma itself or in the interstitial fluid.

Serum is plasma from which fibrinogen and other proteins involved in clotting have been removed as a result of clotting, to be described in Chapter 19.

TABLE 13-1 CONSTITUENTS OF PLASMA

Constituent	Amount/Concentration	Major Functions
Water	93% of plasma weight	Medium for carrying all other constituents
Electrolytes (inorganic):	Total < 1% of plasma weight	Keep H_2O in extracellular compartment; act as buffers; function in membrane excitability
$\quad Na^+$	145 mM	
$\quad K^+$	4 mM	
$\quad Ca^{2+}$	2.5 mM	
$\quad Mg^{2+}$	1.5 mM	
$\quad Cl^-$	103 mM	
$\quad HCO_3^-$	24 mM	
$\quad$ Phosphate (mostly $\quad\quad HPO_4^{2-}$	1 mM	
$\quad SO_4^{2-}$	0.5 mM	
Proteins:	Total = 7.3 g/100 ml (2.5 mM)	Provide nonpenetrating solutes of plasma; act as buffers; bind other plasma constituents (lipids, hormones, vitamins, metals, etc.); clotting factors; enzymes; enzyme precursors; antibodies (immune globulins); hormones
$\quad$ Albumins	4.5 g/100 mL	
$\quad$ Globulins	2.5 g/100 mL	
$\quad$ Fibrinogen	0.3 g/100 mL	Blood clotting
Gases:		
$\quad CO_2$	2 mL/100 mL	No function; a waste product
$\quad O_2$	0.2 mL/100 mL	Oxidative metabolism
$\quad N_2$	0.9 mL/100 mL	No function
Nutrients:		
$\quad$ Glucose and other carbohydrates	100 mg/100 mL (5.6 mM)	
$\quad$ Total amino acids	40 mg/100 mL (2 mM)	
$\quad$ Total lipids	500 mg/100 mL (7.5 mM)	See Chapter 4
$\quad$ Cholesterol	150–250 mg/100 mL (4–7 mM)	
$\quad$ Individual vitamins	0.0001–2.5 mg/100 mL	
$\quad$ Individual trace elements	0.001–0.3 mg/100 mL	
Waste products:		
$\quad$ Urea (from protein)	34 mg/100 mL (5.7 mM)	
$\quad$ Creatinine (from creatine)	1 mg/100 mL (0.09 mM)	
$\quad$ Uric acid (from nucleic acids)	5 mg/100 mL (0.3 mM)	
$\quad$ Bilirubin (from heme)	0.2–1.2 mg/100 mL (0.003–0.018 mM)	
Individual hormones	0.000001–0.05 mg/100 mL	

eral circulation. The appearance of many reticulocytes in blood indicates the presence of unusually rapid erythrocyte production.

Because erythrocytes lack nuclei and organelles, they can neither reproduce themselves nor maintain their normal structure indefinitely. The average life span of an erythrocyte is approximately 120 days, which means that almost 1 percent of the body's erythrocytes are destroyed and must be replaced every day. This amounts to 100 billion cells per day! Erythrocyte destruction occurs in the spleen.

The production of erythrocytes requires the usual nutrients: amino acids, lipids, and carbohydrates. In addition, iron, folic acid, and vitamin B_{12} are essential.

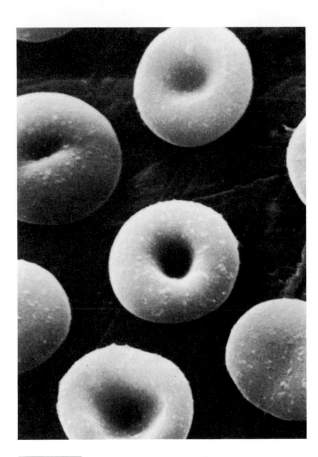

FIGURE 13-2 Scanning electron micrograph of erythrocytes. *(From R. G. Kessel and C. Y. Shih, "Scanning Electron Microscopy in Biology," Springer-Verlag, New York, 1974, p. 265.)*

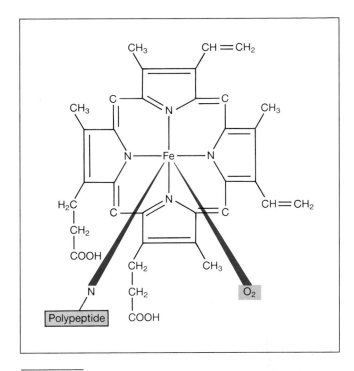

FIGURE 13-3 Heme. Oxygen binds to the iron atom (Fe). Heme attaches to a polypeptide chain by a nitrogen atom to form one subunit of hemoglobin. Four of these subunits bind to each other to make a single hemoglobin molecule.

Iron. As noted above, **iron** is the element to which oxygen binds. Small amounts of iron are lost each day via the urine, feces, sweat, and cells sloughed from the skin. In addition, women lose a similar amount via menstrual blood. In order to remain in iron balance, the amount of iron lost from the body must be replaced by ingestion of iron-containing foods, particularly rich sources of which are meat, liver, shellfish, egg yolk, beans, nuts, and cereals. A significant upset of iron balance can result either in iron deficiency, leading to inadequate hemoglobin production, or in an excess of iron in the body, with serious toxic effects.

The homeostatic control of iron balance resides primarily in the intestinal epithelium, which actively absorbs iron from ingested food. Only a small fraction of ingested iron is absorbed, but what is more important, this fraction is increased or decreased, in a negative-feedback manner, depending upon the state of body iron balance. These fluctuations are mediated by changes in the iron content of the intestinal epithelium: the more iron in the body, the more in the intestinal epithelium, and the less new iron will be absorbed.

A buffer against iron deficiency is the considerable bodily store of iron, mainly in the liver, bound up in a protein called **ferritin**. About 50 percent of the total body iron is in hemoglobin, 25 percent is in heme-containing proteins in other cells of the body, and 25 percent is in liver ferritin. Moreover, the recycling of iron is very efficient (Figure 13-4): As old erythrocytes are destroyed in the spleen, their iron is released into the plasma and bound to an iron-transport protein called **transferrin**. Almost all of this iron is delivered by transferrin to the bone marrow to be incorporated into new erythrocytes. On a much lesser scale, other organs and tissues, some cells of which are continuously dying and being replaced, release iron into the plasma and take up iron from transferrin.

Folic acid and vitamin B₁₂. **Folic acid**, a vitamin found in large amounts in leafy plants, yeast, and liver, is involved in the synthesis of the pyrimidine thymine. It is, therefore, essential for the formation of DNA and, hence, for normal cell division. When this vitamin is not

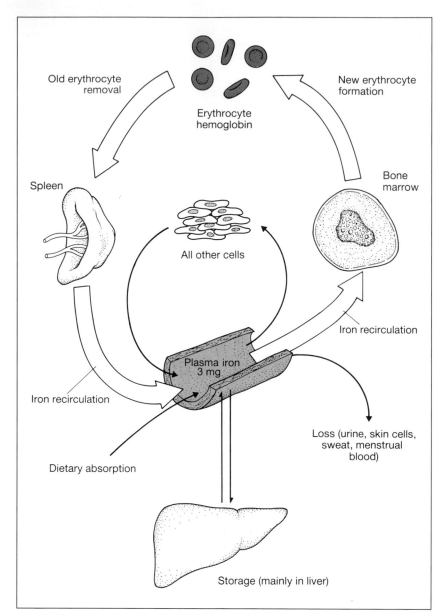

FIGURE 13-4 Summary of iron balance. The thickness of the arrows corresponds approximately to the amount of iron involved. In the steady state, the rate of gastrointestinal iron absorption and the rate of iron loss via urine, skin, and menstrual flow are equal. The rate of absorption is the major homeostatically controlled process. Recirculation of erythrocyte iron is very important because it involves 20 times more iron per day than is absorbed and excreted. (*Adapted from Crosby.*)

present in adequate amounts, impairment of cell division occurs throughout the body but is most striking in rapidly proliferating cells, including erythrocyte precursors. Thus, fewer erythrocytes are produced when folic acid is deficient.

Production of normal erythrocyte numbers also requires extremely small quantities (one-millionth of a gram per day) of a cobalt-containing molecule, **vitamin B$_{12}$,** since this vitamin is required for the action of folic acid. Vitamin B$_{12}$ is not synthesized in the body, and so its supply depends upon dietary intake. It is found only in animal products, and strictly vegetarian diets may be deficient in it. Unlike folic acid, vitamin B$_{12}$ is also

needed for myelin formation, so a variety of neurological symptoms may accompany the poor erythrocyte production when vitamin B$_{12}$ is deficient.

Regulation of erythrocyte production. In a normal person, the total volume of circulating erythrocytes remains remarkably constant due to reflexes that regulate the bone marrow's production of these cells. In the previous section, we stated that iron, folic acid, and vitamin B$_{12}$ must be present for normal erythrocyte production. However, none of these substances constitutes the signal that *regulates* production rate.

The direct control of erythrocyte production (erythropoiesis) is exerted by a hormone called **erythropoietin**, which is secreted into the blood by cells in the kidneys and, to a lesser extent, the liver. Erythropoietin acts on the bone marrow to stimulate the proliferation of precursor cells and their maturation into erythrocytes.

Erythropoietin is normally secreted in relatively small amounts, which stimulates the bone marrow to produce erythrocytes at a rate adequate to replace the usual loss. Erythropoietin secretion rate can be increased markedly above basal values, the stimulus being a decreased oxygen delivery to the kidneys. As a result of the increase in erythropoietin secretion, plasma erythropoietin concentration, erythrocyte production, and the oxygen-carrying capacity of the blood all increase. Therefore, oxygen delivery to the tissues returns toward normal (Figure 13-5).

Testosterone also stimulates the release of erythropoietin. This may account, at least in part, for the higher hemoglobin concentration in men than in women—16 g/dL blood versus 14 g/dL.

The use of erythropoietin synthesized by recombinant DNA techniques holds great medical promise, not only for treating the erythropoietin deficiency of kidney disease but for rapidly stimulating patients with normal kidneys to produce more erythrocytes. For example, such use in patients undergoing surgery should reduce the need for transfusions.

Anemia. Defined as (1) a decrease in the total number of erythrocytes, each having a normal quantity of hemoglobin, or (2) a diminished concentration of hemoglobin per erythrocyte, or (3) a combination of both, **anemia** has a wide variety of causes. These include dietary deficiencies of iron, vitamin B_{12}, or folic acid; bone-marrow failure due to toxic drugs and cancer; excessive blood loss from the body leading to iron deficiency; inadequate secretion of erythropoietin in kidney disease; and excessive destruction of erythrocytes, as in the disease called sickle-cell anemia.

This last disease is due to a genetic mutation that causes substitution of one wrong amino acid in a particular portion of globin. At the low oxygen concentrations existing in many capillaries, the abnormal hemoglobin molecules interact with each other to form fiberlike structures that distort the erythrocyte membrane and cause the cell to form sickle shapes or other bizarre forms. This results both in the blockage of the capillaries, with consequent tissue damage, and in the destruction of the deformed erythrocytes, with consequent anemia.

Laboratory evaluation of anemia starts with analysis of a blood sample for the number of erythrocytes, the average cell size, and the concentration of hemoglobin. Aver-

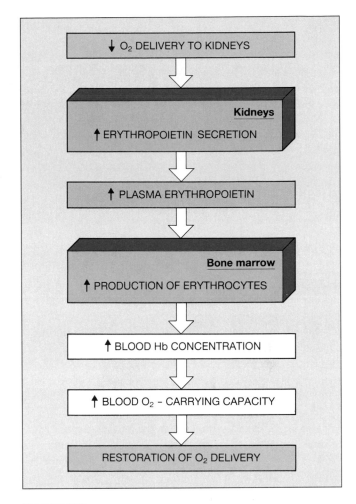

FIGURE 13-5 Reflex by which a decreased oxygen delivery to the kidneys increases erythrocyte production via increased erythropoietin secretion. Situations that trigger this reflex include insufficient pumping of blood by the heart, lung disease, and anemia (a decrease in number of erythrocytes or in hemoglobin concentration).

age cell size is particularly useful since it correlates well with several common types of anemia. For example, **microcytosis** (small cells) most commonly occurs in iron deficiency, **normocytosis** (normal size) accompanies acute blood loss, and **macrocytosis** (large size) is characteristic of vitamin B_{12} or folic acid deficiency. The macrocytosis is due to delayed cell division because of DNA deficiency.

Leukocytes

If one adds appropriate dyes to a drop of blood and examines it under a microscope, the various leukocyte types (Table 13-2) can be seen clearly (Figure 13-6). They are classified according to their structure and affinity for the various dyes.

TABLE 3-2	NUMBERS AND DISTRIBUTIONS OF ERYTHROCYTES, LEUKOCYTES, AND PLATELETS IN NORMAL HUMAN BLOOD

Total erythrocytes = 5,000,000,000 per milliliter of blood

Total leukocytes = 7,000,000 per milliliter of blood

Percent of total leukocytes:
    Polymorphonuclear granulocytes
        Neutrophils 50–70
        Eosinophils 1–4
        Basophils 0.1
    Monocytes 2–8
    Lymphocytes 20–40

Total platelets = 250,000,000 per milliliter of blood

The name **polymorphonuclear granulocytes** refers to the three types of leukocytes with multilobed nuclei and abundant membrane-bound granules. The granules of one group show no dye preference, and the cells are therefore called **neutrophils**. The granules of the second group take up the red dye eosin, thus giving the cells their name **eosinophils**. Cells of the third group have an affinity for basic dyes and are called **basophils**.

A fourth type of leukocyte is the **monocyte**, which is somewhat larger than granulocytes and has a single oval or horseshoe-shaped nucleus and relatively few cytoplasmic granules. The final class of leukocytes is the **lymphocyte**, which contains scanty cytoplasm and, like the monocyte, a relatively large nucleus.

All the leukocyte types are produced in the bone marrow. In addition, many lymphocytes undergo further development and mitosis in tissues outside the bone marrow, as described in Chapter 19. Specific leukocyte functions in the body's defenses are also described in that chapter.

Platelets

The circulating platelets are colorless cell fragments that contain numerous granules and are much smaller than erythrocytes. Platelets are produced when the cytoplasm of large bone marrow cells, termed **megakaryocytes**, become pinched off and enter the circulation. Platelet functions in blood clotting are described in Chapter 19.

Regulation of Blood Cell Production

In children, the marrow of almost all bones produces blood cells. By adulthood, only the bones of the chest, the base of the skull, and the upper limbs remain active. The bone marrow in an adult weighs almost as much as the liver, and it produces cells at an enormous rate.

To reiterate, the bone marrow is the site of production of all blood cells, including the lymphocytes that subsequently undergo mitosis in other tissues. All these cell types are descended from a single population of bone-marrow cells called **pluripotent stem cells**, which are undifferentiated cells capable of giving rise to precursors of any of the individual cell types (Figure 13-7). When a pluripotent stem cell divides, it can replicate itself or become committed to a particular developmental pathway—what governs this "decision" is not known. The first branching yields lymphoid stem cells, which are the precursors of lymphocytes, and myeloid stem cells, the precursors of all the other types. At some point, the proliferating offspring of the myeloid stem cells become committed to differentiate along only one path, for example, into erythrocytes.

What controls the rate of production of each type of blood cell? We have already described the regulatory process for erythrocyte production, which is controlled by the hormone erythropoietin. The production of the leukocyte cell types is stimulated by four other hormones collectively termed **colony-stimulating factors**

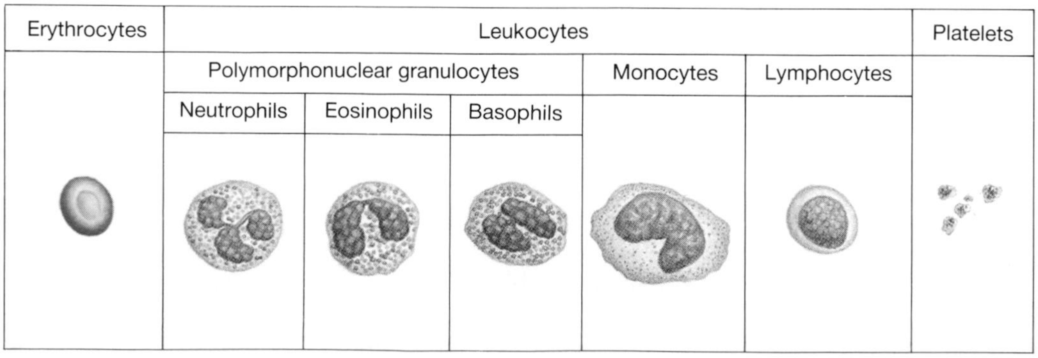

Erythrocytes	Leukocytes					Platelets
	Polymorphonuclear granulocytes			Monocytes	Lymphocytes	
	Neutrophils	Eosinophils	Basophils			

FIGURE 13-6 Normal blood cell types.

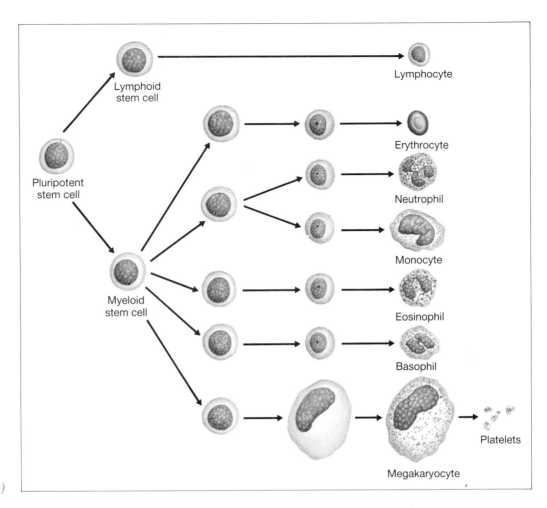

FIGURE 13-7 Production of blood cells by the bone marrow. *(Adapted from Golde and Gasson.)*

Labels in figure: Lymphoid stem cell, Pluripotent stem cell, Myeloid stem cell, Lymphocyte, Erythrocyte, Neutrophil, Monocyte, Eosinophil, Basophil, Megakaryocyte, Platelets

(**CSFs**). For example, granulocyte colony-stimulating factor causes increased production and release of neutrophils by the bone marrow. The physiology of the CSFs is more complex than that of erythropoietin, since these substances are produced by a variety of cells and exert a wide range of effects other than stimulating the production of blood cells. Further description of leukocyte formation is best delayed until Chapter 19 since they are basically part of the body's defenses against infection and cancer.

SECTION B
OVERALL DESIGN OF THE CARDIOVASCULAR SYSTEM

The rapid flow of blood throughout the body is produced by pressures created by the pumping action of the heart. This type of flow is known as **bulk flow** because all constituents of the blood move together. The extraordinary degree of branching of blood vessels assures that almost all cells in the body are within a few cell diameters of at least one of the smallest branches, the capillaries. These distances are small enough for *diffusion* to permit exchanges of nutrients and metabolic end products between capillary blood and the interstitial

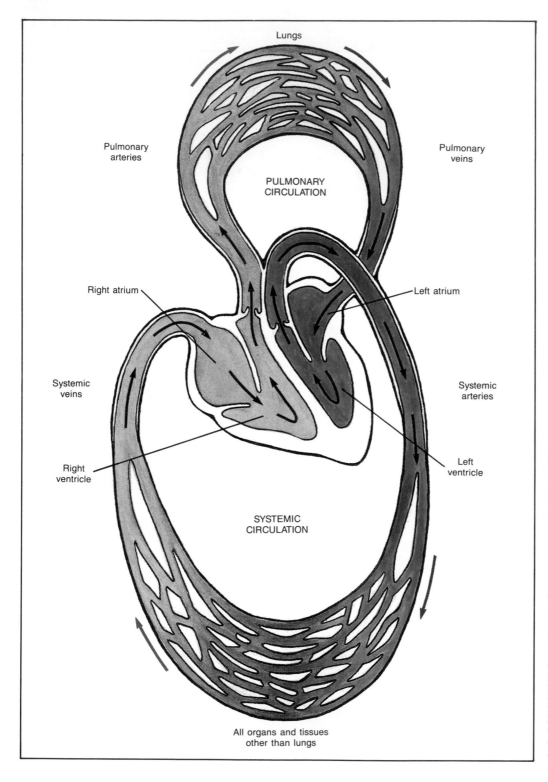

Lungs

Pulmonary
arteries

Pulmonary
veins

PULMONARY
CIRCULATION

Right atrium

Left atrium

Systemic
veins

Systemic
arteries

Right
ventricle

Left
ventricle

SYSTEMIC
CIRCULATION

All organs and tissues
other than lungs

FIGURE 13-8 The systemic and pulmonary circulations. The pulmonary circulation pumps blood through the lungs, and the systemic circulation supplies all other organs and tissues. As depicted by the color change from blue to red, blood is fully oxygenated as it flows through the lungs and then loses some oxygen (red to blue) as it flows through the other organs and tissues. For simplicity, the arteries and veins leaving and entering the heart are depicted as single vessels; in reality, this is true for the arteries but not for the veins (see Figure 13-9).

fluid. Exchanges between the interstitial fluid and cell interior are accomplished by both diffusion and mediated transport.

At any given moment, approximately 5 percent of the total circulating blood is flowing through the capillaries. Yet it is this 5 percent that is performing the ultimate functions of the entire cardiovascular system: the supplying of nutrients and removal of metabolic end products. All other components of the system subserve the overall aim of getting adequate blood flow through the capillaries. This point should be kept in mind as we describe these components.

Cardiovascular physiology as an experimental science began in 1628, when the British physiologist William Harvey demonstrated that the cardiovascular system forms a circle, so that blood pumped out of the heart through one set of vessels returns to the heart via a different set. There are actually two circuits (Figure 13-8), both originating and terminating in the heart, which is divided longitudinally into two functional halves. Each

half contains two chambers: an **atrium** and a **ventricle**. The atrium on each side empties into the ventricle on that side, but there is no direct communication between the two atria or the two ventricles.

Blood is pumped via one circuit (the **pulmonary circulation**) from the right ventricle through the lungs and back into the heart via the left atrium. It is pumped via the other circuit (the **systemic circulation**) from the left ventricle through all the tissues of the body except the lungs and back to the heart via the right atrium. In both circuits, the vessels carrying blood away from the heart are called **arteries** and those carrying blood from the lungs or all other parts of the body (peripheral organs and tissues) back to the heart are called **veins**.

In the systemic circuit, blood leaves the left ventricle via a single large artery, the **aorta** (Figure 13-9). The systemic arteries branch from the aorta, dividing into progressively smaller branches. The smallest arteries form **arterioles** that branch into a huge number of very small, one-cell-thick vessels—the **capillaries**—which

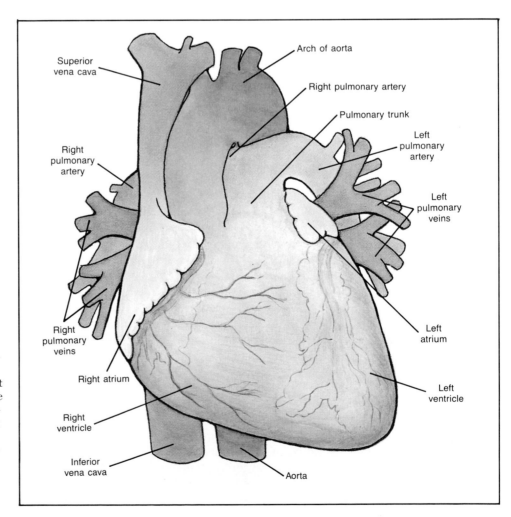

FIGURE 13-9 Blood leaves each of the ventricles via a single artery, the pulmonary trunk from the right ventricle and the aorta from the left ventricle. Because the aorta and pulmonary trunk cross each other before emerging from the heart (Figure 13-13), it looks as if the aorta arises from the right ventricle and the pulmonary trunk from the left ventricle. Blood enters the right atrium via two large veins, the superior vena cava and inferior vena cava. It enters the left atrium via four pulmonary veins.

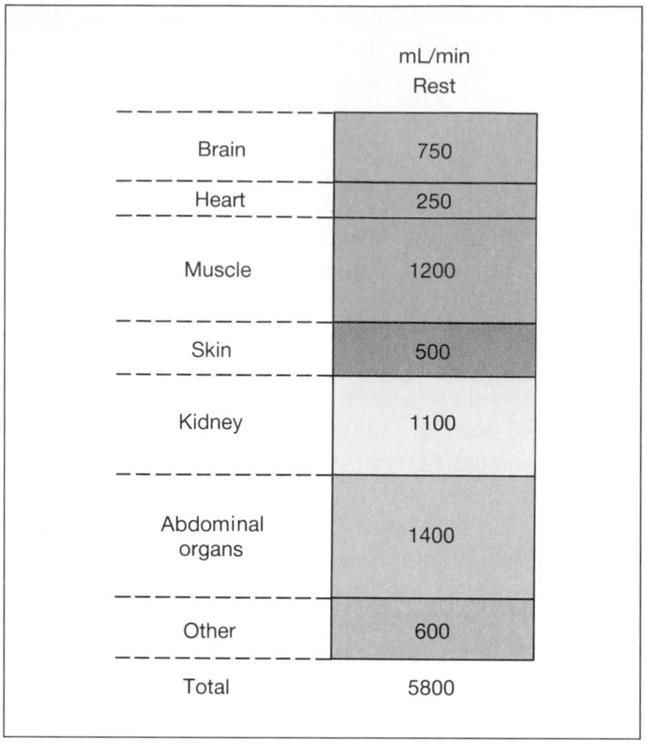

FIGURE 13-10 Distribution of blood flow in the various organs and tissues of the body at rest. (*Adapted from Chapman and Mitchell.*)

unite to form larger and thicker vessels—the **venules**. The arterioles, capillaries, and venules are collectively termed the **microcirculation**.

The venules in the systemic circulation then unite to form larger vessels, the veins. The veins from the various peripheral organs and tissues unite to produce two large veins, the **inferior vena cava**, which collects blood from the lower portion of the body, and the **superior vena cava**, from the upper half of the body. It is via these two veins that blood is returned to the right atrium.

The pulmonary circulation is composed of a similar circuit. Blood leaves the right ventricle via a single large artery, the **pulmonary trunk**, which divides into the two **pulmonary arteries**, one supplying each lung. In the lungs, the arteries continue to branch, ultimately forming capillaries that unite into venules and then veins. The blood leaves the lungs via four **pulmonary veins**, which empty into the left atrium.

As blood flows through the lung capillaries, it picks up oxygen supplied to the adjacent lung air sacs by breathing. Therefore, the blood in the pulmonary veins, left heart, and systemic arteries has a high oxygen content. As this blood flows through the capillaries of tissues and organs throughout the body, some of this oxygen leaves

the blood to enter and be used by cells, resulting in the lower oxygen content of systemic venous blood.

As shown in Figure 13-8, blood can pass from the systemic veins to the systemic arteries only by first being pumped through the lungs. Thus all the blood returning from the body's peripheral organs and tissues via the systemic veins is oxygenated before it is pumped back to them.

It must be emphasized that the lungs receive all the blood pumped by the right heart, whereas each of the peripheral organs and tissues receives only a fraction of the blood pumped by the left ventricle. For reference, the typical distribution of the blood pumped by the left ventricle in an adult at rest is given in Figure 13-10.

Finally, there are several exceptions, notably the liver and kidneys, to the usual anatomical pattern described in this section for the systemic circulation, and these will be presented in the relevant chapters dealing with those organs.

PRESSURE, FLOW, AND RESISTANCE

The last task in this survey of the design of the cardiovascular system is to introduce the concepts of pressure, flow, and resistance. Throughout the system, blood flow F is always from a region of higher pressure to one of lower pressure. The term "pressure" refers to the force exerted by the blood. The pressure exerted by a fluid is often termed a **hydrostatic pressure**. The units for the rate of flow are volume per unit time, usually liters per minute (L/min). The units for the pressure difference (ΔP) driving the flow are millimeters of mercury (mmHg) because historically blood pressure was measured by determining how high a column of mercury could be driven by the blood pressure. It must be emphasized that it is not the absolute pressure at any point in the cardiovascular system that determines flow rate but the *difference* in pressure between the relevant points (Figure 13-11).

Knowing only the pressure difference between two points will not tell you the flow rate, however. For this, you also need to know the **resistance** R to flow, that is, how difficult it is for blood to flow between two points at any given pressure difference. Resistance is the measure of the friction that impedes flow. The basic equation relating these variables is

$$F = \frac{\Delta P}{R}$$

In words, flow rate is directly proportional to pressure difference and inversely proportional to resistance. This equation applies not only to the cardiovascular system

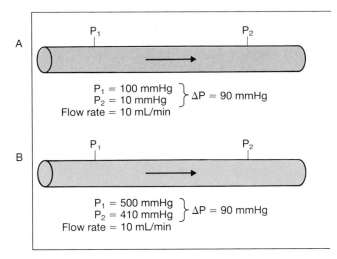

P1 = 100 mmHg
P2 = 10 mmHg } ΔP = 90 mmHg
Flow rate = 10 mL/min

P1 = 500 mmHg
P2 = 410 mmHg } ΔP = 90 mmHg
Flow rate = 10 mL/min

FIGURE 13-11 Flow between two points within a tube is proportional to the pressure difference between the points. The flows in these two identical tubes are the same because the pressure differences are the same.

but to any system in which liquid or air flows through tubes.

Resistance cannot be directly measured but is always calculated from the directly measured F and ΔP. For example, in Figure 13-11 the resistances in both examples can be calculated to be 90 mmHg/10 mL/min = 9 mmHg/mL/min.

This example illustrates how resistance can be *calculated*, but what is it that actually *determines* the resistance?[1] One determinant is the fluid property known as **viscosity**, which is a measure of the friction between adjacent layers of a flowing fluid—the greater the friction, the greater the viscosity. The other determinants of resistance are the tube's length and radius, which determine the amount of friction between the fluid and the tube wall. The following equation defines the contributions of these three determinants:

$$R = \left(\frac{\eta L}{r^4}\right)\left(\frac{8}{\pi}\right)$$

where η = fluid viscosity
L = tube length
r = inside radius of the tube
$8/\pi$ = a constant

In other words, resistance is directly proportional to both the fluid viscosity and tube length and inversely

---

[1] This distinction between how a thing is measured (or calculated) and its determinants may seem confusing to the reader. By standing on a scale, you measure your weight, but your weight is not determined by the scale but rather by how much you eat and exercise, and so on.

proportional to the fourth power of the radius, that is, the radius multiplied by itself four times.

Blood viscosity is not fixed but increases as hematocrit increases, and increases in hematocrit can have important effects on the resistance to flow in certain situations. Under most physiological conditions, however, the hematocrit and, hence, viscosity of blood is relatively constant and does not play a role in the *control* of resistance.

Similarly, since the lengths of the blood vessels remain constant in the body, length is also not a factor in the control of resistance. In contrast, as we shall see, tube radius does not remain constant, and so tube radius—the $1/r^4$ term in our equation—is the most important determinant of changes in resistance. Just how important changes in radius can be is illustrated in Figure 13-12: Increasing the radius of a tube *twofold* decreases

FIGURE 13-12 Effect of tube radius on resistance and flow.

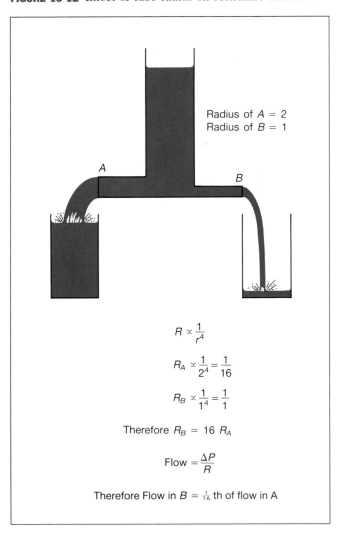

Radius of A = 2
Radius of B = 1

$R \propto \dfrac{1}{r^4}$

$R_A \propto \dfrac{1}{2^4} = \dfrac{1}{16}$

$R_B \propto \dfrac{1}{1^4} = \dfrac{1}{1}$

Therefore $R_B = 16\ R_A$

$\text{Flow} = \dfrac{\Delta P}{R}$

Therefore Flow in $B = \frac{1}{16}$ th of flow in A

TABLE 13-3 THE CARDIOVASCULAR SYSTEM	
Component	**Function**
Heart	
Atria	Chambers through which blood flows from veins to ventricles. Atrial contraction adds to ventricular filling but is not essential.
Ventricles	Chambers whose contractions produce the pressures that drive blood through the pulmonary and systemic vascular systems and back to the heart.
Vascular system	
Arteries	Low-resistance tubes conducting blood to the various organs with little loss in pressure. They act as pressure reservoirs for maintaining blood flow during ventricular relaxation.
Arterioles	Major sites of resistance to flow; responsible for the pattern of blood-flow distribution to the various organs; participate in the regulation of arterial blood pressure.
Capillaries	Sites of nutrient and waste product exchange between blood and tissues.
Venules	Sites of nutrient and waste product exchange between blood and tissues; participate in the regulation of capillary blood pressure.
Veins	Low-resistance conduits for blood flow back to the heart. Their capacity for blood is adjusted so as to facilitate this flow.

its resistance *sixteenfold*. If ΔP is held constant in this example, flow through the tube increases sixteenfold since $F = \Delta P/R$.

This completes our introductory survey of the cardiovascular system. We now turn to a description of its components and their control. In so doing, we might very easily lose sight of the forest for the trees if we do not persistently ask of each section: "How does this component of the circulation contribute to adequate blood flow through the capillaries of the various organs or to an adequate exchange of materials between blood and cells?" Referring to the summary in Table 13-3 as you read the description of each component will help you keep focused on this question.

SECTION C
THE HEART

ANATOMY

The heart is a muscular organ enclosed in a fibrous sac, the **pericardium**, and located in the chest (thorax). The narrow space between the pericardium and the heart is filled with a watery fluid that serves as lubricant as the heart moves within the sac.

The walls of the heart are composed primarily of cardiac-muscle cells and are termed the **myocardium**. The inner surface of the myocardium, that is, the surface in contact with the blood within the cardiac chambers, is lined by a thin layer of cells known as **endothelial cells** or **endothelium**. As we shall see, endothelial cells line not only the heart chambers but the entire vascular system as well.

As noted earlier, the human heart is divided into right and left halves, each consisting of an atrium and a ventricle. Located between the atrium and ventricle in each half of the heart are the **atrioventricular valves (AV valves)**, which permit blood to flow from atrium to ventricle but not from ventricle to atrium (Figure 13-13). The right AV valve is called the **tricuspid valve** and the left the **mitral valve**. The opening and closing of the AV valves is a passive process resulting from pressure differ-

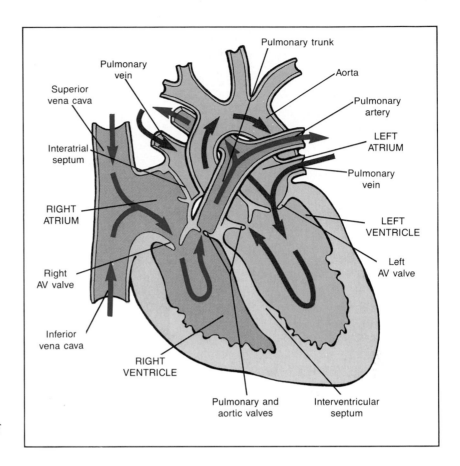

FIGURE 13-13 Diagrammatic section of the heart. The arrows indicate the direction of blood flow.

ences across the valves. When the blood is moving from atrium to ventricle, the valves are pushed open, but when a ventricle contracts, its AV valve is forced closed by the increasing pressure of the blood in the ventricle. Blood is therefore forced into the pulmonary trunk from the right ventricle and into the aorta from the left ventricle, but blood does not move back into the atria.

To prevent the AV valves from being pushed up into the atrium, the valves are fastened to muscular projections (**papillary muscles**) of the ventricular walls by fibrous strands (chordae tendinae). The papillary muscles do *not* open or close the valves. They act only to limit the valves' movements and prevent them from being everted.

The opening of the right ventricle into the pulmonary trunk and that of the left ventricle into the aorta are also regulated by valves, the **pulmonary** and **aortic valves** (Figure 13-13). These valves permit blood to flow into the arteries during ventricular contraction but prevent blood from moving in the opposite direction during ventricular relaxation. Like the AV valves, these valves act in a purely passive manner: Their being open or closed depends upon the pressure difference across them.

There are no valves at the entrances of the venae cavae (plural of vena cava) into the right atrium and of the pulmonary veins into the left atrium. However, atrial contraction pumps very little blood back into the veins because atrial contraction compresses the veins at their sites of entry into the atria, increasing the resistance to backflow. Actually, a little blood is ejected back into the veins, and this accounts for the venous pulse that can often be seen in the neck veins when the atria are contracting.

Figure 13-14 summarizes the path of blood flow through the entire cardiovascular system.

Cardiac Muscle

The cardiac-muscle cells of the myocardium are arranged in layers that are tightly bound together and completely encircle the blood-filled chambers. When the walls of a chamber contract, they come together like a squeezing fist and exert pressure on the blood they enclose.

Cardiac muscle combines properties of both skeletal and smooth muscle (Chapter 11). The cells are striated

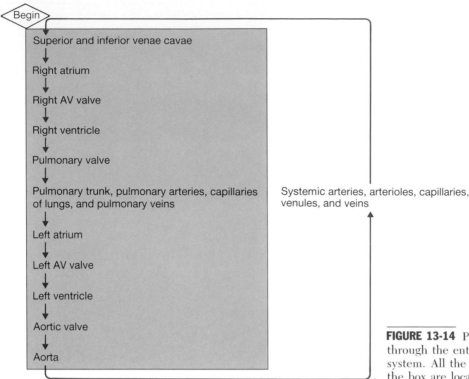

Begin

Superior and inferior venae cavae

↓

Right atrium

↓

Right AV valve

↓

Right ventricle

↓

Pulmonary valve

↓

Pulmonary trunk, pulmonary arteries, capillaries of lungs, and pulmonary veins

↓

Left atrium

↓

Left AV valve

↓

Left ventricle

↓

Aortic valve

↓

Aorta

Systemic arteries, arterioles, capillaries, venules, and veins

FIGURE 13-14 Path of blood flow through the entire cardiovascular system. All the structures within the box are located in the chest.

(Figure 13-15) due to an arrangement of thick myosin and thin actin filaments similar to that of skeletal muscle. Cardiac-muscle cells are considerably shorter than skeletal muscle fibers, however, and have several branching processes. Adjacent cells are joined end to end at structures called intercalated disks, within which are desmosomes that hold the cells together and to which the myofibrils are attached. Adjacent to the intercalated disks are gap junctions, similar to those in many smooth muscles.

Cardiac-muscle cells that contract and produce force constitute 99 percent of the atrial and ventricular muscle fibers. In addition, the myocardium contains specialized cardiac-muscle fibers that are not contractile but are essential for normal heart excitation. They constitute a network known as the **conducting system** of the heart and are in contact with contractile cardiac fibers via gap junctions. The conducting system initiates the heart beat and helps spread the impulse rapidly throughout the heart.

One final point about the cardiac-muscle cells: Certain cells in the atria secrete the family of peptide hormones collectively called atrial natriuretic factor, described in Chapter 15.

Innervation. The heart receives a rich supply of sympathetic and parasympathetic nerve fibers, the latter contained in the vagus nerves. Both sympathetic and

FIGURE 13-15 Electron micrograph of cardiac muscle. Note the striations similar to those of skeletal muscle. The wide horizontal bands are intercalated disks.

Intercalated disk

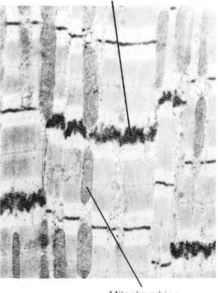

Mitochondrion

parasympathetic postganglionic fibers innervate the cells of the conducting system as well as the contractile cells of the atria and ventricles. The sympathetic postganglionic fibers release primarily norepinephrine, and the parasympathetics release primarily acetylcholine. The receptors for norepinephrine on cardiac muscle are mainly beta-adrenergic, of the beta$_1$ subtype.[2] The hormone epinephrine, from the adrenal medulla, combines with the same receptors as norepinephrine and exerts the same actions on the heart. The receptors for acetylcholine are of the muscarinic type.

Blood supply. The blood being pumped through the heart chambers does not exchange nutrients and metabolic end products with the myocardial cells. They, like those of all other organs, receive their blood supply via arteries that branch from the aorta. The arteries supplying the myocardium are the **coronary arteries**, and the blood flowing through them is termed the **coronary blood flow**. The coronary arteries exit from the very first part of the aorta and lead to a branching network of small arteries, arterioles, capillaries, venules, and veins similar to those in all other organs.

HEARTBEAT COORDINATION

Contraction of cardiac muscle, like that of other muscle types, is triggered by depolarization of the plasma membrane of the cells making up the muscle. As described earlier, myocardial cells are connected to each other by gap junctions. These allow action potentials to spread from one cell to another. Thus, the initial excitation of one myocardial cell eventually results in the excitation of all cells. This initial depolarization normally arises in a small group of conducting-system cells, the **sinoatrial (SA) node**, located in the right atrium near the entrance of the superior vena cava. The action potential then spreads from the SA node throughout the heart in a manner that causes first the atria and then the ventricles to contract.

How does the action potential arise in the SA node, and what is the path and sequence of excitation? To answer these questions, we must first describe cardiac action potentials.

Cardiac Action Potentials

A typical ventricular action potential is illustrated in Figure 13-16A. The membrane permeability changes that underlie it are shown in Figure 13-16B. As in skeletal

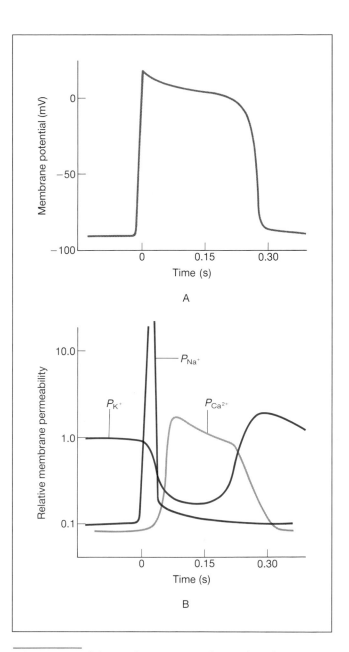

FIGURE 13-16 (A) Membrane potential recording from a ventricular muscle cell. (B) Simultaneously measured permeabilities P to potassium, sodium, and calcium during the action potential of (A). The sodium permeability goes off scale during the spike of the action potential.

muscle cells and neurons, the resting membrane is much more permeable to potassium than to sodium. Therefore, the resting membrane potential is much closer to the potassium equilibrium potential (-90 mV) than to the sodium equilibrium potential ($+60$ mV). Similarly, the depolarizing phase of the action potential is due mainly to a positive-feedback increase in sodium permeability caused by the opening of voltage-gated sodium channels. At almost the same time, the permeability to potassium decreases as potassium channels close, and this also contributes to the membrane depolarization.

---

[2]There are also some alpha-adrenergic receptors, but their physiological significance is not known.

Again as in skeletal muscle cells and neurons, the increased sodium permeability is very transient, since the sodium channels quickly close again. Unlike the case in these other excitable tissues, however, the return of sodium permeability toward its resting value is not accompanied by membrane repolarization. Rather, the membrane remains depolarized at a plateau of about 0 mV. The reasons for this continued depolarization are: (1) Potassium permeability stays below the resting value, and (2) there is a marked increase in the membrane permeability to calcium. The second reason is the more important of the two, and the explanation for it is as follows.

In myocardial cells, the original membrane depolarization causes voltage-gated calcium channels in the plasma membrane to open, which results in a flow of calcium ions down their electrochemical gradient into the cell. These channels are referred to as **slow channels** because there is a delay in their opening. The flow of positive calcium ions into the cell, along with some sodium also entering through the slow channels, just balances the flow of positive potassium charge out of the cell and keeps the membrane depolarized at the plateau value.

Ultimately, repolarization does occur when the permeabilities of calcium, sodium, and potassium all return to their original state, that is, when the slow channels close and the potassium channels reopen.

The membrane potentials of atrial contractile cells are similar in shape to those just described for ventricular cells, but the duration of their plateau phase is shorter.

In contrast, there are extremely important differences between the membrane potentials of contractile myocardial cells and those in the conducting system. Figure 13-17 illustrates the membrane potentials of a cell from the SA node. Note that the resting potential of the SA node cell is not steady but instead manifests a slow depolarization. This gradual depolarization is known as a **pacemaker potential**; it brings the membrane potential to threshold, at which point an action potential occurs. Following the peak of the action potential, the membrane potential repolarizes, and the gradual depolarization begins again.

Thus, the pacemaker potential provides the SA node with **automaticity**, the capacity for spontaneous rhythmic self-excitation. The slope of the pacemaker potential, that is, how quickly the membrane potential changes per unit time, determines how quickly threshold is reached and the next action potential elicited. The inherent rate of the SA node, that is, the rate exhibited in the total absence of any neural or hormonal input to the node, is approximately 100 depolarizations per minute.

What is responsible for the pacemaker potential? The major cause of this gradual depolarization is a progressive, "spontaneous" reduction in membrane permeability of the SA-node cells to potassium, while the permeability to sodium remains relatively constant. Therefore, the inward sodium leak is not balanced by the outward potassium leak, and the membrane slowly depolarizes to threshold.

Other portions of the conducting system are capable of generating pacemaker potentials, but the inherent rate of these other areas is slower than that of the SA node. In contrast, contractile cardiac muscle cells are not normally capable of generating pacemaker potentials.

We can now apply this information concerning cardiac action potentials to understanding the origin of the heartbeat and the spread of excitation throughout the myocardium.

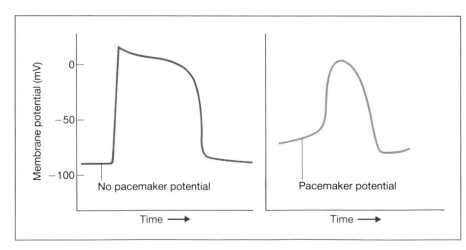

FIGURE 13-17 Comparison of membrane potentials in a ventricular muscle cell (from Figure 13-16) and a sinoatrial- (SA-) nodal cell. The SA-nodal cell's action potential is due to calcium influx.

Sequence of Excitation

To reiterate, the SA node, having the fastest inherent discharge rate of any of the myocardial cells with pacemaker activity, is the normal pacemaker for the entire heart. Its depolarization generates the current that leads to depolarization of all other cardiac-muscle cells, and so its rate of depolarization determines the **heart rate**, the number of times the heart contracts per unit time. Only if the activity of the SA node is depressed or conduction from it is blocked does another portion of the conducting system take over as pacemaker and set the heart rate.

What is the pathway by which the action potential initiated in the SA node spreads throughout the heart, passing from cell to cell by way of gap junctions? The spread throughout the right atrium is mainly from contractile cell to contractile cell. This is also true for the spread from the right atrium to the left atrium as the contractile cells of the two atria contact each other in the central (interatrial) wall they share. The spread is rapid enough that the two atria are depolarized and contract at essentially the same time.

The spread of the action potential to the ventricles is more complicated and involves the rest of the conducting system (Figures 13-18 and 13-19). The link between atrial depolarization and ventricular depolarization is a portion of the conducting system called the **atrioventricular (AV) node**, located at the base of the right atrium. The action potential spreading through the right atrium from contractile cell to contractile cell causes depolariza-

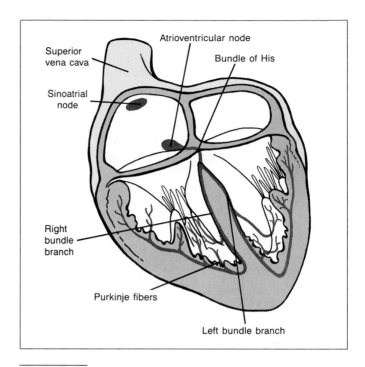

FIGURE 13-18 Conducting system of the heart.

FIGURE 13-19 Sequence of cardiac excitation. Shading denotes areas that are depolarized. Impulse spread from right atrium to left atrium is via the contractile cells where the atria contact each other in the wall shared by the atria. (*Adapted from Rushmer.*)

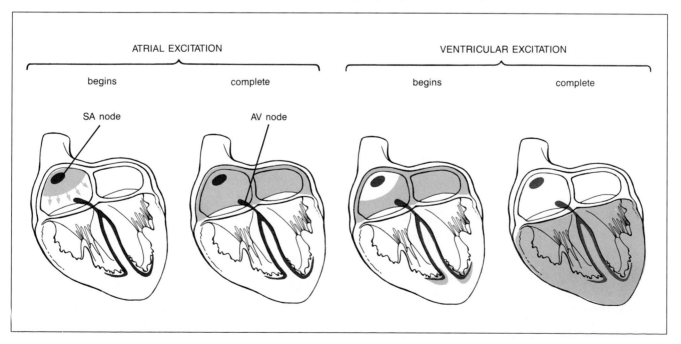

tion of the AV node,[3] and from the AV node the action potential spreads into both ventricles. In this regard, the AV node manifests a particularly important characteristic: The propagation of action potentials through it is relatively slow (requiring approximately 0.1 s) mainly because its cells have small diameters. This delay allows atrial contraction to add additional blood to the ventricles before ventricular excitation occurs.

After leaving the AV node, the impulse enters the wall between the two ventricles (the interventricular septum) via the conducting-system fibers termed the **bundle of His** after its discoverer (pronounced "hiss"). The bundle of His then divides within the septum into **right** and **left bundle branches**, which eventually leave the septum to enter the walls of both ventricles. These fibers in turn branch into a large number of **Purkinje fibers**, large conducting cells that rapidly distribute the impulse throughout much of the ventricles. Finally, the Purkinje fibers make contact with contractile myocardial cells, through which the impulse spreads from cell to cell in the remaining ventricular myocardium.

The rapid conduction along the Purkinje fibers and the diffuse distribution of these fibers cause depolarization of all right and left ventricular cells more or less simultaneously and ensure a single coordinated contraction. Actually, depolarization and contraction begin very slightly earlier in the bottom of the ventricles and spread upward. The result is a more efficient contraction, like squeezing a tube of toothpaste from the bottom up.

It should be emphasized that the AV node and the bundle of His constitute the only electrical link between either of the atria and the ventricles. There are no others because a layer of nonconducting connective tissue, pierced by the bundle, completely separates each atrium from its ventricle. Thus, for example, there can be no impulse transmission from the left atrium to the ventricles. This anatomical pattern ensures that excitation will travel from the SA node to both ventricles only through the AV node, but it also means that malfunction of the AV node may prevent the SA node from influencing the ventricles.

If this occurs, autorhythmic cells in the bundle of His, released from control by the SA node, begin to initiate excitation at their own inherent rate and become the pacemaker for the ventricles. Their rate is quite slow, generally 25 to 40 beats per minute, and it is completely out of synchrony with the atrial contractions, which continue at the normal higher rate of the SA node. Under such conditions, the atria are totally ineffective as pumps

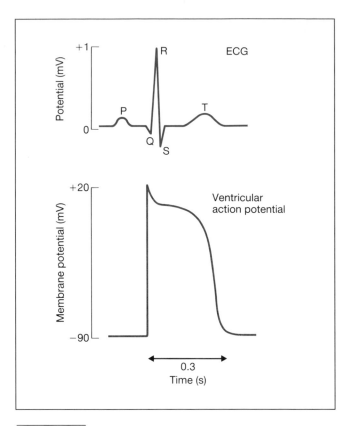

FIGURE 13-20 (Top) Typical electrocardiogram recorded from electrodes connecting the arms. P, atrial depolarization; QRS, ventricular depolarization; T, ventricular repolarization. (Bottom) Ventricular action potential recorded from a single cell. Note the correspondence of the QRS complex with depolarization and the correspondence of the T wave with repolarization.

since they are usually contracting against closed AV valves. Fortunately, atrial pumping, as we shall see, is relatively unimportant for cardiac function except during relatively strenuous exercise.

The Electrocardiogram

The **electrocardiogram** (**ECG** or **EKG**[4]) is primarily a tool for evaluating the *electrical* events within the heart. The action potentials of cardiac-muscle cells can be viewed as batteries that cause charge to move throughout the body fluids. These moving charges—currents, in other words—represent the sum of the action potentials occurring simultaneously in many individual cells and can be detected by recording electrodes at the surface of the skin. Figure 13-20 illustrates a typical normal ECG recorded as the potential difference between the right

---

[3] It has been hypothesized that, in addition to spread along the contractile cells of the right atrium, specialized internodal pathways conduct the impulse directly from the SA node to the AV node. Most experts now believe this is not the case.

[4] The *k* is from the German "kardio" for heart.

and left wrists. The first deflection, the **P wave**, corresponds to atrial depolarization. The second deflection, the **QRS complex**, occurring approximately 0.15 s later, is the result of ventricular depolarization. It is a complex rather than a single deflection, like the P wave, because the paths taken by the wave of depolarization through the thick ventricular walls differ from instant to instant, and the currents generated in the body fluids change direction accordingly. Regardless of its form, for example, the Q and/or S portions may be absent, the deflection is still called a QRS complex. The final deflection, the **T wave**, is the result of ventricular repolarization. Atrial repolarization is usually not evident on the ECG.

A typical clinical ECG makes use of 12 combinations of recording locations on the limbs and chest so as to obtain as much information as possible concerning different areas of the heart. The shapes and sizes of the P wave, QRS complex, and T wave vary with the electrode locations.

To reiterate, the ECG is not a direct record of the changes in membrane potential across cardiac-muscle cells but is rather a measure of the currents generated in the extracellular fluid by these changes. To emphasize this point, Figure 13-20 shows, in addition to an ECG, the simultaneously occurring changes in membrane potential in a single ventricular cell.

Because many myocardial defects alter normal impulse propagation, and thereby the shapes and timing of the waves, the ECG is a powerful tool for diagnosing certain types of heart disease. Figure 13-21 gives one example. It must be emphasized, however, that the ECG provides information concerning only the electrical activity of the heart. Thus, if something is wrong with the heart's mechanical activity, but this defect does not give rise to altered electrical activity, then the ECG will not be of diagnostic value.

Excitation-Contraction Coupling

The depolarization ("excitation") of cardiac-muscle cells triggers their contraction. As described in Chapter 11, the mechanism coupling excitation and contraction in all muscle is an increase in the cytosolic calcium concentration. In cardiac muscle, as in skeletal muscle, calcium combines with the regulator protein, troponin, with the result that inhibition of cross-bridge formation between actin and myosin is eliminated. Also as in skeletal muscle, the major source of the increased cytosolic calcium in cardiac muscle during the action potential is release of calcium from the sarcoplasmic reticulum. A second source is calcium diffusing through calcium channels across the plasma membrane into the cardiac-muscle cell during the action potential.

The amount of calcium entering the cell across the

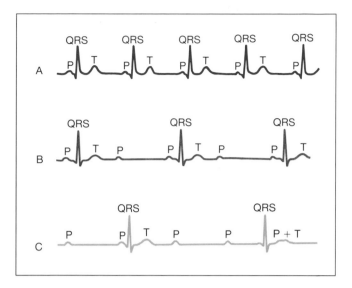

FIGURE 13-21 Electrocardiograms from a healthy person and from two persons suffering from atrioventricular block. (A) A normal ECG. (B) Partial block. Damage to the AV node permits only one-half of the atrial impulses to be transmitted to the ventricles. Note that every second P wave is not followed by a QRS and T. (C) Complete block. There is absolutely no synchrony between atrial and ventricular electrical activities. The ventricles are being driven by a pacemaker in the bundle of His.

plasma membrane is very small compared with the amount simultaneously being released from the sarcoplasmic reticulum, but the former source is crucial for a simple reason: It is this new calcium entering across the plasma membrane that constitutes the signal for release of the much larger amount from the sarcoplasmic reticulum (Figure 13-22).

Contraction ends when the cytosolic calcium concentration is restored to its original extremely low value by active transport of calcium into the sarcoplasmic reticulum. Subsequently, an amount of calcium equal to what entered the cell from the extracellular fluid during excitation is transported out of the cell, so that the total cellular calcium content remains constant.

As we shall see, how much cytosolic calcium concentration increases during excitation is a major determinant of the strength of cardiac-muscle contraction. Cardiac muscle differs greatly in this regard from skeletal muscle, in which the increase in cytosolic calcium occurring during membrane excitation is always adequate to produce maximal "turning-on" of cross bridges by calcium's binding to all troponin sites. In cardiac muscle, the amount of calcium released is usually insufficient to saturate all troponin sites. Thus, the number of active cross bridges, and so the strength of contraction, can be increased if cytosolic calcium is further increased.

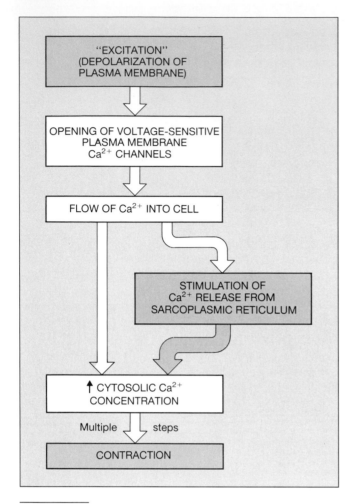

FIGURE 13-22 Excitation-contraction coupling in cardiac muscle. The thick arrow denotes that the sarcoplasmic reticulum is the major source of the increased cytosolic calcium. However, release of calcium from the sarcoplasmic reticulum is triggered by the calcium flow into the cell across the plasma membrane.

Refractory Period of the Heart

Ventricular muscle, unlike skeletal muscle, is incapable of any significant degree of summation of contractions, and this is a very good thing. Imagine that cardiac muscle were able to undergo a prolonged tetanic contraction. During this period no ventricular filling could occur since filling can occur only when the ventricular muscle is relaxed, and the heart would therefore cease to function as a pump.

The inability of the heart to generate such tetanic contractions is the result of the long **refractory period** of cardiac muscle, defined as the period following an action potential during which an excitable membrane cannot be reexcited. As described in Chapter 11, the absolute refractory periods of skeletal muscle are much shorter (1 to 2 ms) than the duration of contraction (20 to 100 ms), and a second contraction can therefore be elicited before the first is over (summation of contractions). In contrast, because of the long plateau, the absolute refractory period of cardiac muscle lasts almost as long as the contraction (250 ms), and the muscle cannot be reexcited in time to produce summation (Figure 13-23).

A common event involving the refractory period is shown in Figure 13-24. In a variety of situations, including, for some people, drinking several cups of coffee or experiencing strong emotion, there is an increased excitability of areas of the heart other than the SA node. When such an area, termed an ectopic focus, initiates an action potential just after the completion of a normal contraction but before the next SA-nodal impulse, a premature excitation wave and contraction occur. As a result, the next normal SA-nodal impulse fires during the refractory period of the premature beat and is not propagated because the myocardial cells are refractory (the SA node still fires because it has a shorter refractory period). The second SA-nodal impulse after the premature contraction is propagated normally. The net result is an unusually long delay between beats. The contraction after the delay is unusually strong (for reasons to be described below) and the person is aware of the heart "skipping a beat." This entire phenomenon is usually benign.

FIGURE 13-23 Relationship between membrane potential changes and contraction in a ventricular-muscle cell. The refractory period lasts almost as long as the contraction.

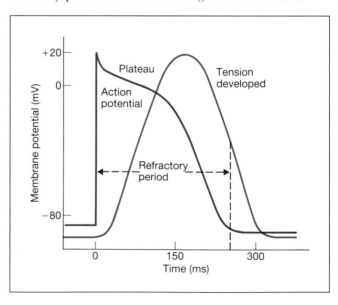

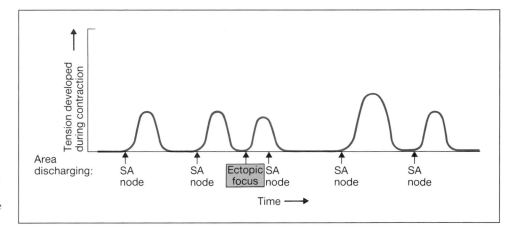

FIGURE 13-24 Effect of the firing of an ectopic focus on the frequency and force of ventricular contraction. The arrows indicate the times at which the SA node or ectopic focus fires.

MECHANICAL EVENTS OF THE CARDIAC CYCLE

The orderly process of depolarization described in the previous sections triggers a recurring **cardiac cycle** of atrial and ventricular contractions and relaxations (Figure 13-25). For orientation, we shall first merely name the parts of this cycle and their key events. Then we shall go through the cycle again, this time describing the pressure and volume changes that cause the events.

The cycle is divided into two major phases, both named for events in the ventricle: the period of ventricular contraction and blood ejection, **systole**, followed by the period of ventricular relaxation, **diastole**, during which the ventricle fills with blood. At an average heart rate of 75 beats/min, each cardiac cycle lasts approximately 0.8 s, with 0.3 s in systole and 0.5 s in diastole.

As illustrated in Figure 13-25, both systole and diastole can be subdivided into several discrete phases. During the very first part of systole, the ventricles are contracting, but all valves in the heart are closed and so no blood can be ejected. This period is termed **isovolumetric ventricular contraction** because the ventricular volume is constant. The ventricular walls are developing tension and squeezing on the blood they encircle, but because the volume of blood in the ventricles is constant and because blood, like water, is essentially incompressible, the ventricular-muscle fibers cannot shorten. Thus, isovolumetric ventricular contraction is analagous to an isometric skeletal-muscle contraction: The muscle develops tension but does not shorten.

Once the rising pressure in the ventricles becomes great enough to open the aortic and pulmonary valves, the **ventricular ejection** period of systole occurs. Blood is forced into the aorta or pulmonary trunk as the con-tracting ventricular-muscle fibers shorten. The volume of blood ejected from a ventricle during systole is termed the **stroke volume**.

During the very first part of diastole, the ventricles relax, but no blood is entering or leaving the ventricles since once again all the valves are closed. Accordingly ventricular volume is not changing, and this period is termed **isovolumetric ventricular relaxation**.[5] The AV valves then open and **ventricular filling** occurs as blood flows in from the atria. Atrial contraction occurs at the very end of diastole, after most of ventricular filling has taken place, and adds to ventricular filling. This is an important point: The ventricle receives blood throughout most of diastole, not just when the atrium contracts. Indeed, in a person at rest, approximately 80 percent of ventricular filling occurs before atrial contraction.

This completes the basic orientation. We now analyze, using Figure 13-26, the pressure and volume changes that occur in the left ventricle and aorta during the cardiac cycle. Events on the right side are described later. Electrical events (ECG) and heart sounds, described in a subsequent section, are at the top of the figure so that their timing can be correlated with the phases of the cycle.

Mid-to-Late Diastole

We begin our analysis at the far left of Figure 13-26 with the events of mid-to-late diastole. The left atrium and ventricle are both relaxed, and atrial pressure (not shown) is very slightly higher than ventricular pressure. Because of the pressure difference, the AV valve is open,

---

[5] Physiologists do not all agree on the dividing line between systole and diastole. As presented here, this dividing line is the point at which ventricular contraction stops. In some definitions, systole includes isovolumetric ventricular relaxation.

SYSTOLE

| ISOVOLUMETRIC VENTRICULAR CONTRACTION | VENTRICULAR EJECTION (Blood flows out of ventricle) |

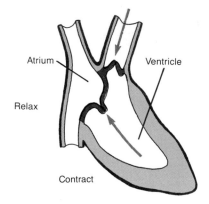

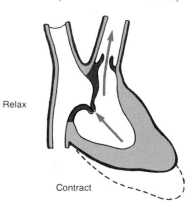

AV valve:	Closed	Closed
Aortic and pulmonary valves:	Closed	Open

DIASTOLE

ISOVOLUMETRIC VENTRICULAR RELAXATION

VENTRICULAR FILLING
Blood flows into ventricle

ATRIAL CONTRACTION

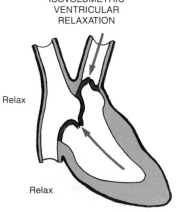

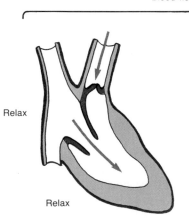

AV valve:	Closed	Open	Open
Aortic and pulmonary valves	Closed	Closed	Closed

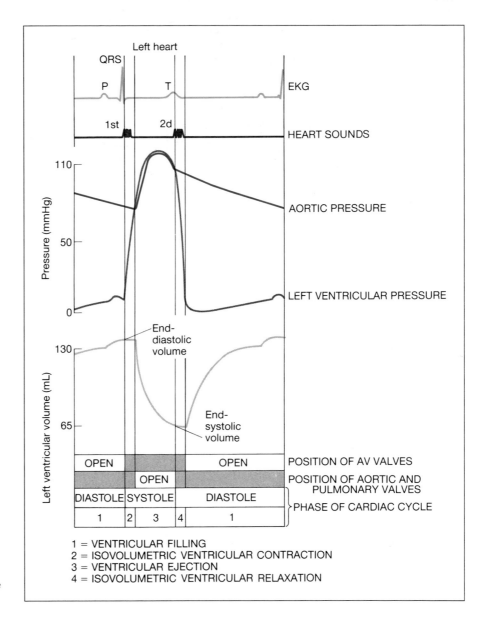

FIGURE 13-26 Summary of events in the left ventricle and aorta during the cardiac cycle.

FIGURE 13-25 (*Opposite*) Divisions of the cardiac cycle. For simplicity, only one atrium and ventricle are shown. The phases of the cycle are identical in both halves of the heart. The direction in which the pressure difference favors flow is denoted by the arrow, but flow can occur only through an open valve.

and blood entering the atrium from the pulmonary veins continues on into the ventricle. Note that during diastole the aortic valve is closed because the aortic pressure is higher than the ventricular pressure.

Throughout diastole, the aortic pressure is slowly falling because blood is moving out of the arteries and through the vascular system. In contrast, ventricular pressure is rising slightly because blood is entering the ventricle, thereby expanding the ventricular volume.

At the very end of diastole, the SA node discharges, the atrium depolarizes (as signified by the P wave of the ECG) and contracts, and a small volume of blood is added to the ventricle (note the small rise in ventricular pressure and blood volume). The amount of blood in the ventricle just prior to systole is called the **end-diastolic volume**.

Systole

From the AV node, the wave of depolarization passes through the ventricle (as signified by the QRS complex of the ECG) and triggers ventricular contraction. Remember that just prior to the contraction the aortic valve is closed and the AV valve is open. As the ventricle contracts, ventricular pressure rises steeply, and almost immediately this pressure exceeds the atrial pressure, closing the AV valve and thus preventing backflow of blood into the atrium. Since the aortic pressure still exceeds the ventricular, the aortic valve remains closed and the ventricle cannot empty despite its contraction.

This brief phase of isovolumetric ventricular contraction ends when the rising ventricular pressure exceeds aortic pressure. The aortic valve opens and ventricular ejection occurs. The ventricular-volume curve shows that ejection is rapid at first and then tapers off. Note that the ventricle does not empty completely. The amount of blood remaining after ejection is called the **end-systolic volume**. Thus:

Stroke volume = end-diastolic volume
$$- \text{ end-systolic volume}$$

You might logically assume that, as soon as ejection begins, ventricular pressure would remain constant, as in an isotonic skeletal-muscle contraction, or even begin to fall, since ventricular volume is decreasing. However, the fact is that ventricular pressure continues to rise during early ejection, as shown in the figure, because the muscle fibers are contracting vigorously enough to elevate the pressure still further despite the diminishing volume it is squeezing upon.

As blood flows into the aorta, the aortic pressure rises along with ventricular pressure. Throughout ejection, only very small pressure differences exist between the ventricle and aorta because the aortic valve opening offers little resistance to flow. Peak ventricular and aortic pressure is reached before the end of ventricular ejection. That is, the pressures start to fall during the last part of systole despite continued ventricular contraction. This is because the rate of ventricular blood ejection during the last part of systole is quite small, as shown by the ventricular-volume curve, and is less than the rate at which blood is leaving the aorta. Accordingly the volume and, therefore, the pressure in the aorta begin to decrease.

Early Diastole

Diastole begins as ventricular contraction and ejection stop and the ventricular muscle relaxes rapidly. Ventricular pressure falls significantly[6] below aortic pressure, and the aortic valve closes. However, at this time, ven-

tricular pressure still exceeds atrial pressure, so that the AV valve also remains closed. This early diastolic phase of isovolumetric ventricular relaxation ends as ventricular pressure falls below atrial pressure, the AV valve opens, and rapid ventricular filling begins.

It is likely that the previous contraction compresses the elastic elements of the ventricle in such a way that the ventricle actually tends to recoil outward once systole is over. This expansion, in turn, lowers ventricular pressure more rapidly than would otherwise occur—and may even create a negative pressure in the ventricle—which enhances filling. Thus, some energy is stored within the myocardium during contraction, and its release during the subsequent relaxation aids filling.

The fact that ventricular filling is almost complete during early diastole is of the greatest importance. It ensures that filling is not seriously impaired during periods when the heart is beating very rapidly, despite a reduction in the duration of diastole and, therefore, total filling time. However, when rates of approximately 200 beats/min or more are reached, filling time does become inadequate, and the volume of blood pumped during each beat is decreased. The significance of this will be described in the section on exercise.

Early ventricular filling also explains why the conduction defects that eliminate the atria as efficient pumps do not seriously impair cardiac function, at least at rest. Many persons lead relatively normal lives for many years despite **atrial fibrillation**—a state in which the atria con-

---

FIGURE 13-27 Pressures in the right ventricle and pulmonary trunk during the cardiac cycle. This figure is done on the same scale as Figure 13-26 to facilitate comparison. The pressure profiles are qualitatively identical in the pulmonary and systemic circuits, but the pressures are much lower in the former.

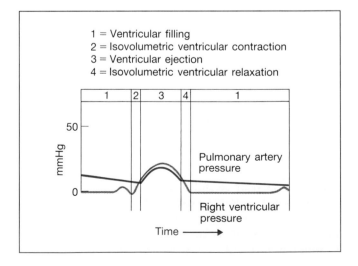

---

[6]The word "significantly" is used in this sentence because, in reality, left ventricular pressure is already slightly less than aortic pressure during late systole, despite the fact that ejection is still occurring. The physical explanation of this pressure "reversal" in late systole is beyond the scope of this book.

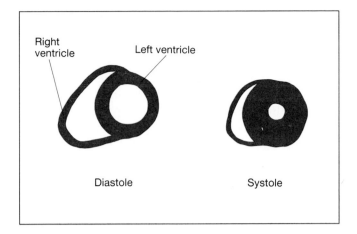

Right ventricle   Left ventricle

Diastole   Systole

FIGURE 13-28 Cross section of the ventricular walls to illustrate the changes that occur during contraction. The view is from above, with the atria removed.

tract in a continuous and completely disordered manner and so fail to serve as effective pumps. Thus, in many respects, the atrium may be conveniently viewed as merely a continuation of the large veins.

Pulmonary Circulation Pressures

The pressure changes in the right ventricle and pulmonary arteries are qualitatively similar to those just described for the left ventricle and aorta (Figure 13-27). There are striking quantitative differences, however. Typical pulmonary artery systolic and diastolic pressures are 24 and 8 mmHg, respectively, compared to systemic arterial pressures of 120/70 mmHg. Similarly, the right ventricular systolic pressure is slightly above 24 mmHg, while the left ventricular systolic pressure is slightly greater than 120 mmHg. Thus, the pulmonary circulation is a low-pressure system, for reasons to be described in a later section. This difference is clearly reflected in the ventricular architecture, the right ventricular wall being much thinner than the left (Figure 13-28). Despite its lower pressure during contraction, however, the right ventricle ejects the same amount of blood as the left.

Heart Sounds

Two sounds, termed **heart sounds**, stemming from cardiac contraction are normally heard through a stethoscope placed on the chest wall. The first sound, a soft low-pitched *lub*, is associated with closure of the AV valves at the onset of systole (Figure 13-26). The second, a louder *dub*, is associated with closure of the pulmonary and aortic valves at the onset of diastole. These sounds, which result from vibrations caused by the closing valves, are perfectly normal, but other sounds, known as **heart murmurs**, are frequently a sign of heart disease.

Murmurs can be produced by blood flowing rapidly in the usual direction through an abnormally narrowed valve (**stenosis**), by blood flowing backward through a damaged leaky valve (**insufficiency**), or by blood flowing between the two atria or two ventricles via a small hole in the wall separating them.

The exact timing and location of the murmur provide the physician with a powerful diagnostic clue. For example, a murmur heard throughout systole suggests a stenotic pulmonary or aortic valve, an insufficient AV valve, or a hole in the interventricular septum. In contrast, a murmur heard during diastole suggests a stenotic AV valve or an insufficient aortic or pulmonary valve.

THE CARDIAC OUTPUT

The volume of blood pumped by *each* ventricle per minute is called the **cardiac output** (**CO**), usually expressed as liters per minute. It is also the volume of blood flowing through *either* the systemic *or* the pulmonary circuit per minute.

The cardiac output is determined by multiplying the heart rate HR—the number of beats per minute—and the stroke volume SV, the blood volume ejected by each ventricle with each beat:

$$CO = HR \times SV$$

Thus, if each ventricle has a rate of 72 beats/min and ejects 70 mL of blood with each beat, the cardiac output is

$$CO = 72 \text{ beats/min} \times 0.07 \text{ L/beat} = 5.0 \text{ L/min}$$

These values are approximately normal for a resting adult. Since, by coincidence, total blood volume is approximately 5 L, this means that essentially all the blood is pumped around the circuit once each minute. During periods of exercise in well-trained athletes, the cardiac output may reach 35 L/min.

The following description of the factors that alter the two determinants of cardiac output—heart rate and stroke volume—applies in all respects to both the right and left heart since stroke volume and heart rate are the same for both under steady-state conditions. It must also be emphasized that heart rate and stroke volume do not always change in the same direction. For example, as we shall see, stroke volume decreases following blood loss while heart rate increases. These changes produce opposing effects on cardiac output.

Control of Heart Rate

Rhythmic beating of the heart at a rate of approximately 100 beats/min will occur in the complete absence of any

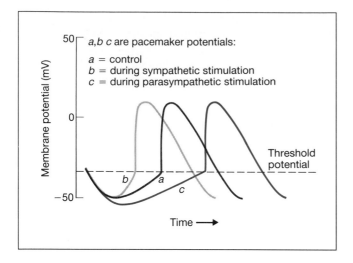

FIGURE 13-29 Effects of sympathetic and parasympathetic nerve stimulation on the slope of the pacemaker potential of an SA-nodal cell. Note that parasympathetic stimulation not only reduces the slope of the pacemaker potential but also causes the membrane potential to be more negative before the pacemaker potential begins. (*Adapted from Hoffman and Cranefield.*)

nervous or hormonal influences on the SA node. This is the inherent autonomous discharge rate of the SA node. The heart rate may be much lower or higher than this, however, since the SA node is normally under the constant influence of nerves and hormones.

As mentioned earlier, a large number of parasympathetic and sympathetic postganglionic fibers end on the SA node. Activity in the parasympathetic (vagus) nerves causes the heart rate to decrease, whereas activity in the sympathetic nerves increases the heart rate. In the rest-

ing state, there is considerably more parasympathetic activity to the heart than sympathetic, and so the normal resting heart rate is well below the inherent rate of 100 beats/min.

Figure 13-29 illustrates how sympathetic and parasympathetic activity influences SA-nodal function. Sympathetic stimulation increases the slope of the pacemaker potential, causing the SA pacemaker cells to reach threshold more rapidly and the heart rate to increase. Stimulation of the parasympathetics has the opposite effect: The slope of the pacemaker potential decreases, threshold is reached more slowly, and heart rate decreases. Parasympathetic stimulation also hyperpolarizes the membrane so that the pacemaker potential starts from a lower value.

How do the neurotransmitters released by the autonomic neurons cause these changes in the pacemaker potential? Acetylcholine, the parasympathetic neurotransmitter, increases the number of open potassium channels, which tends to hold the membrane potential closer to the potassium equilibrium potential. Recall from the section on the SA node that the major cause of the pacemaker potential is the closing of potassium channels. Norepinephrine, the sympathetic neurotransmitter, enhances the rate of pacemaker depolarization mainly by opening channels for sodium and calcium, thereby increasing the flow of positive charge into the cell and hastening its depolarization.

Factors other than the cardiac nerves can also alter heart rate. Epinephrine, the main hormone liberated from the adrenal medulla, speeds the heart by acting on the same beta-adrenergic receptors in the SA node as does norepinephrine. The heart rate is also sensitive to changes in body temperature, plasma electrolyte concentrations, and hormones other than epinephrine.

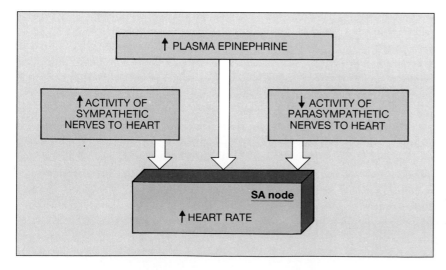

FIGURE 13-30 Major factors that influence heart rate. All effects are exerted upon the SA node. The figure shows how heart rate is increased. Reversal of all the arrows in the boxes would illustrate how heart rate is decreased.

These factors are normally of lesser importance, however. Figure 13-30 summarizes the major determinants of heart rate.

As stated in the section on innervation, sympathetic and parasympathetic neurons innervate other parts of the conducting system as well as the SA node. Thus, sympathetic stimulation not only speeds the SA node but increases conduction through the AV node. Parasympathetic stimulation, in contrast, decreases the rate of spread of excitation along almost the entire conducting system of the heart.

Control of Stroke Volume

The second variable that determines cardiac output is stroke volume, the volume of blood ejected by each ventricle during each contraction. As emphasized earlier, the ventricles never completely empty themselves of blood during contraction. Therefore, a more forceful contraction can produce an increase in stroke volume. Changes in the force of contraction can be produced by a variety of factors, but two are dominant under most physiological conditions: (1) changes in the end-diastolic volume, that is, the volume of blood in the ventricles just prior to contraction, and (2) changes in the magnitude of sympathetic nervous system input to the ventricles.

Relationship between end-diastolic volume and stroke volume: Starling's law of the heart. It is possible to study the characteristics of an isolated heart by using what is called a heart-lung preparation (Figure 13-31). Tubes are placed in the heart and blood vessels of an anesthetized animal so that blood flows from the first part of the aorta into a blood-filled reservoir and from there into the right atrium. The blood is then pumped by the right heart via the lungs into the left heart, from which it is returned to the reservoir.

A key feature of this preparation is that the pressure causing blood flow *into* the heart can be altered simply by raising or lowering the reservoir. This is analogous to altering the systemic venous and right atrial pressures to cause changes in the quantity of blood entering the right ventricle during diastole. Thus when the reservoir is raised, the pressure is raised, and ventricular filling increases, thereby increasing end-diastolic volume. The important finding is that the ventricle, now distended with more blood, responds with a more forceful contraction. The net result is a new steady state in which end-diastolic volume and stroke volume are both increased.

The major explanation of this characteristic of the heart is the length-tension relationship described on page 302. Like skeletal muscle, cardiac muscle contracts more forcefully when, prior to the contraction, it has

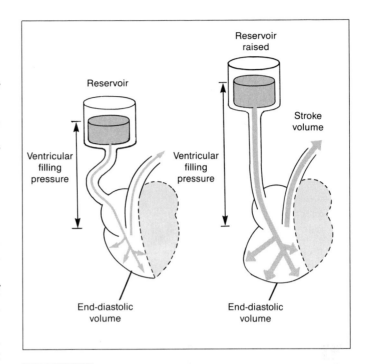

FIGURE 13-31 Experiment for demonstrating the relationship between ventricular end-diastolic volume and stroke volume (Starling's law of the heart). When the reservoir is raised, the pressure causing ventricular filling is increased, and more blood enters the ventricle. This stretches the ventricular muscle, which responds with a stronger contraction.

been stretched over a certain range. End-diastolic volume acts as the preload, determining how stretched the ventricle is just before contraction. The British physiologist Ernest Henry Starling, discovered this relationship between end-diastolic volume and force of contraction, and it is termed **Starling's law of the heart.**

A typical response curve obtained by progressively increasing end-diastolic volume is shown in Figure 13-32. Unlike skeletal muscle, cardiac-muscle length in the resting state is less than that which yields maximal tension during contraction. Therefore, an increase in length from this resting position produces an increase in contractile tension.

The significance of this intrinsic mechanism should be apparent: An increased flow of blood from the veins into the heart (**venous return**) automatically forces an increase in cardiac output by distending the ventricle and increasing stroke volume. One important function of this relationship is maintaining the equality of right and left cardiac outputs. Should the right heart, for example, suddenly begin to pump more blood than the left, the increased blood flow to the left ventricle would automatically produce an equivalent increase in left ventricular

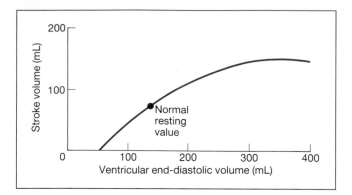

FIGURE 13-32 Relationship between ventricular end-diastolic volume and stroke volume (Starling's law of the heart). The data were obtained by progressively increasing ventricular filling pressure with a heart-lung preparation. The horizontal axis could have been labeled "sarcomere length," and the vertical "contractile force." In other words, this is a length-tension curve.

output. This ensures that blood will not accumulate in the lungs.

The sympathetic nerves. Sympathetic nerves are distributed not only to the SA node and conducting system but to all myocardial cells. The effect of the sympathetic mediator norepinephrine is to increase ventricular **contractility**, defined as the strength of contraction *at any given end-diastolic volume*. Plasma epinephrine also increases myocardial contractility. Thus, the increased force of contraction and stroke volume resulting from sympathetic-nerve stimulation or epinephrine is *independent* of a change in end-diastolic ventricular volume.

FIGURE 13-33 Effects on stroke volume of stimulating the sympathetic nerves to the heart. Stroke volume is increased at any given end-diastolic volume; that is, the sympathetic stimulation has increased ventricular contractility.

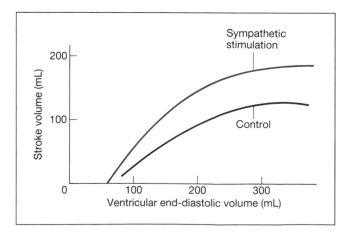

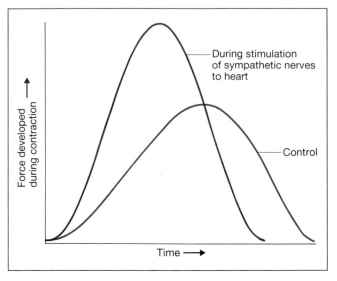

FIGURE 13-34 Effects of sympathetic stimulation on ventricular contraction and relaxation. The dashed portion of each line is during relaxation. Note that both the rate of force development and the rate of relaxation are increased, as is the maximal force developed. All these changes reflect an increased contractility.

Note that a change in contraction force due to increased end-diastolic volume (Starling's law) does *not* reflect increased contractility. Increased contractility is specifically defined as an increased contraction force at any given end-diastolic volume.

The relationship between Starling's law and the cardiac sympathetic nerves as measured in a heart-lung preparation is illustrated in Figure 13-33. The dashed line is the same as the line shown in Figure 13-32 and was obtained by slowly raising ventricular pressure while measuring end-diastolic volume and stroke volume. The solid line was obtained for the same heart during sympathetic-nerve stimulation. Starling's law still applies, but during nerve stimulation the stroke volume is greater at any given end-diastolic volume. In other words, the increased contractility leads to a more complete ejection of the end-diastolic ventricular volume.

Not only does enhanced sympathetic-nerve activity to the myocardium cause the contraction to be more powerful, it also causes both the contraction and relaxation of the ventricles to occur more quickly (Figure 13-34). These latter effects are quite important since, as described earlier, increased sympathetic activity to the heart also increases heart rate. As heart rate increases, the time available for diastolic filling decreases, but the quicker contraction and relaxation induced simultaneously by the sympathetic neurons partially compensate for this problem by permitting a larger fraction of the cardiac cycle to be available for filling.

FIGURE 13-35 Mechanisms by which norepinephrine and epinephrine alter cardiac contractility. Norepinephrine binds to beta$_1$-adrenergic receptors in the plasma membrane of the myocardial cells, and this causes the receptor, via G_s protein, to activate membrane-bound adenylate cyclase. The resulting increase in intracellular cAMP activates cAMP-dependent protein kinase, which phosphorylates at least three proteins: (1) a plasma-membrane protein that controls slow calcium channels; (2) a sarcoplasmic reticulum protein that controls an ATP-dependent calcium-uptake pump; and (3) myosin. These phosphorylated proteins then mediate the observed effects: (1) The plasma-membrane protein causes a greater number of slow calcium channels to be open during excitation, which permits more calcium to enter the cell and bind to troponin, as well as stimulate the sarcoplasmic reticulum to release still more calcium—the result is an increase in the number of actin-myosin cross-bridge attachments, leading to more force development. (2) The sarcoplasmic reticulum protein stimulates the active transport of calcium into the sarcoplasmic reticulum, thereby increasing this organelle's calcium stores and the amount of calcium released into the cytosol during subsequent excitations. The increased rate of calcium pumping from the cytosol back into the sarcoplasmic reticulum after contraction is over also explains why norepinephrine speeds cardiac relaxation. (3) Phosphorylation of myosin causes an increase in the rate at which the cross-bridges cycle. The result is an increase in the velocity of the contraction, that is, the rate of tension development per unit time.

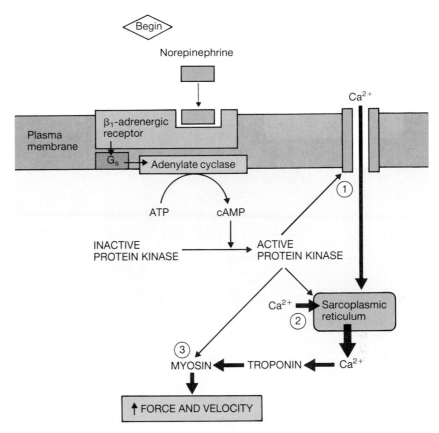

At the level of the contractile apparatus, two processes account for the increased contractility induced by norepinephrine and epinephrine. One is an increase in the number of actin-myosin cross-bridge attachments, resulting from an increased cytosolic calcium concentration. The other is an increase in cross-bridge cycling during a contraction. The cellular mechanisms by which these changes are brought about are summarized in Figure 13-35.

In contrast to the sympathetic nerves, the parasympathetic nerves to the heart normally have only a small effect on ventricular contractility.

Table 13-4 summarizes the effects of the autonomic nerves on cardiac function.

In summary (Figure 13-36), the two major controllers of stroke volume are a mechanism dependent upon changes in end-diastolic volume and a mechanism that is mediated by the cardiac sympathetic nerves and circulating epinephrine and that causes increased ventricular

TABLE 13-4 EFFECTS OF AUTONOMIC NERVES ON THE HEART		
Area Affected	**Sympathetic Nerves**	**Parasympathetic Nerves**
SA node	Increased heart rate	Decreased heart rate
AV node	Increased conduction rate	Decreased conduction rate
Atrial muscle	Increased contractility	Decreased contractility
Ventricular muscle	Increased contractility	Decreased contractility (minor)

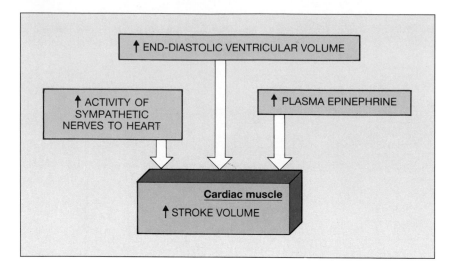

FIGURE 13-36 Major factors that directly influence stroke volume. The figure as drawn shows how stroke volume is increased. A reversal of all arrows in the boxes would illustrate how stroke volume is decreased.

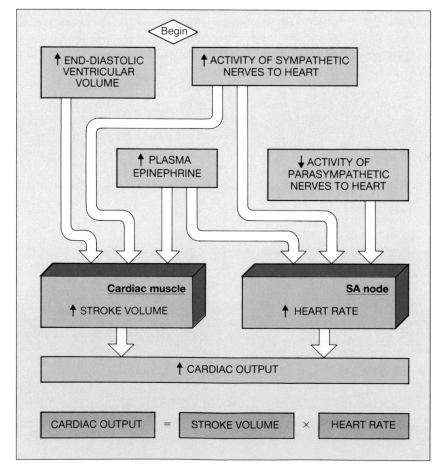

FIGURE 13-37 Major factors determining cardiac output (an amalgamation of Figures 13-30 and 13-36).

contractility. The contribution of each of these two mechanisms in specific physiological situations is described in later sections.[7]

A summary of the major factors that determine cardiac output is presented in Figure 13-37, which combines the information of Figures 13-30 and 13-36.

SECTION D
THE VASCULAR SYSTEM

The functional and structural characteristics of the blood vessels change with successive branching. Yet the entire cardiovascular system, from the heart to the smallest capillary, has one structural component in common: a smooth, single-celled layer of endothelial cells, or endothelium, which lines the inner (blood-contacting) surface of the vessels. Capillaries consist only of endothelium, whereas all other vessels have, in addition, layers of connective tissue and smooth muscle. Endothelial cells have a large number of active functions. These are summarized for reference in Table 13-5,

TABLE 13-5 FUNCTIONS OF ENDOTHELIAL CELLS
1. Serve as a physical lining of heart and blood vessels
2. Secrete endothelium-derived relaxing factors (EDRF), which mediate vascular smooth-muscle responses to many chemical agents and to mechanical force
3. Secrete substances that stimulate angiogenesis (vessel growth)
4. Regulate transport of macromolecules and other substances between plasma and interstitial fluid
5. Regulate platelet clumping, clotting, and anticlotting (Chapter 19)
6. Synthesize active hormones from inactive precursors (Chapter 15)
7. Extract or degrade hormones and other mediators (Chapter 14)
8. Undergo contractile activity, which regulates capillary permeability
9. Influence vascular smooth-muscle proliferation in the disease atherosclerosis

and most of them are described in relevant sections of this chapter.

We have previously described the pressures in the aorta and pulmonary arteries during the cardiac cycle. Figure 13-38 illustrates the pressure changes that occur along the rest of the systemic and pulmonary vascular systems. Text sections below dealing with the individual vascular segments will describe the reasons for these changes in pressure. For the moment, note only that by the time the blood has completed its journey back to the atrium in each circuit, virtually all the pressure originally generated by the ventricular contraction has been dissipated. The reason pressure at any point in the vascular system is less than that at an earlier point is that the blood vessels offer resistance to the flow from one point to the next.

ARTERIES

The aorta and other systemic arteries have thick walls containing large quantities of elastic tissue. Although they also have smooth muscle, arteries can be viewed most conveniently as elastic tubes. Because the arteries have large radii, they serve as low-resistance tubes conducting blood to the various organs. Their second major function, related to their elasticity, is to act as a "pressure reservoir" for maintaining blood flow through the tissues during diastole, as described below.

Arterial Blood Pressure

What are the factors determining the pressure within an elastic container, such as a balloon filled with water? The pressure inside the balloon depends on the volume of water and on how easily the balloon walls can be stretched. If the walls are very stretchable, large quantities of water can be added with only a small rise in pressure. Conversely, the addition of a small quantity of water causes a large pressure rise in a balloon that is difficult to stretch. The term used to denote how easily a

---

[7]A third determinant of stroke volume is the arterial pressure against which the ventricles are pumping during ejection. An increased arterial pressure tends to reduce stroke volume. Just as end-diastolic volume represents the preload for the ventricles, the arterial pressure represents the after-load. This factor is not dealt with in the text since several inherent adjustments reduce the overall influence of after-load on stroke volume.

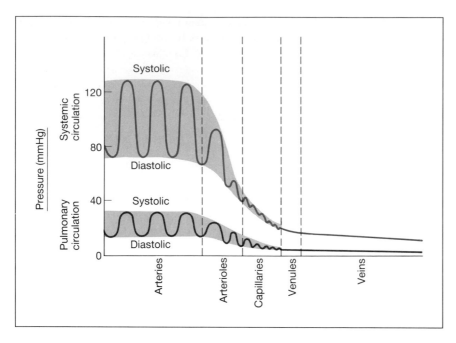

FIGURE 13-38 Pressures in the vascular system.

structure can be stretched is **compliance**. The higher the compliance of a structure, the more easily it can be stretched.

These principles can be applied to an analysis of arterial blood pressure. The contraction of the ventricles ejects blood into the pulmonary and systemic arteries during systole. If a precisely equal quantity of blood were to flow simultaneously out of the arteries, the total volume of blood in the arteries would remain constant and arterial pressure would not change. Such is not the case, however. As shown in Figure 13-39, a volume of blood equal to only about one-third the stroke volume leaves the arteries during systole. The rest of the stroke volume remains in the arteries during systole, distending them and raising the arterial pressure. When ventricular contraction ends, the stretched arterial walls recoil passively, like a stretched rubber band being released, and the arterial pressure continues to drive blood into the arteries during diastole. As blood leaves the arteries, the arterial volume and, therefore, the arterial pressure slowly fall, but the next ventricular contraction occurs while there is still adequate blood in the arteries to stretch them partially. Therefore the arterial pressure does not fall to zero.

The aortic pressure pattern shown in Figure 13-40A is typical of the pressure changes that occur in all the large systemic arteries. The maximum pressure reached during peak ventricular ejection is called **systolic pressure** (**SP**). The minimum pressure occurs just before ventricular ejection begins and is called **diastolic pressure** (**DP**).

FIGURE 13-39 Movement of blood into and out of the arteries during the cardiac cycle. During systole, less blood leaves the arteries than enters from the heart, and so blood is retained in the arteries, stretching their walls. During diastole, the stretched walls recoil passively, and the blood retained in the arteries during the previous systole is driven out of the arteries. The lengths of the arrows denote relative quantities flowing into and out of the arteries and remaining in the arteries during systole.

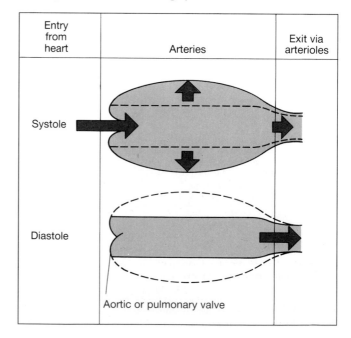

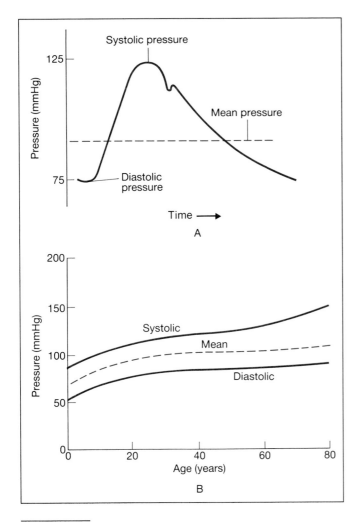

FIGURE 13-40 (A) Typical arterial pressure fluctuations during the cardiac cycle. (B) Changes in arterial pressure with age. *(Adapted from Guyton.)*

Arterial pressure is generally recorded as systolic/diastolic, that is, 125/75 mmHg in our example (see Figure 13-40B for average values in people of different ages).

The difference between systolic pressure and diastolic pressure (125 − 75 = 50 mmHg in the example) is called the **pulse pressure**. It can be felt as a pulsation or throb in the arteries of the wrist or neck with each heart beat. During diastole, nothing is felt over the artery, but the rapid rise in pressure at the next systole pushes out the artery wall, and it is this expansion of the vessel that produces the detectable throb.

Three factors determine the magnitude of the pulse pressure, that is, how much greater systolic pressure is than diastolic: (1) stroke volume, (2) speed at which stroke volume is ejected, and (3) arterial compliance.

The pressure increase produced by a ventricular ejection will be greater if the volume of blood ejected is increased, if the volume is ejected more quickly (as when ventricular contractility is enhanced by sympathetic-nerve stimulation), or if the arteries are less compliant. This last phenomenon occurs in atherosclerosis, a form of "hardening" of the arteries that progresses with age and accounts for the increasing pulse pressure seen so often in older people.

It is evident from Figure 13-40A that arterial pressure is constantly changing throughout the cardiac cycle. The average pressure (**mean arterial pressure, MAP**) in the cycle is not merely the value halfway between systolic and diastolic pressure because diastole usually lasts longer than systole. The true mean arterial pressure can be obtained by complex methods, but for most purposes it is approximately equal to the diastolic pressure plus one-third of the pulse pressure (SP − DP):

$$MAP = DP + \tfrac{1}{3}(SP - DP)$$

In our example, MAP = $75 + \tfrac{1}{3} \times 50 = 92$ mmHg.

The MAP is the most important of the pressures described because it is the pressure driving blood into the tissues averaged over the entire cardiac cycle.

One last point: We can say "arterial" pressure without specifying to which artery we are referring because the aorta and other arteries have such large diameters that they offer only negligible resistance to flow, and the mean pressures are therefore similar everywhere in the arterial tree.

Measurement of Arterial Pressure

Both systolic and diastolic blood pressure are readily measured in human beings with the use of a sphygmomanometer. An inflatable cuff is wrapped around the upper arm, and a stethoscope is placed in a spot on the arm just below the cuff and beneath which the major artery to the lower arm runs.

The cuff is then inflated with air to a pressure greater than systolic blood pressure (Figure 13-41). The high pressure in the cuff is transmitted through the tissue of the arm and completely collapses the artery under the cuff, thereby preventing blood flow to the lower arm. The air in the cuff is then slowly released, causing the pressure in the cuff and on the artery to drop. When cuff pressure has fallen to a value just below the systolic pressure, the artery opens slightly and allows blood flow for this brief time. During this interval, the blood flow through the partially collapsed artery occurs at a very high velocity because of the small opening and the large pressure difference across the opening. The high-velocity blood flow is turbulent and produces vibrations that

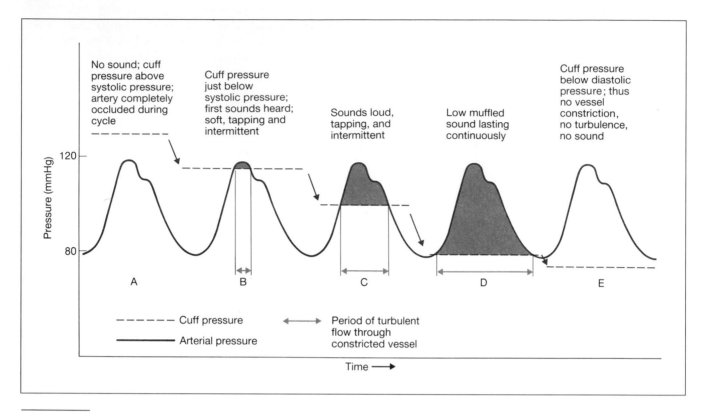

FIGURE 13-41 Sounds heard through a stethoscope while the cuff pressure of a sphygmo-manometer is gradually lowered. Sounds are first heard at systolic pressure, and sounds disappear at diastolic pressure.

can be heard through the stethoscope. Thus, the pressure, measured on the manometer attached to the cuff, at which sounds are first heard as the cuff pressure is lowered is identified as the systolic blood pressure.

As the pressure in the cuff is lowered further, the duration of blood flow through the artery in each cycle becomes longer. When the cuff pressure reaches the diastolic blood pressure, all sound stops, because flow is now continuous and nonturbulent through the open artery. Thus, diastolic pressure is identified as the cuff pressure at which sounds disappear.

It should be clear from this description that the sounds heard during measurement of blood pressure are *not* the same as the heart sounds described earlier, which are due to closing of cardiac valves.

ARTERIOLES

The arterioles play two major roles: (1) They are a major factor in determining mean arterial pressure, and (2) the arterioles in individual organs are responsible for determining the relative blood-flow distribution to those organs. The second function will be de-

scribed in this section, the first in the section on control of blood pressure.

Figure 13-42 illustrates the major principles of blood-flow distribution in terms of a simple model, a fluid-filled tank with a series of compressible outflow tubes. What determines the rate of flow through each exit tube? As stated in Section B of this chapter:

$$F = \frac{\Delta P}{R}$$

Since the driving pressure (the height of the fluid column in the tank) is identical for each tube, differences in flow are completely determined by differences in the resistance to flow offered by each tube. The lengths of the tubes are approximately the same, and the viscosity of the fluid is constant. Therefore, differences in resistance offered by the tubes are due solely to differences in their radii. Obviously, the widest tubes have the greatest flows. If we equip each outflow tube with an adjustable cuff, we can obtain various combinations of flows.

This analysis can now be applied to the cardiovascular system. The tank is analogous to the arteries, which

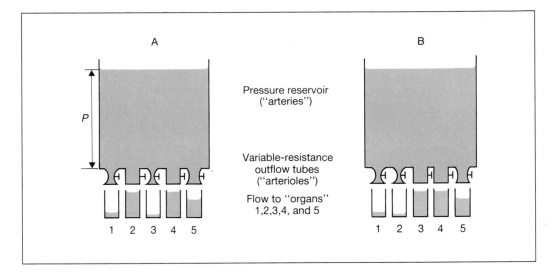

FIGURE 13-42 Physical model of the relationship between arterial pressure, arteriolar radius in different organs, and blood-flow distribution. In A, blood flow is high through tube 2 and low through tube 3, whereas just the opposite is true for B. This shift in blood flow was achieved by constricting tube 2 and dilating tube 3.

serve as a pressure reservoir, the major arteries themselves being so large that they contribute little resistance to flow. The smaller arteries begin to offer some resistance, but it is slight. Therefore, all the arteries of the body can be considered a single pressure reservoir.

The arteries branch within each organ into the arterioles, which are narrow enough to offer considerable resistance to blood flow. Indeed, the arterioles are the major sites of resistance in the vascular tree and are therefore analogous to the outflow tubes in the model. This explains the large decrease in mean pressure—from about 90 to 35 mmHg—as blood flows through the arterioles (Figure 13-38). Flow through the arterioles also eliminates the pulse present upstream in the arteries, leading to a continuous capillary flow.

Like the model's outflow tubes, the arteriolar radii in individual organs are subject to independent adjustment. The blood flow through any organ is given by the following equation, assuming for simplicity that the venous pressure, the other end of the pressure gradient, is zero:

$$F_{organ} = \frac{MAP}{R_{organ}}$$

Since the MAP, the driving force for flow through each organ, is identical throughout the body, differences in flows between organs depend entirely on the relative resistances offered by the arterioles of each organ. Arterioles contain smooth muscle, which can either relax and

cause the vessel radius to increase (**vasodilation**) or contract and decrease the vessel radius (**vasoconstriction**). Thus the pattern of blood-flow distribution depends upon the degree of arteriolar smooth-muscle contraction within each organ and tissue.

Arteriolar smooth muscle possesses a large degree of inherent activity, that is, contraction independent of any neural or hormonal input. Called **myogenic tone**—"tone" in reference to a muscle means degree of muscle contraction—it is responsible for a large portion of the resistance offered by arterioles. What is important, however, is that a variety of physiological factors play upon this smooth muscle, by inducing changes in cytosolic calcium concentration, to either increase or decrease its degree of contraction, thereby altering the vessel's resistance. The mechanisms controlling these changes in arteriolar resistance fall into two general categories: (1) local controls and (2) extrinsic (or reflex) controls.

Local Controls

The term **local controls** denotes mechanisms independent of nerves or hormones by which organs and tissues alter their own arteriolar resistances, thereby self-regulating their blood flows. This self-regulation, termed **autoregulation**, includes the phenomena of active hyperemia, pressure autoregulation, and reactive hyperemia.

Active hyperemia. Most organs and tissues manifest an increased blood flow (**hyperemia**) when their metabolic activity is increased (Figure 13-43). This is termed **active**

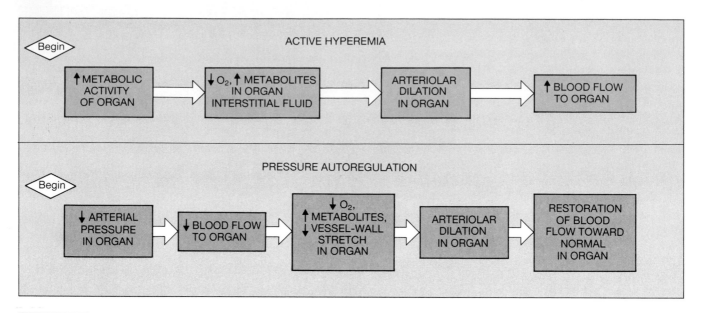

FIGURE 13-43 Local control of organ blood flow. Decreases in metabolic activity or increases in blood pressure would produce changes opposite those shown here.

hyperemia. For example, the blood flow to exercising skeletal muscle increases in direct proportion to the increased activity of the muscle. Active hyperemia is the direct result of arteriolar dilation in the more active organ. The vasodilation does not depend upon the presence of nerves or hormones but is a locally mediated response by which an increased rate of activity in an organ automatically produces an increased blood flow to that organ.

The factors acting upon smooth muscle in active hyperemia to cause it to relax are local chemical changes resulting from the increased metabolic cellular activity. The relative contributions of the various factors implicated vary, depending upon the organs involved and on the duration of the increased activity. Therefore, we shall only name but not quantify some of the chemical changes that occur and appear to be involved: decreased oxygen concentration; increased concentrations of carbon dioxide, hydrogen ion, and metabolites such as adenosine; increased concentration of potassium, perhaps as a result of enhanced potassium movement out of muscle cells during the more frequent action potentials; increased osmolarity resulting from the increased breakdown of high-molecular-weight substances; increased concentrations of eicosanoids (page 151); and increased generation, in several glands, of a peptide known as bradykinin.

Changes in all these variables have been shown to

cause arteriolar dilation under controlled experimental conditions, and they all may contribute to the active-hyperemia response in one or more organs. It must be emphasized once more that all these chemical changes act locally upon the arteriolar smooth muscle, causing it to relax. No nerves or homones are involved.

It should not be too surprising that active hyperemia is most highly developed in heart and skeletal muscle, which show the widest range of normal metabolic activities in the body. It is highly efficient, therefore, that their supply of blood be primarily determined locally.

Pressure autoregulation. In active hyperemia, increased metabolic activity of the tissue or organ is the initial event leading to local vasodilation. However, locally mediated changes in arteriolar resistance may also occur when a tissue or organ suffers a change in its blood supply due to a change in blood pressure (Figure 13-43). The change in resistance is in the direction of maintaining blood flow constant in the face of the pressure change and is therefore termed **pressure autoregulation**. For example, when arterial pressure in an organ is reduced, say, because of a partial occlusion in the artery supplying the organ, local controls cause arteriolar vasodilation, which tends to maintain flow relatively constant.

What is the mechanism of pressure autoregulation? One mechanism is the same metabolic factors described for active hyperemia. When an arterial pressure reduction lowers blood flow to an organ, the supply of oxygen to the organ is diminished and the local oxygen concentration decreases. Simultaneously, the concentrations of carbon dioxide, hydrogen ion, and metabolites all increase because they are not removed by the blood as fast as they are produced, and eicosanoid synthesis is increased by still unclear stimuli. Thus, the local metabolic changes occurring during decreased blood supply at constant metabolic activity are similar to those that occur during increased metabolic activity. This is because both situations reflect an imbalance between blood supply and level of cellular metabolic activity. Note, then, that the vasodilation of active hyperemia and of pressure autoregulation in response to low arterial pressure differ not in their mechanisms—local metabolic factors—but in the event—altered metabolism or altered blood pressure—that brings the mechanism into play.

Pressure autoregulation is not limited to circumstances in which arterial pressure goes down. The opposite events may occur when, for various reasons, arterial pressure increases. The initial increase in flow due to the increase in pressure removes the local vasodilator chemical factors and increases the local concentration of oxygen. This causes the arterioles to constrict, thereby maintaining local flow relatively constant in the face of the increased pressure.

Although our description has emphasized the role of local *chemical* factors in pressure autoregulation, it should be noted that another mechanism may also participate. Arteriolar smooth muscle may respond directly to increased stretch, caused by increased arterial pressure, by contracting to a greater extent. Conversely, decreased stretch, due to decreased arterial pressure, may cause the vascular smooth muscle to decrease its tone. These direct responses of arteriolar smooth muscle to stretch are termed myogenic responses. They are due to stretch-dependent changes in calcium movement into the smooth-muscle cells.

Reactive hyperemia.
When an organ or tissue has had its blood supply completely occluded, a profound transient increase in its blood flow occurs as soon as the occlusion is released. This phenomenon, known as **reactive hyperemia**, is essentially an extreme form of pressure autoregulation. During the period of no blood flow, the arterioles in the affected organ or tissue dilate, owing to the local factors described above. Blood flow is therefore very great through these wide-open arterioles as soon as the occlusion to arterial inflow is removed. It is very easy to observe this phenomenon in the skin simply by inflating a blood pressure cuff around the upper arm to values greater than systolic pressure for a minute or so and then releasing the pressure.

Response to injury.
Tissue injury causes a variety of substances to be released locally from cells or generated from plasma precursors. These substances make arteriolar smooth muscle relax and cause vasodilation in an injured area. This phenomenon, a part of the general process known as inflammation, will be described in detail in Chapter 19.

Extrinsic Controls
Sympathetic nerves. Most arterioles receive a rich supply of sympathetic postganglionic nerve fibers. These neurons release norepinephrine,[8] which combines with alpha-adrenergic receptors on the vascular smooth muscle to cause vasoconstriction.

Note that the receptors for norepinephrine on arterioles are alpha-adrenergic, whereas those for norepinephrine on heart muscle, including the conducting system, are mainly beta-adrenergic. This permits the use of beta-adrenergic antagonists to block the actions of norepinephrine on the heart but not the arterioles, and vice versa for alpha-adrenergic antagonists.

If the nerves to arterioles are virtually all constrictor, how can reflex arteriolar dilation be achieved? Since the sympathetic nerves are seldom completely quiescent but discharge at some finite rate, which varies from organ to organ, they always cause some degree of tonic constriction. Dilation can be achieved by decreasing the rate of sympathetic activity below the basal level.

The skin offers an excellent example of the processes. At room temperature, skin arterioles are already under the influence of a moderate rate of sympathetic discharge. An appropriate stimulus—fear or loss of blood, for example—causes reflex enhancement of this sympathetic discharge, and the arterioles constrict further. In contrast, an increased body temperature reflexly inhibits the sympathetic nerves to the skin, the arterioles dilate, and the skin flushes. This generalization cannot be stressed too strongly: Control of the sympathetic constrictor nerves to arteriolar smooth muscle can accomplish either vasodilation or vasoconstriction.

In contrast to active hyperemia and pressure autoregulation, the primary functions of sympathetic nerves to blood vessels are concerned not with the coordination of local metabolic needs and blood flow but with reflexes. The most common reflex employing these nerves, as we

---

[8]There is one possible exception to the generalization that sympathetic nerves to arterioles release norepinephrine. In some mammalian species, a second group of sympathetic (not parasympathetic) nerves to the arterioles in skeletal muscle releases acetylcholine, which causes arteriolar dilation and increased blood flow. The only known function of these vasodilator fibers is in the response to exercise or stress. Because their existence in human beings is questionable, we shall not deal with them.

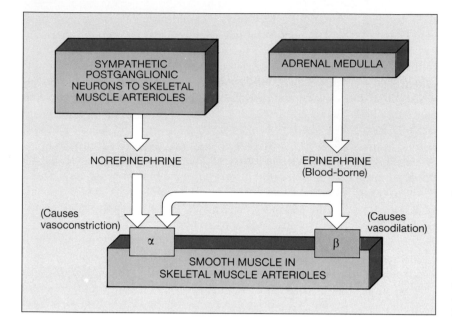

FIGURE 13-44 Effects of sympathetic nerves and plasma epinephrine on the arterioles in skeletal muscle. After its release from neuron terminals, norepinephrine diffuses to the arterioles, whereas epinephrine—a hormone—is blood-borne. Note that activation of alpha-adrenergic receptors and beta-adrenergic receptors produces opposing effects.

shall see, is that which regulates arterial blood pressure by influencing arteriolar resistance throughout the body. Other reflexes redistribute blood flow to achieve a specific function (for example, to increase heat loss from the skin).

Parasympathetic nerves. With but few exceptions—notably the blood vessels supplying the external genitals—there is little important parasympathetic innervation of arterioles. Where such innervation exists, stimulation of the parasympathetic nerves causes vasodilation, mediated by acetylcholine, the parasympathetic neurotransmitter.

Hormones. Epinephrine, like norepinephrine released from sympathetic nerves, can combine with alpha-adrenergic receptors on arteriolar smooth muscle and cause vasoconstriction. The story is more complex, however, because many arteriolar smooth-muscle cells possess beta-adrenergic receptors as well as alpha-adrenergic receptors, and combination of epinephrine with these beta-adrenergic receptors causes the muscle cells to relax rather than contract (Figure 13-44).

In most vascular beds, the existence of beta-adrenergic receptors is of little, if any, importance, since they are greatly outnumbered by the alpha-adrenergic receptors. The arterioles in skeletal muscle are an important exception, however, because they have large numbers of beta adrenergic receptors. Physiological concentrations of circulating epinephrine therefore usually cause dilation in this vascular bed. It must be emphasized that we

are speaking here only of the response to epinephrine, a circulating *hormone*. Stimulation of the norepinephrine-releasing sympathetic *nerves* to skeletal muscle arterioles causes only vasoconstriction.

Another hormone important for arteriolar control is angiotensin II, which directly constricts most arterioles and also increases the activity of the sympathetic nervous system. This peptide is part of the renin-angiotensin system, which will be described fully in Chapter 15.

Yet another important hormone that causes arteriolar constriction is vasopressin, which, as described in Chapter 10, is released into the blood by the posterior pituitary gland. The functions of vasopressin will also be described more fully in Chapter 15.

Endothelial Cells and Vascular Smooth Muscle

It should be clear from the preceding discussion that a large number of substances can induce the contraction or relaxation of vascular smooth muscle. Many of these substances do so by acting directly on the smooth muscle, but others act indirectly via the endothelial cells adjacent to the smooth muscle. Endothelial cells can secrete several substances collectively called **endothelium-derived relaxing factors** (**EDRF**), which cause arteriolar dilation by relaxing vascular smooth muscle. It is now recognized that a number of mediators involved in both reflex and local control of arterioles combine with receptors on endothelial cells and cause the release of EDRF. It then diffuses to the vascular smooth-muscle cells and induces relaxation. Moreover, it is also likely that, under certain circumstances, endothelial cells may secrete vasoconstrictor substances.

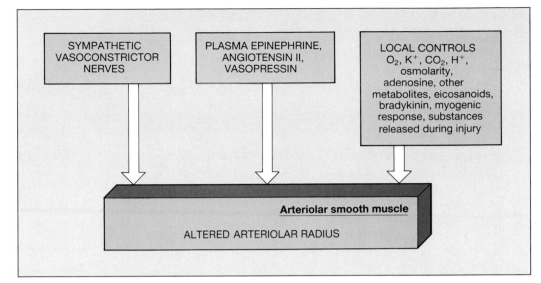

FIGURE 13-45 Major factors affecting arteriolar radius. Endothelium-derived relaxing factor (EDRF) is not shown because it mediates the vasodilation caused by other chemical factors rather than acting as a primary independent input.

Endothelial cells in *arteries* can also secrete EDRF in response to the mechanical shear stress that accompanies high blood-flow rate. The EDRF causes the arterial vascular smooth muscle to relax and the artery to dilate. This "flow-induced arterial vasodilation," which is independent of internal pressure and should be distinguished from *arteriolar* pressure autoregulation, may be important in optimizing the supply of blood to tissues.

Finally, it is likely that endothelium-derived factors are also involved in the long-term structural changes in arterial diameter that occur in response to chronic changes in blood flow to an organ or tissue, as, for example, during growth.

Arteriolar Control in Specific Organs

Figure 13-45 summarizes the factors that determine arteriolar radius. The local metabolic factors and extrinsic reflex controls do not exist in isolation from one another but manifest important interactions. The importance of local and reflex controls varies from organ to organ, and Table 13-6 summarizes the key features of arteriolar control in specific organs.

CAPILLARIES

As mentioned at the beginning of Section B, at any given moment, approximately 5 percent of the total circulating blood is flowing through the capillaries, and it is this 5 percent that is performing the ultimate function of the entire cardiovascular system—the exchange of nutrients and metabolic end products. Some exchange also occurs in the venules, which can be viewed as extensions of capillaries.

The capillaries permeate almost every tissue of the body. Since most cells are no more than 0.01 cm from a capillary, diffusion distances are very small, and exchange is highly efficient. There are an estimated 25,000 miles (42,000 km) of capillaries in an adult, each individual capillary being only about 1 mm long and just wide enough for an erythrocyte to squeeze through.

The essential role of capillaries in tissue function has stimulated many questions concerning how capillaries develop and grow (**angiogenesis**). For example, how do tumors stimulate growth of new capillaries, and what turns on angiogenesis during wound healing? The vascular endothelial cells appear to contain all the information needed to build an entire capillary network by mitosis and cell locomotion, and they are stimulated to do so by a variety of **angiogenic factors** secreted by cells in the vicinity of the endothelial cells. Tumor cells also secrete angiogenic factors.

Anatomy of the Capillary Network

Capillary structure varies considerably from organ to organ, but the typical capillary (Figure 13-46) is a thin-walled tube of endothelial cells one layer thick resting on a basement membrane, without any surrounding smooth muscle or elastic tissue. Some capillaries have a second set of cells adhering to the basement membrane underlying the endothelial cells. These importantly influence the ability of substances to penetrate the capillary wall.

The flat cells that constitute the endothelial tube are not attached tightly to each other but are separated by

TABLE 13-6 SUMMARY OF VASCULAR CONTROL IN SPECIFIC ORGANS

HEART:

High intrinsic tone; oxygen extraction is very high at rest and flow must increase when oxygen consumption increases.

Controlled mainly by local metabolic factors; direct sympathetic influences are minor and normally overridden by local factors.

Vessels are compressed during systole, and so coronary flow occurs mainly during diastole.

SKELETAL MUSCLE:

Controlled by local metabolic factors during exercise.

Sympathetic nerves cause vasoconstriction (mediated by alpha-adrenergic receptors) in reflex response to decreased arterial pressure.

Epinephrine causes vasodilation, via beta-adrenergic receptors, when present in low concentration and vasoconstriction, via alpha-adrenergic receptors, when present in high concentration.

GI TRACT, SPLEEN, PANCREAS, AND LIVER:

Actually two capillary beds partially in series with each other; blood from the capillaries of the GI tract, spleen, and pancreas flows via the portal vein to the liver. In addition, the liver also receives a separate arterial blood supply.

Sympathetic nerves cause vasoconstriction, mediated by alpha-adrenergic receptors, in reflex response to decreased arterial pressure and during stress. The venous constriction causes displacement of a large volume of blood from the liver to the central venous pool.

Active hyperemia occurs following ingestion of a meal and is mediated by local metabolic factors and hormones secreted by the GI tract.

KIDNEYS:

Pressure autoregulation is a major factor.

Sympathetic nerves cause vasoconstriction, mediated by alpha-adrenergic receptors, in reflex response to decreased arterial pressure and during stress. Angiotensin II is also a major vasoconstrictor. These reflexes help conserve sodium and water.

BRAIN:

Excellent pressure autoregulation.

Distribution of blood within the brain is controlled by local metabolic factors.

Vasodilation occurs in response to increased concentration of carbon dioxide in arterial blood.

Influenced very little by the autonomic nervous system.

SKIN:

Controlled mainly by sympathetic nerves, mediated by alpha-adrenergic receptors; reflex vasoconstriction occurs in response to decreased arterial pressure and cold, whereas vasodilation occurs in response to heat.

Substances released from sweat glands and/or as yet uncharacterized nerves also cause vasodilation.

Venous plexus contains large volume of blood which contributes to skin color.

LUNGS:

Very low resistance compared to systemic circulation.

Controlled mainly by gravitational forces and passive physical forces within the lung.

Constriction occurs in response to low oxygen concentration—just opposite to that which occurs in the systemic circulation.

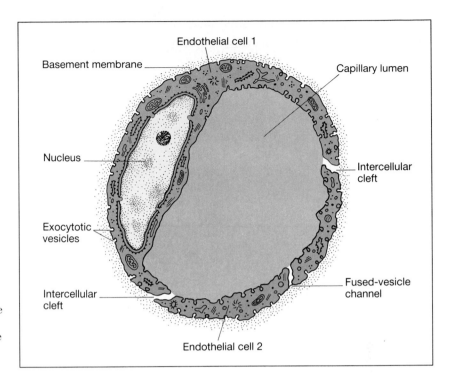

FIGURE 13-46 Capillary cross section. There are two endothelial cells in the figure, but the nucleus of only one is seen because the other is out of the plain of section. (*Adapted from Lentz.*)

narrow water-filled spaces termed **intercellular clefts**. The endothelial cells generally contain large numbers of endocytotic and exocytotic vesicles, and sometimes these fuse to form continuous **fused-vesicle channels** across the cell (Figure 13-46).

Blood flow through capillaries depends very much on the state of the other vessels that constitute the microcir-

culation (Figure 13-47). Thus, vasodilation of the arterioles supplying the capillaries causes increased capillary flow, whereas arteriolar vasoconstriction reduces capillary flow. In addition, in some tissues and organs blood does not enter capillaries directly from arterioles but from vessels, called **metarterioles**, that connect arterioles to venules. Metarterioles are transitional vessels in

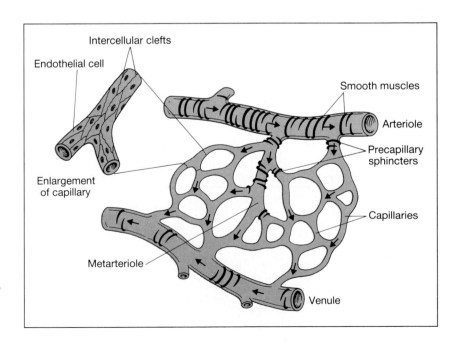

FIGURE 13-47 Diagram of microcirculation. Note the absence of smooth muscle in the capillaries. (*Adapted from Chaffee and Lytle.*)

that they retain, like arterioles, scattered smooth-muscle cells. The site at which a capillary exits from a metarteriole is surrounded by a ring of smooth muscle, the **precapillary sphincter**, which relaxes or contracts in response to local metabolic factors. When contracted, the precapillary sphincter closes the entry to the capillary completely. The more active the tissue, the more precapillary sphincters are open at any moment and the more capillaries in the network are receiving blood. Arterioles may also have precapillary sphincters at the site of capillary exit.

Resistance of the Capillaries

Since capillaries are very narrow, they offer considerable resistance to flow, as evidenced by the pressure decrease along the capillary length in Figure 13-38. Although a single capillary is narrower than an arteriole, the huge total number of capillaries provides such a large cross-sectional area for flow that the *total* resistance of *all* the capillaries is only about 40 percent of that of the arterioles. Moreover, because capillaries lack smooth muscle, their resistance is not altered as that of arterioles is.

Velocity of Capillary Blood Flow

Figure 13-48 illustrates a simple mechanical model of a series of 1-cm-diameter balls being pushed down a single tube that branches into narrower tubes. Although each tributary tube has a smaller cross section than the wide tube, the sum of the tributary cross sections is much

greater than that of the wide tube. Let us assume that in the wide tube each ball moves 3 cm/min. If the balls are 1 cm in diameter and they move two abreast, 6 balls leave the wide tube per minute and enter the narrow tubes, and 6 balls leave the narrow tubes per minute. At what speed does each ball move in the small tubes? The answer is *1 cm/min*.

This example illustrates the following important principle: When a continuous stream moves through consecutive sets of tubes, the velocity of flow decreases as the *sum* of the cross-sectional areas of the tubes increases. This is precisely the case in the cardiovascular system (Figure 13-49). The blood velocity is very great in the aorta, progressively slows in the arteries and arterioles, and then markedly slows as the blood passes through the huge cross-sectional area of the capillaries. The velocity of flow then progressively increases in the venules and veins because the cross-sectional area decreases. The slower flow of blood through the capillaries provides time for the exchange of nutrients and metabolic end products between the blood and tissues.

Diffusion across the Capillary Wall: Exchanges of Nutrients and Metabolic End Products

There are three basic mechanisms by which substances move across the capillary walls in most organs and tissues to enter or leave the interstitial fluid: diffusion, vesicle transport, and bulk flow. Mediated transport constitutes a fourth mechanism in the capillaries of the brain.

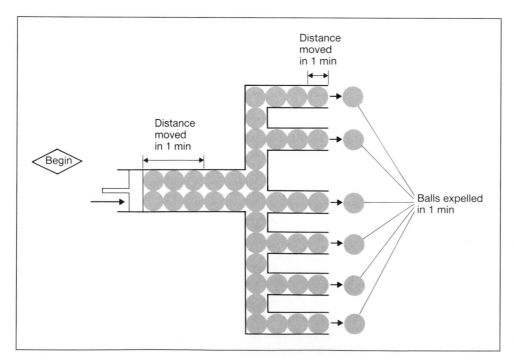

FIGURE 13-48 Relationship between cross-sectional area and flow velocity. The total cross-sectional area of the small tubes is three times greater than that of the large tube. Accordingly, velocity of flow is one-third as great in the small tubes.

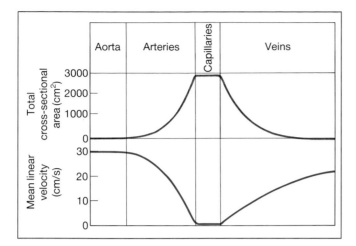

FIGURE 13-49 Relationship between total cross-sectional area and flow velocity in the systemic circulation. *(Adapted from Lytle.)*

In all capillaries, again excluding those in the brain, the first of these mechanisms—diffusion—constitutes the only important means by which net movement of nutrients, oxygen, and metabolic end products, including carbon dioxide, occurs across capillary walls. As described in the next section, there is some movement of these substances by bulk flow, but it is of negligible importance.

The factors determining diffusion rates were described in Chapter 6. Because lipid-soluble substances, including oxygen and carbon dioxide, penetrate cell membranes easily, they pass directly through the plasma membranes of the capillary endothelial cells. In contrast, ions and polar molecules are poorly soluble in lipid and must pass through small water-filled channels in the endothelial lining. The most likely location of these channels is the clefts between adjacent cells.[9] Another set of water-filled channels is provided by the fused-vesicle channels that penetrate the endothelial cells.

The presence of water-filled channels in the capillary walls causes the permeability of ions and small polar molecules to be very high, although still much lower than that of lipid-soluble molecules. The intercellular clefts are, however, too small to allow the passage of proteins, but some protein can diffuse through the fused-vesicle channels, which are larger. Protein may also cross by endocytosis of fluid at the luminal border and movement of the vesicle across the endothelial cell,

with exocytosis of its contents at the interstitial side. Although these various mechanisms do provide a means for protein movement from plasma to interstitial fluid, such movement is quite small in most organs and tissues.

Variations in the size of the water-filled channels from tissue to tissue account for the great differences in the "leakiness" of capillaries in different tissues and organs. At one extreme are the "tight" capillaries of the brain, which have no intercellular clefts, only tight junctions. Therefore even low-molecular-weight, water-soluble substances can gain access to or exit from brain interstitial space only by carrier-mediated transport through the blood-brain barrier.

At the other end of the spectrum are liver capillaries, which have large intercellular clefts as well as holes in the endothelial cells so that even protein molecules can readily pass. This is important because one of the major functions of the liver is the synthesis of plasma proteins and the metabolism of protein-bound substances, both of which must be able to enter or leave the blood.

The capillaries in most organs and tissues lie between these extremes.

What is the sequence of events involved in transfers of materials from capillary to cell? Tissue cells do not exchange material *directly* with blood. The fluid surrounding the cells—the interstitial fluid—acts as intermediary. Therefore, nutrients diffuse first across the capillary wall into the interstitial fluid, from which they gain entry to cells. Conversely, metabolic end products move across the tissue cells' plasma membranes into interstitial fluid, from which they diffuse into the plasma.

Transcapillary diffusion gradients for oxygen, nutrients, and metabolic end products occur as a result of cellular utilization or production of the substance. Let us take two examples: glucose and carbon dioxide in muscle. Glucose is continuously transported from interstitial fluid into the muscle cells by carrier-mediated-transport mechanisms and utilized there. The removal of glucose from interstitial fluid lowers the interstitial-fluid glucose concentration below the glucose concentration in capillary plasma and creates the gradient for diffusion of glucose from the capillary into the interstitial fluid.

Simultaneously, carbon dioxide is continuously produced by muscle cells and diffusing into the interstitial fluid. This causes the carbon dioxide concentration in interstitial fluid to be greater than that in capillary plasma, producing a gradient for carbon dioxide diffusion from the interstitial fluid into the capillary. Note that in both examples metabolism of the substance is the event that ultimately establishes the transcapillary diffusion gradients.

If a tissue is to increase its metabolic rate, it must obtain more nutrients from the blood and eliminate more metabolic end products. One mechanism for

---

[9] Except in the kidney, the basement membrane offers no diffusion barrier, and so we ignore it in this discussion of diffusion.

achieving this is active hyperemia. The second important mechanism is to increase diffusion gradients between plasma and tissue: Increased cellular utilization of oxygen and nutrients lowers their tissue concentrations, while increased production of carbon dioxide and other end products raises their tissue concentrations. In both cases the substance's transcapillary concentration difference is increased, which increases the rate of diffusion.

Bulk Flow across the Capillary Wall: Distribution of the Extracellular Fluid

To reiterate, the exchange of nutrients, oxygen, and metabolic end products across most capillaries occurs almost entirely by diffusion. At the same time these diffusional exchanges are occurring, another completely distinct process is taking place across the capillary: the bulk flow of protein-free plasma. The function of this process is not exchange of materials but rather the distribution of the extracellular fluid. As described in Chapter 1, extracellular fluid comprises the blood plasma and the interstitial fluid. Normally, there is approximately three times more interstitial fluid than plasma, 10 L versus 3 L in a 70-kg person. This distribution is not fixed, and the interstitial fluid functions as a reservoir that can supply fluid to the plasma or receive fluid from it.

As described in the previous section, the capillary wall is highly permeable to water and to almost all plasma solutes, except plasma proteins. Therefore, in the presence of a hydrostatic pressure difference across it, the capillary wall behaves like a porous filter through which protein-free plasma (**ultrafiltrate**) moves by bulk flow from capillary plasma to interstitial fluid through the water-filled channels. The concentrations of all the plasma solutes except protein are virtually the same in the filtering fluid as in plasma.

The magnitude of the bulk flow is directly proportional to the difference between the capillary blood pressure and the interstitial-fluid pressure. Normally, the former is much larger than the latter. Therefore a considerable hydrostatic pressure difference exists to filter protein-free plasma out of the capillaries into the interstitial fluid, the protein remaining behind in the plasma.

Why then does all the plasma not filter out into the interstitial space? The explanation was first elucidated by Starling—the same scientist who expounded the law of the heart that bears his name—and depends upon the principle of osmosis. In Chapter 6 we described how a net movement of water occurs across a semipermeable membrane from a solution of high water concentration to a solution of low water concentration, that is, from a region with a low concentration of solute to which the membrane is impermeable (nonpermeating solute) to a

region of high nonpermeating solute concentration. Moreover, this osmotic flow of water "drags" along with it any dissolved solutes to which the membrane is highly permeable (permeating solute).[10] Thus, a difference in water concentration secondary to different concentrations of nonpenetrating solute on the two sides of a membrane can result in the movement of a solution containing both water and permeating solutes in a manner similar to the bulk flow produced by a hydrostatic pressure difference. Units of pressure can be used in expressing this osmotic flow across a membrane.

This analysis can now be applied to capillary fluid movements. The plasma within the capillary and the interstitial fluid outside it contain large quantities of low-molecular-weight permeating solutes (also termed **crystalloids**), for example, sodium, chloride, or glucose. Since the capillary lining is highly permeable to all these crystalloids, their concentrations in the two solutions are essentially identical.[11] Accordingly, no significant water-concentration difference is caused by the presence of the crystalloids. In contrast, the plasma proteins (also termed **colloids**), being essentially nonpenetrating, have a very low concentration in the interstitial fluid. This difference in protein concentration between plasma and interstitial fluid means that the water concentration of the plasma is lower than that of interstitial fluid, inducing an osmotic flow of water from the interstitial compartment into the capillary. Since the crystalloids in the interstitial fluid move along with the water, osmotic flow of fluid, like flow driven by a hydrostatic pressure difference, does not alter their concentrations in either plasma or interstitial fluid.

A key word in this last sentence is "concentrations." Because the interstitial fluid and plasma have identical crystalloid concentrations to start with, a transfer of either fluid across the capillary wall does not change the crystalloid concentrations in either location. However, the *amount* of water (the *volume*) and the *amount* of crystalloids in the two locations do change. Thus, an increased filtration of fluid from plasma to interstitial fluid increases the volume of the interstitial fluid and decreases the volume of the plasma, even though no changes in crystalloid concentration occur.

In summary (Figure 13-50A), two opposing forces act to move fluid across the capillary wall: (1) The difference

---

[10] The precise physical chemistry of this process remains controversial. Do the water and penetrating solute actually move together or does the water move alone, causing solute concentration differences across the membrane that result in the solute then moving too? In both cases, the final result is the same.

[11] As we have seen, there are small concentration differences for substances that are consumed or produced by the cells. These tend, however, to cancel each other.

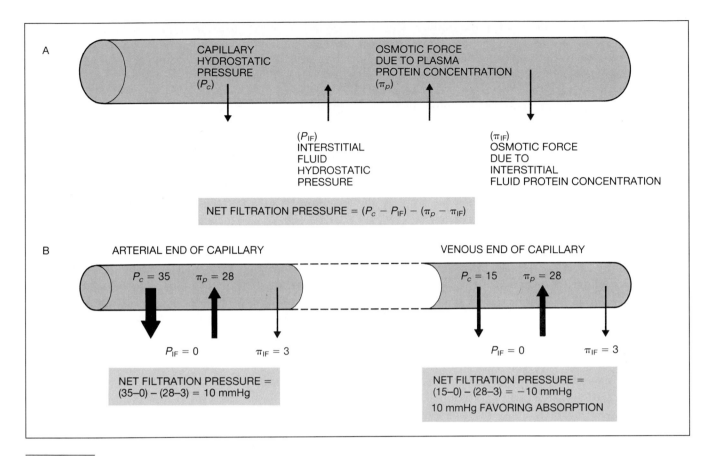

FIGURE 13-50 (A) The four factors determining fluid movement across capillaries. (B) Quantitation of forces causing filtration at the arteriolar end of the capillary and absorption at the venous end. Arrows in B denote magnitude of forces. No arrow is shown for interstitial fluid hydrostatic pressure (P_{IF}) because it is zero.

between capillary blood pressure and interstitial-fluid pressure favors ultrafiltration out of the capillary, and (2) the water-concentration difference between plasma and interstitial fluid, which results from differences in protein concentration, favors the flow of interstitial fluid into the capillary. Accordingly, the movements of fluid depend directly upon four variables: capillary hydrostatic pressure, interstitial hydrostatic pressure, plasma protein concentration, and interstitial fluid protein concentration. These four factors are termed the **Starling forces**.

We may now consider this movement quantitatively in the systemic circulation (Figure 13-50B). Much of the arterial blood pressure has already been dissipated as the blood flows through the arterioles, so that pressure at the beginning of the capillary (the part closest to the arteriole) is about 35 mmHg. Since the capillary also offers resistance to flow, the pressure continuously decreases to approximately 15 mmHg at the end of the

capillary (the part farthest from the arteriole). The interstitial pressure is very low, and we shall assume it to be zero.[12] The plasma protein concentration would produce an osmotic flow of water equivalent to that produced by a hydrostatic pressure of 28 mmHg. The interstitial protein concentration would produce a flow of water equivalent to that produced by a hydrostatic pressure of 3 mmHg. Therefore, the difference in protein concentrations induces a flow of fluid into the capillary equivalent to that produced by a hydrostatic pressure difference of 28 − 3 = 25 mmHg.

Thus, in the beginning of the capillary the hydrostatic pressure difference across the capillary wall (35 mmHg) is greater than the opposing osmotic force (25 mmHg),

---

[12]The exact value of interstitial hydrostatic pressure remains highly controversial. Many physiologists believe that it is not zero but is actually subatmospheric. The outcome of this controversy will not, however, alter the basic concepts being presented here.

and a net movement of fluid out of the capillary (**filtration**) occurs. In the end of the capillary, however, the osmotic force (25 mmHg) is greater than the hydrostatic pressure difference (15 mmHg) and fluid moves into the capillary (**absorption**). The result is that the early and late capillary events tend to cancel each other out. For the aggregate of capillaries in the body, however, there is a small net filtration of approximately 4 L/day. (This number does not include the capillaries in the kidneys.) The fate of this fluid will be described in the section on the lymphatic system.

This analysis of capillary fluid dynamics at the arterial and venous ends of the capillary has been somewhat oversimplified. It is very likely that many individual capillaries manifest only net filtration or net absorption along their entire lengths because the arterioles supplying them are either so dilated or so constricted that a capillary hydrostatic pressure above or below 25 mmHg exists along the entire length of the capillary.

The last paragraph illustrates the very important point that capillary pressure in any vascular bed is subject to physiological regulation, mediated mainly by changes in the resistance of the arterioles in that bed. As shown in Figure 13-51, dilating the arterioles in a particular vascular bed raises capillary pressure because less pressure is lost overcoming resistance between the arteries and the capillaries. Because of the increased capillary pressure, filtration is increased, and more protein-free fluid is lost to the interstitial fluid. In contrast, marked arteriolar constriction produces decreased capillary pressure and, hence, net movement of interstitial fluid into the vascular compartment.

It is this ability to change capillary pressure that determines the distribution of the extracellular fluid volume between the cardiovascular and interstitial compartments. For example, when blood volume is low, as following a hemorrhage, the resulting decrease in capillary pressure causes net movement of interstitial fluid into the capillaries and helps expand the blood volume.

It should be stated again that capillary filtration and absorption play no significant role in the exchange of nutrients and metabolic end products between capillary and tissues. The reason is that the total quantity of a substance, such as glucose or carbon dioxide, moving into or out of a capillary as a result of net bulk flow is extremely small in comparison with the quantities moving by net diffusion.

Finally, this analysis of capillary fluid dynamics has been in terms of the systemic circulation. Precisely the same Starling forces apply to the capillaries in the pulmonary circulation, but the values of the four variables differ. In particular, because the pulmonary circulation is a low-resistance, low-pressure circuit, the normal pulmonary capillary pressure—the major force favoring movement of fluid out of the pulmonary capillaries into the interstitium—is only 15 mmHg.

VEINS

Blood flows from capillaries into venules and then into veins. Some exchange of materials occurs between the interstitial fluid and the venules. Indeed, permeability to macromolecules is often greater for venules than for capillaries, particularly in damaged areas.

The large venules and small veins contain smooth muscle, and constriction of these vessels increases capillary pressure.

The veins outside the chest, the **peripheral veins**, also contain valves that permit flow only toward the heart. Why are these valves necessary if the pressure gradient created by cardiac contraction pushes blood only toward the heart anyway? The answer will be given below.

The veins are the last set of tubes through which blood flows on its way back to the heart. In the systemic circulation the force driving this venous return is the pressure difference between the peripheral veins and the right atrium.[13] Most of the pressure imparted to the blood by the heart is dissipated by resistance as blood flows through the arterioles and capillaries, so that pressure in the first portion of the peripheral veins is only 5 to 10 mmHg. The right atrial pressure is close to 0 mmHg, so that the total driving pressure for flow from the peripheral veins to the right atrium is only 5 to 10 mmHg. This is adequate because of the low resistance to flow offered by the veins, which have large diameters. Thus, a major function of the veins is to act as low-resistance conduits for blood flow from the tissues to the heart.

In addition to their function as low-resistance conduits, the veins perform a second important function: Their diameters are reflexly altered in response to changes in blood volume, thereby maintaining peripheral venous pressure and venous return to the heart. In a previous section, we emphasized that the rate of venous return to the heart is a major determinant of end-diastolic ventricular volume and thereby, stroke volume.

[13] The blood in the right atrium and large veins of the thorax constitute a single "pool"—the central venous pool. The pressure in this pool—the central venous pressure—is easily measured, and it is customary to substitute this pressure for right atrial pressure when dealing with venous return. However, to avoid potential confusion, we have not adopted this convention.

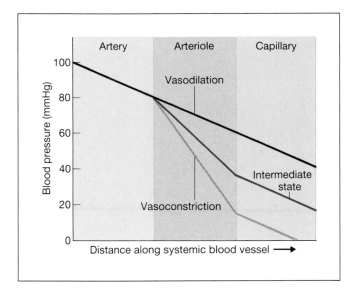

FIGURE 13-51 Effects of arteriolar vasodilation or vasoconstriction in an organ on capillary blood pressure in that organ.

ing skeletal muscle contraction, the veins running through the muscle are partially compressed, which reduces their diameter and forces more blood back to the heart. Now we can describe a major function of the peripheral-vein valves: When the skeletal muscle pump raises venous pressure locally, the valves permit blood flow only toward the heart and prevent flow back toward the tissues (Figure 13-52).

The respiratory pump is somewhat more difficult to visualize. As will be described in Chapter 14, during inspiration of air, the diaphragm descends, pushes on the abdominal contents, and increases abdominal pressure. This pressure increase is transmitted passively to the intraabdominal veins. Simultaneously, the pressure in the thorax decreases, thereby decreasing the pressure in the intrathoracic veins and right atrium. The net effect

Thus, we now see that peripheral venous pressure is an important determinant of stroke volume.

Determinants of Venous Pressure

The factors determining pressure in any elastic tube are, as we know, the volume of fluid within it and the compliance of its wall. Accordingly, total blood volume is one important determinant of venous pressure since, at any given moment, most blood is in the veins. The walls of veins are thinner and much more compliant than those of arteries. Thus, veins can accommodate large volumes of blood with a relatively small increase of internal pressure. Approximately 60 percent of the total blood volume is present in the systemic veins at any given moment, but the venous pressure averages less than 10 mmHg. In contrast, the systemic arteries contain less than 15 percent of the blood, at a pressure of approximately 100 mmHg.

The walls of the veins contain smooth muscle innervated by sympathetic neurons. Stimulation of these neurons releases norepinephrine, which causes contraction of the venous smooth muscle, decreasing the diameter and compliance of the vessels and raising the pressure within them. Increased venous pressure drives more blood out of the veins into the right heart.

Two other mechanisms, in addition to contraction of venous smooth muscle, can increase venous pressure and facilitate venous return. These mechanisms are the **skeletal muscle pump** and the **respiratory pump**. Dur-

FIGURE 13-52 The skeletal muscle pump. During muscle contraction, venous diameter decreases and venous pressure rises. The resulting increase in blood flow can occur only toward the heart because the valves in the veins are forced closed by any backward flow.

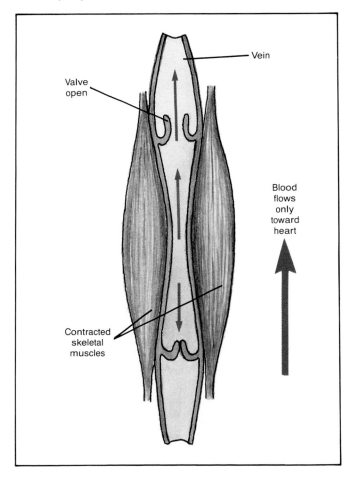

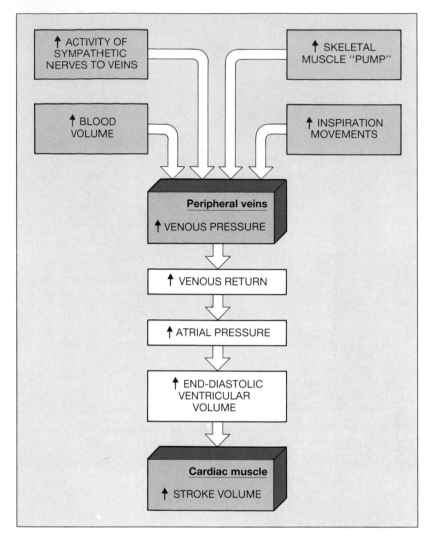

FIGURE 13-53 Major factors determining peripheral venous pressure and, hence, venous return and stroke volume. The figure shows how venous pressure and stroke volume are increased. Reversing the arrows in the boxes would indicate how these can be decreased. The effects of increased inspiration on end-diastolic ventricular volume are actually quite complex, and for the sake of simplicity, they are simply shown as altering venous pressure.

of these pressure changes is to increase the pressure difference between the peripheral veins and the heart. Accordingly, venous return is enhanced during inspiration. The larger the inspiration, the greater the effect. Thus, breathing deeply and frequently, as in exercise, helps blood flow from the peripheral veins to the heart.

One might get the (incorrect) impression from these descriptions that venous return and cardiac output are independent entities. However, any change in venous return, due say to the skeletal muscle pump, almost immediately causes equivalent changes in cardiac output, largely through the operation of Starling's law. Venous return and cardiac output must be identical except for very brief periods of time.

In summary (Figure 13-53), the effects of venous smooth-muscle contraction, the skeletal muscle pump, and the respiratory pump are to facilitate return of blood to the heart, and thereby to enhance cardiac output.

THE LYMPHATIC SYSTEM

The **lymphatic system** is a network of small organs (lymph nodes) and tubes (**lymphatic vessels** or simply "lymphatics") through which **lymph**—a fluid derived from interstitial fluid—flows. The lymphatic system is not part of the cardiovascular system, but it is described in this chapter because its vessels constitute a route for the movement of interstitial fluid to the cardiovascular system (Figure 13-54).

Present in the interstitium of virtually all organs and tissues are numerous **lymphatic capillaries** that are completely distinct from blood-vessel capillaries. Like the latter, they are tubes made of only a single layer of endothelial cells, but they have large water-filled channels that are permeable to all interstitial fluid constituents, including protein. The lymphatic capillaries are the first

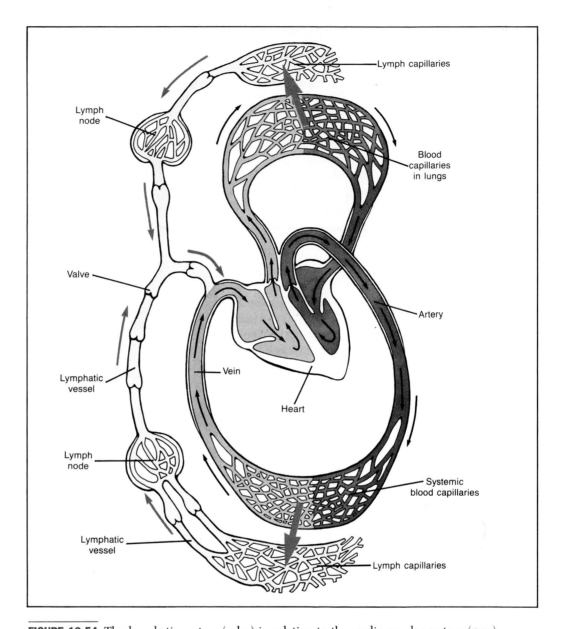

FIGURE 13-54 The lymphatic system (color) in relation to the cardiovascular system (gray). The lymphatic system is a one-way system from interstitial fluid to the cardiovascular system. The excess fluid that filters from the blood capillaries into the interstitial fluid enters the lymph capillaries and flows through the lymphatic vessels and nodes along them. Many lymphatic vessels arise in all systemic organs and the lungs, and each has multiple lymph nodes. All the lymphatic vessels converge into large lymphatic vessels that empty into the systemic circulation via the systemic veins.

of the lymphatic vessels, for unlike the blood-vessel capillaries, no tubes flow into them.

Small amounts of interstitial fluid continuously enter the lymphatic capillaries at rates determined by the interstitial pressure. Now known as lymph, the fluid flows from the lymphatic capillaries into the next set of lymphatic vessels, which converge to form larger and larger lymphatic vessels. Ultimately, the entire network ends in two large lymphatic vessels that drain into veins in the lower neck. Valves at these junctions permit only one-way flow from lymphatic vessel into vein. Thus, the lymphatic vessels carry interstitial fluid to the cardiovascular system. At various points, the lymph flows through lymph nodes, the function of which is described in Chapter 19.

The movement of interstitial fluid to the cardiovascular system via the lymphatics is very important because the amount of fluid filtered out of all the blood-vessel capillaries (excepting those in the kidneys) exceeds that reabsorbed by approximately 4 L each day. This 4 L is returned to the blood via the lymphatic system. In the process, the small amounts of protein that leak out of blood-vessel capillaries into the interstitial fluid are also returned to the cardiovascular system.

Failure of the lymphatic system, due, for example, to occlusion by infectious organisms (as in the disease elephantiasis) allows accumulation of excessive interstitial fluid. The result can be massive swelling of the involved area. The accumulation of interstitial fluid for whatever cause—others are described in the section on heart failure—is termed **edema.**

As will be described in Chapter 16, the lymphatic system also provides the pathway by which fat absorbed from the gastrointestinal tract reaches the blood. The lymphatics also, unfortunately, are often the route by which cancer cells spread from their area of origin to other parts of the body.

Mechanism of Lymph FLow

With no heart to pump it, lymph flow depends largely upon forces external to the lymphatic vessels. These include the same external forces we described for veins— the skeletal muscle and respiratory pumps. Since the lymphatic vessels have valves similar to those in veins, the increased lymphatic pressures produced by these "pumps" cause flow only toward the points at which the lymphatics enter the circulatory system.

In addition to these external forces, the smooth muscle surrounding the lymphatic vessels beyond the lymphatic capillaries exerts a pumplike action by inherent rhythmic contractions. Moreover, this smooth muscle is innervated by sympathetic neurons, and excitation of these neurons in various physiological states such as exercise may contribute to increased lymph flow.

There is also a simple mechanism by which flow of interstitial fluid through the lymphatic vessels back to the cardiovascular system increases when interstitial volume increases, say, as a result of an increased filtration out of the blood-vessel capillaries. The excess interstitial fluid raises interstitial pressure and drives more fluid into the lymphatic capillaries, from which it enters the remaining chain of lymphatic vessels and drains into the cardiovascular system. This mechanism helps prevent edema formation.

SECTION E

INTEGRATION OF CARDIOVASCULAR FUNCTION: REGULATION OF SYSTEMIC ARTERIAL PRESSURE

In Chapter 7 we described the fundamental ingredients of all reflex control systems: (1) the internal-environmental variable being maintained relatively constant; (2) receptors sensitive to changes in this variable; (3) afferent pathways from the receptors; (4) a control center that receives and integrates the afferent inputs; (5) efferent pathways from the control center; (6) effectors "directed" by the efferent pathways to alter their activities. The result is a change in the level of the regulated variable. The control and integration of cardiovascular function will be described in these terms.

The major variable being regulated is the systemic arterial blood pressure. This should not be surprising, since the systemic arterial pressure is the driving force for blood flow through all the organs except the lungs. Maintaining it is, therefore, a prerequisite for assuring adequate blood flow to these organs.

The mean arterial pressure is directly determined by only two factors—the cardiac output and the resistance to flow offered by the vascular system. This relationship can be derived from the basic equation relating flow, pressure, and resistance:

$$F = \frac{\Delta P}{R}$$

Rearranging terms algebraically, we have

$$\Delta P = FR$$

This form of the equation clearly shows that a pressure difference depends upon flow and resistance. Because the systemic vascular system is a continuous series of tubes, this equation holds for the entire system, that

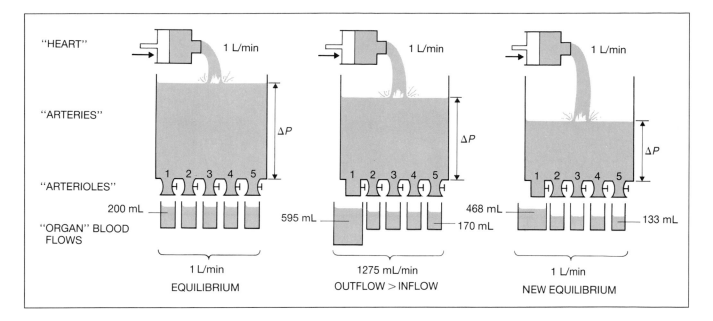

FIGURE 13-55 Dependence of arterial blood pressure upon total arteriolar resistance. Dilating one arteriolar bed affects arterial pressure and organ blood flow if no compensatory adjustments occur. The middle panel is a transient state before the new steady-state occurs.

is, from the very first portion of the aorta to the last portion of the venae cavae just at their entrance to the right atrium. Therefore the ΔP term is mean aortic pressure minus late venae cavae pressure, F is the cardiac output, and R is the **total peripheral resistance,** the sum of the resistances of all the vessels in the systemic vascular system. As described earlier, most of the vascular resistance is supplied by the arterioles, and so we usually equate total peripheral resistance with arteriolar resistance.

Since the pressure in the venae cavae where they empty into the heart is very close to 0 mmHg,

$$\Delta P = \text{mean aortic pressure} - 0$$

or $\qquad \Delta P = \text{mean aortic pressure}$

Since the mean pressure is essentially the same in the aorta and all other large arteries, ΔP is simply equal to mean arterial pressure. Thus the pressure-flow equation for the entire systemic circulation becomes

$$\frac{\text{Mean arterial}}{\text{pressure}} = \frac{\text{cardiac}}{\text{output}} \times \frac{\text{total peripheral}}{\text{resistance}}$$

$$\text{MAP} = \text{CO} \times \text{TPR}$$

That the volume of blood pumped into the arteries

per unit time—the cardiac output—is one of the two direct determinants of mean arterial pressure should come as no surprise. That the resistance of the blood vessels to flow—the TPR—is the other determinant may not be so intuitively obvious, but it can be illustrated by using the model introduced in Figure 13-42.

As shown in Figure 13-55 a pump pushes fluid into a container at the rate of 1 L/min. At steady state, fluid also leaves the container via outflow tubes at a total rate of 1 L/min. Therefore, the height of the fluid column (ΔP), which is the driving pressure for outflow, remains stable. We then disturb the steady state by loosening the cuff on outflow tube 1, increasing its radius, reducing its resistance, and increasing its flow. The total outflow for the system immediately becomes greater than 1 L/min, and more fluid leaves the reservoir than enters from the pump. Therefore the height of the fluid column begins to decrease until a new steady state between inflow and outflow is reached. In other words, at any given pump input, a change in *total* outflow resistance must produce changes in the pressure in the reservoir.

This analysis can be applied to the cardiovascular system by again equating the pump with the heart, the reservoir with the arteries, and the outflow tubes with various arteriolar beds. An analogy to opening outflow tube 1 is exercise: During exercise, the skeletal muscle arterioles dilate, thereby decreasing resistance. If the cardiac

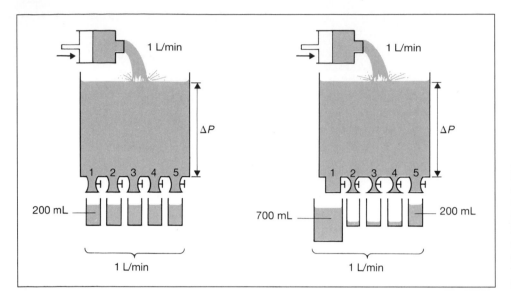

FIGURE 13-56 Compensation for dilation in one bed by constriction in others. When outflow tube 1 is opened, outflow tubes 2 to 4 are simultaneously tightened so that total outflow resistance, total runoff rate, and reservoir pressure all remain constant.

output and the arteriolar diameters of all other vascular beds were to remain unchanged, the increased runoff through the skeletal muscle arterioles would cause a decrease in arterial pressure.

It must be reemphasized that it is the *total* arteriolar resistance that influences arterial blood pressure. The *distribution* of resistances among organs is irrelevant in this regard. Figure 13-56 illustrates this point. On the right, outflow tube 1 has been opened, as in the previous example, while tubes 2 to 4 have been simultaneously tightened. The increased resistance offered by tubes 2 to 4 compensates for the decreased resistance offered by tube 1. Therefore total resistance remains unchanged, and reservoir pressure is unchanged. Total outflow remains 1 L/min, although the distribution of flows is such that flow through tube 1 is increased, that of tubes 2 to 4 is decreased, and that of tube 5 is unchanged.

Applied to the body, this process is analogous to altering the distribution of vascular resistances. When the skeletal muscle arterioles (tube 1) dilate during exercise, the *total* resistance of the system can still be maintained if arterioles were to constrict in other organs, such as the kidneys, gastrointestinal tract, and skin (tubes 2 to 4). In contrast, the brain arterioles (tube 5) remain unchanged, assuring constant brain blood supply. This type of resistance juggling can maintain total resistance only within limits, however. Obviously if tube 1 opens very wide, even total closure of the other tubes could not prevent total outflow resistance from falling. We shall see that this is actually the case during exercise.

The equation MAP = CO × TPR is the fundamental equation of cardiovascular physiology. It provides a way to integrate almost all the information presented in this chapter. For example, we can now explain why pulmonary arterial pressure is much lower than systemic arterial pressure. The cardiac output through the pulmonary and systemic arteries is, of course, the same. Therefore, the pressures can differ only if the resistances differ. Thus, the pulmonary vessels offer much less resistance to flow than the systemic arterioles. In other words, the total pulmonary vascular resistance is lower than the total systemic vascular resistance.

Figure 13-57 presents the grand scheme of factors that determine systemic arterial pressure.[14] None of this information is new, all of it having been presented in previous figures. A change in only a single variable will produce a change in mean arterial pressure by altering either cardiac output or total peripheral resistance. For example, Figure 13-58 illustrates how the decrease in blood volume occurring during hemorrhage leads to a decrease in mean arterial pressure.

Conversely, any deviation in arterial pressure, such as that occurring during hemorrhage, will elicit reflexes so that cardiac output and/or total peripheral resistance will be changed in the direction required to minimize the initial change in arterial pressure.

---

[14] Any model of a system as complex as the circulatory system must, of necessity, be an oversimplification. One basic deficiency in our model is that its chain of causal links is entirely unidirectional (from bottom to top in Figure 13-57) whereas, in fact, there are also important causal interactions in the reverse direction. A reader interested in delving further into these complex interactions and feedback loops should consult the more advanced works listed at the back of the book.

In the short term—seconds to hours—these adjustments are brought about by reflexes termed the baroreceptor reflexes. They utilize mainly changes in the activity of the autonomic nerves supplying the heart and blood vessels, as well as changes in the secretion of the hormones—epinephrine, angiotensin II, and vasopressin—that influence these structures. Over longer time spans, the baroreceptor reflexes become less important, and factors controlling blood volume play a dominant role in determining blood pressure.

BARORECEPTOR REFLEXES

Arterial Baroreceptors

It is only logical that the reflexes that homeostatically regulate arterial pressure originate primarily with arterial receptors that respond to changes in pressure. High in the neck, each of the two major vessels supplying the head, the common carotid arteries, divides into two smaller arteries. At this division, the wall of the artery is thinner than usual and contains a large number of branching, vinelike nerve endings (Figure 13-59). This portion of the artery is called the **carotid sinus.** Its nerve endings are highly sensitive to stretch or distortion. Since the degree of wall stretching is directly related to the pressure within the artery, the carotid sinus serves as a pressure receptor (**baroreceptor**). An area functionally similar to the carotid sinuses is found in the arch of the aorta and is termed the **aortic arch baroreceptor.** The carotid sinuses and aortic arch constitute the **arterial baroreceptors.** Afferent neurons from them travel to the brain and eventually provide input to the neurons of cardiovascular control centers there.

Action potentials recorded in single afferent fibers from the carotid sinus demonstrate the pattern of baroreceptor response (Figure 13-60). In this experiment the pressure in the carotid sinus is artificially controlled so that the pressure is either steady or pulsatile, that is, varying as usual between systolic and diastolic pressure. At a particular steady pressure, for example, 100 mmHg, there is a certain rate of discharge by the neuron. This rate can be increased by raising the arterial pressure, or it can be decreased by lowering the pressure. Thus, the rate of discharge of the carotid sinus is directly proportional to the mean arterial pressure.

If the experiment is repeated using the same mean pressures as before but allowing pressure pulsations, it is found that at any given mean pressure, the larger the pulse pressure, the faster the rate of firing by the carotid sinus. This responsiveness to pulse pressure adds a further element of information to blood pressure regulation since small changes in factors such as blood volume may cause changes in pulse pressure with little or no change in mean pressure.

Other Baroreceptors

The large systemic veins, the pulmonary vessels, and the walls of the heart also contain baroreceptors, most of which function in a manner analogous to the arterial baroreceptors. By keeping brain cardiovascular control centers constantly informed about changes in the systemic venous, pulmonary, atrial, and ventricular pressures, they provide a further degree of regulatory sensitivity. Thus, a slight decrease in atrial pressure may reflexly increase the activity of the sympathetic nervous system even before the change lowers cardiac output and arterial pressure far enough to be detected by the arterial baroreceptors.

The Medullary Cardiovascular Control Center

The primary control center for the baroreceptor reflexes is a diffuse network of highly interconnected neurons in the brainstem medulla called the **medullary cardiovascular center.** The neurons in this center receive input from the various baroreceptors. This input determines the outflow from the center along axons that terminate upon the cell bodies and dendrites of the vagus (parasympathetic) neurons to the heart and the sympathetic neurons to the heart, arterioles, and veins. When the arterial baroreceptors *increase* their rate of discharge, the result is a *decrease* in sympathetic outflow to the heart, arterioles, and veins, and an *increase* in parasympathetic outflow to the heart (Figure 13-61). A decrease in baroreceptor firing rate results in just the opposite pattern.

As parts of the baroreceptor reflexes, angiotensin II generation and vasopressin secretion are also altered so as to help restore blood pressure. Thus, a decreased pressure elicits increased plasma concentrations of both these hormones, which raise arterial pressure by constricting arterioles. For the sake of simplicity, however, we focus in the rest of this chapter only on the sympathetic nervous system when discussing reflex control of arterioles. The roles of angiotensin II and vasopressin will be described in Chapter 15.

Operation of the Baroreceptor Reflex

Our description of the baroreceptor reflex is now complete. If arterial pressure decreases as during a hemorrhage (Figure 13-62), this causes the discharge rate of the carotid sinus and aortic arch baroreceptors to decrease. Fewer impulses travel up the afferent nerves to the medullary cardiovascular center, and this induces: (1) increased heart rate because of increased sympathetic discharge and decreased parasympathetic discharge to

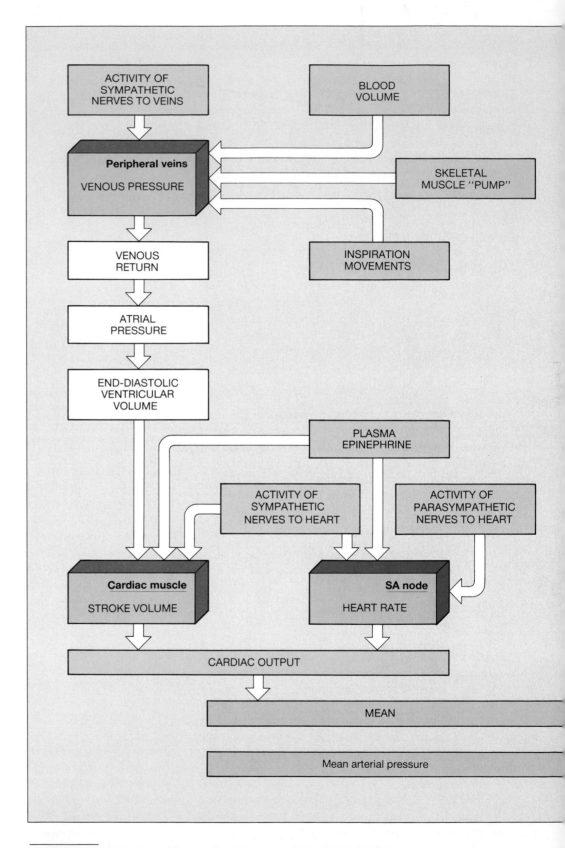

FIGURE 13-57 Summary of factors that determine systemic arterial pressure, an amalgamation of Figures 13-37, 13-45, and 13-53, with the addition of the effect of hematocrit on resistance.

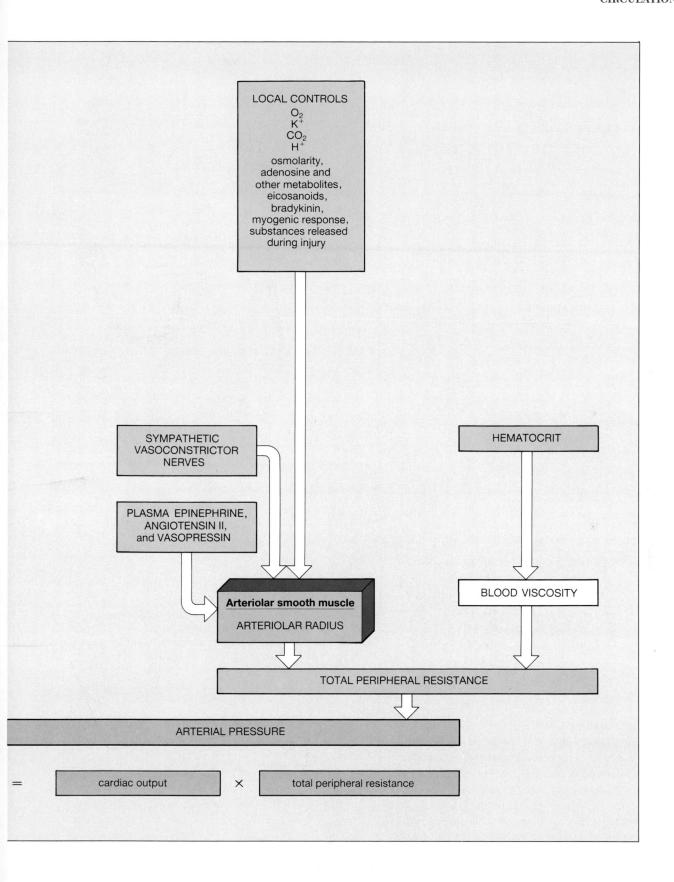

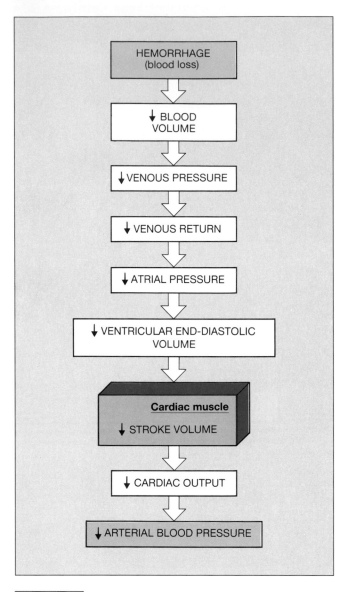

FIGURE 13-58 Sequence of events by which a decrease in blood volume leads to a decrease in mean arterial pressure.

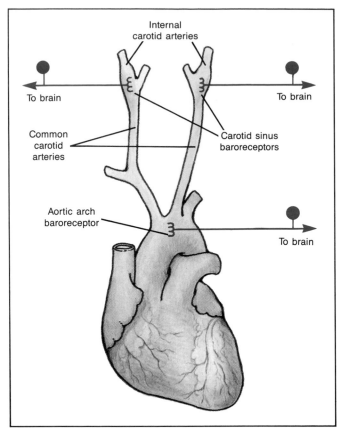

FIGURE 13-59 Locations of arterial baroreceptors.

FIGURE 13-60 Effect of changing mean arterial pressure MAP on the firing of action potentials by afferent neurons from the carotid sinus. This experiment is done by pumping blood through an isolated carotid sinus so as to be able to set the pressure inside it at any value desired.

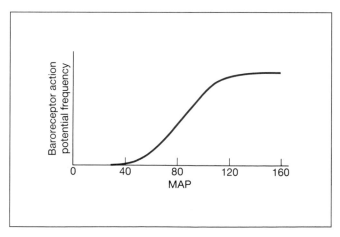

the heart; (2) increased myocardial contractility because of increased sympathetic activity to the heart; (3) arteriolar constriction because of increased sympathetic activity to arterioles and increased plasma concentrations of angiotensin II and vasopressin; and (4) increased venous constriction because of increased sympathetic discharge to veins. The net result is an increased cardiac output (increased heart rate and stroke volume); increased total peripheral resistance (arteriolar constriction), and return of blood pressure toward normal.

An increase in arterial blood pressure for any reason causes *increased* firing of the arterial baroreceptors,

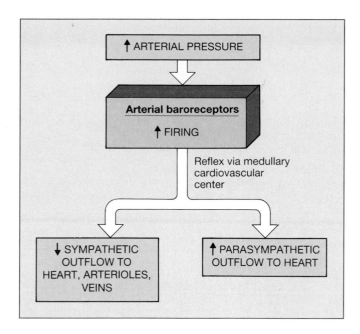

FIGURE 13-61 Neural components of the arterial baroreceptor reflex. When the initial change is a decrease in arterial pressure, all the arrows in the boxes would be reversed.

which reflexly induces a compensatory *decrease* in cardiac output and total peripheral resistance.

Having emphasized the great importance of this baroreceptor reflex, we must now add an equally important qualification. The baroreceptor reflex functions primarily as a *short-term* regulator of arterial blood pressure. It is activated instantly by any blood pressure *change* and attempts to restore blood pressure rapidly toward normal. Yet, if arterial pressure deviates from its normal operating point for more than a few days—recent experiments suggest even less time may be required—the arterial baroreceptors adapt to this new pressure, that is, they have a decreased frequency of action-potential firing at any given pressure. Thus, in patients who have chronically elevated blood pressure, the baroreceptors continue to oppose minute-to-minute changes in blood pressure, but at the higher level.

BLOOD VOLUME AND LONG-TERM REGULATION OF ARTERIAL PRESSURE

The fact that the baroreceptors adapt to prolonged changes in arterial pressure means that the baroreceptor reflex cannot *set* long-term arterial pressure. Many unanswered questions remain regarding long-term pressure regulation. One factor of definite impor-

tance is the blood volume. As described earlier, blood volume is a major determinant of arterial pressure because it influences venous pressure, venous return, end-diastolic volume, stroke volume, and cardiac output. Thus, an increased blood volume increases arterial pressure. But the opposite causal chain also exists—an increased arterial pressure reduces blood volume (more specifically, the plasma component of the blood volume), by increasing the excretion of salt and water by the kidneys.

Figure 13-63 illustrates how these two causal chains constitute a negative-feedback loop that determines both blood volume and arterial pressure. The important point is this: Because arterial pressure influences blood volume but blood volume also influences arterial pressure, blood pressure can stabilize, in the long run, only at a value at which blood volume is also stable. Accordingly, blood volume changes may be the single most important determinant, long-term, of blood pressure.

OTHER CARDIOVASCULAR REFLEXES AND RESPONSES

Stimuli acting upon receptors other than baroreceptors can initiate reflexes that cause changes in arterial pressure. For example, the following stimuli cause an increase in blood pressure: decreased arterial oxygen concentration; increased arterial carbon dioxide concentration; decreased brain blood flow; increased intracranial pressure; and pain originating in the skin (pain originating in the viscera or joints may cause marked decreases in arterial pressure). In addition, many physiological states such as eating and sexual activity are associated with changes in blood pressure.

These changes are triggered by input from receptors or higher brain centers to the medullary cardiovascular center or, in some cases, to pathways distinct from these centers. For example, certain neurons in the cerebral cortex and hypothalamus synapse directly on the sympathetic neurons in the spinal cord, bypassing the medullary center altogether.

Thus, there is a marked degree of flexibility and integration to the control of blood pressure. For example, by stimulating electrically a discrete area of the hypothalamus of an experimental animal, all the usually observed neurally mediated cardiovascular responses to an acute emotional situation can be elicited. Stimulation of other sites elicits cardiovascular changes appropriate to the maintenance of body temperature, feeding, or sleeping. It seems that such outputs are "preprogrammed." The complete pattern can be released by a natural stimulus that initiates the flow of information to the appropriate controlling brain center.

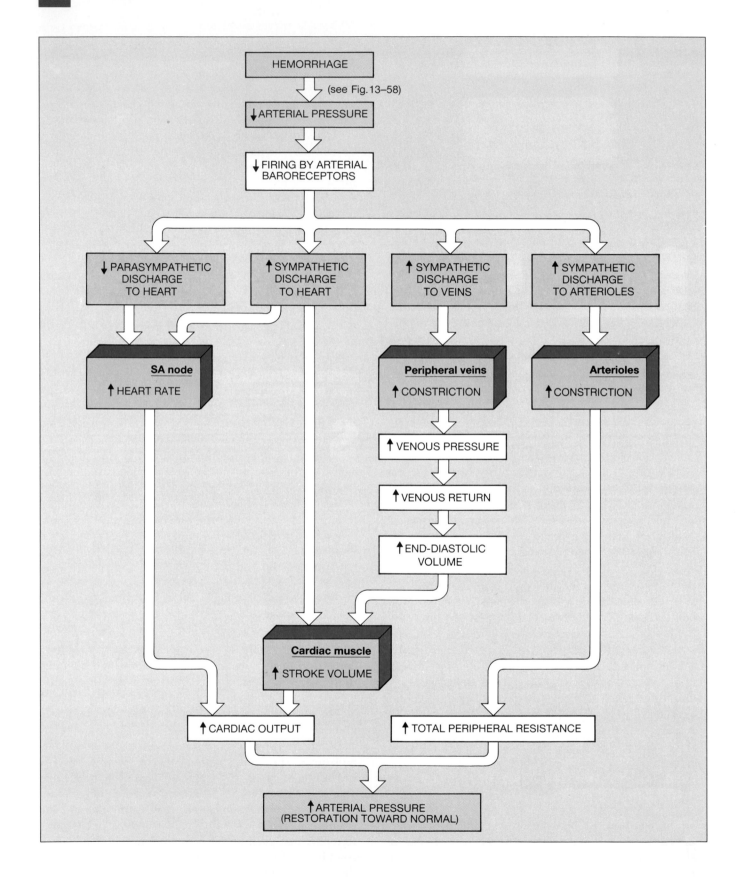

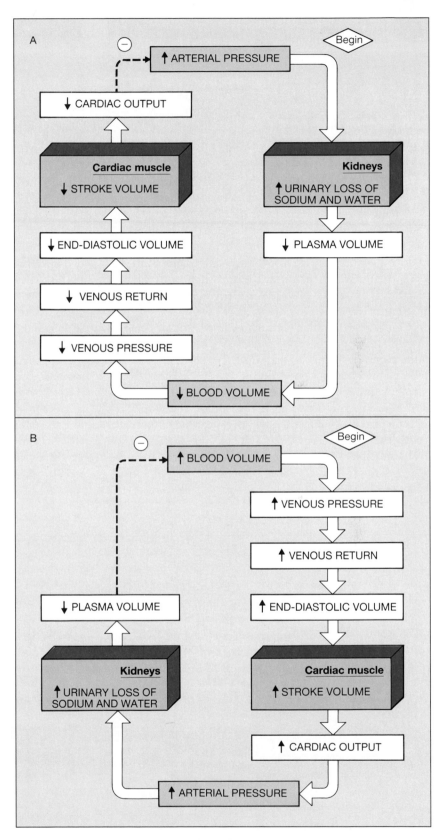

FIGURE 13-63 Causal reciprocal relationships between arterial pressure and blood volume. (A) An increase in arterial pressure due, for example, to an increased cardiac output induces a decrease in blood volume by promoting fluid excretion by the kidneys, which tends to restore arterial pressure to its original value. (B) An increase in blood volume due, for example, to altered kidney function induces an increase in arterial pressure, which tends to restore blood volume to its original value by promoting fluid excretion by the kidneys. Because of these relationships, blood volume is a major determinant of arterial pressure.

FIGURE 13-62 *(Opposite)* Arterial baroreceptor reflex compensation for hemorrhage. Hemorrhage causes a decrease in arterial pressure by causing decreases, in turn, of blood volume, venous pressure, venous return, atrial pressure, ventricular end-diastolic pressure, stroke volume, and cardiac output. The compensatory mechanisms do not restore arterial pressure completely to normal. Beyond the initial decrease in arterial pressure, all arrows signifying increases or decreases are relative to the state immediately following the hemorrhage but before reflex compensation begins. For simplicity, we have not shown the fact that plasma angiotensin II and vasopressin are also reflexly increased and help constrict arterioles.

SECTION F
CARDIOVASCULAR PATTERNS IN HEALTH AND DISEASE

HEMORRHAGE AND OTHER CAUSES OF HYPOTENSION

The term **hypotension** means a low blood pressure, regardless of cause. The decrease in blood volume caused by hemorrhage produces hypotension by the sequence of events shown previously in Figure 13-58. The most serious consequences of hypotension are reduced blood flow to the brain and cardiac muscle.

The immediate counteracting response to hemorrhage is the baroreceptor reflex, previously summarized in Figure 13-62.

Figure 13-64, which shows how four variables change over time when there is a decrease in blood volume, adds a further degree of clarification. The values of factors changed as a direct result of the hemorrhage—stroke volume, cardiac output, and mean arterial pressure—are restored by the baroreceptor reflex *toward*, but not *to*, normal. In contrast, values not altered directly by the hemorrhage but only by the reflex response to hemorrhage—heart rate and peripheral resistance—

are increased above their prehemorrhage values. The increased peripheral resistance results from increases in sympathetic outflow to the arterioles in many vascular beds but not those of the heart and brain. Thus, skin blood flow may decrease markedly because of arteriolar vasoconstriction—this is why the skin becomes cold and pale. Kidney and intestinal blood flow also decrease.

A second important type of compensatory mechanism (one not shown in Figure 13-62) involves the movement of interstitial fluid into capillaries. This occurs because both the drop in blood pressure and the increase in arteriolar constriction decrease capillary hydrostatic pressure, thereby favoring absorption of interstitial fluid (Figure 13-65). Thus, the initial event—blood loss and decreased blood volume—is in large part compensated for by the movement of interstitial fluid into the vascular system. Indeed, 12 to 24 h after a moderate hemorrhage, the blood volume may be restored virtually to normal by this mechanism (Table 13-7). At this time the entire restoration of blood volume is due to expansion of the plasma volume, whereas replacement of the lost erythrocytes requires many days. The altered contribu-

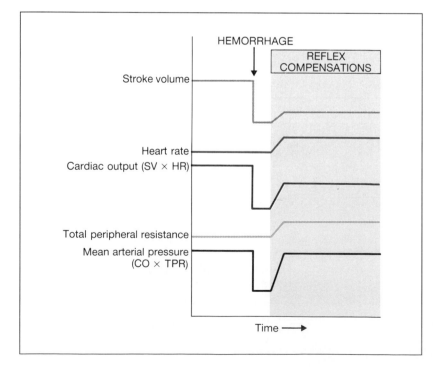

FIGURE 13-64 Four simultaneous graphs showing the time course of cardiovascular effects of hemorrhage. Note that the entire decrease in arterial pressure immediately following hemorrhage is secondary to the decrease in stroke volume and, hence, cardiac output. This figure emphasizes the relativeness of the "increase" and "decrease" arrows of Figure 13-62. All variables shown are increased relative to the state immediately following the hemorrhage, not necessarily to the state prior to the hemorrhage.

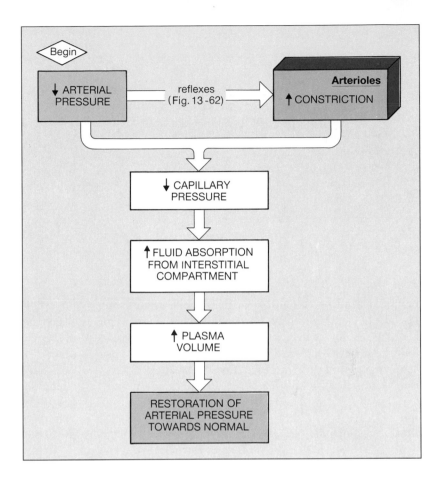

FIGURE 13-65 Mechanisms compensating for blood loss by movement of interstitial fluid into the capillaries.

tions of plasma and erythrocytes to blood volume during this period generally causes no problems.

We must emphasize that absorption of interstitial fluid only *redistributes* the extracellular fluid. Ultimate replacement of the fluid lost involves the control of fluid ingestion and kidney function. Both processes are described in Chapter 15. Replacement of the lost erythrocytes requires stimulation of erythropoiesis by erythropoietin. These replacement processes require days to weeks in contrast to the rapidly occurring reflex compensations described in Figure 13-62.

The early compensatory mechanisms for hemorrhage, the baroreceptor reflexes and fluid absorption, are highly efficient, so that losses of as much as 1.5 L of blood—approximately 30 percent of total blood volume—can be sustained with only slight reductions of mean arterial pressure or cardiac output.

Loss of large quantities of cell-free extracellular fluid rather than whole blood can also cause hypotension. This type of fluid loss may occur via the skin, as in severe sweating or burns, via the gastrointestinal tract, as in diarrhea or vomiting, or via unusually large urinary losses. Regardless of the route, the loss of fluid decreases circulating blood volume and produces symptoms and compensatory cardiovascular changes similar to those seen in hemorrhage.

Hypotension may be caused by events other than blood or fluid loss. One such cause is strong emotion, during which hypotension can cause fainting. Somehow, the higher brain centers involved with emotions inhibit sympathetic activity and enhance parasympathetic activity, resulting in a markedly decreased arterial pressure and brain blood flow. This whole process is usually transient. It should be noted that the fainting that sometimes occurs in a person donating blood is usually due to hypotension brought on by emotion, not the blood loss, since losing 0.5 L of blood will not itself cause serious hypotension in normal persons.

Massive liberation of chemicals that relax arteriolar smooth muscle may also cause hypotension by reducing total peripheral resistance. An important example is the hypotension that occurs during severe allergic responses.

Treatment of patients with hypotension in ways com-

TABLE 13-7 FLUID SHIFTS AFTER HEMORRHAGE

	Normal	Immediately after Hemorrhage	18 h after Hemorrhage	% of Normal Value
Total blood volume, mL	5000	4000 (↓ 20%)	4900	98
Erythrocyte volume, mL	2300	1840 (↓ 20%)	1840	80
Plasma volume, mL	2700	2160 (↓ 20%)	3060	113

monly favored by the uninformed, namely, administering alcohol and covering the person with mounds of blankets, is not appropriate. Both alcohol and excessive body heat cause profound dilation of skin arterioles, thus lowering total peripheral resistance and decreasing arterial blood pressure still further.

Shock

The term **shock** denotes any situation in which there is a generalized decrease in blood flow to the organs and tissues adequate to damage them. As with hypotension, the most common cause of shock is severe hemorrhage. When very large blood losses occur or when compensatory reflex adjustments are inadequate, the decrease in blood flow to the tissues and organs may be large enough to produce shock, in this case called **circulatory shock.** The cardiovascular system, especially the heart, suffers damage if shock is prolonged. As the heart deteriorates, cardiac output declines even more, and shock becomes progressively worse and ultimately irreversible. The person may die even after blood transfusions and other appropriate therapy.

Just as hemorrhage is not the only cause of hypotension and decreased cardiac output, so it is not the only cause of shock. Loss of fluid other than blood, excessive release of vasodilators as in allergy and infection, loss of neural tone to the cardiovascular system, and extensive bodily damage can all lead to severe reductions of tissue blood flow and the positive-feedback cycles culminating in irreversible shock.

THE UPRIGHT POSTURE

One might intuitively think that in a pump-driven circuit with a vertical orientation, gravity would hinder upward flow and facilitate downward flow, but intuition would be wrong in this case. The reason is that gravity acts equally on the outflow and inflow limbs of the circuit, so that the two effects counterbalance each other. Nevertheless, as we shall now explain, gravity does have important effects for another reason—because changes in the circulatory system in going from a lying, horizontal position to a standing, vertical one results in a decrease in the *effective* circulating blood volume. Why this is so requires an understanding of the action of gravity upon the long, continuous columns of blood in the vessels between the heart and the feet.

The pressures we have given in previous sections in this chapter are for an individual in the horizontal position, in which all blood vessels are at approximately the same level as the heart. In this position, the weight of the blood produces negligible pressure. In contrast, when a person is vertical, the intravascular pressure everywhere becomes equal to the pressure generated by cardiac contraction *plus* an additional pressure equal to the weight of a column of blood from the heart to the point of measurement. In an average adult, for example, the weight of a column of blood extending from the heart to the feet amounts to 80 mmHg. In a foot capillary, therefore, the pressure increases from 25 (the pressure resulting from cardiac contraction) to 105 mmHg, the extra 80 mmHg due to the weight of the column of blood.

This increase in pressure due to gravity influences the effective circulating blood volume. The increased hydrostatic pressure that occurs in the legs when a person is quietly standing pushes outward on the highly distensible vein walls, causing marked distension. The result is pooling of blood in the veins; that is, much of the blood emerging from the capillaries simply goes into expanding the veins rather than returning to the heart. Simultaneously, the increase in capillary pressure caused by the gravitational force produces increased filtration of fluid

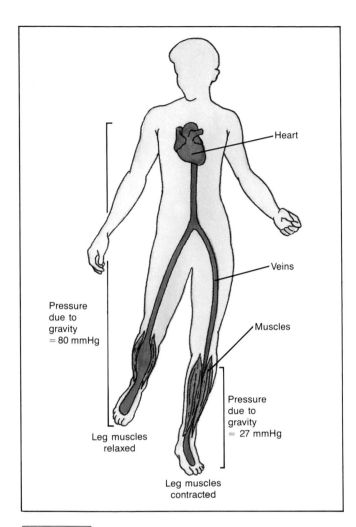

Heart

Veins

Muscles

Pressure
due to
gravity
= 80 mmHg

Pressure
due to
gravity
= 27 mmHg

Leg muscles
relaxed

Leg muscles
contracted

FIGURE 13-66 Role of contraction of the leg skeletal muscles in reducing capillary pressure and filtration in the upright position. The skeletal muscle contraction compresses the veins, causing intermittent complete emptying so that the columns of blood are interrupted. The venous pooling is exaggerated in this figure for purposes of illustration.

out of the capillaries into the interstitial space. This accounts for the fact that our feet swell during prolonged standing. The combined effects of venous pooling and increased capillary filtration reduce the effective circulating blood volume very similarly to the effects caused by a mild hemorrhage. The ensuing decrease in arterial pressure causes reflex compensatory adjustments similar to those shown in Figure 13-62 for hemorrhage.

The reason for specifying that the person in the previous paragraph is standing "quietly" is that the way to minimize the entire sequence described is to contract the leg skeletal muscles. This produces intermittent, emptying of the leg veins so that uninterrupted columns of venous blood from the heart to the feet no longer exist (Figure 13-66). The result is a decrease in both venous

distension and pooling plus a marked reduction in capillary hydrostatic pressure and fluid filtration out of the capillaries.

An example of the importance of this compensation (its absence, really) is soldiers fainting while standing at attention, that is, with minimal contraction of the leg muscles, for long periods of time. Here fainting may be considered adaptive in that venous and capillary pressure changes induced by gravity are eliminated once the person is prone. The pooled venous blood is mobilized, and the filtered fluid is absorbed back into the capillaries. Thus, the wrong thing to do to anyone who has fainted for whatever reason is to hold him or her upright.

EXERCISE

During exercise, cardiac output may increase from a resting value of 5 L/min to a maximal value of 35 L/min in trained athletes. The distribution of this cardiac output during strenuous exercise is illus-

FIGURE 13-67 Distribution of the systemic cardiac output at rest and during strenuous exercise. The values at rest were previously presented in Figure 13-10. (*Adapted from Chapman and Mitchell.*)

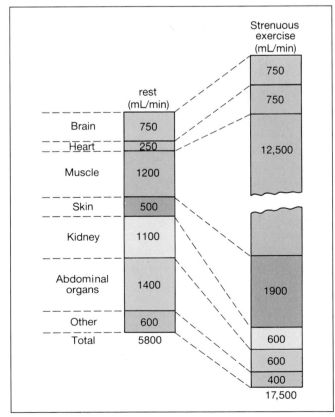

Strenuous
exercise
(mL/min)

rest
(mL/min)

	rest (mL/min)	Strenuous exercise (mL/min)
Brain	750	750
		750
Heart	250	12,500
Muscle	1200	
Skin	500	
Kidney	1100	
Abdominal organs	1400	1900
Other	600	600
Total	5800	600
		400
		17,500

trated in Figure 13-67. As expected, most of the increase in cardiac output goes to the exercising muscles, but there are also increases in flow to skin, required for dissipation of heat, and to heart, required for the additional work performed by the heart in pumping the increased cardiac output. The increases in flow through these three vascular beds are the result of arteriolar dilation in them. In both skeletal and cardiac muscle, the dilation is mediated by local metabolic factors, whereas the dilation in skin is achieved by a decrease in the firing of the sympathetic neurons to the skin. At the same time that arteriolar dilation is occurring in these beds, arteriolar constriction—manifested as decreased blood flow in Figure 13-67—is occurring in the kidneys and gastrointestinal organs, secondary to increased activity of the sympathetic neurons supplying them.

Dilation of arterioles in skeletal muscle, cardiac muscle, and skin arterioles causes a decrease in total peripheral resistance to blood flow. This decrease is partially offset by constriction of arterioles in other organs. Such "resistance juggling," however, is quite incapable of compensating for the huge dilation of the muscle arterioles, and the net result is a marked decrease in total peripheral resistance.

What happens to arterial blood pressure during exercise? As always, the mean arterial pressure is simply the product of cardiac output and total peripheral resistance. During most forms of exercise (Figure 13-68 illustrates the case for mild exercise), the cardiac output tends to increase somewhat more than the total peripheral resistance decreases, so that mean arterial pressure usually increases slightly. Regardless of what happens to mean pressure, however, the pulse pressure increases markedly, partly due to increased stroke volume but mainly because of the quicker ejection of the stroke volume by the sympathetically driven heart.

The cardiac output increase during exercise is caused by greater sympathetic activity and less parasympathetic activity to the heart. The increases in heart rate are usually much greater than those of stroke volume. Note in Figure 13-68 that, in our example, the increased stroke volume occurs without change in end-diastolic ventricular volume. Accordingly, the increased stroke volume in this case cannot be ascribed to Starling's law but is due completely to the increased contractility induced by the cardiac sympathetic nerves.

It would be incorrect to leave the impression that enhanced sympathetic activity to the heart is sufficient to account for the elevated cardiac output that occurs in exercise. The fact is that cardiac output can be increased to high levels only if venous return to the heart is simultaneously facilitated to the same degree. Otherwise, the shortened filling time resulting from the high heart rate

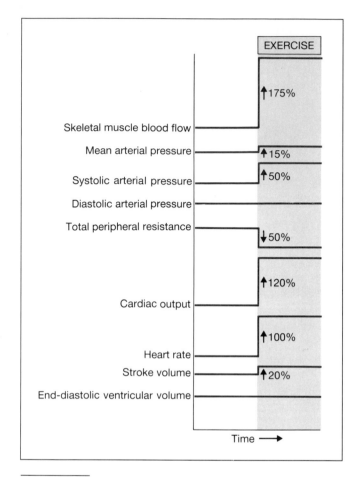

FIGURE 13-68 Summary of cardiovascular changes during mild exercise.

would lower end-diastolic volume and stroke volume (Starling's law). Therefore, factors promoting venous return during exercise are extremely important. They are (1) increased activity of the skeletal muscle pump, (2) increased depth and frequency of inspiration, (3) sympathetically mediated increase in venous tone, and (4) the greater ease of blood flow from arteries to veins through the dilated skeletal muscle arterioles.

During vigorous exercise, in contrast to the mild exercise illustrated in Figure 13-68, these four factors may be so powerful that venous return is enhanced enough to cause an *increase* in end-diastolic ventricular volume. Under such conditions, stroke volume is further enhanced above what it would have been from increased contractility alone.

What are the control mechanisms by which the cardiovascular changes in exercise are elicited? As described previously, dilation of arterioles in skeletal and cardiac muscle once exercise is underway represents active hyperemia secondary to local metabolic factors within the muscle. But what drives the enhanced sympathetic outflow to most other arterioles, the heart, and the veins and the decreased parasympathetic outflow to the heart? It is certainly *not* the arterial baroreceptors since the elevated mean and pulsatile arterial pressure of exercise would, via the baroreceptors, elicit just the opposite pattern of autonomic responses.

Rather, the control of autonomic outflow during exercise offers an excellent example of what we earlier referred to as a "preprogrammed" pattern, modified by continuous afferent input. One or more discrete control centers in the brain are activated during exercise, and descending pathways from these centers to the appropriate autonomic preganglionic neurons elicit the firing pattern typical of exercise. Indeed, these centers begin to "direct traffic" even before the exercise begins, since a person just about to begin exercising already manifests many of the changes in cardiac and vascular function.

Once exercise is underway, chemical and physical changes in the muscle activate receptors in the muscle. Afferent input from these receptors goes to the medullary cardiovascular center and facilitates the output reaching the autonomic neurons from higher brain cen-

ters (Figure 13-69). The result is a further increase in heart rate, myocardial contractility, and vascular resistance in the nonactive organs. Such a system permits a fine degree of matching between cardiac pumping and total oxygen and nutrients required by the exercising muscles.

Table 13-8 summarizes the changes that occur during moderate endurance exercise, that is, exercise that involves large muscle groups for an extended period of time.

Cardiac Output and Training

The limiting factor in endurance exercise is stroke volume, which ultimately limits cardiac output (Figure 13-70). Stroke volume increases with moderate work load, but as work load is progressively increased, stroke volume reaches a maximal value and then decreases. The major factors responsible for this decrease are (1) the very rapid heart rate, which decreases diastolic filling time, and (2) failure of the peripheral factors favoring venous return—muscle pump, respiratory pump, venoconstriction, arteriolar dilation—to elevate venous pressure high enough to maintain adequate ventricular filling during the very short time available. The net result is a

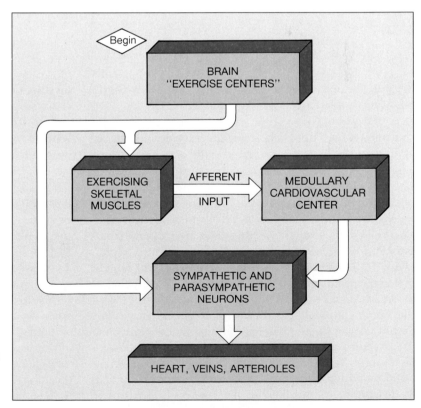

FIGURE 13-69 Control of the autonomic nervous system during exercise. The primary outflow to the sympathetic and parasympathetic neurons is via pathways from "exercise" centers in the brain, pathways parallel to those going to the exercising muscles. Afferent input from chemoreceptors in these muscles also influences the autonomic neurons by way of the medullary cardiovascular center.

TABLE 13-8   CARDIOVASCULAR CHANGES IN MODERATE EXERCISE

Variable	Change	Explanation
Cardiac output	Increases	Heart rate and stroke volume both increase.
Heart rate	Increases	Sympathetic-nerve activity to the SA node increases, and parasympathetic nerve activity decreases.
Stroke volume	Increases	Contractility increases due to increased sympathetic-nerve activity to the ventricular myocardium.
Total peripheral resistance	Decreases	Resistance in heart and skeletal muscles decreases more than resistance in other vascular beds increases.
Mean arterial pressure	Increases	Cardiac output increases more than total peripheral resistance decreases.
Pulse pressure	Increases	Stroke volume and velocity of ejection of the stroke volume increase.
End-diastolic volume	Unchanged	Filling time is decreased by the high heart rates, but this is compensated for by the factors favoring venous return—venoconstriction, skeletal muscle pump, and increased inspiratory movements.
Blood flow to heart and skeletal muscle	Increases	Active hyperemia occurs in both beds, mediated by local metabolic factors.
Blood flow to skin	Increases	Sympathetic nerves to skin vessels are inhibited reflexly by the increase in body temperature.
Blood flow to viscera	Decreases	Sympathetic nerves to the abdominal organs and the kidneys are stimulated.
Blood flow to brain	Unchanged	Autoregulation of brain arterioles maintains constant flow despite the increased mean arterial pressure.

decrease in end-diastolic volume that causes, by Starling's law, a decrease in stroke volume.

A person's maximal endurance work load and oxygen consumption are not fixed at any given value but can be altered by the habitual level of physical activity. For example, prolonged bed rest may decrease maximal exercise capacity of 25 percent, whereas intense long-term physical training may increase it by a similar amount. To be effective, the training must be of an endurance type. We are still uncertain of the most effective relative combinations of intensity and duration, but, as an example, jogging 20 to 30 min three times weekly at 5 to 8 mi/h definitely produces some training effect in most people.

This is manifested at rest as an increased stroke volume and decreased heart rate with no change in cardiac output. At maximal work loads, the trained individual has an increased cardiac output due to an increased maximal stroke volume since maximal heart rate is not altered by training (Figure 13-70). It is unclear how much of this increased maximal stroke volume is due to increased cardiac muscle mass and how much is due to increased

increases in the number of blood vessels in skeletal muscle, which would increase muscle blood flow and venous return. Training also increases blood volume and the concentrations of oxidative enzymes and mitochondria in the exercised muscles.

HYPERTENSION

Hypertension is defined as a chronically increased systemic arterial pressure. The dividing line between normal pressure and hypertension is set at approximately 140/90 mmHg.

Theoretically, hypertension could result from an increase in cardiac output or in total peripheral resistance, or both. In reality, however, the major abnormality in most cases of well-established hypertension is increased total peripheral resistance, caused by abnormally reduced arteriolar radius.

What causes the arteriolar constriction? In only a small fraction of cases is the cause known. For example,

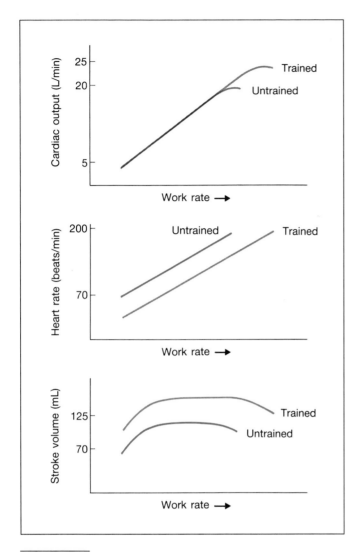

FIGURE 13-70 Changes in stroke volume, heart rate, and cardiac output with increasing work in untrained and trained persons. Note that cardiac output plateaus because stroke volume decreases at maximal work.

diseases that damage a kidney or decrease its blood supply are often associated with **renal hypertension**. The cause of the hypertension may be increased release of renin from the kidney, with subsequent increased generation of the potent vasoconstrictor angiotensin II. However, for more than the 95 percent of the persons with hypertension, the cause of the arteriolar constriction is unknown. Hypertension of unknown cause is called **primary hypertension** (formerly "essential hypertension").

Many hypotheses have been proposed to explain the increased arteriolar constriction. At present, much evidence seems to point to excessive sodium ingestion or retention as a contributing factor in genetically predisposed persons. Although this relationship remains controversial, it is certainly true that many persons with hypertension show a drop in blood pressure after being on low-sodium diets or receiving drugs, termed **diuretics**, that cause increased sodium loss via the urine. Obesity seems to be a definite risk factor for primary hypertension, and weight reduction and exercise are also frequently effective in causing some reduction of blood pressure in overweight sedentary persons with hypertension. Recently, low dietary intake of calcium has been implicated as a possible contributor to primary hypertension, but this, too, remains controversial.

Hypertension causes a variety of problems. One of the organs most affected is the heart. Because the left ventricle in a hypertensive person must chronically pump against an increased arterial pressure, it develops an adaptive increase in muscle mass. In the early phases of the disease, this helps maintain the heart's function as a pump. With time, however, changes in the organization and properties of myocardial cells occur, and these result in diminished contractile function and heart failure. The presence of hypertension also enhances the development of atherosclerosis and heart attacks, occlusion or rupture of a cerebral blood vessel (a stroke), and kidney damage. These diseases are all discussed in subsequent sections.

In addition to diuretics, four other classes of drugs, all of which reduce peripheral resistance, are in common use for treating hypertension: (1) blockers of beta-adrenergic receptors; (2) blockers of calcium channels; (3) blockers of the formation of angiotensin II; and (4) drugs that act either on the brain or outside the brain to antagonize the sympathetic nervous system.

That drugs that block beta-adrenergic receptors *reduce* total peripheral resistance by *dilating* arterioles, should come as a surprise to the reader: Since beta-adrenergic receptors in the vessels mediate dilation, one would deduce that blockage of these receptors should cause arteriolar constriction. Clearly, the beta-adrenergic receptors on the blood vessels are not the critical ones being blocked in the treatment of hypertension, and the explanation of the decrease in resistance is uncertain.

The calcium-channel blockers reduce the entry of calcium into vascular smooth-muscle cells, causing them to contract less strongly.

As will be described in Chapter 15, the final step in the formation of angiotensin II is mediated by an enzyme called angiotensin-converting enzyme. Drugs that block this enzyme therefore reduce the concentration of angiotensin II in plasma.

The antagonists of the sympathetic nervous system reduce sympathetically mediated stimulation of arteriolar smooth muscle.

HEART FAILURE

The heart may fail as a pump for many reasons, for example, pumping against a chronically elevated arterial pressure in hypertension, or structural damage due to decreased coronary blood flow. Regardless of cause, however, the hallmark of **heart failure** is a decreased contractility of the heart. As shown in Figure 13-71, the failing heart shifts downward to a lower Starling curve, that is, to a lower stroke volume at any given end-diastolic volume. How can this be compensated?

For one thing, a reflex increase in sympathetic outflow to the heart helps to increase contractility and also raises heart rate, both of which contribute to a restoration of cardiac output. Unfortunately, the ability of increased sympathetic-nerve activity and plasma epinephrine to drive the heart for long periods is self-limited. Cardiac norepinephrine stores become depleted, and cardiac beta-adrenergic receptors down-regulate, that is, decrease in number.

A second compensation is to increase ventricular end-diastolic volume. The heart in failure is generally distended with blood, as are the veins and capillaries, the major cause being an increase, sometimes massive, in plasma volume. The sequence of events is as follows: Decreased cardiac output causes a decrease in mean and pulsatile arterial pressure. This triggers reflexes, including the baroreceptor reflexes, that induce the kidneys to reduce their excretion of sodium and water. The retained fluid then causes expansion of the extracellular volume, increasing venous pressure, venous return, and end-diastolic ventricular volume and thus tending to restore stroke volume toward normal.

There is, however, an undesirable result of this elevated venous pressure: the formation of edema. Why does an increased venous pressure cause edema? The capillaries, of course, drain via venules into the veins, and so when venous pressure increases, the capillary pressure also increases and causes increased filtration of fluid out of the capillaries into the interstitial fluid. Swelling of the legs and feet is usually prominent, but the engorgement occurs elsewhere as well.

When the left ventricle fails, the edema may be very serious—**pulmonary edema**, which is the accumulation of fluid in the interstitial spaces of the lung or in the air spaces themselves. This impairs gas exchange. The reason for such accumulation is that the relatively ineffective left ventricle fails to pump blood to the same extent as the right ventricle, and so the volume of blood in all the pulmonary vessels increases. The resulting engorgement of pulmonary capillaries raises the capillary pressure above its normally very low value, causing increased filtration out of the capillaries. This situation

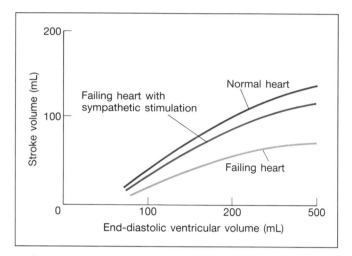

FIGURE 13-71 Relationship between end-diastolic ventricular volume and stroke volume in normal and failing hearts. The normal curve is that shown previously in Figure 13-32. The failing heart can still eject an adequate stroke volume if the sympathetic activity to it is increased or if the end-diastolic volume increases, that is, if the ventricle becomes more distended.

usually worsens at night: During the day, because of the patient's upright posture, fluid accumulates in the legs; then the fluid is slowly absorbed when the patient lies down at night, thus expanding the plasma volume and precipitating an attack of pulmonary edema.

The treatment for heart failure is easily understood in terms of its pathophysiology. The precipitating cause, for example, hypertension, should be corrected if possible. Ventricular contractility can be increased by a drug known as digitalis, which acts by increasing cytosolic calcium in the myocardial cells,[15] and excess fluid can be eliminated by the use of diuretics.

In closing this section, it should be emphasized that there are causes of edema other than heart failure and lymphatic malfunction described earlier. They are all understandable as imbalances in the Starling forces and are listed in Table 13-9.

"HEART ATTACKS" AND ATHEROSCLEROSIS

We have seen that the myocardium does not extract oxygen and nutrients from the blood within the atria and ventricles but depends upon its

---

[15] The mechanism of action of digitalis offers an excellent review of cellular calcium metabolism. Digitalis inhibits Na,K-ATPase in the myocardial plasma membranes, leading to an increase in cytosolic sodium concentration. This decreases the gradient for sodium-calcium exchange across the plasma membranes, thereby decreasing calcium exit and increasing cytosolic calcium concentration.

TABLE 13-9 MAJOR CAUSES OF EDEMA

Physiological Event	Cause of Edema
Increased arterial pressure secondary to increased cardiac output*	Increased capillary pressure, leading to increased filtration.
Local arteriolar dilation, as in exercise or inflammation	Increased capillary pressure, leading to increased filtration.
Increased venous pressure, as in heart failure or venous obstruction	Increased capillary pressure, leading to increased filtration.
Decreased plasma protein concentration, as in liver disease (decreased protein production), kidney disease (loss of protein in the urine), or protein malnutrition	Decreased force for osmotic absorption across capillary. Therefore, *net* filtration is increased.
Increased interstitial fluid protein concentration resulting from increased capillary permeability to protein (as in inflammation)	Decreased force for osmotic absorption across capillary. Therefore, *net* filtration is increased.
Obstruction of lymphatic vessels, as in infection by filaria roundworms (elephantiasis)	Fluid filtered from the blood capillaries into the interstitial compartment is not carried away. Protein also accumulates in the interstitial fluid.

*If arterial pressure is elevated because of increased total peripheral resistance, the capillary pressure may not be elevated because the increased arteriolar resistance causing the increased arterial pressure will prevent most of that increase in pressure from reaching the capillaries.

own blood supply via the coronary arteries. Insufficient coronary blood flow leads to myocardial damage in the affected region and, if severe enough, to death of that portion of the heart. This is a **myocardial infarction** or a **heart attack**.

Of the approximately 1.5 million heart attack victims in the United States each year, three-fourths are admitted to a hospital, where they can be given advanced care. More than 80 percent of these people survive the attack and are discharged. Unfortunately, the other one-fourth of heart attacks occur with so little immediate warning that the victim has no opportunity to reach a hospital before the heart stops. These sudden cardiac deaths are due mainly to **ventricular fibrillation**, an abnormality in impulse conduction resulting in continuous disorganized ventricular contractions that are ineffective in producing flow. (Note that ventricular fibrillation is fatal whereas atrial fibrillation, as described earlier in this chapter, generally causes only minor problems.) Many of these persons can be saved if modern emergency resuscitation procedures are applied immediately after the attack. This treatment is **cardiopulmonary resuscitation (CPR)**, a repeated series of mouth-to-mouth respirations and chest compressions that circulate a small amount of blood to the brain, heart, and other vital

organs when the heart has stopped. CPR is then followed by definitive treatment, including defibrillation, a procedure in which electric current is passed through the heart to try to halt the abnormal electrical activity causing the fibrillation. CPR is a potentially life-saving procedure that is easily learned in a few hours of training at readily available courses sponsored by the Red Cross and other groups.

The symptoms of myocardial infarction include prolonged chest pain, often radiating to the left arm, nausea, vomiting, sweating, weakness, and shortness of breath. Diagnosis is made by ECG changes typical of infarction and by measurement of certain enzymes in plasma. These enzymes are present in cardiac muscle and leak out into the blood when the muscle is damaged. They include particular variants of creatine phosphokinase (CPK) and lactate dehydrogenase (LDH).

The major cause of insufficient coronary flow leading to a heart attack is the presence of atherosclerosis in these vessels. **Atherosclerosis** (sometimes called "hardening of the arteries") is a disease characterized by a thickening of the arterial wall with (1) large numbers of abnormal smooth-muscle cells and (2) deposits of cholesterol and other substances in the portion of the vessel wall closest to the lumen. The mechanisms that initiate this thickening are not clear, but it is known that cigarette smoking, high plasma cholesterol concentration, hypertension, diabetes, and several other factors increase the incidence and the severity of the atherosclerotic process.

The mechanism by which atherosclerosis reduces coronary blood flow is quite simple: The extra muscle cells and various deposits in the wall bulge into the lumen of the vessel and increase resistance to flow. This is usually progressive, often leading ultimately to complete occlusion. Acute coronary occlusion may occur because of (1) sudden formation of a blood clot on the roughened vessel surface, (2) the breaking off of a fragment of blood clot or fatty deposit that then lodges downstream, completely blocking a smaller vessel, or (3) a profound spasm of the vessel's smooth muscle.

If the coronary occlusion is gradual, the heart may remain uninjured because, over time, new accessory vessels supplying the same area of myocardium develop. Many patients experience recurrent short-lived episodes of inadequate coronary blood flow, usually during exertion or emotional tension, before ultimately suffering a heart attack. The pain associated with these episodes is termed **angina pectoris**.

The chronic treatment of coronary artery disease and angina with drugs can be understood in terms of physiological concepts described in this chapter. First, vasodilator drugs such as nitroglycerin help in the following way: They dilate the coronary arteries and the systemic arterioles. The arteriolar effect lowers total peripheral resistance, thereby lowering arterial blood pressure and the work the heart must expend in ejecting blood. Second, beta-adrenergic blocking agents may be used to lower the arterial pressure, but more importantly, they block the effects of the sympathetic nerves on the heart. The result is reduced myocardial work because both heart rate and contractility are reduced. Finally, calcium-channel blockers reduce myocardial work by reducing myocardial contractility. They are also coronary and general vasodilators, and they prevent abnormal electrical rhythms that can lead to ventricular fibrillation. How blockade of calcium channels may achieve each of these effects should be understandable from previous descriptions of the roles of calcium in excitation-contraction coupling in cardiac and smooth muscle.

In the section of Chapter 19 dealing with blood clotting, several other drug therapies for preventing or treating myocardial infarction by opposing clotting will be described.

There are several surgical treatments for coronary artery disease. **Coronary angioplasty** is the passing of a catheter with a balloon at its tip into the occluded artery and then expanding the balloon. This enlarges the lumen by stretching the vessel and breaking up abnormal tissue deposits.

A second surgical technique is **coronary bypass**, in which an area of occluded coronary artery is removed and replaced with a new vessel, usually a vein taken from elsewhere in the patient's body. Tubes made of synthetic materials may also be used. Coronary bypass often produces marked relief of angina and prolongs life in some persons.

The question of whether regular exercise is protective against heart attacks is still controversial, although more and more circumstantial evidence favors this view. Certainly, modest exercise programs induce a variety of changes consistent with a protective effect: (1) increased diameter of coronary arteries; (2) decreased severity of hypertension and diabetes, which are risk factors for atherosclerosis; (3) decreased plasma cholesterol concentration (yet another risk factor) with simultaneous increase in the plasma concentration of a cholesterol-carrying lipoprotein thought to be protective against atherosclerosis; and (4) improved ability to dissolve blood clots. Finally, the results of long-term studies that evaluated the effects of exercise on the incidence of atherosclerosis are also suggestive of some degree of protection.

We do not wish to leave the impression that atherosclerosis attacks only the coronary vessels, for such is not the case. Most arteries of the body are subject to this same occluding process, and wherever the atherosclerosis becomes severe, the resulting symptoms reflect the decrease in blood flow to the specific area. For example, cerebral occlusions (**strokes**) due to atherosclerosis cause brain damage and constitute an important cause of disability and death. Persons with atherosclerotic cerebral vessels may also suffer reversible neurologic deficits, known as transient ischemic attacks (TIA), lasting minutes to hours, without actually experiencing a stroke at the time.

SUMMARY

Section A. Blood

I. Blood is composed of cells (erythrocytes, leukocytes, and platelets) and plasma, the liquid in which the cells are suspended.

II. Plasma contains proteins (albumins, globulins, and fibrinogen), nutrients, metabolic end products, hormones, and mineral electrolytes.

III. Erythrocytes, which make up more than 99 percent of blood cells, contain hemoglobin, an oxygen-binding protein consisting of heme and globin. Oxygen binds to the iron in heme.

   A. Erythrocytes are produced in the bone marrow and destroyed in the spleen.

B. Iron, folic acid, and vitamin B_{12} are essential for erythrocyte formation.

C. Control of erythrocyte production is exerted by the hormone erythropoietin, which is produced by the kidneys in response to low oxygen supply and which stimulates the bone marrow.

D. Anemia can be caused not only by deficiency of iron, vitamin B_{12}, and folic acid but by bone-marrow failure, excessive blood loss, excessive destruction of erythrocytes, and by erythropoietin deficiency.

IV. The leukocytes include three types of polymorphonuclear granulocytes (neutrophils, eosinophils, and basophils), monocytes, and lymphocytes.

V. Platelets are cell fragments essential for blood clotting.

VI. Blood cells are descended from stem cells in the bone marrow.

Section B. Overall Design of the Cardiovascular System

I. The cardiovascular system consists of two circuits: the pulmonary circulation, from the right ventricle to the lungs and then to the left atrium, and the systemic circulation, from the left ventricle to all organs and tissues other than the lungs and then to the right atrium.

II. Arteries carry blood away from the heart, and veins carry blood to the heart.

A. In the systemic circuit, the large artery leaving the left heart is the aorta, and the large veins emptying into the right heart are the superior and inferior venae cavae. The analagous vessels in the pulmonary circulation are the pulmonary trunk and the four pulmonary veins.

B. Arterioles, capillaries, and venules exist between arteries and veins.

III. Flow between two points in the cardiovascular system is directly proportional to the pressure difference between the points and inversely proportional to the resistance: $F = \Delta P/R$.

IV. Resistance is directly proportional to the viscosity of a fluid and to the length of the tube. It is inversely proportional to the fourth power of the tube's radius.

Section C. The Heart

Anatomy

I. The atrioventricular (AV) valves prevent flow from the ventricles back into the atria.

II. The pulmonary and aortic valves prevent backflow from the pulmonary trunk into the right ventricle and from the aorta into the left ventricle.

III. Cardiac-muscle cells are joined by gap junctions that permit current flow from cell to cell.

IV. The myocardium also contains specialized muscle cells that constitute the conducting system of the heart, initiating the heart beat and speeding the wave of excitation.

Heartbeat Coordination

I. Cardiac-muscle cells must undergo action potentials for contraction to occur.

A. The rapid depolarization of the action potential in contractile atrial and ventricular cells is due mainly to a positive-feedback increase in sodium permeability.

B. Following the initial depolarization, the membrane remains depolarized (the plateau phase) almost the entire duration of the contraction because of prolonged entry of calcium into the cell through slow plasma-membrane channels.

II. The SA node generates the current that leads to depolarization of all other cardiac-muscle cells.

A. The SA node manifests a pacemaker potential, which brings its membrane potential to threshold and initiates an action potential.

B. The impulse spreads from the SA node throughout both atria and to the AV node, where a small delay occurs. The impulse then passes, in turn, into the bundle of His, right and left bundle branches, Purkinje fibers, and contractile ventricular fibers.

III. Calcium, mainly released from the sarcoplasmic reticulum, functions as the excitation-contraction coupler in cardiac muscle, as in skeletal muscle, by combining with troponin.

A. The amount of calcium released does not usually saturate all troponin binding sites, and so the number of active cross bridges can be increased if cytosolic calcium is increased still further.

IV. Cardiac muscle cannot undergo summation of contractions because it has a very long refractory period.

Mechanical Events of the Cardiac Cycle

I. The cardiac cycle is divided into systole (ventricular contraction) and diastole (ventricular relaxation).

A. At the onset of systole, ventricular pressure rapidly exceeds atrial pressure, and the AV valves close. The aortic and pulmonary valves are not yet open, however, and so no ejection occurs during this isovolumetric ventricular contraction.

B. When ventricular pressures exceed aortic and pulmonary trunk pressures, the aortic and pulmonary valves open, and ventricular ejection of blood occurs.

C. When the ventricles relax at the beginning of diastole, the ventricular pressures fall significantly below those in the aorta and pulmonary trunk, and the aortic and pulmonary valves close. Because the AV valves are also still closed, no change in ventricular volume occurs during this isovolumetric ventricular relaxation.

D. When the ventricular pressures fall below the pressures in the right atrium and the left atrium, the AV valves open, and the ventricular filling phase of diastole begins.

E. Filling occurs very rapidly at first so that atrial contraction, which occurs at the very end of diastole, usually adds only a small amount of additional blood to the ventricles.

II. The amount of blood in the ventricles just prior to systole is the end-diastolic volume. The volume remaining after ejection is the end-systolic volume, and the volume ejected is the stroke volume.

III. Pressure changes in the systemic and pulmonary circulations have similar patterns, but the pulmonary values are much lower.

IV. The first heart sound is due to the closing of the AV valves, the second to the closing of the aortic and pulmonary valves.

The Cardiac Output

I. The cardiac output is the volume of blood pumped by each ventricle and equals the product of heart rate and stroke volume.

A. Heart rate is increased by the sympathetic nerves to the heart and by epinephrine. It is decreased by the parasympathetic nerves to the heart.

B. Stroke volume is increased by an increase in end-diastolic volume (Starling's law of the heart) and by an increase in contractility due to sympathetic-nerve stimulation or to epinephrine.

Section D. The Vascular System

Arteries

I. The arteries function as low-resistance conduits and as pressure reservoirs for maintaining blood flow to the tissues during ventricular relaxation.

II. The difference between maximal arterial pressure (systolic pressure) and minimal arterial pressure (diastolic pressure) during a cardiac cycle is the pulse pressure.

III. Mean arterial pressure can be estimated as diastolic pressure plus $\frac{1}{3}$ pulse pressure.

Arterioles

I. Arterioles, the major site of resistance to flow in the vascular system, play a major role in determining both mean arterial pressure and the distribution of flows to the various organs and tissues.

II. Arteriolar resistance is determined by local factors and by reflex neural and hormonal input.

A. Local factors that change with the degree of metabolic activity cause the arteriolar vasodilation and increased flow of active hyperemia.

B. Pressure autoregulation, a change in resistance that maintains flow constant in the face of a change in blood pressure, is due to local metabolic factors and to arteriolar myogenic responses to stretch.

C. The sympathetic nerves are the only innervation of most arterioles and cause vasoconstriction via alpha-adrenergic receptors.

D. Epinephrine causes vasoconstriction or vasodilation, depending on the organ or tissue.

E. Angiotensin II and vasopressin cause vasoconstriction.

F. Some vasodilator inputs act by releasing EDRF from endothelial cells.

III. Arteriolar control in specific organs is summarized in Table 13-6.

Capillaries

I. Capillaries are the site of exchange of nutrients and waste products between blood and tissues.

II. Capillary blood flow is determined by the resistance of the arterioles supplying the capillaries and by the number of open precapillary sphincters.

III. Blood flows through the capillaries more slowly than in any other part of the vascular system because of the huge cross-sectional area of the capillaries.

IV. Diffusion is the mechanism by which nutrients and waste products exchange between capillary plasma and interstitial fluid.

A. Lipid-soluble substances move across the entire endothelial wall, whereas ions and polar molecules move through water-filled channels that exist between the cells or as fused-vesicle channels.

B. Plasma proteins move across most capillaries only very slightly, either by diffusion or vesicle transport.

C. The diffusion gradient for a substance across capillaries arises as a result of cell utilization or production of the substance. Increased metabolism increases the diffusion gradient and increases the rate of diffusion.

V. Bulk flow of protein-free plasma or interstitial fluid across capillaries determines the distribution of extracellular fluid between these two fluids.

A. Filtration from plasma to interstitial fluid is favored by the hydrostatic pressure difference between the capillary and the interstitial fluid. Absorption from interstitial fluid to plasma is favored by the plasma protein concentration difference between the plasma and the interstitial fluid.

B. Filtration and absorption do not change the concentrations of crystalloids in the plasma and interstitial fluid because these substances move together with water.

C. There is normally a small excess of filtration over absorption.

Veins

I. Veins serve as low-resistance conduits for venous return.

II. Veins are very compliant and contain most of the blood in the vascular system.

A. Their diameters are reflexly altered by vasoconstriction or vasodilation so as to maintain venous pressure and venous return.

B. The skeletal muscle pump and respiratory pump increase venous pressure locally and enhance venous return. Venous valves permit the pressure to produce only flow toward the heart.

The Lymphatic System

I. The lymphatic system provides a one-way route for movement of interstitial fluid to the cardiovascular system.

II. It returns the excess fluid filtered from the blood-vessel capillaries, as well as the protein that leaks out of the blood-vessel capillaries.

III. Lymph flow is driven by the skeletal muscle pump, the respiratory pump, and contraction of smooth muscle in the larger lymphatic vessels.

Section E. Integration of Cardiovascular Function: Regulation of Systemic Arterial Pressure

I. Mean arterial pressure, the primary regulated variable in the cardiovascular system, equals the product of cardiac output and total peripheral resistance.

II. The factors that determine cardiac output and total peripheral resistance are summarized in Figure 13-57.

Baroreceptor Reflexes

I. The primary baroreceptors are the arterial baroreceptors—the two carotid sinuses and the aortic arch. Nonarterial baroreceptors are located in the systemic veins, pulmonary vessels, and walls of the heart.

II. The firing rates of the arterial baroreceptors are proportional to mean arterial pressure and to pulse pressure.

III. An increase in firing due to an increase in pressure causes, by way of the medullary cardiovascular center, an increase in parasympathetic outflow to the heart and a decrease in sympathetic outflow to the heart, arterioles, and veins. The result is a decrease in cardiac output and total peripheral resistance and, hence, a decrease in mean arterial pressure. The opposite occurs when the initial change is a decrease in arterial pressure.

Blood Volume and Long-term Regulation of Arterial Pressure

I. The baroreceptor reflexes are short-term regulators of arterial pressure but adapt to a maintained change in pressure.

II. The most important long-term regulator of arterial pressure is the blood volume.

Section F. Cardiovascular Patterns in Health and Disease

Hemorrhage and Other Causes of Hypotension

I. The physiological responses to hemorrhage are summarized in Figures 13-62 to 13-65.

II. Hypotension can be caused by loss of fluids other than blood, by strong emotion, and by liberation of vasodilator chemicals.

III. Shock is any situation in which blood flow to the tissues and organs is low enough to cause damage.

The Upright Posture

I. In the upright posture, gravity acting upon unbroken columns of blood reduces venous return by increasing vascular pressures in the veins and capillaries in the limbs.
   A. The increased venous pressure distends the veins, causing venous pooling, and the increased capillary pressure causes increased filtration out of the capillaries.
   B. These effects are minimized by contraction of the skeletal muscles in the legs.

Exercise

I. The cardiovascular changes that occur in endurance-type exercise are illustrated in Figures 13-67 and 13-68.

II. The changes are due to active hyperemia in the exercising skeletal muscles and heart, to increased sympathetic outflow to the heart, arterioles, and veins, and to decreased parasympathetic outflow to the heart.

III. The increase in cardiac output depends not only on the autonomic influences on the heart but on factors that help increase venous return.

IV. Training can increase a person's maximal endurance workload by increasing maximal stroke volume.

Hypertension

I. Hypertension is usually due to increased total peripheral resistance resulting from increased arteriolar constriction.

II. More than 95 percent of hypertension is termed primary in that the cause of the increased arteriolar constriction is unknown.

Heart Failure

I. Heart failure is a decreased cardiac contractility such that cardiac output is inadequate.

II. It leads to fluid retention by the kidneys and formation of edema because of increased capillary pressure.

III. Pulmonary edema can occur when the left ventricle fails.

Heart Attacks and Atherosclerosis

I. Insufficient coronary blood flow can cause damage to the heart.

II. Acute death from a heart attack is usually due to ventricular fibrillation.

III. The major cause of reduced coronary blood flow is atherosclerosis, an occlusive disease of arteries.

IV. Persons may suffer intermittent attacks of angina pectoris without actually suffering a heart attack at the time of the pain.

V. Atherosclerosis can also cause strokes and symptoms of inadequate blood flow in other areas.

REVIEW QUESTIONS

1. Define:

circulatory system	heart
blood vessels	cardiovascular system
vascular system	

Section A. Blood

1. Define:

blood	plasma
erythrocytes	plasma proteins
leukocytes	albumins
platelets	globulins
hematocrit	serum

hemoglobin
heme
globin
bone marrow
reticulocyte
iron
ferritin
transferrin
folic acid
vitamin B_{12}
erythropoietin
anemia

microcytosis
normocytosis
macrocytosis
polymorphonuclear
   granulocytes
neutrophils
eosinophils
basophils
monocytes
lymphocytes
megakaryocytes
pluripotent stem cells
colony-stimulating factors
   (CSFs)

2. Give average values for total blood volume, erythrocyte volume, plasma volume, and hematocrit.

3. Which is the most abundant class of plasma protein?

4. Which is the most abundant plasma solute in terms of millimoles per liter?

5. Describe the structure of hemoglobin.

6. Summarize the production, lifespan, and destruction of erythrocytes.

7. What are the routes of iron gain, loss, and distribution, and how is iron recycled when erythrocytes are destroyed?

8. Describe the control of erythropoietin secretion and the effect of this hormone.

9. State the relative proportions of erythrocytes and leukocytes in blood.

10. Diagram the derivation of the different blood cell lines.

Section B. Overall Design of the Cardiovascular System

1. Define:

bulk flow
atrium
ventricle
pulmonary circulation
systemic circulation
arteries
veins
aorta
arterioles
capillaries

venules
microcirculation
inferior vena cava
superior vena cava
pulmonary trunk
pulmonary arteries
pulmonary veins
hydrostatic pressure
resistance
viscosity

2. State the formula relating flow, pressure difference, and resistance.

3. What are the three determinants of resistance?

Section C. The Heart

1. Define:

pericardium
myocardium
endothelial cells
endothelium
atrioventricular (AV)
   valves

tricuspid valve
mitral valve
papillary muscles
pulmonary valve
aortic valve
conducting system

coronary arteries
coronary blood flow
sinoatrial (SA) node
slow channels
pacemaker potential
automaticity
heart rate
atrioventricular (AV)
   node
bundle of His
right and left bundle
   branches
Purkinje fibers
electrocardiogram (ECG)
P wave
QRS complex
T wave
refractory period (of cardiac
   muscle)
cardiac cycle

systole
diastole
isovolumetric ventricular
   contraction
ventricular ejection
stroke volume
isovolumetric ventricular
   relaxation
ventricular filling
end-diastolic volume
end-systolic volume
atrial fibrillation
heart sounds
heart murmurs
stenosis
insufficiency
cardiac output (CO)
Starling's law of the heart
venous return
contractility

2. List the structures through which blood passes from the systemic veins to the systemic arteries.

3. Contrast and compare cardiac muscle with skeletal and smooth muscle.

4. Describe the autonomic innervation of the heart, including the types of receptors involved.

5. Draw a ventricular action potential. Describe the changes in membrane permeability that underlie the potential changes.

6. Contrast action potentials in contractile cells with SA node action potentials. What is the pacemaker potential due to, and what is its inherent rate? Why is the SA node the normal pacemaker for the entire heart?

7. Describe the spread of excitation from the SA node through the rest of the heart.

8. Draw and label a normal ECG. Relate the P, QRS, and T waves to the ventricular action potential.

9. Describe the sequence of events leading to excitation-contraction coupling in cardiac muscle.

10. Why is the heart incapable of summation of contractions?

11. Draw the pressure changes in the left ventricle and aorta throughout the cardiac cycle. Show when the valves open and close, when the heart sounds occur, and the pattern of ventricular ejection.

12. Contrast the pressures in the right ventricle and pulmonary trunk with those in the left ventricle and aorta.

13. What causes heart murmurs in diastole? In systole?

14. Write the formula relating cardiac output, heart rate, and stroke volume. Give normal values for a resting adult.

15. Describe the effects of the sympathetic and parasympathetic nerves on heart rate. Which is dominant at rest?

16. What are the two major factors influencing force of contraction?

17. Draw a curve illustrating Starling's law of the heart.

18. Describe the effects of the sympathetic nerves on cardiac muscle during contraction and relaxation.

19. Draw a family of curves relating end-diastolic volume and stroke volume during different levels of sympathetic stimulation.

20. Summarize the effects of the autonomic nerves on the heart.

21. Draw a flow diagram summarizing the factors determining cardiac output.

Section D. The Vascular System

1. Define:

compliance
systolic pressure (SP)
diastolic pressure (DP)
pulse pressure
mean arterial pressure
    (MAP)
vasodilation
vasoconstriction
myogenic tone
local controls
autoregulation
hyperemia
active hyperemia
pressure autoregulation
reactive hyperemia
endothelium-derived
    relaxing factors (EDRF)
angiogenesis
angiogenic factors

intercellular clefts
fused-vesicle channels
metarterioles
precapillary sphincter
ultrafiltrate
crystalloids
colloids
Starling forces
filtration
absorption
peripheral veins
skeletal muscle pump
respiratory pump
lymphatic system
lymphatic vessels
lymph
lymphatic capillaries
edema

2. Draw the pressure changes during the cardiac cycle along the systemic and pulmonary vascular systems.

3. What are the two functions of the arteries?

4. What are normal values for systolic, diastolic, and mean arterial pressures? How is mean arterial pressure estimated?

5. What are the three factors that determine pulse pressure?

6. What denotes systolic and diastolic pressure in measurement of arterial pressure with a sphygmomanometer?

7. What are the major sites of resistance in the systemic vascular system?

8. What are two functions of arterioles?

9. Write the formula relating flow through an organ to (1) mean arterial pressure and (2) the resistance to flow offered by that organ.

10. List the chemical factors thought to mediate active hyperemia.

11. Name a mechanism other than chemical factors that may contribute to pressure autoregulation.

12. What is the only autonomic innervation of most arterioles? What are the major adrenergic receptors influenced by these nerves? How can control of sympathetic nerves to arterioles achieve either dilation or constriction?

13. Name three hormones that influence arterioles and describe their effects.

14. Draw a flow diagram summarizing the factors affecting arteriolar radius.

15. What are the relative velocities of flow through the various segments of the vascular system?

16. Contrast diffusion and bulk flow. Which is the mechanism of exchange of nutrients, oxygen, and metabolic end products across the capillary wall?

17. What is the only solute to have significant concentration differences across the capillary wall? How does this difference influence water concentration?

18. What four variables determine flow of fluid across the capillary wall? Give representative values for each of them in the systemic capillaries.

19. How do changes in local arteriolar resistance influence local capillary pressure?

20. What is the relationship between cardiac output and venous return in the steady state? What is the force driving venous return?

21. Contrast the compliances and blood volumes of the veins and arteries.

22. What three factors influence venous pressure?

23. Approximately how much fluid is returned to the blood by the lymphatics each day?

24. Describe the forces that cause lymph flow.

Section E. Integration of Cardiovascular Function

1. Define:

total peripheral
    resistance (TPR)
carotid sinus baroreceptor

aortic arch baroreceptor
arterial baroreceptors
medullary cardiovascular
    center

2. Write the equation relating mean arterial pressure to cardiac output and total peripheral resistance.

3. Why is mean pulmonary arterial pressure lower than mean systemic arterial pressure?

4. Draw a flow diagram illustrating the factors that determine mean arterial pressure.

5. Identify the receptors, afferent pathways, integrating center, efferent pathways, and effectors in the arterial baroreceptor reflex.

6. When the arterial baroreceptors decrease or increase their rate of firing, what changes in automatic outflow and cardiovascular function occur?

7. Describe the role of blood volume in the long-term regulation of arterial pressure.

Section F. Cardiovascular Patterns in Health and Disease

1. Define:

hypotension
shock

circulatory shock
hypertension

renal hypertension
primary hypertension
diuretics
heart failure
pulmonary edema
myocardial infarction
heart attack

ventricular fibrillation
cardiopulmonary
  resuscitation (CPR)
atherosclerosis
angina pectoris
coronary angioplasty
coronary bypass
strokes

2. Draw a flow diagram illustrating the reflex compensation for hemorrhage.

3. What happens to plasma volume and interstitial fluid volume following a hemorrhage?

4. What causes hypotension during a severe allergic response?

5. How does gravity influence effective blood volume?

6. Describe the role of the skeletal muscle pump in decreasing capillary filtration.

7. List the directional changes that occur during exercise for all relevant cardiovascular variables. What are the specific efferent mechanisms that bring about these changes?

8. What factors enhance venous return during exercise?

9. Diagram the control of autonomic outflow during exercise.

10. What is the limiting factor in endurance exercise?

11. What changes in cardiac function occur at rest and during exercise as a result of endurance training?

12. What is the abnormality in most cases of established hypertension?

13. Describe two mechanisms for raising stroke volume in heart failure.

14. How does heart failure lead to edema?

15. Name four risk factors for atherosclerosis.

THOUGHT QUESTIONS

(Answers are given in Appendix A.)

1. A person is found to have a hematocrit of 35 percent. Can you conclude from this that there is a decreased volume of erythrocytes in the blood?

2. Which would cause a greater increase in resistance to flow—a doubling of blood viscosity or a halving of tube diameter?

3. If all plasma-membrane calcium channels in cardiac muscle were blocked with a drug, what would happen to the muscle's action potentials and contraction?

4. A person with a heart rate of 40 has no P waves but normal QRS complexes on the ECG. What is the explanation?

5. A person has a left ventricular systolic pressure of 180 mmHg and an aortic systolic pressure of 110 mmHg. What is the explanation?

6. A person has a left atrial pressure of 20 mmHg and a left ventricular pressure of 5 mmHg during ventricular filling. What is the explanation?

7. A patient is taking a drug that blocks beta-adrenergic receptors. What changes in cardiac function will the drug cause?

8. What is the mean arterial pressure in a person whose diastolic and systolic pressures are, respectively, 160 and 100 mmHg?

9. A person is given a drug that doubles the blood flow to her kidneys but does not change the mean arterial pressure. What must the drug be doing?

10. A blood vessel removed from an experimental animal dilates when exposed to acetylcholine. After the endothelium is scraped from the lumen of the vessel, it no longer dilates in response to this mediator. Explain.

11. A person is accumulating edema throughout the body. Average capillary pressure is 25 mmHg, and lymphatic function is normal. What is the most likely cause of the edema?

12. A person's cardiac output is 7 L/min and mean arterial pressure is 150 mmHg. What is the person's total peripheral resistance?

13. The following data are obtained for an experimental animal before and after a drug. Before: Heart rate = 80 beats/min, and stroke volume = 80 mL/beat. After: Heart rate = 100 beats/min, and stroke volume = 64 mL/beat. Total peripheral resistance remains unchanged. What has the drug done to mean arterial pressure?

14. When, in an experimental animal, the nerves from all the arterial baroreceptors are cut, what happens to mean arterial pressure?

15. What happens to the hematocrit within several hours after a hemorrhage?

14

RESPIRATION

Respiration has two quite different meanings: (1) utilization of oxygen in the metabolism of organic molecules by cells, as described in Chapter 5, and (2) the exchanges of oxygen and carbon dioxide between an organism and the external environment. The second meaning is the subject matter of this chapter.

Our cells obtain most of their energy from chemical reactions involving oxygen. In addition, cells must be

TABLE 14-1   FUNCTIONS OF THE RESPIRATORY SYSTEM

1. Provides oxygen
2. Eliminates carbon dioxide
3. Regulates the blood's hydrogen-ion concentration (pH)
4. Forms speech sounds (phonation)
5. Defends against microbes
6. Influences arterial concentrations of chemical messengers by removing some from pulmonary capillary blood and producing and adding others to this blood
7. Traps and dissolves blood clots

able to eliminate carbon dioxide, the major end product of oxidative metabolism. A unicellular organism can exchange oxygen and carbon dioxide directly with the external environment, but this is obviously impossible for most cells of a complex organism like a human being. Therefore, the evolution of large animals required the development of a specialized system—the respiratory system—to exchange oxygen and carbon dioxide for the entire animal with the external environment.

The **respiratory system** comprises those structures involved in the exchange of gases between the blood and external environment: the lungs (**pulmonary** is the adjectival form of "lungs"), the series of tubes leading to the lungs, and the chest structures responsible for moving air into and out of the lungs during breathing.

In addition to the provision of oxygen and elimination of carbon dioxide, the respiratory system serves other functions as listed in Table 14-1 and discussed in this chapter.

ORGANIZATION OF THE RESPIRATORY SYSTEM

There are two lungs, the right and left, each divided into several lobes. The lungs consist mainly of tiny air-containing sacs called **alveoli** (singular, **alveolus**), which number approximately 150 million per lung and are the sites of gas exchange with the blood. The **airways** are all the tubes through which air flows between the external environment and the alveoli.

Inspiration is the movement of air from the external environment through the airways into the alveoli during breathing. **Expiration** is movement in the opposite direction. An inspiration and an expiration constitute a respiratory cycle. During the entire respiratory cycle, the right ventricle of the heart continuously pumps blood through the capillaries surrounding each alveolus. At rest, in a normal adult, approximately 4 L of environ-

mental air enter and leave the alveoli per minute while 5 L of blood, the entire cardiac output, flows through the pulmonary capillaries. During heavy exercise, the air flow can increase 30 to 40 fold and the blood flow, 5 to 6 fold. At all times, the amounts of air and blood distributed to the individual alveoli and their capillaries must be in the right proportions to each other. The alveolar air and capillary blood are separated from each other by extremely thin membranes, across which oxygen and carbon dioxide diffuse.

The Airways

During inspiration air passes through either the nose (the most common site) or mouth into the **pharynx** (throat), a passage common to the routes followed by air and food (Figure 14-1). The pharynx branches into two tubes, one (the esophagus) through which food passes to the stomach and one, the **larynx**, which is part of the airways. The larynx houses the **vocal cords**, two strong bands of elastic tissue stretched horizontally across its lumen. The flow of air past the vocal cords causes them to vibrate, producing sounds. The nose, mouth, pharynx, and larynx are termed the upper airways.

The larynx opens into a long tube, the **trachea**, which in turn branches into two **bronchi** (singular, **bronchus**), one of which enters each lung. Within the lungs, there are more than 20 generations of branchings, each resulting in narrower, shorter, and more numerous tubes, the names of which are summarized in Figure 14-2. The walls of the trachea and bronchi contain cartilage that gives them their cylindrical shape and supports them. The first airway branches that no longer contain cartilage are termed **bronchioles**. Alveoli first begin to appear, attached to the walls, in the bronchioles called respiratory bronchioles, and their frequency increases in the alveolar ducts (Figure 14-3) until the airways end in grapelike clusters of alveoli.

The airways beyond the larynx can be divided into two zones: (1) the **conducting zone**—from the top of the trachea to the beginning of the respiratory bronchioles—which contains no alveoli and across which gas exchange with the blood does not occur (Table 14-2); and (2) the **respiratory zone**—from the respiratory bronchioles on down—which contains alveoli and across which gas exchange occurs.[1]

The blood vessels supplying the lung generally accompany the airways and also undergo numerous branchings. The smallest of these vessels branch into networks of capillaries that richly supply the alveoli (Figure 14-3).

---

[1] In some terminologies, the respiratory bronchioles are classified as a "transitional zone" between the conducting and respiratory zones.

FIGURE 14-1 Organization of the respiratory

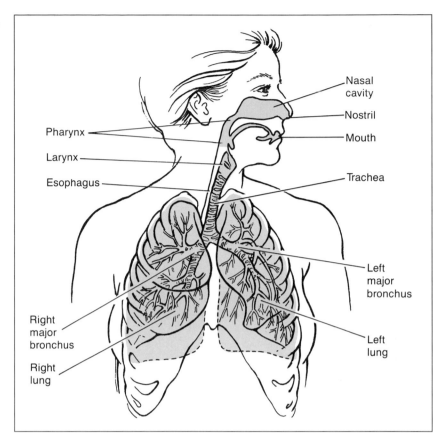

FIGURE 14-2 Airway branching.

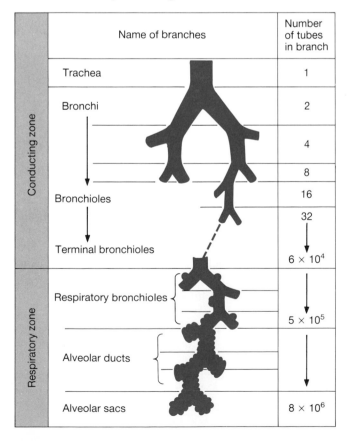

	Name of branches	Number of tubes in branch
Conducting zone	Trachea	1
	Bronchi	2
		4
		8
	Bronchioles	16
		32
	Terminal bronchioles	6×10^4
Respiratory zone	Respiratory bronchioles	5×10^5
	Alveolar ducts	
	Alveolar sacs	8×10^6

The epithelial surfaces of the airways, to the end of the respiratory bronchioles, contain cilia that constantly beat toward the pharynx. They also contain glands and individual epithelial cells that secrete mucus. Particulate matter, such as dust contained in the inspired air, sticks to the mucus, which is continually and slowly moved by the cilia to the pharynx and then swallowed. This mucus escalator is important to keep the lungs clear of particulate matter and the many bacteria that enter the body on dust particles. Ciliary activity can be inhibited by many noxious agents. For example, smoking a single cigarette can immobilize the cilia for several hours. The reduction

TABLE 14-2   FUNCTIONS OF THE CONDUCTING ZONE OF THE AIRWAYS

1. Provides a low-resistance pathway for air flow; resistance is physiologically regulated by changes in contraction of airway smooth muscle and by physical forces acting upon the airways.
2. Defends against microbes and other toxic chemicals and foreign matter; cilia, mucus, and phagocytes perform this function.
3. Warms and moistens the air.
4. Phonates (vocal cords).

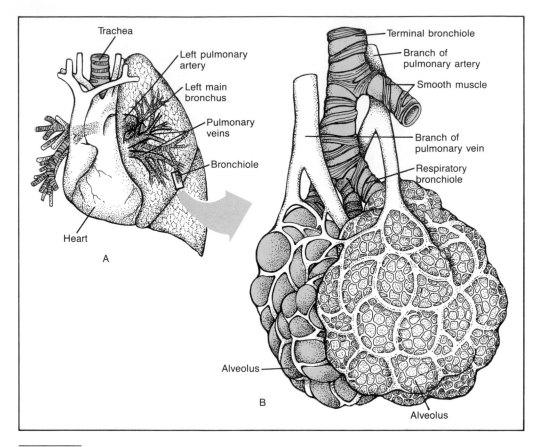

FIGURE 14-3 Relationships between blood vessels and airways. (A) The lung appears transparent so that the relationships can be seen. The airways beyond the bronchiole are too small to be seen. (B) An enlargement of a small section of 14-3A to show the continuation of the airways and the clusters of alveoli at their ends. Virtually the entire lung, not just the surface, consists of such clusters. For clarity, the left cluster of alveoli is shown without its covering capillary network.

in ciliary activity may result in lung infection and/or airway obstruction by stationary mucus. A smoker's early-morning cough is an attempt to clear this obstructive mucus from the airways.

A second protective mechanism is provided by cells, present in the airways and alveoli, that engulf inhaled particles and bacteria and thus keep them from gaining access to other lung cells or from entering the blood. These cells, termed phagocytes, are also injured by cigarette smoke and air pollutants.

Site of Gas Exchange: The Alveoli

The alveoli are tiny hollow sacs whose open ends are continuous with the lumens of the airways (Figures 14-2 and 14-4A). Typically, the air in two alveoli is separated by a single alveolar wall (Figure 14-4A). The air-facing surface(s) of the wall are lined by a continuous layer, one cell thick, of epithelial cells (**type I cells**).

The alveolar walls contain capillaries, the endothelial linings of which are separated from the alveolar epithelial lining only by a basement membrane and a very thin

interstitial space containing interstitial fluid and a loose meshwork of connective tissue (Figure 14-4B). Indeed, in places, the interstitial space may be absent altogether, and the epithelium of the alveolar surface and the endothelium of the capillaries in the wall may fuse. Thus the blood within an alveolar-wall capillary is separated from the air within the alveolus by only an extremely thin barrier (0.2 μm, compared with the 7-μm diameter of an average red blood cell). The total surface area of alveoli in contact with capillaries is approximately 75 m^2 (roughly the size of a tennis court and 80 times greater than the external body surface area). This extensive area, as well as the thinness of the barrier, permits the rapid exchange of large quantities of oxygen and carbon dioxide.

In addition to the type I cells, the alveolar epithelium contains smaller numbers of thicker specialized cells (**type II cells**) (Figure 14-4B) that produce a detergent-like substance, surfactant, to be discussed below. The alveolar walls also contain phagocytes and other connective-tissue cells that function in the lung's defense mech-

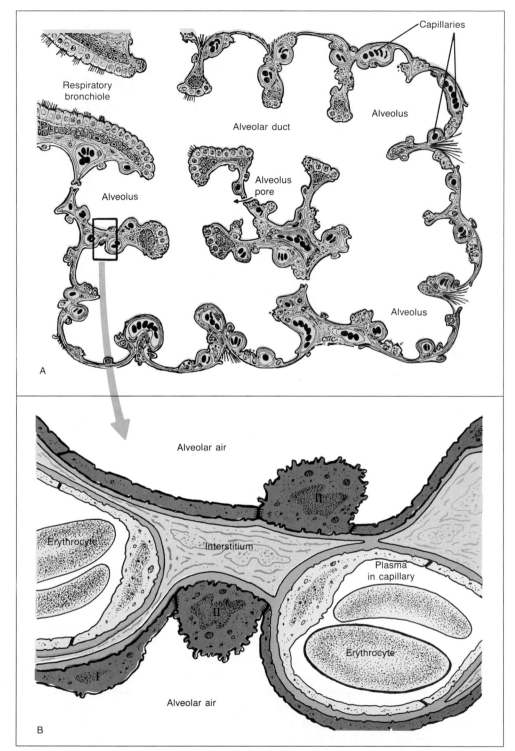

FIGURE 14-4

(A) Cross section through an area of the respiratory zone. There are 18 alveoli, only 4 of which are labeled. Two frequently share a common wall. *(From R. O. Greep and L. Weiss, "Histology," 3d edn., McGraw-Hill Book Company, New York, 1973.)* (B) Schematic enlargement of a portion of an alveolar wall. I = nucleus of a type I cell; II = nucleus of a type II cell. *(Adapted from Gong and Drage.)*

anisms. Finally, the alveolar surfaces in contact with the air are moist.

In some of the alveolar walls there are pores that permit the flow of air between alveoli. This route can be very important when the airway leading to an alveolus is occluded by disease, since some air can still enter the alveolus by way of the pores between it and adjacent alveoli.

Relation of the Lungs to the Thoracic (Chest) Wall

The lungs, like the heart, are situated in the **thorax**, the compartment of the body between the neck and abdomen. Thorax and chest are synonyms, even though common usage often assumes that the "chest" refers only to the front of the thorax. The thorax is a closed compartment, bounded at the neck by muscles and connective

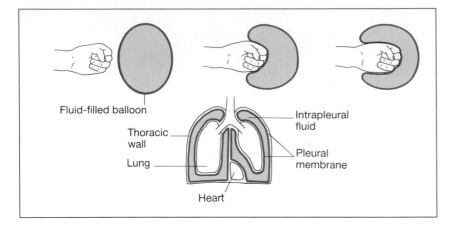

Fluid-filled balloon

Thoracic wall

Lung

Heart

Intrapleural fluid

Pleural membrane

FIGURE 14-5 Relationship of lungs, pleura, and thoracic wall, shown as analogous to pushing a fist into a fluid-filled balloon. Note that there is no communication between the right and left intrapleural fluids. The volume of intrapleural fluid is greatly exaggerated here. It normally consists of an extremely thin layer of fluid between the pleural membrane lining the inner surface of the thoracic wall and that lining the surface of the lungs.

tissue and completely separated from the abdomen by a large dome-shaped sheet of skeletal muscle, the **diaphragm**. The wall of the thorax is formed by the spinal column, the ribs, the breastbone (sternum), and the muscles that lie between the ribs (the **intercostal muscles**). The thoracic wall also contains large amounts of elastic connective tissue.

Each lung is surrounded by a completely closed sac, the **pleural sac**, consisting of a thin sheet of cells called **pleura**. The two pleural sacs, one on each side of the midline, are completely separate from each other. The relationship between a lung and its pleural sac can be visualized by imagining what happens when one pushes a fist into a balloon (Figure 14-5): The arm represents the major bronchus leading to the lung, the fist is the lung, and the balloon is the pleural sac. The fist becomes coated by one surface of the balloon. In addition, the balloon is pushed back upon itself so that its opposite surfaces lie close together. The pleural surface coating the lung (the visceral pleura) is firmly attached to the lung. Similarly, the outer layer (parietal pleura) is attached to and lines the interior thoracic wall and diaphragm. The two layers of pleura are so close to each other that normally they are always in virtual contact, but they are *not* attached to each other. Rather, they are separated by an extremely thin layer of **intrapleural fluid**, the total volume of which is only a few mL.

The intrapleural fluid lubricates the outer surface of the lungs. More important, as we shall see, pressure changes in the intrapleural fluid cause the lung surface and thoracic wall to move in and out together during breathing.

Pulmonary Pressures

An important point for the discussion that follows is that both the lungs and thoracic wall are elastic structures. If they are either stretched or compressed by some force, they will recoil, that is, return to their original sizes and positions when the force is removed.

The volume of the lungs and thorax at the end of a relaxed expiration is determined only by passive elastic forces since no significant contractions of the respiratory muscles are occurring. The volume of air in the lungs at this time is known as the **functional residual capacity**. The following simple experiment reveals that this position reflects a balance point between opposed forces.

In certain kinds of surgery, the chest wall ("chest wall" and "thoracic wall" are synonyms) is opened and the parietal pleura cut, exposing the fluid-filled intrapleural space, but the lung is not cut. Yet the lung on the side of the incision recoils immediately to a smaller volume. Simultaneously, the chest wall on that side moves outward. This experiment shows that as long as the chest wall is intact, some force must be acting to stretch the lung and that this force is eliminated when the chest is opened. The experiment also shows that the intact chest wall is normally partially pulled inward (compressed) by some force resulting from the presence of the lung within it. In other words, the lungs and chest wall would move away from each other, the lungs to recoil inward and the chest wall to pop outward, except that forces exist that keep them from doing so. We now turn to the nature of these forces and how they result in a stable equilibrium when the lung is at functional residual capacity.

Figure 14-6 illustrates the situation that normally exists at the end of an unforced expiration, when no respiratory muscle contraction is occurring and no air is flowing. All pressures in the respiratory system, as in the cardiovascular system, are given relative to **atmospheric pressure**—the pressure of the air surrounding the body (760 mmHg at sea level). The pressure within the lungs

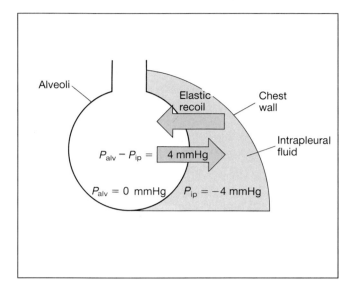

FIGURE 14-6 Alveolar P_{alv}, intrapleural P_{ip}, and transpulmonary $P_{alv} - P_{ip}$ pressures at the end of an unforced expiration. The transpulmonary pressure exactly opposes the elastic recoil of the lung, and the lung volume (and chest wall) remain stable. The lung volume at this time is the functional residual capacity.

(the **alveolar pressure, P_{alv}**) between breaths is 0 mmHg, that is, it is the same as atmospheric pressure. The pressure in the intrapleural fluid, the **intrapleural pressure (P_{ip})**, between breaths is approximately 4 mmHg less than atmospheric pressure, that is, −4 mmHg.[2] There is, therefore, a pressure difference of 4 mmHg [0 − (−4) = 4] across the lung wall. This pressure difference is known as the **transpulmonary pressure**, and it is the force that holds the lungs open, that is, keeps the stretched lungs from collapsing.

At the same time there is also a pressure difference of 4 mmHg keeping the chest wall from moving out: The atmospheric pressure against the outside of the wall is 0 mmHg, and the intrapleural pressure against the inside of the wall is −4 mmHg; the difference is 4 mmHg directed inward and opposing the tendency of the compressed thoracic wall to move outward.

We have so far not described how it is that the intrapleural pressure is subatmospheric. As the lungs (tending to move inward from their stretched position) and the thoracic wall (tending to move outward from its compressed position) move ever so slightly away from each other, there occurs an infinitesimal enlargement of the

---

[2] Physiologists usually express pressures in the respiratory system in cmH$_2$O rather than mmHg; we have chosen, for simplicity, not to do this (for reference, 1 cmH$_2$O = 1.3 mmHg).

fluid-filled intrapleural space between them. But fluid cannot expand the way air can, and so even this tiny enlargement of the intrapleural space drops the intrapleural pressure below atmospheric pressure. In this way, the elastic recoil of both the lung and chest wall creates the subatmospheric intrapleural pressure that keeps them from moving apart more than a tiny amount.

Why the lung collapses when the chest wall is opened should now be apparent. When the chest wall is pierced, atmospheric air rushes through the wound into the intrapleural space, and the intrapleural pressure goes from −4 mmHg to 0 mmHg. The transpulmonary pressure acting to hold the lung open is thus eliminated, and the stretched lung collapses. At the same time, the chest wall moves outward because there is no longer a difference in pressure between the external environment and the intrapleural space to keep the chest wall compressed.

As we shall see in the next section, the magnitudes of the alveolar, intrapleural, and transpulmonary pressures vary during breathing and directly cause the changes in lung size that occur during inspiration and expiration.

VENTILATION AND LUNG MECHANICS

An inventory of steps involved in respiration (Figure 14-7) is provided for orientation before beginning the detailed descriptions of each step.

Ventilation is defined as the exchange of air between the atmosphere and alveoli. Like blood, air moves by bulk flow from a region of high pressure to one of low pressure. We saw on page 360 that bulk flow can be described by the equation

$$F = \frac{\Delta P}{R}$$

That is, flow F is proportional to the pressure difference ΔP between two points and inversely proportional to the resistance R. For air flow into or out of the lungs, the relevant pressures are the atmospheric pressure P_{atm} and the alveolar pressure:

$$F = \frac{P_{atm} - P_{alv}}{R}$$

Air moves into and out of the lungs because the alveolar pressure is made alternately less than and greater than atmospheric pressure (Figure 14-8). These alveolar pressure changes are caused, as we shall see, by changes in the dimensions of the lungs. To describe these changes, we need to learn about one more basic concept. As

FIGURE 14-7 The steps of respiration.

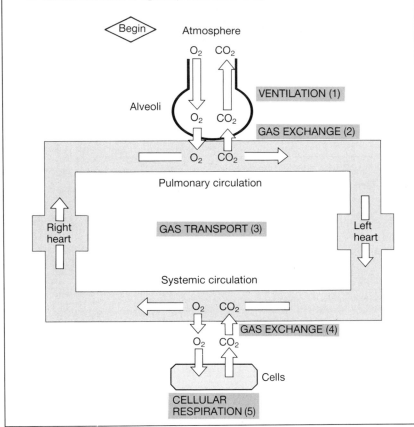

1. **Ventilation**: Exchange of air between atmosphere and alveoli by *bulk flow*
2. Exchange of O_2 and CO_2 between alveolar air and blood in lung capillaries by *diffusion*
3. Transport of O_2 and CO_2 through pulmonary and systemic circulation by *bulk flow*
4. Exchange of O_2 and CO_2 between blood in tissue capillaries and cells in tissues by *diffusion*
5. Cellular utilization of O_2 and production of CO_2

FIGURE 14-8 Relationships required for ventilation. When the alveolar pressure P_{alv} is less than atmospheric pressure P_{atm}, air enters the lungs. Flow F is directly proportional to the pressure difference and inversely proportional to airway resistance R.

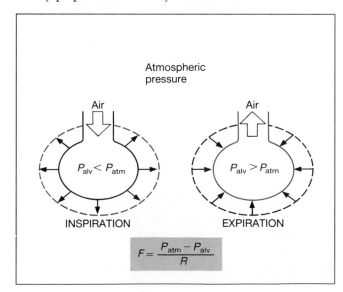

stated by what is known as **Boyle's law** (Figure 14-9), the relationship between the pressure exerted by a fixed number of gas molecules in a container and the volume of the container is as follows: An increase in the volume of the container decreases the pressure of the gas, whereas a decrease in container volume increases the pressure.

Inspiration

Figures 14-10 and 14-11 summarize the events during normal inspiration at rest. Just before the inspiration begins, that is, at functional residual capacity, the respiratory muscles are relaxed, and no air is flowing because the alveolar pressure is atmospheric (0 mmHg). As described earlier, the intrapleural pressure at this point is subatmospheric (-4 mmHg), and the transpulmonary pressure $P_{alv} - P_{ip}$ is 4 mmHg.

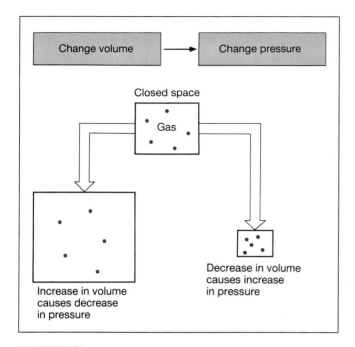

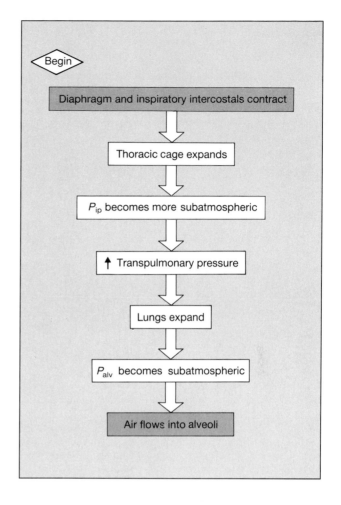

FIGURE 14-9 Boyle's law: The pressure exerted by a constant number of gas molecules in a container is inversely proportional to the volume of the container; that is, PV equals a constant K.

FIGURE 14-10 Sequence of events during inspiration. Figure 14-11 illustrates these events quantitatively.

Inspiration is initiated by the contraction of the diaphragm and the inspiratory intercostal muscles.[3] The diaphragm is the most important inspiratory muscle during normal quiet breathing. When the nerves to it cause it to contract, its dome moves downward into the abdomen, enlarging the thorax. Simultaneously, the nerves to the inspiratory intercostal muscles cause them to contract, leading to an upward and outward movement of the ribs and a further increase in thoracic size. As the thorax enlarges, the thoracic wall moves ever so slightly further away from the lung surface, and the intrapleural fluid pressure becomes even more subatmospheric than it was between breaths. This increases the transpulmonary pressure, which forces the lung to open more.

Thus, when the inspiratory muscles increase the thoracic dimensions, the lungs are also forced to enlarge virtually to the same degree because of the change in intrapleural pressure. The enlargement of the lung causes an increase in the sizes of the alveoli throughout the lung. Therefore, by Boyle's law, the pressure within the alveoli drops to less than atmospheric. This produces the difference in pressure $P_{atm} - P_{alv}$ that causes a bulk flow of air from the atmosphere through the airways into the alveoli. By the end of the inspiration, the pressure in the alveoli again equals atmospheric pressure.[4]

Expiration

Figures 14-11 and 14-12 summarize the sequence of events during expiration. At the end of inspiration, the

---

[3]There is still controversy over just which intercostal muscles are involved in inspiration and expiration, and so here we shall simply refer to them as "inspiratory" and "expiratory" intercostals. It should also be noted that additional muscles, termed the accessory muscles of inspiration, may also be brought into play during labored inspiration. These muscles act both on the upper ribs to enlarge the upper thorax and on the sternum to move it forward and upward.

[4]It should be noted that the analysis of Figure 14-11 treats the lungs as a single alveolus. The fact is that there are significant regional differences in alveolar, intrapleural, and transpulmonary pressures throughout the lungs and thoracic cavity. These differences are due mainly to the effects of gravity. They are of great importance in determining the distribution of inspired air throughout the lung.

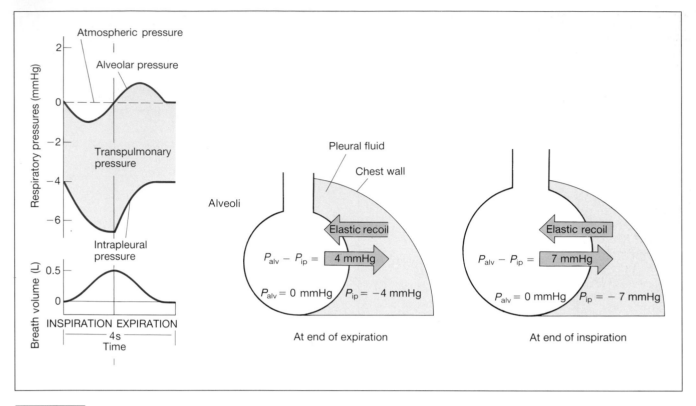

FIGURE 14-11 Summary of alveolar, intrapleural, and transpulmonary pressure changes and air flow during inspiration and expiration of 500 mL of air. Note that normal atmospheric pressure (760 mmHg) has a value of zero on the respiratory pressure scale. Note that the transpulmonary pressure exactly opposes the elastic recoil of the lungs at the end of both inspiration and expiration.

nerves to the diaphragm and inspiratory intercostal muscles cease firing and so these muscles relax. The chest wall and, hence, the lungs *passively* return to their original dimensions. As the lungs shrink, air in the alveoli becomes temporarily compressed so that, by Boyle's law, alveolar pressure exceeds atmospheric. Therefore, air flows from the alveoli through the airways out into the atmosphere. Thus, expiration at rest is completely passive, depending only upon the relaxation of the inspiratory muscles and recoil of the stretched lungs.

Under certain conditions, during exercise, for example, expiration of larger volumes is achieved by contraction of the expiratory intercostal muscles and the abdominal muscles, which *actively* decreases thoracic dimensions. The expiratory intercostal muscles insert on the ribs in such a way that their contraction pulls the chest wall in. Contraction of the abdominal muscles increases intraabdominal pressure and forces the diaphragm up into the thorax.

Lung Compliance

To reiterate, the degree of lung expansion at any instant is proportional to the transpulmonary pressure, that is, the difference between the alveolar pressure and the intrapleural pressure. But just how much any given transpulmonary pressure expands the lung depends upon the stretchability, or compliance, of the lung. **Lung compliance C_L** is defined as the magnitude of the change in lung volume ΔV_L produced by a given change in the transpulmonary pressure:

$$C_L = \frac{\Delta V_L}{\Delta (P_{alv} - P_{ip})}$$

Thus, the higher the compliance, the easier it is to expand the lungs at any given transpulmonary pressure. A low lung compliance means that a greater than normal transpulmonary pressure must be developed across the lung wall to produce a given amount of lung expansion. In other words, when lung compliance is low, intrapleural pressure must be made more subatmospheric than usual during inspiration to achieve lung expansion. This requires more vigorous contractions of the diaphragm and inspiratory intercostal muscles. Thus, the less compliant the lung, the more energy is required for a given amount of expansion. Persons with low lung compliance therefore tend to breathe shallowly and rapidly.

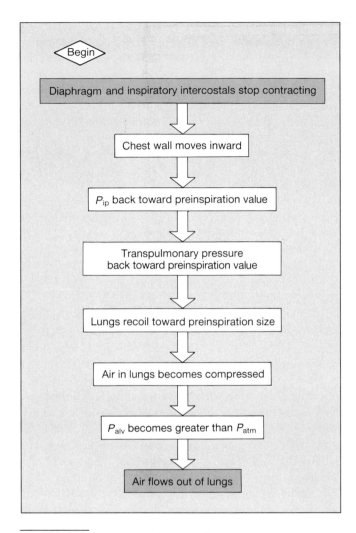

FIGURE 14-12 Sequence of events during expiration.

An abnormally high compliance, as occurs in normal aging and the disease emphysema, also causes problems, as we shall describe in a subsequent section on emphysema.

Determinants of lung compliance. There are two major determinants of lung compliance. One is the stretchability of the lung tissues, particularly its elastic connective tissues. Thus a thickening of the lung tissues decreases lung compliance. However, the single most important determinant of lung compliance is not the stretchability of the lung *tissues* but the surface tension at the air-water interfaces within the alveoli. As noted earlier, the surfaces of the alveolar cells are moist, and so they can be pictured as air-filled sacs lined with water. At an air-water interface, the attractive forces between the water molecules, known as **surface tension**, make the water lining like a stretched balloon that constantly tries to shrink and resists further stretching. Thus, expansion of the lung requires energy not only to stretch the connective tissue of the lung but to overcome the surface tension of the water layer lining the alveoli.

Indeed, the surface tension of pure water is so great that were the alveoli lined with pure water, lung expansion would require exhausting muscular effort and the lungs would tend to collapse. It is extremely important, therefore, that the type II alveolar cells produce a detergent-like phospholipid known as pulmonary **surfactant**, which markedly reduces the cohesive forces between water molecules on the alveolar surface. Therefore, surfactant lowers the surface tension and increases lung compliance (Table 14-3).

The amount of surfactant present depends on the relative rates at which it is being secreted and broken down or reabsorbed. It tends to decrease when breaths are small and constant, even for as little as 30 to 60 minutes. A deep breath stimulates the type II cells to replenish the surfactant. This is why patients who have had thoracic or abdominal surgery and are breathing shallowly because of the pain must be urged to take occasional deep breaths.

A striking example of what occurs when surfactant is deficient is the disease known as **respiratory-distress syndrome of the newborn**, which frequently afflicts premature infants in whom the surfactant-synthesizing cells are too immature to function adequately. Because of low lung compliance, the infant is able to inspire only by the most strenuous efforts, which may ultimately cause complete exhaustion, inability to breathe, lung collapse, and death. Normal maturation of the surfactant-synthesizing apparatus is facilitated by the hormone cortisol, the secretion of which is increased late in pregnancy. Accordingly, administration of cortisol to a pregnant woman who is likely to deliver prematurely provides an important means of preventing this disease.

Airway Resistance

As previously stated, the volume of air that flows into or out of the alveoli per unit time is directly proportional to

TABLE 14-3 SOME IMPORTANT FACTS ABOUT PULMONARY SURFACTANT
1. Pulmonary surfactant is a phospholipid bound to a protein.
2. It is secreted by type II alveolar cells.
3. It lowers surface tension of the water layer at the alveolar surface, which increases lung compliance, that is, makes lungs easier to expand.
4. Its concentration decreases when lung volume is small and constant.

the pressure difference between the atmosphere and alveoli and inversely proportional to the resistance to flow offered by the airways:

$$F = \frac{P_{atm} - P_{alv}}{R}$$

What factors determine airway resistance? Resistance is (1) directly proportional to the magnitude of the frictional interactions between the flowing gas molecules, that is, the viscosity of the air, (2) directly proportional to the length of the airway, and (3) inversely proportional to the fourth power of the airway radius. These factors are similar to those determining resistance in the circulatory system. Resistance in the respiratory tree, just as in the circulatory tree, is largely controlled by the radius of the airways.

The airway radii are normally so large that they offer little resistance to air flow. Air viscosity is also usually negligible, and so is the contribution of airway length. Therefore, resistance to flow is normally so small that very small pressure differences suffice to produce large volumes of air flow. As we have seen (Figure 14-11), the average atmosphere-to-alveoli pressure difference during a normal breath at rest is less than 1 mmHg; yet, approximately 500 mL of air is moved by this tiny difference.

Airway radii and therefore resistance are affected by physical, neural, and chemical factors. One important physical factor is the transpulmonary pressure, which exerts a distending force on the airways just as on the alveoli. This is a major factor keeping the smaller airways—those without cartilage to support them—from collapsing. Because, as we have seen, transpulmonary pressure increases during inspiration, airway radius is larger and airway resistance is smaller when the lungs are expanded than when they are at functional residual capacity.

A second physical factor holding the airways open is the connective-tissue fibers that attach to the airway exteriors and, because of their arrangement, continuously pull outward on the sides of the airways. This is termed lateral traction. Since these fibers become stretched as the lungs expand, they help pull the airways open even more during inspiration. Thus, both the transpulmonary pressure and lateral traction act in the same direction.

Such physical factors also explain why the airways become narrower and airway resistance increases during a forced expiration. Indeed, because of increased airway resistance, there is a limit as to how much one can increase air flow rate during expiration no matter how intense the effort.

Neural regulation of airway size is mediated mainly by parasympathetic neurons, reflex stimulation of which causes contraction of airway smooth muscle and increased resistance. This is the cause of airway constriction, as well as increased mucus secretion, when chemical irritants are inhaled.

There is little, if any, sympathetic innervation of the airways, but epinephrine, the hormone released from the adrenal medulla during exercise or stress, causes airway dilation by acting on beta-adrenergic receptors on the airway smooth muscle.

Histamine, a chemical messenger released locally during allergic responses, causes contraction of airway smooth muscle and increased mucus secretion. The effects of eicosanoids (page 151) on pulmonary airways are also important since the lungs take up, metabolize, and release various members of the eicosanoid family; some are airway constrictors, others are dilators.

The neural and chemical factors listed above may all have a common final pathway in their action—an alteration of smooth-muscle cAMP. Specifically, a *decrease* in cAMP causes *increased* smooth-muscle contraction.

Finally, the airway smooth muscle is highly responsive to carbon dioxide. A high carbon dioxide concentration in the alveoli produces bronchodilation, and a low carbon dioxide bronchoconstriction. The physiological significance of this responsiveness is discussed later. The factors that control airway resistance are summarized in Table 14-4.

One might wonder why we should be concerned with all these factors that *can* influence airway resistance since normally airway resistance is so low that it is no impediment to air flow except perhaps during strenuous exercise. The reason is that, under abnormal circum-

TABLE 14-4   FACTORS CONTROLLING AIRWAY RESISTANCE
Neural and chemical influences on airway smooth muscle:

Airways constricted by:	Airways dilated by:
Histamine	Epinephrine
Parasympathetic nerves	
Decreased carbon dioxide	Increased carbon dioxide
Some eicosinoids	Some eicosinoids
Irritants	

Physical influences:

1. Airways are held open by transpulmonary pressure and lateral traction. They open more during inspiration and may collapse during forced expiration.
2. Airways may be occluded by mucus accumulation.

stances, these factors cause serious changes in airway resistance. Asthma and chronic obstructive pulmonary disease provide important examples.

Asthma. Asthma is a disease characterized by intermittant attacks in which airway smooth muscle contracts, increasing airway resistance. More mucus may also be secreted by the airways, and this mucus may be abnormally thick, leading to a further increase in airway resistance secondary to mucus plugs. Some patients with asthma have a history of allergies, but others do not. The cause of the increased contraction of the airway smooth muscle is not the same in all asthmatic patients, and the chemical mediators described above may all play a role. Accordingly, the bronchodilator drugs used to treat an asthmatic attack can be categorized as follows: drugs that block the action of acetylcholine and histamine; drugs that inhibit the synthesis of eicosanoids; drugs like epinephrine that stimulate beta-adrenergic receptors; drugs that increase concentrations of cAMP in the smooth muscle by blocking phosphodiesterase, the enzyme that mediates the breakdown of cAMP.

Chronic obstructive pulmonary disease. The term **chronic obstructive pulmonary disease** refers to emphysema or chronic bronchitis or a combination of the two. These diseases, which cause severe difficulties not only in breathing but in oxygenation of the blood, are among the major causes of disability and death in the United States. In contrast to asthma, increased smooth-muscle contraction is not the cause of airway obstruction in these diseases.

Emphysema is characterized by destruction of the alveolar walls and consequently a marked enlargement of the alveolar air spaces and loss of pulmonary capillaries. These changes impair gas exchange, as will be described later. In addition, the small airways—those from the terminal bronchioles on down—are reduced in number and have atrophied walls. How all these changes occur is still poorly understood, but it is known that cigarette smoking is an important cause. Air pollution and hereditary factors also play a role in some patients.

The airway obstruction in emphysema is due to collapse of the airways. To understand this, recall from page 438 that two physical factors passively holding the airways open are the transpulmonary pressure and the connective-tissue fibers attached to the airway exteriors. Both of these factors are diminished in emphysema, in association with a high lung compliance, and so the airways collapse. This is why, as mentioned earlier, a high lung compliance is not a good thing.

Chronic bronchitis is characterized by excessive mucus production in the bronchi and chronic inflammatory changes in the small airways. The cause of obstruction is accumulation of mucus in the airways and thickening of the inflamed airways. The same agents listed above—smoking, for example—that cause emphysema also cause chronic bronchitis. That is why the two diseases frequently coexist.

The Heimlich maneuver. The **Heimlich maneuver** is used to aid persons choking on foreign matter caught in the respiratory tract. A sudden increase in abdominal pressure is produced as the rescuer's fists, placed against the victim's abdomen slightly above the navel and well below the tip of the sternum, are pressed into the abdomen with a quick upward thrust (Figure 14-13). The increased abdominal pressure forces the diaphragm up-

---

FIGURE 14-13 The Heimlich maneuver. The rescuer's fists are placed against the victim's abdomen. A quick upward thrust of the fists causes elevation of the diaphragm and a forceful expiration. As this expired air is forced through the trachea and larynx, the foreign object in the airway is expelled.

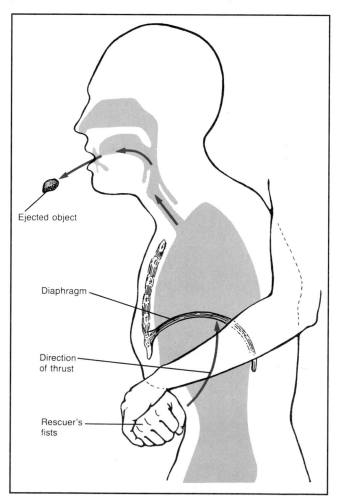

Ejected object

Diaphragm

Direction of thrust

Rescuer's fists

ward into the thorax, reducing thoracic size and, by Boyle's law, increasing alveolar pressure. The forceful expiration produced by the increased alveolar pressure expels the object caught in the respiratory tract.

Lung Volumes and Capacities

The volume of air entering the lungs during a single inspiration is normally approximately equal to the volume leaving on the subsequent expiration and is called the **tidal volume**. During normal quiet breathing, the tidal volume—termed the resting tidal volume—is approximately 500 mL.

After expiration of a resting tidal volume, the lungs still contain air. Recall that this volume is termed the functional residual capacity; it amounts to approximately 2500 mL (Figure 14-14). In other words, at rest, the 500 mL of air inspired with each breath adds to and mixes with 2500 mL of air already in the lungs, and then 500 mL of the total is expired. Through maximal active contraction of the expiratory muscles, it is possible to expire 1500 mL of the 2500 mL remaining after the resting tidal volume has been expired; this additional volume is termed the **expiratory reserve volume**. Even after a maximal active expiration, approximately 1000 mL of air still remains in the lungs and is termed the **residual volume**.

Let us return to the situation in which a person is breathing at rest and see how much inspiration can be increased from this point. The volume of air that can be inspired over and above the resting tidal volume is called the **inspiratory reserve volume** and amounts to approximately 3000 mL of air (Figure 14-14). Thus, someone inspiring maximally after a resting expiration will inspire 500 mL, the resting tidal volume, plus another 3000 mL, the inspiratory reserve volume, for a total of 3500 mL. On the next expiration, this air is all expelled.

A useful clinical measurement is the **vital capacity**, the maximal volume of air that a person can expire, regardless of the time required, after a maximal inspiration. Under these conditions, the person is expiring the resting tidal volume and inspiratory reserve volume just inspired, followed by the expiratory reserve volume (Figure 14-14). In other words, the vital capacity is the sum of these three volumes.

A variant on this method is the **forced vital capacity (FVC)**, in which the person takes a maximal inspiration and then exhales maximally *as fast as possible;* thus the name, "forced" vital capacity. For several reasons the FVC may differ somewhat from the value obtained in an "unforced" vital capacity measurement. But, most important, the apparatus used to measure FVC also measures the volume expired after 1 s (the forced expiratory volume in 1 s, **FEV$_1$**). Normal persons can expire approximately 80 percent of the FVC in 1 s.

FIGURE 14-14 Lung volumes and capacities recorded on a spirometer, an apparatus for measuring inspired and expired volumes. When the subject inspires, the pen moves up; with expiration, it moves down. The capacities are the sums of two or more lung volumes. The lung volumes are the four distinct components of total lung capacity. (Note that residual volume and total lung capacity cannot be measured with a spirometer.)

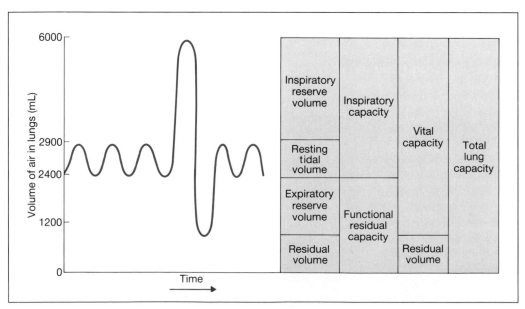

These measurements are useful diagnostic tools. For example, persons with obstructive lung diseases (increased airway resistance) typically cannot expire a normal fraction of the FVC in 1 s because it is difficult to expire air rapidly through the narrowed airways. In contrast to obstructive lung diseases, **restrictive lung diseases** are characterized by normal airway resistance but impaired respiratory movements because of abnormalities in the lung tissue, the pleura, the chest wall, or the neuromuscular machinery. Restrictive lung diseases are characterized by a reduced vital capacity but a normal FEV_1/FVC ratio.

Alveolar Ventilation

The total ventilation per minute, termed the **minute ventilation**, is equal to the tidal volume multiplied by the respiratory rate:

Minute ventilation (mL/min) =
$$\text{tidal volume (mL/breath)} \times \text{respiratory rate (breaths/min)}$$

For example, at rest, a normal person moves approximately 500 mL of air in and out of the lungs with each breath, and takes 10 breaths each minute. The minute ventilation is therefore 500 mL per breath × 10 breaths per minute = 5000 mL of air per minute. However, because of dead space, not all this air is available for exchange with the blood.

Dead space. The airways have a volume of about 150 mL. Exchanges of gases with the blood occur only in the alveoli and not in this 150 mL of the conducting airways. Picture, then, what occurs during expiration of a tidal volume which, in this instance, is of 450 mL: 450 mL of air is forced out of the alveoli and through the airways. Approximately 300 mL of this alveolar air is exhaled at the nose or mouth, but approximately 150 mL still remains in the airways at the end of expiration. During the next inspiration, 450 mL of air flows into the alveoli, but the first 150 mL entering the alveoli is not atmospheric air but the 150 mL left behind from the last breath (Figure 14-15). Thus, only 300 mL of new atmospheric air enters the alveoli during the inspiration. At the end of inspiration, 150 mL of fresh air fills the conducting airways, but no gas exchange with the blood can occur there. At the next expiration, this fresh air will be washed out and again replaced by old alveolar air, thus completing the cycle. The end result is that 150 mL of the 450 mL of atmospheric air entering the respiratory system during each inspiration never reaches the alveoli but is merely moved in and out of the airways. Because these airways do not permit gas exchange with the blood, the space within them is termed the **anatomic dead space**.

Thus the volume of fresh air entering the alveoli during each inspiration equals the tidal volume minus the volume of air in the anatomic dead space. For the previous example:

Tidal volume = 450 mL

Anatomic dead space = 150 mL

Fresh air entering alveoli in one inspiration =

$$450 \text{ mL} - 150 \text{ mL} = 300 \text{ mL}$$

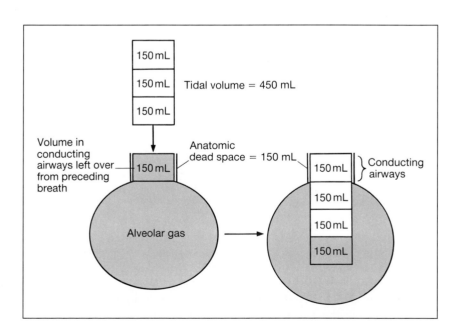

FIGURE 14-15 Effects of anatomic dead space on alveolar ventilation. Anatomic dead space is the volume of the airways.

					Anatomic-dead-space	
Subject	Tidal Volume, mL/Breath	×	Frequency, Breaths/min	= Minute Ventilation, mL/min	Ventilation, mL/min	Alveolar Ventilation, mL/Min
A	150		40	6000	150 × 40 = 6000	0
B	500		12	6000	150 × 12 = 1800	4200
C	1000		6	6000	150 × 6 = 900	5100

TABLE 14-5 EFFECT OF BREATHING PATTERNS ON ALVEOLAR VENTILATION

To determine how much fresh air enters the alveoli per minute, we simply multiply the volume of fresh air entering the alveoli per breath by the breathing frequency (set at 12 breaths/min in our example):

$$300 \text{ mL/breath} \times 12 \text{ breaths/min} = 3600 \text{ mL/min}$$

This total, the volume of fresh air entering the alveoli per minute, is called the **alveolar ventilation**:

Alveolar ventilation =
(tidal volume − dead space) × frequency

The value calculated for alveolar ventilation may be somewhat confusing because it seems to indicate that, in our example, only 3600 mL gas enters and leaves the alveoli in each minute. This is not true—the total is 5400 mL (12×450), but only 3600 mL is fresh air; the rest, of course, is the air from the anatomic dead space.

What is the significance of the anatomic dead space and alveolar ventilation? Since only that portion of inspired air that enters the alveoli, that is, the alveolar ventilation, is useful for gas exchange with the blood, the magnitude of the alveolar ventilation is of much greater significance than is the minute ventilation, as can be demonstrated readily by data in Table 14-5.

In this experiment, subject A breathes rapidly and shallowly, B normally, and C slowly and deeply. Each subject has exactly the same minute ventilation, that is, each is moving the same amount of air in and out of the lungs per minute. Yet, when we subtract the anatomic-dead-space volume from the minute ventilation, we find marked differences in alveolar ventilation. Subject A has no alveolar ventilation and would become unconscious in several minutes, whereas C has a considerably greater alveolar ventilation than B, who is breathing normally. The important deduction to be drawn from this example is that increased *depth* of breathing is far more effective in elevating alveolar ventilation than is an equivalent increase of breathing *rate*. Conversely, a decrease in depth can lead to a critical reduction of alveolar ventilation. This is because a fixed volume of *each* tidal volume goes to the dead space. If the tidal volume decreases, the fraction of the tidal volume going to the dead space increases until, as in subject A, it may represent the entire tidal volume. On the other hand, any increase in tidal volume goes entirely toward increasing alveolar ventilation. These concepts have important physiological implications. Most situations that produce an increased ventilation, such as exercise, reflexly call forth a relatively greater increase in breathing depth than rate.

The anatomic dead space is not the only type of dead space. Some fresh inspired air is not used for gas exchange with the blood even though it reaches the alveoli because some alveoli, for various reasons, have little or no blood supply. This volume of air is known as **alveolar dead space**. It is quite small in normal persons but may be very large in several kinds of lung disease. As we shall see, it is minimized by local mechanisms that match air and blood flows. The sum of the anatomic and alveolar dead space is known as the **total dead space**.

EXCHANGE OF GASES IN ALVEOLI AND TISSUES

We have now completed our discussion of the lung mechanics that produce alveolar ventilation, but this is only the first step in the respiratory process. Oxygen must move across the alveolar membranes into the pulmonary capillaries, be transported by the blood to the tissues, leave the tissue capillaries and enter the extracellular fluid, and finally cross plasma membranes to gain entry into cells. Carbon dioxide must follow a similar path in reverse.

In the steady state, the volume of oxygen that leaves the tissue capillaries and is consumed by the body cells per unit time is exactly equal to the volume of oxygen added to the blood in the lungs during the same time period. Similarly, the rate at which carbon dioxide is produced by the body cells and enters the systemic

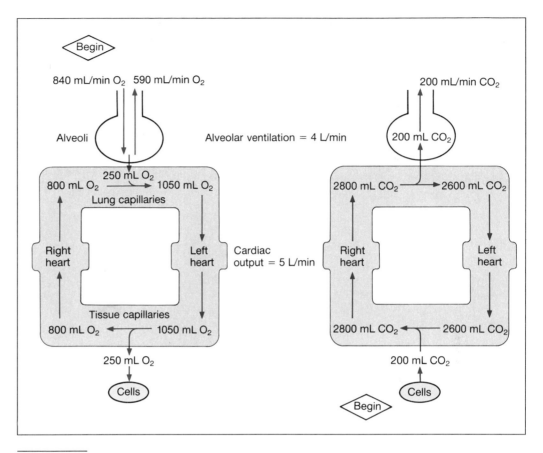

FIGURE 14-16 Summary of oxygen and carbon dioxide exchanges between atmosphere, lungs, blood, and tissues during 1 min. It is assumed that $RQ = 0.8$. Note that the values for oxygen and carbon dioxide in blood are *not* the values per liter of blood but rather the amounts transported per minute in the cardiac output (5 L normally). Thus, the volume of oxygen in 1 L of arterial blood is approximately 1000 mL $O_2 \div 5$ L blood = 200 mL O_2/L.

blood is identical to the rate at which carbon dioxide leaves the blood in the lungs and is expired.

The relative amounts of oxygen consumed by cells and carbon dioxide produced depend primarily upon which nutrients are being used for energy. The ratio CO_2 produced/O_2 consumed is known as the **respiratory quotient (RQ)**. On a mixed diet, the RQ is approximately 0.8, that is, 8 molecules of CO_2 are produced for every 10 molecules of O_2 consumed.[5]

Figure 14-16 presents typical exchange values during 1 min for a person at rest, assuming a cellular oxygen consumption of 250 mL/min, a carbon dioxide production of 200 mL/min, an alveolar ventilation of 4000 mL per minute, and a cardiac output of 5000 mL/min.

Since only 21 percent of atmospheric air is oxygen, the total oxygen entering the alveoli is 21 percent of

4000 mL, or 840 mL per minute. Of this inspired oxygen, 250 mL crosses the alveoli into the pulmonary capillaries, and the remaining 590 mL is exhaled. This 250 mL of oxygen is carried away by 5 L of blood per minute—the cardiac output. Note, however, in Figure 14-16 that blood entering the lungs already contains large quantities of oxygen (the adaptive value of this seeming inefficiency will be described later), to which the new 250 mL is added. The blood then flows from the lungs to the left heart and is pumped by the left ventricle through the tissue capillaries, where 250 mL of oxygen leaves the blood to be taken up and utilized by cells. The quantities of oxygen added to the blood in the lungs and removed in the tissues are identical.

As shown by Figure 14-16, the story reads in reverse for carbon dioxide. There is already a good deal of carbon dioxide in arterial blood; to this is added an additional 200 mL, the amount produced by the cells, as blood flows through tissue capillaries. This additional amount is eliminated as blood flows through the lungs.

[5]The $RQ = 1$ for carbohydrate, 0.7 for fat, and 0.8 for protein.

Blood pumped by the heart carries oxygen and carbon dioxide between the lungs and tissues by bulk flow, but diffusion is responsible for the net movement of these molecules between the alveoli and blood, and between the blood and the cells of the body. Understanding the mechanisms involved in these diffusional exchanges depends upon some basic chemical and physical properties of gases, to which we now turn.

Partial Pressures of Gases

A gas consists of individual molecules constantly in random motion. Since these rapidly moving molecules bombard the walls of any vessel containing them, they exert a pressure against the walls. The magnitude of the pressure is increased by anything that increases the rate of bombardment. The pressure a gas exerts is proportional to (1) the temperature (because heat increases the speed at which molecules move) and (2) the concentration of the gas, that is, the number of molecules per unit volume.

In a mixture of gases, the pressure exerted by each gas is independent of the pressure exerted by the others because gas molecules are normally so far apart that they do not interfere with each other. Since each gas in a mixture behaves as though no other gases were present, the total pressure of the mixture is simply the sum of the individual pressures. These individual pressures, termed **partial pressures**, are denoted by a P in front of the symbol for the gas. For example, the partial pressure of oxygen is represented by P_{O_2}. Net diffusion of a gas will occur from a region where its partial pressure is high to a region where it is low.

Gases such as oxygen and carbon dioxide can dissolve in water or other liquids. Thus, one can refer to the gases as existing in a gas phase or a liquid phase. The term partial pressure is not restricted to the gas phase but also applies to gases dissolved in liquids.

Atmospheric air consists primarily of nitrogen and oxygen with very small quantities of water vapor, carbon dioxide, and inert gases. The sum of the partial pressures of all these gases is termed atmospheric pressure or barometric pressure. It varies in different parts of the world as a result of differences in altitude, but at sea level it is 760 mmHg. Since approximately 21 percent of the molecules in air are oxygen, the P_{O_2} of atmospheric air is 0.21×760 mmHg = 160 mmHg at sea level.

Diffusion of gases in liquids. When a liquid is exposed to air containing a particular gas, molecules of the gas will enter the liquid and dissolve in it. The concentration of the gas in the liquid will be directly proportional to the partial pressure of the gas in the air. This phenomenon reflects the basic definition of pressure.

Suppose, for example, that a closed container contains both water and gaseous oxygen. Oxygen molecules from the gas phase constantly bombard the surface of the water, some entering the water and dissolving. Since the number of molecules striking the surface is directly proportional to the P_{O_2}, the number of molecules entering the water is also directly proportional to P_{O_2}. If the P_{O_2} in the gas phase is higher than the P_{O_2} in the liquid, there will be a net diffusion of oxygen into the liquid until the P_{O_2} is the same in the liquid and gaseous phases.

Conversely, if a liquid containing a dissolved gas at high partial pressure is exposed to a lower partial pressure of that same gas in a gas phase, gas molecules will diffuse from the liquid into the gas phase until the partial pressures in the two phases become equal.

The exchanges *between* gas and liquid phases described in the last two paragraphs are precisely the phenomena occurring between alveolar air and pulmonary capillary blood. In addition, dissolved gas molecules also diffuse *within* a liquid from a region of higher partial pressure to a region of lower partial pressure, an effect that underlies the exchange of gases between cells, extracellular fluid, and capillary blood throughout the body.

The concentration of a gas in a liquid is proportional not only to the partial pressure of the gas but to the solubility of the gas in the liquid; the more soluble the gas, the greater will be its concentration at any given partial pressure. Thus, if a liquid is exposed to two different gases at the same partial pressures, the concentrations of the gases in the liquid at equilibrium are not identical unless their solubilities in that liquid are identical.

With these basic gas properties as the foundation, we can now discuss the diffusion of oxygen and carbon dioxide across alveolar, capillary, and plasma membranes. The pressures of these gases in air and in various sites of the body are given in Figure 14-17 for a resting person at sea level. We start our discussion with the alveolar gas pressures because their values set those of systemic arterial blood.

Alveolar Gas Pressures

Normal alveolar gas pressures are $P_{O_2} = 105$ mmHg, and $P_{CO_2} = 40$ mmHg. (We do not deal with nitrogen, even though it is the most abundant gas in the alveoli, because nitrogen is biologically inert under normal conditions.) Compare these values with the gas pressures in the air being breathed: $P_{O_2} = 160$ mmHg and $P_{CO_2} = 0.3$ mmHg, a value so low that we will simply assume it to be zero. The alveolar P_{O_2} is lower than atmospheric P_{O_2} because some of the oxygen in the air entering the alveoli leaves them to enter the pulmonary capillaries.

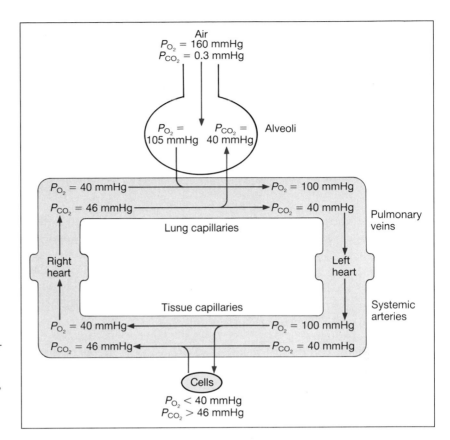

FIGURE 14-17 Partial pressures of carbon dioxide and oxygen in inspired air and various places in the body. Note that the P_{O_2} in the systemic arteries is shown as identical to that in the pulmonary veins; actually, for reasons described in the text, the arterial value should be slightly less, but we have ignored this for the sake of clarity.

Alveolar P_{CO_2} is higher than atmospheric P_{CO_2} because carbon dioxide enters the alveoli from the pulmonary capillaries.

The factors that determine the value of alveolar P_{O_2} are (1) the P_{O_2} of atmospheric air, (2) the rate of cellular oxygen consumption, and (3) the alveolar ventilation. At any particular atmospheric P_{O_2}, it is the *ratio* of oxygen consumption to alveolar ventilation that determines alveolar P_{O_2}—the *higher* the ratio, the *lower* the alveolar P_{O_2}. The story is simpler for P_{CO_2} since atmospheric P_{CO_2} is essentially zero. Alveolar P_{CO_2} is therefore determined only by the ratio of carbon dioxide production to alveolar ventilation—the *higher* the ratio, the *higher* the alveolar P_{CO_2}. Although there are equations for calculating the alveolar gas pressures from these variables, we will describe the interactions in an intuitive qualitative manner.

First, breathing air having a low P_{O_2}, as at high altitude, results in a low alveolar P_{O_2} since fewer molecules of oxygen are present in the atmospheric air flowing through the alveoli. In contrast, administration of a high-oxygen gas mixture to a person increases alveolar P_{O_2}.

Next, consider the effects of changing only the alveolar ventilation on the alveolar gas pressures (Figure 14-18). Suppose that the alveolar ventilation is larger than the usual 4 L/min because a resting person is deliberately breathing rapidly and deeply, but that the cellular oxygen utilization is unchanged at 250 mL/min. Under these conditions, the 250 mL of oxygen that leaves the alveoli each minute to enter the blood and be supplied to cells is a smaller fraction of the oxygen supplied by the alveolar ventilation. Therefore, steady-state alveolar P_{O_2} is high, that is, more closely approximates atmospheric air than usual. At the same time, the alveolar P_{CO_2} is low because the still normal amount of carbon dioxide (200 mL) entering the alveoli from the blood is diluted more than usual by the larger alveolar ventilation.

In contrast, a reduced alveolar ventilation in a person with a normal cellular oxygen utilization and carbon dioxide production will result in a lower alveolar P_{O_2} and a higher alveolar P_{CO_2}.

Finally, consider the effects of increasing cellular oxygen utilization and carbon dioxide production at constant alveolar ventilation. In this case, a larger fraction of the oxygen in the alveolar ventilation will leave the alveoli to enter the pulmonary capillaries; therefore, alveolar P_{O_2} will decrease. At the same time, the larger amount of carbon dioxide entering the alveoli from the blood will be diluted less than usual by the alveolar ventilation; therefore, the alveolar P_{CO_2} will increase.

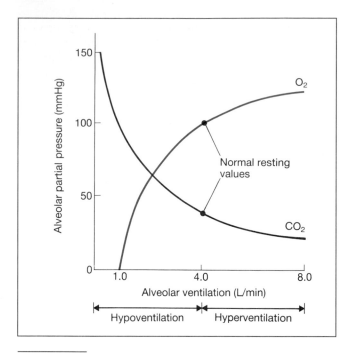

FIGURE 14-18 Effects of increasing or decreasing alveolar ventilation on alveolar partial pressures in a person having a constant metabolic rate (cellular oxygen consumption and carbon dioxide production). Note that alveolar P_{O_2} approaches zero when alveolar ventilation is about 1 L/min. At this point all the oxygen entering the alveoli crosses into the blood, leaving virtually no oxygen in the alveoli.

Several terms can now be defined. **Hypoventilation** is a decrease in ventilation without a similar decrease in oxygen consumption and carbon dioxide production. The result is that alveolar P_{O_2} decreases and alveolar P_{CO_2} increases. **Hyperventilation** is increased ventilation without a similar increase in oxygen consumption and carbon dioxide production. The result is that alveolar P_{O_2} increases and alveolar P_{CO_2} decreases. **Hyperpnea** is increased ventilation in exact proportion to an increased oxygen consumption and carbon dioxide production. Accordingly, the alveolar gas pressures do not change.

Alveolar-Blood Gas Exchange

The blood that enters the pulmonary capillaries is, of course, systemic venous blood pumped to the lungs via the pulmonary arteries. Having come from the tissues, it has a high P_{CO_2} (46 mmHg) and a low P_{O_2} (40 mmHg) (Table 14-6). The differences in the partial pressures of oxygen and carbon dioxide on the two sides of the alveolar-capillary membrane result in the net diffusion of oxygen from alveoli to blood and of carbon dioxide from blood to alveoli. As this diffusion occurs, the capillary blood P_{O_2} rises and its P_{CO_2} falls. The net diffusion of

TABLE 14-6	NORMAL GAS PRESSURES IN SYSTEMIC BLOOD	
	Venous	**Arterial**
P_{O_2}	40 mmHg	100 mmHg
P_{CO_2}	46 mmHg	40 mmHg

these gases ceases when the capillary partial pressures become equal to those in the alveoli.

In a normal person, the rates at which oxygen and carbon dioxide diffuse are so rapid and the blood flow through the capillaries so slow that complete equilibrium is reached well before the end of the capillaries (Figure 14-19). Only during the most strenuous exercise, when blood flows through the lung capillaries very rapidly, is there insufficient time for complete equilibration.

Thus, the blood that leaves the pulmonary capillaries to return to the heart and be pumped into the systemic arteries has essentially the same P_{O_2} and P_{CO_2} as alveolar air. Accordingly, the factors described in the previous section (atmospheric P_{O_2}, cellular oxygen consumption and carbon dioxide production, and alveolar ventilation) determine the alveolar gas pressures, which then determine the systemic arterial gas pressures. Actually, systemic arterial P_{O_2} is normally 5 to 10 mmHg lower than alveolar P_{O_2}. The major reason for this is described in the

FIGURE 14-19 Complete equilibration of blood P_{O_2} with alveolar P_{O_2} along the length of the pulmonary capillaries.

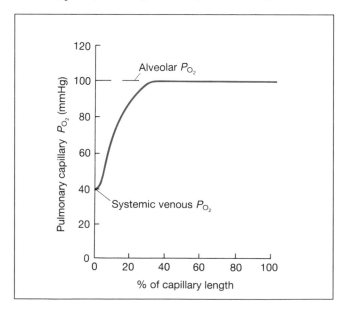

section that immediately follows on matching of ventilation and blood flow.

Factors other than the partial-pressure differences also normally influence net diffusion rates of oxygen and carbon dioxide between alveolar air and pulmonary capillary blood. One such additional factor is the surface area available for diffusion. Many of the pulmonary capillaries are normally closed at rest. During exercise, these capillaries open and receive blood, thereby increasing the surface area for diffusion and enhancing gas exchange. The mechanism by which this occurs is a simple physical one; the pulmonary circulation at rest is at such a low pressure that the pressure in many capillaries is inadequate to keep the capillaries open, but the increased cardiac output of exercise raises pulmonary vascular pressures, which opens these capillaries.

The diffusion of gases between alveoli and capillaries may be impaired in a number of ways, resulting in inadequate gas exchange. For one thing, the surface area available for diffusion may be decreased. In lung infections or pulmonary edema, for example, some of the alveoli may become filled with fluid. Diffusion may also be impaired if the alveolar walls become thickened, as, for example, in the disease (of unknown cause) called diffuse interstitial fibrosis. In both these cases, there may be significant reduction of oxygen supply to the blood, particularly during exercise when the time for equilibration is reduced, but no impairment of carbon dioxide elimination. The reason that carbon dioxide diffusion remains adequate is that carbon dioxide is much more soluble in water than is oxygen and so more carbon dioxide molecules are available for diffusion.

Matching of Ventilation and Blood Flow in Alveoli

The major disease-induced cause of inadequate gas exchange between alveoli and pulmonary-capillary blood is not diffusion impairment. Rather, it is the mismatching of the air supply and blood supply in individual alveoli.

The lungs are composed of approximately 300 million discrete alveoli, each capable of receiving carbon dioxide from, and supplying oxygen to, the pulmonary-capillary blood. To be most efficient, the right proportion of alveolar air flow (ventilation) and capillary blood flow (perfusion) should be available to each alveolus. Even in normal persons there are factors, the most important of which is gravity, that cause ventilation and perfusion to be out of proportion to each other in parts of the lung. This is termed **ventilation-perfusion inequality.**

In disease states, regional changes in lung compliance, airway resistance, and vascular resistance can cause marked ventilation-perfusion inequalities. The most striking example of this is emphysema. First, because of the destruction of lung tissue as well as the obstruction of airways, some areas of lung may receive large amounts of air while others receive little or none. Second, there is also a maldistribution of blood flow because capillaries are lost when the alveolar walls are destroyed.

What are the effects of ventilation-perfusion inequalities on gas exchange? In areas receiving too much ventilation relative to blood flow, the alveolar P_{O_2} will be high and the alveolar P_{CO_2} will be low. In areas receiving too little ventilation relative to their blood flows, the alveolar P_{O_2} will be low, and the P_{CO_2} high. These differences tend to cancel each other out with regard to the alveolar gas pressures averaged over the entire lung. However, for complex reasons, the presence of both types of regions in the lungs do *not* cancel out each other's effects so far as overall oxygenation of the blood is concerned. The result is that when the blood from the entire lung is mixed in the pulmonary veins, it has a lower P_{O_2} than the average alveolar P_{O_2}. This blood, of course, will become the systemic arterial blood, and so the arterial blood flowing to the tissues also has a lower P_{O_2} than the average alveolar P_{O_2}.

There is enough ventilation-perfusion inequality in normal people to account for arterial P_{O_2} being 5 to 10 mmHg less than alveolar P_{O_2}. In disease states the ventilation-perfusion inequalities may be so great that the arterial P_{O_2} becomes low enough to interfere seriously with the delivery of oxygen to the tissues.

Carbon dioxide elimination is also impaired by ventilation-perfusion inequalities, but, again for complex reasons, not to the same degree as oxygen uptake. Nevertheless, severe ventilation-perfusion inequalities in disease can cause a significant elevation of arterial P_{CO_2}.

There are two local homeostatic responses within the lungs to minimize the mismatching of ventilation and blood flow. One operates on the airways to alter the distribution of ventilation, the other on the blood vessels to alter blood-flow distribution. If an alveolus is receiving an excess of air relative to its blood supply, the P_{CO_2} in the alveolus will be low. The low P_{CO_2} causes the smooth muscle of the airway supplying the alveolus to contract, thereby reducing air flow. By this local mechanism, ventilation can be better matched to blood supply (Figure 14-20).

The second mechanism for matching air flow and blood flow is the control of pulmonary vascular smooth muscle by oxygen. A decreased P_{O_2} in the alveoli causes vasoconstriction of the small pulmonary blood vessels, analogous to arterioles in the systemic circulation, supplying the alveolus, and an increased P_{O_2} causes vasodilation. Note that this local effect of oxygen on pulmonary blood vessels is precisely the opposite of that exerted on systemic arterioles. How does this effect of oxygen on

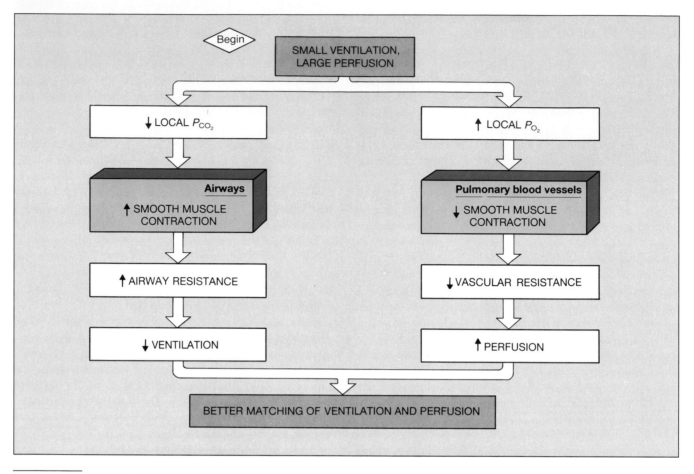

FIGURE 14-20 Local matching of ventilation and blood flow. High local P_{CO_2} affects airway resistance, and low local P_{O_2} affects pulmonary vascular resistance. These events occur simultaneously. The opposite events occur in areas with large air flow and small blood flow; such areas have low local P_{CO_2} and high local P_{O_2}.

pulmonary vessels provide matching of air and blood supplies? Recall the example above in which an alveolus has a large ventilation and small blood flow. The alveolus will therefore have a high P_{O_2}. This causes dilation of the blood vessels supplying it, which increases blood supply to match the high air supply (Figure 14-20).

Gas Exchange in the Tissues

As the systemic arterial blood enters capillaries throughout the body, it is separated from the interstitial fluid only by the thin capillary wall, which is highly permeable to both oxygen and carbon dioxide. The interstitial fluid, in turn, is separated from intracellular fluid by cells' plasma membranes, which are also quite permeable to oxygen and carbon dioxide. Metabolic reactions occurring within cells are constantly consuming oxygen and producing carbon dioxide. Therefore, as shown in Figure 14-17, intracellular P_{O_2} is lower and P_{CO_2} higher

than in blood. As a result, there is a net diffusion of oxygen from blood into cells and a net diffusion of carbon dioxide from cells into blood. In this manner, as blood flows through systemic capillaries, its P_{O_2} decreases and its P_{CO_2} increases. This accounts for the systemic venous blood values shown in Figure 14-17.

The mechanisms that enhance diffusion of oxygen and carbon dioxide between cells and blood when a tissue increases its metabolic activity were discussed on page 393.

In summary, the supply of new oxygen to the alveoli and the consumption of oxygen in the cells create P_{O_2} gradients that produce net diffusion of oxygen from alveoli to blood in the lungs and from blood to cells in the rest of the body. Conversely, the production of carbon dioxide by cells and its elimination from the alveoli via expiration create P_{CO_2} gradients that produce net diffusion of carbon dioxide from cells to blood in the rest of the body and from blood to alveoli in the lungs.

TRANSPORT OF OXYGEN IN BLOOD

Table 14-7 summarizes the oxygen content of systemic arterial blood (we shall henceforth refer to systemic arterial blood simply as arterial blood). Each liter contains the number of oxygen molecules equivalent to 200 mL of pure gaseous oxygen at atmospheric pressure. The oxygen is present in two forms: (1) dissolved in the plasma and erythrocyte water and (2) reversibly combined with hemoglobin molecules.

The amount of oxygen dissolved in blood is directly proportional to the P_{O_2} of the blood. Because oxygen is relatively insoluble in water, only 3 mL an be dissolved in 1 L of blood at the normal arterial P_{O_2} of 100 mmHg. The other 197 mL of oxygen in a liter of arterial blood, more than 98 percent of the total oxygen, is carried in the erythrocytes reversibly bound to hemoglobin.

As described on page 352, hemoglobin is a protein composed of four polypeptide chains, each of which contains a single atom of iron. It is the iron atom to which a molecule of oxygen binds. Thus each hemoglobin molecule can bind four molecules of oxygen. However, the equation for the reaction between oxygen and hemoglobin is usually written in terms of a single polypeptide-heme chain of a hemoglobin molecule:

$$O_2 + Hb \rightleftharpoons HbO_2 \qquad (14\text{-}1)$$

This chain can exist in one of two forms—**deoxyhemoglobin** (Hb) and **oxyhemoglobin** (HbO$_2$). In a blood sample containing many hemoglobin molecules, the fraction of all the hemoglobin in the form of oxyhemoglobin is expressed as the percent hemoglobin saturation:

$$\text{Percent saturation} = \frac{\text{amount of O}_2 \text{ on Hb}}{\text{maximal possible amount}} \times 100$$

For example, if 40 percent of all the hemoglobin in a blood sample is oxyhemoglobin, the sample is said to be 40 percent saturated.

TABLE 14-7 OXYGEN CONTENT OF SYSTEMIC ARTERIAL BLOOD

1 liter (L) arterial blood contains:

	3 mL	O$_2$ physically dissolved
	197 mL	O$_2$ bound to hemoglobin
Total	200 mL	O$_2$

Cardiac output = 5 L/min

O$_2$ carried to tissues/min = 5 L/min × 200 mL O$_2$/L
               = 1000 mL O$_2$/min

What factors determine the percent hemoglobin saturation? By far the most important is the blood P_{O_2}. Thus, the P_{O_2} determines not only the concentration of oxygen dissolved in the blood but also the amount that binds to hemoglobin.

Effect of P_{O_2} on Hemoglobin Saturation

From inspection of Equation 14-1 and the law of mass action, one can see that raising the blood P_{O_2} should increase the combination of oxygen with hemoglobin. The quantitative relationship between these variables is shown in Figure 14-21, which is called an **oxygen-hemoglobin dissociation curve**. (The term dissociate means "to separate," in this case oxygen from hemoglobin; it could just as well have been called an oxygen association curve.) The curve is an S-shaped curve with a steep slope between 10 and 60 mmHg P_{O_2} and a flat portion between 70 and 100 mmHg P_{O_2}. Thus, the extent to which hemoglobin combines with oxygen increases very rapidly from 10 to 60 mmHg so that, at a P_{O_2} of 60 mmHg, 90 percent of the total hemoglobin is combined with oxygen. From this point on, a further increase in P_{O_2} produces only a small increase in oxygen binding.

The importance of this plateau at higher P_{O_2} values is as follows. Many situations, including high altitudes and

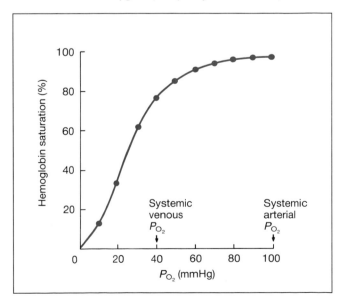

FIGURE 14-21 Oxygen-hemoglobin dissociation curve. This curve applies to blood at 37°C and a normal arterial hydrogen-ion concentration. The vertical axis could also have plotted oxygen carriage, in milliliters of oxygen. At 100 percent saturation, the amount of hemoglobin in normal blood carries 200 mL of oxygen. (*Adapted from Comroe.*)

cardiac or pulmonary disease, are characterized by a moderate reduction of alveolar and therefore arterial P_{O_2}. Even if the P_{O_2} fell from the normal value of 100 to 60 mmHg, the total quantity of oxygen carried by hemoglobin would decrease by only 10 percent since hemoglobin saturation is still close to 90 percent at a P_{O_2} of 60 mmHg. The plateau therefore provides an excellent safety factor in the supply of oxygen to the tissues.

The flatness of the upper portion of the curve also explains another fact: In a normal person at sea level, raising the alveolar (and therefore the arterial) P_{O_2} either by hyperventilating or by breathing 100 percent oxygen adds very little additional oxygen to the blood. A small additional amount dissolves, but because hemoglobin is already almost completely saturated with oxygen at the normal arterial P_{O_2} of 100 mmHg, it simply cannot pick up any more oxygen when the P_{O_2} is elevated beyond this point. This applies only to normal people at sea level. If the person initially has a low arterial P_{O_2} because of lung disease or high altitude, then there would be a great deal of deoxyhemoglobin initially present in the arterial blood. Therefore, raising the alveolar and, thereby, the arterial P_{O_2} would result in significantly more oxygen carriage.

We now retrace our steps and reconsider the movement of oxygen across the various membranes, this time including hemoglobin in our analysis. It is essential to recognize that the oxygen bound to hemoglobin does not contribute directly to the P_{O_2} of the blood. Only dissolved oxygen does so. Therefore, oxygen diffusion is governed only by the dissolved portion, a fact that permitted us to ignore hemoglobin in discussing transmembrane partial-pressure gradients. However, the presence of hemoglobin plays a critical role in determining the *total amount* of oxygen that will diffuse, as illustrated by a simple example (Figure 14-22).

Two solutions separated by a semipermeable membrane contain equal quantities of oxygen, the gas pressures are equal, and no net diffusion occurs. Addition of hemoglobin to compartment B destroys this equilibrium because much of the oxygen combines with hemoglobin. Despite the fact that the total quantity of oxygen in compartment B is still the same, the number of dissolved oxygen molecules has decreased. Therefore, the P_{O_2} of compartment B is less than that of A, and net diffusion of oxygen occurs from A to B. At the new equilibrium, the oxygen pressures are once again equal, but almost all the oxygen is in compartment B and is combined with hemoglobin.

Let us now apply this analysis to capillaries of the lung and tissue (Figure 14-23). The plasma and erythrocytes entering the lungs have a P_{O_2} of 40 mmHg. As we can see from Figure 14-21, hemoglobin saturation at this P_{O_2}

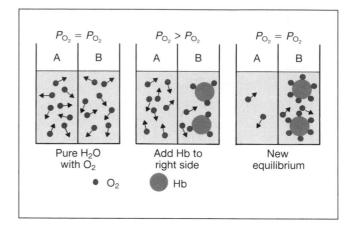

FIGURE 14-22 Effect of added hemoglobin on oxygen distribution between two compartments containing a fixed number of oxygen molecules and separated by a semipermeable membrane. At the new equilibrium, the P_{O_2} values are again equal to each other but lower than before the hemoglobin was added. However, the total oxygen, in other words, that dissolved plus that combined with hemoglobin, is now much higher on the right side of the membrane. (*Adapted from Comroe.*)

is 75 percent. Oxygen then diffuses from the alveoli, because of its higher P_{O_2} (100 mmHg), into the plasma. This increases plasma P_{O_2} and induces diffusion of oxygen into the erythrocytes, elevating erythrocyte P_{O_2} and causing increased combination of oxygen and hemoglobin. The vast preponderance of the oxygen diffusing into the blood from the alveoli does not remain dissolved but combines with hemoglobin. Therefore, the blood P_{O_2} normally remains less than the alveolar P_{O_2} until hemoglobin is virtually 100 percent saturated. Thus the diffusion gradient favoring oxygen movement into the blood is maintained despite the very large transfer of oxygen.

In the tissue capillaries, the procedure is reversed: As the blood enters the capillaries, plasma P_{O_2} is greater than interstitial fluid P_{O_2} and net oxygen diffusion occurs out of the capillary from the plasma. Plasma P_{O_2} is now lower than erythrocyte P_{O_2}, and oxygen diffuses out of the erythrocyte into the plasma. The lowering of erythrocyte P_{O_2} causes the dissociation of some of the oxygen from hemoglobin, thereby liberating oxygen. Simultaneously, the oxygen that had diffused into the interstitial fluid is moving into cells along the concentration gradient generated by cellular utilization of oxygen. The net result is a transfer of large quantities of oxygen from hemoglobin into cells purely by diffusion.

The fact that, under resting conditions, hemoglobin is still 75 percent saturated as the blood leaves the tissue capillaries underlies an important automatic mechanism

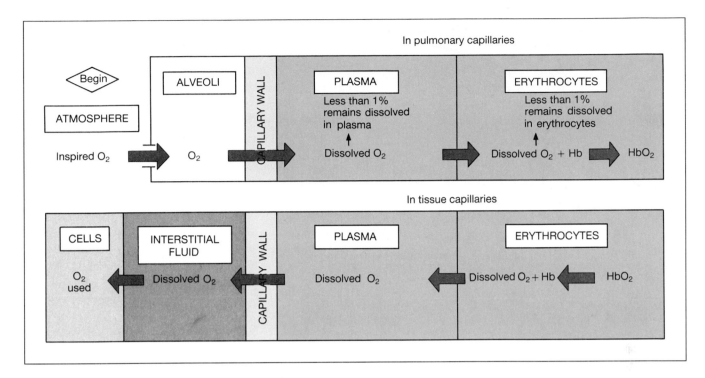

FIGURE 14-23 Oxygen movement in lungs and tissues. Movement of inspired air into the alveoli is by bulk flow; all movement across membranes is by diffusion.

by which cells can obtain more oxygen whenever they increase their activity. An exercising muscle consumes more oxygen, thereby lowering its tissue P_{O_2}. This increases the blood-to-tissue P_{O_2} gradient and the diffusion of oxygen from blood to cell. In turn, the resulting reduction of erythrocyte P_{O_2} causes additional dissociation of hemoglobin and oxygen. Because of these gradient changes, an exercising muscle can extract almost all the oxygen from its blood supply. This process is so effective because it takes place in the steep portion of the oxygen-hemoglobin dissociation curve. Of course, an increased blood flow to the muscles also contributes greatly to the increased oxygen supply.

Effects of Blood P_{CO_2}, H$^+$ Concentration, Temperature, and DPG Concentration on Hemoglobin Saturation

At any given P_{O_2}, a variety of other factors influence the degree of hemoglobin saturation: blood P_{CO_2}, H$^+$ concentration, temperature, and the concentration of a substance—DPG—produced by the erythrocytes. As illustrated in Figure 14-24, an increase in any of these factors causes the dissociation curve to shift to the right, which means that, at any given P_{O_2}, hemoglobin has less affinity for oxygen. In contrast, a decrease in any of these factors causes the dissociation curve to shift to the left, which means that, at any given P_{O_2}, hemoglobin has a greater affinity for oxygen.

The effects of increased P_{CO_2}, H$^+$ concentration, and temperature are continuously exerted on the blood in tissue capillaries because each of these factors is higher in tissue-capillary blood than in arterial blood: The P_{CO_2} is increased because of the carbon dioxide entering the blood from the tissues. For reasons to be described later, the H$^+$ concentration is elevated because of the elevated P_{CO_2}. The temperature is increased because of the heat produced by tissue metabolism. Therefore, hemoglobin exposed to this elevated blood P_{CO_2}, H$^+$ concentration, and temperature as it passes through the tissue capillaries has its affinity for oxygen decreased, and therefore, it gives up even more oxygen than it would have if the decreased tissue-capillary P_{O_2} had been the only operating factor.

The more active a tissue is, the greater is its P_{CO_2}, H$^+$ concentration, and temperature. This causes hemoglobin to release more oxygen, at any given P_{O_2}, during passage through the tissue's capillaries, thereby providing the more active cells with additional oxygen.

What is the mechanism by which these factors influence hemoglobin's affinity for oxygen? Carbon dioxide

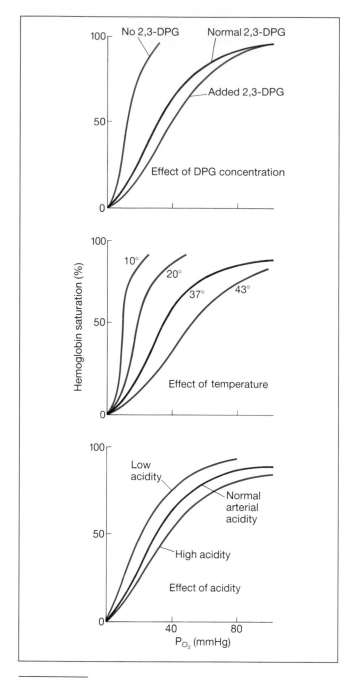

FIGURE 14-24 Effects of 2,3-DPG concentration, temperature, and acidity on the relationship between P_{O_2} and hemoglobin saturation. *(Adapted from Comroe.)*

What is the mechanism by which these factors influence hemoglobin's affinity for oxygen? Carbon dioxide and hydrogen ions do so by combining with the hemoglobin and altering its molecular configuration. Thus, these effects are a form of allosteric modulation, described on page 56. An elevated temperature also decreases hemoglobin's affinity for oxygen by altering its molecular configuration.

Erythrocytes contain large quantities of the substance **2,3-diphosphoglycerate (DPG)**, which is present in only trace amounts in other mammalian cells. DPG, which is produced by the erythrocytes during glycolysis, binds reversibly with hemoglobin, causing it to have a lower affinity for oxygen (Figure 14-24). Therefore, the effect of increased DPG is to shift the oxygen-hemoglobin dissociation curve to the right, the same as the effect exerted by an increased temperature, P_{CO_2}, or H^+ concentration. The net result is that whenever DPG levels are increased, there is enhanced unloading of oxygen from hemoglobin as blood flows through the tissues. Such an increase in DPG concentration is triggered by a variety of conditions associated with inadequate oxygen supply to the tissues and helps to maintain oxygen delivery.

TRANSPORT OF CARBON DIOXIDE

As noted earlier, carbon dioxide is much more soluble in water than is oxygen, and so more dissolved carbon dioxide than dissolved oxygen is carried in blood. Even so, only a small amount of blood carbon dioxide is transported in this way.

Some of the carbon dioxide molecules, after dissolving in the blood, can also react reversibly with the amino groups of proteins, particularly hemoglobin, to form **carbamino compounds**:

$$RNH_2 + CO_2 \rightleftharpoons RNHCOOH$$

For simplicity, this reaction with hemoglobin is often written as:

$$CO_2 + Hb \rightleftharpoons HbCO_2 \qquad (14\text{-}2)$$

But most of the carbon dioxide molecules, after dissolving in the blood, are converted to bicarbonate:

$$CO_2 + H_2O \underset{\text{anhydrase}}{\overset{\text{Carbonic}}{\rightleftharpoons}} H_2CO_3 \rightleftharpoons HCO_3^- + H^+ \quad (14\text{-}3)$$

Carbonic acid   Bicarbonate

The first reaction in Equation 14-3 is rate-limiting and is catalyzed by the enzyme **carbonic anhydrase**, which is present in the erythrocytes but not the plasma. Carbonic acid is converted to bicarbonate and hydrogen ion as fast as it is produced.

Because carbon dioxide undergoes these various reactions in blood, it is customary to add up the amounts of dissolved carbon dioxide, bicarbonate, and carbon dioxide in carbamino form and call this sum the total carbon dioxide of the blood. Normally, this amounts to approxi-

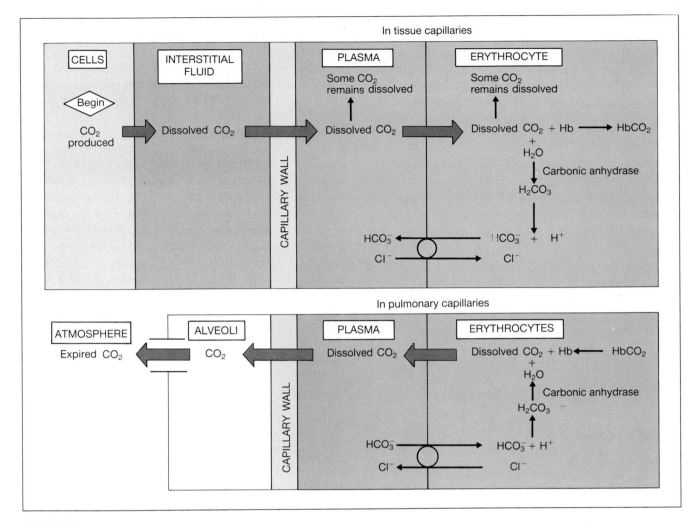

FIGURE 14-25 Summary of CO_2 movement. All movement across membranes is by diffusion. Note that most of the CO_2 entering the blood in the tissues ultimately is converted to HCO_3^-. This occurs almost entirely in the erythrocytes because the carbonic anhydrase is located there, but most of the HCO_3^- then moves out of the erythrocytes into the plasma in exchange for chloride ions.

mately 520 mL per liter of arterial blood. Bicarbonate is by far the most abundant of the three contributors, constituting approximately 90 percent of the total carbon dioxide of arterial blood.

Metabolism, in a resting person, generates about 200 mL of carbon dioxide per minute. When arterial blood flows through tissue capillaries, this volume of carbon dioxide diffuses from the tissues into the blood. Since cardiac output is 5 L/min, each liter of systemic venous blood contains 40 mL (that is, 200 mL/5 L) more carbon dioxide than the systemic arterial blood does. Let us review, with the addition of a few more important

facts, the reactions that this carbon dioxide undergoes after it enters the blood (Figure 14-25).

1. A small fraction (10 percent) of the carbon dioxide entering the blood remains physically dissolved in the plasma and red blood cells.
2. Another fraction (30 percent) reacts directly with hemoglobin to form the carbamino compound $HbCO_2$. This process is facilitated because much of the oxyhemoglobin becomes deoxyhemoglobin as blood flows through the tissues, and deoxyhemoglo-

bin reacts more readily with carbon dioxide than does oxyhemoglobin.

3. Most of the carbon dioxide (60 percent) is converted to bicarbonate and hydrogen ions (Equation 14-2).[6] This occurs mainly in the erythrocytes because they contain large quantities of the carbonic anhydrase required for the reactions to go quickly. Once formed, most of the bicarbonate moves out of the erythrocytes via a 1:1 exchange with chloride ions and is carried in the plasma. The reactions described in Equation 14-3 also explain why, as mentioned earlier, the H^+ concentration in tissue-capillary blood and systemic venous blood is higher than that of the arterial blood and increases as metabolic activity increases. The fate of these hydrogen ions will be discussed in the next section.

Just the opposite events occur as systemic venous blood flows through the lung capillaries (Figure 14-25). Because the blood P_{CO_2} is higher than alveolar P_{CO_2}, a net diffusion of CO_2 from blood into alveoli occurs. This loss of carbon dioxide from the blood lowers the blood P_{CO_2} and drives these reactions to the left: HCO_3^- and H^+ combine to give H_2CO_3, which then dissociates to CO_2 and H_2O. Similarly, $HbCO_2$ generates Hb and free

---

[6]The 60 percent figure might seem to conflict with the previous statement that bicarbonate constitutes 90 percent of the carbon dioxide content of arterial blood. There is no conflict, however. The *new* carbon dioxide added to blood in tissue capillaries does not distribute among carbamino and bicarbonate in the same proportions as those already present in *arterial blood*. The reason is that the concentration of deoxyhemoglobin is much higher in the *tissue-capillary blood* than arterial blood and is available for binding carbon dioxide to form carbamino.

FIGURE 14-26 Binding of hydrogen ions by hemoglobin as blood flows through tissue capillaries. This reaction is facilitated because deoxyhemoglobin, formed as oxygen comes off hemoglobin, has a greater affinity for hydrogen ions than does oxyhemoglobin. For this reason, Hb and HbH are both abbreviations for deoxyhemoglobin.

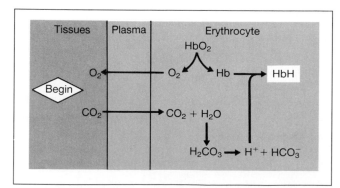

CO_2. Normally, as fast as this CO_2 is generated from HCO_3^- and H^+ and from $HbCO_2$, it diffuses into the alveoli. In this manner, all the CO_2 delivered into the blood in the tissues now is delivered into the alveoli, from which it is eliminated from the body by expiration.

TRANSPORT OF HYDROGEN IONS BETWEEN TISSUES AND LUNGS

To repeat, as blood flows through the tissues, a fraction of oxyhemoglobin loses its oxygen to become deoxyhemoglobin while, simultaneously, a large quantity of carbon dioxide enters the blood and undergoes, primarily in the erythrocytes, the reactions that generate bicarbonate and hydrogen ions. What happens to these hydrogen ions? Deoxyhemoglobin has a much greater affinity for H^+ than does oxyhemoglobin, and so it binds most of the hydrogen ions (Figure 14-26). Indeed, deoxyhemoglobin is often abbreviated HbH rather than Hb to denote its binding of H^+. In effect, the reaction is: $HbO_2 + H^+ \rightleftharpoons HbH + O_2$.

In this manner only a small number of the hydrogen ions generated in the blood remain free. This explains why the acidity of venous blood is only slightly greater than that of arterial blood. The pH of arterial blood is 7.4 and that of venous blood is 7.36. Remember that pH goes down as H^+ concentration goes up.

As the venous blood passes through the lungs, all these reactions are reversed. Deoxyhemoglobin becomes converted to oxyhemoglobin and, in the process, releases the hydrogen ions it had picked up in the tissues. The hydrogen ions react with bicarbonate to give carbon dioxide and water, and the carbon dioxide diffuses into the alveoli to be expired. Normally all the hydrogen ions that are generated from the reaction of carbon dioxide and water in the tissue capillaries recombine with bicarbonate to form carbon dioxide and water in the pulmonary capillaries. Therefore, none of these hydrogen ions appear in the arterial blood.

But what if the person is hypoventilating or has a lung disease that prevents normal elimination of carbon dioxide? Not only would arterial P_{CO_2} rise as a result but so would arterial H^+ concentration. Increased arterial H^+ concentration due to carbon dioxide retention is termed **respiratory acidosis**. Conversely, hyperventilation would lower the arterial values of both P_{CO_2} and H^+ concentration—a **respiratory alkalosis**.

In the course of describing the transport of oxygen, carbon dioxide, and H^+ in blood, we have presented multiple factors that influence the binding of these substances by hemoglobin. They are all summarized in Table 14-8.

TABLE 14-8 EFFECTS OF VARIOUS FACTORS ON HEMOGLOBIN
The affinity of hemoglobin for oxygen is decreased by: 1. Increased hydrogen-ion concentration 2. Increased P_{CO_2} 3. Increased temperature 4. Increased DPG concentration
The affinity of hemoglobin for both hydrogen ions and carbon dioxide is decreased by increased P_{O_2}.

CONTROL OF RESPIRATION

Neural Generation of Rhythmic Breathing

Like cardiac muscle, the inspiratory muscles normally contract rhythmically, that is, alternately contract and relax. The origins of these contractions, however, are quite different. Certain cardiac-muscle cells have automaticity, and the nerves to the heart merely alter this inherent rate but are not required for cardiac contraction. In contrast, the diaphragm and intercostal muscles consist of skeletal muscle, which cannot contract unless stimulated to do so by nerves. Thus, breathing depends entirely upon cyclical respiratory-muscle excitation by the motor nerves to the diaphragm and the intercostal muscles. Destruction of these nerves or the areas from which they originate, as in poliomyelitis, for example, results in paralysis of the respiratory muscles and death, unless some form of artificial respiration can be instituted.

Inspiration is initiated by a burst of action potentials in the nerves to the inspiratory muscles. Then, abruptly, the action potentials cease, the inspiratory muscles relax, and expiration occurs as the elastic lungs recoil. In situations when expiration is facilitated by contraction of expiratory muscles, the nerves to these muscles, which were quiescent during inspiration, begin firing during expiration.

By what mechanism are nerve impulses to the respiratory muscles alternately increased and decreased? Automatic control of this neural activity resides primarily in neurons in the medulla, the same area of brain that contains the major cardiovascular control centers. In various parts of the medulla, clusters of neurons discharge in synchrony with inspiration and cease discharging during expiration. These neurons are called **medullary inspiratory neurons**. They provide, through either direct or interneuronal connections, the rhythmic input to the motor neurons innervating the inspiratory muscles.

There are two components to the question of what determines the discharge pattern of the medullary inspiratory neurons: (1) What factors are essential for the generation of rhythmical firing? (2) What factors modify the timing of the bursts and their intensity? This section deals only with the first question; the subsequent section, with the second.

Just how the brain generates alternating cycles of firing and quiescence in the medullary inspiratory neurons remains controversial. It is possible that the inspiratory neurons, or neurons that synapse upon them, are pacemaker neurons, that is, that they have inherent automaticity and rhythmicity. However, most experts believe that even though such pacemakers might play a role, a much more complex model is required to explain the on-off cycle of the neuronal output.

The medullary inspiratory neurons also receive a rich synaptic input from neurons in various areas of the pons, the part of the brainstem just above the medulla. This input modulates the output of the medullary inspiratory neurons and may help terminate inspiration by inhibiting them.

Another cutoff signal for inspiration comes from **pulmonary stretch receptors**, which lie in the airway smooth-muscle layer and are activated by lung inflation. Action potentials in the afferent nerve fibers from them travel to the brain and inhibit the medullary inspiratory neurons. This is known as the Hering-Breur reflex. Thus feedback from the lungs helps to terminate inspiration. However, the threshold of these receptors in human beings is very high, and this pulmonary stretch receptor reflex plays a role in setting respiratory rhythm only under conditions of very large tidal volumes, as in exercise.

One last point about the medullary inspiratory neurons: They are quite sensitive to depression by drugs, especially barbiturates and morphine, and death from an overdose of these drugs is often secondary to a cessation of ventilation.

Control of Ventilation by P_{O_2}, P_{CO_2}, and H^+ Concentration

Respiratory rate and tidal volume are not fixed but can be altered over a wide range. For simplicity, we shall describe the control of total ventilation without discussing whether rate or depth makes the greater contribution to the change.

There are many inputs to the medullary inspiratory neurons, but the most important for the involuntary control of ventilation are from peripheral chemoreceptors and central chemoreceptors (Table 14-9).

The **peripheral chemoreceptors**, located high in the neck at the bifurcation of the common carotid arteries and in the thorax on the arch of the aorta (Figure 14-27),

TABLE 14-9	STIMULI FOR THE CENTRAL AND PERIPHERAL CHEMORECEPTOR

Peripheral chemoreceptors—that is, carotid bodies and aortic bodies—respond to changes in the *arterial blood*. They are stimulated by:

1. decreased P_{O_2}
2. increased hydrogen-ion concentration
3. increased P_{CO_2}, mainly via associated changes in hydrogen-ion concentration

Central chemoreceptors—that is, located in the medulla—respond to changes in the *brain interstitial fluid*. They are stimulated by increased P_{CO_2}, which increases hydrogen-ion concentration.

FIGURE 14-27 Location of the carotid and aortic bodies. Note that the carotid body is quite close to the carotid sinus, the major arterial baroreceptor. Both right and left common carotid bifurcations contain a carotid sinus and a carotid body.

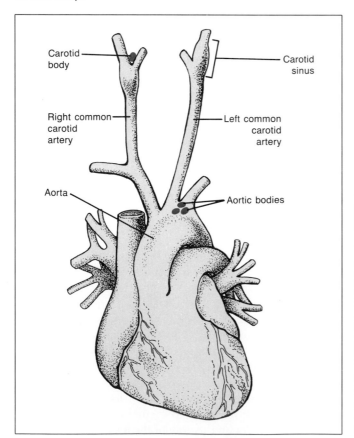

are called the **carotid and aortic bodies**. In both locations they are quite close to, but distinct from, the arterial baroreceptors described in Chapter 13. Of the two groups of peripheral chemoreceptors, the carotid bodies are by far the more important. They are composed of epithelial-like cells and neuron terminals in intimate contact with the arterial blood. Afferent nerve fibers arising from these terminals pass to the brainstem, where they synapse ultimately with the medullary inspiratory neurons. The peripheral chemoreceptors are sensitive to changes in the arterial P_{O_2}, H^+ concentration, and P_{CO_2}.

The **central chemoreceptors** are located in the medulla and provide synaptic input to the medullary inspiratory neurons. They respond to changes in the H^+ concentration of the brain's extracellular fluid. As we shall see, such changes occur particularly when the blood P_{CO_2} is altered.

The stimuli for the central and peripheral chemoreceptors are summarized in Table 14-9.

Control by P_{O_2}. Figure 14-28 illustrates an experiment in which healthy subjects breath low-P_{O_2} gas mixtures for several minutes. The experiment is performed in such a way that keeps arterial P_{CO_2} constant so that the pure effects of changing only P_{O_2} can be studied. Little increase in ventilation is observed until the oxygen content of the inspired air is reduced enough to lower arterial P_{O_2} to 60 mmHg. Beyond this point, any further reduction in arterial P_{O_2} causes a marked reflex hyperventilation.

FIGURE 14-28 The effect on ventilation of breathing low-oxygen mixtures. The arterial P_{CO_2} was maintained at 40 mmHg throughout the experiment.

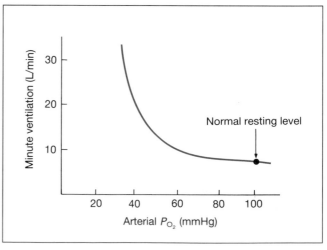

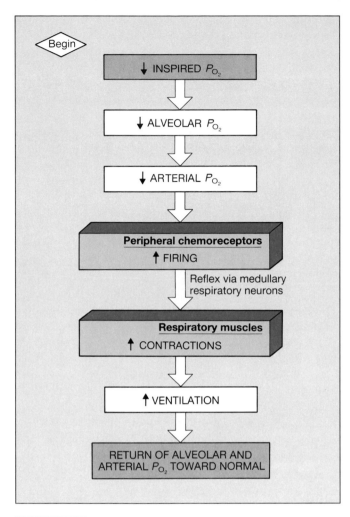

FIGURE 14-29 Sequence of events by which a low arterial P_{O_2} causes hyperventilation, which maintains alveolar (and hence arterial) P_{O_2} at a value higher than would exist if the ventilation had remained unchanged.

This reflex is mediated by the peripheral chemoreceptors (Figure 14-29). The low arterial P_{O_2} increases the rate at which the receptors discharge, resulting in an increased number of action potentials traveling up the afferent nerve fibers and stimulating the medullary inspiratory neurons. The resulting increase in ventilation provides more oxygen to the alveoli and minimizes the drop in alveolar and arterial P_{O_2} produced by the low-P_{O_2} gas mixture.

It may seem surprising that we are so insensitive to smaller reductions of arterial P_{O_2}, but look again at the oxygen-hemoglobin dissociation curve (Figure 14-21). Total oxygen carriage by the blood is not really reduced very much until the arterial P_{O_2} falls below 60 mmHg. Therefore, increased ventilation would not result in very much more oxygen being added to the blood until that point is reached.

To reiterate, the peripheral chemoreceptors respond to decreases in arterial P_{O_2}, as occurs in lung disease or exposure to high altitude. However, the peripheral chemoreceptors are *not* stimulated in situations in which there are modest reductions in the oxygen content of the blood but no change in arterial P_{O_2}. An example of this is anemia (page 355), where there is a decrease in the amount of hemoglobin present in the blood (Chapter 13) but no decrease in arterial P_{O_2}, because the concentration of dissolved oxygen in the blood is normal. This same analysis holds true when total oxygen delivery is reduced by the presence of carbon monoxide, a gas that has an extremely high affinity for the oxygen-binding sites in hemoglobin and reduces the amount of oxygen combined with hemoglobin by competing for these sites. Since carbon monoxide does not affect the amount of oxygen that can dissolve in blood, the arterial P_{O_2} is unaltered, and no increase in peripheral chemoreceptor output occurs.

Control by P_{CO_2}.   That an increase in arterial P_{CO_2} plays an important role in the control of ventilation can be demonstrated by a simple experiment in which subjects breath air containing variable quantities of carbon dioxide. The presence of carbon dioxide in the inspired air causes an elevation of alveolar P_{CO_2} and thereby an elevation of arterial P_{CO_2}. As can be seen in Figure 14-30,

FIGURE 14-30 Effects on respiration of increasing arterial P_{CO_2} achieved by adding carbon dioxide to inspired air.

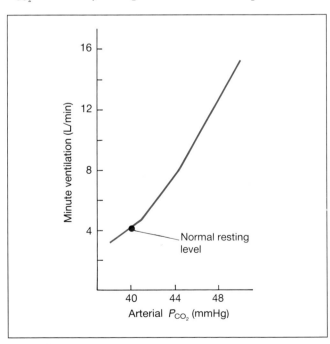

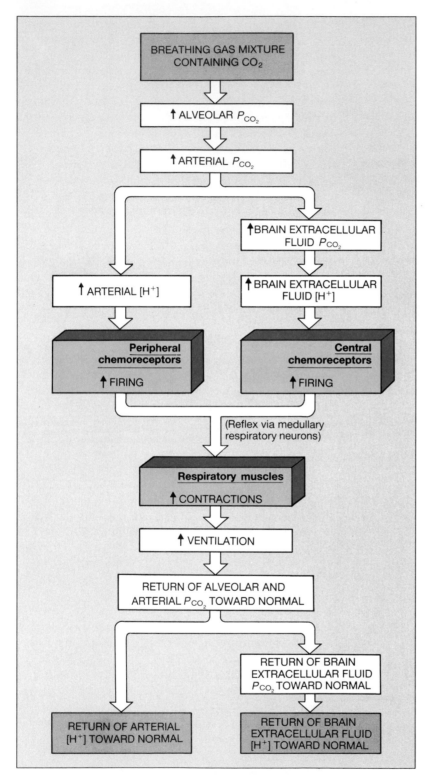

FIGURE 14-31 Extracellular-fluid hydrogen-ion concentration is maintained relatively constant via regulation of ventilation. Note that the peripheral chemoreceptors are stimulated by an *increase* in H$^+$ concentration, whereas they are stimulated by a *decrease* in P_{O_2} (Figure 14-29).

increased P_{CO_2} markedly stimulates ventilation, an increase of 2 to 5 mmHg in alveolar P_{CO_2} causing a 100 percent increase in ventilation.

Thus, unlike the case for lowered P_{O_2}, even the slightest change in arterial P_{CO_2} is associated with a significant reflex change in ventilation. Experiments like this have documented that small changes in arterial P_{CO_2} are resisted by the reflex mechanisms controlling ventilation to a greater degree than are equivalent changes in arterial P_{O_2}. An increase in arterial P_{CO_2} due, say, to lung disease, results in stimulation of ventilation to promote the elimination of carbon dioxide. Conversely, a decrease in arterial P_{CO_2} below normal levels removes some of the stimulus for ventilation, thereby reducing ventilation and allowing metabolically produced carbon dioxide to accumulate and return the P_{CO_2} to normal. In this manner, the arterial P_{CO_2} is stabilized at the normal value of 40 mmHg.

The pathways by which changes in the arterial P_{CO_2} result in changes in ventilation include both the peripheral and central chemoreceptors. Before looking at the specific reflexes, we must emphasize that the ability of changes in arterial P_{CO_2} to control ventilation reflexly is largely due to associated changes in H^+ concentration (Equation 14-3).[7]

Figure 14-31 summarizes the pathways by which increased arterial P_{CO_2} stimulates ventilation. For one thing, the peripheral chemoreceptors are stimulated by the increased arterial H^+ concentration resulting from the increased P_{CO_2}. At the same time, because carbon dioxide diffuses rapidly across the membranes separating capillary blood and brain tissue, the increase in arterial P_{CO_2} causes a rapid identical increase in brain extracellular fluid P_{CO_2}. This increased P_{CO_2} increases H^+ concentration, which stimulates the central chemoreceptors. Inputs from both peripheral and central chemoreceptors stimulate the medullary inspiratory neurons to increase ventilation. The end result is a return of arterial and brain extracellular fluid P_{CO_2} and H^+ concentration toward normal.

Of the two sets of receptors involved in this reflex, the central chemoreceptors are the more important, accounting for about 70 percent of the stimulus to ventilation. Contrast this to the low-oxygen reflex, which is mediated only by the peripheral chemoreceptors.

The effects of increased P_{CO_2} and decreased P_{O_2} not only exist as independent inputs to the medulla but manifest synergistic interactions as well. Acute ventilatory response to combined low P_{O_2} and high P_{CO_2} is con-

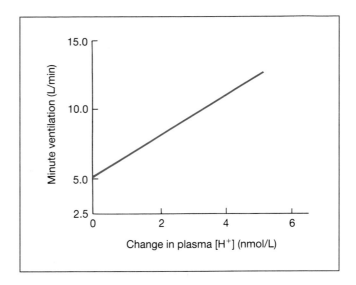

FIGURE 14-32 Changes in ventilation in response to an elevation of plasma hydrogen-ion concentration, produced by the administration of lactic acid. (*Adapted from Lambertsen.*)

siderably greater than the sum of the individual responses.

Throughout this section, we have described the *stimulatory* effects of carbon dioxide on ventilation via *reflex* input to the medulla, but *very* high levels of carbon dioxide have just the opposite effect and may be lethal. This is because such concentrations of carbon dioxide act *directly* on the medulla to *inhibit* the respiratory neurons by an anesthetic-like effect. Symptoms caused by increased blood P_{CO_2} include severe headaches, restlessness, and dulling or loss of consciousness. The accompanying increased H^+ concentration also causes serious problems.

Control by changes in arterial H^+ concentration not due to altered carbon dioxide. There are many normal and pathological situations in which a change in arterial H^+ concentration is due to some cause other than a primary change in P_{CO_2}. These are termed **metabolic acidosis** when H^+ concentration is increased and **metabolic alkalosis** when it is decreased. In such cases, the peripheral chemoreceptors play the major role in altering ventilation.

For example, increased addition of lactic acid to the blood, as in strenuous exercise, causes hyperventilation almost entirely by stimulation of the peripheral chemoreceptors (Figures 14-32 and 14-33). The central chemoreceptors are only minimally stimulated in this case because brain H^+ concentration is increased to only a small extent, at least early on, by the hydrogen ions generated

---

[7] An increased P_{CO_2} exerts some direct stimulation of the peripheral chemoreceptors, but this is very small compared to its indirect effects via changes in hydrogen-ion concentration.

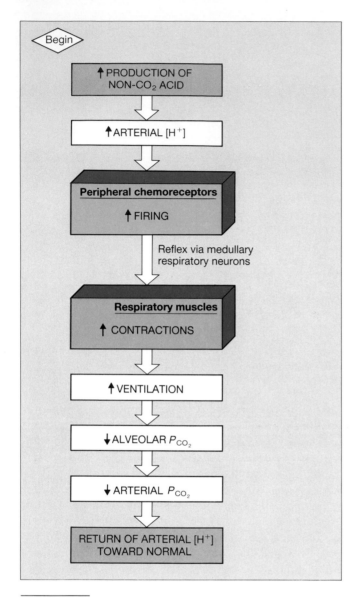

Begin

↑ PRODUCTION OF
NON-CO$_2$ ACID

↑ ARTERIAL [H$^+$]

Peripheral chemoreceptors
↑ FIRING

Reflex via medullary
respiratory neurons

Respiratory muscles
↑ CONTRACTIONS

↑ VENTILATION

↓ ALVEOLAR P_{CO_2}

↓ ARTERIAL P_{CO_2}

RETURN OF ARTERIAL [H$^+$]
TOWARD NORMAL

FIGURE 14-33 Reflexly induced hyperventilation minimizes the change in arterial hydrogen-ion concentration when acids are produced in excess in the body. Note that under such conditions, arterial P_{CO_2} is reflexly reduced below its normal value.

from the lactic acid. This is because hydrogen ions penetrate the blood-brain barrier very slowly. Contrast this with how easily carbon dioxide penetrates the blood-brain barrier and changes brain H$^+$ concentration.

The converse of the above situation is also true: When arterial H$^+$ concentration is lowered by any means other than by a reduction of P_{CO_2} (for example, by loss of hydrogen ions from the stomach in vomiting), ventilation is

reflexly depressed because of decreased peripheral chemoreceptor output.

The adaptive value such reflexes have in regulating arterial H$^+$ concentration should be evident (Figure 14-33). The hyperventilation induced by a metabolic acidosis lowers arterial P_{CO_2}, which lowers arterial H$^+$ concentration back toward normal. Similarly, hypoventilation induced by a metabolic alkalosis results in an elevated arterial P_{CO_2} and a restoration of H$^+$ concentration toward normal. Notice that when a change in arterial H$^+$ concentration due to some acid unrelated to carbon dioxide influences ventilation via the peripheral chemoreceptors, P_{CO_2} is displaced from normal. This is a reflex that regulates arterial H$^+$ concentration at the expense of changes in arterial P_{CO_2}.

Figure 14-34 summarizes the control of ventilation by P_{O_2}, P_{CO_2}, and H$^+$ concentration.

Control of Ventilation during Exercise

During exercise, the alveolar ventilation may increase as much as twentyfold. On the basis of our three variables—P_{O_2}, P_{CO_2}, and H$^+$ concentration—it might seem easy to explain the mechanism that induces this increased ventilation. Unhappily, such is not the case.

Increased P_{CO_2} as the stimulus? It would seem logical that, as the exercising muscles produce more carbon dioxide, blood P_{CO_2} would increase. This is true for systemic venous blood but not for arterial blood. Therefore, the reflex mechanisms by which increased arterial P_{CO_2} increases ventilation are not brought into play, and increased P_{CO_2} does not seem to be the signal for increased ventilation during exercise.

The reader might well be wondering why arterial P_{CO_2} does not increase during exercise. Recall two facts from the section on alveolar gas pressures: (1) alveolar P_{CO_2} sets arterial P_{CO_2}, and (2) alveolar P_{CO_2} is determined by the *ratio* of carbon dioxide production and alveolar ventilation. During exercise, the alveolar ventilation increases in exact proportion to the increased carbon dioxide production (hyperpnea), and so alveolar and therefore arterial P_{CO_2} do not change. Indeed, in very strenuous exercise, the alveolar ventilation increases relatively more than carbon dioxide production (hyperventilation), and thus alveolar and systemic arterial P_{CO_2} actually decrease (Figure 14-35).

Decreased P_{O_2} as the stimulus? The story is similar for oxygen: Systemic venous P_{O_2} decreases during exercise, but alveolar P_{O_2} and, hence, systemic arterial P_{O_2} usually remain unchanged because cellular oxygen consumption and alveolar ventilation increase in exact proportion to each other (Figure 14-35). The hyperventila-

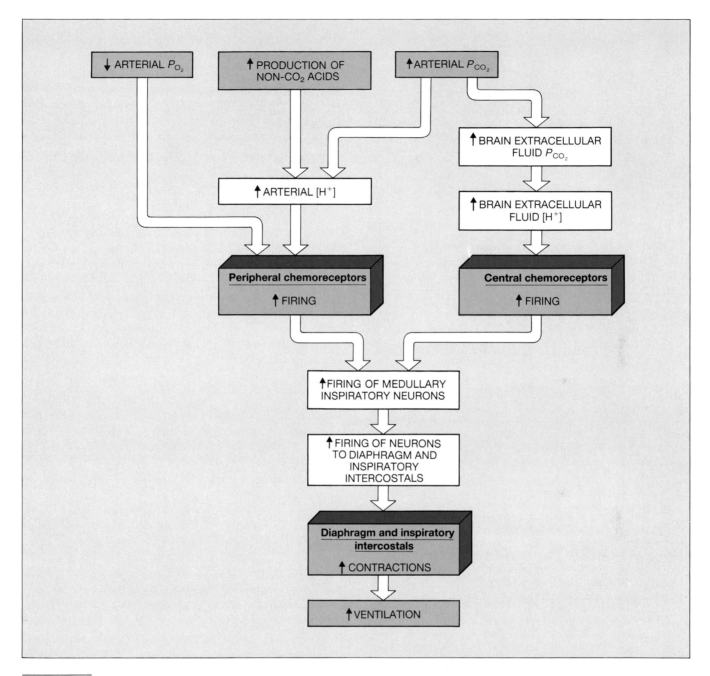

FIGURE 14-34 Summary of chemical inputs that stimulate ventilation. When arterial P_{O_2} increases or when P_{CO_2} or hydrogen-ion concentration decreases, ventilation is reflexly decreased. This is a combination of Figures 14-29, 14-31, and 14-33.

tion of very strenuous exercise actually raises alveolar P_{O_2}, just as it lowers alveolar P_{CO_2}, but for a variety of reasons, the increase in alveolar P_{O_2} is not accompanied by an increase in arterial P_{O_2}.[8]

---

[8] Indeed, especially in highly conditioned athletes, the arterial P_{O_2} may decrease somewhat during strenuous exercise, and may also contribute to stimulation of ventilation.

Increased H+ as the stimulus? Since the arterial P_{CO_2} does not change during moderate exercise and decreases in severe exercise, there is no accumulation of excess H^+ resulting from carbon dioxide accumulation. But during strenuous exercise, there *is* an increase in arterial H^+ concentration (Figure 14-35) for quite a different reason: generation and release of lactic acid into the blood. This change in H^+ concentration is partly responsible for stimulating the hyperventilation of severe exercise.

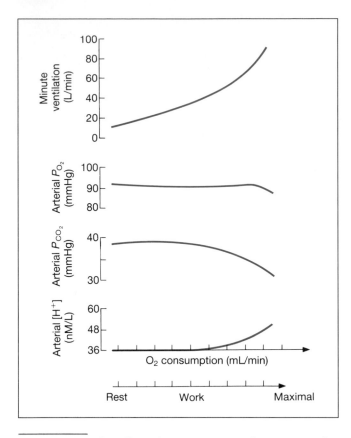

FIGURE 14-35 The effect of exercise on ventilation, arterial gas pressures, and hydrogen-ion concentration. (*Adapted from Comroe.*)

FIGURE 14-36 Ventilation changes during exercise. Note (1) the abrupt increase at the onset of exercise and (2) the equally abrupt but larger decrease at the end of exercise.

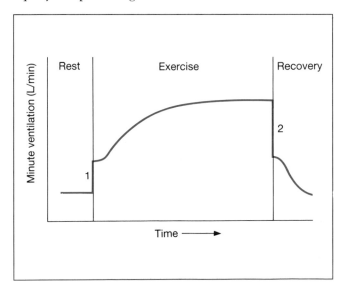

We are left with the fact that we do not know what input stimulates ventilation during exercise. Our big three—decreased arterial P_{O_2}, increased arterial P_{CO_2}, and increased arterial H^+ concentration—appear presently to be inadequate, but many physiologists still believe that they will ultimately be shown to be the critical inputs. These scientists reason that the constancy of arterial P_{CO_2} during moderate exercise is strong evidence that ventilation is, in fact, controlled by P_{CO_2}. In other words, if P_{CO_2} were not the major controller, how else could it remain unchanged in the face of the marked increase in carbon dioxide production?

Moreover, the problem is even more complicated. As shown in Figure 14-36, there is an abrupt increase—within seconds—in ventilation at the onset of exercise and an equally abrupt decrease at the end. Clearly, these changes occur too rapidly to be explained by alteration of chemical constituents of the blood. They may represent a conditioned, or learned, response.

Other factors.   Many receptors in joints and muscles are stimulated by the physical movements that accompany muscle contraction. It is quite likely that afferent pathways from these receptors play a significant role in stimulating respiration during exercise. Thus, the mechanical events of exercise help to coordinate alveolar ventilation with the metabolic requirements of the tissues.

An increase in body temperature also frequently occurs as a result of increased physical activity and contributes to stimulation of alveolar ventilation. Body temperature also increases in many situations other than exercise, and when it does, it stimulates ventilation. Thus, people with fever show increased ventilation.

Epinephrine and norepinephrine are also respiratory stimulants. The increased plasma concentrations of these catecholamines with exercise probably contribute to the increased ventilation.

Finally, as seems to be the case for control of the cardiovascular system during exercise, branches come off the axons descending from the brain to motor neurons supplying exercising muscles, and these branches influence the neurons that control the respiratory muscles.

Figure 14-37 summarizes various factors that influence ventilation during exercise.

Other Ventilatory Responses

Protective reflexes.   A group of responses protect the respiratory system against irritant materials. Most familiar are the cough and the sneeze reflexes, which originate in receptors located between airway epithelial cells.

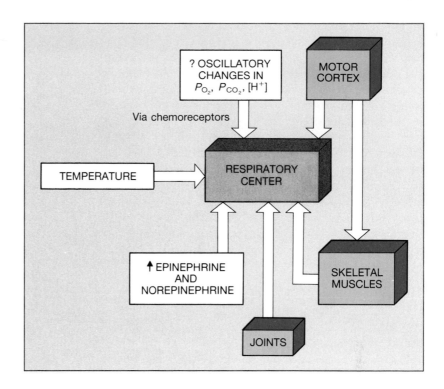

FIGURE 14-37 Summary of factors that stimulate ventilation during exercise.

When these receptors are stimulated, the medullary respiratory neurons cause reflexly a deep inspiration and a violent expiration. In this manner, particles can be literally exploded out of the respiratory tract. The cough reflex is inhibited by alcohol, which may contribute to the susceptibility of alcoholics to choking and pneumonia.

Another example of a protective reflex is the immediate cessation of respiration which is frequently triggered when noxious agents are inhaled.

Pain and emotion.   Painful stimuli anywhere in the body can produce reflex stimulation of the respiratory neurons. Emotional states are also often accompanied by marked alterations of respiration, as evidenced by the rapid breathing that characterizes fright, anxiety, and many similar emotions. In addition, movement of air in or out of the lungs is required for such involuntary expressions of emotion as laughing and crying. In such situations, descending pathways from higher brain centers are the controlling input.

Voluntary control of breathing.   Although we have discussed in detail the involuntary nature of most respiratory reflexes, it is quite obvious that considerable voluntary control of respiratory movements exists. Voluntary control is accomplished by descending pathways from the cerebral cortex to the motor neurons of the respiratory muscles. This voluntary control of respiration cannot be maintained when the involuntary stimuli, such as an elevated P_{CO_2} or H^+ concentration, become intense. An example is the inability to hold one's breath for very long.

The opposite of breath holding—deliberate hyperventilation—lowers alveolar and arterial P_{CO_2} and increases P_{O_2}. Unfortunately, swimmers sometimes voluntarily hyperventilate immediately before a race to be able to hold their breath longer during the race. We say "unfortunately" because the low P_{CO_2} may still permit breath holding at a time during the race when the exertion is lowering the arterial P_{O_2} to levels that can cause unconsciousness and lead to drowning.

Besides the obvious forms of voluntary control, respiration must also be controlled during such complex actions such as speaking and singing.

Breathing during sleep.   Ventilation normally decreases during sleep more than the decrease in the body's cellular oxygen consumption decreases. Accordingly, the arterial P_{O_2} decreases and the arterial P_{CO_2} increases.

This normal tendency for ventilation to decrease during sleep is profoundly and dangerously exaggerated in some persons, who suffer from sleep apnea, recurrent periods—up to 500 per night—during which breathing ceases completely for abnormally long periods of time, often 1 to 2 min or more. This disorder may be caused by a reduced sensitivity of the peripheral chemoreceptors occurring cyclically during certain phases of sleep (Chapter 20). Sleep apnea is being studied intensively because it may be one of the causes of what is called the **sudden-infant-death syndrome**, the disease in which an infant who had seemed to be in good health is found dead in its crib.

Finally, the function of that most common of respiratory movements associated not with sleep but sleepiness—the yawn—is not known.

HYPOXIA

Hypoxia is defined as a deficiency of oxygen at the tissue level. There are many potential causes of hypoxia, but they can be classed in four general categories: (1) **hypoxic hypoxia** (also termed **hypoxemia**), in which the arterial P_{O_2} is reduced; (2) **anemic hypoxia**, in which the arterial P_{O_2} is normal but the total oxygen *content* of the blood is reduced because of inadequate numbers of erythrocytes, deficient or abnormal hemoglobin, or competition for the hemoglobin molecule by carbon monoxide; (3) **ischemic hypoxia**, in which blood flow to the tissues is too low; and (4) **histotoxic hypoxia**, in which the quantity of oxygen reaching the tissues is normal but the cell is unable to utilize the oxygen because a toxic agent—cyanide, for example—has interfered with the cell's metabolic machinery.

The primary causes of hypoxic hypoxia in disease are listed in Table 14-10. Exposure to the reduced P_{O_2} of high altitude also causes hypoxia but is not a "disease." The brief summaries in Table 14-10 provide a review of many of the key aspects of respiratory physiology and pathophysiology described in this chapter.

It must be emphasized that, as described in the table, some of the diseases that produce hypoxia also produce carbon dioxide retention. Treating only the oxygen deficit by administering oxygen may constitute inadequate therapy because it does nothing about the carbon dioxide retention.[9]

---

[9] Indeed, such therapy may be dangerous. The primary respiratory drive in such patients is the hypoxia since for several reasons the response to an increased P_{CO_2} may be lost in chronic situations. The administration of pure oxygen may cause such persons to stop breathing entirely.

TABLE 14-10 CAUSES OF A DECREASED ARTERIAL P_{O_2} (HYPOXIC HYPOXIA) IN DISEASE
1. Hypoventilation may be caused by (a) a defect anywhere along the respiratory control pathway, from the medulla through the respiratory muscles, (b) by severe thoracic cage abnormalities, and (c) by major obstruction of the upper airway. The hypoxemia of hypoventilation is always accompanied by an increased arterial P_{CO_2}.
2. Diffusion impairment results from thickening of the alveolar membranes or a decrease in their surface area. In turn, it causes failure of equilibration of blood P_{O_2} with alveolar P_{O_2}. Often it is apparent only during exercise. Arterial P_{CO_2} is either normal, since carbon dioxide diffuses more readily than oxygen, or reduced, if the hypoxemia reflexly stimulates ventilation.
3. Shunt results from blood bypassing ventilated alveoli in passing from the right heart to the left heart. The most common causes are cardiac defects that permit blood to flow directly from the right heart to the left heart. Arterial P_{CO_2} generally does not rise since the effect of the shunt on it is counterbalanced by the increased ventilation reflexly stimulated by the hypoxemia.
4. Ventilation-perfusion inequality is by far the most common cause of hypoxemia. It occurs in chronic obstructive lung diseases and many other lung diseases. Arterial P_{CO_2} may be normal or increased, depending upon how much ventilation is reflexly stimulated.

Acclimatization to High Altitude

We will go into some detail on this topic because of its inherent interest but also because the responses to high altitude are essentially the same as the responses to hypoxia from any cause. Thus, a person with severe hypoxia from lung disease may show many of the same changes—increased hematocrit, for example—as a high-altitude sojourner.

Atmospheric pressure progressively decreases as altitude increases. Thus, at the top of Mt. Everest (approximately 29,000 ft or 9000 m), the atmospheric pressure is 253 mmHg (recall that it is 760 mmHg at sea level). The air is still 21 percent oxygen, which means that the P_{O_2} is 53 mmHg, that is, 0.21×253 mmHg. Obviously, the alveolar and arterial P_{O_2} must decrease as one ascends unless pure oxygen is breathed.

The effects of oxygen lack vary from one individual to another, but most persons who ascend rapidly to altitudes above 10,000 ft experience some degree of mountain sickness. This disorder consists of breathlessness,

TABLE 14-11	ACCLIMATIZATION TO THE HYPOXIA OF HIGH ALTITUDE

1. The peripheral chemoreceptors stimulate ventilation.
2. Erythropoietin, secreted by the kidneys, stimulates erythrocyte synthesis, resulting in increased erythrocyte and hemoglobin content of blood.
3. DPG increases and shifts the hemoglobin dissociation curve to the right, facilitating oxygen unloading in the tissues.
4. Increases in capillary density, mitochondria, and muscle myoglobin occur, all of which increase oxygen transfer.

heart palpitations, headache, nausea, fatigue, and impairment of mental processes. Much more serious is the appearance, in some persons, of life-threatening pulmonary edema.[10]

Over the course of several days, the symptoms of mountain sickness disappear, although maximal physical capacity remains reduced. Acclimatization to high altitude is achieved by the compensatory mechanisms listed below. The highest villages permanently inhabited by people are in the Andes at 19,000 ft (5700 m). These villagers work quite normally, and the only major precaution they take is that the women come down to lower altitudes during late pregnancy.

In response to hypoxia, oxygen supply to the tissues is maintained in the four ways summarized in Table 14-11. It should be noted that the DPG effect is not always adaptive and may be maladaptive. This is because, at very high altitudes, a right shift in the curve impairs oxygen *loading* in the lungs, an effect that outweighs any benefit from facilitation of *unloading* in the tissues. Indeed, many high-altitude species such as the llama have a left-shifted curve, indicating an adaptive value for this factor.

NONRESPIRATORY FUNCTIONS OF THE LUNGS

The lungs have a variety of functions in addition to their roles in gas exchange and regulation of H^+ concentration. Most notable are the influences they have on the arterial concentrations of a large number of biologically active substances. Many substances—neurotransmitters and paracrines, for example—re-

leased locally into interstitial fluid may diffuse into capillaries and thus make their way into the systemic venous system. The lungs partially or completely remove some of these substances from the blood and thereby prevent them from reaching other locations in the body via the arteries. The cells that perform this function are the endothelial cells lining the pulmonary capillaries.

In contrast, the lungs may also produce and add new substances to the blood. Some of these substances play local regulatory roles within the lungs, but if produced in large enough quantity, they may diffuse into the pulmonary capillaries and be carried to the rest of the body. For example, inflammatory responses (Chapter 19) in the lung may lead, via excessive release of potent chemicals such as histamine, to profound alterations of systemic blood pressure or flow. In at least one case, the lungs contribute a hormone, angiotensin II, to the blood (Chapter 15).

Finally, the lungs also act as a "sieve" that traps and dissolves small blood clots generated in the systemic circulation, thereby preventing them from reaching the systemic arterial blood where they could occlude blood vessels in other organs.

SUMMARY

Organization of the Respiratory System

I. The respiratory system comprises the lungs, the airways leading to them, and the chest structures responsible for movement of air into and out of them.
   A. The conducting zone of the airways consists of the trachea, bronchi, and terminal bronchioles.
   B. The respiratory zone of the airways consists of the alveoli, which are the sites of gas exchange, and the airways to which the alveoli are attached.
   C. The lungs and interior of the thorax are covered by pleura between the surfaces of which is an extremely thin layer of intrapleural fluid.

II. The volume of air in the lungs at the end of an unforced expiration is the functional residual capacity (FRC). At this volume, only passive forces of the lungs and chest wall are acting.
   A. At FRC, the lungs are stretched and are attempting to recoil, whereas the chest wall is compressed and attempting to move outward. This creates a subatmospheric intrapleural pressure.
   B. The difference between the intrapleural pressure and the alveolar pressure is the transpulmonary pressure, which is the force acting to hold the lungs open.

Ventilation and Lung Mechanics

I. Bulk flow of air between the atmosphere and alveoli is proportional to the difference between the atmospheric and alveolar pressures and inversely proportional to the airway resistance: $F = (P_{atm} - P_{alv})/R$.

---

[10] The cause of the pulmonary edema is still unclear. It may be due to increased permeability to protein and/or increased hydrostatic pressure in the pulmonary capillaries.

II. During inspiration, the contractions of the diaphragm and inspiratory intercostal muscles increase the volume of the thoracic cage.
   A. This makes intrapleural pressure more subatmospheric, increases transpulmonary pressure, and causes the lungs to expand.
   B. This expansion initially makes alveolar pressure subatmospheric, which creates the pressure difference between atmosphere and alveoli to drive air flow into the lungs.
III. During expiration, the inspiratory muscles cease contracting, allowing the lungs and chest wall to recoil passively toward their original size.
   A. This initially compresses the alveolar air, raising alveolar pressure above atmospheric pressure and driving air out of the lungs.
   B. In forced expirations, the contraction of expiratory intercostal muscles and abdominal muscles actively decreases thoracic dimensions.
IV. Lung compliance is determined by the stretchability of the lung elastic connective tissues and the surface tension of the fluid lining the alveoli. The latter is greatly reduced, and compliance increased, by surfactant, produced by the type II cells of the alveoli.
V. Airway resistance determines how much air flows into the lungs at any given pressure difference between atmosphere and alveoli. The factors that determine airway resistance are summarized in Table 14-4.
VI. The vital capacity is the sum of resting tidal volume, inspiratory reserve volume, and expiratory reserve volume. The volume expired during the first second of a forced vital capacity (FVC) measurement is the FEV_1 and normally averages 80 percent of FVC.
VII. Minute ventilation is the product of tidal volume and respiratory rate; alveolar ventilation is the product of (tidal volume − anatomic dead space) and respiratory rate.

Exchange of Gases in Alveoli and Tissues

I. In the steady state, the net volumes of oxygen and carbon dioxide exchanged in the lungs per unit time are equal to the net volumes exchanged in the tissues.
   A. Typical volumes per minute are 250 mL for oxygen consumption and 200 mL for carbon dioxide production.
   B. Exchange of these gases in lungs and tissues is by diffusion, due to differences in partial pressures.
II. Normal alveolar gas pressure for oxygen is 105 mmHg and for carbon dioxide, 40 mmHg.
   A. At any given inspired P_{O_2}, the ratio of oxygen consumption to alveolar ventilation determines alveolar P_{O_2}—the higher the ratio, the lower the alveolar P_{O_2}.
   B. The higher the ratio of carbon dioxide production to alveolar ventilation, the higher the alveolar P_{CO_2}.
III. Average value at rest for systemic venous P_{O_2} is 40 mmHg and for P_{CO_2}, 46 mmHg.
IV. As systemic venous blood flows through the pulmonary capillaries, there is net diffusion of oxygen from alveoli to blood and of carbon dioxide from blood to alveoli.
   A. By the end of the pulmonary capillaries, the blood gas pressures have become equal to those in the alveoli.
   B. Therefore, the P_{O_2} and P_{CO_2} in the systemic arterial blood are virtually identical to those in the alveoli.
V. Inadequate gas exchange between alveoli and pulmonary capillaries may occur when the capillary surface area is decreased, when the alveolar walls thicken, and when there are ventilation-perfusion inequalities.
   A. The latter are minimized by local intrapulmonary homeostatic responses in which the low P_{O_2} constricts the pulmonary blood vessels and the high P_{CO_2} dilates the airways.
   B. The major consequence of significant uncompensated mismatching is that the systemic arterial P_{O_2} is considerably lower than the alveolar P_{O_2}.
VI. In the tissues, net diffusion of oxygen occurs from blood to cells, and net diffusion of carbon dioxide from cells to blood.

Transport of Oxygen in the Blood

I. Each liter of systemic arterial blood normally contains 200 mL of oxygen, more than 98 percent of which is bound to hemoglobin, and the rest of which is dissolved.
II. The major determinant of the degree to which hemoglobin is saturated with oxygen is blood P_{O_2}.
   A. Hemoglobin is almost 100 percent saturated at the normal systemic arterial P_{O_2} of 105 mmHg. The fact that saturation is already more than 90 percent at a P_{O_2} of 60 mmHg permits relatively normal uptake of oxygen by the blood even when alveolar P_{O_2} is moderately reduced.
   B. Hemoglobin is 75 percent saturated at the normal systemic venous P_{O_2} of 40 mmHg. Thus, only 25 percent of the oxygen has dissociated from hemoglobin and entered the tissues.
III. The affinity of hemoglobin for oxygen is decreased by an increase in P_{CO_2}, hydrogen-ion concentration, and temperature. All these conditions exist in the tissues and facilitate the dissociation of oxygen from hemoglobin.
IV. The affinity of hemoglobin for oxygen is also decreased by erythrocyte DPG, which increases in situations associated with inadequate oxygen supply and helps maintain oxygen delivery to the tissues.

Transport of Carbon Dioxide by the Blood

I. Each liter of systemic arterial blood contains approximately 550 mL of "total" carbon dioxide, 90 percent of which is bicarbonate.
II. When carbon dioxide molecules diffuse from the tissues into the blood, 10 percent remains dissolved in plasma and erythrocytes, 30 percent combines in the erythrocytes with deoxyhemoglobin to form carbamino compounds, and 60 percent combines in the erythrocytes with water to form carbonic acid, which then dissociates to yield bicarbonate and hydrogen ions. Most of the bicarbonate then diffuses out of the erythrocytes into the plasma in exchange for chloride ions.
III. As venous blood flows through lung capillaries, P_{CO_2} de-

creases because of diffusion of carbon dioxide out of the blood into the alveoli, and all these reactions are reversed.

Transport of Hydrogen Ions between Tissues and Lungs

I. Most of the hydrogen ions generated in the erythrocytes from carbonic acid during blood passage through tissue capillaries bind to deoxyhemoglobin because deoxyhemoglobin formed as oxygen unloads from oxyhemoglobin has a high affinity for hydrogen ions.

II. The binding of hydrogen ions to deoxyhemoglobin is reversed as the blood flows through the lung capillaries and the ions combine with bicarbonate to yield carbon dioxide and water.

Control of Respiration

I. Breathing depends upon cyclical inspiratory muscle excitation by the nerves to the diaphragm and intercostal muscles. This neural activity is triggered by the medullary inspiratory neurons.

II. The most important inputs to the medullary inspiratory neurons for the involuntary control of minute ventilation are from the peripheral chemoreceptors—the carotid and aortic bodies—and the central chemoreceptors.

III. Ventilation is reflexly stimulated, via the peripheral chemoreceptors, by a decrease in arterial P_{O_2}, but only when the decrease is large.

IV. Ventilation is reflexly stimulated, via both the peripheral and central chemoreceptors, when the arterial P_{CO_2} goes up even a slight amount. The stimulus for this reflex is not the increased P_{CO_2} itself but the concomitant increased hydrogen-ion concentration in arterial blood and brain extracellular fluid.

V. Ventilation is also stimulated, mainly via the peripheral chemoreceptors, by an increase in arterial hydrogen-ion concentration resulting from causes other than an increase in P_{CO_2}. The result of this reflex is to restore hydrogen-ion concentration toward normal by lowering P_{CO_2}.

VI. Ventilation is reflexly inhibited by an increase in arterial P_{O_2} and by a decrease in arterial P_{CO_2} or hydrogen-ion concentration.

VII. During moderate exercise, ventilation increases in exact proportion to metabolism, but the signals causing this are not known. During very strenuous exercise, ventilation increases more than metabolism.

   A. The proportional increases in ventilation and metabolism during moderate exercise cause the arterial P_{O_2}, P_{CO_2}, and hydrogen-ion concentration to remain unchanged.

   B. Arterial hydrogen-ion concentration increases during very strenuous exercise because of increased lactic acid production. This increase accounts for some of the hyperventilation seen in that situation.

VIII. Ventilation is also controlled by reflexes originating in airway receptors, by painful and emotional stimuli, and by conscious intent.

Hypoxia

I. The four types of hypoxia are listed in Table 14-10.

II. During exposure to hypoxia, as at high altitude, oxygen supply to the tissues is maintained by the four responses listed in Table 14-11.

Nonrespiratory Functions of the Lungs

I. The lungs influence arterial blood concentrations of biologically active substances by removing some from systemic venous blood and adding others to systemic arterial blood.

II. The lungs also act as sieves that dissolve small clots formed in the tissues.

REVIEW QUESTIONS

1. Define:

respiration (two definitions)	Heimlich maneuver
respiratory system	tidal volume
pulmonary	expiratory reserve volume
alveoli	residual volume
airways	inspiratory reserve volume
inspiration	vital capacity
expiration	forced vital capacity (FVC)
pharynx	FEV_1
larynx	restrictive lung diseases
vocal cords	minute ventilation
trachea	anatomic dead space
bronchi	alveolar ventilation
bronchioles	alveolar dead space
conducting zone	total dead space
respiratory zone	respiratory quotient (RQ)
type I cells	partial pressures
type II cells	hypoventilation
thorax	hyperventilation
diaphragm	hyperpnea
intercostal muscles	ventilation-perfusion
pleural sac	inequality
pleura	deoxyhemoglobin
intrapleural fluid	oxyhemoglobin
functional residual capacity	oxygen-hemoglobin
atmospheric pressure	dissociation curve
alveolar pressure (P_{alv})	2,3-diphosphoglycerate
intrapleural pressure (P_{ip})	(DPG)
transpulmonary pressure	carbamino compounds
ventilation	carbonic anhdrase
Boyle's law	respiratory acidosis
lung compliance (C_L)	respiratory alkalosis
surface tension	medullary inspiratory
surfactant	neurons
respiratory-distress	pulmonary stretch receptors
syndrome of the newborn	peripheral chemoreceptors
asthma	carotid bodies
chronic obstructive	aortic bodies
pulmonary disease	central chemoreceptors
emphysema	metabolic acidosis
chronic bronchitis	metabolic alkalosis

sudden-infant-death
   syndrome
hypoxia
hypoxic hypoxia

hypoxemia
anemic hypoxia
ischemic hypoxia
histotoxic hypoxia

2. List the functions of the respiratory system.

3. At rest, how many liters of air and blood flow through the lungs per minute?

4. Describe four functions of the conducting portion of the airways.

5. At functional residual capacity, in what directions are the lungs and chest wall tending to move? What prevents them from doing so?

6. What are normal values for intrapleural pressure, alveolar pressure, and transpulmonary pressure at the end of an unforced expiration?

7. Which respiration steps occur by diffusion and which by bulk flow?

8. Write the equation relating air flow into or out of the lungs to atmospheric pressure, alveolar pressure, and airway resistance.

9. Describe the sequence of events that causes air to move into the lungs during inspiration and expiration. Diagram the changes in intrapleural pressure and alveolar pressure.

10. What factors determine lung compliance? Which is most important?

11. How does surfactant increase lung compliance?

12. How is airway resistance influenced by air viscosity, airway length, and airway radii?

13. List the physical, nervous, and chemical factors that alter airway resistance.

14. Contrast the causes of increased airway resistance in asthma, emphysema, and chronic bronchitis.

15. What distinguishes lung capacities, as a group, from lung volumes?

16. State the formula relating minute ventilation, tidal volume, and respiratory rate. Give representative values for each at rest.

17. State the formula for calculating alveolar ventilation. What is an average value for alveolar ventilation?

18. State typical values for oxygen consumption, carbon dioxide production, and cardiac output at rest. How much oxygen (in milliliters per liter) is present in systemic venous and systemic arterial blood?

19. The concentration of a gas in a liquid is proportional to what two factors?

20. State the alveolar gas pressures for oxygen and carbon dioxide in a normal person at rest.

21. What factors determine alveolar gas pressures?

22. What is the mechanism of gas exchange between alveoli and pulmonary capillaries? In normal persons at rest, what are the gas pressures at the end of the pulmonary capillaries, relative to those in the alveoli?

23. Why does thickening of alveolar membranes impair oxygen movement but have little effect on carbon dioxide exchange?

24. What is the major result of ventilation-perfusion inequalities throughout the lungs? What two homeostatic responses minimize mismatching?

25. What generates the diffusion gradients for oxygen and carbon dioxide in the tissues?

26. In what two forms is oxygen carried in the blood? What are the normal quantities (in milliliters per liter) in each form in arterial blood?

27. Draw an oxygen-hemoglobin dissociation curve. Put in the points that represent systemic venous and systemic arterial blood. What is the adaptive importance of the plateau?

28. Would breathing pure oxygen cause a large increase in oxygen carriage by the blood in a normal person? In a person with a low alveolar P_{O_2}?

29. Describe the effects of increased P_{CO_2}, H^+ concentration, and temperature on the oxygen-hemoglobin dissociation curve. How are these effects adaptive for oxygen unloading in the tissues?

30. Describe the effects of increased DPG on the oxygen-hemoglobin dissociation curve. Under what conditions does an increase in DPG occur?

31. Draw figures showing the reactions carbon dioxide undergoes entering the blood in the tissue capillaries and leaving the blood in the alveoli. What fractions are contributed by dissolved carbon dioxide, bicarbonate, and carbamino?

32. What happens to most of the hydrogen ions formed in the erythrocytes from carbonic acid? What happens to blood H^+ concentration as blood flows through tissue capillaries?

33. What are the effects of P_{O_2} on carbamino formation and H^+ binding by hemoglobin?

34. In what area of the brain does automatic control of rhythmic respirations reside?

35. Describe the function of the pulmonary stretch receptors.

36. What changes stimulate the peripheral chemoreceptors? The central chemoreceptors?

37. Why does moderate anemia or carbon monoxide exposure not stimulate the peripheral chemoreceptors?

38. Is respiratory control more sensitive to changes in arterial P_{O_2} or P_{CO_2}?

39. Describe the pathways by which increased arterial P_{CO_2} stimulates ventilation. Which pathway is most important?

40. Describe the pathway by which a change in arterial H^+ concentration independent of altered carbon dioxide influences ventilation. What is the adaptive value of this reflex?

41. What happens to arterial P_{O_2}, P_{CO_2}, and H^+ concentration during moderate and severe exercise? List other factors that may stimulate ventilation during exercise.

42. List four general causes of hypoxic hypoxia.

43. Describe two general ways in which the lungs can alter the concentrations of substances other than oxygen, carbon dioxide, and H^+ in the arterial blood.

THOUGHT QUESTIONS

(*Answers are given in Appendix A.*)

1. At the end of a normal expiration, a person's FRC is 2 L, his alveolar pressure is 0 mmHg, and his intrapleural pressure is −4 mmHg. He then inhales 800 mL, and at the end of inspiration the alveolar pressure is 0 mmHg and the intrapleural pressure is −8 mmHg. Calculate this person's lung compliance.

2. A patient has an inability to produce surfactant. In order to inhale a normal tidal volume, will her intrapleural pressure have to be more or less subatmospheric during inspiration, relative to a normal person?

3. A patient is being artificially ventilated by a machine during surgery at a rate of 20 breaths/min and a tidal volume of 250 mL/breath. Assuming a normal anatomic dead space of 150 mL, is this patient receiving an adequate alveolar ventilation?

4. Why must a person floating on the surface of the water and breathing through a snorkel increase his tidal volume and/or breathing frequency if alveolar ventilation is to remain normal?

5. A normal person breathing room air voluntarily increases her alveolar ventilation twofold and continues to do so until new steady-state alveolar gas pressures for oxygen and carbon dioxide are reached. Are the new values higher or lower than normal?

6. A person has an alveolar P_{O_2} of 105 mmHg and an arterial P_{O_2} of 80 mmHg. Could hypoventilation, say, due to respiratory muscle weakness, produce these values?

7. A person's alveolar membranes have become thickened enough to moderately decrease the rate at which gases diffuse across them at any given partial-pressure differences. Will this person necessarily have a low arterial P_{O_2} at rest? During exercise?

8. A person is breathing 100 percent oxygen. How much will the oxygen content (in milliliters per liter of blood) of the arterial blood increase compared to when the person is breathing room air?

9. Which of the following have higher values in systemic venous blood than in systemic arterial blood: plasma P_{CO_2}, erythrocyte P_{CO_2}, plasma bicarbonate concentration, erythrocyte bicarbonate concentration, plasma hydrogen-ion concentration, erythrocyte hydrogen-ion concentration, erythrocyte carbamino concentration, erythrocyte chloride concentration, plasma chloride concentration?

10. If the spinal cord were severed where it joins the brainstem, what would happen to respiration?

11. The peripheral chemoreceptors are denervated in an experimental animal, and the animal then breathes a gas mixture containing 10 percent oxygen. What changes occur in the animal's ventilation? What changes occur when this denervated animal is given a mixture of air containing 21 percent oxygen and 5 percent carbon dioxide to breath?

12. Patients with severe, uncontrolled diabetes mellitus produce large quantities of certain organic acids. Can you predict their ventilation pattern and whether their arterial P_{O_2} and P_{CO_2} increase or decrease?

CHAPTER

15

THE KIDNEYS
AND REGULATION
OF WATER AND
INORGANIC IONS

This chapter deals with the kidneys and how they help regulate water and inorganic ions. Most people assume that the kidneys' sole job is to eliminate wastes and poisons from the body, but that is only one of their tasks. Indeed, their prime function is to balance the body's water and inorganic ions so as to maintain stable concentrations of these substances in the extracellular fluid, that is, the internal environment. Altering urinary excretion is the means to this balancing end.

Regulation of the total-body balance of any substance can be studied in terms of the balance concept described in Chapter 7. Theoretically, a substance can appear in the body either as a result of ingestion or as a product of metabolism. On the loss side, a substance can be excreted from the body or can be metabolized. Therefore, if the quantity of any substance in the body is to be maintained at a constant level over a period of time, the total amounts ingested and produced must equal the total amounts excreted and metabolized.

For water and hydrogen ions, all four possible pathways apply. However, balance is simpler for the mineral electrolytes such as sodium and potassium: Since they are neither synthesized nor metabolized by cells, total-body balance is a function of only ingestion and excretion.

Reflexes that alter excretion, specifically excretion via the urine, constitute the major mechanisms that regulate the body balances of water and many of the inorganic ions determining the properties of the extracellular fluid. The extracellular concentrations of these ions are given on page 352. We will first describe how the kidneys work in general and then apply this information to how they process specific substances—sodium, water, potassium, and so on—and participate in reflexes that regulate these substances. It should be noted that the kidneys are not the major regulators of *all* inorganic substances. In particular, the body balances of many of the trace elements, such as zinc and iron, are regulated mainly by control of gastrointestinal absorption of the element or by control of its excretion in the bile by the liver.

SECTION A

BASIC PRINCIPLES OF RENAL PHYSIOLOGY

FUNCTIONS OF THE KIDNEYS

The kidneys process blood by removing substances from it and, in a few cases, by adding substances to it. In so doing, they perform a variety of functions, as summarized in Table 15-1. As noted earlier, the major function is to regulate the water content, mineral composition, and acidity of the body by excreting each substance in an amount adequate to achieve total-body balance and maintain normal concentrations in the extracellular fluid.

A second renal (the adjective denoting kidney) function is the excretion of metabolic **waste products**, so termed because they serve no function. These include **urea** from protein, **uric acid** from nucleic acids, **creatinine** from muscle creatine, the end products of hemoglobin breakdown, which give urine much of its color, and many others. Some of these waste products, for example, urea, are relatively harmless, but others are toxic, and their accumulation during periods of renal malfunction accounts for some of the disordered body functions typical of severe kidney disease. The identity of these "toxins" is still uncertain.

TABLE 15-1   FUNCTIONS OF THE KIDNEYS
1. Regulation of water and inorganic-ion balance
2. Removal of metabolic waste products from the blood and their excretion in the urine
3. Removal of foreign chemicals from the blood and their excretion in the urine
4. Secretion of hormones: A. Erythropoietin, which controls erythrocyte production (Chapter 13)   B. Renin, which controls formation of angiotensin, which influences blood pressure and sodium balance (this chapter)   C. 1,25-dihydroxyvitamin D_3, which influences calcium balance (this chapter)

The kidneys have another general excretory function: the elimination of some foreign chemicals—drugs, pesticides, food additives, and so on—and their metabolites. These substances are removed from the blood by the kidneys and excreted in the urine.

Finally, in addition to their excretory functions, the kidneys act as endocrine glands, secreting at least three substances that are components of hormonal systems: erythropoietin (Chapter 13), renin, and the active form of vitamin D. These last two hormones are discussed in this chapter.

STRUCTURE OF THE KIDNEYS AND URINARY SYSTEM

The kidneys are paired organs that lie in the back of the abdominal wall but not actually in the abdominal cavity. They are retroperitoneal, meaning just behind the lining—the peritoneum—of this cavity.

Each kidney is composed of approximately 1 million functional units bound together by small amounts of connective tissue containing blood vessels, nerves, and lymphatics. One such unit, or **nephron**, is shown in Figure 15-1. Each nephron consists of (1) an initial component called the **glomerulus**, which forms a protein-free filtrate of blood that passes into (2) a **tubule** that processes the filtrate as it flows through the tubule before exiting the kidneys as urine. Let us describe the anatomy of these two structures, beginning with the glomerulus.

Systemic arterial blood enters each kidney via a renal artery, which then divides into progressively smaller branches. Each of the smallest arteries gives off, at right angles to itself, a series of arterioles, the **afferent arterioles**, each of which conducts blood to a compact tuft of capillaries called the **glomerular capillaries** (Figure 15-2). The capillary tuft protrudes into a balloon-like hollow capsule—**Bowman's capsule**. The combination of the glomerular capillaries and Bowman's capsule constitutes the glomerulus.[1]

One way of visualizing the glomerulus is to imagine a loosely clenched fist—the glomerular capillary tuft—punched into a balloon—Bowman's capsule. The part of Bowman's capsule in contact with the glomerular capillaries becomes pushed inward but does not make contact with the opposite side of the capsule. Accordingly, a fluid-filled space, **Bowman's space**, exists within the capsule. Blood in the glomerular capillaries is separated from the fluid in Bowman's space by a filtration barrier consisting of three layers: the single-celled capillary endothelium, the single-celled epithelial lining of Bowman's capsule, and the noncellular proteinaceous layer

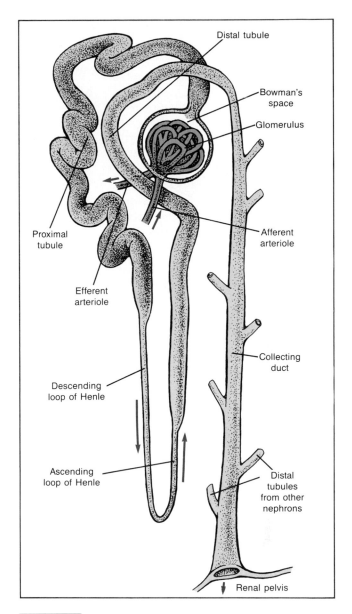

FIGURE 15-1 Basic structure of a nephron. The glomerulus consists of the glomerular capillaries and Bowman's capsule. Between the ascending loop of Henle and the distal tubule is a very short tubular segment, called the macula densa.

of basement membrane between the endothelium and epithelium.

Extending out from Bowman's capsule is the nephron tubule, the lumen of which is continuous with Bowman's space. Throughout its course, the tubule is composed of a single layer of epithelial cells resting on a basement membrane. The epithelial cells differ in structure and function along the tubule's length, and four major divi-

---

[1]There is no complete agreement as to whether the glomerulus should refer only to the capillaries or to the capillaries plus Bowman's capsule. However, the latter is presently the more common usage.

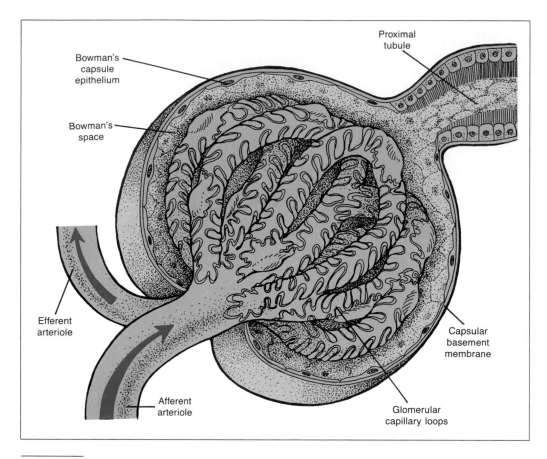

Bowman's
capsule
epithelium

Bowman's
space

Efferent
arteriole

Afferent
arteriole

Proximal
tubule

Capsular
basement
membrane

Glomerular
capillary loops

FIGURE 15-2 Anatomy of the glomerulus.

sions are recognized (Figure 5-1).[2] The segment of the tubule that drains Bowman's capsule is the **proximal tubule**. The next portion of the tubule is the **loop of Henle**, which is a sharp hairpin-like loop consisting of a descending limb coming from the proximal tubule and an ascending limb leading to the next tubular segment, the **distal tubule**. Until the distal tubule, each tubule is completely separate from its neighbors, but then distal tubules from several adjacent nephrons join to form **collecting ducts**, the fourth of the tubular segments. Distal tubules from still other nephrons drain into the collecting ducts formed by these unions as the ducts course toward the kidney's central cavity, the **renal pelvis**.

There are important regional differences in the kidney (Figure 15-3). The outer portion, the **renal cortex**, contains all the glomeruli, the proximal and distal tubules, and the outer portions of the loops of Henle and collecting ducts. The loops of Henle extend from the

cortex down into the inner portion, the **renal medulla**, through which also course the collecting ducts on their way to the renal pelvis. The loops and collecting ducts are parallel to each other and interspersed in the medulla.

The renal pelvis is continuous with the **ureter**, the tube that passes from the kidney to the **urinary bladder**, where urine is temporarily stored and from which it is eliminated during urination via the **urethra**, the tube leading from the bladder to the external world (Figure 15-4).

To return to the kidney blood vessels: The glomerular capillaries of each tuft, instead of leading to veins, recombine to form another arteriole, the **efferent arteriole**, through which blood leaves the glomerulus. Each efferent arteriole soon divides into a second set of capillaries, the **peritubular capillaries**, which branch profusely to form a network surrounding the tubule and then eventually rejoin to form the veins by which blood leaves the kidney.

One additional anatomical detail involving both the tubule and the arterioles must be mentioned. Just before the ascending limb of the loop of Henle becomes the

---

[2]This subdivision of the tubule into four segments is highly simplified, since each of these segments can be further subdivided into multiple distinct functional components. A description of each of these components is beyond the scope of this book, however, and so we shall deal with the four overall segments as though each were homogenous.

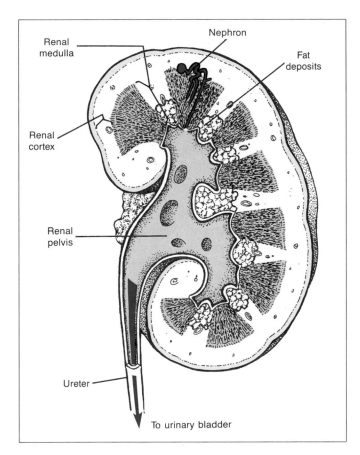

FIGURE 15-3 Section of a human kidney. For clarity, the nephron illustrated to show nephron orientation is not in scale—its outline would not be clearly visible without a microscope. The outer kidney, the cortex, has a granular appearance because it contains all the glomeruli. It also contains all the proximal tubules, distal tubules, and upper portions of both the loops of Henle and collecting ducts. The inner kidney, the medulla, contains the remainder of the loops of Henle and the collecting ducts. These medullary structures run parallel to each other, giving the medulla a striped appearance. The collecting ducts drain into the renal pelvis.

distal tubule, it passes between the arterioles supplying its glomerulus (Figure 15-1). This very short segment of tubule is the **macula densa**. The wall of the afferent arteriole at this point contains secretory cells known as **granular cells**. The combination of macula densa and granular cells is known as the **juxtaglomerular apparatus (JGA)**. The JGA serves several functions to be described later: (1) The granular cells secrete the hormone renin, which is involved in blood pressure regulation and sodium balance; and (2) the macula densa functions as a sensor of tubular fluid flow and/or composition in local homeostatic responses that regulate both the secretion of renin and the rate of fluid filtration by the glomeruli.

BASIC RENAL PROCESSES

To reiterate, blood flows from the afferent arterioles into the glomerular capillaries and thence into the efferent arterioles. Urine formation begins with **glomerular filtration**, the bulk flow of protein-free plasma from the glomerular capillaries through the glomerular membranes—capillary endothelium, basement membrane, and epithelium of Bowman's capsule—and into Bowman's capsule. This capsular fluid, now called the **glomerular filtrate**, contains all the substances, except protein, present in plasma and in virtually the same concentrations they have in plasma. The filtrate contains almost no protein because the glomerular membranes restrict the movement of such high-molecular-weight substances.

The urine that eventually enters the renal pelvis is quite different from the glomerular filtrate because, as the filtrate flows from Bowman's capsule through the tubule, its composition is altered. This change occurs by two general processes, tubular reabsorption and tubular secretion (Figure 15-5). The tubule is at all points intimately associated with the peritubular capillary net-

FIGURE 15-4 Urinary system in a woman. The urine, formed by the kidneys, flows from the kidneys through the ureters into the bladder, from which it is eliminated via the urethra. In the male, the urethra courses through the penis (Chapter 18).

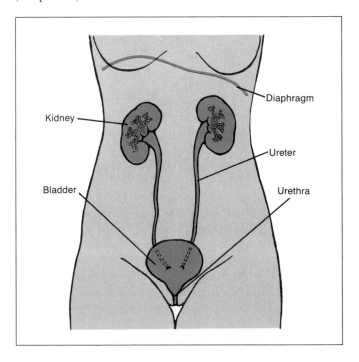

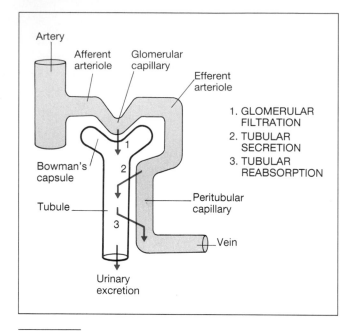

Artery

Afferent arteriole

Glomerular capillary

Efferent arteriole

1. GLOMERULAR FILTRATION
2. TUBULAR SECRETION
3. TUBULAR REABSORPTION

Bowman's capsule

Tubule

Peritubular capillary

Vein

Urinary excretion

FIGURE 15-5 The three basic components of renal function. This figure is to illustrate only the directions of reabsorption and secretion, not specific sites or order of occurrence. Depending on the particular substance, reabsorption and secretion can occur at various sites along the tubule.

work, a relationship that permits transfer of materials between peritubular blood and the lumen of the tubule. When the direction of transfer is from tubular lumen to peritubular capillary plasma, the process is called **tubular reabsorption**, or simply reabsorption. Movement in the opposite direction, that is, from peritubular plasma to tubular lumen, is called **tubular secretion**, or simply secretion.

Tubular reabsorption and tubular secretion denote only the direction of the transfer, not the mechanism, which is diffusion for some substances and mediated transport for others. In the latter case the transport processes are in the plasma membranes of the tubular epithelial cells, and this explains the common use of the phrases "the tubule secretes" and "the tubule reabsorbs."

The most common relationships between the three basic renal processes—glomerular filtration, tubular reabsorption, and tubular secretion—are shown in the hypothetical situation of Figure 15-6. Plasma containing three low-molecular-weight substances, X, Y, and Z, enters the glomerular capillaries, and approximately 20 percent of the plasma is filtered into Bowman's capsule. The filtrate, which contains X, Y, and Z in the same concentrations as in the plasma remaining in the capillaries,

enters the proximal tubule and begins its flow through the rest of the tubule. Simultaneously, the remaining 80 percent of the plasma, with its X, Y, and Z, leaves the glomerular capillaries via the efferent arteriole and enters the peritubular capillaries.

The tubule can secrete 100 percent of the peritubular-capillary X into the tubular lumen but cannot reabsorb X. Thus, by the combination of filtration and tubular secretion, all the plasma that originally entered the renal artery is cleared of substance X, which leaves the body via the urine.

The tubule can reabsorb Y and Z. The amount of Y reabsorption is small, so that most of the filtered material is not reabsorbed and escapes from the body. But for Z the reabsorptive mechanism is so powerful that all the filtered Z is transported back into the plasma. Therefore no Z is lost from the body. Hence, for Z the processes of filtration and reabsorption have canceled each other out, and the net result is as though Z had never entered the kidney.

For each substance in plasma, a particular combination of filtration, tubular reabsorption, and tubular secretion applies. The critical point is that, for many substances, the rates at which the processes proceed are subject to physiological control. What is the effect, for example, if the Y filtration rate were increased by some physiological mechanism or its reabsorption rate decreased? Either change means that more Y would be lost from the body via the urine. By triggering such changes in filtration, reabsorption, or secretion whenever the plasma concentration of a substance changes from normal, homeostatic mechanisms can regulate the substance's plasma concentration.

Glomerular Filtration

Composition of the filtrate. As stated above, the glomerular filtrate, that is, the fluid within Bowman's capsule, is essentially protein free and contains all other substances present in plasma in virtually the same concentrations as in plasma. The only exceptions to the last part of this generalization are certain low-molecular-weight substances that would otherwise be filterable but are bound to proteins and therefore not filtered. For example, half of the plasma calcium and virtually all of the plasma fatty acids are bound to protein.

We must point out the reason for use of the term "essentially protein free" in the previous paragraph. In reality, there is a very small amount of protein in the filtrate, since the glomerular membranes are not perfect sieves for protein. Normally, this filtered protein is completely removed from the tubule so that virtually no protein appears in the final urine. However, in diseased kidneys, the glomerular membranes may become much more permeable to protein and/or the tubules may lose

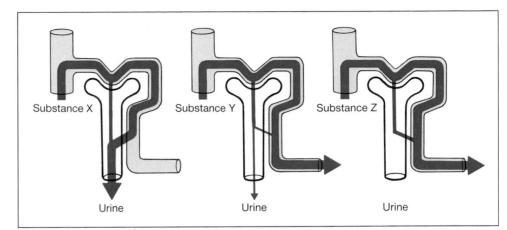

FIGURE 15-6 Renal handling of three hypothetical substances X, Y, and Z. X is filtered and secreted but not reabsorbed. Y is filtered, and a fraction is then reabsorbed. Z is filtered and is completely reabsorbed.

their ability to remove it from the tubule. In either case, protein will appear in the urine.

Forces involved in filtration. Glomerular filtration, like filtration across any capillary, is a bulk-flow process. As described in Chapter 13 (page 394), capillary filtration is determined by opposing forces: The hydrostatic pressure difference across the capillary wall favors filtration, while the protein concentration difference across the wall creates an osmotic force that opposes filtration. This also applies to the glomerular capillaries, as summarized in Figure 15-7.

The pressure of the blood in the glomerular capillaries averages 55 mmHg.[3] This is higher than in other capillaries of the body because the nephron afferent arterioles, having relatively large diameters, offer less resistance than most arterioles, and so more of the arterial pressure is transmitted to the capillaries. This glomerular capillary hydrostatic pressure is a force favoring filtration.

The fluid in Bowman's capsule exerts a hydrostatic pressure of 15 mmHg, and this opposes filtration into the capsule. A second opposing force results from the presence of protein in the glomerular capillary plasma and its virtual absence in Bowman's capsule. This unequal distribution of protein causes the water concentration of the plasma to be less than that of the fluid in Bowman's capsule. This difference in water concentration favors an osmotic flow of fluid—water plus all the low-molecular-weight solutes—from Bowman's capsule into the glomerular capillary, a flow equivalent to that produced by a pressure difference of 30 mmHg.

---

[3]Glomerular capillary pressure cannot be directly measured in humans, and this figure is the best educated guess from work in experimental animals.

Thus, as can be seen from Figure 15-7, the **net glomerular filtration pressure**, that is, the algebraic sum of the three relevant forces, is approximately 10 mmHg. This pressure initiates urine formation by forcing an es-

---

FIGURE 15-7 Forces involved in glomerular filtration. The symbol π denotes an osmotic force due to differences in solute and hence water concentration. The difference in solute concentration is due entirely to the presence of protein in plasma but not in Bowman's space. The filtrate is essentially protein-free plasma and averages 180 L/day.

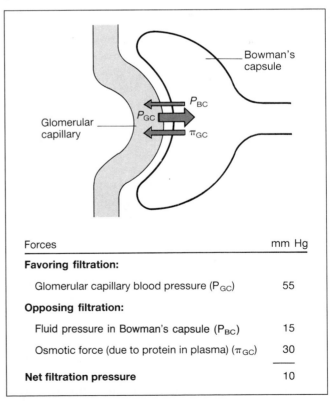

Forces	mm Hg
Favoring filtration:	
Glomerular capillary blood pressure (P_{GC})	55
Opposing filtration:	
Fluid pressure in Bowman's capsule (P_{BC})	15
Osmotic force (due to protein in plasma) (π_{GC})	30
Net filtration pressure	10

sentially protein-free filtrate of plasma through the glomerular membranes into Bowman's capsule and thence down the tubule into the renal pelvis. The glomerular membranes serve only as a filtration barrier and have no active, that is, energy-requiring, function.

Rate of glomerular filtration.   The volume of fluid filtered from the glomerular capillaries into Bowman's capsule per unit time is known as the **glomerular filtration rate (GFR)**. The glomerular capillaries are so much more permeable to fluid than, say, a muscle or skin capillary, that the net filtration pressure of 10 mmHg causes massive filtration. In a 70-kg person, the average volume filtered into Bowman's capsule is 180 L/day! Contrast this figure with the net filtration of fluid across all the other capillaries in the body—4 L/day, as described in Chapter 13.

There are several implications of this remarkable filtration rate. First, in order to form such a huge volume of filtrate, the kidneys must receive a large share of the cardiac output. Each moment the kidneys receive 20 to 25 percent of the blood pumped by the left ventricle even though the combined weights of these two organs is only about 1 percent of body weight. Second, when we recall that the total volume of plasma in the cardiovascular system is approximately 3 L, it follows that the entire plasma volume is filtered by the kidneys some 60 times a day. This opportunity to process such large volumes of plasma enables the kidneys to regulate the constituents of the internal environment and to excrete large quantities of waste products.

It is possible to measure the total amount of any nonprotein substance (assuming also that the substance is not bound to protein) filtered into Bowman's capsule by multiplying the GFR by the plasma concentration of the substance. This amount is called the **filtered load** of the substance. For example, if the GFR is 180 L/day and plasma glucose concentration is 1 g/L, then 180 L/day × 1 g/L = 180 g/day is the filtered load of glucose.

Once we know the filtered load of the substance, we can compare it to the amount of the substance excreted and tell whether the substance undergoes net tubular reabsorption or net secretion. Whenever the quantity of a substance excreted in the urine is less than the filtered load, tubular reabsorption must have occurred. Conversely, if the amount excreted in the urine is greater than the filtered load, tubular secretion must have occurred. Such measurements have played an important role in determining which substances undergo tubular reabsorption or tubular secretion.

Tubular Reabsorption

Many filterable plasma components are either absent from the urine or present in smaller quantities than were filtered at the glomerulus. This fact alone is sufficient to prove that these substances undergo tubular reabsorption. An idea of the magnitude and importance of these reabsorptive mechanisms can be gained from Table 15-2, which summarizes data for a few plasma components that undergo filtration and reabsorption.

The values in Table 15-2 are typical for a normal person on an average diet. There are at least three important conclusions to be drawn from this table: (1) The filtered loads are enormous, generally larger than the amounts of the substances in the body. For example, the body contains about 40 L of water, but the volume of water filtered each day is, as we have seen, 180 L. (2) Reabsorption of waste products is relatively incomplete, for example, only 44 percent in the case of urea, so that large fractions of their filtered loads are excreted in the urine. (3) Reabsorption of most useful plasma components, for example, water, inorganic ions, and organic nutrients, is relatively complete so that the amounts excreted in the urine represent very small fractions of the filtered loads.

In this last regard, an important distinction should be made between reabsorptive processes that can be controlled physiologically and those that cannot. The reabsorption rates of many, but not all, organic nutrients, for example, glucose, are always very high and are not physiologically regulated, and so the filtered loads of these substances are normally completely reabsorbed, none appearing in the urine. For these substances, like substance Z in our earlier example, it is as though the kidneys do not exist since the kidneys do not eliminate them from the body at all. Therefore the kidneys do not help *regulate* the plasma concentrations of these substances, that is, minimize changes from their operating points. Rather, the kidneys merely maintain whatever plasma concentrations already exist, generally the result of hormonal regulation of nutrient metabolism (Chapter 17).

In contrast, the reabsorptive rates for water and many ions, although also very high, are regulatable. Consider

TABLE 15-2 AVERAGE VALUES FOR SEVERAL COMPONENTS THAT UNDERGO FILTRATION AND REABSORPTION

Substance	Amount Filtered per Day	Amount Excreted per Day	Percent Reabsorbed
Water, L	180	1.8	99.0
Sodium, g	630	3.2	99.5
Glucose, g	180	0	100
Urea, g	54	30	44

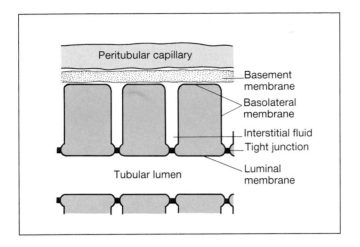

FIGURE 15-8 Diagrammatic representation of tubular epithelium. The basement membrane of the tubule is a homogeneous proteinaceous structure that plays no significant role in transport and will not be shown in subsequent figures illustrating transport in this chapter.

what happens when a person drinks a lot of water. Within 1 to 2 h, all the excess water has been excreted in the urine, chiefly, as we shall see, as the result of decreased tubular reabsorption of water, and water balance is restored to normal. The critical point is that the rates at which water and the inorganic ions are reabsorbed, and, therefore, the rates at which they are excreted, are subject to physiological control.

Tubular reabsorption is a process fundamentally different from glomerular filtration. The latter occurs completely by bulk flow, with water and all low-molecular-weight solutes moving together. In contrast, there is relatively little bulk flow across the epithelial cells of the tubule from lumen to interstititial fluid, so that reabsorption is not by mass movement. Rather, the reabsorption of some substances is by diffusion, while that of others requires more or less distinct mediated transport systems. The phrase "more or less distinct" denotes the fact that, as we shall see, the reabsorption of different substances is often linked.

Reference to Figure 15-8 highlights the fact that, except for substances that can diffuse across the tight junctions between cells, tubular reabsorption requires movement of the substance across several membranes. To describe this, we need to define some terms concerning the plasma membrane of the tubular epithelial cells. The portion of the plasma membrane facing the lumen is the **luminal membrane**. Beginning at the tight junctions and constituting the plasma membrane of the sides and base of the cell is the **basolateral membrane**.

To be reabsorbed through the cell, a substance must first cross the luminal membrane, then diffuse through the cytosol of the cell, and finally cross the basolateral membrane into the interstitial fluid, from which it crosses the basement membrane and peritubular capillary endothelium to gain entry into the capillaries. The mechanisms of this type of epithelial transport were described on page 131 and should be reviewed at this time. Suffice it to point out here that if the movement across either the luminal membrane or the basolateral membrane is active, then the entire process is termed active. Movement across the basement membrane and capillary endothelium into the peritubular capillaries is by a combination of diffusion and bulk flow. This does not contradict our earlier statement that movement across the *tubular epithelial cells* from tubular lumen to interstitial fluid is *not* by bulk flow.[4]

T_m-limited transport mechanisms. Many of the active reabsorptive systems in the renal tubule have a limit, termed a **transport maximum (T_m)**, to the amounts of material they can transport per unit time, because the membrane proteins responsible for the transport become saturated. An important example is the active-transport process for glucose. As noted earlier, normal persons do not excrete glucose in their urine because all filtered glucose is reabsorbed. However, if the plasma glucose concentration and, thus, the filtered load become very high, glucose will appear in the urine because the tubules cannot reabsorb the entire filtered load. This occurs in people with diabetes mellitus, a disease in which the hormonal control of plasma glucose concentration is defective (Chapter 17).

The pattern described for glucose is also true for a large number of other organic substances. For example, most amino acids and water-soluble vitamins are filtered in large amounts each day, but almost all these filtered molecules are reabsorbed. If for some reason the plasma concentration becomes high enough, however, reabsorption of the filtered load will not be as complete, and the substance will appear in larger amounts in the urine. Thus, persons ingesting large quantities of vitamin C manifest progressive increases in their plasma concentrations of vitamin C until the filtered load exceeds the tubular reabsorptive T_m for this substance and any additional ingested vitamin C is excreted in the urine.

Reabsorption by diffusion. Urea reabsorption provides an example of passive reabsorption by diffusion.

---

[4] Bulk flow occurs from interstitial fluid into peritubular capillaries because the balance of hydrostatic and osmotic forces acting across these capillaries favors movement in this direction. The hydrostatic pressure in the capillary is low, and the protein concentration is high.

An analysis of urea concentrations in the tubule will help elucidate the mechanism. The urea concentration in the fluid within Bowman's capsule is the same as that in the plasma. Then, as this fluid flows through the proximal tubule, water reabsorption occurs (by mechanisms to be described later), increasing the concentrations of tubular-fluid solutes that, like urea, are not simultaneously being reabsorbed by active transport. The result is that the tubular-fluid urea concentration becomes greater than the peritubular-plasma urea concentration. Accordingly, urea diffuses down this concentration gradient from tubular lumen to peritubular capillary. Urea reabsorption is thus a passive process dependent upon the reabsorption of water. As noted earlier, this process causes reabsorption of about 50 percent of the filtered urea.[5]

Reabsorption by diffusion is also of considerable importance for many foreign chemicals. The plasma membranes of the tubular epithelium, like all plasma membranes, are primarily lipid, and so lipid-soluble substances can penetrate them fairly readily. Recall from Chapter 6 that one of the major determinants of lipid solubility is the polarity of a molecule—the less polar, the more lipid-soluble. Many drugs and environmental pollutants are nonpolar and therefore highly lipid-soluble. This makes their excretion from the body via the urine difficult since they are filtered at the glomerulus but then reabsorbed by diffusion as water reabsorption causes their intratubular concentrations to increase.

The body does have a way of making these substances more excretable, however. The liver transforms them to more polar metabolites that, because of their reduced lipid solubility, diffuse across the tubular wall poorly and are therefore excreted.

Tubular Secretion

Tubular secretory processes, by which substances move from peritubular capillaries into the tubular lumen, constitute a second pathway into the tubule, the first being glomerular filtration. The most important substances secreted by the tubules are hydrogen ions and potassium, and the mechanisms by which they are secreted are described in the sections on these substances.

The kidney is also able to secrete a large number of organic ions, some of which are normally occurring metabolites such as choline and creatinine, while others are foreign chemicals—penicillin, for example. These sub-

stances diffuse from the peritubular capillaries into the interstitial fluid outside the basolateral membranes of the tubular epithelial cells. They are then actively transported across the basolateral membrane and into the cell, followed by exit across the luminal membrane and into the tubular lumen.

Metabolism by the Tubules

Although renal physiologists have traditionally listed glomerular filtration, tubular reabsorption, and tubular secretion as the three basic renal processes, a fourth process—metabolism by the cells of the tubule—is of considerable importance for many substances. The cells of the renal tubules are able to synthesize certain substances, notably ammonia, which are then added to the lumen fluid and excreted. Why the cells would make something just to have it excreted will be made clear when we discuss the role of ammonia. The cells are also capable of catabolizing certain organic substances—peptides, for example—taken up from either the tubular lumen or peritubular capillaries, thus eliminating them from the body as surely as if they had been excreted in the urine.

MICTURITION

The composition of the urine is not significantly altered after it leaves the collecting ducts, and so the renal pelves (plural of pelvis), ureters, bladder, and urethra simply serve as plumbing. Urine flow through the ureters to the bladder occurs continuously, propelled by contractions of the ureteral wall smooth muscle. The urine is stored in the bladder and intermittently ejected during urination, termed **micturition**.

The bladder is a balloon-like chamber, with walls of smooth muscle. The smooth muscle at the neck of the bladder is sometimes called the internal urethral sphincter. It is not a distinct muscle, however, but rather the last portion of the bladder and the first portion of the urethra. When the bladder is relaxed, the outlet of the bladder is closed. When the bladder either actively contracts or is passively distended, the outlet is pulled open by changes in bladder shape.

In an infant, micturition is basically a local spinal reflex, and its place and time are determined by the volume of urine in the bladder, acting via this reflex. The bladder wall contains stretch receptors whose afferent fibers enter the spinal cord and stimulate the parasympathetic nerves supplying the bladder smooth muscle. These nerves cause bladder contraction. When the bladder contains only small amounts of urine, its internal pressure is low, there is little stimulation of the bladder

---

[5]You might logically ask why there should be *any* reabsorption of a waste product like urea. Put in another way, is there any adaptive value in not excreting the entire filtered load of this substance? The answer is probably no, and the partial reabsorption simply reflects an inevitable consequence of the easy diffusibility of urea through plasma membranes.

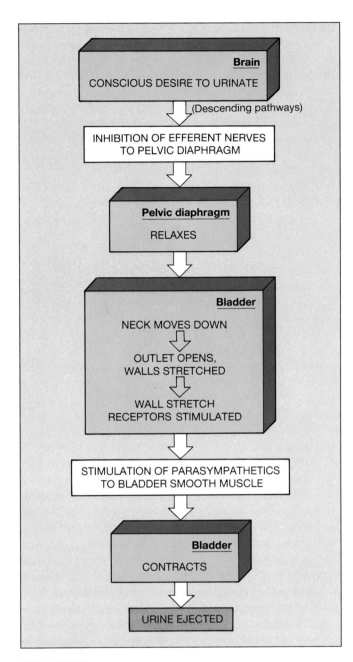

FIGURE 15-9 Voluntary control of micturition. The opening of the bladder outlet permits bladder contraction to produce emptying. Not shown is the fact that the conscious desire to urinate is usually initiated, when the bladder becomes distended with urine, by input from the same bladder stretch receptors that then participate in the response by triggering increased parasympathetic input to the bladder. Thus, these receptors play a dual role in micturition.

stretch receptors, and the parasympathetics are relatively quiescent. As the bladder fills with urine, it becomes distended, and the stretch receptors are stimulated, thereby reflexly eliciting stimulation of the para-

sympathetic neurons and contraction of the bladder. This contraction pulls open the bladder outlet and simultaneously produces the pressure required to cause urine flow through the urethra.

Voluntary control of micturition, learned during childhood, involves the skeletal muscles of the **pelvic diaphragm**, which forms the floor of the pelvis and helps support the lower bladder. Voluntary relaxation of the pelvic diaphragm allows the neck of the bladder to move downward. This tends to open the bladder outlet while simultaneously stretching the walls of the bladder and eliciting reflex bladder contractions via the parasympathic nerves. If the increased pressure resulting from contraction of the bladder wall is insufficient to force urine into the urethra, the pressure can be further increased by voluntary contraction of the abdominal muscles and the respiratory diaphragm. In contrast, micturition can be stopped voluntarily simply by contracting the pelvic diaphragm. Figure 15-9 summarizes the voluntary control of micturition.

In an adult with damage to the central nervous system that interrupts the descending pathways mediating voluntary control of the pelvic diaphragm, micturition once again becomes the pure spinal reflex it was in infancy. Another situation sometimes associated with loss of bladder control and involuntary bladder emptying is fear or other strong emotion, acting via descending pathways to the bladder's innervation.

SECTION B
REGULATION OF SODIUM AND WATER BALANCE

TOTAL-BODY BALANCE AND INTERNAL DISTRIBUTION OF SODIUM AND WATER

Table 15-3 summarizes total-body water balance. These are average values, which are subject to considerable normal variation. The two sources of body water are metabolically produced water, resulting largely from the oxidation of organic nutrients, and water ingested in liquids and so-called solid food (a rare steak is approximately 70 percent water). There are four sites from which water is lost to the external environment: skin, respiratory passageways, gastrointestinal tract, and urinary tract. Menstrual flow constitutes a fifth potential source of water loss in women. The loss of water by evaporation from the cells of the skin and the lining of respiratory passageways is a continuous process, often

TABLE 15-3 DAILY WATER GAIN AND LOSS IN ADULTS	
Intake: Drunk	1200 mL
In food	1000 mL
Metabolically produced	350 mL
Total	2550 mL
Output: Insensible loss (skin and lungs)	900 mL
Sweat	50 mL
In feces	100 mL
Urine	1500 mL
Total	2550 mL

referred to as **insensible water loss**, because the person is unaware of its occurrence. Additional water can be made available for evaporation from the skin by the production of sweat. Normal gastrointestinal loss of water in feces is generally quite small but can be severe in diarrhea. Gastrointestinal loss can also be large in vomiting.

Table 15-4 is a summary of total-body balance for sodium chloride. The excretion of sodium and chloride via the skin and gastrointestinal tract is normally quite small but may increase markedly during severe sweating, vomiting, or diarrhea. Hemorrhage can also result in the loss of large quantities of both salt and water.

Under normal conditions, as can be seen from Tables 15-3 and 15-4, salt and water losses exactly equal salt and water gains, and no net change of body salt and water occurs. As we shall see, this matching of losses and gains is primarily the result of regulation of urinary loss, which can be varied over an extremely wide range. For example, urinary water excretion can vary from approximately 0.4 L/day to 25 L/day, depending upon whether one is lost in the desert or participating in a beer-drinking contest. Similarly, some persons ingest 20 to 25 g of sodium chloride per day, whereas a person on a low-salt diet may ingest only 50 mg. The normal kidney can readily alter its excretion of salt over this range so as to match loss with gain.

The reflexes that achieve balance of salt in the body operate mainly on sodium, and the renal processing of chloride is usually coupled to that of sodium. Accordingly, we shall have little more to say about chloride even though it is the most abundant anion in the extracellular fluid.

So far we have been dealing only with total-body balance. In order to understand the reflexes that control sodium and water excretion, however, one must also know how these substances are distributed between the extracellular and intracellular fluids. Figure 15-10 summarizes this information for water. As noted in Chapter 2, water is the most abundant substance in the body, accounting for approximately 60 percent of total-body weight. Approximately two-thirds of the water (28 L) is inside cells, and the remaining one-third (14 L) is in the extracellular fluid—the interstitial fluid and the plasma. Water moves freely across most plasma membranes and so always equilibrates by osmosis between the intracellular and extracellular compartments according to how much total solute is in each.

This is a good place to review several concepts described in Chapter 6. The total solute concentration of a solution is known as its osmolarity. But since solute concentration and water concentration are two sides of the same coin, osmolarity is also a "reverse" measure of water concentration: The higher the osmolarity of a solution, the lower its water concentration. Accordingly, there is net movement of water from a region of low osmolarity to one of high osmolarity.

Sodium is distributed quite differently from water. Very little is inside cells because plasma-membrane Na, K-ATPase pumps actively transport this ion out of cells (Chapter 6). Approximately 50 percent of body sodium is in the extracellular fluid, and the rest is in bone, a component that we shall ignore in this discussion because movement of sodium into or out of bone is not a closely regulated process. Sodium is by far the major solute of the extracellular fluid, and so it is the major contributor to this fluid's osmolarity.

These facts can now be used to explain a very important generalization: The volume of extracelluar fluid depends primarily upon the amount of sodium in the body. Let us take an example. If one ingests a large amount of sodium, it is absorbed from the intestinal tract into the blood and distributes throughout the extracellular fluid. This raises the osmolarity of the extracellular fluid and pulls water out of cells, increasing the extracellular volume while decreasing the intracellular volume. As we shall see, the increased osmolarity also makes one thirsty. Finally, the increased osmolarity causes the kidneys to retain water.

When sodium is lost from the body, analogous but opposite events occur, that is, the volume of extracellular fluid decreases.

BASIC RENAL PROCESSES FOR SODIUM AND WATER

Being of low molecular weight and not bound to protein, sodium and water are both freely filterable at the glomerulus. They both undergo considerable reabsorption—normally more than 99 percent (Table 15-2)—but no secretion. Most renal energy utili-

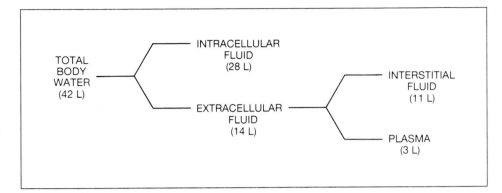

FIGURE 15-10 Distribution of body water. The volumes are for a normal 70-kg person. Total-body water averages approximately 60 percent of body weight.

zation goes to accomplish this enormous reabsorptive task. The mechanisms for reabsorption of these substances can be summarized by two generalizations: (1) Sodium reabsorption is a primary active process, and (2) water reabsorption is by osmosis and is dependent upon sodium reabsorption.

Primary Active Sodium Reabsorption

The reabsorption of sodium and water occurs largely by the mechanisms of epithelial transport described in Chapter 6. The key feature is the primary active transport of sodium, via Na, K-ATPase pumps in the epithelial-cell basolateral membrane, out of the cells and into the interstitial fluid (Figure 15-11). This active transport keeps the intracellular concentration of sodium low. Also the cell interior is electrically negative with respect to the outside. Therefore, there exist both a chemical difference and an electrical difference to move sodium out of the lumen into the tubular epithelial cells. This movement across the luminal membrane is either by diffusion through sodium channels or by carrier-mediated downhill transport, the specific mechanism varying from segment to segment of the tubule depending upon which protein channels and/or transporters are present in their luminal membranes.

When sodium moves through channels, it moves alone, but the carrier-mediated pathways cotransport or countertransport (Chapter 6, page 119) a large number

of other substances (Figure 15-12) with the sodium. These substances, therefore, undergo secondary active reabsorption or secretion. For this reason, sodium reabsorption is critical not only for retention of this ion but for the renal processing of many solutes.

For example, the cotransport of glucose with sodium into proximal tubular epithelial cells raises the intracel-

FIGURE 15-11 Mechanism of sodium movement from lumen to interstitial fluid during reabsorption. The sizes of the letters for Na^+ and K^+ denote high and low concentrations of these ions. Movement of Na^+ across the luminal membrane can occur through sodium channels, as shown in the figure, or as cotransport and countertransport with other substances, as shown in Figure 15-12. In either case, movement across the luminal membrane is passive. It is caused by the low epithelial-cell cytosolic sodium concentration produced by the active Na, K-ATPase sodium pumps in that cell's basolateral membrane. These pumps are located all along the basolateral membrane to the tight junctions, not just in the basal portion.

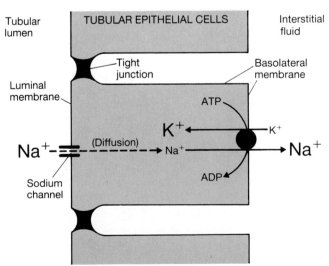

TABLE 15-4 DAILY SODIUM CHLORIDE INTAKE AND LOSS	
Intake: Food	10.5 g
Output: Sweat	0.25 g
Feces	0.25 g
Urine	10.00 g
Total output	10.5 g

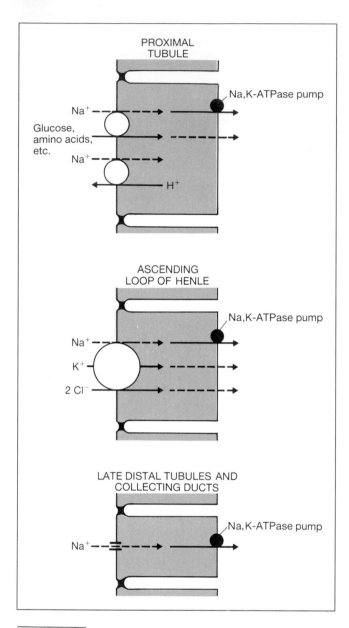

FIGURE 15-12 Sodium reabsorption in several nephron segments is coupled to reabsorption or secretion of other substances. The carrier-mediated transport systems that move sodium across the luminal membrane in the proximal tubule and ascending loop of Henle produce only downhill (passive) movement of sodium across this membrane. However, they simultaneously achieve secondary active cotransport of glucose, amino acids, potassium, and chloride into the cell or countertransport of hydrogen ion into the lumen. The cotransported substances then move passively across the basolateral membrane into the interstitial fluid and thence into peritubular capillaries. The origin of the hydrogen ion countertransported with sodium is described later in the chapter.

lular glucose concentration, which drives the downhill movement of glucose across the basolateral membrane. The net result is secondary active reabsorption of glucose. The luminal-membrane proteins that mediate glucose-sodium cotransport are found only in the proximal tubule, and so this is the only site of glucose reabsorption. Cotransport with sodium accounts for the reabsorption of a large number of other organic substances, for example, amino acids and lactate, by the proximal tubule.

In contrast, the epithelial cells of the ascending limb of the loop of Henle possess luminal-membrane proteins that cotransport sodium and chloride, achieving the secondary active reabsorption of chloride in this nephron segment. This system also cotransports potassium so that it is termed a Na,K, 2Cl cotransporter.

To take another example, hydrogen ions are countertransported with sodium in the proximal tubule, that is, they are transported from the cell into the lumen as sodium moves from lumen to cell, and this constitutes a major mechanism for the secondary active secretion of hydrogen ion by this nephron segment.

To reiterate, sodium moves passively from the lumen into the cell and is actively transported by Na, K-ATPase pumps into the interstitial fluid. The last step in the reabsorptive process, as is the case for all reabsorbed substances, is the movement of interstitial fluid, by bulk flow, into the peritubular capillaries.

Coupling of Water Reabsorption to Sodium Reabsorption

How does active sodium transport lead to passive water transport? The movement of sodium from the tubular lumen to the interstitium across the epithelial cells lowers the osmolarity, that is, raises the water concentration, of the luminal fluid. It simultaneously raises the osmolarity, that is, lowers the water concentration, of the interstitial fluid adjacent to the epithelial cells. The difference in water concentration between lumen and interstitum causes net osmosis of water from the lumen across the tubular cells' plasma membranes and/or tight junctions into the interstitial fluid. From there, water, sodium, and everything else in the interstitial fluid move together by bulk flow into peritubular capillaries as the final step in reabsorption (Figure 15-13).

Water reabsorption can occur, however, only if the tubular epithelium is permeable to water. No matter how large its concentration gradient, water cannot cross an epithelium impermeable to it. The water permeability of the proximal tubule is always very high, and so water molecules are reabsorbed almost as rapidly as sodium ions. As a result, the proximal tubule always reabsorbs sodium and water in the same proportions and the fluid leaving the proximal tubule, just like the fluid en-

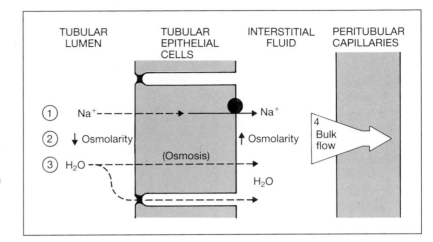

FIGURE 15-13 Coupling of water and sodium reabsorption. (1) Reabsorption of sodium creates (2) a difference in osmolarity between lumen and interstitial fluid, which causes (3) the osmosis of water in the same direction either through the cell or across the tight junctions. (4) Movement of both solute and water from interstitial fluid into peritubular capillaries occurs by bulk flow.

tering it from Bowman's capsule, is isosmotic to plasma—300 mOsmol/L. We will make use of this fact in a while. About two-thirds of the filtered sodium and water are reabsorbed by the proximal tubule.

Earlier we pointed out that lipid-soluble substances like urea are reabsorbed by diffusion since water reabsorption increases their luminal concentrations above that of plasma. Accordingly the high rate of water reabsorption in the proximal tubule causes considerable reabsorption of these solutes as well. The proximal tubule is also relatively permeable to chloride and so the increase in luminal chloride concentration produced by water reabsorption drives passive chloride reabsorption, just like passive urea reabsorption, in this nephron segment.

The water permeabilities of the nephron segments beyond the proximal tubules vary considerably. We will talk about those of the loops of Henle in the next section. The water permeability of the late distal tubules and collecting ducts can be high or low because it is subject to physiological control. The reabsorption of sodium and water, therefore, can be dissociated from each other in these nephron segments, that is, as sodium is reabsorbed, water sometimes is also reabsorbed and sometimes not.

The major determinant of this controlled permeability, and hence of water reabsorption in these segments, is a peptide hormone secreted by the posterior pituitary and known as **antidiuretic hormone (ADH)** or **vasopressin**.[6] ADH stimulates production of cyclic AMP in the epithelial cells of late distal tubules and collecting ducts, which leads to the appearance, in the luminal membranes of these segments, of proteins that function as water channels. Accordingly, in the presence of a high plasma concentration of ADH, the water permeability of the late distal tubules and collecting ducts is very great, water reabsorption is maximal, and the final urine volume is small—less than 1 percent of the filtered water. In the absence of ADH, the water permeability of these segments is very low, and very little water is reabsorbed from these sites.[7] Therefore, a large volume of water remains behind in the tubule to be excreted in the urine, which is hyposmotic, that is, has an osmolarity much lower than that of plasma. This increased urine excretion resulting from low ADH is termed **water diuresis**.

The disease **diabetes insipidus**, which is distinct from diabetes mellitus or "sugar diabetes," illustrates what happens when the ADH system is disrupted. Persons with this disease have lost the ability to produce ADH, usually as a result of damage to the hypothalamus. Thus, the permeability to water of the late distal tubules and collecting ducts is low and unchanging regardless of the state of the body fluids, and diabetes insipidus is characterized by a constant water diuresis—as much as 25 L/day. In most cases, the flow can be restored to normal by the administration of ADH.

From what has been said so far about sodium-water coupling, one might logically, but wrongly, conclude that, even in the presence of maximal amounts of ADH, the kidneys would be unable to excrete a hyperosmotic urine, that is, one with an osmolarity greater than that of plasma. For this to occur, relatively more water must be reabsorbed than sodium, but how can this happen when

---

[6] Throughout this chapter, we use the name "antidiuretic hormone" in keeping with the action of this hormone on the kidney. Its other name, "vasopressin," which is becoming its "official" label, reflects the fact, noted in Chapter 13, that this hormone can also exert a direct vasoconstrictor action on arterioles, resulting in elevated total peripheral resistance and, thereby, increased arterial blood pressure.

[7] What makes the plasma membranes of these tubular segments, in contrast to most plasma membranes, so impermeable to water in the absence of the water channels induced by ADH is not known.

water reabsorption is always secondary to sodium reabsorption? The answer is given in the next section.

The ability of the kidneys to produce hyperosmotic urine is a major determinant of one's ability to survive without water. The human kidney can produce a maximal urinary concentration of 1400 mOsmol/L, almost five times the osmolarity of plasma, which is 300 mOsmol/L. The urea, sulfate, phosphate, other waste products, and ions excreted each day amount to approximately 600 mOsmol. Therefore, the minimal volume of urine water in which this mass of solute can be dissolved equals:

$$\frac{600 \text{ mOsmol/day}}{1400 \text{ mOsmol/L}} = 0.444 \text{ L/day}$$

This volume of urine is known as the **obligatory water loss**. The loss of this minimal volume of urine contributes to dehydration when a person is deprived of water intake.

Urine Concentration: The Countercurrent Multiplier System

Urinary concentration takes place as the tubular fluid flows through the collecting ducts coursing through the medulla toward the renal pelvis. The interstitial fluid surrounding these ducts is very hyperosmotic, and in the presence of ADH water therefore diffuses out of the ducts into the interstitial fluid and thence into the blood vessels of the medulla. The key question is: How did the medullary interstitial fluid become hyperosmotic?

The complex process that sets up this interstitial hyperosmolarity is called the **countercurrent multiplier system** and takes place in the loops of Henle. These loops, like the collecting ducts, extend into the medulla. The fluid in a loop flows first in one direction—down the descending limb—and then in the opposite direction—up the ascending limb; thus the name "countercurrent."

First let us explain the principle of countercurrent multiplication with a model that has some features of the loop of Henle but is simpler to visualize (Figure 15-14). Then we will move to the more complex real-life situation. Our model is a hairpin-loop tube through which flows a sodium chloride solution that enters the loop at an osmolarity of 300 mOsmol/L. The ascending limb actively cotransports sodium and chloride into the descending limb and is relatively impermeable to water so that relatively little water can follow the salt. The pump can achieve a maximal osmolarity difference of 200 mOsmol/L across the loop. Let us look at what occurs under conditions of flow, simplifying the analysis by assuming that fluid flow and ion pumping occur in discontinuous out-of-phase steps. Think of it as a series of freeze-frames in a movie we are viewing, or perhaps as a series of snapshots of the loop.

We begin (A) by allowing the entire loop to fill with fluid having an osmolarity of 300 mOsmol/L. Now we stop the flow and allow the pumps all along the ascending loop to operate so that the osmolarity in the entire ascending limb falls to 200 mOsmol/L and that of the descending limb rises to 400 mOsmol/L (B). Now we restart flow (C): new 300-mOsmol/L fluid enters the top

FIGURE 15-14 Model to illustrate the basic principle of countercurrent multiplication. The ascending limb actively transports sodium chloride into the descending limb. The transport process and the flow of fluid through the loop actually occur continuously but have been separated into discontinuous steps for purposes of illustration.

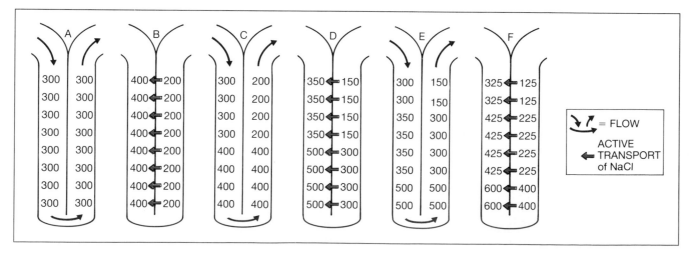

of the descending limb and pushes the entire column down, around, and up, so that an equal volume of 400-mOsmol/L fluid is forced out of the ascending limb. Again flow is stopped and the pumps begin operating until once again they have established a 200-mOsmol/L difference across the tubule (D). The sequence is repeated once more (E and F).

The critical point emerges after this last cycle is completed: Even though only a 200-mOsmol/L difference exists across the loop at each level, a 375-mOsmol/L difference exists from the top of the system to the bottom. In other words, the 200-mOsmol/L difference at each level has been *multiplied* by the countercurrent system. With continued small cycles of the system, the multiplication would have been much greater than the extra 175 mOsmol/L shown here.

We can now apply this model to the loop of Henle. The fluid entering the descending limb from the proximal tubule is, as we have seen, 300 mOsmol/L, the same as plasma. The *ascending* limb actively cotransports sodium and chloride and is relatively impermeable to water, so that little water follows the salt. Now, however, comes the major difference between the previous model and the loop of Henle: The pumps in the ascending limb do not transport sodium and chloride into the descending limb but rather into the interstitial fluid surrounding the limbs.

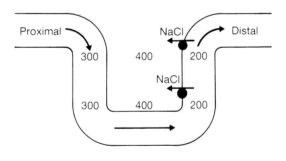

In contrast, the *descending* limb does not pump sodium chloride and is highly permeable to water. Therefore, there is a net diffusion of water out of the descending limb into the more concentrated interstitial fluid until the osmolarities inside this limb and in the interstitial fluid are again equal. The interstitial osmolarity is maintained at 400 mOsmol/L during this equilibration because the ascending limb continues to pump sodium chloride to maintain the 200 mOsmol/L difference between it and the interstitium.

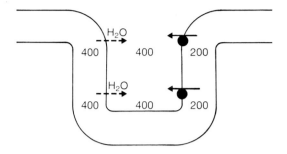

Thus, the osmolarities of the descending limb and interstitial fluid become equal, and both are 200 mOsmol/L higher than that of the ascending limb. This is the essence of the system: The loop countercurrent multiplier causes the interstitial fluid of the medulla to become concentrated. It is this hyperosmolarity that will draw water out of the collecting ducts and concentrate the urine.

When we now allow flow in this system, as in our simpler model, the result is the same (Figure 15-15): Multiplication of the osmolarity difference from 200 mOsmol/L at any given level to a much higher value—1400 mOsmol/L in this case—at the bend in the loop. It should be emphasized that the active sodium chloride transport mechanism in the ascending limb is the essential component of the entire system. Without it, the countercurrent flow would have no effect whatever on loop and interstitial concentrations.

Now we have our concentrated interstitial fluid, but we must still follow the fluid from the loop through the distal tubule and into the collecting duct (Figure 15-15). The countercurrent multiplier system concentrated the descending-loop fluid, but then it immediately rediluted it so that the fluid entering the distal tubule is actually more dilute than the plasma. Then, in the presence of ADH, which permits water movement across the tubular epithelium, the fluid in the late distal tubules reequilibrates with peritubular plasma by losing water until it becomes isosmotic to plasma in the peritubular capillaries, that is, until it is once again at 300 mOsmol/L. This fluid then enters and flows along the collecting ducts.

Under the influence of ADH, the collecting ducts are highly permeable to water, which diffuses out of the collecting ducts into the interstitial fluid as a result of the high-osmolarity set up there by the loop countercurrent multiplier system. This water then enters the medullary capillaries and is carried out of the kidneys by the venous blood. This water reabsorption occurs all along the lengths of the collecting ducts so that the fluid at the end of the collecting ducts has essentially the same osmolarity as the interstitial fluid surrounding the bend in the

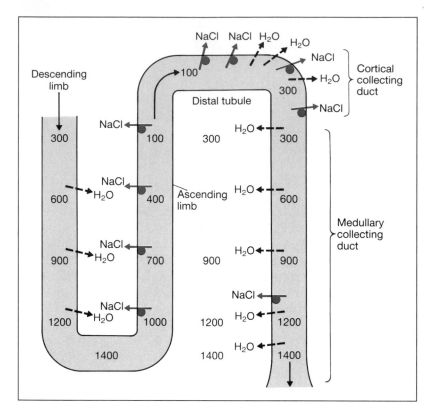

FIGURE 15-15 Operation of the renal countercurrent multiplier system in the formation of hyperosmotic urine.

loops, that is, at the bottom of the medulla. By this means, the final urine is hyperosmotic.[8] By retaining as much water as possible, the kidneys compensate for a bodily water deficit.

In contrast, when plasma ADH concentration is low, the late distal tubules and collecting ducts become relatively impermeable to water. Therefore, fluid in the late distal tubules does not reequilibrate with peritubular plasma, and, even more important, the high medullary interstitial osmolarity set up by the loop is ineffective in inducing water movement out of the collecting ducts. As a result, a large volume of hyposmotic urine is excreted, thereby compensating for a bodily water excess.

RENAL SODIUM REGULATION

In normal persons, urinary sodium excretion is reflexly increased when there is a sodium excess in the body and reflexly decreased when there is a sodium deficit.

[8] We have presented here only the absolutely essential components of the countercurrent multiplier system. It is actually much more complex and includes vascular components, nonhomogeneity of the ascending loop, and a contribution from urea as well as sodium and chloride. The interested reader can consult the suggested readings for this chapter.

These reflexes are so precise that total-body sodium normally varies by only a few percent despite a wide range of sodium intakes and the sporadic occurrence of large losses via the skin and gastrointestinal tract.

Since sodium is freely filterable at the glomerulus and actively reabsorbed but not secreted, the amount of sodium excreted in the urine represents the resultant of two processes:

Sodium excreted = sodium filtered
$$- \text{ sodium reabsorbed}$$

It is possible to reflexly adjust sodium excretion by changing either or both variables. Thus, when total-body sodium decreases for any reason, sodium excretion is reflexly decreased below normal levels by lowering the GFR and raising sodium reabsorption.

Most of these reflexes are initiated by various cardiovascular baroreceptors, such as the carotid sinus. As described in Chapter 13, baroreceptors respond to pressure changes within the cardiovascular system and initiate reflexes that regulate these pressures. The reason that regulation of cardiovascular pressures by baroreceptors simultaneously achieves regulation of total-body sodium is that these variables are closely correlated. As described on page 482, changes in total-body sodium result in similar changes in extracellular volume. Now, since extracellular volume comprises plasma volume and interstitial volume, plasma volume is also significantly affected by total-body sodium. We saw in Chapter 13 that plasma volume is, in turn, an important determinant of the pressures in the veins, cardiac chambers, and arteries. Thus, the chain linking total-body sodium to cardiovascular pressures is completed: Low total-body sodium causes low cardiovascular pressures, which, via baroreceptors, initiate reflexes that (1) restore the cardiovascular pressures via direct actions on the cardiovascular system (Chapter 13) and (2) simultaneously lower GFR and increase sodium reabsorption. These latter events decrease sodium excretion, thereby retaining sodium in the body. Increases in total-body sodium have the reverse reflex effects.

The reader may be surprised that no mention has been made of receptors that monitor plasma sodium concentration. Such receptors do exist but are unimportant compared to the baroreceptors in controlling sodium excretion.

Control of GFR

Figure 15-16 summarizes the major mechanisms by which a lower total-body sodium, as caused by diarrhea, for example, elicits a decrease in GFR. The key event—a reduced glomerular capillary pressure—occurs both as a direct consequence of a lowered arterial pressure in the kidneys and, more importantly, as a result of baroreceptor reflexes acting on the afferent arterioles. Note that the latter are simply the basic baroreceptor reflexes described in Chapter 13, where it was pointed out that a decrease in cardiovascular pressures causes reflex vasoconstriction in many areas of the body.

Conversely, an increased GFR is elicited when an increased total-body sodium causes increased plasma volume, and this increased GFR contributes to the increased renal sodium loss that returns extracellular volume to normal.[9]

Control of Sodium Reabsorption

So far as long-term regulation of sodium excretion is concerned, the control of sodium reabsorption is more important than the control of GFR. The major factor determining the rate of tubular sodium reabsorption is aldosterone.

Aldosterone and the renin-angiotensin system. The adrenal cortex produces a steroid hormone, **aldosterone**,

which stimulates sodium reabsorption by the late distal tubules and collecting ducts. When this hormone is absent, a person will excrete 15 g of sodium per day (equivalent to 35 g of sodium chloride), whereas excretion may be virtually zero when aldosterone is present in large quantities.[10] In a normal person, the amounts of aldosterone produced and sodium excreted lie somewhere between these extremes, varying with the amount of sodium ingested.

Aldosterone, like other steroids, acts by inducing the synthesis of proteins, in this case proteins involved in sodium transport. For example, it stimulates the production of proteins that function as sodium channels in the luminal membrane. By this same mechanism, aldosterone also stimulates sodium transport into the blood from the lumens of both the large intestine and the ducts carrying fluid from the sweat glands and salivary glands. In this manner, less sodium is lost in the feces and from the surface of the skin in sweat.

Aldosterone secretion is controlled, in part, by reflexes involving the kidneys. As noted earlier, the granular cells in the walls of afferent arterioles in the juxtaglomerular apparatus synthesize and secrete into the blood an enzyme known as **renin**. This enzyme then splits off a small polypeptide, **angiotensin I**, from a large plasma protein, **angiotensinogen** (Figure 15-17). Angiotensin I then undergoes further cleavage to form **angiotensin II**. This conversion is mediated by an enzyme known as **converting enzyme**, which is found in very high concentration on the luminal surface of capillary endothelial cells, particularly those in the lungs. Thus, most conversion of angiotensin I to angiotensin II occurs as blood flows through the lungs. Angiotensin II is a potent stimulator of aldosterone secretion and constitutes a major input to the adrenal cortex controlling the production and release of this hormone.

What determines the plasma concentration of angiotensin II? Angiotensinogen is synthesized by the liver and is always present in the blood, and converting enzyme is also always present in amounts adequate to convert all angiotensin I to angiotensin II. Therefore, the rate-limiting factor in angiotensin II formation is the concentration of plasma renin, which, in turn, depends upon the rate of renin secretion by the kidneys. The critical question now becomes: What controls renin secretion?

There are at least three distinct inputs to the renin-secreting granular cells: (1) the renal sympathetic

---

[9]In addition to being acted upon by these reflex controls of GFR, the kidney has the ability to self-regulate its own GFR (and blood flow). These mechanisms are described in the suggested readings for this chapter.

[10]Aldosterone controls the reabsorption of only about 2 percent of the filtered load of sodium and yet is the single most important input. This is because reabsorption of most of the other 98 percent occurs automatically, that is, without reflex control.

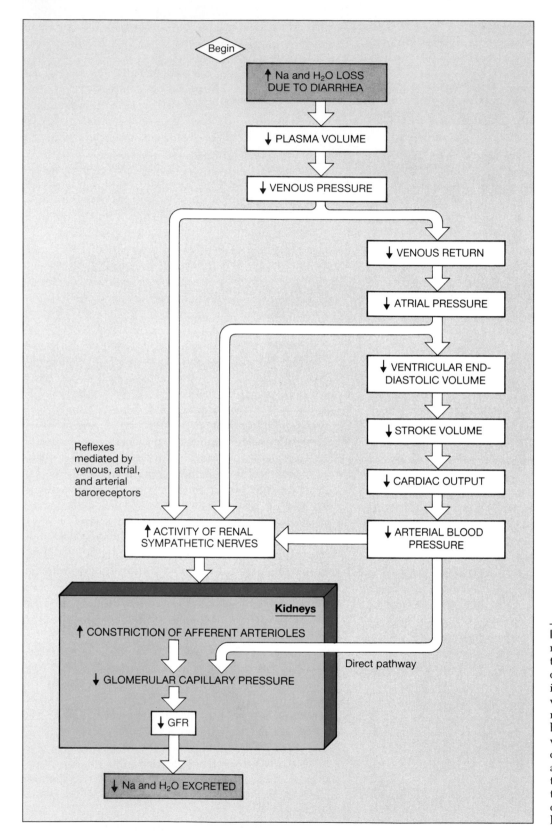

FIGURE 15-16 Direct and reflex pathways by which the GFR and hence sodium and water excretion is decreased when plasma volume decreases. The reflex pathway, initiated by baroreceptors in large veins, the atria, and the carotid sinuses and aortic arch, is much more effective in lowering GFR than is the direct effect on the kidneys of the lowered arterial pressure.

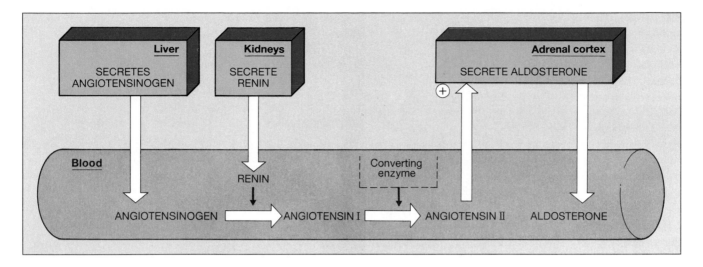

FIGURE 15-17 Summary of the renin-angiotensin-aldosterone system. Converting enzyme is located on the surface of capillary endothelial cells, particularly in the lungs. Renin is the rate-limiting factor in the system.

nerves, (2) the intrarenal baroreceptor, and (3) the macula densa. It is not yet possible to assign quantitative roles to each of them. That the renal sympathetic nerves constitute one important input makes excellent sense, teleologically, since a reduction in body sodium and plasma volume lowers blood pressure and, via baroreceptors external to the kidneys, triggers increased sympathetic discharge to the granular cells, stimulating renin release (Figure 15-18). This sets off the hormonal chain of events that increases sodium reabsorption and restores sodium balance and plasma volume toward normal.

The other two inputs for controlling renin release—the intrarenal baroreceptor and the macula densa—are totally contained within the kidneys and require no external neuroendocrine input. The renin-secreting granular cells are themselves pressure-sensitive, and so they function as **intrarenal baroreceptors**: When renal arterial pressure decreases, as would occur when extracellular volume is down, these cells secrete more renin (Figure 15-18). Thus, normally, the granular cells respond simultaneously to the combined effects of sympathetic input, triggered by baroreceptors external to the kidneys, and to their own pressure sensitivity.

The second completely internal input to the granular cells is via the macular densa, which senses sodium or chloride concentration in the tubular fluid, a decreased salt concentration causing increased renin secretion. For several reasons, macula densa sodium and chloride concentrations tend to decrease when a person's arterial

pressure is decreased, and so this input also signals for increased renin release at the same time that the sympathetic nerves and intrarenal baroreceptor are doing so.

By helping to regulate sodium balance and, thereby, plasma volume, the renin-angiotensin system contributes to the control of blood pressure. However, this is not the only way in which it influences arterial pressure. Recall from Chapter 13 (page 388) that angiotensin II is a potent constrictor of arterioles and that this effect on peripheral resistance increases arterial pressure. Abnormal increases in the activity of the renin-angiotensin system (as occurs, for example, when a renal artery becomes partially occluded) can contribute to the development of hypertension (high blood pressure) via both the sodium-retaining and arteriole-constricting effects. It should not be surprising, therefore, that certain persons suffering from hypertension are treated with drugs that block the renin-angiotensin system. One such group of drugs blocks the action of converting enzyme, thus preventing the generation of angiotensin II.

Atrial natriuretic factor. Although aldosterone is the most important controller of sodium reabsorption, a variety of other factors also play roles. One of these is the peptide hormone known as **atrial natriuretic factor (ANF)**, which is synthesized and secreted by cells in the cardiac atria. ANF acts on the kidneys to inhibit sodium reabsorption (Figure 15-19). It also inhibits the secretion of both renin and aldosterone, which also results in less sodium reabsorption. As would be predicted, the secre-

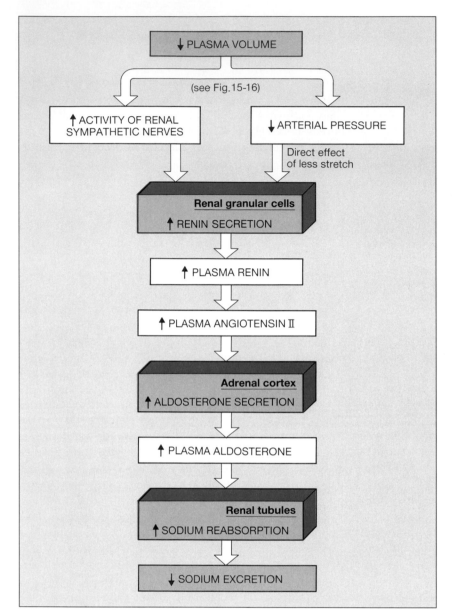

FIGURE 15-18 Pathway by which decreased plasma volume leads to increased sodium reabsorption and hence decreased sodium excretion. The renal sympathetic nerves stimulate the granular cells, which also function as intrarenal baroreceptors responding directly to the decreased stretch produced by the lowered arterial pressure. The third signal, not shown in the figure, for stimulating renin secretion is via the macula densa.

tion of ANF is increased when there is an excess of sodium in the body, the stimulus being an increase in atrial distention.

This completes our survey of the control of sodium excretion, which depends upon the control of two renal variables—the GFR and sodium reabsorption. The latter is controlled by the renin-angiotensin-aldosterone hormone system and by other factors, including atrial natriuretic factor. The reflexes that control both GFR and sodium reabsorption are essentially blood pressure regulating reflexes since they are most frequently initiated by changes in arterial or venous pressures.

RENAL WATER REGULATION

Water excretion, like sodium excretion, is the difference between the volume of water filtered (the GFR) and the volume reabsorbed. Accordingly, the baroreceptor-initiated GFR-controlling reflexes described in the previous section tend to have the same effects on water excretion as on sodium excretion. As in the case of sodium, however, the major regulated determinant of water excretion is not GFR but, instead, the rate of water reabsorption. As we have seen, this is determined by ADH, and so total-body water is regu-

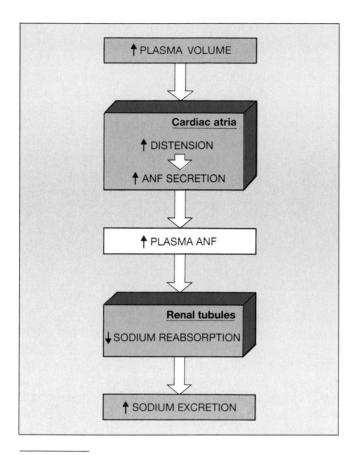

FIGURE 15-19 Atrial natriuretic factor in the direct control of sodium reabsorption and hence sodium excretion. In addition, ANF inhibits the secretion of renin and of aldosterone and this also leads to decreased sodium reabsorption.

lated mainly by reflexes that alter the secretion of this hormone.

As described in Chapter 10, ADH is produced by a discrete group of hypothalamic neurons whose axons terminate in the posterior pituitary, from which ADH is released into the blood. The most important of the inputs to these neurons are from baroreceptors and osmoreceptors.

Baroreceptor Control of ADH Secretion

We have seen that a decreased extracellular volume, due, say, to diarrhea or hemorrhage, reflexly calls forth, via the renin-angiotensin system, an increased aldosterone secretion. The new fact is that decreased extracellular volume also triggers increased ADH secretion. This increased ADH increases the water permeability of the late distal tubules and collecting ducts, more water is reabsorbed and less is excreted, and so water is retained in the body to help stabilize the extracellular volume.

This reflex is mediated by neural input to the ADH-

secreting cells from several baroreceptors, particularly a group located in the left atrium (Figure 15-20). The baroreceptors decrease their rate of firing with the decreased atrial blood pressure that occurs when blood volume decreases. When fewer impulses are transmitted from the baroreceptors via afferent neurons and ascending pathways to the hypothalamus, the result is increased ADH secretion. Conversely, increased atrial pressure causes more firing by the baroreceptors, resulting in a decrease in ADH secretion.

Osmoreceptor Control of ADH Secretion

We have seen how changes in extracellular volume simultaneously elicit reflex changes in the excretion of *both* sodium *and* water. This is adaptive since the situations causing extracellular volume alterations are very often associated with loss or gain of both sodium and water in approximately proportional amounts. In contrast, changes in total-body water in which no change in total-body sodium occurs are compensated for by altering water excretion without altering sodium excretion.

The major change caused by water loss or gain out of proportion to sodium loss or gain is a change in osmolarity of the body fluids. The reason this is a key point is that, under conditions of predominantly water gain or loss, the receptors that initiate the reflexes controlling

FIGURE 15-20 Baroreceptor pathway by which ADH secretion is decreased when plasma volume is increased. The greater plasma volume raises left atrial pressure, which stimulates the atrial baroreceptors and inhibits ADH secretion.

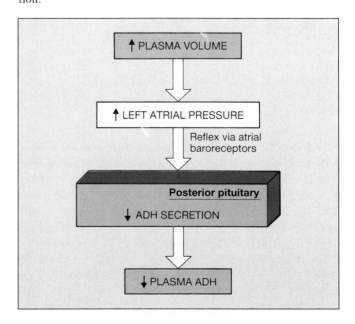

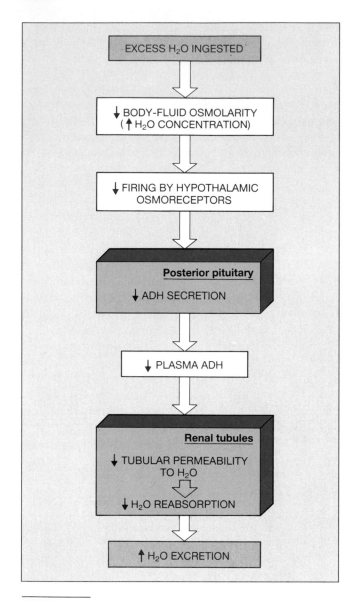

FIGURE 15-21 Osmoreceptor pathway by which ADH secretion is lowered and water excretion raised when excess water is ingested.

ADH secretion are **osmoreceptors**, receptors responsive to changes in osmolarity (how they work is unknown). These osmoreceptors are located in the hypothalamus.

Take, as an example, a person drinking 2 L of beer, which contains little sodium or any other solute, in a short time. The excess water lowers the body-fluid osmolarity, which reflexly inhibits ADH secretion via the hypothalamic osmoreceptors (Figure 15-21). As a result, water permeability of the late distal tubules and collecting ducts becomes very low, water is not reabsorbed from these segments, and a large volume of hyposmotic

urine is excreted. In this manner, the excess water is eliminated.

At the other end of the spectrum, when the osmolarity of the body fluids is increased, say, because of water deprivation, ADH secretion is reflexly increased via the osmoreceptors, water reabsorption by the late distal tubules and collecting ducts is increased, and a very small volume of highly concentrated urine is excreted. In this manner, the kidneys retain water in excess of solute and reduce the body-fluid osmolarity toward normal.

We have now described two afferent pathways controlling the ADH-secreting hypothalamic cells, one from baroreceptors and one from osmoreceptors. To add to the complexity, these cells receive synaptic input from many other brain areas, so that ADH secretion, and therefore urine volume and concentration, can be altered by pain, fear, and a variety of other factors. ADH secretion can also be altered by drugs. For example, alcohol is a powerful inhibitor of ADH release, and this probably accounts for much of the increased urine volume produced following the ingestion of alcohol, a urine volume well in excess of the volume of beverage consumed. All these effects are usually short-lived, however, and should not obscure the generalization that ADH secretion is determined primarily by the states of extracellular volume and osmolarity.

A SUMMARY EXAMPLE: THE RESPONSE TO SWEATING

Figure 15-22 shows the factors that control renal sodium and water excretion in response to severe sweating. Sweat is a hyposmotic solution containing mainly water, sodium, and chloride. Therefore, sweating causes both a decrease in extracellular volume and an increase in body-fluid osmolarity. The renal retention of water and sodium helps to compensate for the water and salt lost in the sweat.

THIRST AND SALT APPETITE

Now we turn to the other component of any balance, control of intake. Deficits of salt and water must eventually be compensated by ingestion of these substances.

The subjective feeling of thirst, which drives one to obtain and ingest water, is stimulated both by a lower extracellular volume and a higher plasma osmolarity (Figure 15-23). Note that these are precisely the same two changes that stimulate ADH production, and the osmoreceptors and atrial baroreceptors that control

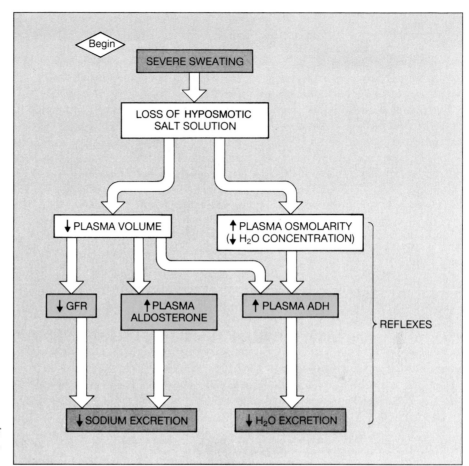

FIGURE 15-22 Pathways by which sodium and water excretion are decreased in response to severe sweating. This figure is an amalgamation of Figures 15-16, 15-18, and the reverse of 15-20 and 15-21.

ADH secretion are identical to those for thirst. The brain centers that receive input from these receptors and mediate thirst are located in the hypothalamus, very close to those areas that produce ADH.

Another influencing factor is angiotensin II, which stimulates thirst by a direct effect on the brain. Thus, the renin-angiotensin system helps regulate not only sodium balance but water balance as well and constitutes one of

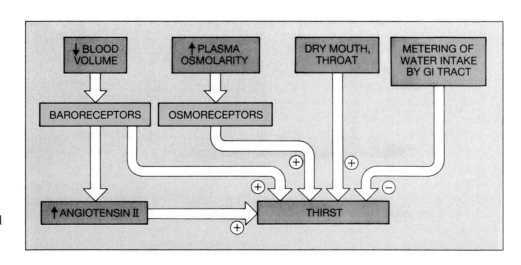

FIGURE 15-23 Inputs reflexly controlling thirst. Psychosocial factors and conditioned responses are not shown.

the pathways by which thirst is stimulated when extracellular volume is decreased.

There are still other pathways controlling thirst. For example, dryness of the mouth and throat causes profound thirst, which is relieved by merely moistening them. Animals such as the camel—and people, to a lesser extent—that have been markedly dehydrated will rapidly drink just enough water to replace their previous losses and then stop. What is amazing is that when they stop, the water has not yet had time to be absorbed from the gastrointestinal tract into the blood. Some kind of "metering" of the water intake by the gastrointestinal tract has occurred, but its nature remains a mystery.

Although the analog of thirst for sodium, **salt appetite,** is an important component of sodium homeostasis in most mammals, its contribution to everyday sodium homeostasis in normal human beings is probably slight. On the other hand, people certainly like salt, as manifested by almost universally large intakes of sodium whenever it is cheap and readily available. Thus, the average American intake of salt is 10 to 15 g/day, despite the fact that human beings can survive quite normally on less than 0.5 g/day. Present evidence suggests that a large salt intake may be an important contributor to the pathogenesis of hypertension in certain persons.

SECTION C
REGULATION OF POTASSIUM, CALCIUM, AND HYDROGEN ION

The total-body balances and extracellular concentrations of these three important ions are closely regulated by the kidneys, acting in concert with other mechanisms.

sium losses via the sweat and gastrointestinal tract are normally quite small, although large quantities can be lost during vomiting or diarrhea. The control of renal function is the major mechanism by which body potassium is regulated.

POTASSIUM REGULATION

Potassium is, as we have seen, the most abundant intracellular ion. Although only 2 percent of the total-body potassium is in the extracellular fluid, the potassium concentration in this fluid is extremely important for the function of excitable tissues, notably nerve and muscle. Recall (Chapter 8) that the resting-membrane potentials of these tissues are directly related to the relative intracellular and extracellular potassium concentrations. An increase in extracellular potassium concentration depolarizes plasma membranes, triggering action potentials and often rendering the membrane inexcitable afterward. In the heart this may cause serious arrhythmias. Conversely, lowering the external potassium concentration hyperpolarizes plasma membranes. Therefore, early manifestations of potassium depletion are weakness of skeletal muscles, because of decreased firing of action potentials, and abnormalities of cardiac muscle conduction and rhythm.

Total-body potassium balance is analogous to that of sodium in that the normal person remains in balance by daily excreting an amount of potassium in the urine equal to the amount ingested minus the amounts eliminated in the feces and sweat. Also like sodium, potas-

Renal Regulation of Potassium

Potassium is freely filterable at the glomerulus. The amounts excreted in the urine are generally a small fraction of the quantity filtered, thus establishing the existence of tubular potassium reabsorption . Under certain conditions, however, the excreted quantity may actually exceed that filtered, thus establishing the existence of tubular potassium secretion. Thus, potassium can undergo net reabsorption or net secretion.

Normally, most of the filtered potassium is reabsorbed regardless of changes in body potassium balance. In other words, the *reabsorption* of potassium does not seem to be controlled so as to achieve potassium homeostasis. Therefore, changes in potassium *excretion* are due mainly to changes in potassium *secretion*, which occurs in the late distal tubules and the collecting ducts. Thus, during potassium depletion, when the homeostatic response is to minimize potassium loss, there is no significant potassium secretion in these portions of the nephron, and only the small amount of filtered potassium that escaped reabsorption is excreted. In all other situations, to this same small amount of unreabsorbed potassium is added a variable amount of secreted potassium necessary to maintain balance (Figure 15-24). Therefore, in describing the homeostatic control of potassium excretion, we may focus on only the factors that alter the rate of

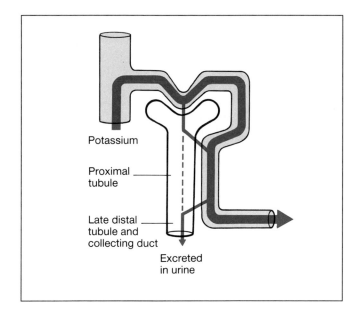

FIGURE 15-24 Simplified model of the basic renal handling of potassium. Net reabsorption always occurs in the proximal portions of the nephron, but the distal portions of the nephron can show either net reabsorption or net secretion. Usually, net distal secretion occurs, as in this figure. In a person who is potassium-depleted, however, net distal reabsorption occurs.

tubular potassium secretion in the distal portions of the nephron.

The initial step in potassium secretion is the active potassium transport, by Na, K-ATPase pumps, across the basolateral membrane into the tubular epithelial cells. This is the same process that, as illustrated in Figure 15-11, transports sodium out of the cell into the interstitial fluid. The fate of the potassium that enters the cell was not shown in that figure since we were concentrating on sodium at that time. In the late distal tubules and collecting ducts, there are potassium channels in the luminal membrane and therefore much of this potassium diffuses out of the cell down its concentration gradient into the lumen to complete the secretory process. (In contrast, there are no such luminal channels in the proximal tubules but there are in the basolateral membrane, and so the potassium pumped into the cell simply diffuses right back out across the basolateral membrane.)

One of the most important factors altering potassium secretion is aldosterone, which in addition to stimulating sodium reabsorption, increases potassium secretion. The reflex by which an excess or deficit of potassium controls aldosterone production is completely different from the reflex initiated by changes in extracellular volume. The latter constitutes the complex pathway discussed earlier

involving renin and angiotensin. The former is much simpler (Figure 15-25): The aldosterone-secreting cells of the adrenal cortex are sensitive to the potassium concentration of the extracellular fluid bathing them. Thus, an increased intake of potassium leads to an increased extracellular potassium concentration, which in turn directly stimulates aldosterone production by the adrenal cortex. The resulting increased plasma aldosterone concentration increases potassium secretion and thereby eliminates the excess potassium from the body.

Conversely, a lowered extracellular potassium concentration decreases aldosterone production and thereby reduces potassium secretion. Less potassium than usual is excreted in the urine, thus helping to restore the normal extracellular concentration.

The control and renal tubular effects of aldosterone are summarized in Figure 15-26. The fact that a single hormone—aldosterone—regulates both sodium and potassium excretion raises the question of potential conflicts between homeostasis of the two ions. For example, if a person were sodium-deficient and therefore secreting large amounts of aldosterone, the potassium-secret-

FIGURE 15-25 Pathway by which an increased potassium intake induces greater potassium excretion mediated by aldosterone.

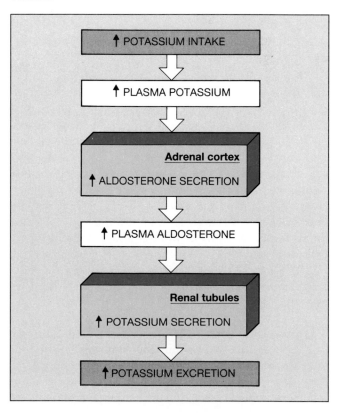

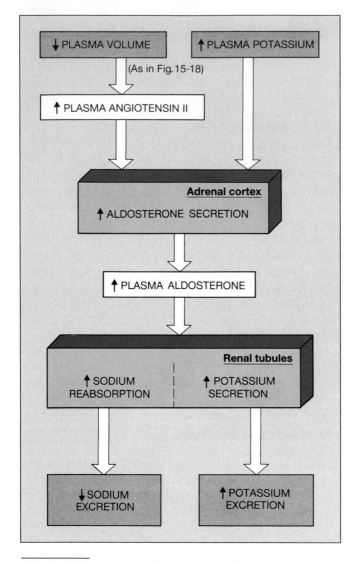

FIGURE 15-26 Summary of the control of aldosterone and its effects on sodium reabsorption and potassium secretion.

ing effects of this hormone would tend to cause some potassium loss even though potassium balance were normal to start with. Normally, such conflicts cause only minor imbalances because there are a variety of other counteracting controls of sodium and potassium excretion.

CALCIUM REGULATION

Extracellular calcium concentration is also normally held relatively constant, the requirement for precise regulation stemming primarily from the effects of calcium on neuromuscular excitability. A low calcium concentration increases the excitability of nerve and

muscle cell membranes, so that persons with low plasma calcium suffer from hypocalcemic tetany, characterized by skeletal muscle spasms. Hypercalcemia is life threatening, too, in that it causes cardiac arrhythmias as well as depressed neuromuscular excitability. These effects reflect calcium's ability to bind to plasma-membrane proteins that function as ion channels. The binding alters the open or closed state of the channels. This effect of calcium on membranes is totally distinct from its role as an excitation-contraction coupler.

Effector Sites for Calcium Homeostasis

Earlier sections on sodium, potassium, and water homeostasis were concerned almost entirely with the *renal* handling of these substances. It was possible to do so for several reasons: (1) Although internal exchanges between extracellular fluid and such storage sites as bone and intracellular fluid are of some importance for these three substances, the major homeostatic controls act via the kidneys; and (2) intestinal absorption of these substances, that is, movement of ingested substance from the intestinal lumen into the blood, approximates 100 percent under normal circumstances and is not a major controlled variable. Neither of these statements holds true for calcium homeostasis, and so this section must deal not only with the renal handling of calcium but with the other two major effector sites for calcium homeostasis—bone and the gastrointestinal tract.

Bone. Approximately 99 percent of total-body calcium is contained in bone, which is primarily a framework of organic molecules, produced by bone cells, upon which calcium phosphate crystals are deposited. Contrary to popular belief, bone is not a fixed, unchanging tissue but is constantly being remodeled and, what is more important, can either withdraw calcium from extracellular fluid or deposit it there.

Gastrointestinal tract. Under normal conditions a considerable amount of ingested calcium is not absorbed from the intestine and simply leaves the body along with the feces. Accordingly, control of the active-transport system that moves this ion from intestinal lumen to blood can result in large increases or decreases in the amount of calcium absorbed.

Kidneys. The kidneys handle calcium by filtration and reabsorption. In addition, as we shall see, the renal handling of phosphate also plays a role in the regulation of extracellular calcium.

Parathyroid Hormone

All three of the effector sites described above are subject, directly or indirectly, to control by a protein hor-

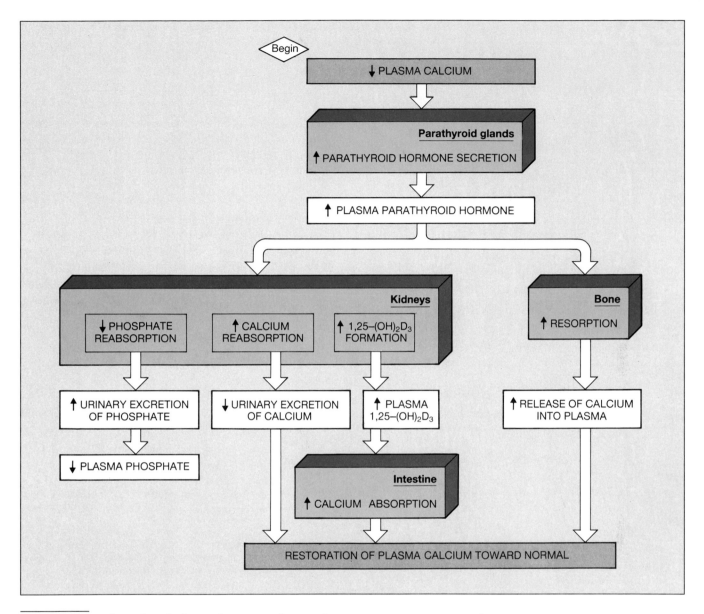

FIGURE 15-27 Reflexes by which a reduction in plasma calcium concentration is restored toward normal via the actions of parathyroid hormone. [See Figure 15-28 for a description of 1,25-$(OH)_2D_3$.]

mone called **parathyroid hormone**, produced by the parathyroid glands. These glands are embedded in the surface of the thyroid gland but are distinct from it. Parathyroid hormone production is controlled by the plasma calcium concentration. Decreased plasma calcium concentration stimulates parathyroid hormone secretion, and an increased plasma calcium concentration does just the opposite. Extracellular calcium concentration acts directly upon the parathyroid glands without any intermediary hormones or nerves.

Parathyroid hormone exerts at least four distinct ef-

fects influencing calcium homeostasis (Figure 15-27):

1. It increases the movement of calcium from bone into extracellular fluid, making available this large store of calcium for the regulation of extracellular calcium concentration.
2. It stimulates the activation of vitamin D (see below), and this latter hormone then increases intestinal absorption of calcium.
3. It increases renal tubular calcium reabsorption, thus decreasing urinary calcium excretion.

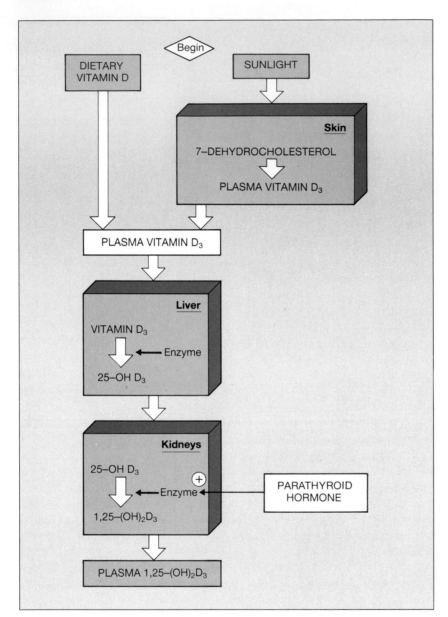

FIGURE 15-28 Metabolism of vitamin D to the active form: $1,25\text{-}(OH)_2D_3$. The kidney enzyme that mediates the final step is activated by parathyroid hormone. The major action of $1,25\text{-}(OH)_2D_3$ is to stimulate the absorption of calcium from the gastrointestinal tract.

4. It reduces the tubular reabsortion of phosphate, thus raising its urinary excretion and lowering its extracellular concentration.

The adaptive value of the first three is straightforward: They all tend to increase extracellular calcium concentration, thus compensating for the decreased concentration that originally stimulated parathyroid hormone secretion. The adaptive value of the fourth effect for calcium regulation requires further explanation as follows.

Parathyroid hormone causes both calcium and phosphate to be released from bone into the extracellular fluid. If the extracellular phosphate concentration were to increase as a result of this release, further movement of calcium from bone would be retarded because a high extracellular phosphate causes calcium and phosphate to be deposited on bone and other tissues. Therefore it is adaptive that parathyroid hormone also decreases tubular reabsorption of phosphate, thus permitting the excess phosphate to be eliminated in the urine.

Vitamin D

The term **vitamin D** denotes a group of closely related chemicals that regulate absorption of ingested calcium by the intestine. **Vitamin D$_3$** is formed by the action of ultraviolet radiation (from sunlight, usually) on a substance—7-dehydrocholesterol—found in skin. A second form of vitamin D is that ingested in food. Because of clothing and decreased out-of-doors living, people are often dependent upon this dietary source of vitamin D, and for this reason, it was classified as a vitamin. The form of vitamin D found in food differs only trivially in structure from vitamin D_3, and no distinction will be made between them in the following description.

Vitamin D_3 is metabolized by addition of hydroxyl groups, first in the liver, then in the kidneys (Figure 15-28). The end result of these changes is **1,25-dihydroxy-vitamin D_3** [abbreviated **1,25-$(OH)_2D_3$**], the active form of vitamin D. It should be clear from this description that 1,25-$(OH)_2D_3$ is really a hormone, not a vitamin, since it is made in the body.

The major action of 1,25-$(OH)_2D_3$ is to stimulate the intestine to actively absorb the calcium ingested in food. Thus, the major event in vitamin D deficiency is decreased intestinal calcium absorption, resulting in decreased plasma calcium. When this occurs in children, the newly formed bone protein matrix fails to be calcified normally because of the low plasma calcium, leading to the disease **rickets.** The soft and easily fractured bones become distorted, and the child with rickets typically is severely bowlegged due to the effect of weight bearing on the legs.

The blood concentration of 1,25-$(OH)_2D_3$ is subject to physiological control. The major control point is the addition of the second hydroxyl group, which occurs in the kidneys. The enzyme catalyzing this step is stimulated by parathyroid hormone. Thus, as we have seen, a low plasma calcium concentration stimulates the secretion of parathyroid hormone, which in turn, enhances the production of 1,25-$(OH)_2D_3$, and both hormones contribute to the restoration of the plasma calcium toward normal.

HYDROGEN-ION REGULATION

Most metabolic reactions are highly sensitive to the hydrogen-ion concentration of the fluid in which they occur. This sensitivity is due to the influence on enzyme function exerted by the hydrogen ion, which changes the shapes of these molecules. Accordingly, the hydrogen-ion concentration of the extracellular fluid is closely regulated. At this point the reader should review the section on hydrogen ions and acidity in Chapter 2.

Sources of Hydrogen-Ion Gain or Loss

The regulation of total-body hydrogen-ion balance can be viewed in the same way as the balance of any other ion, that is, as the matching of gains and losses. Table

TABLE 15-5   SOURCES OF HYDROGEN-ION GAIN OR LOSS
Gain
1. Generation of hydrogen ions from CO_2
2. Production of phosphoric and sulfuric acids from the metabolism of protein and other organic molecules
3. Gain of hydrogen ions due to loss of bicarbonate in diarrhea
4. Gain of hydrogen ions due to loss of bicarbonate in the urine
Loss
1. Loss of hydrogen ions in vomitus
2. Loss of hydrogen ions in the urine

15-5 summarizes the major routes for gains and losses of hydrogen ion. Normally, the major route for gain is the metabolic generation of hydrogen ions from the following sources.

First, as described in Chapter 14, a huge quantity of CO_2 is generated daily as the result of oxidative metabolism and yields hydrogen ions via the reactions:

$$CO_2 + H_2O \xrightarrow{\text{Carbonic anhydrase}} H_2CO_3 \rightarrow HCO_3^- + H^+ \quad (15\text{-}1)$$

This source does not normally constitute a net gain of hydrogen ions, however, since all the hydrogen ions generated via these reactions during passage of blood through the tissues are reincorporated into water when the reactions are reversed during passage of blood through the lungs (Chapter 14). Net retention of CO_2, however, as in hypoventilation, will result in a net gain of hydrogen ions. Conversely, net loss of CO_2, as in hyperventilation, causes net elimination of hydrogen ions.

The body also produces acids, both organic and inorganic, from sources other than CO_2. These include phosphoric acid and sulfuric acid, generated during the catabolism of proteins and other organic molecules containing sulfur and phosphorus, as well as lactic acid and other organic acids. In the United States, where the diet is high in protein, people normally have a net daily production of 40 to 80 mmol of these acids.

A third potential source of net bodily gain or loss of hydrogen ion is gastrointestinal secretions leaving the body. Vomitus contains a high concentration of hydrogen ions and so constitutes a source of net loss. In contrast, the other gastrointestinal secretions are alkaline, that is, contain very little hydrogen ion but a concentration of bicarbonate higher than exists in plasma. Loss of these fluids, as in diarrhea, constitutes, in essence, a bodily gain of hydrogen ions. This is a very important point: Given the mass action relationship shown in Equation 15-1, the loss of a bicarbonate ion from the body has virtually the same net result as gaining a hydrogen ion because loss of the bicarbonate causes the reaction to be driven to the right, thereby generating a hydrogen ion. Similarly, the *gain* of a bicarbonate by the body has virtually the same net result as *losing* a hydrogen ion.

Finally, the urine constitutes the fourth source of net hydrogen ion gain or loss. As is the case for the other inorganic ions described in this chapter, the renal excretion of hydrogen ion is regulated so as to achieve a stable balance and hence maintain a relatively stable hydrogen-ion concentration in the body fluids. Thus, the kidneys normally excrete the 40 to 80 mmol of hydrogen ion generated by the average American diet. The kidneys also adjust their excretion of hydrogen ion to compensate for any net retention or elimination of CO_2, for any increase in the metabolic production of hydrogen ions, and for any increased loss of hydrogen ion or bicarbonate via the gastrointestinal tract. Before turning to these renal mechanisms, we describe the buffering of hydrogen ions in the body.

Buffering of Hydrogen Ions in the Body

The idea that hydrogen-ion regulation involves the same kind of input-output balancing as does that of sodium and other ions is easily obscured by the phenomenon of **buffering**, the reversible binding of hydrogen ions by anions when the hydrogen-ion concentration changes. The general form of buffering reactions is:

$$\text{Buffer}^- + \text{H}^+ \rightleftharpoons \text{HBuffer} \qquad (15\text{-}2)$$

HBuffer is a weak acid in that it can exist as the undissociated molecule (HBuffer) or can dissociate to Buffer$^-$ plus H^+. When H^+ concentration increases for whatever reason, the reaction is forced to the right, and more H^+ is bound by Buffer$^-$ to form HBuffer. For example, when H^+ concentration is increased because of increased production of lactic acid, some of the hydrogen ions combine with the body's buffers, and the hydrogen-ion concentration is lowered below what it otherwise would have been. Conversely, when H^+ concentration decreases, Equation 15-2 proceeds to the left, and H^+ is released from HBuffer. In this manner, the body buffers stabilize H^+ concentration against changes in either direction.

Between their generation in the body and their elimination, most hydrogen ions are buffered by extracellular and intracellular buffers. The normal extracellular-fluid pH of 7.4 corresponds to a hydrogen-ion concentration of only 0.00004 mmol/L. Without buffering, the daily turnover rate of the 40 to 80 mmol of H^+ produced by our diet would cause huge changes in body-fluid hydrogen-ion concentration.

The only important *extracellular* buffer is the CO_2/ HCO_3^- system summarized in Equation 15-1. The major *intracellular* buffers are phosphates and proteins. An example is hemoglobin, as described in Chapter 14, page 454.

Renal Regulation of Extracellular Hydrogen-Ion Concentration

Buffering minimizes changes in hydrogen-ion concentration but does not actually eliminate hydrogen ions from the body or retain them. This is the function of the kidneys. They regulate body-fluid hydrogen-ion concentration in two major ways: (1) by altering their excretion of hydrogen ion, and (2) by altering their reabsorption and hence excretion of the bicarbonate filtered at the glomeruli. To understand why bicarbonate reabsorption is so important, recall the point made earlier that the loss of a bicarbonate ion from the body (in this case via the urine) has the same effect on body-fluid hydrogen-ion concentration as the gain of a hydrogen ion. Accordingly, if the kidneys did not reabsorb the large amount of filtered bicarbonate, the body would become very acidic, just as though a large amount of hydrogen ion had been added to it.

Both hydrogen-ion excretion and bicarbonate reabsorption are controlled homeostatically so as to compensate for changes in body-fluid hydrogen-ion concentration. Thus, when the blood is more acid than normal, say, because of increased hydrogen-ion production in the body, the kidneys excrete the additional hydrogen ions and simultaneously reabsorb all filtered bicarbonate. In contrast, if the blood is less acid than normal, say, because of persistent vomiting, the kidneys excrete little, if any, hydrogen ion and simultaneously permit some filtered bicarbonate to go unabsorbed, thereby raising the blood hydrogen-ion concentration back toward normal.

Hydrogen-ion excretion and bicarbonate reabsorption both are achieved by a single mechanism—tubular secretion of hydrogen ions. Let us look first at how such secretion achieves hydrogen-ion excretion. Virtually all the hydrogen ions excreted in the urine enter the tubule via secretion since the hydrogen-ion concentration in plasma is so small that only negligible amounts are filtered. The secretory process, which occurs in most nephron segments and which is illustrated in Figure 15-29, may seem odd at first because the hydrogen ions to be secreted into the tubule are generated *inside* the tubular cells rather than coming directly from the blood. In the cells, carbon dioxide and water combine in reac-

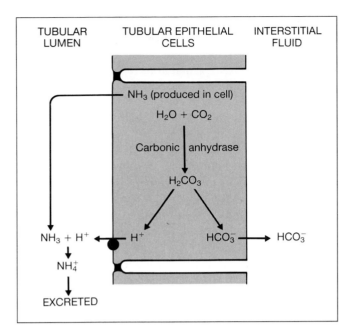

FIGURE 15-29 Tubular secretion of hydrogen ion. Once in the lumen, the hydrogen ion can combine with ammonia (NH_3) synthesized by the tubule and can be excreted. Alternatively, it can combine with filtered phosphate (not shown in the figure) or filtered bicarbonate. Note that for each hydrogen ion formed in the cell and secreted in the lumen, a bicarbonate ion is also formed in the cell and moves into the blood.

creased extracellular hydrogen-ion concentration lasting for more than 1 to 2 days stimulates the rate of tubular-cell NH_3 production. This extra NH_3 provides the additional luminal buffer required for combination with and excretion of the larger number of secreted hydrogen ions.

This completes our description of how hydrogen-ion secretion results in hydrogen-ion excretion. Now we turn to how this same secretory process achieves bicarbonate reabsorption. First, look again at Figure 15-29 and recall that a new bicarbonate appears in the blood every time a hydrogen ion is secreted. Now look at Figure 15-30. The difference between this figure and the previous one is that the secreted hydrogen ion, upon entering the lumen, combines with a filtered bicarbonate rather than ammonia (or some other buffer) and the bicarbonate is converted to carbon dioxide. The overall result is that an "old" filtered bicarbonate disappears

tions identical to Equation 15-1 and catalyzed by intracellular carbonic anhydrase. The resulting hydrogen ion is actively transported across the luminal membrane into the tubular lumen while the resulting bicarbonate is transported across the basolateral membrane and enters the blood. The net result is the same as if the secreted hydrogen ion had come from the blood rather than being generated in the tubular cell.

Once in the lumen, the hydrogen ion to be excreted combines with one of several buffers, and it is in this form that it is excreted. The major urinary buffers are HPO_4^{2-} and ammonia (NH_3):

$$HPO_4^{2-}+H^+ \rightarrow H_2PO_4^-$$
$$NH_3 + H^+ \rightarrow NH_4^+$$

The HPO_4^{2-} reaches the tubular fluid in the glomerular filtrate. The NH_3 is formed in the tubular cells by the deamination of certain amino acids supplied by the blood. From the cells the NH_3 diffuses into the lumen, where it combines with H^+ to form NH_4^+ (Figure 15-29).

An important feature of this system is that an in-

FIGURE 15-30 Reabsorption of bicarbonate. This is achieved by hydrogen-ion secretion identical to that shown in Figure 15-29. The hydrogen ion in the lumen combines with a filtered bicarbonate ion and generates carbon dioxide and water. Thus the filtered bicarbonate disappears at the same time that the bicarbonate formed in the cell enters the blood. Therefore, even though the two bicarbonate ions are different in origin, there has been, in effect, a reclaiming (reabsorption) of the filtered bicarbonate ion. In the proximal tubule, but not in later nephron segments, the secreted hydrogen ions are countertransported across the luminal membrane (Figure 15-12) with sodium. Therefore, a sodium ion is reabsorbed along with a bicarbonate ion.

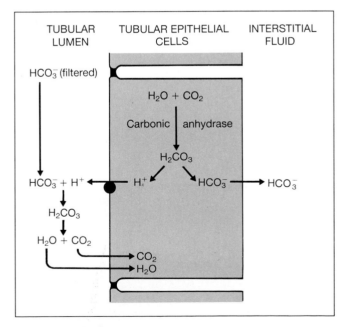

while a "new" bicarbonate—the one resulting from the intracellular events—appears in the blood. The overall result is the same as if the filtered bicarbonate had been reabsorbed.

In actuality, the events of Figures 15-29 and 15-30 occur simultaneously so that a rate of hydrogen-ion secretion adequate to achieve a high rate of hydrogen-ion excretion also usually achieves complete reabsorption of filtered bicarbonate. As noted above, this is the renal compensation for an elevated blood hydrogen-ion concentration. Similarly, low rates of hydrogen-ion excretion and incomplete bicarbonate reabsorption occur together, and this is the renal compensation for decreased blood acidity. The factors that regulate hydrogen-ion secretion in these situations so as to achieve homeostasis are beyond the scope of this book.[11]

Classification of Disordered Hydrogen-Ion Concentration

Acidosis refers to any situation in which the hydrogen-ion concentration of arterial blood is elevated; **alkalosis** denotes a reduction. It should be evident that acidosis and alkalosis are the results of an imbalance between hydrogen-ion gain and loss, and all such situations fit into two distinct categories (Table 15-6): (1) **respiratory acidosis** or **alkalosis**; (2) **metabolic acidosis** or **alkalosis**.

As its name implies, the respiratory category results either from failure of the lungs to eliminate carbon dioxide as fast as it is produced (respiratory acidosis) or from elimination of carbon dioxide faster than it is produced (respiratory alkalosis). As described earlier, the imbalance of arterial hydrogen-ion concentrations in such cases is completely explainable in terms of mass action. Thus, the hallmark of a respiratory acidosis is elevated arterial P_{CO_2} and hydrogen ion: that of respiratory alkalosis is a reduction in both.

Metabolic acidosis or alkalosis includes all situations other than those in which the primary problem is respiratory. Some common causes of metabolic acidosis are excessive production of lactic acid during severe exercise or hypoxia, or of ketone bodies in uncontrolled diabetes mellitus or fasting (both described in Chapter 17). Metabolic acidosis can also result from excessive loss of bicarbonate, as in diarrhea. A frequent cause of metabolic alkalosis is persistent vomiting, with its associated loss of hydrogen ions as HCl from the stomach.

What is the arterial P_{CO_2} in metabolic acidosis or alkalosis? Since, by definition, metabolic acidosis and alkalosis must be due to something other than excess retention or loss of carbon dioxide, one might have predicted that arterial P_{CO_2} would be unchanged, but such is not the case. As described in Chapter 14, the elevated hydrogen-ion concentration associated with the metabolic acidosis reflexly stimulates ventilation and lowers arterial P_{CO_2}. By mass action this helps restore the hydrogen-ion concentration toward normal. Conversely, a person with metabolic alkalosis will reflexly have ventilation inhibited. The result is a rise in arterial P_{CO_2} and, by mass action, an associated restoration of hydrogen-ion concentration toward normal. To reiterate, the carbon dioxide changes in metabolic acidosis and alkalosis are not the *cause* of the acidosis or alkalosis but rather are compensatory reflex responses to primary nonrespiratory abnormalities.

---

[11]An added complexity is that bicarbonate may be secreted to some degree, as well as reabsorbed.

TABLE 15-6	CHANGES IN THE ARTERIAL CONCENTRATIONS OF HYDROGEN ION AND CARBON DIOXIDE IN ACID-BASE DISORDERS			
Primary Disorder	H^+	CO_2	Cause of CO_2 Change	
Respiratory acidosis	↑	↑	Primary abnormality	
Respiratory alkalosis	↓	↓		
Metabolic acidosis	↑	↓	Reflex ventilatory compensations	
Metabolic alkalosis	↓	↑		

SECTION D
DIURETICS AND KIDNEY DISEASE

DIURETICS

Drugs used clinically to increase the volume of urine excreted are known as **diuretics**. Such agents act by inhibiting the reabsorption of sodium, along with chloride and/or bicarbonate, resulting in increased excretion of these ions. Since water reabsorption is dependent upon sodium reabsorption, water reabsorption is also reduced, resulting in increased water excretion.

Diuretics are among the most commonly used medications. For one thing, they are used to treat diseases characterized by renal retention of salt and water. In normal persons, the mechanisms for controlling total-body sodium are so precise that sodium balance does not vary by more than a few percent despite marked changes in dietary intake or in losses due to sweating, vomiting, or diarrhea. In several types of diseases, however, sodium balance becomes deranged by the failure of the kidneys to excrete sodium normally. Sodium excretion may fall virtually to zero despite continued sodium ingestion, and the person may retain huge quantities of sodium and water, leading to abnormal expansion of the extracellular fluid and formation of edema.

The most common example of this phenomenon is congestive heart failure (page 416). A person with a failing heart manifests (1) a decreased GFR and (2) increased aldosterone secretion, both of which contribute to the virtual absence of sodium from the urine. The net result is extracellular volume expansion and formation of edema. The sodium-retaining responses are triggered by the lower cardiac output—a result of cardiac failure—and the resulting decrease in arterial pressure.

Another common use of diuretics is in the treatment of hypertension (page 418). It is still unclear why the decrease in body sodium and water resulting from the diuretic-induced excretion of these substances brings about arteriolar dilation and a lowering of the blood pressure.

As summarized in Table 15-7, diuretics are classified according to the mechanisms by which they inhibit ion reabsorption. Except for one category, the potassium-sparing diuretics, all diuretics not only increase sodium excretion but also cause increased potassium excretion as well, an unwanted side-effect. One member of the exceptional category blocks the action of aldosterone by competing for this hormone's receptors in the tubule.

TABLE 15-7   CLASSES OF DIURETICS

Class	Mechanism	Major Site Affected
Carbonic anhydrase inhibitors	Inhibits secretion of hydrogen-ion: this causes less reabsorption of sodium and bicarbonate.	Proximal tubules
Loop diuretics	Inhibits Na, K, 2Cl cotransport in luminal membrane.	Ascending loops of Henle
Thiazides	Inhibits Na, Cl cotransport in luminal membrane.	Early distal tubules
Potassium-sparing diuretics	Inhibits action of aldosterone.	Late distal tubules and collecting ducts
	Blocks sodium channels in luminal membrane.	Same

Since aldosterone stimulates both sodium reabsorption and potassium secretion, antagonizing its actions results in increased sodium excretion but decreased potassium excretion. The other members of the potassium-sparing group act by blocking sodium channels in the luminal membrane of the collecting duct cells. They lower potassium *excretion* because potassium *secretion* is normally coupled to sodium reabsorption in this nephron segment.

KIDNEY DISEASE

The term "kidney disease" is no more specific than "car trouble" since many diseases affect the kidneys. Bacteria, allergies, congenital defects, stones, tumors, and toxic chemicals are some possible sources of kidney damage. Obstruction of the urethra or a ureter may cause injury due to a buildup of pressure and may predispose the kidneys to bacterial infection.

Disease can attack the kidneys at any age. Experts estimate that there are at present more than 3 million undetected cases of kidney infection in the United States and that 25,000 to 75,000 Americans die of kidney failure each year.

Although many diseases of the kidney are self-limited and produce no permanent damage, others progress if untreated. The end stage of progressive diseases, regardless of the nature of the damaging agent—bacteria, toxic chemical, and so on—is a shrunken, nonfunctioning kidney. Similarly, the symptoms of profound renal malfunction are relatively independent of the damaging agent and are collectively known as **uremia**, literally "urine in the blood."

The severity of uremia depends upon how well the impaired kidneys are able to preserve the constancy of the internal environment. Assuming that the person continues to ingest a normal diet containing the usual quantities of nutrients and electrolytes, what problems arise? The key fact to keep in mind is that kidney destruction markedly reduces the number of functioning nephrons. Accordingly, the many substances that gain entry to the tubule entirely or primarily by filtration are filtered and therefore excreted in diminished amounts. In addition, the excretion of potassium and hydrogen ions is impaired because there are too few nephrons capable of normal tubular secretion. The buildup of all these substances in the blood causes many of the problems of uremia.

The remarkable fact is how large the safety factor is in renal function. In general, the kidneys are still able to perform their regulatory function quite well as long as 10 percent of the nephrons are functioning. This is because the remaining nephrons undergo alterations in function—filtration, reabsorption, and secretion—so as to compensate for the missing nephrons. For example, each remaining nephron increases its rate of potassium secretion so that the total amount of potassium excreted by the kidneys can be maintained at normal levels. The limits of regulation are restricted, however. To use potassium as our example again, if the person with severe renal disease were to go on a diet high in potassium, the remaining nephrons might not be able to secrete enough potassium to prevent potassium retention.

To take another example, the diseased kidney is unable to concentrate the urine normally, which should not be surprising given the complex interactions required for normal functioning of the countercurrent multiplier system. Whereas water balance can be maintained by diseased kidneys when the person is ingesting reasonable quantities of water, the kidneys may not be able to reduce urine volume adequately to avoid dehydration if the person is water deprived.

Other problems arise in uremia because of abnormal secretion of the hormones produced by the kidneys. Thus, decreased secretion of erythropoietin results in anemia. Decreased ability to form $1,25\text{-}(OH)_2D_3$ results in deficient absorption of calcium from the gastrointestinal tract, with a resulting decrease in plasma calcium and inadequate bone calcification. Both of these hormones are now available for administration to patients with uremia.

The problem with renin, the third of the renal hormones, is rarely too little secretion but rather too much secretion by the granular cells of the damaged kidneys. The result is increased plasma angiotensin II concentration and the development of hypertension.

Creatinine Clearance and Plasma Creatinine

There are many tests of renal function, but the most useful single piece of information for clinicians is the GFR. Since this value is the sum of the fluid filtered by all the individual nephrons, a decrease in GFR generally reflects a decrease in the numbering of functioning nephrons. The GFR can be closely approximated in patients by performing a **creatinine clearance** as follows. The waste product creatinine produced by muscle is filtered by the glomeruli and does not undergo reabsorption. If we assume for the moment that creatinine does not undergo tubular secretion, then the entire mass of creatinine *excreted* per unit time must be equal to the mass *filtered* over the same time period. Therefore, if we simply divide the mass of creatinine excreted in the urine per unit time by the person's plasma creatinine concentration, we obtain the volume of fluid filtered per unit time, that is, the GFR:

$$\frac{\text{Creatinine excreted}}{\text{Plasma creatinine concentration}} = \text{GFR}$$

$$\frac{\text{mg/min}}{\text{mg/mL}} = \text{mL/min}$$

The term "creatinine clearance" reflects the fact that the GFR is the volume of plasma "cleared" of its creatinine per unit time. Unfortunately, there actually is some creatinine secreted as well as filtered, and so the value for GFR obtained by this method is too high. But it is fairly close to the true GFR since the amount of creatinine secreted is normally quite small compared to the amount filtered. When more exact measures of the GFR are required, a foreign polysaccharide called *inulin* (not insulin), which is neither reabsorbed nor secreted, can be infused intravenously and the inulin clearance measured.

A simpler but cruder estimate of whether a person's GFR is decreased is provided by the plasma creatinine concentration alone. Since the excretion of creatinine depends upon its filtration, a decrease in GFR will result in creatinine retention in the body and a proportionately elevated plasma creatinine concentration.

Hemodialysis, Peritoneal Dialysis, and Transplantation

As described above, failing kidneys reach a point when they can no longer excrete water and ions at rates that maintain body balances of these substances, nor can they excrete waste products as fast as they are produced. Dietary alterations can minimize these problems, for example, by lowering potassium intake and thereby reducing the amount of potassium to be excreted, but such alterations cannot eliminate the problems. The techniques used to replace the kidneys' excretory functions are hemodialysis and peritoneal dialysis. The general term dialysis means to separate substances using a membrane.

The artificial kidney is an apparatus that utilizes a process termed **hemodialysis** to remove excess substances from the blood. During hemodialysis, blood is pumped from one of the patient's arteries through tubing that is bathed by a large volume of fluid. The tubing then conducts the blood back into the patient by way of a vein. The tubing is generally made of cellophane that is highly permeable to most solutes but relatively impermeable to protein and completely impermeable to blood cells—characteristics quite similar to those of capillaries. The bath fluid is a salt solution with ionic concentrations similar to or lower than those in normal plasma. As blood flows through the tubing, the concentrations of nonprotein plasma solutes tend to reach diffusion equilibrium with those of the solutes in the bath fluid. For example, if the plasma potassium concentration of the patient is above normal, potassium diffuses out of the blood across the cellophane tubing and into the bath fluid. Similarly, waste products and excesses of other substances also diffuse into the bath and thus are eliminated from the body.

Patients with acute reversible renal failure may require hemodialysis only for days or weeks. Patients with chronic irreversible renal failure require treatment for the rest of their lives, however, unless they can receive a renal transplant. Such patients undergo hemodialysis several times a week, often at home.

Another way of removing excess substances from the blood is **peritoneal dialysis**, which uses the lining of the person's own abdominal cavity (peritoneum) as a dialysis membrane. Fluid is injected, via a needle inserted through the abdominal wall, into this cavity and allowed to remain there for hours, during which solutes diffuse into the fluid from the person's blood. The dialysis fluid is then removed by reinserting the needle and is replaced with new fluid. This procedure can be performed several times daily by a patient who is simultaneously doing normal activities.

The treatment of choice for most patients with permanent renal failure is kidney transplantation. Rejection of the transplanted kidney by the recipient's body is a potential problem with transplants (Chapter 19), but great strides have been made in reducing the frequency of rejection. Many people who might benefit from a transplant cannot receive one because a kidney is not available. Presently, the major source of kidneys for transplanting is recently deceased persons, and improved public understanding should lead to many more persons giving permission in advance to have their kidneys and other organs used following their death.

SUMMARY

SECTION A: Basic Principles of Renal Physiology

Functions and Structure of the Kidneys

I. The kidneys regulate the water and ionic composition of the body, excrete waste products and foreign chemicals, and secrete three hormones—renin, 1,25-dihydroxyvitamin D_3, and erythropoietin. The first two functions are accomplished by continuous processing of the plasma.

II. Each nephron in the kidneys consists of a glomerulus and a tubule.

   A. Each glomerulus comprises a capillary tuft, which is supplied by an afferent arteriole, and a Bowman's capsule, into which the tuft protrudes.

   B. The tubule extends out from Bowman's capsule and is subdivided into four segments: proximal tubule, loop of Henle, distal tubule, and collecting duct. Distal tubules from different nephrons merge to form the collecting

ducts, and the latter empty into the renal pelvis, from which urine flows through ureters to the bladder.
  C. An efferent arteriole leaves the glomerular capillary tuft and branches into peritubular capillaries, which supply the tubule.

Basic Renal Processes
  I. The three basic renal processes are glomerular filtration, tubular reabsorption, and tubular secretion. In addition, the kidneys synthesize and/or catabolize certain substances.
  II. Urine formation begins with glomerular filtration—approximately 180 L/day—of essentially protein-free plasma into Bowman's capsule.
    A. Glomerular filtrate contains all the plasma substances, other than protein and substances bound to protein, in virtually the same concentrations as in plasma.
    B. Glomerular filtration is driven by the hydrostatic pressure in the glomerular capillaries and is opposed by both the hydrostatic pressure in Bowman's capsule and the osmotic force due to the proteins in the glomerular capillary plasma.
  III. As the filtrate moves through the tubules, certain substances are reabsorbed into the peritubular capillaries.
    A. Tubular reabsorption rates are generally very high for nutrients, ions and water, but are lower for waste products.
    B. Reabsorption may occur by carrier-mediated mechanisms or by diffusion.
    C. Many of the carrier-mediated systems manifest transport maximums, so that when the filtered load of a substance to be transported becomes very high, large amounts of the substance will escape reabsorption and appear in the urine.
    D. Diffusion occurs for some substances to which the tubular epithelium is permeable, because water reabsorption creates tubule-interstitium diffusion gradients for them.
  IV. Tubular secretion, movement from the peritubular capillaries into the tubules, is a pathway in addition to glomerular filtration for a substance to gain entry to the tubule.

Micturition
  I. Micturition occurs when bladder distention stimulates stretch receptors that trigger spinal reflexes leading to parasympathetically mediated contraction of the bladder smooth muscle.
  II. Voluntary control is exerted chiefly by stimulating or inhibiting the nerves to the pelvic diaphragm.

Section B. Regulation of Sodium and Water Balance
Total-Body Balance and Internal Distribution of Sodium and Water
  I. The body gains water via ingestion and internal production, and it loses water via urine, the gastrointestinal tract, and evaporation from the skin and respiratory tract (as insensible loss and sweat).

  II. For both water and sodium, the major homeostatic control point for maintaining stable balance is renal excretion.
  III. Approximately two-thirds of the body water is intracellular and one-third is extracellular. The amount of sodium in the body determines the extracellular volume.

Basic Renal Processes for Sodium and Water
  I. Sodium is freely filterable at the glomerulus, and its reabsorption is a primary active process dependent upon Na,K-ATPase pumps in the basolateral membranes of the tubular epithelium.
  II. Sodium reabsorption also drives, by cotransport, the secondary active reabsorption of glucose, amino acids, and chloride and, by countertransport, the secretion of hydrogen ions.
  III. Sodium reabsorption creates an osmotic gradient across the tubule, which drives water reabsorption by osmosis.
  IV. Water reabsorption from the late distal tubules and collecting ducts depends on the presence of the posterior pituitary hormone antidiuretic hormone (ADH), which increases the permeability of these segments to water.
  V. A large volume of dilute urine is produced when plasma ADH concentration and, therefore, water reabsorption is low.
  VI. A small volume of concentrated urine is produced by the renal countercurrent multiplier system when plasma ADH concentration is high.
    A. The active transport of sodium chloride by the ascending loop of Henle causes a progressive concentration of the interstitial fluid of the medulla.
    B. This concentrated interstitium causes water to move out of the collecting ducts, made highly permeable to water by ADH. The result is concentration of the collecting duct fluid and the urine.

Renal Sodium Regulation
  I. Sodium excretion is the difference between the amount of sodium filtered and the amount reabsorbed.
  II. GFR, and therefore filtered load of sodium, are controlled by baroreceptor reflexes. Decreased vascular pressures cause decreased baroreceptor firing and increased sympathetic outflow to the renal arterioles, resulting in vasoconstriction and decreased GFR.
  III. The major control of tubular sodium reabsorption is the adrenal cortical hormone aldosterone, which stimulates sodium reabsorption in the late distal tubules and collecting ducts.
  IV. The renin-angiotensin system is one of the major controllers of aldosterone secretion.
    A. Renin, the rate-limiting variable in the renin-angiotensin system, is secreted by the granular cells of the kidney juxtaglomerular apparatus and catalyzes, in the blood, the formation of angiotensin I from circulating angiotensinogen produced by the liver. Angiotensin I is then cleaved to yield angiotensin II, which acts on the adrenal cortex to stimulate aldosterone secretion.
    B. When extracellular volume decreases, renin secretion is stimulated by three inputs: (1) stimulation of the

renal sympathetic nerves to the granular cells by baroreceptor reflexes; (2) pressure decreases sensed by the granular cells, themselves; and (3) a signal generated by low sodium or chloride concentration in the lumen of the macula densa.

V. Atrial natriuretic factor, secreted by cells in the atria in response to atrial distention, inhibits sodium reabsorption.

Renal Water Regulation

I. Water excretion is the difference between the amount of water filtered and the amount reabsorbed.

II. GFR regulation via the baroreceptor reflexes plays some role in regulating water excretion, but the major control is via ADH-control of water reabsorption.

III. ADH secretion by the posterior pituitary is controlled by baroreceptors, particularly those in the left atrium, and by osmoreceptors in the hypothalamus.
A. Via the baroreceptor reflexes, a low extracellular volume stimulates ADH secretion, and a high extracellular volume inhibits it.
B. Via the osmoreceptors, a high body-fluid osmolarity stimulates ADH secretion and a low osmolarity inhibits it.

Thirst and Salt Appetite

I. Thirst is stimulated by a variety of inputs, including baroreceptors, osmoreceptors, and angiotensin II.

II. Salt appetite is not of major regulatory importance in people.

Section C. Regulation of Potassium, Calcium, and Hydrogen Ion

Potassium Regulation

I. A person remains in potassium balance by excreting an amount of potassium in the urine equal to the amount ingested minus the amounts lost in the feces and sweat.

II. Potassium is freely filterable at the glomerulus and undergoes both reabsorption and secretion, the latter being the major controlled variable.

III. When body potassium is increased, aldosterone secretion is stimulated by an increased plasma concentration of potassium, and the increased aldosterone then stimulates potassium secretion.

Calcium Regulation

I. Plasma calcium concentration is maintained by control of urinary excretion, gastrointestinal absorption, and movement into and out of bone.

II. Parathyroid hormone increases plasma calcium concentration by influencing these processes.
A. It stimulates tubular reabsorption of calcium, movement of calcium out of bone, and formation of the hormone 1,25-dihydroxyvitamin D_3, which stimulates calcium absorption by the intestine.
B. It also inhibits the tubular reabsorption of phosphate, and the lowered plasma phosphate facilitates calcium movement out of bone.

III. Vitamin D_3 is formed in the skin or ingested and then is metabolized, under stimulation by parathyroid hormone, to the active form, 1,25-dihydroxyvitamin D_3, in the kidneys.

Hydrogen-Ion Regulation

I. Total-body balance of hydrogen ions is the result of both metabolic production of these ions and net gains or losses via the gastrointestinal tract and urine. A stable balance is achieved by regulation of urinary losses.

II. Buffering is a means of minimizing changes in hydrogen-ion concentration by combining these ions reversibly with anions such as bicarbonate and intracellular proteins.

III. The kidneys not only excrete hydrogen ions but reabsorb bicarbonate.
A. Both processes require tubular hydrogen-ion secretion in a process catalyzed by carbonic anhydrase.
B. The excreted hydrogen ions are in combination with urinary buffers, notably ammonia produced and secreted by the tubules.

IV. Acid-base disorders are categorized as respiratory or metabolic.
A. Respiratory acidosis is due to retention of carbon dioxide, and respiratory alkalosis to excessive elimination of carbon dioxide.
B. All other causes of acidosis or alkalosis are termed metabolic and reflect gain or loss, respectively, of hydrogen ions from a source other than carbon dioxide.

Section D. Diuretics and Kidney Disease

I. Diuretics inhibit reabsorption of sodium and water, thereby enhancing their excretion. Different diuretics act on different nephron segments (Table 15-7).

II. Many of the symptoms of uremia—general renal malfunction—are due to retention of substances because of reduced GFR and, in the case of potassium and hydrogen-ion, reduced secretion. Other symptoms are due to inadequate secretion of the renal hormones.
A. GFR can be approximated by determining the creatinine clearance. The plasma creatinine concentration, alone, is used as an indicator of GFR changes.
B. Either hemodialysis or peritoneal dialysis can be used chronically to eliminate the water, ions, and waste products retained during uremia.

REVIEW QUESTIONS

Section A. Basic Principles of Renal Physiology

1. Define:

waste products	tubule
urea	afferent arteriole
uric acid	glomerular capillaries
creatinine	Bowman's capsule
nephron	Bowman's space
glomerulus	proximal tubule

loop of Henle
distal tubule
collecting ducts
renal pelvis
renal cortex
renal medulla
ureter
urinary bladder
urethra
efferent arteriole
peritubular capillaries
macula densa
granular cells
juxtaglomerular apparatus
  (JGA)

glomerular filtration
glomerular filtrate
tubular reabsorption
tubular secretion
net glomerular filtration
  pressure
glomerular filtration rate
  (GFR)
filtered load
luminal membrane
basolateral membrane
transport maximum (T_m)
micturition
pelvic diaphragm

2. What are the functions of the kidneys?

3. What three hormones do the kidneys secrete?

4. Fluid flows through what sequence of structures from the glomerulus to the bladder? Blood flows through what structures from the renal artery to the renal vein?

5. What are the three basic renal processes that lead to the formation of urine?

6. How does the composition of the glomerular filtrate compare with that of plasma?

7. Describe the forces that determine the magnitude of the GFR. What is a normal value for GFR?

8. Contrast the mechanisms of reabsorption for glucose and urea. Which one shows a T_m?

9. Diagram the sequence of events leading to micturition in infants and in adults.

Section B. Regulation of Sodium and Water Balance

1. Define:

insensible water loss
antidiuretic hormone
  (ADH)
vasopressin
water diuresis
diabetes insipidus
obligatory water loss
countercurrent multiplier
  system
aldosterone

renin
angiotensin I
angiotensinogen
angiotensin II
converting enzyme
intrarenal baroreceptors
atrial natriuretic factor
  (ANF)
osmoreceptors
salt appetite

2. What are the sources of water gain and loss in the body? What are they for sodium?

3. Describe the distribution of water and sodium between intracellular and extracellular fluids.

4. What is the relationship between body sodium and extracellular fluid?

5. What is the mechanism of reabsorption of sodium, and how is the reabsorption of other solutes coupled to it?

6. What is the mechanism of water reabsorption, and how is it coupled to sodium reabsorption?

7. What is the effect of ADH on the renal tubules, and what are the sites affected?

8. Describe the characteristics of the two limbs of the loop of Henle with regard to their transport of sodium, chloride, and water.

9. Diagram the osmolarities in the two limbs of the loop of Henle, medullary interstitium, distal tubule, and collecting duct in the presence of ADH. What happens to the collecting duct values in the absence of ADH?

10. What two factors determine how much sodium is excreted per unit time?

11. Diagram the sequence of events by which a decrease in blood pressure leads to a decreased GFR.

12. List the sequence of events leading from increased renin secretion to increased aldosterone secretion.

13. What are the three inputs controlling renin secretion?

14. Diagram the sequence of events leading from a decreased atrial pressure or from an increased plasma osmolarity to an increased secretion of ADH.

15. What are the stimuli for thirst?

Section C. Regulation of Potassium, Calcium, and Hydrogen Ion

1. Define:

parathyroid hormone
vitamin D
vitamin D_3
1,25-dihydroxyvitamin D_3
  [1,25-$(OH)_2D_3$)]
rickets
buffering

acidosis
alkalosis
respiratory acidosis
respiratory alkalosis
metabolic acidosis
metabolic alkalosis

2. Which of the basic renal processes apply to potassium, and which of them is the controlled process?

3. Diagram the reflex leading from increased plasma potassium to increased potassium excretion.

4. What are the two major controls of aldosterone secretion, and what are this hormone's major actions?

5. What is the stimulus controlling the secretion of parathyroid hormone, and what are this hormone's four major effects?

6. Describe the formation and action of 1,25-$(OH)_2D_3$. How does parathyroid hormone influence the production of this hormone?

7. What are the sources of gain and loss of hydrogen ions in the body?

8. List the body's major buffer systems.

9. How does the secretion of hydrogen ions occur and what happens to them once they are in the lumen? How does hydrogen-ion secretion achieve bicarbonate reabsorption?

10. Classify the four types of acid-base disorders according to plasma hydrogen-ion concentration, bicarbonate concentration, and P_{CO_2}.

Section D. Diuretics and Kidney Disease

1. Define:

diuretics   hemodialysis

uremia   peritoneal dialysis

creatinine clearance

2. What information is supplied by a measurement of the creatinine clearance, and how does this differ from simply measuring plasma creatinine concentration?

THOUGHT QUESTIONS

(Answers are given in Appendix A.)

1. Substance T is present in the urine. Does this prove that it is filterable at the glomerulus?

2. Substance V is not normally present in the urine. Does this prove that it is neither filtered nor secreted?

3. The concentration of glucose in plasma is 100 mg/100 mL, and the GFR is 125 mL/min. How much glucose is filtered per minute?

4. A person is found to be excreting abnormally large amounts of a particular amino acid. From the theoretical description of T_m-limited reabsorptive mechanisms in the text, list several possible causes.

5. The concentration of urea in urine is always much higher than the concentration in plasma. Does this mean that urea is secreted?

6. If a drug that blocked the reabsorption of sodium were taken, what would happen to the reabsorption of water, urea, chloride, glucose, and amino acids and to the secretion of hydrogen ions?

7. Compare the changes in GFR and renin secretion occurring in response to a moderate hemorrhage in two persons: one taking a drug that blocks the sympathetic nerves to the kidneys and the other not taking such a drug.

8. If a person were taking a drug that completely inhibited converting enzyme, what would happen to aldosterone secretion when the person went on a low-sodium diet?

9. In the steady state, is the amount of sodium chloride excreted daily in the urine by a normal person ingesting 12 g of sodium chloride per day (A) 12 g/day or (B) less than 12 g/day? Explain.

10. A person who has suffered a head injury seems to have recovered but is thirsty all the time. What do you think might be the cause?

11. A patient has a tumor in the adrenal cortex that continuously secretes large amounts of aldosterone. What effects does this have on the total amount of sodium and potassium in her body?

16

THE DIGESTION AND ABSORPTION OF FOOD

The **gastrointestinal system** (Figure 16-1) includes the **gastrointestinal tract** (mouth, pharynx, esophagus, stomach, small intestine, large intestine, and rectum) and the glandular organs (salivary glands, liver, gallbladder, and pancreas) that are not part of the tract but secrete substances into it via ducts connecting the organs to the tract.

The gastrointestinal system serves to transfer organic molecules, salts, and water from the external environment to the body's internal environment, where they can be distributed to cells by the circulatory system.

The adult gastrointestinal tract is a tube approximately 15 ft long, running through the body from mouth to anus. The lumen of the tract, like the hole in a dough-

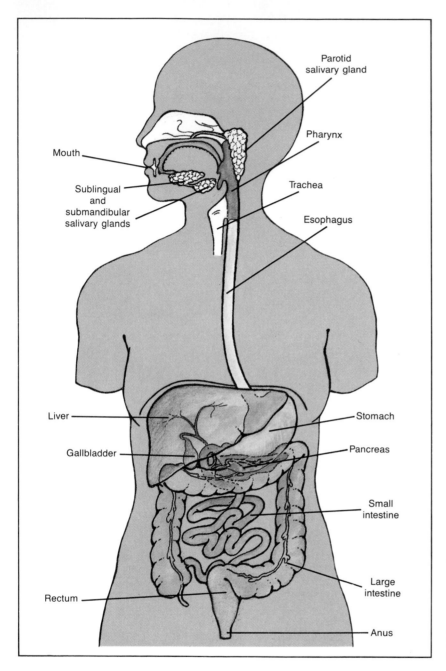

FIGURE 16-1 Anatomy of the gastrointestinal system.

nut, is continuous with the external environment, which means that its contents are technically outside the body. This fact is relevant to understanding some of the tract's properties. For example, the large intestine is inhabited by millions of bacteria, most of which are harmless and even beneficial in this location. However, if the same bacteria enter the blood, as may happen, for example, in the case of a ruptured appendix, they may be extremely harmful and even lethal.

Most food is taken into the mouth as large particles containing many macromolecules, such as proteins and polysaccharides, that are unable to cross the wall of the gastrointestinal tract. Before ingested food can be absorbed, it must be dissolved and broken down into much smaller molecules. This dissolving and breaking-down process—**digestion**—is accomplished by the action of hydrochloric acid secreted by the stomach, bile secreted by the liver, and a variety of digestive enzymes that are released by the system's exocrine glands, a process termed **secretion**.

The molecules produced by digestion then move from the lumen of the gastrointestinal tract across a layer of epithelial cells and enter the blood or the lymph—**absorption**.

While digestion, secretion and absorption are taking place, contractions of smooth muscles in the gastrointestinal tract wall mix the luminal contents with the various secretions and move them through the tract from mouth to anus—**motility**.

The functions of the gastrointestinal system can be described in terms of these four processes—digestion, secretion, absorption, and motility (Figure 16-2)—and the mechanisms controlling them.

The gastrointestinal system is designed to maximize absorption, and within certain wide limits it will absorb as much as is ingested. With a few important exceptions, the gastrointestinal system does not *regulate* the amount of nutrients absorbed or their concentration in the internal environment. This is primarily the function of the kidneys, as we saw in Chapter 15, and a number of endocrine systems, as will be described in Chapter 17.

Small amounts of certain metabolic end products are excreted via the gastrointestinal tract, but the elimination of most waste products from the internal environment is achieved by the lungs and kidneys. Thus, the elimination of waste products is only a minor function of the gastrointestinal system. The material—feces—leaving the system at the end of the gastrointestinal tract consists almost entirely of bacteria and ingested material that was not digested and absorbed, that is, material that was never actually part of the internal environment.

OVERVIEW: FUNCTIONS OF THE GASTROINTESTINAL ORGANS

The gastrointestinal tract begins with the **mouth**, and digestion starts there with chewing, which breaks up large pieces of food into smaller particles that can be swallowed without choking. **Saliva**, secreted by the **salivary glands** (Figure 16-3), located in the head, drains into the mouth though a short series of ducts. Saliva moistens and lubricates the food particles prior to swallowing. It contains the carbohydrate-digesting enzyme **amylase**, which partially digests polysaccharides. A third function of saliva is to dissolve some of the food molecules. Only in the dissolved state can these molecules react with chemoreceptors in the mouth, giving rise to the sensation of taste.

The next segments of the tract, the **pharynx** and **esophagus**, contribute nothing to digestion but provide the pathway by which ingested food and drink reach the stomach. The muscles in the walls of these segments control swallowing.

FIGURE 16-2 Four processes carried out by the gastrointestinal tract: digestion, secretion, absorption, and motility.

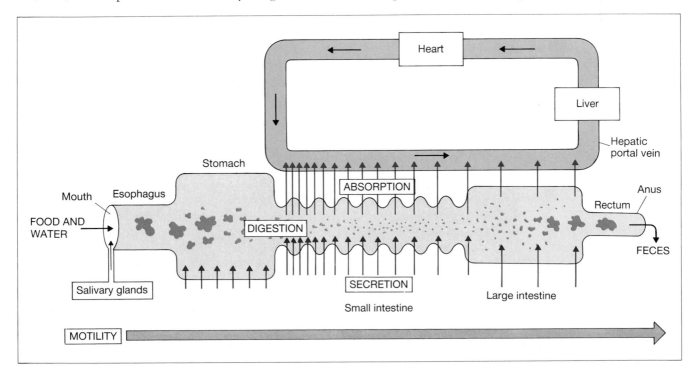

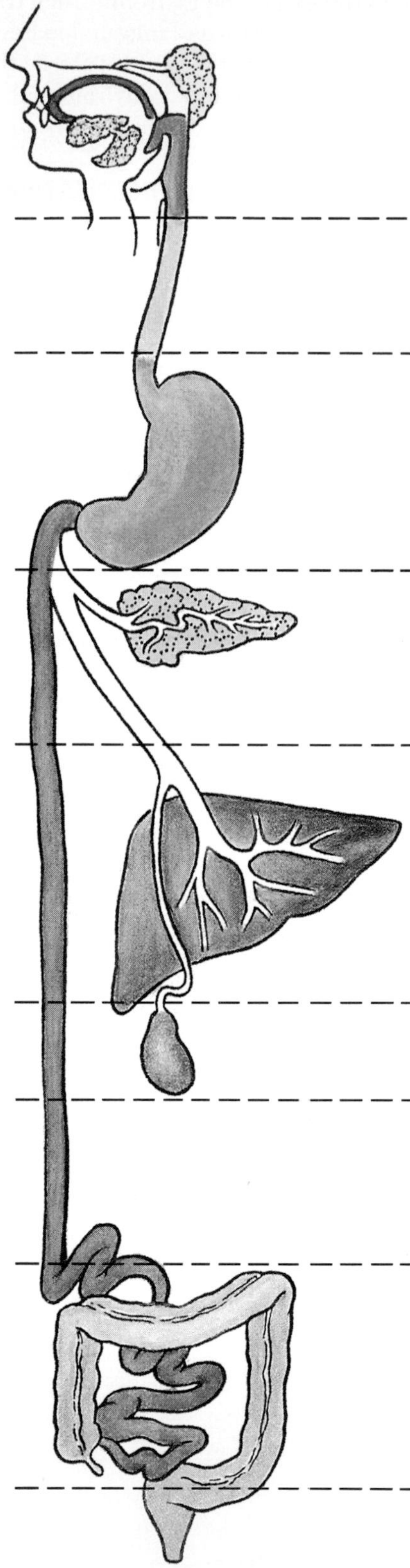

ORGAN	EXOCRINE SECRETIONS	FUNCTIONS
Mouth and pharynx		Chewing (mechanical digestions); initiation of swallowing reflex
Salivary glands	Salt and water	Moisten food
	Mucus	Lubrication
	Amylase	Polysaccharide-digesting enzyme
Esophagus		Move food to stomach by peristaltic waves
	Mucus	Lubrication
Stomach		Store, mix, and dissolve food; regulate emptying of dissolved food into small intestine
	HCl	Solubilization of food particles; kill microbes
	Pepsin	Protein-digesting enzyme
	Mucus	Lubricate and protect epithelial surface
Pancreas		Secretion of enzymes and bicarbonate; also has nondigestive endocrine functions
	Enzymes	Digest carbohydrates, fats, proteins, and nucleic acids
	Bicarbonate	Neutralize HCl entering small intestine from stomach
Liver		Secretion of bile; many other nondigestive functions
	Bile salts	Solubilize water-insoluble fats
	Bicarbonate	Neutralize HCl entering small intestine from stomach
	Organic waste products and trace metals	Elimination in feces
Gallbladder		Store and concentrate bile between meals
Small intestine		Digestion and absorption of most substances; mixing and propulsion of contents
	Enzyme	Food digestion
	Salt and water	Maintain fluidity of luminal contents
	Mucus	Lubrication
Large intestine (colon)		Storage and concentration of undigested matter; mixing and propulsion of contents
	Mucus	Lubrication
Rectum		Defecation

FIGURE 16-3 Functions of the gastrointestinal organs.

The **stomach** is a saclike organ, located between the esophagus and the small intestine, whose functions are to store, dissolve, and partially digest food, and to deliver its contents to the small intestine in amounts optimal for digestion and absorption. The glands lining the stomach's wall secrete a strong acid, **hydrochloric acid**, and several protein-digesting enzymes collectively known as **pepsin**.

Hydrochloric acid's primary function is to dissolve the particulate matter in food. The acid environment in the stomach lumen alters the ionization of polar molecules, especially proteins, disrupting the extracellular network of connective-tissue proteins that form the structural framework of tissues (food particles). The proteins and polysaccharides released by hydrochloric acid's dissolving action are partially digested in the stomach by both pepsin and amylase, the latter from the salivary glands, into smaller molecular fragments. The one component of food that is not dissolved by acid is fat. The fat released from food particles in the stomach aggregates into large water-insoluble globules.

Hydrochloric acid also kills bacteria that enter along with food. This process is not 100 percent effective, and some bacteria survive to take up residence and multiply in the intestinal tract, particularly the large intestine.

The digestive actions of the stomach reduce food particles to a solution, known as **chyme**, which contains molecular fragments of proteins and polysaccharides and droplets of fat. None of these digestion products can cross the epithelium of the stomach wall, and thus little absorption of organic nutrients occurs in this organ.

Digestion's final stages and most absorption occur in the next section of the tract, the **small intestine**. Here molecules of intact or partially digested carbohydrate, fat, and protein are broken down by enzymes into monosaccharides, fatty acids, and amino acids. The products of digestion cross the epithelial cells and enter the blood and/or lymph. Vitamins, minerals, and water, which do not require enzymatic digestion, are also absorbed in the small intestine.

The small intestine is a 1.5-in-diameter tube leading from the stomach to the large intestine. Its 9-ft length is divided into three segments: An initial short segment, the **duodenum**, is followed by the **jejunum** and then by the longest segment, the **ileum**. Normally, most absorption occurs in the first quarter of the small intestine, in the duodenum and jejunum. Thus, the small intestine has a considerable functional reserve, making it almost impossible to exceed its absorptive capacity even when very large quantities of food are ingested.

Several major glands—the pancreas and liver—secrete substances that flow via ducts into the duodenum of the small intestine. The **pancreas**, a large elongated gland located behind the stomach, has both endocrine and exocrine functions, but only the latter are directly involved in gastrointestinal function. Similarly, the **liver**, a large gland located in the upper right portion of the abdomen, has a large variety of functions, as described in various chapters, but we will be concerned here with only its functions that are directly related to the gastrointestinal tract.

The exocrine portion of the pancreas secrets (1) digestive enzymes specific for each class of organic molecule and (2) a fluid rich in bicarbonate ions. These secretions enter the small intestine through a duct leading from the pancreas to the duodenum. The high acidity of the chyme coming from the stomach would inactivate the pancreatic enzymes in the small intestine if the acid were not neutralized by the bicarbonate ions in the pancreatic fluid.

Because fat is insoluble in water, the fat droplets entering the small intestine from the stomach must first be dissolved before they can be effectively digested and absorbed in a water environment. Fat is solubilized by the action of **bile salts**, one of the components of the fluid secreted by the liver—**bile**. In addition to bile salts, bile contains bicarbonate ions, cholesterol, and a number of organic waste products. In the liver, bile is secreted into small ducts that join to form the single duct that empties into the duodenum. Between meals, secreted bile is stored in a small sac underneath the liver, the **gallbladder**, that concentrates the bile by absorbing salts and water. During a meal, the smooth muscles in the gallbladder wall contract, causing a concentrated bile solution to be injected via the bile duct into the duodenum.

In the small intestine, monosaccharides and amino acids are absorbed by specific carrier-mediated transport processes in the plasma membranes of the intestinal epithelial cells, whereas fatty acids enter these cells by diffusion. Most mineral ions are actively absorbed, and water diffuses passively down osmotic gradients. Digestion and absorption have been largely completed by the middle portion of the small intestine.

The motility of the small intestine, brought about by the smooth muscles lining it, mixes the luminal contents with the various secretions, brings the contents into contact with the epithelial surface where absorption takes place, and slowly advances the luminal material toward the large intestine. Since most substances are absorbed in the early portion of the small intestine, only a small volume of water, salt, and undigested material is passed on to the next segment, the **large intestine** (or **colon**), which temporarily stores the undigested material (some of which is acted upon by bacteria) and concentrates it by absorbing salt and water. Contractile activities in the

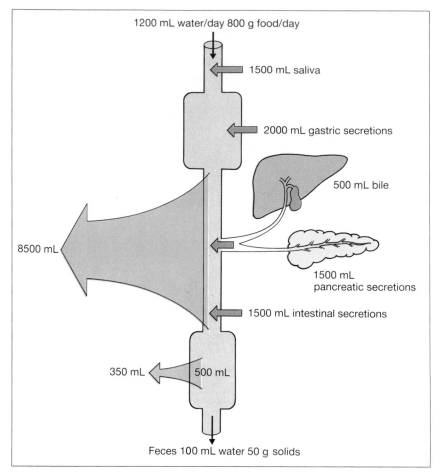

1200 mL water/day 800 g food/day

1500 mL saliva

2000 mL gastric secretions

500 mL bile

8500 mL

1500 mL pancreatic secretions

1500 mL intestinal secretions

350 mL   500 mL

Feces 100 mL water 50 g solids

FIGURE 16-4 Average amounts of food and fluid ingested, secreted, absorbed, and excreted from the gastrointestinal tract daily.

final segment of the gastrointestinal tract, the **rectum** and associated sphincter muscles, eliminate the feces—**defecation**.

Although the average adult consumes about 800 g of food and 1200 mL of water per day, this is only a fraction of the material entering the lumen of the gastrointestinal tract. An additional 7000 mL of fluid from salivary glands, stomach, pancreas, liver, and the intestines (Figure 16-4) is secreted into the tract each day. Only about 100 mL of water is lost in the feces, the rest being absorbed into the blood. Almost all the salts in the secreted fluids are also absorbed into the circulation. The secreted enzymes are themselves digested and the resulting amino acids absorbed.

This completes our overview of the gastrointestinal system. Since its major task is digestion and absorption, we begin our more detailed description with these processes. Subsequent sections of the chapter will then describe, organ by organ, the regulation of the secretions and motility that produce the optimal conditions for digestion and absorption. A prerequisite for all this physiology, however, is a knowledge of the structure of the gastrointestinal tract wall.

STRUCTURE OF THE GASTROINTESTINAL TRACT WALL

From the mid-esophagus to the anus, the wall of the gastrointestinal tract has the general structure illustrated in Figure 16-5. Most of the tube's luminal surface is highly convoluted, a feature that greatly increases the surface area available for absorption. From the stomach on, this surface is covered by a single layer of epithelial cells linked together along the edges of their luminal surfaces by tight junctions.

Included in this epithelial layer are exocrine cells that secrete mucus and endocrine cells that release hormones into the blood. Invaginations of the epithelium into the underlying tissue form tubular exocrine glands that secrete mucus, acid, enzymes, water, and ions.

Just below the epithelium is a layer of connective tissue, the lamina propria, through which pass small blood vessels, nerve fibers, and lymphatic ducts. The lamina propria is separated from underlying tissues by a thin layer of smooth muscle, the muscularis mucosa. The combination of these three layers—the epithelium, lam-

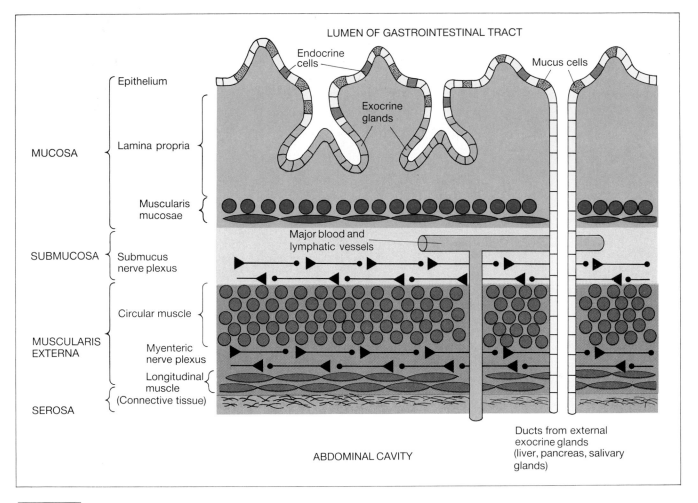

FIGURE 16-5 Structure of the gastrointestinal wall in longitudinal section. Not shown are the smaller blood vessels, neural connections between the two nerve plexuses, and neural terminations on muscles and glands.

ina propria, and muscularis mucosa—is called the **mucosa**.

Beneath the mucosa is a second connective-tissue layer, the **submucosa**, containing a network of nerve cells, the **submucus plexus**, and blood and lymphatic vessels whose branches penetrate into both the overlying mucosa and the underlying layers of smooth muscle called the **muscularis externa**. Contractions of these muscles provide the forces for moving and mixing the gastrointestinal contents. The muscularis externa has two layers: (1) a relatively thick inner layer of **circular muscle**, whose fibers are oriented in a circular pattern around the tube such that contraction produces a narrowing of the lumen, and (2) a thinner outer layer of **longitudinal muscle**, whose contraction shortens the tube. Between these two layers is a second network of nerve cells known as the **myenteric plexus**.

Finally, surrounding the outer surface of the tube is a layer of connective tissue called the **serosa**. Thin sheets of connective tissue connect the serosa to the abdominal wall, supporting the gastrointestinal tract in the abdominal cavity. At various points along the tube, the secretions of exocrine glands lying outside the tract (the salivary glands, liver, and pancreas) are delivered to the lumen of the tract via ducts.

The area available for nutrient absorption in the small intestine is greatly increased by the extensive folding of intestinal surface. Extending from the surface are finger-like projections known as **villi** (Figure 16-6). The surface of each villus is covered with a single layer of epithelial cells whose surface membranes form small projections called **microvilli** (Figure 16-7). The combination of folded mucosa, villi, and microvilli increases the small intestine's surface area about 600-fold over that

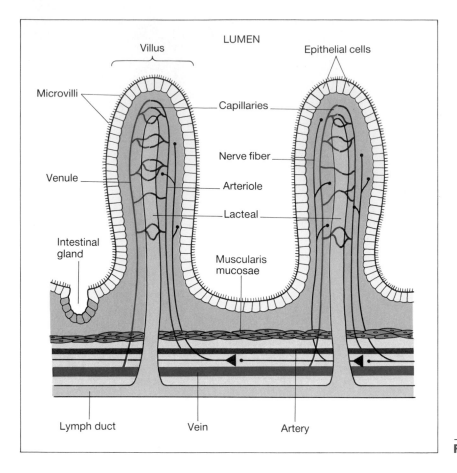

LUMEN

Villus

Epithelial cells

Microvilli

Capillaries

Nerve fiber

Venule

Arteriole

Lacteal

Intestinal gland

Muscularis mucosae

Lymph duct

Vein

Artery

FIGURE 16-6 Structure of intestinal villi.

of a flat-surfaced tube having the same length and diameter. The human small intestine's surface area is about 300 m², the area of a tennis court.

Epithelial surfaces in the gastrointestinal tract are continuously being replaced by new epithelial cells. In the small intestine, new cells arise by mitosis of cells at the base of the villi. These cells differentiate as they migrate to the top of the villus, replacing older cells that disintegrate and are discharged into the intestinal lumen. These disintegrating cells release into the lumen intracellular enzymes that then contribute to the digestive process. About 17 billion epithelial cells are replaced each day, and the entire epithelium of the small intestine is replaced approximately every 5 days. It is because of this rapid cell turnover that the lining of the intestinal tract is so susceptible to damage by agents that inhibit cell division, such as radiation or anticancer drugs.

The center of each intestinal villus is occupied both by a single blind-ended lymphatic vessel termed a **lacteal** and by a capillary network (Figure 16-6). As we will see, most of the fat absorbed in the small intestine enters the lacteals while other absorbed nutrients enter the blood capillaries. The venous drainage from the intesti-

nal villi, as well as from the large intestine, pancreas, and portions of the stomach does not empty directly into the vena cava but passes first via the **hepatic portal vein** into the liver. There it flows through a second capillary network before leaving the liver to return to the heart. Thus, material absorbed into the intestinal capillaries can be processed by the liver before entering the general circulation.

DIGESTION AND ABSORPTION

Carbohydrate

Carbohydrate intake per day ranges from about 250 to 800 g in a typical American diet. About two-thirds of this carbohydrate is the plant polysaccharide starch and the remainder is mostly the disaccharides sucrose (table sugar) and lactose (milk sugar) (Table 16-1). Only small amounts of monosaccharides are normally present in the diet. Cellulose and certain other complex polysaccharides found in vegetable matter cannot be broken down by the enzymes in the small intestine. They are passed on to the large intestine, where they are metabolized by bacteria.

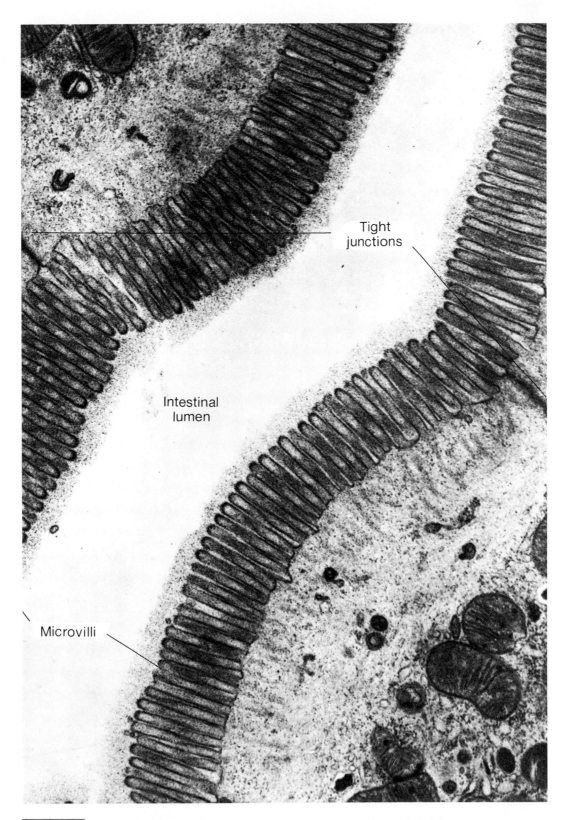

Tight
junctions

Intestinal
lumen

Microvilli

FIGURE 16-7 Microvilli on the surface of intestinal epithelial cells. [*From D. W. Fawcett, J. Histochem. Cytochem. 13:75–91 (1965). Courtesy of Susumo Ito.*]

TABLE 16-1 CARBOHYDRATES IN FOOD		
Class	**Examples**	**Made up of**
Polysaccharides	Starch	Glucose
	Cellulose	Glucose
	Glycogen	Glucose
Disaccharides	Sucrose	Glucose-fructose
	Lactose	Glucose-galactose
	Maltose	Glucose-glucose
Monosaccharides	Glucose	
	Fructose	
	Galactose	

Starch is partially digested by salivary amylase in the upper part of the stomach before the enzyme is destroyed by gastric acid, and digestion is continued in the small intestine by pancreatic amylase. The products formed by the amylases, along with ingested sucrose and lactose, are broken down into monosaccharides—glucose, galactose and fructose—by enzymes located in the plasma membranes of the small intestinal epithelial cells. These monosaccharides are then transported across the intestinal epithelium into the blood. Fructose crosses the epithelium by facilitated diffusion, while glucose and galactose undergo secondary active transport coupled to sodium (page 119).

Following a normal meal, most of the ingested carbohydrate is digested and absorbed within the first 20 percent of the small intestine. The carbohydrate transport capacity of the small intestine, unlike that of the kidney, is practically impossible to saturate.

Protein

A daily intake of about 40 to 50 g of protein is required by a normal adult to supply essential amino acids and replace the amino-acid nitrogen converted to urea. (A typical American diet contains about 125 g of protein.) In addition to dietary protein, a large amount of protein, in the form of enzymes and mucus, is secreted into the gastrointestinal tract or enters it via the disintegration of epithelial cells. Most of this protein is broken down into amino acids and absorbed by the small intestine.

Proteins are broken down to peptide fragments in the stomach by pepsin and in the small intestine by **trypsin** and **chymotrypsin**, which are secreted by the pancreas. These fragments are further digested to free amino acids by **carboxypeptidase** secreted by the pancreas and by **aminopeptidase**, located in the luminal epithelial membranes of the small intestine. These last two enzymes split off amino acids from the carboxyl and amino ends of the peptide chains, respectively.

The free amino acids then undergo secondary active transport, coupled with sodium, across the intestinal wall. Short chains of two or three amino acids are also actively absorbed.[1] This is in contrast to carbohydrate absorption, in which disaccharides are not absorbed. As with carbohydrates, protein digestion and absorption are largely completed in the early portion of the small intestine.

Small amounts of intact proteins are able to cross the intestinal epithelium and gain access to the interstitial fluid. They do so by a combination of endocytosis and exocytosis. Protein absorptive capacity is much greater in infants than in adults, and antibodies (proteins involved in the immunological defense system of the body) secreted in the mother's milk can be absorbed by the infant, providing some immunity until the infant can produce its own antibodies.

Fat

Fat intake ranges from about 25 to 160 g/day in a typical American diet; most is in the form of triacylglycerols. Fat digestion occurs almost entirely in the small intestine. The major digestive enzyme in this process is pancreatic **lipase**, which catalyzes the splitting of bonds linking fatty acids to the first and third carbon atoms of glycerol, producing two free fatty acids and a monoglyceride as products:

$$\text{Triacylglycerol} \xrightarrow{\text{Lipase}} \text{monoglyceride} + 2 \text{ fatty acids}$$

As noted earlier, the triacylglycerols entering the small intestine from the stomach are insoluble in water and are aggregated into large lipid droplets. Since only the lipids at the surface of these droplets are accessible to the water-soluble lipase, digestion would proceed very slowly without the solubilizing action of bile. Furthermore, the products of lipase action (fatty acids and monoglycerides) are themselves insoluble in water.

These problems are circumvented by bile salts, which increase the rate of fat digestion and absorption in two ways: (1) By a process known as **emulsification**, they prevent large lipid droplets from aggregating into still larger ones, and (2) they combine with the fatty acids and monoglycerides produced by lipase at the droplet surface to form very small, water-soluble aggregates known as **micelles**. In the absence of bile salts and, hence, the resulting lipid emulsification and micelle formation, fat diges-

---

[1] Amino acids are actually absorbed at a greater rate when present in small peptides than as free amino acids. This has practical applications when adequate levels of nutrients must be supplied to patients with various GI disorders.

tion and absorption occur so slowly that much ingested fat passes on to the large intestine and is excreted in the feces.

Bile salts are formed in the liver from cholesterol and are amphipathic molecules having a polar and a nonpolar surface (Figure 16-8). The nonpolar side of the steroid ring associates with other nonpolar surfaces, leaving the polar side exposed at the water surface. Mechanical agitation in the intestine breaks up the large fat globules, and the resulting droplets become coated with bile salts. Because of the negative charge on the bile salts at the surface, the droplets repel each other and so do not aggregate. The resulting suspension of lipid droplets, each about 1 μm in diameter, is known as an **emulsion** (Figure 16-9).

Although digestion is speeded up by emulsification, absorption of the insoluble products of the lipase reaction would be very slow if it were not for the second action of bile salts, the formation of micelles, which are similar in structure to emulsion droplets, but are much smaller. Micelles consist of bile salts, fatty acids, monoglycerides, and phospholipids, all clustered together with the polar ends of each molecule oriented toward the

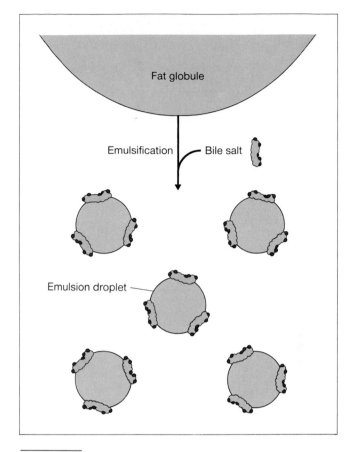

FIGURE 16-9 Emulsification of fat by bile salts.

FIGURE 16-8 Structure of bile salts. (A) Chemical formula of glycocholic acid, one of several bile salts secreted by the liver (polar groups in color). (B) Three-dimensional structure of a bile salt, showing its polar and nonpolar surfaces.

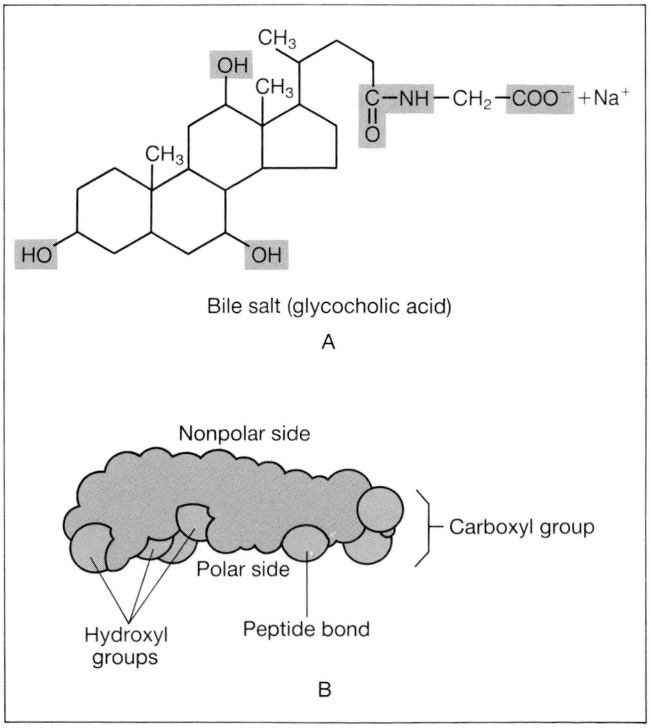

micelle's surface and the nonpolar portions of these amphipathic molecules forming the micelle's core.

How do micelles increase absorption? Although free fatty acids and monoglycerides have an extremely low solubility in water, a few molecules do exist in solution and are free to diffuse across plasma membranes. The fatty acids and monoglycerides in the micelles are in equilibrium with those free in solution, molecules being continuously exchanged between the two states. As the concentration of free lipids falls because of their diffusion out of the lumen, more lipids shift out of the micelles into the free phase (Figure 16-10). Thus, the micelles provide a means of storing the insoluble fat digestion products so that they are available, through equilibrium with the free lipids, for absorption. Note that it is not the micelle that is absorbed but rather the lipids that are free in solution.

Although fatty acids and monoglycerides enter the epithelial cells from the intestinal lumen, it is triacylglycerol that is released on the other side of the cell into the interstitial fluid. Thus, during their passage through the epithelial cells, fatty acids and monoglycerides are

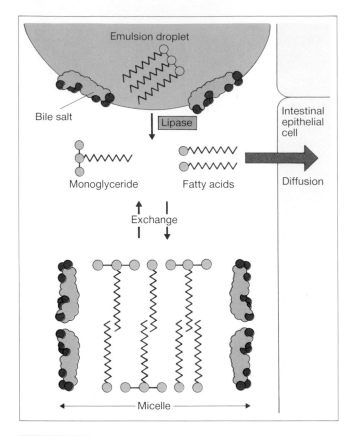

FIGURE 16-10 The products of fat digestion by lipase are held in solution in the micellar state, combined with bile salts. The micellar contents rapidly exchange with the free products in aqueous solution, which are able to diffuse into intestinal epithelial cells.

resynthesized into triacylglycerols. This occurs in the agranular endoplasmic reticulum, where the enzymes for triacylglycerol synthesis are located. Within this organelle, the resynthesized fat aggregates into small droplets coated with an amphipathic protein that performs an emulsifying function similar to that of a bile salt.

The exit of this fat droplet from the cell follows the same pathway as does a secreted protein (page 129). Vesicles containing the droplet pinch off the endoplasmic reticulum, are processed through the Golgi apparatus, and eventually fuse with the plasma membrane, releasing the fat droplet into the interstitial fluid. These small, extracellular fat droplets are known as **chylomicrons**. Chylomicrons contain, in addition to triacylglycerols, other lipids (including phospholipids, cholesterol, and fat-soluble vitamins) that have been absorbed by the same process that led to fatty acid and monoglyceride absorption.

The chylomicrons pass into the lacteals rather than into the capillaries. The chylomicrons cannot enter the capillaries because a basement membrane (an extracellular polysaccharide layer) at the outer surface of the capillary provides a barrier to the relatively large chylomicrons. The lacteals do not have basement membranes, and thus the chylomicrons can pass through the lacteal wall into the lymph. As everywhere else in the body, the lymph from the small intestine eventually empties into systemic veins. In the next chapter we describe how the lipids in the chylomicrons, circulating in the blood, are made available to the cells of the body.

Figure 16-11 summarizes the pathway taken by fat in moving from the intestinal lumen into the lymphatic system.

Vitamins

Most vitamins undergo little enzymatic modification during digestion and absorption. Digestion releases the vitamins from food particles, transferring them to a soluble form in which they can be absorbed.

The fat-soluble vitamins—A, D, E, and K—follow the pathway for fat absorption just described. They are solubilized in micelles. Thus, any interference with the secretion of bile or the action of bile salts in the intestine decreases the absorption of the fat-soluble vitamins.

Most water-soluble vitamins are absorbed by diffusion or carrier-mediated transport. However, one—vitamin B_{12}—is a very large, charged molecule. In order to be absorbed, vitamin B_{12} must first bind to a protein, known as **intrinsic factor**, secreted by the acid-secreting cells in the stomach. The resulting complex then binds to specific sites on the epithelial cells in the lower portion of the ileum, where vitamin B_{12} is absorbed. As described on page 354, vitamin B_{12} is required for erythrocyte formation and prevention of anemia. This form of anemia may occur when the stomach has been removed (as, for example, to treat ulcers or gastric cancer) or fails to secrete intrinsic factor. Since the absorption of vitamin B_{12} occurs in the lower part of the ileum, removal of this segment because of disease can also result in anemia.

Water and Minerals

Water is the most abundant substance in chyme. Approximately 9000 mL of ingested and secreted fluid enters the small intestine each day, but only 500 mL is passed on to the large intestine, since 95 percent of the fluid is absorbed in the small intestine. The epithelial membranes are very permeable to water. Therefore, net water diffusion occurs across the epithelium whenever a water concentration difference is established by the active absorption of solutes (page 131).

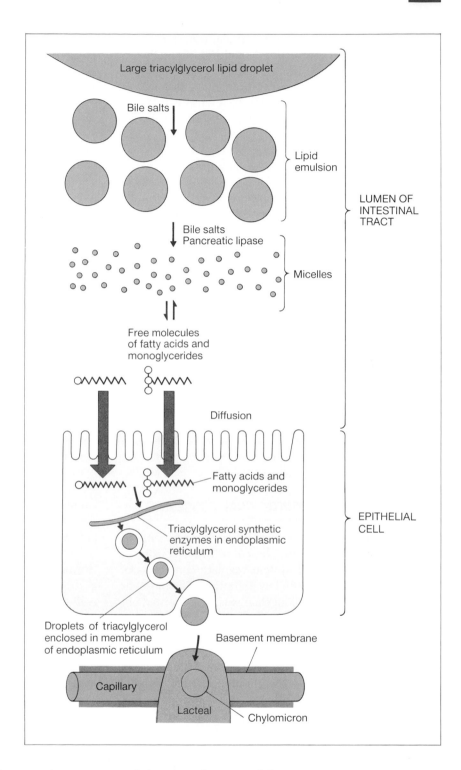

Large triacylglycerol lipid droplet

Bile salts

Lipid emulsion

Bile salts
Pancreatic lipase

Micelles

Free molecules
of fatty acids and
monoglycerides

Diffusion

Fatty acids and
monoglycerides

Triacylglycerol synthetic
enzymes in endoplasmic
reticulum

LUMEN OF
INTESTINAL
TRACT

EPITHELIAL
CELL

Droplets of triacylglycerol
enclosed in membrane
of endoplasmic reticulum

Basement membrane

Capillary

Lacteal

Chylomicron

FIGURE 16-11 Summary of fat absorption across the walls of the small intestine.

Sodium ions account for most of the actively transported solutes because they constitute the most abundant solute in chyme. Sodium absorption is a primary active process, using the Na,K-ATPase pumps in a manner similar to that for renal tubular sodium reabsorption. Other minerals present in smaller concentrations, such as potassium, magnesium, and calcium, are also absorbed, as are trace elements, such as iron, zinc, and

iodide. Consideration of the transport processes associated with all these is beyond the scope of this book, and we shall briefly consider as an example the absorption of only one of them—iron.

Only about 10 percent of ingested iron is absorbed into the blood each day. Iron ions are actively transported into intestinal epithelial cells, where most of them are incorporated into ferritin, the protein-iron complex that functions as an intracellular iron store (page 353). The absorbed iron that does not bind to ferritin is released on the blood side and bound to the plasma protein, transferrin, in which form it circulates throughout the body. Most of the iron bound to ferritin is released back into the intestinal lumen when the intestinal cells disintegrate at the tip of the villus.

Iron absorption depends on the body's iron content. When body stores are ample, the amount of iron bound to ferritin increases, which reduces the amount available for release to the blood. When the body stores drop, as, for example, when there is a loss of hemoglobin during hemorrhage, the amount of iron bound to intestinal ferritin decreases, increasing the iron available for release into the blood. The feedback pathways that control the formation of intestinal ferritin are unknown.

Iron absorption also depends on the type of food ingested. This is because iron binds to many negatively charged ions in food that can retard its absorption. For example, iron in ingested liver is much more absorbable than is iron in egg yolk since the latter contains phosphates that bind the iron to form an insoluble complex.

The absorption of iron is typical of most trace metals in several respects: (1) Cellular storage proteins and plasma carrier proteins are involved, and (2) the control of absorption is the major mechanism for the homeostatic control of the body's content of the trace metal.

REGULATION OF GASTROINTESTINAL PROCESSES

Unlike control systems that regulate variables in the internal environment, the control mechanisms of the gastrointestinal system regulate conditions in the lumen of the tract. With few exceptions like those for trace metals, these control mechanisms are governed not by the nutritional state of the body, but rather by the volume and composition of the luminal contents.

Basic Principles

Gastrointestinal reflexes are initiated by a relatively small number of luminal stimuli: (1) distension of the wall by the luminal contents; (2) chyme osmolarity (total solute concentration); (3) chyme acidity; and (4) the concentrations of specific digestion products—monosaccharides, fatty acids, peptides, and amino acids. These stimuli act on receptors located in the wall of the tract—mechanoreceptors, osmoreceptors, and chemoreceptors—to trigger reflexes that influence the effectors—the muscle layers in the wall of the tract and the exocrine glands that secrete substances into its lumen.

Neural regulation. The gastrointestinal tract has its own local nervous system, known as the **enteric nervous system**, in the form of two nerve networks, the myenteric plexus and the submucous plexus (Figure 16-5). There neurons either synapse with other neurons in the plexus or end near smooth muscles and glands. Many axons leave the myenteric plexus and synapse with neurons in the submucous plexus, and vice versa, so that neural activity in one plexus influences the activity in the other. Moreover, stimulation at one point in the plexus can lead to impulses that are conducted both up and down the tract. Thus, stimuli in the upper part of the small intestine may affect smooth-muscle and gland activity in the stomach as well as in the lower part of the intestinal tract. Many of the receptors mentioned earlier are, in fact, parts of plexus neurons.

Nerve fibers from both the sympathetic and parasympathetic branches of the autonomic nervous system enter the intestinal tract and synapse with neurons in both plexuses. Via these pathways, the CNS can influence the motor and secretory activity of the gastrointestinal tract. In addition, there are intratract neural reflexes that are independent of the CNS. Thus, two types of neural reflex arcs exist (Figure 16-12): **short reflexes** from receptors through the nerve plexuses to effector cells, and **long reflexes** from receptors to the CNS by way of afferent nerves and back to the nerve plexuses and effector cells by way of the autonomic nerve fibers. Some controls are mediated solely by short reflexes or by long reflexes, whereas others use both.

Finally, it should be noted that not all neural reflexes are initiated by signals within the tract. The sight or smell of food and the emotional state of an individual can have significant effects on the gastrointestinal tract that are mediated by the CNS via autonomic neurons.

Hormonal regulation. The hormones that control the gastrointestinal system are secreted mainly by endocrine cells scattered throughout the epithelium of the stomach and small intestine; that is, these cells are not clustered into discrete organs like the thyroid or adrenal glands. One surface of each endocrine cell is exposed to the lumen of the gastrointestinal tract. At this surface, various chemical substances in the chyme stimulate the cell to release its hormones from the opposite, blood side of the cell. Although some of these hormones can also be detected in the lumen and may therefore be acting locally as paracrines, most of the gastrointestinal hormones reach their target cells via the circulation.

Several dozen substances are currently being investigated as possible gastrointestinal hormones, but only four—**secretin, cholecystokinin (CCK), gastrin,** and **glucose insulinotropic peptide (GIP)**[2]—have met all the criteria for hormones. They, as well as several candidate

---

[2]GIP was originally called gastric inhibitory peptide.

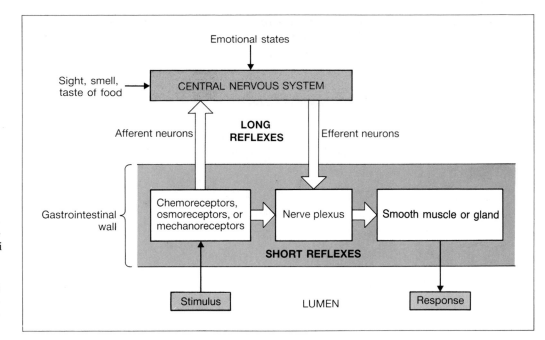

FIGURE 16-12 Long and short neural reflex pathways initiated by stimuli in the gastrointestinal tract. The long reflexes utilize neurons that link the central nervous system to the gastrointestinal tract.

hormones, also exist in the CNS and in gastrointestinal plexus neurons, where they may be functioning as neurotransmitters or neuromodulators.

Table 16-2, which summarizes the major characteristics of the four established GI hormones, not only serves as a reference for future discussions but also illustrates several generalizations: (1) Each hormone participates in a feedback control system that regulates some aspect of the luminal environment, and (2) each hormone affects more than one type of target cell.

These two generalizations can be illustrated by CCK. The presence of fatty acids in the small intestine triggers CCK secretion from the duodenum into the blood, and circulating CCK then stimulates the secretion from the pancreas of digestive enzymes, including lipase, which digests fat. CCK also causes the gallbladder to contract, delivering to the intestine the bile salts required for fat digestion and absorption. As fat is digested and absorbed, the stimulus (fatty acids in the lumen) for CCK release is removed.

In many cases, a single effector cell contains receptors for more than one hormone, as well as receptors for neurotransmitters and paracrines, with the result that a variety of inputs can affect the cell's response. One example of such interactions is the phenomenon known as potentiation. CCK strongly stimulates pancreatic enzyme secretion, whereas the hormone secretin is a weak stimulus for enzyme secretion. In the presence of secretin, however, CCK stimulates the secretion of pancreatic enzymes more strongly than would be predicted by the sum of the individual stimulatory effects of CCK and

secretin. Thus, secretin potentiates the effect of CCK. One of the consequences of potentiation is that small changes in the plasma concentration of a particular hormone can have considerable effects on the actions of other gastrointestinal hormones.

In addition to their stimulation (or in some cases inhibition) of effector-cell functions, the gastrointestinal hormones also have tropic (growth-promoting) effects on various tissues, including the gastric and intestinal mucosa and the exocrine portions of the pancreas.

Phases of gastrointestinal control. The neural and hormonal control of the gastrointestinal system is, in large part, divisible into three phases—cephalic, gastric, and intestinal—according to stimulus location.

The **cephalic phase** is initiated when receptors in the head (*cephalic*, head) are stimulated by sight, smell, taste, and chewing as well as by various emotional states. The efferent pathways for these reflexes are parasympathetic fibers, most of which are in the vagus nerves, and sympathetic fibers. These fibers activate neurons in the gastrointestinal nerve plexuses, which in turn affect secretory and contractile activity.

Three stimuli in the stomach initiate the reflexes that constitute the **gastric phase** (*gastric*, stomach) of regulation—distension, low acidity, and peptides formed during the digestion of ingested protein. The responses to these stimuli are mediated by short and long neural reflexes and by gastrin release.

Finally, the **intestinal phase** is initiated by stimuli in the intestinal tract—distension, high acidity, osmolarity,

TABLE 16-2   PROPERTIES OF GASTROINTESTINAL HORMONES

	Gastrin	CCK	Secretin	GIP
Structure	Peptide	Peptide	Peptide	Peptide
Endocrine cell location	Antrum of stomach	Small intestine	Small intestine	Small intestine
Stimuli for hormone release	Amino acids, peptides in stomach; parasympathetic nerves	Amino acids, fatty acids in small intestine	Acid in small intestine	Glucose, fat in small intestine
Stimuli inhibiting hormone release	Acid in stomach lumen			
Target-Cell Responses				
Stomach				
Acid secretion	Stimulates		Inhibits	
Antrum contraction	Stimulates		Inhibits	
Pancreas				
Bicarbonate secretion		Potentiates secretin's actions	Stimulates	
Enzyme secretion		Stimulates	Potentiates CCK's actions	
Insulin secretion				Stimulates
Liver				
Bicarbonate secretion		Potentiates secretin's actions	Stimulates	
Gallbladder contraction		Stimulates		
Sphincter of Oddi		Relaxes		
Small intestine motility	Stimulates ileum; inhibits ileocecal sphincter			
Large intestine motility	Stimulates mass movement			
Growth of	Stomach and small intestine	Exocrine pancreas	Exocrine pancreas	

and various digestive products. Like the gastric phase, the intestinal phase is mediated by both long and short reflexes and by gastrointestinal hormones, in this case secretin, CCK, and GIP.

It must be emphasized that each of these phases is named for the site at which the various stimuli initiate the reflex and not for the sites of effector activity. Each phase is characterized by efferent output to virtually all organs in the gastrointestinal tract. These phases do not occur in temporal sequence except at the very beginning of a meal. Rather, during ingestion and the much longer absorptive period, reflexes characteristic of all three phases may be occurring simultaneously.

Keeping in mind the neural and hormonal mechanisms available for regulating gastrointestinal activity, we can now examine the specific contractile and secretory processes that occur in each segment of the gastrointestinal system.

Mouth, Pharynx, and Esophagus

Chewing.   Chewing is controlled by the somatic nerves to the skeletal muscles of the mouth and jaw. In addition to the voluntary control of these muscles, rhythmic chewing motions are reflexly activated by the pressure of food against the gums, hard palate at the roof of the mouth, and tongue. Activation of these mechanoreceptors leads to reflexive inhibition of the muscles holding the jaw closed. The resulting relaxation of the jaw reduces the pressure, leading to a new cycle of contraction and relaxation.

Although chewing prolongs the subjective pleasure of taste, it does not appreciably alter the rate at which the food will be digested and absorbed from the small intestine. On the other hand, attempting to swallow a large particle of food can lead to choking if the particle lodges over the trachea, blocking the entry of air into the lungs. A number of preventable deaths occur each year from choking, the symptoms of which are often confused with those of a heart attack so that no attempt is made to remove the obstruction from the airway. The Heimlich maneuver, described on page 439, can often dislodge the obstructing particle from the airways.

Saliva.   Saliva is secreted by three pairs of exocrine glands: the **parotid**, the **submandibular**, and the **sublingual** (Figure 16-1). The major salivary proteins are the enzyme amylase and the **mucins**, which when mixed with water form the highly viscous solution known as mucus.

The secretion of saliva is controlled by both sympathetic and parasympathetic neurons. Unlike their antagonistic activity in most organs, however, both systems stimulate salivary secretion, the parasympathetics producing the greater response. In the absence of ingested material, a low rate of salivary secretion keeps the mouth moist. In the presence of food, salivary secretion increases markedly. This reflex response is initiated by chemoreceptors and pressure receptors in the walls of the mouth and tongue.

Swallowing.   Swallowing is a complex reflex initiated when pressure receptors in the walls of the pharynx are stimulated by food or drink forced into the rear of the mouth by the tongue. These receptors send afferent impulses to the **swallowing center** in the brainstem medulla, which coordinates the swallowing process via efferent fibers to the muscles in the pharynx, larynx, and esophagus, as well as to the respiratory muscles.

As the ingested material moves into the pharynx, the soft palate is elevated and lodges against the back wall of the pharynx, preventing food from entering the nasal cavity (Figure 16-13B). Impulses from the swallowing center inhibit respiration, raise the larynx, and close the glottis (the area around the vocal cords at the beginning of the trachea), keeping food from moving into the trachea. As the tongue forces the food further back into the pharynx, the food tilts a flap of tissue, the epiglottis, backward to cover the closed glottis (Figure 6-13C).

The next stage of swallowing occurs in the esophagus, the foot-long tube that passes through the thoracic cavity, penetrates the diaphragm, which separates the thoracic cavity from the abdominal cavity, and joins the stomach a few centimeters below the diaphragm. Skeletal muscles surround the upper third of the esophagus, smooth muscles the lower two-thirds

As was described in Chapter 14, the pressure in the thoracic cavity is 4 to 10 mmHg less than atmospheric, and this subatmospheric pressure is transmitted across the thin wall of the esophagus to the lumen. In contrast,

FIGURE 16-13 Movements of food through the pharynx and upper esophagus during swallowing.

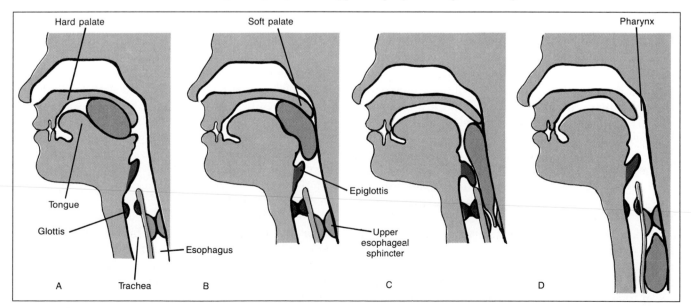

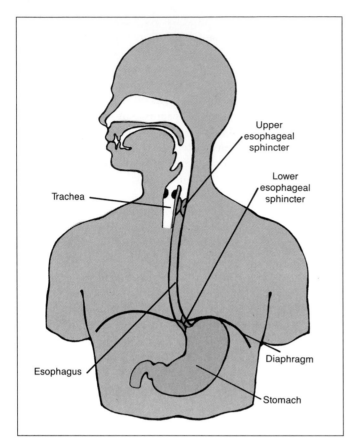

Upper esophageal sphincter

Lower esophageal sphincter

Trachea

Diaphragm

Esophagus

Stomach

FIGURE 16-14 Location of upper and lower esophageal sphincters.

the luminal pressure at the beginning of the esophagus is equal to atmospheric pressure, and the pressure at the opposite end of the esophagus in the stomach is slightly greater than atmospheric. These pressure differences would tend to force air (from above) and stomach contents (from below) into the esophagus. This does not occur, however, because both ends of the esophagus are normally closed by the contraction of sphincter muscles. Skeletal muscles surround the esophagus just below the pharynx, forming the **upper esophageal sphincter**, whereas the smooth muscles in the last portion form the **lower esophageal sphincter** (Figure 16-14).

The esophageal phase of swallowing begins with relaxation of the upper esophageal sphincter. Immediately after the food has passed, the sphincter closes, the glottis opens, and breathing resumes. Once in the esophagus, the food is moved toward the stomach by a progressive wave of muscle contractions that proceeds along the esophagus, compressing the lumen and forcing the food ahead of it. Such waves of contraction in the muscle layers surrounding a tube are known as **peristaltic waves**.

One esophageal peristaltic wave takes about 9 s to reach the stomach.

Swallowing can occur even while a person is upside down since it is not primarily gravity but the peristaltic wave that moves the food to the stomach. The lower esophageal sphincter opens and remains relaxed throughout the period of swallowing, allowing the arriving food to enter the stomach. After the food has passed, the sphincter closes, resealing the junction between the esophagus and the stomach.

Swallowing is an example of a reflex in which multiple responses occur in a temporal sequence determined by the pattern of synaptic connections between neurons in the coordinating center. Since both skeletal and smooth muscles are involved, the swallowing center must direct efferent activity in both somatic nerves (to skeletal muscle) and autonomic nerves, specifically parasympathetic fibers (to smooth muscle). Simultaneously, afferent fibers from receptors in the esophageal wall send to the swallowing center information that can alter the efferent activity. For example, if a large food particle does not reach the stomach during the initial peristaltic wave, the distension of the esophagus by the particle activates receptors that initiate reflexes causing repeated waves of peristaltic activity (**secondary peristalsis**).

The ability of the lower esophageal sphincter to maintain a barrier between the stomach and the esophagus is aided by the fact that its last portion lies below the diaphragm and is subject to the same abdominal pressures as is the stomach. In other words, if the pressure in the abdominal cavity is raised, for example, during cycles of respiration or by contraction of the abdominal muscles, the pressures on both the stomach contents and the terminal segment of the esophagus are raised together, preventing the formation of a pressure difference that could force the stomach contents into the esophagus.

During pregnancy the growth of the fetus not only increases the pressure on the abdominal contents but also pushes the terminal segment of the esophagus through the diaphragm into the thoracic cavity. The sphincter is therefore no longer assisted by changes in abdominal pressure. Accordingly, during the last half of pregnancy there is a tendency for increased abdominal pressures to force some of the stomach contents up into the esophagus. The hydrochloric acid from the stomach irritates the esophageal walls, causing contractile spasms of the smooth muscle that produces pain known as **heartburn** (because the pain appears to be located over the heart). Heartburn often subsides in the last weeks of pregnancy as the uterus descends lower into the pelvis prior to delivery, decreasing the pressure on the abdominal organs. Heartburn can also occur in anyone after a large meal, which can raise the pressure in the stomach enough to force acid into the esophagus.

Stomach

Glands in the thin-walled upper portions of the stomach, the **body** and **fundus** (Figure 16-15), secrete mucus, hydrochloric acid, and the enzyme precursor pepsinogen. The lower portion of the stomach, the **antrum**, has a much thicker layer of smooth muscle. The glands in this region secrete little acid, but they do contain the endocrine cells that secrete the hormone gastrin.

The epithelial layer lining the stomach invaginates into the mucosa, forming numerous tubular glands (Figure 16-16). The cells at the opening of these glands secrete mucus, whereas lining the walls of the glands are **parietal cells**[3], which secrete acid and **chief cells**, which secrete pepsinogen. Thus, each of the three major gastric exocrine secretions—mucus, acid, and pepsinogen—is secreted by a distinct cell type.

HCl secretion.

The stomach secretes about 2 L of hydrochloric acid per day. The concentration of hydrogen ions in the stomach lumen may reach 150 mM, 3 million times greater than the concentration in the blood.

A primary H^+-ATPase in the luminal membrane of the parietal cells pumps hydrogen ions into the stomach's lumen (Figure 16-17). As hydrogen ions are secreted into the lumen, bicarbonate ions are being secreted on the opposite side of the cell into the blood, lowering the acidity in the venous blood from the stomach.

---

[3] Parietal cells are also known as oxyntic cells.

During a meal, the rate of HCl secretion increases markedly. Table 16-3 summarizes the many factors controlling this secretion. First, during the cephalic phase, the message for increased acid secretion reaches the stomach by way of parasympathetic nerves that synapse on neurons in the stomach-wall nerve plexus. Axons from these enteric neurons terminate near parietal cells, releasing acetylcholine, which stimulates acid secretion. Some of the enteric neurons also end near gastrin-releasing cells in the antrum, and their stimulation releases gastrin, which in turn stimulates acid secretion by parietal cells.

Once food has reached the stomach, the gastric phase ensues in response to a variety of intragastric stimuli—distension, low acidity, and peptides derived from protein digestion. Distension produces its effects through receptors located in the stomach wall that activate both long and short reflexes leading to the simulation of parietal-cell acid secretion. In contrast, peptides and acid in the stomach's lumen exert their effects on acid secretion via gastrin.

The greater the protein content of a meal, the greater the amount of acid secretion. This occurs for two reasons. First, peptides formed by the digestive action of pepsin stimulate the release of gastrin. The second reason is more complicated but is related to the fact that a high acid concentration *inhibits* the release of gastrin and hence inhibits gastrin-stimulated acid secretion. Before food enters the stomach, there is a low rate of acid secretion, but the H^+ *concentration* in the lumen is high because there are few buffers present to bind the

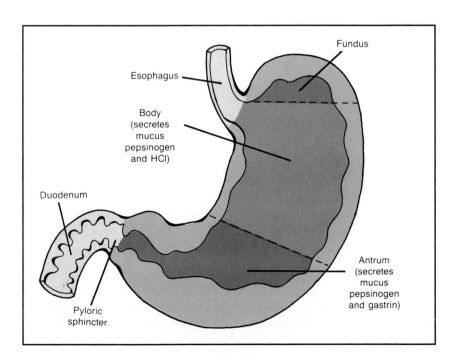

FIGURE 16-15 The three regions of the stomach: fundus, body, and antrum.

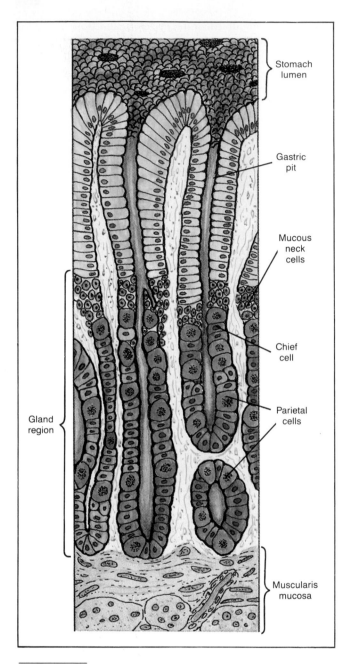

Stomach
lumen

Gastric
pit

Mucous
neck
cells

Chief
cell

Gland
region

Parietal
cells

Muscularis
mucosa

FIGURE 16-16 Gastric glands in the body of the stomach.

This sequence provides a negative-feedback control that regulates luminal acid concentration.

Ingested substances other than protein can also stimulate acid secretion when they reach the stomach. For example, caffeine, a substance found in coffee, tea, chocolate, and cola drinks, stimulates gastrin release and thus increases acid secretion. Alcohol, contrary to popular belief, has little direct effect on gastric acidity in humans.

We now come to the intestinal phase of acid secretion control, the phase in which stimuli in the early portion of the small intestine reflexly influence acid secretion by the stomach. Excess acidity in the duodenum triggers reflexes that inhibit gastric acid secretion. This is beneficial for the following reason. The digestive activity of enzymes and bile salts in the small intestine is strongly inhibited by acid solutions, and so it is essential that the chyme entering the small intestine from the stomach not contain so much acid that it cannot be rapidly neutralized by the bicarbonate-rich fluids simultaneously secreted into the intestine by the liver and pancreas.

Duodenal chyme acidity is not the only factor triggering reflex inhibition of gastric acid secretion. Distension, hypertonic solutions, and solutions containing amino acids, fatty acids, or monosaccharides also do so. Thus, the extent to which acid secretion is inhibited during the intestinal phase varies, depending upon the volume and composition of the meal, but the net result is the same—balancing the secretory activity of the stomach with the digestive and absorptive capacities of the small intestine.

The inhibition of gastric secretion during the intestinal phase is mediated by short and long reflexes and by hormones that inhibit either parietal-cell acid secretion or gastrin secretion. The hormone or hormones released by the intestinal tract that inhibit stomach activity are called **enterogastrones**, a general term. A portion of enterogastrone activity may be due to secretin and CCK, but additional unidentified hormones may be involved.

Acid secretion can also be altered by several paracrines, especially histamine and prostaglandins. **Histamine**, released from the stomach mucosa, can produce a marked increase in parietal-cell acid secretion (indeed, histamine is often administered clinically to measure the maximal acid-secreting capacity of the stomach), but its physiological role in the normal control of acid secretion remains unclear.

Pepsin secretion. Pepsin is secreted by chief cells in an inactive precursor form known as **pepsinogen**. The high acidity in the stomach lumen alters the shape of pepsinogen, exposing its active site so that this site can act on other pepsinogen molecules to break off a small

hydrogen ions. The high H$^+$ concentration inhibits gastrin release. The protein in food is an excellent buffer so that as food enters the stomach, the concentration of free hydrogen ions drops as the hydrogen ions bind to the proteins. This decrease in acidity removes the inhibition of gastrin secretion and, thereby, of acid secretion. The more protein in a meal, the greater the buffering of acid, the more gastrin secreted, and the more acid secreted.

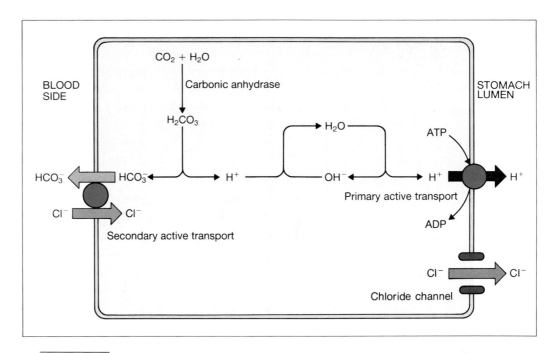

FIGURE 16-17 Secretion of hydrochloric acid by parietal cells. The hydrogen ions secreted into the lumen by primary active transport are derived from the breakdown of water molecules, leaving hydroxyl ions (OH^-) behind. These hydroxyl ions are neutralized by combination with other hydrogen ions generated by the reaction between carbon dioxide and water, a reaction catalyzed by the enzyme carbonic anhydrase, which is present in high concentrations in parietal cells. The bicarbonate ions formed by this reaction move out of the parietal cell on the blood side, in exchange for chloride ions.

chain of amino acids from their ends. This cleavage converts pepsinogen to a fully active form, pepsin (Figure 16-18). The activation of pepsin is thus an autocatalytic, positive-feedback process.

The synthesis and secretion of pepsinogen, followed by its intraluminal activation to pepsin, provides an example of a process that occurs with other proteolytic enzymes in the gastrointestinal tract. Because the en-

TABLE 16-3 CONTROL OF HCl SECRETION DURING A MEAL		
Stimuli	Pathways to the Parietal Cells	Result
Cephalic phase 　Sight 　Smell 　Taste 　Chewing	Parasympathetic nerves, gastrin	↑ HCl secretion
Gastric contents (gastric phase) 　Distension 　↓ H^+ concentration 　↑ Peptides	Long and short neural reflexes, gastrin	↑ HCl secretion
Intestinal contents (intestinal phase) 　Distension 　↑ H^+ concentration 　↑ Osmolarity 　↑ Nutrient concentrations	Long and short neural reflexes, secretin, CCK, and other unspecified hormones	↓ HCl secretion

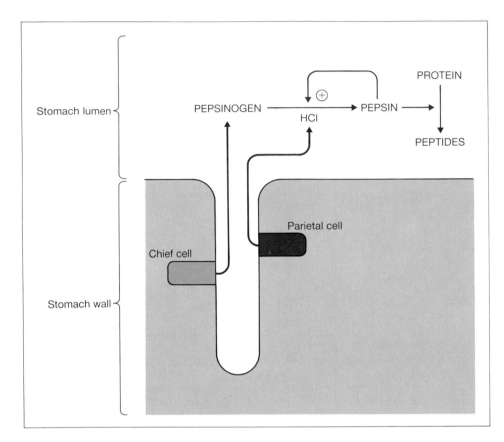

FIGURE 16-18 Conversion of pepsinogen to pepsin in the stomach.

zymes are synthesized in an inactive form, the intracellular proteins that could be substrates for the enzymes are protected from digestion, thus preventing the destruction of the enzyme-synthesizing cell.

Pepsin is active only in the presence of a high H^+ concentration. It becomes inactive, therefore, when it enters the small intestine, where the hydrogen ions are neutralized by the bicarbonate ions secreted into the small intestine. Since high acidity is required for both pepsin activation and pepsin activity, it is appropriate that, as we have seen, peptide fragments produced by pepsin activity are a stimulus for acid secretion.

Only a fraction of the total protein in a meal is digested by pepsin. Even in the absence of pepsin, as occurs in some pathological conditions, protein can be completely digested by the enzymes in the small intestine.

The primary pathway for stimulating pepsinogen secretion is input to the chief cells from the nerve plexuses. During the cephalic, gastric, and intestinal phases, most of the factors that stimulate or inhibit acid secretion exert the same effect on pepsinogen secretion. Thus pepsinogen secretion parallels acid secretion.

Gastric Motility. The empty stomach has a volume of only about 50 mL, and the diameter of its lumen is only slightly larger than that of the small intestine. When a meal is swallowed, however, the smooth muscle in the fundus and body relaxes prior to the arrival of the food, allowing the stomach's volume to increase to as much as 1.5 L with little increase in pressure. This **receptive relaxation** is mediated by the parasympathetic nerves and coordinated by the swallowing center in the brain.

As in the esophagus, the stomach's primary contractile activity produces peristaltic waves. Each wave begins in the body of the stomach and produces only a ripple as it proceeds toward the antrum, a contraction too weak to produce much mixing of the luminal contents with acid and pepsin. As the wave approaches the larger mass of wall muscle surrounding the antrum, however, it produces a more powerful contraction that both mixes the luminal contents and closes the **pyloric sphincter**, a ring of smooth muscle and connective tissue between the antrum and the duodenum (Figure 16-19). The pyloric sphincter is normally relaxed and closes only upon arrival of a peristaltic wave. As a consequence of sphincter closing, only a small amount of chyme is expelled into the duodenum with each wave, and most of the antral contents are forced backward toward the body of the stomach.

What is responsible for producing these gastric peristaltic waves? Their (three-per-minute) rhythm is gener-

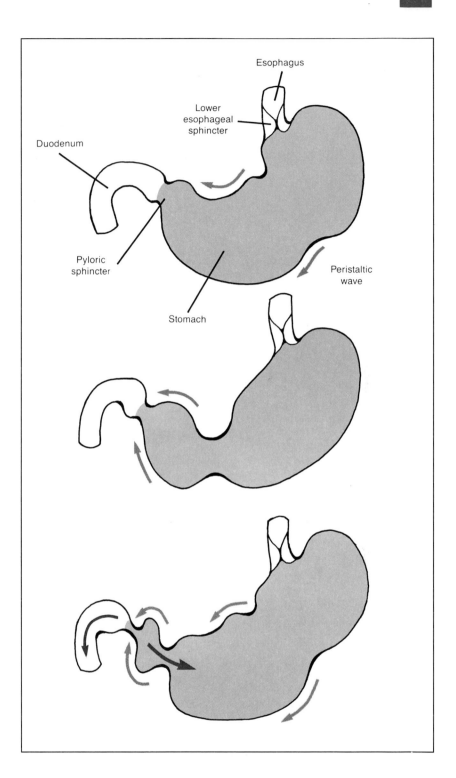

FIGURE 16-19 Peristaltic waves passing over the stomach force a small amount of luminal material into the duodenum. When a strong peristaltic wave arrives at the antrum, the pressure on the antral contents is increased, but the antral contraction also closes the pyloric sphincter, with the result that the antral contraction forces most of the antral contents back into the body of the stomach. This contributes to the mixing of the antral contents.

ated by pacemaker cells in the longitudinal muscle layer that undergo spontaneous depolarization-repolarization cycles (slow waves) known as the **basic electrical rhythm**. These slow waves are propagated through gap junctions along the stomach's longitudinal muscle layer and also induce similar slow waves in the overlying circular muscle layer. In the absence of neural or hormonal input, however, these depolarizations are too small to cause the muscle membranes to reach threshold and fire the action potentials required to elicit contractions. Excitatory neurotransmitters and hormones act upon the muscle to depolarize the membrane, thereby bringing it closer to threshold. Action potentials are then generated at the peak of the slow wave cycle (Figure 16-20), and

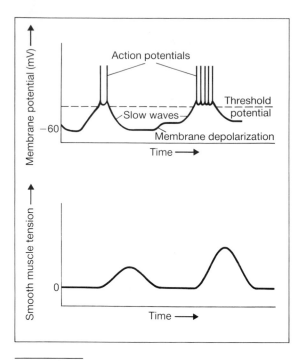

FIGURE 16-20 Slow wave oscillations in the membrane potential of gastric smooth-muscle fibers trigger bursts of action potentials at the wave peak when threshold potential is reached. Membrane depolarization brings the slow wave close to threshold increasing the action potential frequency and thus the force of smooth-muscle contraction.

the number of spikes fired with each wave determines the strength of the elicited muscle contraction.

Thus, whereas the frequency of contraction is determined by the basic electrical rhythm and remains essentially constant, the force of contraction and therefore the amount of gastric emptying per contraction are determined by neural and hormonal input to the antral smooth muscle.

The initiation of these reflexes depends upon the contents of both the stomach and small intestine. First, all the factors previously discussed that regulate gastrin release also alter gastric motility indirectly since gastrin increases the force of antral contractions. Distension of the stomach also increases the force of antral contractions through long and short reflexes triggered by mechanoreceptors in the stomach. Therefore, the larger a meal, the faster the stomach's initial emptying rate, and as the volume of the stomach decreases, so do the force of gastric contractions and the rate of emptying.

In contrast, distension of the duodenum or the presence of fat, acid, or hypertonic solutions in its lumen all inhibit gastric emptying (Figure 16-21). Fat is the most

potent of the chemical stimuli. These responses to intestinal stimuli are mediated by enterogastrones and by long and short neural reflexes.

Autonomic nerve fibers to the stomach can be activated independently of the reflexes originating in the stomach and duodenum and can influence gastric motility. Decreased parasympathetic or increased sympathetic activity inhibits motility. Via these pathways, pain and emotions such as sadness, depression, and fear tend to decrease motility, whereas aggression or anger tend to increase it. These relationships are not always predictable, however, and different people show different gastrointestinal responses to apparently similar emotional states.

Vomiting.   Vomiting is the forceful expulsion of the contents of the stomach and upper intestinal tract through the mouth. Like swallowing, vomiting is a complex reflex coordinated by a region in the brainstem medulla, in this case known as the **vomiting center**. Neural input to this center from receptors in many different regions of the body can initiate the vomiting reflex. For example, excessive distension of the stomach or small intestine, various substances acting upon chemoreceptors in the intestinal wall or in the brain, increased pressure within the skull, rotating movements of the head (motion sickness), intense pain, and tactile stimuli applied to the back of the throat can all initiate vomiting.

What is the adaptive value of this reflex? Obviously, the removal of ingested toxic substances before they can be absorbed is of benefit. Moreover, the nausea that usually accompanies vomiting may have the adaptive value of conditioning the individual to avoid the future ingestion of foods containing such toxic substances. Why other types of stimuli, such as those producing motion sickness, have become linked to the vomiting center is not clear.

Vomiting is usually preceded by increased salivation, sweating, increased heart rate, pallor, and feelings of nausea—all characteristic of a general sympathetic nervous system discharge in response to stress. The events leading to vomiting begin with a deep inspiration, closure of the glottis, and elevation of the soft palate. The abdominal muscles then contract, raising the abdominal pressure, which is transmitted to the stomach's contents. The lower esophageal sphincter relaxes, and the high abdominal pressure forces the contents of the stomach into the esophagus. This initial sequence of events can occur repeatedly without expulsion via the mouth and is known as **retching**. Vomiting occurs when the abdominal contractions become so strong that the increased intrathoracic pressure forces the contents of the esophagus through the upper esophogeal sphincter.

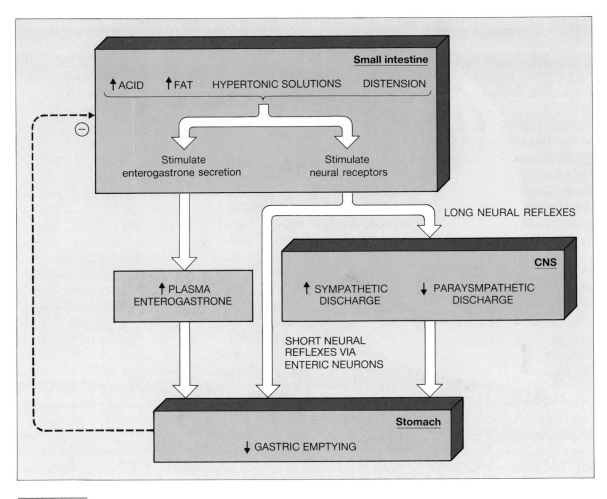

FIGURE 16-21 Intestinal-phase pathways inhibiting gastric emptying.

Vomiting is also accompanied by strong contractions in the upper portion of the small intestine, contractions that tend to force some of the intestinal contents back into the stomach from which they can be expelled. Thus, some bile may be present in the vomitus.

Excessive vomiting can lead to large losses of the water and salts that normally would be absorbed in the small intestine. This can result in severe dehydration, upset the body's salt balance, and produce circulatory problems due to a decrease in plasma volume. The loss of gastric acid lowers blood acidity.

Pancreatic Secretions

The exocrine portion of the pancreas secretes bicarbonate ions and a number of digestive enzymes into ducts that converge into the pancreatic duct, the latter joining the common bile duct from the liver just before entering the duodenum (Figure 16-22). The enzymes are secreted by cells at the base of the exocrine glands, whereas bicarbonate ions are secreted by the cells lining the early portion of the pancreatic ducts.

The mechanism of bicarbonate secretion is somewhat analogous to that of hydrochloric acid secretion by the stomach, except that the directions of hydrogen-ion and bicarbonate-ion movement are reversed. Hydrogen and bicarbonate ions are formed in the duct lumen by the action of carbonic anhydrase located on the luminal membranes of the duct cells. The hydrogen ions are then actively transported into the duct cells and released into the blood, whereas the bicarbonate ions remain in the duct lumen.

The amount of bicarbonate secreted by the pancreas is approximately equal to the amount of acid secreted by the stomach. Therefore, the amount of acid released into the blood by the pancreas is normally equal to the amount of bicarbonate released into the blood by the stomach, the net result being little change in the acidity of the blood returning to the heart (Figure 16-23). How-

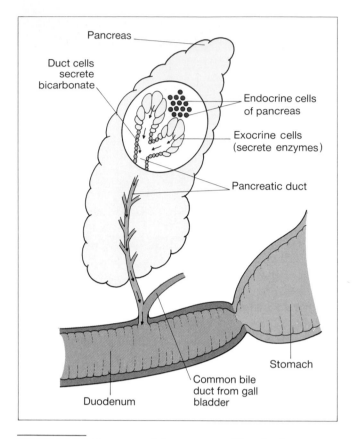

LUMEN OF
GASTROINTESTINAL
TRACT

HCO_3^- ← Stomach parietal cell → H^+

$+$ $+$

H^+ ← Pancreatic duct cells and liver → HCO_3^-

H_2CO_3 H_2CO_3

$CO_2 + H_2O$ $CO_2 + H_2O$

FIGURE 16-22 Structure of the pancreas. The gland areas are shown greatly enlarged relative to the entire pancreas and ducts.

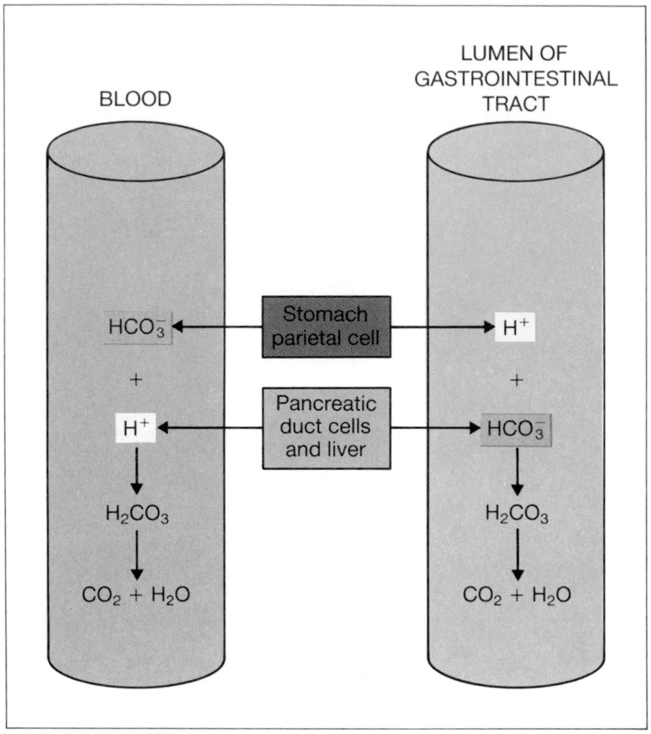

FIGURE 16-23 Acid secreted by the stomach is neutralized in the lumen of the small intestine by the bicarbonate secreted by the pancreas and liver. Bicarbonate released into the blood by the stomach is neutralized by the acid released into the blood by the pancreas and liver.

ever, the loss of large quantities of bicarbonate ions from the intestinal tract during periods of prolonged diarrhea leads to a net accumulation of acid in the blood, just as loss of acid from the stomach by vomiting leads to a net alkalinization of the blood.

The pancreatic enzymes digest fat, polysaccharides, proteins, and nucleic acids to fatty acids, sugars, amino acids, and nucleotides, respectively. A partial list of these enzymes and their activities is given in Table 16-4. Most of the proteolytic enzymes are secreted in inactive forms and then activated in the duodenum by other enzymes[4]. A key step in this process is mediated by **enterokinase**, which is embedded in the luminal plasma membrane of the intestinal epithelium. It is a proteolytic enzyme that splits off a peptide from pancreatic **trypsinogen**, forming the active enzyme trypsin. Trypsin is a proteolytic enzyme, and once activated, it activates the

other pancreatic proteolytic enzyme precursors by splitting off peptide fragments (Figure 16-24). This function is in addition to trypsin's role in digesting ingested protein.

Pancreatic secretion increases during a meal, due mainly to stimulation by the hormones secretin and CCK (Table 16-2). Secretin is the primary stimulant for bicarbonate secretion, whereas CCK stimulates enzyme secretion. (As noted earlier on page 527, each of these hormones has a potentiating effect on the activity of the other.)

Since the function of pancreatic bicarbonate is to neutralize the acid entering the duodenum from the stomach, it is appropriate that the major stimulus for secretin release is acid in the duodenum. As the acid in the duodenum is neutralized, the stimulus for secretin release is decreased, and less bicarbonate is secreted by the pancreas. In analogous fashion, since CCK stimulates the secretion of digestive enzymes, including those for fat and protein digestion, it is appropriate that the stimuli for its release are fatty acids and amino acids in the duodenum. Thus, the organic nutrients in the small intes-

---

[4] Several of the pancreatic enzymes, such as amylase and lipase, are secreted in their active forms but require additional factors, such as ions and bile salts encountered in the intestinal lumen, to produce maximal activity.

TABLE 16-4 PANCREATIC ENZYMES		
Enzyme	**Substrate**	**Action**
Trypsin, chymotrypsin	Proteins	Breaks peptide bonds in proteins to form peptide fragments
Carboxypeptidase	Proteins	Splits off terminal amino acid from carboxyl end of protein
Lipase	Fat	Splits off two fatty acids from triaylglycerols, forming free fatty acids and monoglycerides
Amylase	Polysaccharide	Splits polysaccharides into glucose and maltose
Ribonuclease, deoxyribonuclease	Nucleic acids	Splits nucleic acids into free mononucleotides

tine initiate, via hormonal reflexes, the secretions involved in their own digestion[5].

Although about 75 percent of the pancreatic exocrine secretions are controlled by stimuli arising from the intestinal phase of digestion, cephalic and gastric stimuli, by way of the parasympathetic nerves to the pancreas, also play a role. Thus, the taste of food or the distension of the stomach by food, will lead to increased pancreatic secretion. Figures 16-25 and 26 summarize the main factors controlling pancreatic secretion.

---

[5] As the amount of trypsin in the small intestine increases, the rate of CCK release decreases, and thus pancreatic enzyme secretion decreases. How trypsin initiates this negative-feedback control of pancreatic enzyme secretion is currently unknown.

Bile Secretion

Bile is secreted by liver cells into a number of small ducts, the bile canaliculi (Figure 16-27), which converge to form the common hepatic duct (Figure 16-28). A small sac branches from this duct on the underside of the liver, forming the gallbladder, which stores and concentrates the bile. Contraction of the gallbladder ejects the concentrated bile into the common bile duct, which is a continuation of the common hepatic duct, in response to various gastrointestinal stimuli. The gallbladder can be surgically removed without impairing bile secretion by the liver or its flow into the intestinal tract. In fact, many animals that secrete bile do not have a gallbladder.

Bile contains six major ingredients: (1) bile salts; (2) cholesterol; (3) lecithin (a phospholipid); (4) bicarbonate

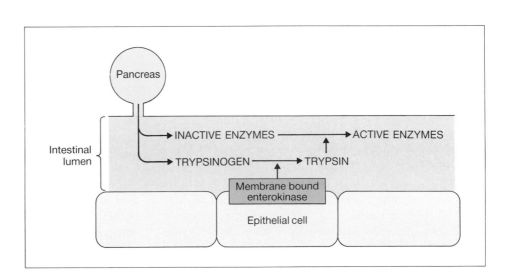

FIGURE 16-24 Activation of pancreatic enzyme precursors in the small intestine.

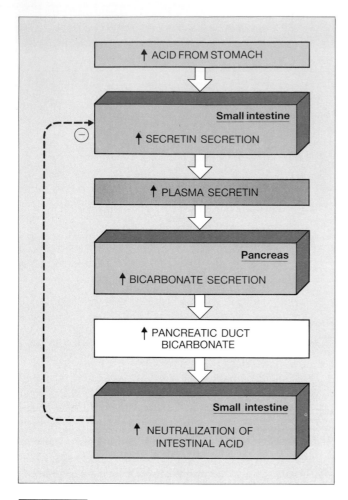

FIGURE 16-25 Hormonal regulation of pancreatic bicarbonate secretion.

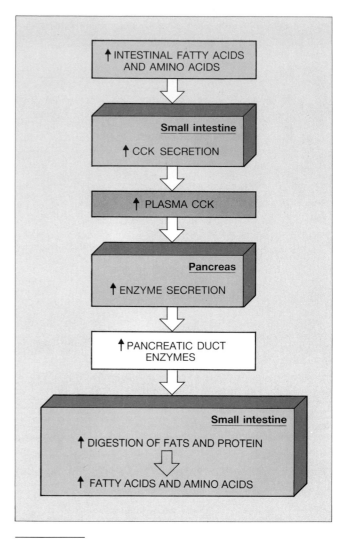

FIGURE 16-26 Hormonal regulation of pancreatic enzyme secretion.

ions and other salts; (5) bile pigments and small amounts of other end products of organic metabolism; and (6) trace metals. The first three ingredients are synthesized in the liver and solubilize fat in the small intestine. Bicarbonate ions help neutralize acid in the duodenum, and the fifth and sixth categories represent substances excreted from the body in the feces.

The **bile pigments** are substances formed when the heme portions of hemoglobin are broken down during the destruction of old or damaged erythrocytes in the spleen. The predominant bile pigment is **bilirubin**. Liver cells extract bilirubin from the blood and actively secrete it into the bile. It is bilirubin that gives bile its yellow color. After entering the intestinal tract via the bile, bilirubin is modified by bacterial enzymes to form the brown pigments that give feces their characteristic color. Some of the bile pigments are absorbed into the plasma during their passage through the intestinal tract

and are eventually excreted in the urine, giving urine its yellow color.

From the standpoint of gastrointestinal function, the most important components of bile are the bile salts, which are necessary to digest and absorb fats. During the digestion of a fatty meal, most of the bile salts entering the intestinal tract are reabsorbed in the ileum (the last segment of the small intestine) and returned to the liver via the portal vein, where they are once again secreted into the bile. This "recycling" pathway from the intestine to the liver and back to the intestine via the bile duct is known as the **enterohepatic circulation** (Figure 16-29). A small amount of the bile salts are not recycled, however, but are lost in the feces. The liver synthesizes new bile salts to replace them.

Cholesterol secretion into the bile is one of the mech-

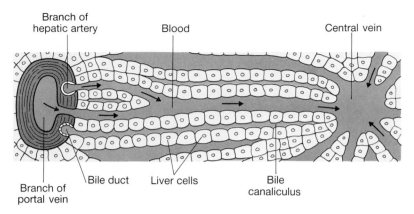

FIGURE 16-27 A small section of the liver showing location of bile canaliculi and ducts with respect to blood and liver cells. (*Adapted from Kappas and Alvares.*)

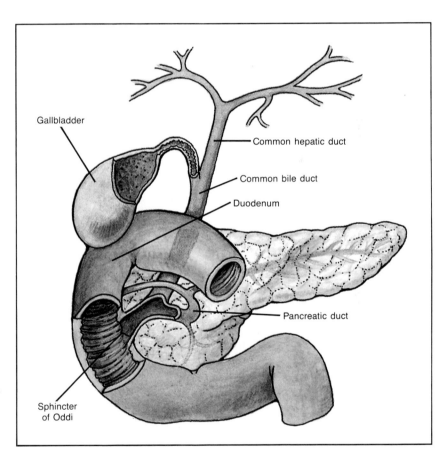

FIGURE 16-28 Bile ducts from liver converge to form the common hepatic duct, from which branches the duct leading to the gallbladder. Beyond this branch the common hepatic duct becomes the common bile duct. The common bile duct and the pancreatic duct converge and empty their contents into the duodenum at the sphincter of Oddi.

anisms by which cholesterol homeostasis is maintained (Chapter 17). Cholesterol is insoluble in water, and its solubility in the bile is achieved by its incorporation into micelles by bile salts. The presence of lecithin in the bile increases the amount of cholesterol that can be solubilized in micelles.

Like pancreatic secretions, the components of the bile are secreted by two different cell types. The bile salts, cholesterol, lecithin, and bile pigments are secreted by **hepatocytes** (liver cells), whereas most of the

bicarbonate-rich salt solution is secreted by the epithelial cells lining the bile ducts. Since bicarbonate ions in the bile help to neutralize acid in the duodenum, it is appropriate that the salt solution secreted by the bile ducts, just like that secreted by the pancreas, is stimulated by secretin in response to the presence of acid in the duodenum.

Secretin does not, however, stimulate the secretion of bile salts by hepatocytes. Bile salt secretion is controlled by the concentration of bile salts in the blood—the

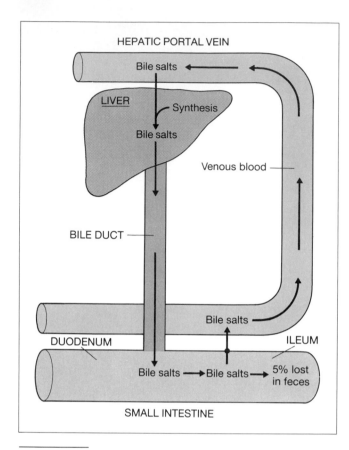

FIGURE 16-29 Enterohepatic circulation of bile salts and regulation of bile bicarbonate secretion.

Shortly after the beginning of a fatty meal, the sphincter of Oddi relaxes, and the smooth muscles in the wall of the gallbladder contract, discharging concentrated bile into the duodenum. The signal for gallbladder contraction and sphincter relaxation is the intestinal hormone CCK—appropriately so, since a major stimulus for this hormone's release is the presence of fat in the duodenum. (It is from this ability to cause contraction of the gallbladder that cholecystokinin received its name: *chole*, bile; *cysto*, bladder; *kinin*, to move). Figure 16-30 summarizes the factors controlling the entry of bile into the small intestine.

Small Intestine

Secretions. In addition to the chyme entering the small intestine from the stomach, liver, and pancreas, approximately 2000 mL of water is secreted from blood to lumen by the walls of the small intestine each day. Still larger volumes are simultaneously moving in the opposite direction—from the lumen into the blood—so that normally there is an overall net absorption of water from the small intestine.

One reason for water movement into the lumen is that the chyme entering the small intestine from the stomach may be hypertonic because of a high concentra-

FIGURE 16-30 Regulation of bile entry into the small intestine.

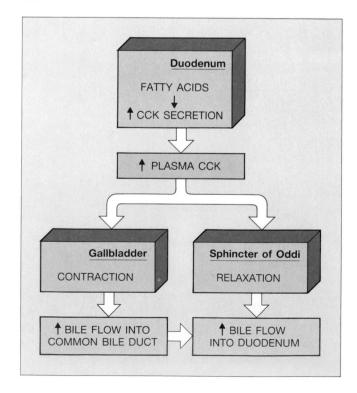

greater the plasma concentration of bile salts, the greater their secretion into the bile. Between meals, most of the secreted bile enters the gallbladder and becomes concentrated. Only a small amount of bile salts enters the intestine, and, therefore, little is absorbed, and the plasma concentration of bile salts is low, that is, the enterohepatic cycling of bile salts is minimal. During a meal, the gallbladder contracts, emptying concentrated bile into the small intestine. The higher plasma concentration of absorbed bile salts leads to an increased rate of bile salt secretion.

Although, as we have seen, bile secretion is greatest during and just after a meal, some bile is always being secreted by the liver. Surrounding the bile duct at the point where it enters the duodenum is a ring of smooth muscle known as the **sphincter of Oddi**. When this sphincter is closed, the bile secreted by the liver is shunted into the gallbladder where sodium is actively transported from the bile into the blood. As solute is pumped out of the bile, water follows by osmosis. This absorption of water concentrates the unabsorbed bile salts, bile pigments, and cholesterol in the gallbladder.

tion of solutes in the meal and because digestion breaks down large molecules into many more small molecules. This hypertonicity causes the osmotic movement of water from the isotonic plasma into the lumen.

As we have seen, a hypertonic solution in the duodenum is one of the stimuli inhibiting gastric emptying. This reflex prevents large amounts of hypertonic fluid from entering the duodenum. Otherwise, large net movements of water from the blood into the intestine would occur, with a resulting decrease in plasma volume and impairment of the circulation.

In the description above, water movement is the result of events in the lumen. In addition, the intestinal epithelium itself secretes a number of mineral ions into the lumen, and water moves with these ions. The continuous disintegration of the intestinal epithelium and the secretion of mucus provide additional sources of material entering the lumen of the small intestine.

Motility.   In contrast to the peristaltic waves that sweep over the stomach, the most common motion of the small intestine during a meal is a stationary contraction and relaxation of intestinal segments with little apparent net movement toward the large intestine (Figure 16-31). Each contracting segment is only a few centimeters long,

and the contraction lasts a few seconds. The chyme in the lumen of a contracting segment is forced both up and down the intestine. This rhythmical contraction and relaxation of the intestine, known as **segmentation**, produces a continuous division and subdivision of the intestinal contents, thoroughly mixing the chyme in the lumen and bringing it into contact with the intestinal wall.

These segmenting movements are initiated by electrical activity generated by pacemaker cells in the longitudinal smooth muscle. Like the slow waves in the stomach, this basic electrical rhythm produces oscillations in smooth-muscle membrane potential that, if threshold is reached, trigger action potentials causing muscle contraction. The frequency of segmentation is set by the frequency of the intestinal basic electrical rhythm, but unlike the stomach, which normally has a single rhythm, the intestinal rhythm varies along the length of the intestine, each successive region having a slightly lower frequency than the one above. For example, segmentation in the duodenum occurs at a frequency of about 12 contractions/min, whereas in the terminal portion of the ileum the rate is only 9 contractions/min. Segmentation does produce, therefore, a slow migration of the intestinal contents toward the large intestine because the contraction frequency is greater in the upper portion of the small intestine than in the lower, forcing more chyme downward, on the average, than upward.

The intensity of segmentation can be altered by hormones, the enteric nervous system, and autonomic nerves—parasympathetic activity increases the force of contraction, and sympathetic stimulation decreases it. As is true for the stomach, these inputs produce changes in the force of smooth-muscle contraction but do not significantly change their frequency. Following a meal, distension of the small intestine increases, which by both long and short reflexes increases the force of segmenting contractions, increasing the degree of mixing.

After most of a meal has been absorbed, the segmenting contractions cease and are replaced by a pattern of peristaltic activity known as the **migrating motility complex**. Each peristaltic wave travels a short distance along the small intestine and then dies out. The site at which the waves are initiated slowly migrates from the duodenum down the small intestine. Thus, during the postabsorptive period, large regions of the small intestine show no signs of contractile activity, neither segmentation nor peristalsis. By the time the migrating motility complex reaches the end of the ileum, new waves are beginning in the duodenum and the process is repeated. Upon the arrival of a new meal in the stomach, the migrating motility complex ceases and is replaced by segmentation.

The peristaltic waves sweep into the large intestine

FIGURE 16-31 Segmentation movements of the small intestine. The small arrows indicate the movements of the luminal contents

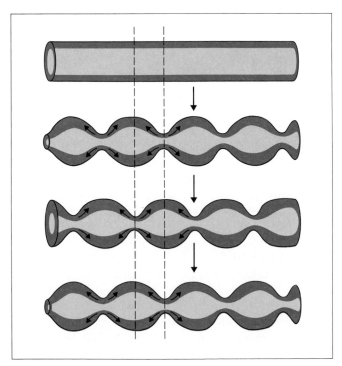

any undigested material still remaining in the small intestine and also prevent bacteria from remaining in the small intestine long enough to grow and multiply. In diseases in which there is an aberrant migrating motility complex, bacterial overgrowth in the small intestine can become a major problem.

The patterns of contractile activity during segmentation and peristalsis are coordinated by the enteric nervous system. However, the control mechanisms that produce the change from one pattern to the other are unknown.

The contractile activity in various regions of the small intestine can be altered by reflexes initiated at different points along the gastrointestinal tract. Segmentation intensity in the ileum increases during periods of gastric emptying, and this is known as the **gastroileal reflex**. Also, large distensions of the intestine, injury to the intestinal wall, and various bacterial infections in the intestine lead to a complete cessation of motor activity, the **intestino-intestinal reflex**.

Most of these reflexes are mediated by long reflexes. Gastrin may play a role in the gastroileal reflex since gastrin increases the ileal motility and relaxes the iliocecal sphincter, the junction between the ileum and the beginning of the large intestine.

A person's emotional state can also affect the contractile activity of the intestine and, therefore, the rate of chyme propulsion and mixing. Fear tends to decrease motility whereas hostility increases it, although these intestinal responses vary greatly in different individuals. These responses are mediated by autonomic nerves.

As much as 500 mL of air may be swallowed during a meal. Most of this air travels no further than the esophagus, from which it is eventually expelled by belching. Some of the air reaches the stomach, however, and is passed on to the intestines, where its percolation through the chyme as the intestinal contents are mixed produces gurgling sounds that are often quite loud.

Large Intestine

The large intestine is a tube 2.5 in in diameter and about 4 ft long. Its first portion, the **cecum**, forms a blind-ended pouch from which extends the **appendix**, a small finger-like projection having no known function (Figure 16-32). The colon consists of three relatively straight segments—the ascending, transverse, and descending portions. The terminal portion of the descending colon is S shaped, forming the sigmoid colon, which empties into a relatively straight segment, the **rectum**, which is the last part of the intestinal tract.

Although the large intestine has a greater diameter than the small intestine, it is about half as long, its surface is not convoluted, and its mucosa lacks villi. There-

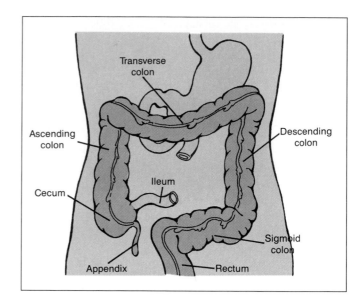

FIGURE 16-32 The large intestine (colon) and rectum.

fore, its epithelial surface area is far less than that of the small intestine. Only about 4 percent of the material entering the gastrointestinal tract is absorbed by the colon each day, mainly ions and water. The secretions of the colon are scanty, lack digestive enzymes, and consist mostly of mucus. The primary function of the large intestine is to store and concentrate fecal material prior to defecation.

Chyme enters the colon through the ileocecal sphincter. This sphincter is normally closed, but after a meal, when the gastroileal reflex increases ileal contractions, it relaxes each time the terminal portion of the ileum contracts, allowing chyme to enter the large intestine. Distension of the colon, on the other hand, produces a reflex contraction of the sphincter, preventing fecal material from entering the small intestine. About 500 mL of chyme enters the colon from the small intestine each day. This material is derived largely from the secretions of the lower small intestine since most of the ingested food has been absorbed before reaching the large intestine.

The primary absorptive process in the large intestine is the active transport of sodium from lumen to blood, with the accompanying osmotic reabsorption of water. If fecal material remains in the large intestine for a long time, almost all of the water is absorbed, leaving behind dry fecal pellets. There is normally a small net movement of potassium into the colon lumen, but severe depletion of total-body potassium can result when large volumes of fluid are excreted in the feces.

The large intestine also absorbs some of the products (vitamins, for example) synthesized by the bacteria inhabiting this region. Although this source of vitamins generally provides only a small part of the normal daily requirement, it may make a significant contribution when dietary vitamin intake is low.

Other bacterial products include gas (**flatus**), which is a mixture of nitrogen and carbon dioxide, with small amounts of the flammable gases hydrogen, methane, and hydrogen sulfide. Bacterial fermentation of undigested polysaccharides produces gas in the colon at the rate of about 400 to 700 mL/day. Certain foods, beans, for example, contain large amounts of carbohydrates that cannot be digested by intestinal enzymes but are readily metabolized by bacteria in the large intestine, producing large amounts of gas.

Motility and defecation. Contractions of the circular smooth muscle in the colon produce a segmentation motion with a rhythm considerably slower than that in the small intestine. A contraction may occur only once every 30 min. Because of this slow movement, material entering the colon from the small intestine remains for some 18 to 24 h, providing time for bacteria to grow and multiply. Three to four times a day, generally following a meal, a wave of intense contraction, known as a **mass movement,** spreads rapidly over the colon toward the rectum. This usually coincides with the gastroileal reflex. Unlike a peristaltic wave, in which the smooth muscle at each point relaxes after the wave of contraction has passed, the smooth muscle of the colon remains contracted for some time after a mass movement.

The **anus,** the exit from the rectum, is normally closed by the **internal anal sphincter,** which is composed of smooth muscle, and the **external anal sphincter,** composed of skeletal muscle under voluntary control. The sudden distension of the walls of the rectum produced by the mass movement of fecal material into it initiates the defecation reflex, which is mediated primarily by the external nerves to the terminal end of the large intestine. The reflex response consists of a contraction of the rectum, relaxation of the internal anal sphincter, contraction of the external anal sphincter, and increased peristaltic activity in the sigmoid colon. Eventually a pressure is reached in the rectum that triggers relaxation of the external anal sphincter, however, allowing the feces to be expelled.

The conscious urge to defecate, mediated by stretched mechanoreceptors, accompanies the initial distension of the rectum. Brain centers can, however, via descending pathways to somatic nerves, override the afferent input from the defecation reflex, thereby keeping the external sphincter closed and allowing a person to delay defecation. If defecation does not occur, the smooth muscle in the walls of the rectum relax and the urge to defecate subsides until the next mass movement

propels more feces into the rectum, increasing its volume and again initiating the defecation reflex. As with most skeletal-muscle activity, voluntary control of the external anal sphincter is learned. Thus a child gradually learns to control its anal sphincter. Spinal cord damage can lead to a loss of voluntary control over defecation.

Defecation is normally assisted by a deep inspiration, followed by closure of the glottis and contraction of the abdominal and thoracic muscles, producing an increase in abdominal pressure that is transmitted to the contents of the large intestine and rectum. This maneuver also causes a rise in intrathoracic pressure, which leads to a transient rise in blood pressure followed by a fall in pressure as the venous return to the heart is decreased. The cardiovascular stress resulting from excessive strain during defecation may precipitate a stroke or heart attack in individuals with cardiovascular disease.

About 150 g of feces, consisting of about 100 g of water and 50 g of solid material, is normally eliminated each day. The solid matter is mostly bacteria, undigested polysaccharides, bile pigments, and small amounts of salts.

PATHOPHYSIOLOGY OF THE GASTROINTESTINAL TRACT

Since the end result of gastrointestinal function is the absorption of nutrients, salts, and water, most malfunctions of this organ system affect either the nutritional state of the body or its salt and water content. The following provide a few familiar examples of disordered gastrointestinal function.

Ulcers

Considering the high concentration of acid and pepsin secreted by the stomach, it is natural to wonder why the stomach does not digest itself. Several factors protect the walls of the stomach from being digested. The surface of the mucosa is lined with cells that secrete a slightly alkaline mucus, which forms a thin layer over the luminal surface. Both the protein content of mucus and its alkalinity neutralize hydrogen ions in the immediate area of the epithelium. Thus, mucus forms a chemical barrier between the highly acid contents of the lumen and the cell surface. In addition, the tight junctions between the epithelial cells lining the stomach restrict the diffusion of hydrogen ions into the underlying tissues. Finally, these epithelial cells are replaced every few days by new cells arising by the division of cells within the gastric pits.

Yet, in about 10 percent of the U.S. population, these protective mechanisms are inadequate, and erosions (**ulcers**) of the epithelial surface occur. Ulcers can occur not only in the stomach but also in the lower part of the esophagus and in the duodenum. Indeed, duodenal ulcers are about 10 times more frequent than gastric ulcers. Damage to blood vessels in the tissues underlying the ulcer may cause bleeding into the gastrointestinal lumen. On occasion, the ulcer may penetrate the entire wall, resulting in leakage of the luminal contents into the abdominal cavity.

Ulcer formation involves breaking the mucosal barrier and exposing the underlying tissue to the corrosive action of acid and pepsin, but it is not clear what produces the initial damage to the barrier. Many factors may be involved—genetic susceptibility, drugs, alcohol, bile salts, and an excessive secretion of acid and pepsin are just some of the contributing factors.

Although acid is essential for ulcer formation, it is not necessarily the primary factor that breaks the mucosal barrier. Many patients with ulcers have normal or even subnormal rates of acid secretion. Regardless of the cause of barrier breakdown, reducing the level of acid and pepsin secretion either by cutting the parasympathetic (vagus) nerves to the stomach or by the use of drugs that inhibit acid secretion promotes healing and tends to prevent the reoccurrence of ulcers.

Despite popular notions that ulcers are due to emotional stress and despite the existence of a pathway (the parasympathetic nerves) for mediating stress-induced increases in acid secretion, the role of stress in producing ulcers remains unclear. Once the ulcer has been formed, however, emotional stress can aggravate it by increasing acid secretion.

Gallstones

As described earlier, bile contains not only bile salts but cholesterol and phospholipids, which are water-insoluble and are maintained in soluble form in the bile as micelles. When the concentration of cholesterol in the bile becomes too high in relation to the concentrations of phospholipid and bile salts, cholesterol will crystallize out of solution, forming **gallstones.** This can occur when the liver secretes excessive amounts of cholesterol or when the bile becomes concentrated in the gallbladder. Although cholesterol gallstones are the most frequently encountered gallstone in the Western world, the precipitation of bile pigments can occasionally be responsible for gallstone formation.

If a gallstone is small, it may pass through the common bile duct into the intestine with no complications. A larger stone may become lodged in the opening of the gallbladder, causing contractile spasms of the smooth muscle and pain. A more serious complication arises when a gallstone lodges in the common bile duct, thereby preventing bile from entering the intestine. The

absence of bile in the intestine decreases the rate of fat digestion and absorption, so that approximately half of ingested fat is not digested and passes on to the large intestine and eventually appears in the feces. Furthermore, bacteria in the large intestine convert some of this fat into fatty acid derivatives that alter salt and water movements, leading to a net flow of fluid into the large intestine. The result is diarrhea and increased fluid loss.

The buildup of pressure in a blocked bile duct inhibits further secretion of bile. As a result, bilirubin, which is normally secreted into the bile from the blood, accumulates in the blood and diffuses into tissues, where it produces the yellowish coloration of the skin and eyes known as **jaundice.** It should be noted that bile-duct occlusion is not the only cause of jaundice; bilirubin accumulation can also occur if hepatocytes are damaged by liver disease and therefore fail to secrete bilirubin into the bile.

Since the duct from the pancreas joins the common bile duct just before it enters the duodenum, a gallstone that becomes lodged at this point prevents both bile and pancreatic secretions from entering the intestine. This results in failure both to neutralize acid and to digest most organic nutrients, not just fat, adequately. The end result is severe nutritional deficiencies.

Why some individuals develop gallstones and others do not is still unclear. Women, for example, have about twice the incidence of gallstone formation as men, and Native Americans have a very high incidence compared with other ethnic groups in America.

Lactose Intolerance

Lactose is the major carbohydrate in milk. It cannot be absorbed directly but must first be digested into its components—glucose and galactose—which are readily absorbed by active transport. Lactose is digested by the enzyme **lactase,** which is embedded in the plasma membranes of intestinal epithelial cells. Lactase is present at birth, but in approximately 25 percent of white Americans and in most northern Europeans its concentration then declines when the child is between 18 and 36 months old. In the absence of lactase, lactose is not digested and remains in the small intestine. Since the absorption of water requires prior absorption of solute to provide an osmotic gradient, the unabsorbed lactose prevents some of the water from being absorbed. This lactose-containing fluid is passed on to the large intestine, where bacteria digest the lactose. They then metabolize the released monosaccharides, producing large quantities of gas (which distends the colon, producing pain) and organic products that inhibit active ion absorption. This causes fluid movement into the lumen of the large intestine, producing diarrhea. The response to milk ingestion by adults whose lactase levels have diminished during development varies from mild discomfort to severely dehydrating diarrhea, according to the volume of milk and milk products ingested and the amount of lactase present in the intestine.

Constipation and Diarrhea

Many people have a mistaken belief that, unless they have a bowel movement every day, the absorption of toxic substances from fecal material in the large intestine will somehow poison them. Attempts to identify such toxic agents in the blood following prolonged periods of fecal retention have been unsuccessful. There appears to be no physiological necessity for having bowel movements at frequent intervals. Whatever maintains a person in a comfortable state is physiologically adequate, whether this means a bowel movement after every meal, once a day, or only once a week.

On the other hand, there often are symptoms—headache, loss of appetite, nausea, and abdominal distension—that may arise when defecation has not occurred for several days or even weeks, depending on the individual. These symptoms of **constipation** are caused not by toxins but by distension of the rectum. In addition, the longer that fecal material remains in the large intestine, the more water is absorbed and the harder and drier the feces become, making defecation more difficult and sometimes painful. Thus, constipation tends to promote constipation.

Decreased motility of the large intestine is the primary factor causing constipation. This often occurs in the elderly or may result from damage to the colon's enteric nervous system. Emotional stress can also decrease the motility of the large intestine.

One of the factors increasing colonic motility, and thus opposing the development of constipation, is distension of the large intestine. A variety of plant products, including cellulose and other complex polysaccharides collectively known as **dietary fiber**, provide a natural laxative in ingested food. Bran, cabbage, and carrots have a relatively high dietary fiber content. These substances are not digested by the enzymes in the small intestine and are passed on to the large intestine, where their bulk produces distension and thereby increases motility.

Laxatives, which increase the frequency or ease of defecation, act through a variety of mechanisms. Some, such as mineral oil, simply lubricate the feces, making defecation easier and less painful. Others contain magnesium and aluminum salts, which are poorly absorbed and therefore lead to water retention in the intestinal tract. Still others, such as castor oil, stimulate the motil-

ity of the colon and alter ion transport across the wall, thus indirectly affecting water movement.

Excessive use of laxatives in attempting to maintain a preconceived notion of regularity leads to a decreased responsiveness of the colon to normal defecation-promoting signals. In such cases, a long period without defecation may occur following cessation of laxative intake, appearing to confirm the necessity of taking laxatives to promote regularity.

Diarrhea is characterized by an increased water content in the feces and an increased frequency of defecation. Diarrhea results from decreased fluid absorption, increased fluid secretion, or both. The increased motility that accompanies diarrhea probably does not cause the diarrhea (by decreasing the time available for fluid absorption) but rather is the result of the distension produced by increased luminal fluid.

A number of bacterial, protozoan, and viral diseases of the intestinal tract cause diarrhea. In many cases they do so by releasing substances that either alter ion-transport processes in the epithelial membranes or damage the epithelial barriers by direct penetration into the mucosa. As described earlier, the presence of unabsorbed solutes in the lumen, resulting from decreased digestion or absorption, also results in retained fluid and diarrhea.

The major consequences of severe diarrhea are decreased blood volume and the disturbances in electrolyte and acid-base homeostasis resulting from the excessive loss of ions such as sodium, potassium and bicarbonate.

SUMMARY

1. The gastrointestinal system transfers organic nutrients, minerals, and water from the external environment to the internal environment. The four processes used to accomplish this function are (1) digestion, (2) secretion, (3) absorption, and (4) motility.
   A. The system is designed to maximize the absorption of most nutrients, not to regulate the amount absorbed.
   B. It does not play a major role in the removal of waste products from the internal environment.

Overview: Functions of the Gastrointestinal Organs

I. The names and functions of the gastrointestinal organs are summarized in Figure 16-3.
II. Each day the gastrointestinal tract secretes about 3.5 times more fluid into the lumen than is ingested. Only about 1 percent of the luminal contents are excreted in the feces.

Structure of the Gastrointestinal Tract

I. The structure of the wall of the gastrointestinal tract is summarized in Figure 16-5.

A. The area available for absorption in the small intestine is greatly increased by the folding of the intestinal wall, the presence of villi, and the microvilli on the surface of the epithelial cells.

B. The epithelial cells lining the intestinal tract are continuously replaced by new cells arising by mitosis.

C. The venous blood from the small intestine, containing absorbed nutrients other than fats, passes to the liver via the hepatic portal vein before returning to the heart. Fat is absorbed into the lymphatic vessels (lacteals) in the middle of each villus.

Digestion and Absorption

I. Starch is digested by amylases secreted by the salivary glands and pancreas, and the resulting products, as well as ingested disaccharides, are digested to monosaccharides by enzymes in the luminal membranes of epithelial cells in the small intestine.

A. The monosaccharides are absorbed by secondary active transport.

B. Some polysaccharides, such as cellulose, cannot be digested and pass to the large intestine, where they are metabolized by bacteria.

II. Proteins are broken down into small peptides and amino acids, which are absorbed by secondary active transport in the small intestine.

A. The breakdown step to peptides is catalyzed by pepsin in the stomach and by the pancreatic enzymes trypsin and chymotrypsin in the small intestine.

B. In the small intestine, peptides are broken down into amino acids by reactions catalyzed by pancreatic carboxypeptidase and intestinal aminopetidase.

III. The digestion and absorption of fat by the small intestine requires mechanisms that solubilize the fat and its digestion products.

A. Large fat globules leaving the stomach are emulsified in the small intestine by bile salts secreted by the liver.

B. Lipase from the pancreas digests fat at the surface of the emulsion droplets forming fatty acids and monoglycerides.

C. These water-insoluble products of lipase action, when combined with bile salts, form micelles that are in equilibrium with the free molecules.

D. Free fatty acids and monoglycerides diffuse across the luminal membranes of epithelial cells, within which they are enzymatically recombined to form triacylglycerol, which is released as chylomicrons from the blood side of the cell by exocytosis.

E. The released chylomicrons enter lacteals in the intestinal villi and pass, by way of the lymphatic system, to the cardiovascular system.

IV. Fat-soluble vitamins are absorbed by the same pathway used for fat absorption. Most water-soluble vitamins are absorbed in the small intestine by diffusion or carrier-mediated transport. Vitamin B_{12} is absorbed in the ileum after combining with intrinsic factor secreted by the stomach.

V. Water is absorbed from the small intestine by osmosis following the active absorption of solutes, primarily sodium chloride.

Regulation of Gastrointestinal Processes

I. Most gastrointestinal reflexes are initiated by luminal stimuli: (1) distension, (2) osmolarity, (3) acidity, and (4) digestion products.

A. Neural reflexes are mediated by short reflexes in the enteric nervous system and by long reflexes involving afferent and efferent neurons to and from the CNS.

B. Gastrin, secretin, CCK, and GIP are secreted by endocrine cells scattered throughout the epithelium of the stomach (gastrin) and small intestine (secretin, CCK, and GIP). Their properties are summarized in Table 16-2.

C. The three phases of gastrointestinal regulation—cephalic, gastric, and intestinal—are named after the location of the stimulus that initiates the response.

II. Chewing breaks up food into smaller particles suitable for swallowing but is not essential for the eventual digestion and absorption of food.

III. Salivary secretion is stimulated by food in the mouth acting via chemo- and pressure receptors. Both sympathetic and parasympathetic stimulation increases salivary secretion.

IV. Food moved into the pharynx by the tongue initiates swallowing, which is coordinated by the swallowing center in the medulla.

A. Food is prevented from entering the trachea by inhibition of respiration and by closure of the glottis.

B. The upper esophageal sphincter relaxes as food is moved into the esophagus and then closes.

C. Food is moved through the esophagus toward the stomach by peristaltic waves. The lower esophageal sphincter remains open throughout swallowing.

D. If food does not reach the stomach with the first peristaltic wave, distension of the esophagus will initiate secondary peristalsis.

V. The factors controlling acid secretion by parietal cells in the stomach are summarized in Table 16-3.

VI. Pepsinogen, secreted by the gastric chief cells in response to most of the same reflexes that control acid secretion, is converted to the active proteolytic enzyme pepsin in the stomach's lumen by acid and by pepsin itself.

VII. Peristaltic waves sweeping over the stomach become stronger in the antrum, where most mixing occurs. With each wave, only a small portion of the stomach contents are expelled into the small intestine through the pyloric sphincter.

A. The basic electrical rhythm generated by stomach smooth muscle determines peristaltic wave frequency. Contraction strength depends on the frequency of action potentials triggered by the slow waves.

B. The larger a meal, the faster the stomach empties. Distension of the small intestine, and fat, acid, or hypertonic solutions in the intestinal lumen inhibit gastric emptying.

VIII. Vomiting is coordinated by the vomiting center in the medulla. Contraction of abdominal muscles force the contents of the stomach into the esophagus (retching), and if the contractions are strong enough, they force the contents of the esophagus through the upper esophageal sphincter into the mouth (vomiting).

IX. The exocrine portion of the pancreas secretes digestive enzymes and bicarbonate ions, which reach the duodenum through the pancreatic duct.

A. The bicarbonate ions neutralize acid entering the small intestine from the stomach.

B. Most proteolytic enzymes, including trypsin, are secreted by the pancreas in inactive forms. Trypsin is activated by enterokinase located on the luminal membranes of the small intestine and in turn activates the other inactive enzymes.

C. The hormone secretin, released from the small intestine in response to increased luminal acidity, stimulates bicarbonate secretion, while CCK released from the small intestine in response to the products of fat and protein digestion stimulates pancreatic enzyme secretion.

X. The major ingredients in bile are bile salts, cholesterol, lecithin, bicarbonate ions, bile pigments, and trace metals.

A. Bilirubin, a breakdown product of hemoglobin, is absorbed from the blood by the liver and secreted into the bile.

B. After reaching the small intestine, bile salts are absorbed from the ileum and travel by the hepatic portal vein back to the liver, where they are again secreted into the bile. The liver also synthesizes new bile salts to replace those lost in the feces.

C. The greater the bile salt concentration in the hepatic portal blood, the greater the rate of bile secretion.

D. Secretin stimulates bicarbonate secretion by the cells lining the bile ducts in the liver.

E. Bile is concentrated in the gallbladder by the absorption of salt and water.

F. Following a meal, the release of CCK from the small intestine causes the gallbladder to contract and the sphincter of Oddi to relax, thereby injecting concentrated bile into the intestine.

XI. In the small intestine, the digestion of polysaccharides and proteins increases the osmolarity of the luminal contents, producing an osmotic flow of water into the lumen.

A. Hypertonic solutions in the lumen inhibit gastric emptying.

B. Small amounts of salt, bicarbonate, and water are secreted by the small intestine. However, there is net absorption of these substances.

XII. Intestinal motility is coordinated by the enteric nervous system and modified by long and short reflexes and hormones.

A. During and shortly after a meal, the intestinal contents are mixed by segmenting movements of the intestinal wall.

B. After most of the food has been digested and absorbed, segmentation is replaced by the migrating motility complex that sweeps the undigested material into the large intestine by a series of peristaltic waves.

XIII. The primary function of the large intestine is to store and concentrate fecal matter prior to defecation.

A. Water is absorbed from the large intestine secondary to the active absorption of salt, leading to the concentration of fecal matter.

B. Flatus is produced by bacterial fermentation of undigested polysaccharides.

C. Three to four times a day, mass movements in the large intestine move its contents into the rectum.

D. Distension of the rectum initiates defecation, which is assisted by a forced expiration against a closed glottis.

E. Defecation can be voluntarily controlled through somatic nerves to the skeletal muscles of the external anal sphincter.

Pathophysiology of the Gastrointestinal Tract

I. The factors that normally prevent breakdown of the mucosal barrier (ulcers) are: (1) the secretion of an alkaline mucus, (2) tight junctions between epithelial cells, and (3) rapid replacement of epithelial cells.

II. Precipitation of cholesterol or, less often, bile pigments in the gallbladder forms gallstones that can block the exit of the gallbladder or bile ducts, leading to: (1) decreased digestion and absorption of fat and (2) accumulation of bile pigments in the blood and tissues—jaundice.

III. In the absence of lactase, lactose cannot be digested, and its presence in the small intestine retains water, leading to diarrhea and increased flatus production.

IV. Constipation is primarily the result of decreased colonic motility. The symptoms of constipation are produced by the overdistension of the rectum, not by the absorption of bacterial toxic products.

V. Diarrhea is caused primarily by decreased fluid absorption, increased fluid secretion, or both.

REVIEW QUESTIONS

1. Define:

gastrointestinal system	amylase
gastrointestinal tract	pharynx
digestion	esophagus
secretion	stomach
absorption	hydrochloric acid
motility	pepsin
feces	chyme
mouth	small intestine
saliva	duodenum
salivary glands	jejunum
	ileum

pancreas
liver
bile salts
bile
gallbladder
large intestine
colon
rectum
defecation
mucosa
submucosa
submucus plexus
muscularis externa
circular muscle
longitudinal muscle
myenteric plexus
serosa
villi
microvilli
lacteal
hepatic portal vein
trypsin
chymotrypsin
carboxypeptidase
aminopeptidase
lipase
emulsification
micelles
emulsion
chylomicrons
intrinsic factor
enteric nervous system
short reflexes
long reflexes
secretin
cholecystokinin (CCK)
gastrin
glucose insulinotropic
    peptide (GIP)
cephalic phase
gastric phase
intestinal phase
parotid gland
submandibular gland
sublingual gland
mucins

swallowing center
upper esophageal sphincter
lower esophageal sphincter
peristaltic waves
secondary peristalsis
heartburn
body of stomach
fundus
antrum
parietal cells
chief cells
enterogastrones
histamine
pepsinogen
receptive relaxation
pyloric sphincter
basic electrical rhythm
vomiting center
retching
enterokinase
trypsinogen
bile pigments
bilirubin
enterohepatic circulation
hepatocytes
sphincter of Oddi
segmentation
migrating motility complex
gastroileal reflex
intestino-intestinal reflex
cecum
appendix
rectum
flatus
mass movement
anus
internal anal sphincter
external anal sphincter
ulcers
gallstones
jaundice
lactase
constipation
dietary fiber
diarrhea

2. List the four processes used in accomplishing the functions of the gastrointestinal system.

3. List the primary functions performed by each of the organs in the gastrointestinal system.

4. Approximately how much fluid is secreted into the gastrointestinal tract each day, compared with the amount of food and drink ingested? How much of this appears in the feces?

5. What structures are responsible for the large surface area of the small intestine?

6. Where does the venous blood go after leaving the small intestine?

7. Identify the enzymes involved in carbohydrate digestion and the mechanism of carbohydrate absorption in the small intestine.

8. List three ways in which proteins or their digestion products can be absorbed from the small intestine.

9. How do bile salts solubilize fats?

10. What is the role of micelles in fat absorption?

11. Describe the movement of fat across the wall of the small intestine.

12. How does the absorption of fat-soluble vitamins differ from that of water-soluble vitamins?

13. Specify two conditions that may lead to failure to absorb vitamin B_{12}.

14. How is water absorbed in the small intestine?

15. List the four types of stimuli that initiate most gastrointestinal reflexes.

16. Describe the location of the enteric nervous system and its role in both short and long reflexes.

17. State the major functions of the four gastrointestinal hormones.

18. Describe the neural reflexes leading to increased salivary secretion.

19. Describe the sequence of events that occur during swallowing.

20. List the cephalic, gastric and intestinal phase stimuli that stimulate or inhibit acid secretion by the stomach.

21. Describe the function of gastrin and the factors controlling its secretion.

22. By what mechanism is pepsinogen converted to pepsin in the stomach?

23. Describe the factors that control gastric emptying.

24. Describe the process of vomiting.

25. Describe the mechanisms controlling pancreatic bicarbonate and enzyme secretion.

26. How are pancreatic enzymes activated in the small intestine?

27. List the major constituents of bile.

28. Describe the recycling of bile salts by the enterohepatic circulation.

29. What determines the rate of bile secretion by the liver?

30. Describe the effects of secretin and CCK on the bile ducts and gallbladder.

31. What causes water to move from the blood to the lumen of the duodenum following gastric emptying?

32. Describe the type of intestinal motility found during and shortly after a meal and the type found several hours after a meal.

33. Describe the production of flatus by the large intestine.

34. Describe the factors that initiate and control defecation.

35. Why is the stomach wall normally not digested by the acid and digestive enzymes in the lumen?

36. What are the consequences of the blocking of the common bile duct with a gallstone?

37. What are the consequences of the failure to digest lactose in the small intestine?

38. Distinguish the factors that cause constipation from those that produce diarrhea.

THOUGHT QUESTIONS

(Answers are given in Appendix A.)

1. If the salivary glands were unable to secrete amylase, what effect would this have on starch digestion?

2. Milk or a fatty snack taken prior to the ingestion of alcohol decreases the rate of intoxication. What is the mechanism of this effect?

3. What are some of the consequences of removing a large portion of the stomach?

4. A patient brought to a hospital after a period of prolonged vomiting has an elevated heart rate, decreased blood pressure, and below-normal blood acidity. Explain these symptoms in terms of the consequences of excessive vomiting.

5. Can fat be digested and absorbed in the absence of bile salts? Explain.

6. How might damage to the lower portion of the spinal cord affect defecation?

7. One of the procedures used in the treatment of ulcers is vagotomy, the surgical cutting of the vagus (parasympathetic) nerves to the stomach. By what mechanism will this procedure help ulcers to heal and decrease the incidence of new ulcers?

CHAPTER

17

REGULATION OF ORGANIC METABOLISM, GROWTH, AND ENERGY BALANCE

Chapter 4 introduced the concepts of energy and of organic metabolism at the level of the individual cell. This chapter now deals with a variety of topics that are concerned in one way or another with those same concepts but for the entire body. First we describe how the metabolic pathways for carbohydrate, fat, and protein are controlled so as to provide a continuous source of energy to the various tissues and organs. The next topic is how the metabolic changes that underlie bodily growth occur and are controlled. Finally, we describe the determinants of total-body energy balance in terms of energy intake and output and how the body temperature—a measure of the heat content of the body—is regulated.

SECTION A

CONTROL AND INTEGRATION OF CARBOHYDRATE, PROTEIN, AND FAT METABOLISM

EVENTS OF THE ABSORPTIVE AND POSTABSORPTIVE STATES

Mechanisms have evolved for survival during alternating periods of plenty and fasting. We speak of two functional states or periods: the **absorptive state**, during which ingested nutrients are entering the blood from the gastrointestinal tract, and the **postabsorptive state**, during which the gastrointestinal tract is empty of nutrients, and energy must be supplied by the body's own stores. Since an average meal requires approximately 4 h for complete absorption, our usual three-meal-a-day pattern places us in the postabsorptive state during the late morning and afternoon and almost the entire night.

During the absorptive period, some of the ingested nutrients supply the energy needs of the body, and the remainder are added to the body's energy stores, to be called upon during the next postabsorptive period. Total-body energy stores are adequate for the average person to easily withstand a fast of many weeks.

Figures 17-1 and 17-2 summarize the major pathways to be described in this chapter. Although they may appear formidable at first glance, they should give little difficulty after we have described the component parts,

and they should be referred to constantly during the following discussion.

Absorptive State

We shall assume, for this discussion, an average meal containing all three of the major nutrients—carbohydrate, protein, and fat, the carbohydrate constituting most of the meal's energy content (calories). Recall from Chapter 16 that carbohydrate and protein enter the blood supplying the gastrointestinal tract primarily as monosaccharides and amino acids, respectively. The blood leaves the gastrointestinal tract to go directly to the liver by way of the hepatic portal vein, allowing the liver to alter the composition of the blood before it returns to the heart to be pumped to the rest of the body. In contrast to carbohydrate and amino acids, fat is absorbed, as triacylglycerols in chylomicrons, into the lymph, which drains into the systemic venous system. Thus, the liver does not get first crack at absorbed fat.

Absorbed glucose. Some of the carbohydrate absorbed from the gastrointestinal tract is galactose and fructose, but since they are either converted to glucose by the liver or enter essentially the same metabolic pathways as does glucose, we shall simply refer to these sugars as glucose.

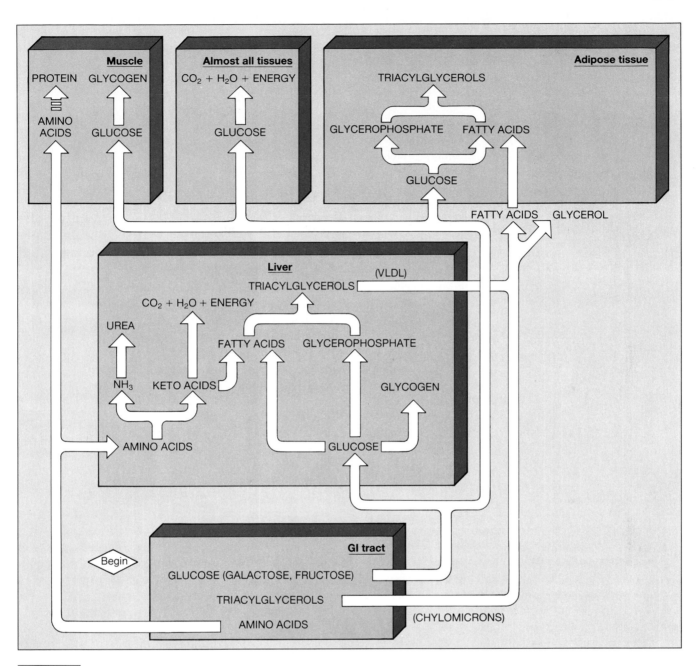

FIGURE 17-1 Major metabolic pathways of the absorptive state. Glycerophosphate is the form of glycerol used to synthesize triacylglycerol. The glycerol formed by the action of lipoprotein lipase in the adipose-tissue capillaries is not taken up by the adipose-tissue cells but circulates to the liver for further metabolism. All arrows between boxes denote carriage of the substance via the blood. VLDL = very low density lipoproteins.

As shown in Figure 17-1, much of the absorbed glucose enters the liver cells. This is a very important point: During the absorptive period there is net uptake of glucose by the liver. We shall describe the fate of this glucose in a moment.

Of the absorbed glucose that does *not* enter liver cells, a very large fraction enters most other body cells and is catabolized to carbon dioxide and water, providing the energy for ATP formation. Indeed, and this is another key point, glucose is the body's major energy source during the absorptive state.

Another fraction of the absorbed glucose that does not enter liver cells enters skeletal muscle and is stored there as the polysaccharide glycogen. Yet another fraction enters adipose-tissue cells, where it is transformed to fat (triacylglycerols). The importance of glucose as a precursor of fat cannot be overemphasized, since glucose is the precursor for both the glycerophosphate and the fatty acids that are linked together in the synthesis of triacylglycerols.

To return to the absorbed glucose that enters the liver: As shown in Figure 17-1, it is either stored as glycogen, as in skeletal muscle, or transformed to glycerophosphate and fatty acids, which are then used to synthesize triacylglycerols, just as in adipose tissue. Some of this fat synthesized from glucose in the liver is stored there, but most is packaged, along with specific proteins, into molecular aggregates of lipids and proteins. These aggregates are secreted by the liver cells and enter the blood. They are called **very low density lipoproteins (VLDL)** because they contain much more fat than protein, and fat is less dense than protein. The synthesis of VLDL by liver cells occurs by processes similar to those for synthesis of chylomicrons by intestinal mucosal cells, as described on page 524.

Once in the bloodstream, VLDL complexes, being quite large, do not readily penetrate capillary walls. Instead, their triacylglycerols are hydrolyzed to glycerol (not glycerophosphate) and fatty acids by the enzyme **lipoprotein lipase** located on the capillary endothelium surface in contact with the blood. The capillaries in adipose tissue are particularly rich in lipoprotein lipase, and the fatty acids generated by this enzyme's action diffuse across the capillary wall and into the adipose-tissue cells. There they combine with glycerophosphate supplied, as we have seen, by glucose, to form triacylglycerols once again. Thus, most of the fatty acids in the VLDL triacylglycerol synthesized from glucose by the *liver* end up being stored in triacylglycerol in *adipose tissue*. The glycerol released by lipoprotein lipase is not taken up by the adipose-tissue cells but recirculates to the liver.

To summarize, the major fates of glucose during the absorptive phase are utilization for energy, storage as glycogen in liver and skeletal muscle, and storage as fat in adipose tissue. Of the two types of storage, fat is quantitatively the more important.

Absorbed triacylglycerols. To reiterate, almost all ingested fat is absorbed as chylomicrons into the lymph, which flows into the systemic circulation. The biochemical processing of these chylomicron triacylglycerols in plasma is quite similar to that just described for VLDL produced by the liver. The fatty acids of plasma chylomicrons are released, mainly in adipose tissue capillaries, by action of endothelial lipoprotein lipase, and the released fatty acids leave the capillaries to enter adipose-tissue cells and combine with glycerophosphate to form triacylglycerols.

Thus, there are three major sources of the fatty acids found in adipose-tissue triacylglycerol: (1) glucose that enters adipose tissue and is converted to fatty acids, (2) glucose that is converted in the liver to VLDL triacylglycerols, which are transported via the blood to the adipose tissue, and (3) ingested fat transported to adipose tissue in chylomicrons.

This description has emphasized the *storage* of ingested fat. For simplicity, we have not shown in Figure 17-1 that a fraction of the ingested fat is not stored but is oxidized during the absorptive state by various organs to provide energy. The relative amounts of carbohydrate and fat used for energy during the absorptive period depend largely on the content of the meal.

Absorbed amino acids. A minority of the absorbed amino acids enter liver cells. They are used to synthesize a variety of proteins, including liver enzymes and plasma proteins, or they are converted to carbohydrate-like intermediates known as **keto acids** by removal of the amino group (deamination, page 99). The amino groups are used to synthesize urea, which is excreted by the kidneys. The keto acids generated by deamination are *not* converted to glucose during the absorptive phase, unlike the situation to be described for the postabsorptive phase. However, they can enter the Krebs tricarboxylic acid cycle and be catabolized to provide energy for the liver cells, or they can be converted to fatty acids, thereby participating in fat synthesis by the liver.

Most ingested amino acids are not taken up by the liver cells but enter other cells (Figure 17-1), where they may be synthesized into protein. We have simplified the diagram by showing "nonliver" amino acid uptake only by muscle because muscle contains the largest amount of body protein. It should be emphasized, however, that all cells require a constant supply of amino acids for protein synthesis and participate in the dynamics of protein metabolism.

To reiterate, there is net synthesis of protein during the absorptive phase. This process is represented by the dashed line in the muscle box in Figure 17-1 to call attention to an important fact: A minimal supply of ingested amino acids is essential in order to maintain normal protein stores, but excess amino acids are not *stored* as protein, in the sense that glucose is stored as glycogen or that both glucose and fat are stored as fat. Therefore, eating large amounts of protein does not, in itself, cause large increases in body protein. Rather, the excess amino acids are merely converted to carbohydrate or fat. This discussion does not apply to growing children, who manifest a continuous increase of body protein, or to adults who are actively building body mass as, for example, by weight lifting.

We do not wish to leave the impression that amino

acids are all either synthesized into protein or converted into carbohydrate and fat. There are many other metabolic fates of amino acids throughout the body. For example as we have seen, amino acids are used to synthesize the thyroid hormones and to supply the ammonia used as a urinary buffer by the kidneys.

Summary. During the absorptive period: (1) Energy is provided primarily by absorbed carbohydrate, (2) there is net uptake of glucose by the liver, (3) carbohydrates, fats, and proteins in excess of those utilized for energy (or protein synthesis in the case of the last nutrient) are stored mostly as fat, (4) glycogen constitutes a quantitatively less important storage form for excess ingested carbohydrate, and (5) there is net synthesis of body proteins.

Postabsorptive State

As the absorptive period ends, net synthesis of glycogen, fat, and protein ceases, and net catabolism of all these substances begins to occur. The overall significance of these events can be understood in terms of the essential problem during the postabsorptive period: No glucose is being absorbed from the intestinal tract, yet the plasma glucose concentration must be maintained because the brain normally utilizes only glucose for energy. Too low a plasma glucose concentration can result in alterations of neural activity ranging from subtle impairment of mental function to coma and even death.

The events that maintain plasma glucose concentration fall into two categories: (1) reactions that provide sources of blood glucose and (2) glucose sparing (fat utilization).

Sources of blood glucose.   The sources of blood glucose during the postabsorptive period (Figure 17-2) are as follows.

1. **Glycogenolysis**, the hydrolysis of glycogen stores to glucose, occurs in the liver. The glucose then leaves the liver and enters the blood. Hepatic glycogenolysis, a rapidly occurring event, is the first line of defense in maintaining plasma glucose concentration. Its role, however, is relatively short-lived for the following reason: After the absorptive period is completed, the normal liver contains less than 100 g of glycogen; at 4 kcal/g, the subsequent breakdown of all this glycogen provides 400 kcal, enough to fulfill an average person's total caloric need for only 4 h.
2. Muscle (and to a lesser extent, other tissues) contains approximately the same amount of glycogen as the liver, and this glycogen can serve, by glycogenolysis, as a source of glucose. A complication arises, however, because muscle, unlike liver, lacks the enzyme

necessary to form glucose from the glucose 6-phosphate formed during glycogenolysis (page 95). The glucose 6-phosphate undergoes glycolysis within the muscle to yield pyruvate and lactate. These substances are liberated into the blood, circulate to the liver, and are converted into glucose, which can then leave the liver cells to enter the blood. Thus, muscle glycogen contributes to the blood glucose indirectly via the liver.

3. The catabolism of triacylglycerols yields glycerol and fatty acids, a process termed **lipolysis**. The glycerol and fatty acids then enter the blood. The liver extracts the glycerol and converts it to glucose. Thus, a source of glucose during the postabsorptive period is the glycerol released when adipose-tissue triacylglycerol is broken down.
4. Protein is by far the major source of blood glucose during a fast of more than a few hours. Large quantities of protein in muscle and, to a lesser extent, other tissues can be catabolized without serious cellular malfunction. There are, of course, limits to this process, and continued protein loss ultimately means functional disintegration, sickness, and death. Before this point is reached, however, protein breakdown can supply large quantities of amino acids that enter the blood and are picked up by the liver, which converts them, via the keto acid pathway, to glucose.

Synthesis of glucose from pyruvate, lactate, glycerol, and amino acids is known as **gluconeogenesis**, that is, new formation of glucose. During a 24-h fast, this process provides approximately 180 g of glucose. The liver is not the only organ capable of gluconeogenesis; the kidneys also perform gluconeogenesis but mainly during a prolonged fast. At the end of such a fast, the kidneys may be contributing as much glucose as the liver.

Glucose sparing (fat utilization).   The 180 g of glucose per day produced by the liver and kidneys during fasting supplies 720 kcal: 180 g/day $\times$ 4 kcal/g = 720 kcal/day. As described later in this chapter, normal total energy expenditure equals 1500 to 3000 kcal/day. Accordingly, gluconeogenesis cannot supply all the body's energy needs. The following essential adjustment must therefore take place during the transition from absorptive to postabsorptive state: Most organs and tissues markedly reduce their glucose catabolism and increase their fat utilization, the latter becoming the major energy source. This metabolic adjustment, termed **glucose sparing**, "spares" the glucose produced by the liver for use by the nervous system.

The essential step in this adjustment is lipolysis, the catabolism of adipose-tissue triacylglycerol, which liberates glycerol and fatty acids into the blood. We described

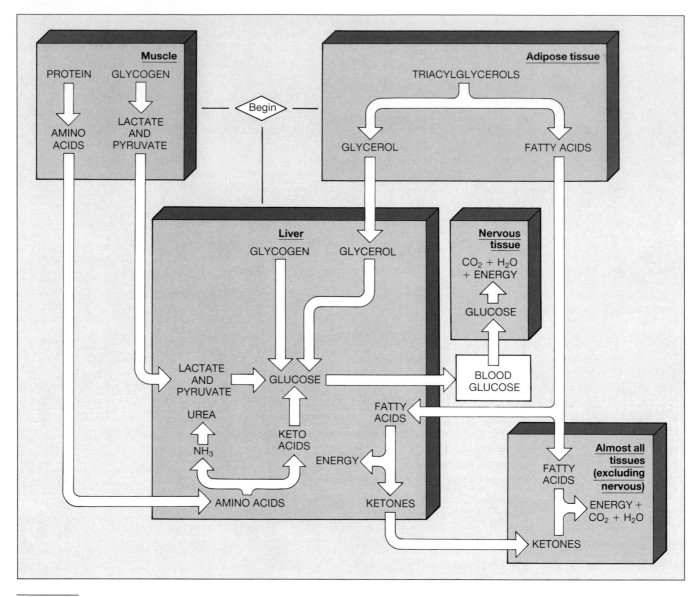

FIGURE 17-2 Major metabolic pathways of the postabsorptive state. The central focus is regulation of the blood glucose concentration. All arrows between boxes denote carriage of the substance via the blood.

lipolysis in the previous section in terms of its importance in providing glycerol to the liver for conversion to glucose. Now, we focus on the liberated fatty acids, which circulate in combination with plasma albumin. (Despite this binding to protein, they are known as free fatty acids in that they are "free" of glycerol.) The circulating fatty acids are picked up by almost all tissues, excluding the nervous system. They enter the Krebs cycle by way of acetyl CoA and are catabolized to carbon dioxide and water (Chapter 5), thereby providing energy.

The liver also catabolizes fatty acids to acetyl CoA, but the liver is unique in that most of the acetyl CoA it forms from fatty acids during the postabsorptive state does not enter the Krebs cycle but is processed into three compounds collectively called **ketones** (or ketone bodies). The reader should distinguish ketones from keto acids, which are metabolites of amino acids. The ketones are released into the blood and provide an important energy source during the postabsorptive phase for the many tissues capable of oxidizing them via the Krebs cycle. One of the ketones is acetone, some of which is exhaled and accounts for the distinctive breath odor of persons undergoing prolonged fasting or, as we shall see, suffering from severe untreated diabetes mellitus.

The net result of fatty acid and ketone utilization during fasting is provision of energy for the body and sparing of glucose for the brain. Moreover, an important change in brain metabolism with prolonged starvation also occurs. Many areas of the brain are capable of utilizing ketones for energy, and so as these substances begin to build up in the blood after the first few days of a fast, the brain begins to utilize them, as well as glucose, for its energy source. The survival value of this phenomenon is very great; When the brain greatly reduces its glucose requirement by utilizing ketones, much less protein need be broken down to supply the amino acids for gluconeogenesis. Accordingly, the protein stores will last longer, and the ability to withstand a long fast without serious tissue disruption is enhanced.

The combined effects of glycogenolysis, gluconeogenesis, and the switch to fat utilization are so efficient that, after several days of complete fasting, the plasma glucose concentration is reduced by only a few percent. After 1 month, it is decreased only 25 percent.

Summary. In the postabsorptive state: (1) Glycogen, fat, and protein syntheses are curtailed and net breakdown occurs; (2) glucose is formed in the liver (and, later, in the kidneys also) from its own glycogen, and by gluconeogenesis from lactate, pyruvate, glycerol, and amino acids; (3) this glucose is released into the blood, but its utilization for energy is greatly reduced in most tissues; (4) lipolysis releases adipose-tissue fatty acids into the blood, and the oxidation of these fatty acids and of ketones produced from them by the liver provides most of the body's energy supply; and (5) the brain continues to use glucose but also starts using ketones as they build up in the blood.

ENDOCRINE AND NEURAL CONTROL OF THE ABSORPTIVE AND POSTABSORPTIVE STATES

We now turn to the endocrine and neural factors that control and integrate these metabolic pathways and transformations. As before, the reader should constantly refer to Figures 17-1 and 17-2. We shall focus primarily on the following questions, summarized in Figure 17-3, raised by the previous discussion: (1) What controls the shift from the net anabolism of protein, glycogen, and triacylglycerol to net catabolism in going from the absorptive phase to the postabsorptive phase? (2) What induces primarily glucose utilization for energy during the absorptive phase but fat utilization during the postabsorptive phase? (3) What drives net glucose uptake by the liver during the absorptive phase

but gluconeogenesis and glucose release during the postabsorptive phase?

The most important controls of these transitions from feasting to fasting and vice versa are two pancreatic hormones—insulin and glucagon, the hormone epinephrine from the adrenal medulla, and the sympathetic nerves to liver and adipose tissue. Insulin and glucagon are peptides secreted by the **islets of Langerhans**, clusters of endocrine cells in the pancreas. Appropriate histological techniques reveal four distinct types of islet cells, termed A (or alpha), B (or beta), D, and F cells, each of which secretes a different hormone. The **B cells** are the source of insulin, and the **A cells** of glucagon. There are two other pancreatic hormones, somatostatin and pancreatic polypeptide, secreted by the D cells and the F cells, respectively, but the functions of these two pancreatic hormones are not known.

Insulin

Insulin is the most abundant hormone secreted by the islets and is the single most important controller of organic metabolism. It acts directly or indirectly on most tissues of the body. For simplicity, insulin's many actions are often divided into two broad categories—metabolic effects on carbohydrate, lipid, and protein synthesis and growth-promoting effects on DNA synthesis, mitosis, and cell differentiation. In general, the metabolic effects occur rapidly after a physiological rise in insulin concentration, whereas the growth-promoting effects require more time (hours and days) to become manifest after exposure to elevated insulin concentrations.

This section deals with the metabolic effects, which are so important and widespread that an injection of insulin into a fasting person duplicates the absorptive-state pattern of Figures 17-1 and 17-3. Conversely, persons suffering from insulin deficiency manifest the postabsorptive pattern of Figures 17-2 and 17-3. From these statements alone, one might predict, correctly, that insulin secretion is stimulated during eating and inhibited during fasting. Although insulin acts on many cells, we shall deal mainly with the major targets—muscle, adipose tissue, and liver.

The metabolic effects of insulin are the direct result of changes in (1) the target-cell plasma membrane's capacity to transport glucose and amino acids and/or (2) enzyme function. The biochemical sequences of events leading to these changes are initiated when insulin binds to plasma-membrane receptors on its target cells.

Insulin receptors and signal transduction. Insulin, secreted from the pancreas and transported by the blood, combines with specific receptors on the outer

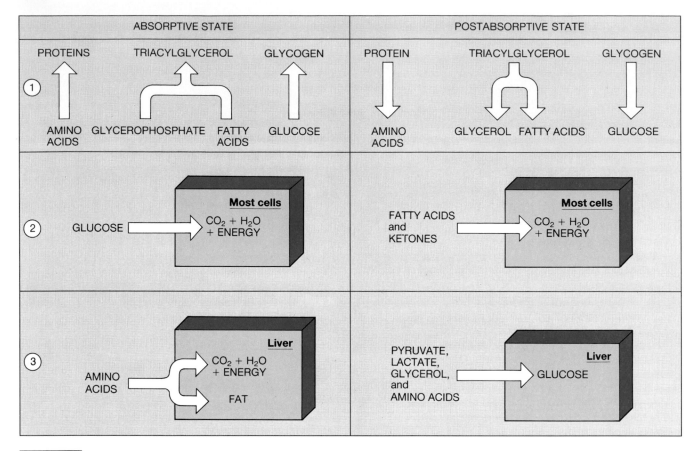

ABSORPTIVE STATE	POSTABSORPTIVE STATE

FIGURE 17-3 Summary of critical points in transition from absorptive state to postabsorptive state. The phrase "absorptive state" could be replaced with "actions of insulin," and "postabsorptive state" with "results of decreased insulin."

surface of the plasma membrane of its target cells. As noted on page 162, the insulin-activated receptor is a protein kinase enzyme that catalyses its own phosphorylation (using ATP). This phosphorylation further activates the receptor's kinase activity. How the activated receptor then influences plasma-membrane transport and intracellular enzyme activity, that is, what constitutes the signal transduction mechanism (Chapter 7), is still unclear. Insulin, however, is definitely one hormone that does *not* utilize the adenylate cyclase-cAMP second-messenger system (Chapter 7). Indeed, it inhibits adenylate cyclase, thereby decreasing intracellular cyclic AMP concentration. Whatever the signal transduction mechanisms turn out to be, it is through them that insulin's effects on membrane transport and enzyme function are brought about (Table 17-1 and Figure 17-4).

Insulin's effects on membrane transport. Glucose enters most cells by the carrier-mediated mechanism that we described as facilitated diffusion on page 116. Insulin acts on this transport process in many cells, particularly muscle and adipose tissue, to increase the movement of glucose into the cell. Insulin does not act on the transport process in liver or brain, but we shall see that another of its actions does, nevertheless, result in increased glucose diffusion into liver cells.

How does insulin increase facilitated diffusion in muscle, adipose tissue, and other cells? It causes, via its (unknown) signal transduction mechanism, glucose transport proteins stored in the cytoplasm to be inserted in the plasma membrane. Greater glucose entry into cells then increases this sugar's availability for all the reactions in which it participates—glycolysis, glycogen synthesis, and synthesis of fatty acids and glycerophosphate that then combine to form triacylglycerol—all key events of the absorptive phase.

Insulin also stimulates the active transport of amino acids into cells, thereby making more amino acids avail-

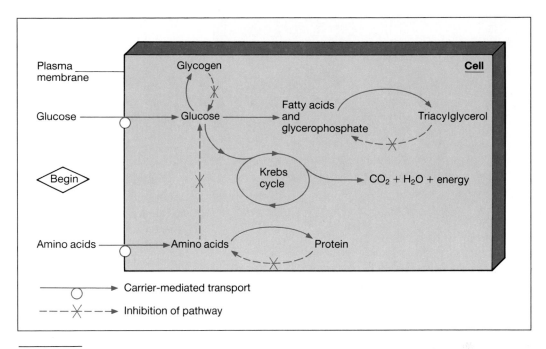

FIGURE 17-4 Major metabolic effects of insulin (see also Table 17-1). Each solid arrow represents a process enhanced by insulin, whereas a dashed arrow denotes a reaction inhibited by insulin. Conversely, when insulin concentration is low, the activity of the dashed pathways is enhanced and that of the solid pathways is decreased. The nontransport effects are mediated by insulin-dependent enzymes. Not all these effects are exerted upon all cells.

able for protein synthesis, the remaining key event of the absorptive phase.

Insulin's effects on enzymes.

Insulin favors glycolysis and net synthesis of glycogen, triacylglycerol, and protein not only by its effects on membrane transport but by altering the activities or concentrations of many intracellular enzymes involved in the metabolic pathways of these substances in muscle, adipose tissue, and liver.

Glycogen synthesis in muscle provides an excellent example (Figure 17-4): Increased glucose transport stimulates glycogen synthesis by increasing glucose availability, but in addition, insulin increases the activity of the enzyme catalyzing the rate-limiting step in glycogen synthesis and inhibits the enzyme that catalyzes glycogen catabolism. Thus, insulin favors glucose transformation to glycogen in muscle by a triple-barreled effect! In liver, the stimulation of glycogen synthesis is only double-barreled since, as noted, glucose transport is not stimulated in liver.

At the same time that net glycogen synthesis is being stimulated in liver, insulin inhibits almost all the critical liver enzymes that catalyze gluconeogenesis. The net result is that insulin abolishes glucose release by the liver. You might think, by putting together this information and the absence of an insulin effect on membrane transport in liver, that the insulin-stimulated liver neither releases nor takes up glucose. The fact is, however, that the insulin-stimulated liver takes up glucose. Insulin causes this uptake not by stimulating a membrane transport process for glucose, as it does in muscle and adipose tissue, but rather by increasing the activity of the enzyme that catalyzes the first step in glucose metabolism—its phosphorylation. This keeps the cytosolic glucose concentration low, resulting in a large concentration gradient favoring glucose movement from plasma to cytosol. Other enzymes important for glycolysis, both in liver and elsewhere, are also activated by insulin.

The situation for triacylglycerol storage is analogous to that for carbohydrate. As we saw in the previous section, insulin stimulates the facilitated diffusion of glucose into adipose-tissue cells, and the glucose provides the precursors for triacylglycerol synthesis. Simultaneously, insulin increases the activity of adipose-tissue enzymes that catalyze fatty acid synthesis, and most important, it inhibits the intracellular lipase that catalyzes triacylglycerol breakdown (Figure 17-4). Again there is a triple-

TABLE 17-1 METABOLIC EFFECTS OF INSULIN

Carbohydrate metabolism
  Stimulation of glucose uptake by cells (T)*
  Stimulation of glycolysis (E)*
  Stimulation of glycogen synthesis (E)
  Inhibition of glycogen catabolism (E)
  Inhibition of gluconeogenesis (E)
     Net result: Decrease in plasma glucose
             concentration; increases in glucose
             utilization and glycogen storage
           Net glucose uptake, rather than
             release, by liver
Lipid metabolism
  Stimulation of triacylglycerol synthesis (E)
  Inhibition of triacylglycerol catabolism (E)
  Stimulation of endothelial-cell lipoprotein lipase (E)
     Net result: Decreases in plasma concentrations of
             glycerol and free fatty acids
           Net fat storage and decreased
             utilization of fat for energy
Protein metabolism
  Stimulation of amino acid uptake by cells (T)
  Stimulation of protein synthesis (E)
  Inhibition of protein catabolism (E)
     Net result: Decreased plasma concentrations of
             amino acids
           Net protein anabolism

*T denotes an effect on a transport process; E, an effect on enzymes.

barreled effect favoring, in this case, triacylglycerol storage in adipose tissue. Actually there is a four-barreled effect, for insulin also stimulates endothelial-cell lipoprotein lipase, which catalyzes the breakdown of plasma triacylglycerols and releases fatty acids that can then enter the adipose-tissue cells and be built into triacylglycerol there. The reader must remember to distinguish lipoprotein lipase, which is on the surface of endothelial cells and promotes fat *storage*, from the completely different intracellular lipase that mediates lipolysis, thereby favoring fat *mobilization*. And similarly for protein: Insulin stimulates uptake of amino acids by target cells while simultaneously increasing the activity of some of the ribosomal enzymes that mediate the synthesis of protein from these amino acids and inhibiting the enzymes that mediate protein catabolism.

The metabolic effects of insulin are summarized in Table 17-1 and Figure 17-4.

Effects of decreases in plasma insulin concentration.
A decrease in plasma insulin concentration, by eliminating or reducing its effects, will bring about just the oppo-

site results: (1) Net protein catabolism due to less entry of amino acids into cells, decreased activity of the enzymes mediating protein synthesis, and less inhibition of the enzymes mediating protein catabolism; (2) net glycogen catabolism due to less glucose entry into cells, decreased activity of the enzyme mediating glycogen synthesis, and less inhibition of the enzyme mediating glycogen catabolism; (3) net fat catabolism due to less entry of glucose into adipose-tissue cells, decreased activity of the enzymes mediating fat synthesis, and less inhibition of the enzyme mediating lipolysis; (4) decreased glycolysis due to less entry of glucose into cells and decreased activity of the glycolytic enzymes; (5) increased utilization of fatty acids and ketones due to increased availability of these substances secondary to increased fat catabolism; and (6) gluconeogenesis and release of glucose by the liver due to less inhibition of the gluconeogenic enzymes.

Thus, the metabolic pathways acted upon by insulin are in a dynamic state, capable of proceeding in either direction. For this reason, energy metabolism can be shifted from the absorptive to the postabsorptive pattern merely by lowering plasma insulin concentration via reflex lowering of its secretion rate.

We have now given a major part of the answers to the three questions posed on page 559 in that a decrease in plasma insulin concentration is mainly responsible for the metabolic transformations that occur in going from the absorptive phase to the postabsorptive phase: Thus, 1, 2, and 3 in the earlier paragraph explain the shift from net anabolism to net catabolism, 4 and 5 explain the shift from glucose utilization to fatty acid and ketone utilization for energy, and 6 explains the appearance of gluconeogenesis. With the next meal, the plasma insulin concentration increases due to reflexly increased insulin secretion and reverses all these transformations, that is, restores the absorptive phase metabolic profile.

Control of insulin secretion. The most important control of insulin secretion is the glucose concentration of the blood flowing through the pancreas, a simple system requiring no participation of nerves or of other hormones. An increase in blood glucose concentration stimulates insulin secretion, whereas a decrease inhibits secretion. The feedback nature of this system is shown in Figure 17-5: The rise in plasma glucose concentration following a meal stimulates insulin secretion, and the insulin induces rapid entry of glucose into cells as well as cessation of glucose output by the liver. These effects reduce the blood concentration of glucose, thereby removing the stimulus for insulin secretion, which returns to its previous level.

There are also numerous insulin-secretion controls

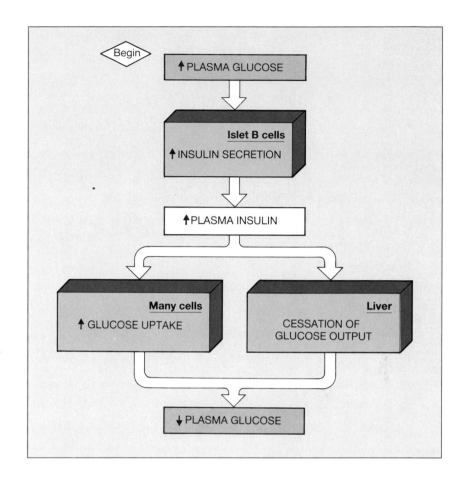

FIGURE 17-5 Negative-feedback nature of plasma glucose control over insulin secretion.

other than plasma glucose concentration (Figure 17-6). One is the plasma concentration of certain amino acids, an elevated amino acid concentration causing enhanced insulin secretion. This is another negative-feedback control: Amino acid concentrations increase after a meal, especially a high-protein meal, and the increased plasma insulin stimulates cell uptake of these amino acids.

There are also important hormonal controls over insulin secretion. For example, one or more of the hormones secreted by the gastrointestinal tract in response to eating stimulate the release of insulin. This provides an "anticipatory" (or feedforward) component to glucose regulation during ingestion of a meal, that is, insulin secretion will rise earlier and to a greater extent than it would if plasma glucose were the only controller.

Finally, the autonomic neurons to the islets of Langerhans also influence insulin secretion. Activation of the parasympathetic neurons, which occurs during ingestion of a meal, stimulates secretion of insulin and constitutes a second type of anticipatory regulation. In contrast, activation of the sympathetic neurons to the islets or an increase in plasma epinephrine concentration, both of which occur during exercise or stress, inhibits insulin

secretion. The significance of this relationship will be described later in this chapter.

To repeat, insulin unquestionably plays the primary role in controlling the metabolic adjustments required for feasting or fasting, but other hormonal and neural factors also play significant roles. They all oppose the action of insulin in one way or another and are known as **glucose-counterregulatory controls**. Of these, the most important is glucagon.

Glucagon

As noted earlier, **glucagon** is the peptide hormone produced by the A cells of the pancreatic islets. Glucagon and glucagon-like substances are also secreted by cells in the lining of the gastrointestinal tract, but the function of these substances is unclear.

Glucagon binds to receptors in the plasma membrane of its target cells, leading to the activation of adenylate cyclase and generation of cyclic AMP in the cells. Cyclic AMP, in turn, activates a protein kinase that phosphorylates enzymes that mediate the effects of glucagon. All these effects are opposed to those of insulin (Figure 17-7): (1) Increased glycogen breakdown in liver (but,

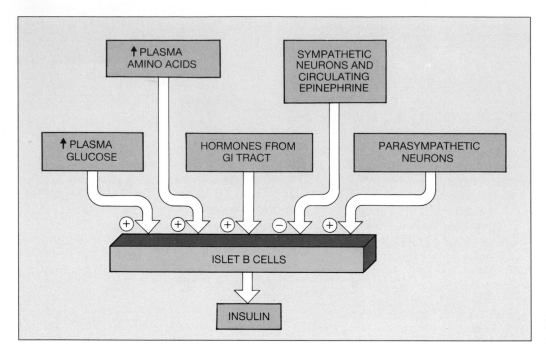

FIGURE 17-6 Control of insulin secretion.

interestingly, not in skeletal muscle); (2) increased lipolysis; (3) increased gluconeogenesis by liver; and (4) synthesis of ketones by the liver. Thus, the overall results of glucagon's effects are to increase the plasma concentrations of glucose, fatty acids, glycerol, and ketones, all of which are important for the postabsorptive period.

From a knowledge of these effects, one would logically suppose that glucagon secretion should increase during the postabsorptive period and prolonged fasting, and such is the case. The stimulus for glucagon secretion is a decreased plasma glucose concentration (**hypoglycemia**). The adaptive value of such a reflex is obvious. A decreasing plasma glucose concentration induces increased release of glucagon, which, by its effects on metabolism, serves to restore normal blood glucose levels by glycogenolysis and gluconeogenesis, while at the same time supplying fatty acids and (if the fast is prolonged) ketones for cell utilization. Conversely, an increased plasma glucose concentration inhibits glucagon's secretion, thereby helping to return plasma glucose level toward normal.[1]

Thus far, the story is quite uncomplicated. The B and A cells of the pancreatic islets constitute a push-pull system for regulating plasma glucose by producing two hormones—insulin and glucagon—whose actions and major controlling input are just the opposite of each other. However, we must now point out a complicating feature: A second major control of glucagon secretion is the plasma amino acid concentration acting directly on the A cells, and in this regard the effect is identical to, rather than opposite that for insulin. Glucagon secretion, like insulin secretion, is strongly stimulated by a rise in plasma amino acid concentration such as occurs following a protein-rich meal.

Thus, during absorption of a carbohydrate-rich meal containing little protein, there occurs an increase in insulin secretion alone, caused by the rise in plasma glucose. During absorption of a low-carbohydrate, high-protein meal, glucagon levels also increase along with insulin, under the influence of the increased plasma amino acid concentration. The usual meal is somewhere between these extremes and is accompanied by a rise in insulin level and a relatively small increase in glucagon since the simultaneous increases in blood glucose and amino acids tend to counteract each other so far as glucagon secretion is concerned.

What is the adaptive value of the fact that a high-protein meal stimulates glucagon secretion? Imagine what might occur were glucagon not part of the response to a high-protein meal. Insulin secretion would be in-

---

[1]The mechanism by which changes in plasma glucose concentration act on the islets to alter glucagon secretion are not known. It has been proposed that the A cells are controlled not by glucose, per se, but by insulin, that is, that an increase in insulin directly inhibits glucagon secretion whereas a decrease in insulin removes the inhibition, resulting in increased secretion. Pancreatic somatostatin may also inhibit glucagon secretion.

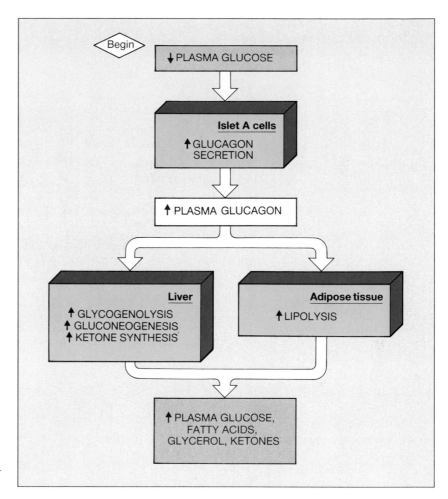

FIGURE 17-7 Negative-feedback nature of plasma glucose control over glucagon secretion.

creased by the amino acids, but, since little carbohydrate was ingested and therefore available for absorption, the increase in plasma insulin could cause a marked and sudden drop in plasma glucose. In reality, the rise in glucagon secretion simultaneously caused by the amino acids permits the plasma-glucose-raising effects of this hormone to counteract the plasma-glucose-lowering actions of insulin, and the net result is a stable plasma glucose. Thus, a high-protein meal virtually free of carbohydrate can be absorbed with little change in plasma glucose concentration despite a marked increase in insulin secretion.

The secretion of glucagon, like that of insulin, is controlled not only by the plasma concentrations of glucose and amino acids but also by neural and hormonal inputs to the islets (Figure 17-8). For example, the sympathetic nerves stimulate glucagon secretion—just the opposite of their effect on insulin secretion. The adaptive significance of this relationship for exercise and stress will be described subsequently.

Epinephrine and Sympathetic Nerves to Liver and Adipose Tissue

Both epinephrine, the major hormone secreted by the adrenal medulla, and the sympathetic nerves to liver and adipose tissue help control plasma glucose. As described in Chapter 10, epinephrine release is controlled entirely by the preganglionic sympathetic fibers to the adrenal medulla.

Epinephrine inhibits insulin secretion and stimulates glucagon secretion. In addition, this hormone also affects nutrient metabolism directly (Figure 17-9). Its direct effects include stimulation of (1) glycogenolysis by both liver and skeletal muscle, (2) gluconeogenesis, and (3) lipolysis, as well as (4) inhibition of glucose uptake by skeletal muscle. Activation of the sympathetic nerves to liver and adipose tissue elicits essentially the same responses by these organs as does circulating epinephrine.

Thus, enhanced sympathetic nervous system activity exerts effects on organic metabolism that are opposite to

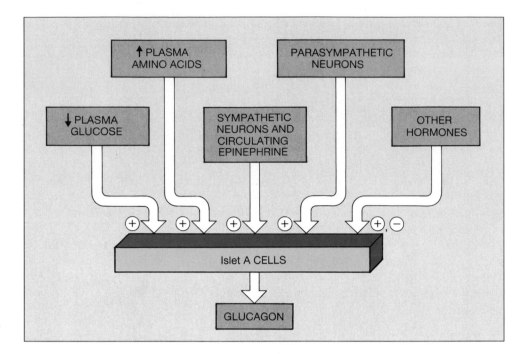

FIGURE 17-8 Control of glucagon secretion. Note that the sympathetic and parasympathetic nerves to the A cells both stimulate glucagon secretion.

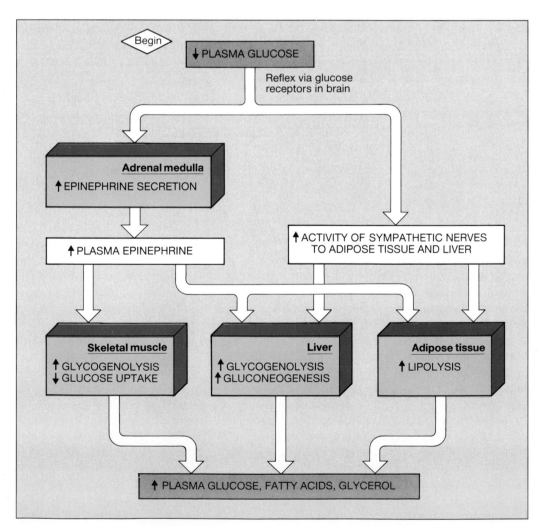

FIGURE 17-9 Participation of the sympathetic nervous system in the response to a low plasma glucose concentration. Glucose receptors in the liver (not shown) may also participate in this reflex.

those of insulin and similar to those of glucagon—increased plasma concentrations of glucose, glycerol, and fatty acids.

As might be predicted from these effects, hypoglycemia leads reflexly to increases in both epinephrine secretion and sympathetic-nerve activity to liver and adipose tissue. This is the same stimulus that, as described above, leads to increased secretion of glucagon, although the receptors and pathways are totally different. When the plasma glucose concentration decreases, glucose receptors in the brain and the liver initiate the reflexes that lead to increased activity in the sympathetic pathways to the adrenal medulla, liver, and adipose tissue. The adaptive value of the response is the same as that for the glucagon response to hypoglycemia: Blood glucose returns toward normal and fatty acids are supplied for cell utilization.

How important is the sympathetic nervous system in the acute compensatory response to hypoglycemia? The present view is that this system is less important than is glucagon but nevertheless does contribute to bringing plasma glucose back toward normal.

In contrast to its increase during *acute* hypoglycemia, sympathetic nervous system activity decreases during prolonged fasting or the ingestion of low-calorie diets; the adaptive significance of this change is discussed later in this chapter.

Other Hormones

In addition to the three hormones already described in this section, there are many others—growth hormone, cortisol, thyroid hormones, sex steroids, vasopressin, and angiotensin—that have various effects on organic metabolism. However, the secretion of these hormones is not primarily keyed to the transitions between the absorptive and postabsorptive states.[2] Instead, it is controlled by other factors, and these hormones are, for the most part, involved in homeostatic processes described elsewhere in this book. However, several of the effects of cortisol and growth hormone can, under certain circumstances, importantly influence plasma glucose concentration and so warrant description here.

Cortisol.   Cortisol, the major glucocorticoid produced by the adrenal cortex, plays an important "permissive" role in the adjustments to fasting. We have described how fasting is associated with stimulation of both gluconeogenesis and lipolysis; however, neither of these critical metabolic transformations occurs to the usual degree

in a person deficient in cortisol. In other words, plasma cortisol level need not *rise* during fasting and usually does not. There must be, however, a basal level of cortisol in the blood in order to "permit" other hormonal changes—decreased insulin, increased glucagon and epinephrine—to be fully effective in eliciting gluconeogenesis and lipolysis. Somehow, the presence of even small amounts of cortisol in the blood maintains the concentrations of the key liver and adipose-tissue enzymes required for gluconeogenesis and lipolysis. Therefore, in response to fasting, persons with cortisol deficiency develop hypoglycemia serious enough to interfere with brain function.

Cortisol can play more than a permissive role. We have stated that this hormone does not normally increase during fasting and does not itself trigger the metabolic responses to fasting. Whenever its plasma concentration does increase, however, as occurs during stress (Chapter 19), cortisol elicits many metabolic events ordinarily associated with fasting (Table 17-2). Clearly, here is another hormone, in addition to glucagon and epinephrine, that can exert actions opposite to those of insulin. Indeed, persons with very high plasma levels of cortisol due either to abnormally high secretion or to cortisol administration for medical reasons (Chapter 19) develop symptoms similar to those seen in persons with insulin deficiency.

Growth hormone.   The primary physiological effects of growth hormone are to stimulate both growth and protein anabolism. Compared to these effects, those it exerts on carbohydrate and lipid metabolism are quite minor under most physiological situations. Nonetheless, as is true for cortisol, either severe deficiency or marked excess of growth hormone does produce significant abnormalities of lipid and carbohydrate metabolism. Growth hormone's effects on these nutrients, but not on protein, are similar to those of cortisol and are opposite to those of insulin. Growth hormone (1) stimulates lipol-

---

[2] Some people do show a small burst of growth hormone secretion several hours after a meal, but it is unlikely that this change contributes significantly to postabsorptive adaptations.

TABLE 17-2   EFFECTS OF CORTISOL ON ORGANIC METABOLISM
(A) Basal concentrations are permissive for stimulation of gluconeogenesis and lipolysis in the postabsorptive state.
(B) Increased plasma concentrations cause:
(1) Increased protein catabolism
(2) Increased gluconeogenesis
(3) Decreased glucose uptake by cells
(4) Increased triacylglycerol breakdown
Net result: Increased plasma concentrations of amino acids, glucose, and free fatty acids

TABLE 17-3 SUMMARY OF GLUCOSE-COUNTERREGULATORY CONTROLS

	Glucagon	Epinephrine	Cortisol	Growth Hormone
Glycogenolysis	X	X		
Gluconeogenesis	X	X	X	X
Lipolysis	X	X	X	X
Blockade of glucose uptake		X	X	X

Note: All the processes listed on the left—glycogenolysis, gluconeogenesis, lipolysis, and blockade of glucose uptake—are opposed to insulin's actions and are stimulated by one or more of the glucose-counterregulatory hormones in the table. An X indicates that the hormone stimulates the process; no X indicates that the hormone has no effect on the process. Epinephrine stimulates glycogenolysis in both liver and skeletal muscle, whereas glucagon does so only in liver.

ysis in adipose tissue, (2) increases gluconeogenesis by the liver, and (3) decreases glucose uptake by many peripheral tissues.

Summary

To a great extent insulin may be viewed as the "hormone of plenty." Its secretion and plasma concentration are increased during the absorptive period and decreased during postabsorption, and these changes are adequate to cause most of the metabolic changes associated with these periods. In addition, opposed in various ways to insulin's effects are the actions of four major glucose-counterregulatory controls—glucagon, epinephrine and the sympathetic nerves to liver and adipose tissue, cortisol, and growth hormone (Table 17-3). Glucagon and the sympathetic nervous system are activated during the postabsorptive period and definitely play roles in preventing hypoglycemia, glucagon being the more important. The rates of secretion of cortisol and growth hormone are not usually coupled to the absorptive-postabsorptive pattern; nevertheless, their presence in the blood at basal concentrations is necessary for normal adjustment of lipid and carbohydrate metabolism to the postabsorptive period, and excessive amounts of either hormone causes abnormally elevated plasma glucose concentrations.

FUEL HOMEOSTASIS IN EXERCISE AND STRESS

During exercise large quantities of fuels must be mobilized to provide the energy required for muscle contraction. As described in Chapter 11, these fuels include plasma glucose and fatty acids as well as the muscle's own glycogen.

The plasma glucose used during exercise is supplied by the liver, both by breakdown of its glycogen stores and by gluconeogenesis—conversion of pyruvate, lactate, glycerol, and amino acids into glucose. The glycerol is made available by a marked increase in adipose-tissue lipolysis with resultant release of glycerol and fatty acids into the blood, the fatty acids serving along with glucose as a fuel source for the exercising muscle.

What happens to blood glucose concentration during exercise (Figure 17-10)? It changes very little in short-term, mild to moderate exercise and may increase slightly with strenuous short-term activity. However, during prolonged exercise, more than 90 min, plasma glucose concentration decreases, sometimes by as much as 25 percent. These relatively minor changes in blood glucose in the presence of marked increases in glucose utilization clearly demonstrate that glucose output by the liver must increase approximately in proportion to the increased utilization, at least until the later stages of exercise when it begins to lag somewhat.

The metabolic profile seen in an exercising individual—increases in hepatic glucose production, triacylglycerol breakdown, and fatty acid utilization—is similar to that seen in a fasting person, and the controls are also the same. Exercise is characterized by a fall in insulin secretion and a rise in glucagon secretion (Figure 17-10), plus increased activity of the sympathetic nervous system. In addition, cortisol and growth hormone are also secreted in increased amounts. All these events tend to increase plasma glucose, balancing the glucose-lowering effect caused by increased muscle glucose utilization.

What triggers the stimulation of glucagon secretion and the inhibition of insulin secretion during exercise? One signal during prolonged exercise is the modest decrease in plasma glucose that occurs as the muscles con-

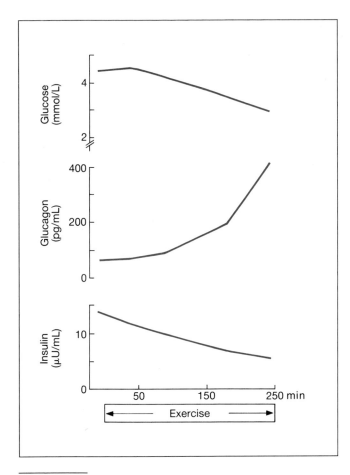

FIGURE 17-10 Plasma concentrations of glucose, glucagon, and insulin during prolonged moderate exercise at a fixed intensity. (*Adapted from Felig and Wahren.*)

sume glucose. This is the same signal that controls the secretion of these hormones in fasting. A second input controlling insulin and glucagon secretion is increased circulating epinephrine and enhanced activity of the sympathetic neurons supplying the pancreatic islets, mediated by the central nervous system as part of the neural response to exercise. Thus, the increased sympathetic nervous system activity characteristic of exercise not only contributes directly to fuel mobilization by acting on the liver and adipose tissue but contributes indirectly by inhibiting the secretion of insulin and stimulating that of glucagon.

One component of the response to exercise is quite different from fasting. In exercise, glucose uptake and utilization by the muscles is increased, whereas in fasting, it is markedly reduced. How is it that, during exercise, the movement of glucose into muscle can remain high in the presence of reduced plasma insulin and increased plasma concentrations of epinephrine, cortisol,

and growth hormone, all of which decrease glucose uptake by skeletal muscle? An as yet unidentified local chemical change in the muscle is the cause.

Exercise is not the only situation characterized by the neuroendocrine profile of decreased insulin and increased glucagon, sympathetic activity, cortisol, and growth hormone. The body reacts in the same way to a variety of nonspecific stresses, both physical and emotional. The adaptive value of these neuroendocrine responses to stress is that the resulting shifts in metabolism prepare the body for exercise ("fight or flight") in the face of real or threatened injury. In addition, the amino acids liberated by catabolism of body protein stores, because of decreased insulin and increased cortisol, not only provide energy via gluconeogenesis but also constitute a potential source of amino acids for tissue repair should injury occur. The subject of stress and the body's responses to it are further described in Chapter 19.

DIABETES MELLITUS

The name diabetes, meaning "syphon" or "running through," was used by the Greeks over 2000 years ago to describe the striking urinary volume excreted by people suffering from this disease. Mellitus, meaning "sweet," distinguishes this urine from the large quantities of nonsweet ("insipid") urine produced by persons suffering from vasopressin deficiency. As described in Chapter 15, the latter disorder is known as diabetes insipidus, and the unmodified word "diabetes" is often used as a synonym for diabetes mellitus.

Diabetes can be due to a deficiency of insulin or to a hyporesponsiveness to insulin, for it is not one but several diseases with different causes. Classification of these diseases rests on how much insulin the person is secreting and whether therapy requires administration of insulin. In **type 1 (insulin-dependent) diabetes**, the hormone is completely or almost completely absent from the islets of Langerhans and the plasma, and therapy with insulin is essential (this protein hormone cannot be given orally because gastrointestinal enzymes would digest it). In **type 2 (insulin-independent) diabetes**, the hormone is often present in plasma at near-normal or even above-normal levels, and therapy often does not require administration of insulin.

Type 1 is the less common, affecting 10 percent of diabetic patients. It is due to the total or near-total destruction of the pancreatic B cells. The cause of the B-cell destruction seems to be attack by the body's own defense mechanisms (autoimmune disease, page 681) in genetically susceptible persons. The triggering events for this autoimmune response are unknown, although viral infection is a strong candidate in some cases.

Because of their insulin deficiency, untreated patients with type 1 diabetes always have elevated plasma glucose concentrations, both because glucose fails to enter cells normally and because the liver continues to make and release glucose. Since a low insulin concentration results in the metabolic profile characteristic of the postabsorptive state, the changes (other than the increased plasma glucose) seen in untreated type 1 diabetes represent a caricature of this state: marked glycogenolysis, gluconeogenesis, and lipolysis, with resultant elevation of plasma glucose, glycerol, fatty acids, and ketones.

If extreme, these metabolic changes culminate in the acute life-threatening emergency called **diabetic ketoacidosis** (Figure 17-11). Some of the problems are due to the effects a markedly elevated plasma glucose produces on renal function. In Chapter 15, we pointed out that a normal person does not excrete glucose because all glucose filtered at the glomerulus is reabsorbed by the tubules. However, the elevated plasma glucose of diabetes may so increase the filtered load of glucose that the maximum tubular reabsorptive capacity is exceeded and large amounts of glucose may be excreted. For the same reasons, large amounts of ketones may also appear in the urine. These urinary losses aggravate the situation by depleting the body of nutrients and leading to weight loss. Far worse, however, is the effect of these solutes on sodium and water excretion. In Chapter 15, we saw how tubular water reabsorption is a passive process induced by active solute reabsorption. In diabetes, the osmotic force exerted by unreabsorbed glucose and ketones holds water in the tubule, thereby preventing its reabsorption. Sodium reabsorption is also retarded, and the net result is marked excretion of sodium and water, which can lead, by the sequence of events shown in Figure 17-11, to hypotension, brain damage, and death.

The other serious abnormality in diabetic ketoacidosis is the increased plasma hydrogen-ion concentration caused by the accumulation of ketones, two of which are weak acids. This increased hydrogen-ion concentration causes brain dysfunction that can contribute to the development of coma and death.[3]

Diabetic ketoacidosis is seen only in patients with untreated type 1 diabetes, that is, those with almost total inability to secrete insulin. However, 90 percent of diabetics are in the type 2 category and never develop metabolic derangements severe enough to go into diabetic ketoacidosis. Type 2 diabetes is a disease mainly of overweight, sedentary adults. Given the earlier mention of progressive weight loss as a symptom of diabetes, it may seem contradictory that many diabetics are overweight. The paradox is resolved when one realizes that patients with type 2 diabetes, in contrast to those with type 1, do not suffer enough glucose excretion in the urine to cause weight loss.

How is it that type 2 diabetes is characterized by near-normal or even above-normal plasma levels of insulin? The problem is hyporesponsiveness to insulin, termed **insulin resistance**. Insulin's target cells do not respond normally to the circulating insulin because of alterations either in the insulin receptors or some intracellular process occurring after receptor activation. An insufficient number of insulin receptors per target cell seems to be one of the causes of the insulin resistance in type 2 diabetics who are overweight.

The sequence of events coupling obesity to a diminished number of insulin receptors may be as follows (Figure 17-12). Because insulin secretion is increased during food absorption, any person—diabetic or not—who chronically overeats secretes, on the average, increased amounts of insulin. Over time, the resulting elevation of plasma insulin induces a reduction (downregulation) in the number of insulin receptors. Thus, insulin itself is responsible for the decrease in target-cell responsiveness (insulin resistance), producing a higher plasma glucose concentration at any given plasma insulin level. The tendency toward a higher plasma glucose concentration is small in nondiabetic overeaters because the islet cells respond by secreting enough additional insulin to get the job done despite the reduction in available receptors. In contrast, the diabetes-prone person may also secrete additional insulin but not enough to prevent significant hyperglycemia.

The insulin hyporesponsiveness secondary to overeating can usually be completely reversed if the person simply reduces his or her caloric intake. As the number of receptors returns to normal—this begins even while the person is still overweight as long as caloric intake is below expenditure—so do the responses of target cells to insulin. Thus dietary control, without any other therapy, is frequently sufficient to eliminate the elevated blood glucose of obese type 2 diabetes. An exercise program also is useful since the number of insulin receptors is increased by frequent endurance exercise, independent of changes in body weight.[4]

---

[3]The acidosis also leads, in a positive-feedback manner, to worsening of the diabetes and the acidosis because there is a decrease in the affinity of receptors for insulin.

---

[4]This is one reason that patients with type 1 diabetes require *less* insulin to keep glucose normal when they are physically active than when they are sedentary. Another reason mentioned earlier is that local changes occur during exercise that make the muscle less dependent on insulin.

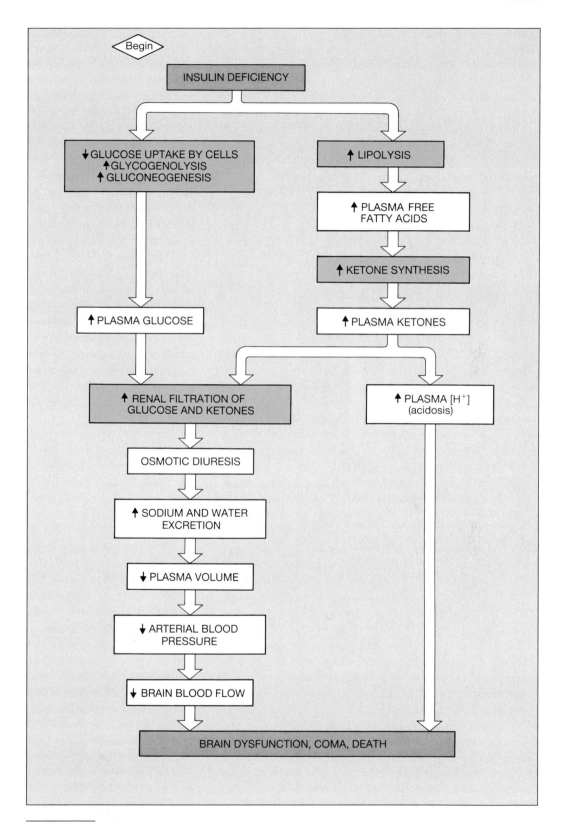

FIGURE 17-11 Effects of severe untreated insulin deficiency (diabetic ketoacidosis).

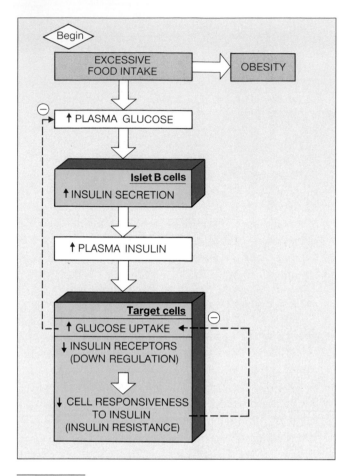

FIGURE 17-12 Postulated mechanism by which chronic over-eating leads to chronically elevated plasma insulin and diminished cell responsiveness to insulin. Note that a chronically increased plasma insulin level contributes, via down-regulation, to a negative-feedback control over cell responsiveness to insulin. This insulin resistance reduces the ability of insulin to lower plasma glucose concentration.

Given that even untreated type 2 diabetics do not usually develop the severe metabolic and urinary derangements typical of untreated type 1 diabetes, one might wonder why treatment is so important in type 2 diabetes. Unfortunately, persons with either type tend to develop a variety of chronic abnormalities, including atherosclerosis, small-vessel and nerve disease, susceptibility to infection, and blindness. The factors responsible are still unclear, but it is quite likely that an elevated plasma glucose contributes directly or indirectly to some, if not all, of these abnormalities. Accordingly the more normal the plasma glucose, the less likely it is that complications will arise.

This discussion of diabetes has focused on insulin, but it is now clear that the hormones that elevate plasma glucose concentration may contribute to the severity of diabetes. Glucagon seems particularly important in this regard. Since glucagon secretion is inhibited by an elevated plasma glucose level, one would expect to find a low plasma glucagon concentration in diabetic persons. However, most diabetic patients, for unknown reasons, have plasma glucagon concentrations that are either increased or unchanged. This absolute or relative glucagon excess contributes to the metabolic dysfunction typical of diabetes. Thus, if glucagon secretion is eliminated in experimental animals with type 1 diabetes, the massive hepatic overproduction of glucose and ketones usually observed in this type of diabetes does not occur, despite the total absence of insulin.

Finally, as we have seen, all the systems that raise plasma glucose concentration are activated during stress, which explains why stress exacerbates the symptoms of diabetes. Since diabetic ketoacidosis, itself, constitutes a severe stress, a positive-feedback cycle is triggered in which marked lack of insulin induces ketoacidosis, which elicits activation of the glucose-counterregulatory systems, which worsens the ketoacidosis.

HYPOGLYCEMIA AS A CAUSE OF SYMPTOMS

As we have seen, "hypoglycemia" means a low plasma glucose concentration. Plasma glucose concentration can drop to very low values during fasting in persons with several types of disorders. This is termed fasting hypoglycemia, and the relatively uncommon disorders responsible for it can be understood in terms of the regulation of blood glucose concentration. They include (1) an excess of insulin, due to an insulin-producing tumor, a drug that stimulates insulin secretion, or the taking of too much insulin by a diabetic, and (2) a defect in one or more of the glucose-counterregulatory systems, for example, inadequate glycogenolysis and/or gluconeogenesis due to liver disease, glucagon deficiency, or cortisol deficiency.

Fasting hypoglycemia causes many symptoms, some of which—hunger, increased heart rate, tremulousness, weakness, nervousness, and sweating—are accounted for by activation of the sympathetic nervous system caused reflexly by the hypoglycemia. Other symptoms, such as headache, confusion, uncoordination, and slurred speech are direct consequences of too little glucose reaching the brain. More serious brain effects, including convulsions and coma, can occur if the plasma glucose concentration goes low enough.

In contrast, low plasma glucose concentration has not been shown to routinely produce either acute or chronic symptoms of fatigue, lethargy, loss of libido, depression, or many other symptoms for which the lay press frequently holds it responsible. Despite this, a large number of persons who suffer from such symptoms have been told or have assumed that the symptoms are due to hypoglycemia. Few of these people have ever had their blood glucose concentrations measured at the time of the symptoms, and the blood sugar is most often within the normal range in those cases where measurements have been made. For all these reasons, most experts believe that most of the symptoms popularly ascribed to hypoglycemia have other causes.

REGULATION OF PLASMA CHOLESTEROL

In the previous sections, we described the flow of lipids to and from adipose tissue in the form of fatty acids and triacylglycerols complexed with proteins. One very important lipid—**cholesterol**—was not mentioned earlier because it, unlike the fatty acids and triacylglycerols, serves not as a metabolic fuel but rather as a precursor for plasma membranes, bile salts, steroid hormones, and other specialized molecules. A large number of studies have been devoted to the factors that govern the plasma cholesterol concentration because high plasma cholesterol enhances the development of atherosclerosis, the arterial thickening that leads to heart attacks, strokes, and other forms of cardiovascular damage (page 418).

Based on studies of the relationship between plasma cholesterol levels and these diseases, recent recommendations from the National Institutes of Health call a total plasma cholesterol below 200 mg/dL (1 dL = 100 mL) "desirable," 200 to 239 mg/dL "borderline high," and 240 mg/dL or greater "high."

The story is more complicated, however, since cholesterol circulates in the plasma as part of various lipoprotein complexes. These include chylomicrons (page 524) VLDL (page 556), **low-density lipoproteins (LDL)**, and **high-density lipoproteins (HDL)**. LDL are the main cholesterol carriers, and they deliver cholesterol to almost all cells, which use them for their structural and metabolic requirements. LDL bind to plasma-membrane receptors specific for a protein component of the LDL, and the LDL are taken up by the cell. In contrast, HDL serve as acceptors of cholesterol from various tissues. They promote the removal of cholesterol from cells and its secretion into the bile by the liver. For these reasons, LDL cholesterol is "bad" cholesterol with regard to enhancing the deposition of cholesterol in arterial walls, whereas HDL cholesterol is "good" cholesterol.

The best single indicator of the likelihood of developing atherosclerotic heart disease is, therefore, not *total* plasma cholesterol but rather the *ratio* of plasma LDL-cholesterol to plasma HDL-cholesterol. A ratio of 5:1 is associated with the average American risk, which is quite high; a ratio of 3.5:1 or less corresponds to half the average risk, whereas a ratio of 9:1 doubles the average risk. Cigarette smoking, a known risk factor for heart attacks, lowers plasma HDL, whereas regular exercise increases it.

Plasma cholesterol concentration changes considerably with age. At birth, the value is approximately 50 mg/100 mL, but it then rises very rapidly to reach 165 mg/100 mL by the age of 2. Then there is relatively little change until late adolescence, at which time, in men, the mean value begins a progressive rise to reach a peak of 245 mg/100 mL in the 50s (Figure 17-13). For unknown reasons (estrogen may play a role), women show a smaller rise until after menopause, after which the values rise to or above male levels. This lower plasma cholesterol may explain why premenopausal women have so much less atherosclerotic cardiovascular disease than men. After menopause, the rates become similar.

A schema for cholesterol metabolism is illustrated in Figure 17-14. Dietary cholesterol comes from animal sources, egg yolk being by far the richest in this lipid; a single egg contains about 250 mg of cholesterol. The average daily intake of cholesterol per person by Americans is 550 mg. Some of this ingested cholesterol is not absorbed into the blood, however, but simply moves the length of the gastrointestinal tract to be excreted in the feces. A second source of cholesterol is synthesis within the body. The liver and cells lining the gastrointestinal tract are major producers of cholesterol, most of which enters the blood. In addition, almost all cells can synthesize some of the cholesterol required for their own plasma membranes, but most cannot do so in adequate amounts and depend upon receiving cholesterol from the blood.

Now for the other side of cholesterol balance—the pathways for net loss from the body. Because of recycling processes, these pathways appear in Figure 17-14 to be more complex than they really are. First of all, some plasma cholesterol is picked up by liver cells and enters the bile, which flows into the intestinal tract. Here it is treated much like ingested cholesterol, some being absorbed back into the blood and the remainder being excreted in the feces. Second, much of the cholesterol picked up by the liver cells is metabolized into bile acids (Chapter 16). After their production by the liver, these bile acids flow through the bile duct into the small intestine. Most are then reclaimed by absorption back into the blood across the wall of the lower small intes-

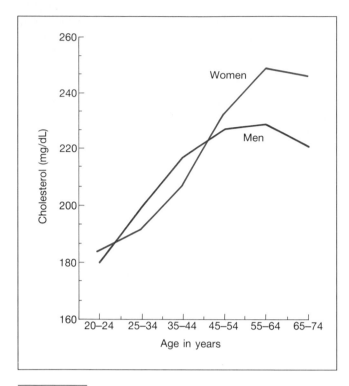

FIGURE 17-13 Average plasma cholesterol concentrations in healthy American adults.

tine, with those escaping absorption being excreted in the feces. This is an important point: The drain on plasma cholesterol represented by its conversion to bile acids depends upon how much of these bile acids are ultimately lost in the feces.

Thus, the liver is the center of the cholesterol universe. At one and the same time, it may be synthesizing new cholesterol, picking it up from the blood and secreting it into the bile, and transforming it into bile acids. The homeostatic control mechanisms that keep plasma cholesterol relatively constant operate on all of these processes, but the single most important response involves cholesterol production: The synthesis of cholesterol by the liver is inhibited whenever dietary cholesterol is increased. This is because cholesterol inhibits the enzyme critical for cholesterol synthesis by the liver.

Thus, as soon as the plasma cholesterol level starts rising because of increased cholesterol ingestion, hepatic synthesis is inhibited, and the plasma concentration remains close to its original value. Conversely, when dietary cholesterol is reduced and plasma cholesterol begins to fall, hepatic synthesis is stimulated (released from inhibition), and this increased production opposes any further fall. This homeostatic control of synthesis is the major reason why it is difficult to alter plasma cholesterol very much in either direction by altering only dietary cholesterol.

Thus far, the relative constancy of plasma cholesterol has been emphasized. However, there are environmen-

FIGURE 17-14 Metabolism of cholesterol.

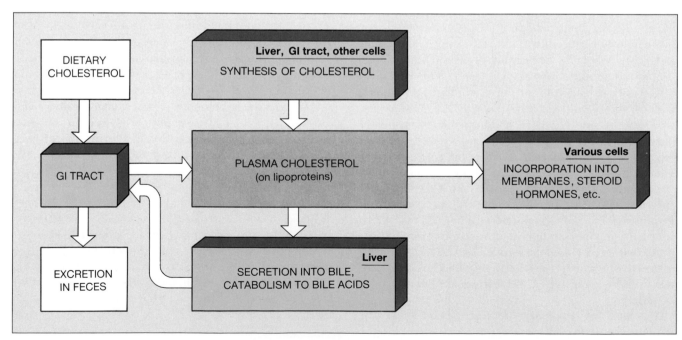

tal and genetic factors that can alter plasma cholesterol concentrations considerably by influencing one or more of the metabolic pathways for cholesterol. Perhaps the most intensively studied of these factors are the quantity and type of dietary fatty acids. Ingesting saturated fatty acids, the dominant fatty acids of animal fat, raises plasma cholesterol. Eating either polyunsaturated fatty acids, the dominant plant fatty acids, or monounsaturated fatty acids such as those in olive or peanut oil, lowers plasma cholesterol. The fatty acids exert their ef-

fects by altering cholesterol synthesis, excretion, and catabolism. A variety of drugs now in common use also are capable of lowering plasma cholesterol by influencing one or more of the metabolic pathways or by interfering with the reabsorption of bile acids.

Thus, dietary changes, in combination with drug therapy if necessary, can result in significant lowering of plasma cholesterol concentration and a decrease in the LDL:HDL ratio, with important consequences for reducing cardiovascular disease.

SECTION B
CONTROL OF GROWTH

Growth is a complex process influenced by genetics, endocrine function, and a variety of environmental factors, including nutrition and the presence of infection. The process involves cell multiplication and net protein synthesis throughout the body, but a person's height is determined specifically by bone growth.

that linear growth is not necessarily correlated with the rates of growth of specific organs. The pubertal growth spurt lasts several years in both sexes but is greater in boys. This, plus the fact that boys have a longer period for prepubertal growth, because they begin puberty approximately 2 years later than girls, accounts for the differences in height between men and women.

BONE GROWTH

Bone is a living tissue consisting of a protein matrix upon which calcium salts, particularly calcium phosphates, are deposited. The cells responsible for laying down this matrix are **osteoblasts**. A growing long bone is divided, for descriptive purposes, into the ends, or **epiphyses**, and the shaft. The portion of each epiphysis that is in contact with the shaft is a plate of actively proliferating cartilage, the **epiphyseal growth plate** (Figure 17-15). Osteoblasts at the shaft edge of the epiphyseal growth plate convert the cartilaginous tissue at this edge to bone while new cartilage is simultaneously being laid down in the plate by cells called **chondrocytes**. In this manner, the epiphyseal growth plate remains intact (indeed, actually widens) and is gradually pushed away from the bony shaft as the latter lengthens. Linear bone growth can continue as long as the epiphyseal growth plates exist but ceases when the plates are themselves ultimately converted to bone as a result of hormonal influences at puberty. This is known as **epiphyseal closure** and occurs at different times in different bones. Accordingly, a person's "bone age" can be determined by x-raying the bones and determining which ones have undergone epiphyseal closure.

As shown in Figure 17-16, children manifest two periods of rapid body growth, one during the first two years of life and the second during puberty. Note in this figure

ENVIRONMENTAL FACTORS INFLUENCING GROWTH

Adequacy of nutrient supply and freedom from disease are the primary environmental factors influencing growth. Lack of sufficient amounts of any of the essential amino acids, essential fatty acids, vitamins, or minerals interferes with growth. Total protein and sufficient nutrients to provide energy must also be adequate. No matter how much protein is ingested, growth cannot be normal if the intake of energy-providing nutrients is too low since the protein is simply catabolized for energy.

The growth-inhibiting effects of malnutrition can be seen at any time of development but are most profound when they occur very early in life. Thus, maternal malnutrition may cause growth retardation in the fetus. Since low birth weight is strongly associated with increased infant mortality, prenatal malnutrition causes increased numbers of prenatal and early postnatal deaths. Moreover, irreversible stunting of brain development may be caused by prenatal malnutrition.

Malnutrition during infancy can also cause stunting of both brain development and total-body growth. The individual seems locked in to a younger developmental age. On the other hand, the final height a child reaches cannot be increased beyond the genetically determined maximum by the child eating more than adequate vita-

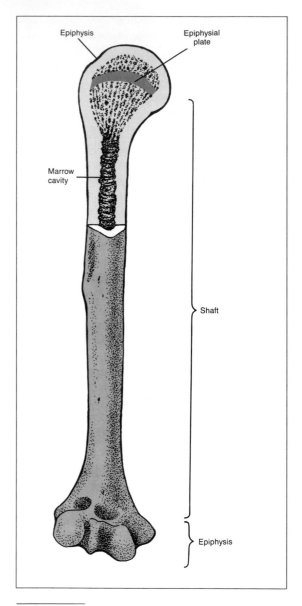

FIGURE 17-15 Anatomy of a long bone during growth.

mins, protein, or total calories. Overfeeding produces obesity, not growth.

Sickness can also stunt growth, but if the illness is temporary, the recovered child manifests a remarkable growth spurt ("catch-up" growth) that rapidly brings him or her up to the normal growth curve.

HORMONAL INFLUENCES ON GROWTH

The hormones most important to human growth are growth hormone, thyroid hormones, insulin, testosterone, and estrogens, all of which exert widespread effects. Other hormones—ACTH, TSH, prolactin, and the pituitary gonadotropins FSH and LH—selectively influence the growth and development of their target organs: the adrenal cortex, thyroid gland, breasts, and gonads, respectively.

In addition to all these hormones, there is a group of peptide **growth factors** (for example, nerve growth factors), each of which is highly effective in stimulating differentiation and/or mitosis of certain cell types. There are also peptide **growth-inhibiting factors** that modulate growth by inhibiting mitosis in specific tissues. These growth factors and growth-inhibiting factors are usually produced by multiple cell types rather than by discrete endocrine glands. Indeed, many of them are produced and released in the immediate vicinity of their sites of action and so are categorized as paracrines and autocrines. Some, however, may enter the blood and function as hormones. In any case, their secretion may be controlled by other hormones.

The physiology of the growth factors and growth-inhibiting factors is important not just for understanding normal growth control but also because these factors may be involved in the development of cancer. For example, one of the cancer-associated oncogenes (page 72)

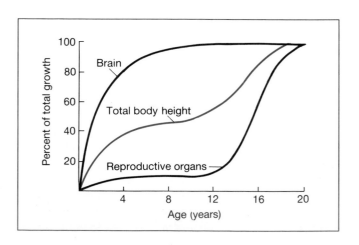

FIGURE 17-16 Relative growth in brain, total-body height (a measure of long bone growth), and reproductive organs. Note that brain growth is nearly complete by age 5, whereas maximal height and reproductive-organ size is not reached until the late teens.

has been shown to code for a version of the receptor for epidermal growth factor that is always in the activated state even in the absence of the growth factor. This activated receptor imparts a continuous growth signal to the cells containing it.

Closely related to the growth factors described above, and sometimes classified with them, are several other peptide hormones—erythropoietin (Chapter 13) and thymosin and interleukin-1 (Chapter 19)—which influence blood-cell differentiation and mitosis.

The various hormones and growth factors do not all stimulate growth at the same times. For example, fetal growth is largely independent of the major hormones that stimulate growth during childhood—growth hormone, the thyroid hormones, and the sex steroids—but is dependent upon insulin.

Growth Hormone

Although growth hormone, secreted by the anterior pituitary, has little or no effect on fetal growth, it is the single most important stimulus for postnatal growth. Its major growth-promoting effect is stimulation of mitosis in its many target tissues. Thus, growth hormone promotes bone lengthening by stimulating the maturation and mitosis of the chondrocytes in the epiphyseal plates, thereby continuously widening the plates and providing more cartilaginous material for bone formation.

Growth hormone excess during childhood produces **giantism**, whereas deficiency produces **dwarfism**. However, when excess growth hormone is secreted in adults after epiphyseal closure, it cannot lengthen the bones further, but it does produce the disfiguring bone thickening and overgrowth of other organs known as **acromegaly**.

Importantly, growth hormone exerts its mitosis-stimulating (mitogenic) effect not *directly* on cells but rather *indirectly* through the mediation of a chemical messenger whose synthesis and release are induced by growth hormone. This messenger is called **insulin-like growth factor 1 (IGF-1)** (also known as **somatomedin C**). Under the influence of growth hormone, IGF-1 is secreted by the liver into the blood, which carries it to its target cells. Thus, IGF-1 is categorized as a hormone. But IGF-1 is secreted by many cells other than those in the liver, and it acts in these sites as a paracrine or autocrine.

Current concepts of how growth hormone and IGF-1 interact on the epiphyseal plates are as follows: (1) Growth hormone stimulates the chondrocyte precursor cells (prechondrocytes) and/or young differentiating chondrocytes in the epiphyseal plates to differentiate into chondrocytes; (2) during this differentiation, the cells begin both to secrete IGF-1 and to become responsive to IGF-1; (3) IGF-1 then acts locally to stimulate the differentiating chondrocytes to undergo mitosis.

It is likely that a similar interplay between growth hormone and IGF-1 underlies the ability of growth hormone to stimulate growth in target tissues other than bone; that is, growth hormone stimulates precursor cells to differentiate and to secrete IGF-1, and IGF-1 stimulates the newly differentiated cells to undergo mitosis.

The importance of IGF-1 mediating the major growth-promoting effect of growth hormone is illustrated by the fact that Pygmies have been found to have normal or high plasma concentrations of growth hormone but very low concentrations of IGF-1. Thus, a genetic defect in the ability to produce IGF-1 in response to growth hormone probably accounts for their short stature.

The secretion and activity of IGF-1 can be influenced by several hormones other than growth hormone and by the nutritional status of the individual. For example, malnutrition during childhood inhibits the production of IGF-1 even though plasma growth hormone concentration is elevated and should be stimulating IGF-1 secretion.

In addition to its specific growth-promoting effect on mitosis via IGF-1, growth hormone directly stimulates protein synthesis in various tissues and organs. It does this by increasing amino acid uptake by cells as well as the synthesis of RNA and ribosomes. All these events are essential for protein synthesis.[5] This anabolic effect on protein metabolism facilitates the ability of tissues and organs to enlarge. Table 17-4 summarizes the multiple effects of growth hormone, all of which have been described in this chapter.

---

[5]The similarity of growth hormone's effect on protein synthesis to that exerted by insulin may seem puzzling when one recalls the previously described antiinsulin effects of growth hormone on carbohydrate and lipid metabolism. It has been postulated that these varied actions of growth hormone may be mediated by different receptors responding to different amino acid sequences on the large growth hormone molecule.

TABLE 17-4 MAJOR EFFECTS OF GROWTH HORMONE

1. Promotes growth: Induces precursor cells to differentiate and secrete insulin-like growth factor 1 (IGF-1), which stimulates mitosis
2. Stimulates protein synthesis
3. Antiinsulin effects: Stimulates lipolysis and gluconeogenesis; inhibits glucose uptake

The hypothalamic hormones controlling growth hormone secretion were described in Chapter 10 (Figure 10-24). Briefly, growth hormone secretion is stimulated by growth hormone-releasing hormone (GHRH) and inhibited by hypothalamic somatostatin. Growth hormone secretion occurs in episodic bursts and manifests a striking diurnal rhythm. During most of the day, there is little or no growth hormone secreted, although bursts may be elicited by certain stimuli, including stress,[6] hypoglycemia, or exercise. In contrast, 1 to 2 h after a person falls asleep, one or more larger, prolonged bursts of secretion may occur. The total 24-h secretion rate of growth hormone, almost all occurring during sleep, is highest during adolescence—the period of most rapid growth—next highest in children, and lowest in adults.

The availability of large quantities of human growth hormone produced by recombinant-DNA technology has greatly facilitated the treatment of children with short stature due to deficiency of growth hormone. Because growth hormone is relatively species specific, humans cannot use growth hormone from animals other than primates (which would have to be sacrificed to obtain it), and so the only previous significant source for medical use was growth hormone extracted from the anterior pituitaries of donated human cadavers.

Controversial at the present time is the question of whether growth hormone should be used to enhance growth in short children who do not exhibit evidence of growth hormone deficiency. Also controversial is the use of growth hormone by athletes to increase muscle mass through this hormone's stimulating effect on protein synthesis.

Thyroid Hormones

The thyroid hormones (TH)—thyroxine and triiodothyronine—are essential for normal growth because they are permissive for the growth-promoting effects of growth hormone. Accordingly, infants and children with deficient thyroid function manifest retarded growth due to slowed bone growth. Because epiphyseal closure is delayed, bone age is decreased. Growth can be restored to normal by administration of physiological quantities of thyroid hormone.

Quite distinct from its growing-promoting effect, TH also plays a crucial role in development of the central nervous system during fetal life and the first few months after birth. Hypothyroid infants (cretins) are mentally retarded, a defect that can be completely repaired by early treatment with TH. However, if the infant is un-

treated for more than several months, the developmental failure is largely irreversible.

This effect on brain *development* must be distinguished from other stimulatory effects TH exerts on the nervous system throughout life, not just during infancy. A hypothyroid person exhibits sluggishness and poor mental function, and these effects are completely reversible at any time with administration of TH. Conversely, a person with excessive secretion of TH is jittery and hyperactive.

Insulin

It should not be surprising that adequate amounts of insulin are necessary for normal growth since insulin is, in all respects, an anabolic hormone. Its stimulatory effects on amino acid uptake and protein synthesis are particularly important in this regard.

In addition to this general anabolic effect, however, insulin exerts specific growth-promoting effects on cell differentiation and mitosis during fetal life. Insulin may also exert some growth-promoting effect during childhood since it can stimulate the secretion of IGF-1 and can, itself, combine with and activate IGF-1 receptors.

Sex Hormones

As will be described in Chapter 18, sex hormone secretion (testosterone in the male and estrogen in the female) begins in earnest at about the age of 8 to 10 and progressively increases to reach a plateau over the next 5 to 10 years. A normal pubertal growth spurt, which reflects growth of the long bones, requires this increased production of the sex hormones. One major effect of the sex steroids is to stimulate the secretion of growth hormone, and the sex steroids are probably responsible, at least in part, for the large increase in growth hormone secretion during puberty. However, the growth-promoting effects of the sex steroids probably also involve other mechanisms not yet elucidated.

Unlike growth hormone, the sex hormones ultimately stop bone growth by inducing epiphyseal closure. The dual effects of the sex hormones explains the pattern seen in adolescence—rapid lengthening of the bones culminating in complete cessation of growth for life.

In addition to these dual effects on bone, testosterone, but not estrogen, exerts a direct anabolic effect on protein synthesis in many nonreproductive organs and tissues of the body. This accounts, at least in part, for the increased muscle mass of men, compared with women.

This also is why, like growth hormone, testosterone-like agents termed anabolic steroids are used by athletes in an attempt to increase their muscle mass and strength. However, it remains controversial as to how much these agents, in men who are not testosterone-

---

[6] In contrast, during childhood severe prolonged psychosocial stress may inhibit growth hormone secretion and lead to diminished growth.

TABLE 17-5 MAJOR HORMONES INFLUENCING GROWTH	
Hormone	**Principal Actions**
Growth hormone	Major stimulus of postnatal growth: Induces precursor cells to differentiate and secrete insulin-like growth factor 1 (IGF-1), which stimulates mitosis Stimulates protein synthesis
Insulin	Stimulates fetal growth ? Stimulates postnatal growth by stimulating secretion of IGF-1 and directly activating IGF-1 receptors Stimulates protein synthesis
Thyroid hormones	Permissive for growth hormone's actions Promote development of the central nervous system
Testosterone	Stimulates growth at puberty, at least in part by stimulating the secretion of growth hormone Causes eventual epiphyseal closure Stimulates protein synthesis
Estrogen	Stimulates growth at puberty, at least in part by stimulating the secretion of growth hormone Causes eventual epiphyseal closure
Cortisol	Inhibits growth

deficient to start with, really increase muscle mass or improve athletic performance. Moreover, the steroids generally used have multiple toxic side-effects. Although these drugs definitely have a positive effect on protein synthesis in women, they produce masculinization as well as the same toxic side-effects as in men.

Cortisol

Cortisol, the major hormone secreted by the adrenal cortex in response to stress, has potent *antigrowth* effects. It inhibits DNA synthesis and stimulates protein catabolism in many organs, and it inhibits bone growth. In children, the elevation in plasma cortisol that accompanies infections and other stresses is, at least in part, responsible for the retarded growth that occurs with illness.

This completes our survey of the major hormones that affect growth. Their actions are summarized in Table 17-5.

COMPENSATORY GROWTH

We have dealt thus far only with growth during childhood. During adult life, a specific type of regenerative organ growth, known as **compensatory growth**, can occur in many human organs. For example, after the surgical removal of one kidney, the cells of the other begin to manifest increased mitosis, and the kidney ultimately grows until its total mass approaches the initial mass of the two kidneys combined. The causes of this compensatory growth are not known.

SECTION C
SUMMARY OF LIVER FUNCTIONS

In this chapter and previous chapters we have described a large number of functions served by the liver. Although a few more functions remain to be described in future chapters, this is a convenient place to provide, in Table 17-6, a comprehensive reference list of liver functions.

TABLE 17-6 SUMMARY OF LIVER FUNCTIONS

A. Endocrine functions
   1. In response to growth hormone, secretes insulin-like growth factor 1 (IGF-1), which promotes growth by stimulating mitosis in various tissues, including bone (Chapter 17).
   2. Contributes to the activation of vitamin D (Chapter 15).
   3. Forms triiodothyronine (T_3) from thyroxine (T_4) (Chapter 10).
   4. Secretes angiotensinogen, which is acted upon by renin to form angiotensin (Chapter 15).
   5. Metabolizes hormones (Chapter 10).
B. Clotting functions
   1. Produces many of the plasma clotting factors, including prothrombin and fibrinogen (Chapter 19).
   2. Produces bile salts, which are essential for the gastrointestinal absorption of vitamin K, which is, in turn, needed for production of the clotting factors (Chapter 19).
C. Plasma proteins: Synthesizes and secretes plasma albumin (Chapter 13), acute phase proteins (Chapter 19), binding proteins for steroid hormones (Chapter 10) and trace elements (Chapter 13), lipoproteins (Chapter 17), and other proteins mentioned elsewhere in this table.
D. Digestive functions via bile production and secretion (Chapter 16)
   1. Synthesizes and secretes bile acids, which are necessary for adequate digestion and absorption of fats.
   2. Secretes into the bile a bicarbonate-rich solution of inorganic ions, which helps neutralize acid in the duodenum.
E. Organic metabolism (Chapter 17)
   1. Converts plasma glucose into glycogen and triacylglycerols during absorptive period.
   2. Converts plasma amino acids to fatty acids, which can be incorporated into triacylglycerols, during absorptive period.
   3. Synthesizes triacylglycerols and secretes them as lipoproteins during absorptive period.
   4. Produces glucose from glycogen and other sources (gluconeogenesis) during postabsorptive period and releases the glucose into the blood.
   5. Converts fatty acids into ketones during fasting.
   6. Produces urea, the major end product of amino acid (protein) catabolism, and releases it into the blood.
F. Cholesterol metabolism (Chapter 17)
   1. Synthesizes cholesterol and releases it into the blood.
   2. Secretes plasma cholesterol into the bile.
   3. Converts plasma cholesterol into bile acids.
G. Excretory and degradative functions
   1. Secretes bilirubin and other bile pigments into the bile (Chapter 16).
   2. Excretes, via the bile, many endogenous and foreign organic molecules as well as trace metals (Chapter 19).
   3. Biotransforms many endogenous and foreign organic molecules (Chapter 19).

SECTION D

REGULATION OF TOTAL-BODY ENERGY BALANCE

BASIC CONCEPTS OF ENERGY EXPENDITURE AND CALORIC BALANCE

The breakdown of organic molecules liberates the energy locked in their molecular bonds. This is the energy cells use to perform the various forms of biological work—muscle contraction, active transport, and molecular synthesis. As described in Chapter 5, the first law of thermodynamics states that energy can be neither created nor destroyed but can be converted from one form to another. Thus, internal energy liberated ΔE during breakdown of an organic molecule can either appear as heat H or be used for performing work W:

$$\Delta E = H + W$$

During metabolism, about 60 percent of the energy released from organic molecules appears immediately as heat, and the rest is used for work. The energy used for work must first be incorporated into molecules of ATP, the subsequent breakdown of which serves as the immediate energy source for the work. It is essential to realize that the body is not a heat engine since it is totally incapable of converting heat to work, but the heat released in its chemical reactions is valuable for maintaining body temperature.

Biological work can be divided into two general categories: (1) **external work**—movement of external objects by contracting skeletal muscles and (2) **internal work**—all other forms of work, including skeletal muscle activity not used in moving external objects. As just stated, much of the energy liberated from nutrient catabolism appears immediately as heat. What may not be obvious is that all internal work, too, is ultimately transformed to heat except during periods of growth. Several examples will illustrate this essential point.

1. Internal work is performed during cardiac contraction, but this energy appears ultimately as heat generated by the resistance to blood flow offered by the blood vessels.
2. Internal work is performed during secretion of HCl by the stomach and $NaHCO_3$ by the pancreas, but this work appears as heat when the H^+ and HCO_3^- combine in the small intestine.
3. The internal work performed during synthesis of a plasma protein is recovered as heat during the inevitable catabolism of the protein since with few exceptions all bodily constituents are constantly being built up and broken down. During periods of net synthesis of protein, fat, and other molecules, however, energy is stored in the bonds of these molecules and does not appear as heat.

Thus, the total energy liberated when organic nutrients are catabolized by cells may be transformed into body heat, appear as external work, or be stored in the body in the form of organic molecules. The **total energy expenditure** of the body is therefore given by the equation:

Total energy expenditure = heat produced by the body + external work done + energy stored

Metabolic Rate

The unit for energy is the **kilocalorie (kcal)**[7], which is the amount of heat required to heat one liter of water one degree centigrade. Total energy expenditure per unit time is called the **metabolic rate**, which can be measured directly or indirectly. In either case, the measurement is much simpler if the person is fasting and at rest. Total energy expenditure then becomes equal to heat production since energy storage and external work are eliminated.

The direct method of measuring metabolic rate is simple to understand but difficult to perform. The subject is placed in a chamber known as a calorimeter, and heat production is measured by the temperature changes in water flowing through the calorimeter (Figure 17-17). This is an excellent method in that it measures heat production directly, but calorimeters are found in only a few research laboratories. Accordingly, a simple, indirect method has been developed for widespread use.

---

[7] In the field of nutrition, the three terms "Calorie" (with a capital C), "large calorie," and "kilocalorie" are synonyms.

FIGURE 17-17 Direct method for measuring metabolic rate. The water flowing through the calorimeter carries away the heat produced by the person's body. The amount of heat produced is calculated from the total volume of water and the difference between inflow and outflow temperatures.

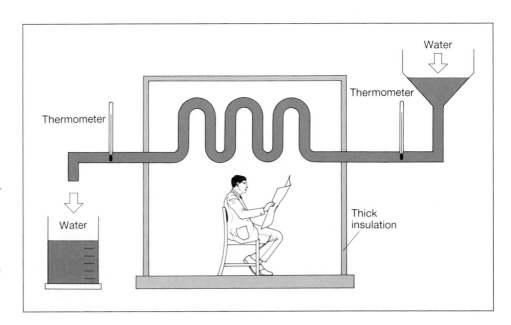

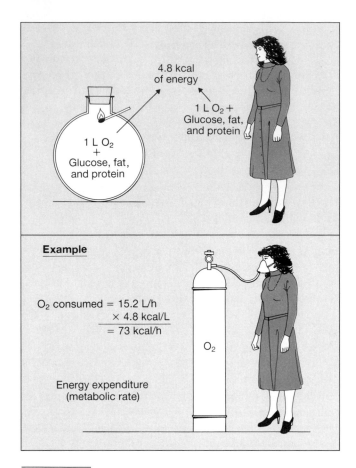

4.8 kcal
of energy

1 L O₂ +
Glucose, fat,
and protein

1 L O₂
+
Glucose, fat,
and protein

Example

O₂ consumed = 15.2 L/h
    × 4.8 kcal/L
    = 73 kcal/h

O₂

Energy expenditure
(metabolic rate)

FIGURE 17-18 Indirect method for measuring metabolic rate. When 1 L of oxygen is utilized in the oxidation of organic nutrients, approximately 4.8 kcal of energy is liberated as heat. In this example, the metabolic rate is (15.2 L of O_2/h) × (4.8 kcal/L) = 73 kcal/h, a typical value for a normal, fasted, resting adult. This rate of energy expenditure is approximately equal to that of a 100-W bulb.

In the indirect procedure (Figure 17-18), one simply measures the subject's oxygen uptake per unit time by measuring total ventilation and the oxygen content of both inspired and expired air. From this value one calculates heat production based on the fundamental principle that the energy liberated by the catabolism of foods inside the body must be the same as that liberated when the foods are catabolized to the same products outside the body. We know precisely how much heat is liberated when 1 L of oxygen is utilized in the oxidation of fat, protein, or carbohydrate outside the body. This same quantity of heat must be produced when 1 L of oxygen is utilized in the body. Fortunately, we do not need to know precisely which type of nutrient is being oxidized internally because the quantities of heat produced per liter of oxygen utilized are reasonably similar for the oxi-

dation of fat, carbohydrate, and protein, and average 4.8 kcal/L of oxygen. When more exact calculations are required, additional techniques are used to estimate the relative quantity of each nutrient.

Determinants of Metabolic Rate

Basal metabolic rate. Since many factors cause the metabolic rate to vary (Table 17-7), the test used clinically to evaluate it specifies certain standardized conditions and measures what is known as the **basal metabolic rate (BMR)**. The subject is at mental and physical rest in a room at comfortable temperature and has not eaten for at least 12 h. These conditions are arbitrarily designated "basal," even though the metabolic rate during sleep may be less than the BMR. For the following discussion, it must be emphasized that the term BMR can be applied to a person's metabolic rate only when the specified conditions are met. Thus, a person who has recently eaten or is exercising has a metabolic rate but not a *basal* metabolic rate.

BMR is often termed the "metabolic cost of living," and most of the BMR is expended, as might be imagined, by the heart, liver, kidneys, and brain. The magnitude of the BMR is related not only to physical size but to age and sex as well. The growing child's BMR, expressed on a per-weight basis, is considerably higher than the adult's, in part because the child expends a great deal of energy in net synthesis of new tissue. On the other end of the age scale, the BMR gradually decreases with advancing age. A woman's BMR is generally less than that of a man, even taking into account size differences, but increases markedly during pregnancy

TABLE 17-7   SOME FACTORS AFFECTING METABOLIC RATE
Age
Sex
Height, weight, and body surface area
Growth
Pregnancy, menstruation, lactation
Infection or other disease
Body temperature
Recent ingestion of food
Prolonged alteration in amount of food intake
Muscular activity
Emotional state
Sleep
Environmental temperature
Circulating levels of various hormones, especially epinephrine and thyroid hormone

and lactation. The greater demands upon the body by infection or other disease generally increase BMR. However, when a person suffers wasting because of infection, BMR may decrease below normal.

Thyroid hormones. The thyroid hormones are the single most important determinant of BMR at any given size, age, and sex. Indeed, the BMR test was once commonly used for evaluation of thyroid hormone status but has now been superseded by more specific tests. TH increases the oxygen consumption and heat production of most body tissues, a notable exception being the brain. This ability to increase BMR is termed a **calorigenic effect**. The mechanism of TH's calorigenic effect is presently uncertain, but by means of it, the plasma concentration of TH is the major factor that "sets" the body's BMR.

Long-term excessive TH, as in persons with hyperthyroidism, induces a host of effects secondary to the calorigenic effect. For example, the increased metabolic demands markedly increase hunger and food intake; the greater intake frequently remains inadequate to meet the metabolic needs, and net catabolism of protein and fat stores leads to loss of body weight. Also the greater heat production activates heat-dissipating mechanisms (skin vasodilation and sweating), and the person suffers from marked intolerance to warm environments. In contrast, the hypothyroid individual complains of cold intolerance.

The calorigenic effect of TH is only one of a bewildering variety of effects exerted by these hormones. With one exception—facilitation of the activity of the sympathetic nervous system, the functions of the thyroid hormones have all been described earlier in this chapter and are listed for reference in Table 17-8. The control of the secretion of these hormones, as well as their metabolism, is described in Chapter 10.

Epinephrine. Epinephrine is another hormone that exerts a calorigenic effect. This effect may be related to the hormone's stimulation of glycogen and triacylglycerol catabolism, since ATP splitting and energy liberation occur in both the breakdown and the subsequent resynthesis of these molecules. Regardless of the mechanism, when epinephrine secretion by the adrenal medulla is stimulated, the metabolic rate rises. This probably accounts for part of the greater heat production associated with emotional stress, although increased muscle tone is also contributory.

Food-induced thermogenesis. The ingestion of food acutely increases the metabolic rate by 10 to 20 percent for a few hours after eating. This effect is known as **food-induced thermogenesis**. Ingested protein gives the greatest effect, carbohydrate and fat, less. Most of the increased heat production is secondary to the processing of the absorbed nutrients by the liver, not to the energy expended by the gastrointestinal tract in digestion and

---

TABLE 17-8 FUNCTIONS OF THE THYROID HORMONES (TH)

(1) Required for normal maturation of the nervous system in the fetus and infant
   Deficiency: Mental retardation (cretinism)

(2) Required for normal bodily growth because they facilitate the effects of growth hormone
   Deficiency: Deficient growth in children

(3) Required for normal alertness and reflexes at all ages
   Deficiency: Mentally and physically slow and lethargic
   Excess: Restless, irritable, anxious, wakeful

(4) Major determinant of the rate at which the body produces heat during the basal metabolic state.
   Deficiency: Low BMR, sensitivity to cold, decreased food appetite
   Excess: High BMR, sensitivity to heat, increased food appetite, increased catabolism of nutrients

(5) Facilitates the activity of the sympathetic nervous system by stimulating the synthesis of one class of receptors (beta receptors) for epinephrine and norepinephrine
   Excess: Symptoms similar to those observed with activation of the sympathetic nervous system (for example, increased heart rate)

TABLE 17-9 ENERGY EXPENDITURE DURING DIFFERENT TYPES OF ACTIVITY FOR A 70-KG (154-LB) PERSON	
Form of activity	Energy kcal/h
Lying still, awake	77
Sitting at rest	100
Typewriting rapidly	140
Dressing or undressing	150
Walking on level, 4.3 km/h (2.6 mi/h)	200
Bicycling on level, 9 km/h (5.5 mi/h)	304
Walking on 3 percent grade, 4.3 km/h (2.6 mi/h)	357
Sawing wood or shoveling snow	480
Jogging, 9 km/h (5.3 mi/h)	570
Rowing, 20 strokes/min	828

absorption. It is to avoid the contribution of food-induced thermogenesis that BMR is performed in the postabsorptive state.

To reiterate, food-induced thermogenesis is the *rapid* increase in energy expenditure in response to ingestion of a meal. As we shall see, *prolonged* alterations in food intake (either increased or decreased total calories) also have significant effects on metabolic rate but are not termed "food-induced thermogenesis."

Muscle activity. The factor that can increase metabolic rate most is altered skeletal muscle activity. Even minimal increases in muscle tone significantly increase metabolic rate, and strenuous exercise may raise heat production more than fifteenfold (Table 17-9). Thus, depending on the degree of physical activity, total energy expenditure may vary for a normal young adult from a value of approximately 1500 kcal/24 h to more than 7000 kcal/24 h (for a lumberjack). Changes in muscle activity also account, in part, for the changes in metabolic rate that occur during sleep (decreased muscle tone), during exposure to a low environmental temperature (increased muscle tone and shivering), and with strong emotions.

Total-Body Energy Balance

The laws of thermodynamics dictate that the total energy expenditure (metabolic rate) of the body must equal total energy intake. We have already identified the ultimate forms of energy expenditure: internal heat production, external work, and net molecular synthesis (energy storage). The source of input is the energy contained in ingested food. Therefore:

$$\text{Energy from food intake} = \text{internal heat produced} + \text{external work} + \text{energy stored}$$

Our equation includes no term for loss of fuel from the body via excretion of nutrients because, in normal persons, only negligible losses occur via the urine, feces, and as sloughed hair and skin. In certain diseases, however, the most important being diabetes, urinary losses of organic molecules may be quite large and would have to be included in the equation.

Let us now rearrange the equation to focus on energy storage:

$$\text{Energy stored} = \text{energy from food intake} - (\text{internal heat produced} + \text{external work})$$

Thus, whenever energy intake differs from the sum of internal heat produced and external work, changes in energy storage occur, that is, the total-body energy content increases or decreases (Table 17-10).

Total-body energy content in adults is usually regulated around a relatively constant operating point. Theoretically, this constancy could be achieved through (1) adjusting metabolic energy expenditure to compensate for randomly determined food intake or (2) adjusting food intake to changing metabolic expenditures or (3) both. There is no question that (2) is the major means of regulating total-body energy content.

Nevertheless, there is some contribution of (1), that is, adjustment of metabolic expenditure when food intake changes. A striking experiment documenting this phenomenon was the reduction of caloric intake in normal male volunteers from 3492 kcal/day to 1570 kcal/day for 24 weeks. These men were soldiers performing considerable physical activity, and weight loss was initially very rapid and ultimately averaged 24 percent of original body weight. However, the important fact is that the weights did stabilize at this point, that is, energy balance was reestablished despite the continued caloric intake of only 1570 kcal/day. Clearly, metabolic rate must have decreased to the same value. The decrease was due partly to a reduction in physical activity as the men became apathetic and reluctant to engage in any activity, but BMR was also decreased.[8]

Experiments have also been done at the other end of the spectrum, to see whether prolonged deliberate overfeeding of volunteers would induce an increased meta-

---

[8] A decreased total body mass contributed to the decrease in BMR (a smaller body utilizes less energy), but there was also a decrease in metabolism by the liver and other organs out of proportion to their changes in weight. The mechanism of this latter effect was a decrease in the activity of the sympathetic nervous system.

TABLE 17-10   THREE POSSIBLE STATES OF ENERGY BALANCE

State	Result
1. Energy intake = (internal heat production + external work)	Body energy content remains constant (body fat content remains constant).
2. Energy intake > (internal heat production + external work)	Body energy content increases (body fat content increases).
3. Energy intake < (internal heat production + external work)	Body energy content decreases (body fat content decreases).

bolic rate adequate to reestablish caloric balance despite the continued high-calorie diet. In many subjects, a compensatory increase in metabolic rate did, in fact, occur, and the weight gains of these subjects were less than that expected from the degree of overfeeding.

To summarize these studies, after several weeks of dietary alteration in either direction, the resulting deviations in an individual's total-body energy content trigger significant, although generally relatively small, counteracting changes in metabolism. This helps explain why some dieters lose 10 lb or so of fat fairly easily and then become stuck at a plateau, losing additional weight at a much slower rate. It may also help explain why some very thin people have difficulty trying to gain much weight.

We now turn to the control of food intake, the more important mechanism for keeping body energy content constant.

Control of Food Intake

The control of food intake can be analyzed in the same way as any other biological control system, that is, in terms of its various components—regulated variable, receptor sensitive to this variable, afferent pathway, and so on. As our previous description emphasized, the variable being maintained relatively constant in this system is total-body energy content. Here we have a problem similar to that described in Chapter 15 for total-body sodium balance: What kind of receptors could possibly detect the *total*-body content of a particular variable, in this case, calories?

It is very unlikely that any such receptors exist, and so, instead, the system depends on signals that are intimately correlated to food intake and total energy storage. These include the plasma concentrations of glucose and the hormones that regulate organic metabolism. These inputs are thought to constitute **satiety signals** that cause the person to cease feeling hungry and set the time period before hunger returns once again.

Figure 17-19 presents one hypothesized model for

satiety: Plasma glucose concentration and the rate of cellular glucose utilization rise during eating as the ingested carbohydrate is digested and absorbed. Detection of this increase by glucose receptors in the brain could constitute a signal to those portions of the brain that control eating behavior and lead to cessation of hunger. Conversely, in the postabsorptive period, plasma glucose and glucose utilization decrease, the signal to the brain glucose receptors would be removed, and the individual once again would become hungry.[9]

Similarly, changes in plasma concentrations of the hormones that regulate carbohydrate and fat metabolism

---

[9]Despite extensive evidence from other species that such glucose receptors exist for the control of food appetite, it is not certain that they are present in humans.

FIGURE 17-19 Hypothesized mechanism by which plasma glucose acts as a satiety signal to shut off eating.

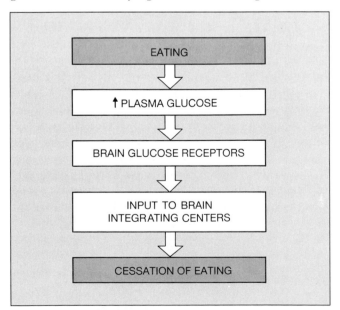

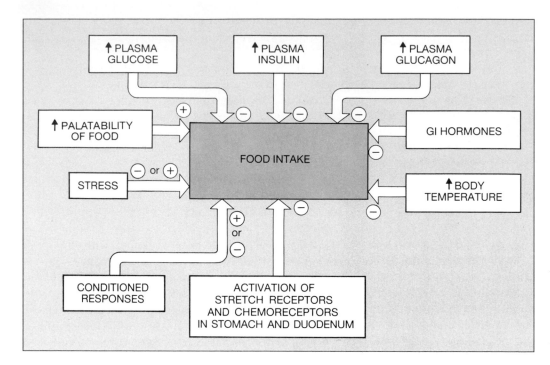

FIGURE 17-20 Inputs controlling food intake. The minus signs denote hunger suppression, and the plus signs denote hunger stimulation.

could be important satiety signals to the brain's integrating centers. For example, insulin, which increases during food absorption, has been shown to suppress hunger. Glucagon, which may also increase after consumption of a high-protein or mixed-nutrient meal, also suppresses hunger.

A nonchemical correlate of food ingestion—body temperature—is yet another possible satiety signal. The increase in metabolic rate induced by eating tends to raise body temperature slightly, and this probably constitutes a signal inhibitory to hunger.

All the satiety signals described above reflect events occurring as a result of nutrient *absorption*, but there are also important signals initiated by the presence of food within the gastrointestinal tract. These include neural signals triggered by stimulation of both stretch receptors and chemoreceptors in the stomach and duodenum, as well as by several of the hormones (cholecystokinin, for example) released from the stomach and duodenum during eating.

Thus far we have been dealing with those inputs that would help to control meal size and frequency. To be effective over long periods of time in regulating total caloric intake and matching it to energy expenditure, however, the magnitude of these signals must reflect not merely meal-eating patterns but the total-body energy content itself. The identities of such long-term regulators remain unknown.

It is presently not possible to quantify the contributions of the many signals postulated to control food in-take—summarized in Figure 17-20—and to describe the location and nature of the receptors stimulated by each. Moreover, there appear to be multiple brain areas, in addition to the hypothalamus, that integrate all these afferent inputs and cause the individual either to feel hungry or not. The ways in which these various areas interact remain to be determined.

Although food intake is very likely controlled by the reflex input from some combination of glucose receptors, thermoreceptors, and so on, it also is strongly influenced by the reinforcement, both positive and negative, of such things as smell, taste, and texture. For example, rats given a "cafeteria-style" choice of foods highly palatable to people overeat by 70 to 80 percent and soon become obese. An analogous (but reverse) experiment has shown that obese people lose weight when they must obtain all their food from a monotonous diet delivered via a tube in response to lever pressing. Thus, the behavioral concepts of reinforcement, drive, and motivation, to be described in Chapter 20, must be incorporated into any comprehensive theory of food-intake control. So must factors such as stress. For example, repeated tail-pinching will cause rats to overeat, and this response can be blocked by drugs that antagonize endogenous opiates.

Obesity

Obesity has been defined, by a 1985 Consensus Development Conference at the National Institutes of Health, as an excess of body fat resulting in a significant impairment of health from a variety of diseases, notably hypertension, atherosclerotic heart disease, and diabetes. The conference concluded that a level of 20 percent or more above "desirable" body weight qualifies as obesity and is associated with sufficient risk to health to justify clinical intervention.

How does one determine desirable weight? The most commonly used ideal or desirable weight tables are those put out by the Metropolitan Life Insurance Company and are based on mortality data (Table 17-11). However, there is considerable controversy over the fact that the tables issued in 1983 indicate that weights approximately 12 to 14 pounds higher than those of the 1959 tables are safe and in fact may be associated with *lower* mortality risk. Most experts believe that the 1959 tables are more accurate predictors of health consequences but that achieving the weights defined by the 1983 tables would be a major and desirable advance in health for the millions of people in the United States and other countries who now exceed those weights.

In any case, the presently preferred simple method for assessing degree of obesity is the **body mass index** (**BMI**), calculated as weight in kilograms divided by the square of height in meters. For example, a 70-kg person with a height of 180 cm would have a BMI of 21.6 (70/1.8^2). This method correlates quite well with body-fat measurement techniques such as skinfold thickness. Normal BMI ranges for adults are 19 to 25; a BMI of more than 27.8 for men and 27.3 for women is defined as obesity.

The bad news about the causes of obesity is that hereditary influences are quite strong. For example, identical twins who have been separated soon after birth and raised in different households manifest strikingly similar body weights as adults. The good news is that studies of animal models and human populations indicate that environmental factors, which should, in theory, be alterable, can also play important roles. One example is the experiment described earlier in which normally nonobese rats became obese when given a diet of snack foods from the supermarket. Human studies indicate that the prevalence of obesity in industrialized countries is related to social conditions, economic circumstances, or both. For example, in the United States obesity is six times more common among members of the lowest socioeconomic class than among those of the highest class. In nonindustrialized countries, this relationship does not hold because the poor cannot obtain adequate food.

The methods and goals of treating obesity are presently undergoing extensive rethinking. An increase in fat must be due to an excess of food intake over metabolic rate, and low-calorie diets have long been the mainstay of therapy. However, it is now clear that such diets alone have limited effectiveness in many obese people partly because, as described earlier, the person's metabolic rate drops, sometimes falling low enough to prevent further weight loss on as little as 1000 calories a day. Related to this, studies have also shown that many obese people get fat or stay fat on a caloric intake no greater than, and sometimes less than, the amount consumed by people of normal weight. These persons must either be less active than normal or have lower basal metabolic rates. Finally, at least half of definitely obese people—those who are more than 20 percent overweight—who try to diet down to desirable weights suffer medically, physically, and psychologically.

Exercise has proven to be a valuable element in any weight-loss program. The exercise itself utilizes calories, though depressingly few, and usually does not stimulate appetite in obese persons. More importantly, exercise partially offsets the tendency for metabolic rate to decrease during long-term caloric restriction and weight loss. Also, the combination of exercise and diet causes the person to lose a relatively larger proportion of fat to protein than with diet alone.

Despite considerable gloom over the inadequacy of present therapy for obesity, most experts believe that new basic-research techniques will soon provide the information needed to develop rational therapy with drugs and other interventions. For example, cloning of the gene that makes certain strains of mice obese should open the way to determining how this "obesity gene" acts.

TABLE 17-11 METROPOLITAN LIFE INSURANCE "DESIRABLE-WEIGHT" TABLE (1959), MEDIUM FRAME			
Men		Women	
Height	Weight, lb	Height	Weight, lb
5'2"	114–126	4'9"	94–106
5'3"	117–129	4'10"	97–109
5'4"	120–132	4'11"	100–112
5'5"	123–136	5'0"	103–115
5'6"	127–140	5'1"	106–118
5'7"	131–145	5'2"	109–122
5'8"	135–149	5'3"	112–126
5'9"	139–153	5'4"	116–131
5'10"	143–158	5'5"	120–135
5'11"	147–163	5'6"	127–140
6'0"	151–168	5'7"	128–143
6'1"	155–173	5'8"	132–147
6'2"	160–178	5'9"	136–151
6'3"	165–183	5'10"	140–155

Note: Height is without shoes, and weight is without clothes.

Finally, as an exercise in energy balance, let us calculate how rapidly a person can expect to lose weight on a reducing diet. Suppose a person whose steady-state metabolic rate per 24 h is 2000 kcal goes on a 1000 kcal/day diet. How much of the person's own body fat will be required to supply this additional 1000 kcal/day? Since fat contains 9 kcal/g:

$$\frac{1000 \text{ kcal/day}}{9 \text{ kcal/g}} = 111 \text{ g/day, or } 777 \text{ g/week}$$

Approximately another 77 g of water is lost from the adipose tissue along with this fat (adipose tissue is 10 percent water), so that the grand total for 1 week's loss equals 854 g, or 1.8 lb. Thus, during a long-term diet, the person can reasonably expect to lose approximately this amount of weight per week, assuming no decrease in metabolic rate occurs. Actually, the amount of weight loss during the first week will probably be considerably greater since a large amount of water may be lost early in the diet, particularly when the diet contains little carbohydrate. This early loss, which is associated with depletion of glycogen stores in liver and skeletal muscle, is really of no value so far as elimination of excess fat is concerned, and often underlies the claims made for fad diets.

Anorexia Nervosa

Anorexia nervosa is a disorder that almost exclusively affects adolescent girls and young women. The typical patient becomes pathologically afraid of gaining weight and reduces her food intake so severely that she may die of starvation.

It is not known whether the cause of anorexia nervosa is primarily psychological or biological. There are many other abnormalities associated with it—loss of menstrual periods, low blood pressure, low body temperature, altered secretion of many hormones—but it is not clear whether these are simply the result of starvation or whether they represent signs, along with the eating disturbances, of primary hypothalamic malfunction.

REGULATION OF BODY TEMPERATURE

Animals capable of maintaining their body temperatures within very narrow limits are termed **homeothermic**. The relatively constant and high body temperature frees biochemical reactions from fluctuating with the external temperature. However, the maintenance of a relatively high body temperature imposes a requirement for precise regulatory mechanisms since further large elevations of temperature cause nerve malfunction and protein denaturation. Some people suffer convulsions at a body temperature of 41°C (106°F), and 43°C is the absolute limit for survival of most people.

Several important generalizations about normal human body temperature should be stressed at the outset: (1) Oral temperature averages about 0.5 C° less than rectal, which is generally used as an estimate of internal

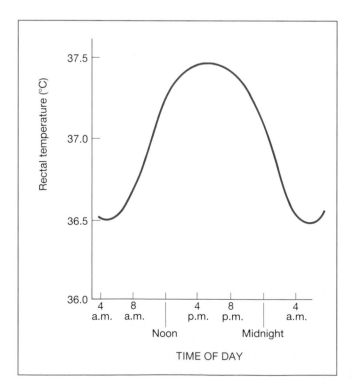

FIGURE 17-21 Circadian changes in core (measured as rectal) body temperature in normal males and in normal females in the first half of their menstrual cycle (*Adapted from Scales et al.*).

temperature; thus, not all parts of the body have the same temperature. (2) Internal temperature varies several degrees in response to activity pattern and external temperature. (3) There is a characteristic circadian fluctuation (Figure 17-21), so that temperature is lowest during sleep and highest during the awake state even if the person remains relaxed in bed. (4) An added variation in women is a higher temperature during the last half of the menstrual cycle.

If temperature is viewed as a measure of heat "concentration," temperature regulation can be studied by our usual balance methods. In this case, the total heat content of the body is determined by the net difference between heat produced and heat lost. Maintaining a constant body temperature implies that, overall, heat production must equal heat loss. Both these variables are subject to precise physiological control.

Temperature regulation offers a classic example of a biological control system, whose generalized components are shown in Figure 17-22. The balance between heat production and heat loss is continuously being disturbed, either by changes in metabolic rate (exercise being the most powerful influence) or by changes in the external environment that alter heat loss or gain. The resulting changes in body temperature that are detected by thermoreceptors in the skin and bodily interior reflexly change the output of the effectors so that heat production and/or loss are changed and body temperature is restored toward normal.

Heat Production

The basic concepts of heat production were described on page 78. Recall that heat is produced by virtually all chemical reactions occurring in the body and that the cost-of-living metabolism by all organs sets the basal level of heat production, which can be increased mainly as a result of skeletal muscular contraction or the action of several hormones.

Changes in muscle activity.
Changes in muscle activity constitute the major control of heat production for temperature regulation. The first muscle changes in response to cold are a gradual and general increase in skeletal muscle contraction. This may lead to shivering, the characteristic muscle response to cold, which consists of oscillating rhythmical muscle tremors occurring at the rate of about 10 to 20 per second. During shivering, the efferent motor nerves to the skeletal muscles are controlled by descending pathways under the primary control of the hypothalamus. So effective are these contractions that body heat production may be increased severalfold within seconds to minutes. Because no external work is performed by shivering, all the energy liberated by the metabolic machinery appears as internal heat

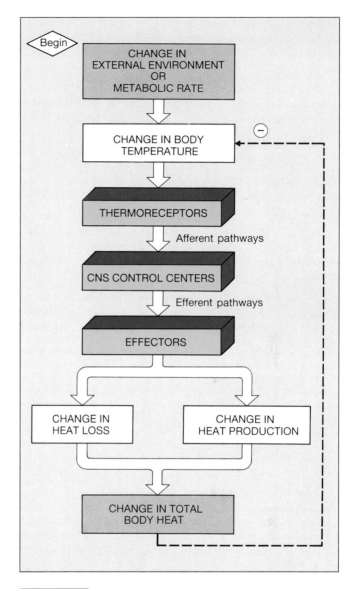

FIGURE 17-22 Generalized pathway for temperature regulation. The external environment influences body temperature by altering heat loss or gain. The determinants of metabolic rate are given in Table 17-7.

and is known as **shivering thermogenesis**. People also use their muscles for voluntary heat-production activities such as foot stamping and hand clapping.

Thus far, our discussion has focused primarily on the muscle response to cold; the opposite reactions occur in response to heat. Muscle tone is reflexly decreased and voluntary movement is also diminished. These attempts to reduce heat production are relatively limited, however, both because muscle tone is quite low to start with and because an increased body temperature acts directly on cells to increase metabolic rate.

Nonshivering thermogenesis. In most experimental animals, chronic cold exposure induces an increase in metabolic rate that is not due to increased muscle activity and is termed **nonshivering thermogenesis**. Its cause is an increased secretion of epinephrine, with some contribution by thyroid hormone as well. Nonshivering thermogenesis is quite minimal, if present at all, in adult human beings. It does occur in infants.

Heat-Loss Mechanisms

The surface of the body exchanges heat with the external environment by radiation, conduction, and convection (Figure 17-23) and by evaporation of water.

Radiation, conduction, and convection. **Radiation** is the process by which the surfaces of all objects constantly emit heat in the form of electromagnetic waves. The rate of emission is determined by the temperature of the radiating surface. Thus, if the body surface is

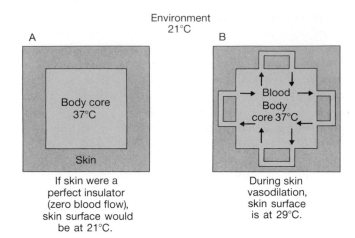

FIGURE 17-24 Relationship of skin's insulating capacity to skin blood flow. (A) If skin were a perfect insulator—in other words, if there were zero blood flow (a situation never reached in a normal person even during maximal vasoconstriction)—the temperature of its outer surface would equal that of the external environment. (B) The skin blood vessels dilate, and the increased blood flow carries heat to the body surface. In this way, the insulating capacity of the skin is reduced, and its surface temperature becomes intermediate between that of the core and that of the external environment.

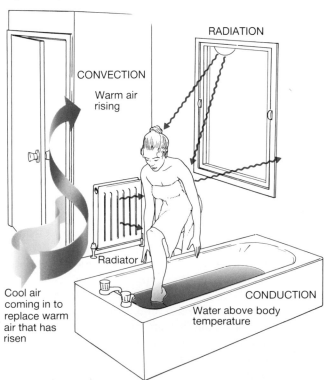

FIGURE 17-23 Mechanisms of heat transfer. In radiation, heat is transferred by electromagnetic waves. In conduction, heat moves by direct transfer of thermal energy from molecule to molecule. Here, the thermal energy from the water molecules is being transferred to the molecules of the foot. In convection, warm air moves away from the body to be replaced by cooler air.

warmer than the *average* of the various surfaces in the environment, net heat is lost from the body, the rate being directly dependent upon the temperature difference between the surfaces.

Conduction is the gain or loss of heat by transfer of thermal energy during collisions between adjacent molecules. In essence, heat is "conducted" from molecule to molecule. The body surface loses or gains heat by conduction through direct contact with cooler or warmer substances, including the air or water.

Convection is the process whereby conductive heat loss or gain is aided by movement of the air or water next to the body. For example, air next to the body is heated by conduction, moves away, and carries off the heat just taken from the body. The air that moved away is replaced by cool air, which in turn follows the same pattern. Convection is always occurring because warm air is less dense and therefore rises, but it can be greatly facilitated by external forces such as wind or fans. Thus, convection aids conductive heat exchange by continuously maintaining a supply of cool air.

In the absence of convection, conduction would be important only in such unusual circumstances as immersion in cold water. Henceforth we shall also imply convection when we use the term "conduction." Because of the great importance of air movement in aiding heat loss, attempts have been made to quantify the cooling effect of combinations of air speed and temperature. The most useful tool is called the wind-chill index.

It is convenient to view the body as a central core surrounded by a shell consisting of skin and subcutaneous tissue; we shall refer to this complex outer shell simply as skin. It is the temperature of the central core that is being regulated at approximately 37°C. As we shall see, the temperature of the outer surface of the skin changes markedly.

If the skin were a perfect insulator, no heat would ever be lost from the core. The temperature of the outer skin surface would equal the environmental temperature (except during direct exposure to the sun), and net conduction and radiation would be zero. The skin is not a perfect insulator, however, and so the temperature of its outer surface generally is somewhere between that of the external environment and that of the core.

The skin's effectiveness as an insulator is subject to physiological control by a change in the blood flow to it. The more blood reaching the skin from the core, the more closely the skin's temperature approaches that of the core. In effect, the blood vessels diminish the insulating capacity of the skin by carrying heat to the surface (Figure 17-24) to be lost to the external environment. These vessels are controlled largely by vasoconstrictor sympathetic nerves,[10] the firing rate of which is increased in response to cold and decreased in response to heat.[11] Certain areas of skin participate much more than others in these vasomotor responses, and so skin temperatures vary with location.

Three *behavioral* mechanisms for altering heat loss by radiation and conduction remain to be described: changes in surface area, in clothing, and in choice of surroundings. Curling up into a ball, hunching the shoulders, and similar maneuvers in response to cold reduce the surface area exposed to the environment, thereby decreasing heat loss by radiation and conduction. In human beings, clothing is also an important component of temperature regulation, substituting for the insulating effects of feathers in birds and fur in other mammals. The outer surface of the clothes forms the true "exterior" of the body surface. The skin loses heat directly to the air space trapped by the clothes, which in turn pick up heat from the inner air layer and transfer it to the external environment. The insulating ability of clothing is determined primarily by the thickness of the trapped air layer.

Clothing is important not only at low temperatures but at very high temperatures. When the environmental temperature is greater than body temperature, radiation and conduction favor heat *gain* rather than heat loss. People therefore insulate themselves against such temperatures by wearing clothes. The clothing, however, must be loose so as to allow adequate movement of air to permit evaporation (see below). White clothing is cooler since it reflects more radiant energy, which dark colors absorb. Contrary to popular belief, loose-fitting light-colored clothes are far more cooling than going nude during direct exposure to the sun.

The third behavioral mechanism for altering heat loss is to seek out warmer or colder surroundings, as, for example by moving from a shady spot to the sunlight. Raising or lowering the thermostat of a house or turning on an air conditioner also fits this category.

Evaporation. Water **evaporation** from the skin and membranes lining the respiratory tract is the other major process for loss of body heat. A very large amount of energy—600 kcal/L—is required to transform water from the liquid to the gaseous state. Thus, whenever water vaporizes from the body's surface, the heat required to drive the process is conducted from the surface, thereby cooling it.

Even in the absence of sweating, there is loss of water by diffusion through the skin, which is not waterproof. A similar amount is lost from the respiratory lining during expiration. These two losses are known as **insensible water loss** and amount to approximately 600 mL/day in human beings. Evaporation of this water accounts for a significant fraction of total heat loss. In contrast to this passive water loss, sweating requires the active secretion of fluid by **sweat glands** and its extrusion into ducts that carry it to the skin surface.

Production of sweat is stimulated by the sympathetic nerves. Sweat is a dilute solution containing sodium chloride as its major solute. There are an estimated 2.5 million sweat glands spread over the adult human body, and sweating rates of over 4 L/h have been reported. The evaporation of 4 L of water would eliminate almost 2400 kcal from the body!

It is essential to recognize that sweat must evaporate in order to exert its cooling effect. The most important factor determining evaporation is the water vapor concentration of the air, that is, the relative humidity. The discomfort suffered on humid days is due to the failure of evaporation. The sweat glands continue to secrete, but the sweat simply remains on the skin or drips off, rather than evaporating.

---

[10]There is also a population of sympathetic neurons to the skin whose neurotransmitters (as yet unidentified) cause active vasodilation. This response is intimately involved with sweating.

[11]In addition, the affinity of alpha-adrenergic receptors for norepinephrine, the sympathetic neurotransmitter, is increased directly by cold, and so the ability of any given rate of nerve firing to cause vasoconstriction is increased when the skin is cold.

TABLE 17-12 SUMMARY OF EFFECTOR MECHANISMS IN TEMPERATURE REGULATION

Desired Effect	Mechanism
	Stimulated by Cold
Decrease heat loss	1. vasoconstriction of skin vessels
	2. reduction of surface area (curling up, etc.)
	3. behavioral response (put on warmer clothes, raise thermostat setting, etc.)
Increase heat production	1. increased muscle tone
	2. shivering and increased voluntary activity
	3. increased secretion of thyroid hormone and epinephrine
	4. increased food appetite
	Stimulated by Heat
Increase heat loss	1. vasodilation of skin vessels
	2. sweating
	3. behavioral response (put on cooler clothes, turn on fan, etc.)
Decrease heat production	1. decreased muscle tone and voluntary activity
	2. decreased secretion of thyroid hormone and epinephrine
	3. decreased food appetite

Summary of Effector Mechanisms in Temperature Regulation

Table 17-12 summarizes the effector mechanisms regulating temperature, none of which is an all-or-none response but a graded, progressive increase or decrease in activity. Changes in skin blood flow alone can regulate body temperature, by altering heat loss, over a range of environmental temperatures (approximately 25 to 30°C or 75 to 86°F) known as the **thermoneutral zone**. At temperatures lower than this, even maximal vasoconstriction cannot prevent heat loss from exceeding heat production, and the body must increase its heat production to maintain temperature. At environmental temperatures above the thermoneutral zone, even maximal vasodilation cannot eliminate heat as fast as it is produced, and another heat-loss mechanism—sweating—is therefore brought strongly into play. Indeed, at environmental temperatures above that of the body, heat is actually added to the body by radiation and conduction, and evaporation is the sole mechanism for heat loss. A person's ability to tolerate such temperatures is determined by the humidity and by the maximal sweating rate. For example, when the air is completely dry, a person can tolerate a temperature of 130°C (255°F) for 20 min or longer, whereas very moist air at 46°C (115°F) is bearable for only a few minutes.

Brain Centers Involved in Temperature Regulation

Via descending pathways, neurons in the hypothalamus and other brain areas control the output of motor neurons to skeletal muscle for muscle tone and for shivering. They also control sympathetic neurons to skin arterioles for vasoconstriction and dilation, to sweat glands, and to the adrenal medulla. When thyroid hormone is a component of the response to cold, the temperature-regulating centers also control the output of hypothalamic thyrotropin-releasing hormone (page 272).

Afferent Input to the Integrating Centers

The final component of temperature-regulating systems to be described is really the first component, the afferent input. Obviously these temperature-regulating reflexes require receptors capable of detecting changes in body temperature. There are two groups of receptors, one in the skin (peripheral thermoreceptors) and the other in deeper body structures (central thermoreceptors).

Peripheral thermoreceptors. In the skin and certain mucous membranes are two populations of nerve endings, one stimulated by a lower and the other by a higher range of temperatures. These **peripheral thermorecep-**

tors are often termed cold and warm receptors.[12] Information from these receptors is transmitted via the afferent neurons and ascending pathways to the hypothalamus and other integrating areas, which respond with appropriate efferent output. In this manner, the firing of cold receptors stimulates heat-producing and heat-conserving mechanisms. These receptors also account for one's ability to identify a hot or cold area of the skin. The skin thermoreceptors provide feedforward information in thermoregulatory reflexes.

Central thermoreceptors. The skin thermoreceptors alone would be highly inefficient regulators of body temperature for the simple reason that it is the core temperature, not the skin temperature, that is being regulated. **Central thermoreceptors** in the hypothalamus, spinal cord, abdominal organs, and other internal locations have synaptic connections with the same integrating centers in the hypothalamus (and other brain areas) that receive input from the skin thermoreceptors.

This completes our survey of the temperature-regulating pathways, which are summarized in Figure 17-25.

Temperature Acclimatization

Acclimatization to heat. Changes in sweating onset, volume, and composition determine people's chronic adaptation to high temperatures. A person newly arrived in a hot environment has a poor ability to do work initially; body temperature rises, and severe weakness and illness may occur. After several days, there is a great improvement in work tolerance, with little increase in body temperature, and the person is said to have acclimatized to the heat (see page 144 for a discussion of the concept of acclimatization). Body temperature does not rise as much because sweating begins sooner and the volume of sweat produced is greater.

There is also an important change in the composition of the sweat, namely, a marked reduction in its sodium concentration. This adaptation, which minimizes the loss of sodium from the body via the sweat, is due to increased secretion of the mineralocorticoid hormone aldosterone. The sweat-gland secretory cells produce a solution with a sodium concentration similar to that of plasma, but some of the sodium is absorbed back into the blood as the secretion flows along the sweat-gland ducts toward the skin surface. Aldosterone stimulates this absorption in a manner identical to its stimulation of sodium reabsorption in the renal tubules. The reflexes that activate the renin-angiotensin-aldosterone system were described in Chapter 15.

Acclimatization to cold. Cold acclimatization has been less studied than heat acclimatization because of the difficulty of subjecting persons to total-body cold stress sufficient to produce acclimatization. The best-studied group was the Korean women who, wearing only cotton garments, used to dive for shellfish and edible seaweed in the middle of winter when the sea water temperature was as low as 10°C. Moreover, the fact that the acclimatization characteristics, to be described, disappeared within several years after the women began using wet suits is strong evidence that these characteristics really represented acclimatization, not genetic differences.

The major components of their acclimatization were: (1) an increase in metabolic rate; (2) an increase in the insulating ability of a given amount of skin fat; and (3) an ability to withstand a colder water temperature without shivering.

A different kind of adaptation is exhibited by Eskimos and various groups of fishermen who manifest less vasoconstriction of finger blood vessels during exposure of the hands to cold water. This is an adaptation that permits these persons to do work with their hands despite intense cold. The Eskimos and fishermen, unlike the diving women, normally expose only their hands to the cold and use warm clothing to protect the rest of the body.

Fever

The term **hyperthermia** denotes an elevation of body temperature regardless of cause. The specific type of hyperthermia termed **fever** is due to a "resetting of the thermostat" in the hypothalamus (Figure 17-26), that is, fever is a hyperthermia in which the person still regulates body temperature in response to heat or cold but at a higher set point. The most common cause of fever is infection, but physical trauma and possibly other situations can also induce fever.

The onset of fever during infection is frequently gradual, but it is most striking when it occurs rapidly in the form of a chill. The brain thermostat is suddenly raised, the person feels cold, and marked vasoconstriction and shivering occur. The person also curls up and puts on more blankets. This combination of decreased heat loss and increased heat production serves to drive body temperature up to the new set point, where it stabilizes. It will continue to be regulated at this new value until the thermostat is reset to normal and the fever "breaks." The person then feels hot, throws off the covers, and manifests profound vasodilation and sweating.

What is the basis for the thermostat resetting? A chemical originally called **endogenous pyrogen** (**EP**) is released from monocytes and macrophages in the presence of infection or inflammation. EP acts upon the ther-

---

[12] In one sense, these are misleading terms since cold is not a separate entity but only a lesser degree of warmth.

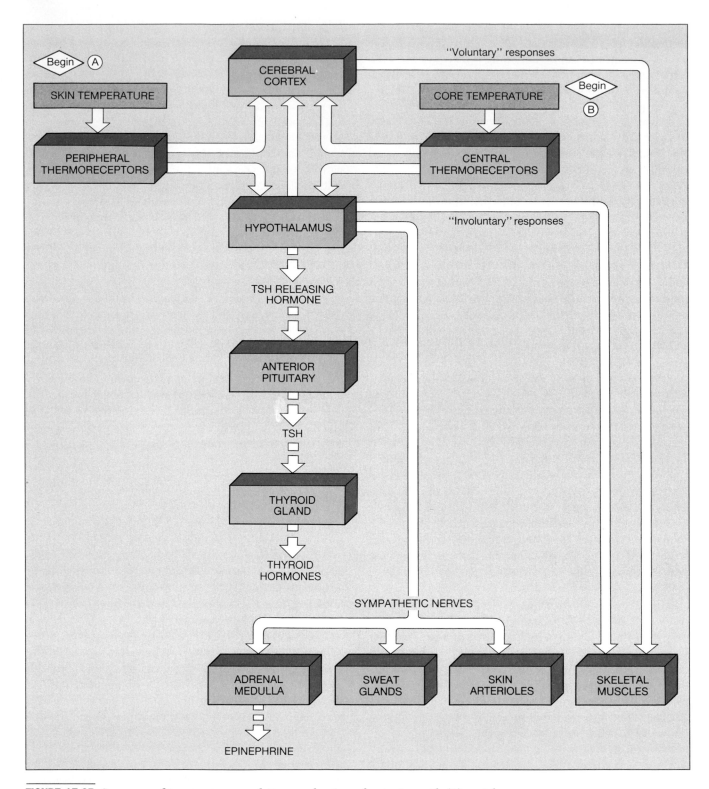

FIGURE 17-25 Summary of temperature-regulating mechanisms, beginning with (A) peripheral thermoreceptors and (B) central thermoreceptors. The dashed arrow indicates hormonal pathways, which are probably of minor importance in human beings. The solid arrows denote neural pathways. For simplicity, other nonhypothalamic integrating areas are not shown.

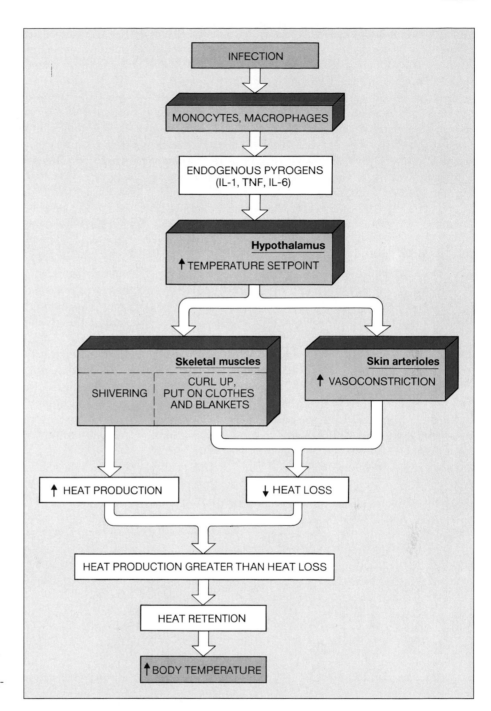

FIGURE 17-26 Pathway by which infection causes fever. IL-1 = interleukin 1; TNF = Tumor necrosis factor; IL-6 = interleukin 6.

moreceptors in the hypothalamus (and perhaps other brain areas), altering their rate of firing and their input to the integrating centers. The action of EP is mediated via local release of prostaglandins, which then directly alter central thermoreceptor function. Aspirin reduces fever by inhibiting prostaglandin synthesis.

Terminology can be confusing in this field. EP was assumed to be a single substance, which was ultimately identified as the specific peptide **interleukin 1 (IL-1)**. However, it is now clear that there are two forms of IL-1 (termed alpha and beta) and that these are not the only pyrogenic peptides released from monoctyes and macrophages. Others include **tumor necrosis factor** and **interleukin 6** (Chapter 19), and so these substances belong, with IL-1, in the general category of endogenous pyrogens. In addition to their effects on temperature, IL-1 and the other endogenous pyrogens have many other effects—to be described in Chapter 19—that have the common denominator of enhancing resistance to infection and promoting the healing of damaged tissue.

One would expect fever, which is such a consistent concomitant of infection, to play some important protective role, and most evidence suggests that such is the case. Increased body temperature stimulates a large number of the body's defensive responses to infection. The likelihood that fever is a beneficial response raises as yet unanswered questions concerning the use of aspirin and other drugs to suppress fever during infection. It must be emphasized that these questions apply to the usual modest fevers. There is no question that an extremely high fever can be harmful, particularly in its effects on the central nervous system, and must be vigorously opposed with drugs and other forms of therapy.

Other Causes of Hyperthermia

To reiterate, fever is a hyperthermia caused by an elevation of thermal set point. There are a variety of situations other than a change in set point in which hyperthermia occurs.

Exercise. The most common cause of hyperthermia in normal people is sustained exercise, during which body temperature rises and is maintained as long as the exercise continues. It is possible that a small fraction of this

rise represents a new set point, and some IL-1 may be released during sustained exercise, perhaps because of slight tissue damage. However, most of the rise in body temperature is not a fever but is simply a physical consequence of the internal heat generated by the exercising muscles. As shown in Figure 17-27, heat production rises immediately during the initial stage of exercise and exceeds heat loss, causing heat storage in the body and a rise in core temperature. This rise in core temperature triggers reflexes, via the central thermoreceptors, for increased heat loss—increased skin blood flow and sweating, and the discrepancy between heat production and heat loss starts to diminish but does not disappear. Therefore core temperature continues to rise, albeit more slowly. Ultimately, core temperature will be high enough to drive, via the central thermoreceptors, the heat-loss reflexes at a rate such that heat loss equals heat production, and core temperature stabilizes at this elevated value despite continued exercise.

Heat exhaustion and heat stroke. **Heat exhaustion** is a state of collapse, often taking the form of fainting, due to hypotension brought on by (1) depletion of plasma volume secondary to sweating and (2) extreme dilation of skin blood vessels. Thus, decreases in both cardiac output and peripheral resistance contribute to the hypotension. Heat exhaustion occurs as a direct consequence of the activity of heat-loss mechanisms, and because these mechanisms have been so active, the body temperature is only modestly elevated. In a sense, heat exhaustion is a safety valve that, by forcing cessation of work in a hot environment when heat-loss mechanisms are overtaxed, prevents the larger rise in body temperature that would precipitate the far more serious condition of heat stroke.

In contrast to heat exhaustion, **heat stroke** represents a complete breakdown in heat-regulating systems so that body temperature keeps going up and up. It is an extremely dangerous situation, characterized by collapse, delirium, seizures, or prolonged unconsciousness—all due to marked elevation of body temperature. It almost always occurs in association with exposure to or overexertion in hot and humid environments. In some persons, particularly the elderly, heat stroke may appear with no apparent prior period of severe sweating, but in most cases, it comes on as the endstage of prolonged untreated heat exhaustion. Exactly what triggers the transition to heat stroke is not clear—impaired circulation to the brain due to dehydration is one factor—but the striking finding is that even in the face of a rapidly rising body temperature, the person fails to sweat. This sets off a positive-feedback situation in which the rising body temperature directly stimulates metabolism, that is, heat production, which further raises body temperature.

FIGURE 17-27 Thermal changes during exercise. Heat loss is reflexly increased, and when it once again equals heat production, core temperature stabilizes.

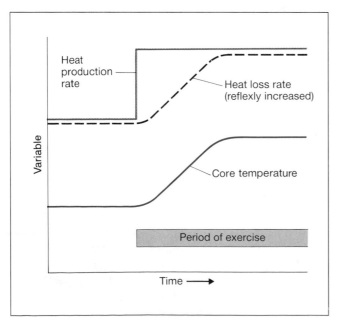

A recent finding of clinical importance is that some of the commonly used tranquilizer drugs interfere with neurotransmitters in hypothalamic thermoregulatory centers, and some people, particularly the elderly, using these drugs are very prone to heat stoke.

SUMMARY

Section A. Control and Integration of Carbohydrate, Protein, and Fat Metabolism

Events of the Absorptive and Postabsorptive States

I. During absorption, energy is provided primarily by absorbed carbohydrate, and net synthesis of glycogen, triacylglycerol, and protein occurs.
   A. Some absorbed carbohydrate not used for energy is converted to glycogen, mainly in the liver and skeletal muscle, but most is converted to triacylglycerol in adipose tissue and the liver. The latter organ releases its triacylglycerol in very low density lipoproteins, the fatty acids of which are picked up by adipose tissue.
   B. The fatty acids of some absorbed triacylglycerol are used for energy, but most are rebuilt into fat in adipose tissue.
   C. Absorbed amino acids drive net protein synthesis, but excess amino acids are converted to carbohydrate and fat.
   D. There is net uptake of glucose by the liver.
II. In the postabsorptive state, blood glucose is maintained by a combination of glucose production by the liver and a switch from glucose utilization to fatty acid and ketone utilization by most tissues.
   A. Synthesis of glycogen, fat, and protein is curtailed, and net breakdown of these molecules occurs.
   B. The liver forms glucose first by glycogenolysis of its own glycogen and then by gluconeogenesis from lactate and pyruvate (from breakdown of muscle glycogen), glycerol (from adipose-tissue lipolysis), and especially amino acids (from protein catabolism).
   C. Glycolysis is decreased, and most of the body's energy supply comes from the oxidation of fatty acids released by adipose-tissue lipolysis and of ketones produced from fatty acids by the liver.
   D. The brain continues to use glucose, but also starts using ketones as they build up in the blood.

Endocrine and Neural Control of the Absorptive and Postabsorptive States

I. The major hormones secreted by the pancreatic islets of Langerhans are insulin by the B cells and glucagon by the A cells.
II. Insulin is the most important hormone controlling metabolism.
   A. It stimulates glucose and amino acid entry into cells and glycolysis, suppresses glucose production by the liver, stimulates the synthesis of glycogen, fat, and protein, and inhibits their catabolism.
   B. The major stimulus for insulin secretion is an increased plasma glucose concentration, but secretion is also influenced by many other factors summarized in Figure 17-6.
III. Glucagon, epinephrine, cortisol, and growth hormone all exert effects on carbohydrate and lipid metabolism that are opposed to those of insulin. They all raise plasma concentrations of glucose, glycerol, and fatty acids.
   A. Glucagon stimulates glycogenolysis by the liver, gluconeogenesis, lipolysis, and ketone synthesis.
   B. The major stimulus for glucagon secretion is hypoglycemia, but secretion is also stimulated by plasma amino acids and by both the sympathetic and parasympathetic nerves to the islets.
   C. Epinephrine released from the adrenal medulla in response to hypoglycemia stimulates glycogenolysis in liver and muscle, gluconeogenesis, and lipolysis and blocks glucose uptake by muscle. The sympathetic nerves to liver and adipose tissue exert effects similar to epinephrine.
   D. Cortisol is permissive for gluconeogenesis and lipolysis and, in higher concentrations, stimulates gluconeogenesis and blocks glucose uptake. These last two effects are also exerted by growth hormone.

Fuel Homeostasis in Exercise and Stress

I. During exercise, the muscles use as their energy sources plasma glucose, plasma fatty acids, and their own glycogen.
   A. Glucose is produced by the liver, and fatty acids are provided by adipose-tissue lipolysis.
   B. The changes in plasma insulin, glucagon, and epinephrine are similar to those that occur during the postabsorptive period and are mediated mainly by the sympathetic nervous system.
II. Stress causes metabolic and hormonal changes similar to those of exercise.

Diabetes Mellitus

I. Type 1 diabetes is due to absolute insulin deficiency and can lead to diabetic ketoacidosis.
II. Type 2 diabetes is usually associated with obesity, which causes insulin resistance, and a normal or even elevated plasma insulin concentration.

Regulation of Plasma Cholesterol

I. Plasma cholesterol, a precursor for synthesis of plasma membranes, bile acids, and steroid hormones, is carried mainly by low-density lipoproteins, which deliver it to cells; high-density lipoproteins carry cholesterol from cells to the liver. The ratio of LDL/HDL correlates with the incidence of coronary heart disease.
II. Cholesterol synthesis is mainly by the liver and varies inversely with ingested cholesterol.
III. The liver also secretes cholesterol into the bile and converts cholesterol to bile acids.

Section B. Control of Growth

Bone Growth

I. A bone lengthens as osteoblasts at the shaft edge of the epiphyseal growth plates convert cartilage to bone while new cartilage is being laid down in the plate.

II. Growth ceases when the plates are completely converted to bone.

Environmental Factors Influencing Growth

I. The major environmental factors influencing growth are nutrition and disease.

II. Malnutrition during in utero life and infancy may produce irreversible stunting.

Hormonal Influences on Growth

I. Growth hormone is the major stimulus of postnatal growth.
   A. It stimulates the release of IGF-1 from the liver and many other cells, and IGF-1 then acts locally or as a hormone to stimulate mitosis.
   B. Growth hormone also acts directly on cells to stimulate protein synthesis.
   C. Growth hormone secretion occurs mainly during sleep and is highest during adolescence.

II. The thyroid hormones stimulate brain development during infancy and bone growth during childhood and adolescence.

III. Insulin stimulates growth mainly during in utero life.

IV. Testosterone and estrogen stimulate bone growth during adolescence but also cause epiphyseal closure. Testosterone also stimulates protein synthesis.

V. Cortisol, in high concentration, inhibits growth and stimulates protein catabolism.

Section C. Summary of Liver Functions (Table 17-6)

Section D. Regulation of Total-Body Energy Balance

Basic Concepts of Energy Expenditure and Caloric Balance

I. The energy liberated during a chemical reaction appears either as heat or work.

II. Total energy expenditure = heat produced + external work done + energy stored.

III. Metabolic rate is influenced by the many factors summarized in Table 17-7.

IV. Basal metabolic rate is increased by the thyroid hormones and epinephrine.

V. Energy storage, as fat, can be positive or negative when metabolic rate is less than or greater than, respectively, the energy content of ingested food.
   A. Energy storage is regulated mainly by reflex adjustment of food intake to metabolic rate.
   B. In addition, metabolic rate increases or decreases, to some extent, when food intake is chronically increased or decreased, respectively.

VI. Food appetite is controlled by the many factors summarized in Figure 17-20.

VII. Obesity, the result of an imbalance between food intake and metabolic rate, increases the incidence of many diseases.

Regulation of Body Temperature

I. Body temperature, measured as rectal temperature, shows a circadian rhythm, being highest during the day and lowest at night.

II. Body temperature is regulated by altering heat production and/or heat loss so as to change total-body heat content.
   A. Heat production is altered by increasing muscle tone, shivering, and voluntary activity.
   B. Heat-loss mechanisms include radiation, conduction, convection, and evaporation of water from the body surface.

III. Heat loss by radiation, conduction, and convection depends on the difference in temperature between the skin surface and the environment.
   A. In response to cold, skin temperature is decreased by decreasing skin blood flow through reflex stimulation of the sympathetic nerves to the skin. In response to heat, skin temperature is increased by inhibiting the nerves.
   B. Behavioral responses such as putting on more clothes also influence heat loss.

IV. Evaporation of water occurs all the time as insensible loss from the skin and respiratory lining. Additional water for evaporation is supplied by sweat, stimulated by the sympathetic nerves to the sweat glands.

V. Increased heat production is essential for temperature regulation at environmental temperatures below the thermoneutral zone, and sweating is essential at temperatures above this zone.

VI. The hypothalamus and other brain areas are the integrating centers for temperature-regulating reflexes.

VII. Both peripheral and central thermoreceptors participate in these reflexes.

VIII. Temperature acclimatization to the heat is achieved by an earlier onset of sweating, an increased volume of sweat, and a decreased sodium concentration of the sweat.

IX. The most common causes of hyperthermia are fever and exercise.
   A. Fever is due to a resetting of the temperature set point so that heat production is increased and heat loss is decreased in order to raise body temperature to the new set point and keep it there. The stimulus is endogenous pyrogens, probably including IL-1, TNF, and IL-6.
   B. The hyperthermia of exercise is not due primarily to a changed set point but to the increased heat produced by the muscles.

REVIEW QUESTIONS

Section A. Control and Integration of Carbohydrate, Protein, and Fat Metabolism

1. Define:

absorptive state
postabsorptive state
very low density
   lipoproteins (VLDL)
lipoprotein lipase
keto acids
glycogenolysis
lipolysis
gluconeogenesis
glucose sparing
ketones
islets of Langerhans
B cells
A cells
insulin

glucose-counterregulatory
   controls
glucagon
hypoglycemia
diabetes mellitus (type 1,
   insulin-dependent; and
   type 2, insulin-
   independent)
diabetic ketoacidosis
insulin resistance
cholesterol
low-density lipoproteins
   (LDL)
high-density lipoproteins
   (HDL)

2. Using a diagram, summarize the events of the absorptive period.

3. In what two organs does major glycogen storage occur?

4. How do the liver and adipose tissue metabolize glucose during the absorptive period?

5. How does adipose tissue metabolize absorbed triacylglycerol, and what are the three major sources of the fatty acids in adipose-tissue triacylglycerol?

6. What happens to the absorbed amino acids when an excess of protein is ingested?

7. Using a diagram, summarize the events of the postabsorptive period. Include the four sources of blood glucose and the pathways leading to ketone formation.

8. Distinguish between the roles of glycerol and free fatty acids during fasting.

9. List the metabolic effects of insulin on membrane transport and enzyme function, and describe the net results of these actions. What effects occur when plasma insulin concentration decreases?

10. List five inputs controlling insulin secretion, and state the physiological significance of each.

11. List four metabolic effects of glucagon and their consequences.

12. List three inputs controlling glucagon secretion, and state the physiological significance of each.

13. List four metabolic effects of epinephrine and the sympathetic nerves to the liver and adipose tissue, and state the net results of each.

14. List the permissive effects of cortisol and the effects that occur when plasma cortisol concentration increases.

15. List three effects of growth hormone on carbohydrate and lipid metabolism.

16. Which hormones stimulate gluconeogenesis? Glycogenolysis in liver? Glycogenolysis in skeletal muscle? Lipolysis? Blockade of glucose uptake?

17. Describe how plasma glucose, insulin, glucagon, and epinephrine levels change during exercise or stress. What causes the changes in the concentrations of the hormones?

18. Describe the metabolic disorders of severe type 1 diabetes.

19. How does obesity contribute to type 2 diabetes?

20. Hypersecretion of which hormones can induce a diabetic state?

21. Using a diagram, describe the sources of cholesterol gain and loss. Include three roles of the liver in cholesterol metabolism, and state the controls over these processes.

22. What are the effects of saturated and unsaturated fatty acids on plasma cholesterol?

23. What is the significance of the LDL cholesterol/HDL cholesterol ratio?

Section B. Control of Growth

1. Define:

osteoblasts
epiphyses
epiphyseal growth plate
chondrocytes
epiphyseal closure
growth factors
growth-inhibiting factors

giantism
dwarfism
acrogemaly
insulin-like growth factor 1
   (IGF-1, somatomedin C)
compensatory growth

2. Describe the process by which bone is lengthened.

3. What are the effects of malnutrition on growth?

4. List the major hormones that control growth.

5. Describe the relationship between growth hormone and IGF-1, and the roles of each in growth.

6. What are effects of growth hormone on protein synthesis?

7. What is the status of growth hormone secretion at different stages of life?

8. State the effects of the thyroid hormones on growth and development.

9. Describe the effects of testosterone on growth, cessation of growth, and protein synthesis. Which of these effects are shared by estrogen?

10. What is the effect of cortisol on growth?

Section D. Regulation of Total-Body Energy Balance

1. Define:

external work
internal work
total energy expenditure
kilocalorie
metabolic rate
basal metabolic rate (BMR)
calorigenic effect

food-induced thermogenesis
satiety signals
obesity
body mass index (BMI)
anorexia nervosa
homeothermic
shivering thermogenesis

nonshivering thermogenesis	central thermoreceptors
radiation	hyperthermia
conduction	fever
convection	endogenous pyrogen (EP)
evaporation	interleukin 1 (IL-1)
insensible water loss	tumor necrosis factor (TNF)
sweat glands	interleukin 6 (IL-6)
thermoneutral zone	heat exhaustion
peripheral thermoreceptors	heat stroke

2. State the formula relating total energy expenditure, heat produced, external work, and energy storage.

3. What two hormones alter basal metabolic rate?

4. State the equation for total-body energy balance. Describe the three possible states of balance with regard to energy storage.

5. What happens to basal metabolic rate after a person has either lost or gained weight?

6. List five satiety signals.

7. List three beneficial effects of exercise in a weight-loss program.

8. Compare and contrast the four mechanisms for heat loss.

9. Describe the control of skin blood vessels during exposure to cold or heat.

10. With a diagram, summarize the reflex responses to heat or cold. What are the dominant mechanisms for temperature regulation in the thermoneutral zone and in temperatures below and above this range?

11. What changes are exhibited by a heat-acclimatized person?

12. Summarize the sequence of events leading to a fever, and contrast this to the sequence leading to hyperthermia during exercise.

THOUGHT QUESTIONS

(Answers are given in Appendix A.)

1. What happens to the triacylglycerol concentrations in the plasma and in adipose tissue after administration of a drug that blocks the action of lipoprotein lipase?

2. A resting, unstressed person has increased plasma concentrations of free fatty acids, glycerol, amino acids, and ketones. What situations might be responsible, and what additional plasma measurement would distinguish between them?

3. A normal volunteer is given an injection of insulin. The plasma concentrations of which hormones increase?

4. What happens to the plasma insulin concentration of a normal volunteer who has been given an injection of epinephrine?

5. If the sympathetic preganglionic fibers to the adrenal medulla were cut in an animal, would this eliminate the sympathetically mediated component of increased gluconeogenesis and lipolysis during exercise? Explain.

6. A patient with type 1 diabetes suffers a broken leg. Would you advise this person to increase or decrease his dosage of insulin?

7. A person has a defect in the ability of her small intestine to absorb bile acids. What effect will this have on plasma cholesterol concentration?

8. A well-trained athlete is found to have a moderately elevated plasma cholesterol concentration. What additional measurement would you advise this person to have done?

9. A full-term newborn infant is abnormally small. Is this most likely due to deficient growth hormone, deficient thyroid hormones, or deficient in utero nutrition?

10. Why might the administration of androgens to stimulate growth in a small 12-year-old male turn out to be counterproductive?

11. Without measuring any plasma hormones, how would you distinguish between thyroid deficiency and growth hormone deficiency as the cause of poor growth in a 1-year-old?

12. What are the sources of heat loss for a person immersed up to the neck in a 40°C bath?

13. Lizards can regulate their body temperatures only through behavioral means. Can you predict what they do when they are injected with bacteria?

REPRODUCTION

efore we begin detailed descriptions of male and female reproductive systems, it is worthwhile to summarize some general terminology and concepts. This brief description is for orientation, and the specifics of these processes will be dealt with in subsequent sections.

The primary reproductive organs are known as the **gonads**, the **testes** (singular *testis*) in the male and the **ovaries** in the female. In both sexes, the gonads serve dual functions: (1) production of the reproductive cells, termed **gametes**—**spermatozoa** (singular *spermatozoan*, usually shortened to **sperm**) by males and **ova** (singular

ovum) by females; and (2) secretion of particular steroid hormones, often termed **sex hormones**—**testosterone** by the male and **estrogen** and **progesterone** by the female.

The systems of ducts through which the sperm or ova are transported and the glands lining or emptying into the ducts are termed the **accessory reproductive organs**. In the female the breasts are also usually included in this category. The **secondary sexual characteristics** comprise the many external differences—hair distribution and body contours, for example—between males and females. The secondary sexual characteristics are not directly involved in reproduction.

Reproductive function is largely controlled by a chain of hormones (Figure 18-1). The first hormone in the chain is **gonadotropin-releasing hormone (GnRH)**, secreted by neuroendocrine cells in the hypothalamus and reaching the anterior pituitary via the hypothalamo-pituitary portal blood vessels (Chapter 10). Once in the anterior pituitary, GnRH stimulates the release of the **pituitary gonadotropins**—**follicle-stimulating hormone (FSH)** and **luteinizing hormone (LH)**. These two protein hormones were named for their effects in the female, but their molecular structures are the same in both sexes.[1] They act together upon the gonads, the result being production of gametes (**gametogenesis**) and sex hormone secretion. In turn, the sex hormones exert many effects on all portions of the reproductive system, including the gonads from which they come, and other parts of the body, as well. In addition, the sex hormones and one other gonadal hormone—inhibin—exert feedback effects on the secretion of GnRH, FSH, and LH. These controls of reproductive function change markedly during a person's lifetime and may be divided into the stages summarized in Table 18-1.

FIGURE 18-1 General pattern of reproduction control in both males and females. GnRH, like all hypothalamic releasing hormones, reaches the anterior pituitary via the hypothalamo-pituitary portal vessels (Chapter 10). The arrow within the gonads denotes the fact that the sex hormones act locally, as paracrines, to stimulate gamete formation. Not shown in this introductory figure but described later is another hormone, inhibin, which is secreted by the gonads.

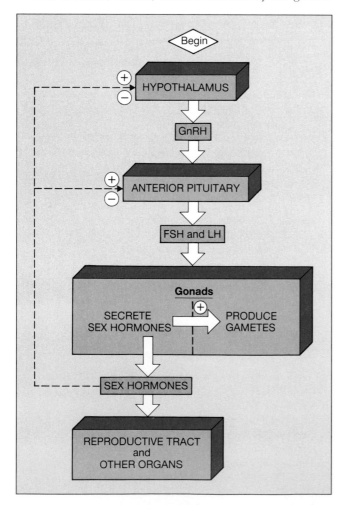

GENERAL PRINCIPLES OF GAMETOGENESIS

The populations of cells that give rise to the gametes are known as **germ cells**. Gametogenesis involves both mitosis of these cells and the type of cell division known as meiosis. Because the general principles of gametogenesis are essentially the same in males and females, they are treated in this section, and features specific to the male or female are described later.

The first stage in gametogenesis is proliferation of the primordial germ cells by mitosis. Recall from our discussion of mitosis on page 69 that each nucleated human cell, except the gametes, contains 46 chromosomes, 23 from each parent. Each maternal chromosome in a cell

---

[1] In the male, LH is sometimes called interstitial cell-stimulating hormone, ICSH.

TABLE 18-1 STAGES IN THE CONTROL OF REPRODUCTIVE FUNCTION

1. During the initial stage, which begins during fetal life and ends in the first year of life (infancy), the gonadotropins and gonadal sex hormones are secreted at relatively high levels.

2. From infancy to puberty, the secretion rates of these hormones are very low, and reproductive function in general is quiescent.

3. Beginning at puberty, hormonal secretion rates increase markedly, being stable in men but varying greatly in women during the menstrual cycle, and this ushers in the period of active reproduction.

4. Finally, reproductive function diminshes later in life as the gonads become less responsive to the gonadotropins, the ability to reproduce ceasing entirely in women.

has a corresponding paternal chromosome that contains the same type of genetic information in its DNA. For example, both chromosomes of a particular pair of corresponding maternal and paternal chromosomes contain genes for eye color. The two corresponding chromosomes in such a pair are said to be homologous to each other. In mitosis all the dividing cell's 46 chromosomes are replicated, and each of the two cells, termed daughter cells, resulting from the division receives a full set of 46 chromosomes—23 pairs of homologous chromosomes—identical to those of the original cell. Thus each daughter cell is identical to the original cell.

Mitosis of primordial germ cells, the first stage of gametogenesis, provides a supply of identical germ cells for the next stages. The timing of mitotic activity in germ cells differs greatly in females and males. In the female, mitosis of germ cells occurs exclusively during the individual's embryonic existence. In the male, some mitosis occurs in the embryo to generate the population of primitive germ cells present at birth, but mitosis really begins in earnest at puberty and usually continues throughout life.

The second stage of gametogenesis is **meiosis**, in which each resulting gamete receives only 23 chromo-

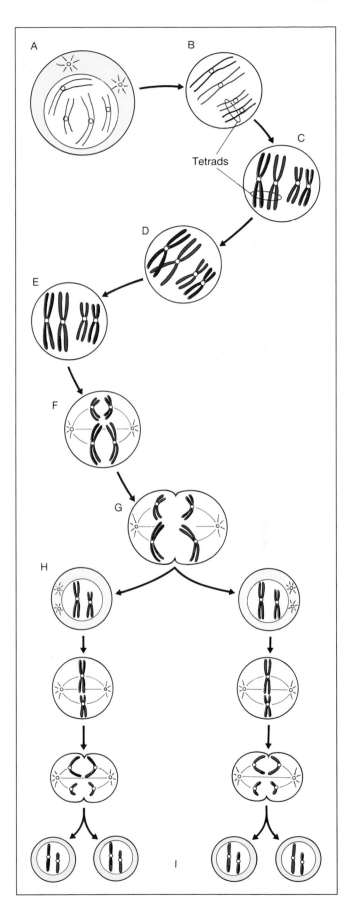

FIGURE 18-2 Stages of meiosis in a generalized germ cell. For simplicity, the initial cell (A), which is in interphase, is given only four chromosomes rather than 46, the human number. The letters are keyed to descriptions in the text. (*Adapted from Patten and Carlson.*)

somes from a 46-chromosome germ cell, one chromosome from each homologous pair. In other words, the male member of any given homologous chromosome pair from the original cell ends up in one daughter cell and the female member in the other daughter cell. Because a mature sperm and ovum each has only 23 chromosomes, their union at fertilization results in a cell with a full complement of 46 chromosomes.

Let us see how meiosis works, using Figure 18-2 in which the letters are keyed to the text. Meiosis consists of two cell divisions in succession, and the events preceding the first meiotic division are identical to those preceding a mitotic division (the material on mitosis on page 69 should be reviewed at this point for terminology). During the interphase preceding meiosis, the DNA of the chromosomes was replicated. Thus the resting interphase cell still has 46 chromosomes, but each chromosome consists of two identical strands of DNA termed sister chromatids and joined together by a centromere (A).

As the first meiotic division begins, homologous chromosomes, each consisting of two identical sister chromatids, come together and line up point for point along their entire lengths. Thus, 23 four-chromatid groupings, called tetrads, are formed (B). The sister chromatids of each chromosome condense into thick rodlike structures and become highly visible (C). Then, within a tetrad, corresponding segments of homologous chromosomes come to overlap one another (D). At these points, portions of the homologous chromosomes break off and exchange with each other in a process known as **crossing over** (E). Thus, crossing over results in recombination of genes on homologous chromosomes.

Following crossing over, the tetrads line up in the center of the cell so that, for any given tetrad, the maternal sister chromatids are on one side of the equatorial pole and the paternal on the other (F). The orientation of the tetrads on the equator is random, by which we mean sometimes the maternal portion points "north" and sometimes the paternal portion points "north." The cell now divides, with the two maternal sister chromatids of any tetrad going to one daughter cell and the two paternal sister chromatids going to the other (G). Because of the random orientation at the equator, it is extremely unlikely that all 23 maternal chromosomes would end up in one cell and all 23 paternal chromosomes in the other. Thus, the dividing cell's maternal and paternal chromosomes, partially scrambled because of crossing over, are distributed randomly to the two daughter cells. The daughter cells now contain 23 chromosomes, each still consisting of two sister chromatids.

The second division of meiosis occurs without any further replication of DNA. The two chromatids of each chromosome separate and move apart into the new daughter cells (H). The daughter cells resulting from the second meiotic division, therefore, contain 23 one-chromatid chromosomes (I).

To summarize, meiosis produces daughter cells having only 23 chromosomes, and two events during the first meiotic division contribute to the enormous genetic variability of the daughter cells: (1) crossing over and (2) the random distribution of maternal chromatid pairs and paternal chromatid pairs between the two daughter cells—over 8 million (2^{23}) different combinations of maternal and paternal chromosomes can result during this division.

SECTION A
MALE REPRODUCTIVE PHYSIOLOGY

ANATOMY

The male reproductive system includes the two testes, the system of ducts that store and transport sperm to the exterior, the glands that empty into these ducts, and the penis. The duct system, glands, and penis constitute the male accessory reproductive organs.

The testes are suspended outside the body in the **scrotum**, which is an outpouching of the abdominal wall and is divided internally into two sacs, one for each testis. During embryonic development, the testes are located in the abdomen, but during the seventh month of

intrauterine development, they descend into the scrotum. This descent is essential for normal sperm production during adulthood, since sperm formation requires a temperature lower than normal internal body temperature. The testes are kept cooler than this by air circulating around the scrotum and by a heat-exchange mechanism in the blood vessels supplying the testes. In contrast to spermatogenesis, testosterone secretion can occur normally at internal body temperature, and so failure of testes descent does not impair testosterone secretion.

The sites of sperm formation (**spermatogenesis**) in the testes are the many tiny convoluted **seminiferous tu-**

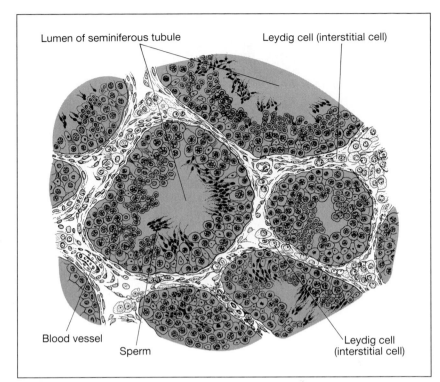

Lumen of seminiferous tubule   Leydig cell (interstitial cell)

Blood vessel

Sperm

Leydig cell
(interstitial cell)

FIGURE 18-3 Cross-section of an area of testis. The colored areas are the seminiferous tubules, the sites of sperm production. Mature sperm from each tubule are found in the tubular lumen. This low-power magnification is only for general orientation, and the positions of the other cells in the tubules are shown in a later figure (Figure 18-8). The tubules are separated from each other by interstitial space that contains Leydig cells and blood vessels.

bules (Figure 18-3), the combined length of which is 250 m. Each seminiferous tubule is bounded by a basement membrane and a layer of smooth-muscle-like cells. The latter cells are responsible for peristaltic movements of the tubules. In the center of each tubule is a fluid-filled lumen. The remainder of the tubule contains germ cells and another cell type, to be described later, called Sertoli cells. The endocrine **Leydig cells** (also called **interstitial cells**), which secrete testosterone, lie in small connective-tissue spaces between the tubules. Thus, the sperm-producing and testosterone-producing functions of the testes are carried out by different structures.

Sperm, following their formation, move through the duct system leading from the seminiferous tubules to the urethra, the tube that courses from the urinary bladder through the penis. First, the seminiferous tubules from different areas of a testis unite to form a network of interconnected tubes, the rete testis (Figure 18-4). Small ducts termed efferent ductules leave the rete testis, pierce the fibrous covering of the testis, and empty into a single duct within a structure called the **epididymis** (plural *epididymides*). The epididymis is loosely attached to the outside of the testis, and the duct of the epididymis is so convoluted that, when straightened out at dissection, it measures 6 m. In turn, the epididymis draining each testis (we shall follow the common practice of short-

ening the term "duct of the epididymis" to simply "epididymis") leads to a **vas (ductus) deferens**, a large thick-walled tube. The vas deferens and the blood vessels and nerves supplying the testis are bound together in the **spermatic cord**, which passes through a slitlike passage, the inguinal canal, in the abdominal wall.

After entering the abdomen, the two vas deferens—one from each side—course to the back of the bladder base (Figure 18-5) and become the **ejaculatory ducts**. Just at this transition, two large glands, the **seminal vesicles**, which lie behind the bladder, drain into the two ductus deferens. The ejaculatory ducts then enter the substance of the **prostate gland** and join the urethra, coming from the bladder. The prostate gland is a single donut-shaped gland below the bladder and surrounding the upper part of the urethra. The urethra leaves the prostate gland to enter the penis. The paired **bulbourethral glands**, lying below the prostate, drain into the urethra just after it leaves the prostate.

The prostate gland and seminal vesicles secrete the bulk of the fluid in which ejaculated sperm are suspended. This fluid, plus the sperm cells, constitute **semen**, the sperm contributing only a few percent of the total volume. The glandular secretions contain a large number of different chemical substances, including nutrients, buffers for protecting the sperm against the

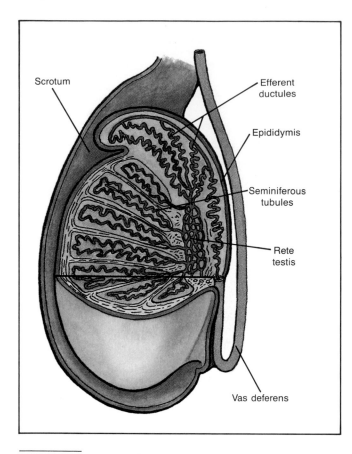

FIGURE 18-4 Section of a testis. The upper portion of the testis has been removed to show its interior. On the right, the scrotal sac covering the epididymis and vas deferens has also been removed to show these structures.

acidic vaginal secretions, and prostaglandins.[2] The function of the prostaglandins in semen is still not clear. The bulbourethral glands contribute a small volume of mucoid secretions.

In addition to providing a route for sperm from the seminiferous tubules to the exterior, several of the duct-system segments perform additional functions to be described in the section on sperm transport.

SPERMATOGENESIS

The various stages of spermatogenesis are summarized in Figure 18-6. The undifferentiated germ cells, which are termed **spermatogonia** (singular *spermatogonium*) begin to divide mitotically at puberty.

---

[2]So-named because it was originally thought that the prostaglandins in semen were produced by the prostate rather than the actual site—the seminal vesicles.

The daughter cells of this first division then divide and so on for a specified number of division cycles so that a clone of spermatogonia are produced from each original spermatogonium. The morphology of the daughter cells produced at each mitotic division differs slightly from that of the parent cell, that is, some differentiation occurs as well as mitosis. The cells that result from the final mitotic division and differentiation in the series are called **primary spermatocytes**, and these are the cells that will undergo the first meiotic division of spermatogenesis.

It should be emphasized that if all the cells in the clone produced by each original spermatogonium were to follow this pathway, the spermatogonia would disappear, that is, would all be converted to primary spermatocytes. This does not occur because, at an early point, one of the cells of each clone "drops out" of the mitosis-differentiation cycle and reverts to being a primitive spermatogonium that, at a later time, will enter into its own full sequence of divisions. In turn, one cell of the clone it produces will do likewise, and so on. Thus, the supply of spermatogonia does not decrease.

Each primary spermatocyte increases markedly in size and undergoes the first meiotic division to form two **secondary spermatocytes**, each of which in turn undergoes the second meiotic division into two **spermatids**. Thus, each primary spermatocyte, containing 46 chromosomes, gives rise to four spermatids, each containing 23 chromosomes.

The final phase of spermatogenesis is the differentiation of the spermatids into spermatozoa. This process involves extensive cell remodeling, including elongation, but no further cell divisions. The head of a sperm (Figure 18-7) consists almost entirely of the nucleus, containing the DNA that bears the sperm's genetic information. The tip of the nucleus is covered by the **acrosome**, a protein-filled vesicle containing several enzymes that play an important role in the sperm's penetration of the ovum. Most of the tail is a flagellum—a group of contractile filaments that produce whiplike movements capable of propelling the sperm at a velocity of 1 to 4 mm/min. The sperm's mitochondria form the midpiece of the tail and provide the energy for the sperm's movement.

In any small segment of each seminiferous tubule, spermatogenesis proceeds in a regular sequence. For example, at any given time, virtually all the primary spermatocytes in one portion of the tubule are undergoing division, whereas in an adjacent segment, all the secondary spermatocytes may be dividing. The entire process, from primary spermatocyte to sperm, takes approximately 64 days. The normal human male manufactures approximately thirty million sperm per day.

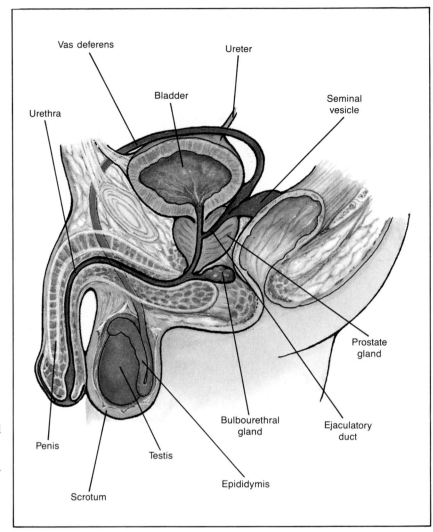

Vas deferens

Ureter

Bladder

Seminal
vesicle

Urethra

Prostate
gland

Penis

Bulbourethral
gland

Ejaculatory
duct

Testis

Scrotum

Epididymis

FIGURE 18-5 Anatomic organization of the male reproductive tract. This figure shows the testis, epididymis, vas deferens, ejaculatory duct, seminal vesicle, and bulbourethral gland on only one side of the body, but they are all paired structures. They and the prostate gland and penis constitute the accessory reproductive organs in the male. The bladder and a ureter are shown for orientation but are not part of the reproductive tract. Once the ejaculatory ducts join the urethra in the prostate, the urinary and reproductive tracts have merged.

Thus far, we have described spermatogenesis without regard to its orientation within the seminiferous tubules or the participation of a second seminiferous-tubule cell type, the **Sertoli cell**, with which the developing germ cells are intimately associated. As noted earlier, each seminiferous tubule is surrounded by a basement membrane. Each Sertoli cell extends from the basement membrane all the way to the lumen in the center of the tubule and is joined to adjacent Sertoli cells by means of tight junctions (Figure 18-8). Thus, the Sertoli cells form an unbroken ring around the outer circumference of the seminiferous tubule, and the tight junctions divide the tubule into two compartments—a basal compartment between the basement membrane and the tight junctions, and a central, or adluminal, compartment, beginning at the tight junctions and including the lumen.

This arrangement has several very important results: (1) The ring of interconnected Sertoli cells forms a **blood-testis barrier** that prevents the movement of many chemicals (proteins, other charged organic molecules, and ions) from blood into the lumen of the seminiferous tubule;[3] and (2) different stages of spermatogenesis take place in different compartments and, hence, different environments.

Mitosis of spermatogonia to yield primary spermatocytes takes place entirely in the basal compartment. The primary spermatocytes then move through the tight junctions of the Sertoli cells, which open in front of them

[3]The basement membrane of the tubule may also contribute to the barrier but is permeable to some substances that cannot pass the Sertoli cell tight junctions.

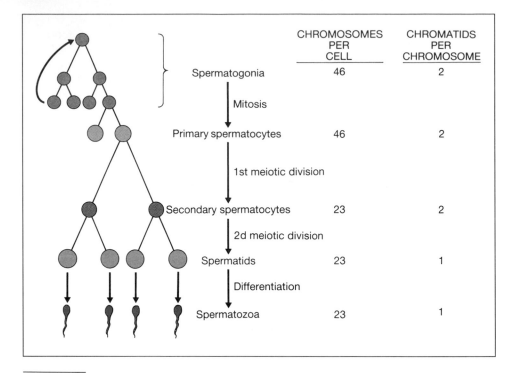

	CHROMOSOMES PER CELL	CHROMATIDS PER CHROMOSOME
Spermatogonia	46	2
Mitosis		
Primary spermatocytes	46	2
1st meiotic division		
Secondary spermatocytes	23	2
2d meiotic division		
Spermatids	23	1
Differentiation		
Spermatozoa	23	1

FIGURE 18-6 Summary of spermatogenesis beginning at puberty. Each spermatogonium yields, by mitosis, a clone of spermatogonia. For simplicity, the figure shows only two such cycles, with a third mitotic cycle generating two primary spermatocytes. Each primary spermatocyte then yields two secondary spermatocytes, each of which yields two spermatids, each of which contains 23 chromosomes and differentiates into a mature spermatozoan. Thus, each primary spermatocyte yields four sperm. The arrow from one of the spermatogonia back to an original spermatogonium denotes the fact that one cell of the clone does not go on to generate primary spermatocytes but reverts to being an undifferentiated spermatogonium that gives rise to a new clone.

while at the same time new tight junctions form behind them, to gain entry into the central compartment. In this central compartment, the meiotic divisions of spermatogenesis occur, and the spermatids are remodeled into spermatozoa while contained in recesses formed by invaginations of the Sertoli cell plasma membranes. When sperm formation is completed, the Sertoli cell cytoplasm around the sperm retracts, and the sperm is released into the lumen to be bathed by the luminal fluid.

Sertoli cells serve as the route by which nutrients reach developing germ cells, and they also secrete most of the fluid of the tubule lumen. This fluid has a highly characteristic ionic composition and contains androgen-binding protein, which binds testosterone secreted by the Leydig cells and crossing the blood-testis barrier to enter the tubule.

The Sertoli cells also have important endocrine functions summarized in Table 18-2 and discussed in later sections.

TRANSPORT OF SPERM

From the seminiferous tubules, the sperm pass through the rete testis and efferent ductules into the epididymis and thence into the vas deferens. The vas deferens and the portion of the epididymis closest to it serve as a storage reservoir for sperm.

Movement of the sperm as far as the epididymis results from the pressure created both by the continuous formation of fluid by the Sertoli cells back in the seminiferous tubules and by peristalsis of the tubules. The

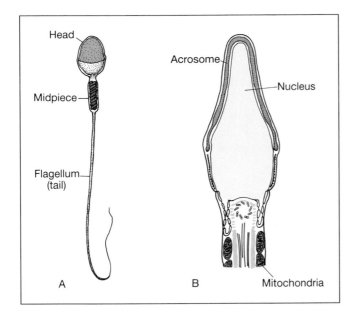

FIGURE 18-7 (A) Diagram of a human mature sperm. (B) A close-up of the head. The acrosome contains enzymes required for fertilization of the ovum.

sperm themselves are nonmotile at this time. During passage through the epididymis, which takes about 12 days, the sperm undergo a final maturation process that consists of changes in size, shape, and metabolic properties.

FIGURE 18-8 Relation of the Sertoli cells and germ cells. The Sertoli cells form a ring around the entire tubule. For convenience of presentation, the various stages of spermatogenesis are shown as though the germ cells move down a line of adjacent Sertoli cells. In reality, all stages beginning with any given spermatogonium take place between the same two Sertoli cells. Spermatogonia (A and B) are found only in the basal compartment (between the tight junctions of the Sertoli cells and the basement membrane of the tubule). After several mitotic cycles (A to B), the spermatogonia (B) give rise to primary spermatocytes (C). Each of the latter crosses a tight junction, enlarges (D), and divides into two secondary spermatocytes (E), which divide into spermatids (F), which differentiate into spermatozoa (G). This last step involves loss of cytoplasm by the spermatids. (*Adapted from Tung.*)

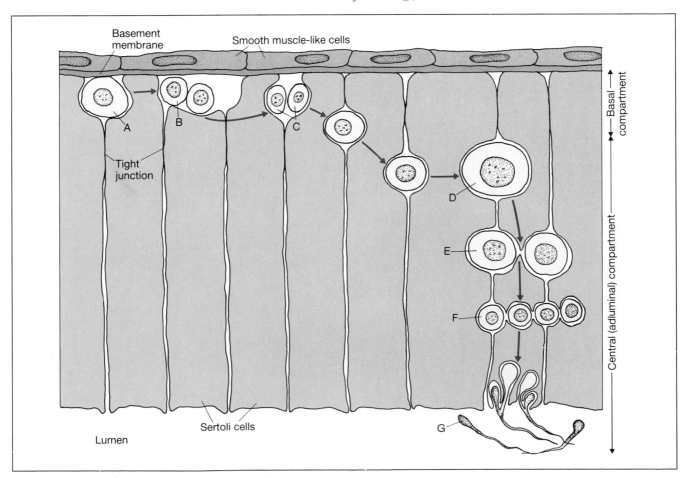

TABLE 18-2 FUNCTIONS OF SERTOLI CELLS

1. Provide blood-testis barrier to chemicals
2. Nourish developing sperm
3. Secrete luminal fluid, including androgen-binding protein
4. Are stimulated by testosterone and FSH to secrete chemical messengers that stimulate sperm production and maturation
5. Secrete the protein hormone inhibin, which inhibits FSH secretion
6. Phagocytize defective sperm

Another event that occurs during passage through the epididymis is a hundred-fold concentration of the sperm by fluid absorption from the lumen of the epididymis. Therefore, as the sperm pass from the end of the epididymis into the vas deferens, they are a densely packed mass whose transport is no longer a result of fluid movement but is due to peristaltic contractions of the smooth muscle in the epididymis and vas deferens. The absence

of a large quantity of fluid accounts for the fact that **vasectomy**—surgical tieing-off and removal of a segment of each vas deferens—does not cause the accumulation of much fluid behind the tied-off point. The sperm do build up, however, and these are removed mainly by phagocytosis. Vasectomy has no effect on testosterone secretion.

The next step in sperm transport is ejaculation, usually preceded by erection, which permits entry of the penis into the vagina.

Erection

The penis becoming rigid—**erection**—is a vascular phenomenon that can be understood by the structure of the penis. This organ consists almost entirely of three cylindrical vascular cords running its entire length. Normally the blood vessels supplying the cords are constricted so that the cords contain little blood and the penis is flaccid. During sexual excitation, the inflow vessels dilate, the cords become engorged with blood at high pressure, and the penis becomes rigid. Moreover, as the cords expand, the veins emptying them are passively compressed, thus contributing to the engorgement. This entire process occurs rapidly, complete erection sometimes taking only 5 to 10 s.

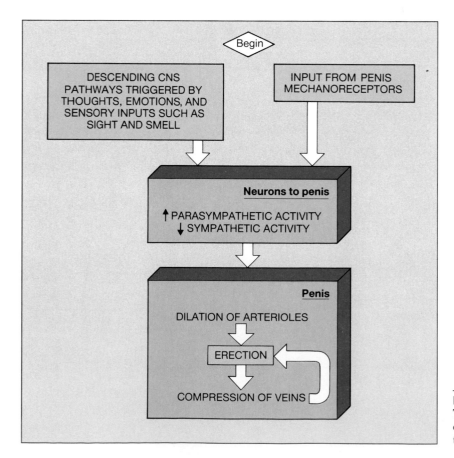

FIGURE 18-9 Reflex pathways for erection. The reflex can be initiated by mechanoreceptors in the penis and/or by input from the brain.

The vascular dilation is accomplished by stimulation of the parasympathetic nerves and inhibition of the sympathetic nerves to the arterioles of the penis (Figure 18-9). This is one of the few cases of direct parasympathetic control over high-resistance blood vessels.

Which receptors and afferent pathway initiate these reflexes? The primary tactile input comes from highly sensitive mechanoreceptors in the genital region, particularly in the head of the penis. The afferent fibers carrying the impulses synapse in the lower spinal cord on interneurons, which trigger the efferent outflow.

It must be stressed, however, that higher brain centers, via descending pathways, may exert profound stimulatory or inhibitory effects upon the autonomic neurons to the arterioles of the penis. Thus, mechanical stimuli from areas other than the penis as well as thoughts, emotions, sight, and odors, can induce erection in the complete absence of penile stimulation. Conversely, failure of erection (**impotence**) may frequently be due to psychological factors acting through descending pathways.

The ability of alcohol to inhibit erection is probably due to its effects on higher brain centers.

Ejaculation

The discharge of semen from the penis—**ejaculation**—is also basically a spinal reflex, the afferent pathways from penile mechanoreceptors being identical to those described for erection. When the level of stimulation produces sufficient summation of synaptic potentials, there is elicited a patterned automatic sequence of efferent discharge that can be divided into two phases: (1) The smooth muscles of the epididymis, ductus deferens, ejaculatory ducts, prostate, and seminal vesicles contract as a result of sympathetic stimulation, emptying the sperm and glandular secretions into the urethra (**emission**); and (2) the semen (average volume = 3 mL, containing 300 million sperm) is then expelled from the urethra by a series of rapid contractions of the urethral smooth muscle as well as the skeletal muscle at the base of the penis.

During ejaculation, the sphincter at the base of the bladder is closed so that sperm cannot enter the bladder nor can urine be expelled from it.

The rhythmical muscular contractions that occur during ejaculation are associated with intense pleasure and many systemic physiological changes, the entire event being termed an **orgasm**. A marked skeletal muscle contraction occurs throughout the body, and there is a large increase in heart rate and blood pressure. This is followed by the rapid onset of muscular and psychological relaxation.

Once ejaculation has occurred, there is a latent period during which a second erection is not possible. The latent period is quite variable but may last from minutes to hours.

As is true of erection, premature ejaculation, failure to ejaculate, or lack of generalized orgasm at the time of ejaculation can be the result of influence by higher brain centers.

HORMONAL CONTROL OF MALE REPRODUCTIVE FUNCTIONS

As mentioned earlier, male reproductive function is controlled by a series of hormones: (1) A single gonadotropin releasing hormone[4] from the hypothalamus stimulates the secretion of both gonadotropins—follicle-stimulating hormone and luteinizing hormone—from the anterior pituitary; (2) FSH and LH stimulate the gonads, the result being the production of mature sperm and the secretion of testosterone; and (3) testosterone exerts a wide variety of effects both on the reproductive system, including stimulation of spermatogenesis, and on other organs or tissues.

Each link in this hormonal chain is essential, for malfunction of either the hypothalamus or the anterior pituitary can result in failure of testosterone secretion and spermatogenesis as surely as if the gonads themselves were diseased.

GnRH, FSH, and LH

Recall from Chapter 10 that the hypothalamic cells that produce releasing hormones are neurons that secrete the hormones into the hypothalamo-pituitary portal vessels as a result of action potentials generated in the neurons. Thus, GnRH is a neurohormone. In a normal adult man, the GnRH-secreting neuroendocrine cells are thought to fire a burst of action potentials approximately every 2 h, secreting GnRH at these times, with virtually no secretion in between. It is not yet known what underlies the rhythmical nature of this so-called pulse generator. This pattern of GnRH secretion is important because the anterior pituitary will not respond to GnRH if its plasma concentration remains constant over time. The reason is that down-regulation of GnRH receptors occurs under such circumstances. Accordingly, when exogenous GnRH is administered to patients, it must be given episodically. As we shall see, both the amplitude and the frequency of the GnRH pulses are under physiological control in both men and women.

---

[4]There is some evidence that two other hypothalamic substances can control the gonadotropins, one that selectively releases LH and one that selectively inhibits release of LH. However, their physiological roles, if any, remain conjectural.

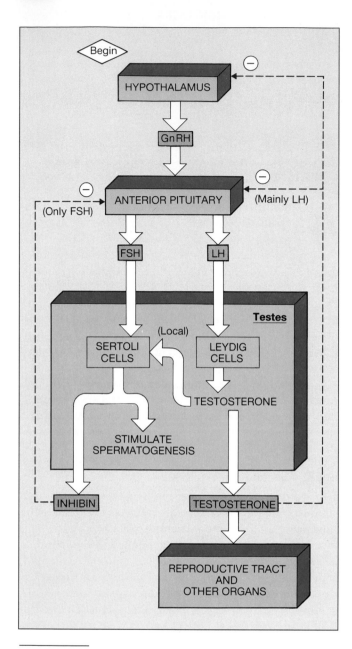

FIGURE 18-10 Summary of hormonal control of male reproductive function. GnRH reaches the anterior pituitary via the hypothalamo-pituitary portal vessels. Note that FSH acts only on the Sertoli cells, and its secretion is inhibited by inhibin, a protein hormone secreted by the Sertoli cells. LH acts only on the Leydig cells, and its secretion is inhibited by testosterone, the hormone secreted by the Leydig cells. Testosterone, acting locally, is essential for spermatogenesis.

The GnRH reaching the anterior pituitary during each periodic pulse triggers the release from the anterior pituitary of both LH and FSH, although not necessarily in equal amounts, as we shall see. Accordingly, systemic plasma concentrations of FSH and LH also show rhythmical episodic changes—rapid increases during the pulse followed by slow decreases over the next 90 min or so as the hormones are slowly removed from the plasma.

There is a clear separation of actions of FSH and LH within the testes (Figure 18-10). FSH acts on the Sertoli cells to stimulate spermatogenesis and other Sertoli cell functions, whereas LH acts on the Leydig cells to stimulate testosterone secretion. It must be emphasized that, as shown in Figure 18-10, despite the absence of any *direct* effect of LH on spermatogenesis, this hormone exerts an essential *indirect* effect because it stimulates testosterone secretion and testosterone is required for spermatogenesis.

Exactly how the Sertoli cells, in response to FSH, stimulate spermatogenesis is not clear, but they are known to secrete at least one paracrine chemical messenger, called spermiogenesis growth factor, that influences germ-cell division and differentiation in the seminiferous tubules.

The last components of the hypothalamo-pituitary control of male reproduction that remain to be discussed are the negative feedbacks exerted by testicular hormones. That such inhibition exists is evidenced by the fact that castration (surgical removal of the gonads) results in marked increases in the secretion of both LH and FSH.

Testosterone inhibits LH secretion in two ways (Figure 18-10): (1) It acts on the hypothalamus to decrease the frequency of GnRH bursts, thereby resulting in less GnRH reaching the pituitary over any given period of time; and (2) it acts on the anterior pituitary to cause less LH secretion, but not less FSH secretion, in response to any given level of GnRH. This direct action on the pituitary helps explain why testosterone inhibits LH secretion so much more than it does FSH secretion.

How does the presence of functioning testes reduce FSH secretion, if not mainly via testosterone? The inhibitory signal, exerted directly on the anterior pituitary, is the protein hormone **inhibin** secreted by the Sertoli cells (Figure 18-10). That the Sertoli cells are the source of feedback inhibition of FSH secretion makes sense since, as pointed out above, the facilitatory effect of FSH on spermatogenesis is exerted via the Sertoli cells. Thus, these cells are in all ways the link between FSH and spermatogenesis.

Despite these complexities, one should not lose sight of the fact that the total amounts of GnRH, LH, FSH, and testosterone secreted and of sperm produced are relatively constant from day to day in the adult male.[5] This is completely different from the large cyclical swings of activity so characteristic of female reproductive processes.

Testosterone

Only the testes produce significant amounts of testosterone. As pointed out in Chapter 10, other steroids with actions similar to those of testosterone are produced by the adrenal cortex and are, along with testosterone, collectively known as **androgens**. The adrenally produced androgens, however, are much less potent than testosterone and are unable to maintain testosterone-dependent functions should testosterone secretion be decreased or eliminated by disease or castration. Accordingly, the only androgen we shall discuss in our description of male reproductive function is testosterone. The effects of testosterone are summarized in Table 18-3 and described below.

Spermatogenesis.

To reiterate, FSH, acting via Sertoli cells, is required for normal spermatogenesis. So are adequate amounts of testosterone, and sterility is an invariable result of testosterone deficiency.[6] Recall that the Leydig cells, which secrete testosterone, lie in the interstitial spaces between the seminiferous tubules. The stimulatory effects of testosterone on spermatogenesis are exerted locally by the hormone, acting as a paracrine, in this case, moving from the Leydig cells into the seminiferous tubules. There, testosterone enters Sertoli cells, and it is via these cells that it facilitates spermatogenesis.

It must be emphasized that although testosterone is required for spermatogenesis, testosterone production does not depend upon spermatogenesis. In other words, testosterone deficiency produces sterility by interrupting spermatogenesis, but interference with the function of the seminiferous tubules does not alter normal testosterone production by the Leydig cells.

---

[5] However, this is not to say that the interplay between GnRH, LH, FSH, testosterone, and inhibin results in absolutely unchangeable levels of these hormones. The GnRH-secreting cells in the hypothalamus receive much synaptic input from other neurons, some excitatory and some inhibitory. This input can cause changes in the frequency at which the GnRH pulses occur and, hence, changes in the average rates at which FSH, LH, and testosterone are secreted. For example, watching sexually arousing movies causes a marked increase in the secretion of all these hormones.

[6] It is an unsettled question whether the hormonal requirements for maintenance of spermatogenesis in adult men differ from those needed for its initiation at puberty. It is possible that once FSH and testosterone, acting together, have initiated spermatogenesis at puberty, only testosterone is required for maintaining it. This question has important implications for the development of a male contraceptive.

TABLE 18-3   EFFECTS OF TESTOSTERONE IN THE MALE
1. Is essential for spermatogenesis (acts via Sertoli cells)
2. Induces differentiation of male accessory reproductive organs and maintains their function
3. Induces male secondary sex characteristics
4. Stimulates protein anabolism, bone growth, and cessation of bone growth
5. Maintains sex drive and (?) enhances aggressive behavior
6. Decreases GnRH secretion via an action on the hypothalamus and therefore decreases LH and FSH secretion.
7. Directly inhibits LH secretion, via an action on the anterior pituitary

Accessory reproductive organs.

The morphology and function of the entire male duct system, glands, and penis depend upon testosterone. Following castration in the adult, all the accessory reproductive organs decrease in size, the glands markedly reduce their secretion rates, and the smooth-muscle activity of the ducts is diminished. Erection and ejaculation may be deficient. These defects disappear upon the administration of testosterone.

Secondary sex characteristics and growth.

Virtually all the male secondary sex characteristics are testosterone-dependent. For example, a male castrated before puberty does not develop a beard or either underarm or pubic hair. Other testosterone-dependent secondary sexual characteristics are deepening of the voice resulting from growth of the larynx, thick secretion of the skin oil glands (this predisposes to acne), and the masculine pattern of fat distribution. Testosterone also stimulates muscle and bone growth but ultimately shuts off the latter, as described in Chapter 17, by causing closure of the bones' epiphyseal plates. Interestingly, testosterone is necessary for expression of the gene that determines baldness.

Behavior.

Testosterone is essential in males for the development of sex drive at puberty. It also plays an important role in maintaining sex drive in the adult male, although men often remain sexually active, albeit usually at a reduced level, for years after castration.

A controversial question is whether testosterone influences human behavior in addition to sex, that is, are there any other inherent male-female behavioral differences, or are any observed differences all socially conditioned? It has proven very difficult to answer such questions with respect to human beings, but there is little doubt that behavioral differences based on gender do exist in other mammals. For example, aggression is clearly greater in males and is testosterone-dependent.

Mechanism of action. Like other steroid hormones, testosterone crosses plasma membranes readily and, within target cells, combines with specific receptors for it. In the cell's nucleus it influences the transcription of certain genes into messenger RNA. The result is a change in the rate at which the target cell synthesizes the proteins for which these genes code. It is the changes in the concentrations of these proteins that underlie the cell's overall response to the hormone. For example, testosterone induces in the prostate gland increased synthesis of enzymes that catalyze the formation of the gland's secretions.

In Chapter 10 we mentioned that hormones sometimes must undergo transformation in their target cells in order to be most effective, and this is true of testosterone in many of its target cells, particularly those of the male reproductive tract. In these cells, after its entry into the cytoplasm, testosterone undergoes an enzyme-mediated conversion to another steroid, **dihydrotestosterone**, and it is mainly this molecule that then combines with androgen receptors and induces effects.

Quite startling is the fact that, in still other target cells, notably neurons in certain areas of the brain, testosterone is transformed not to dihydrotestosterone but to the estrogen, estradiol, which then combines with estrogen receptors to exert its effect. Thus, a male sex hormone must first be transformed to a female sex hormone to act on these neurons.

Prolactin

Another anterior pituitary hormone, **prolactin**, has significant functions in men (again, as we shall see, its name reflects its long-recognized major function in women). It potentiates the stimulatory effects of LH on Leydig cells and of testosterone on many of its target cells.

SECTION B
FEMALE REPRODUCTIVE PHYSIOLOGY

ANATOMY

The female reproductive system includes the two ovaries and the female reproductive tract—two uterine tubes, a uterus, and a vagina. These structures are also termed the **female internal genitalia** (Figure 18-11A and B). In the female, unlike the male, the urinary and reproductive duct systems are entirely separate from each other.

The ovaries are almond-sized organs in the upper pelvic cavity, one on each side of the uterus. The ends of the **uterine tubes** (also known as oviducts or fallopian tubes) are not directly attached to the ovaries but open into the abdominal cavity close to them. The opening of each uterine tube is funnel-shaped and surrounded by long, finger-like projections (the fimbriae) lined with ciliated epithelium. The other ends of the uterine tubes are attached to the uterus and empty directly into its cavity. The **uterus** is a hollow thick-walled muscular organ lying between the urinary bladder and rectum. It is the source of bleeding during menstruation and it houses the fetus during pregnancy. The lower portion of the uterus is the **cervix**. A small opening in the cervix leads to the **vagina**, the canal leading from the uterus to the outside.

The **female external genitalia** (Figure 18-12) include the mons pubis, labia majora and minora, clitoris, vestibule of the vagina, and vestibular glands. The term **vulva** is another name for all these structures. The mons pubis is the rounded fatty prominence over the junction of the pubic bones. The labia majora, the female analogue of the scrotum,[7] are two prominent skin folds that form the outer lips of the vulva. The labia minora are small skin folds lying between the labia majora. They surround the urethral and vaginal openings, and the area thus enclosed is the vestibule, into which the vestibular glands empty. The vaginal opening lies behind that of the urethra. Partially overlying the vaginal opening is a thin fold of mucous membrane, the hymen. The **clitoris**, the female analogue of the penis, is an erectile structure located at the top of the vulva.

Unlike the continuous sperm production of the male, the maturation and release of the female gamete, the ovum, is cyclic. This cyclic pattern is also true for the

---

[7]The term "analogue" or "homologue" used frequently in this chapter does not imply that one structure is dominant and its analogue somehow an inferior imitation. It simply means that the two structures are derived embryologically from the same source and/or have similar functions.

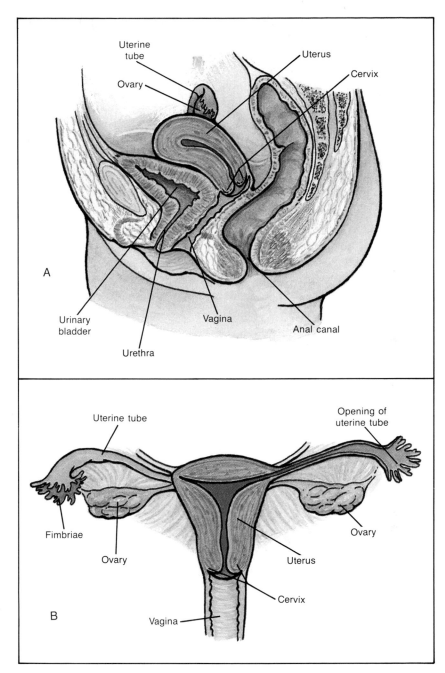

FIGURE 18-11 Female reproductive system. (A) Section through a female pelvis. (B) Diagram showing the continuity between the organs of the reproductive duct system—uterine tubes, uterus, and vagina.

function and structure of virtually the entire female reproductive system. In human beings, these cycles are called **menstrual cycles**. Their phases can be named for events occurring in either the uterus, the source of menstrual bleeding (**menstruation**), or in the ovaries. We shall first describe the ovarian events because changes in hormone secretion by the ovaries completely underlie the uterine events that ultimately cause bleeding. The length of a menstrual cycle varies considerably from woman to woman, and in any particular woman, averaging about 28 days. The first day of bleeding is termed day 1.

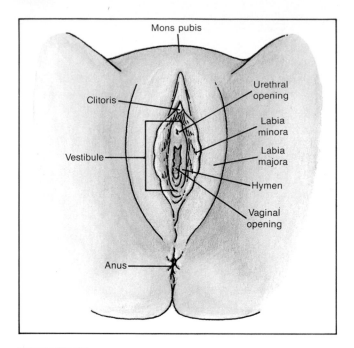

FIGURE 18-12 Female external genitalia.

OVARIAN FUNCTION

The ovary, like the testis, serves a dual purpose: (1) **oogenesis**, the production of gametes—the ova; and (2) secretion of the female sex hormones, estrogen and progesterone. Prior to ovulation, the gametogenic and endocrine functions take place in a specific structure—the follicle—and after ovulation the follicle differentiates into a corpus luteum, which has only an endocrine function. For comparison, recall that in the testes the production of gametes and the secretion of sex steroids take place in different compartments—in the seminiferous tubules and in the Leydig cells, respectively.

This section will simply describe the events that occur in the ovary during a menstrual cycle, and the next section will replay these events, adding the controls.

Oogenesis

At birth, a female's ovaries contain an estimated total of 1 million ova, and no new ones appear after birth. (As will be described in a subsequent section, the ovum, like the sperm, is referred to by different names at different stages of development, but for simplicity, we shall often simply use the term "ovum.") Thus, in marked contrast to the male, the newborn female already has all the germ cells she will ever have. Only a few, perhaps 400, are

destined to reach full maturity during her active reproductive life. All the others degenerate at some point in their development so that few remain by the time a woman reaches approximately 50 years of age. One result of this developmental pattern is that the ova that are released near age 50 are 30 to 35 years older than those ovulated just after puberty. It has been suggested that certain defects more common among children of older women are the result of aging changes in the ovum.

During early fetal development, the primitive germ cells, or **oogonia** (singular *oogonium*), a term analogous to spermatogonia in the male, undergo numerous miotic divisions (Figure 18-13). At some point in fetal life, the oogonia cease dividing, and from this point on no new germ cells are generated. In the fetus, all the oogonia then develop into **primary oocytes** (analogous to primary spermatocytes), which begin a first meiotic division by replicating their DNA. They do not, however, complete the division. Accordingly, all the germ cells present at birth are primary oocytes containing 46 chromosomes, each with two sister chromatids. The cells are said to be in a state of meiotic arrest.

This state continues until puberty and the onset of renewed activity in the ovaries. Indeed, the first meiotic division is completed only just before an oocyte is about to be released from the ovary. This division is analogous to the division of the primary spermatocyte, and each daughter cell receives 23 chromosomes, each with two chromatids. In this division, however, one of the two daughter cells, the **secondary oocyte**, retains virtually all the cytoplasm. The other, termed the **first polar body**, is very small and adheres to the secondary oocyte. Thus, the primary oocyte, which is already as large as the mature ovum will be, passes on to the secondary oocyte half of its chromosomes but almost all of its nutrient-rich cytoplasm.

The second meiotic division occurs, in a uterine tube, after ovulation but only if the secondary oocyte is fertilized, that is, penetrated by a sperm. As a result of this division the daughter cells each receive 23 chromosomes, each with a single chromatid. Once again, one daughter cell, now termed a **mature ovum** (note that because of the timing, a "mature ovum" can also be termed a "fertilized ovum"), retains nearly all the cytoplasm, whereas the other, termed the **second polar body**, is very small and nonfunctional. Interestingly, the first polar body also undergoes a second meiotic division at the same time the secondary oocyte does. However, as far as is known, neither of these resulting cells nor the second polar body is fertilized, and they all eventually disintegrate.

The net result of oogenesis is that each primary oocyte produces only one fertilizable ovum (Figure 18-13).

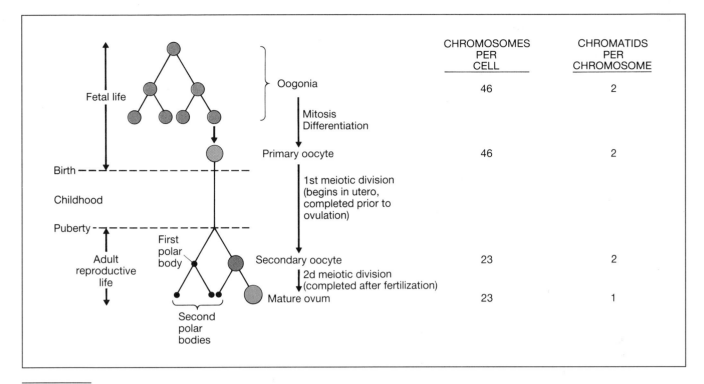

		CHROMOSOMES PER CELL	CHROMATIDS PER CHROMOSOME
Oogonia		46	2
Mitosis Differentiation			
Primary oocyte		46	2
1st meiotic division (begins in utero, completed prior to ovulation)			
Secondary oocyte		23	2
2d meiotic division (completed after fertilization)			
Mature ovum		23	1

FIGURE 18-13 Summary of oogenesis. Compare with the male pattern of Figure 18-6. The secondary oocyte is ovulated and does not complete its meiotic division unless it is penetrated (fertilized) by a sperm. Thus, it is a semantic oddity that the ovum is not termed "mature" until after fertilization occurs. Note that each primary oocyte yields only one secondary oocyte, which can yield only one mature ovum.

In contrast, each primary spermatocyte produces four viable spermatozoa.

Follicle Growth

Throughout their life in the ovaries, the oocytes exist in structures known as **follicles**. In the adult ovary, most of these follicles are either **primordial follicles** or, the next stage of development, **primary follicles**. Both of these follicle types consist of one primary oocyte surrounded by a single layer of cells called **granulosa cells**, the primary follicles differing from the primordial follicles only in the shape of the granulosa cells.

Further development from the primary follicle stage (Figure 18-14) is characterized by an increase in size of the oocyte and a proliferation of the granulosa cells. The oocyte becomes separated from the granulosa cells by a thick layer of material, the **zona pellucida**, secreted by the granulosa cells and oocyte. Despite the presence of the zona pellucida, however, the inner layer of granulosa cells remains intimately associated with the oocyte by means of cytoplasmic processes that traverse the zona pellucida and form gap junctions with the oocyte.

Through these gap junctions, nutrients and chemical messengers are passed to the oocyte.

The follicle grows as new cell layers are formed, not only from mitosis of the original granulosa cells but as a result of the differentiation of specialized ovarian connective-tissue cells surrounding the granulosa cells. Thus, the follicle now consists of the oocyte, layers of granulosa cells surrounding the oocyte, and outer layers of cells known as the **theca**.

When the follicle reaches a certain diameter, a fluid-filled space, the **antrum**, begins to form in the midst of the granulosa cells as a result of fluid they secrete. By the time the antrum begins to form, the oocyte has reached full size. From this point on, most of the follicle's enlargement is due to the expanding antrum.

The progression from primordial follicle to primary follicle to the preantral and small-antral stages occurs continuously in the adult ovary, so that at any moment there are always present a sizable number of preantral follicles and a few small-antral follicles. Then, at the beginning of each menstrual cycle, several of these follicles begin to develop into larger antral follicles. About 1

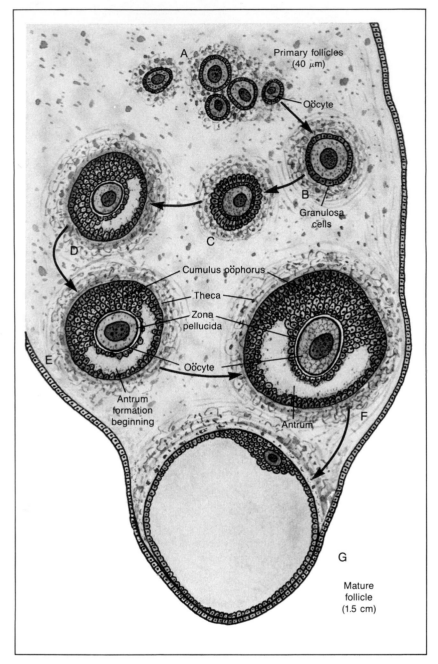

A — Primary follicles (40 μm)

Oöcyte

Granulosa cells

B

C

D

Cumulus oöphorus

Theca

Zona pellucida

Oöcyte

Antrum

E

Antrum formation beginning

F

G

Mature follicle (1.5 cm)

FIGURE 18-14 Development of a human oocyte and ovarian follicle.

week into the cycle, a selection process occurs: Only the largest follicle, the **dominant follicle**, continues to develop, and the others that had begun to enlarge during that week degenerate, a process termed atresia. In other words, once a follicle has matured beyond a certain point, it is committed to go on and ovulate its oocyte or, more commonly, to become atretic. However, on occasion (1 to 2 percent of all cycles), two or more follicles reach maturity, and more than one oocyte may be ovu-

lated. This is the commonest cause of multiple births. In such cases the siblings are fraternal, not identical.

Ultimately, the oocyte is surrounded by granulosa cells that project from the inner follicle wall into the antrum and are termed the cumulus oophorous. The mature follicle (also termed a Graafian follicle) becomes so large (diameter about 1.5 cm) that it balloons out on the surface of the ovary. **Ovulation** occurs when the walls of the follicle and ovary at this site rupture and the oocyte, surrounded by its tightly adhering zona pellucida and several layers of granulosa cells is carried out of the ovary and onto the ovarian surface by the antral fluid. All this happens on approximately day 14 of the cycle.

Formation of the Corpus Luteum

After the mature follicle discharges its antral fluid and oocyte, its remnant in the ovary collapses around the antrum and undergoes a rapid transformation. The cells enlarge greatly, newly formed capillaries move into the interior, and the entire glandlike structure is now known as the **corpus luteum**. If the discharged oocyte, now in the uterine tube, is not fertilized, the corpus luteum reaches its maximum development within approximately 10 days and then rapidly degenerates and is catabolized. As we shall see, it is the disappearance of the corpus luteum that leads to menstruation and the beginning of a new menstrual cycle.

In terms of ovarian function, therefore, the menstrual cycle may be divided into two phases approximately equal in length and separated by ovulation (Figure 18-15): (1) the **follicular phase**, during which a single mature follicle and secondary oocyte develop; and (2) the **luteal phase**, beginning after ovulation and lasting until the demise of the corpus luteum.

Ovarian Hormones

Just as "androgen" refers to a group of hormones with similar actions, the term "estrogen" denotes not a single specific hormone but rather a group of steroid hormones that have similar effects on the female reproductive tract. These include **estradiol**, estrone, and estriol, of which the first is the major estrogen secreted by the ovaries. It is common to refer to any of them simply as estrogen, and we shall follow this practice. Estrogen is secreted by the granulosa cells and, following ovulation, by the corpus luteum.

Progesterone, another ovarian steroid hormone, is secreted in very small amounts by the granulosa cells just prior to ovulation, but its major source is the corpus luteum.

Two peptide hormones are also produced by the ovaries, inhibin by the granulosa cells and relaxin by the corpus luteum.

The actions of all these ovarian hormones will be described later in this chapter when we take up the control of reproduction.

CONTROL OF OVARIAN FUNCTION

The basic factors controlling ovarian function are analogous to the controls described for testicular function in that they constitute a hormonal series made up of GnRH, the anterior pituitary gonadotropins FSH and LH, and gonadal sex hormones—estrogen and progesterone in the female.[8] As in the male, the entire sequence of controls depends upon the secretion of GnRH from hypothalamic neuroendocrine cells in episodic pulses. In the female, however, the frequency of these pulses and hence the total amount of GnRH secreted during a 24-h period changes in a patterned manner over the course of the menstrual cycle. So does the responsiveness of both the anterior pituitary to GnRH and of the ovaries to FSH and LH.

For purposes of orientation, let us look first, in Figure 18-16, at the patterns of plasma hormone concentrations in arterial plasma during a normal menstrual cycle. GnRH is not shown since its important concentration is

---

[8] It is now recognized that these hormones are not the exclusive regulators of ovarian function. Several other hormones and growth factors (for example, insulin and insulin-like growth factors) may play important but still poorly understood roles. Some of the substances are produced by granulosa cells and may function as autocrines.

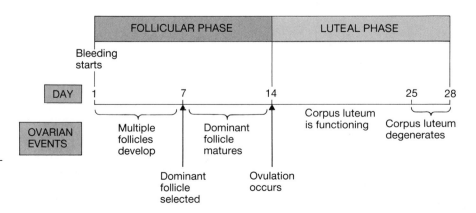

FIGURE 18-15 Summary of ovarian events during a menstrual cycle. The first day of the cycle, the onset of menstrual bleeding, is named for a uterine event even when ovarian events are used to denote the cycle phases.

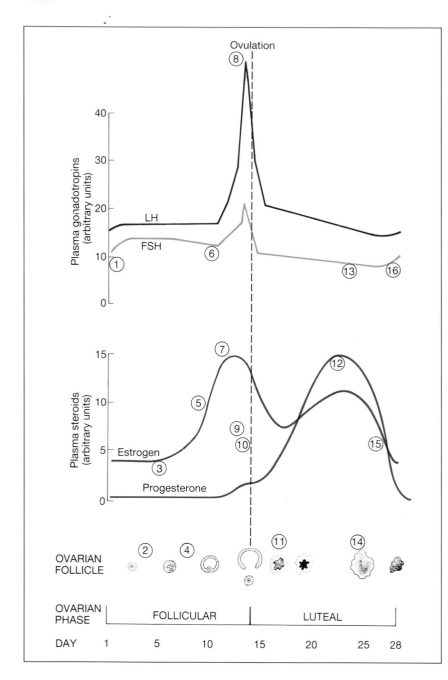

Ovulation
8

FIGURE 18-16 Summary of plasma hormone concentrations and ovarian events during the menstrual cycle. The events marked by the circled numbers are described later in the text and are listed in this legend to provide a summary, the arrows in the legend denoting causality. There are no arrows preceding numbers 4 and 14 because the causes of these events are not yet known. (1) FSH and LH secretion increase (because plasma estrogen concentration is low and exerting little negative feedback). → (2) Multiple follicles begin to enlarge and secrete estrogen. → (3) Plasma estrogen concentration remains stable and then begins to rise. (4) One follicle becomes dominant (? cause) and secretes very large amounts of estrogen. → (5) Plasma estrogen level increases markedly. → (6) FSH secretion and plasma FSH concentration decrease, and (7) plasma estrogen reaches levels high enough to exert a positive feedback on gonadotropin secretion. → (8) An LH surge is triggered. → (9) The oocyte completes its first meiotic division and cytoplasmic maturation while the follicle secretes less estrogen accompanied by some progesterone, (10) ovulation occurs, and (11) the corpus luteum forms and begins to secrete large amounts of both estrogen and progesterone. → (12) Plasma estrogen and progesterone increase. → (13) FSH and LH secretion is inhibited, and their plasma concentrations progressively fall. (14) The corpus luteum begins to degenerate (? cause) and decrease its hormone secretion. → (15) Plasma estrogen and progesterone concentrations fall. → (16) FSH and LH secretion begins to increase, and a new cycle begins.

in the hypothalamo-pituitary blood, not in the arterial blood, and such measurements are not available for humans.

Note that FSH is slightly elevated in the early part of the follicular phase and then steadily decreases throughout the remainder of the cycle except for a small midcycle peak. LH is constant during most of the follicular phase but then shows a large midcycle rise—the **LH surge**—peaking approximately 18 h before ovulation, followed by a rapid return to presurge values and then a further slow decline during the luteal phase.

The estrogen pattern is more complex. After remaining fairly low and stable for the first week, it rises rapidly during the second week. Estrogen then starts falling shortly after LH has peaked. This is followed by a second rise due to secretion by the corpus luteum, and finally, a rapid decline during the last days of the cycle. The progesterone pattern is simplest of all: Virtually no progesterone is secreted by the ovaries during the follicular phase until just prior to ovulation, but very soon after ovulation, the developing corpus luteum begins to secrete large amounts of progesterone, and from this point the progesterone pattern is similar to that for estrogen.

The following discussion will explain how these changes are all interrelated to yield a self-cycling pattern. The numbers in Figure 18-16 are keyed to the text.

Follicle Development and Estrogen Secretion during the Early and Middle Follicular Phase

As mentioned earlier, there are always a number of preantral and small-antral follicles in the adult ovary. The cells of such follicles then acquire receptors for FSH and LH, and further development requires stimulation by these hormones. The increase in gonadotropin secretion that occurs as one cycle ends and the next begins (number 1 in Figure 18-16) causes the plasma concentrations of FSH and LH to become high enough to stimulate this further development (2). (The explanation of what causes gonadotropin secretion to increase at this time is best delayed until we discuss the events at the end of a cycle.) During the next week or so, there is a separation of function of FSH and LH on the follicle: FSH acts on the granulosa cells, and LH acts on the theca cells.

FSH stimulates the granulosa cells to multiply and produce estrogen, and it also stimulates formation and enlargement of the antrum. Some of the estrogen produced diffuses into the blood and maintains a relatively stable plasma concentration (3), but much remains in the follicle and, along with FSH, stimulates the granulosa

cells to increase still further their proliferation and their production of estrogen.[9]

The granulosa cells, however, require help to produce estrogen because they lack the ability to produce the androgens that are the precursors for estrogen. This aid is supplied by the theca cells. As shown in Figure 18-17, LH acts upon the theca cells, stimulating them not only to proliferate but to synthesize androgens. The androgens diffuse into the granulosa cells and are converted to estrogen. Thus, the secretion of estrogen by the granulosa cells requires the interplay of both types of follicle cells and both pituitary gonadotropins.

At this point it is worthwhile to emphasize the similarities the two follicle cell types bear, during this period of the cycle, to cells of the testes: The granulosa cell is analogous to the Sertoli cell in that it controls the microenvironment in which the germ cell develops and matures, and it is stimulated by both FSH and the major gonadal sex hormone. The theca cell is analagous to the Leydig cell in that it produces androgens and is stimulated to do so by LH.

By the beginning of the second week, one follicle has become dominant (4 in Figure 18-16), and it becomes the major producer of estrogen. So much estrogen is secreted that the plasma concentration of this steroid begins to rise markedly (5).

---

[9]The evidence from experimental studies has documented that granulosa cells are under the combined control of FSH and estrogen in several species. However, very recent studies in monkeys were unable to find estrogen receptors in these cells and so have raised the question as to whether estrogen helps control granulosa cells in this species. Definitive evidence is not yet available for humans concerning this very important question, and we have chosen to present the time-honored concept of dual control.

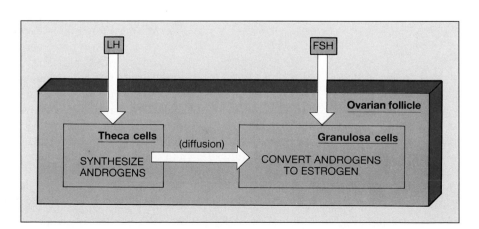

FIGURE 18-17 Control of estrogen secretion during the early and middle follicular phase.

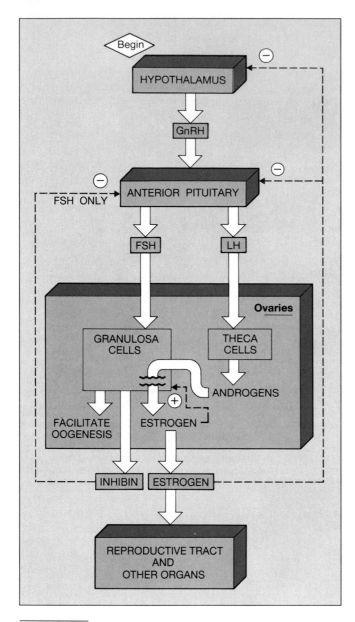

FIGURE 18-18 Summary of hormonal control of ovarian function during the early and middle follicular phases. Compare with the analogous pattern of the male (Figure 18-10). Inhibin is a protein hormone that inhibits FSH secretion.

What about hormonal feedbacks throughout this part of the cycle? For about the first 11 days of the cycle, plasma estrogen exerts a negative-feedback inhibition over the secretion of gonadotropins (Table 18-4 and Figure 18-18). At the range of plasma concentrations existing throughout these days, estrogen acts on the anterior pituitary to reduce the amount of FSH and LH secreted in response to any given amount of GnRH. It may also

act on the hypothalamus to decrease the amplitude of GnRH pulses and, hence, the total amount of GnRH secreted over any time period.

Therefore, as expected from this negative feedback, the plasma concentration of FSH begins to fall as the follicular phase continues and estrogen starts to increase (6 in Figure 18-16). The granulosa cells also secrete the protein hormone inhibin, which, as in the male, preferentially inhibits secretion of FSH (Figure 18-18 and Table 18-4).

It is the decrease in FSH during the second week that causes the regression of all enlarging follicles except the dominant one. There is simply not enough FSH to sustain them. Of course, the next question is why the dominant follicle is not affected in the same way by the decrease in FSH. There is presently no agreed-upon explanation for the ability of the dominant follicle to prosper even though plasma FSH is down.[10]

LH Surge and Ovulation

The inhibitory effect estrogen exerts on the pituitary and hypothalamus occurs only when plasma estrogen concentration is relatively *low*, as during the early and middle follicular phases. In contrast, high plasma concentrations of estrogen for 1 to 2 days, as occurs during the estrogen peak of the late follicular phase (7 in Figure 18-16), act upon the pituitary to *enhance* the sensitivity of LH-releasing mechanisms to GnRH (Figure 18-19). The high estrogen may also stimulate *increased* secretion of GnRH by the hypothalamus. These effects are termed the positive-feedback effects of estrogen. The net result is that, as estrogen secretion rises rapidly during the late follicular phase, its blood concentration eventually becomes high enough to cause an outpouring of LH—the LH surge (8 in Figure 18-16). The small increase in the plasma concentration of progesterone that exists during the late follicular phase may facilitate the estrogen-induced LH surge by an effect on the hypothalamus.

The midcycle surge of LH is perhaps the single most decisive event of the entire menstrual cycle, for it induces ovulation. One more important piece of information is needed to understand how it does so. We emphasized in the previous section that, during the early and middle follicular phases, FSH acts on the granulosa cells and LH acts on the theca cells. As the dominant follicle matures, however, this situation changes. LH receptors, induced by FSH with the assistance of estrogen, begin to appear in large numbers on the granulosa cells, and this

---

[10]One hypothesis is that the estrogen secreted by the dominant follicle induces an increased number of FSH receptors in this follicle, thereby increasing the follicle's responsiveness to FSH.

TABLE 18-4 SUMMARY OF MAJOR FEEDBACK EFFECTS OF ESTROGEN, PROGESTERONE, AND INHIBIN

1. Estrogen, in *low* plasma concentrations, causes the anterior pituitary to secrete less FSH and LH in response to GnRH and also may inhibit the hypothalamic neurons that secrete GnRH. Result: negative-feedback inhibition of FSH and LH secretion.

2. Estrogen, in *high* plasma concentrations, causes the anterior pituitary cells to secrete more LH (and FSH) in response to GnRH and also may stimulate the hypothalamic neurons that secrete GnRH. Result: positive-feedback stimulation of the LH surge.

3. The small rise in progesterone that occurs in the late follicular phase may act on the hypothalamus to stimulate GnRH secretion. Result: facilitation of the estrogen-induced LH surge.

4. High plasma concentrations of progesterone, in the presence of estrogen, inhibit the hypothalamic neurons that secrete GnRH. Result: negative-feedback inhibition of FSH and LH secretion and prevention of LH surges.

5. Inhibin acts on the pituitary to inhibit the secretion of FSH.

---

sets the stage for the LH surge to act upon the granulosa cells to cause the events (9 in Figure 8-16), presented in Table 18-5, that culminate in ovulation (10).

The function of the granulosa cells in mediating the effects of the LH surge is the last in the series of these cells' functions described in this chapter. They are all summarized in Table 18-6.

The Corpus Luteum

The LH surge not only induces ovulation by the mature follicle but also triggers the reactions that transform into a corpus luteum the remaining granulosa cells of that follicle in the ovary (11). Even though its plasma concentration becomes relatively low, LH continues to stimulate the corpus luteum throughout the luteal phase.

During its short life in the nonpregnant woman, the corpus luteum secretes large quantities of progesterone and estrogen (12). In the presence of estrogen, the high plasma concentration of progesterone acts on the hypothalamus to suppress the secretion of GnRH by reducing the frequency of GnRH pulses (Table 18-4). The decrease in GnRH secretion results in a decrease in the secretion of the gonadotropins by the pituitary. Accordingly, during the luteal phase of the cycle, plasma concentrations of the gonadotropins, particularly FSH, are very low (13), which explains why no new follicles can develop beyond the preantral or small-antral stage during this second half of the cycle.

FIGURE 18-19 In the late follicular phase, the dominant follicle secretes large amounts of estrogen, which act on the hypothalamus and anterior pituitary to cause an LH surge. The increased plasma LH then triggers both ovulation and corpus luteum formation. These actions of LH are mediated via the granulosa cells.

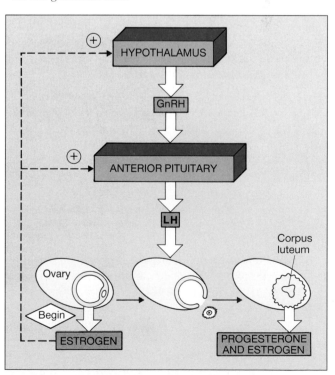

TABLE 18-5 EFFECTS OF THE LH SURGE ON OVARIAN FUNCTION

1. The primary oocyte completes its first meiotic division and undergoes cytoplasmic changes that prepare the ovum for implantation after fertilization. These LH effects on the oocyte are mediated by messengers from the granulosa cells.

2. A marked increase in antrum size and in blood flow to the follicle occurs.

3. The granulosa cells begin secreting progesterone and decrease their secretion of estrogen, which accounts for the midcycle drop in plasma estrogen concentration and the small rise in plasma progesterone just before ovulation.

4. Enzymes and prostaglandins are synthesized that break down the follicular-ovarian membranes. These weakened membranes rupture, allowing the oocyte and its surrounding granulosa cells to be carried out onto the surface of the ovary.

5. The granulosa cells remaining in the ovary after ovulation are transformed into the corpus luteum, which begins to secrete progesterone and estrogen.

Recall from page 619 that the corpus luteum degenerates and disappears within 2 weeks if pregnancy does not occur (14). What causes this? The most likely hypothesis at present is that the corpus luteum "self-destructs" after 10 to 14 days by producing large quantities of a prostaglandin that interferes locally with the corpus luteum's activities.

With degeneration of the corpus luteum, plasma progesterone and estrogen concentrations decrease (15). GnRH, FSH, and LH secretions increase (16 and 1) as a result of being relieved from the inhibiting effects of high concentrations of progesterone, and a new group of follicles is stimulated to mature. The cycle begins anew.

TABLE 18-6 FUNCTIONS OF GRANULOSA CELLS

1. Nourish oocyte

2. Secrete antral fluid

3. Are the site of action for estrogen and FSH in the control of follicle development during early and mid-follicular phases

4. Secrete estrogen from the androgens reaching them from the theca cells

5. Secrete inhibin, which inhibits FSH secretion via an action on the pituitary

6. Are the site of action for LH induction of changes in the oocyte and follicle culminating in ovulation and formation of the corpus luteum

UTERINE CHANGES IN THE MENSTRUAL CYCLE

Because we have been describing ovarian function, we have so far referred to the phases of the menstrual cycle in terms of ovarian events—follicular and luteal phases, separated by ovulation. However, the phases of the menstrual cycle can also be named in terms of uterine events (Figure 18-20). Day 1 is, as noted earlier, the first day of menstrual bleeding, and the entire period of menstruation is known as the **menstrual phase**, which is generally about 3 to 5 days in a typical 28-day cycle. The menstrual flow then ceases, and the glandular epithelium (**endometrium**) lining the uterine cavity begins to thicken as it regenerates. This period of growth, the **proliferative phase**, lasts for the 10 days or so between cessation of menstruation and occurrence of ovulation. Soon after ovulation, the endometrium begins to secrete various substances, and so the part of the menstrual cycle between ovulation and the onset of the next menstruation is called the **secretory phase**.

Thus, the ovarian follicular phase includes the uterine menstrual and proliferative phases, whereas the ovarian luteal phase is the same as the uterine secretory phase.

The uterine changes during a menstrual cycle reflect the effects of varying plasma concentrations of estrogen and progesterone (Figure 18-20). During the proliferative phase, a rising estrogen level stimulates growth of both the endometrium and the underlying uterine smooth muscle (**myometrium**). In addition, it induces the synthesis of receptors for progesterone on endometrial cells. Then, following ovulation and formation of the

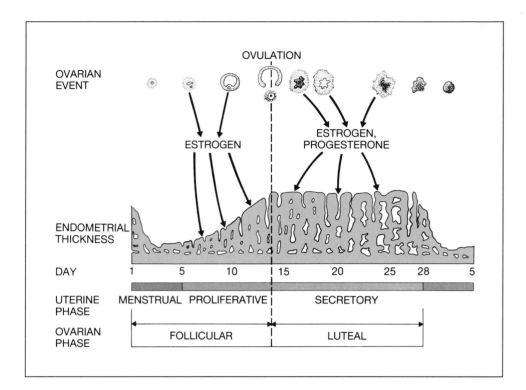

OVULATION

OVARIAN
EVENT

ESTROGEN

ESTROGEN,
PROGESTERONE

ENDOMETRIAL
THICKNESS

DAY   1   5   10   15   20   25   28   5

UTERINE
PHASE   MENSTRUAL PROLIFERATIVE   SECRETORY

OVARIAN
PHASE   FOLLICULAR   LUTEAL

FIGURE 18-20 Relationships between ovarian and uterine changes during the menstrual cycle.

corpus luteum, during the secretory phase, progesterone acts upon this estrogen-primed endometrium to convert it to an actively secreting tissue. Its glands become coiled and filled with glycogen, the blood vessels become more numerous, and various enzymes accumulate in the glands and connective tissue. All these changes make the endometrium a hospitable environment for an embryo.

Progesterone also inhibits myometrial contractions, in large part by opposing the stimulatory actions of estrogen and locally generated prostaglandins. This helps ensure that the embryo, once it arrives in the uterus, will not be swept out by uterine contractions before it can implant in the wall.

Estrogen and progesterone also have important effects on the mucus secreted by the cervix. Under the influence of estrogen alone, this mucus is abundant, clear, and nonviscous. All these characteristics are most pronounced at ovulation and allow sperm deposited in the vagina to move easily through the mucus on their way to the uterus and thence to the uterine tubes. In contrast, progesterone, present in significant concentrations only after ovulation, causes the mucus to become thick and sticky—in essence a "plug" that prevents sperm or bacteria from entering the uterus from the vagina. The antibacterial blockade protects the fetus should conception have occurred. The adaptive value of the block to sperm movement into the uterus during early conception is unclear.

The fall in plasma progesterone and estrogen levels that results from degeneration of the corpus luteum deprives the highly developed endometrium of its hormonal support. The immediate result is profound constriction of the uterine blood vessels, which leads to diminished supply of oxygen and nutrients. Disintegration starts, and the entire lining, except for a thin, deep layer that will regenerate the endometrium in the next cycle, begins to slough. Also, the uterine smooth muscle begins to undergo rhythmical contractions.

Both the vasoconstriction and the uterine contractions are mediated by prostaglandins produced by the endometrium. The major cause of menstrual cramps, **dysmenorrhea**, is overproduction of these prostaglandins, leading to excessive uterine contractions. Prostaglandin's effects on smooth muscle elsewhere in the body also account for the systemic symptoms—nausea, vomiting, and headache—that sometimes accompany the cramps.

After the initial period of vascular constriction, the endometrial arterioles dilate, resulting in hemorrhage through the weakened capillary walls. The menstrual flow consists of this blood mixed with endometrial debris. Typical blood loss per menstrual period is about 50 to 150 mL.

TABLE 18-7	SUMMARY OF THE MENSTRUAL CYCLE
Day(s)	**Major Events**
1–5	Estrogen and progesterone are low because the previous corpus luteum has completely regressed. *Therefore:* (a) Endometrial lining sloughs. (b) Secretion of FSH and LH is released from inhibition, and their plasma concentrations rise. *Therefore:* Several follicles are stimulated to enlarge.
7	Dominant follicle is selected.
7–12	Plasma estrogen rises because of secretion by the dominant follicle. *Therefore:* Endometrium is stimulated to proliferate.
12–13	LH surge is induced by high plasma estrogen. *Therefore:* (a) Oocyte is induced to complete its first meiotic division and undergo cytoplasmic maturation. (b) Follicle is stimulated to secrete lytic enzymes and prostaglandin.
14	Ovulation is caused by follicular enzymes and prostaglandin.
15–25	Corpus luteum forms and, under influence of LH, secretes estrogen and progesterone, so plasma concentrations of these hormones increase. *Therefore:* (a) Secretory endometrium develops. (b) Secretion of FSH and LH is inhibited, lowering their plasma concentrations. *Therefore:* No new follicles develop.
25–28	Corpus luteum degenerates. *Therefore:* Plasma estrogen and progesterone concentrations decrease. *Therefore:* Endometrium begins to slough at conclusion of day 28, and a new cycle begins.

This completes our description of the menstrual cycle, the major events of which are summarized in Table 18-7.

OTHER EFFECTS OF ESTROGEN AND PROGESTERONE

The uterine effects of estrogen and progesterone represent only one set of the many exerted by these hormones, all of which are listed in Table 18-8. The effects of estrogen in the female are analogous to those of testosterone in the male in that estrogen influences the gonads, the accessory sex organs, and the secondary sex characteristics. As described above, estrogen is required for follicle and ovum maturation. Estrogenic stimulation maintains the uterus, uterine tubes, vagina, the external genitalia, and the breasts. Estrogen is also responsible for the female body-hair distribution and the general female body configuration: narrow shoulders, broad hips, and the characteristic female fat deposition in the hips, abdomen and breasts. Estrogen does not have testosterone's anabolic effect on skeletal muscle, but, as described in Chapter 17, it does cause bone growth as well as eventual cessation of growth by closure of the bones' epiphyseal plates.

As is true for testosterone, estrogen acts on the cell nucleus, and its biochemical mechanism of action is at the level of gene transcription.

Progesterone is present in significant amounts only during the luteal phase of the menstrual cycle, and its

TABLE 18-8 EFFECTS OF FEMALE SEX STEROIDS

Estrogen

1. Stimulates ovary and follicle growth.
2. Stimulates growth of smooth muscle and epithelial linings of reproductive tract. In addition:
   a. Uterine tubes: Increases contractions and ciliary activity.
   b. Uterus: Increases myometrial contractions. Stimulates secretion of abundant, clear cervical mucus. Prepares endometrium for progesterone's actions by inducing progesterone receptors.
   c. Vagina: Increases "cornification" (layering of epithelial cells).
3. Stimulates external genitalia growth.
4. Stimulates breast growth, particularly ducts and fat deposition.
5. Stimulates female body configuration development: narrow shoulders, broad hips, female fat distribution.
6. Stimulates a more-fluid sebaceous gland secretion ("antiacne" effect).
7. Stimulates development of female pubic hair pattern (growth of pubic and axillary hair is androgen-stimulated).
8. Stimulates bone growth and ultimate cessation of bone growth (closure of epiphyseal plates); protects against osteoporosis.
9. Vascular effects (deficiency produces "hot flashes").
10. Has feedback effects on hypothalamus and anterior pituitary (Table 18-4).
11. Stimulates retention of fluid.
12. Stimulates prolactin secretion but inhibits prolactin's milk-inducing action on the breasts.
13. ? Protects against atherosclerosis.

Progesterone

1. Stimulates endometrial glands to secrete.
2. Induces thick, sticky cervical mucus.
3. Decreases contractions of uterine tubes and myometrium.
4. Decreases vaginal cornification.
5. Stimulates breast growth, particularly glandular tissue.
6. Inhibits milk-inducing effects of prolactin.
7. Has feedback effects on hypothalamus and anterior pituitary (Table 18-4).

effects are less widespread than those of estrogen, the endometrial changes being the most prominent. Progesterone also exerts important effects on the breasts, the uterine tubes, and the uterine smooth muscle, the significance of which will be described later. Progesterone also causes what is called "cornification"—an increase in layering—of the cells lining the vagina, and the microscopic examination of some of these cells provides an indicator that ovulation has or has not occurred. Note that in this regard, and several others in Table 18-8, progesterone exerts an "antiestrogen effect," probably by inhibiting the synthesis of estrogen receptors.

Another indicator that ovulation has occurred is a small rise (approximately 0.5 C°) in body temperature that usually occurs at this time and persists throughout the luteal phase. This change is probably due to an action of progesterone on temperature regulatory centers in the brain.

In closing this section, brief mention should be made of the **premenstrual syndrome (PMS)** because it is often ascribed to estrogen excess or progesterone deficiency. In fact, however, the cause of this syndrome is not known. PMS is a cluster of physical and psychological symptoms that occur during the week or so preceding menstruation and that are relieved by the onset of menstruation (it is completely distinct from dysmenorrhea).

The symptoms may include painful or swollen breasts, bloating, headache, backache, depression, anxiety, irritability, and other behavioral and motor changes. These symptoms may occur singly or in varying combinations. Estimates of the incidence of this syndrome depend on what criteria are used and vary from as low as 2 to 5 percent of adult women to as high as 30 to 40 percent. Research efforts to understand the pathophysiology of PMS are still at an early stage, and no specific therapies are presently available.

ANDROGENS IN WOMEN

Testosterone is not a uniquely male hormone but is found in very low concentration in the blood of normal women as a result of production by the ovaries and adrenal glands as well as peripheral conversion of adrenal androgens. Of greater importance is the fact that androgens other than testosterone are found in quite significant concentrations in the blood of women. The sites of production are mainly the adrenal glands, which contribute these same androgens in the male.

In contrast to their lack of significance in the male, adrenal androgens do play several important roles in the female, notably maintenance of sexual drive (see below). In several disease states, the female adrenals may secrete abnormally large quantities of androgen, which produce virilism: The female fat distribution disappears, a beard appears along with the male body-hair distribution, the voice lowers in pitch, the skeletal muscle mass enlarges, the clitoris enlarges, and the breasts diminish in size.

FEMALE SEXUAL RESPONSE

The female response to sexual intercourse is characterized by marked vasocongestion and muscular contraction in many areas of the body. For example, increasing sexual excitement is associated with vascular engorgement of the breasts and erection of the nipples, resulting from contraction of muscle fibers in them. The clitoris, which is endowed with a rich supply of sensory nerve endings, also becomes erect due to vascular congestion. During intercourse, the vaginal epithelium becomes highly congested and secretes a mucus-like lubricant.

In addition to intense pleasure, there are many physical correlates of orgasm in the female as in the male: There is a sudden increase in skeletal muscle activity involving almost all parts of the body; the heart rate and blood pressure increase; the female counterpart of male genital contraction is a transient rhythmical contraction of the vagina and uterus. Orgasm seems to play no essential role in assuring fertilization since conception can occur in the absence of an orgasm.

A question related to the female sexual response is sex drive. Incongruous as it may seem, sexual desire in adult women is probably more dependent upon androgens, secreted by the adrenal glands, than estrogen, and sex drive is maintained beyond the menopause when estrogen levels become very low.

This completes our survey of normal reproductive physiology in the nonpregnant female. In weaving one's way through this maze, it is all too easy to forget the prime function subserved by this entire system, namely, reproduction. Accordingly, we must now return to the ovum we left free on the surface of the ovary, obtain a sperm for it, and carry the fertilized ovum through pregnancy and delivery.

PREGNANCY

Following their ejaculation into the vagina, sperm live approximately 48 h. After ovulation, the ovum remains fertile for 10 to 15 h. The net result is that for pregnancy to occur, sexual intercourse must be performed no more than 48 h before or 15 h after ovulation. These are only average figures, and there is probably considerable variation in the survival time of both sperm and ovum.

Ovum Transport

At ovulation, the ovum is extruded from the ovary, and its first mission is to gain entry into a uterine tube. The fimbriae at the end of the uterine tubes are lined with ciliated epithelium. At ovulation, the fimbriae smooth muscle causes them to pass over the ovary while the cilia beat in waves toward the interior of the duct. These ciliary motions sweep the ovum into the uterine tube as it emerges onto the ovarian surface.

Once in the uterine tube, the ovum moves rapidly for several minutes, propelled by cilia and by contractions of the tube's smooth-muscle coating. The muscular contractions soon diminish, and ovum movement, driven almost entirely by the cilia, becomes so slow that the ovum takes about 4 days to reach the uterus. Thus, if fertilization is to occur, it must do so in the uterine tube because of the short life span of the unfertilized ovum.

Sperm Transport

Within a minute or so after intercourse, some sperm can be detected in the uterus. The act of intercourse itself provides some impetus for transport of sperm out of the vagina through the cervix into the uterus because of the fluid pressure of the ejaculate and the pumping action of the penis in the vagina during ejaculation. In addition, beating of the cilia on the inner surface of the cervix probably wafts the sperm toward the uterus. Passage through the cervical mucus by the swimming sperm is dependent on the estrogen-induced changes in consistency of the mucus described earlier. Transport of the sperm the length of the uterus and into the uterine tube probably is via the sperm's own propulsions and the currents set up by the beating of uterine cilia. The possible role of uterine smooth-muscle contractions remains unclear.

The mortality rate of sperm during the trip is huge. Of the several hundred million deposited in the vagina, only a few hundred reach the uterine tube. This is one of the major reasons there must be so many sperm in the ejaculate for fertilization to occur.

Sperm Capacitation and Activation

In addition to aiding transport of sperm, the female reproductive tract exerts a second critical effect on them, namely, the conferring upon them of the ability to fertilize an ovum. Although, as we have mentioned, sperm "mature" during their stay in the epididymis, they are still not able to fertilize the ovum until they have resided in the female tract for several hours.

The first event of this two-stage process conferring fertilizability upon the sperm is known as **sperm capacitation**. It involves the stripping off from the sperm's surface of a layer of glycoprotein molecules originally acquired in the epididymis. Next, the capacitated sperm undergo, in the immediate vicinity of an ovum, multiple changes collectively known as **sperm activation**: (1) Anatomical rearrangements at the sperm's surface expose to the exterior the enzymes in the acrosome; (2) the previously regular wavelike beats of the sperm's tail are replaced by a more whiplike action that propels the sperm forward in strong lurches; and (3) the sperm's plasma membrane becomes altered so that it is capable of fusing with the surface membrane of the ovum.

Entry of the Sperm into the Ovum

Fertilization is the fusion of a sperm and ovum. The sperm, after moving between the granulosa cells still surrounding the ovum, binds to the zona pellucida. This is a species-specific binding between sperm receptors in the zona pellucida's outer surface and complementary proteins in the sperm plasma membrane. The highly motile sperm then move through the zona pellucida with the aid of acrosomal proteolytic enzymes.

Once through the zona pellucida, the sperm plasma membrane makes contact with the ovum plasma mem-

brane and fuses with it. The sperm then slowly passes into the ovum cytoplasm, usually losing its tail in the process. This penetration is achieved not by the sperm's motility but as a result of contractile elements in the ovum, which draw the sperm in. After penetration by the sperm, the ovum completes its second meiotic division over the next few hours, and the one daughter cell with practically no cytoplasm—the second polar body—is extruded and disintegrates. The two sets of chromosomes—23 from the ovum and 23 from the sperm—each become surrounded by distinct membranes and are known as pronuclei, which migrate to the center of the cell. During this period of a few hours, the DNA of the chromosomes in both pronuclei is replicated, the pronuclear membranes break down, the cell is ready to undergo a mitotic division, and fertilization is complete.

Viability of the fertilized ovum (**zygote**) depends upon stopping the entry of additional sperm. The mechanism of this "block to polyspermy" is as follows: The initial fusion of the sperm and ovum plasma membranes triggers a reaction in which secretory vesicles located around the ovum's periphery release their contents, by exocytosis, into the space between the plasma membrane and the zona pellucida. Some of these molecules are enzymes that enter the zona pellucida and cause both hardening of the zona pellucida and inactivation of its sperm-binding sites. This prevents additional sperm from binding to or moving through the zona.

Fertilization, the major events of which are summarized in Figure 18-21, also triggers activation of ovum enzymes required for the ensuing cell divisions and embryogenesis.

The zygote is now ready to begin its development as it continues its passage, propelled by cilia, down the uterine tube to the uterus. If fertilization had not occurred, the ovum would have slowly disintegrated and been phagocytized by cells lining the uterus. Rarely, a fertilized ovum remains in the uterine tube and embeds itself in the tube wall. Such tubal pregnancies cannot succeed because of lack of space for the fetus to grow, and surgery will be necessary, unless there is spontaneous abortion, because of a risk of maternal hemorrhage. Even more rarely a fertilized ovum may be expelled into the abdominal cavity where implantation may take place and proceed to term, the infant being delivered surgically. Tubal and abdominal pregnancies are termed **ectopic pregnancies**.

Early Development, Implantation, and Placentation

During the leisurely 3- to 4-day passage through the uterine tube, the zygote undergoes a number of cell divisions, a process known as **cleavage**. Identical or monozygotic twins result when, at some point during cleav-

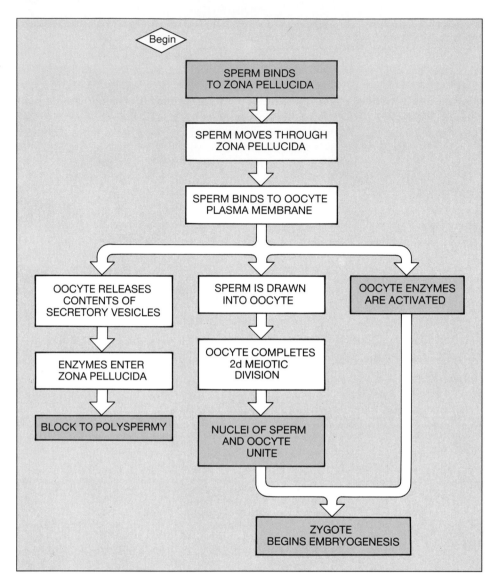

FIGURE 18-21 Events leading to fertilization and the beginning of embryogenesis.

age, the dividing cells become completely separated into two independently growing cell masses. In contrast, as we have seen, dizygotic (fraternal) twins result from two ova being ovulated and fertilized.

After reaching the uterus, the cluster of cells floats free in the intrauterine fluid, from which it receives nutrients, for approximately three days, all the while undergoing further cell divisions. The ball of cells develops into the structure known as a **blastocyst**, consisting of an outer layer of cells, the **trophoblast**, an **inner cell mass**, and a central fluid-filled cavity (Figure 18-22). During subsequent development, the inner cell mass will give

rise to the developing human—termed an **embryo**[11] during the first two months and a **fetus** after that—and some of the membranes associated with it. The trophoblast will surround the embryo and fetus throughout development and be involved in its nutrition as well as in the secretion of several important hormones.

The period during which the blastocyst forms corre-

---

[11] Terminology can be confusing here. Sometimes embryo is used to refer to everything to which the zygote gives rise, whereas, as here, it denotes only the developing individual and not the membranes associated with the individual.

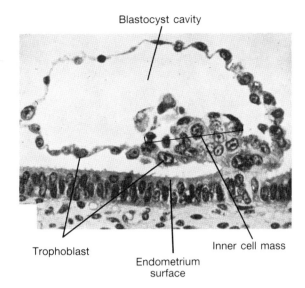

Blastocyst cavity

Trophoblast

Endometrium surface

Inner cell mass

FIGURE 18-22 Photomicrograph showing the beginning implantation of a 9-day monkey blastocyst. [*From C. H. Heuser and G. L. Streeter, Carnegie Contrib. Embryol. 29:15 (1941).*]

sponds with days 14 to 21 of the typical menstrual cycle. During this period the uterine lining is being prepared by estrogen and progesterone, secreted by the corpus luteum, to receive the blastocyst. By approximately the

twenty-first day of the cycle, that is, 7 days after ovulation, **implantation**—the embedding of the blastocyst into the endometrium—begins. The trophoblast cells are quite sticky, particularly in the region overlying the inner cell mass, and it is this portion of the blastocyst that adheres to the endometrium upon contact and initiates implantation.

The initial contact between blastocyst and endometrium induces rapid proliferation of the trophoblast, the cells of which penetrate between endometrial cells. The endometrium, too, is undergoing changes at the site of contact: increases in vascular permeability, edema, and compositional changes in the intercellular matrix and in the endometrial cells. Implantation is soon completed (Figure 18-23), and the nutrient-rich endometrial cells provide the metabolic fuel and raw materials required for early growth of the embryo.

This simple nutritive system, however, is adequate to provide for the embryo only during the first few weeks, when it is very small. The structure taking over this function is the **placenta**, a combination of interlocking fetal and maternal tissues that serves as the organ of exchange between mother and fetus for the remainder of the pregnancy. The fetal portion of the placenta is supplied by the outermost membrane of the trophoblast,

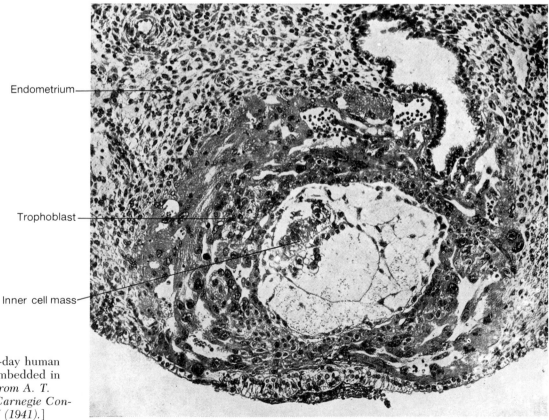

Endometrium

Trophoblast

Inner cell mass

FIGURE 18-23 Eleven-day human embryo completely embedded in the uterine lining. [*From A. T. Hertig and J. Rock, Carnegie Contrib. Embryol. 29:127 (1941).*]

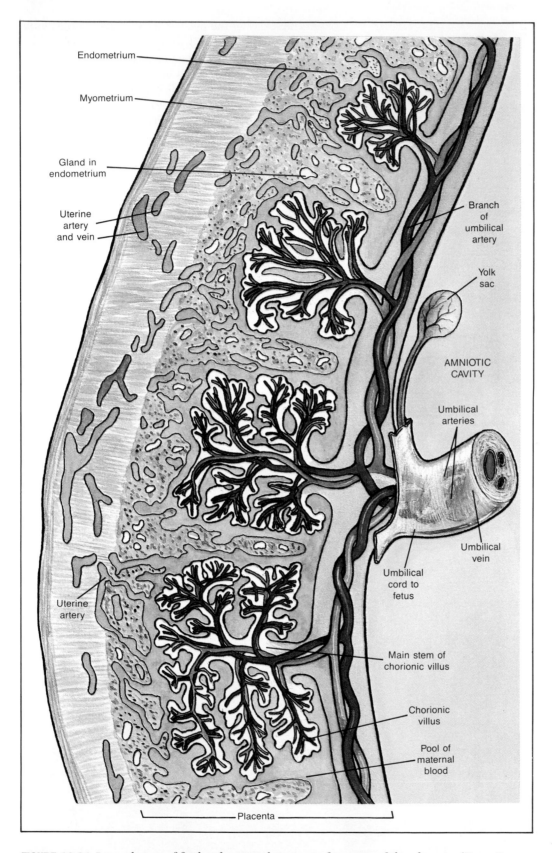

FIGURE 18-24 Interrelations of fetal and maternal tissues in formation of the placenta. (*From B. M. Patten and B. M. Carlson, "Human Embryology," 4th edn., McGraw-Hill, New York, 1985.*)

termed the **chorion**, and the maternal portion by the endometrium underlying the chorion. Finger-like projections, **chorionic villi**, extend from the chorion into the endometrium (Figure 18-24). The villi contain a rich network of capillaries. The endometrium around the villi is altered by enzymes secreted from the cells of the villi so that each villus comes to be completely surrounded by a pool, or sinus, of maternal blood.

The maternal blood enters these placental sinuses via the uterine artery, percolates through them, and then exits via the uterine veins. Simultaneously, blood flows from the fetus into the capillaries of the chorionic villi via the **umbilical arteries** and out of the capillaries back to the fetus via the **umbilical vein**. The umbilical vessels are contained in the **umbilical cord**, a long ropelike structure that connects the fetus to the placenta.

Five weeks after implantation, the placenta has become well established, the fetal heart has begun to pump blood, and the entire mechanism for nutrition of the fetus and excretion of its waste products is in operation. Waste products move from blood in the fetal capillaries across the cells lining the villi into the maternal blood, and nutrients move in the opposite direction. Many substances, such as oxygen and carbon dioxide, move by simple diffusion, whereas others are carried by medi-

ated-transport mechanisms in the lining. It must be emphasized that there is an *exchange of materials* between the two blood streams but no actual *mingling* of the fetal and maternal blood.

Meanwhile, a space called the **amniotic cavity** has formed between the inner cell mass and the trophoblast. The membrane lining the cavity is derived from the inner cell mass and called the **amnion** (or **amniotic sac**). It eventually fuses with the inner surface of the chorion so that only a single combined membrane surrounds the fetus. The fluid in the amniotic cavity, the **amniotic fluid**, resembles that of fetal extracellular fluid, and it buffers mechanical disturbances and temperature variations.

The fetus, floating in the amniotic cavity and attached by the umbilical cord to the placenta (Figure 18-25), develops into a viable infant during the next 8 months. (Description of intrauterine development is beyond the scope of this book but can be found in any standard textbook of human embryology.) Note in Figure 18-25 that only the amnionic sac separates the fetus from the uterine lumen.

Amniotic fluid can be sampled (**amniocentesis**) as early as the fourth month by inserting a needle in the amniotic cavity. A large number of genetic diseases can

FIGURE 18-25 The uterus at (A) 3, (B) 5, and (C) 8 weeks after fertilization. Embryos and their membranes are drawn to actual size. (*From B. M. Pattern and B. M. Carlson, "Human Embryology," 3rd edn., McGraw-Hill, New York, 1968.*)

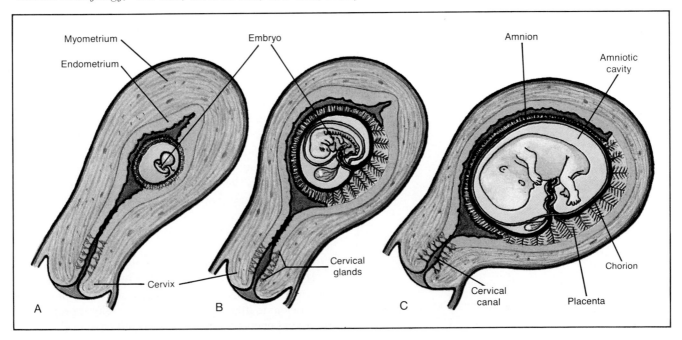

be diagnosed by the finding of certain chemicals either in the fluid or in cells suspended in the fluid. The chromosomes of these cells can also be examined for diagnosis of certain disorders as well as to determine the sex of the fetus.

Another technique for fetal diagnosis presently under investigation is that of **chorionic villus sampling**. This technique, which can be performed as early as the beginning of the third month of pregnancy, involves obtaining tissue from a chorionic villus of the placenta. The risks and reliability of this technique remain to be fully determined. Finally, a third technique for fetal diagnosis is that of ultrasound, which provides a "picture" of the fetus without the use of x-rays.

A point of importance is that the developing embryo and fetus is subject to considerable influence by a host of factors (noise, radiation, chemicals, viruses, and so on) to which the mother may be exposed. For example, drugs taken by the mother can reach the fetus via the placenta and impair fetal growth and development. In this regard, it must be emphasized that aspirin, alcohol, and the chemicals in cigarette smoke are very potent agents. For example, when mothers ingest aspirin within 5 days of delivery, many of their offspring have transient bleeding tendencies because of aspirin's inhibitory effects on blood clotting (Chapter 19).

FIGURE 18-26 Urinary excretion of estrogen, progesterone, and chorionic gonadotropin during pregnancy. Urinary excretion rates are an indication of blood concentrations of these hormones.

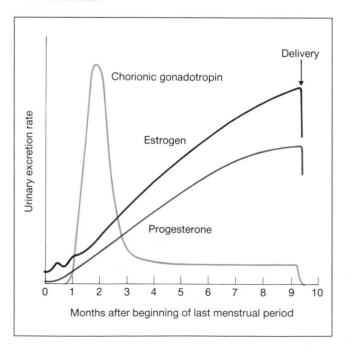

Maternal nutrition is also crucial for the fetus. Malnutrition during pregnancy retards fetal growth and results in infants with higher-than-normal death rates, reduced growth after birth, and an increased incidence of learning disabilities.

Hormonal and Other Changes during Pregnancy

Throughout pregnancy, plasma concentrations of estrogen and progesterone remain high (Figure 18-26). Estrogen stimulates growth of the uterine muscle mass, which will eventually supply the contractile force needed to deliver the fetus. Progesterone inhibits uterine motility so that the fetus is not expelled prematurely. During approximately the first 2 months of pregnancy, almost all the estrogen and progesterone are supplied by the extremely active corpus luteum.

Recall that if pregnancy had not occurred, this gland-like structure would have degenerated within 2 weeks after its formation. The persistence of the corpus luteum during pregnancy is due to a hormone called **chorionic gonadotropin (CG)**, which starts to be secreted by the trophoblast cells almost immediately after they have started their endometrial invasion. The CG gains entry to the maternal blood via the endometrial blood vessels of the placenta. This protein hormone is very similar to but not identical to LH, and it maintains the corpus luteum and strongly stimulates steroid secretion by it. Thus, the signal that preserves the corpus luteum comes from the embryonic tissues—recall that the trophoblast derives from the embryo, not the mother's tissues.

The detection of CG in urine or plasma is the basis of most pregnancy tests. The secretion of CG increases rapidly during early pregnancy, reaching a peak at 60 to 80 days after the end of the last menstrual period (Figure 18-26). It then falls just as rapidly, so that by the end of the third month it has reached a low, but definitely detectable, level that remains relatively constant for the duration of the pregnancy. Associated with this falloff of CG secretion, the placenta begins to secrete large quantities of estrogen and progesterone, and the very marked increases in blood steroids during the last 6 months of pregnancy are due almost entirely to the placental secretion. The corpus luteum remains, but its contribution is dwarfed by that of the placenta. Indeed, removal of the ovaries during the last 7 months of pregnancy has no effect at all upon the pregnancy, whereas removal during the first 2 months causes immediate abortion because the endometrium disintegrates due to a lack of estrogen and progesterone.

Recall that secretion of GnRH and, hence, of LH and FSH is powerfully inhibited by progesterone in the presence of estrogen. Since both these steroid hormones are present in high concentration throughout pregnancy, the blood concentrations of the pituitary gonadotropins remain extremely low. By this means, further follicle development and ovulation are eliminated for the duration of the pregnancy.

An important and clinically useful aspect of placental steroid secretion is that the placenta has the enzymes required for the synthesis of progesterone but not those for the complete synthesis of the major estrogen of pregnancy, **estriol**. The enzymes the placenta lacks, however, are present in special cells in the fetal adrenal cortex. Therefore, the placenta and fetal adrenals, working

together, with intermediates transported between them via the fetal circulation, produce the estriol required to maintain the pregnancy. Measurement of this hormone in maternal blood provides a means for monitoring the well-being of the fetus.

The placenta produces not only steroids and CG but many other hormones as well. Some of these are identical to hormones normally produced by other endocrine glands whereas others are different. One different hormone has effects very similar to those of both prolactin

TABLE 18-9 MATERNAL RESPONSES TO PREGNANCY

	Response
Placenta	Secretion of estrogen, progesterone, chorionic gonadotropin, placental lactogen, and other hormones.
Anterior pituitary	Increased secretion of prolactin and ACTH. Secretes very little FSH and LH.
Adrenal cortex	Increased secretion of aldosterone and cortisol.
Posterior pituitary	Increased secretion of vasopressin.
Parathyroids	Increased secretion of parathyroid hormone.
Kidneys	Increased secretion of renin, erythropoietin, and 1,25-dihydroxyvitamin D_3. Retention of salt and water. *Cause:* Increased aldosterone, vasopressin, and estrogen
Breasts	Enlarge and develop mature glandular structure. *Cause:* Estrogen, progesterone, prolactin, and placental lactogen
Blood volume	Increases. *Cause:* Erythrocyte volume is increased by erythropoietin and plasma volume by salt and water retention.
Calcium balance	Positive. *Cause:* Increased parathyroid hormone and 1,25-dihydroxyvitamin D_3.
Body weight	Increases by average of 12.5 kg, 60 percent of which is water.
Circulation	Cardiac output increases, total peripheral resistance decreases (vasodilation in uterus, skin, breasts, GI tract, and kidneys), mean arterial pressure stays constant.
Respiration	Hyperventilation (arterial P_{CO_2} decreases).
Organic metabolism	Metabolic rate increases. Plasma glucose, gluconeogenesis, and fatty acid mobilization all increase. *Cause:* Hyporesponsiveness to insulin due to insulin antagonism by placental lactogen and cortisol.
Appetite and thirst	Increase.
Nutritional RDAs	Increase.

and growth hormone. This hormone, **placental lactogen**, may play important roles in the mother: maintaining a positive protein balance, mobilizing fats for energy, stabilizing plasma glucose at relatively high levels to meet the needs of the fetus, and facilitating development of the breasts.

Some of the numerous other physiological changes, hormonal and nonhormonal, in the mother during pregnancy are summarized in Table 18-9.

One additional word about fluid balance during pregnancy: Some women retain abnormally great amounts of fluid and manifest both protein in the urine and hypertension. These are the symptoms of the disease known as **toxemia of pregnancy** or eclampsia, a condition that can usually be controlled by salt restriction. All attempts to determine the factors responsible for the disease have failed.

Parturition

A normal human pregnancy lasts approximately 40 weeks, although many babies are born 1 to 2 weeks earlier or later. Safe survival of premature infants is now possible at about the twenty-seventh week of pregnancy, but treatment of these infants often requires heroic efforts. Delivery of the infant, followed by the placenta, the entire process being termed **parturition**, is produced by strong rhythmical contractions of the myometrium. Actually, weak and infrequent uterine contractions begin at approximately 30 weeks and gradually increase in both strength and frequency. During the last month, the entire uterine contents shift downward so that the baby is brought into contact with the cervix. In over 90 percent of births, the baby's head is downward and will act as the wedge to dilate the cervical canal when labor begins (Figure 18-27).

This dilation can occur because, during the last few weeks of pregnancy, the cervix becomes soft and flexible, which makes the cervical canal much more extensible. The cervical softening (often termed "ripening") is largely the result of a breakup of collagen fibers. These biochemical changes are mediated by a variety of messengers, including estrogen, prostaglandins, and the polypeptide hormone, **relaxin**, secreted mainly by the corpus luteum. Its contribution to cervical softening is the major function of relaxin.

When labor begins in earnest, the uterine contractions become coordinated and quite strong (although usually painless at first) and occur at approximately 10- to 15-min intervals. The contractions begin in the upper portion of the uterus and sweep downward. This coordination is made possible by the fact that the uterine smooth-muscle cells are connected to each other by gap junctions. These junctions increase markedly in number near the end of pregnancy.

At the onset of labor or before, the amniotic sac ruptures and the amniotic fluid escapes through the vagina.

As the contractions increase in intensity and frequency, the cervical canal is gradually forced open to a maximum diameter of approximately 10 cm. Until this point, the contractions have not moved the fetus out of the uterus but have served only to dilate the cervix. Now the contractions move the fetus through the cervix and vagina. At this time the mother, by bearing down to increase abdominal pressure, can help the uterine contractions to deliver the baby. The umbilical vessels and placenta are still functioning, so that the baby is not yet on its own, but within minutes of delivery both the umbilical vessels and the placental vessels completely constrict, stopping blood flow to the placenta. The entire placenta becomes separated from the underlying uterine wall, and a wave of uterine contractions delivers the placenta as the **afterbirth**.

Ordinarily, parturition proceeds automatically from beginning to end and requires no significant medical intervention. In a small percentage of cases, however, the position of the baby or some maternal defect can interfere with normal delivery. The head-first position of the fetus is important for several reasons: (1) If the baby is not oriented head first, another portion of its body is in contact with the cervix and is generally a far less effective wedge. (2) Because of the head's large diameter compared with the rest of the body, if the body were to go through the cervical canal first, the canal might obstruct the passage of the head, leading to problems when the partially delivered baby attempts to breath. (3) If the umbilical cord becomes caught between the canal wall and the baby's head or chest, mechanical compression of the umbilical vessels can result. Despite these potential problems, however, most babies who are not oriented head first are born normally.

What mechanisms control the events of parturition? Let us consider a set of fairly well established facts:

1. The smooth-muscle cells of the myometrium have inherent rhythmicity and are capable of autonomous contractions that are facilitated as the muscle is stretched by the growing fetus.
2. The autonomic neurons to the uterus are of little importance in parturition since anesthetizing them does not interfere with delivery.
3. Progesterone exerts a powerful inhibitory effect upon uterine contractions by decreasing the sensitivity to estrogen, the hormone oxytocin, and prostaglandins, all of which stimulate uterine contractions. Shortly before delivery, the ratio of progesterone to estrogen often decreases, mainly because of increased estrogen synthesis.
4. The pregnant uterus near term secretes a prostaglan-

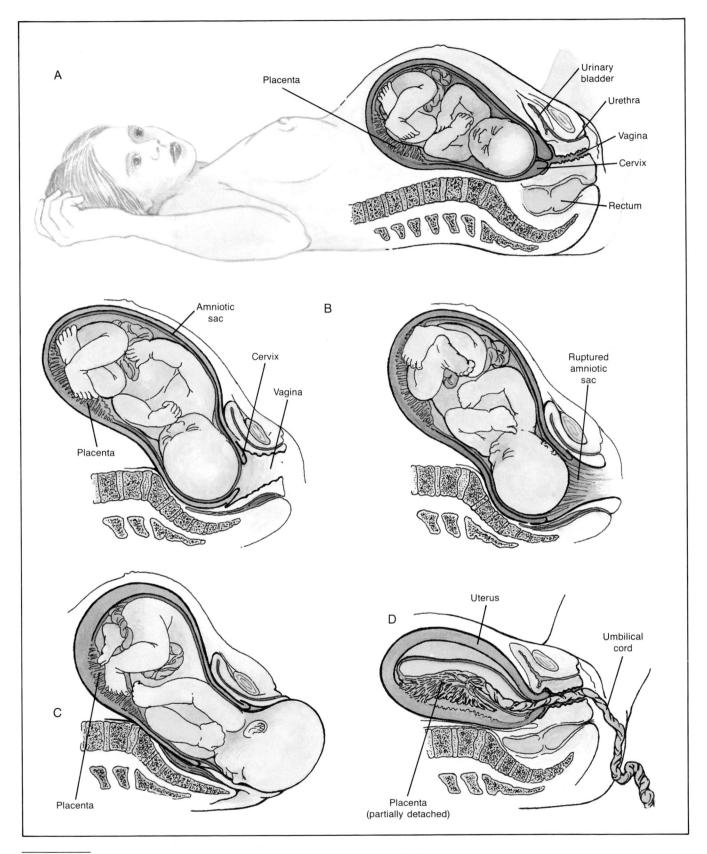

FIGURE 18-27 Stages of parturition. (A) Parturition has not yet begun. (B) The cervix has dilated. (C) The fetus is moving through the cervical canal and vagina. (D) The placenta is coming loose from the uterine wall preparatory to its expulsion.

din that is a profound stimulator of uterine smooth-muscle contraction. An increase in the release of this substance has been demonstrated during labor.

5. **Oxytocin**, one of the hormones released from the posterior pituitary, is an extremely potent uterine-muscle stimulant. It not only acts directly on uterine smooth muscle but stimulates it to synthesize prostaglandins. Oxytocin is reflexly released as a result of afferent input to the hypothalamus from receptors in the uterus, particularly the cervix. During pregnancy the concentration of oxytocin receptors in the uterus markedly and progressively increases as a result of estrogen stimulation and, possibly, uterine distention near term.

These facts can now be put together in a unified pattern, as shown in Figure 18-28. Once started, the uterine contractions exert a positive-feedback effect upon themselves via both local facilitation of inherent uterine contractions and reflex stimulation of oxytocin secretion. But what *starts* the contractions? The most tenable hypothesis at present involves both oxytocin and prostaglandin: Even though there is no dramatic change in circulating oxytocin just prior to the onset of labor, the increased number of oxytocin receptors in the myometrium makes the uterus so sensitive to oxytocin that contractions become coordinated and strong. The decrease in progesterone to estrogen ratio further increases responsiveness to oxytocin. Simultaneously, the increased sensitivity to oxytocin results in an increased oxytocin-induced local synthesis of uterine prostaglandin, which further stimulates uterine contractions.

The action of prostaglandin on parturition is the last in a series of prostaglandin effects on the female reproductive system we have described. They are summarized in Table 18-10.

Lactation

The secretion of milk by the breasts, or **mammary glands**, is termed **lactation**. The breasts contain ducts that branch all through the tissue and converge at the nipples (Figure 18-29). These ducts arise in saclike glands called **alveoli**, the same term used to denote the lung air sacs. The breast alveoli, which secrete milk, look like bunches of grapes with stems terminating in the ducts. The alveoli and the ducts immediately adjacent to them are surrounded by specialized contractile cells called **myoepithelial cells**.

Before puberty, the breasts are small with little internal glandular structure. With the onset of puberty in females, the increased estrogen causes a marked enhancement of duct growth and branching but relatively little development of the alveoli, and much of the breast enlargement at this time is due to fat deposition. Progesterone secretion also commences at puberty during the luteal phase of each cycle, and this hormone contributes to breast growth.

During each menstrual cycle, the breasts undergo fluctuations in association with the changing blood concentrations of estrogen and progesterone, but these changes are small compared with the marked breast enlargement that occurs during pregnancy as a result of the stimulatory effects of high plasma concentrations of estrogen, progesterone, prolactin, and placental lactogen. This last hormone, as described earlier, is secreted by the placenta, whereas prolactin is secreted by the anterior pituitary. Under the influence of all these hor-

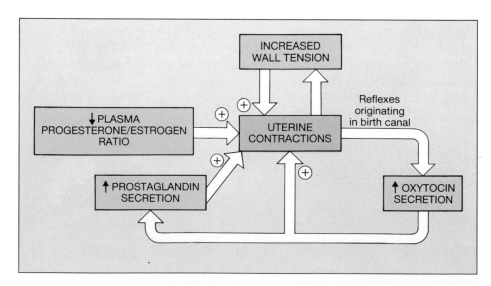

FIGURE 18-28 Factors stimulating uterine contractions during parturition. Note the positive-feedback nature of several of the inputs. Not shown in the figure is the important fact that estrogen (or a decrease in the progesterone to estrogen ratio) increases the number of oxytocin receptors in the myometrium.

TABLE 18-10 SOME EFFECTS OF PROSTAGLANDINS* ON THE FEMALE REPRODUCTIVE SYSTEM

Site of Production	Action	Result
Late-antral follicle	Stimulates production of lytic enzymes	Rupture of follicle
Corpus luteum	Interferes with corpus luteum's hormone secretion and function	Death of corpus luteum
Uterus	Constricts blood vessels in endometrium	Onset of menstruation
	Causes changes in endometrial blood vessels and cells early in pregnancy	Facilitates implantation
	Increases contraction of myometrium	Helps initiate both menstruation and parturition

*The term "prostaglandins" is used loosely here to include all the eicosanoids (Chapter 7).

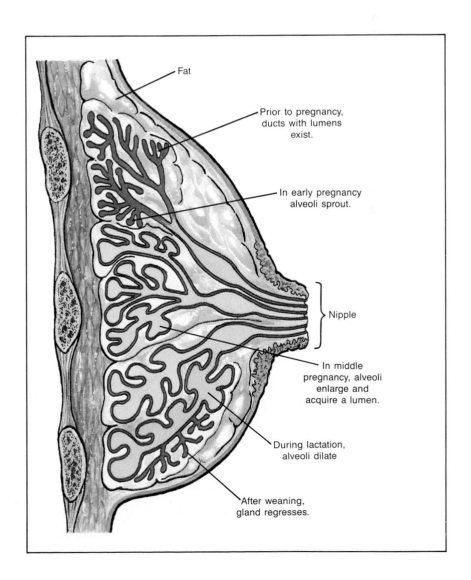

FIGURE 18-29 Anatomy of the breast. (*Adapted from Elias et al.*)

mones, both the ductal and the alveolar structures become fully developed.

As pointed out in Chapter 10, prolactin secretion by the anterior pituitary is controlled by several hypothalamic hormones, at least one of which—**prolactin inhibiting hormone (PIH)**—is inhibitory and one—**prolactin releasing hormone (PRH)**—stimulatory. Under the dominant influence of PIH, which has been identified as the chemical messenger dopamine, prolactin secretion is low prior to puberty. It then increases considerably at puberty in girls but not in boys. This increase is caused by the increased plasma estrogen concentration that occurs at this time, for estrogen acts directly on the anterior pituitary to stimulate prolactin secretion. During pregnancy, there is a marked further increase in prolactin secretion due to stimulation by the elevated plasma estrogen, beginning at about 8 weeks and rising throughout the remainder of the pregnancy.

Prolactin is the single most important hormone promoting milk production. Yet despite the fact that prolactin is elevated and the breasts are markedly enlarged and fully developed as pregnancy progresses, there is no secretion of milk. This is because estrogen and progesterone, in large concentrations, prevent milk production by inhibiting this particular action of prolactin on the breasts. Thus, although estrogen causes an increase in the secretion of prolactin and acts with it in promoting breast growth and differentiation, it, along with progesterone, is antagonistic to prolactin's ability to induce milk secretion. Delivery removes the source—the placenta—of the large amounts of sex steroids and, thereby, the inhibition of milk production.

Following parturition, *basal* prolactin secretion decreases from its peak late-pregnancy levels and after several months may actually return to prepregnancy levels even though the mother continues to nurse. Superimposed upon this basal level, however, are large secretory bursts of prolactin during each nursing period. The episodic pulses of prolactin are the signals to the breasts for maintenance of milk production, which ceases several days after the mother completely stops nursing her infant but continues uninterrupted for years if nursing is continued.

The reflexes mediating the prolactin bursts are initiated by afferent input to the brain from nipple receptors stimulated by suckling. It is still unclear whether this input's major effect is to inhibit the hypothalamic neurons that release dopamine or to stimulate the neurons that secrete PRH (Figure 18-30).

One other reflex process is essential for nursing. Milk is secreted into the lumen of the alveoli, but the infant cannot suck the milk out of the alveoli. It must first be moved into the ducts, from which it can be sucked. This

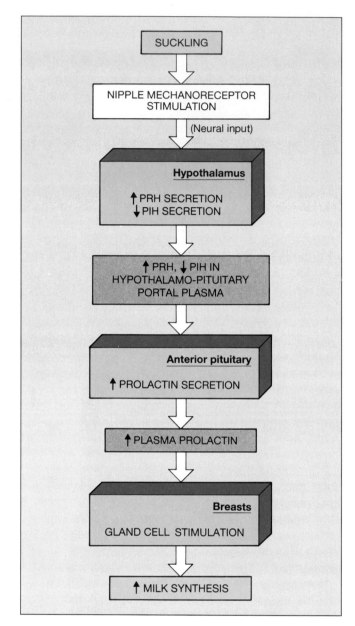

FIGURE 18-30 Control of prolactin secretion during nursing. Neural input from nipple mechanoreceptors inhibits the hypothalamic cells that secrete dopamine (PIH) and probably also stimulates the hypothalamic cells that secrete PRH.

movement is called the **milk ejection reflex** (formerly called milk let-down) and is accomplished by contraction of the myoepithelial cells surrounding the alveoli. The contraction is under the control of oxytocin, which is reflexly released by suckling (Figure 18-31), just like prolactin. Higher brain centers can also exert an important influence over oxytocin release: A nursing mother

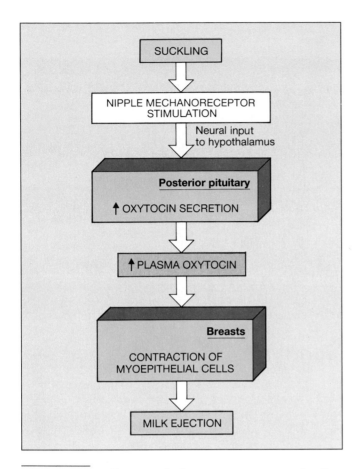

FIGURE 18-31 Reflex control of oxytocin secretion and milk ejection reflex. This reflex is triggered by the same stimulus—suckling acting via nipple mechanoreceptors—that triggers prolactin secretion.

may actually leak milk when she hears her baby cry or even thinks about nursing.

Milk contains four major constituents: water, protein, fat, and the carbohydrate lactose (milk sugar). The alveolar cells must be capable of extracting the raw materials—amino acids, fatty acids, glycerol, glucose, and so on—from the blood and building them into the higher-molecular-weight substances. Although prolactin is the single most important hormone controlling these synthetic processes, insulin, growth hormone, cortisol, and still other hormones also participate.

Another neuroendocrine reflex triggered by suckling and mediated, in part, by prolactin is inhibition of the hypothalamic-pituitary-ovarian chain at a variety of steps, with resultant block of ovulation. If suckling is continued at high frequency, ovulation can be delayed for years. When supplements are added to the baby's diet and the frequency of suckling is decreased, how-

ever, most women will resume ovulation even though they are continuing to lactate. Failure to recognize this fact may result in an unplanned pregnancy.

Fertility Control

More than half of all American women between the ages of 15 and 44 use some form of fertility control (**contraception**) (Table 18-11). Some—vaginal diaphragms, spermicides, and condoms—prevent sperm from reaching the ovum. [In addition, condoms significantly reduce the risk of sexually transmitted diseases (STDs) such as syphillis, gonorrhea, AIDS, and herpes.] In contrast, the **oral contraceptives** are based on the fact that estrogen and progesterone can inhibit pituitary gonadotropin release, thereby preventing ovulation. One type is a combination of a synthetic estrogen and a progesterone-like substance (a progestagen or progestin). Another type is the so-called minipill, which contains only the progesterone-like substance.[12] In actuality, the oral contraceptives, particularly the minipill, do not always prevent ovulation, yet still are effective because they have multiple antifertility effects. In other words, the hormonal milieu required for normal pregnancy is such that these exogenous steroids interfere with many of the steps between intercourse and implantation of the blastocyst. For example, the progestagen component inhibits the estrogen-induced proliferation of the endometrium, making it inhospitable for implantation. In addition, the progestagen affects the composition of the cervical mucus, preventing passage of sperm through the cervix.

Taken correctly, oral contraceptives are almost 100 percent effective. Although potentially lethal side-effects, such as intravascular clotting, have been reported, they occur in only a very small number of women who are over 35 and smoke or who have a previous history of blood-clotting disorders. Indeed, the fatality rate is considerably less than that associated with pregnancy. Moreover, oral contraceptives have been shown to protect against cancer of both the ovaries and the uterine lining, the latter the third most common cancer among women in the United States, and against several other diseases as well. Very recent data, however, have raised the question of whether oral contraceptives increase the risk of breast cancer.

Another effective contraceptive is the **intrauterine device (IUD)**. Placing one of these small objects in the uterus prevents pregnancy, probably by somehow inter-

---

[12]The combination oral contraceptives are given for a certain number of days and then stopped, which produces menstruation. This is done to avoid the continued buildup of the endometrium and so-called "breakthrough bleeding." The minipill is taken daily with no break. Menstruation generally occurs normally but may be more irregular than prior to pill use.

TABLE 18-11   SUMMARY OF COMPLETELY REVERSIBLE CONTRACEPTIVE METHODS*

Type	Estimated Effectiveness	Advantages	Disadvantages	Comments
Birth-control pill (oral contraceptive)	98% (combination) 97% (mini)	Most effective reversible contraceptive. Results in lighter, more regular periods. Protects against cancer of the ovaries and uterine lining. Decreases risk of pelvic inflammatory disease, fibrocystic breast disease, and benign ovarian cysts.	Minor side-effects similar to early pregnancy (nausea, breast tenderness, fluid retention) during first 3 months of use. Major complications (blood clots, hypertension) may occur in smokers and those over 35. Controversial as to whether breast cancer is increased. Must be taken on a regular daily schedule.	Combination types contain both synthetic estrogen and progestagen. Minipill contains only progestagen and may produce irregular bleeding. Available by prescription only.
Intrauterine device (IUD)	95%	Once inserted, usually stays in place. Remains effective for a year.	May cause bleeding and cramping. Increased risk of pelvic inflammatory disease. If pregnancy occurs, increased risk that it may be ectopic. Must check for placement after each period. Requires annual replacement.	Only one type currently available in the U.S. Available by prescription only.
Condom (rubber, prophylactic, sheath)	90%	Protects against sexually transmitted diseases, including AIDS and herpes. May protect against cervical cancer.	Must be applied immediately before intercourse. Rare cases of allergy to rubber. May break. Blunting of sensation.	More effective when the woman uses a spermicide.
Vaginal spermicide	70–80% (used alone)	Available over the counter as jellies, foam, creams, and suppositories.	Messiness. Must be applied no more than 1 h before intercourse.	Best results occur when used with a condom or diaphragm.
Diaphragm	80–98% (with spermicide)	Can be inserted up to 2 h before intercourse.	Increased risk of urinary tract infection. Rare cases of allergy to rubber.	Must be used with spermicide. Available by prescription only and must be fitted.
Vaginal sponge	85–90%	Easy to use because spermicide is self-contained. May be inserted as much as (but no more than) 24 h prior to intercourse.	May be hard to remove; may fragment. May irritate vaginal lining. Higher failure rate in women who have given birth.	Must be left in for 6 h after intercourse.

*Coitus interruptus (withdrawal) and periodic abstinence from sexual intercourse (rhythm method) are not included because of their high failure rate. (Table adapted from University of California, Berkeley, *Wellness Letter*, 1987.)

fering with the endometrial preparation for acceptance of the blastocyst. Unfortunately, the intrauterine device also is associated with a small incidence of various adverse effects, and all but one brand have been removed from the market in the United States.

Another widely used fertility control method is the rhythm method, in which couples abstain from sexual intercourse near the time of ovulation. Unfortunately, it is difficult to time ovulation precisely even with laboratory techniques. For example, the small rise in body temperature or change in cervical mucus and vaginal epithelium, all of which are indicators of ovulation, occur only *after* ovulation. This problem, combined with the marked variability of the time of ovulation in many women explains why this technique has a high failure rate.

Considerable research is also being done on new methods for terminating early pregnancy, using either prostaglandins or antagonists of progesterone, and on a possible male contraceptive. For example, a drug that blocked the secretion or action of FSH but not LH would prevent spermatogenesis but not testosterone secretion.

Finally, surgical sterilization—tubal ligation in women and vasectomy in men—is widely available and effective but is often irreversible.

The other side of the coin of fertility prevention is the problem of unwanted infertility. Approximately 12 percent of the married couples in the United States are infertile. There are many reasons—some known, some unknown—for infertility. Careful investigation of infertile couples frequently permits diagnosis and therapy of the basic problem.

In some cases when the basic defect cannot be treated, it can be circumvented by the technique of in vitro fertilization. With this technique, the ovum is removed, via a very small abdominal incision, from the ovary immediately before ovulation and placed in a dish for several days with sperm from the woman's husband. After the fertilized ovum has developed into a cluster of two to eight cells, it is transferred to the uterus. After a few days the cluster of cells implants. This technique was first used only for women whose oviducts were absent or damaged, but its use has now been broadened to include other disorders, including unexplained infertility.

SECTION C
THE CHRONOLOGY OF REPRODUCTIVE FUNCTION

This section treats a variety of topics that have to do, in one way or another, with sequential changes in reproductive development or function. Thus, genetic inheritance, which sets the sex of the individual—**sex determination**, is established at the moment of fertilization. This is followed by **sex differentiation**, the multiple processes in which development of the reproductive system occurs in the fetus. Then there is the maturation of the system at puberty and the eventual decline that occurs with aging.

SEX DETERMINATION

One's sex is determined by genetic inheritance of two chromosomes called the **sex chromosomes**. The larger of the sex chromosomes is called the **X chromosome** and the smaller, the **Y chromosome**. Males possess one X and one Y, whereas females have two X chromosomes. Thus, the genetic difference between male and female is simply the difference in one chromosome.

The reason for the approximately equal sex distribution of the population should be readily apparent: The ovum can contribute only an X chromosome, whereas half of the sperm produced during meiosis are X and half are Y. When the sperm and ovum join, 50 percent should have XX and 50 percent XY. Interestingly, however, sex ratios at birth are not exactly 1:1. Rather, for unclear reasons, there tends to be a slight preponderance of male births.

An easy method exists for determining whether a person's cells contain two X chromosomes, the normal female pattern. When two X chromosomes are present, only one functions and the nonfunctional X chromosome condenses to form a nuclear mass, termed the **sex chromatin**, that is readily observable with a light microscope (scrapings from the cheek mucosa are a convenient source of cells to be examined). The single X chromosome in male cells rarely condenses to form sex chromatin.

A more exacting technique for determining sex-chromosome composition employs tissue-culture visualization of all the chromosomes—a **karyotype**. This technique has revealed a group of genetic sex abnormalities characterized by such unusual chromosomal combinations as XXX, XXY, X, and others. The end result of such combinations is usually the failure of normal anatomical and functional sexual development.

SEX DIFFERENTIATION

It is not surprising that persons with abnormal genetic endowment manifest abnormal sexual development, but careful study has also revealed people with normal chromosomal combinations but abnormal sexual appearance and function. In these people, sex differentiation has been abnormal, and their appearance may even be at odds with their genetic sex, that is, the presence of XX or XY chromosomes.

It will be important to bear in mind during the following description one essential generalization: The genes directly determine only whether the individual will have testes or ovaries. All the rest of sex differentiation depends upon the presence or absence of the genetically determined gonads.

Differentiation of the Gonads

The male and female gonads derive embryologically from the same site in the body. Until the sixth week of intrauterine life, there is no differentiation of this site. In the genetic male (in other words, if the Y chromosome is present), the testes begin to develop during the seventh week. The critical messenger (**testes determining factor**) that induces testes formation is coded for by a single gene on the Y chromosome. In the absence of a Y chromosome, testes do not develop and, instead, ovaries begin to develop at about 11 weeks.

Differentiation of Internal and External Genitalia

So far as its internal duct system and external genitalia are concerned, the fetus is capable of developing into either sex. Prior to the functioning of the fetal gonads, the primitive reproductive tract includes a double genital duct system—**Wolffian ducts** and **Müllerian ducts**—and a common opening for the genital ducts and urinary system to the outside. Normally, most of the reproductive tract develops from only one of these duct systems. In the male, the Wolffian ducts persist and the Müllerian ducts regress, whereas in the female, the opposite happens. The external genitalia in the two sexes and the vagina do not develop from these duct systems but from other structures at the body surface.

Which of the two duct systems and types of external genitalia develop depends on the presence or absence of fetal testes. If testes are present, they secrete (1) testosterone, from the Leydig cells under stimulation by chorionic gonadotropin secreted by the placenta, and (2) a protein hormone called **Müllerian inhibiting hormone (MIH)**, from the Sertoli cells. MIH causes the Müllerian duct system to degenerate, and testosterone causes the Wolffian ducts to differentiate into epididymis, ductus deferens, ejaculatory duct, and seminal vesicle. Exter-

nally and somewhat later, under the influence of testosterone, after conversion to dihydrotestosterone, a penis forms and the tissue near it fuses to form the scrotum, into which the testes will ultimately descend, stimulated to do so by both MIH and testosterone (Figure 18-32A).

In contrast, the female fetus, lacking testes and thus testosterone and MIH, develops uterine tubes and a uterus from the Müllerian system as well as a vagina and female external genitalia from the other structures (Figure 18-32B). A female gonad need not be present for the female organs to develop.

There is a variety of conditions in which normal sex differention does not occur. For example, in the syndrome known as testicular feminization (or androgen insensitivity syndrome), the person's genetic endowment is XY and testes are present, but he has female external genitalia, a vagina, and no duct system at all. The problem causing this is a lack of androgen receptors due to a genetic defect. The fetal testes differentiate as usual, and they secrete both MIH and testosterone. MIH causes the Müllerian ducts to regress, but the inability of the Wolffian ducts to respond to testosterone also causes them to regress, and so no duct system develops. The tissues that give rise to external genitalia (and the vagina, in the female) are also unresponsive to testosterone and so female external genitalia and a vagina develop rather than male structures.

Sexual Differentiation of the Central Nervous System

In humans and other primates it appears that there are no inherent male-female differences in the ability of the hypothalamus to secrete GnRH in response to neuronal or hormonal inputs. For example, administration of large amounts of estrogen to castrated male monkeys elicit LH surges indistinguishable from those shown by females.

The situation may be different for sexual behavior in that qualitative differences in the brain may be formed during development: Genetic female monkeys given testosterone during late fetal life manifest evidence of masculine sex behavior (mounting, for example) as adults.

Related to this is the question of whether exposure to androgens during fetal existence causes development of other behavior patterns in addition to sex. Again, for other primates, the answer seems a clear-cut *yes*. For example, the female monkey offspring given testosterone during fetal development manifest a high degree of male-type play behavior.

The evidence for human beings is scanty. Perhaps the best-studied group has been women whose mothers were given, during pregnancy, synthetic hormones not recognized to be androgenic at the time. At birth, as would be predicted from our description of sex differentiation, they had malelike external genitalia, including a

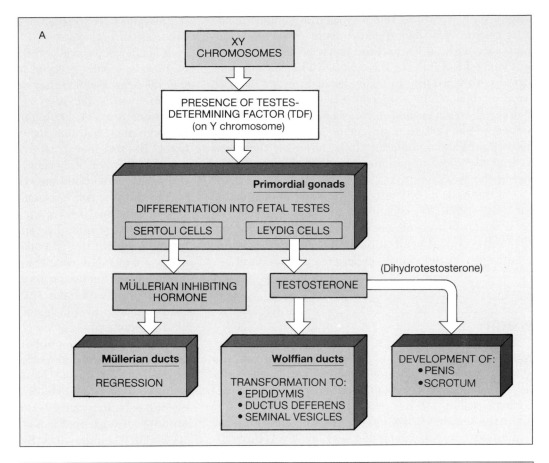

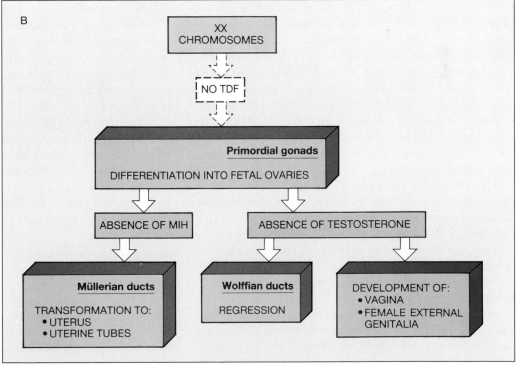

FIGURE 18-32 Sex differentiation. (A) Male. (B) Female.

fused empty scrotum, but normal ovaries and female duct system. The abnormalities were corrected surgically very early in life and the behavior of the subjects was studied carefully for the next 5 to 15 years and compared with a control group matched in every possible way. Certain of the subjects' behaviors differed from those of the girls in the control group. For example, the subjects took part in more rough-and-tumble outdoor activities. When studied as adults, however, there was little evidence of important behavioral differences between the two groups.

On the basis of this study and several others like it, the present tentative conclusions are that, although prenatal sex hormones influence certain forms of behavior, the most powerful factors ultimately determining gender identity are the individual's experience and socialization.

PUBERTY

Puberty is the period, usually occurring sometime between the ages of 10 and 14, during which the reproductive organs mature and reproduction becomes possible.

In the female, GnRH, the pituitary gonadotropins, and estrogen are all secreted at very low levels throughout childhood. Accordingly, follicle maturation and menstrual cycles do not occur, the female accessory sex organs remain small and nonfunctional, and there are minimal secondary sex characteristics. The onset of puberty is occasioned by an alteration in brain function that raises secretion of GnRH. This releasing hormone in turn stimulates secretion of pituitary gonadotropins, which stimulate follicle development and estrogen secretion. Estrogen, in addition to its critical role in follicle development, induces the striking changes in the accessory sex organs and secondary sex characteristics associated with puberty.

The picture for the male is analogous to that for the female. An increased GnRH secretion at puberty causes increased secretion of pituitary gonadotropins, which stimulate the seminiferous tubules and testosterone secretion. Testosterone, in addition to its critical role in spermatogenesis, induces the pubertal changes in the accessory reproductive organs, secondary sex characteristics, and sexual drive.

The mechanism of the brain change that results in increased GnRH secretion at puberty remains unknown. One proposed candidate for the trigger is a decrease in secretion of the hormone melatonin by the pineal gland, but there is no convincing evidence for this hypothesis. Whatever the mechanism, the process is not abrupt but develops over several years, as evidenced by slowly rising plasma concentrations of the gonadotropins and testosterone or estrogen.

It should be recognized that the maturational events of puberty usually proceed in an orderly sequence but that the ages at which they occur may vary among individuals. In girls, the onset of breast development is usually the first event, beginning at an average age of 11, although pubic hair may, on occasion, appear first. **Menarche**, the first menstrual period, is a later event (average of 12.3 years) and occurs almost invariably after the peak of the total-body growth spurt has passed.

The age at which menarche occurs is influenced by a variety of factors, one of the most important being the girl's amount of body fat. The girl must have at least a minimum or threshold amount of body fat in order to begin and maintain normal menstrual cycles. The mechanism underlying this relationship between body fat and reproductive function is still unclear but seems to involve signals to the hypothalamic areas controlling GnRH secretion.

This relationship is not limited to puberty but continues throughout reproductive life: Women who become extremely lean due to dieting or exercise frequently cease ovulating and having menstrual cycles. Such a phenomenon must have been adaptive during evolution of our ancestors in that it assured that females could conceive only when they had the energy stores to complete a pregnancy successfully—it takes 50,000 to 80,000 kilocalories to produce an infant and another 500 to 1000 kilocalories a day for lactation.

In boys, the first sign of puberty is accelerated growth of testes and scrotum. Pubic hair appears a trifle later, and axillary and facial hair still later. Acceleration of penis growth begins on the average at 13 years and is complete by 15.

Some children with brain tumors or other lesions of the hypothalamus may undergo precocious puberty, that is, sexual maturation at an unusually early age, sometimes within the first 5 years of life. The youngest mother on record gave birth to a full-term, healthy infant by Caesarean section (abdominal incision) at 5 years, 8 months.

MENOPAUSE

Fertility in women peaks in their midtwenties and declines gradually after the age of 30. Around the age of 50, on the average, menstrual cycles become less regular and ovulation may often fail to occur because not enough estrogen is secreted to trigger an LH surge. Ultimately, the cycles cease entirely, and this cessation is known as the **menopause**. The phase of life beginning with menstrual irregularity and culminating in menopause is known as the **climacteric**, the counterpart of puberty. It involves numerous physical and emotional changes as sexual maturity gives way to cessation of reproductive function.

Menopause and the irregular function leading to it are caused by ovarian failure. The ovaries lose their ability to respond to the gonadotropins, partly because of a decreasing number of follicles and partly because the remaining follicles are hyporesponsive. That the hypothalamus and anterior pituitary are functioning normally is evidenced by the fact that the gonadotropins are secreted in greater amounts in response to the decreasing plasma estrogen levels.

Although some ovarian secretion of estrogen generally continues beyond the menopause, as does peripheral conversion of adrenal androgens to estrogen, plasma estrogen gradually diminishes until it is inadequate to maintain the estrogen-dependent tissues. The breasts and genital organs gradually atrophy to a large degree. Marked decreases in bone mass and strength, termed **osteoporosis**, may occur because of bone resorption and can result in bone fractures. Sexual drive frequently stays the same and may even increase. The hot flashes so typical of menopause result from dilation of the skin arterioles, causing a feeling of warmth and marked sweating, but why estrogen deficiency causes this is unknown.

Many of the symptoms associated with menopause can be reduced by the administration of estrogen. The desirability of such administration is controversial, however, because of the fact that estrogen, under some circumstances, increases the risk of developing uterine cancer. This risk may be ameliorated, however, by concurrent treatment with a progestagen.

Another aspect of menopause is its relationship with cardiovascular diseases. Women have much less hypertension and atherosclerosis than men until after the menopause, but then the incidence becomes similar in both sexes.

Changes in the male reproductive system with aging are less drastic. Once testosterone and pituitary gonadotropin secretions are initiated at puberty, they continue, at least to some extent, throughout adult life. There is a steady decrease, however, in testosterone secretion, beginning at about the age of 40, which apparently reflects slow deterioration of testicular function and, as in the female, failure to respond to the pituitary gonadotropins. Along with the decreasing testosterone levels, both sex drive and capacity diminish, although many men continue to be fertile in their seventies and eighties.

With aging, some men manifest increased emotional problems, such as depression, and this is sometimes referred to as "male menopause." It is not clear, however, what role hormone changes play in this phenomenon.

SUMMARY

General Principles of Gametogenesis

I. The first stage of gametogenesis is mitosis of primordial germ cells.

II. This is followed by meiosis, a sequence of two cell divisions resulting in each gamete receiving 23 chromosomes.

III. Crossing over and random distribution of maternal and paternal chromatids to the daughter cells cause genetic variability in the gametes.

Section A. Male Reproductive Physiology

Spermatogenesis

I. The male gonads, the testes, produce sperm in the seminiferous tubules and secrete testosterone from the Leydig cells.

II. The meiotic divisions of spermatogenesis result in each sperm containing 23 chromosomes, compared to the original 46 of the spermatogonia.

III. The developing germ cells are intimately associated with the Sertoli cells, which perform many functions, as summarized in Table 18-2.

Transport of Sperm

I. From the seminiferous tubules, the sperm pass through the epididymis, where they mature and are concentrated.

II. The epididymis and vas deferens store the sperm, and the seminal vesicles and prostate secrete the bulk of the semen.

III. Erection of the penis occurs because of vascular engorgement accomplished by relaxation of the arterioles and passive occlusion of the veins.

IV. Ejaculation includes emission—emptying of semen into the urethra—followed by expulsion of the semen from the urethra

Hormonal Control of Male Reproductive Function

I. Hypothalamic GnRH stimulates the anterior pituitary to secrete FSH and LH, which then act on the testes: FSH on the Sertoli cells to stimulate spermatogenesis and inhibin secretion, and LH on the Leydig cells to stimulate testosterone secretion.

   A. Testosterone exerts a negative-feedback inhibition on both the hypothalamus and the anterior pituitary to reduce LH secretion.

   B. Inhibin exerts a negative-feedback inhibition on FSH secretion.

II. Testosterone, acting locally on the Sertoli cells, is essential for maintaining spermatogenesis.

III. Testosterone also maintains the accessory reproductive organs and male secondary sex characteristics and stimulates growth of muscle and bone. In many of its target cells it must first undergo transformation to dihydrotestosterone or (in the brain) to estradiol.

Section B. Female Reproductive Physiology

Ovarian Function

I. The female gonads, the ovaries, produce ova and secrete estrogen, progesterone, inhibin, and relaxin.

II. The two meiotic divisions of oogenesis result in each ovum having 23 chromosomes, in contrast to the 46 of the original oogonia.

III. The follicles surrounding the ova consist of inner layers of granulosa cells and outer layers of theca cells.

IV. At the beginning of each menstrual cycle, several follicles begin to develop into antral follicles, but soon only the largest (dominant) follicle continues its development to full maturity and ovulation.

V. Following ovulation the remaining cells of that follicle are transformed into a corpus luteum, which lasts about 10 to 14 days if pregnancy does not occur.

VI. The menstrual cycle can be divided, according to ovarian events, into a follicular phase and a luteal phase, which are approximately 14 days each and separated by ovulation.

Control of Ovarian Function

I. During the early and middle follicular phases, FSH stimulates the granulosa cells to proliferate and secrete estrogen, and LH stimulates the theca cells to proliferate and produce the androgens that the granulosa cells use to make estrogen. During this period, estrogen exerts a negative feedback on the hypothalamus and anterior pituitary to inhibit the secretion of GnRH and the gonadotropins.

II. During the late follicular phase, plasma estrogen becomes high enough to elicit a surge of LH, which then causes, via the granulosa cells, completion of the oocyte's first meiotic division and cytoplasmic maturation, ovulation, and formation of the corpus luteum.

III. During the luteal phase, under the influence of small amounts of LH, the corpus luteum secretes progesterone and estrogen. Regression of the corpus luteum results in a cessation of the secretion of these hormones.

Uterine Changes in the Menstrual Cycle

I. The ovarian follicular phase is equivalent to the uterine menstrual and proliferative phases, the first day of menstruation being the first day of the cycle. The ovarian luteal phase is equivalent to the uterine secretory phase.

A. Menstruation occurs when the plasma progesterone and estrogen levels fall as a result of the regression of the corpus luteum.

B. During the proliferative phase, estrogen stimulates growth of the endometrium and myometrium and causes the cervical mucus to be readily penetrable by sperm.

C. During the secretory phase, progesterone converts the estrogen-primed endometrium to a secretory tissue and makes the cervical mucus relatively impenetrable to sperm.

Other Effects of Estrogen and Progesterone

I. The many effects of estrogen and progesterone are summarized in Table 18-8.

Pregnancy

I. After ovulation, the ovum is swept into a uterine tube, where a sperm, having undergone capacitation and activation in the female tract, fertilizes it.

II. Following fertilization the ovum undergoes its last meiotic division, and the nuclei of the ovum and sperm fuse. Reactions in the ovum block penetration by other sperm and trigger cell division and embryogenesis.

III. The zygote undergoes cleavage, becoming a blastocyst that implants in the endometrium on approximately day 7 after ovulation.

A. The trophoblast gives rise to the fetal part of the placenta, whereas the inner cell mass develops into the embryo proper.

B. Although they do not mix, fetal blood and maternal blood both flow through the placenta, exchanging gases, nutrients, and waste products.

C. The fetus is surrounded by amniotic fluid in the amniotic sac.

IV. The progesterone and estrogen required to maintain the uterus during pregnancy come from the corpus luteum for the first two months, their secretion stimulated by chorionic gonadotropin produced by the trophoblast.

V. During the last 7 months of pregnancy, the corpus luteum is not important because the placenta itself produces large amounts of progesterone and estrogen.

VI. The high levels of progesterone, in the presence of estrogen, inhibit the secretion of GnRH and, thereby, that of the gonadotropins, so that menstrual cycles are eliminated.

VII. Parturition occurs by rhythmical contractions of the uterus, which first dilate the cervix and then move the infant, followed by the placenta, through the birth canal. The contractions are stimulated by a decrease in the progesterone to estrogen ratio (progesterone inhibits and estrogen stimulates contractions), by oxytocin released from the posterior pituitary in a reflex triggered by uterine mechanoreceptors, and by uterine prostaglandins.

VIII. The breasts develop markedly during pregnancy as a result of the combined influences of estrogen, progesterone, prolactin, and placental lactogen.

A. Prolactin secretion is stimulated during pregnancy by estrogen acting on the anterior pituitary, but milk is not synthesized because high concentrations of estrogen and progesterone inhibit the milk-producing action of prolactin on the breasts.

B. As a result of the suckling reflex, large bursts of prolactin and oxytocin occur during nursing, the oxytocin causing milk ejection.

Section C. The Chronology of Reproductive Function

Sex Determination and Sex Differentiation

I. Sex is determined by the two sex chromosomes: Males are XY and females are XX.

II. A gene on the Y chromosome codes for a protein whose action leads to the development of testes. In the absence of a Y chromosome, testes do not develop and ovaries do instead.

III. When a functioning male gonad is present to secrete testosterone and MIH, a male reproductive tract and external genitalia develop. In the absence of testes, the female system develops.

Puberty

I. At puberty, the hypothalamic-anterior-pituitary-gonadal chain of hormones becomes active due to a change in brain function that permits increased secretion of GnRH.
II. In girls the first sign of puberty is either the beginning of breast development or appearance of pubic hair. In boys it is acceleration of growth of testes and scrotum.

Menopause

I. Fertility in women peaks in their mid twenties and declines gradually after age 30.
II. Around the age of 50, a woman's menstrual periods become less regular and ultimately disappear—the menopause.
   A. The cause of menopause is a decrease in the number of ovarian follicles and their hyporesponsiveness to the gonadotropins.
   B. The symptoms of menopause are largely due to the marked decrease in plasma estrogen concentration.
III. Men show a steady decrease in testosterone secretion after age 40 but generally no complete cessation of reproductive function.

REVIEW QUESTIONS

1. Define:

gonads
testes
ovaries
gametes
spermatozoa
sperm
ova
ovum
sex hormones
testosterone
estrogen
progesterone
accessary reproductive organs

secondary sexual characteristics
gonadotropin-releasing hormone (GnRH)
pituitary gonadotropins
follicle-stimulating hormone (FSH)
luteinizing hormone (LH)
gametogenesis
germ cells
meiosis
crossing over

2. Describe the stages of gametogenesis and how meiosis results in genetic variability.

Section A. Male Reproductive Physiology

1. Define:

scrotum
spermatogenesis
seminiferous tubules
Leydig cells
interstitial cells
epididymis

vas (ductus) deferens
spermatic cord
ejaculatory ducts
seminal vesicles
prostate gland
bulbourethral glands

semen
spermatogonia
primary spermatocytes
secondary spermatocytes
spermatids
acrosome
Sertoli cell
blood-testis barrier
vasectomy

erection
impotence
ejaculation
emission
orgasm
inhibin
androgens
dihydrotestosterone
prolactin

2. Describe the sequence of events leading from spermatogonia to sperm.

3. List the functions of the Sertoli cells.

4. Describe the path taken by sperm from the seminiferous tubules to the urethra.

5. State the roles of the prostate gland, seminal vesicles, and bulbourethral glands in the formation of semen.

6. Describe the neural control of erection and ejaculation.

7. Diagram the hormonal chain controlling the testes. Contrast the effects of FSH and LH.

8. What are the feedback controls from the testes to the hypothalamus and pituitary?

9. List the effects of testosterone on accessory reproductive organs, secondary sex characteristics, growth, protein metabolism, and behavior.

Section B. Female Reproductive Physiology

1. Define:

female internal genitalia
uterine tubes
uterus
cervix
vagina
female external genitalia
vulva
clitoris
menstrual cycles
menstruation
oogenesis
oogonia
primary oocytes
secondary oocytes
first polar body
mature ovum
second polar body
follicles
primordial follicles
primary follicles
granulosa cells
zona pellucida
theca
antrum
dominant follicle
ovulation
corpus luteum
follicular phase
luteal phase

estradiol
LH surge
menstrual phase
endometrium
proliferative phase
secretory phase
myometrium
dysmenorrhea
premenstrual syndrome (PMS)
sperm capacitation
sperm activation
fertilization
zygote
ectopic pregnancy
cleavage
blastocyst
trophoblast
inner cell mass
embryo
fetus
implantation
placenta
chorion
chorionic villi
umbilical arteries
umbilical vein
umbilical cord
amniotic cavity

amnion
amniotic sac
amniotic fluid
amniocentesis
chorionic villus sampling
chorionic gonadotropin
  (CG)
estriol
placental lactogen
toxemia of pregnancy
parturition
relaxin
afterbirth

oxytocin
mammary glands
lactation
alveoli
myoepithelial cells
prolactin inhibiting
  hormone (PIH)
prolactin releasing hormone
  (PRH)
milk ejection reflex
contraception
oral contraceptives
intrauterine device (IUD)

2. Draw the female reproductive tract.

3. Describe the various stages from oogonium to mature ovum.

4. Describe the progression from a primordial follicle to a dominant follicle.

5. Name four hormones produced by the ovaries and name the cells that produce them.

6. What are the analogies between granulosa cells and Sertoli cells, and between theca cells and Leydig cells?

7. Diagram the changes in plasma concentrations of estrogen, progesterone, LH, and FSH during the menstrual cycle.

8. List the effects of FSH and LH on the follicle.

9. Describe the effects of estrogen on gonadotropin secretion during the early, middle, and late follicular phases.

10. List the effects of the LH surge on the ovum and the follicle.

11. What are the effects of the sex steroids on gonadotropin secretion during the luteal phase?

12. Describe the hormonal control of the corpus luteum and its history in a nonpregnant cycle.

13. What happens to the sex steroids and the gonadotropins as the corpus luteum degenerates?

14. Compare the phases of the menstrual cycle according to uterine and ovarian events.

15. Describe the effects of estrogen and progesterone on the endometrium, cervical mucus, and myometrium.

16. Describe the uterine events associated with menstruation.

17. List the effects of estrogen on the accessory sex organs and secondary sex characteristics.

18. List the effects of progesterone on the breasts and body temperature.

19. What are the source and effects of androgens in women?

20. How does the ovum get from the ovary to a uterine tube?

21. What are the mechanisms for the movement of sperm and ova in the uterine tubes, and where does fertilization normally occur?

22. Describe the events that occur during fertilization.

23. How many days after ovulation does implantation occur, and in what stage is the embryo at that time?

24. Describe the components of the placenta and the mechanisms of exchange between maternal and fetal blood.

25. State the sources of estrogen and progesterone during different stages of pregnancy. What is the dominant estrogen of pregnancy, and how is it produced?

26. What is the state of gonadotropin secretion during pregnancy, and what is the cause?

27. What anatomical feature permits coordinated contractions of the myometrium?

28. Describe the mechanisms and messengers that contribute to parturition.

29. List the effects of prostaglandins on the female reproductive system.

30. Describe the development of the breasts after puberty and during pregnancy, and list the hormonal effects responsible.

31. Describe the effects of hypothalamic releasing hormones and estrogen on prolactin secretion during pregnancy. Diagram the suckling reflex for prolactin release.

32. Which is the major hormone responsible for stimulating milk production, and why is there no milk production during pregnancy?

33. Diagram the milk ejection reflex.

Section C: The Chronology of Reproductive Function

1. Define:

sex determination
sex differentiation
sex chromosomes
X chromosome
Y chromosome
sex chromatin
karyotype
testes determining factor
Wolffian ducts

Müllerian ducts
Müllerian inhibiting
  hormone (MIH)
puberty
menarche
menopause
climacteric
osteoporosis

2. State the genetic difference between males and females and a method for identifying genetic sex.

3. Describe the sequence of events, the timing, and the control of the development of the gonads and the internal and external genitalia.

4. What is the state of gonadotropin and sex hormone secretion before puberty?

5. What is the state of estrogen and gonadotropin secretion after menopause?

6. List the changes that occur after menopause.

THOUGHT QUESTIONS

(Answers are given in Appendix A.)

1. What symptom will be common to a person whose Leydig cells have been destroyed and a person whose Sertoli cells have been destroyed? What symptom will not be common?

2. What sexual problem might a man have who was taking a drug that blocks cholinergic receptors?

3. A man who is sterile (unable to produce sperm capable of causing fertilization) is found to have no evidence of demasculinization, an increased blood concentration of FSH and a normal plasma concentration of LH. What is the most likely basis of his sterility?

4. If you were a scientist trying to develop a male contraceptive acting on the anterior pituitary, would you try to block the secretion of FSH or that of LH? Explain the reason for your choice.

5. A 30-year-old man has very small muscles, a sparse beard, and a high-pitched voice. His plasma concentration of LH is elevated. Explain the likely cause of all these findings.

6. There are disorders of the adrenal cortex in which excessive amounts of androgens are produced. If this occurs in a woman, what will happen to her menstrual cycles?

7. Women with inadequate secretion of GnRH are often treated for their sterility with drugs that mimic the action of this hormone. Can you suggest a possible reason that such treatment often is associated with multiple births?

8. Which of the following would be a signal that ovulation was soon to occur: the cervical mucus becoming thick and sticky; an increase in body temperature; a marked rise in plasma LH?

9. The absence of what phenomenon would interfere with the ability of sperm obtained by masturbation to fertilize an ovum in a test tube?

10. If a woman 7 months pregnant is found to have a marked decrease in plasma estriol but no change in plasma progesterone, what would you conclude?

11. What types of drugs might you work on if you were trying to develop one to stop premature labor?

12. If a genetic male failed to produce MIH during in utero life, what would the result be?

13. Could the symptoms of menopause be treated by injections of FSH and LH?

CHAPTER

19

DEFENSE MECHANISMS OF THE BODY

SECTION A
IMMUNOLOGY: DEFENSES AGAINST FOREIGN MATTER

Immunology is the study of the physiological responses by which the body destroys or neutralizes foreign matter, both living and nonliving, as well as cells of its own that have become altered in certain ways. Immune responses protect against infection by **microbes**—viruses, bacteria, fungi, and other parasites—and against entry of nonmicrobial foreign matter. They also destroy cancer cells that arise in the body, a function known as **immune surveillance**, and worn-out or damaged body cells such as old erythrocytes.

Immune responses can be classified into two categories: nonspecific and specific, which interact with each other.[1] **Nonspecific immune responses** nonselectively protect against foreign substances or cells without having to recognize their specific identities. **Specific immune responses** depend upon recognition of the substance or cell to be attacked.

Before introducing the cells that participate in immune responses, let us first look at the major microbes we shall be concerned with in this chapter—bacteria and viruses. **Bacteria** are unicellular organisms that have an outer coating, the cell wall, in addition to a plasma membrane. Bacteria can damage tissues at their sites of invasion or release, into the extracellular fluid, toxins that are carried by the blood and disrupt physiological functions in other parts of the body.

Viruses are essentially nucleic acids surrounded by a protein coat. Unlike bacteria, which can carry out metabolic activity and replicate independent of other cells, viruses lack both the enzyme machinery for energy production and the ribosomes essential for protein synthesis. Thus, they cannot multiply by themselves but must "live" inside other cells whose biochemical apparatus they make use of. The viral nucleic acid directs the host cell to synthesize the proteins required for viral replication, with the required nucleotides and energy sources also being supplied by the host cell.

The effect of viral habitation and replication within a cell depends upon the type of virus. Some viruses, after entering a cell, multiply rapidly, kill the cell, and then move on to other cells. Others replicate inside their host cells very slowly, and the viral nucleic acid may even become associated with the cell's own DNA molecules, replicating along with them and being passed on to the daughter cells during cell division. Such a virus may remain in the cell or its offspring for many years and then suddenly begin to multiply rapidly. As we shall see, cells infected with such "slow" viruses may be damaged by the body's own defense mechanisms turned against the cells because they are no longer recognized as "self." Finally, certain viruses cause transformation of their host cells into cancer cells.

CELLS MEDIATING IMMUNE RESPONSES

The cells that carry out immune responses are collectively termed the **immune system**, but they do not constitute a "system" in the sense of anatomically connected organs like the gastrointestinal or urinary systems. Rather, they are a diverse collection of cells found both in the blood and in organs and tissues throughout the body. The major cell types are listed in Table 19-1, along with their sites of production. Because of the large number of cells and the far larger number of chemical messengers that participate in immune responses, a miniglossary defining those cells and messengers discussed in this chapter is given at the end of Section A (Table 19-13, page 684).

The anatomy of the **leukocytes**, their origin in the bone marrow, and their distribution in blood were described on pages 356 to 357 and should be reviewed at this time. Unlike erythrocytes, the leukocytes use the blood mainly for transportation and leave the circulatory system to enter the tissues and function there. In this regard, the lymphocytes are the most complex of the leukocytes, and we shall delay until a subsequent section the description of the various classes and subclasses of lymphocytes, their origins, migrations, and major organs of residence—the lymphoid organs.

Plasma cells are not a distinct cell line but differentiate from a particular set of lymphocytes (the B lymphocytes) during immune responses. Despite their name, plasma cells are not usually found in the blood but rather in the tissues in which they differentiated from lymphocytes.

Macrophages are large cells found in virtually all the organs and tissues, their structures varying somewhat from location to location. They are derived from mono-

---

[1]Sometimes the nonspecific immune responses are not classified as immune responses. We believe, however, that the classification used in this chapter is useful because of the close interactions between nonspecific and specific responses.

TABLE 19-1   CELLS MEDIATING IMMUNE RESPONSES

Name	Site Produced
Leukocytes (white blood cells)	
Neutrophils	Bone marrow
Basophils	Bone marrow
Eosinophils	Bone marrow
Monocytes	Bone marrow
Lymphocytes	Bone marrow, thymus, and
B cells	peripheral lymphoid organs
T cells	
Cytotoxic T cells	
Helper T cells	
Suppressor T cells	
"Third population"*	
NK cells	
Plasma cells	Peripheral lymphoid organs; differentiate from B lymphocytes during immune responses
Macrophages	Almost all tissues and organs; differentiate from monocytes
Mast cells	Almost all tissues and organs; differentiate from basophils

*This term refers to a non-B and non-T population of lymphocytes that includes NK cells and another poorly understood subset not discussed in this chapter.

cytes that pass out of blood vessels to enter the tissues and become transformed into macrophages. By the time of birth, this process has already supplied the tissues with a large number of tissue macrophages that occupy fixed positions, but the migration of monocytes continues throughout life. In keeping with one of their major functions, the engulfing of particles, including microbes, macrophages are strategically placed where they will encounter their targets. For example, they are found in large numbers in the various epithelia in contact with the external environment, and, in several organs, they line the vessels through which blood or lymph flows. Some macrophages can leave their tissues and be carried by the lymph to other locations.

Mast cells are also found scattered throughout the organs and tissues. They are derived from differentiation of basophils that have left the blood vessels and have entered the interstitial fluid. Their most striking anatomical feature is a very large number of secretory vesicles, and they secrete many locally acting chemical messengers.

The functions of all these cells are briefly listed in Table 19-2 for reference and will be described in subsequent sections.

NONSPECIFIC IMMUNE RESPONSES

The nonspecific immune responses, those that protect against foreign matter without having to recognize its identity, include external barriers to invasion, like skin, and the response to injury known as inflammation.

External Anatomic and Chemical Barriers

The body's first lines of defense against infection are the barriers offered by surfaces exposed to the external environment. Very few microorganisms can penetrate the intact skin, and the sweat, sebaceous, and lacrymal glands all secrete antimicrobial chemicals. The mucus secreted by the epithelial linings of the respiratory and upper gastrointestinal tracts also contains antimicrobial chemicals, but more important, mucus is sticky. Particles that adhere to it are prevented from entering the blood. They are either swept by ciliary action up into the pharynx and then swallowed, as occurs in the upper respiratory tract, or are engulfed by macrophages in the various linings.

TABLE 19-2 MAJOR FUNCTIONS OF CELLS MEDIATING IMMUNE RESPONSES

Neutrophils	1. Phagocytosis. 2. Release chemicals involved in inflammation (vasodilators, chemotaxins, etc.).
Basophils	1. Have functions in blood similar to those of mast cells in tissues. 2. Enter tissues and are transformed into mast cells.
Eosinophils	Destroy parasitic worms.
Monocytes	1. Have functions in blood similar to those of macrophages in tissues. 2. Enter tissues and are transformed into macrophages.
B cells	1. Initiate antibody-mediated immune responses by binding specific antigens to their plasma-membrane receptors, which are antibodies. 2. During activation, are transformed into plasma cells, which secrete antibodies.
Cytotoxic T cells	Cell-mediated immunity, that is, bind to antigens on plasma membrane of target cells (mainly virus-infected cells and cancer cells) and directly destroy the cells.
Helper T cells	Secrete protein messengers (lymphokines) that activate B, cytotoxic T, and NK cells and convert macrophages into effector macrophages.
Suppressor T cells	Inhibit B and cytotoxic T cells.
NK cells	Same functions as cytotoxic T cells.
Plasma cells	Secrete antibodies.
Macrophages	1. Phagocytosis. 2. Process and present antigens to lymphocytes. 3. Secrete protein messengers (monokines) involved in inflammation, activation of helper T cells, and systemic responses to infection or injury (the acute phase response).
Mast cells	Release histamine and other chemicals involved in inflammation.

Other specialized surface barriers are the hairs at the entrance to the nose, the cough and sneeze reflexes, and the acid secretion of the stomach, which kills microbes. Finally, a major barrier to infection are the many relatively innocuous microbes normally found on the skin and other linings exposed to the external environment. These microbes suppress the growth of other potentially more dangerous ones.

Inflammation

Inflammation is the body's local response to infection or injury. It can be elicited by a variety of injurious agents—cold, heat, and trauma, for example—not just by microbes, but the response is relatively stereotyped regardless of cause. The function of inflammation is to destroy or inactivate the foreign invaders and/or set the stage for tissue repair.

The key actors in inflammation are **phagocytes**. This term denotes any cell capable of **phagocytosis**, the process whereby particulate matter is engulfed and destroyed inside the phagocyte. The most important phagocytes are neutrophils, monocytes, and macrophages.

In this section, we describe inflammation as it occurs in a nonspecific response. We shall see later that inflammation is an important component of many specific immune responses also, the difference being that in the latter the inflammation becomes amplified and made more effective.

The sequence of local events in a typical nonspecific inflammatory response to a bacterial infection, due say to cutting oneself with a bacteria-covered knife, is summarized in Table 19-3. Although our example is for an infection, many of the same events occur in response to tissue damage or the presence of a nonliving foreign body even when no infection is present.

The familiar manifestations of inflammation are local redness, swelling, heat, and pain. The events of inflammation that underlie these manifestations are induced and regulated by a large number of chemical mediators of varying origins, as summarized in Table 19-4. Based on origin, they fall into two general categories: (1) substances released into the extracellular fluid from cells that either already exist in the infected area or enter it during inflammation; and (2) peptides generated in the

---

TABLE 19-3 SEQUENCE OF EVENTS IN A LOCAL INFLAMMATORY RESPONSE TO BACTERIA

1. Initial entry of bacteria into tissue
2. Vasodilation of the microcirculation in the infected area, leading to increased blood flow
3. Marked increase in protein permeability of the capillaries and venules in the infected area, with resulting diffusion of protein and filtration of fluid into the extracellular fluid, causing swelling
4. Exit of neutrophils and, later, of monocytes from the capillaries and venules into the extracellular fluid of the infected area
5. Destruction of bacteria in the tissue either through phagocytosis or by mechanisms not requiring prior phagocytosis
6. Tissue repair

---

infected area by the enzymatic splitting of proteins that circulate in the plasma. An example of the first category is **histamine**, which is released from mast cells. An example of the second is the generation of **kinins** from the plasma protein kininogen.

Any given event, vasodilation for example, may be induced by multiple mediators, and any given mediator may induce more than one event. For example, histamine is a potent inducer of steps 2 to 4 of Table 19-3.

Note in Table 19-4 that phagocytes—neutrophils, monocytes, and macrophages—release important inflammatory mediators. Two of the protein mediators, **interleukin 1 (IL-1)** and **tumor necrosis factor (TNF)**, secreted by monocytes and macrophages will keep

recurring in this chapter because they play many roles, both local and hormonal, in the body's defenses (they were described in Chapter 17 as mediators of fever). All protein messengers secreted by monocytes and macrophages are collectively termed **monokines**.

Let us now go step by step through the process summarized in Table 19-3, assuming that the bacterial infection is localized to the tissue just beneath the skin and that the bacteria have not entered the blood. If entry into the blood or lymph were to occur, then similar inflammatory responses would take place in any other tissue or organ invaded by the blood-borne or lymph-borne bacteria (the latter case will be described in some detail in the section on specific immune responses). Also, microbes can be phagocytized in the blood itself by neutrophils and monocytes.

Vasodilation and increased permeability to protein. In response to the tissue injury and microbial entry, chemical mediators are released into the extracellular fluid from local cells or generated there from proteins in the plasma. These mediators dilate most of the microcirculation vessels in the damaged area. They also cause the local capillaries to become quite permeable to proteins by inducing the capillary endothelial cells to contract, opening spaces between them. The adaptive value of these vascular changes is twofold: (1) Increased blood flow to the inflamed area increases the delivery of phagocytic leukocytes and plasma proteins crucial for immune responses, and (2) increased capillary permeability to protein ensures that the relevant plasma proteins—many of which are normally restrained by the capillary membranes—can gain entry to the inflamed area.

TABLE 19-4 SOME IMPORTANT LOCAL INFLAMMATORY MEDIATORS

Mediator	Source
Kinins	Plasma proteins
Complement	Plasma proteins
Products of blood clotting	Plasma proteins
Histamine	Mast cells
Eicosanoids	Many cells
Platelet-activating factor	Many cells
Monokines (e.g., interleukin 1 and tumor necrosis factor)	Monocytes and macrophages
Lysosomal enzymes and oxygen-derived substances	Neutrophils and macrophages
Antibodies*	Lymphocytes
Lymphokines* (e.g., gamma interferon)	Lymphocytes

*Play a role only in specific immune responses.

As described in Chapter 13, page 396, the vasodilation and increased permeability to protein cause net filtration of plasma into the interstitial fluid and the formation of edema. This accounts for the swelling, which is simply a consequence of the changes in the microcirculation and has no known adaptive value of its own.

Chemotaxis. Within 30 to 60 min of the onset of inflammation, circulating neutrophils begin to stick to the inner surface of the endothelium of the capillaries and venules in the infected area. Then, a narrow projection of the neutrophil is inserted into the space between two endothelial cells and the entire neutrophil squeezes through the capillary wall and into the interstitium (**neutrophil exudation**). In this way, huge numbers of neutrophils migrate into the inflamed area and move toward the microbes.

The entire process, beginning with attachment to the endothelium, is known as **chemotaxis**, and any chemical mediator that can cause it is termed a **chemotaxin**. Several of the inflammatory mediators listed in Table 19-4 function as chemotaxins.

Movement of leukocytes from the blood into the damaged area is not limited to neutrophils. Monocytes follow later and once in the tissue undergo the anatomical and functional changes that transform them to macrophages. Meanwhile, some of the macrophages already present in the tissue may become active.

Phagocytosis. The initial step in phagocytosis is contact between the surfaces of the phagocyte and microbe. Such contact is not itself always sufficient to cause firm attachment and trigger engulfment, however, particularly with the many bacteria that are surrounded by a thick polysaccharide capsule. As we shall see, chemical factors produced by the body can bind the phagocyte to the microbe and markedly enhance phagocytosis. Any substance that does this is known as an **opsonin**, from the Greek word that means "to prepare for eating."

Following contact, the phagocyte engulfs the microbe

FIGURE 19-1 Phagocytosis. The foreign particulate substance is taken into the cell by endocytosis, and a membrane-bound phagosome is formed inside the phagocyte. The phagosome then merges with a lysosome, which brings together the digestive enzymes of the lysosome and the contents of the phagosome. After digestion has taken place in the phagolysosome, the end products are released to the outside of the cell by exocytosis or used by the cell for its own metabolism.

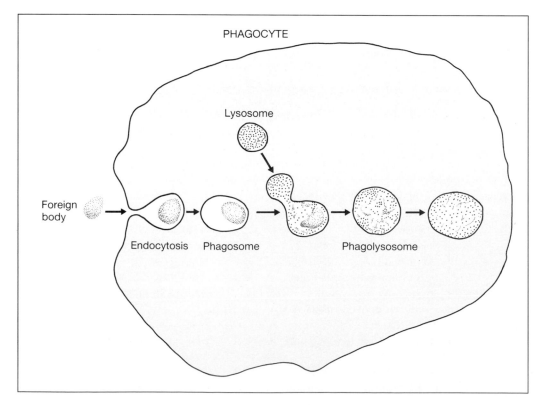

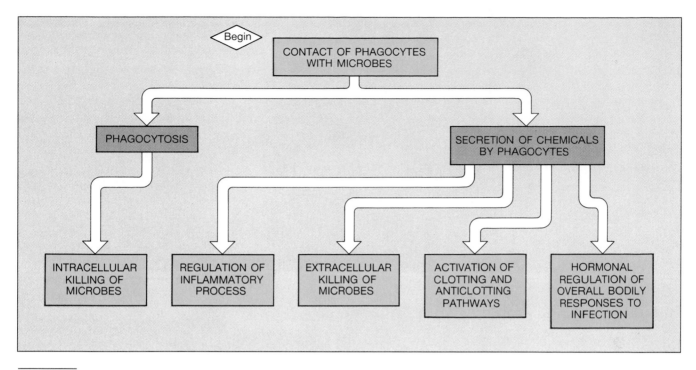

FIGURE 19-2 Role of phagocytes in nonspecific immune responses.

by endocytosis (Figure 19-1), and the microbe-containing sac formed in this step is called a **phagosome**. The microbe remains in the phagosome, a layer of plasma membrane separating it from the phagocyte's cytosol. The phagosome membrane makes contact with one of the phagocyte's lysosomes, which are filled with a variety of hydrolytic enzymes, the membranes of the two structures fuse, and the combined vesicles are now called the **phagolysosome**. Inside the phagolysosome, the microbe's macromolecules are broken down by the lysosomal enzymes. In addition, enzymes found in the phagolysosome membrane produce hydrogen peroxide and other oxygen derivatives all of which are extremely destructive to macromolecules.

Contact of phagocytes with microbes not only triggers phagocytosis but induces the phagocytes to secrete a large number of substances into the extracellular fluid (Figure 19-2). Some of these substances function as inflammatory mediators. Thus, a positive feedback occurs in which the more phagocytes enter the area and encounter microbes, the more inflammatory mediators are released to bring in more phagocytes. Other substances released by the phagocytes activate both the clotting and anticlotting pathways (described in Section C of this chapter). Still others, specifically those secreted by macrophages, enter the blood and exert widespread effects on the body. We will delay description of these systemic (as opposed to local) effects until a subsequent section.

Finally, the phagocytes can also secrete their antimicrobial hydrolytic enzymes and oxygen derivatives into the extracellular fluid, where these chemicals can destroy the microbes (and, sometimes, also damage surrounding normal tissue, as we shall see). Thus, phagocytes can also kill microbes without prior phagocytosis.

Extracellular killing by complement. The family of proteins known as **complement** provides another means for killing microbes without prior phagocytosis. Certain of the complement proteins always circulate in the blood in an inactive state. Upon activation of one of the group in response to infection or damage (the stimuli will be described in a moment), there occurs a cascade in which active molecules are generated in the extracellular fluid of the infected area from inactive precursors that have entered from the blood. Since this system consists of at least 20 distinct proteins, it is extremely complex, and we shall identify the roles of only a few of the individual proteins.

Five of the active proteins generated in the cascade form a complex, the **membrane attack complex (MAC)**, which imbeds itself in the microbial plasma membrane. In this manner, channels are created in the membrane, making it leaky. Water and salts enter the cell, disrupting the intracellular ionic environment and killing the microbe.

In addition to supplying a means for direct killing of

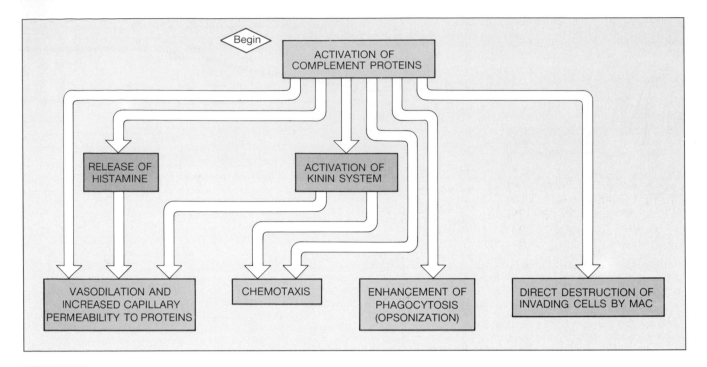

FIGURE 19-3 Functions of complement proteins.

microbes, the complement system serves other important functions in inflammation (Figure 19-3). Some of the activated complement molecules along the cascade cause both vasodilation and increased capillary permeability to protein, in part by stimulating release of histamine from mast cells. Some are chemotaxins and so stimulate neutrophil exudation and movement toward the microbes. Component C_{3b} is an opsonin—the only important one in nonspecific inflammatory responses—and attaches the phagocyte to the microbe (Figure 19-4).

How is the complement sequence initiated during nonspecific inflammation? As we shall see later, antibodies, a class of proteins secreted by lymphocytes, are required to activate the first protein (C_1) in the full sequence known as the classical complement pathway, but lymphocytes are not involved in the nonspecific inflammation we are presently describing. There is, however, an **alternate complement pathway**, one that is not antibody-dependent and bypasses C_1. The alternate pathway plugs in at approximately the middle of the cascade, with the formation of C_{3b}, and is triggered nonspecifically by certain carbohydrates on the microbial surface. Not all microbes have these molecules and so not all will trigger the alternate pathway.

Tissue repair. The final stage of inflammation is tissue repair. Depending upon the tissue involved, multiplication of organ-specific cells may or may not occur during

tissue repair. For example, liver cells multiply but neurons do not. In addition, fibroblasts (a type of connective-tissue cell) in the damaged area divide rapidly and begin to secrete large quantities of collagen. These events, like the others of inflammation, are brought about by chemical mediators.

FIGURE 19-4 Function of complement C_{3b} as an opsonin. One portion of this complement molecule binds nonspecifically to the surface of the microbe, whereas another portion binds to specific receptor sites for it on the plasma membrane of the phagocyte. The structures are not drawn to scale.

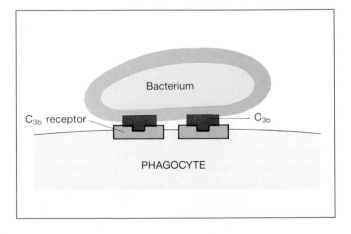

The end result of infection or damage anywhere in the body may be either complete repair, with or without a scar, or formation of a granuloma. A **scar** is a permanent mass of connective tissue that replaces normal tissue (this should not be confused with a scab, which is the temporary crust that forms over a skin wound during healing). A **granuloma** is a structure that develops when the inflammation has been caused by certain microbes, such as the bacteria causing tuberculosis, that are engulfed by phagocytes but survive within them. It also occurs when the inflammatory agent is a nonmicrobial substance (shrapnel, for example) that cannot be digested by the phagocytes. A granuloma is a ball of numerous layers of phagocytic-type cells, the central ones of which contain the offending material. The granuloma, embedded in the tissue, is usually surrounded by a fibrous capsule. Thus, a person may harbor live tuberculosis-producing bacteria for many years and show no ill effects as long as the microbes are contained within a granuloma and not allowed to escape.

Another mechanism for containing the spread of microbes, especially when tissue breakdown during inflammation is severe and the microbes cannot all be phagocytized, is formation of an **abscess**, a bag of pus—microbes, leukocytes, and liquified debris—walled off by fibroblasts and collagen. An abscess must be drained surgically, for it will not be absorbed spontaneously.

SPECIFIC IMMUNE RESPONSES

Lymphocytes mediate specific immune responses. As shown in Table 19-1, there are multiple types of lymphocytes, and our first task is to describe how these cells arise and the organs and tissues in which they come to reside.

Lymphoid Organs and Lymphocyte Types

Lymphoid organs. Like all leukocytes, lymphocytes circulate in the blood, but the great majority of lymphocytes are not in the blood but rather in a group of organs and tissues collectively termed the **lymphoid organs**. These are subdivided into the primary lymphoid organs and the peripheral lymphoid organs.

The **primary lymphoid organs** are the bone marrow and the thymus, which supply the peripheral lymphoid organs with mature lymphocytes already programmed to perform their functions. The bone marrow and thymus are not normally sites in which specific immune responses occur.

The **peripheral lymphoid organs** are the lymph nodes, spleen, tonsils, and the lymphocyte accumulations in the linings of the intestinal, respiratory, genital, and urinary tracts. It is in the peripheral lymphoid organs that lymphocytes are stimulated to participate in specific immune responses.

A distinction must be made between the "lymphoid organs" and the "lymphatic system," described in Chapter 13 (page 399). The latter is a network of lymphatic vessels and the lymph nodes found along these vessels. Of all the lymphoid organs, only the lymph nodes belong to the lymphatic system. There are no anatomical links, other than via the cardiovascular system, between the various lymphoid organs. Let us look briefly at these structures, excepting that of the bone marrow, which was described in Chapter 13.

The **thymus** lies in the upper part of the chest. Its size varies with age, being relatively large at birth and continuing to grow until puberty, when it gradually atrophies and is replaced by fatty tissue. Prior to its atrophy, the thymus consists mainly of lymphocytes that will eventually migrate via the blood to the peripheral lymphoid organs. It also contains endocrine cells that secrete a group of hormones, known collectively as **thymosin** (or thymopoietin), which exert a still poorly understood regulatory effect on the peripheral lymphocytes of thymic origin.

Recall from Chapter 13 that the fluid flowing along the lymphatic vessels is called lymph and that it is interstitial fluid that has entered the lymphatic capillaries and is being routed to the large lymphatic vessels that drain into veins in the neck. During this trip, the lymph flows through **lymph nodes** scattered along the vessels. Lymph, therefore, is the route by which lymph-node cells encounter the materials that trigger their immune responses. Each node is a honeycomb of sinuses (Figure 19-5) lined by macrophages, with large clusters of lymphocytes between the sinuses. Like other peripheral lymphoid organs, lymph nodes also contain cells that are distinguishable from macrophages but carry out macrophage-like functions. For simplicity, these cells are included in our use of the term macrophages.

The **spleen** is the largest of the lymphoid organs and lies in the left part of the abdominal cavity between the stomach and the diaphragm. In essence, the spleen is to the circulating blood what the lymph nodes are to the lymph. Blood percolates through the vascular meshwork of the spleen's interior, and large collections of lymphocytes and macrophages are found in the spaces of the meshwork. The macrophages of the spleen, in addition to interacting with lymphocytes, also phagocytize aging or dead erythrocytes.[2]

The **tonsils** are a group of small rounded organs in the pharynx. They are filled with lymphocytes and macro-

---

[2] In the fetus, the spleen is an important organ for forming all types of blood cells, but in the adult only lymphocytes are formed there.

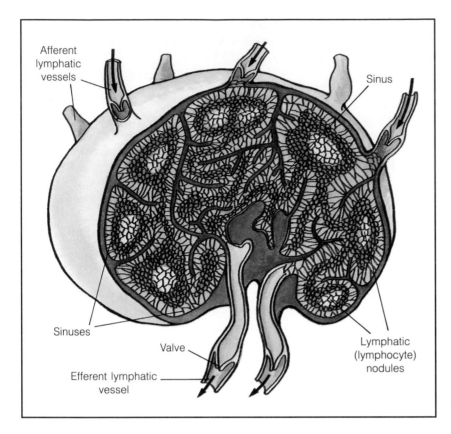

FIGURE 19-5 Anatomy of a lymph node.

phages and have openings ("crypts") to the surface of the pharynx. Their lymphocytes respond to infectious agents that arrive by way of ingested food as well as inspired air. Similarly the lymphocytes in the linings of the various tracts exposed to the external environment respond to infectious agents that penetrate into these linings from the lumen of the tract.

We have emphasized that the bone marrow and thymus supply lymphocytes to the peripheral lymphoid organs. Most of the lymphocytes in the peripheral organs are not, however, cells that originated in the primary lymphoid organs. The explanation is that, once in the peripheral organ, a lymphocyte can undergo mitosis to produce additional identical lymphocytes that in turn undergo mitosis and so on. In other words, all lymphocytes are *descended* from mature ancestors that were produced in the bone marrow or thymus, but may not themselves have arisen in these organs.

Finally, we must describe the source of the lymphocytes in blood. Some are cells on their way from the bone marrow or thymus to the peripheral organs but the vast majority are cells that are participating in lymphocyte traffic between the peripheral lymphoid organs, blood, lymph, and all the tissues of the body. Lymphocytes from the peripheral lymphoid organs constantly enter the lymph and are carried, via lymphatic vessels, to the blood. Simultaneously, some blood lymphocytes are pushing through the endothelium of blood-vessel capillaries or venules all over the body to enter the interstitial fluid. From there, they move into lymphatic capillaries and along the lymphatic vessels to lymph nodes. They may then leave the lymphatic vessels to take up residence in the node. This recirculation is going on all the time, not just during an infection. It greatly increases the likelihood that any given lymphocyte will encounter the target it is specifically programmed to recognize.

Lymphocyte types and their origins. There are multiple populations and subpopulations of lymphocytes (Table 19-1 and 19-2). The **B lymphocytes**, or simply **B cells,** mature in the bone marrow and then are carried by the blood to the peripheral lymphoid organs. This overall process of maturation and migration continues throughout a person's life. All generations of lymphocytes that subsequently arise from these cells by mitosis in the peripheral lymphoid organs will be identical to the parent cells, that is, will also be B cells.

In contrast to the B cells, other lymphocytes leave the bone marrow in an immature state during fetal and early neonatal life. They are carried to the thymus and mature in this organ before heading on to the peripheral lymphoid organs. They constitute the second major class of lymphocytes, the **T lymphocytes** or **T cells.** Production of T cells by the thymus is completed early in life, but like B cells, T cells also undergo mitosis in peripheral lymphoid organs, the offspring being identical to the original T cells.

Importantly, the maturation process that occurs in the thymus is not identical for all T cells, and there exist, therefore, multiple subsets of T cells. Based on their functions, they are termed **cytotoxic, helper,** and **suppressor T cells.**[3]

In addition to the B and T cells, there is another broad class of lymphocytes often termed the "third population." The origin of these cells is still unclear. This class has several subsets, but we shall deal only with the one called **natural killer cells** (**NK cells**).

The functions of all these lymphocyte types will be described in subsequent sections and are summarized in Table 19-2.

Antigens

Unlike the nonspecific defense mechanisms, lymphocytes must recognize the specific foreign matter to be attacked. Any foreign molecule that can trigger a specific immune response is termed an **antigen.** Most antigens are either proteins or very large polysaccharides. The term antigen does not denote a specific structure in the way anatomical terms like "microtubule" or "integral membrane protein" do. Rather it is a functional term, that is, any molecule, no matter what its function may otherwise be, that can induce a specific immune response is, by definition, an antigen. It is the ability of lymphocytes to distinguish one antigen from another that confers specificity upon the immune responses in which they participate.

In some cases the antigen is an individual molecule not part of a cell. A common example is ragweed pollen, the antigen that causes the specific immune response we know as hay fever. In other cases the antigen is part of the surface of a cell—microbe, virus-infected bodily cell, tumor cell, or transplanted cell—that appears in the body. The reason for saying "appears in" rather than "enters" is that, as we shall see, bodily cells, themselves, produce the "foreign" molecules that act as antigens on

---

[3]Another way to categorize T cells is not by function but rather by the presence of one of two proteins, called T_4 and T_8, in the plasma membrane of the cell: T_8 cells include cytotoxic and suppressor T cells, whereas T_4 cells are the helper T cells. Also, another designation coming into use for these cells uses the term CD rather than T; thus CD_4 and CD_8.

tumor cells and virus-infected cells. The molecules are foreign in the sense that they are not present in normal cells.

Antigen processing. In most specific immune responses, an antigen must be processed in the body before lymphocytes can interact with it. Macrophages are able to perform this task, and they form, therefore, a link between nonspecific and specific responses. After a microbe or noncellular antigen has been phagocytized by a macrophage in a nonspecific response, it is partially broken down to fragments that will be capable of reaction with specific receptors on the surfaces of lymphocytes. The antigen, if a protein, may also be unfolded inside the macrophage to expose internal amino acid sequences important for recognition by the lymphocyte receptors. The macrophages then shuttle these processed molecules, which still retain the capacity to act as antigens, to their plasma membrane surfaces, where they are accessible to the lymphocytes. In future discussions, we will use the term antigen to denote both these processed antigens and the original antigen.

Several other cell types related to macrophages can also perform this processing and "presenting" function. An example is the keratinocyte, the most numerous cell in the outer layers of the skin. Accordingly, the more general term **antigen-presenting cell** is often used to include all such cells, including macrophages.

Categories of Specific Immune Responses

As we shall see in a moment, there are two broad categories of specific immune responses and a variety of variations on a theme within each category. In following the details of all these responses, it is easy to lose sight of the forest because of the trees, so before dealing with the details, we present the basic pattern common to them.

A typical specific immune response can be divided into three stages: (1) antigen encounter and recognition by the lymphocytes, (2) lymphocyte activation, and (3) the attack.

1. A lymphocyte programmed to "recognize" a specific antigen encounters it, and the antigen becomes bound to receptors specific for that antigen on the surface of the lymphocyte. This binding is the physicochemical meaning of the word "recognize." Accordingly, the ability of lymphocytes to distinguish one antigen from another is determined by the nature of the receptors the lymphocytes have. Encounter and recognition may take place anywhere in the body since the lymphocytes are continuously moving about but most commonly occur in the peripheral lymphoid organs.

2. A lymphocyte that has combined with antigen undergoes cycles of mitotic divisions, and the progeny differentiate into cells that serve multiple functions, depending upon the lymphocyte type. This process of mitosis and differentiation, termed **lymphocyte activation**, usually occurs at the site of encounter and recognition. Some lymphocyte types, once activated, will direct or carry out the next step, the attack, while others will influence both the activation and function of these "attack" cells.

3. The activated cells launch an attack against all antigens of the kind that initiated the immune response. The result of the attack is to neutralize noncellular antigens and to destroy antigen-bearing cells wherever they may be in the body. It is essential to understand that, theoretically, it takes only one or two antigen molecules to initiate an immune response that will then attack all of the other antigens of that kind in the body. The nature of the attack varies depending upon the type of lymphocyte involved.

With this framework in mind, we now turn to the two broad categories of immune responses: (1) **antibody-mediated** or **humoral**, and (2) **cell-mediated**. These names denote the methods of attack utilized by the populations of lymphocytes that mediate the responses: B cells in the first category and both cytotoxic T cells and NK cells in the second (Figure 19-6).

Antibodies are proteins that are both present in the plasma membranes of B cells and secreted by them—to be more precise, secreted by the plasma cells into which B cells differentiate after they are activated. Via the blood,[4] the secreted antibodies travel all over the body to reach antigens of the kind that stimulated the immune response, combine with the antigens, and then direct an

---

[4]This explains why antibody-mediated immune responses are also called "humoral"; this word, though derived from the ancient belief concerning four bodily humors, has come to denote communication by way of chemical messengers in the blood, antibodies in this case.

FIGURE 19-6 Summary of roles of B, cytotoxic T, and helper T cells in immune responses. The two other less well understood lymphocyte populations are suppressor T cells, which inhibit B cells and cytotoxic T cells, and NK cells, which function similarly to cytotoxic T cells.

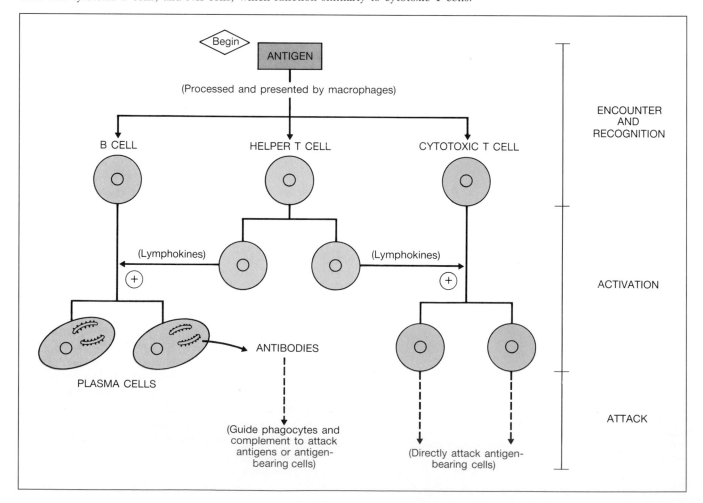

attack (by phagocytes and complement, as we shall see) that eliminates the antigens or the cells bearing them.

Antibodies belong to the family of proteins known as **immunoglobulins**. Each antibody molecule is composed of four interlinked polypeptide chains (Figure 19-7). The two long chains are called heavy chains, and the two short ones, light chains. There are five major classes of immunoglobulins, determined by the amino acid sequences in the heavy chains. The classes are designated by the letters A, D, E, G, and M after the symbol Ig (for immunoglobulin).

As illustrated in Figure 19-7, antibodies have a "stem," called the **Fc** portion and comprising the lower half of the two heavy chains, and two "prongs," each containing one **antigen-binding site** (also termed antibody-combining site)—the amino acid sequences that bind antigen. The amino acid sequences of the Fc portion are identical for all antibodies of the same class. In contrast, the amino acid sequences of the antigen-binding sites vary from antibody to antibody in a given class. Thus, each class contains many thousands of unique antibodies, each capable of combining with only one specific antigen or, in some cases, several antigens closely related in structure.

The second category of specific immune responses is cell-mediated immunity, in which cytotoxic T cells and/or NK cells travel to the location of cells bearing, on their surface, antigens of the kind that initiated the im-mune response and directly kill them, via secreted chemicals, without the intervention of antibody.

It is worth reemphasizing the important geographic difference in antibody-mediated and cell-mediated responses. In the former case, the lymphocytes remain in whatever location the recognition and activation steps occurred and send their antibodies forth, via the blood, to seek out antigens or antigen-bearing cells identical to those that triggered the response. In the cell-mediated responses, the lymphocytes themselves must enter the blood and seek out the targets. It should not be surprising, therefore, that the great majority of blood lymphocytes are T lymphocytes.

Although antibody-mediated and cell-mediated responses are often triggered by the same targets, several generalizations concerning their roles hold: (1) Antibody-mediated responses, carried out by B cells, have an extremely wide diversity of targets and are the major defense against bacteria, viruses, and other microbes in the extracellular fluid, and against toxic molecules; and (2) cell-mediated killing, carried out by cytotoxic cells and NK cells, is directed against a more limited number of targets, specifically the body's own cells that have become cancerous or infected with viruses.

The explanation of these different roles is that the receptors of each lymphocyte population have general characteristics that differ from those of the others and dictate the kinds of antigen with which lymphocytes in that population can combine during the recognition step. (We are talking here of receptor characteristics common to an entire class of lymphocyte, the B cells, for example, not the characteristics unique to each lymphocyte within that class.)

We have now assigned roles to the B, cytotoxic T, and NK cells. What roles do the other two classes of T cells perform? As their name implies, the helper T cells help activate both B cells and cytotoxic T cells. Thus, helper T cells participate in both humoral and cell-mediated immune responses. Indeed, these helper cells are essential for the production of antibodies, except in the case of a small number of antigens.

FIGURE 19-7 Antibody structure. The Fc portions are the same for all antibodies of a particular class. Each "prong" contains a single antigen-binding site. The links between chains represent disulfide bonds.

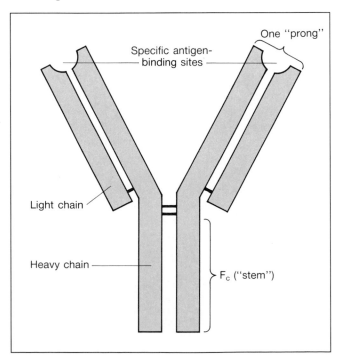

TABLE 19-5   CLASSIFICATION OF SPECIFIC IMMUNE RESPONSES

1. **Antibody-mediated** or **humoral:** Mediated by antibodies secreted by plasma cells, which arise from activated B cells. Constitute major protection against bacteria and viruses in the extracellular fluid.
2. **Cell-mediated:** Mediated by cytotoxic T cells and natural killer cells. Constitute major defense against intracellular viruses and cancer cells.

Both (1) and (2) are facilitated by helper T cells and inhibited by suppressor T cells.

Helper T cells go through the usual first two stages of the immune response in that they must combine with antigen and then undergo activation. Once activated, however, they do not direct an attack against antigens but rather secrete protein messengers that act on the B and cytotoxic T cells.

Suppressor T cells inhibit the function of both B cells and cytotoxic T cells. Suppressor T cells are still poorly understood, and thus we shall have relatively little to say about them in this chapter. Suffice it to emphasize that they seem to function, in a negative-feedback fashion, to dampen strong immune responses.

One last note on terminology: Protein chemical messengers secreted by monocytes and macrophages are termed monokines, and those secreted by lymphocytes are **lymphokines**. However, because some proteins are secreted by both monocytes/macrophages and lymphocytes, as well as several other cell types, the term **cytokine** is often used to include all protein messengers involved in immune responses.

Figure 19-6 and Table 19-5 summarize the basic interactions between B, cytotoxic T, and helper T cells presented in this section. The last pieces of information we need before describing various complete specific immune responses concern the characteristics of the B-, T-, and NK-cell receptors.

Lymphocyte Receptors

B-cell receptors. Each B cell always displays on its plasma membrane copies of the antibody it is capable of producing. This surface antibody acts as the receptor for the antigen specific to it.

Any given B cell or clone of identical B cells produces antibodies with unique antigen-binding sites. Therefore, the body has had to arm itself with millions of different B cells in order to ensure that antibodies will exist that are specific for the vast number of different antigens the organism *might* encounter during its lifetime. The antibody that any given B cell and all of its offspring can produce was determined during the cell's maturation in the bone marrow. Diversity arose as the result of a unique series of rearrangements of the multiple genes that code for the variable regions of the antibodies.

T-cell receptors. Since T cells do not produce antibodies, T-cell surface receptors for antigen are not antibodies. Rather, they are two-chained proteins that, like antibodies, have specific regions that differ from one T cell to another. As in B-cell development, multiple gene transformations occurred during T-cell maturation, leading to a large repertoire of distinct T cells—distinct in that each cell and its offspring possesses receptors of a single specificity. For T cells, this maturation occurs during their residence in the thymus.

In addition to their structural differences, the B- and T-cell receptors differ in a much more important way: The T-cell receptor cannot combine with antigen unless the antigen becomes complexed with certain of the body's own plasma-membrane proteins. The T-cell receptor combines with the entire complex of antigen and bodily protein.

The plasma-membrane proteins that must be complexed with the antigen for T-cell recognition to occur are a group of proteins coded for by genes known as the **major histocompatibility complex** (**MHC**) and therefore called **MHC proteins**.[5] Since no two persons other than identical twins have the same MHC genes, no two persons have the same MHC proteins on the plasma membranes of their cells.

There are two classes of MHC proteins: I and II. Class I proteins are found on the surface of virtually all cells of a person's body, excluding erythrocytes. Class II proteins are found only on the surface of macrophages and a few other cell types, including B cells.

The MHC proteins are often termed restriction elements since the ability of a T cell's receptor to recognize an antigen is restricted to situations in which the antigen is complexed with an MHC protein. The different subsets of T cells do not all have the same MHC requirements (Table 19-6): Cytotoxic and suppressor T cells require antigen to be associated with Class I proteins, whereas helper T cells require Class II proteins.

At this point you might well be asking: How do antigens, which are foreign, end up on the surface of the body's own cells complexed with MHC proteins? The answer will be given for each class in subsequent sections.

NK-cell receptors. NK-cell receptors show no MHC restriction, and it is unclear how is it that they have targets similar to those of the cytotoxic T cells. Of considerable importance is the fact that any given NK-cell receptor is much less specific than most lymphocyte receptors and can bind to a broad spectrum of antigens.

This completes our framework for understanding specific immune responses. We can now present typical responses for each system from beginning to end.

Antibody-Mediated Immune Responses

A classical antibody-mediated response is that which results in the destruction of bacteria. The sequence of events is summarized in Table 19-7.

---

[5]The term "histocompatibility" means "tissue compatibility," and it refers to the fact that how likely it is that a tissue transplant from one person to another will take is often evaluated by seeing how closely the two persons' MHC proteins match.

TABLE 19-6   MHC RESTRICTION OF THE LYMPHOCYTE RECEPTORS

Cell Type	MHC Restriction
B	None
Helper T	Class II, found only on macrophages and a few other cell types, including B cells
Cytotoxic T and suppressor T	Class I, found on all nucleated cells of the body
NK	None

Antigen processing and recognition. The story starts in the same way as for nonspecific responses, with the bacteria penetrating one of the body's linings in contact with the external environment and entering the interstitial fluid. There, they can be phagocytized nonspecifically by a tissue macrophage, yielding a variety of bacterial antigens, including some from the bacteria's surface. The macrophage may remain in the lining and perform its presenting function there. Alternatively, the macrophage may move into a lymphatic vessel, carry the bacterial antigens to a lymph node, enter the node, and present the antigens there. A third possibility is for unphagocytized bacteria to enter the lymphatic system or bloodstream at their sites of bodily entry and be removed from the lymph or blood by the macrophages that line the sinuses of the lymph nodes and the spleen, respectively.

Thus, the initial encounter between macrophage-processed antigen and B cell and the binding of antigen by the B cell may occur simultaneously in more than one peripheral lymphoid organ. Importantly, the chances of an antigen-specific lymphocyte encountering the antigen in any of these locations is enhanced by the fact that so many lymphocytes are continuously recirculating between the organs, as described earlier.

One more point about antigen presentation: We have emphasized the role of macrophages and macrophage-like cells in performing this function, but B cells are capable of binding certain unprocessed, unpresented antigens, a fact to which we will return.

With few exceptions, the binding of antigen is not, by itself, adequate to trigger activation of the B cell. An additional requirement is signals in the form of lymphokines released into the interstitial fluid by helper T cells near the antigen-bound B cells.

B-cell activation: The role of helper T cells. What causes helper T cells to secrete their protein messen-

TABLE 19-7   SEQUENCE OF EVENTS IN ANTIBODY-MEDIATED IMMUNITY AGAINST BACTERIA

1. In peripheral lymphoid organs, bacterial antigen, usually processed by macrophages, binds to specific receptors on the plasma membrane of B cells.

2. Simultaneously, macrophages present to helper T cells processed antigen complexed to MHC Class II proteins on the macrophages and secrete IL-1, which acts on the helper T cells.

3. In response, the helper T cells secrete IL-2, which activates the helper T cells to proliferate and secrete other lymphokines that activate the antigen-bound B cells to proliferate and differentiate into plasma cells. Certain B cells differentiate into memory cells rather than plasma cells.

4. The plasma cells secrete antibodies specific for the antigen that initiated the response, and the antibodies circulate all over the body via the blood.

5. Antibodies combine with antigen on the surface of the bacteria anywhere in the body.

6. Presence of antibody bound to antigen facilitates phagocytosis of the bacteria by neutrophils and macrophages and activates the complement system, which further enhances phagocytosis and also directly kills the bacteria by making their membranes leaky.

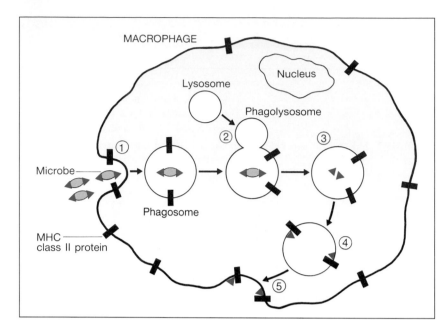

MACROPHAGE

Nucleus

Lysosome

Phagolysosome

① ② ③

Microbe

④

Phagosome

MHC
class II protein

⑤

FIGURE 19-8 Sequence of events by which processed antigen becomes inserted into macrophage plasma membrane in association with MHC Class II protein. The numbers are keyed to the text.

gers? The key point is that at the same time the B cell is binding antigen, the helper T cell is binding, via its receptor, the complex of antigenic fragments and Class II MHC proteins on the surface of a macrophage or other antigen presenting cell. Thus, macrophages present both uncomplexed antigen, which the B cell binds, and complexed antigen, which the helper T cell binds.

How does this complex arise? The processing of antigen by a macrophage begins with phagocytosis of a free antigen or of an antigen-bearing cell like our bacterium. The membrane of the phagosome formed during phagocytosis contains Class II MHC proteins, as did the plasma membrane from which it pinched off [(1) in Figure 19-8]. As usual, a lysosome merges with the phagosome to form a phagolysosome (2), within which the antigen is broken down (3). Some of the antigenic fragments bind to Class II proteins in the phagolysosomal membrane (4). One molecule of antigenic fragment becomes bound to one molecule of Class II MHC protein so that the antigen-MHC protein becomes a single complex.

The phagolysosome then migrates to the macrophage surface and inserts in the membrane (5). It is to this complex displayed on the surface of the macrophage that a helper T cell of the appropriate specificity binds via its receptor. In other words, the T cell attaches to the macrophage via the MHC-antigenic-fragment complex.

To repeat, macrophages simultaneously present antigen to both B and helper T cells in peripheral lymphoid organs. The two presentations can be performed by different nearby macrophages, not necessarily by a single one. Meanwhile, the macrophage is secreting a monokine mentioned in the section on inflammation—interleukin 1. IL-1 stimulates the helper T cell to en-

large and to secrete a lymphokine termed **interleukin 2 (IL-2)**. This messenger acts as an autocrine to stimulate the helper T cell to undergo activation. The cell undergoes mitotic cycles to form a clone of activated helper T cells, and these cells then release several other lymphokines (Figure 19-9).

It is these additional lymphokines that activate the antigen-bound B cells, which undergo mitotic cycles and differentiate into plasma cells.

Thus, as shown in Figure 19-9, we are dealing with a series of protein messengers interconnecting the various cell types, the helper T cells serving as the central coordinator: The macrophage releases IL-1, which acts on the helper T cell to stimulate release of IL-2, which stimulates the helper T cell to multiply, the activated progeny then releasing still other lymphokines that activate antigen-bound B cells.

A recent important discovery is that the roles of macrophages in the recruitment and activation of helper T cells can also be served by B cells. They, like macrophages, possess Class II MHC molecules on their membranes and can process and present antigen to helper T cells. B cells can also secrete IL-1. When one combines this information with the fact, mentioned earlier, that B cells can sometimes bind unprocessed antigen, it becomes apparent that macrophages are not always required for antibody-mediated immune responses (Figure 19-10).

Antibody secretion. The most striking aspect of the transformation of activated B cells into plasma cells is a marked expansion of the cytoplasm, which consists almost entirely of the granular endoplasmic reticulum

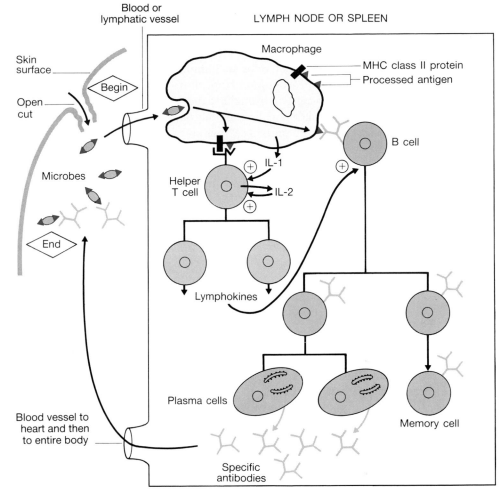

FIGURE 19-9 Sequence of events by which a bacterial infection leads to antibody synthesis in peripheral lymphoid organs and the secreted antibodies travel by the blood to the site of infection where they bind to bacteria of the type that induced the response. The processed bacterial antigen is presented to both a B cell and a helper T cell by macrophages, but the same macrophage need not present to both cells. Not shown is the fact that the B cell can sometimes also combine with unprocessed antigen.

found in cells that manufacture proteins for export. Plasma cells produce thousands of antibody molecules per second before they die in a day or so.[6]

Some of the B-cell progeny do not differentiate into plasma cells but rather become **memory cells** (Figure 19-9), ready to respond rapidly and effectively should the antigen ever reappear at a future time.

After their synthesis, antibodies are released from the plasma cell and enter the extracellular fluid. They may then enter the blood, which carries them from the peripheral lymphoid organ in which recognition and activation occurred to all tissues and organs of the body. At sites of infection, the antibodies leave the blood (recall that nonspecific inflammation had already made capillaries leaky at these sites) and combine with the type of

FIGURE 19-10 Case in which B-cell activation does not require macrophages. The B cell, itself, performs the macrophage functions of processing and presenting antigen to a helper T cell as well as secreting the IL-1 needed to activate the T cell.

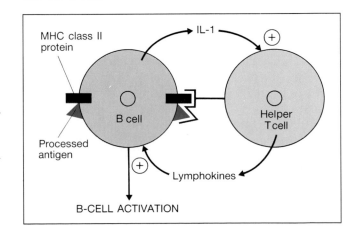

---

[6] What shuts off antibody formation? For one thing, elimination of the foreign material causing the immune response will do so. In addition, it is likely that the helper T cells, in addition to triggering B-cell proliferation and differentiation, may also perform the same functions for suppressor T cells, which then act on the B cells to dampen their activity.

bacterial surface antigen that had initiated the immune response (Figure 19-9). These antibodies will then direct the attack against the bacteria to which they are now bound.

Thus, antibodies play two distinct roles in immune responses: (1) During the initial recognition step, those on the surface of B cells bind to antigen brought to them; and (2) those secreted by the plasma cells seek out and bind to bacteria bearing the same antigens, "marking" them as the targets to be attacked.

The attack: Effects of antibodies. We mentioned earlier that there are five major classes of antibodies. The most abundant are the **IgG** antibodies, also commonly called **gamma globulins**. They and **IgM** provide the bulk of specific immunity against bacteria and viruses. **IgE** antibodies mediate allergic responses and participate in defenses against multicellular parasites. **IgA** antibodies are secreted by the linings of the gastrointestinal, respiratory, and genitourinary tracts. IgA antibodies generally do not circulate but act locally in the linings or on their surfaces. They are also secreted by the mammary glands and so are the major antibodies in milk. The functions of **IgD** are still unclear.

Thus, the different classes have different overall functions. However, the basic mechanisms by which they exert their effects follow common principles: The presence of antigen-antibody complexes triggers events that profoundly amplify inflammation. In other words, a common denominator of antibody-mediated immune mechanisms is that they enhance the vasodilation, vascular permeability to protein, neutrophil exudation, phagocytosis, and extracellular killing of microbes already started in a nonspecific way.

Activation of complement system. The single most important mechanism by which the presence of antigen-antibody complexes enhances inflammation is activation of the complement system. As described on page 659, the complement system, which directly kills microbes by its membrane attack complex and also enhances virtually every process in inflammation, including phagocytosis, is activated in the nonspecific inflammatory response via the alternate complement pathway. However, the presence of antibody attached to antigen increases participation of the complement system because the antibody-antigen complex is a powerful activator of the first step in the **classical complement pathway**. The first molecule in this pathway C_1 binds to complement receptors on the Fc portions of an antibody that has combined with antigen (Figure 19-11). This binding of C_1 activates the enzymatic portions of this complement molecule, thereby initiating the entire classical pathway.

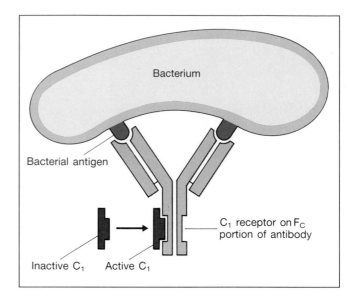

FIGURE 19-11 Initiation of classical complement pathway. The enzymatic ability of C_1 is activated by its binding to the Fc portion of an antibody, itself bound to specific antigen. The subsequent events, not shown in the figure, are: Active C_1 activates the next complement molecule in the pathway while still attached to the antibody; the membrane attack complex formed at the end of cascade kills the bacterium. The structures in the figure are not drawn to scale.

It is important to note that C_1 binds not to the unique antigen-binding sites in the antibody's prongs but rather to binding sites in the Fc portions. Since the latter are the same in virtually all antibodies of the IgG and IgM classes, the complement molecules will bind to *any* antigen-bound antibodies belonging to these classes. In other words, there is only one set of complement molecules, and once activated, they do essentially the same thing, regardless of the specific identity of the invader.

To summarize, the function of the antibodies is to identify invading cells as foreign by combining with specific antigens on the invading cell's surface. The complement system is subsequently activated when the first complement molecules in the sequence combine with this antigen-bound antibody, and the end product of this cascade, the MAC, does the actual destroying of the invaders by making their membranes leaky. Other activated complement molecules enhance phagocytosis and the steps in inflammation leading to it.

Direct enhancement of phagocytosis. Activation of complement is not the only mechanism by which antibodies enhance phagocytosis. Merely the presence of an IgG antibody attached to antigen on the microbe's surface has some enhancing effect on phagocytosis. In other words, antibodies can act directly as opsonins in addition to their indirect action via complement.

The mechanism is analogous to that for complement in that the antibody links the phagocyte to the antigen. As shown in Figure 19-12, the phagocyte has membrane receptors that bind to the Fc portion of antibodies. This linkage promotes attachment of the antigen to the phagocyte and the triggering of phagocytosis.

Direct neutralization of bacterial toxins and viruses. Bacterial toxins can act as antigens to induce antibody production. The antibodies then combine with the toxins, thus preventing the interaction of the toxin with susceptible cell-membrane sites. Since each antibody has two binding sites for combination with antigen, chains of antibody-antigen complexes are formed (Figure 19-13), and these chains are then phagocytized.

A similar binding process is the major antibody-mediated mechanism for eliminating viruses in the extracellular fluid. Certain of the viral surface molecules serve as antigens, and the antibodies produced against them combine with them, preventing attachment of the virus to plasma membranes of potential host cells. This prevents the virus from entering cells. As with bacterial toxins, chains of antibody-viral complexes are formed and are phagocytized.

Active and passive humoral immunity. We have been discussing antibody formation without regard to the course of events in time. The response of the antibody-producing machinery to invasion by a foreign antigen varies enormously, depending upon whether the machinery has previously been exposed to that antigen. Antibody production responds slowly over several days to the first contact with a microbial antigen, but any subsequent infection by the same invader elicits an immediate and marked outpouring of additional specific antibody (Figure 19-14). This response, which is mediated by the memory cells described earlier, confers a greatly enhanced resistance toward subsequent infection with that particular microorganism. Resistance built up as a result of the body's contact with microorganisms and their toxins or other antigenic components is known as **active immunity**.

Until this century, the only way to develop active immunity was to suffer an infection, but now the injection of microbial derivatives in **vaccines** is used. The material in the vaccine may be small quantities of living or dead microbes, small quantities of toxins, or harmless antigenic molecules derived from the microorganism or its toxin. The general principle is always the same: Exposure of the body to the agent results in the induction of the memory cells required for rapid, effective response to possible future infection by that particular organism.

A second kind of immunity, known as **passive immunity**, is simply the direct transfer of actively formed antibodies from one person (or other animal) to another, the recipient thereby receiving preformed antibodies. Such transfers normally occur between mother and fetus since IgG is selectively transported across the placenta. Breast-fed children also receive antibodies in the mother's milk. These are important sources of protection for

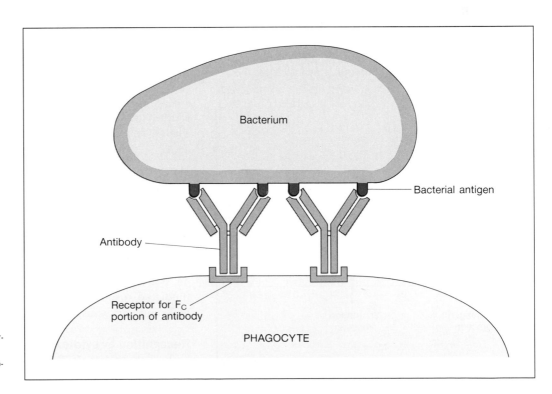

FIGURE 19-12 Direct enhancement of phagocytosis by antibody. The antibody links the phagocyte to the bacterium.

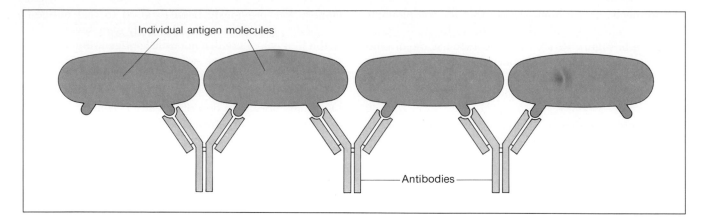

FIGURE 19-13 A chain of interlocking antigen-antibody complexes.

the infant during the first months of life, when the antibody-synthesizing capacity is relatively poor.

The same principle is used clinically when specific antibodies or pooled gamma globulin is given to a person exposed to or suffering from certain infections, such as measles, hepatitis, or tetanus. The protection afforded by this transfer of antibodies is relatively short-lived, usually lasting only a few weeks. The procedure is not without danger since the injected antibodies, often of nonhuman origin, may themselves serve as antigens, eliciting antibody production by the recipient and possibly severe allergic responses.

Summary. We can now summarize the interplay between nonspecific and specific humoral immune mecha-

FIGURE 19-14 Rate of antibody production following initial contact with an antigen and subsequent contact with the same antigen.

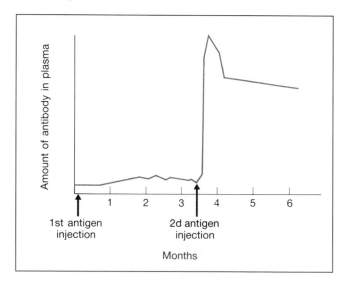

nisms in resisting a bacterial infection. When we encounter particular bacteria for the first time, *nonspecific* defense mechanisms resist their entry and, if entry is gained, attempt to eliminate them by phagocytosis and nonphagocytic killing. Simultaneously, the bacterial antigens induce the specific B-cell clones to differentiate into plasma cells capable of antibody production. If the nonspecific defenses are rapidly successful, these slowly developing *specific* immune responses may never play an important role. If the nonspecific responses are only partly successful, the infection may persist long enough for significant amounts of antibody to be produced. The presence of antibody leads to both enhanced phagocytosis and direct destruction of the foreign cells. All subsequent encounters with that type of bacteria will be associated with the same sequence of events, with the crucial difference that the specific responses may be brought into play much sooner and with greater force; that is, the person may have active immunity against that type of bacteria.

Cell-Mediated Immune Responses: Defenses against Virus-Infected and Cancer Cells

The second broad category of specific immune responses is cell-mediated immunity. In humoral immunity, as we just saw, the antigen-recognition step is carried out by a B cell whereas the destruction of the antigen-bearing invader is achieved either by phagocytes or by complement circulating in the blood. In contrast, both the recognition and the killing steps in cell-mediated immunity are carried out by a single cell, either a cytotoxic T cell or an NK cell. Figure 19-15 summarizes a typical cell-mediated response triggered by viral infection of body cells and serves as a guide for the rest of this section.

Recognition by cytotoxic and NK cells. As mentioned earlier, the cytotoxic T-cell receptor binds specific anti-

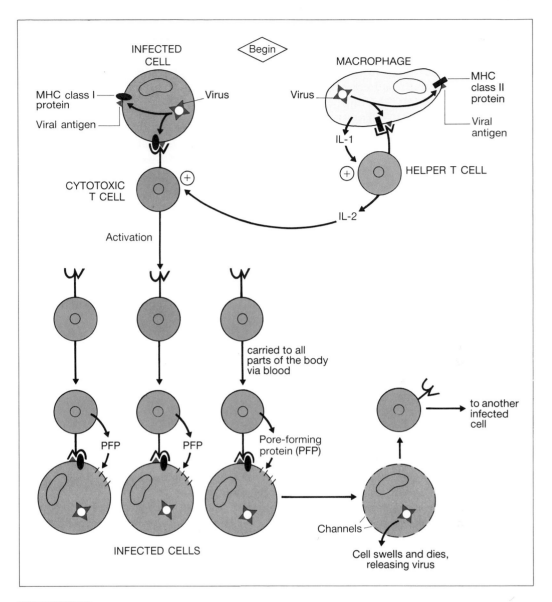

FIGURE 19-15 Cell-mediated killing of virus-infected cells by cytotoxic T cells. Specific cytotoxic T cells bind to viral antigen complexed to MHC Class I protein on an infected cell. Simultaneously, helper T cells have bound to processed viral antigens on macrophages (in association with the macrophage's MHC Class II proteins), have become activated, and are secreting IL-2. Under stimulation by this lymphokine, the cytotoxic T cells proliferate, and many enter the blood to reach other cells infected with the same virus. They bind to the infected cell's antigen-MHC Class I complexes and kill by releasing pore-forming protein molecules, which insert into the membrane of the infected cells and make them leaky. The infected cell swells and dies, releasing virus into the extracellular fluid, where it can be attacked by antibodies. The cytotoxic T cell is not damaged in the process but can move on to another infected cell. Natural killer cells function in the same way except that their attachment to target cells is not MHC restricted. For clarity, we have not shown in the figure the fact that helper T cells also secrete gamma interferon, which enhances the killing abilities of cytotoxic T cells (and NK cells).

gen complexed with the body's own Class I MHC proteins. This MHC restriction explains why cytotoxic T cells recognize virus-infected cells and cancer cells (as opposed to cells that have no connection with the body's own cells). Let us look at each situation in turn.

Once a virus has taken up residence inside a host cell,

the virus itself is not available for triggering immune responses. However, some of the proteins that the viral nucleic acid causes the host cell to manufacture end up not in new viruses but rather inserted in the plasma membrane of the host cell complexed with one of the cell's own Class I MHC proteins. Recall from page 666 that, in contrast to MHC Class II proteins, Class I proteins are produced by all of a person's cells and so this complex can be formed in any virus-infected cell. The complex serves as the binding site for antigen-specific cytotoxic T cells.

Most cancer cells also have one or more membrane proteins different from those of other body cells. This is because cancer arises as a result of genetic changes induced by chemicals, radiation, viruses, or other factors in previously normal body cells. The altered genes specify the malignant traits of the cell and the new "foreign" plasma-membrane protein. A complex of this antigen and one of the cell's own Class I MHC proteins provides the binding site for antigen-specific cytotoxic T cells.

What about NK cells? As noted earlier, their receptors show no MHC restriction, and it is unclear why they have targets similar to those of the cytotoxic T cells.

Helper T cells and lymphocyte activation in cell-mediated immunity.

Helper T cells facilitate cell-mediated immunity in ways similar to those in antibody-mediated immunity. They are brought into play when macrophages phagocytize viruses or cancer cells and then process and present the antigens, in association with the macrophage's MHC Class II proteins, to the helper T cells.

In response to binding the antigen and to stimulation by IL-1 secreted by the participating macrophages, the helper T cells release IL-2. As we have seen, this lymphokine acts as an autocrine to stimulate the proliferation of helper T cells. We now add the fact that the IL-2 secreted by helper T cells also activates cytotoxic T and NK cells, stimulating them to proliferate and to be able to attack their targets more vigorously.

Why is proliferation important if a cytotoxic T or NK cell has already found its target? There is rarely just one infected cell during an infection or one cancer cell. By increasing the pool of cytotoxic T and NK cells capable of recognizing the particular antigen, proliferation increases the likelihood that the other infected or cancer cells will be encountered by one of the killer cells.

In addition to IL-2, helper T cells secrete at least one other lymphokine, **gamma interferon**, that promotes the functions of cytotoxic T and NK cells. Gamma interferon belongs to a family of proteins collectively called interferon, which we shall describe in a subsequent section.

Activated NK cells also secrete gamma interferon, and so stimulate themselves in an autocrine manner.

The attack in cell-mediated immunity.

There are several mechanisms of target-cell killing in cell-mediated immunity, but the best-defined one is as follows. The cytotoxic T or NK cell releases, by exocytosis, the contents of its secretory vesicles into the extracellular space between itself and the target cell to which it is bound. These vesicles contain molecules of a protein, **pore-forming protein**, that then insert into the target cell's membrane like staves in a barrel and form channels through the membrane. In this manner, pore-forming protein causes the attacked cell to become leaky and burst. The fact that pore-forming protein is released into an enclosed space between the tightly attached attacking cell and the target assures that innocent bystander cells will not be killed since pore-forming protein is not at all specific.

Thus, humoral immunity, which uses circulating complement, and cell-mediated killing, which uses pore-forming protein released by the lymphocyte itself, can cause the death of their targets in the same way—by channel formation. Indeed, pore-forming protein and the proteins of complement's membrane attack complex are very similar in structure.

Effector Macrophages in Specific Immune Responses

In addition to its effects on cytotoxic T and NK cells, gamma interferon also has important effects on macrophages. Under the influence of this protein messenger (and probably other lymphokines as well) secreted by helper T cells in specific immune responses, macrophages undergo changes to become **effector macrophages** (Figure 19-16). These macrophages can eliminate their targets much more effectively and rapidly than ordinary macrophages. Most important, the effector macrophage is particularly good at eliminating cancer cells and cells infected with viruses or other microbes. Thus, these kinds of target cells are attacked not only by cytotoxic T cells and NK cells but by effector macrophages, as well.

The effector macrophage is primed to secrete large amounts of monokines. One of these, tumor necrosis factor (TNF), mentioned earlier as an important inflammatory mediator, has been shown to destroy tumors (hence its name) in vitro and in experimental animals. For this reason TNF is undergoing clinical testing in patients with terminal cancer.

Because the targets they attack are similar to the targets of cytotoxic T and NK cells, we have placed the description of effector macrophages after the section on cell-mediated immunity. You should recognize, how-

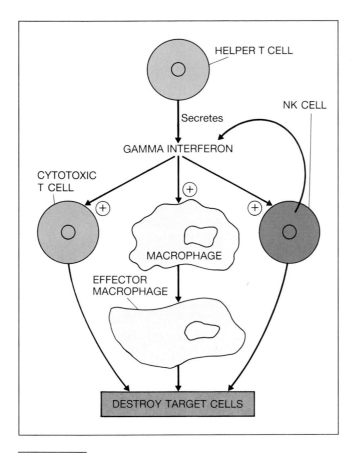

FIGURE 19-16 Role of gamma interferon, secreted by helper T cells and NK cells, in stimulating the killer ability of cytotoxic T cells, NK cells, and macrophages. Note that gamma interferon functions as an autocrine for NK cells since it is secreted by these cells and acts upon them.

TABLE 19-8 SOME IMPORTANT ACTIONS OF LYMPHOKINES SECRETED BY HELPER T CELLS

1. Interleukin 2 exerts several effects on T-cell populations:
   a. It acts as an autocrine to stimulate its cell of origin, helper T cells, to proliferate and secrete other lymphokines.
   b. It stimulates the proliferation and differentiation of cytotoxic T cells and NK cells.
 2. Several lymphokines stimulate B-cell proliferation and differentiation into antibody-secreting plasma cells.
 3. Gamma interferon transforms macrophages into effector macrophages, which are highly effective in killing target cells.

ever, that macrophages can become effector macrophages during antibody-mediated immunity as well since helper T cells secrete gamma interferon in both antibody-mediated and cell-mediated immune responses.

In the course of this chapter, we have mentioned several lymphokines secreted by helper T cells. Table 19-8 summarizes this information.

Summary of Defenses against Viral Infection

In various sections we have touched on the body's defenses against viruses. This section provides a summary and adds one new component—the ability of interferon to inhibit viral replication. The "Comment" column of Table 19-9 shows that host defense mechanisms against viruses operate in two spheres: (1) when the virus is in the extracellular fluid and (2) after the virus has taken up residence within the body's cells.

Antibody-mediated responses can function against viruses only when the viruses are in the extracellular fluid. In contrast, the defense against intracellular virus is mainly cell-mediated, by cytotoxic T and NK cells. In order to attack a virus once it has entered a cell, the T cell or NK cell must destroy the invaded cell. Effector macrophages also perform a similar function. With destruction of the host cell comes release of the viruses, which can then be directly neutralized by circulating antibody. Generally, only a few host cells must be sacrificed in this way, but once viruses have had a chance to replicate and spread from cell to cell, so many virus-infected host cells may be killed by the body's own defenses that organ malfunction may occur.

Interferon. In an earlier section we described how gamma interferon, secreted by helper T and NK cells, stimulates cytotoxic T cells, NK cells, and macrophages to be more potent killer cells. As is so often the case in immunology, this mediator has other effects as well. It is a member of a larger family of proteins, the other two members being alpha and beta interferons, collectively termed **interferon**. All three interferons nonspecifically inhibit viral replication inside host cells. This effect is quite distinct from the effect of gamma interferon on macrophages and other killer cells.

In response to infection by a virus, several different cell types, not just helper T cells, secrete one or more of the interferons into the extracellular fluid. Interferon then binds to plasma-membrane receptors on nearby cells, whether they are infected or not (Figure 19-17). It also enters the circulation and reaches cells at far-removed sites. Thus cells that can synthesize interferon provide it to cells that cannot.

How does interferon prevent viral replication? Its binding to the plasma membrane triggers the synthesis of several enzymes by the cell. If the cell is infected or

TABLE 19-9 SUMMARY OF HOST RESPONSES TO VIRUSES

	Main Cells Involved	Comment on Action
Nonspecific Responses		
Anatomic barriers	Body surface linings	Physical barrier; antiviral chemicals
Inflammation	Tissue macrophages	Phagocytosis of extracellular virus
Specific Responses		
Humoral immunity	Plasma cells derived from B cells secrete antibodies that perform the actions listed to the right	Neutralize virus and thus prevent entry to cell Activate complement, which leads to enhanced phagocytosis of extracellular virus
Cell-mediated immunity	Cytotoxic T and NK cells secrete chemicals that perform the actions listed to the right	Destroy host cell and thus induce release of virus so that it can be phagocytized
Effector macrophages	Macrophages activated by gamma interferon secreted by helper T cells	Destroy virus-infected cells
Interferon	Multiple cell types after viruses enter them	Nonspecifically prevents viral replication inside host cells

eventually becomes infected, these enzymes block the synthesis of proteins the virus requires for replication. Interferon is not specific. Many, but not all, viruses induce interferon synthesis, and interferon in turn can inhibit the multiplication of many kinds of viruses.

Rejection of Tissue Transplants

Although antibodies play some role, the cell-mediated immune system is mainly responsible for the recognition and destruction of tissue transplants, called grafts. This is termed **graft rejection**.

Except in the case of identical twins, the Class I MHC proteins on the cells of a graft differ from the recipient's. So are the Class II molecules present on the macrophages present in the graft (recall that virtually all organs and tissues have macrophages). Accordingly, the MHC

FIGURE 19-17 Role of interferon in preventing viral replication. (A) An infected body cell secretes interferon (cells capable of doing so are helper T cells, NK cells, and several other cell lines), which enters the interstitial fluid and blood and binds to interferon receptors on adjacent or far-removed cells (B). This induces synthesis of proteins (C) that inhibit viral replication should virus enter the cell (D).

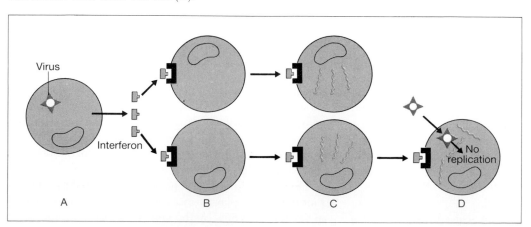

proteins of both classes are recognized as foreign, and the cells bearing them are destroyed by cytotoxic T cells with the aid of helper T cells.[7]

Some of the tools aimed at reducing graft rejection are radiation and drugs that kill actively dividing lymphocytes and thereby decrease the T-cell population. The drug, cyclosporin, which does not kill lymphocytes but rather alters their function, has proved extremely effective in reducing graft rejection. The major action of cyclosporin is to block the production of IL-2 and other lymphokines by helper T cells. This eliminates a critical signal for proliferation of the cytotoxic T cells.

SYSTEMIC MANIFESTATIONS OF INFLAMMATION AND INFECTION

We have thus far described the *local* aspects of inflammation and specific responses to infection. In addition, there are many *systemic* responses to infection, that is, responses of organs and tissues distant from the site of infection. It is natural to think of these responses as part of the disease, but the fact is that most actually represent adaptive responses to the disease.

Probably the single most common and striking systemic sign of infection is fever. Present evidence suggests that fever is often beneficial, in that an increase in body temperature enhances many of the protective responses described in this chapter. On page 596, we discussed the implications of this fact for the taking of drugs that reduce fever.

Decreases in the plasma concentrations of iron and zinc occur in response to infection and are due to changes in the uptake and/or release of these trace elements by liver, spleen, and other tissues. The decrease in plasma iron concentration has adaptive value since bacteria require a high concentration of iron to multiply. The role of the decrease in zinc is not known. The decrease in appetite characteristic of infection may also deprive the invading organisms of nutrients they require to proliferate.

Another adaptive response to infection is the secretion by the liver of a host of proteins known collectively as **acute phase proteins**. These proteins exert a large array of effects on the inflammatory process, immune cell function, and tissue repair.

Another response to infection, release of neutrophils

---

[7]There is an interesting problem here. To be recognized by the host's cytotoxic T cells, the graft's MHC proteins should theoretically be complexed with the host's MHC proteins on host antigen presenting cells. Whether this really occurs or whether foreign MHC proteins are recognized without such complexing is not clear.

and monocytes by the bone marrow, is of obvious value. Finally, there is release of amino acids from muscle, and these amino acids provide the building blocks for the synthesis of proteins required to fight the infection and for tissue repair.

All these responses and many others not listed are elicited by one or more of the monokines released from stimulated macrophages. Thus, IL-1, TNF, and another monokine, termed **interleukin 6** (**IL-6**), all of which serve local roles in immune responses, also serve as hormones to elicit far-flung responses such as fever (Figure 19-18). Several lymphokines and protein messengers secreted by other cell types also participate.

For example, **colony-stimulating factors**, which stimulate bone marrow to produce more neutrophils and monocytes (Chapter 13, page 356), are secreted not only by macrophages and lymphocytes but by endothelial cells and fibroblasts. A cascade is involved here since these last two cell types are stimulated to do so by IL-1 and TNF.

This is a good place to review the various functions of macrophages in immune responses (Table 19-10).

FACTORS THAT ALTER THE BODY'S RESISTANCE TO INFECTION

There are many factors that determine the body's capacity to resist infection. We offer here only a few examples.

Nutritional status is extremely important, and protein-calorie malnutrition is, worldwide, the single greatest contributor to decreased resistance to infection. The role that excesses or deficits of individual nutrients other than protein play in altering resistance to infection is now receiving much study, and no simple conclusions can be made. For example, severe iron deficiency predisposes to infection, but so does an excess of iron.

A preexisting disease, infectious or noninfectious, can also predispose the body to infection. People with diabetes, for example, suffer from a propensity to numerous infections, at least partially explainable on the basis of defective leukocyte function. Moreover, any injury to a tissue lowers its resistance, perhaps by altering the chemical environment or interfering with blood supply.

It is also likely, although this point remains controversial, that a person's state of mind can influence his or her resistance to infection. The physiological links between state of mind and resistance remain to be determined, but the entire question has been given great impetus by the emerging field of psychoneuroimmunology. Until recently, the tendency was to treat immune responses, on the one hand, and the nervous and endocrine sys-

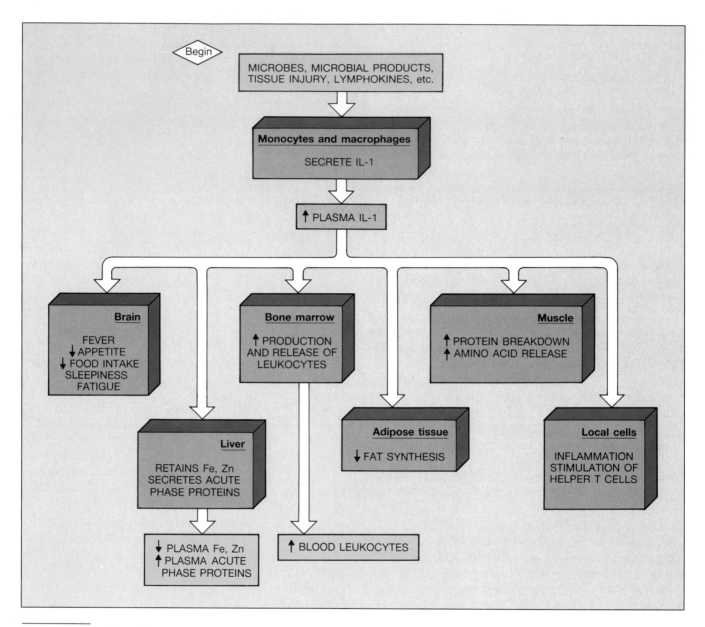

FIGURE 19-18 Effects of IL-1. Except for the local effects on inflammation and helper T cells, all other effects are probably exerted by *blood-borne* IL-1 acting as a hormone. Many of these same effects are exerted by tumor necrosis factor, IL-6, and perhaps other cytokines as well.

tems, on the other, as entirely separate. It is now clear, however, that lymphoid tissue is innervated and that the cells mediating immunity possess receptors for many neurotransmitters and hormones. Conversely, many of the cytokines released by immune cells have important effects on the brain and endocrine system. Moreover, lymphocytes secrete several of the same hormones produced by endocrine glands. It seems certain, therefore, that the nervous and endocrine systems interact with the immune system in important ways. For example, it has

been shown that the production of antibodies can be altered by psychological conditioning in much the same way as can other bodily functions.

Resistance to infection will be impaired if one of the basic resistance mechanisms itself is deficient, as, for example, in persons having a congenital deficiency of plasma gamma globulin, that is, failure to synthesize antibodies. These persons experience frequent and sometimes life-threatening infections that can be prevented by regular replacement injections of gamma globulin.

TABLE 19-10 ROLE OF MACROPHAGES IN IMMUNE RESPONSES
1. In nonspecific inflammation, phagocytize particulate matter, including microbes. Also secrete antimicrobial chemicals and protein messengers (monokines) that function as local inflammatory mediators. The inflammatory monokines include IL-1 and TNF.
2. Process and present antigen to lymphocytes.
3. Participate in activation of lymphocytes: Secrete interleukin 1 (IL-1), which stimulates helper T cells to secrete interleukin 2 (IL-2), which then activates both helper T cells and cytotoxic T cells.
4. During specific immune responses, perform same killing and inflammation-inducing functions as in (1) but are more efficient because antibodies act as opsonins and because the cells are transformed into effector macrophages by a lymphokine, gamma interferon, secreted by helper T cells.
5. IL-1, TNF, and another monkine, IL-6, mediate many of the systemic responses to infection or injury.

Another congenital defect is combined immunodeficiency, an absence of both B and T cells. If untreated, infants with this disorder usually die within their first year of life from overwhelming infections. This condition can be cured by a bone-marrow transplantation, which supplies both B cells and cells that will migrate to the thymus and become T cells.

An environmentally induced decrease in the production of leukocytes is also an important cause of lowered resistance, as, for example, in patients given drugs specifically to inhibit rejection of tissue or organ transplants.

The total quantity of leukocytes circulating is not necessarily critical. Patients with leukemia, for example, may have tremendous numbers of blood neutrophils or monocytes, but these cells are almost all immature or otherwise incapable of normal function. Such patients are extremely prone to infection. Reduced functional activity of lymphocytes is also seen in the elderly and may account for their decreased resistance.

AIDS

The most striking example of the lack of a basic resistance mechanism is the disease called **acquired immune deficiency syndrome** (**AIDS**). This disease is caused by the **human immunodeficiency virus** (**HIV**), which incapacitates the immune system. HIV belongs to the retrovirus family, which are unusual in that their nucleic acid core is RNA rather than DNA. Retroviruses possess an enzyme that, once the virus is inside a host cell, transcribes the virus's RNA into DNA, which is then integrated into the host cell's chromosomes. The association is permanent; every time the host cell reproduces, it also reproduces the retrovirus DNA. The virus may remain dormant inside the cell for years. Thus, individuals known to be infected with HIV, based on the finding of antibodies against HIV in their blood, will show no symptoms until the virus becomes active and undergoes rapid replication insde the cell, resulting in the cell's death. What causes the virus to become active is not known.

Unfortunately, the cells that HIV preferentially enters and resides in are helper T cells, so the active disease causes depletion of these cells, with catastrophic consequences. Without adequate numbers of normally functioning helper T cells, B cells cannot make antibodies and cytotoxic T cells cannot function fully in cell-mediated immunity. Thus the AIDS patient dies from infections and cancers that ordinarily would be readily handled by the immune system.

AIDS was first described in 1981, but it has since reached epidemic proportions, with nearly 71,000 cases reported in the United States alone by August 1988; 56 percent of these people had already died. It is estimated, based on the finding of antibody against HIV in the blood, that 1 to 2 million more Americans are infected with the virus but have no symptoms of AIDS as yet. This is an important point: One must distinguish between the presence of the actual disease—AIDS—and asymptomatic infection with HIV. We do not know what fraction of infected persons will ultimately develop AIDS.

The transmission of HIV is known to occur through transfer of contaminated blood or blood products from one person to another, during sexual intercourse with an infected partner, and by transmission from an infected mother to her child. This explains why AIDS is concentrated in several high-risk groups: (1) Patients who had received frequent blood transfusions and thus had a high likelihood of having received infected blood during one of the transfusions, (2) persons who use intravenous drugs, (3) sexually active men who are homosexual or both homosexual and heterosexual (the frequency of transmission among lesbians is very low), (4) the heterosexual partners of individuals infected with HIV, and (5) babies born to women infected with HIV.

Transfusion of blood is no longer a significant source of AIDS in this country since blood is now tested and discarded if found to be positive for the presence of antibody to HIV.[8] Infection with HIV among intravenous drug users is related to exposure to contaminated blood through the practice of sharing needles. The major reason that homosexual men are a high-risk group is that anal intercourse is often associated with small tears in blood vessels, which allows HIV-containing semen from an infected person to pass into the blood of the partner. Transmission from infected mother to offspring can occur either across the placenta during pregnancy or via breast milk.

There is no question that HIV can be transmitted during heterosexual intercourse, but at present the incidence of such transmission is relatively low in this country—approximately 4 percent of all AIDS cases are presently attributed to heterosexual transmission, with a 1:3.5 ratio of male to female cases. It is clear that heterosexual transmission can be male-to-female or female-to-male and that most cases occur during vaginal intercourse.

The pattern seen in the United States, in which most cases of AIDS are homosexual males or intravenous drug users, is typical of many parts of the world in which AIDS is prevalent. In contrast, most cases of AIDS in sub-Saharan Africa occur among heterosexuals, and the reasons for this are not completely clear. One hypothesis is that the high incidence of other venereal diseases in that part of the world makes transmission during heterosexual intercourse much more likely because of the presence of open genital sores that permit entry of HIV from infected semen.

There is no evidence that HIV is transmitted via saliva, tears, sweat, feces, urine, vomit, insects, toilet seats, or swimming pools. Even prolonged close contact with persons infected with HIV does not lead to infection in the absence of the transmission routes described above.

Evidence indicates that the use of condoms reduces the risk of transmission during sexual intercourse. The ultimate hope for prevention of AIDS is development of a vaccine. For a variety of reasons related to the nature of the virus and the fact that it infects a cell crucial for immune responses, this will not be an easy task. Drug therapy is also difficult given that, among the few existing antiviral drugs, few show any effect against HIV. Several drugs are presently being evaluated including one that blocks reverse transcription of the HIV RNA into DNA inside the host cells.

Antibiotics

Finally, we should at least mention the most important of the external agents we employ in altering resistance to bacteria, the **antibiotics**, such as penicillin. Unfortunately, there are no antibiotics known to be effective against viruses.

Antibiotics exert a wide variety of effects, but the common denominator is interference with the synthesis of one or more of the bacterium's essential macromolecules. Antibiotics must not be used indiscriminately. For one thing, they may exert toxic effects on the body's cells as well as the bacteria. A second reason for judicious use is the problem of drug resistance. Most large bacterial populations contain a few mutants that are not sensitive to the drug. These few are capable of multiplying into large populations resistant to the effects of that particular antibiotic. Perhaps even more important, resistance can be transferred from one microbe directly to another previously nonresistant microbe by means of chemical agents ("resistance factors") passed between them. A third reason for the judicious use of antibiotics is that these agents may actually contribute to a new infection by eliminating certain species of relatively harmless bacteria that ordinarily prevent growth of more dangerous ones.

ALLERGY (HYPERSENSITIVITY REACTIONS)

A certain portion of the population is capable of acquiring specific immune reactivity to environmental antigens such as dusts, pollens, and food constituents. Despite the relative harmlessness of most of these antigens, subsequent exposure to them elicits an immune attack that produces distressing symptoms and, often, outright bodily damage. This phenomenon, known as **allergy** or **hypersensitivity**, is in essence immunity gone wrong, for the response is inappropriate to the stimulus.

Depending upon the antigen, allergic responses may be due to activation of either the antibody-mediated or

---

[8]There is an extremely slight possibility that a blood donor might have become infected with HIV so recently that the antibodies would not have become evident at the time of blood giving. This possibility is further minimized by careful questioning of prospective donors to try to determine whether they might be in a high-risk group.

cell-mediated systems. In the latter cases—the skin rash after contact with poison ivy is an example—the inflammatory response is a T-cell-mediated type. Because it takes several days to develop, it is known as **delayed hypersensitivity**. In contrast, antibody-mediated hypersensitivity responses are usually very rapid in onset and are termed **immediate hypersensitivity**. They involve a group of antibodies and a sequence of events different from the classical antibody-mediated response to bacteria.

Immediate Hypersensitivity

In immediate hypersensitivity, initial exposure to the antigen leads to some antibody synthesis but, more important, to the production of memory cells that mediate active immunity. Upon reexposure, the antigen elicits a more powerful antibody response. So far, none of this is unusual, but the difference is that the particular antigens that elicit allergic reactions stimulate the production of IgE antibodies.[9] Upon their release from plasma cells, these antibodies circulate to various parts of the body and become attached to connective-tissue mast cells (Figure 19-19). When subsequently an antigen combines with the mast-cell IgE, this triggers release of the mast cell's secretory vesicles, which contain histamine, other vasoactive chemicals, and chemotaxins. In addition to releasing these preformed inflammatory mediators stored in mast-cell vesicles, the binding of antigen to IgE triggers the mast cell to synthesize prostaglandins and leukotrienes as well as a substance called platelet-activating factor. All these mediators then initiate a local inflammatory response.

Thus, the symptoms of antibody-mediated allergy reflect the various effects of these mediators and the body site in which the antigen-IgE-mast-cell combination occurs. For example, when a previously sensitized person inhales ragweed pollen, the antigen combines with IgE on mast cells in the respiratory passages. The mediators released cause increased mucus secretion, increased blood flow, swelling of the epithelial lining, and contraction of the smooth muscle surrounding the airways. Thus, there follow the symptoms of congestion, running nose, sneezing, and difficulty in breathing that characterize hayfever.

Allergic symptoms may be localized to the site of entry of the antigen. If very large amounts of the chemicals released by the mast cells enter the circulation, however, systemic symptoms may result and cause severe hypotension and bronchiolar constriction. This sequence of events (**anaphylactic shock**) can cause death due to circulatory and respiratory failure. It can be elicited in some sensitized people by the antigen in a single bee sting.

The very rapid component of immediate hypersensitivity often proceeds to a late-phase reaction lasting many hours or days, during which large numbers of leukocytes, particularly eosinophils, migrate into the inflamed area. The chemotaxins that attract them are mediators released by the mast cells. The eosinophils, once in the area, secrete mediators that prolong the inflammation and sensitize the tissues so that less stimulation is needed the next time to evoke a response.

The therapy of allergic reactions is aimed at blocking the release or action of the various mediators causing the inflammatory response. Some examples are: (1) antihistamines; (2) adrenal corticosteroids, which in large doses block arachidonic acid metabolism by inhibiting phospholipase A_2 and hence preventing formation of the prostaglandins and leukotrienes (Chapter 7, page 151); (3) nonsteroidal antiinflammatory drugs, which block cyclooxygenase and so prevent synthesis of prostaglandins; and (4) the frequent injection of minute amounts of the specific antigen, which may activate suppressor T cells capable of inhibiting specific IgE antibody synthesis and/or may stimulate the production of IgG antibodies that combine with the offending antigen and prevent it from combining with IgE.

Given the inappropriateness of most immediate hypersensitivity responses, how did such a system evolve? The normal physiological function of the IgE/mast-cell/eosinophil pathways is to repel invasion by multicellular parasites. The mediators released by the mast cells stimulate the inflammatory response against the parasites, and the eosinophils serve as the major killer cells against them. How this system also came to be inducible by harmless agents is not clear.

AUTOIMMUNE DISEASE

Given the huge diversity of recognition sites for antigens on lymphocytes, why does the immune system not attack the body's own cells? This may seem a strange question given our definition of antigens as foreign molecules. But how do the lymphocytes "know" that a protein is a "self" protein rather than a foreign one? It is thought that during in utero and early postnatal life, the lymphocytes bearing receptors that could bind self proteins and thus induce immune responses against them are destroyed in the thymus. It is also possible that some of these lymphocytes are inhib-

---

[9] Production of IgE requires the participation of a specific group of helper T cells. It is likely that most people have few of these cells and so do not develop immediate hypersensitivity reactions.

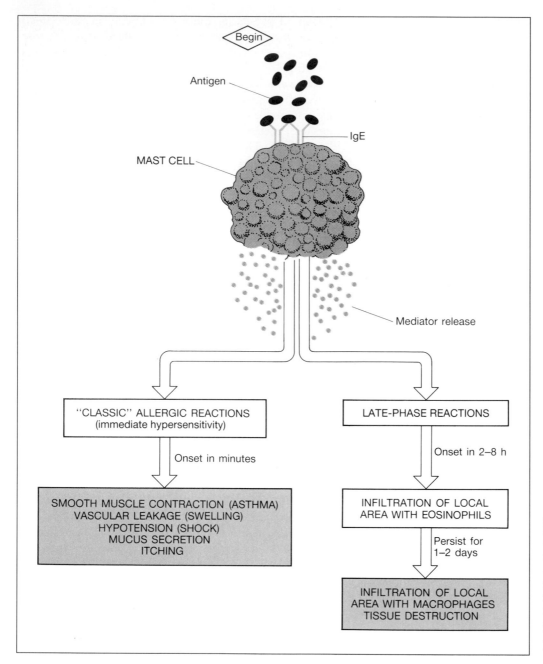

FIGURE 19-19 Sequence of events in an antibody-mediated allergic response. The mediators include substances stored in vesicles, such as histamine, and substances synthesized upon stimulation, such as leukotrienes.

ited, rather than killed, and that suppressor T cells play a role in this process.

Unfortunately, the body does, in fact, sometimes suffer antibody-mediated or cell-mediated attack against its own tissues, the result being cell damage or alteration of function. A growing number of human diseases are being recognized as **autoimmune** in origin. Examples are: multiple sclerosis, in which myelin is attacked; myesthenia gravis, in which the receptors for acetylcholine on skeletal muscle are the target; rheumatoid arthritis, in which joints are damaged; and type I diabetes, in which the insulin-producing cells of the pancreas are destroyed. There are multiple causes for the body's failure to recognize its own cells, as summarized in Table 19-11.

This table focuses on reasons for the appearance of antibodies or killer lymphocytes against the body's own cells. However, autoimmune damage may also be brought about in several other quite different ways. An overzealous inflammatory response—for example, too much generation of complement or release of inflammatory mediators and toxic chemicals from macrophages— may cause both damage to adjacent normal tissues and

TABLE 19-11 SOME POSSIBLE CAUSES OF AUTOIMMUNE ATTACK

1. Normal body proteins may be altered by combination with drugs or environmental chemicals. This leads to an attack on the cells bearing the now-"foreign" protein.
2. In cell-mediated immune attacks on virus-infected bodily cells, so many cells may be destroyed that disease results.
3. Genetic mutations in the body's cells may yield new proteins that serve as antigens.
4. The body may encounter microbes whose antigens are so close in structure to certain of the body's own proteins that the antibodies or cytotoxic T lymphocytes produced against these microbial antigens also attack cells bearing the self-proteins.
5. A deficiency of suppressor T cells could contribute to the development of autoimmunity.

potentially lethal systemic responses. For example, release of large amounts of IL-1 and tumor necrosis factor in response to an infection with certain types of bacteria can cause profound vasodilation throughout the body, precipitating hypotension and shock. Moreover, prolonged secretion of tumor necrosis factor can cause marked catabolism of fat and protein so as to lead to general wasting of the body. In other words, it is not the bacteria themselves that cause the shock and the general wasting but rather the mediators released in response to the bacteria.

Thus, the various mediators of immunity are a double-edged sword: In usual amounts they are essential for normal resistance, but in excessive amounts they can cause illness. Thus, at the same time that some scientists are studying how to enhance resistance to microbes and cancer by *administering* mediators such as IL-1 and TNF, other scientists are developing drugs to *block* the action of these same mediators in situations associated with excessive secretion.

Another important question in the field of autoimmunity concerns the relationship between mother and fetus: Why does the mother's immune system not reject the fetus, since half of the fetal proteins are, of course, paternal and therefore foreign to the mother? Several possible mechanisms are currently undergoing investigation.

TRANSFUSION REACTIONS AND BLOOD TYPES

Transfusion reactions are a special example of tissue rejection, one that illustrates the fact that antibodies rather than cytotoxic T cells can sometimes be the major factor in the destruction of nonmicrobial cells. Erythrocytes do not have MHC proteins, but they do have other membrane proteins that can function as antigens when exposed to another person's blood. The most important of these are designated A, B, and O. These proteins are inherited, and persons having a gene for A and/or B will express the A and/or B antigen, respectively. Persons with neither the gene for A nor the one for B will express the O antigen. Accordingly, the possible blood types are A, B, AB, and O.

If a typical pattern of antibody induction were followed, one would expect that a type A person would develop antibodies against type B cells only if the B cells were introduced into the body. However, what is atypical of this system is that even without initial exposure, type A persons always have anti-B antibodies in their plasma. Similarly, type B persons have plasma anti-A antibodies. Type AB persons have neither anti-A nor anti-B antibody, and type O persons have both. Anti-O antibodies do not exist. Because these antierythrocyte antibodies exist without prior exposure, they are called

TABLE 19-12 SUMMARY OF ABO BLOOD-TYPE INTERACTIONS

Recipient	Donor	Compatible ?
A	O, A	yes
B	O, B	yes
AB	O, A, B, AB	yes
O	O	yes
A	AB, B	no
B	AB, A	no
AB		
O	AB, A, B	no

TABLE 19-13 A MINIGLOSSARY OF CELLS AND CHEMICAL MEDIATORS INVOLVED IN IMMUNE FUNCTIONS

Cells

Antigen-presenting cell (accessory cell): Cells, such as macrophages, that process and present antigen to lymphocytes.

B cells: Lymphocytes that, upon activation, proliferate and differentiate into antibody-secreting plasma cells; provide major defense against bacteria, viruses in the extracellular fluid, and toxins; can function as antigen-presenting cells for helper T cells.

Cytotoxic T cells: The class of T lymphocytes that, upon activation by specific antigen, directly attack the cell bearing that type of antigen; with NK cells and effector macrophages, are major killers of virus-infected cells and tumor cells; bind antigen associated with MHC Class 1 proteins.

Effector macrophages: Macrophages whose killing ability has been enhanced by lymphokines, particularly gamma interferon.

Eosinophils: Granulocytic leukocytes involved in allergic responses and destruction of parasites.

Helper T cells: The class of T cells that play a stimulatory role in the production of antibodies and in the function of cytotoxic T cells and NK cells; bind antigen associated with MHC Class II proteins.

Lymphocytes: The type of leukocyte responsible for specific immune defenses; categorized mainly as B cells and T cells (NK cells constitute another category).

Macrophages: Cell type that (1) functions as a phagocyte, (2) processes and presents antigen to lymphocytes, and (3) secretes monokines involved in inflammation, activation of lymphocytes, and the systemic acute phase response to infection or injury.

Mast cells: Cell type derived from basophils; releases inflammatory mediators in immediate allergy.

Memory cells: B cells that differentiate during an initial infection and respond rapidly during a subsequent exposure to the same antigen.

Monocytes: A type of leukocyte; leaves the bloodstream and is transformed into a macrophage; has functions similar to those of macrophages.

Natural killer (NK) cells: Class of lymphocytes that bind to cells bearing foreign antigens without MHC restriction and kill them directly.

Neutrophils: Granulocytic leukocytes that function as phagocytes and also release chemicals involved in inflammation.

Plasma cells: Cells that differentiate from activated B lymphocytes and that secrete antibodies.

Suppressor T cells: The class of T cells that inhibit antibody production and cytotoxic T cell function.

T cells: Lymphocytes derived from precursors that differentiated in the thymus; see cytotoxic T cells, helper T cells, and suppressor T cells.

Chemical Mediators

Acute phase proteins: Group of proteins secreted by the liver during systemic response to injury or infection; stimulus for their secretion is IL-1 and other cytokines.

Antibodies: Immunoglobulins that function as antigen receptors on B cells and are secreted by plasma cells; secreted antibodies combine with the type of antigen that stimulated their production and direct an attack against the antigen or cell bearing it.

C_1: The first protein in the classical complement pathway.

C_3: The first protein activated in the alternate complement pathway.

Chemotaxin: A general name given to any chemical mediator that stimulates chemotaxis of neutrophils or other phagocytes.

natural antibodies. It is thought that these antibodies are induced early in life by antigens very similar to A and B on the surface of intestinal bacteria.

With this information as background, we can predict what happens when a type A person is given type B blood. There are two incompatibilities: (1) The recipi-ent's anti-B antibody causes the transfused cells to be attacked; and (2) the anti-A antibody in the transfused plasma causes the recipient's cells to be attacked. The latter is generally of little consequence, however, be-cause the transfused antibodies become so diluted in the recipient's plasma that they are of no consequence. It is

TABLE 19-13 (continued)

Complement: A group of plasma proteins that, upon activation, kill microbes directly and facilitate the various steps of the inflammatory process, including phagocytosis; the "classical complement pathway" is triggered by antigen-antibody complexes, whereas the "alternate pathway" can operate independently of antibody.

Cytokines: General term for protein messengers secreted either by macrophages and monocytes (monokines) or by lymphocytes (lymphokines).

Eicosanoids: General term for products of arachidonic acid metabolism (prostaglandins, prostacyclin, thromboxanes, leukotrienes); function as important inflammatory mediators.

Gamma interferon: See interferon.

Histamine: An inflammatory mediator secreted mainly by mast cells; acts on microcirculation to cause vasodilation and increased permeability to protein.

IgA: The class of antibodies secreted by the lining of the body's various "tracts."

IgD: A class of antibodies whose function is unknown.

IgE: The class of antibodies that mediate immediate hypersensitivity ("allergy") and resistance to parasites.

IgG: The most abundant class of plasma antibodies.

IgM: A class of antibody that, along with IgG, provides the bulk of specific humoral immunity against bacteria and viruses.

Immunoglobulin (Ig): A synonym for antibody; the five major classes are IgA, IgD, IgE, IgG, and IgM.

Interferon: Class of proteins that nonspecifically inhibit viral replication; in addition, gamma interferon stimulates the killing ability of cytotoxic T cells, NK cells, and macrophages.

Interleukin 1 (IL-1): Protein secreted by macrophages (and other cells) that activates helper T cells, exerts many inflammatory effects, and mediates many of the systemic acute phase responses including fever.

Interleukin 2 (IL-2): Protein secreted by activated helper T cells that causes helper T cells, cytotoxic T cells, and NK cells to proliferate.

Interleukin 6 (IL-6): Protein secreted by macrophages (and other cells) that mediates many of the acute phase responses.

Kinins: Peptides split from kininogens in inflamed areas and that facilitate the vascular changes associated with inflammation; they also activate neuronal pain receptors.

Leukotrienes: A class of eicosanoids that function as inflammatory mediators.

Lymphokines: Protein messengers secreted by lymphocytes.

Membrane attack complex (MAC): Group of complement proteins that form channels in the surface of a microbe, making it leaky and killing it.

Monokines: Protein messengers secreted by macrophages and monocytes.

Natural antibodies: Antibodies to the erythrocyte antigens of the A or B type not present in person's body; these antibodies are present without prior exposure to the antigen.

Opsonin: General name given to any chemical mediator that promotes phagocytosis.

Platelet-activating factor: Mediator synthesized and released by mast cells that stimulates inflammation.

Pore-forming protein: Protein secreted by cytotoxic T and NK cells that forms channels in the plasma membrane of the target cell, making it leaky and killing it; structure and function are similar to MAC in the complement system.

Prostaglandins: A class of eicosanoids that function as inflammatory mediators.

Tumor necrosis factor: Monokine that kills target cells outright, stimulates inflammation, and mediates many of the systemic acute phase responses.

the destruction of the transfused cells by the recipient's antibodies that produces the problems. The range of possibilities is shown in Table 19-12. It should be evident from this table why type O people are frequently called "universal donors" whereas type AB people are "universal recipients." These terms are misleading, however, since besides those of the ABO type, there are a host of other erythrocyte membrane proteins that can act as antigens and plasma antibodies against them. Therefore, except in dire emergency, the blood of donor and recipient must be carefully matched.

Another group of erythrocyte membrane antigenic

proteins of medical importance is the **Rh factor**, so called because it was first studied in rhesus monkeys. The Rh system follows the classic immunity pattern in that no-one develops anti-Rh antibodies unless exposed to Rh-type cells, usually termed Rh-positive cells, from another person. Although this can be a problem if an Rh-negative person, that is, one whose cells have no Rh antigen, is subjected to multiple transfusions with Rh-positive blood, its major importance is in the mother-fetus relationship. When an Rh-negative mother carries an Rh-positive fetus, some of the fetal erythrocytes may cross the placental barriers into the maternal circulation, inducing her to synthesize anti-Rh antibodies. This occurs mainly during separation of the placenta at delivery. Therefore, a first Rh-positive pregnancy rarely offers any danger to the fetus since delivery occurs before the antibodies are made by the mother. In future pregnancies, however, these anti-Rh antibodies will already be present in the mother and can cross the placenta to attack the erythrocytes of an Rh-positive fetus. The risk increases with each Rh-positive pregnancy as the mother becomes more and more sensitized.

Fortunately, Rh disease can be prevented by giving any Rh-negative mother gamma globulin against Rh erythrocytes within 72 h after she has delivered an Rh-positive infant. These antibodies bind to the antigenic sites on any Rh erythrocytes that might have entered the mother's blood during delivery and prevent them from inducing antibody synthesis by the mother. The administered antibodies are eventually catabolized.

SUMMARY

Table 19-13 presents a summary of immune mechanisms in the form of a miniglossary of cells and chemical mediators involved in immune responses. All the material in this table has been covered in this chapter.

SECTION B
METABOLISM OF FOREIGN CHEMICALS

The body is exposed to a huge number of nonmicrobial foreign chemicals, "foreign" in the sense that they are not normally found in the body. Many are products of the natural world, but many are manmade. There are now more than 10,000 foreign chemicals being commercially synthesized, and over 1 million have been synthesized at one time or another. They find their way into the body either because they are purposely administered as drugs, or simply because they are in the air, water, and food we use.

As described in Section A of this chapter, foreign materials can induce inflammation and specific immune responses. These defenses are directed mainly against foreign cells, however, and although noncellular foreign chemicals can also elicit certain of these responses, as in allergy, for example, such immune responses do not constitute the major defense mechanisms against foreign chemicals. Rather, molecular alteration (**biotransformation**) and excretion do.

The body's metabolism of foreign chemicals is summarized in Figure 19-20. First, the chemical gains entry to the body through the gastrointestinal tract, lungs, skin, or placenta in the case of a fetus. Once in the blood, the chemical may become bound reversibly to plasma proteins or to erythrocytes. Such binding lowers its free concentration and thereby its ability to alter cell function. It may accumulate in storage depots—for example, DDT in fat tissue—or it may undergo enzyme-mediated biotransformation in various organs and tissues, as described in a later section. The metabolites resulting from biotransformation may enter the blood and are subject to the same fates as the parent molecules. Finally, the foreign chemical and its metabolites may be eliminated from the body in the urine, expired air, skin secretions, or feces.

The blood concentration of any foreign chemical is determined by the interplay of all these metabolic pathways. For example, kidney function and biotransformation both tend to decrease with age, and this explains why the same dose of a drug often produces much higher blood concentrations in elderly people than in young people.

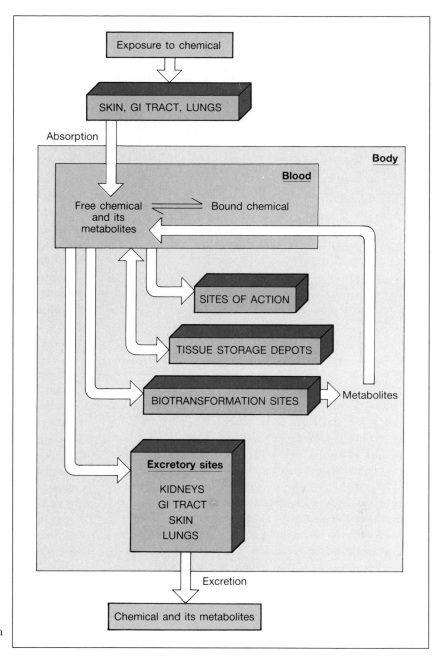

FIGURE 19-20 Metabolic pathways for foreign chemicals.

ABSORPTION

In practice, most foreign molecules move through the lining of some portion of the gastrointestinal tract fairly readily, either by diffusion or by carrier-mediated transport. This should not be surprising since the gastrointestinal tract evolved to favor absorption of the wide variety of nutrient molecules in the environment. The nonnutrient chemicals are the beneficiaries of these relatively nondiscriminating transport mechanisms.

The lung alveoli are highly permeable to most foreign chemicals and therefore offer an easy entrance route for those that are airborne. Lipid solubility is all-important for entry through the skin, so that this route is of little importance for charged molecules but can be used by oils, steroids, and other lipids.

The penetration of the placental membranes by foreign chemicals is important since the effects of environmental agents on the fetus during critical periods of development may be quite marked and, in many cases, irreversible. Diffusion across the placenta is an important mechanism for lipid-soluble substances, and carrier-mediated systems, which evolved for the carriage of endogenous nutrients, may be usurped by foreign chemicals to gain entry into the fetus.

STORAGE SITES

The major storage sites for foreign chemicals are cell proteins, bone, and fat. The chemical bound to cell proteins or bone or dissolved in the fat is in equilibrium with the free chemical in the blood, so that an increase in blood concentration causes more movement into storage, up to the point of saturation. Conversely, as the chemical is eliminated from the body and its blood concentration falls, movement occurs out of storage sites.

These storage sites are a source of protection, but it sometimes happens that the storage sites accumulate so much chemical that they become damaged.

EXCRETION

As described in Chapter 15, to appear in the urine, a chemical must either be filtered through the glomerulus or secreted across the tubular epithelium. Glomerular filtration is a bulk-flow process so that all low-molecular-weight substances in plasma undergo filtration. Accordingly, there is considerable filtration of most foreign chemicals, except those bound to plasma proteins or erythrocytes. In contrast, tubular secretion is by discrete transport processes, and many foreign chemicals—penicillin, for example—utilize the mediated-transport systems available for naturally occurring substances.

Once in the tubular lumen, via either filtration or tubular secretion, the foreign chemical will still not be excreted if it is reabsorbed across the tubular epithelium into the blood. As the filtered fluid moves along the renal tubules, molecules that are lipid-soluble passively diffuse along with reabsorbed water through the tubular epithelium and back into the blood. The net result is that little is excreted in the urine, and the chemical is retained in the body. If these chemicals could be transformed into more polar and therefore less lipid-soluble molecules, their passive reaborption from the tubule would be retarded and they would be excreted more readily. This type of transformation is precisely what occurs in the liver, as described in the next section.

An analogous problem exists for foreign molecules secreted in the bile. Many of these substances, having reached the lumen of the small intestine, are absorbed back into the blood, thereby escaping excretion in the feces. This cyclic enterohepatic circulation is described in Chapter 16.

BIOTRANSFORMATION

The metabolic alteration of foreign molecules occurs mainly in the liver but to some extent also in kidney, skin, placenta, and other organs. A large number of distinct enzymes and pathways are involved, but the common denominator of most of them is that they transform chemicals into more polar, less lipid-soluble substances. One consequence of this transformation is that the chemical may be rendered less toxic, but this is not always so. The second, more important, consequence is that the chemical's tubular reabsorption is diminished and urinary excretion facilitated. Similarly, for substances handled by biliary secretion, gut absorption of the metabolite is less likely so that fecal excretion is also enhanced.

The hepatic enzymes that perform these transformations are called the **microsomal enzyme system (MES)** and are located mainly in the smooth endoplasmic reticulum. One of the most important facts about this enzyme system is that it is easily inducible, that is, the number of these enzymes can be greatly increased by exposure to a chemical that acts as a substrate for the system. For example, chronic overuse of alcohol results in an increased rate of alcohol catabolism because of the induction of the microsomal enzymes. This accounts for much of the tolerance to alcohol, that is, the fact that increasing doses must be taken to achieve a given magnitude of effect.

The hepatic biotransformation mechanisms vividly demonstrate how an adaptive response may, under some circumstances, turn out to be maladaptive. These enzymes all too frequently "toxify" rather than "detoxify" a drug or pollutant. In fact, many foreign chemicals are quite nontoxic until the liver enzymes biotransform them. Of particular importance is the fact that many chemicals that cause cancer do so only after biotransformation. For example, a major component of cigarette smoke is transformed by the MES into a carcinogenic compound. Persons with a particularly highly inducible MES have a 20- to 40-fold higher risk of developing lung cancer from smoking.

The MES can also cause problems in another way because it evolved primarily not to defend against foreign chemicals, which were much less prevalent during our evolution, but rather to metabolize endogenous substances, particularly steroids and other lipid-soluble molecules. Therefore, the induction of these enzymes by a drug or pollutant increases metabolism not only of that drug or pollutant but of the endogenous substances as well. The result is a decreased concentration in the body of the normal substance, resulting in its possible deficiency.

Another fact of importance concerning the MES is that whereas certain chemicals induce it, others inhibit it. The presence of such chemicals in the environment could have deleterious effects on the system's capacity to protect against those chemicals it transforms. Just to illustrate how complex this picture can be, note that any chemical that inhibits the microsomal enzyme system may actually confer protection against other chemicals that must undergo transformation in order to become toxic.

SECTION C
HEMOSTASIS: THE PREVENTION OF BLOOD LOSS

Elimination of bleeding is known as **hemostasis**. We shall discuss the probable sequence of events when small vessels—arterioles, capillaries, venules—are damaged because they are the most common source of bleeding in everyday life and because the hemostatic mechanisms are most effective in dealing with such injuries. In contrast, the bleeding from a medium or large artery is not usually controllable by the body. Venous bleeding is less dangerous because veins have low blood pressure. Indeed, the drop in pressure induced by raising the bleeding part above the heart level may stop hemorrhage from a vein. In addition, if the venous bleeding is into the tissues, the accumulaton of blood may increase interstitial pressure enough to eliminate the pressure gradient required for continued blood loss. Accumulation of blood in the tissues can occur from bleeding from any vessel type and is termed a hematoma.

When a blood vessel is severed or otherwise injured, the immediate inherent response of its smooth muscle is to constrict. This shortlived response slows the flow of blood in the affected area. In addition, this constriction presses the opposed endothelial surfaces of the vessel together, and this contact induces a stickiness capable of keeping them "glued" together against high pressures.

Permanent closure of the vessel by constriction and contact stickiness occurs only in the very smallest vessels of the microcirculation, however, and the staunching of bleeding ultimately is dependent upon two other interdependent processes that occur in rapid succession: (1) platelet plug formation; and (2) blood coagulation. Platelets are so involved in both processes that a deficiency or abnormality of platelets is a common cause of abnormal bleeding.

FORMATION OF A PLATELET PLUG

The involvement of platelets in hemostasis requires their adhesion to a surface. Platelets have a propensity for adhering to surfaces, but they do not adhere to the normal endothelial cells lining the blood vessels. However, injury to a vessel disrupts the endothelium and exposes the underlying connective-tissue collagen molecules—fibrous proteins that constitute a major component of the interstitial matrix. Platelets adhere to collagen, and this attachment triggers them to release the contents of their secretory vesicles ("granules"), which contain a variety of chemical agents. Many of these agents, including adenosine diphosphate (ADP), then induce changes in the surface of the platelets, causing new platelets to adhere to the old ones. Termed **platelet aggregation**, this self-perpetuating (positive-feedback) process rapidly creates a **platelet plug** inside the vessel (Figure 19-21).

The adherence of platelets to a damaged vessel wall is facilitated by **von Willebrand factor (vWF)**, a plasma protein secreted by endothelial cells. This protein forms a bridge between the vessel wall and the first layers of platelets: It binds to collagen, and then platelets bind to it. Because, as we shall see, platelet aggregation is also essential for clotting, an absence or abnormality of vWF (von Willebrand's disease) is a common cause of inadequate clotting.[10]

Chemical agents in the platelets' secretory granules

---

[10]vWF serves a second function in clotting: It acts as a binding protein for one of the other proteins—factor VIII—involved in clotting and described in the section on clotting.

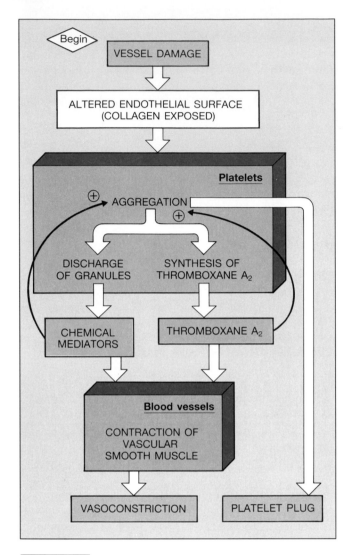

FIGURE 19-21 Sequence of events leading to vasoconstriction and formation of a platelet plug following damage to a blood-vessel wall. Note the two positive feedbacks in the pathways.

are not the only stimulator of platelet aggregation. In addition, adherence of the platelets to collagen causes the conversion of arachidonic acid in the platelet plasma membrane to **thromboxane A_2**, a member of the eicosanoid family (Chapter 7, page 151). This substance powerfully stimulates platelet aggregation and secretion of platelet granules.

The platelet plug can completely seal breaks in blood-vessel walls. Its effectiveness is further enhanced by another property of platelets—contraction. Platelets contain a very high concentration of contractile proteins, which are stimulated to contract in aggregated platelets.

This results in a compression and strengthening of the platelet plug.

While the plug is being built up and compacted, the vascular smooth muscle in the vessels around it is being stimulated to contract (Figure 19-21), thereby decreasing the blood flow to the area and the pressure within the damaged vessel. This vasoconstriction is also the result of platelet activity, for it is mediated by thromboxane A_2 and by several chemicals contained in the secreted platelet granules.

Why does the platelet plug not continuously expand, spreading away from the damaged endothelium along normal endothelium in both directions? One important reason involves the ability of the adjacent vessel wall to make the eicosanoid known as **prostacyclin** (also termed prostaglandin I_2, **PGI_2**). Whereas stimulated *platelets* possess the enzymes that produce thromboxane A_2 from arachadonic acid, normal *endothelial cells* contain a different enzyme, one that converts intermediates formed from arachidonic acid not to thromboxane A_2 but to PGI_2 (Figure 19-22). PGI_2 is a profound inhibitor of platelet aggregation.

To reiterate, the platelet plug is built up extremely rapidly and is the primary sealer of breaks in vessel walls. In the next section, we shall see that platelet aggregation is also absolutely essential for the next, more slowly occurring, hemostatic event—blood coagulation. In turn, certain of the chemicals liberated in the coagulation process act as yet additional stimulators of platelet aggregation and granule release.

BLOOD COAGULATION: CLOT FORMATION

Blood coagulation (**clotting**) is the transformation of blood into a solid gel termed a **clot** or **thrombus**. Clotting is the dominant hemostatic defense. It occurs around the original platelet plug when the plasma protein fibrinogen is converted to fibrin.

Fibrinogen is a large, rod-shaped protein produced by the liver and always present in the plasma of normal persons. At a site of vessel damage, the enzyme **thrombin** catalyzes a reaction in which several polypeptides are split from the fibrinogen molecules. The still large fibrinogen remnants then bind to each other to form the protein polymer known as **fibrin** (Figure 19-23). The fibrin, initially a loose mesh of interlacing strands, is rapidly stabilized and strengthened by enzymatically mediated formation of covalent cross-linkages. This chemical linking is catalyzed by an enzyme in plasma known as **factor XIII**, which is converted from an inactive form to an active form by thrombin. Thus thrombin catalyzes the

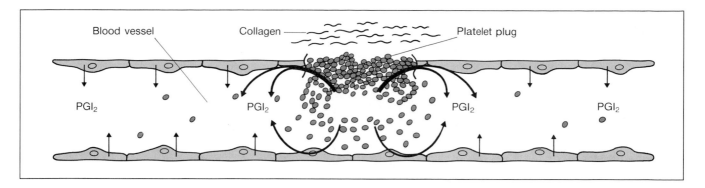

FIGURE 19-22 Normal endothelium adjacent to a platelet plug produces PGI_2 from platelet and endothelial-cell arachidonic acid, and this eicosanoid inhibits platelet aggregation. This prevents spread of platelet aggregation from the damaged site.

formation not only of fibrin but also of active factor XIII, which will stabilize the fibrin network.[11]

In the process of clotting, many erythrocytes and other cells are trapped in the fibrin meshwork, but it must be emphasized that the essential component of the

---

[11] In addition to its crucial role in clotting, fibrinogen also participates in the formation of platelet plugs. It does so by forming bridges between platelets. The binding sites for fibrinogen on the platelet plasma membrane somehow become exposed as platelet aggregation proceeds.

FIGURE 19-23 Scanning electron micrograph of an erythrocyte enmeshed in fibrin. (*Courtesy of Elia Kairinen, Gillett Research Institute.*)

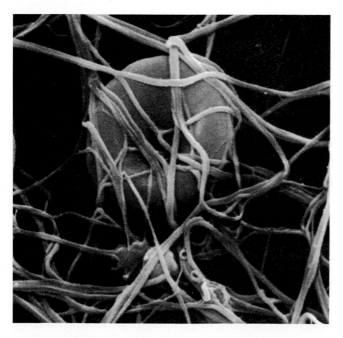

clot is fibrin, and clotting can occur in the absence of all cells except platelets. The function of the clot is to support and reinforce the platelet plug and to solidify blood that remains in the wound channel.

Fibrinogen is always present in the blood, but thrombin is normally absent and is formed only when the stimulus for clotting occurs. Circulating in the blood is an inactive precursor of thrombin called **prothrombin**. At the site of blood-vessel damage, prothrombin is enzymatically converted to thrombin, which then enzymatically converts fibrinogen to fibrin and activates factor XIII (Figure 19-24).

We have now only pushed the essential question one step farther back: Where does the enzyme that catalyzes the conversion of the prothrombin to thrombin come from? The answer is that this enzyme is always present in the plasma in an inactive form and is converted to its

FIGURE 19-24 Generation of thrombin and its effects on fibrinogen and factor XIII.

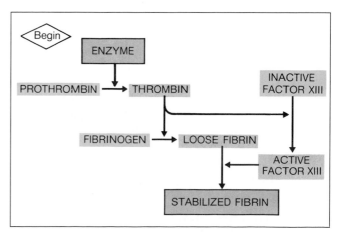

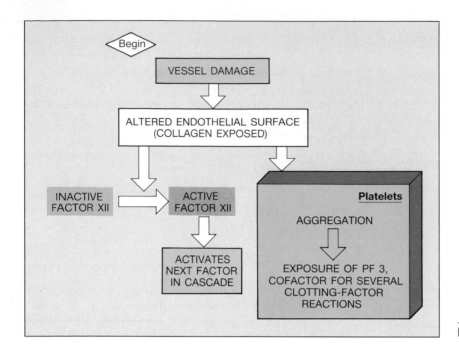

FIGURE 19-25 Initial events in clotting.

active form by yet another enzyme, which itself was activated by another enzyme. Thus, we are dealing with a cascade of plasma proteins, each normally an inactive proteolytic enzyme until activated by the previous one in the sequence. Ultimately, the final enzyme in the sequence is activated and, in turn, catalyzes the conversion of prothrombin to thrombin. The first factor in the sequence is called **factor XII**, and, as will be described in subsequent sections, it has several other important functions in addition to initiating clotting.

But we still have not answered the question of what *initiates* clotting, that is, what activates the first enzyme (factor XII) in the catalytic sequence? The answer is contact of factor XII with a damaged vessel surface, specifically with the collagen fibers underlying the damaged endothelium.[12] Recall that this is precisely the same stimulus that triggers platelet aggregation (Figure 19-25). This fact is important because several of the sequential enzyme activation steps in the clotting pathway require a particular phospholipid, called **PF₃**, exposed in the plasma membranes of the aggregated platelets. Thus, platelets are essential for normal blood clotting.

Figure 19-26 summarizes the intravascular sequence of events leading to blood clotting. It is known as the

intrinsic clotting pathway because everything necessary for it is in ("intrinsic" to) the blood. Calcium is required at various steps in the clotting cascade. However, calcium concentration in the blood can never get so low as to cause clotting defects because death would have occurred from some other cause before such low concentrations were reached.

Two of the circulating clotting factors—**VIII** and **V**—do not form links in the cascade chain but serve as cofactors in two of the cascade reactions—those leading to active factor X and to thrombin, respectively. Another important fact concerning factors VIII and V is that their activation is stimulated by thrombin. Thus, thrombin not only catalyzes the formation of fibrin from fibrinogen and the activation of factor XIII but also helps activate two of the cofactors in the cascade leading to its own formation.

Moreover, not shown in the figure is yet another action of thrombin—it profoundly stimulates platelet aggregation and thereby the exposure of platelet phospholipid. Thus, once thrombin formation has begun, the overall reactions leading to clotting occur very rapidly, owing to the positive-feedback effects of thrombin (stimulation of platelet aggregation and activation of factors VIII and V) on its own generation. The top of Table 19-14 presents a summary of the procoagulant effects of thrombin.

The value of having such a complex cascade system for clotting lies in the amplification gained at many steps, that is, in the manyfold increase in the number of active molecules produced. (Recall from Chapter 7 that an

---

[12]Why does blood coagulate when it is taken from the body and put in a glass tube? This has nothing whatever to do with exposure to air, as popularly supposed, but happens because the glass surface induces the same activation of factor XII and aggregation of platelets as does a damaged vessel surface. A silicone coating markedly delays clotting by reducing the activating effect of the glass surface.

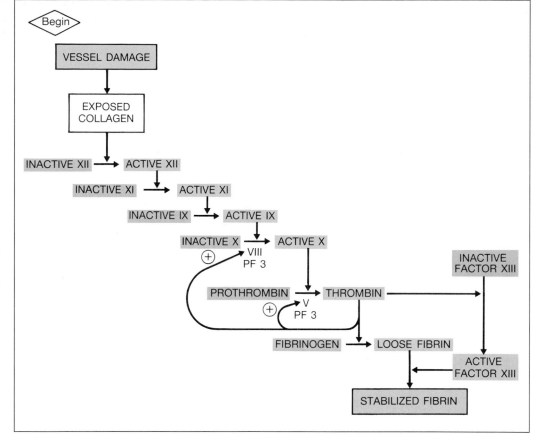

FIGURE 19-26 Intrinsic clotting pathway. For the sake of clarity, the roles of calcium are not shown. Note the two positive-feedback effects of thrombin (activating factors VIII and V) on its own formation. You might think that factors IX and X were accidentally transposed in this figure; such is not the case; the order of activation really is XI, IX, X.

analogous cascade of enzyme activations was described for the cyclic AMP-protein-kinase system.) There is at least one disadvantage in such a cascade, however, in that a single defect, either genetic or environmentally induced, anywhere in the system can block the entire cascade, thereby interfering with clot formation. For example, the disease **hemophilia**, in which excessive bleeding occurs, is usually due to a genetic absence of factor VIII.

In addition to the intrinsic clotting pathway there is an **extrinsic clotting pathway**, which also leads to thrombin formation and clotting (Figure 19-27). It is termed "extrinsic" because it utilizes a protein—**tissue factor**—from outside of the blood, specifically in the interstitial compartment. Tissue factor (also called tissue thromboplastin) forms an active complex with another protein, **factor VII**. This complex then plugs into the intrinsic clotting pathway by activating factors IX and X. Thus, exposure of blood to tissue factor bypasses the first steps in the intrinsic clotting sequence. Since such exposure

TABLE 19-14 ACTIONS OF THROMBIN	
Procoagulant	1. Cleaves fibrinogen to fibrin 2. Activates clotting factors VIII, V, and XIII 3. Stimulates platelet aggregation.
Anticoagulant	1. Activates protein C, which deactivates clotting factors VIII and V and inactivates an inhibitor of tissue plasminogen activator 2. Directly inactivates an inhibitor of tissue plasminogen activator

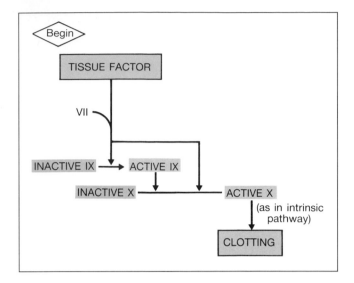

FIGURE 19-27 Extrinsic clotting pathway. From factor X on, the extrinsic pathway is identical to the intrinsic pathway.

usually occurs when a blood vessel is damaged, this pathway goes into action right along with the intrinsic pathway. The existence of two converging pathways tends to reduce any clotting problems due to a defect in the early steps of either.

During infection, the clotting pathways are also frequently triggered by chemicals secreted into the interstitial fluid by leukocytes and macrophages. The resulting interstitial fibrin clots may block further spread of the bacteria.

The liver plays several important indirect roles in clotting (Figure 19-28), and persons with liver disease frequently have serious bleeding problems. First, the liver is the site of production for many of the plasma clotting factors. Second, the bile salts produced by the liver (Chapter 16) are important for normal intestinal absorption of the lipid-soluble substance **vitamin K**, which the liver requires to produce the normal structures of prothrombin and factors VII, IX, and X. Thus, the liver produces bile salts, which help in the absorption of vitamin K, which helps in the liver's production of clotting factors.

ANTICLOTTING SYSTEMS

Factors that Oppose Clot Formation

There are at least three different mechanisms that oppose clot formation once under way. The first of these mechanisms—PGI_2—was described earlier as the eicosanoid that is generated by endothelial cells and inhibits platelet aggregation, a process essential for subse-

quent clot formation. The other two, named protein C (*not* protein kinase C, page 162) and antithrombin III, inhibit clotting directly, thereby helping to terminate this positive-feedback process and preventing it from spreading excessively.

The initiator of the second inhibitory mechanism is, perhaps not surprisingly, thrombin. As illustrated in Figure 19-29, thrombin activates a plasma protein, **protein C**, which then deactivates factors VIII and V. Thus, thrombin directly activates factors VIII and V but also deactivates them indirectly via protein C.

The third naturally occurring brake on clot formation is a plasma protein called **antithrombin III**, which inactivates thrombin. To do so, antithrombin III requires the participation of **heparin**, a substance that is present on the surface of endothelial cells and binds antithrombin III. Patients with genetic deficiency in the ability to form antithrombin III suffer from hypercoagulability and excessive intravascular clotting.

The Fibrinolytic System

Protein C and antithrombin III halt clot formation. The system to be described now dissolves a clot *after* it is formed.

FIGURE 19-28 Roles of the liver in clotting.

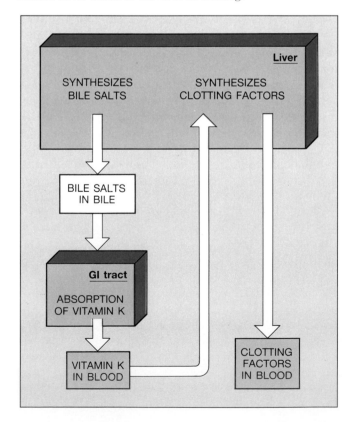

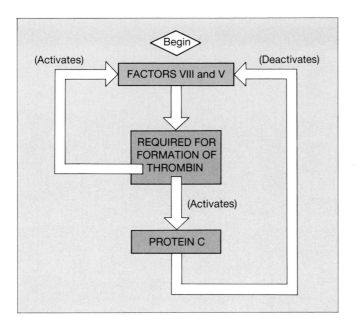

FIGURE 19-29 Thrombin activates factors VIII and V directly but deactivates them indirectly via protein C, exerting both stimulatory and inhibitory effects on its own formation.

A fibrin clot is not designed to last forever. It is a transitory device until permanent repair of the vessel occurs. The **fibrinolytic** (or **thrombolytic**) **system** is the principal effector of clot removal. The physiology of this system (Figure 19-30) is analogous to that of the clotting system: It constitutes a plasma proenzyme, **plasminogen**, which can be activated to the active enzyme **plasmin** by protein **plasminogen activators**. Once formed, plasmin digests fibrin, thereby dissolving the clot.

The fibrinolytic system is proving to be every bit as complicated as the clotting system, with multiple types of plasminogen activators and pathways for generating them, as well as several inhibitors of these plasminogen activators. In describing how this system can be set into motion, we restrict our discussion to one example—the particular plasminogen activator known as **tissue plasminogen activator** (**t-PA**), which is produced by endothelial cells and circulates in the blood. During clotting, both plasminogen and t-PA bind to fibrin and become incorporated throughout the clot. The binding of t-PA to fibrin is crucial because t-PA is a very weak enzyme in the absence of fibrin. The presence of fibrin profoundly increases the ability of t-PA to generate plasmin from plasminogen. Thus, fibrin is an important initiator of the fibrinolytic process that leads to its own dissolution.

One more example of the many interactions between the clotting and fibrinolytic systems is as follows: Not only is t-PA a weak enzyme in the absence of fibrin, but there is a normal circulating inhibitor of t-PA. During clotting, thrombin inactivates this inhibitor in the clot (Figure 19-31 and Table 19-14). Elimination of the inhibitor increases the ability of t-PA to dissolve the clot.

Finally, it should be mentioned that plasminogen activators are normally involved not only in fibrinolysis but in a host of other physiological processes, including tissue repair, ovulation, fertilization, and cell proliferation.

The secretion of t-PA is the last of the various anticlotting functions exerted by endothelial cells that we have mentioned in this chapter. These are summarized in Table 19-15.

ANTICLOTTING DRUGS

Various drugs are used clinically to prevent or reverse clotting, and a brief description of their actions serves as a review of key clotting mechanisms. One of the most common uses of these drugs is in the prevention and treatment of myocardial infarction (heart attacks, page 418), which is often the result of abnormal clotting in an atherosclerotic coronary vessel.

Aspirin inhibits the cyclo-oxygenase enzyme in the pathway for formation of thromboxane formation (Chapter 7, page 151). Since thromboxane, produced by the platelets, is important for platelet aggregation, aspirin reduces both platelet aggregation and the ensuing coagulation. A recent study has demonstrated that the taking of one aspirin every other day for several years by men with no previous history of heart disease reduced the incidence of heart attacks by 50 percent.

Coumarin is a drug that interferes with the action of vitamin K, which is required for synthesis of clotting factors by the liver.

Heparin, the naturally occurring cofactor for antithrombin III, can also be administered as a drug.

FIGURE 19-30 Basic fibrinolytic system. There are many different plasminogen activators and many different pathways for bringing them into play.

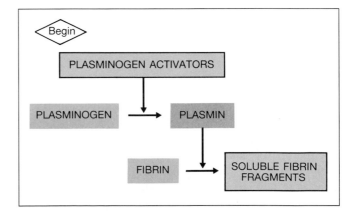

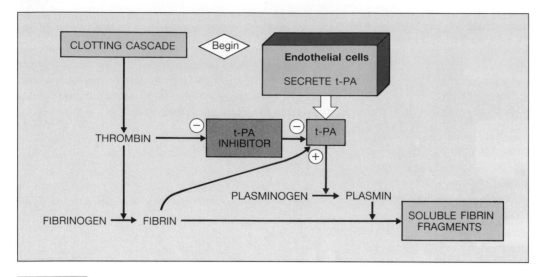

FIGURE 19-31 Fibrinolytic pathway involving tissue plasminogen activator t-PA. Fibrin activates t-PA directly, and thrombin does so indirectly by inhibiting an inhibitor of t-PA. This inhibition of the inhibitor is mediated both by thrombin itself and by protein C, activated by thrombin.

TABLE 19-15 ANTICLOTTING ROLES OF ENDOTHELIAL CELLS	
Action	**Result**
Synthesize and release PGI_2	PGI_2 inhibits platelet aggregation
Bind thrombin, which then activates protein C	Active protein C inactivates clotting factors VIII and V and inactivates an inhibitor of tissue plasminogen activator
Secrete tissue plasminogen activator	Tissue plasminogen activator catalyzes the formation of plasmin, which dissolves clots
Display heparin-like molecules on the luminal surfaces of their plasma membranes	Heparin binds antithrombin III, and this molecule then inactivates thrombin.

In contrast to aspirin, coumarin, and heparin, all of which prevent clotting, the fourth type of drug—plasminogen activators—dissolves a clot once formed. One such drug is called streptokinase. Its administration, either intravenously or directly into an occluded coronary artery, as soon as possible after myocardial infarction often results in reopening of the artery. Recently, t-PA, produced by recombinant-DNA techniques, has been used for this purpose with even greater success.

SECTION D
RESISTANCE TO STRESS

Much of this book has been concerned with the body's response to stress in its broadest meaning of an environmental change that must be adapted to if health and life are to be maintained. Thus, any change in external temperature, water intake, and so on sets into motion mechanisms designed to prevent a significant change in some physiological variable. In this section, however, we describe the basic stereotyped re-

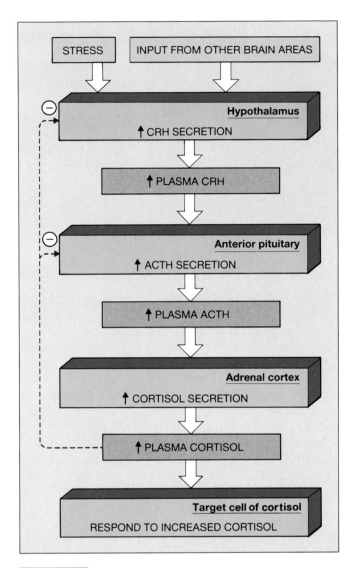

FIGURE 19-32 Pathway by which stressful stimuli elicit increased cortisol secretion. Additional stimuli, not shown in the figure, for ACTH release are vasopressin, epinephrine, and several cytokines released from immune cells.

sponse to **stress** in the more limited sense of noxious or potentially noxious stimuli. These stimuli comprise an immense number of situations, including physical trauma, prolonged exposure to cold, prolonged heavy exercise, infection, shock, decreased oxygen supply, pain, fright, and other emotional stresses.

It is obvious that the overall response to cold exposure is very different from that to infection or fright, but in one respect the response to all these situations is the same: Invariably, secretion of the glucocorticoid hormone **cortisol** by the adrenal cortex is increased. Indeed, the term stress has come to mean to physiologists any event that elicits increased cortisol secretion. Activity of the sympathetic nervous system, including release of the

hormone **epinephrine** from the adrenal medulla, is also usually increased in stress.

The increased cortisol secretion of stress is mediated mainly by the hypothalamus-anterior pituitary system described in Chapter 10. As illustrated in Figure 19-32, neural input to the hypothalamus induces secretion of **corticotropin releasing hormone** (**CRH**), which is carried by the hypothalamo-pituitary portal vessels to the anterior pituitary and stimulates **adrenocorticotropic hormone** (**ACTH**) release. The ACTH, in turn, circulates to the adrenal cortex and stimulates cortisol release.

It has recently become clear that the secretion of ACTH and therefore of cortisol is stimulated by several hormones in addition to hypothalamic CRH. These include vasopressin and epinephrine, both of which are usually increased in stress. But the most interesting recent finding is that the monokine interleukin 1 and at least one lymphokine also stimulate ACTH secretion. These cytokines provide a means for eliciting a classic stress response when the immune system is stimulated. The possible significance of this relationship for immune function is described below.

FUNCTIONS OF CORTISOL IN STRESS

The major effects of increased cortisol during stress are summarized in Table 19-16. The effects on organic metabolism, as described in Chapter 17, are to mobilize fuels—to increase the plasma concentrations of amino acids, glucose, glycerol, and free fatty acids. These effects are ideally suited to meet a stressful situation. First, an animal faced with a potential threat is usually forced to forego eating, and these metabolic changes are essential for survival during fasting. Second, the amino acids liberated by catabolism of body protein stores not only provide a source of glucose, via gluconeogenesis, but also constitute a potential source of amino acids for tissue repair should injury occur.

A few of the medically important implications of these cortisol-induced effects on organic metabolism are as follows: (1) Any patient ill or subjected to surgery catabolizes considerable quantities of body protein; (2) a diabetic who suffers an infection requires more insulin than usual; and (3) a child subjected to severe stress of any kind manifests retarded growth.

Cortisol has important effects during stress other than those on organic metabolism. It enhances vascular reactivity, that is, increases the ability of vascular smooth muscle to contract in response to stimuli such as norepinephrine. A patient with insufficient cortisol faced with even a moderate stress may develop hypotension, due primarily to a marked decrease in total peripheral resistance.

TABLE 19-16 EFFECTS OF INCREASED PLASMA CORTISOL CONCENTRATION DURING STRESS

1. Effects on organic metabolism
   (a) Stimulation of protein catabolism
   (b) Stimulation of liver uptake of amino acids and their conversion to glucose (gluconeogenesis)
   (c) Inhibition of glucose uptake and oxidation by many body cells ("insulin antagonism") but not by the brain
   (d) Stimulation of triacylglycerol catabolism in adipose tissue, with release of glycerol and fatty acid into the blood
2. Enhanced vascular reactivity, that is, increased ability to maintain vasoconstriction in response to norepinephrine and other stimuli
3. Unknown protective effects against the damaging influences of stress
4. ? Inhibition of inflammation and specific immune responses

---

As denoted by item 3 in Table 19-16, we still do not know the other reasons increased cortisol is so important for the body's optimal response to stress, that is, for its ability to resist the damaging influences of stress.

Effect 4 in the table is controversial. It stems from the known fact that administration of very large amounts of cortisol profoundly reduces the inflammatory response to injury or infection. We mentioned earlier that such doses of cortisol block the production of the eicosanoids, important mediators of inflammation in allergy. Such doses also inhibit the release or action of other inflammatory mediators and hence can block almost every step of inflammation—vasodilation, increased vascular permeability, and phagocytosis. Cortisol can also reduce the number of circulating lymphocytes and decrease both antibody production and the activity of cytotoxic T cells. Because of these effects, cortisol is an invaluable tool in the treatment of allergy, arthritis, other inflammatory diseases, and graft rejection.

These antiinflammatory effects are classified among the various *pharmacological* effects of cortisol because such large doses are required to achieve them. It has usually been assumed by physiologists that these effects do not occur at the usual plasma concentrations of cortisol. Now, however, that view is being questioned, and it may well be that the plasma concentrations achieved during stress are, in fact, antiinflammatory. Thus, the increased cortisol typical of infection or trauma might be exerting a dampening effect on the body's immune responses, protecting against possible damage from excessive inflammation.

This would also explain the significance of the fact, mentioned earlier, that several cytokines, including interleukin 1, stimulate the secretion of ACTH and thereby cortisol. Such stimulation might be part of a negative-feedback system in which the increased cortisol then partially blocks the inflammatory processes in which the cytokines participate.

One more point about the ACTH-cortisol system. Until recently, it was assumed that the only role of ACTH in stress was to stimulate the secretion of cortisol. It is now known, however, that ACTH, independent of its stimulation of cortisol secretion, is one of the peptides that facilitates learning and memory (Chapter 20). Thus, it may well be that the rise in ACTH secretion induced by psychosocial stress helps one to cope with the stress by facilitating the learning of appropriate responses to it.

FUNCTIONS OF THE SYMPATHETIC NERVOUS SYSTEM IN STRESS

Activation of the sympathetic nervous system during stress is often termed the **fight-or-flight response**, and the name is appropriate. A list of the major effects of increased general sympathetic activity, including secretion of epinephrine, almost constitutes a guide on how to meet emergencies in which physical activity may be required and bodily damage may occur. Most of these actions have been discussed in other sections of the book, and they are listed in Table 19-17 with little or no comment. Actions (6) and (7) have not been mentioned before; they reflect, respectively, stimulation by epinephrine of the brain respiratory centers and of platelet aggregation.

The adaptive value of these responses in a fight-or-flight situation is obvious. But what purpose do they serve in the psychosocial stresses so common to modern life, when neither fight nor flight is appropriate? A question yet to be answered is whether these effects, if prolonged, might enhance the development of certain diseases, particularly atherosclerosis and hypertension. For example, one can easily imagine that the increased blood

TABLE 19-17 ACTIONS OF THE SYMPATHETIC NERVOUS SYSTEM, INCLUDING EPINEPHRINE SECRETED BY THE ADRENAL MEDULLA, IN STRESS
1. Increased hepatic and muscle glycogenolysis (provides a quick source of glucose)
2. Increased breakdown of adipose tissue triacylglycerol (provides a supply of glycerol for gluconeogenesis and of fatty acids for oxidation)
3. Decreased fatigue of skeletal muscle
4. Increased cardiac output secondary to increased cardiac contractility and heart rate
5. Shunting of blood from viscera to skeletal muscles by means of vasoconstriction in the former beds and vasodilation in the latter
6. Increased ventilation
7. Increased coagulability of blood

lipid concentration and cardiac work could contribute to the former disease. Considerable work remains to be done to evaluate such possibilities.

OTHER HORMONES RELEASED DURING STRESS

Other hormones that are usually released during many kinds of stress are aldosterone, vasopressin (ADH), growth hormone, and glucagon. Insulin secretion is usually decreased. The increases in vasopressin and aldosterone ensure the retention of water and sodium within the body, an important adaptation in the face of potential losses by hemorrhage or sweating. Vasopressin also stimulates the secretion of ACTH, as we have seen, and may also influence learning. As described in Chapter 17, the overall effects of the changes in growth hormone, glucagon, and insulin are, like those of cortisol and epinephrine, to mobilize energy stores.

This list of hormones whose secretion rates are altered by stress is by no means complete. It is likely that the secretion of almost every known hormone may be influenced by stress. For example, prolactin and thyroid hormone are often increased, whereas the pituitary gonadotropins and the sex steroids are decreased. The adaptive significance of many of these changes is unclear.

The secretion of endorphin and B-lipotropin is also increased during stress. As described in Chapter 10, these substances are secreted from the anterior pituitary along with ACTH. Endorphin is a potent endogenous opiate, and its possible role in mediating both relief of pain and mood alterations in stress is a subject of great interest. Moreover, endorphin has many other effects as well; for example, it decreases appetite.

Finally, we mention once again the potential role of interleukin-1 and other monokines secreted by stimulated macrophages. To the extent that any particular stress causes significant tissue damage, that stress will trigger release of these proteins, which exert the far-reaching protective effects described earlier. A question for future study is whether either psychological stresses or physical stresses that cause little, if any, tissue damage can also induce release of monokines, perhaps via hormonal input to the macrophages.

SUMMARY

Section A. Immunology: Defenses against Foreign Material

Cells Mediating Immune Responses

I. Immune responses may be nonspecific, in which the identity of the target is not recognized, or specific, in which it is.

II. The cells of the immune system are leukocytes (neutrophils, eosinophils, basophils, monocytes, and lymphocytes), plasma cells, macrophages, and mast cells. The leukocytes use the blood for transportation but function mainly in the tissues.

Nonspecific Immune Responses

I. External barriers to infection are the skin, antimicrobial chemicals in glandular secretions, the linings of the respiratory and GI tracts, and the cilia of these linings.

II. Inflammation, the local inflammatory response to injury or infection, includes vasodilation, increased vascular permeability to protein, phagocyte chemotaxis, destruction of the invader via phagocytosis or extracellular killing, and tissue repair.

   A. The mediators controlling these processes, summarized in Table 19-4, are either released from cells in the area or generated locally from plasma proteins.

B. The cells that function as phagocytes are the neutrophils, monocytes, and macrophages. They also secrete many mediators.

C. One group of inflammatory mediators—the complement family of plasma proteins, activated during nonspecific inflammation by the alternate pathway—not only stimulates many of the steps of inflammation but mediates extracellular killing via its membrane attack complex.

D. The end result of infection or tissue damage is complete repair, with or without a scar, or formation of a granuloma.

Specific Immune Responses

I. Lymphocytes mediate specific immune responses.

II. The lymphoid organs are categorized as primary (bone marrow and thymus) or peripheral (lymph nodes, spleen, tonsils, and lymphocyte collections in the linings of the body's tracts).

A. The primary lymphoid organs are the sites of maturation of lymphocytes that will then be carried to the peripheral lymphoid organs, which are the major sites of lymphocyte mitosis and specific immune responses.

B. Lymphocytes undergo a continuous recirculation between the peripheral lymphoid organs, lymph, blood, and all the body's organs and tissues.

III. The three populations of lymphocytes are B, T, and NK cells.

A. B cells mature in the bone marrow and are carried to the peripheral lymphoid organs; there, additional B cells arise by mitosis.

B. T cells leave the bone marrow in an immature state, are carried to the thymus and undergo maturation there. These cells are then carried to the peripheral lymphoid organs, and new T cells arise from them by mitosis.

C. The origin of NK cells is unknown.

IV. Specific immune responses occur in three stages.

A. A lymphocyte programmed to recognize a specific antigen encounters it, usually after the antigen has been processed by a macrophage, and binds the antigen to plasma-membrane receptors specific for the antigen.

B. The lymphocyte is activated to undergo mitosis and further differentiation.

C. The activated cells launch an attack against antigens of that kind all over the body.

V. There are two broad categories of specific immune responses.

A. Antibody-mediated immunity—the major defense against bacteria and viruses in the extracellular fluid—is carried out by antibodies secreted by plasma cells into which B cells differentiate upon activation. Antibodies are composed of four interlocking polypeptide chains. The variable regions of the antibodies are the sites that bind antigen.

B. Cell-mediated immunity—the major defense against virus-infected cells and cancer cells—is carried out by cytotoxic T cells and NK cells.

C. Helper T cells stimulate B, cytotoxic T, and NK cells, whereas suppressor T cells inhibit them.

VI. B-cell receptors are copies of the specific antibody the cell is capable of producing. Any given B cell or clone of B cells produces antibodies with a unique antigen-binding site.

VII. T-cell receptors are not antibodies but do have specific antigen-binding sites that differ from T cell to T cell.

A. The T-cell receptor binds antigen only when the antigen is complexed to one of the body's own MHC proteins on the plasma membrane of a macrophage or other antigen-bearing cell.

B. Class I MHC proteins are found on all nucleated cells of the body whereas Class II MHC proteins are found only on macrophages, B cells, and several other antigen-presenting cells. Cytotoxic T cells require antigen to be complexed to Class I proteins, whereas helper T cells require Class II proteins.

VIII. In humoral immunity, the membrane receptors of a B cell bind antigen, and at the same time a helper T cell also binds antigen in association with a Class II MHC protein on a macrophage or other presenting cell.

A. The helper T cell, activated both by the antigen and by IL-1 secreted by the presenting cell, secretes IL-2, which then causes the helper T cell to proliferate into a clone of cells that secrete other lymphokines.

B. These lymphokines then stimulate the antigen-bound B cell to proliferate and differentiate into plasma cells, which secrete antibodies. Some of the activated B cells become memory cells, which are responsible for active immunity.

C. There are five major classes of secreted antibodies: IgG, IgM, IgA, IgD, and IgE. The first two are the major antibodies against bacterial and viral infection.

D. The secreted antibodies are carried throughout the body by the blood and combine with antigen. The antigen-antibody complex enhances the inflammatory response mainly by activating the complement system. Complement proteins mediate many steps of inflammation, act as opsonins, and directly kill antibody-bound cells via the membrane attack complex.

E. Antibodies of the IgG class also act directly as opsonins.

F. Antibodies also neutralize toxins and extracellular viruses.

IX. In cell-mediated immunity, the major targets of which are virus-infected cells and cancer cells, a cytotoxic T cell binds, via its membrane receptor, to cells bearing antigen in association with a Class I MHC protein. NK-cell binding of these antigens is independent of MHC proteins and is relatively nonspecific.

A. The cytotoxic T or NK cell then releases pore-forming protein, which kills the attached target cell by making it leaky.

B. Helper T cells are also activated in cell-mediated responses and secrete IL-2, which helps activate cytotoxic T or NK cells.

X. Helper T cells in both humoral- and cell-mediated immune responses also secrete gamma interferon, which stimulates cytotoxic T cells and NK cells and transforms macrophages into more effective killer cells against microbe-infected cells and cancer cells.

XI. Defenses against viruses include antibodies, which can neutralize them; cell-mediated responses, which kill the host cell in which they are residing; and interferon, which nonspecifically stimulates the production of proteins that inhibit viral replication.

XII. Rejection of tissue transplants is initiated by MHC proteins on the transplanted cells and is mediated mainly by cytotoxic T cells.

XIII. Systemic manifestations of infection—the acute phase response—are summarized in Figure 19-18. The major mediators of these responses are IL-1, tumor necrosis factor, and I1-6.

Factors that Alter the Body's Resistance to Infection

I. The body's capacity to resist infection is influenced by nutritional status, the presence of other diseases, psychological factors, and the intactness of the immune system.

II. AIDS is caused by a retrovirus that destroys helper T cells and therefore reduces the ability to resist infection and cancer.

III. Antibiotics interfere with the synthesis of macromolecules by bacteria.

Allergy (Hypersensitivity) Reactions

I. Allergic reactions are classified as delayed (mediated by cytotoxic T cells) or immediate (mediated by IgE antibodies).

II. In immediate hypersensitivity, antigen binds to IgE antibodies that are bound to mast cells. The mast cells then release inflammatory mediators such as histamine that produce the symptoms of the allergy.

III. The late phase of immediate hypersensitivity is mediated by eosinophils.

Autoimmune Disease

I. The body does not normally attack its own tissues because clones of lymphocytes bearing receptors against proteins found in the body are either destroyed early in life or suppressed.

II. Under abnormal circumstances, antibodies or killer cells can attack normal tissues and produce autoimmune disease.

III. Excessive inflammation and immune-complex disease can also cause destruction of normal body tissues.

Transfusion Reactions and Blood Types

I. Transfused erythrocytes will be destroyed if the recipient has natural antibodies against antigens—type A or type B—on the cells.

II. Antibodies against Rh-positive erythrocytes can be produced following exposure of an Rh-negative person to such cells.

Section B. Metabolism of Foreign Chemicals

I. The concentration of a foreign chemical in the body depends upon degree of exposure to the chemical, its rate of absorption across the GI tract, lung, or placenta, and its rates of storage, biotransformation, and excretion.

II. Biotransformation occurs in the liver and other tissues and is mediated by multiple enzymes, notably the microsomal enzyme system (MES). Its major function is to make lipid-soluble substances more polar (less lipid-soluble), thereby decreasing renal tubular reabsorption and increasing excretion.

A. This system can be induced or inhibited by the chemicals it processes and by other chemicals.

B. It detoxifies some chemicals but toxifies others, notably carcinogens.

Section C. Hemostasis: Prevention of Blood Loss

I. The initial response to blood-vessel damage is vasoconstriction as well as sticking together of the opposed endothelial surfaces.

II. The next events are formation of a platelet plug followed by blood coagulation (clotting).

Formation of a Platelet Plug

I. Platelets adhere to exposed collagen in a damaged vessel and release the contents of their secretory vesicles.

A. These substances enhance further platelet aggregation.

B. This process is also enhanced by vonWillebrand factor, secreted by the endothelial cells, and by thromboxane A_2, produced by the platelets.

C. Contractile elements in the platelets compress and strengthen the plug.

II. The platelet plug does not spread along normal endothelium because the latter secretes prostacyclin, which inhibits platelet aggregation.

Blood Coagulation: Clot Formation

I. Blood is transformed into a solid gel when, at the site of vessel damage, plasma fibrinogen is converted into fibrin molecules, which bind to each other to form a mesh.

II. This reaction is catalyzed by the enzyme thrombin, which also activates factor XIII, a plasma protein that stabilizes the fibrin meshwork.

III. The formation of thrombin from the plasma protein prothrombin is the end result of a cascade of reactions in which an inactive plasma protein is activated and then enzymatically activates the next protein in the series.

A. In the intrinsic clotting pathway, the cascade begins with the activation of factor XII by contact with collagen underlying a damaged vessel.

B. In the extrinsic clotting pathway, the cascade begins when tissue factor forms an active complex with plasma factor VII, and this complex activates factors IX and X, plugging into the intrinsic pathway at this point.

C. Both calcium and a platelet phospholipid, PF_3, are required at several steps in the cascade.

D. Two of the factors involved in clotting, factors VIII and V, are activated in a positive-feedback manner by thrombin.

IV. Vitamin K is required by the liver for normal production of clotting factors.

Anticlotting Systems

I. Clotting is opposed by protein C and antithrombin III.
A. Protein C is activated by thrombin and inactivates factors VIII and V.
B. Antithrombin III binds to heparin on endothelial cells and inactivates thrombin.

II. Clots are dissolved by the fibrinolytic system, in which the plasma proenzyme, plasminogen, is activated by plasminogen activators to plasmin, which digests fibrin.
A. Tissue plasminogen activator is secreted by endothelial cells and is activated by fibrin in a clot.
B. Thrombin inactivates a plasma inhibitor of tissue plasminogen activator.

Section D. Resistance to Stress

I. Classic responses to stress, whether physical or psychological, are increased secretion of cortisol from the adrenal cortex and activation of the sympathetic nervous system, including release of epinephrine by the adrenal medulla.

II. The functions of these responses, summarized in Tables 19-16 and 19-17, can be viewed both as a preparation for fight-or-flight and for coping with new situations.

III. Other hormones released during stress include aldosterone, vasopressin, glucagon, growth hormone, and prolactin. Insulin secretion is usually decreased.

lymphoid organs	IgG
primary lymphoid organs	gamma globulins
peripheral lymphoid organs	IgM
thymus	IgE
thymosin	IgA
lymph nodes	IgD
spleen	classical complement
tonsils	pathway
B lymphocytes	active immunity
B cells	vaccines
T lymphocytes	passive immunity
T cells	gamma interferon
cytotoxic T cells	pore-forming protein
helper T cells	effector macrophages
suppressor T cells	interferon
natural killer (NK) cells	graft rejection
antigen	acute phase proteins
antigen presenting cell	interleukin 6 (IL-6)
lymphocyte activation	colony-stimulating factors
antibody-mediated	acquired immune
immunity	deficiency syndrome
humoral immunity	(AIDS)
cell-mediated immunity	human immunodeficiency
antibodies	virus (HIV)
immunoglobulins	antibiotics
Fc	allergy
antigen-binding site	hypersensitivity
lymphokines	delayed hypersensitivity
cytokines	immediate hypersensitivity
major histocompatibility	anaphylactic shock
complex (MHC)	autoimmune disease
MHC proteins	natural antibodies
interleukin 2 (IL-2)	Rh factor
memory cells	

REVIEW QUESTIONS

Part A. Immunology: Defenses against Foreign Matter

1. Define:

immunology	kinins
microbes	interleukin 1 (IL-1)
immune surveillance	tumor necrosis factor (TNF)
nonspecific immune	monokines
responses	neutrophil exudation
specific immune responses	chemotaxis
bacteria	chemotaxin
viruses	opsonin
immune system	phagosome
leukocytes	phagolysosome
plasma cells	complement
macrophages	membrane attack complex
mast cells	(MAC)
inflammation	alternate complement
phagocytes	pathway
phagocytosis	granuloma
histamine	abscess

2. What are the major cells of the immune system and their general functions?

3. Describe the major anatomic and biochemical barriers to infection.

4. Name the three cell types that function as phagocytes.

5. List the sequence of events in an inflammatory response and describe each.

6. Name the major inflammatory mediators and their sources.

7. What triggers the alternate pathway for complement activation? What roles does complement play in inflammation and cell killing?

8. Name the lymphoid organs. Contrast the functions of the bone marrow and thymus with that of the peripheral lymphoid organs.

9. Name the various populations and subpopulations of lymphocytes and state their roles in specific immune responses.

10. Contrast the major targets of humoral immunity and cell-mediated immunity.

11. How do the Fc and combining site portions of antibodies differ?

12. What are the differences between B-cell receptors and T-cell receptors? Between cytotoxic T-cell receptors and helper T-cell receptors?

13. Diagram the sequence of events in a B-cell-mediated response, including the role of helper T cells, interleukin 1, and interleukin 2.

14. Contrast the general functions of the different antibody classes.

15. How does complement activation get triggered in the classical complement pathway, and how does complement "know" what cells to attack?

16. Name two ways in which the presence of antibodies enhances phagocytosis.

17. Compare and contrast the origins, receptors, and attack methods of cytotoxic T cells and NK cells.

18. Diagram the sequence of events by which a cell-mediated response is activated and leads to the killing of target cells. Include the roles of helper T cells, interleukin 1, and interleukin 2.

19. What are the interactions between helper T cells, NK cells, gamma interferon, effector macrophages, and tumor necrosis factor?

20. What is the major cell type involved in graft rejection?

21. Contrast the extracellular and intracellular phases of immune responses to viruses, including the role of interferon.

22. List the systemic responses to infection or injury and the mediators responsible for them.

23. What factors influence the body's resistance to infection?

24. What is the major defect in AIDS, and what causes it?

25. Diagram the sequences of events in immediate hypersensitivity.

26. Make a table of blood-type compatibilities and incompatibilities.

Section B. Metabolism of Foreign Chemicals

1. Define:

biotransformation
microsomal enzyme system (MES)

2. Why is the urinary excretion of lipid-soluble substances generally very low?

3. What are two functions of biotransformation mechanisms?

4. What are two ways in which activation of biotransformation mechanisms may actually cause malfunction?

Section C. Hemostasis

1. Define:

hemostasis
platelet aggregation
platelet plug
von Willebrand factor (vWF)
thromboxane A_2
prostacyclin (PGI_2)
blood coagulation
clotting
clot
thrombus
fibrinogen
thrombin
fibrin

factor XIII
prothrombin
factor XII
phospholipid PF_3
intrinsic clotting pathway
factor VIII
factor V
hemophilia
extrinsic clotting pathway
tissue factor
factor VII
vitamin K
protein C
antithrombin III
heparin
fibrinolytic (thrombolytic) system
plasminogen
plasmin
plasminogen activators
tissue plasminogen activator (t-PA)

2. Describe the sequence of events leading to platelet aggregation and the formation of a platelet plug. What helps keep this process localized?

3. Diagram the intrinsic clotting pathway.

4. What is the role of platelets in clotting?

5. List the procoagulant effects of thrombin.

6. Diagram the extrinsic clotting pathway. At what point do the intrinsic and extrinsic pathways merge?

7. Describe the roles of the liver and vitamin K in clotting.

8. Diagram the fibrinolytic system.

9. How does fibrin help initiate the fibrinolytic system?

10. List the anticoagulant roles of thrombin.

Section D. Resistance to Stress

1. Define:

stress
cortisol
epinephrine
corticotropin releasing hormone (CRH)
adrenocorticotropic hormone (ACTH)
fight-or-flight response

2. Diagram the CRH-ACTH-cortisol pathway.

3. List the functions of cortisol in stress.

4. List the major effects of activation of the sympathetic nervous system during stress.

5. List four hormones other than those listed in previous questions that increase during stress and one that decreases.

THOUGHT QUESTIONS

(Answers are given in Appendix A.)

1. If an individual failed to develop a thymus because of a genetic defect, what would happen to humoral and cell-mediated immunity?

2. What abnormalities would a person with a neutrophil deficiency display? A person with a monocyte deficiency?

3. An experimental animal is given a drug that blocks phagocytosis. Would this drug prevent the animal's immune system from killing foreign cells via the complement system?

4. If the Fc portion of a patient's antibodies were abnormal, what effects could this have on humoral immunity?

5. Would you predict patients with AIDS to develop fever in response to an infection? Explain.

6. A patient with symptoms of hyperthyroidism is found to have circulating antibodies against the receptors for the thyroid hormones. Can you deduce the cause of the hyperthyroidism?

7. Barbiturates and alcohol are normally metabolized by the MES. Can you deduce why the actions of an administered barbiturate last for a shorter time than normal in persons who chronically consume large quantities of alcohol?

8. If factor XII were absent because of a genetic defect, would clotting be eliminated?

CONSCIOUSNESS AND BEHAVIOR

---

he term consciousness includes two distinct concepts: **conscious experiences** and **states of consciousness**. The first concept refers to those things a person is aware of—thoughts, feelings, perceptions, ideas, dreams, reasoning—during any of the states of consciousness. A person's state of consciousness, that is, whether awake, asleep, drowsy, and so on, is defined in two ways: (1) by behavior, covering the spectrum from maximum attentiveness to coma, and (2) by the pattern of brain activity that can be recorded electrically. This record is the **electroencephalogram** (EEG), which portrays the electrical-potential difference between two points on the surface of the scalp.

STATES OF CONSCIOUSNESS

he EEG is such an important tool in identifying the different states of consciousness that we begin with it.

Electroencephalogram

When neurons are active, their activity is indicated by the electric signals known as graded potentials and action potentials (page 191). The electrical activity that is going on in the brain's neurons, particularly those near the surface, can be recorded from the outside of the head. Electrodes, which are wires attached to the head

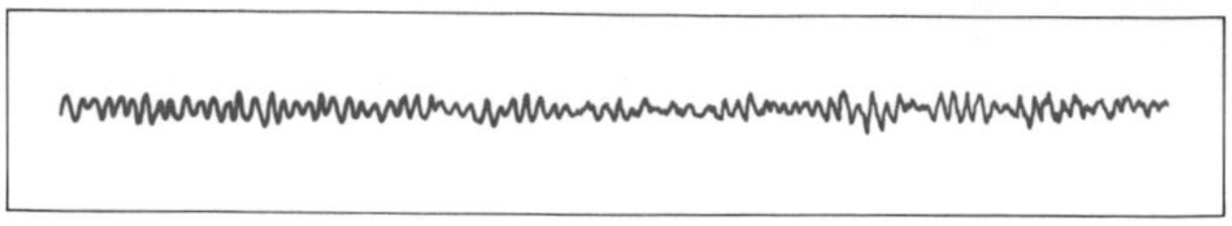

FIGURE 20-1 EEG patterns are wavelike.

by a salty paste that conducts electricity, pick up the electric signals from the head and transmit them to a machine that transforms them into the EEG. Thus, the EEG patterns are the result of the varying currents in the neurons that underlie the recording electrode.

While we often think of neural activity in terms of action potentials, they contribute little to the EEG, except in the unusual circumstances when the action potentials in a large group of neurons are synchronized. Rather, the EEG is largely due to graded potentials, in this case summed postsynaptic potentials in the brain neurons.

EEG patterns, such as that shown in Figure 20-1, are waves, albeit complex ones, and can be described in two ways: (1) The wave's amplitude indicates how much electrical activity is going on beneath the recording electrode at any given time. The amplitude is measured in volts, or, because the amplitudes are so small, in microvolts. Thus, one can speak of the voltage of an EEG to mean the wave's amplitude (Figure 20-2). The amplitude may range from 0.5 to 100 μV. Note that EEG amplitudes are about 100 times smaller than an electrocardiogram, or EKG, and about 1000 times smaller than an action potential. (2) The wave's frequency indicates how often the wave cycles from its maximal to its minimal amplitude and back. The frequency is measured in hertz, or cycles per second, and may vary from 1 to 30 Hz.

FIGURE 20-2 Characteristics of a wave. The wave amplitude is also called the wave voltage. The frequency of this wave pattern is 2 cycles per second (Hz).

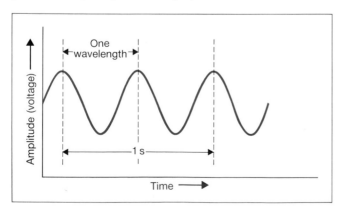

TABLE 20-1	THE FOUR MAJOR FREQUENCIES OF THE EEG		
Wave	Frequency Range, Hz		Corresponding State of Consciousness
Delta	0.5–4		Sleep stages 3–4
Theta	4–8		Sleep stages 3–4
Alpha	8–13		Awake, relaxed, eyes closed
Beta	13–25		Awake, alert

Changes in EEG patterns are correlated with changes in behavior. These patterns will be described in the next two sections. Four major frequency ranges are found in EEG patterns (Table 20-1). In general, lower EEG frequencies indicate less responsive behaviors, such as sleep, whereas higher frequencies indicate arousal.

The cause of the wavelike nature, or rhythmicity, of the EEG is not certain. It is currently thought that clusters of neurons in the thalamus are the rhythm generators that provide a fluctuating output through nerve fibers leading from the thalamus to the cortex.

The EEG is a useful clinical tool because the normal patterns are altered over brain areas that are diseased or damaged. **Epilepsy** is a common neurological disease associated with distinctive high amplitude (up to 1000 μV) wave patterns, known as spikes and waves (Figure 20-3). This disease is also associated with stereotyped changes in behavior that vary according to the part of the brain affected and can include a temporary loss of consciousness.

The Waking State

Behaviorally, the waking state is far from homogeneous, comprising the infinite variety of things one can be

FIGURE 20-3 Spike and wave pattern in the EEG of a patient during an epileptic seizure.

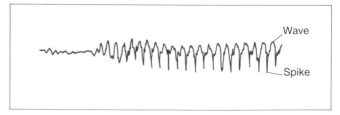

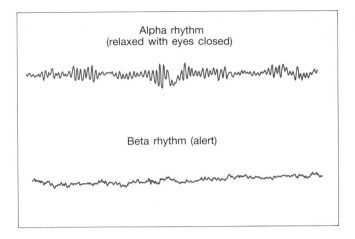

FIGURE 20-4 The (A) alpha and (B) beta rhythms of the EEG.

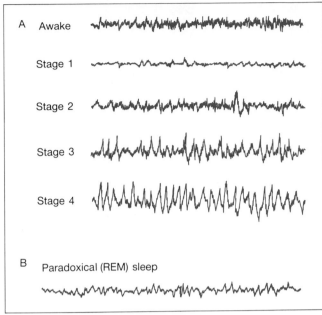

FIGURE 20-5 The EEG record of a person (A) passing from the awake state to deep sleep (stage 4) and (B) during paradoxical (REM) sleep.

doing. The most prominent EEG wave pattern of an awake, relaxed adult whose eyes are closed is a slow oscillation of approximately 10 Hz, known as the **alpha rhythm** (Figure 20-4). The alpha rhythm is associated with decreased levels of attention, and when alpha rhythms are being generated, subjects commonly report that they feel relaxed and happy. However, people who normally experience much alpha rhythm have not been shown to be psychologically different from people with lower levels.

When people are attentive to an external stimulus or are thinking hard about something, the alpha rhythm is replaced by lower-amplitude, higher-frequency oscillations, the **beta rhythm** (Figure 20-4). This transformation is known as **EEG arousal** and is associated with the act of paying attention to a stimulus rather than with the act of perception. For example, if people open their eyes in a completely dark room and try to see, EEG arousal occurs even though the people are able to perceive nothing. With decreasing attention to repeated stimuli, the EEG pattern reverts to the alpha rhythm.

Sleep

The EEG pattern changes profoundly in sleep. As a person becomes increasingly drowsy and finally falls asleep, the alpha rhythm gradually shifts toward slower-frequency theta and delta wave patterns (Figure 20-5A). These EEG changes are accompanied by complex behavioral, psychological, and physiological events.

Neurophysiologists recognize two stages of sleep—slow-wave sleep and paradoxical sleep. The initial phase of sleep, called **slow-wave sleep,** is divided into four stages, each successive stage having an EEG pattern characterized by a slower frequency and higher voltage (amplitude) than the preceding one (Figure 20-5A and

Table 20-2). Initially there is considerable tone in the postural muscles, but muscle tone progressively declines during slow-wave sleep. The eyes begin slow, rolling movements until they finally stop in stage 4 (deep sleep) with the eyes turned upward. If the sleeper is awakened during stage 4 sleep, he or she rarely reports dreaming.

Sleep always begins with the progression just described, from stage 1 to stage 4. This progression takes 30 to 45 min, and then the process reverses itself, the EEG ultimately resuming the low-voltage, high-frequency pattern characteristic of the alert, awake state. Instead of the person waking, however, the behavioral characteristics of sleep continue at this time. Skeletal-muscle tone, decreased to some extent during slow-wave sleep, is now markedly reduced except in the eye muscles, where rapid bursts of saccade-like eye movements occur (page 240). Rapid eye movement is so characteristic of this state of sleep that the state is often called **REM** (rapid eye movement) **sleep.** It is also called **paradoxical sleep** because the sleeper is difficult to arouse despite having an EEG that is characteristic of the alert, awake state (Figure 20-5B). When awakened during REM sleep, subjects report 80 to 90 percent of the time that they have been dreaming. This is true even in people who do not remember dreaming when they awaken later spontaneously.

If uninterrupted, sleep continues in this cyclic fash-

TABLE 20-2 SLEEP-WAKEFULNESS STAGES

Stage	Behavior	EEG
Relaxed wakefulness	Awake, relaxed with eyes closed.	Mainly alpha rhythm at 10 Hz. Changes to beta rhythm in response to internal or external stimuli.
Relaxed drowsiness	Fatigued, tired, or bored; eyelids may narrow and close; head may start to droop; momentary lapses of attention and alertness. Sleepy but not asleep.	Decrease in alpha wave amplitude and frequency.
Slow-wave sleep Stage 1	Light sleep; easily aroused by moderate stimuli or even by neck muscle jerks triggered by muscle stretch receptors as head nods; continuous lack of awareness.	Alpha waves reduced in frequency, amplitude, and percentage of time present; gaps in alpha rhythm filled with delta and theta activity.
Stage 2	True sleep; further lack of sensitivity to activation and arousal.	Alpha waves replaced by random waves of greater amplitude.
Stages 3 and 4	Deep sleep; in stage 4, activation and arousal occur only with vigorous stimulation; when awakened, person does not report dreaming.	Much delta and theta activity; predominant delta in 4.
Paradoxical (REM) sleep	Deepest sleep; greatest relaxation and difficulty of arousal; begins 50–90 min after sleep onset; episodes are repeated every 60–90 min, each episode lasting 10 min; dreaming occurs; rapid eye movements behind closed eyelids; marked increase in brain O_2 consumption.	EEG resembles that of alert awake state.

ion, beginning with slow-wave sleep and punctuated at regular intervals by episodes of REM sleep. Continuous recordings of adults show that the average total sleep period comprises four or five such cycles, each lasting 90 to 100 min (Figure 20-6). Slow-wave sleep constitutes about 80 percent of the total sleeping time in adults, and paradoxical sleep about 20 percent. The time spent in paradoxical sleep increases toward the end of an undisturbed night.

Although adults spend about one-third of their time sleeping, physiologists know little about the functions served by it. During the sleep cycle, there are major changes throughout the body. During slow-wave sleep, there are pulsatile releases of growth hormone (page 276) and gonadotropins from the pituitary and decreases in blood pressure, heart rate, and respiratory rate. Based on these changes, some researchers speculate that slow-wave sleep may serve as a time of the body's rest and metabolic restoration.

REM sleep is associated with an increase and irregularity in blood pressure, heart rate, and respiratory rate. Moreover, erection of the penis, engorgement of the clitoris, and twitches of the facial or limb muscles may occur (despite the generalized lack of skeletal-muscle tone). The correlation between dreaming and REM sleep indicates that the mind is highly active at this time.

Whereas slow-wave sleep is seen as a relatively primitive rest state, REM sleep is thought to be an active functional state of the brain. For example, REM sleep may allow for the expression, through dreams, of concerns in the "subconscious" and for the long-term chemical and structural changes that the brain must undergo to make learning and memory possible.

Neural Substrates of States of Consciousness

Periods of sleep and being awake alternate about once a day, that is, they manifest a circadian rhythm consisting typically of 8 h asleep and 16 h awake. This basic rhythm is controlled by the biological-clock functions of the hypothalamic suprachiasmatic nucleus (page 238). Within the sleep portion of this circadian cycle, slow-wave sleep and REM sleep alternate, as we have seen.

A likely explanation of the sleep-wake rhythms involves two interacting systems in the brainstem, one an

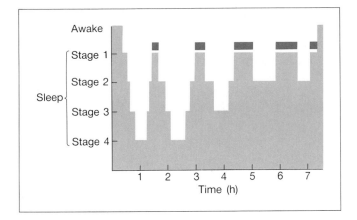

FIGURE 20-6 A typical night's sleep in an average young adult. The heavy lines indicate periods of paradoxical sleep, which are characterized by stage 1 EEG patterns and the presence of rapid eye movements. *(Adapted from Nicholi.)*

arousal system and the other (with several components) a sleep-producing system (Figure 20-7). Thus, neurons in some brainstem regions are most active during waking and are inhibited during sleep, while neurons in other regions show the opposite activity pattern. A separate cluster of neurons is responsible for the cyclical imposition of REM sleep upon slow-wave sleep. The neurotransmitters norepinephrine, serotonin, dopamine, and

FIGURE 20-7 Brainstem structures involved in arousal, paradoxical sleep, and slow-wave sleep. *(Adapted from Jouvet.)*

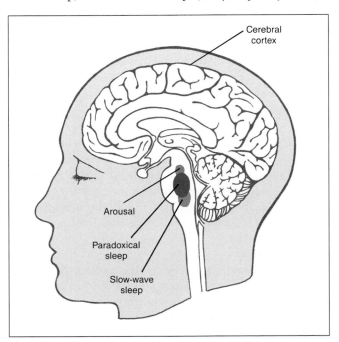

acetylcholine all play a role in the alternating states of consciousness.

The changes in activity of the cortex, as reflected in EEG changes during waking and the two phases of sleep, are mediated by pathways passing from the brainstem to neurons in the thalamus and then from the thalamic neurons to the cortex. The EEG of REM sleep, even though it is similar to the awake EEG, is mediated by different pathways, which include the norepinephrine-releasing neurons of the locus ceruleus.

Skeletal-muscle activity decreases with the transition from waking to slow-wave sleep because central motor commands are dramatically lowered during this time. Then, during REM sleep, most muscle activity becomes essentially impossible because the motor neurons are actively inhibited by pathways descending from the reticular formation. An exception to this generalized motor inhibition during REM sleep is, as mentioned earlier, the system of eye muscles that sweep the eyes rapidly back and forth.

Afferent pathways or other brain centers can override the periodic inhibition of the arousal system that occurs during the sleep portion of the cycle. This blocking of inhibition can maintain activity in the arousal systems high enough to keep one awake or to interrupt sleep. In fact, the waking mechanisms seem to be more easily activated than those causing sleep. An example familiar to all parents is that it is much easier to arouse a sleeping child than to get an alert, attentive child to sleep.

In addition to the transmitters mentioned above, there are a number of other sleep-inducing chemical substances, such as delta-sleep inducing peptide (DSIP) and substance S, found in blood, urine, cerebrospinal fluid, and brain tissue. It is not known if these substances trigger slow-wave sleep or paradoxical sleep.

Coma and Brain Death

There have been great advances in the last 100 years in the development of resuscitation techniques. For example, it is now fairly common for artificial respiration to supply oxygen to the lungs and for cardiac massage to restore a nonfunctional circulation. However, these techniques may result in a body with adequate cardiovascular and respiratory function but with an inactive nervous system—in other words, a living body that contains a dead brain—because the brain is particularly susceptible to oxygen lack and is likely to suffer damage even though resuscitative techniques have restored the functioning of the heart and lungs.

The term **coma** describes a severe or total decrease in mental function due to structural or physiological impairment of the brain. The EEG of a person in a coma is illustrated in Figure 20-8. A person in a coma is characterized by a sustained loss in the capacity for arousal

FIGURE 20-8 The EEG of a person in a coma.

even in response to vigorous stimulation. There is no expression of any mental function, the eyes are closed, and sleep-wake cycles disappear. But is a person in a coma dead or alive?

The question "When is a person actually dead?" often has urgent medical, legal, and social consequences. For example, with the advent of organ transplantation and the need for viable tissues, it became imperative to know how soon transplant organs could be taken from a "dead" person.

Brain death describes the irreversible loss of function of all neural structures above the spinal cord such that the brain no longer functions and has no possibility of functioning again. Brain death is widely accepted by doctors and lawyers as the criterion for death.

The problem now becomes practical: How does one know when a person in a coma is considered brain dead? There is general agreement that the criteria listed in Table 20-3, if met, denote brain death. Notice that the

cause of a coma must be known because comas due to drug poisoning are usually reversible. Also, the criteria specify that there be no evidence of functioning neural tissues above the spinal cord because fragments of spinal reflexes may remain for several hours or longer after the brain is dead. The criterion for lack of spontaneous respiration (apnea) can be difficult to test because once a patient is on a respirator, it is of course inadvisable to remove him or her for the 10-min test because of the danger of further brain damage due to oxygen lack. Therefore apnea is diagnosed if there is no spontaneous attempt to fight the respirator, that is, the patient's reflexes do not drive respiration at a rate or depth different from those of the respirator.

CONSCIOUS EXPERIENCES

Having run the gamut of the states of consciousness from the awake, alert state to coma and brain death, we turn now to the conscious experiences during the awake state. We begin with the subject of how we pay attention because we are aware of (that is, we experience) only those ideas, perceptions, and actions to which we pay some degree of attention. For instance, I am aware of the weight of my glasses resting on my nose only when I think about it. But it is believed that the amount of attention that can be paid to different events at any one time is limited, thus limiting the number of things we can consciously experience at one time.

Directed Attention

The term **directed attention** means the avoidance of distraction by irrelevant stimuli while seeking out and focusing on stimuli that are important. An example familiar to students is ignoring distracting events in a busy library while studying there.

Presentation of a novel stimulus to a relaxed subject showing an alpha EEG pattern causes EEG arousal (if the subject is already concentrating on something, the EEG is already in the beta rhythm). If the stimulus has meaning for the individual, behavioral changes occur. The person stops what he or she is doing and looks around, listening intently and orienting toward the stimulus source. This behavior is called the **orienting response**. If the person is concentrating hard and is not distracted by the novel stimulus, the orienting response does not occur.

For attention to be directed only toward stimuli that are meaningful, the nervous system must evaluate the importance of the incoming sensory information. A decision that the stimulus is irrelevant results in a progressive decrease in response to the repeated stimulus (**ha-**

TABLE 20-3 CRITERIA FOR BRAIN DEATH

1. The nature and duration of the coma must be known.
  a. Known structural damage to brain or irreversible systemic metabolic disease.
  b. No chance of drug intoxication, especially paralyzing or sedative drugs.
  c. No sign of brain function for 6 h in cases of known structural cause and when no drugs or alcohol are involved; otherwise, 12 to 24 h without signs of brain function plus a negative drug screen.
2. Cerebral and brainstem function are absent.
  a. No response to painful stimuli administered above the spinal cord.
  b. Pupils unresponsive to light.
  c. No eye movement in response to ice-water stimulation of the vestibular reflex.
  d. Apnea (no spontaneous breathing) for 10 min.
  e. Systemic circulation may be intact.
  f. Purely spinal reflexes may be retained.
3. Supplementary (optional) criteria.
  a. Flat EEG (wave amplitudes less than 2 μV).
  b. Responses absent in vital brainstem structures.
  c. No cerebral circulation.

bituation). For example, when a loud bell is sounded for the first time, it may evoke an orienting response because the person might be frightened at or curious about the novel stimulus. After several ringings, however, the person makes progressively less response and eventually may ignore the bell altogether. An extraneous stimulus of another type or the same stimulus at a different intensity restores the original response (**dishabituation**).

Habituation is not due to receptor fatigue or adaptation. Rather, it involves a depression of synaptic transmission possibly related to a prolonged inactivation of calcium channels in the presynaptic axon terminals. Such inactivation results in a decreased calcium influx during depolarization and, hence, a decrease in the amount of neurotransmitter released in response to action potentials.

Neural mechanisms for directed attention. The locus ceruleus, a brainstem nucleus (Figure 20-9) that projects to the parietal cortex and to many other parts of the central nervous system as well, is strongly implicated in directed attention. The system of fibers leading from the locus ceruleus determine which brain area is to gain temporary predominance in the ongoing stream of the conscious experience.

The activity of neurons in the locus ceruleus increases with sensory stimulation and with increased attention. Norepinephrine, the transmitter released by these neurons, decreases the background electrical activity in the brain so that signals from the sensory systems are clearer, that is, the weak signals are suppressed and the vigorous signals are enhanced so the difference between them is increased. Thus, neurons of the locus ceruleus seem to improve information processing during directed attention.

Neural Mechanisms for Conscious Experiences

Finding the neural substrate of conscious experience begins with the question "Which neural actions are specifically required for conscious experiences to occur?" Certainly, there can be neural responses to stimulation without any accompanying conscious experience of the event. For example, stimulation of the cerebral cortex surface exposed during neurosurgery can produce large neural responses but no awareness of the event by a conscious, alert subject. Another example, stimulation of peripheral nerves does not always lead to conscious sensation. From these observations, we conclude that specific kinds of neural activity must be present in the brain if conscious sensory experiences are to occur.

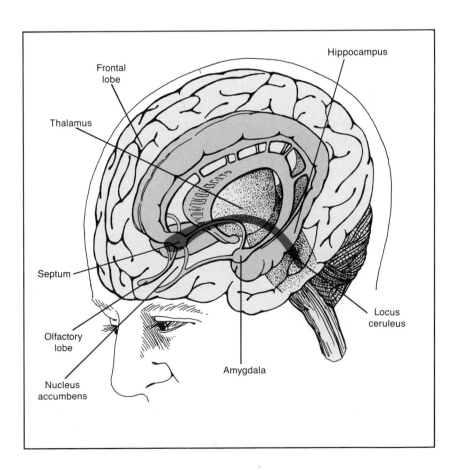

FIGURE 20-9 Brain structures involved in emotion, motivation, and the affective disorders. Dark blue indicates region of medial forebrain bundle. (*Redrawn from Bloom et al.*)

All conscious experiences are popularly attributed to the workings of the "mind," a word that conjures up the image of a nonneural "me," a phantom interposed between afferent and efferent impulses, with the implication that mind is something more than neural activity. Although a good definition of mind is not possible at this time, mind includes such actions as thinking, perceiving, making decisions, feeling, and imagining. Thus, we can conclude that mind is a "process" that gives rise to and includes conscious experiences. The truth of the matter is, however, that physiologists have only a minimal understanding of the mechanisms that give rise to mind or to conscious experiences.

A few things are clear. Conscious experiences and the brain are not the same thing. Conscious experience (and, for that matter, mind) is a process, an action, that is a *function* of the machinery of the brain. Thus, without the neurophysiological operations of the machinery of the brain, conscious experiences and the other functions we attribute to the mind would not exist.

It is most likely that conscious experience does *not* arise because of the molecular or membrane properties of a single neuron or even because of the combined activity of neurons in such large brain centers as the thalamus or limbic system. It is now becoming clear that chemistry and physics are not the sciences that will lead to an understanding of conscious experiences, even though they do lead to an understanding of the basic operations of the brain's machinery. Rather, it seems that conscious experiences can be approached only mathematically at the level of the interactions of networks of neurons.

The best guess now available is that mind and conscious experience are processes that occur during the flow of electric signals (action potentials and graded potentials) among the extremely complicated networks that link the vast numbers of interacting neurons in the brain. In fact, the discouraging possibility exists that even if the interacting-networks theory of conscious experiences is right, our abilities to map the networks and understand their interrelations will fall far short of the goal of understanding the neural basis of mind since an immense number of units are involved.

Not only must certain neural activity be present for conscious experiences to occur, the neural activity must be present for a minimal period of 500 ms. For example, there is a substantial delay between the first electrical response of the brain to sensory stimulation and the conscious experience of the stimulation.

A general hypothesis has been proposed that many if not all conscious experiences require a minimum time of cortical activation. If cortical activation occurs for less than this minimum time, mental operations that determine the meaning and emotional value of an event still occur, but they are unconscious. In fact, a major determinant of whether an event will be conscious or unconscious is the duration of the appropriate neural activity.

A great deal of mental activity occurs during this processing time, but we are unaware of it, that is, it is unconscious. The results of this unconscious processing can nevertheless influence the experiences, thoughts, and actions that occur during conscious experience. It is believed, in fact, that most neural operations, including many perceptions, memories, judgments, problem solving, and discriminative responses to stimulation, proceed unconsciously. Most of us can recall times when, "aha!", we knew how to solve a problem, even though we had not been consciously working on it. On the other hand, the unconscious mental activity that precedes conscious experience can shut down (repress, in psychological terms) components of the mental processing and prevent them from reaching conscious experience. For instance, you may "forget" to call your mother because, at a subconscious level, you are afraid she will confront you with your failure to write her.

While examining the basis of conscious experiences, it is interesting to study those of persons undergoing periods of sensory deprivation. Student volunteers lived 24 h a day in as complete isolation as possible—even to the extent that their movements were greatly restricted. External stimuli were almost completely absent, and stimulation of the body surface was relatively constant. At first, the students slept excessively, but soon they began to be disturbed by vivid hallucinations that sometimes became so distorted and intense that the students refused to continue the experiment. The neural bases of these hallucinations are poorly understood, but it has been suggested that patterns of neural activity corresponding to those normally elicited by peripheral stimuli may be generated when varied sensory input is absent and that conscious experience is not solely dependent upon the senses.

There seems to be an optimal amount of afferent stimulation necessary for the maintenance of the normal, awake consciousness. Levels of stimulation greater or less than this optimal amount can lead to trances, hypnotic states, hallucinations, "highs," or other altered states of consciousness. In fact, alteration of sensory input is commonly used to induce such experiences intentionally.

MOTIVATION AND EMOTION

otivation is a factor in most, if not all, behaviors, while emotions accompany many of our conscious experiences.

Motivation

Those processes responsible for the goal-directed quality of behavior are the **motivations** for that behavior. **Primary motivated behavior** is related directly to homeostasis, that is, the maintenance of a relatively unchanging internal environment, an example being putting on a sweater when one is cold. In such homeostatic goal-directed behavior, specific bodily needs are satisfied, the word "needs" having a physical or chemical correlate. Thus, in our example the correlate of need is a drop in body temperature, and the correlate of need satisfaction is the return of the body temperature to normal. The neurophysiologic integration of much homeostatic goal-directed behavior has been discussed earlier (thirst and drinking, Chapter 15; food intake and temperature regulation, Chapter 17; reproduction, Chapter 18).

In many kinds of behavior, however, the relation between the behavior and the primary goal is indirect. For example, the selection of a particular sweater on the basis of style has little if any apparent relation to homeostasis. The motivation in this case is called **secondary**. Much of human behavior fits this latter category and is influenced by habit, learning, experience, and emotions. These factors as well as the actual homeostatically related "needs" are thought to determine the degree of motivation, or drive, behind a particular behavior.

Concepts inseparable from motivation are those of reward and punishment. Rewards are things that organisms work for or things that make the behavior that leads to them occur more often. They are related to what is called **appetitive motivation** in that rewards may be said to satisfy appetites, for example, as eating satisfies hunger and drinking satisfies thirst. Punishments are the opposite and are associated with **aversive motivations** that lead to behaviors in which an organism tries to escape from a painful or life-threatening situation, avoid a situation that in the past has proved to be harmful, or remove by aggression a potential source of harm. Aversive motivation can be learned, a fact some parents try to utilize when, for example, they punish a child for running into the street, but some aversive motivations are innate, such as those experienced in response to painful stimuli, restraint, or suffocation.

Much of the available information concerning the neural substrates of motivation has been obtained by studying the effects of **brain self-stimulation**. In this technique, an unanesthetized experimental animal (or a human undergoing neurosurgery for other reasons) regulates the rate at which electric stimuli are delivered through electrodes implanted in discrete brain areas. The animal is placed in a box containing a lever it can press (Figure 20-10). If no stimulus is delivered to the brain when the bar is pressed, the animal usually presses it occasionally at random.

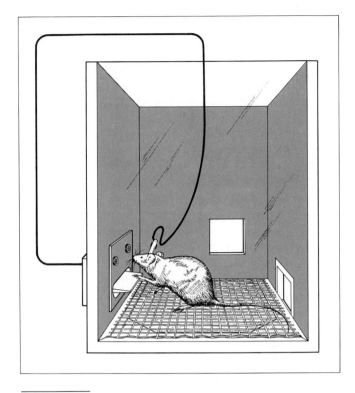

FIGURE 20-10 Apparatus for self-stimulation experiments. *(Adapted from Olds.)*

If, in contrast, a stimulus is delivered to the brain as a result of the bar press, a different behavior occurs, depending on the location of the electrodes. If the animal increases the bar-pressing rate above the level of random presses in the absence of a stimulus, the electric stimulus is by definition rewarding. If the animal decreases the press rate below the random level, the stimulus is punishing.

Thus, the rate of bar pressing with the electrode in different brain areas is taken to be a measure of the effectiveness of the reward or punishment. Different pressing rates were found in different brain regions. Regions that are part of appetitive motivational systems lead to increased rates of bar pressing, and regions that are part of aversive motivational systems lead to decreased rates.

Brain self-stimulation of the lateral regions of the hypothalamus serves as a positive reward. Animals with electrodes in these areas have been known to bar-press to stimulate their brains 2000 times per hour continuously for 24 h until they drop over from exhaustion! In fact, electric stimulation of the lateral hypothalamus is more rewarding than external rewards in that hungry rats, for example, often ignore available food for the sake of electrically stimulating their brains at that location.

One relatively nonspecific neural system that passes through the lateral hypothalamus and is strongly implicated in reward is the **medial forebrain bundle,** a large tract of ascending and descending axons, some long, passing from one end of the brain to the other, and some short, extending only a few millimeters. Fibers in this tract affect virtually every level of the brain, but they have a particularly strong influence on the hypothalamus. Interestingly, axons of locus ceruleus neurons, mentioned earlier in the context of directed attention, constitute a significant portion of the fibers in the medial forebrain bundle.

Although the rewarding sites are more densely packed in the lateral hypothalamus than anywhere else in the brain and animals will bar-press at higher rates when the electrodes are implanted there, self-stimulation can be obtained from a large number of other brain areas. It seems that self-stimulation in these brain areas triggers such appetitive behaviors as feeding, drinking, and sexual behavior. Consistent with this is the fact that the animal's rate of self-stimulation in some areas increases when the animal is deprived of food. In other areas, it is decreased by castration and restored by the administration of sex hormones. Thus, neurons controlling homeostatic goal-directed behavior are themselves intimately involved in the reinforcing effects of reward and punishment.

Chemical mediators. Two catecholamines, norepinephrine and dopamine, are transmitters in the pathways that mediate the brain reward systems and appetitive motivation. For this reason, drugs that increase synaptic activity in the catecholamine pathways—amphetamines, for example—increase self-stimulation rates. Conversely, drugs such as chlorpromazine hydrochloride, an antipsychotic agent, which lower activity in the catecholamine pathways, decrease self-stimulation.

In addition to the catecholamine pathways, a system of neurons activated by morphine and enkephalin is involved in motivation. The catecholamines and enkephalin are also, as we shall see, implicated in pathways subserving learning. This is not unexpected since rewards and punishments are believed to constitute the incentives for learning.

Emotion

Emotions are related to motivation in that they lend strength or intensity to an action. For purposes of this discussion, we distinguish two aspects of emotion: (1) **inner emotions**, such as the feelings of fear, love, anger, joy, anxiety, hope, and so on, that are entirely within a person, and (2) the outward expressions and displays of **emotional behavior**. Emotional behavior includes such complex behaviors as attack and such simple actions as laughing, sweating, crying, or blushing. Emotional be-

havior is achieved by integrated activity of the autonomic and somatic efferent systems and provides an outward sign that an inward emotion has occurred.

The two aspects of emotion are served by different parts of the brain, and these aspects can occur independently in certain diseases and in experimental situations. Inner emotions involve the cerebral cortex and various areas of the limbic system, whereas emotional behavior involves one part of the limbic system—the hypothalamus—and the brainstem. Emotional behavior, unlike inner emotions, can be studied reasonably well because it includes responses of the autonomic, endocrine, and motor systems that are easily measured.

Stimulation of limbic structures other than the hypothalamus causes a wide variety of complex emotional behaviors that receive their final coordination by the hypothalamus. For example, stimulation of one area of the limbic system caused an experimental animal to approach the researcher as though expecting a reward, whereas stimulation of a second area caused the animal to stop the behavior it was performing, as though fearing punishment. Stimulation of a third region caused the animal to arch its back, puff out its tail, hiss, snarl, bare its claws and teeth, flatten its ears, and strike. Simultaneously, its heart rate, blood pressure, respiration, salivation, and concentrations of plasma epinephrine and fatty acids all increased. Clearly, this behavior typified that of an enraged or threatened animal.

Moreover, the animal's behavior could be changed from savage to docile and back again simply by altering different areas of the limbic system. Damage to regions of the septum (Figure 20-9) produced vicious rage in a tame animal, and this could be counteracted by destruction of the amygdala, a group of nuclei in the tip of the temporal lobe (Figure 20-9), which we shall encounter again in our discussion of learning. Moreover, some animals with damage in the amygdala manifest bizarre sexual behavior in which they attempt to mate with animals of other species and females assume male positions and attempt to mount other animals.

Limbic areas have been stimulated in awake human beings undergoing neurosurgery. These patients reported vague feelings of fear or anxiety during periods of stimulation to certain areas (even though they were not told when the current was on). Stimulation of other areas induced pleasurable sensations that the subjects found hard to define precisely.

The models presently available to explain inner emotions and emotional behavior are far from complete. It seems that nonhypothalamic parts of the limbic system receive information from cortical association areas, particularly those in the frontal lobe, and send output directly to the hypothalamus. Thus, the limbic system serves as the route by which information about the emotional meaning of an external stimulus (for example, whether it is threatening or friendly), including information gleaned from memory and understanding, is passed to the hypothalamus. The hypothalamus then integrates the endocrine, autonomic, and even some of the motor activities that form appropriate emotional behavior.

In addition to its role in elaborating the conscious experience of emotional feelings, the cerebral cortex provides the neural mechanisms that direct the motor responses to the external event during emotional behavior, for example, to approach or avoid a situation. Moreover, it is forebrain structures, including the cerebral cortex, that account for the modulation, direction, understanding, or even inhibition of emotional behaviors.

ALTERED STATES OF CONSCIOUSNESS

The state of consciousness may be different from the commonly experienced wakefulness, drowsiness, and so on. Other, more bizarre situations, such as those occurring with hypnosis, mind-altering drugs, and certain diseases, are referred to as altered states of consciousness.

Schizophrenia

One of the diseases that induces altered states of consciousness is **schizophrenia**. Schizophrenia is a family of disorders rather than a single disease, but the general symptoms include altered motor behavior, perceptual distortions, disturbed thinking, altered mood, and abnormal interpersonal behavior. The motor behavior can range from total immobilization (catatonia) to wild, purposeless activity. Perceptual distortions can include hallucinations, particularly auditory ones, such as hearing voices or hearing one's own thoughts out loud. The disturbed thinking can include delusions, which are illogical thoughts and beliefs that are false or improbable but cannot be changed by contrary evidence or argument, for example, the belief that one has been chosen for a special mission or is persecuted by others.

Schizophrenia is a fairly common disease. It can appear gradually, often during adolescence in an otherwise bright and normal person, or it can occur suddenly after a single precipitating event. Despite extensive research, the causes of schizophrenia remain unclear, although it is generally accepted that there is a genetic component to the disease. Whereas 1 percent of the general population will be diagnosed as having the disease, 10 to 16 percent of people having a schizophrenic parent will themselves become schizophrenic.

It is very likely that there is something biochemically different about people with this disease. The most widely accepted explanation for schizophrenia is the **dopamine hypothesis**, which suggests that increased activity in the dopaminergic system in the brain at least partially explains the symptoms of the disease. This hypothesis is supported by the fact that virtually all drugs that alter schizophrenia influence transmission at dopamine-mediated synapses. For example, the symptoms are made worse by amphetamine-like drugs, which are dopamine agonists, and the most useful of the antipsychotic drugs used in treating schizophrenia block dopamine receptors.

Excessive dopamine activity could theoretically occur because of malfunction of any of the steps in the synthesis, storage, release, receptor activation, reuptake, or metabolism of the neurotransmitter, but abnormalities in receptors on the postsynaptic neurons in the dopamine pathways are most suspect.

Abnormalities in other neurotransmitters and neuromodulators, including serotonin, histamine, acetylcholine, the endorphins, and norepinephrine have also been implicated in schizophrenia.

The Affective Disorders: Depressions and Manias

Affect is a term used in psychology to indicate feeling or emotion. The **affective disorders** are those characterized by serious, prolonged disturbances of mood. They include the **depressions**, the **manias**, and swings between these two states, the so-called **bipolar affective disorders**. In the depressive disorders, the prominent feature is a pervasive sadness, loss of interest or pleasure, and feelings of worthlessness. The essential features of the manias are elated mood, sometimes with euphoria (that is, an exaggerated sense of well-being), overconfidence, and irritability. Along with schizophrenia, these affective disorders form the major psychiatric illnesses today.

Norepinephrine is the major neurotransmitter implicated in the affective disorders, although acetylcholine, GABA, and serotonin may be involved as well. In major depression the locus ceruleus shows increased activity. This nucleus contains over half the cells in the brain that use norepinephrine as their neurotransmitter. Axons of these cells branch extensively and innervate every major region of the central nervous system.

Much of the evidence supporting the involvement of norepinephrine in the affective disorders is circumstantial, coming from measurements of the levels of norepinephrine and its metabolites in the cerebrospinal fluid and blood plasma. When a person is in the midst of a major depression, norepinephrine levels are increased. Following the administration of antidepressant drugs, the levels return to baseline. Here one encounters the kind of difficulty met when trying to analyze the cause of a mental disorder: Do the changes in levels of norepinephrine and its metabolites reflect disturbances that are the cause of the disease, or do they reflect the altered behaviors and mental states resulting from it?

Surprisingly, drugs used to combat depressions generally enhance the effects of norepinephrine. For example, the antidepressant drugs classed as monoamine oxidase inhibitors interfere with an enzyme responsible for norepinephrine breakdown, and the tricyclic antidepressant drugs, such as Elavil and Pamelor, interfere with norepinephrine uptake by presynaptic endings. In both cases, the result is *increased* concentration of norepinephrine at the synapses.

The explanation for the unexpected effectiveness of these drugs in depression is that the locus ceruleus neurons, in addition to using norepinephrine as a neurotransmitter at their axon endings, have receptors for this substance on their own cell bodies. The receptors on the cell bodies are inhibitory, so that drugs that increase the availability of norepinephrine to these receptors cause a decrease in firing of the locus ceruleus fibers throughout the brain.

The drug most effective in treating bipolar affective disorder is lithium carbonate. It is highly specific, normalizing both manic and depressing moods and slowing down thinking and motor behavior without causing sedation. In addition, it decreases the severity of the swings between mania and depression that occur in the bipolar disorders and, in some cases, it is even effective in depression not associated with manias. Lithium decreases the release of norepinephrine from presynaptic terminals and enhances its uptake, which fits nicely with the hypothesis that there is too much norepinephrine at excitatory norepinephrine receptors in the affective disorders. It does not explain, however, why lithium is occasionally effective in treatment of the depressions when its actions are opposite those of the tricyclic antidepressants. The confusing patterns of drug effectiveness in these diseases contributes to the poor understanding of their causes.

Psychoactive Drugs, Tolerance, and Addiction

In the previous sections, we mentioned several drugs used to combat altered states of consciousness. These as well as other psychoactive drugs are also used as "street drugs" in a deliberate attempt to elevate mood (euphorigens) and produce unusual states of consciousness ranging from meditational states to hallucinations.

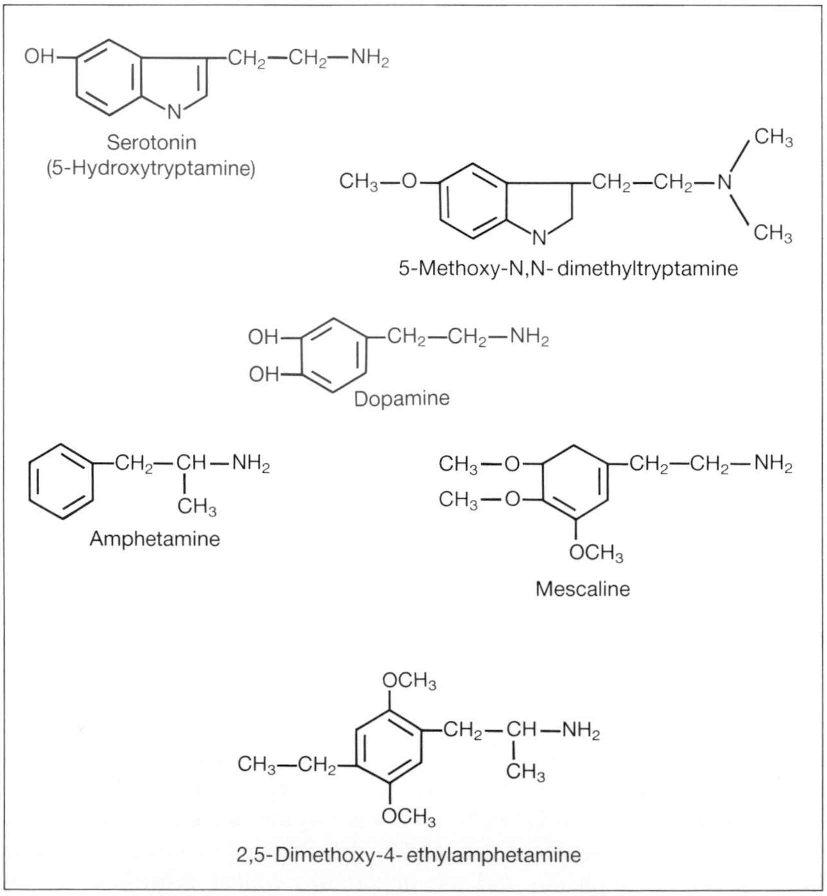

FIGURE 20-11 Molecular similarities between neurotransmitters (blue type) and some euphorigens (black type). At high doses, these euphorigens can cause hallucinations.

Virtually all the psychoactive drugs exert their actions either directly or indirectly by altering neurotransmitter-receptor interactions. As mentioned on page 210, psychoactive drugs are often chemically similar to neurotransmitters such as serotonin, dopamine (Figure 20-11), and norepinephrine, and they interact with the receptors activated by these transmitters.

Tolerance to a drug occurs when increasing doses of that drug are required to achieve effects that initially occurred in response to a smaller dose, that is, it takes more drug to do the same job. Tolerance may develop because the presence of the drug stimulates the synthesis, especially in the liver microsomal system, of those enzymes that degrade it. As drug concentrations increase, so do the concentrations of the enzymes that degrade it. Thus, more drug must be administered to produce the same plasma concentrations of the drug and hence the same initial effect.

Other postulated causes for tolerance have nothing to do with drug degradation but result from the drug's action at synapses. The drug's effects may add to those of the normally occurring neurotransmitter, thereby producing an increased response that, by feedback mechanisms, decreases the release of the neurotransmitter. Let us use enkephalin and morphine as an example.

Normally, enkephalin, which is a neurotransmitter in central nervous system pathways involved both in pain inhibition and in a sense of well-being, is continually released, at least to some degree, by those neurons that use it as their neurotransmitter. Accordingly, the enkephalin receptors are always exposed to a certain amount of stimulation (Figure 20-12). When morphine, an opiate drug that reacts with some of the receptors normally activated by enkephalin, is present, it binds to receptors not already occupied by enkephalin and thereby increases the analgesic-euphorigenic effects of the enkephalin pathways. As the enkephalin receptors are stimulated more and more strongly, a feedback system acts on the presynaptic neuron, thereby decreasing the synthesis and release of enkephalin. The postsynaptic receptors, no longer receiving their usual amount of enkephalin, can therefore accept more morphine. In other words, to regain the previous analgesic-euphorigenic effects of full receptor activation, increased amounts of morphine are required.

This type of reasoning can also explain the physical symptoms associated with cessation of drug use, that is, **withdrawal.** When the drug is stopped, the receptors receive for some time neither the drug nor the normal neurotransmitter because the synthesis and release of the neurotransmitter had been inhibited by the previous prolonged drug use. Withdrawal symptoms disappear as neuronal transmitter release returns to normal levels.

Drug dependence, the now preferred term for addic-

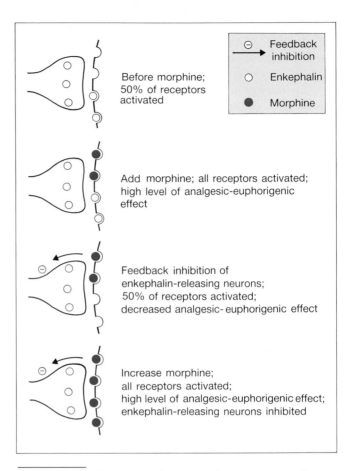

FIGURE 20-12 Hypothesized steps in the production of morphine dependence in a person

tion, has two facets that may occur together or independently: (1) a psychological dependence that is experienced as a craving for the drug and inability to stop using the substance at will, and (2) a physical dependence that causes unpleasant symptoms with cessation of drug use. Actually, drug dependence has not been defined to everyone's satisfaction. Regardless, the definition that is ultimately agreed upon should include the following concepts: (1) drug dependence is an abnormal response to the exposure to alcohol or certain other drugs; (2) it is characterized by the tendency toward progressive increase in consumption of the drug; and (3) it is characterized by a persistent tendency to relapse to use of the substance even after abstinence has been achieved and withdrawal symptoms are no longer in evidence. Thus, because of point 3, a person does not have to be using drugs to be drug dependent. Note that different substances have different powers of causing dependence. Thus, just because you can drink or not drink beer as you choose, it does not follow that you will be able to use or not use cocaine or heroin as you choose.

The physical basis for dependence and the parts of the brain involved in it are not fully known. The brain reward system, mentioned earlier in the contexts of motivation and emotion, allows a person to experience pleasure in response to pleasurable events or in response to certain drugs. Apparently, drugs that lead to dependence in some people act on these neural pathways to increase drug-taking behavior. The nucleus accumbens, a structure that lies in the lower part of the forebrain and is part of the reward system (Figure 20-9), and the neurotransmitter dopamine have been most strongly implicated. While some researchers believe that people take drugs primarily to feel pleasure, there has been a recent revival of the 1950s concept that once drug dependence has developed, people take drugs primarily to avoid the discomfort of withdrawal.

Alcohol: An example.   Medically, the central nervous system depressants, such as alcohol, are used to produce sedation, sleep, and anesthesia. Nonmedicinally, these drugs are used for relaxation and relief from anxiety, to produce a mild euphoria, and to promote sleep. With repeated usage, they produce tolerance and both the physical and psychological forms of dependence.

Excess ingestion of alcohol causes a variety of central nervous system symptoms, for example, poor coordination, sluggish reflexes, emotional instability, and out-of-character behavior such as belligerence and aggression in a person typically shy, retiring, and mild-mannered. Sometimes, amnesia for events that happen during the intoxicated period (blackout) occurs.

The degree of intoxication relates to body size, type of beverage consumed, rapidity of drinking, tolerance, the presence or absence of food in the stomach, and many other factors. Individual response to alcohol consumption varies greatly, but signs of intoxication are almost always present at blood levels of 2000 mg/L, unconsciousness usually occurs when blood levels exceed 4000 mg/L, and death frequently occurs at concentrations above 5000 mg/L. A blood alcohol level of 1000 mg/L, or 0.1 g/dL, is the legal upper limit for driving a car in most states. Unconsciousness usually occurs before the person can drink enough to die, but the rapid consumption of large amounts of alcohol can cause death either by depression of the medullary respiratory center or by vomiting followed by aspiration of the vomitus into the respiratory tract.

The length of the intoxication period ranges from several hours to 8 to 12 h or longer after alcohol consumption stops. More than 95 percent of the alcohol consumed is metabolized in the liver by oxidation to acetaldehyde and coenzyme A (page 89), and this occurs at a relatively slow rate. For example, it takes 5 to 6 h for the blood of an average-sized person to be cleared of the alcohol in 120 mL (4 oz) of whisky or 1.2 L (1.25 qt) of beer, and it takes up to 14 h for the ability to make accurate skilled maneuvers and sound judgments to return.

Withdrawal symptoms following cessation of prolonged, heavy drinking are tremors, nausea and vomiting, a dry mouth, headache and muscular weakness, nightmares, and feelings of anxiety, guilt, and irritability. Hallucinations and seizures can also occur. Withdrawal symptoms disappear in 5 to 7 days.

Alcohol produces a variety of central nervous system effects, and it is unclear which, if any, are related to the drug's behavioral effects. It is generally accepted that alcohol interacts with plasma membranes of neurons not by way of receptors but by causing nonspecific changes in the membranes' lipids and proteins. These nonspecific changes seem, however, to cause specific indirect changes in membrane receptors that are coupled to adenylate cyclase or ion channels. In particular, alcohol increases the movement of chloride ions through GABA receptors, thereby increasing the effectiveness of the brain's major inhibitory neurotransmitter.

The benzodiazepines, such as Librium and Valium, and the barbiturates also interact with GABA receptors. This probably explains why drugs of these two classes and alcohol produce cross tolerance and cross dependence. Because of such interactions, the withdrawal symptoms suffered by a person who is addicted to alcohol, for example, can be relieved by taking either benzodiazepines or barbiturates.

LEARNING AND MEMORY

Learning is the acquisition and storage of information as a consequence of experience. It is measured by an increase in the likelihood of a particular behavioral response to a stimulus. Generally, rewards or punishments, as mentioned earlier, are crucial ingredients of learning, as is contact with and manipulation of the environment. **Memory** is the relatively permanent storage form of the learned information.

Neural circuits in the central nervous system connect the sensory reception of the new information to be learned with the behavioral response. For example, if an animal learns to blink its eye in response to a tone, neural circuits must in some way connect the auditory pathways from the ear to the motorneurons controlling the eye muscles. Some molecular or cellular change has to take place somewhere along these neural circuits if learning is to be achieved. This change is called the **memory trace.** It is important to distinguish between the entire neural circuit necessary to demonstrate learning, which usually involves a sensory and a motor component, and the memory trace, which is the part of that circuit that shows the training-induced changes. Just where the memory trace occurs and what form it takes are the paramount questions in the study of learning, but unfortunately, the answers to these questions are still disappointingly vague. We shall discuss them after a brief description of memory.

Memory

Depending on how long a memory lasts, it is designated as either **working memory,** a short-term memory that lasts seconds to hours, or **long-term memory,** which lasts days to years. The process of transferring memories from the working to the long-term form is called **memory consolidation** (Figure 20-13).

Working memory serves as the initial depository of information and consists of the set of things we are paying attention to at any one time. Working memory has a limited capacity and has been said to contain no more than seven meaningful pieces of data at a time. (The series 149217761990 counts as 12 pieces, but arranged in a more meaningful way—1492, 1776, and 1990— it counts as only 3.)

Conditions such as coma, deep anesthesia, electroconvulsive shock, and insufficient blood supply to the brain, all of which interfere with the electrical activity of the brain, also interfere with working memory. Thus, it is assumed that working memory exists in the form of graded or action potentials. Working memory is interrupted when a person becomes unconscious from a blow on the head and cannot remember anything that happened for about 30 min before the blow, so-called **retrograde amnesia.** The loss of consciousness in no way interferes with memories of experiences that were learned before the period of amnesia. Working memory is also susceptible to external interference, such as occurs when one attempts to learn conflicting information.

After their existence in working memory, memories may either fade away or be transferred to long-term memory. Which of these two events happens depends on factors such as attention, motivation, and the presence of various hormones. It is as though, for memory consolidation to occur, a "fix signal" must follow the event that is to be remembered. Such a signal is probably not specific for a particular event but rather may indicate "whatever just happened, remember it." Thus, the signal for memory consolidation may be likened to the fixation step in photographic developing in that the latent image on the film will fade rapidly unless a chemical fixative is applied. The same fixative is used for photographs of any subject and itself contains no specific information.

The physiological "fix" signal may be hormonal, since experiments show that, after a training session, the administration of various hormones, including epinephrine, ACTH, vasopressin, and beta-endorphin, influences the retention of the learned experience. These hormones are normally released in response to stressful or even mildly stimulating experiences, suggesting that the hormonal consequences of our experiences regulate the memory consolidation of these experiences.

Behavioral investigations and common experience indicate that memories of past events and well-learned behavior patterns normally can have very long life spans. Unlike working memory, long-term memory has a seemingly unlimited capacity because people's memories are never so full that they cannot learn something new. Long-term memory can survive deep anesthesia, trauma, or electroconvulsive shock, all of which disrupt the normal patterns of neural conduction in the brain.

Memories can be consolidated very rapidly, in some situations after just one trial, and they can be retained over extended periods during which most molecules of brain neurons have been renewed many times. Information can be retrieved from long-term stores after long periods of disuse, and the common notion that memory, like muscle, atrophies with lack of use is not always true. In fact, many memories become more vivid and less easily forgotten with time.

Long-term memories seem to be of two kinds: procedural and declarative. **Procedural memories** are those involved in remembering *how* to do something, and **de-**

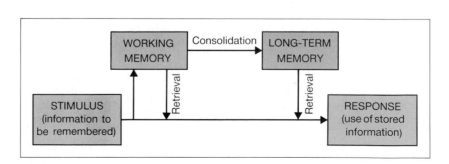

FIGURE 20-13 Events involved in learning and retrieving information.

clarative memories are involved in remembering *what* something is or what happened. Persons who suffer from severe deficits in declarative memory but have intact procedural memory can, for example, learn and retain motor skills, but they report no memory of having done so. One case study describes a pianist who learned a new piece to accompany a singer at a concert but had no recollection of the event the following morning.

The Location of Memory

One difficulty in locating the site of a memory trace is that learning in some form or other seems to be a general property of the entire nervous system. Nevertheless, most researchers agree that memory traces for specific types of tasks are localized and not widely distributed throughout the brain.

The structures currently thought to be most involved in the memory trace are the hippocampus and amygdala (Figure 20-9), which are both parts of the limbic system, as well as the cerebellum and the cerebral cortex. For example, the hippocampus is essential for memory consolidation but it is unnecessary for adding new information to working memory or retaining memories once they are consolidated. Moreover, pathways from the septal region of the forebrain (Figure 20-9) to the hippocampus degenerate in people with Alzheimer's disease, a condition marked by serious memory deficits (page 208).

Certain regions of the cerebellum are necessary for learning discrete, meaningful movements, and the amygdala seems to be necessary for learning to avoid pain. The cerebral cortex is probably essential for memories involving language and complex spatial information and for its role in helping focus the attention of the learner on the task at hand. Memory traces need not be localized to a single cell, however, and within localized regions of the brain, traces may involve a number of neurons, parallel circuits, and feedback loops.

The Neural Basis of Learning and Memory

As we learned earlier, formation of a memory trace involves cellular or molecular changes in those neurons that make up the trace. The changes that occur during memory consolidation are caused by electrical activity in the neurons during working memory. This activity, in turn, causes brief alterations in synaptic effectiveness (page 205). Strengthening the effectiveness of existing synapses in the memory trace may occur by changes in ion-channel permeabilities. Changes in ion permeabilities, particularly Ca^{2+} permeability, alter the amount of transmitter released from presynaptic terminals in response to action potentials in the terminal and can be long lasting. Such changes occur in invertebrate systems

during memory formation, but it is not known if similar events occur in the human brain.

Researchers seeking a biochemical basis for long-term memory began by looking for changes in brain-cell DNA that could be correlated with learning because DNA is the only type of molecule stable in the brain over the many years that some memories are known to last. No changes in DNA occur with learning, however, but there is an increased synthesis of messenger RNA and proteins. There is also growing evidence that certain brain neurons have increased metabolism of specific proteins during memory consolidation.

Two of the opioid peptides, enkephalin and endorphin, seem to interfere with learning and memory, particularly when the lesson involves a painful stimulus. Their inhibitory effect on learning may occur in one of two ways: (1) They may simply decrease the emotional (fear, anxiety) component of the painful experience associated with the learning situation, thereby decreasing the motivation necessary for learning to occur, or (2) they may inhibit the norepinephrine pathways that facilitate learning.

In contrast to working memory, long-term memory was improved after significant anatomical changes in the brain had been induced. Experiments show that exposing subjects to an enriched environment improved long-term memory. The enriched environment—one containing toys, social situations, physical challenges, and so on—induces dramatic anatomic changes in the subjects' brain, particularly the cerebral cortex, hippocampus, and cerebellum. Compared to the brains of animals raised in isolation (an impoverished environment), the brains of subjects in an enriched environment had more neuroglial cells, a more complex branching of neurons' dendrites, an increase in the spines on nerve-cell processes, which serve as sites for synapses, and a change in the synapse structure. This ability of neural tissue to change because of its activation is known as **plasticity**.

When the isolation animals and enriched animals were later tested and compared in learning ability, the enriched animals learned faster. When the animals' brains were examined, a direct correlation was found between learning ability and the brain structural changes induced by the enriched environment: the more capable of learning, the greater the brain changes. Thus, there are indications that structural changes may be associated with learning.

Because the early studies of this type were conducted in young animals, the problem of explaining learning in adults remained. However, it has recently been shown that plasticity occurs even when the enriched environment begins at middle or old age. Thus, it is clear that the patterns by which neurons are connected during youth are not permanent and unchanging as was once supposed.

CEREBRAL DOMINANCE AND LANGUAGE

The use of language and its development clearly depends on brain structure. To the casual observer, the two cerebral hemispheres appear to be symmetrical, but each has anatomical, chemical, and functional specializations. For example, the left hemisphere is superior at producing language—the conceptualization of what one wants to say or write, the neural control of the act of speaking or writing, and recent verbal memory—in 90 percent of the population.

The left hemisphere is specialized for controlling rapid changes in muscle groups, as in the production of speech sounds, and for perceiving events with rapidly changing temporal patterns, as in listening to someone talking. Understanding of both spoken and written language also usually depends on the left hemisphere although the difference between the hemispheres may not be as marked for language comprehension as it is for language production.

Within the left cerebral cortex, different areas are related to specific aspects of language (Figure 20-14). Areas in the frontal, parietal, and temporal lobes contain

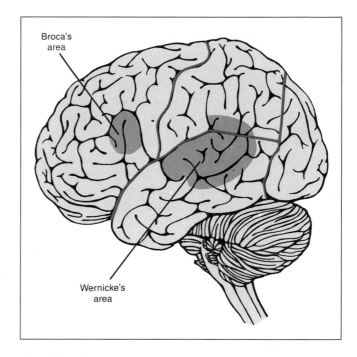

FIGURE 20-15 Areas found clinically to be involved in the comprehension (Wernicke's area) and motor (Broca's area) aspects of language.

FIGURE 20-14 Areas involved in the comprehension (light blue), memory (dark blue), and motor (gray) aspects of language. *(Adapted from Ojemann.)*

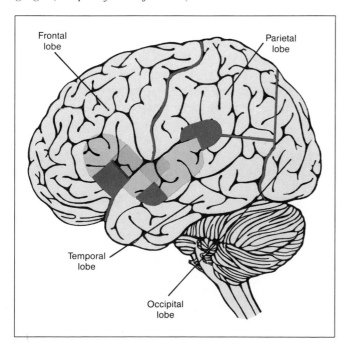

mechanisms involved in the motor aspects of language output, but some of these same mechanisms are also essential for decoding speech sounds. Thus, there seem to be common mechanisms underlying both language production and comprehension.

Surrounding these areas in the left frontal, parietal, and temporal lobes are regions of cortex related to certain individual functions of language, such as the grammatical aspects, naming objects, and the identification of word meaning. And surrounding these are cortical areas essential for verbal memory. Parts of thalamus and hippocampus also play a role in verbal memory.

Such neural specialization is demonstrated by the **aphasias**, specific language defects not due to mental retardation or paralysis of the muscles involved in speech. Thus, in most people, damage to the left cerebral hemisphere, but not the right, interferes with the capacity for language, and damage to different areas of the left cerebral hemisphere affects language differently.

Despite the fact that current research is identifying common areas of left cortex for both language production and understanding, the aphasias are seen clinically as divided into two main types. Damage to the temporal region, known as **Wernicke's area** (Figure 20-15) results in aphasias that are more closely related to conceptualization—the persons cannot understand spoken or written language even though their hearing and vision are unimpaired. In contrast, damage to **Broca's area**, the language areas in frontal cortex, results in expressive aphasias—the persons are unable to carry out the coordinated respiratory and oral movements necessary for language even though they can move their lips and tongue. They understand spoken language and know what they want to say but are unable to speak. Although Wernicke and Broca aphasics are still distinguished, both types exhibit deficits broad enough to support the overlap of language comprehension and production mechanisms seen experimentally.

The anatomical and physiological asymmetries in the two hemispheres are present at birth but are fairly flexible in early years of life. For example, after accidental damage to the left hemisphere of children under the age of 2, adequate (though perhaps not equal) language develops in the intact right hemisphere. In these children, however, defects occur in those functions in which the right hemisphere normally plays a role, that is, language develops at the expense of usual right-hemisphere functions, such as spatial localization.

Even if the left hemisphere is traumatized in children after the onset of language development, functional language ability is reestablished in the right hemisphere after variable periods of loss. The transfer of language functions to the right hemisphere becomes rapidly worse as the age at which the damage occurs increases, so that after the early teens, language development in the right hemisphere is interfered with permanently.

The dramatic change in the teens in the possibility of learning language, or the ease of learning a second language, may be related to the fact that the brain attains its structural, biochemical, and functional maturity at that time. Apparently, with brain maturation, language functions are irrevocably assigned, and for people with language specialization in the left hemisphere, the utilization of language propensities of the right hemisphere is no longer possible.

Just as the left hemisphere is dominant for events that occur in sequences over time such as those seen in language production, the right hemisphere is thought to be specialized for processing information that is treated as a unified block, where occurrence over time is less important, such as the perception of faces and other three-dimensional objects.

Memories are handled differently in the two hemispheres, too, with verbal memories more apt to be associated with the left hemisphere and nonverbal memories, for example, visual patterns, with the right. Even the emotional responses of the two hemispheres seem to be different. For example, when electroconvulsive shock is administered in the treatment of depression, better effects are often obtained when the electrodes are placed over the right hemisphere. The two sides of the brain also differ in their sensitivity to psychoactive drugs.

CONCLUSION

Until recently it was thought that most complex mental acts were handled almost exclusively by the cerebral cortex. Now it is understood that mental tasks are performed not by any single area of the brain but by the operation of many basic units, each of which is localized to a specific part of the brain. Because many basic units function even in simple mental tasks, widely distributed areas of the brain are involved, and their function must be orchestrated in the performance of each task. In other words, even though mental acts seem intuitively to be single acts, they are not. Instead, each act is composed of many individual information-processing events.

The performance of any one brain region is shared or influenced by structures in other areas. For example, cortex and the subcortical regions—particularly limbic and reticular systems—form a highly interconnected system in which many parts contribute to a particular mental task, and both hemispheres are involved. In fact, the nervous system is so abundantly interconnected that it is difficult to know where any particular subsystem begins or ends.

In the early seventeenth century, the brain's mode of operation was compared to that of a clock. When computers became widely used, the brain was compared to a computer. Then the brain was compared to a hologram, a photographic process that records specially processed light waves, rather than the image of an object. These widely divergent analogies only emphasize how little we know of how the brain really functions, and this remains one of the most baffling questions in science.

SUMMARY

Consciousness includes conscious experiences and states of consciousness. The electroencephalogram provides one means of defining the state of consciousness.

States of Consciousness

I. Electric currents in the cerebral cortex due predominately to summed postsynaptic potentials are recorded as the EEG.
   A. Slower EEG waves correlate with less responsive behaviors.
   B. Rhythm generators in the thalamus are probably responsible for the wavelike nature of the EEG.
II. Alpha rhythms and, during EEG arousal, beta rhythms characterize the EEG of an awake person.
III. Slow-wave sleep progresses from stage 1 (faster, lower-amplitude waves) through stage 4 (slower, higher-amplitude waves), followed by an episode of REM (paradoxical) sleep. There are generally five of these cycles per night.

IV. In the brainstem, a sleep-producing system and an arousal system interact to produce sleep-wake cycles.

V. Brain structures involved in directed attention, particularly the locus ceruleus, are thought to determine which areas of the brain gain temporary predominance in the ongoing stream of conscious experience.

Conscious Experiences

I. It is believed that conscious experiences, which depend on directed attention, occur because of activity in interacting neuron networks.

II. In order for conscious experiences to occur, neural activity must be present in the brain for a minimum period of time.

III. Neural processes that result in perceptions, judgments, problem-solving ability, and discriminative responses to stimuli can occur even though conscious experiences do not develop.

Motivation and Emotion

I. Behaviors that satisfy homeostatic needs are primary motivated behaviors. Behavior not related to homeostasis is a result of secondary motivation.
 A. Repetition of a behavior indicates appetitive motivation. Avoidance of a behavior indicates aversive motivation.
 B. Lateral portions of the hypothalamus are involved in appetitive motivation.
 C. Norepinephrine, dopamine, and enkephalin are transmitters in the brain pathways that mediate appetitive motivation and reward.

II. Two aspects of emotion, inner emotions and emotional behavior, can be distinguished.
 A. Different brain areas mediate these two aspects of emotion, but both aspects work together so that the behaviors truly represent the inner emotions.
 B. The limbic system integrates inner emotions and behavior.

Altered States of Consciousness

I. Schizophrenia is probably due, at least in part, to hyperactivity in brain dopaminergic systems.

II. The affective disorders are caused, at least in part, by disturbances in transmission at norepinephrine-mediated synapses in the brainstem.

III. Psychoactive drugs, which are often chemically related to neurotransmitters, result in tolerance, withdrawal, and drug dependence. Tolerance and withdrawal may result from synaptic feedback mechanisms that keep the transmitter-receptor interaction at a constant level.

Learning and Memory

I. Motivations (rewards and punishments) are generally crucial to learning.
 A. Working memory has limited capacity, is short term, and depends on functioning brain electrical activity.
 B. After registering in working memory, facts either fade away or are consolidated in long-term memory, depending on attention, motivation, and various hormones.
 C. Long-term memory, of which there are two forms—procedural and declarative—seems to have an unlimited capacity and to be independent of brain electrical activity.

II. Memory traces are localized, discrete brain areas containing cellular or molecular changes specific to different memories.

III. Learning is enhanced after subjects are exposed to enriched environments. The enhancement is presumably due to the brain neural and chemical development that follows such exposure.

Cerebral Dominance and Language

I. The two cerebral hemispheres differ anatomically, chemically, and functionally. In 90 percent of the population, the left hemisphere is superior at producing language and in performing other tasks that require rapid changes over time.

II. After damage to the dominant hemisphere, some language function can be acquired by the opposite hemisphere—the younger the patient, the greater the transfer of function.

Conclusion

I. Many brain units are involved in the performance of even simple mental tasks.

II. Each unit is localized to a specific brain area, but, because many units are involved, widely distributed brain areas take part in mental tasks.

REVIEW QUESTIONS

1. Define:

conscious experiences	appetitive motivation
states of consciousness	aversive motivations
electroencephalogram (EEG)	brain self-stimulation
	medial forebrain bundle
epilepsy	inner emotions
alpha rhythm	emotional behavior
beta rhythm	schizophrenia
EEG arousal	dopamine hypothesis
slow-wave sleep	affective disorders
REM sleep	depressions
paradoxical sleep	manias
coma	bipolar affective disorders
brain death	tolerance
directed attention	withdrawal
orienting response	drug dependence
habituation	learning
dishabituation	memory
motivations	memory trace
primary motivated behavior	working memory
	long-term memory
secondary motivated behavior	memory consolidation
	retrograde amnesia

procedural memories   aphasias
declarative memories   Wernicke's area
plasticity   Broca's area

2. State the two criteria used to define one's state of consciousness.

3. What type of neural activity is recorded as the EEG?

4. Draw EEG records that show alpha and beta rhythms, the stages of slow-wave sleep, and REM sleep; indicate the characteristic wave frequencies of each.

5. Using four characteristics, distinguish slow-wave sleep from REM sleep.

6. Briefly describe the neural mechanisms that determine the states of consciousness.

7. Name the criteria used to distinguish brain death from coma.

8. Describe the orienting response as a form of directed attention.

9. State two criteria that determine whether or not an experience will be conscious.

10. List the effects that unconscious processes can have on mental experiences.

11. Distinguish primary from secondary motivated behavior.

12. Distinguish appetitive from aversive emotions and explain how rewards and punishments are related to them.

13. Explain what brain self-stimulation can tell about appetitive and aversive emotions and rewards and punishments.

14. Name two neurotransmitters that mediate the brain reward systems.

15. Distinguish inner emotions from emotional behavior. Name the brain areas involved in each.

16. Briefly describe the role of the limbic system in emotions.

17. Name the neurotransmitters involved in schizophrenia and the affective disorders.

18. Describe a mechanism that could explain tolerance and withdrawal.

19. Describe the time course of alcohol metabolism, and name the chemicals that result from this process.

20. Distinguish working memory from long-term memory, and explain the relationship between them.

21. Name three factors that influence memory consolidation.

22. Explain how we can say that basic neural units are "strictly localized to specific parts of the brain" yet say that "mental tasks are not performed by any single area of the brain."

THOUGHT QUESTIONS

(Answers are given in Appendix A.)

1. Explain why patients given drugs to treat parkinsonism (page 338) sometimes develop symptoms similar to schizophrenia.

2. Explain how clinical observations of persons with various aphasias help physiologists understand the neural basis of language.

Chapter 4

4-1. A drug could inhibit acid secretion by: (1) binding to the membrane sites that inhibit acid secretion, which would produce the same effect as the body's compounds, that is, inhibit acid secretion; or (2) binding to membrane proteins that normally lead to acid secretion but without triggering acid secretion, thereby preventing the body's natural compounds from binding (competition); or (3) having an allosteric effect upon the binding sites, which would increase the affinity of the sites that bind inhibitors or decrease the affinity of those sites that bind stimulating compounds.

4-2. The reason for a lack of insulin effect could be either a decrease in the number of available binding sites to which insulin can bind or a decrease in the affinity of the binding sites for insulin so that less insulin is bound. A third possibility, which does not involve the process of insulin binding, would be a defect in the way in which the binding site triggers a cell response once it has bound insulin.

4-3. An increase in the concentration of compound A will lead to a decrease in the concentration of compound H by the route shown below. Sequential activations and inhibitions of proteins of this general type are frequently encountered in physiological control systems.

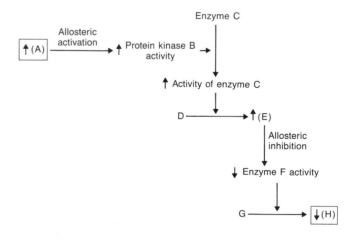

4-4. (A) Acid secretion could be increased to 20 mM/h by: (1) increasing the concentration of compound X from 2 pM

to 8 p*M*, thereby increasing the number of binding sites occupied; or (2) increasing the affinity of the binding site for compound X, thereby increasing the amount bound without changing the concentration of compound X. (B) Increasing the concentration of compound X from 18 to 28 p*M* will not increase acid secretion because, at 18 p*M*, all the binding sites are occupied (the system is saturated), and there are no further binding sites available.

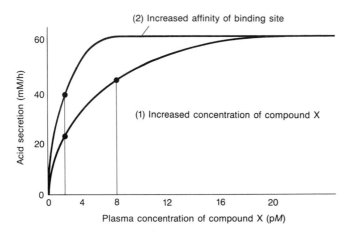

4-5. Phosphoprotein phosphatase removes the phosphate group from proteins that have been covalently modulated by protein kinase. Without phosphoprotein phosphatase, the protein could not return to its unmodulated state and would remain in its activated state. The ability to decrease as well as increase protein activity is essential to the regulation of physiological processes.

4-6. Nucleotide bases in DNA pair, A-T and G-C. Given the base sequence of one DNA strand as:

A-G-T-G-C-A-A-G-T-C-T

the corresponding strand of DNA would be:

T-C-A-C-G-T-T-C-A-G-A

and the sequence in mRNA transcribed from the given strand would be:

U-C-A-C-G-U-U-C-A-G-A

[Recall that uracil (U) replaces thymine (T) in RNA.]

4-7. The triplet code G-T-A for the amino acid histidine will be transcribed into mRNA as C-A-U, and the anticodon in tRNA corresponding to C-A-U will be G-U-A.

4-8. If the gene for a protein containing 100 amino acids was only composed of the triplet exon code words, the gene would be 300 nucleotides in length—a triplet of 3 nucleotides codes for one amino acid. However, because of the presence of intron segments, which account for 75 to 90 percent of nucleotide in a gene, the gene would be between 1000 and 3000 nucleotides

long. Thus, the exact size of a gene cannot be determined from knowing the number of amino acids in the protein coded by the gene.

4-9. Tubulin is the protein that polymerizes to form microtubules. The tubulin monomers become linked together in a spiral that forms the walls of the hollow microtubule. Without microtubules to form the spindle apparatus, the chromosomes will not separate during mitosis.

4-10. A drug that inhibits the replication of DNA will inhibit mitosis since a duplicate set of chromosomes is necessary for cell division. Since one of the characteristics of cancer cells is their ability to undergo unlimited division, a drug that inhibits DNA replication will inhibit the growth of cancer cells. Unfortunately, such drugs will also inhibit the division of normal cells as well, and their use must be carefully monitored to balance the damage done to normal tissues against the inhibition of tumor growth.

Chapter 5

5-1. The reactant molecules have a combined energy content of 55 + 93 = 148 kcal/mol, and the combined energy content of the products is 62 + 87 = 149. Thus, the energy content of the products exceeds that of the reactants by 1 kcal/mol, and this amount of energy must be added to A and B to form the products C and D.

The reaction is a reversible reaction since the difference in energy content between the reactants and products is small, 1 kcal/mol. When the reaction reaches chemical equilibrium, there will be a slightly higher concentration of reactants than products.

5-2. The rate of the reaction can be increased to rate X in two ways: (1) The substrate concentration could be increased to B. (2) The enzyme affinity could be increased by covalent or allo-

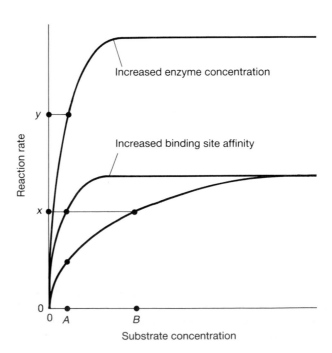

steric modulation, without changing the substrate concentration.

The rate of the reaction can be increased to rate Y in only one way—by increasing the concentration of the enzyme. Increasing either the substrate concentration or the affinity of the enzyme will not increase the rate of the reaction beyond its maximal rate, which, in the absence of an increase in enzyme concentration, is less than rate Y.

5-3. The maximum rate at which the end product E can be formed is 5, the rate of the slowest—the rate-limiting—reaction in the pathway.

5-4. No. Under normal conditions, the concentration of oxygen at the level of the mitochondria is sufficient to saturate the enzyme that combines oxygen with hydrogen to form water. In the electron-transport chain, the rate-limiting reactions depend on the available concentrations of ADP and P_i which are combined to form ATP.

If ADP and P_i concentrations are low, the rate of oxygen consumption will be low and will not be increased by increasing the oxygen concentration. In certain types of lung and circulatory diseases in which the transfer of oxygen from normal air to the blood and the delivery of the blood to the tissues is impaired, the concentration of oxygen at the level of the mitochondria can become rate limiting, and increasing the concentration of oxygen in the inhaled air will result in the delivery of more oxygen to the tissues and increase the rate of ATP formation.

5-5. During starvation, in the absence of ingested glucose, the body's stores of glycogen are rapidly depleted. Glucose, which is the major fuel used by the brain, must now be synthesized from other types of molecules. The major source of this newly formed glucose comes from the breakdown of proteins to amino acids and their conversion to glucose. To a lesser extent, the glycerol portion of fat is converted to glucose. The fatty acid portion of fat cannot be converted to glucose.

5-6. Since the Krebs cycle will function only during aerobic conditions, the catabolism of fat is dependent on the presence of oxygen. Fatty acids are broken down to acetyl coenzyme A during beta oxidation and acetyl coenzyme A enters the Krebs cycle to be converted to carbon dioxide. In the absence of oxygen, acetyl coenzyme A cannot be broken down to carbon dioxide and the increased concentration of acetyl coenzyme A will inhibit the further breakdown of fatty acids.

5-7. Since the liver is the site at which ammonia is converted to urea, diseases that damage the liver can lead to an increase in the blood levels of ammonia, which is especially toxic to nerve cells. Note that it is not the liver that produces the ammonia. The ammonia is formed in most cells during the oxidative deamination of amino acids and then travels to the liver via the blood.

Chapter 6

6-1. (a) During diffusion, net flux always occurs from high to low concentration. Thus it will be from 2 to 1 in case A and from 1 to 2 in case B. (b) At equilibrium the concentrations of solute in the two compartments will be equal: 4 M in case A

and 31 M in case B. (c) Both will reach diffusion equilibrium at the same rate since the difference in concentration across the membrane is the same in each case, 2 M; (3 − 5) = −2, and (32 − 30) = 2. The two one-way fluxes will be much larger in B than in A, but the net flux has the same magnitude in both cases, although oriented in opposite directions.

6-2. The ability of one amino acid to decrease the flux of a second amino acid across a cell membrane is an example of the competition of two molecules for the same binding site, as explained in Chapter 4. The same binding site on the carrier protein that binds alanine can also bind leucine. The higher the concentration of alanine, the greater the number of carrier-binding sites occupied by it, and the fewer available for binding leucine. Thus less leucine will be moved into the cell.

6-3. The net transport will be out of the cell, in the direction from the higher-affinity site on the intracellular surface to the lower-affinity site on the extracellular surface. More molecules will be bound to the carrier on the high-affinity side of the membrane, and thus more will be moving out of the cell than into it, until the concentration in the extracellular fluid becomes large enough so that the number of bound carriers at the extracellular surface is equal to the number of bound carriers at the intracellular surface.

6-4. Although ATP is not used directly in secondary active transport, it is necessary for the primary active transport of sodium out of cells. Since it is the sodium concentration gradient across the plasma membrane that provides the energy for most secondary active-transport systems, a decrease in ATP production will decrease active sodium transport, leading to a decrease in the sodium concentration gradient and thus to a decrease in secondary active transport.

6-5. The solution with the greatest osmolarity will have the lowest water concentration. The osmolarities are

$$
\begin{aligned}
\text{A.} \quad & 20 + 30 + (2 \times 150) + (3 \times 10) = 380 \text{ m}Osm \\
\text{B.} \quad & 10 + 100 + (2 \times 20) + (3 \times 50) = 300 \text{ m}Osm \\
\text{C.} \quad & 100 + 200 + (2 \times 10) + (3 \times 20) = 380 \text{ m}Osm \\
\text{D.} \quad & 30 + 10 + (2 \times 60) + (3 \times 100) = 460 \text{ m}Osm
\end{aligned}
$$

Solution D thus has the lowest water concentration. (Recall that NaCl forms two ions in solution and $CaCl_2$, three.)

Solutions A and C are isosmotic since they have the same osmolarity.

6-6. Initially the osmolarity of compartment 1 is $(2 \times 200) + 100 = 500$ mOsm and that of 2 is $(2 \times 100) + 300 = 500$ mOsm. The two solutions are thus isosmotic, and there is no difference in water concentration across the membrane. Since the membrane is permeable to urea but not to NaCl, urea will diffuse from compartment 2 into 1, producing an increase in the osmolarity of 1, lowering its water concentration and thus causing a net flow of water into 1, increasing its volume. This net movement of urea and water into compartment 1 will continue until (a) urea reaches diffusion equilibrium at a concentration of 200 mM in each compartment and (b) the movement of water has lowered the concentration of NaCl in 1 to 150 mM and raised that of NaCl in compartment 2 to 150 mM. Note that the same movement of water would have occurred if there were no urea present in either compartment.

It is only the concentration of nonpenetrating solutes, NaCl in this case, that determines the direction of net flow, regardless of the concentration of any penetrating solutes that are present.

6-7. The osmolarities and nonpenetrating-solute concentrations are:

Solution	Osmolarity, mOsm	Nonpenetrating Solute Concentration, mM
A	$(2 \times 150) + 100 = 400$	$2 \times 150 = 300$
B	$(2 \times 100) + 150 = 350$	$2 \times 100 = 200$
C	$(2 \times 200) + 100 = 500$	$2 \times 200 = 400$
D	$(2 \times 100) + 50 = 250$	$2 \times 100 = 200$

Only the concentration of nonpenetrating solutes, NaCl in this case, will determine the change in cell volume. Since the intracellular concentration of nonpenetrating solutes is 300 mOsm, solution A will produce no change in cell volume, solutions B and D will cause cells to swell since these solutions have a higher water concentration than the intracellular fluid, and solution C will cause cells to shrink because it has a lower water concentration than the intracellular fluid.

6-8. Solution A is isotonic—having the same concentration of nonpenetrating solutes as intracellular fluid (300 mM). Solution A is also hyperosmotic since its total osmolarity is greater than 300 mOsm, as are solutions B and C. Solution B is also hypotonic since its concentration of nonpenetrating solutes is less than 300 mM. Solution C is hypertonic since its concentration of nonpenetrating solutes is greater than 300 mM. Solution D is hypotonic (less than 300 mM of nonpenetrating solutes) and also hyposmotic (having a total osmolarity of less than 300 mOsm).

6-9. Exocytosis is triggered by an increase in intracellular calcium concentration. Calcium ions can be actively transported out of cells by a secondary countertransport carrier that is coupled to the entry of sodium ions on the same carrier. If the intracellular concentration of sodium ions were increased, the sodium concentration gradient across the membrane would be decreased, and this would decrease the secondary active transport of calcium out of the cell. This would lead to an increase in intracellular calcium concentration, which would trigger increased exocytosis.

Chapter 7

7-1. 4.4 mM. If you answered 8 mM, you ignored the fact that plasma potassium concentration is homeostatically regulated so that the doubling of input will lead to negative-feedback reflexes that oppose an equivalent increase in plasma concentration. If you answered 4 mM, you ignored the fact that homeostatic control systems cannot *totally* prevent changes in the regulated variable when a perturbation occurs. Thus, there must be *some* rise in plasma potassium concentration in this situation (to serve as the error signal for the compensating re-

flexes), and this is consistent with the answer 4.4 mM. (The actual rise would have to be experimentally determined. There is no way you could have predicted that the rise would be 10 percent. All you could predict is that it would neither double nor stay absolutely unchanged.)

7-2. No. There may in fact be one, but there is another possibility—that the altered skin blood flow in the cold represents an *acclimatization* undergone by each Eskimo during his or her lifetime as a result of performing such work repeatedly.

7-3. Patient A's drug very likely acts to block phospholipase A_2, whereas patient B's drug blocks lipoxygenase. (See Figure 7-10.)

7-4. The chronic loss of exposure of the heart's receptors to norepinephrine causes an up-regulation of this receptor type, that is, more receptors in the heart for norepinephrine. The drug, being an agonist of norepinephrine, that is, able to bind to norepinephrine's receptors and activate them, is now more effective since there are more receptors for it to combine with.

7-5. None. Since you are told that all six responses are mediated by the cAMP system, then blockade of any of the steps listed in the question would eliminate all six of the responses. This is because the cascade for all six responses is identical from the receptor through the formation of cAMP and activation of cAMP protein kinase. Thus, the drug must be acting at a point beyond this kinase, for example, at the level of the phosphorylated protein mediating this response.

7-6. Not in most cells, since there are other physiological mechanisms by which signals impinging on the cell can increase cytosolic calcium concentration. These include IP_3-induced release of calcium from the endoplasmic reticulum and voltage-sensitive calcium channels.

7-7. Cyclic AMP itself. All three substances activate one or more protein kinases.

Chapter 8

8-1. Little change in the resting membrane potential would occur when the pump first stops because the pump's direct contribution to charge separation (recall that the pump is electrogenic) is very small. With time, however, the membrane potential would depolarize progressively toward zero because the concentration gradients, which depend on the Na,K-ATPase pumps and which give rise to the diffusion potentials that constitute most of the membrane potential, would have run down.

8-2. The resting potential would decrease, that is, become less negative, because the concentration gradient causing net diffusion of this positively charged ion out of the cell would be smaller. The action potential would fire more easily, that is, with smaller stimuli, because the resting potential would be closer to threshold. It would repolarize more slowly because repolarization depends on net potassium diffusion from the cell and the concentration gradient driving this diffusion is lower. Also, the afterhyperpolarization would be smaller.

8-3. The hypothalamus was probably damaged. It plays a critical role in appetite, thirst, and sexual capacity.

8-4. The drug probably blocks cholinergic muscarinic receptors. These receptors on effector cells mediate the actions of the parasympathetic nerves. Therefore, the drug would remove the slowing effect of these nerves on the heart, allowing the heart to speed up. Blocking their effect on the salivary glands would cause the dry mouth. The drug is not blocking cholinergic nicotinic receptors because the skeletal muscles are not affected.

8-5. Since the membrane potential of the cells in this question depolarizes, that is, becomes less negative, when chloride channels are blocked, one can predict that there was net chloride diffusion into the cells through these channels prior to the drug. Therefore, one can also predict that this passive inward movement was being exactly balanced by active transport of chloride out of the cells.

8-6. Without acetylcholinesterase, acetylcholine would remain bound to the receptors and all the autonomic actions normally caused by acetylcholine would be accentuated. Thus, there would be marked narrowing of the pupil, airway constriction, stomach cramping and diarrhea, sweating, salivation, slowing of the heart, and fall in blood pressure. On the other hand, in skeletal muscles, which must repolarize after excitation in order to be excited again, there would be weakness, fatigue, and finally inability to contract. In fact, poisoning by high doses of cholinesterase inhibitors occurs because of paralysis of the muscles that are used in respiration. Low doses of these compounds are used therapeutically.

8-7. These potassium channels, which open after a short delay following the initiation of an action potential, increase potassium diffusion out of the cell, hastening repolarization, and they account for the increased potassium permeability that causes the afterhyperpolarization. Therefore, the action potential would be broader, returning to its resting level more slowly, and the afterhyperpolarization would be absent.

Chapter 9

9-1. a. Block synaptic transmission in the pathways that convey information about pain to the brain. For example, if substance P is the neurotransmitter at the central endings of the nociceptor afferent fibers, give a drug that blocks substance P receptors.

b. Cut the dorsal root at the level of entry of the nociceptor fibers to prevent transmission of their action potentials into the central nervous system.

c. Give a drug that activates receptors in the descending pathways that block transmission of the incoming or ascending pain information.

d. Stimulate the neurons in these same descending pathways to increase their blocking activity (stimulation produced analgesia or, possibly, acupuncture).

e. Cut the ascending pathways that transmit information from the nociceptor afferents.

f. Deal with the emotions, attitudes, memories, and so on, to decrease the sensitivity to the pain.

g. Stimulate nonpain, low-threshold afferent fibers to block transmission through the pain pathways (TENS).

h. Block transmission in the afferent nerve with a local anesthetic such as procaine hydrochloride.

Chapter 10

10-1. Epinephrine falls to very low levels during rest and fails to increase during stress. The sympathetic preganglionics are the only major control of the adrenal medulla.

10-2. The increased concentration of binding protein causes more TH to be bound, thereby lowering the plasma concentration of *free* TH. This causes less negative-feedback inhibition of TSH by the anterior pituitary, and the increased TSH causes the thyroid to secrete more TH until the free concentration has returned to normal. The end result is an increased *total* plasma TH—most bound to the protein—but a normal free TH. There is no hyperthyroidism because it is only the free concentration that exerts effects on TH's target cells.

10-3. Destruction of the anterior pituitary or hypothalamus. These symptoms reflect the absence of, in order: growth hormone; the gonadotropins; and ACTH (the symptom is due to the resulting decrease in the secretion of cortisol). One cannot tell from these data alone whether the problem is primary hyposecretion of the anterior-pituitary hormones or secondary hyposecretion because the hypothalamus is not secreting releasing hormones normally.

10-4. Vasopressin and oxytocin, that is, the posterior-pituitary hormones. The anterior-pituitary hormones would not be affected because the influence of the hypothalamus on these hormones is exerted not by connecting nerves but via the releasing hormones in the portal vascular system.

10-5. The secretion of both GH and TSH would increase. Somatostatin, coming from the hypothalamus, normally exerts an inhibitory effect on the secretion of these two hormones.

10-6. Norepinephrine and many other neurotransmitters are released by neurons that terminate on the hypothalamic neurons that secrete the releasing hormones. Therefore, manipulation of these neurotransmitters will alter the secretion of the releasing hormones and, thereby, the anterior-pituitary hormones.

10-7. The high dose of the cortisol-like substance inhibits the secretion of ACTH by feedback inhibition of (1) hypothalamic corticotropin-releasing hormone and (2) the response of the anterior pituitary to this releasing hormone. The lack of ACTH causes the adrenal to atrophy and decrease its secretion of cortisol.

10-8. The hypothalamus. The low basal TSH indicates that either the pituitary is defective or that it is receiving inadequate stimulation (TRH) from the hypothalamus. If the thyroid, itself, had been defective, basal TSH would have been elevated because of less negative-feedback inhibition by TH. The TSH increase in response to TRH shows that the pituitary is capable of responding to a stimulus and so is unlikely to be defective. Therefore, the problem is that the hypothalamus is secreting too little TRH.

Chapter 11

11-1. Under resting conditions, the myosin in the thick filaments has already bound and hydrolyzed a molecule of ATP, resulting in an energized molecule of myosin (M*·ADP·P$_i$). Since ATP is necessary to detach the myosin cross bridge from actin at the end of cross-bridge movement, the absence of ATP will result in rigor mortis, in which the cross bridges become bound to actin but do not detach, leaving myosin bound to actin (A·M).

11-2. The transverse tubule conducts the muscle action potential from the plasma membrane into the interior of the fiber, where it can trigger the release of calcium from the sarcoplasmic reticulum. If the transverse tubules were not attached to the plasma membrane, an action potential could not be conducted to the sarcoplasmic reticulum and there would be no release of calcium to initiate contraction, even though an action potential can be generated in the plasma membrane of the muscle fiber.

11-3. As a muscle shortens, the thick and thin filaments slide past each other, and thus the amount of overlap changes, changing the number of cross bridges that can bind to actin and thus the maximal force of contraction. This gives rise to the length-tension relation. As the sarcomere length becomes shorter than the optimal length, the maximum tension that can be generated decreases. With a light load, the muscle will continue to shorten until its maximal tension just equals the load. No further shortening is possible since at shorter sarcomere lengths the tension would be less than the load and the load would have the effect of stretching the sarcomere. The heavier the load, the less the distance shortened before entering the isometric state since maximum tension is decreasing as the muscle shortens below l_o.

11-4. Maximum tension is produced when the fiber is (1) stimulated by a frequency of action potentials that is high enough to produce a maximal tetanic tension and (2) at its optimum length l_o, where the thick and thin filaments have maximal overlap and thus provide the greatest number of cross bridges for tension production.

11-5. Moderate tension—for example, 50 percent of maximal tension—is accomplished by recruiting sufficient numbers of motor units to produce this degree of tension. (If all motor units were the same size, which they are not, this would mean recruiting 50 percent of the motor units in the muscle.) If activity is maintained at this level for prolonged periods, some of the active fibers will begin to fatigue and their contribution to the total tension will decrease. The same level of total tension can be maintained, however, by recruiting additional motor units as some of them fatigue. At this point, one might have 50 percent of the fibers active, and 25 percent fatigued and 25 percent still unrecruited. Eventually the whole muscle will fatigue when all the fibers have fatigued and there are no additional motor units to recruit.

11-6. The oxidative motor units, both fast and slow, will be affected by a decrease in blood flow since they depend on blood flow to provide both the fuel—glucose and fatty acids—and the oxygen required to metabolize the fuel. The fast-glycolytic motor units will be little affected by the decreased blood flow since they rely predominantly on internal stores of glycogen, which is anaerobically metabolized by glycolysis. Note

that there are many more blood vessels in the region of oxidative fibers than around glycolytic fibers.

11-7. Two factors lead to recovery of muscle force: (1) Some new fibers can be formed by the fusion and development of undifferentiated satellite cells. This will replace some, but not all, of the fibers that were damaged. (2) Some of the restored force is the result of the hypertrophy of the surviving fibers. The undamaged fibers must produce more force to move a given load because of the loss of supporting fibers in the accident. These fibers undergo increased synthesis of actin and myosin, resulting in increases in fiber diameter and thus their force of contraction.

11-8. In the absence of extracellular calcium ions, skeletal muscle contracts normally in response to a muscle action potential because the calcium required to trigger contraction comes entirely from the sarcoplasmic reticulum within the muscle fibers. If the motor neuron to the muscle is stimulated in a calcium-free medium, however, the muscle will not contract because the influx of calcium from the extracellular fluid into the motor nerve terminal is necessary to trigger the release of acetylcholine that in turn triggers an action potential in the muscle.

The response of smooth muscle in a calcium-free medium depends on the type of smooth muscle. Smooth muscles that depend primarily on calcium released from the sarcoplasmic reticulum will behave like skeletal muscle and respond to direct stimulation but not to nerve stimulation in the absence of extracellular calcium. Smooth muscles that rely on the influx of calcium from the extracellular fluid to trigger contraction will fail to contract in response to both types of stimuli.

11-9. The simplest model to explain the experimental observations is that upon stimulation, a neurotransmitter is released that binds to receptor sites on the membranes of smooth-muscle cells that trigger contraction. The substance released is not acetylcholine and does not act on ACh receptors. Presumably it has its own membrane receptors. (Not ruled out by these observations is the possibility that the neurotransmitter acts upon some nonmuscle cells that in turn release a substance that acts upon the smooth muscle to produce contraction.)

Action potentials in the parasympathetic nerves are essential for initiating contraction. If the nerves are prevented from generating action potentials by blocking their voltage-sensitive sodium channels, there is no response to stimulation. However, the muscle still responds to the nonmetabolized form of ACh because smooth-muscle action potentials depend on opening voltage-sensitive calcium channels, not voltage-sensitive sodium channels. ACh is the neurotransmitter released from most, but not all, parasympathetic endings. The observation that the muscle can be contracted by injecting a nonmetabolized form of ACh indicates that the smooth muscle probably has muscarinic receptors for ACh, and this response can be blocked by a drug that specifically blocks ACh at the muscarinic receptors. When muscarinic receptors are blocked, however, stimulation of the parasympathetic nerves still produces a contraction, providing the evidence that some substance other than ACh is producing contraction. Either the postganglionic parasympathetic nerve endings release a substance other than ACh, or the ending releases ACh, which acts on nicotinic receptors (that are not blocked by muscarinic drugs) located on a neuron that then releases the neurotransmitter that acts upon the smooth muscle.

Chapter 12

12-1. None. The gamma motor neurons are important in preventing the muscle-spindle stretch receptors from going slack, but the knee jerk is tested when the extensor muscles in the thigh, and the spindle muscle fibers within them, are stretched and the stretch receptors are responsive.

12-2. The efferent pathway of the reflex arc (the alpha motor neurons) would not be activated, the effector organ (the skeletomotor muscle fibers) would not be activated, and there would be no reflex response.

12-3. The drawing must have excitatory synapses on the motor neurons of both ipsilateral extensor and ipsilateral flexor muscles. Thus, there will be no reciprocal inhibition.

12-4. A toxin that interferes with the inhibitory synapses on motor neurons would leave unbalanced the normal excitatory input to those neurons. Thus, the motor neurons would fire excessively, which would result in increased muscle contraction. This is exactly what happens in lockjaw as a result of the toxin produced by the tetanus bacillus.

Chapter 13

13-1. No. Decreased erythrocyte volume is certainly one possible explanation, but there is a second: The person might have a normal erythrocyte volume but an abnormally increased plasma volume. Convince yourself of this by writing the hematocrit equation as erythrocyte volume/(erythrocyte volume + plasma volume).

13-2. A halving of tube radius. Resistance is directly proportional to blood viscosity but inversely proportional to the *fourth power* of tube radius.

13-3. The plateau of the action potential and the contraction would be absent. You might think that contraction would persist since most calcium in excitation-contraction coupling in the heart comes from the sarcoplasmic reticulum. However, the signal for the release of this calcium is the calcium entering across the plasma membrane.

13-4. The SA node is not functioning, and the ventricles are being driven by a pacemaker in the vicinity of the A-V node.

13-5. The person has a narrowed aortic valve. Normally, the resistance across the aortic valve is so small that there is only a tiny pressure difference between left ventricle and aorta during ventricular ejection. In this case, the large pressure difference indicates that resistance across the valve must be very high.

13-6. This question is analogous to question 5 in that the large pressure difference across a valve while the valve is open indicates an abnormally narrowed valve—in this case the left A-V valve.

13-7. Decreased heart rate and contractility. These are effects mediated by the sympathetic nerves on the beta-adrenergic receptors in the heart.

13-8. 120 mmHg. $MAP = DP + \frac{1}{3}$ pulse pressure.

13-9. The drug must have caused the arterioles in the kidneys to dilate enough to reduce their resistance by 50 percent. Blood flow to an organ is determined by mean arterial pressure and that organ's resistance to flow.

13-10. The experiment suggests that acetylcholine causes vasodilation by releasing EDRF from endothelial cells.

13-11. A low plasma protein concentration. Capillary pressure is, if anything, lower than normal and so cannot be causing the edema.

13-12. 15 mmHg/L/min. TPR = MAP/CO.

13-13. Nothing. Cardiac output and TPR have remained unchanged, and so their product, MAP, has also remained unchanged. This question emphasizes that MAP depends on cardiac output but not upon the combination of heart rate and stroke volume that produces the cardiac output.

13-14. It increases. There are a certain number of impulses traveling up the nerves from the arterial baroreceptors. When these nerves are cut, the number of impulses reaching the medullary cardiovascular center goes to zero just as it would physiologically if the mean arterial pressure were to decrease markedly. Accordingly, the medullary cardiovascular center responds to the absent impulses by reflexly increasing arterial pressure.

13-15. It decreases. The hemorrhage causes no immediate change in hematocrit since erythrocytes and plasma are lost in the same proportion. As interstitial fluid starts entering the capillaries, however, it expands the plasma volume and decreases hematocrit. (This is too soon for any new erythrocytes to be synthesized.)

Chapter 14

14-1. 200 mL/mmHg.

Lung compliance

$$= \frac{\Delta \text{ lung volume}}{\Delta (P_{alv} - P_{ip})}$$

$$= \frac{800 \text{ mL}}{[0 - (-8)] \text{ mmHg} - [0 - (-4)] \text{ mmHg}}$$

$$= \frac{800 \text{ mL}}{4 \text{ mmHg}} = 200 \text{ mL/mmHg}$$

14-2. More subatmospheric than normal. A decreased surfactant level causes the lungs to be less compliant, that is, more difficult to expand. Therefore, a greater transpulmonary pressure $(P_{alv} - P_{ip})$ is required to expand them a given amount.

14-3. No.

Alveolar ventilation
$$= (\text{tidal volume} - \text{dead space}) \times \text{breathing rate}$$
$$= (250 \text{ mL} - 150 \text{ mL}) \times 20 \text{ breaths/min}$$
$$= 2000 \text{ mL/min}$$

whereas normal alveolar ventilation is approximately 4000 mL/min.

14-4. The volume of the snorkel constitutes an additional dead space, and so total pulmonary ventilation must be increased if alveolar ventilation is to remain constant.

14-5. The alveolar P_{O_2} will be higher than normal, and the alveolar P_{CO_2} will be lower. If you do not understand why, review the factors that determine the alveolar gas pressures.

14-6. No. Hypoventilation reduces arterial P_{O_2} but only because it reduces the alveolar P_{O_2}. That is, in hypoventilation, both alveolar and arterial P_{O_2} are decreased to essentially the same degree. In this problem, alveolar P_{O_2} is normal, and so the person is not hypoventilating. The low arterial P_{O_2} must therefore represent a defect that causes a discrepancy between alveolar P_{O_2} and arterial P_{O_2}. Possibilities include impaired diffusion, a shunting of blood from the right side of the heart to the left through a hole in the heart wall, and mismatching of airflow and blood flow in the alveoli.

14-7. Not at rest, if the defect is not too severe. Recall that equilibration of alveolar air and pulmonary capillary blood is normally so rapid that it occurs well before the end of the capillaries. Therefore, even though diffusion may be retarded, as in this problem, there may still be enough time for equilibration to be reached. In contrast, the time for equilibration is decreased during exercise, and failure to equilibrate is much more likely to occur, resulting in a lowered arterial P_{O_2}.

14-8. Only a few percent (specifically, from approximately 200 mL O_2/L blood to approximately 215 mL/O_2/L). The reason the increase is so small is that almost all the oxygen in blood is carried bound to hemoglobin, and hemoglobin is almost 100 percent saturated at the arterial P_{O_2} achieved by breathing room air. The high arterial P_{O_2} achieved by breathing 100 percent oxygen does cause a directly proportional increase in the amount of oxygen *dissolved* in the blood, but this still remains a small fraction of the total oxygen in the blood. Review the numbers given in the text.

14-9. All except plasma chloride concentration. The reasons are all given in the text.

14-10. It would cease. Respiration depends upon descending input from the medulla to the nerves supplying the diaphragm and the inspiratory intercostal muscles.

14-11. The 10 percent oxygen mixture will markedly lower alveolar and thus arterial P_{O_2}, but no increase in ventilation will occur because the reflex response to hypoxia is initiated solely by the peripheral chemoreceptors. The 5 percent carbon dioxide mixture will markedly increase alveolar and arterial P_{CO_2}, and a large increase in ventilation will reflexly be elicited via the central chemoreceptors. The increase will not be as large as in a normal animal because the peripheral chemoreceptors do play a role, albeit minor, in the reflex response to elevated P_{CO_2}.

14-12. These patients have profound hyperventilation, with marked increases in both the depth and rate of ventilation. The stimulus, mainly via the peripheral chemoreceptors, is the marked increase in their arterial hydrogen-ion concentration due to the acids produced. The hyperventilation causes an increase in their arterial P_{O_2} and a decrease in their arterial P_{CO_2}.

Chapter 15

15-1. No. It is a possibility, but there is another: Substance T may be secreted by the tubules.

15-2. No. It is a possibility, but there is another: Substance V may be filtered and/or secreted, but the V entering the lumen via these routes may be completely reabsorbed.

15-3. 125 mg/min. The amount of any substance filtered per unit time is given by the product of the GFR and the filterable plasma concentration of the substance, in this case 125 mL/min × 100 mg/100 mL.

15-4. The plasma concentration might be so high that the T_m for the amino acid is exceeded, and so all the filtered amino acid is not reabsorbed. A second possibility is that there is a specific defect in the tubular transport for this amino acid. A third possibility is that some other amino acid is present in the plasma in high concentration and is competing for reabsorption.

15-5. No. Urea is filtered and then partially reabsorbed. The reason its concentration in the tubule is higher than in plasma is that relatively more water is reabsorbed than urea. Therefore, the urea in the tubule becomes concentrated. Despite the fact that urea concentration in the urine is greater than in the plasma, the amount excreted is less than the filtered load, that is, reabsorption has occurred.

15-6. They would all be decreased. The transport of all these substances is coupled, in one way or another, to that of sodium.

15-7. GFR would not go down as much and renin secretion would not go up as much as in a person not receiving the drug. The sympathetic nerves are a major pathway for both responses during hemorrhage.

15-8. There would be little, if any, increase in aldosterone secretion. The major stimulus for increased aldosterone secretion is angiotensin II, but this substance is formed from angiotensin I by the action of converting enzyme, and so blockade of this enzyme would block the entire pathway.

15-9. (B) Urinary excretion in the steady state must be less than ingested sodium chloride by an amount equal to that lost in the sweat and feces. This is normally quite small, less than 1 g/day, so that urine excretion in this case equals approximately 11 g/day.

15-10. If the hypothalamus had been damaged, there might be inadequate secretion of ADH. This would cause loss of a large volume of urine, which would tend to dehydrate the person and make her thirsty. Of course, the area of the brain involved in thirst might itself have suffered damage.

15-11. Because aldosterone stimulates sodium reabsorption and potassium secretion, there will be total-body retention of sodium and loss of potassium. Interestingly, the person in this situation actually retains very little sodium because urinary sodium excretion returns to normal despite the continued presence of the high aldosterone. One hypothesized explanation for this is that GFR and atrial natriuretic hormone both go up.

Chapter 16

16-1. If the salivary glands fail to secrete amylase, the undigested starch that reaches the small intestine will be digested by the amylase secreted by the pancreas. Thus, starch digestion is not affected by the absence of salivary amylase.

16-2. Alcohol can be absorbed across the stomach wall, but absorption is much more rapid from the small intestine with its larger surface area. Ingestion of foods containing fat release enterogastrones from the small intestine, and these hormones inhibit gastric emptying and thus prolong the time alcohol spends in the stomach before reaching the small intestine. This delay decreases the rate at which alcohol enters the blood. Milk, contrary to popular belief, does not "protect" the lining of the stomach by coating it with a fatty layer. Rather it is the fat content of milk that decreases the absorption of alcohol by decreasing the rate of gastric emptying.

16-3. When large portions of the stomach are removed because of gastric cancer, for example, the storage function of the stomach is lost as well as the control of gastric emptying. When a meal is ingested, it passes rapidly into the small intestine, where the digestion of macromolecules increases the osmolarity of the luminal fluid, causing water to move into the lumen from the blood. The increased volume of fluid in the duodenum triggers vomiting. Furthermore, the fluid loss from the blood may produce a sufficient decrease in blood volume to produce the symptoms of shock. To overcome these problems, patients without a stomach must eat an increased number of very small meals.

16-4. Vomiting results in the loss of fluid and acid from the body. The fluid comes from the luminal contents of the stomach and duodenum, most of which was secreted by the gastric glands, pancreas, and liver and thus is derived from the blood. The cardiovascular symptoms of this patient are the result of the decrease in blood volume that accompanies vomiting. The secretion of acid by the stomach produces an equal number of bicarbonate ions, which are released into the blood. Normally these bicarbonate ions would be neutralized by hydrogen ions released by the pancreas when bicarbonate ions are secreted into the small intestine to neutralize the acid from the stomach. Because gastric acid is lost during vomiting, the pancreas is not stimulated to secrete bicarbonate, and the corresponding hydrogen ions that would be formed are not available to neutralize the bicarbonate released by the stomach. As a result, the acidity of the blood decreases.

16-5. Fat can be digested and absorbed in the absence of bile salts but in greatly decreased amounts. Without emulsification of fat by bile salts, only the fat at the surface of large lipid droplets is available to pancreatic lipase, and the rate of fat digestion is very slow. Without the formation of micelles with the aid of bile salts, the products of fat digestion become dissolved in the large lipid droplets, where they are not readily available for diffusion into the epithelial cells. In the absence of bile salts, only about 50 percent of the ingested fat is digested and absorbed. The undigested fat is passed on to the large intestine, where the bacteria produce compounds that increase colonic motility and promote the secretion of fluid into the lumen of the large intestine, leading to diarrhea.

16-6. Damage to the lower portion of the spinal cord produces a loss of voluntary control over defecation due to disruption of the somatic nerves to the skeletal muscle of the external anal sphincter. When fecal matter distends the rectum, it initiates the defecation reflex, but the damage to the somatic nerves leaves the external sphincter in an uncontracted state. Under these conditions, defecation occurs whenever the rectum becomes distended.

16-7. Vagotomy decreases the secretion of acid by the stomach. Impulses in the parasympathetic nerves directly stimulate acid secretion by the parietal cells as well as the release of gastrin, which in turn stimulates acid secretion. Impulses in the vagal nerves are increased during both the cephalic and gastric phases of digestion. Decreasing the amount of acid secreted by vagotomy decreases irritation of existing ulcers, which promotes healing, and decreases the probability of forming new ulcers.

Chapter 17

17-1. The concentration in plasma would increase, and the concentration in adipose tissue would decrease. Lipoprotein lipase cleaves plasma triacylglycerols, so its blockade would decrease the rate at which these molecules were cleared from plasma and would decrease the availability of the fatty acids in them for synthesis of intracellular triacylglycerols. However, this would only reduce but not eliminate such synthesis, since the adipose-tissue cells could still synthesize their own fatty acids from glucose.

17-2. The person might be a type I diabetic or might be a normal fasting person. Plasma glucose would be increased in the first case but decreased in the second. Plasma insulin concentration would not be useful because it would be decreased in both cases. The fact that the person was resting and unstressed was specified because severe stress or exercise could also produce the plasma changes specified in the question. Plasma glucose would be increased in the former and decreased in the latter.

17-3. Glucagon, epinephrine, and growth hormone. The insulin will produce hypoglycemia, which then induces reflex increases in these hormones.

17-4. It may increase, remain approximately constant, or decrease. This is the result of opposing effects: Epinephrine will increase the plasma glucose, which will *stimulate* insulin release, but at the same time epinephrine acts directly on the B cells to *inhibit* insulin release.

17-5. It might reduce it but not eliminate it. The sympathetic effects on organic metabolism during stress are mediated not only by circulating epinephrine but by the sympathetic nerves to the liver (glycogenolysis and gluconeogenesis), adipose tissue (lipolysis), and to the islets (inhibition of insulin secretion and stimulation of glucagon secretion).

17-6. Increase. The stress of the accident will elicit increased activity of all the glucose-counterregulatory controls, and will, therefore necessitate more insulin to oppose these influences.

17-7. It will lower plasma cholesterol concentration. Bile acids are formed from cholesterol, and losses of these bile acids in the feces will be replaced by synthesis of new ones from cholesterol. Chapter 16 describes how bile acids are normally absorbed from the small intestine so that very few of those secreted into the bile are normally lost from the body.

17-8. Plasma concentration of HDL. It is the ratio of LDL cholesterol to HDL cholesterol that best correlates with the development of atherosclerosis, that is, HDL cholesterol is "good" cholesterol. The answer to this question would have been the same regardless of whether the person was an athlete or not, but the question was phrased this way to emphasize that persons who exercise generally have increased HDL cholesterol.

17-9. In utero malnutrition. Neither growth hormone nor the thyroid hormones influences in utero growth.

17-10. The androgens stimulate growth but also cause the ultimate cessation of growth by closing the epiphyseal plates. Therefore, there might be a rapid growth spurt in response to the androgens but a subsequent premature cessation of growth. Estrogens exert similar effects.

17-11. Thyroid deficiency in a 1-year-old would be accompanied by mental retardation, and other symptoms such as cold intolerance, whereas short stature would be the only significant defect in a pure growth hormone deficiency.

17-12. Heat loss from the head, mainly via convection and sweating, is the major route for loss under these conditions. The rest of the body is *gaining* heat by conduction, and sweating is of no value in the rest of the body because it cannot evaporate. Heat is also lost via the expired air (insensible loss), and some people actually begin to pant under such conditions. The rapid shallow breathing increases air flow and heat loss without causing hyperventilation.

17-13. They seek out warmer places, if available, so that their body temperature increases, that is, they use behavior to develop a fever. This is excellent evidence that the hyperthermia of infection is a fever, that is, a set-point change.

Chapter 18

18-1. Sterility due to lack of spermatogenesis would be the common symptom. The Sertoli cells are essential for spermatogenesis, and so is testosterone produced by the Leydig cells. The person with Leydig cell destruction, but not the person with Sertoli cell destruction, would also have symptoms of testosterone deficiency.

18-2. Impotence. Stimulation of the parasympathetic nerves to the vessels of the penis is essential for erection, and acetylcholine is the neurotransmitter released by these nerves.

18-3. Impaired function of the seminiferous tubules, notably of the Sertoli cells. The increased plasma FSH concentration is due to the lack of negative-feedback inhibition of FSH secretion by inhibin, itself secreted by the Sertoli cells. The Leydig cells seem to be functioning normally in this person since the lack of demasculinization and the normal plasma LH indicate normal testosterone secretion.

18-4. FSH secretion. Since FSH acts on the Sertoli cells and LH acts on the Leydig cells, the answer to this question is

essentially the same as that for question 1—sterility would result in either case, but the loss of LH would also cause an undesirable elimination of testosterone and its effects.

18-5. These findings are all due to testosterone deficiency. You would also expect to find that the testes and penis were small.

18-6. They will be eliminated. The androgens act on the hypothalamus to inhibit the secretion of GnRH and on the pituitary to inhibit the response to GnRH. The result is inadequate secretion of gonadotropins and therefore inadequate stimulation of the ovaries. In addition to the loss of menstrual cycles, the woman will suffer some degree of masculinization of the secondary sex characteristics because of the combined effects of androgen excess and estrogen deficiency.

18-7. Such treatment may cause so much secretion of FSH that multiple dominant follicles are stimulated to develop simultaneously and be ovulated during the LH surge.

18-8 An increased plasma LH. The other two are due to increased plasma progesterone and so do not occur until *after* ovulation and formation of the corpus luteum.

18-9. The absence of sperm capacitation. When test-tube fertilization is performed, special techniques are used to induce capacitation.

18-10. The fetus is in difficulty. The placenta produces progesterone entirely on its own, whereas estriol requires participation of the fetus, specifically, the fetal adrenal cortex.

18-11. Prostaglandin antagonists, oxytocin antagonists, and drugs that lower cytosolic calcium concentration. You may not have thought of the last category since calcium in not mentioned in this context in the text, but as in all muscle, calcium is the immediate cause of contraction in the myometrium.

18-12. This person would have normal male external genitals and testes, although the testes may not have descended fully, but would also have some degree of development of uterine tubes, a uterus, and a vagina. These female structures would tend to develop because no MIH was present to cause degeneration of the Müllerian duct system.

18-13. No. These two hormones are already elevated in menopause, and the problem is that the ovaries are unable to respond to them with estrogen secretion. Thus, the treatment must be with estrogen, itself.

Chapter 19

19-1. Both would be impaired because T cells would not differentiate. The absence of cytotoxic T cells would impair cell-mediated immunity, and the absence of helper T cells would impair both types of immunity.

19-2. Neutrophil deficiency would impair nonspecific inflammatory responses to bacteria. Monocyte deficiency, by causing macrophage deficiency, would impair both nonspecific inflammation and specific immune responses.

19-3. The drug might reduce but would not eliminate the action of complement, since this system destroys cells directly (via the membrane attack complex) as well as by facilitating phagocytosis.

19-4. Antibodies would bind normally to antigen but might not be able either to activate complement or to act as opsonins. The reason for these defects is that the sites to which both complement C_1 and phagocytes bind are located in the Fc portion of antibodies.

19-5. They do develop fever, although often not to the same degree as normal. They can do so because IL-1 and other cytokines secreted by macrophages cause fever, whereas the defect in AIDS is failure of helper T-cell function.

19-6. This person is suffering from an autoimmune attack against the receptors. The antibodies formed then bind to the receptors and activate them just as the thyroid hormones would have.

19-7. Alcohol over time induces a high level of MES activity, which causes administered barbiturate to be metabolized more rapidly than normal.

19-8. No. Indeed, it might be virtually unaffected because the extrinsic clotting pathway does not require this factor.

Chapter 20

20-1. Dopamine is depleted in the basal ganglia of people having Parkinson's disease, and they are given dopamine agonists, usually L-dopa. This treatment raises dopamine levels in other parts of the brain, however, where the dopamine levels were previously normal. Schizophrenia is associated with increased brain dopamine levels, and symptoms of this disease appear when dopamine levels are high. The converse therapeutic problem can occur during the treatment of schizophrenics with dopamine-lowering drugs when the symptoms of Parkinson's disease sometimes show as a side-effect.

20-2. Experiments done on anesthetized animals often involve either stimulating a brain part to see the effects of increased neuronal activity or damaging ("lesioning") an area to observe resulting deficits. Such experiments on animals, which lack the complex language mechanisms of people, cannot help with language studies. Diseases sometimes mimic these two experimental situations, and behavioral studies of the resulting language deficits in people with aphasia, coupled with study of the brains after death, have provided a wealth of information.

ENGLISH
AND
METRIC
UNITS

	English	Metric
Length	1 foot = 0.305 meter 1 inch = 2.54 centimeters	1 meter = 39.37 inches 1 centimeter (cm) = 1/100 meter 1 millimeter (mm) = 1/1000 meter 1 micrometer (μm) = 1/1000 millimeter 1 nanometer (nm) = 1/1000 micrometer *[1 angstrom (Å) = 1/10 nanometer]
Mass	†1 pound = 433.59 grams 1 ounce = 27.1 grams	1 kilogram (kg) = 1000 grams = 2.2 pounds 1 gram (g) = 0.037 ounce 1 milligram (mg) = 1/1000 gram 1 microgram (μg) = 1/1000 milligram 1 nanogram (ng) = 1/1000 microgram 1 picogram (pg) = 1/1000 nanogram
Volume	1 gallon = 3.785 liters 1 quart = 0.946 liter	1 liter = 1000 cubic centimeters = 0.264 gallon 1 liter = 1.057 quarts 1 milliliter (ml) = 1/1000 liter 1 microliter (μl) = 1/1000 milliliter

*The angstrom unit of length is not a true metric unit, but has been included because of its frequent use, until recently, in the measurement of molecular dimensions.

†A pound is actually a unit of force, not mass. The correct unit of mass in the English system is the slug, while the newton is the metric unit of force. When we write 1 kg = 2.2 pounds, this means that one kilogram of *mass* will have a *weight* under standard conditions of *gravity* at the Earth's surface of 2.2 pounds *force*.

EQUATIONS

I. The **Nernst equation** describes the equilibrium potential for any ion species, that is, the electric potential necessary to balance a given ionic concentration gradient across a membrane so that the net passive flux of the ion is zero. The Nernst equation is:

$$E = \frac{RT}{zF} \ln \frac{C_o}{C_i}$$

where E = equilibrium potential for the particular ion in question

C_i = intracellular concentration of the ion

C_o = extracellular concentration of the ion

z = valence of the ion (+1 for sodium and potassium, −1 for chloride)

R = gas constant [8314.9 J/(kg · mol · K)]

T = absolute temperature (temperature measured on the Kelvin scale: degrees centigrade + 273)

F = Faraday (the quantity of electricity contained in 1 mol of electrons: 96,484.6 C/mol of charge)

$\ln$ = logarithm taken to the base e

sodium, and chloride (and other ions if they are in sufficient concentrations) and on their relative permeability properties. It is used to calculate the value of the membrane potential when the potential is determined by more than one ion. The Goldman equation is:

$$V_m = \frac{RT}{F} \ln \frac{P_K \times K_o + P_{Na} \times Na_o + P_{Cl} \times Cl_i}{P_K \times K_i + P_{Na} \times Na_i + P_{Cl} \times Cl_o}$$

where V_m = membrane potential

R = gas constant [8314.9 J/(kg · mol · K)]

T = absolute temperature (temperature measured on the Kelvin scale: degrees centigrade + 273)

F = Faraday (the quantity of electricity contained in 1 mol of electrons: 96,484.6 C/mol of charge)

$\ln$ = logarithm taken to the base e

P_K, P_{Na}, and P_{Cl} = membrane permeabilities for potassium, sodium, and chloride, respectively

K_o, Na_o, and Cl_o = extracellular concentrations of potassium, sodium, and chloride, respectively

K_i, Na_i, and Cl_i = intracellular concentrations of potassium, sodium, and chloride, respectively

A band one of transverse bands making up repeating striations of cardiac and skeletal muscle; composed of thick filaments and located in middle of sarcomere

A cell glucagon-secreting cell of pancreatic islet of Langerhans; also called alpha cell

abscess (AB-sess) microbes, leukocytes, and liquified tissue debris walled off by fibroblasts and collagen; thus, a bag of pus

absolute refractory period time during which an excitable membrane cannot generate an action potential in response to any stimulus

absorption movement of materials across an epithelial layer from body cavity or compartment toward the blood

absorptive state period during which nutrients enter bloodstream from gastrointestinal tract

accessory reproductive organ duct through which sperm or ova are transported, or a gland emptying into such a duct (in the female the breasts are usually included)

acclimatization (ah-climb-ah-ti-ZA-shun) environmentally induced improvement in functioning of a genetically based physiological system

accommodation adjustment of eye for viewing various distances by changing lens's shape

acetone (ASS-ih-tone) ketone body produced from acetyl CoA during prolonged fasting or untreated severe diabetes mellitus

acetyl coenzyme A (acetyl CoA) (ASS-ih-teal koh-EN-zime A, koe-A) metabolic intermediate that transfers acetyl groups to Krebs cycle and various synthetic pathways

acetyl group CH_3CO-

acetylcholine (ACh) (ass-suh-teal-KOH-lean) neurotransmitter released by pre- and postganglionic parasympathetic neurons, somatic neurons, and some CNS neurons

acetylcholinesterase (ass-suh-teal-koh-lin-ES-ter-ase) enzyme that breaks down acetylcholine into acetic acid and choline

ACh *see* acetylcholine

acid molecule capable of releasing a hydrogen ion; solution having H^+ concentration greater than that of pure water, that is, pH less than 7; *see also* strong acid, weak acid

acidity concentration of free, unbound hydrogen ions in a solution; the higher the H^+ concentration, the greater the acidity

acidosis (ass-ih-DOUGH-sis) any situation in which arterial H^+ concentration is elevated; *see also* metabolic acidosis, respiratory acidosis

acquired immune deficiency syndrome (AIDS) disease caused by *human immunodeficiency virus* (HIV) and characterized by profound inability to resist many infections; major deficit is lack of helper T cells

acrosome (AK-roh-sohm) cytoplasmic vesicle containing digestive enzymes and located at head of a sperm

ACTH *see* adrenocorticotropic hormone

actin (AK-tin) globular contractile protein to which myosin cross bridges bind; located in muscle thin filaments and in microfilaments of cytoskeleton

action potential electric signal propagated by nerve and muscle cells; an all-or-none reversal of membrane polarity; has a threshold and refractory period and is conducted without decrement

activation *see* lymphocyte activation

activation energy energy necessary to disrupt existing chemical bonds during a chemical reaction

active hyperemia (hi-per-EE-me-ah) increased blood flow through a tissue due to increased metabolic activity

active immunity resistance to reinfection acquired by contact with microorganisms, their toxins, or other antigenic material; *compare* passive immunity

active site region of enzyme to which substrate binds

active transport an energy-requiring carrier-mediated transport system that can move molecules across a membrane against an electrochemical gradient; *see also* primary active transport, secondary active transport

acute (ah-CUTE) lasting a relatively short time; *compare* chronic

acute phase protein one of a group of proteins secreted by liver during systemic response to injury or infection

adaptation (evolution) a biologic characteristic that favors survival in a particular environment; (neural) decrease in action-potential frequency in a neuron despite constant stimulus

adenine (A) (ADD-ah-neen) purine base of DNA and RNA

adenosine diphosphate (ADP) (ah-DEN-oh-seen die-FOS-fate) two-phosphate product of ATP breakdown

adenosine monophosphate (AMP) monophosphate derivative of ATP; nucleotide in RNA

adenosine triphosphate (ATP) major molecule that transfers energy from metabolism to cell functions during its breakdown to ADP and P_i

adenylate cyclase (ah-DEN-ah-late SIGH-klase) enzyme that catalyzes transformation of ATP to cyclic AMP

ADH (antidiuretic hormone) *see* vasopressin

adipocyte (ah-DIP-oh-site) cell specialized

for triacylglycerol synthesis and storage; fat cell

adipose tissue (ADD-ah-poss) tissue composed largely of fat-storing cells

ADP *see* adenosine diphosphate

adrenal cortex (ah-DREE-nal KOR-tex) endocrine gland that forms outer shell of each adrenal gland; secretes mainly cortisol, aldosterone, and androgens; *compare* adrenal medulla

adrenal gland one of a pair of endocrine glands above each kidney; each gland consists of outer *adrenal cortex* and inner *adrenal medulla*

adrenal medulla (mah-DULL-ah) endocrine gland that forms inner core of each adrenal gland; main secretion is epinephrine; *compare* adrenal cortex

adrenaline (ah-DREN-ah-lyn) British name for epinephrine

adrenergic (ad-ren-ER-jik) pertaining to norepinephrine or epinephrine; compound that acts like norepinephrine or epinephrine

adrenocorticotropic hormone (ACTH) (adren-oh-kor-tih-koh-TROH-pik) polypeptide hormone secreted by anterior pituitary; stimulates adrenal cortex to secrete cortisol; also called corticotropin

aerobic (air-OH-bik) in presence of oxygen

affective disorder mania, depression, or bipolar disorder; any disease characterized by serious, prolonged disturbances of mood

afferent (AF-er-ent) carrying toward

afferent arteriole vessel in kidney that carries blood from artery to glomerulus

afferent neuron neuron whose cell body lies outside CNS; carries information from receptors at its peripheral endings to CNS

afferent pathway component of reflex arc that transmits information from receptor to integrating center

affinity strength with which ligand binds to its binding site

afterbirth placenta and associated membranes expelled from uterus after delivery of infant

agonist (AG-ah-nist) chemical messenger that binds to receptor and triggers cell's response; often refers to drug that mimics action of chemical normally in the body

AIDS *see* acquired immune deficiency syndrome

airway tube through which air flows between external environment and lung alveoli

albumin (al-BU-min) most abundant group of plasma proteins

aldosterone (al-doe-stir-OWN or al-DOS-stirown) mineralocorticoid steroid hormone secreted by adrenal cortex; regulates electrolyte balance

alkaline having H^+ concentration lower than that of pure water, that is, having a pH greater than 7

alkalosis (alk-ah-LOW-sis) any situation in which arterial blood H^+ concentration is reduced; *see also* metabolic alkalosis, respiratory alkalosis

all or none event that occurs maximally or not at all

allergy acquired, specific immune reactivity to environmental antigens involving IgE antibodies; hypersensitivity

allosteric modulation (al-low-STAIR-ik) control of protein binding-site properties by modulator molecules that bind to regions of the protein other than the binding site altered by them

allosteric protein protein whose binding-site characteristics are subject to allosteric modulation

alpha-adrenergic receptor plasma-membrane receptor for epinephrine and norepinephrine that, when activated, opens a calcium channel in the membrane; *compare* beta-adrenergic receptor

alpha-gamma coactivation simultaneous, or near simultaneous, activation of alpha and gamma motor neurons

alpha helix coiled conformation of polypeptide chain found in many proteins

alpha motor neuron motor neuron that innervates skeletomotor muscle fibers

alpha rhythm prominent 10-Hz oscillation in the electroencephalograms of awake, relaxed adults with their eyes closed

alternate complement pathway sequence for complement activation that bypasses first steps in classical pathway and is not antibody dependent

alveolar dead space (al-VEE-oh-lar) volume of inspired air that reaches alveoli but cannot undergo gas exchange with blood

alveolar pressure air pressure in pulmonary alveoli

alveolar ventilation volume of atmospheric air entering alveoli each minute

alveolus (al-VEE-oh-lus) thin-walled, airfilled "outpocketing" from terminal air passageways in lungs; cell cluster at end of duct in secretory gland

Alzheimer's disease degenerative brain disease that is most common cause of declining intellectual function in late life

amine hormone (ah-MEAN) hormone derived from amino acid tyrosine; includes thyroid hormone, epinephrine, and norepinephrine

amino acid (ah-MEAN-oh) molecule containing amino group, carboxyl group, and side chain attached to a carbon atom; molecular subunit of protein

amino group $-NH_2$

aminoacyl-tRNA synthetase (ah-MEAN-ohACE-il tee-RNA SIN-the-tase) general name for 20 enzymes, each of which catalyzes covalent linkage of a specific amino acid to its particular tRNA during protein synthesis

aminopeptidase (ah-mean-oh-PEP-tih-dase) one of a family of enzymes located in the intestinal epithelial membrane; breaks peptide bond at amino end of polypeptide

ammonia NH_3; produced during amino acid breakdown; converted in liver to urea

amnesia memory loss; *see also* retrograde amnesia

amniotic membrane (am-nee-AHT-ik) membrane surrounding fetus in utero

AMP *see* adenosine monophosphate

amphetamine drug that increases transmission at catecholamine-mediated synapses in brain

amphipathic molecule (am-fuh-PATH-ik) molecule containing polar or ionized groups at one end and nonpolar groups at the other

amplitude height, how much, magnitude of change

amylase (AM-ih-lase) enzyme that partially breaks down polysaccharides

anabolic steroid (an-ah-BOL-ik STEAR-oid) testosterone-like agent that increases protein synthesis

anabolism (an-NAB-oh-lizm) cellular synthesis of organic molecules

anaerobic (an-ih-ROW-bik) in absence of oxygen

analgesia (an-al-JEE-zee-ah) removal of pain

anatomic dead space space in respiratory tract airways whose walls do not permit gas exchange with blood

androgen (AN-dro-jen) any chemical with testosterone-like actions

anemia (ah-KNEE-me-ah) reduction in total blood hemoglobin

anemic hypoxia (ah-KNEE-mik hi-POK-seeah) hypoxia with normal arterial O_2 pressure but reduced total blood oxygen content

ANF *see* atrial natriuretic factor

angina pectoris (an-JI-nah PEK-tor-iss) chest pain associated with inadequate blood flow to heart muscle

angiotensin (an-gee-oh-TEN-sin) angiotensin II

angiotensin I peptide generated in plasma by renin's action on angiotensinogen

angiotensin II hormone formed by enzymatic action on angiotensin I; stimulates aldosterone secretion from adrenal cortex, vascular smooth-muscle contraction, and thirst

angiotensinogen (an-gee-oh-ten-SIN-oh-jen) plasma protein precursor for angiotensin I

anion (AN-eye-on) negatively charged ion; *compare* cation

antagonist (muscle) muscle whose action opposes intended movement; (drug) molecule that competes with another for a receptor and binds to the receptor but does not trigger the cell's response

anterior toward or at the front

anterior pituitary anterior portion of pituitary gland; synthesizes, stores, and releases various hormones including ACTH, GH, TSH, prolactin, FSH, and LH

antibody (AN-tih-bah-dee) immunoglobulin that both functions as antigen receptor on B cell and is secreted by plasma cell; combines with type of antigen that stimulated its production; directs attack against antigen or cell bearing it; *see also* natural antibody

antibody-mediated immune response *see* humoral immune response

anticodon (an-tie-KOE-don) three-nucleotide sequence in tRNA able to base-pair with complementary codon in mRNA during protein synthesis

antidiuretic hormone (ADH) (an-tie-die-your-ET-ik) *see* vasopressin

antigen (AN-tih-jen) any foreign molecule that stimulates a specific immune response

antihistamine (an-tie-HISS-tah-mean) chemical that blocks histamine action

antithrombin III (an-tie-throm-bin THREE) plasma anticlotting protein that inactivates thrombin

antrum (AN-trum) (gastric) lower portion of stomach, that is, region closest to pyloric sphincter; (ovarian) fluid-filled cavity in maturing ovarian follicle

aorta (a-OR-tah) largest artery in body; carries blood from left ventricle of heart to thorax and abdomen

aortic arch baroreceptor (a-OR-tik) *see* arterial baroreceptor

aortic body chemoreceptor chemoreceptor located near aortic arch; sensitive to arterial blood O_2 pressure and H^+ concentration

aortic valve valve between left ventricle of heart and aorta

aphasia (ah-FAY-see-ah) specific language deficit not due to mental retardation or muscular weakness

apnea (AP-nee-ah) cessation of respiration

appendix small fingerlike projection from cecum of large intestine

appetitive motivation goal-directed quality of behavior that is related to rewards

aqueous (A-kwee-us) watery; prepared with water

arachidonic acid (ah-rak-ah-DON-ik) polyunsaturated fatty acid precursor of eicosanoids

arrhythmia (a-RYTH-me-ah) any variation from normal heartbeat rhythm

arterial baroreceptor nerve endings sensitive to stretch or distortion produced by arterial blood pressure changes; located in carotid sinus or aortic arch arteries; also called the carotid sinus and aortic arch baroreceptors

arteriole (are-TEER-ee-ole) blood vessel between artery and capillary, surrounded by smooth muscle; primary site of vascular resistance

artery (ARE-ter-ee) thick-walled, elastic vessel that carries blood away from heart to arterioles

ascending limb portion of Henle's loop of renal tubule leading to distal convoluted tubule

aspartate (ah-SPAHR-tate) a major excitatory neurotransmitter in CNS; ionized form of the amino acid aspartic acid

association cortex *see* cortical association areas

asthma (AS-muh) disease characterized by severe airway constriction and plugging of the airways with mucus

atherosclerosis (ath-er-oh-skluh-ROW-sis) disease characterized by thickening of arterial walls with abnormal smooth-muscle cells, cholesterol deposits, and connective tissue; results in narrowing of vessel lumen

atmospheric pressure air pressure surrounding the body (760 mmHg at sea level)

atom smallest unit of matter that has unique chemical characteristics; has no net charge; combines with other atoms to form molecules

atomic mass relative value that indicates an atom's mass relative to the mass of other types of atoms, based on the assignment of a value of 12 to carbon

atomic nucleus dense region, consisting of protons and neutrons, at center of atom

atomic number number of protons in nucleus of atom

ATP *see* adenosine triphosphate

ATPase enzyme that breaks down ATP to ADP and inorganic phosphate; *see* Na^+, K^+−ATPase, Ca^{2+}−ATPase

atrial natriuretic factor (ANF) (nat-rye-your-ET-ik) peptide hormone that is secreted by atrial cells in response to atrial distension and causes decreased renal sodium reabsorption

atrioventricular (AV) node (a-tree-oh-ven-TRICK-you-lar) region at base of right atrium near interventricular septum, containing specialized cardiac-muscle cells through which electrical activity must pass to go from atria to ventricles

atrioventricular (AV) valve valve between atrium and ventricle of heart; AV valve on right side of heart is the *tricuspid valve*, and that on left is the *mitral valve*

atrium (ATE-ree-um) chamber of heart that receives blood from veins and passes it on to ventricle on same side of heart

atrophy (AT-row-fee) wasting away; decrease in size; *see also* disuse atrophy, denervation atrophy

auditory (AW-dih-tor-ee) pertaining to sense of hearing

auditory cortex region of cerebral cortex that receives nerve fibers from auditory pathways

autocrine (AW-toe-crin) chemical messenger that is secreted into extracellular fluid and acts upon cell that secreted it

autoimmune disease (aw-toe-im-MUNE) disease produced by antibody-mediated or cell-mediated attack against body's own cells, which results in damage or alteration of cell function

automaticity (aw-toe-mah-TISS-ih-tee) capable of self-excitation

autonomic nervous system (aw-toe-NAHM-ik) component of efferent division of peripheral nervous system that consists of sympathetic and parasympathetic subdivisions; innervates cardiac muscle, smooth muscle, and glands; *compare* somatic nervous system

autoregulation (aw-toe-reg-you-LAY-shun) ability of an individual organ to control (self-regulate) its vascular resistance independent of neural and hormonal influence

AV node *see* atrioventricular node

AV valve *see* atrioventricular valve

aversive motivation goal-directed quality of behavior that is related to avoidance, escape, or aggression

axon (AXE-own) extension from neuron cell body; propagates action potentials away from cell body; also called a nerve fiber

axon-axon synapse *see* presynaptic synapse

axon terminal end of axon; forms synaptic or neuroeffector junction with postjunctional cell

axon transport process involving intracellular filaments by which materials are moved from one end of axon to other

B cell (immune system) lymphocyte that, upon activation, proliferates and differentiates into antibody-secreting plasma cell; (endocrine system) insulin-secreting cell in pancreatic islet of Langerhans; also called beta cell

B lymphocyte *see* B cell

bacteria unicellular organisms that have outer cell wall and a plasma membrane

balance concept if quantity of any substance in body is to be maintained constant over time, amount ingested and produced must equal amount excreted and consumed

barometric pressure *see* atmospheric pressure

baroreceptor receptor sensitive to pressure and to rate of change in pressure; *see also* arterial baroreceptor, intrarenal baroreceptor

basal (BAY-sul) resting level

basal ganglia nuclei deep in cerebral hemispheres that code and relay information associated with control of body movements; specifically, caudate nucleus, globus pallidus, and putamen

basal metabolic rate (BMR) metabolic rate when one is at mental and physical rest but not sleeping, at comfortable temperature, and fasted at least 12 h; also called metabolic cost of living

base (acid-base) any molecule that can combine with H⁺; (nucleotide) molecular ring of carbon and nitrogen, which, with a phosphate group and a sugar, constitutes a nucleotide; *see also* purine base, pyrimidine base

basement membrane thin proteinaceous layer of extracellular material upon which epithelial and endothelial cells sit

basic electrical rhythm spontaneous depolarization-repolarization cycles of pacemaker cells in longitudinal smooth-muscle layer of stomach and intestines; coordinates repetitive muscular activity of GI tract

basilar membrane (BAS-ih-lar) membrane that separates cochlear duct and scala tympani in inner ear; supports organ of Corti

basolateral membrane (bay-so-LAH-ter-al) side of epithelial cell facing away from lumen; also called serosal or blood side of cell

basophil (BAY-so-fill) polymorphonuclear granulocytic leukocyte whose granules stain with basic dyes; enters tissues and becomes mast cell

beta-adrenergic receptor (BAY-ta ad-ren-ER-jik) plasma-membrane receptor for epinephrine and norepinephrine that utilizes cAMP second-messenger system; *compare* alpha-adrenergic receptor

beta oxidation (ox-ih-DAY-shun) series of reactions that transfers hydrogens from fatty acid breakdown to oxidative phosphorylation for ATP synthesis

beta rhythm low, fast EEG oscillations in alert, awake adults who are paying attention to (or thinking hard about) something

bicarbonate (by-CAR-bah-nate) HCO₃⁻

bile fluid secreted by liver; contains bicarbonate, bile salts, cholesterol, lecithin, bile pigments, metabolic end products, and certain trace metals

bile canaliculi (can-al-IK-you-lee) small ducts adjacent to liver cells into which bile is secreted

bile duct carries bile from liver and gallbladder to small intestine

bile pigment colored substance, derived from breakdown of heme group or hemoglobin, secreted in bile

bile salt one of a family of steroid molecules secreted in bile by the liver; promotes solubilization and digestion of fat in small intestine

bilirubin (bill-eh-RUE-bin) yellow substance resulting from heme breakdown; excreted in bile as a bile pigment

binding site region of protein to which a specific ligand binds

biogenic amine (by-oh-JEN-ik ah-MEAN) one of family of neurotransmitters having basic formula R−NH₂; includes dopamine, norepinephrine, epinephrine, serotonin, and histamine

"biological clock" cells that function rhythmically in absence of apparent external stimulation

biotransformation (by-oh-trans-for-MAY-shun) alteration of foreign molecules by an organism's metabolic pathways

bipolar affective disorder affective disorder characterized by mood swings between mania and depression

bipolar cell retinal neuron postsynaptic to rod or cone

bladder *see* urinary bladder

blastocyst (BLAS-toe-cyst) early embryonic stage consisting of ball of developing cells surrounding central cavity

blood-brain barrier group of anatomical barriers and transport systems that controls kinds of substances entering brain extracellular space from blood and their rates of entry

blood coagulation (koh-ag-you-LAY-shun) blood clotting

blood sugar glucose

blood-testis barrier barrier that limits chemical movements between blood and lumen of seminiferous tubules

blood type blood classification according to presence of A, B, or O antigens

BMR *see* basal metabolic rate

body mass index (BMI) method for assessing degree of obesity calculated as weight in kilograms divided by square of height in meters

bone marrow highly vascular, cellular substance in central cavity of some bones; site of erythrocyte, leukocyte, and platelet synthesis

Bowman's capsule blind sac leading to tubular component of kidney nephron

Boyle's law (boils) pressure of a fixed amount of gas in a container is inversely proportional to container's volume

bradykinin (braid-ee-KI-nin) protein formed by action of the enzyme kallikrein on precursor; dilates vessels, increases capillary permeability, and probably stimulates pain receptors

brainstem brain subdivision consisting of medulla oblongata, pons, and midbrain and located between spinal cord and forebrain

Broca's area (BRO-kahz) region of left frontal lobe associated with speech production

bronchiole (BRON-key-ole) small division of a bronchus

bronchus (BRON-kus) large-diameter air passage that enters lung; located between trachea and bronchioles

buffer weak acid or base that can exist in undissociated (Hbuffer) or dissociated (H⁺ + buffer) form

buffering reversible hydrogen-ion binding by anions when H⁺ concentration changes; tends to minimize changes in acidity of a solution when acid is added or removed

bulbourethral gland (bul-bo-you-WREATH-ral) one of paired glands in male that secretes fluid components of semen into the urethra

bulk flow movement of fluids or gases from region of higher pressure to one of lower pressure

C₁ first protein activated in classical complement pathway

C₃ first protein activated in alternate complement pathway

C₃b complement protein that functions as an opsonin

Ca²⁺ calcium ion

Ca²⁺−ATPase primary active-transport carrier protein for calcium in plasma membrane and in several organelle membranes

calmodulin (kal-MOD-you-lin) intracellular calcium-binding protein that mediates many of calcium's second-messenger functions

calorie (cal) unit of heat-energy measurement; amount of heat needed to raise temperature of 1 g of water 1 C°; one large calorie (Cal or kcal), used in nutrition, equals 1000 calories

calorigenic effect (kah-lor-ih-JEN-ik) increase in metabolic rate caused by epinephrine or thyroid hormones

cAMP *see* cyclic AMP

cancer uncontrolled cell growth

candidate hormone substance suspected of being a hormone but not yet proven to be one

capacitation *see* sperm capacitation

capillary smallest blood vessel type

carbamino compound (car-bah-ME-no) compound resulting from combination of carbon dioxide and protein amino groups, particularly in hemoglobin

carbohydrate substance composed of carbon, hydrogen, and oxygen according to general formula $C_n(H_2O)_n$, where n is any whole number

carbon monoxide CO; gas that reacts with hemoglobin and decreases blood oxygen-carrying capacity

carbonic acid (car-BAHN-ik) H_2CO_3; an acid formed from H_2O and CO_2

carbonic anhydrase (an-HI-drase) enzyme that catalyzes the reaction $CO_2 + H_2O \rightarrow H_2CO_3$

carboxyl group (car-BOX-il) —COOH

carboxypeptidase (car-box-ee-PEP-tih-dase) enzyme secreted into small intestine by exocrine pancreas as precursor, procarboxypeptidase; breaks peptide bond at carboxyl end of protein

carcinogen (car-SIN-oh-jen) any agent that can induce cancerous transformation of cells

cardiac (CAR-dee-ak) pertaining to the heart

cardiac cycle one contraction-relaxation sequence of heart

cardiac muscle heart muscle

cardiac output blood volume pumped by each ventricle per minute (not total output pumped by both ventricles)

cardiovascular center neuron cluster in brainstem medulla that serves as a major integrating center for reflexes affecting heart and blood vessels

cardiovascular system heart and blood vessels

carotid (ka-RAH-tid) pertaining to two major arteries (carotid arteries) in neck that convey blood to head

carotid body chemoreceptor chemoreceptor near main branching of carotid artery; sensitive to blood O_2 pressure and H^+ concentration

carotid sinus dilatation of internal carotid artery just above main carotid branching; location of carotid baroreceptors

carotid sinus baroreceptor *see* arterial baroreceptor

carrier integral membrane protein capable of combining with specific molecules, enabling them to pass through membrane

cascade (kas-KADE) multiplicative sequence of events in which the number of reaction products increases at one or more steps

catabolism (kah-TAH-bowl-ism) cellular breakdown of organic molecules

catalyst (CAT-ah-list) substance that accelerates chemical reactions but does not itself undergo any net chemical change during the reaction

catechol-O-methyl transferase (CAT-eh-cole oh METH-il) enzyme that degrades catecholamine neurotransmitters

catecholamine (cat-eh-COLE-ah-mean) dopamine, epinephrine, or norepinephrine, all of which have similar chemical structures

cation (CAT-eye-on) ion having net positive charge; *compare* anion

CCK *see* cholecystokinin

cecum (SEE-come) dilated pouch at beginning of large intestine into which open the ileum, colon, and appendix

cell adhesion molecule (CAM) glycoprotein important in nervous system development

cell body in cells with long extensions, the part that contains the nucleus

cell differentiation *see* differentiation

cell-mediated immune response type of specific immune response mediated by cytotoxic T and NK cells; major defense against intracellular viruses and cancer cells

cell organelle (or-guh-NELL) membrane-bound compartment, nonmembranous particle, or filament that performs specialized functions in cell

cellulose polysaccharide composed of glucose subunits; found in plant cells

center of gravity point in a body at which body mass is in perfect balance; if the body

were suspended from a string attached to this point, there would be no movement

central chemoreceptor receptor in brainstem medulla that responds to H^+ concentration changes of brain extracellular fluid

central nervous system (CNS) brain plus spinal cord

central pattern generator cluster of brainstem or spinal neurons that activates skeletal muscle in cyclic manner for repetitive behavior

central thermoreceptor temperature receptor in hypothalamus, spinal cord, abdominal organ, or other internal location

centriole (SEN-tree-ole) small cytoplasmic body having nine fused sets of microtubules; participates in nuclear and cell division

cephalic phase (seh-FAHL-ik) (of gastrointestinal control) initiation of the neural and hormonal reflexes regulating gastrointestinal functions by stimulation of receptors in head, that is, cephalic receptors—sight, smell, taste, and chewing

cerebellum (ser-ah-BELL-um) brain subdivision lying behind forebrain and above brainstem; deals with muscle movement control

cerebral cortex (SER-ah-brul or sah-REE-brul) cellular layer covering the cerebrum

cerebrospinal fluid (CSF) (sah-ree-broh-SPY-nal) fluid that fills cerebral ventricles and the subarachnoid space surrounding brain and spinal cord

cerebrum (SER-ah-brum or sah-REE-brum) frontmost part of brain, which, with diencephalon, forms the forebrain

cervix (SIR-vix) lower portion of uterus; cervical opening connects uterine and vaginal lumens

CG *see* chorionic gonadotropin

cGMP *see* cyclic GMP

channel small passage in plasma membrane formed by integral membrane proteins and through which certain small-diameter molecules and ions can diffuse; *see also* receptor-operated channel, voltage-sensitive channel

charge *see* electric charge

chemical bond interaction between electric forces of subatomic particles of adjacent atoms; holds atoms together in a molecule

chemical element specific type of atom

chemical equilibrium *see* dynamic equilibrium

chemical reaction breaking of some chemical bonds and formation of new ones, which changes one type of molecule to another; *see also* reversible reaction, irreversible reaction

chemical specificity *see* specificity

chemical synapse (SIN-apse) synapse at which neurotransmitters released by one neuron diffuse across an extracellular gap to influence a second neuron's activity

chemoreceptor afferent nerve ending (or cells associated with it) sensitive to concentrations of certain chemicals

chemotaxin (key-moh-TAX-in) any chemical that causes chemotaxis

chemotaxis (key-moh-TAX-iss) movement of cells, particularly phagocytes, in a specific direction in response to a chemical stimulus

chief cell gastric gland cell that secretes pepsinogen, precursor of pepsin

cholecystokinin (CCK) (koh-lee-sis-toe-KI-nin) peptide hormone secreted by duodenum that regulates gastric motility and secretion, gall bladder contraction, and pancreatic enzyme secretion; possible satiety signal

cholesterol particular steroid molecule; precursor of steroid hormones and bile salts and a component of plasma membranes

cholinergic (koh-lyn-ER-jik) pertaining to acetylcholine; a compound that acts like acetylcholine

chondrocyte (KON-droh-site) cell type that forms new cartilage

chorionic gonadotropin (CG) (kor-ee-ON-ik go-NAD-oh-troh-pin) protein hormone secreted by trophoblastic cells of blastocyst and placenta; maintains secretory activity of corpus luteum during first 3 months of pregnancy

choroid plexus (KOR-oid) highly vascular epithelial structure lining portions of cerebral ventricles; responsible for much of cerebrospinal fluid formation

chromatid (KROME-ah-tid) one of two identical strands of chromatin resulting from DNA duplication during mitosis or meiosis

chromatin (KROM-ih-tin) combination of DNA and nuclear proteins that is the principle component of chromosomes

chromophore retinal component of a photopigment

chromosome highly coiled, condensed form of chromatin formed in cell nucleus during mitosis and meiosis

chronic (KRON-ik) persisting over a long time; *compare* acute

chylomicron (kye-low-MY-kron) small droplet consisting of lipids and protein that is released from intestinal epithelial cells into the lacteals during fat absorption

chyme (kyme) solution of partially digested food in stomach and intestinal lumens

chymotrypsin enzyme secreted by exocrine pancreas; breaks certain peptide bonds in proteins and polypeptides

cilia (SILL-ee-ah) hairlike projections from specialized epithelial cells; sweep back and forth in a synchronized way to propel material along epithelial surface

circadian (sir-KAY-dee-an) recurring once approximately every 24 h

circular muscle smooth-muscle layer in

stomach and intestinal walls and having muscle fibers circumferentially oriented around these organs

citric acid (SIT-rik) six-carbon intermediate in Krebs cycle

citric acid cycle *see* Krebs cycle

classical complement pathway antibody-dependent system for activating complement

clearance volume of plasma from which a particular substance has been completely removed in a given time

climacteric (kli-MAK-ter-ik) phase involving physical and emotional changes as sexual maturity gives way to cessation of reproductive function

clitoris (KLIT-or-iss) small body of erectile tissue in female external genitalia; homologous to penis

clone genetically identical molecules or organisms

CNS central nervous system

CO₂ carbon dioxide

coactivation *see* alpha-gamma coactivation

cochlea (COCK-lee-ah) inner ear; fluid-filled spiral-shaped compartment that contains cochlear duct

cochlear duct (COCK-lee-er) fluid-filled membranous tube that extends length of inner ear, dividing it into compartments; contains organ of Corti

code word three-nucleotide sequence in DNA that signifies a given amino acid in a synthesized protein

codon (KOH-don) three-base sequence in mRNA that corresponds to given code word in DNA

coenzyme (koh-EN-zime) organic cofactor; generally serves as a carrier that transfers atoms or small molecular fragments from one reaction to another; is not consumed in the reaction and can be reused

coenzyme A (koh-en-zime A) coenzyme that transfers acetyl groups from one reaction to another

cofactor (KOH-fact-or) organic or inorganic substance that binds to specific region of an enzyme and is necessary for the enzyme's activity

collagen (COLL-ah-jen) strong, fibrous protein that functions as extracellular structural element in connective tissue

collecting duct portion of renal tubules between distal tubules and renal pelvis

colloid (KAH-loid) large molecule, mainly protein, to which capillaries are relatively impermeable

colon (KOH-lon) large intestine, specifically that part extending from cecum to rectum; *see also* sigmoid colon

colony-stimulating factor (**CSF**) collective term for four hormones that stimulate leukocyte production

color blindness defect in color vision often due to absence of one or even two cone photopigments

coma deep, prolonged unconsciousness not fulfilling all the criteria for brain death, thus, there may be spontaneous movement and some intact reflexes

"command" neuron neuron or one of a group of neurons whose activity initiates series of neural events resulting in a voluntary action

commissure (KOM-iss-sure) bundle of nerve fibers linking right- and left-brain halves

compensatory growth type of regeneration present in many organs after tissue damage

competition ability of similar molecules to combine with the same receptor

complement (KOM-plih-ment) group of plasma proteins that, upon activation, kill microbes directly and facilitate the inflammatory process, including phagocytosis

compliance stretchability; *see also* lung compliance

concentration amount of material per unit volume of solution

concentration gradient gradation in concentration that occurs between two regions having different concentrations

conducting system network of cardiac-muscle fibers specialized to conduct electrical activity to different areas of heart

conducting zone air passages that extend from top of trachea to beginning of respiratory bronchioles and have walls too thick for gas exchange between air and blood

conduction heat exchange by transfer of thermal energy during collisions of adjacent molecules

conductor material having low resistance to charge flow

cone one of two retinal receptor types for photic energy; gives rise to color vision

conformation three-dimensional shape of a molecule

congestive heart failure set of signs and symptoms associated with decreased contractility of heart and engorgement of heart, veins, and capillaries

connective-tissue cell cell specialized to form extracellular elements that connect, anchor, and support body structures

conscious experience things of which a person is aware; thoughts, feelings, perceptions, ideas, and reasoning during any state of consciousness

consciousness *see* conscious experience, state of consciousness

contractility (kon-trak-TIL-ity) force of heart contraction that is independent of fiber length

contraction operation of the tension-generating process in a muscle

contraction time time between a muscle action potential and development of peak twitch tension by the muscle

contralateral on the opposite side of the body

control system collection of interconnected components that keeps a physical or chemical parameter within predetermined range of values

convection (kon-VEK-shun) process by which air or water next to a warm body is heated by conduction, moves away, and is replaced by colder air that, in turn, follows the same cycle

convergence (neuronal) many presynaptic neurons synapsing upon one postsynaptic neuron; (of eyes) turning of eyes inward, that is, toward nose, to view near objects

converting enzyme enzyme that catalyzes removal of two amino acids from angiotensin I to form angiotensin II

core temperature temperature of inner body

cornea (KOR-nee-ah) transparent structure covering front of eye; forms part of eye's optical system and helps focus an object's image on retina

cornification (kor-nif-ih-KAY-shun) process by which outer epithelial layer is converted to dead, horny cells

coronary blood flow blood flow to heart muscle

corpus callosum (KOR-pus kal-LOW-sum) wide band of nerve fibers connecting the two cerebral hemispheres; a brain commissure

corpus luteum (LOO-tee-um) ovarian structure formed from the follicle after ovulation; secretes estrogen and progesterone

cortex (CORE-tex) outer layer of organ; *see also* adrenal cortex, cerebral cortex; *compare* medulla

cortical association area region of parietal and frontal cerebral cortex; receives input from various sensory types, memory stores, and so on, and performs further perceptual processing

corticobulbar pathway (kor-tih-koh-BUL-bar) descending pathway having its neuron cell bodies in cerebral cortex; its axons pass without synapsing to region of brainstem motor neurons

corticospinal pathway descending pathway having its neuron cell bodies in cerebral cortex; its axons pass without synapsing to region of spinal motor neurons; also called the pyramidal tract

corticosteroid (kor-tih-koh-STEAR-oid) steroid produced by adrenal cortex or drug that resembles one

corticotropin-releasing hormone (**CRH**) (kort-ih-koh-TROH-pin) hypothalamic hormone that stimulates ACTH (corticotropin) secretion by anterior pituitary

cortisol (KORT-ih-sol) main glucocorticoid hormone secreted by adrenal cortex; regulates various aspects of organic metabolism

cotransmitter chemical messenger released with a neurotransmitter from synapse or neuroeffector junction

cotransport form of secondary active transport in which net movement of actively transported substance and "downhill" movement of molecule supplying the energy are in the same direction

countercurrent multiplier system mechanism associated with loops of Henle that creates in renal medulla a region having high interstitial-fluid osmolarity

countertransport form of secondary active transport in which net movement of actively transported molecule is in direction opposite to "downhill" movement of molecule supplying the energy

covalent bond (koh-VAY-lent) chemical bond between two atoms in which each atom shares one of its electrons with the other

covalent modulation alteration of a protein's shape and therefore its function by the covalent binding of various chemical groups to it

CP creatine phosphate

cranial nerve one of 24 peripheral nerves (12 pairs) that joins brainstem or forebrain

creatine phosphate (KREE-ah-tin) molecule that transfers phosphate and energy to ADP to generate ATP

creatinine (kree-AT-ih-nin) waste product derived from muscle creatine

creatinine clearance plasma volume from which creatinine is removed by renal filtration per unit time; approximates glomerular filtration rate

CRH corticotropin-releasing hormone

critical period time during development when a system is most readily influenced by environmental factors

cross bridge myosin projection extending from a muscle thick filament and capable of exerting force on the thin filament, causing the filaments to slide past each other

cross-bridge cycle sequence of events between one binding of a cross bridge to actin, its release, and reattachment

crossed-extensor reflex increased activation of extensor muscles contralateral to a limb flexion

crossing over process in which segments of maternal and paternal chromosomes exchange with each other during chromosomal pairing in meiosis

crystalloid low-molecular-weight solute

CSF *see* cerebrospinal fluid or colony-stimulating factor

cumulus oophorous (KEW-mew-lus oh-oh-FOR-us) granulosa cells surrounding ovum where ovum projects into ovarian follicle antrum

current movement of electric charge; in biological systems, this is achieved by ion movement

cutaneous (cue-TAY-nee-us) pertaining to skin

cyclic AMP (cAMP) cyclic 3′,5′-adenosine monophosphate; cyclic nucleotide that serves as a second messenger for many chemical messengers

cyclic AMP-dependent protein kinase (KI-nase) enzyme that is activated by cyclic AMP and then phosphorylates specific proteins, thereby altering their activity

cyclic GMP (cGMP) cyclic 3′,5′-guanosine monophosphate; cyclic nucleotide that acts as second messenger in some cells

cyclooxygenase (cy-klo-OX-ah-jen-ase) enzyme that acts on arachidonic acid and leads to production of prostaglandins, prostacyclin, and thromboxane

cytochrome (SIGH-toe-krome) one of a series of enzymes that couples energy to ATP formation during oxidative phosphorylation

cytokine (SIGH-toe-kine) general term for protein messenger secreted by macrophages and monocytes (*monokine*) or by lymphocytes (*lymphokine*)

cytokinesis (sigh-toe-kin-EE-sis) stage of cell division during which cytoplasm divides to form two new cells

cytoplasm (SIGH-toe-plasm) region of cell outside the nucleus

cytosine (C) (SIGH-toe-seen) pyrimidine base in DNA and RNA

cytoskeleton cytoplasmic filamentous network associated with cell shape and movement

cytosol (SIGH-toe-sol) intracellular fluid that surrounds cell organelles and nucleus

cytotoxic T cell (sigh-toe-TOX-ik) T lymphocyte that, upon activation by specific antigen, directly attacks a cell bearing that type of antigen; major killer of virus-infected and tumor cells

D cell somatostatin-secreting cell of pancreatic islet of Langerhans

DAG diacylglycerol

daughter cell one of the two new cells formed when a cell divides

dead space volume of inspired air that cannot be exchanged with blood; *see also* anatomic dead space, alveolar dead space, total dead space

deamination (dee-am-in-A-shun) removal of amino (—NH₂) group from a molecule

decremental decreasing in amplitude

defecation (deaf-ih-KAY-shun) expulsion of feces from rectum

dehydroepiandrosterone (dee-hydro-epi-andro-stir-OWN) androgen hormone secreted by adrenal cortex

delta wave slow 0.5- to 4-Hz oscillation of the electroencephalogram; associated with sleep

dendrite (DEN-drite) highly branched extension of neuron cell body; receives synaptic input from other neurons

denervation atrophy (AT-row-fee) decrease in size of muscle fiber whose nerve supply is destroyed

dense body cytoplasmic structure to which thin filaments of a smooth-muscle fiber are anchored

deoxyhemoglobin (Hb) (dee-ox-see-HE-mo-glo-bin) hemoglobin not combined with oxygen; reduced hemoglobin

deoxyribonucleic acid (DNA) (dee-ox-see-rye-bow-new-CLAY-ik) nucleic acid that stores and transmits genetic information; consists of double strand of nucleotide subunits that contain deoxyribose

depolarize to change membrane potential value toward zero, so that cell interior becomes less negative

depression disturbance in mood characterized by pervasive sadness or loss of interest or pleasure

descending limb (of Henle's loop) segment of renal tubule into which proximal tubule drains

desmosome (DEZ-moe-zome) cell junction that holds two cells together; consists of plasma membranes of adjacent cells separated by a 20-nm extracellular space filled with a cementing substance

diabetes *see* diabetes mellitus

diabetes insipidus (die-ah-BE-tees in-SIP-ih-des) disease due to defective control of urine concentration by ADH; marked by great thirst and excretion of a large volume of dilute urine

diabetes mellitus (MEL-ih-tus) disease in which plasma glucose control is defective because of insulin deficiency or decreased target-cell response to insulin

diabetic ketoacidosis (die-ah-BET-ik KEY-toe ass-ih-DOE-sis) acute life-threatening emergency in type I diabetes; characterized by increased plasma glucose and ketones, marked urinary losses, and metabolic acidosis

diacylglycerol (DAG) (die-ace-ill-GLIS-er-all) second messenger that activates protein kinase C, which then phosphorylates a large number of other proteins

dialysis (die-AL-ih-sis) process of altering the concentration of substances in the blood by using concentration differences between plasma and a bathing solution separated by a semipermeable membrane

diaphragm (DIE-ah-fram) dome-shaped skeletal-muscle sheet that separates the abdominal and thoracic cavities; principal muscle of respiration

diastole (die-ASS-toe-lee) period of cardiac cycle when ventricles are not contracting

diastolic pressure (die-ah-STAL-ik) minimum blood pressure during cardiac cycle

diencephalon (die-en-SEF-ah-lon) core of anterior part of brain; lies beneath cerebral hemispheres and contains thalamus and hypothalamus

differentiation (dif-fer-en-she-A-shun) process by which cells acquire specialized structural and functional properties

diffusion (dif-FU-shun) random movement of molecules from one location to another because of random thermal molecular motion; net diffusion always occurs from a region of higher concentration to a region of lower concentration

diffusion equilibrium state during which diffusion fluxes in opposite directions are equal, that is, the net flux = 0

diffusion potential voltage difference created by net diffusion of ions

digestion process of breaking down large particles and high-molecular-weight substances into small molecules

dihydrotestosterone (die-hi-droh-tes-TOS-ter-own) steroid formed by enzyme-mediated alteration of testosterone; active form of testosterone in certain of its target cells

1,25-dihydroxyvitamin D₃ (1-25-die-hi-DROX-ee-vie-tah-min DEE-3) hormone that is formed by kidneys and is active form of vitamin D

2,3-diphosphoglyceraldehyde (DPG) (2-3-die-fos-foe-gliss-er-AL-dah-hide) substance produced by erythrocytes during glycolysis; binds reversibly to hemoglobin, causing it to release oxygen

disaccharide (die-SAK-er-ide) carbohydrate molecule composed of two monosaccharides

disinhibition removal of inhibition from a neuron, thereby allowing its activity to increase

dissociation separation from

distal (DIS-tal) farther from reference point; *compare* proximal

distal tubule portion of kidney tubule between loop of Henle and collecting duct

disuse atrophy (AT-row-fee) decrease in size of muscle fibers that are not used for a long time

diuresis (die-u-REE-sis) increased urine excretion

diuretic (die-u-RET-ik) substance that inhibits fluid reabsorption in renal tubule, thereby increasing urine excretion

diurnal (die-URN-al) daily; occurring in a 24-h cycle

divergence (die-VER-gence) (neuronal) one presynaptic neuron synapsing upon many postsynaptic neurons; (of eyes) turning of eyes outward to view distant objects

dl deciliter; 1/10 L

DNA deoxyribonucleic acid

DNA polymerase enzyme that, during DNA replication, forms new DNA strand by joining together nucleotides already base-paired with an existing DNA strand

dopamine (DOPE-ah-mean) catecholamine neurotransmitter; precursor of epinephrine and norepinephrine

dopamine hypothesis theory that schizophrenia is due to increased activity in dopaminergic pathways

dorsal (DOR-sal) toward or at the back

dorsal root a group of afferent nerve fibers that enters the dorsal region of spinal cord

double bond two covalent chemical bonds formed between same two atoms; symbolized by =

double helix molecular conformation in which two strands of molecules are coiled around each other; conformation of DNA

down-regulation decrease in number of target-cell receptors for a given messenger in response to a chronic high concentration of that messenger

DPG *see* 2,3-diphosphoglyceraldehyde

dual innervation (in-ner-VAY-shun) innervation of an organ or gland by both sympathetic and parasympathetic nerve fibers

ductus deferens *see* vas deferens

duodenum (due-oh-DEE-num) first portion of small intestine (between stomach and jejunum)

dynamic equilibrium rates of forward and reverse components of a chemical reaction are equal, and no net change in reactant or product concentration occurs

eardrum *see* tympanic membrane

ECG *see* electrocardiogram

ectopic focus (ek-TOP-ik) region of heart other than SA node that assumes role of cardiac pacemaker

ectopic pregnancy implantation and development of a fetus at a site other than the uterus

edema (eh-DEE-mah) accumulation of excess fluid in interstitial space

EDV *see* end-diastolic volume

EEG *see* electroencephalogram

EEG arousal transformation of EEG pattern from alpha to beta rhythm during increased levels of attention

effector (ee-FECK-tor) cell or cell collection whose change in activity constitutes the response in a control system

effector macrophage macrophage whose killing ability has been enhanced by lymphokines, particularly gamma interferon

efferent (EF-er-ent) carrying away from

efferent arteriole renal vessel that conveys blood from glomerulus to peritubular capillaries

efferent neuron neuron that carries information away from CNS

efferent pathway component of reflex arc that transmits information from integrating center to effector

eicosanoid (eye-KOH-sah-noid) general term for products of arachadonic acid metabolism (prostaglandins, prostacyclin, thromboxanes, and leukotrienes); function as paracrines or autocrines

ejaculation (ee-jak-you-LAY-shun) discharge of semen from penis

ejaculatory duct (ee-JAK-you-lah-tory) continuation of vas deferens after it is joined by seminal vesicle duct; joins urethra in prostate gland

EKG *see* electrocardiogram

electric charge particle having excess positivity or negativity

electric force force that causes particles to move toward regions having an opposite charge and away from regions having a like charge

electric potential (or electric potential difference) *see* potential

electric signal graded potential or action potential

electric synapse (SIN-apse) synapse at which local currents resulting from electrical activity flow between two neurons through gap junctions joining them

electrocardiogram (ECG, EKG) (ee-lek-tro-CARD-ee-oh-gram) recording at skin surface of the electric currents generated by cardiac-muscle action potentials

electrochemical gradient force determining direction and magnitude of net charge movement; combination of electrical and chemical gradients

electrode (ee-LEK-trode) probe used to stimulate electrically, or record from, a tissue

electroencephalogram (EEG) (eh-lek-tro-en-SEF-ah-low-gram) brain electrical activity as recorded from scalp

electrogenic pump (elec-tro-JEN-ik) active-transport system that directly separates electric charge, thereby producing a potential difference

electrolyte (ee-LEK-tro-lite) substance that dissociates into ions when in aqueous solution

electromagnetic radiation radiation composed of waves with electric and magnetic components

electron (ee-LEK-tron) subatomic particle that carries one unit of negative charge

electrostatic pertaining to electric charges at rest

embryo (EM-bree-oh) organism during early stages of development; in human beings, the first 2 months of intrauterine life

emission (ee-MISH-un) movement of male genital duct contents into urethra prior to ejaculation

emotion *see* inner emotion, emotional behavior

emotional behavior outward expression and display of inner emotions

emphysema (em-fah-SEEM-ah) lung disease characterized by alveolar wall destruction and consequent impairment of gas exchange

emulsification (eh-mul-sih-fih-KAY-shun) fat-solubilizing process in which large lipid droplets are broken into smaller ones

end-diastolic volume (EDV) (die-ah-STAH-lik) amount of blood in ventricle just prior to systole

end-plate potential (EPP) depolarization of motor end plate of skeletal-muscle fiber in response to acetylcholine; initiates action potential in muscle plasma membrane

end-product inhibition inhibition of a metabolic pathway by final product's action upon allosteric site on an enzyme (usually the rate-limiting enzyme) in the pathway

end-systolic volume (ESV) (sis-TAH-lik) amount of blood remaining in ventricle after ejection

endocrine gland (EN-doe-krin) group of cells that secretes into the extracellular space hormones that then diffuse into bloodstream; also called a ductless gland

endocrine system all the body's hormone-secreting glands

endocytosis (en-doe-sigh-TOE-sis) process in which plasma membrane folds into the cell forming small pockets that pinch off to produce intracellular, membrane-bound vesicles; *see also* phagocytosis

endogenous pyrogen (EP) (en-DAHJ-en-us PIE-row-jen) one of a family of monokines (tumor necrosis factor, interleukin 1, and interleukin 6) that acts in the brain to cause fever

endometrium (en-dough-ME-tree-um) glandular epithelium lining uterine cavity

endoplasmic reticulum (en-doe-PLAS-mik ree-TIK-you-lum) cell organelle that consists of interconnected network of membrane-bound branched tubules and flattened sacs; two types are distinguished: *granular*, with ribosomes attached, and *agranular*, which is smooth-surfaced

endorphin (en-DOR-fin) any of the "endogenous-morphine" family of peptides; functions as neurotransmitter at synapses activated by opiate drugs and as paracrine and hormone

β-endorphin powerful opioid peptide secreted by anterior pituitary; synthesized as part of pro-opiomelanocortin and cosecreted with ACTH; *see also* endorphin

endothelium (en-doe-THEE-lee-um) thin layer of cells that lines heart cavities and blood vessels

endothelium-derived relaxing factors (EDRF) substances secreted by vascular endothelium that relax vascular smooth muscle and cause arteriolar dilation

energy ability to produce change; measured by amount of work performed during a given change

enkephalin (en-KEF-ah-lyn) peptide neurotransmitter at some synapses activated by opiate drugs; an endorphin

enteric nervous system (en-TAIR-ik) neural network residing in and innervating walls of gastrointestinal tract

enterogastrone (en-ter-oh-GAS-trone) one of a family of hormones released by intestinal tract in response to fat in lumen; inhibits stomach activity

enterohepatic circulation (en-ter-oh-hih-PAT-ik) reabsorption of bile salts (and other substances) from intestines, passage to liver (via hepatic portal vein), and secretion back to intestines (via bile)

enterokinase (en-ter-oh-KI-nase) an enzyme in luminal plasma membrane of intestinal epithelial cells; converts pancreatic trypsinogen to trypsin

entrainment (en-TRAIN-ment) adjusting biological rhythm to environmental cues

enzymatic activity rate at which an enzyme converts substrate to product

enzyme (EN-zime) protein catalyst that accelerates specific chemical reactions but does not itself undergo net change during the reaction

eosinophil (ee-oh-SIN-oh-fil) polymorphonuclear granulocytic leukocyte whose granules take up red dye eosin; involved in allergic responses and parasite destruction

epididymis (ep-ih-DID-eh-mus) portion of male reproductive duct system located between seminiferous tubules and vas deferens

epiglottis (ep-ig-GLOT-iss) thin cartilage flap that folds down, covering trachea, during swallowing

epilepsy high-amplitude, spike-and-wave neuronal activity that spreads to adjacent neural tissue; can give rise to convulsions

epinephrine (ep-ih-NEF-rin) hormone secreted by adrenal medulla and involved in regulation of organic metabolism; a catecholamine neurotransmitter; also called adrenaline

epiphysis (eh-PIF-ih-sis) end of long bone

epiphyseal closure (eh-PIF-ih-see-al) conversion of epiphyseal growth plate to bone

epiphyseal growth plate actively proliferating cartilage near bone ends; region of bone growth

epithelial cell (ep-ih-THEE-lee-al) cell at surface of body or hollow organ; specialized to secrete or absorb ions and organic molecules; with other epithelial cells, forms an *epithelium*

epithelial transport molecule movement from one extracellular compartment across epithelial cells into a second extracellular compartment

epithelium (ep-ih-THEE-lee-um) tissue that covers all body surfaces, lines all body cavities, and forms most glands

EPP *see* end-plate potential

EPSP *see* excitatory postsynaptic potential

equilibrium (ee-qua-LIB-rium) no net change occurs in a system; requires no energy

equilibrium potential voltage gradient across a membrane that is equal in force but opposite in direction to concentration force affecting a given ion species

erection stiffening of the penis or clitoris due to vascular congestion

error signal steady-state difference between level of regulated variable in a control system and set point for that variable

erythrocyte (eh-RITH-row-site) red blood cell

erythropoiesis (eh-rith-row-poy-EE-sis) erythrocyte formation

erythropoietin (eh-rith-row-POY-ih-tin) hormone secreted mainly by kidney cells; stimulates red-blood-cell production

esophagus portion of digestive tract that connects throat (pharynx) and stomach

essential amino acid amino acid that cannot be formed by the body at all (or at rate adequate to meet metabolic requirements) and must be obtained from diet

essential nutrient substance required for normal or optimal body function but synthesized by the body either not at all or in amounts inadequate to prevent disease

estradiol (es-tra-DIE-ol) steroid hormone of estrogen family; major ovarian estrogen

estriol (ES-tree-ol) steroid hormone of estrogen family; major estrogen secreted by placenta during pregnancy

estrogen group of steroid hormones that have effects similar to estradiol on female reproductive tract

ESV *see* end-systolic volume

euphorigen (you-FOR-ih-jen) drug that elevates mood

excitability ability to produce electric signals

excitation-contraction coupling mechanism in muscle fibers linking plasma-membrane depolarization with cross-bridge force generation

excitatory postsynaptic potential (EPSP) (post-sin-NAP-tic) depolarizing graded potential in postsynaptic neuron in response to activation of excitatory synaptic endings

excitatory synapse (SIN-apse) synapse that, when activated, increases likelihood that postsynaptic neuron will undergo action potentials or increases frequency of existing action potentials

excretion appearance of a substance in urine or feces

exocrine gland (EX-oh-krin) cluster of epithelial cells specialized for secretion and having ducts that lead to an epithelial surface

exocytosis (ex-oh-sigh-TOE-sis) intracellular vesicle fuses with plasma membrane, the vesicle opens, and its contents are liberated into the extracellular fluid

exon (EX-on) gene region containing code words for an amino acid sequence in a protein

expiration (ex-pur-A-shun) movement of air out of lungs

expiratory reserve volume (ex-PIE-ruh-tor-ee) volume of air that can be exhaled by maximal contraction of expiratory muscles after a normal expiration

extension straightening a joint

extensor muscle muscle whose activity straightens a joint

external anal sphincter ring of skeletal muscle around lower end of rectum

external environment environment surrounding external surfaces of an organism

external genitalia (jen-ih-TAY-lee-ah) (female) mons pubis, labia majora and minora, clitoris, vestibule of the vagina, and vestibular glands; (male) penis and scrotum

external work movement of external objects by skeletal-muscle contraction

extracellular fluid fluid outside cell; the interstitial fluid and plasma

extrafusal fiber *see* skeletomotor fiber

extrapyramidal system *see* multineuronal pathway

extrinsic (ex-TRIN-sik) coming from outside

extrinsic clotting pathway formation of fibrin clots by pathway using tissue factor from interstitium

facilitated diffusion (fah-SIL-ih-tay-ted) carrier-mediated transport system that moves molecules from high to low concentration across a membrane; energy not required

facilitation (fah-sil-ih-TAY-shun) general depolarization of a neuron when excitatory synaptic input exceeds inhibitory input

factor XII initial factor in sequence of reactions that results in blood clotting

factor XIII plasma enzyme that catalyzes cross-link formation between fibrin molecules to strengthen blood clot

farsighted vision defect that occurs because eyeball is too short for the lens so that near objects are focused behind the retina

fast fiber skeletal muscle fiber that contains myosin having high ATPase activity

fat mobilization increased breakdown of triacylglycerols and release of glycerol and fatty acids into blood

fat-soluble vitamin a vitamin that is soluble in nonpolar solvents and insoluble in water; vitamin A, D, E, or K

fatty acid carbon chain with carboxyl group at one end through which chain can be linked to glycerol to form triacylglycerol; *see also* polyunsaturated fatty acid, saturated fatty acid, unsaturated fatty acid

Fc portion "stem" part of antibody

feces (FEE-sees) material expelled from large intestine during defecation

feedback characteristic of control systems in which output response influences input to system; *see also* negative feedback, positive feedback

feedforward aspect of some control systems that allows system to anticipate changes in a regulated variable

ferritin (FAIR-ih-tin) iron-binding protein that stores iron in body

fertilization union of sperm and ovum

fetus (FEE-tus) period of human development from second month of intrauterine life until birth

fever increased body temperature due to setting of "thermostat" of temperature-regulating mechanisms at higher-than-normal level

fibrillation (fib-rih-LAY-shun) rapid, unsynchronized cardiac-muscle contractions that prevent effective pumping of blood

fibrin (FYE-brin) protein polymer resulting from enzymatic cleavage of fibrinogen; can turn blood into gel (clot)

fibrinogen (fye-BRIN-oh-jen) plasma protein precursor of fibrin

fibrinolytic system (fye-brin-oh-LIT-ik) plasma-enzyme cascade that breaks down clots; also called thrombolytic system

fight-or-flight response activation of sympathetic nervous system during stress

filtered load amount of any substance filtered from renal glomerular capillaries into Bowman's capsule

filtration movement of essentially protein-free plasma out across capillary walls due to a pressure gradient across the wall

first messenger extracellular chemical messenger or electric message that arrives at a cell's plasma membrane

first-order neuron *see* primary afferent

flatus (FLAY-tus) intestinal gas or gas expelled through anus

flexion (FLEK-shun) bending a joint

flexion reflex bending of those joints that withdraw an injured part away from a painful stimulus

flexor muscle muscle whose activity bends a joint

fluid mosaic model (mo-ZAY-ik) cell membranes consist of proteins embedded in bimolecular lipid that has the physical properties of a fluid, allowing membrane proteins to move laterally within it

flux movement of a substance across a surface in a unit of time; *see also* net flux

folic acid (FOE-lik) vitamin of B-complex group; essential for nucleotide formation

follicle *see* ovarian follicle, primary follicle

follicle-stimulating hormone (FSH) protein hormone secreted by anterior pituitary in males and females; a gonadotropin

follicular phase (foe-LIK-you-lar) that portion of menstrual cycle during which a follicle and ovum develop to maturity prior to ovulation

forced vital capacity (FVC) maximal expiration as fast as possible following maximal inspiration

forebrain large, anterior brain subdivision consisting of right and left cerebral hemispheres (the cerebrum) and diencephalon

fovea (FO-vee-ah) area near center of retina where cones are most concentrated; gives rise to most acute vision

free-running rhythm cyclical activity driven by biological clock in absence of environmental cues

frequency number of times an event occurs per unit time

frontal lobe region of anterior cerebral cortex where motor cortex, Broca's speech center, and some association cortex are located

fructose (FRUK-tose) five-carbon sugar; present in sucrose (table sugar)

FSH *see* follicle-stimulating hormone

functional residual capacity lung volume after relaxed expiration

functional site binding site on allosteric protein, which, when activated, carries out protein's physiological function

functional unit organ subunit; all subunits of a given organ have similar structural and functional properties

fused-vesicle channel endocytotic or exocytotic vesicles that have fused to form a continuous water-filled channel through capillary endothelial cell

G protein plasma-membrane regulatory protein that responds to an activated receptor and, in turn, interacts with membrane ion channels or enzymes; for the G proteins that regulate adenylate cyclase, subscript "i" (G_i *protein*) denotes inhibitory action, and "s" (G_s *protein*) denotes stimulatory action.

GABA *see* gamma-aminobutyric acid

galactose (gah-LAK-tose) six-carbon monosaccharide; present in lactose (milk sugar)

gallbladder small sac under the liver; concentrates bile and stores it between meals; contraction of gallbladder ejects bile into small intestine

gallstone precipitate of cholesterol (and occasionally other substances) in gallbladder or common bile duct

gamete (GAM-eat) germ cell or reproductive cell; sperm in male and ovum in female

gametogenesis (gam-ee-toe-JEN-ih-sis) gamete production

gamma-aminobutyric acid (GABA) major inhibitory neurotransmitter in CNS

gamma globulin immunoglobulin G (IgG), most abundant class of plasma antibodies

gamma interferon *see* interferon

gamma motor neuron small motor neuron that controls spindle (intrafusal) muscle fibers

ganglion (GANG-glee-on) (pl. ganglia); generally reserved for cluster of neuron cell bodies outside CNS

ganglion cell retinal neuron that is postsynaptic to bipolar cells; axons of ganglion cells form optic nerve

gap junction intercellular junction that allows ions and small molecules to flow between cytoplasms of adjacent cells; the two adjacent plasma membranes are joined by small tubes

gastric (GAS-trik) pertaining to the stomach

gastric phase (of gastrointestinal control) initiation of neural and hormonal gastrointestinal reflexes by stimulation of stomach wall

gastrin (GAS-trin) peptide hormone secreted by antral region of stomach; stimulates gastric acid secretion

gastroileal reflex (gas-tro-ILL-ee-al) reflex increase in contractions of ileum during gastric emptying

gastrointestinal system (gas-tro-in-TES-tin-al) gastrointestinal tract plus salivary glands, liver, gallbladder, and pancreas

gastrointestinal tract mouth, pharynx, esophagus, stomach, and small and large intestines

GDP guanosine diphosphate

gene unit of hereditary information; portion of DNA containing information required to determine a protein's amino acid sequence

genetic code sequence of nucleotides in gene, three nucleotides indicating the location of one amino acid in protein specified by that gene

germ cell cell that gives rise to male or female gametes (the sperm and ova)

GFR *see* glomerular filtration rate

GH *see* growth hormone

GI gastrointestinal

gland group of epithelial cells specialized for secretion; *see also* endocrine gland, exocrine gland

glial cell (GLEE-al) nonneuronal cell in CNS; helps regulate extracellular environment of CNS; also called neuroglial cell

globin (GLOW-bin) polypeptide chains of hemoglobin molecule

globulin (GLOB-you-lin) one of a family of proteins found in blood plasma

glomerular filtration (glow-MER-you-lar) movement of an essentially protein-free plasma from renal glomerular capillaries into Bowman's capsule

glomerular filtration rate (GFR) volume of fluid filtered from renal glomerular capillaries into Bowman's capsule per unit time

glomerulus (glow-MER-you-lus) structure at beginning of kidney nephron consisting of vascular component and Bowman's capsule

glottis opening between vocal cords and surrounding area through which air passes

glucagon (GLUE-kah-gahn) peptide hormone secreted by A cells of pancreatic islets of Langerhans; leads to rise in plasma glucose

glucocorticoid (glue-koh-KORT-ih-coid) steroid hormone produced by adrenal cortex and having major effects on glucose metabolism

gluconeogenesis (glue-koh-nee-o-JEN-ih-sis) formation of glucose by the liver from pyruvate, lactate, glycerol, or amino acids

glucose major monosaccharide (carbohy-drate) in the body; a six-carbon sugar, $C_6H_{12}O_6$; also called blood sugar

glucose-counterregulatory control neural or hormonal factor that opposes insulin's actions; glucagon, epinephrine and sympathetic nerves to liver and adipose tissue, cortisol, and growth hormone

glucose insulinotropic peptide (GIP) intestinal hormone formerly called gastric inhibitory peptide; stimulates insulin secretion in response to glucose and fat in small intestine

glucose-6-phosphate (FOS-fate) first intermediate in glycolytic pathway

glucose sparing switch from glucose to fat utilization by most cells during postabsorptive state

glutamate (GLU-tah-mate) formed from the amino acid glutamic acid; a major excitatory CNS neurotransmitter

glycerol (GLIS-er-ol) three-carbon carbohydrate; forms backbone of triacylglycerol

glycerophosphate (glis-er-o-FOS-fate) three-carbon molecule; combines with free fatty acids in triacylglycerol synthesis

glycine (GLY-seen) an amino acid; a neurotransmitter at some inhibitory synapses in CNS

glycogen (GLY-koh-jen) highly branched polysaccharide composed of glucose subunits; major carbohydrate storage form in body

glycogenolysis (gly-koh-jen-NOL-ih-sis) glycogen breakdown to glucose

glycolysis (gly-KAHL-ih-sis) metabolic pathway that breaks down glucose to two molecules of pyruvic acid (aerobically) or two molecules of lactic acid (anaerobically)

glycolytic fiber skeletal muscle fiber that has a high concentration of glycolytic enzymes and large glycogen stores

glycoprotein protein containing covalently linked carbohydrate groups

GnRH *see* gonadotropin-releasing hormone

Golgi apparatus (GOAL-gee) cell organelle consisting of membranes and vesicles, usually near nucleus; processes newly synthesized proteins for secretion and distribution to other organelles

Golgi tendon organ tension-sensitive mechanoreceptor ending of afferent nerve fiber; wrapped around collagen bundles in tendon

gonad (GO-nad) gamete-producing reproductive organ, that is, testes in male and ovaries in female

gonadotropic hormone (go-nad-oh-TROH-pik) hormone secreted by anterior pituitary; controls gonadal function; FSH or LH

gonadotropin-releasing hormone (GnRH) hypothalamic hormone that controls LH and FSH secretion by anterior pituitary in males and females

graded potential membrane potential change of variable amplitude and duration that is conducted decrementally; has no threshold or refractory period

gradient (GRAY-dee-ent) continuous increase or decrease of a variable over distance

gram atomic mass amount of element in grams equal to the numerical value of its atomic mass

granular cell cell in renal afferent arteriole wall that secretes renin; part of juxtaglomerular apparatus

granuloma (gran-you-LOW-mah) mass of layered phagocytic-type cells, the central one of which contains a microbe or other potentially harmful substance

granulosa cell (gran-you-LOW-sah) cell that surrounds ovum and antrum in ovarian follicle; secretes estrogen and inhibin

gray matter area of brain and spinal cord that appears gray in unstained specimens and consists mainly of cell bodies and unmyelinated portions of nerve fibers

ground substance material surrounding cells and fibers in connective tissue

growth cone specialized enlargement at tip of growing neuron process

growth factor one of a group of peptides that is highly effective in stimulating mitosis and/or differentiation of certain cell types

growth hormone (GH) peptide hormone secreted by anterior pituitary; stimulates somatomedin release; enhances body growth by acting on carbohydrate and protein metabolism

growth hormone release-inhibiting hormone *see* somatostatin

growth hormone releasing hormone (GRH) hypothalamic hormone that stimulates growth hormone secretion by anterior pituitary

growth-inhibiting factor one of a group of peptides that modulates growth by inhibiting mitosis in specific tissues

GTP *see* guanosine triphosphate

guanine (G) (GWA-neen) purine base in DNA and RNA

guanosine triphosphate (GTP) (GWA-no-seen tri-FOS-fate) energy-transporting molecule similar to ATP except that it contains the base guanine rather than adenine

H zone one of transverse bands making up striated pattern of cardiac and skeletal muscle; light region that bisects A band

habituation (hah-bit-you-A-shun) reversible decrease in response strength upon repeatedly administered stimulation

hair cell mechanoreceptor in organ of Corti and vestibular apparatus

Hb *see* reduced hemoglobin

HbO$_2$ *see* oxyhemoglobin

H$_2$CO$_3$ carbonic acid

HDL *see* high-density lipoprotein

heart attack damage to or death of cardiac muscle; also called myocardial infarction

heartburn pain that seems to occur in region of heart but is due to pain receptors in esophageal wall stimulated by acid refluxed from stomach

heart murmur heart sound caused by turbulent blood flow through narrowed or leaky valves or through hole in interventricular or interatrial septum

heart rate number of heart contractions per minute

heart sound noise that results from vibrations due to closure of atrioventricular valves (first heart sound) or pulmonary and aortic valves (second heart sound)

heat exhaustion state of collapse due to hypotension because of plasma volume depletion, secondary to sweating, and extreme skin blood-vessel dilation; thermoregulatory centers still function

heat stroke positive-feedback situation in which heat gain exceeds heat loss, which causes body temperature to become so high that brain thermoregulatory centers do not function, allowing temperature to rise even higher

Heimlich maneuver (HIME-lik) forceful elevation of diaphragm produced by rescuer's fist against choking victim's abdomen; causes sudden sharp increase in alveolar pressure to expel obstructing material that is causing choking

helper T cell T cell that enhances antibody production and cytotoxic T- and NK-cell function

hematocrit (he-MAT-oh-krit) percentage of blood volume occupied by erythrocytes

heme iron-containing organic molecule bound to each of the four polypeptide chains of hemoglobin or to cytochromes

hemoglobin (HE-mo-glo-bin) protein composed of four polypeptide chains, heme, and iron; located in erythrocytes and transports most blood oxygen

hemoglobin saturation percent of oxygen binding sites in hemoglobin combined with oxygen

hemophilia (he-moh-FILL-ee-ah) hereditary disorder in which a clotting factor is absent, resulting in excessive bleeding

hemorrhage (HEM-or-age) bleeding

hemostasis (he-mo-STAY-sis) stopping blood loss from a damaged vessel

Henle's loop *see* loop of Henle

heparin (HEP-ah-rin) anticlotting agent found on endothelial-cell surfaces; binds antithrombin III; an anticoagulant drug

hepatic (hih-PAT-ik) pertaining to the liver

hepatic portal vein vein that conveys blood from capillaries in the intestines and portions of the stomach and pancreas to capillaries in the liver

hertz (Hz) cycles per second; measure used for wave frequencies

high-density lipoprotein plasma lipid-protein aggregate having low proportion of lipid; promotes removal of cholesterol from cells

hippocampus (hip-oh-KAM-pus) portion of limbic system associated with learning and emotions

histamine (HISS-tah-mean) inflammatory chemical messenger secreted mainly by mast cells; monoamine neurotransmitter

histotoxic hypoxia (hiss-toe-TOKS-ik hi-POK-see-ah) hypoxia in which cell cannot utilize oxygen because a toxic agent has interfered with its metabolic machinery

homeostasis (home-ee-oh-STAY-sis) relatively stable condition of extracellular fluid that results from regulatory system actions

homeostatic control system (home-ee-oh-STAT-ik) control system that consists of interconnected components and keeps a physical or chemical parameter of internal environment relatively constant

homeothermic (home-ee-oh-THERM-ik) capable of maintaining body temperature within very narrow limits

homologous (ho-MAHL-oh-gus) corresponding in origin, structure, and position

hormone chemical messenger synthesized by specific endocrine gland in response to certain stimuli and secreted into the blood, which carries it to target cells

HPO$_4^{2-}$ phosphate

HR *see* heart rate

humoral immune response immune response mediated by antibodies; major protection against bacteria and viruses in extracellular fluid

hydrochloric acid (hi-dro-CLOR-ik) HCl; strong acid secreted into stomach lumen by parietal cells

hydrogen bond weak chemical bond between two molecules or parts of the same molecule, in which negative region of one polarized substance is electrostatically attracted to a positively polarized hydrogen atom in the other

hydrogen ion (EYE-on) H$^+$; single proton; H$^+$ concentration of a solution determines its acidity

hydrogen peroxide H$_2$O$_2$; chemical produced by phagosome and highly destructive to the macromolecules within it

hydrolysis (hi-DRAHL-ih-sis) breaking of chemical bond with addition of elements of water (—H and —OH) to the products formed; also called a hydrolytic reaction

hydrophilic (hi-dro-FILL-ik) attracted to, and easily dissolved in, water

hydrophobic (hi-dro-FOE-bik) not attracted to, and insoluble in, water

hydrostatic pressure (hi-dro-STAT-ik) pressure exerted by fluid

hydroxyl group (hi-DROX-il) —OH

hyper- too much

hypercalcemia increased plasma calcium

hyperemia (hi-per-EE-me-ah) increased blood flow; *see also* active hyperemia

hyperglycemia (hi-per-glye-SEE-me-ah) plasma glucose concentration increased above normal levels

hyperosmotic (hi-per-oz-MAH-tik) having total solute concentration greater than plasma

hyperpnea (hi-PERP-nee-ah) increased ventilation in exact proportion to increased oxygen consumption and carbon dioxide production

hyperpolarize to change membrane potential so cell interior becomes more negative than its resting state

hypersensitivity *see* allergy

hypertension chronically increased arterial blood pressure

hyperthermia increased body temperature regardless of cause

hypertonia (hi-per-TOE-nee-ah) abnormally high muscle tone

hypertonic (hi-per-TAH-nik) containing a higher concentration of effectively membrane-impermeable solute particles than cells contain

hypertrophy (hi-PER-tro-fee) enlargement of a tissue or organ due to increased cell size rather than increased cell number

hyperventilation increased ventilation without similar increase in O$_2$ consumption or CO$_2$ production

hypo- too little

hypocalcemic tetany (hi-po-kal-SEE-mik TET-ah-nee) skeletal-muscle spasms due to a low extracellular calcium concentration

hypoglycemia (hi-po-glye-SEE-me-ah) low blood sugar (glucose) concentration

hypoosmotic (hi-po-oz-MAH-tik) having total solute concentration less than that of plasma

hypotension low blood pressure

hypothalamic releasing hormone (hi-po-thah-LAM-ik) hormone released from hypothalamic neurons into hypothalamo-pituitary portal vessels to control release of anterior pituitary hormone

hypothalamus (hi-po-THAL-ah-mus) brain region below thalamus; responsible for integration of many basic behavioral patterns involving correlation of neural and endocrine functions, especially those concerned with regulation of internal environment

hypotonia (hi-po-TOE-nee-ah) abnormally low muscle tone

hypotonic (hi-po-TAH-nik) containing a lower concentration of effectively membrane-impermeable solute particles than cells contain

hypoventilation decrease in ventilation without similar decrease in O$_2$ consumption or CO$_2$ production

hypoxia (hi-POK-see-ah) deficiency of oxygen at tissue level; *see also* hypoxic hypoxia, anemic hypoxia, ischemic hypoxia, histotoxic hypoxia

hypoxic hypoxia (hi-POK-sik hi-POK-see-ah) hypoxia due to decreased arterial P_{O_2}; also called hypoxemia

Hz *see* hertz

I band one of transverse bands making up repeating striations of cardiac and skeletal muscle; located between A bands of adjacent sarcomeres and bisected by Z line

IgA class of antibodies secreted by, and acting locally in, lining of gastrointestinal, respiratory, and urinary tracts

IgD class of antibodies whose function is unknown

IgE class of antibodies that mediates immediate hypersensitivity (allergy) and resistance to parasites

IGF-1 *see* insulin-like growth factor

IgG gamma globulin; most abundant class of antibodies

IgM class of antibodies that, along with IgG, provides major specific humoral immunity against bacteria and viruses

IL *see* interleukin

ileocecal sphincter (ill-ee-oh-SEE-kal) ring of smooth muscle separating small and large intestines, that is, ileum and cecum

ileum (ILL-ee-um) final, longest segment of small intestine

immune response *see* nonspecific immune response, specific immune response

immune surveillance (sir-VAY-lence) recognition and destruction of cancer cells that arise in body

immunity physiological mechanisms that allow body to recognize materials as foreign or abnormal and to neutralize or eliminate them; *see also* active immunity, passive immunity

immunodeficiency (im-mun-oh-dee-FISH-en-see) absence of B or T cells or, in the case of *combined immunodeficiency,* both

immunoglobulin (**Ig**) (im-mun-oh-GLOB-you-lin) synonym for antibody; five classes are IgG, IgA, IgD, IgM, and IgE

implantation (im-plan-TAY-shun) event during which fertilized ovum becomes embedded in uterine wall

inferior vena cava (VEE-nah CAVE-ah) large vein that carries blood from lower half of body to right atrium of heart

inflammation (in-flah-MAY-shun) local response to injury or infection characterized by local swelling, pain, heat, and redness

inhibin (in-HIB-in) protein hormone secreted by seminiferous-tubule Sertoli cells and ovarian follicles; inhibits FSH secretion

inhibitory postsynaptic potential (**IPSP**) hyperpolarizing graded potential that arises in postsynaptic neuron in response

to activation of inhibitory synaptic endings upon it

inhibitory synapse (SIN-apse) synapse that, when activated, decreases likelihood that postsynaptic neuron will fire an action potential (or decreases frequency of existing action potentials)

initial segment first portion of axon plus part of cell body where axon joins

inner ear cochlea; contains organ of Corti

inner emotion emotional feelings that are entirely within a person

innervate to supply with nerves

inorganic pertaining to substances that do not contain carbon; *compare* organic

inorganic phosphate (FOS-fate) $H_2PO_4^-$ or HPO_4^{2-}; phosphate that is not linked to an organic molecule

inositol trisphosphate (IP_3) (in-OS-ih-tol tris-FOS-fate) second messenger that causes calcium release from endoplasmic reticulum into cytosol

insensible water loss water loss of which a person is unaware, that is, loss by evaporation from skin and respiratory passages

inspiration air movement from atmosphere into lungs

inspiratory muscles those skeletal muscles whose contraction contributes to inspiration

inspiratory neuron neuron in brainstem medulla that fires in synchrony with inspiration and ceases firing during expiration

inspiratory reserve volume maximal air volume that can be inspired above resting tidal volume

insulin (IN-suh-lin) peptide hormone secreted by B cells of pancreatic islets of Langerhans; has metabolic and growth-promoting effects; stimulates glucose and amino-acid uptake by most cells and stimulates protein, fat, and glycogen synthesis

insulin-like growth factor one of a group of peptides that have growth-promoting effects

insulin-like growth factor 1 (**IGF-1**) insulin-like growth factor that mediates mitosis-stimulating effect of growth hormone on bone and possibly other tissues; also known as somatomedin C

insulin resistance hyporesponsiveness of insulin's target cells to circulating insulin due to altered insulin receptors or intracellular processes

integral protein protein that is embedded in membrane lipid layer; may span entire membrane or be located at only one surface

integrating center cells that receive one or more signals, compare them to a physiological set point, and send out appropriate response; also called an *integrator*

intercalated disk (in-TER-kah-lay-ted) thickened end of cardiac-muscle cell where cell is connected to adjacent cell

intercellular cleft a narrow water-filled space between capillary endothelial cells

intercostal muscle (in-ter-KOS-tal) skeletal muscle that lies between ribs and whose contraction causes rib cage movement during breathing

interferon (in-ter-FEAR-on) class of proteins that nonspecifically inhibit viral replication; *gamma interferon* also stimulates killing ability of cytotoxic T cells, NK cells, and macrophages

interleukin 1 (**IL-1**) (in-ter-LEW-kin) protein secreted by macrophages and other cells that activates helper T cells, exerts many inflammatory effects, and mediates many of the systemic acute phase responses, including fever

interleukin 2 (**IL-2**) protein secreted by activated helper T cells that causes activated helper T, cytotoxic, and NK cells to proliferate

interleukin 6 protein secreted by macrophages and other cells; mediates many of the acute phase responses

internal anal sphincter smooth-muscle ring around lower end of rectum

internal environment extracellular fluid (interstitial fluid and plasma)

internal genitalia (jen-ih-TAY-lee-ah) (female) ovaries, uterine tubes, uterus, and vagina

internal urethral sphincter (you-REETH-ral) part of smooth muscle of urinary bladder wall that opens and closes the bladder outlet

internal work energy-requiring activities in body; *see also* work

interneuron neuron whose cell body and axon lie entirely in CNS

interphase period of cell-division cycle between end of one division and beginning of next

interstitial cell (in-ter-STISH-al) (of testis) *see* Leydig cell

interstitial-cell stimulating hormone (**ICSH**) *see* luteinizing hormone

interstitial fluid extracellular fluid surrounding tissue cells; excludes plasma, which is extracellular fluid surrounding blood cells

interstitium (in-ter-STISH-ium) interstitial space; fluid-filled space between tissue cells

interventricular septum (in-ter-ven-TRIK-you-lar) partition in heart separating right and left ventricles

intestinal phase (of gastrointestinal control) initiation of neural and hormonal gastrointestinal reflexes by stimulation of intestinal-tract walls

intestino-intestinal reflex reflex cessation of contractile activity of intestines in response to various stimuli in intestine

intracellular fluid fluid in cells; cytosol plus fluid in cell organelles, including nucleus

intrafusal fiber *see* spindle fiber

intrapleural fluid (in-tra-PLUR-al) thin fluid film in thoracic cavity between pleura lining the inner wall of thoracic cage and pleura covering lungs

intrapleural pressure pressure in pleural space; also called intrathoracic pressure

intrarenal baroreceptor pressure-sensitive granular cell of afferent arteriole, which responds to decreased renal arterial pressure by secreting renin

intrathoracic pressure *see* intrapleural pressure

intrinsic (in-TRIN-sik) situated entirely within a part

intrinsic clotting pathway intravascular sequence of fibrin formation initiated by factor XII

intrinsic factor glycoprotein secreted by parietal cells in stomach and necessary for absorption of vitamin B_{12} in the ileum

intron (IN-trahn) region of noncoding base sequences in gene

inversely proportional relationship in which, as one factor increases by a given amount, the other decreases by a proportional amount

ion (EYE-on) atom or small molecule containing unequal number of electrons and protons and, therefore, carrying a net positive or negative electric charge

ionic bond (eye-ON-ik) strong electrical attraction between two oppositely charged ions

ionization (eye-on-ih-ZAY-shun) process of removing electrons from or adding them to an atom or small molecule to form an ion

IP$_3$ *see* inositol trisphosphate

ipsilateral (ip-sih-LAT-er-al) on the same side of the body

IPSP *see* inhibitory postsynaptic potential

iris ringlike structure surrounding pupil of eye

irreversible reaction chemical reaction that releases large quantities of energy and results in almost all the reactant molecules being converted to product; *compare* reversible reaction

ischemia (iss-KEY-me-ah) reduced blood supply

ischemic hypoxia (iss-KEY-mik hi-POK-see-ah) too little blood flow to tissues to deliver adequate amounts of oxygen

islet of Langerhans (EYE-let of LAN-ger-hans) cluster of pancreatic endocrine cells; different islet cells secrete insulin, glucagon, somatostatin, and pancreatic polypeptide

isometric contraction (ice-so-MET-rik) contraction of muscle under conditions in which it develops tension but does not change length

isosmotic (ice-oz-MAH-tik) having the same osmotic pressure, thus the same solute-particle concentration

isotonic (ice-oh-TAH-nik) containing the same number of effectively nonpenetrating solute particles as cells; *see also* isotonic contraction

isotonic contraction contraction of a muscle under conditions in which load on the muscle remains constant but muscle shortens

isovolumetric ventricular contraction (iso-vol-you-MET-rik) early phase of systole when atrioventricular and aortic valves are closed and ventricular volume does not change

isovolumetric ventricular relaxation early phase of diastole when atrioventricular and aortic valves are closed

isozyme (EYE-so-zime) one of two or more enzymes that catalyze the same reaction but have slightly different structures and therefore slightly different affinities for substrate

jaundice (JAWN-dis) condition in which the skin is yellowish due to bilirubin buildup as a result of the failure of bilirubin excretion into the bile by the liver

jejunum (je-JEW-num) segment of small intestine between duodenum and ileum

juxtaglomerular apparatus (**JGA**) (jux-tah-glow-MER-you-lar) renal structure consisting of macula densa and granular cells; site of renin secretion and sensor for renal homeostatic responses

k_p *see* membrane permeability constant

kcal *see* kilocalorie

KCl potassium chloride

keto acid (KEY-toe) molecule formed by removal of amino group from amino acid; contains carbonyl (—CO—) and carboxyl (—COOH) group

ketone (KEY-tone) product of fatty acid metabolism that accumulates in blood during starvation and in untreated diabetes mellitus; acetoacetic acid, acetone, or β-hydroxybutyric acid; also called ketone body

kilocalorie (**kcal**) (KILL-oh-kal-ah-ree) amount of heat required to heat 1 L water 1 C°; calorie used in nutrition; also called Calorie and large calorie

kinesthesia (kin-ess-THEE-zee-ah) sense of movement derived from joint position and movement

kininogen (ki-NIN-oh-jen) protein from which the kinins are generated

kinin (KI-nin) peptide split from kininogen in inflamed area; facilitates vascular changes associated with inflammation; may also stimulate pain receptors

Krebs cycle mitochondrial metabolic pathway that utilizes fragments derived from carbohydrate, protein, and fat breakdown and produces carbon dioxide, hydrogen, and small amounts of ATP; also called *tricarboxylic acid cycle* or *citric acid cycle*

L-dopa dopamine precursor; administered to patients with parkinsonism

lactase (LAK-tase) small-intestine enzyme that breaks down lactose (milk sugar) into glucose and galactose

lactate *see* lactic acid

lactation (lak-TAY-shun) production and secretion of milk by mammary glands

lacteal (lak-TEAL) blind-ended lymph vessel in center of each intestinal villus

lactic acid (LAK-tic) three-carbon molecule formed by glycolytic pathway in absence of oxygen; dissociates to form lactate and hydrogen ions

lactose (LAK-tose) disaccharide composed of glucose and galactose; also called milk sugar

lactose intolerance inability to digest lactose because of lack of intestinal lactase; leads to accumulation of large amounts of gas and fluid in large intestine, which causes pain and diarrhea

larynx (LAIR-inks) part of air passageway between pharynx and trachea; contains the vocal cords

latent period (LAY-tent) period lasting several milliseconds between action-potential initiation in a muscle fiber and beginning of mechanical activity

lateral position farther from the midline

lateral inhibition method of refining sensory information in afferent neurons and ascending pathways whereby the fibers inhibit each other, the most active fibers causing greatest inhibition of adjacent fibers

lateral sac enlarged region at end of each sarcoplasmic reticulum segment; adjacent to transverse tubule

law of mass action maxim that an increase in reactant concentration causes a chemical reaction to proceed in direction of product formation; the opposite occurs with decreased reactant concentration

LDL *see* low-density lipoprotein

lecithin (LESS-ih-thin) a phospholipid

lengthening contraction contraction as a load pulls a muscle to a longer length despite opposing forces generated by the active cross bridges

lens adjustable part of eye's optical system, which helps focus object's image on retina

leukocyte (LEW-ko-site) white blood cell

leukotriene (LOO-koe-treen) type of eicosanoid formed by action of lipoxygenase

Leydig cell (LIE-dig) testosterone-secreting endocrine cell that lies between seminiferous tubules of testes; also called interstitial cell

LH *see* luteinizing hormone

LH surge large rise in plasma luteinizing-hormone concentration due to increased secretion by anterior pituitary about day 13 of menstrual cycle

lidocaine (LIE-doh-kain) local anesthetic that

works by preventing increase in sodium permeability required for action-potential generation

ligand (LIE-gand) any molecule or ion that binds to protein surface by noncovalent bonds

ligand-operated channel *see* receptor-operated channel

ligase (LIE-gase) enzyme that forms recombinant DNA by linking ends of two DNA fragments previously split by a restriction enzyme

limbic system (LIM-bik) interconnected brain structures in cerebrum; involved with emotions and learning

lipase (LIE-pase) *see* pancreatic lipase, lipoprotein lipase

lipid (LIP-id) molecule composed primarily of carbon and hydrogen and characterized by insolubility in water

lipid bilayer a sheet consisting of two layers of phospholipid

lipolysis (lie-POL-ih-sis) triacylglycerol breakdown

lipoprotein (lip-oh-PRO-teen) lipid aggregate that is partially coated by protein; involved in lipid transport in blood

lipoprotein lipase capillary endothelial enzyme that hydrolyzes triacylglycerol to glycerol and fatty acids

β-lipotropin (bay-tah lip-oh-TRO-pin) endorphin peptide; synthesized as part of pro-opiomelanocortin by anterior pituitary and cosecreted with ACTH; upon cleavage gives rise to beta-endorphin

lipoxygenase (lie-POX-ih-jen-ase) enzyme that acts on arachidonic acid and leads to leukotriene formation

load force acting on a muscle due to weight of an object

local current flow movement of positive ions toward more negative membrane region, and simultaneous movement of negative ions in opposite direction

local homeostatic response (home-ee-oh-STAH-tik) response acting in immediate vicinity of a stimulus, without nerves or hormones, and having net effect of counteracting stimulus

local potential small, graded potential difference between two points that is conducted decrementally and has no threshold or refractory period

local spinal reflex reflex whose afferents, efferents, and integrating center are located in only a few spinal cord segments

locus ceruleus (sih-RUE-lee-us) brainstem nucleus that projects to many brain parts and is implicated in directed attention

long-loop negative feedback inhibition of anterior pituitary and/or hypothalamus by hormone secreted by another endocrine gland

long-term memory memory that has a relatively long lifespan

loop of Henle (HEN-lee) hairpin-like segment of kidney tubule; situated between proximal and distal tubules

low-density lipoprotein (lip-oh-PRO-teen) protein-lipid aggregate that is major carrier of plasma cholesterol to cells

lower esophageal sphincter smooth muscle of lowest 4 cm of esophagus; can act as sphincter, closing off esophageal opening into the stomach

lumen (LOO-men) space in hollow tube or organ

luminal membrane (LOO-min-al) portion of plasma membrane facing the lumen; also called apical or mucosal membrane

lung compliance (come-PLY-ance) change in lung volume caused by given change in transpulmonary pressure; the greater the lung compliance, the more stretchable the lung wall

luteal phase (LOO-tee-al) last half of menstrual cycle; corpus luteum is active ovarian structure

luteinizing hormone (LH) (LOO-tea-en-izing HOR-mone) peptide gonadotropin hormone secreted by anterior pituitary; increase in female at mid menstrual cycle initiates ovulation; also called *interstitial cell-stimulating hormone* (ICSH) in male

lymph (limf) fluid in lymphatic vessels

lymph node small organ, containing lymphocytes, located along lymph vessel; site of lymphocyte formation and storage and immune reactions

lymphatic system (lim-FAT-ik) network of vessels that conveys lymph from tissues to blood, and lymph nodes along these vessels

lymphocyte (LIMF-oh-site) type of leukocyte that is responsible for specific immune defenses; mainly B cells and T cells

lymphocyte activation mitosis and differentiation of lymphocytes

lymphoid organ (LIMF-oid) lymph node, spleen, thymus, tonsil, or aggregate of lymphoid follicles; *see also* primary lymphoid organ, peripheral lymphoid organ; *compare* lymphatic system

lymphokine (LIMF-oh-kine) protein chemical messengers secreted by lymphocytes

lysosome (LIE-so-zome) membrane-bound cell organelle containing digestive enzymes that break down bacteria and large molecules that have entered cell and damaged components of cell

M line one of transverse bands making up repeating striations of cardiac and skeletal muscle; thin dark band in middle of sarcomere

MAC *see* membrane attack complex

macrophage (MAK-ro-faje) cell that phagocytizes, processes and presents antigen to lymphocytes, and secretes monokines involved in inflammation, activation of lymphocytes, and systemic acute phase response to infection or injury; *see also* effector macrophage

macula densa (MAK-you-lah DEN-sah) portion of renal tubule where loop of Henle joins distal tubule; component of juxtaglomerular apparatus

major histocompatibility complex (MHC) group of genes that code for *major histocompatibility complex proteins*, which are important for immune function

malignant (mah-LIG-nant) tending to become worse and result in death; opposite of benign

mammary gland milk-secreting gland in breast; also used synonomously with breast

mass movement contraction of large segments of colon; propels fecal material into rectum

mast cell cell drived from basophil; does not circulate in blood; releases histamine and other chemicals involved in inflammation

maximal tubular capacity (T_m) maximal rate of mediated transport of substance across renal tubule wall

mean arterial pressure (MAP) average blood pressure during cardiac cycle; approximately diastolic pressure plus one-third pulse pressure

mechanoreceptor (meh-CAN-oh-re-sep-tor) sensory receptor that responds preferentially to mechanical stimuli such as bending, twisting, or compressing

medial forebrain bundle neural fibers lateral to hypothalamus that pass between forebrain and brainstem; contains fibers from locus ceruleus and influences widespread brain areas

median eminence (EM-ih-nence) region at base of hypothalamus containing capillary tufts into which hypothalamic releasing and inhibiting hormones are secreted from neuron terminals

mediate (MEE-dee-ate) bring about

mediated transport movement of molecules across membrane by binding to protein carrier; characterized by specificity, competition, and saturation; includes facilitated diffusion and active transport

medulla (meh-DULL-ah or meh-DUEL-ah) innermost portion of an organ; *compare* cortex

medullary cardiovascular center *see* cardiovascular center

medullary inspiratory neuron *see* inspiratory neuron

megakaryocyte (meg-ah-CARE-ee-oh-site) large bone-marrow cell that gives rise to platelets

meiosis (my-OH-sis) process of cell division leading to gamete (sperm and ova) formation; daugher cells receive only half the chromosomes present in original cell

melatonin (mel-ah-TOE-nin) candidate hormone secreted by pineal gland; suspected role in puberty onset and control of body rhythms

membrane structural barrier composed of lipids and proteins; provides selective barrier to molecule and ion movement and structural framework to which enzymes, fibers, and ligands are bound

membrane attack complex (MAC) group of complement proteins that form channels in microbe surface and destroy microbe

membrane potential voltage difference between inside and outside of cell

memory *see* working memory, long-term memory

memory cell B cell that differentiates during an initial infection and responds rapidly during subsequent exposure to same antigen

memory consolidation processes by which memory is transferred from working to long-term form

memory trace neural substrate of memory

menarche (MEN-ark-ee) onset, at puberty, of menstrual cycling

meninges (men-IN-gees) protective membranes that cover brain and spinal cord

menopause (MEN-oh-pause) cessation of menstrual cycling

menstrual cycle (MEN-stru-al) cyclic rise and fall in female reproductive hormones and processes

menstrual flow blood mixed with debris from disintegrating uterine wall

menstruation (men-stroo-A-shun) flow of menstrual fluid from uterus; also called menstrual period

MES *see* microsomal enzyme system

messenger RNA (mRNA) ribonucleic acid that transfers genetic information from DNA to ribosome

metabolic acidosis (met-ah-BOL-ik ass-ih-DOE-sis) acidosis due to any cause other than accumulation of carbon dioxide

metabolic alkalosis (al-kah-LOW-sis) alkalosis resulting from any cause other than excessive respiratory removal of carbon dioxide

metabolic end product waste product from a metabolic reaction or series of reactions

metabolic pathway sequence of enzyme-mediated chemical reactions by which molecules are synthesized and broken down in cells

metabolic rate total-body energy expenditure per unit time

metabolism (meh-TAB-ol-izm) aggregate of chemical reactions that occur in a living organism

metabolite (meh-TAB-oh-lite) substance produced by metabolism

metabolize change by chemical reactions

metarteriole (MET-are-tier-ee-ole) blood vessel that directly connects arteriole and venule

methyl group —CH_3

Mg^{2+} magnesium ion

MHC protein plasma-membrane protein coded for by the major histocompatibility gene complex; restricts T-cell receptor's ability to combine with antigen on cell; categorized as Class I and Class II

micelle (MY-cell) soluble cluster of amphipathic molecules in which molecules' polar regions line surface and nonpolar regions are oriented toward center; formed from fatty acids, 2-monoglycerides, and bile salts during fat digestion in small intestine

microbe bacterium, virus, fungus, or other parasite

microcirculation combination of arterioles, capillaries, and venules

microfilament rodlike cytoplasmic protein filament that forms major component of cytoskeleton

microsomal enzyme system (MES) (my-kro-ZOME-al) enzymes, found in smooth endoplasmic reticulum of liver cells, that transform molecules into more polar, less lipid-soluble substances

microtubule tubular cytoplasmic filament that provides internal support for cells; allows change in cell shape and organelle movement in cell

microvilli (my-kro-VIL-eye) small finger-like projections from epithelial-cell surface; greatly increase surface area of cell; characteristic of epithelium lining small intestine and kidney nephrons

micturition (mik-tu-RISH-un) urination

middle-ear cavity air-filled space in temporal bone: contains three ear bones that conduct sound waves from tympanic membrane to cochlea

migrating motility complex pattern of peristaltic waves that pass over small segments of intestine after absorption of meal

MIH *see* Müllerian inhibitory hormone

milk ejection reflex process by which milk is moved from mammary gland alveoli into ducts, from which it can be sucked; due to oxytocin; formerly called milk let-down

milliliter (mL) (MIL-ih-lee-ter) volume equal to 0.001 L

millimole (MIL-lee-mole) concentration equal to 0.001 mole

millivolt (mV) (MIL-ih-volt) electric potential equal to 0.001 V

mineral inorganic substance, that is, without carbon; major minerals in body are calcium, phosphorus, potassium, sulfur, sodium, chloride, and magnesium

mineralocorticoid (min-er-al-oh-KORT-ih-coid) steroid hormone produced by adrenal cortex that has major effect on sodium and potassium balance; major mineralocorticoid is aldosterone

minute ventilation total ventilation per minute; equals tidal volume times respiratory rate

mitochondrion (my-toe-KON-dree-un) rod-shaped or oval cytoplasmic organelle that produces most of cell's ATP; site of Krebs cycle and oxidative-phosphorylation enzymes

mitogenic (my-toe-JEN-ik) stimulates mitosis

mitosis (my-TOE-sis) process in cell division in which DNA is duplicated and an identical set of chromosomes is passed to each daughter cell

mitotic spindle structure that appears during division of cell nucleus and consists of microtubules (*spindle fibers*) that pass from one side of cell to other

mitral valve (MY-tral) valve between left atrium and left ventricle of heart

modulation *see* allosteric modulation and covalent modulation

modulator molecule ligand that, by acting at an allosteric regulatory site, alters properties of other binding sites on a protein and thus regulates its functional activity

molarity (mo-LAR-ih-tee) number of moles of solute per liter of solution

mole (also mol) number of molecules of substance; number of moles = weight in grams/molecular mass

molecular mass sum of atomic masses of all atoms in molecule

molecule chemical substance formed by linking atoms together

monoamine (mah-no-ah-MEAN) class of neurotransmitters having the structure of R—NH_2, where R is molecule remainder; by convention, excludes peptides and amino acids

monoamine oxidase enzyme that inactivates catecholamine neurotransmitters and serotonin

monocyte (MAH-no-site) type of leukocyte; leaves bloodstream and is transformed into a macrophage

monoglyceride (mah-no-GLISS-er-ide) glycerol linked to one fatty acid side chain

monokine (MAH-no-kine) any protein messenger secreted by a monocyte or macrophage

monosaccharide (mah-no-SAK-er-ide) carbohydrate consisting of one sugar molecule, which generally contains five or six carbon atoms

monosynaptic reflex (mah-no-sih-NAP-tik) reflex in which the afferent neuron directly activates the motor neuron

motivation *see* primary motivated behavior, secondary motivated behavior, appetitive motivation, aversive motivation

motor having to do with muscles and movement

motor control hierarchy the brain areas having a role in skeletal-muscle control are rank-ordered in three functional groups

motor cortex strip of cerebral cortex along posterior border of frontal lobe; gives rise to many axons descending in corticospinal and multineuronal pathways; *see also* primary motor cortex

motor end plate specialized region of muscle-cell plasma membrane that lies directly under axon terminal of a motor neuron

motor neuron efferent neuron that innervates skeletal muscle

motor neuron pool all the motor neurons for a given muscle

motor potential electrical activity that can be recorded over motor cortex about 50 to 60 ms before a movement begins

motor system those CNS parts that contribute to control of skeletal-muscle movements

motor unit motor neuron plus the muscle fibers it innervates

mRNA *see* messenger RNA

mucin (MU-sin) protein that, when mixed with water, forms mucus

mucosa (mu-KO-sah) three layers of gastrointestinal tract wall nearest lumen, that is, *epithelium, lamina propria,* and *muscularis mucosa*

mucus highly viscous solution secreted by mucous membranes

Müllerian duct (mul-AIR-ee-an) part of embryo that, in a female, develops into reproductive system ducts, but in a male, degenerates

Müllerian inhibiting hormone (MIH) protein secreted by fetal testes that causes Müllerian ducts to degenerate

multineuronal pathways pathways made up of chains of neurons functionally connected by synapses; specifically, descending motor pathways that synapse in basal ganglia and other subcortical nuclei and in the brainstem; only final neuron of chain reaches region of motor neurons; extrapyramidal system; also called *multisynaptic pathways*

multiunit smooth muscle smooth muscle that exhibits little, if any, propagation of electrical activity from fiber to fiber and whose contractile activity is closely coupled to its neural input

muscarinic receptor (mus-cur-IN-ik) acetylcholine receptor that responds to the mushroom poison muscarine; located on smooth muscle, cardiac muscle, some CNS neurons, and glands

muscle number of muscle fibers bound together by connective tissue

muscle fatigue decrease in mechanical response of muscle with prolonged stimulation; *compare* psychological fatigue

muscle fiber muscle cell

muscle-spindle stretch receptor capsule-enclosed arrangement of afferent nerve fiber endings in skeletal muscle; sensitive to stretch

muscle tension force exerted by a contracting muscle on an object

muscle tone degree of resistance of muscle to passive stretch

mutation (mu-TAY-shun) any change in base sequence of DNA that changes genetic information

mV *see* millivolt

myasthenia gravis (my-as-THEE-nee-ah GRAH-vis) autoimmune neuromuscular disease associated with skeletal-muscle weakness and fatigue; due to destruction of skeletal-muscle receptors for acetylcholine

myelin (MY-ah-lin) insulating material covering axons of many neurons; consists of layers of myelin-forming cell plasma membrane wrapped around axon

myenteric plexus (my-en-TER-ik PLEX- us) nerve cell network between circular and longitudinal muscle layers in esophagus, stomach, and intestinal walls

myo- (MY-oh) pertaining to muscle

myoblast (MY-oh-blast) embryological cell that gives rise to muscle fibers

myocardial infarction *see* heart attack

myocardium (my-oh-CARD-ee-um) cardiac muscle, which forms heart walls

myoepithelial cell (my-oh-ep-ih-THEE-lee-al) specialized contractile cell around certain exocrine glands; contraction forces gland's secretion through ducts

myofibril (my-oh-FI-bril) thick or thin contractile filament in cytoplasm of striated muscle; myofibril clusters are arranged in repeating sarcomere pattern along longitudinal axis of muscle

myogenic (my-oh-JEN-ik) originating in muscle

myoglobin (my-oh-GLOW-bin) muscle-fiber protein that binds oxygen

myometrium (my-oh-ME-tree-um) uterine smooth muscle

myosin (MY-oh-sin) contractile protein that forms thick filaments in muscle fibers

myosin ATPase enzymatic site on globular head of myosin that catalyzes ATP breakdown to ADP and P_i, releasing the chemical energy used to produce force of muscle contraction

myosin light-chain kinase smooth-muscle protein kinase; when activated by Ca-calmodulin, phosphorylates myosin

NaCl sodium cholride

Na$^+$, K$^+$−ATPase pump primary active-transport carrier protein that splits ATP and releases energy that is used to transport sodium out of cell and potassium into cell

natriuretic hormone (nat-rye-yur-ET-ik) hormone that is secreted by atrium of heart; inhibits sodium reabsorption in renal tubule

natural antibody antibody to erythrocyte antigens A or B; are present without prior exposure to antigen

natural killer (NK) cell type of lymphocyte that binds relatively nonspecifically to cells bearing foreign antigens and kills them directly; no MHC restriction

nearsighted vision defect because eyeball is too long for lens, so that images of distant objects are focused in front of retina

negative balance substance loss from body exceeds gain and total amount in body decreases; also used for physical parameters such as body temperature and energy; *compare* postive balance

negative feedback aspect of control systems in which system's response opposes input to system; *compare* positive feedback

nephron (NEF-ron) functional unit of kidney; has vascular and tubular component

nerve group of many nerve fibers traveling together in peripheral nervous system

nerve cell cell specialized to initiate, integrate, and conduct electric signals; also called neuron

nerve fiber *see* axon

nerve growth factor peptide that stimulates growth and differentiation of some neurons

net amount remaining after opposing quantities are subtracted from each other; final amount

neuroeffector junction "synapse" between a neuron and muscle or gland cell

neuroglia *see* glial cell

neurohormone chemical messenger that is released by a neuron and travels in bloodstream to its target cell

neuromodulator chemical messenger that acts on neurons, usually by a second-messenger system, to alter response to a neurotransmitter

neuron (NUR-on) *see* nerve cell

neuropeptide family of at least 50 neurotransmitters composed of two or more amino acids; often functions as chemical messengers in nonneural tissues

neurotransmitter chemical messenger used by neurons to communicate with each other or with effectors

neutrophil (NEW-tro-fil) polymorphonuclear granulocytic leukocyte whose granules show preference for neither eosin nor basic dyes; functions as phagocyte and releases chemicals involved in inflammation

neutrophil exudation (ex-ooh-DAY-shun) amoeba-like movement of neutrophils from capillary lumen to tissue extracellular space

NH$_3$ ammonia

NH$_4^+$ ammonium ion

nicotinic receptor (nik-oh-TIN-ik) acetylcholine receptor that responds to nicotine; primarily, receptors at motor end plate and on postganglionic autonomic neurons

nociceptor (NO-sih-sep-tor) sensory receptor whose stimulation causes pain

node of Ranvier (RAHN-vee-a) space between adjacent myelin-forming cells along myelinated axon where axonal plasma membrane is exposed to extracellular fluid

nonpolar molecule molecule containing predominantly chemical bonds in which electrons are shared equally between atoms; has few polar or ionized groups

nonspecific ascending pathway chain of synaptically connected neurons in CNS that are activated by sensory units of several different types; signals general information; *compare* specific ascending pathway

nonspecific immune response response that nonselectively protects against foreign material without having to recognize its specific identity

norepinephrine (nor-ep-ih-NEF-ryn) catecholamine neurotransmitter released at most sympathetic postganglionic endings, from adrenal medulla, and in many CNS regions

Novocaine local anesthetic that works by preventing increased sodium permeability required to produce action potential

nuclear envelope double membrane surrounding cell nucleus

nuclear pore opening in nuclear envelope through which molecular messengers pass between nucleus and cytoplasm

nuclear protein complex of nuclei acid and protein

nucleic acid nucleotide polymer in which phosphate of one nucleotide is linked to the sugar of the adjacent one; stores and transmits genetic information; includes DNA and RNA

nucleolus (new-KLEE-oh-lus) densely staining nuclear structure containing DNA that codes for rRNA; site of ribosome assembly

nucleotide (NEW-klee-oh-tide) molecular subunit of nucleic acid; a purine or pyrimidine base, sugar, and phosphate

nucleus (NEW-klee-us) (pl. nuclei) (cell) large membrane-bound organelle that contains cell's DNA; (neural) cluster of neuron cell bodies in CNS

nucleus accumbens (ah-COME-bens) lower forebrain structure that is part of reward system

obstructive lung disease disease characterized by narrowing or blocking of airways; *see also* restrictive lung disease

occipital lobe posterior region of cerebral cortex where primary visual cortex is located

off **response** increase in cell activity upon stimulus removal; decrease in cell's background activity in response to stimulation

Ohm's law (omz) current *I* is directly proportional to voltage *E* and inversely proportional to resistance *R* such that $I = E/R$

olfactory (ol-FAK-tor-ee) pertaining to sense of smell

olfactory mucosa (mu-KO-sah) mucous membrane in upper part of nasal cavity containing receptors for sense of smell

olfactory nerve cranial nerve I; relays information about sense of smell

oligodendroglia (oh-lih-go-DEN-dro-glee-ah) type of glial cell; responsible for myelin formation in CNS

on **response** neuron or receptor response to stimulus onset

oncogene (ON-ko-jean) altered genes associated with cancer

oogenesis (oh-oh-JEN-ih-sis) gamete production in female

oogonium (oh-oh-GO-nee-um) primitive ovum that, upon mitotic division, gives rise to additional oogonia or to primary oocyte

operating point steady-state value maintained by homeostatic control system; also called set point

opioid peptide (OH-pea-oid) leu- or met-enkephalin, β-endorphin, or nalorphin

opponent color cell neuron whose activity is increased by input from cones of one color sensitivity but decreased by cones having a different sensitivity

opsin (OP-sin) protein component of photopigment

opsonin (op-SO-nin) any substance that binds a microbe to a phagocyte and promotes phagocytosis

optic nerve cranial nerve II; relays information about vision

optimal length *(l_o)* length at which muscle fiber develops maximal tension

organ collection of tissues joined in structural unit to serve common function

organ of Corti (KOR-tee) structure in inner ear capable of transducing sound-wave energy into action potentials

organ system organs that together serve an overall function

organelle *see* cell organelle

organic pertaining to carbon-containing substances; *compare* inorganic

orgasm (OR-gazm) inner emotions and systemic physiological changes that mark apex of sexual intercourse; usually accompanied in the male by ejaculation

orienting response behavior in response to a novel stimulus, that is, the person stops what he or she is doing, looks around, listens intently, and turns toward stimulus

osmolarity (oz-mo-LAIR-ih-tee) total solute concentration of a solution; measure of water concentration in that the higher the solution osmolarity, the lower the water concentration

osmoreceptor (OZ-mo-ree-sep-tor) neural receptor that responds to changes in osmolarity of surrounding fluid

osmosis (oz-MO-sis) net diffusion of water across a selective barrier from region of low solute concentration (high water concentration) to region of high solute concentration (low water concentration)

osmotic pressure (oz-MAH-tik) pressure that must be applied to a solution on one side of a membrane to prevent osmotic flow of water across the membrane from a compartment of pure water

osteoblast (OS-tee-oh-blast) cell type responsible for laying down protein matrix of bone

otolith (OH-toe-lith) calcium carbonate "stone" in gelatinous mass of utricle and saccule of vestibular apparatus

oval window membrane-covered opening between middle-ear cavity and scala vestibuli of inner ear

ovarian follicle (oh-VAR-ee-an FOL-ih-kle) ovum and its encasing granulosa and theca cells prior to ovulation

ovary (OH-vah-ree) gonad in female

oviduct *see* uterine tube

ovulation (ov-you-LAY-shun) release of ovum, surrounded by its zona pellucida and cumulus, from ovary

ovum (pl. ova) gamete of female

oxidation (ox-ih-DAY-shun) combining, or causing a substance to combine, with oxygen

oxidative (OX-ih-day-tive) using oxygen

oxidative deamination (dee-am-ih-NAY-shun) reaction in which an amino group from an amino acid is replaced by oxygen to form a keto acid

oxidative fiber muscle fiber that has numerous mitochondria and, therefore, a high capacity for oxidative phosphorylation

oxidative phosphorylation (fos-for-ih-LAY-shun) process by which energy derived from reaction between hydrogen and oxygen to form water is transferred to ATP during its formation

oxyhemoglobin (ox-see-HE-mo-glow-bin) HbO_2; hemoglobin combined with oxygen

oxyntic cell *see* parietal cell

oxytocin (ox-see-TOE-sin) peptide hormone synthesized in hypothalamus and released from posterior pituitary; stimulates mammary glands to release milk and uterus to contract

P wave component of electrocardiogram reflecting atrial depolarization

pacemaker neurons that set rhythm of biological clocks independent of external cues; any nerve or muscle cell that has an inherent autorhythmicity and determines activity pattern of other cells

pacemaker potential spontaneous depolarization to threshold of some nerve and muscle cells' plasma membranes

pacinian corpuscle (pah-SIN-ee-an) mechanoreceptor specialized to respond to vibrating stimuli

pair to join together

pancreas (PAN-kree-us) gland in abdomen near stomach; connected by duct to small intestine; contains endocrine gland cells, which secrete insulin, glucagon, and somatostatin into the bloodstream, and exocrine gland cells, which secrete digestive enzymes and bicarbonate into intestine

pancreatic lipase (LIE-pase) enzyme secreted by exocrine pancreas; acts on triacylglycerols to form 2-monoglycerides and free fatty acids

paracrine (PEAR-ah-krin) chemical messenger that exerts its effects on tissues near its secretion site; by convention, excludes neurotransmitters

paradoxical sleep sleep state associated with small, rapid EEG oscillations, complete loss of tone in postural muscles, and dreaming; also called rapid-eye-movement, or REM, sleep

parasympathetic nervous system (pear-ah-sim-pah-THET-ik) portion of autonomic nervous system whose preganglionic fibers leave CNS from brainstem and sacral portion of spinal cord; most of its postganglionic fibers release acetylcholine; *compare* sympathetic nervous system

parasympathomimetic (pear-ah-sim-path-oh-mih-MET-ik) chemical that produces effects on an organ similar to those produced by stimulating parasympathetic nerves to the organ

parathormone *see* parathyroid hormone

parathyroid gland one of four parathyroid-hormone secreting glands on thyroid gland surface

parathyroid hormone (**PTH**) peptide hormone secreted by parathyroid glands; regulates calcium and phosphate concentration of internal environment; also called parathormone

parietal cell (pah-RYE-ih-tal) gastric gland cell that secretes hydrochloric acid and intrinsic factor; also called oxyntic cell

parietal lobe region of cerebral cortex containing sensory cortex and some association cortex

Parkinson's disease disease characterized by tremor, rigidity, and delay in initiation of movement; due in part to deficit of dopamine in basal ganglia

parotid (pah-ROT-id) one of the three pairs of salivary glands

partial pressure that part of total gas pressure due to molecules of one gas species; measure of concentration of a gas in a gas mixture

parturition (par-tu-RISH-un) birth; delivery of infant and placenta

passive immunity resistance to infection resulting from direct transfer of antibodies or sensitized T cells from one person (or animal) to another; *compare* active immunity

pathway group of nerve fibers in CNS

pelvic diaphragm (PEL-vik DIE-ah-fram) sheet of skeletal muscle that forms floor of pelvis and helps support abdominal and pelvic viscera

pepsin (PEP-sin) family of several protein-digesting enzymes formed in the stomach; breaks protein down to peptide fragments

pepsinogen (pep-SIN-oh-jen) inactive precursor of pepsin; secreted by chief cells of gastric mucosa

peptide (PEP-tide) short polypeptide chain; by convention, having less than 50 amino acids

peptide bond polar covalent chemical bond

$$\overset{\displaystyle O}{\underset{\displaystyle H}{-C-N-}}$$ joining two amino acids; forms

protein backbone

percent hemoglobin saturation *see* hemoglobin saturation

perception understanding of objects and events of external world that we acquire from neural processing of sensory information

perfusion capillary blood flow

pericardium (pear-ah-CAR-dee-um) connective-tissue sac surrounding heart

period *see* menstruation

peripheral chemoreceptor carotid or aortic body; responds to changes in blood O_2 and CO_2 pressures and H^+ concentration

peripheral lymphoid organ (LIM-foid) lymph node, spleen, tonsil, or lymphocyte accumulation in gastrointestinal, respiratory, urinary, or reproductive tracts

peripheral nervous system nerve fibers extending from CNS

peripheral protein protein that is bound to surface of plasma membrane and does not associate with lipid bilayer

peripheral thermoreceptor cold or warm receptor in skin or certain mucous membranes

peristaltic wave (per-ih-STALL-tik) progressive wave of muscle contraction that proceeds along wall of a tube, compressing the tube and causing its contents to move

peritoneum (per-ih-toe-NEE-um) membrane lining abdominal and pelvic cavities and covering organs there

peritubular capillary capillary closely associated with tubular component of kidney nephron

permissiveness situation whereby small quantities of one hormone are required in order for a second hormone to exert its full effects

peroxisome (per-OX-ih-zome) cell organelle that destroys certain toxic products of oxidative reactions

PF₃ ecosinoid exposed in membranes of aggregated platelets; important in activation of several enzymes in clot formation

PGI₂ phospholipid that inhibits platelet aggregation in blood clotting; also called prostacyclin

pH expression of a solution's acidity; negative logarithm to base 10 of H^+ concentration; pH decreases as acidity increases

phagocyte (FAH-go-site) any cell capable of phagocytosis

phagocytosis (fah-go-sigh-TOE-sis) engulfment of particles by a cell followed by the particles' digestion

phagolysosome (fah-go-LIE-so-zome) cell organelle formed by fusion of a phagosome and lysosome; destroys engulfed material

phagosome (FAH-go-zome) membrane-enclosed sac formed when a phagocyte engulfs particulate matter

pharmacological effect effect produced by much larger amounts of hormone than are normally present

pharynx (FARE-inks) throat; passage common to routes taken by food and air

phasic (FASE-ik) intermittent; *compare* tonic

phosphate group (FOS-fate) —PO_4^{2-}

phosphatidylinositol bisphosphate (**PIP₂**) (fos-fa-tid-il-in-OS-ih-tol bis-FOS-fate) plasma membrane phospholipid that forms inositol trisphosphate and diacylglycerol when catalyzed by phospholipase C

phosphodiesterase (fos-foe-die-ES-ter-ase) enzyme that catalyzes cyclic AMP breakdown to AMP

phospholipase A₂ (fos-fo-LIE-pace A-two) enzyme that splits arachidonic acid from plasma-membrane phospholipid

phospholipase C receptor-controlled plasma-membrane enzyme that catalyzes phosphatidyl bisphosphate breakdown to inositol trisphosphate and diacylglycerol

phospholipid (fos-foe-LIP-id) lipid subclass similar to triacylglycerol except that a phosphate group (—PO_4^{2-}) and small nitrogen-containing molecule are attached to third hydroxyl group of glycerol; major component of cell membranes

phosphoprotein phosphatase (FOS-fah-tase) enzyme that removes phosphate from protein, restoring protein's original shape

phosphoric acid (fos-FOR-ik) acid generated during catabolism of phosphorus-containing compounds; dissociates to form inorganic phosphate and hydrogen ions

phosphorylation (fos-for-ah-LAY-shun) addition of phosphate group to an organic molecule

photopigment light-sensitive molecule altered by absorption of photic energy of certain wavelengths; consists of opsin bound to a chromophore

photoreceptor receptor sensitive to light (photic energy)

physiology branch of biology dealing with the mechanisms by which the body functions

pitch degree of how high or low a sound is perceived

pituitary (pih-TWO-ih-tair-ee) endocrine gland that lies in bony pocket below hypothalamus; includes anterior pituitary and posterior pituitary

pituitary gonadotropin *see* gonadotropic hormone

placenta (plah-SEN-tah) interlocking fetal and maternal tissues that serve as organ of molecular exchange between fetal and maternal circulations

placental lactogen (plah-SEN-tal LAK-toe-jen) hormone that is produced by placenta and has effects, particularly in the mother, similar to growth hormone

plasma (PLAS-muh) liquid portion of blood; component of extracellular fluid

plasma cell cell that differentiates from activated B lymphocytes and secretes antibodies

plasma membrane membrane that forms outer surface of cell and separates cell's contents from extracellular fluid

plasma protein albumins, globulins, and fibrinogen

plasmin (PLAZ-min) proteolytic enzyme able to decompose fibrin and, thereby, to dissolve blood clots

plasminogen (plaz-MIN-oh-jen) inactive precursor of plasmin

plasminogen activator any protein that activates the proenzyme plasminogen

plasticity (plas-TISS-ih-tee) ability of neural tissue to change its responsiveness to stimulation because of its past history of activation

platelet (PLATE-let) cell fragment present in blood; plays several roles in blood clotting

platelet activating factor mediator secreted by mast cells, stimulates inflammation

platelet aggregation (ag-reh-GAY-shun) positive-feedback process resulting in platelets sticking together

pleura (PLUR-ah) thin cellular sheet attached to thoracic cage interior (*parietal pleura*) and, folding back upon itself, is attached to lung surface (*visceral pleura*); forms two enclosed *pleural sacs* in thoracic cage

plexus *see* myenteric plexus, submucous plexus

pluripotent stem cells (plur-ih-poe-tent STEM) single population of bone-marrow cells from which all blood cells are descended

polar body small cell resulting from unequal distribution of cytoplasm during division of primary or secondary oocyte

polar covalent bond covalent chemical bond in which two electrons are shared unequally between two atoms; atom to which the electrons are drawn becomes slightly negative and other atom slightly positive

polar molecule molecule containing polar covalent bonds; part of molecule to which electrons are drawn becomes slightly negative and region from which they are drawn becomes slightly positive; molecule is soluble in water

polarized (PO-lar-ized) having two electric poles, one negative and one positive

polymer (POL-ih-mer) large molecule formed by linking together smaller similar subunits

polymorphonuclear granulocyte (pol-ee-morf-oh-NUK-lee-er GRAN-you-low-site) subclass of leukocytes; *see also* eosinophil, basophil, neutrophil

polypeptide (pol-ee-PEP-tide) polymer consisting of amino acid subunits joined by peptide bonds; also called peptide and protein

polysaccharide (pol-ee-SAK-er-ide) large carbohydrate formed by linking monosaccharide subunits together

polysynaptic reflex (pol-ee-sih-NAP-tik) reflex employing one or more interneurons in its reflex arc

polyunsaturated fatty acid fatty acid that contains more than one double bond

pore-forming protein protein secreted by cytotoxic and NK cells; forms channel in plasma membrane of target cell and destroys it

portal vessel blood vessel that links two capillary networks

positive balance gain of substance exceeds loss, and amount of that substance in body increases; *compare* negative balance

positive feedback aspect of control systems in which an initial disturbance sets off train of events that increases the disturbance even further; *compare* negative feedback

postabsorptive state (post-ab-SORP-tive) period during which nutrients are not present in gastrointestinal tract and energy must be supplied by body's endogenous stores

posterior toward or at the back

posterior pituitary portion of pituitary from which oxytocin and vasopressin are released

postganglionic (post-gang-glee-ON-ik) autonomic-nervous-system neuron or nerve fiber whose cell body lies in ganglion and whose axon terminals form neuroeffector junctions; conducts impulses away from ganglion toward periphery; *compare* preganglionic

postsynaptic neuron (post-sin-NAP-tik) neuron that conducts information away from a synapse

postsynaptic potential local potential that arises in postsynaptic neuron in response to activation of synapses upon it; *see also* excitatory postsynaptic potential, inhibitory postsynaptic potential

postural reflex reflex that maintains or restores upright, stable posture

potential (or potential difference) voltage difference between two points

potentiation (po-ten-she-A-shun) presence of one agent enhances response to a second agent such that final response is greater than the sum of the two individual responses

precapillary sphincter (SFINK-ter) smooth-muscle ring around capillary where it exits from thoroughfare channel or arteriole

preganglionic autonomic-nervous-system neuron or nerve fiber whose cell body lies in CNS and whose axon terminals lie in a ganglion; conducts action potentials from CNS to ganglion; *compare* postganglionic

presbyopia (prez-bee-OH-pea-ah) vision impairment due to lens stiffening; makes accommodation difficult

pressure autoregulation ability of individual arteries and arterioles to alter their resistance in response to changing blood pressure so that relatively constant blood flow is maintained

presynaptic neuron (pre-sin-NAP-tic) neuron that conducts action potentials toward a synapse

presynaptic synapse relation between two neurons in which axon terminal of one neuron ends on axon terminal of second neuron; action potentials in first neuron alter neurotransmitter release from second, altering effectiveness of synapse second neuron makes with a third

PRH *see* prolactin releasing hormone

primary active transport active transport in which chemical energy is transferred directly from ATP to carrier protein

primary afferent afferent neuron; also called first-order neuron

primary cortical receiving area region of cerebral cortex where specific ascending pathways end; somatosensory, visual, auditory, or taste cortex

primary follicle (FAH-lick-el) cell cluster in ovaries that consists of one primary oocyte surrounded by single layer of granulosa cells

primary hypersecretion increased hormone secretion due to overfunction of secretory gland

primary hypertension hypertension of unknown cause

primary hyposecretion decreased hormone secretion due to underfunction of secretory gland

primary motivated behavior behavior related directly to achieving homeostasis

primary oocyte (OH-oh-site) female germ cell that undergoes first meiotic division to form secondary oocyte and polar body

primary spermatocyte (sper-MAH-toe-site) male germ cell derived from spermatogonia; undergoes meiotic division to form two secondary spermatocytes

primary visual cortex first part of visual cortex to be activated by visual pathways

process long extension from neuron cell body

product molecule formed in enzyme-catalyzed chemical reaction

progestagen (pro-JES-tah-jen) progesterone-like substance

progesterone (pro-JES-ter-own) steroid hormone secreted by corpus luteum and placenta; stimulates uterine-gland secretion, inhibits uterine smooth-muscle contraction, and stimulates breast growth

program related sequence of neural activity preliminary to motor act

prohormone peptide precursor from which are cleaved one or more active peptide hormones

prolactin (pro-LAK-tin) peptide hormone secreted by anterior pituitary; stimulates milk secretion by mammary glands

prolactin release-inhibiting hormone (PIH) dopamine, which serves as hypothalamic hormone to inhibit prolactin secretion by anterior pituitary

prolactin releasing hormone (PRH) one or more hypothalamic hormones that stimulate prolactin release from anterior pituitary

proliferative phase (pro-LIF-er-ah-tive) stage of menstrual cycle between menstruation and ovulation during which endometrium repairs itself and grows

promoter specific nucleotide sequence at beginning of gene; determines which of the paired strands of DNA is transcribed into mRNA

pro-opiomelanocortin (pro-oh-pee-oh-mel-an-oh-KOR-tin) large protein precursor for ACTH, endorphins, and several other hormones

propagation (prop-ah-GAY-shun) transmission by reproduction of self

proprioception (pro-pree-oh-SEP-shun) sense of bodily movement and position in space

prostacyclin *see* PGI₂

prostaglandin (pros-tah-GLAN-din) one of a group of eicosanoids that function mainly as paracrine or autocrine

prostate gland (PROS-tate) large gland encircling urethra in the male; secretes fluid into urethra

protein large polymer consisting of one or more sequences of amino acid subunits joined by peptide bonds

proteinaceous (pro-teen-A-shus) pertaining to protein

protein binding site *see* binding site

protein C plasma protein that inhibits clotting

protein kinase (KI-nase) one of family of enzymes that phosphorylates certain other proteins by transferring to them a phosphate group from ATP

protein kinase C enzyme that phorphorylates certain proteins when activated by diacylglycerol

proteolytic (pro-tee-oh-LIT-ik) breaks down protein

prothrombin (pro-THROM-bin) inactive precursor of thrombin; produced by liver and normally present in plasma

proton (PRO-tahn) positively charged subatomic particle

proximal (PROX-sih-mal) nearer; closer to reference point; *compare* distal

proximal tubule first tubular component of a nephron after Bowman's capsule

psychoactive drug drug that affects the mind or behavior

psychological dependence craving for substance use and inability to stop at will; *see also* physical dependence

psychological fatigue mental factors that cause individuals to stop muscle activity even though muscles can still contract

PTH *see* parathyroid hormone

puberty attainment of sexual maturity when conception becomes possible; as commonly used, refers to 3 to 5 years of sexual development that culimnates in sexual maturity

pulmonary (PULL-mah-nair-ee) pertaining to lungs

pulmonary circulation circulation through lungs; portion of cardiovascular system between pulmonary trunk, as it leaves the right ventricle, and pulmonary veins, as they enter the left atrium

pulmonary edema (ed-DEE-mah) fluid accumulation in lung interstitium and air sacs

pulmonary stretch receptor afferent nerve ending lying between airway smooth muscle cells and activated by lung inflation

pulmonary surfactant *see* surfactant

pulmonary trunk large artery that carries blood from right ventricle of heart

pulmonary valve valve between right ventricle of heart and pulmonary trunk

pulmonary ventilation *see* minute ventilation

pulse pressure difference between systolic and diastolic arterial blood pressures

pupil opening in iris of eye through which light passes to reach retina

purine base (PURE-ene) double-ring, nitrogen-containing subunit of nucleotide; adenine or guanine

Purkinje fiber (purr-KIN-jee) specialized myocardial cell that constitutes part of conducting system of heart; conveys excitation from bundle branches to ventricular muscle

pyloric sphincter (pie-LOR-ik) ring of smooth muscle between stomach and small intestine

pyramidal tract *see* corticospinal pathway

pyrimidine base (pur-RIM-ih-dean) single-ring, nitrogen-containing subunit of nucleotide; cytosine, thymine, or uracil

pyrogen any substance that causes fever

pyruvate (PIE-rue-vate) anion formed when pyruvic acid loses a hydrogen ion

pyruvic acid (pie-RUE-vik) three-carbon intermediate in glycolytic pathway that, in absence of oxygen, forms lactic acid or, in presence of oxygen, enters Krebs cycle

QRS complex component of electrocardiogram corresponding to ventricular depolarization

R— in chemical formula, signifies remaining portion of molecule

radiation emission of heat from object's surface in form of electromagnetic waves

rapid-eye-movement sleep *see* paradoxical sleep

rate-limiting enzyme enzyme in metabolic pathway most easily saturated with substrate; determines rate of entire metabolic pathway

rate-limiting reaction slowest reaction in metabolic pathway; catalyzed by rate-limiting enzyme

reactant (re-AK-tent) molecule that enters a chemical reaction; called the *substrate* in enzyme-catalyzed reactions

reaction *see* chemical reaction

reactive hyperemia (hi-per-EE-me-ah) transient increase in blood flow following release of occlusion of blood supply

readiness potential electrical activity recorded over parts of brain during "decision-making period," about 0.8 s before a movement begins

receptive field (of neuron) part of body, which, if stimulated, results in activity in that neuron

receptive relaxation reflex decrease in smooth-muscle tension in walls of hollow organs in response to distension

receptor (in sensory system) specialized peripheral ending of afferent neuron, or separate cell intimately associated with it, that detects changes in some aspect of environment; (in chemical communication) specific protein-binding site in plasma membrane or interior of target cell with which a chemical messenger combines to exert its effects

receptor activation change in receptor conformation caused by combination of messenger with receptor

receptor-operated channel plasma-membrane channel that is opened or closed following chemical-messenger binding to a receptor

receptor potential graded potential that arises in afferent-neuron ending, or a specialized cell intimately associated with it, in response to stimulation

reciprocal inhibition inhibition of motor neurons activating those muscles whose contraction would oppose the intended movement

recognition binding of antigen to receptor on lymphocyte surface specific for that antigen

recombinant DNA (re-KOM-bih-nent) DNA formed by joining portions of two DNA molecules previously fragmented by a restriction enzyme

recruitment activation of additional cells in response to increased stimulus strength

rectum short segment of large intestine between sigmoid colon and anus

red muscle fiber muscle fiber having high oxidative capacity and large amount of myoglobin

reflex (REE-flex) biological control system linking stimulus with response and mediated by a reflex arc

reflex arc neural or hormonal components that mediate a reflex; usually includes receptor, afferent pathway, integrating center, efferent pathway, and effector

reflex response final change due to action of stimulus upon reflex arc; also called effector response

refractory period (re-FRAK-tor-ee) time during which an excitable membrane does not respond to a stimulus that normally causes response; *see also* absolute refractory period, relative refractory period

regulatory site site on protein that interacts with modulator molecule; alters functional-site properties

relative refractory period time during which excitable membrane will produce action potential only to a stimulus of greater strength than the usual threshold strength

relaxin (re-LAX-in) polypeptide hormone secreted mainly by corpus luteum; softens cervix prior to parturition

releasing hormone hormone secreted by hypothalamic neuron; controls hormone release by anterior pituitary

REM sleep (rem) *see* paradoxical sleep

renal (REE-nal) pertaining to kidneys

renal pelvis large cavity at base of each kidney; receives urine from collecting ducts and empties it into ureter

renin (REE-nin) enzyme secreted by kidneys; catalyzes splitting off of angiotensin I from angiotensinogen in plasma

replicate (REP-lih-kate) duplicate

repolarize return transmembrane potential to its resting level

residual volume air volume remaining in lungs after maximal expiration

resistance hinderance to movement through a particular substance or tube

respiration (cellular) oxygen utilization in metabolism of organic molecules; (respiratory system) oxygen and carbon dioxide exchange between organism and external environment

respiratory acidosis increased arterial H^+ concentration due to carbon dioxide retention

respiratory alkalosis decreased arterial H^+ concentration when carbon dioxide elimination from lungs exceeds its production

respiratory distress syndrome disease of premature infants in whom surfactant-producing cells do not function adequately

respiratory "pump" effect on venous return of changing intrathoracic and intraabdominal pressures associated with respiration

respiratory quotient *(RQ)* (KWO-shunt) ratio of carbon dioxide produced to oxygen consumed during metabolism

respiratory rate number of breaths per minute

respiratory system structures involved in gas exchange between blood and external environment, that is, lungs, tubes leading to lungs, and chest structures responsible for breathing

respiratory zone portion of airways from beginning of respiratory bronchi to alveoli; contains alveoli across which gas exchange occurs

resting membrane potential voltage difference between inside and outside of cell in absence of excitatory or inhibitory stimulation; also called resting potential

restriction element *see* MHC protein

restriction nuclease (NEW-clee-ase) bacterial enzyme that splits DNA into fragments, acting at different loci in the two DNA strands

restrictive lung disease disease characterized by normal airway resistance but impaired respiratory movement; *compare* obstructive lung disease

retching strong involuntary attempt to vomit but without anything coming up

reticular activating system (ree-TIK-you-ler) neurons in brainstem reticular formation and its thalamic extension whose activity is concerned with alertness and direction of attention to selected events

reticular formation extensive neuron network extending through brainstem core; receives and integrates information from many afferent pathways and from other CNS regions

retina thin layer of neural tissue lining back of eyeball; contains receptors for vision

retinal (ret-in-AL) form of vitamin A that forms chromophore component of photopigment

retrograde amnesia (RET-row-grade am-NEE-jha) loss of memory for events immediately preceding a memory-disturbing trauma such as a blow to the head

retrovirus (RET-ro-vi-rus) virus with RNA core and enzyme capable of transcribing that RNA to DNA, which is then incorporated into host's DNA

reversible reaction chemical reaction in which reactants are converted to products and, simultaneously, products are converted to reactants; *compare* irreversible reaction

reward system brain structures, including locus ceruleus and nucleus accumbens, that provide pleasurable inner emotions in response to certain behaviors and drugs

Rh factor group of erythrocyte plasma-membrane antigens that may (Rh+) or may not (Rh−) be present

rhodopsin (ro-DOP-sin) photopigment in rods

rhythm method contraceptive technique in which couples refrain from sexual intercourse near time of ovulation

ribonucleic acid (RNA) (rye-bo-new-CLAY-ik) single-stranded nucleic acid involved in transcription of genetic information and translation of that information into protein structure; contains the sugar ribose; *see also* messenger RNA, ribosomal RNA, and transfer RNA

ribosomal RNA (rRNA) (rye-bo-ZOME-al) type of RNA used in ribosome assembly; becomes part of ribosome

ribosome (RYE-bo-zome) cytoplasmic particle that mediates linking together of amino acids to form proteins; attached to endoplasmic reticulum as *bound ribosome*, or suspended in cytoplasm as *free ribosome*

rickets disease in which new bone matrix is inadequately calcified due to 1,25-dihydroxyvitamin D_3 deficiency

rigidity (rih-JID-ih-tee) hypertonia with normal muscle reflexes; *compare* spasticity

rigor mortis (RIG-or MORE-tiss) stiffness of skeletal muscles after death resulting from ATP loss

RNA *see* ribonucleic acid

RNA polymerase (POL-ih-mer-ase) enzyme that forms RNA by joining together appropriate nucleotides after they have base-paired to DNA

RNA processing enzymatic removal of intron sequences from newly formed RNA

rod one of two receptor types for photic energy; contains the photopigment rhodopsin

round window membrane-covered opening between scala tympani of inner ear and the middle-ear cavity

RQ *see* respiratory quotient

rRNA *see* ribosomal RNA

SA node *see* sinoatrial node

saccade (sah-KADE) short jerking eyeball movement

saliva watery solution of salts and proteins, including mucins and amylase, secreted by salivary glands

sarcomere (SAR-ko-mere) repeating structural unit of myofibril; composed of thick and thin filaments; extends between two adjacent Z lines

sarcoplasmic reticulum (sar-ko-PLAS-mik re-TIK-you-lum) endoplasmic reticulum in muscle fiber; site of storage and release of calcium ions

satiety signal (sah-TIE-ih-tee) input to food-control centers that causes hunger to cease and sets time period before hunger returns

saturated fatty acid fatty acid whose carbon atoms are linked by single covalent bonds

saturation degree to which protein-binding sites are occupied by ligands

scala tympani (SCALE-ah TIM-pah-nee) fluid-filled inner-ear compartment that receives sound waves from basilar membrane and transmits them to round window

scala vestibuli (ves-TIB-you-lee) fluid-filled inner-ear compartment that receives sound waves from oval window and transmits them to basilar membrane and cochlear duct

schizophrenia (skit-zo-FREE-nee-ah) disease, or family of diseases, characterized by altered motor behavior, distorted perceptions, disturbed thinking, altered mood, and abnormal interpersonal behavior

Schwann cell nonneural cell that forms myelin sheath in peripheral nervous system

scrotum (SKRO-tum) sac that contains testes and epididymides

second messenger intracellular substance that increases as result of combination of extracellular chemical messenger (the "first" messenger) with plasma-membrane receptor; serves as relay from plasma membrane to intracellular biochemical machinery, where it alters some aspect of cell's function

secondary active transport active transport in which energy released during transmembrane movement of one substance from higher to lower concentration is transferred to the simultaneous movement of another substance from lower to higher concentration

secondary motivated behavior behavior not directed toward achieving homeostasis

secondary peristalsis (pear-ih-STALL-sis) esophageal peristaltic waves not immediately preceded by pharyngeal phase of swallow

secondary sexual characteristics external differences between male and female not directly involved in reproduction

secondary spermatocyte (sper-MAH-toe-site) male germ cell derived from primary spermatocyte as a result of the first meiotic division

secretin (SEEK-reh-tin) peptide hormone secreted by upper small intestine; stimulates pancreas to secrete bicarbonate into small intestine

secretion (sih-KREE-shun) elaboration and release of organic molecules, ions, and water by cells in response to specific stimuli

secretory phase (SEEK-rih-tor-ee) that stage of menstrual cycle following ovulation during which secretory type of endometrium develops

secretory vesicle membrane-bound vesicle produced by Golgi apparatus; contains protein to be secreted by cell

section cut surface; slice

segmentation (seg-men-TAY-shun) series of stationary rhythmic contractions and relaxations of rings of intestinal smooth muscle; mixes intestinal contents

semen (SEE-men) sperm-containing fluid of male ejaculate

semicircular canal passage in temporal bone; contains sense organs for equilibrium and movement

seminal vesicle one of pair of glands in males that secrete fluid into vas deferens

seminiferous tubule (sem-ih-NIF-er-ous) tubule in testis in which sperm production occurs; contains Sertoli cells

semipermeable membrane (sem-eye-PER-me-ah-ble) membrane permeable to some substances but not to others

sensor first component of control system; detects specific environmental changes; also called receptor

sensory deprivation (dep-rih-VA-shun) experimental situation in which subject is isolated as completely as possible from sensory stimulation; includes restriction of movements

sensory information information that originates in stimulated sensory receptors

sensory system parts of nervous system that receive, conduct, or process information that leads to perception of a stimulus

sensory unit afferent neuron plus receptors it innervates

sensorimotor cortex (sen-sor-ee-MO-tor) areas of cerebral cortex that play a role in skeletal-muscle control, including primary motor, somatosensory, and parts of parietal-lobe association cortex and the premotor area

septum (SEP-tum) forebrain area anterior to hypothalamus; involved in emotional behavior

serosa (sir-OH-sah) connective-tissue layer surrounding outer surface of stomach and intestines

serotonin (sair-oh-TOE-in) monoamine neurotransmitter; paracrine in blood platelets and digestive tract; also called 5-hydroxytryptamine, or 5-HT

Sertoli cell (sir-TOE-lee) cell intimately associated with developing germ cells in seminiferous tubule; creates "blood-testis barrier," secretes fluid into seminiferous tubule, and mediates hormonal effects on tubule

serum (SEER-um) blood plasma from which fibrinogen and other clotting proteins have been removed as result of clotting

sex chromatin (CHROM-ah-tin) nuclear mass not usually found in cells of males; consists of condensed X chromosome

sex chromosome X or Y chromosome

sex determination genetic basis of individual's sex, XY determining male and XX, female

sex differentiation (dif-er-en-she-A-shun) development of male or female reproductive organs

sex hormone estrogen, progesterone, testosterone, or related hormones

short-loop negative feedback inhibition of hypothalamus by an anterior pituitary hormone

sickle-cell anemia disease in which an amino acid in hemoglobin is abnormal, and at low oxygen concentrations erythrocytes assume sickle shapes or other bizarre forms that block capillaries

sigmoid colon (SIG-moid) S-shaped terminal portion of colon

signal sequence initial portion of newly synthesized protein (if protein is destined for secretion)

signal transduction information relay from plasma-membrane receptor to cell's response mechanism; see also transduction

single-unit smooth muscle smooth muscle that responds to stimulation as single unit because gap junctions join fibers, allowing electrical activity to pass from cell to cell

sinoatrial (SA) node (sigh-no-A-tree-al) region in right atrium of heart containing specialized cardiac-muscle cells that depolarize spontaneously faster than other heart cells; determines heart rate

sister chromatid (CHROM-ah-tid) one of two identical DNA threads joined together during meiosis

skeletal muscle striated muscle attached to bones or skin and responsible for skeletal movements and facial expression; controlled by somatic nervous system

skeletal-muscle "pump" pumping effect of contracting skeletal muscles on blood flow through underlying vessels

skeletomotor fiber primary skeletal-muscle fiber, as opposed to modified fibers in muscle spindle

sleep see paradoxical sleep, slow-wave sleep

sleep center neuron cluster in brainstem whose activity periodically opposes the awake state and induces cycling of slow-wave and paradoxical sleep

sliding filament mechanism process of muscle contraction in which shortening occurs by thick and thin filaments sliding past each other

slow channel voltage-sensitive calcium channel in myocardial-cell plasma membrane; opens, after a short delay, upon depolarization

slow fiber muscle fiber whose myosin has low ATPase activity

slow virus virus that replicates very slowly; may become associated with cell's DNA and be passed to daughter cells upon cell division

slow-wave sleep sleep state associated with large, slow EEG waves and considerable postural-muscle tone but not dreaming

smooth muscle nonstriated muscle that surrounds hollow organs and tubes; controlled by autonomic nervous system, hormones, and paracrines; see also single-unit smooth muscle, multiunit smooth muscle

sodium inactivation turning off of increased sodium permeability at action potential peak

soft palate (PAL-et) nonbony region at back of roof of mouth

solute (SOL-ute) substances dissolved in a liquid

solution liquid (solvent) containing dissolved substances (solutes)

solvent liquid in which substances are dissolved

somatic (so-MAT-ik) pertaining to the body; related to body's framework or outer walls, including skin, skeletal muscle, tendons, and joints

somatic nervous system component of efferent division of peripheral nervous system; innervates skeletal muscle; *compare* autonomic nervous system

somatic receptor neural receptor that responds to mechanical stimulation of skin or hairs and underlying tissues, rotation or bending of joints, temperature changes, or painful stimuli

somatomedin C *see* insulin-like growth factor

somatosensory cortex (so-mat-oh-SEN-so-ree) strip of cerebral cortex in parietal lobe in which nerve fibers transmitting somatic sensory information synapse

somatostatin (SS) (so-mat-oh-STAT-in) hypothalamic hormone that inhibits growth hormone and TSH secretion by anterior pituitary; possible neurotransmitter; also in stomach and pancreatic islets

somatotropin *see* growth hormone

sound wave air disturbance due to variations between regions of high air-molecule density (*compression*) and low density (*rarefaction*)

spasm (SPAH-zm) sustained involuntary muscle contraction

spasticity (spaz-TISS-ih-tee) hypertonia with increased responses to motor reflexes; *compare* rigidity

spatial summation (SPAY-shul) adding together effects of simultaneous inputs to different places on a neuron to produce potential change greater than that caused by single input

specific ascending pathway chain of synaptically connected neurons in CNS, all activated by sensory units of same type

specific immune response response that depends upon recognition of specific foreign material for reaction to it; *see also* cell-mediated immunity, humoral immunity

specificity selectivity; ability of binding site to react with only one, or a limited number of, types of molecules

sperm male gamete; also called spermatozoa

sperm activation changes undergone by sperm in vicinity of ovum so sperm is capable of fertilizing ovum

sperm capacitation (kah-pas-ih-TAY-shun) process by which sperm in female reproductive tract gains ability to fertilize ovum

spermatid (SPER-mah-tid) immature sperm

spermatogenesis (sper-mah-toe-JEN-ih-sis) sperm formation

spermatogonium (sper-mah-toe-GO-nee-um) undifferentiated germ cell that gives rise to primary spermatocyte

spermatozoon *see* sperm

sphincter (sfink-ter) smooth-muscle ring that surrounds a tube, closing tube as muscle contracts

sphincter of Oddi (OH-dee) smooth-muscle ring surrounding bile duct at its entrance into duodenum

sphygmomanometer (sfig-mo-mah-NOM-eh-ter) device consisting of inflatable cuff and pressure gauge for measuring arterial blood pressure

spinal nerve one of 86 (43 pairs) peripheral nerves that join spinal cord

spinal reflex reflex whose afferent and efferent components are in spinal nerves; can occur in absence of brain control

spindle fiber (muscle) modified skeletal-muscle fiber in muscle spindle; also called intrafusal fibers; (mitosis) microtubule that connects chromosome to centriole during mitosis

spleen largest lymphoid organ; located between stomach and diaphragm

stable balance net loss of substance from body equals net gain, and body's amount of substance neither increases nor decreases; *compare* positive balance, negative balance

stapes (STAY-peas) third middle-ear bone; transmits sound waves to scala vestibuli of inner ear

starch moderately branched plant polysaccharide composed of glucose subunits

Starling force factor that determines direction of fluid movement across capillary wall

Starling's law of the heart within limits, increased end-diastolic volume of heart (increased muscle-fiber length) increases force of cardiac contraction

state of consciousness degree of mental alertness; that is, whether awake, drowsy, asleep, and so on

steady state no net change occurs; continual energy input to system is required, however, to prevent net change; *compare* equilibrium

steroid (STEER-oid) lipid subclass; molecule consists of four interconnected carbon rings to which polar groups may be attached

STH *see* growth hormone

stimulus detectable change in environment

stress environmental change that must be adapted to if health and life are to be maintained; event that elicits increased cortisol secretion

stretch receptor afferent nerve ending that is depolarized by stretching; *see also* muscle-spindle stretch receptor

stretch reflex monosynaptic reflex, mediated by muscle-spindle stretch receptor, in which muscle stretch causes contraction of that muscle

striated muscle (STRI-ate-ed) muscle having transverse banding pattern due to repeating sarcomere structure; *see also* skeletal and cardiac muscle

stroke brain damage due to blood stoppage because of occlusion or rupture of cerebral vessel

stroke volume blood volume ejected by a ventricle during one heartbeat

strong acid acid that ionizes completely to form hydrogen ions and corresponding anions when dissolved in water; *compare* weak acid

subcortical nucleus neuron cluster deep in brain; includes basal ganglia

sublingual gland (sub-LING-wal) salivary gland under tongue

submandibular gland (sub-man-DIB-you-lar) salivary gland in lower jaw; formerly called submaxillary gland

submucosa (sub-mu-KO-sah) connective-tissue layer under mucosa in gastrointestinal tract

submucous plexus (sub-MU-kus PLEX-us) nerve-cell network in submucosa of esophageal, stomach, and intestinal walls

substance P neurotransmitter released by afferent neurons in pain pathway as well as other sites

substrate (SUB-straight) reactant in enzyme-mediated reaction

substrate phosphorylation (fos-for-ih-LAY-shun) direct transfer of phosphate group from metabolic intermediate to ADP to form ATP

subsynaptic membrane (sub-sih-NAP-tik) that part of postsynaptic neuron's plasma membrane under synaptic knob

subthreshold potential (sub-THRESH-old) depolarization less than threshold potential

subthreshold stimulus stimulus capable of depolarizing membrane but not by enough to reach threshold

sucrose (SUE-krose) disaccharide composed of glucose and fructose; also called table sugar

sulfate SO_4^{2-}

sulfhydryl group (sulf-HI-dral) —SH

sulfuric acid (sulf-YOUR-ik) acid generated during catabolism of sulfur-containing compounds; dissociates to form sulfate and hydrogen ions

summation (sum-MAY-shun) increase in muscle tension or shortening in response to rapid, repetitive stimulation relative to single twitch

superior vena cava (VEE-nah CAVE-ah) large vein that carries blood from upper half of body to right atrium of heart

supersensitivity increased responsiveness of target cell to given messenger due to up-regulation

suppressor T cell T cell that inhibits antibody production and cytotoxic T-cell function

suprathreshold stimulus (soup-ra-THRESH-old) any agent capable of depolarizing membrane more than its threshold potential, that is, closer to zero

surface tension attractive forces between water molecules at surface resulting in net force that acts to reduce surface area

surfactant (sir-FACT-ant) detergent-like phospholipid produced by pulmonary alveolar cells; reduces surface tension of fluid film lining alveoli

sympathetic nervous system portion of autonomic nervous system whose preganglionic fibers leave CNS at thoracic and lumbar portions of spinal cord; *compare* parasympathetic

sympathetic trunk one of paired chains of interconnected sympathetic ganglia that lie on either side of vertebral column

sympathomimetic (sym-path-oh-mih-MET-ik) produces effects similar to sympathetic nervous system

synapse (SIN-apse) anatomically specialized junction between two neurons where electrical activity in one neuron influences excitability of second; *see also* chemical synapse, electric synapse, excitatory synapse, inhibitory synapse

synaptic cleft narrow extracellular space separating pre- and postsynaptic neurons at chemical synapse

synergistic muscle (sin-er-JIS-tik) muscle whose action aids intended motion

systemic circulation (sis-TEM-ik) circulation from left ventricle through all organs except lungs and back to heart

systole (SIS-toe-lee) period of ventricular contraction

systolic pressure (sis-TAHL-ik) maximum arterial blood pressure during cardiac cycle

T_m *see* maximal tubular capacity

T_3 *see* triiodothyronine

T_4 *see* thyroxin

T cell lymphocyte derived from precursor that differentiated in thymus; *see also* cytotoxic T cell, helper T cell, suppressor T cell

T lymphocyte *see* T cell

t tubule *see* transverse tubule

T wave component of electrocardiogram corresponding to ventricular repolarization

target cell cell influenced by a certain hormone

taste bud sense organ that contains chemoreceptors for taste

tectorial membrane (tek-TOR-ee-al) structure in organ of Corti in contact with receptor-cell hairs

teleology (teal-ee-OL-oh-gee) explanation of events in terms of ultimate purpose served by them

template (TEM-plit) pattern

temporal lobe region of cerebral cortex where primary auditory cortex and Wernicke's speech center is located

temporal summation membrane potential produced as two or more inputs, occurring at different times, are added together; potential change is greater than that caused by single input

tendon (TEN-don) collagen fiber bundle that connects muscle to bone and transmits muscle contractile force to the bone

tension force; *see also* muscle tension

termination code word three-nucleotide sequence in DNA that signifies gene end

testes determining factor chemical messenger, coded for on Y chromosome, that induces testes formation in male

testis (TES-tiss) (pl. testes) gonad in male

testosterone (tes-TOS-ter-own) steroid hormone produced in interstitial cells of testes; major male sex hormone; essential for spermatogenesis and maintains growth and development of reproductive organs and secondary sexual characteristics of male

tetanus (TET-ah-nus) maintained mechanical response of muscle to high-frequency stimulation; the disease lockjaw

tetrad (TET-rad) grouping of two homologous chromosomes, each with its sister chromatid, during meiosis

TH *see* thyroid hormone

thalamus (THAL-ah-mus) subdivision of diencephalon; integrating center for sensory input on its way to cerebral cortex; also contains motor nuclei

theca (THEE-kah) cell layer that surrounds ovarian-follicle granulosa cells; formed from follicle and connective-tissue cells

thermogenesis (ther-mo-JEN-ih-sis) heat generation

thermoneutral zone temperature range over which changes in skin blood flow alone can regulate body temperature

thermoreceptor sensory receptor for temperature and temperature changes, particularly in low (*cold receptor*) or high (*warm receptor*) range

theta wave (THAY-tah) slow 4- to 8-Hz oscillation of electroencephalogram; associated with sleep

thick filament 12- to 18-nm myosin filament in muscle cell

thin filament 5- to 8-nm filament in muscle cell; consists of actin, troponin, and tropomyosin

"third population" non-B and non-T lymphocytes; includes NK cells

thoracic cavity (thor-ASS-ik) chest cavity

thoracic wall chest wall

thorax (THO-raks) closed body cavity between neck and diaphragm; contains lungs, heart, thymus, large vessels, and esophagus; also called the chest

threshold (THRESH-old) (or threshold potential) membrane potential to which excitable membrane must be depolarized to initiate an action potential

threshold stimulus stimulus capable of depolarizing membrane to threshold

thrombin (THROM-bin) enzyme that catalyzes conversion of fibrinogen to fibrin

thrombolytic system *see* fibrinolytic system

thrombosis (throm-BO-sis) clot formation in body

thromboxane (throm-BOX-ain) eicosanoid closely related to prostaglandins; synthesized from arachidonic acid

thromboxane A_2 thromboxane that, among other effects, stimulates platelet aggregation in blood clotting

thrombus (THROM-bus) blood clot

thymine (T) (THIGH-mean) pyrimidine base in DNA but not RNA

thymosin (THIGH-mo-sin) group of hormones secreted by thymus; also called *thymopoietin*

thymus (THIGH-mus) lymphoid organ in upper part of chest; site of T lymphocyte differentiation and thymosin secretion

thyroglobulin (thigh-ro-GLOB-you-lin) large protein to which thyroid hormones bind in thyroid gland; storage form of thyroid hormones

thyroid gland paired endocrine gland in neck; secretes thyroid hormones and calcitonin

thyroid hormones (TH) collective term for amine hormones released from thyroid gland, that is thyroxine (T_4) and triiodothyronine (T_3)

thyroid-stimulating hormone (TSH) glycoprotein hormone secreted by anterior pituitary; induces secretion of thyroid hormone; also called thyrotropin

thyrotropin (thigh-roe-TRO-pin) *see* thyroid-stimulating hormone

thyrotropin-releasing hormone (TRH) hypothalamic hormone that stimulates thyrotropin and prolactin secretion by anterior pituitary

thyroxine (T_4) (thigh-ROCKS-in) tetraiodothyronine; iodine-containing amine hormone secreted by thyroid gland

tidal volume air volume entering or leaving lungs with single breath during any state of respiratory activity

tight junction cell junction in epithelial tissues; extends around cell and restricts molecule diffusion through space between cells

tissue aggregate of differentiated cells of similar type united in performance of particular function; also denotes general cellular fabric of a given organ

tissue factor extravascular enzyme capable of initiating clot formation; also called tissue thromboplastin

tissue plasminogen activator (t-PA) plasma

protein produced by endothelial cells; after binding to fibrin, activates the proenzyme plasminogen

tissue thromboplastin *see* tissue factor

tolerance condition in which increasing drug doses are required to achieve effects that initially occurred in response to a smaller dose

tone *see* muscle tone

tonic (TAH-nik) continuous activity; *compare* phasic

tonsil one of several small lymphoid organs in pharynx

total carbon dioxide sum total of dissolved carbon dioxide, bicarbonate, and carbamino-CO_2

total dead space alveolar and anatomic dead space; *see also* alveolar dead space, anatomic dead space

total energy expenditure sum of external work done plus heat produced plus any energy stored by body

total peripheral resistance (TPR) total resistance to flow in systemic blood vessels from beginning of aorta to end of venae cavae

toxemia of pregnancy (tox-EE-me-ah) disease occurring in pregnant women and associated with fluid retention, urinary protein, hypertension, and possibly convulsions; also called eclampsia

toxin (TOX-sin) poison

t-PA *see* tissue plasminogen activator

TPR *see* total peripheral resistance

trace element mineral present in body in extremely small quantities

trachea (TRAY-key-ah) single airway connecting larynx with bronchi

tract large, myelinated nerve-fiber bundle in CNS

transamination (trans-am-ih-NAY-shun) reaction in which an amino-acid amino group (—NH₂) is transferred to a keto acid, the keto acid thus becoming an amino acid

transcription formation of mRNA containing, in linear sequence of its nucleotides, genetic information in specific gene; first stage of protein synthesis

transduction process by which stimulus energy is transformed into a response

transepithelial transport *see* epithelial transport

transfer RNA (tRNA) type of RNA; different tRNAs combine with different amino acids and with codon on mRNA specific for that amino acid, thus arranging amino acids in sequence to form specified protein

transferrin (trans-FAIR-in) iron-binding protein carrier for iron in plasma

translation assembly of amino acids in correct order during protein synthesis, according to genetic instructions in mRNA; occurs on ribosomes

transmural pressure pressure difference exerted on the two sides of a wall

transport maximum (T_m) upper limit to

amount of material that carrier-mediated transport can move across a membrane

transpulmonary pressure difference between alveolar and intrapleural pressures; force that holds lungs open

transverse tubule (t tubule) tubule extending from striated-muscle plasma membrane into the fiber, passing between opposed sarcoplasmic-reticulum segments; conducts muscle action potential into muscle fiber

TRH *see* thyrotropin-releasing hormone

triacylglycerol (tri-ace-il-GLISS-er-ol) subclass of lipids composed of glycerol and three fatty acids; also called fat, neutral fat, or triglyceride

tricarboxylic acid cycle *see* Krebs cycle

tricuspid valve (try-CUS-pid) valve between right atrium and right ventricle of heart

tricyclic antidepressant drug that interferes with reuptake of norepinephrine and serotonin by presynaptic endings, thereby increasing amount of neurotransmitter in synaptic cleft

triglyceride *see* triacylglycerol

triiodothyronine (T₃) (try-eye-oh-doe-THIGH-ro-neen) iodine-containing amine hormone secreted by thyroid gland

triplet code three-base sequence in DNA and RNA that specifies particular amino acid

tRNA *see* transfer RNA

trophoblast (TRO-fo-blast) outer layer of blastocyst; gives rise to placental tissues

tropic (TRO-pic) growth-promoting

tropic hormone hormone that stimulates the secretion of another hormone

tropomyosin (tro-po-MY-oh-sin) regulatory protein capable of reversibly covering binding sites on actin; associated with muscle thin filaments

troponin (tro-PO-nin) regulatory protein bound to actin and tropomyosin of striated-muscle thin filaments; site of calcium binding that initiates contractile activity

trypsin (TRIP-sin) enzyme secreted into small intestine by exocrine pancreas as precursor trypsinogen; breaks certain peptide bonds in proteins and polypeptides

trypsinogen (trip-SIN-oh-jen) inactive precursor of trypsin; secreted by exocrine pancreas

tryptophan (TRIP-toe-fan) essential amino acid; serotonin precursor

TSH *see* thyroid-stimulating hormone

tubular reabsorption transfer of materials from kidney-tubule lumen to peritubular capillaries

tubular secretion transfer of materials from peritubular capillaries to kidney-tubule lumen

tumor necrosis factor (TNF) (neh-CROW-sis) monokine that kills cells, stimulates inflammation, and mediates many systemic acute phase responses

twitch mechanical response of muscle to single action potential

tympanic membrane (tim-PAN-ik) membrane stretched across end of ear canal; also called ear drum

type I diabetes diabetes in which insulin is completely or almost completely absent; also called insulin-dependent diabetes

type II cell pulmonary-alveolar cell that produces surfactant

type II diabetes diabetes in which insulin is present at near normal or even above normal levels, but the tissues do not respond optimally to it; also called insulin-independent diabetes

tyrosine (TIE-ro-seen) amino acid; precursor of catecholamines and thyroid hormones

tyrosine kinase protein kinase that is part of a receptor and is activated upon ligand binding to the receptor; phosphorylates specifically tyrosine portion of proteins

ulcer an erosion, or sore, as in stomach or intestinal wall

ultrafiltrate (ul-tra-FILL-trate) essentially protein-free fluid formed from plasma as it is forced through capillary walls by pressure gradient

umbilical vessel (um-BIL-ih-kal) artery or vein transporting blood between fetus and placenta

unsaturated fatty acid fatty acid containing one or more double bonds

up-regulation increase in number of target-cell receptors for given messenger in response to chronic low extracellular concentration of that messenger; *see also* supersensitivity; *compare* down-regulation

upper esophageal sphincter (ih-sof-ih-JEE-al SFINK-ter) skeletal-muscle ring surrounding esophagus just below pharynx that, when contracted, closes entrance to esophagus

upper motor neuron neuron of the motor cortex or descending pathway

uracil (YOUR-ah-sil) pyrimidine base; present in RNA but not DNA

urea (you-REE-ah) major nitrogenous waste product of protein breakdown and amino acid catabolism

uremia (you-REE-me-ah) general term for symptoms of profound kidney malfunction

ureter (YUR-ih-ter) tube that connects renal pelvis to urinary bladder

urethra (you-REE-thrah) tube that connects urinary bladder to outside of body

uric acid (YUR-ik) waste product derived from nucleic acid catabolism

urinary bladder thick-walled sac composed of smooth muscle; stores urine prior to urination

uterine tube (YOU-ter-in) one of two tubes that carries ovum from ovary to uterus; also called fallopian tube, oviduct

uterus (YOU-ter-us) hollow organ in pelvic

region of females; houses fetus during pregnancy; also called womb

V *see* volt

vagina (vah-JI-nah) canal leading from uterus to outside of body; also called birth canal

vagus nerve (VAY-gus) cranial nerve X; major parasympathetic nerve

valence (VAY-lence) number of hydrogen atoms (or its equivalent) that an ion can hold in combination or displace in a reaction

van der Waals force (walls) weak attracting force between nonpolar regions of molecules

varicosity (ver-ih-KOS-ih-tee) swollen region of axon; contains neurotransmitter-filled vesicles; analogous to presynaptic ending

vas deferens (vahz DEF-er-enz) one of paired male reproductive ducts that connect epididymis of testis to urethra; also called ductus deferens

vasectomy (vay-SEK-toe-me) cutting and tying off of both ductus ("vas") deferens, which results in sterilization of male without loss of testosterone

vasoconstriction (vay-zoh-kon-STRIK-shun) decrease in blood-vessel diameter due to vascular smooth-muscle contraction

vasodilation (vaz-oh-die-LAY-shun) increase in blood-vessel diameter due to vascular smooth-muscle relaxation

vasopressin (vaz-oh-PRES-sin) peptide hormone synthesized in hypothalamus and released from posterior pituitary; increases water permeability of kidneys' collecting ducts; also called antidiuretic hormone (ADH)

vein any vessel that returns blood to heart

vena cava (VEE-nah CAVE-ah) (pl. venae cavae) one of two large veins that returns systemic blood to heart; *see also* superior vena cava, inferior vena cava

venous return (VR) blood volume flowing to heart per unit time

ventilation air exchange between atmosphere and alveoli; alveolar air flow

ventral (VEN-tral) toward or at the front of body

ventral root one of two groups of efferent fibers that leave ventral side of spinal cord

ventricle (VEN-trick-il) cavity, as in *cerebral ventricle* or *heart ventricle*

venule (VEEN-ule) small vessel that carries blood from capillary network to vein

very low density lipoprotein (VLDL) (lip-oh-PRO-teen) lipid-protein aggregate having high proportion of fat

vesicle (VES-ik-il) small, membrane-bound organelle

vestibular apparatus *see* vestibular system

vestibular receptor hair cell in semicircular canal, utricle, or saccule

vestibular system sense organ in temporal bone of skull; consists of three semicircular canals, a utricle, and a saccule; also called vestibular apparatus, sense organ of balance

villi (VIL-eye) finger-like projections from highly folded surface of small intestine; covered with single layered epithelium

virilism (VIR-il-ism) development of masculine physical characteristics in a woman

virus nucleic acid core surrounded by protein coat; lacks enzyme machinery for energy production and ribosomes for protein synthesis; thus, cannot survive or reproduce except inside other cells whose biochemical apparatus it uses; *see also* slow viruses

viscera (VISS-er-ah) organs in thoracic and abdominal cavities

viscoelastic (viss-ko-ee-LAS-tik) having viscous and elastic properties

viscosity (viss-KOS-ih-tee) property of fluid that makes it resist flow

visual field that part of world being viewed at a given time

vital capacity maximal amount of air that can be expired, regardless of time required, following maximal inspiration

vitalism (VI-tal-ism) view that life processes require a "life force" rather than physicochemical processes alone

vitamin organic molecule that is required in trace amounts for normal health and growth but is not manufactured by metabolic pathways and must be supplied by diet; classified as *water-soluble* (vitamins C and the B complex) and *fat-soluble* (vitamins A, D, E, and K)

VLDL *see* very low density lipoprotein

vocal cord one of two elastic-tissue bands stretched across laryngeal opening and caused to vibrate by air movement past them, producing sounds

volt (V) unit of measurement of electric potential between two points

voltage measure of potential of separated electric charges to do work; measure of electric force between two points

voltage-sensitive channel cell-membrane ion channel that is opened or closed by electric signals

vomiting center neurons in brainstem medulla that coordinate vomiting reflex

von Willebrand factor (vWF) (von-VILL-ih-brand) plasma protein secreted by endothelial cells; facilitates adherence of platelets to damaged vessel wall

vulva (VUL-vah) female external genitalia; the mons pubis, labia majora and minora, clitoris, vestibule of the vagina, and vestibular glands

Wallerian degeneration after injury of neuron, degeneration of those parts separated from cell body by the injury

wavelength distance between two wave peaks in oscillating medium

weak acid acid whose molecules do not completely ionize to form hydrogen ions when dissolved in water; *compare* strong acid

white matter portion of CNS that appears white in unstained specimens and contains primarily myelinated nerve fibers

white muscle fiber muscle fiber lacking appreciable amounts of myoglobin

withdrawal (with-DRAW-al) physical symptoms (usually unpleasant) associated with cessation of drug use

withdrawal reflex *see* flexion reflex

Wolffian duct (WOLF-ee-an) part of embryonic duct system that, in male, remains and develops into reproductive-system ducts, but in female, degenerates

work measure of energy required to produce physical displacement of matter; *see also* external work, internal work

working memory short-term, limited-capacity memory storage process serving as initial depository of information from which memory may be transferred to long-term storage or forgotten

X chromosome *see* sex chromosome

Xylocaine *see* lidocaine

Y chromosome *see* sex chromosome

Z line structure running across myofibril at each end of striated muscle sarcomere; anchors one end of thin filaments

zona pellucida (ZO-nah peh-LEW-sih-dah) thick, clear layer separating ovum from surrounding granulosa cells

APPENDIX

E

REFERENCES FOR FIGURE ADAPTATIONS

Bernhard, S.: "The Structure and Function of Enzymes," Benjamin, New York, 1968.

Berne, R. M.: "Physiology," 2d edn., Mosby, St. Louis, 1988.

Bloom, F. E., A. Lazerson, and L. Hofstadter: "Brain, Mind, and Behavior." Freeman, New York, 1985.

Carlson B. M.: "Patten's Foundations of Embryology," 5th edn., McGraw-Hill, New York, 1988.

Chaffee, E. E., and I. M. Lytle: "Basic Physiology and Anatomy," 4th edn., Lippincott, Philadelphia, 1980.

Chapman, C. B., and J. H. Mitchell: *Sci. Amer.*, May 1965.

Comroe, J. H.: "Physiology of Respiration," Year Book, Chicago, 1965.

Crosby, W. H.: Hemochromatosis: Current concepts and management. *Hosp. Prac.* February 1987.

Curtis, B. A., S. Jacobson, and E. M. Marcus: "An Introduction to the Neurosciences," Saunders, Philadelphia, 1972.

Davis, H., and H. R. Silverman: "Hearing and Deafness," Holt, Rinehart, and Winston, New York, 1970.

Davson, H.: "The Eye," Vol. 1, 2nd edn., Academic Press, New York, 1969.

Elias, H., J. E. Pauly, and E. R. Burns: "Histology and Human Microanatomy," 4th edn., Wiley, New York, 1978.

Felig, P., and J. Wahren: *N. Eng. J. Med.*, **293**:1078 (1975).

Ganong, W. F.: "Review of Medical Physiology," 4th edn., Lange, Los Altos, Calif., 1969.

Gardner, E.: "Fundamentals of Neurology," 5th edn., Saunders, Philadelphia, 1968.

Goldberg, N. D.: *Hosp. Prac.*, May 1974.

Golde, D. W., and Gasson, J. C.: Hormones that stimulate the growth of blood cells. *Sci. Amer.*, July 1988.

Gregory, R. L.: "Eye and Brain: The Psychology of Seeing," McGraw-Hill, New York, 1966.

Guyton, A. C.: "Functions of the Human Body," 3d edn., Saunders, Philadelphia, 1969.

Hedge, G. A., H. D. Colby, and R. L. Goodman: "Clinical Endocrine Physiology," Saunders, Philadelphia, 1987.

Hoffman, B. F., and P. E. Cranefield: "Electrophysiology of the Heart," McGraw-Hill, New York, 1960.

Hubel, D. H., and T. N. Wiesel: *J. Physiol.*, **154**:572 (1960).

Hudspeth, A. J.: *Sci. Amer.* January 1983.

Jouvet, M.: *Sci. Amer.*, February 1967.

Kandel, E. R., and J. H. Schwartz: "Principles of Neural Science," 2nd edn., Elsevier/North-Holland, New York, 1985.

Kappas, A., and A. P. Alvares: *Sci. Amer.*, June 1975.

Kruger, D. T., and J. B. Martin: *N. Eng. J. Med.*, **304**:876, 1981.

Kuffler, W. W., J. G. Nicholls, and A. R. Martin: "From Neuron to Brain," 2nd edn., Sinauer Associates, Sunderland, Mass., 1984.

Lambersten, C. J.: in P. Bard (ed.), "Medical Physiological Psychology, 11th edn., Mosby, St. Louis, 1961.

Landauer, T. K.: "Readings in Physiological Psychology," McGraw-Hill, New York, 1967.

Lehninger, A. L.: "Biochemistry," Worth Publishers, New York, 1970.

Lentz, T. L.: "Cell Fine Structure," Saunders, Philadelphia, 1971.

Little, R. C.: "Physiology of the Heart and Circulation," 2d edn., Year Book, Chicago, 1981.

Maxwell, D. J., and M. Rethely: *TINS*, **10**:117 (1987).

Mollon, J. D.: *Ann., Rev. Psych.*, **33**:41 (1982).

Nieuwenhuys, R., J. Voogd, and C. van Huizen: "The Human Central Nervous System: A Synopsis and Atlas," 2nd edn., Springer-Verlag, New York, 1981.

Ojemann, J.: *Behav. Brain Sci.*, **6**:199 (1983).

Oldenhoff, W. H.: *Hosp. Prac.*, **17**:143 (1982).

Olds, J.: *Sci. Am.*, October 1956.

Pitts, R. F.: "Physiology of the Kidney and Body Fluids," 2d edn., Year Book, Chicago, 1968.

Rasmussen, A. T.: "Outlines of Neuroanatomy," 2d edn., Wm. C. Brown, Dubuque, Ia., 1943.

Routtenberg, A.: *Sci. Am.*, November 1978.

Rushmer, R. F.: "Cardiovascular Dynamics," 2d edn., Saunders, Philadelphia, 1961.

Singer, S. J., and G. L. Nicholson: *Science*, **175**:720 (1972).

Smith, H. W.: "The Kidney," Oxford University Press, New York, 1951.

Snyder, S. H.: "Drugs and the Brain," Freeman, New York, 1986.

Steinberger, E., and W. O. Nelson: *Endocrinology*, **56**:429 (1955).

Tung, K.: Immunopathology and male infertility, *Hosp. Prac.*, June 1988.

von Bekesy, G.: *Sci. Amer.*, August 1957.

Wang, C. C., and M. I. Grossman: *Am. J. Physiol.*, **164**:527 (1951).

Wersall, J., L. Gleisner, and P. G. Lundquist: in A. V. S. de Reuck and J. Knight (eds.), "Myostatic, Kinesthetic, and Vestibular Mechanisms," Ciba Foundation Symposium, Little, Brown, Boston, 1967.

Wurtsman, R. J.: *Sci. Amer.*, January 1985.

SUGGESTED READING

Like all scientists, physiologists report the results of their experiments in scientific journals. Approximately 1300 such journals in the life sciences publish 300,000 papers each year. Every physiologist must be familiar with the ever-increasing number of articles in his or her own specific field of research.

Another type of scientific writing is the review article, a summary and synthesis of the relevant research reports on a specific subject. Such reviews, which are published in many different journals, are extremely useful. Also, the American Physiological Society has a massive program of publishing comprehensive reviews in almost all fields of physiology. These reviews are gathered into a series of ongoing volumes known collectively as the "Handbook of Physiology," Another important series is the *Annual Review of Physiology*, which publishes yearly reviews of the most recent research articles in specific fields and provides the most rapid means of keeping abreast of developments in physiology.

Another type of review that is of value to the nonexpert is the *Scientific American* article. Written in a manner usually intelligible to the layperson, these articles, in addition to reviewing a topic, often present experimental data so that the reader can obtain some insight into how the experiments were done and the data interpreted. Many articles published in *Scientific American* can be obtained individually as inexpensive offprints from W. H. Freeman and Company, San Francisco, CA, 94104, which also publishes collections of *Scientific American* articles in a specific area, for example, psychobiology. Two other magazines that frequently publish excellent reviews of physiology for the nonexpert are *New England Journal of Medicine* and *Hospital Practice*.

Another level of scientific writing is the monograph, an extended review of a subject area broader than the review articles described above. Monographs are usually in book form and generally do not attempt to cover all the relevant literature on the topic but instead analyze and synthesize the most important data and interpretations. At the last level is the textbook, which deals with a much larger field than the monograph or review article. Its depth of coverage of any topic is accordingly much less complete and is determined by the audience to which it is aimed.

We have restricted our suggested readings to textbooks and monographs because they are usually available in college and university libraries. Their bibliographies will serve as a further entry into the scientific literature. We list first a group of books that cover large areas of physiology.

BOOKS COVERING WIDE AREAS OF PHYSIOLOGY

Cell Physiology

Alberts, Bruce, Dennis Bray, Julian Lewis, Martin Raff, Kenith Roberts, and James D. Watson: "Molecular Biology of the Cell," Garland, New York, 1983.

Darnell, James, Harvey Lodish, and David Baltimore: "Molecular Cell Biology," Scientific American Books, New York, 1986.

DeRobertis, E. D. P., and E. M. F. DeRobertis, Jr.: "Essentials of Cell and Molecular Biology," Lea & Febiger, Philadelphia, 1987.

Macleod, Alexander, and Karol Sikora: "Molecular Biology and Human Disease," Blackwell Scientific Publications, St. Louis, Mo., 1984.

Organ System Physiology

Berne, R. M., and M. N. Levy: "Physiology," 2nd edn., Mosby, St. Louis, 1988.

Ganong, W. F.: "Review of Medical Physiology," 13th edn., Lange, Los Altos, Calif., 1987.

Guyton, A. C.: "Textbook of Medical Physiology," 7th edn., Saunders, Philadelphia, 1986.

Mountcastle, V. B. (ed.): "Medical Physiology, 14th edn., Mosby, St. Louis, 1980.

Selkurt, E. E. (ed.): "Physiology," 5th edn., Little, Brown, Boston, 1984.

Anatomy

Basmajian, J. V.: "Grant's Method of Anatomy," 11th edn., Williams & Wilkins, Baltimore, 1988.

Carlson, B. M.: "Patten's Foundations of Embryology," 5th edn., McGraw-Hill, New York, 1988.

Clemente, C. D.: "Gray's Anatomy of the

Human Body," 30th American edn., Lea & Febiger, Philadelphia, 1984.

Hollinshead, W. H., and C. Rosse: "Textbook of Anatomy," 4th edn., Harper & Row, New York, 1985.

Woodburne, R. T.: "Essentials of Human Anatomy," 7th edn., Oxford University Press, New York, 1983.

Pathophysiology

Frohlich, Edward D. (ed.): "Pathophysiology," Lippincott, Philadelphia, 1983.

Porth, Carol: "Pathophysiology," 2nd edn., Lippincott, Philadelphia, 1986.

Sodeman, William A., and T. M. Sodeman, Jr.: "Pathologic Physiology Mechanisms of Disease," 7th edn., Saunders, Philadelphia, 1985.

SUGGESTIONS FOR INDIVIDUAL CHAPTERS

Chapter 1

Bernard, C.: "An Introduction to the Study of Experimental Medicine," Dover, New York, 1957 (paperback).

Brooks, C. McC., and P. F. Cranefield (eds.): "The Historical Development of Physiological Thought," Hafner, New York, 1959.

Butterfield, H.: "The Origins of Modern Science." Macmillan, New York, 1961 (paperback).

"The Excitement and Fascination of Science: A Collection of Autobiographical and Philosophical Essays," Annual Reviews, Palo Alto, Calif., 1966.

Fulton, J. F., and L. C. Wilson (eds.): "Selected Readings in the History of Physiology," 2d edn., Thomas, Springfield, Il., 1966.

Harvey, W.: "On the Motion of the Heart and Blood in Animals," Gateway, Henry Regnery, Chicago, 1962 (paperback).

Leake, Chauncey: "Some Founders of Physiology" American Physiological Society, Washington, D. C., 1961.

Chapter 2

Creighton, Thomas E.: "Proteins: Structures and Molecular Principles," Freeman, New York, 1984.

Lehninger, Albert L.: "Principles of Biochemistry," Worth, New York, 1982.

Smith, Emil L., Robert L. Hill, I. Robert Lehman, Robert J. Lefkowitz, Philip Handler, and Abraham White: "Principles of Biochemistry," 7th edn., McGraw-Hill, New York, 1983.

Stryer, Lubert: "Biochemistry," 3d edn., Freeman, San Francisco, 1989.

Chapter 3

Berns, Michael W.: "Cells," 2d edn., Holt, Rinehart, and Winston, New York, 1983.

Bloom, W., and D. W. Fawcett: "A Textbook of Histology," 11th edn., Saunders, Philadelphia, 1986.

Fawcett, D. W.: "The Cell," 2d edn., Saunders, Philadelphia, 1981.

Schliwa, M.: "The Cytoskeleton: An Introductory Survey," Springer-Verlag, New York, 1986.

Weiss, Leon, et al. (eds.): "Cell and Tissue Biology: A Textbook of Histology," 6th edn., Urban & Schwarzenberg, Baltimore, 1988.

Chapter 4

Alberts, Bruce, Dennis Bray, Julian Lewis, Martin Raff, Kenith Roberts, and James D. Watson: "Molecular Biology of the Cell," Garland, New York, 1983.

Darnell, James, Harvey Lodish, and David Baltimore: "Molecular Cell Biology," Scientific American Books, New York, 1986.

Creighton, Thomas E.: "Proteins: Structures and Molecular Principles," Freeman, New York, 1984.

Lloyd, David, Robert K. Poole, and Steven W. Edwards: "The Cell Division Cycle," Academic, New York, 1982.

Ruddon, Raymond W.: "Cancer Biology," 2d edn., Oxford University Press, New York, 1987.

Szekely, M.: "From DNA to Protein: The Transfer of Genetic Information," Macmillan, New York, 1980.

Watson, James D., Nancy H. Hopkins, Jeffrey W. Roberts, Joan Argetsinger Steitz, and Alan M. Weiner: "The Molecular Biology of the Gene," 4th edn., Benjamin, New York, 1987.

Watson, James D., John Tooze, and David T. Kurtz: "Recombinant DNA: A Short Course," Freeman, New York, 1983.

Chapter 5

Baker, Jeffrey J. W., and Garland E. Allen: "Matter, Energy, and Life," 4th edn., Addison-Wesley, Reading, Mass., 1982.

Fersht, A.: "Enzyme Structure and Mechanism," 2d edn., Freeman, New York, 1985.

Lehninger, Albert L.: "Principles of Biochemistry," Worth, New York, 1982.

Newsholme, E. A., and A. R. Leech: "Biochemistry for Medical Science," Wiley, New York, 1983.

Smith, Emil L., Robert L. Hill, I. Robert Lehman, Robert J. Lefkowitz, Philip Handler, and Abraham White: "Principles of Biochemistry," 7th edn., McGraw-Hill, New York, 1983.

Stryer, Lubert: "Biochemistry," 3d edn., Freeman, San Francisco, 1989.

Wills, Eric D.: "Biochemical Basis of Medicine," Wright, Bristol, 1985.

Chapter 6

Alberts, Bruce, Dennis Bray, Julian Lewis, Martin Raff, Kenith Roberts, and James D. Watson: "Molecular Biology of the Cell," Garland, New York, 1983.

Andreoli, T. E., J. F. Hoffman, and D. D. Fanestil (eds.): "Membrane Physiology," Plenum, New York, 1978.

Darnell, James, Harvey Lodish, and David Baltimore: "Molecular Cell Biology," Scientific American Books, New York, 1986.

Finkelstein, Alan: "Water Movement Through Lipid Bilayers, Pores, and Plasma Membranes: Theory and Reality," Wiley, New York, 1987.

Harrison, Roger, and George G. Lunt: "Biological Membranes," 2d edn., Halsted, New York, 1980.

Weissman, Gerald, and Robert Claiborne (eds.): "Cell Membranes: Biochemistry, Cell Biology, and Pathology," Hospital Practice Publishing, New York, 1976.

Chapter 7

Bernard, C.: "An Introduction to the Study of Experimental Medicine," Dover, New York, 1957 (paperback).

Cannon, W. B.: "The Wisdom of the Body," Norton, New York, 1939 (paperback).

Goodman, Maurice: "Basic Medical Endocrinology," Raven, New York, 1988.

Langley, L. L. (ed.): "Homeostasis: Origin of the Concept," Dowden, Hutchinson, and Ross, Stroudsburg, Pa., 1973.

Slonin, N. B. (ed.): "Environmental Physiology," Mosby, St. Louis, 1974.

Tepperman, T.: "Metabolic and Endocrine Physiology," 5th edn., Year Book, Chicago, 1987.

Williams, R. W. (ed.): "Textbook of Endocrinology," 7th edn., Saunders, Philadelphia, 1987.

Chapter 8

Adelman, G.: "Encyclopedia of Neuroscience," Vols. 1 and 2, Birkhauser, Boston, 1987.

Angevine, J. B., Jr., and C. W. Cotman: "Principles of Neuroanatomy," Oxford University Press, New York, 1981.

Kandel, E. R., and J. H. Schwartz: "Principles of Neural Science," 2nd edn., Elsevier/North-Holland, New York, 1985.

Kuffler, S. W., J. G. Nicholls, and A. Robert Martin: "From Neuron to Brain," 2d edn., Sinauer Associates, Sunderland, Mass., 1984.

Matzke, H. A., and F. M. Foltz: "Synopsis of Neuroanatomy," 4th edn., Oxford University Press, 1983.

McGeer, P. L., J. C. Eccles, and E. G. McGeer: "Molecular Neurobiology of the Mammalian Brain," 2nd edn., Plenum, New York, 1987.

Shepherd, G. M.: "Neurobiology," 2nd edn., Oxford University Press, New York, 1987.

Thompson, R. F.: "The Brain: An Introduction to Neuroscience," Freeman, 1985.

Chapter 9

Adelman, G.: "Encyclopedia of Neuroscience," Vols. 1 and 2, Birkhauser, Boston, 1987.

Kandel, E. R., and J. H. Schwartz: "Principles of Neural Science," 2nd edn., Elsevier/North-Holland, New York, 1985.

Kuffler, S. W., J. G. Nicholls, and A. Robert Martin: "From Neuron to Brain," 2d edn., Sinauer Associates, Sunderland, Mass., 1984.

Ottoson, D.: "Physiology of the Nervous System," Oxford University Press, New York, 1983.

Shepherd, G. M.: "Neurobiology," 2nd edn., Oxford University Press, New York, 1987.

Chapter 10

Frigley, M. J., and W. G. Luttge: "Human Endocrinology: An Interactive Text," Elsevier, New York, 1982.

Goodman, Maurice: "Basic Medical Endocrinology," Raven, New York, 1988.

Hedge, G. A., H. D. Colby, and R. L. Goodman: "Clinical Endocrine Physiology," Saunders, Philadelphia, 1987.

Tepperman, J.: "Metabolic and Endocrine Physiology," 5th edn., Year Book, Chicago, 1987.

Williams, R. W. (ed.): "Textbook of Endocrinology," 7th edn., Saunders, Philadelphia, 1986.

Chapter 11

Bulbring, Edith, Alison F. Brading, Allan W. Jones, and Tadao Tomita (eds.): "Smooth Muscle," University of Texas Press, Austin, 1981.

Hoyle, C.: "Muscles and Their Neural Con-

trol," Wiley-Interscience, New York, 1983.

Keynes, R. D., and D. J. Aidley: "Nerve and Muscle," Cambridge University Press, Cambridge, 1981.

Peachy, Lee D., and Richard H. Adrian (eds.): "Handbook of Physiology: Section 10, Skeletal Muscle," American Physiological Society, Bethesda, Md., 1983.

Shepard, Roy J.: "Exercise Physiology," Decker, Philadelphia, 1987.

Squire, John: "The Structural Basis of Muscle Contraction," Plenum, New York, 1981.

Squire, John M.: "Muscle Design and Disease," Benjamin/Cummings, Menlo Park, Calif., 1986.

Twarog, Betty M., Rhea J. C. Levine, and Maynard M. Dewey (eds.): "Basic Biology of Muscles: A Comparative Approach," Raven, New York, 1982.

Walton, John: "Disorders of Voluntary Muscle," 5th edn., Churchill Livingstone, New York, 1988.

Woledge, Roger C., Nancy A. Curtin, and Earl Homsher: "Energetic Aspects of Muscle Contraction," Academic, New York, 1985.

Chapter 12

Brooks, V. B.: "The Neural Basis of Motor Control," Oxford, New York, 1986.

Evarts, E. V., S. P. Wise, and D. Bousfield (eds.): "The Motor System in Neurobiology," Elsevier, New York, 1985.

Kandel, E. R., and J. H. Schwartz: "Principles of Neural Science," 2nd edn., Elsevier/North-Holland, New York, 1985.

Kuffler, S. W., J. G. Nicholls, and A. Robert Martin: "From Neuron to Brain," 2d edn., Sinauer Associates, Sunderland, Mass., 1984.

Shepherd, G. M.: "Neurobiology," 2nd edn., Oxford University Press, New York, 1987.

Chapter 13

Beck, W. S.: "Hematology," MIT Press, Cambridge, 1985.

Berne, R. M., and M. N. Levy: "Cardiovascular Physiology," 4th edn., Mosby, St. Louis, 1981.

Mohrman, D. E., and L. J. Heller: "Cardiovascular Physiology," 2nd edn., McGraw-Hill, New York, 1986.

Honig, C. R.: "Modern Cardiovascular Physiology," 2nd edn., Little, Brown, Boston, 1986.

Little, R. C.: "Physiology of the Heart and Circulation," 3rd edn., Year Book, Chicago, 1985.

Smith, J. J., and J. P. Kampine: "Circulatory Physiology: The Essentials," 2nd edn., Williams & Wilkins, Baltimore, 1985.

Chapter 14

Davenport, H. W.: "The ABC of Acid-Base Chemistry," 7th edn., University of Chicago Press, Chicago, 1978.

Levitzky, M. G.: "Pulmonary Physiology," 2nd edn., McGraw-Hill, New York, 1986.

Mines, A. H.: "Respiratory Physiology," 2nd edn., Raven, New York, 1987.

Murray, J. F.: "The Normal Lung," 2nd edn., Saunders, Philadelphia, 1985.

Slonim, N. B., and L. H. Hamilton: "Respiratory Physiology," 5th edn., Mosby, St. Louis, 1986.

West, J. B.: "Respiratory Physiology: The Essentials," 3rd cdn., Williams & Wilkins, Baltimore, 1985.

Chapter 15

Brenner, B. M. and F. C. Rector (eds.): "The Kidney," 3rd edn., Saunders, Philadelphia, 1986.

Davenport, H. W.: "The ABC of Acid-Base Chemistry," 7th edn., University of Chicago Press, Chicago, 1978.

Marsh, D. T.: "Renal Physiology," Raven, New York, 1983.

Rose, B. D.: "Clinical Physiology of Acid-Base and Electrolyte Disorders," 2nd edn., McGraw-Hill, New York, 1984.

Smith, H. W.: "From Fish to Philosopher," Anchor Books, Doubleday, Garden City, N.Y., 1961 (paperback).

Sullivan, L. P., and J. J. Grantham: "Physiology of the Kidney," 2nd edn., Lea and Febiger, Philadelphia, 1982.

Valtin, H.: "Renal Function," 2d edn., Little, Brown, Boston, 1983.

Vander, A. J.: "Renal Physiology," 3rd edn., McGraw-Hill, New York, 1985.

Chapter 16

Brooks, Frank P. (ed.): "Gastrointestinal Pathophysiology," 2nd edn., Oxford University Press, New York, 1978.

Davenport, Horace W.: "Physiology of the Digestive Tract," 5th edn., Year Book Medical Publishers, Chicago, 1982.

Granger, D. Neil, James A. Barrowman, and Peter R. Kvietys: "Clinical Gastrointestinal Physiology," Saunders, Philadelphia, 1985.

Johnson, L. R. (ed.): "Gastrointestinal Physiology," 3rd edn., Mosby, St. Louis, 1985.

Johnson, Leonard R., James Christensen, Eugene D. Jacobson, and John H. Walsh (eds.): "Physiology of the Gastrointestinal Tract," 2nd edn., Raven, New York, 1987 (two volumes).

Chapter 17

Adolph. E. F.: "Physiology of Man in the Desert," Interscience, New York, 1947.

Daughty, W. H.: "Endocrine Control of Growth," Elsevier, New York, 1982.

Falkner, F., and Tanner, J. M. (eds.): "Human Growth," Plenum, New York, 1986.

Goodman, Maurice: "Basic Medical Endocrinology," Raven, New York, 1988.

Horvath, S. M., and M. K. Yousef (eds.): "Environmental Physiology: Aging, Heat, and Altitude," Elsevier, New York, 1981.

Kluger, M. J.: "Fever: Its Biology, Evolution, and Function," Princeton University Press, Princeton, 1979.

Tepperman, J.: "Metabolic and Endocrine Physiology," 5th edn., Year Book, Chicago, 1987.

Vander, A. J.: "Nutrition, Stress, and Toxic Chemicals: An Approach to Environmental Health Controversies," University of Michigan Press, Ann Arbor, 1981.

Williams, R. W. (ed.): "Textbook of Endocrinology," 7th edn., Saunders, Philadelphia, 1986.

Chapter 18

Carlson, B. M.: "Patten's Foundations of Embryology," 5th edn., McGraw-Hill, New York, 1988.

Goodman, Maurice: "Basic Medical Endocrinology," Raven, New York, 1988.

Johnson, M., and B. Everett: "Essential Reproduction," 2nd edn., Blackwell Scientific Publications, Oxford, 1985.

Money, J., and A. E. Ehrhardt: "Man and Woman, Boy and Girl: The Differentiation and Dimorphism of Gender Identity from Conception to Maturity," Johns Hopkins University Press, Baltimore, 1973 (paperback).

Tepperman, J.: "Metabolic and Endocrine Physiology," 5th edn., Year Book, Chicago, 1987.

Williams, R. W. (ed.): "Textbook of Endocrinology," 7th edn., Saunders, Philadelphia, 1986.

Yen, S. S. C., and Jaffee, R. B.: "Reproductive Endocrinology," 2nd edn., Saunders, Philadelphia, 1986.

Chapter 19

Asterita, M. E.: "The Physiology of Stress," Sciences, 1985.

Barrett, J. T.: "Textbook of Immunology," 5th edn., Mosby, St. Louis, 1988.

Beck, W. S.: "Hematology," MIT Press, Cambridge, 1985.

Clark, W. R.: "The Experimental Foundation of Modern Immunology," 3rd edn., Wiley, 1987.

Roit, I.: "Essential Immunology," 6th edn., Blackwell, St. Louis, 1988.

Vander, A. J.: "Nutrition, Stress, and Toxic Chemicals: An Approach to Environmental Health Controversies," University of Michigan Press, Ann Arbor, 1981.

Weiner, H.: "Psychobiology and Human Disease," Elsevier, New York, 1977.

Chapter 20

Bloom, F. E., A. Lazerson, and L. Hofstadter: "Brain, Mind, and Behavior," Freeman, New York, 1985.

Kandel, E. R., and J. H. Schwartz: "Principles of Neural Science," 2nd edn., Elsevier/North-Holland, New York, 1985.

Kuffler, S. W., J. G. Nicholls, and A. Robert Martin: "From Neuron to Brain." 2d edn., Sinauer Associates, Sunderland, Mass., 1984.

Shepherd, G. M.: "Neurobiology," 2nd edn., Oxford University Press, New York, 1987.

Snyder, S. H.: "Drugs and the Brain," Freeman, New York, 1986.